JOHN WILLARD RIPPON, Ph.D.

Associate Professor of Medicine
The Pritzker School of Medicine
The University of Chicago
Chicago, Illinois

Second Edition

Medical Mycology

The Pathogenic Fungi
and
The Pathogenic Actinomycetes

W. B. SAUNDERS COMPANY   **1982**

Philadelphia   London   Toronto   Mexico City   Rio de Janeiro   Sydney   Tokyo

IO.4

W. B. Saunders Company   West Washington Square
                           Philadelphia, PA   19105

                           1 St. Anne's Road
                           Eastbourne, East Sussex BN21 3UN, England

                           1 Goldthorne Avenue
                           Toronto, Ontario M8Z 5T9, Canada

                           Apartado 26370 — Cedro 512
                           Mexico 4, D.F., Mexico

                           Rua Coronel Cabrita, 8
                           Sao Cristovao Caixa Postal 21176
                           Rio de Janeiro, Brazil

                           9 Waltham Street
                           Artarmon, N.S.W. 2064, Australia

                           Ichibancho, Central Bldg., 22-1 Ichibancho
                           Chiyoda-Ku, Tokyo 102, Japan

Library of Congress Cataloging in Publication Data

Rippon, John Willard.

Medical mycology.

Includes bibliographies and indexes.

1. Medical mycology.   I. Title. [DNLM: 1. Actinomycetales
   infections.   3. Fungi.   4. Mycoses. WC 540 R593m]

RC117.R5 1982   616.9'69   81–51073

ISBN 0–7216–7586–7   AACR2

Photograph facing the title page: A "jungle" of *Candida albicans* on corn meal Tween agar.

Medical Mycology: The Pathogenic Fungi and The Pathogenic Actinomycetes   ISBN   0-7216-7586-7

Last digit is the print number: 9   8   7   6   5   4   3   2   1

Dedicated to Antonio González Ochoa

When I was a student, my late teacher and friend, Professor Ester Meyer, praised *el maestro* as an exemplary investigator, teacher, clinician, and gentleman. In my later professional association with him, I found these qualities not merely true but magnified.

Note to the Reader

A set of three color filmstrips on *Medical Mycology* is available from the publisher. The filmstrips are for individual use and teaching purposes, and nicely complement this textbook.

The three filmstrips (A, B, and C) consist of a total of 220 separate frames. For your convenience, the slide numbers and their corresponding text figure numbers are listed here.

Slide No.	Figure No.	Slide No.	Figure No.
A 1	2–1	A 32	8–18
A 2	2–2	A 33	8–19, *A*
A 3	2–5	A 34	8–20
A 4	2–8	A 35	8–22
A 5	2–11	A 36	8–23
A 6	3–1	A 37	8–24
A 7	3–2	A 38	8–25
A 8	3–8	A 39	8–27
A 9	4–1	A 40	8–28
A 10	4–2	A 41	8–29
A 11, *A, B, C*	5–1, *A*; 5–2, *A, B*	A 42	8–30
A 12	5–3, *A*	A 43	8–31, *A*
A 13	5–4	A 44	8–32
A 14	5–6, *A*	A 45	8–33
A 15	5–8	A 46	8–34
A 16	5–9	A 47	8–36
A 17	7–1	A 48	8–37
A 18	7–5	A 49	8–38
A 19	7–6	A 50	8–42
A 20	7–7	B 1	8–43, *A*
A 21	7–10	B 2, *A, B*	8–43, *B*
A 22	8–7	B 3	8–43, *C*
A 23	8–8	B 4	8–43, *D*
A 24	8–10	B 5, *A, B*	8–53, *A, B*
A 25	8–11	B 6	8–53, *C*
A 26	8–12	B 7	8–53, *D*
A 27	8–13	B 8	8–53, *E*
A 28	8–14	B 9	8–52
A 29	8–15	B 10, *A, B, C, D*	8–56, *A, B, C, D*
A 30	8–17	B 11	8–58
A 31	8–16	B 12	8–46

Slide No.	Figure No.	Slide No.	Figure No.
B 13	8–60	C 7	15–12
B 14	7–44	C 8	15–15
B 15	20–2	C 9	15–16
B 16	20–4	C 10	16–3
B 17	20–7, A; 20–8, B	C 11	17–4
B 18	20–14	C 12	17–6
B 19	20–15, C	C 13	17–7
B 20	20–15, D	C 14	17–8
B 21	20–20, A	C 15	17–10
B 22	20–21	C 16	17–15
B 23	20–22	C 17	17–22
B 24	21–6	C 18	19–3
B 25	21–8	C 19	19–6
B 26	21–10	C 20	19–12
B 27	21–11	C 21	19–13
B 28	21–14, A; 22–1, B	C 22	23–3
B 29	9–2	C 23	23–6
B 30	9–3	C 24	23–7
B 31	9–4	C 25	23–8
B 32	9–14	C 26	23–12
B 33	10–2	C 27	23–15
B 34	10–3	C 28	23–16
B 35	10–4	C 29	23–19
B 36, A, B, C, D	10–7, 8, 9, 10	C 30	23–21
B 37	10–12, A	C 31	25–2
B 38	10–14	C 32	25–3
B 39	11–2, A	C 33	25–4
B 40	12–2	C 34	25–8
B 41	18–5	C 35	25–9
B 42	18–6	C 36	25–12, A
B 43	18–7	C 37	26–1
B 44	18–13	C 38	26–2
B 45	18–15	C 39	26–8
B 46	18–21	C 40	26–10
B 47	18–23	C 41	26–7, C, D
B 48	15–2, F	C 42	27–1
B 49	15–5, A	C 43	27–2
B 50	15–6	C 44	27–3
C 1	15–4	C 45	27–4
C 2	15–7, A	C 46	27–5
C 3	15–7, B	C 47	27–7
C 4	15–9, B	C 48	27–11
C 5	15–9, A	C 49	27–12
C 6	15–11	C 50	28–1

Preface to the Second Edition

"Tempus fugit." This adage of life also applies to medical mycology. Since the publication of the first edition there have been many and significant changes in many areas of fungal disease. This has necessitated significant revision in almost all sections of this book and complete rewriting in many areas. As reflected in the Chapter Six "introduction to the science of mycology," our understanding of taxonomy has been greatly altered by research on methods of conidiation in the fungi. Our understanding of the basis for the varied manifestations of clinical disease has also changed. This is nowhere more apparent than in perhaps the most famous mycotic infection, histoplasmosis. The interaction of this parasite of low virulence with the cellular defenses of the individual host with all its nuances causes vastly different disease to occur from one patient to another. An attempt has been made to categorize and synthesize this new information into useful clinical classification schemes. In the area of antimycotic therapy, some significant additions to our roster of useful drugs are now appearing. The imidazoles have promise of greatly altering our ability to counteract fungal invasion. The new drug ketoconazole is being found useful in chronic *Candida* infections, paracoccidioidomycosis, and perhaps in the entire spectrum of the mycoses. Its great advantage is that it can be given orally. Some of the established imidazoles such as miconazole are found efficacious in *Pseudallescheria* and other infections that had no useful treatment previously. Also, flucytosine is now being noted as the drug of choice in chromoblastomycosis and other infections by dematiaceous fungi. It is expected that other, more tolerable antimycotic agents for systemic disease will be forthcoming. These topics, in addition to new insights into basic mechanisms of fungal pathogenicity, have made this an exciting decade in medical mycology. An overview of all these advances and conceptual changes is included in the present text.

I would like to thank the many people who have assisted me in preparing this second edition. Particular gratitude goes to Daila Shefner, who has thoroughly gone over this manuscript, translated my spelling into that used in the English language, and kept an eye on the grammar and syntax. Great thanks go to J. W. "Bill" Carmichal and Mike McGinnis for keeping my taxonomy reasonably straight and to my mentor in medical mycology, Li Ajello, for his many suggestions, corrections, cultures, photographs and pieces of information. Although we do not always agree, our conversations are always lively and provocative. I am very grateful to Janet Gallup for the pictures of the laboratory contaminants used in Part Five of the book and to all others whose photographs appear herein. The sources of all photographs are acknowledged where known. However in the interchange of teaching material among colleagues the original source of an illustration is sometimes lost. For this I apologize.

Basically the same format has been used in this current edition as in the first. References are concentrated in the late 1970's for current information unless previous ones are of historical interest, establish a significant point, or remain the best review of a particular topic.

JOHN WILLARD RIPPON
University of Chicago

Preface

The purpose of this volume is threefold: (1) to provide a basic presentation of the subject matter of medical mycology; (2) to present an overview of research efforts concerning both the fungal pathogen and its host; and (3) to serve as a reference source for both historical and current literature of importance in the development of the field. In this way it is hoped that this text will be found useful to students at all levels—the professional medical mycologist, the investigator in this and related disciplines, and the clinician and pathologist who may be concerned with fungous diseases only occasionally or whose practice involves frequent encounters with the mycoses. There is presently a renewed interest in mycotic diseases, particularly in the so-called opportunistic infections, and in basic research on host-parasite interactions. For this reason, it was felt that a text should be designed to include these recent developments in the field of medical mycology and to provide a fundamental understanding of the diseases and the organisms that cause them. It is hoped that this text will provide such an expanded view of the subject.

The history and development of medical mycology are characterized by several stages. In the days of Gruby, Malmsten, and Schoenlein, around 1840, there was a wave of excitement in this new field, and investigators were determining the fungal etiology of several dermatologic diseases, such as the tineas and thrush. A few years later, the sister field of bacteriology began to overshadow clinical mycology because of the trail-blazing work of Koch and Pasteur, and the study of fungal disease went into its first decline. The critical investigator was replaced by the dilettante, and in that period many papers appeared describing dozens of new fungal pathogens (generally contaminants) and attributing a fungal etiology to everything from warts and acne to psoriasis and pemphigus.

In the early 1900's, this dermatologic and mycologic chaos was brought into order temporarily by a dermatologist who was also a scientist. Sabouraud's monumental work on the tineas cleared the air, and he set forth the principles of medical mycology: careful observation of the disease, critical evaluation of its etiology, and a cautious, unbiased approach to therapy. Soon after this, however, there was another period of decline and confusion. There were by then some 130 synonyms for *Candida albicans* and descriptions of more than 300 dermatophyte species in 40 genera. Beginning in the late 1930's, however, trained mycologists, such as Conant and Emmons, began working in the field and placed the nomenclature of the etiologic agent and evaluation of the disease syndrome on a scientific basis. Conant's lucid *Manual of Clinical Mycology* became the bible of the discipline and was later joined by the works of Emmons and Lacaz. These works clearly defined and described the pathogenic fungi as well as the diseases they produce and remain standard texts in the field.

Beginning in the early 1960's, a new aspect of medical mycology has emerged and gained increasing importance. This is the growing incidence of opportunistic fungous infections. To a large extent, this has paralleled advances in medicine involving the treatment of neoplasms, collagen diseases, and other debilitations as well as the progress made in organ transplantation. Such techniques prolong the life of the patient but abrogate normal host defenses and frequently allow saprophytic species to proliferate and invade, thereby causing disease.

In this volume I have generally approached the subject matter along traditional lines used in previous textbooks of medical mycology. In addition, I have tried to emphasize the growing importance of opportunistic infections and to delineate the differences between pathogenic and opportunistic infecting fungi and the diseases they elicit.

Another area of great importance in medical mycology is the increased basic research aimed at elucidating the physiology, biochemistry, and mechanisms of pathogenicity of the fungi in addition to the responses of the host when challenged by infection. An overview of some of this work is included in the appropriate chapters. This is not intended to be a review of all investigations in a particular field but only an indication of some areas in which active research is being carried out. Sections are also included concerning both natural and experimental infections in animals. I have tried to compare and contrast the diseases as they occur in man and animals, and in this way I hope that the text will be useful in the study of both human and veterinary mycology. In addition, there are chapters on other important aspects of medical mycology, such as the pharmacology of antimycotic agents, allergic diseases, intoxications, the genetics of pathogenic fungi, and taxonomy.

In choosing the references to be included in the various chapters, I have elected to pick those that illustrate two aspects of the subject. The first group includes the historically significant papers that first described a disease or some important facet of it. It has been very rewarding and sometimes amusing to read original literature on a particular disease. Papers up to 1960 that are still pertinent or whose content has not been updated by more recent investigations or reviews have been included. The second criterion for choice of a paper was its recentness and significance in the field. Most of the references cited have appeared since 1960. These papers not only include new information but also review previous publications and thus act as a bibliography for those interested in a particular subject.

I am profoundly and deeply grateful for the assistance of Dr. Josephine Morello. She has patiently transformed (translated) the language of my manuscript into readable English. This has been done after many hours of writing, rewriting, and debating, leading to deletion and compromise. The reader should be aware that if he finds this text at all readable it is because of her arduous efforts.

I also wish to thank my board of consultants, Libero Ajello, E. S. McDonough, Angela Restrepo, Shirley McMillen, and Howard Larsh, mycologists; Sharon Thomsen, Francis Straus, and Philip Graff, pathologists; Francisco Battistini and Allan L. Lorincz, dermatologists; John Fennessy, radiologist; and Nicholas J. Gross and Frederick C. Kittle, specialists in diseases of the chest. I would also like to thank Martha Berliner, P. Kulkavni, C. Satyanarayana, F. Mariet, S. Banerjee, and F. Pifano, among others, for the use of illustrations.

Special thanks go to my friend and associate in many research projects, Edward D. Garber, who has contributed the chapter on genetics and taxonomy.

The maps were prepared by Robert Williams and the drawings by Robert Williams, Charles Wellek, and John Rippon. A final thank you is given to Daila Shefner for proofreading this manuscript.

JOHN WILLARD RIPPON

Contents

Introduction to Medical Mycology

The ability of fungi to cause disease appears to be an accidental phenomenon. With the exception of a few dermatophytes, pathogenicity among the molds is not necessary for the maintenance of dissemination of the species. Further, the fungi that are able to cause disease seem to do so because of some peculiar trait of their metabolism not shared by taxonomically similar species. Thus the survival and growth of fungi at the elevated temperature of the body, the reduced oxidation-reduction environment of tissue, and the ability to overcome the host's defense mechanisms set apart these few species from the great numbers of saprophytic and plant pathogenic fungi.

Among the best examples of this transient adaptation to invasion and growth within tissue are the dimorphic fungi. In nature they grow as soil saprophytes, usually in a restricted ecologic niche, producing mycelium and conidia similar to other fungi. However, when their conidia are inhaled or gain entrance by other means into man or other animals, the organisms are able to adapt and grow in this unnatural environment. In so doing, drastic changes occur in their morphology, metabolism, cell wall content and structure, enzyme systems, and methods of reproduction. If the host's defenses are unable to counteract the organism — and this is rare — the infection leads to the death of both host and parasite. Infection for the fungus is a blind alley and is not contagious to other hosts or generally of use in disseminating the species. Such diseases include histoplasmosis, blastomycosis, coccidioidomycosis, and, to a degree, sporotrichosis. The previously mentioned also represent primary pathogens among the fungi in that they are able to cause disease in a normal healthy host, provided that sufficient numbers are present in the infecting dose. In debilitated hosts, their course of infection and disease is exaggerated.

The second major group of fungus diseases are the dermatophytoses. This is a closely related group of organisms with the ability to utilize keratin and to establish a kind of equilibrium, albeit transitory, with the host. In the soil are numerous related keratolytic species, most of which seldom, if ever, are recovered from clinical disease. Others, especially the anthropophilic species, are universal agents of ringworm and probably no longer have a significant soil reservoir, depending on human or fomite contact for transmission. These are among the commonest infectious agents of man. The dermatophytes are in a sense specialized saprophytes, as they do not invade living tissue, utilizing only the dead cornified appendages of the host, such as hair, skin, nails, fur, and feathers. Dermatophytosis may be considered a colonization of cornified structure by such organisms. The clinical disease caused by the dermatophytes is, for the most part, a toxic and allergic reaction by the host to the presence of the organisms and its by-products.

A group of diseases referred to as the subcutaneous mycoses include chromoblastomycosis, mycetoma, sporotrichosis, and entomophthoromycosis. These are organisms of limited invasive ability, gaining entrance to the body by traumatic implantation, and may take years to produce a noticeable disease, requiring that much time to adapt to the tissue environment. The clinical course is a chronic progressive one and very slow to develop. Mycetoma or madura foot and chromoblastomycosis are clinical entities

1

that may have a number of different species as etiologic agents. Mycetoma in particular may be caused by a diverse group of bacterial and fungal organisms that are totally unrelated to one another, but the clinical disease elicited is similar. The organisms are all soil saprophytes of regional epidemiology.

There remains a large category of opportunistic fungous infections. Included are diseases that are manifest almost exclusively in patients debilitated by some other cause and whose normal defense mechanisms are impaired. Formerly, some of these diseases and the fungal infections were in almost constant association such as mucormycosis with diabetes, candidiasis with hypoparathyroidism and other endocrine disturbances, and aspergillosis with chronic lung disease. In present-day medicine, the advent of cytotoxic drugs, long-term steroid treatment, and immunosuppressive agents has markedly increased the number and severity of diseases in this category. The diverse array of organisms being isolated from these cases emphasizes that probably all fungi may be considered potential pathogens when normal defenses are sufficiently abrogated. Fungi are particularly remarkable for their ability to adapt and propagate in a wide variety of environmental situations; thus their invasion of debilitated patients is not surprising.

Also included in this book will be a brief discussion of conditions arising from ingestion or inhalation of fungal products. These include allergic manifestations to conidia, such as farmer's lung and coniosporosis; toxic reactions due to consumption of infected food products, such as aflatoxin; and, of universal interest, mushrooms and mushroom poisoning.

HISTORY

Medical mycology is a science with traditions as any other specialty of human endeavor. For this reason, some bacterial diseases are included in this text. For many years actinomycetes were considered a "link" between bacteria and fungi. The diseases they produce were chronic granulomatous diseases similar to true mycotic infections. Morphologically, physiologically, and biochemically the actinomycetes are bacteria. Further, they are sensitive to antibacterial antibiotics to which fungi are insensitive, but not to antifungal drugs. Because of similarities in clinical disease, however, the pathogenic actino-

mycetes are treated along with eumycotic infections.

The discovery of the causal relation of certain fungi to infectious disease precedes by several years the pioneer work of Pasteur and Koch with pathogenic bacteria. Schoenlein and Gruby studied the fungus of the scalp infection favus (*Trichophyton schoenleinii*) in 1839, and in the same year Lagenbeck described the yeastlike organism of thrush (*Candida albicans*). Robert Remak had described favus earlier, but his work was ignored. Gruby even isolated the fungus of favus and produced the disease by inoculating a healthy subject, thus fulfilling Koch's postulates before Koch formulated them. Prior to this, Bassi described the fungal etiology of muscardine of silkworms (*Beauveria bassiana*).

In spite of its earlier beginnings, medical mycology was soon overshowed by bacteriology and has never received as much attention, though some of the fungous diseases (dermatophytoses) are among the more common infections of man. This is perhaps attributable to the relatively benign nature of the common mycoses, the rarity of the more serious ones, and the difficulty of differentiation of these structurally complex forms which, in a practical sense, sets them off sharply from the bacteria.

A great impetus was given the study of fungous diseases by the careful work of perhaps the most famous name in the field of medical mycology — Raymond Sabouraud. The publication of the classic work on dermatophyte infections, "Les Teignes," was a model of scientific observation. Unfortunately, his followers were not as careful, and the literature became cluttered with numerous synonyms for almost every fungus infection and with a fungal etiology for almost every human disease. There are over 100 names given for the yeastlike organism, *Candida albicans*. Some true fungous infections (histoplasmosis and coccidioidomycosis) were at first described as caused by protozoan parasites, but the works of Ophuls, Brumpt, Gilchrist, and Smith (most of them dermatologists) delineated their true nature and the extent of their epidemiology. A group of Latin American scientist-clinicians is responsible for a large portion of our present knowledge. This group includes González Ochoa, F. Almeida, Mackinnon, and others. The terrible confusion in nomenclature was finally brought into order by the work of the outstanding mycologists, Norman Conant and Chester Emmons. Present research is aimed toward improved diagnostic techniques, specific serologic tests for fungous diseases, accurate taxonomy, and new and improved chemotherapeutic agents. Moreover, efforts are being made to elucidate the mechanisms of fungal pathogenicity. The field of medical mycology is indeed fortunate, now, to have a number of competent investigators directing their attention to the problems of this discipline.

PATHOGENIC FUNGI

Today we recognize some 175 "pathogens" among the approximately 100,000 species of fungi — about 20 that may cause systemic infections, about 20 which are regularly isolated from cutaneous infections, and a dozen that are associated with severe, localized, subcutaneous disease. In addition, there is a long list of opportunistic organisms that may cause disease in the debilitated patient. A few of the diseases discussed in this book, e.g., actinomycosis, candidiasis, and pityriasis versicolor, are caused by endogenous organisms, i.e., species that are part of the normal flora of man; all other fungous and actinomycetous infections are exogenous in origin.

The criterion of pathogenicity is one of the poorest that can be used in differentiation of microorganisms. Because pathogenicity is variable and difficult to determine, by its use parasitic microorganisms are grouped together that are, in fact, much more closely related to certain free-living forms than they are to one another. The superficial nature of pathogenicity as a differential characteristic is nowhere better illustrated than among the fungi. The pathogenic forms which constitute the subject matter of medical mycology form a heterogeneous group that includes some of the actinomycetes, certain molds and moldlike fungi, and a number of yeasts and yeastlike organisms. As stated before, very few are primary pathogens, i.e., able to produce disease in a healthy host, and infection is a blind alley for the organism. With the possible exception of perhaps a half dozen anthropophilic dermatophytes, none of this group is an obligate parasite, and most are misplaced soil saprophytes. From a general biological point of view, then, the pathogenicity of certain fungi is of very minor significance; from the point of view of the parasitized host, man, it is of considerably greater interest. Thus, for lack of a better definition, a fungal pathogen will denote an organism

regularly isolated from a given disease process; rarely, isolated organisms from a variety of clinical circumstances will be considered opportunistic organisms (Table 1).

Mechanisms of Pathogenesis

The mechanisms of fungal pathogenesis have been the subject of considerable basic research. The two major physiologic barriers to fungal growth within tissue are temperature and redox potential. Most fungi are mesophilic and have an optimal growth range considerably below the temperature of the human body. Similarly, the majority of fungi are saprobic and their enzymatic pathways function more efficiently at the oxidation-reduction potential of nonliving substrates than at the relatively more reduced state of living metabolizing tissue. In addition, the body has a highly efficient set of cellular defenses to combat fungal proliferation. Thus, the mammalian species have remained rather free of obligate parasitism by the fungi throughout evolutionary history. Even the so-called "true pathogenic fungi" that are able to grow and proliferate at the temperature and redox potential of the human body are of relatively low virulence. Thousands of cases of infection or infestation by *Histoplasma* or *Coccidioides* occur each year, but only a few result in a frank disease process. It must be concluded that spontaneous resolution is the norm in such infections and that any disease elicited by any fungus is an "opportunistic infection"; some defense of the host was insufficient when challenged. For the fungi of even lesser virulence (*Cryptococcus, Aspergillus, Mucor,* and so forth), the defect in the host's defense must be even greater. These organisms, though thermotolerant, are less metabolically efficient in living tissue and thus require greater advantage (a greater host defect) in order to adjust to this environment and proliferate. Fungal species and strains that are not thermotolerant, cannot adjust to the tissue environment and cannot withstand even a debilitated host's defenses are thus unable to invade and cause disease. Conversely, the basic mechanism of fungal pathogenicity is its ability to adapt to the tissue environment and temperature and withstand the lytic activity of the host's cellular defenses.

A wide spectrum of adaptability exists among the various agents of the human mycoses. Within the group termed the "true

Table 1 *Clinical Types of Fungous Infections**

Type	*Disease*	*Causative Organism*
Superficial infections	Pityriasis versicolor	*Malassezia furfur*
	Piedra	*Trichosporon beigelii* (white)
		Piedraia hortae (black)
Cutaneous infections	Ringworm of scalp, glabrous skin, nails	Dermatophytes (*Microsporum* sp., *Trichophyton* sp., *Epidermophyton* sp.)
	Candidiasis of skin, mucous membranes, and nails	*Candida albicans* and related species
Subcutaneous infections	Chromoblastomycosis	*Fonsecaea pedrosoi* and related forms
	Mycotic mycetoma	*Pseudallescheria boydii. Madurella mycetomatis*, etc.
	Entomorphthoromycosis	*Basidiobolus ranarum*
		Conidiobolus coronatus
	Rhinosporidiosis	*Rhinosporidium seeberi*
	Lobomycosis	*Loboa loboi*
	Sporotrichosis	*Sporothrix schenckii*
Systemic infections	Pathogenic fungous infections	
	Histoplasmosis	*Histoplasma capsulatum*
	Blastomycosis	*Blastomyces dermatitidis*
	Paracoccidioidomycosis	*Paracoccidioides brasiliensis*
	Coccidioidomycosis	*Coccidioides immitis*
	Opportunistic fungous infections	
	Histoplasmosis	*Cryptococcus neoformans*
	Aspergillosis	*Aspergillus fumigatus*, etc.
	Mucormycosis	*Mucor* sp., *Absidia* sp., *Rhizopus* sp., *Rhizomucor* sp.
	Candidiasis (systemic)	*Candida albicans*
	Pseudallescheriasis	*Pseudallescheria boydii*
Miscellaneous and rare mycoses and algosis	Phaeohyphomycosis	*Wangiella dermatitidis, Phialophora* sp., etc.
	Hyphomycosis	
	Algosis	*Paecilomyces* sp., *Beauveria* sp., *Scopulariopsis* sp. etc.
	Basidiomycosis	*Prototheca* sp.
		Schizophyllum commune, Coprinus sp., etc.

pathogenic" fungi, the ability to adapt to a tissue environment is quite marked and is expressed as thermal dimorphism. The organisms grow as mycelial fungi producing conidia at room temperature, but at 37°C they are transformed to a completely different morphologic state, a so-called parasitic phase. In three of the true pathogenic fungi, budding, yeastlike cells are formed (*Histoplasma, Blastomyces,* and *Paracoccidioides*); in the fourth *(Coccidioides),* an endosporulating spherule is formed. In the parasitic state the metabolic rate of the organism increases several fold and different sets of metabolic pathways are favored. This results in a rapidly growing, rapidly multiplying organism whose cell wall structure, carbohydrate content, lipid composition, and RNA aggregates are completely different from those of the organism growing at lower temperatures. These new entities are quite susceptible to phagocytosis and killing by mammalian macrophages, however. Therefore most infections are aborted and resolve spontaneously. Blastomycosis may be an exception to this.

As already noted, the other group of systemic infecting agents, the "opportunists," require a greater defect in the normal defenses to become established. They do not demonstrate thermal dimorphism; they are temperature tolerate, however, and strains isolated from human infection are metabolically more active at 37°C and in the redox potential of human tissue than are isolants of the same species from soil.

A gamut of adaptability is also seen among the agents of the subcutaneous mycoses. These agents are of low virulence capacity and require mechanical introduction of the organism into tissue (traumatic implantation) before they can manifest a disease process. In *Sporothrix schenckii,* the most virulent of the low virulent organisms, transformation to a yeastlike stage accompanies this adaptation to tissue growth, another case of thermal dimorphism. With the agents of chromoblastomycosis and mycetoma, much time is required for this adaptation, and a different morphological form is finally produced. This is the planate-dividing "sclerotic cell" in chromoblastomycosis and the microcolonies massing to form grains (presumably resistant to the action of phagocytic cells) in mycetoma.

The disease process in dermatophytosis is completely different. Two important considerations are apparent here. The first is that living tissue is not invaded; the organism is simply colonizing the keratinized *stratum corneum* because of its keratinolytic ability. The "disease" is the result of the reaction of the host to percutaneous absorption of the metabolic products of the fungus. The second important factor is that in this group are the only fungi that have evolved a dependency on human or animal infection for the survival and dissemination of the species. Thus there are anthropophilic and zoophilic as well as soil inhabiting dermatophytes. The anthropophilic species have evolved into a life cycle in which the fungus elicits the least possible irritation or reaction when it is grown on its particular host (man). This ensures a longer infection and increases the chance of producing more conidia or infectious units that can be transferred to a new host. The same is true of zoophilic dermatophytes and their particular hosts. When a zoophilic species infects a human host, however, a much more severe disease is produced. In general this inflammatory reaction quickly aborts the infection. Some anthropophilic species have evolved a particular association with particular population groups or races of man. The mechanism of this "host recognition" in dermatophytes is unknown.

The entire field of medical mycology offers a fascinating study in the interplay of the prospective host and the prospective parasite. A few species, such as the anthropophilic dermatophytes, are quite successful as parasites. In the others we see varying degrees of success in trying to utilize the tissue environment of man and animals as substrate for growth.

The fungi are structurally complex, showing a bewildering variety of reproductive structures associated with sexual and asexual processes, in addition to vegetative nonreproductive elements and hyphal structures. Their differentiation into genera, species, and varieties is made, in large part, on a morphologic basis—especially the morphology of the reproductive structures. In contrast to bacteria, physiologic and immunologic characteristics of fungi are usually of minor or no importance for purposes of differentiation or identification (except in

yeasts). The biochemistry of the fungi has been extensively investigated, with the elucidation of decomposition occurring in nature, as well as the application of fungi to industrial processes, e.g., antibiotic production, alcohol and citric acid formation, and so on. Present knowledge of the respiratory mechanisms of the cell as well as of catabolic and anabolic pathways of metabolism is derived in large part from the studies of yeasts. However, identification of the fungi is still based on the morphology of the organism and demands contemplative observation.

IMMUNITY AND SEROLOGY

Natural immunity to fungous diseases is very high. Infection therefore depends on exposure to a sufficient inoculum size of the organism, and the general resistance of the host. This is well demonstrated by the two general categories of systemic fungous diseases. In the first, infection occurs in the patient who is in a particular endemic area and who inhales sufficient numbers of conidia. The majority of such infections are asymptomatic or resolve quickly. This is followed by a specific resistance to reinfection. Only rarely does the infection become serious. On the other hand, opportunistic infections are caused by universally present organisms of very low virulence. Establishment of disease depends entirely on lowered resistance of the host. If the patient recovers from such an infection, no specific resistance is noted.

As far as can be ascertained at present, humoral antibodies play little or no part in our ability to contain a fungous infection. Cellular defenses appear to be the only efficacious resistance to invasion by fungi. This is well illustrated by the types of infections exhibited by patients with the various forms of lymphomatous disease or with genetic defects of leukocytic function. In patients who have defects or dyscrasias of their T-cell lymphocytes, opportunistic fungous infections are common. If the defect is only in the B cells, fungous disease is uncommon, whereas bacterial infections, particularly those caused by grampositive cocci, are frequent. The importance of cellular defense mechanisms is also reflected in the pathologic manifestations evoked by infection by the organism. The pathogenic fungi in the normal host induce a pyogenic reaction, followed by a granulomatous reaction. The response elicited by invasion by opportunistic fungi is necrotic and suppurative. The host is deficient and cannot contain the organism. As stated before, all fungous infections are opportunistic. The normal, healthy, well-nourished adult is resistant to all fungi, except in cases of overwhelming exposure. Slight debilitation, in some cases transient, may afford the opportunity for establishment of disease. These subtle differences in host defenses are a fascinating area of research that is just now being exploited. Such factors, and they appear to be numerous, determine who will have a fungous infection, either pathogenic or opportunistic, the severity of the disease, and the final outcome of the host-parasite interaction. Though some differences in the virulence of different strains of fungi have been found, these appear to be of minor importance.

Fungi serve as poor antigens. For this rea-

son, in most cases, serology is often of little importance as a diagnostic or prognostic aid. Complement-fixing antibodies, precipitins, and so forth, when present, may be used to evaluate the status of disease in a few cases. However, there are many cross-reactions, conflicting opinions on interpretation, and no standardization of antigens, so that the usefulness of these procedures is limited. Standardization of antigens and the use of immunodiffusion procedures in a few diseases are the subjects of much research presently and offer many advantages over the older complement-fixation procedures. Specific fluorescent antibody procedures are also useful in most diseases.

Hypersensitivity and allergy to fungi are manifested in various ways. Frank allergy may simply be due to inhalation of the conidia of fungi. This may induce asthma, asthmalike

Table 2 *Tissue Reactions in Fungous Diseases**

Chronic inflammatory reactions
  Lymphocytes, plasma cells, neutrophils, and fibroblasts, occasionally giant cells
  Rhinosporidium seeberi
  Entomophthoromycosis

Pyogenic reactions
  Acute or chronic, suppurative neutrophilic infiltrate
  Actinomyces israelii: sulphur granules, also lipid-laden peripheral histiocytes
  Nocardia asteroides
  Acute aspergillosis
  Acute candidiasis

Mixed pyogenic and granulomatous reactions
  Neutrophilic infiltration and granulomatous reaction, lymphocytes, plasma cells
  Blastomyces dermatitidis
  Paracoccidioides brasiliensis
  Coccidioides immitis: neutrophils, especially at broken spherule
  Sporothrix schenckii: organism rarely seen in tissue
  Chromoblastomycosis: chronic pyogenic inflammation; epithelioid cell nodules and giant cells
  Mycetoma; in addition, may be large foamy giant cells similar to xanthoma

Pseudoepitheliomatous hyperplasia
  Following chronic inflammation in skin (hyperplasia of epidermal cells, hyperkeratosis, elongation of rete ridges)
  Blastomyces dermatitidis
  Paracoccidioides brasiliensis
  Chromoblastomycosis
  Coccidioides immitis

Histiocytic granuloma
  Histiocytes frequently with intracellular organisms, sometimes becoming multinucleate giant cells
  Histoplasma capsulatum
  Meningeal *Cryptococcus neoformans*

Granuloma with caseation
  Granulomatous reaction, Langhans' giant cells (L.G.C.), central necrosis
  Histoplasma capsulatum
  Coccidioides immitis

Granuloma "sarcoid" type
  Nonnecrotizing
  Cryptococcus neoformans
  Occasionally *Histoplasma capsulatum*

Fibrocaseous pulmonary granuloma; "tuberculoma"
  Histoplasma capsulatum: thick fibrous wall surrounding epithelioid and L.G.C. organisms in soft center, often calcification
  Coccidioides immitis: Thin fibrous wall, rarely calcified
  Cryptococcus neoformans: poorly defined

Thrombotic arteritis
  Thrombosis, purulent coagulative necrosis, invasion of vessels
  Aspergillosis
  Mucormycosis

Fibrosis
  Proliferating fibroblasts, deposition of collagen; may resemble keloid
  Loboa loboi

Sclerosing foreign body granuloma
  In paranasal sinuses or following viral infection
  Aspergillus sp., bizarre hyphae in giant cells

*A Gram stain is used for actinomycosis, nocardiosis, actinomycotic mycetoma, and candidiasis; otherwise a periodic acid-Schiff (PAS) stain is recommended. Methenamine silver stains are useful for both actinomycotic and eumycotic organisms.

diseases, and even fibrosis and consolidation of lungs. Allergies may also manifest as "ids" or eruptions associated with cutaneous infections due to dermatophytes or *Candida* sp. Allergic reactions, such as erythema nodosum, may be part of the primary infection of a systemic disease, as in coccidioidomycosis and histoplasmosis. Hypersensitivity, as demonstrated by a positive skin reaction (e.g., to coccidioidin or histoplasmin), is often associated with good resistance to infection or reinfection.

PATHOLOGY

The tissue response of the host to the offending agent varies widely with the variety of organism. In dermatophyte infections, erythema is generally produced and is probably a response to the irritation caused by the organism or by its products of metabolism. Occasionally, severe inflammation, followed by scar tissue and keloid formation, will occur. In organisms that invade living tissue, such as those responsible for subcutaneous and systemic disease, a rather uniform pyogenic response is generally elicited. This usually gives way to a variety of chronic disease processes, which are listed in Table 2.

THERAPY

In general, antibacterial antibiotics are not effective in treating fungous diseases, and the few widely used antifungal agents are similarly ineffective against bacteria. For this reason it is very important to establish a diagnosis of the fungal agent early in the course of the disease. It is not unusual to find patients on long-term antitubercular therapy, when, at last, it was found that a fungus was responsible for their disease.

In the preantibiotic era, about the only treatment for systemic mycoses was supportive therapy of the patient. Clinical cures can now be obtained for most of these mycotic diseases. Cutaneous infections were treated with various combinations of keratolytic and antimicrobial agents, such as Castellani's paint and Whitfield's ointment. Modifications of these are still useful today in dermatologic practice. Sulfur ointments and salicylic acid remain the treatments of choice for some of these diseases. However, a variety of new specific drugs are now useful for treatment of these infections.

Only a few useful antifungal agents are presently available for the therapy of fungous diseases. They generally have the twin drawbacks of being insoluble or only partially soluble in water and variably toxic in therapeutic dosage. One group of such substances, formed by various species of *Streptomyces,* is the polyene antibiotics, which are effective against certain pathogenic fungi. They have the general structure $(CH{=}CH)_n$, and of the dozen or so which have been described, nystatin and amphotericin B have proved to be the most useful. Nystatin is a diene-tetraene polyene derived from *Streptomyces noursei,* and amphotericin B is a heptene derived from *S. nodosus.* Both contain the diaminomethyl deoxypentose and mycosamine. Nystatin also contains a lactone structure. Both appear to act against the cell membrane of the fungus, causing leakage of potassium and other metabolites. Both are relatively insoluble in water or saline. Nystatin is effective in topical *Candida* infections and occasionally in keratitis and otomycosis of other fungal origin, but not in deep mycoses or dermatophyte infections. Amphotericin B is effective against the deep mycoses, e.g., blastomycosis, cryptococcosis, histoplasmosis, coccidioidomycosis, and systemic candidiasis, but its usefulness is somewhat limited by toxicity. However, this agent, given in dextrose solution to prevent precipitation, offers about the only effective treatment for these diseases. New drugs, such as pimaricin from *S. natalensis,* are now receiving clinical trials. Pimaricin has now become established as the drug of choice in mycotic keratitis.

A second group of antifungal agents that have received considerable attention recently are the imidazoles. These include miconazole, clotrimazole, and ketoconazole. They are regularly used as topical agents and have the advantage of being broad spectrum (i.e., antifungal, antiyeast and antibacterial), although some skin sensitizing problems do exist. Many problems have occurred, however, when these drugs have been used systemically; these include high relapse rate (miconazole) and high level of toxicity (clotrimazole). Ketoconazole is now undergoing clinical trials. Miconazole, though not useful in all systemic infections, appears to be the drug of choice now in paracoccidioidomycosis, pseudallescheriasis, and perhaps infections by dematiaceous fungi.

Other new drugs for specific diseases are

regularly appearing, such as 5-fluorocytosine for cryptococcosis and chromoblastomycosis, pimaricin for keratitis, and so forth; these will be discussed in separate chapters. Potassium iodide, an old treatment for deep mycoses, is still the treatment of choice for sporotrichosis and entomophthoromycosis.

Another widely used antibiotic, griseofulvin, produced by *Penicillium griseofulvum,* is an effective chemotherapeutic agent in the dermatophytoses, but not in the deep mycoses. It has been found to be 7-chloro-2′,4,6-trimethyoxy-6′-methylspirobenzofuran-2 (3HO, 1′-[2]-cyclohexane)-3,4′-dione. It may be applied topically but is of questionable efficacy by this route. When given orally, it is absorbed and later concentrated in the keratin-containing structures. It causes abnormal multiple branching of the fungus and interferes with protein and nucleic acid synthesis. The latter probably is the most important mode of action. This antibiotic was isolated in 1939 by Raistrick but was of no interest at that time because it had no antibacterial activity. It was used for the treatment of some fungous infections of plants, but its chemotherapeutic activity in mammalian dermatophyte infection was not exploited until the work of Gentles and Blank and Roth in 1959, as discussed in the chapters on dermatophytes and pharmacology. Since then, it has been increasingly more widely and successfully used for the treatment of the dermatophytoses. Again, its usefulness is limited by solubility problems and difficulty of absorption. The latter is partially alleviated by administration following a fatty meal and use of the ultramicronized product. Long periods of treatment are often necessary with this drug, and the relapse rate is very high. Another antidermatophyte drug that has gained wide acceptance is tolnaftate. It is used topically and sometimes combined with orally administered griseofulvin. It is not effective in hair and nail infections. Haloprogin and thiabendazole are also useful in some types of dermatophytic infections. Dihydroxystilbamidine is a drug that was widely used some years ago for a variety of fungous infections. Today it still has a definite place in the treatment of some forms of blastomycosis but otherwise has been replaced by more effective therapeutic agents.

A great deal of research is presently underway to develop more effective and less toxic antifungal agents. Because they are bacteria, the pathogenic actinomycetes are susceptible to antibacterial chemotherapy.

LABORATORY MYCOLOGY

The isolation of fungi is a relatively easy procedure; the identification and determination of significance are much more difficult. Unlike pathogenic bacteria, fungi are not nutritionally fastidious, literally growing on wet cement. A simple source of organic nitrogen and carbohydrate will suffice. The most commonly employed medium in medical mycology is Sabouraud's agar. This contains beef extract and dextrose at a pH of 5.6, which discourages bacterial growth. It is not necessarily true that fungi grow best on this media (they do not), but it is traditional to use it. Further, the colonial morphology noted, studied, described, and taught is that of the fungus growing on Sabouraud's media. Colony morphology varies greatly with the medium on which the organism is grown. Therefore, it is recommended that this media continue to be used for standardized gross morphologic appraisal. In present-day use, the medium contains added antimicrobial agents to make it more selective for the common fungal pathogens. Antibacterial antibiotics, such as Aureomycin or Chloromycetin to reduce bacterial overgrowth and Actidione (cycloheximide) to retard the growth of many nonpathogenic fungi, are added. The latter antibiotic, derived from *Streptomyces griseus,* is a general inhibitor of eukaryotic protein synthesis, having no effect on prokaryotic cells. For some unknown reason (probably relative uptake), most pathogenic fungi, dermatophytes, and deep-infecting fungi are not inhibited by it, though most soil saprophytes, such as *Penicillium, Aspergillus,* and *Alternaria,* are inhibited by it. Some pathogenic fungi (such as *Cryptococcus,* for which it was once used in therapy) are also inhibited, however, and antibiotic-free media must be used when these are suspected. The opportunistic fungal pathogens, such as *Aspergillus, Mucor,* and *Fusarium,* are also sensitive to Actidione. Since the antibacterial antibiotics inhibit actinomycetes, nonselective bacteriologic media must be used in their isolation.

Clinical specimens of all types may be planted on the media and incubated for growth. Since most fungi grow better at room temperature, this is the temperature most commonly utilized for primary isolation. Generally, tubes should be observed from four to six weeks or more, because most fungal pathogens grow very slowly.

The identification of the isolated fungus is

an exercise in contemplative observation. Classification in mycology is based on morphologic attributes of the fungus. Only familiarity by repeated study of the various organisms will lead to confidence in identification. There is no easy way. The procedure for identification is to accumulate the set of characteristics of an unknown and then determine its identity. This involves (1) the gross morphology of the colony or thallus (the obverse), its color, texture, topography, and rate of growth; (2) the reverse, for presence or absence of characteristic pigment; and (3) the microscopic morphology, its size, shape, topography and arrangement of conidia, types of hyphal appendages, and hyphal modifications. In addition to these, one must be familiar with strain variations that are common and numerous, and variations influenced by culture media, age of growth, temperature, and perhaps the phase of the moon.

The handling of the isolated growth for identification differs depending on the category of disease from which the specimen is obtained. Dermatophytes are planted on slide cultures, as described in Part Five, to observe their microscopic morphology. Scraping the culture for a wet mount is to be discouraged, as conidia arrangement is an important characteristic for identification. Yeasts are planted on special medium for conidia production, assimilation, and fermentation profiles. Growth from deep-infecting fungi in the mycelial stage is handled with great care, as the conidia are infectious. *Histoplasma* and *Coccidioides* are quite dangerous and should be examined and manipulated only in a sterile hood or biological safety cabinet. The yeast or parasitic stages of dimorphic fungi are not infectious and may be examined with the usual precautions.

After the organism has been isolated and identified, its significance must be determined. Of course, a single colony of a known pathogen, such as *Histoplasma, Blastomyces,* and *Coccidioides,* is to be considered significant. With other organisms, significance may depend on where they were isolated. In skin lesions or internal organs, the presence of *Candida albicans* may be significant; in the buccal cavity, rectum, or vagina, however, it may represent normal flora. In the latter instances, the numbers of the organisms isolated are important. In the case of opportunistic infections, such as *Aspergillus, Fusarium,* and *Mucor,* the significance of mere presence is much greater. Any fungus isolated from a sterilely obtained specimen from a closed body cavity or organ should be regarded as a possible etiologic agent of disease. On the other hand, a single colony of *Aspergillus* from a sputum culture, wound swab, or other nonsterile specimen usually represents a contaminant. Here again, numbers isolated are important. If the same organism is repeatedly isolated and in relatively large numbers, then the organism should be considered as an etiologic agent of disease. This is particularly true in debilitated patients in whom infection with opportunistic fungi, such as *Aspergillus* and *Mucor,* is increasingly common.

There are three other procedures which may be used as adjuncts to the diagnosis of disease of fungal origin. The first is demonstration of the organism in biopsy or tissue sections. Presently many specific stains exist for the histopathologic examination and demonstration of fungi. The staining is often difficult to carry out, and control slides must always be run. The presence of a few fungal organisms in tissue is most easily seen in the Gomori-Grocott chromic acid methenamine silver method. Detail of structure for ease of identification is best seen using the Gridley or Hotchkiss-McManus modification of the periodic acid-Schiff stain. These stains are useful for fungi but do not generally stain actinomycetes. Demonstration of *Nocardia, Streptomyces, Actinomyces,* and so forth, since they are bacteria, is easy using the Brown-Brenn modification of the Gram stain or the methenamine stain.

Serologic procedures are often very useful in diagnosis and prognosis of patients with some of the systemic fungus diseases. They are of little value in dermatophyte infection or subcutaneous infections. In systemic infections, precipitin titers usually appear first, followed by complement-fixing titers. The disappearance or persistence of the latter is correlated with the progress of the patient, and serial tests are very useful in determining prognosis. These procedures are somewhat complicated and are best done by an experienced serologist. Much expertise and special equipment are also required for application of the fluorescent antibody techniques. These may be very useful for the demonstration of specific fungi in tissue, sputum, and exudate and may be the only procedures to establish a diagnosis in the event cultures fail or were not taken.

A relatively simple procedure coming into greater use in the laboratory is immunodiffusion. Through the use of an Ouchterlony plate, patients' sera are placed in one well and antigen in the other. The presence or absence of precipitin lines and the numbers of lines may indicate the etiologic agent. In some diseases, precipitin patterns have been correlated with presence or absence of active disease in contrast to past infection. As more specific antigens are produced, this procedure will become increasingly useful in the diagnosis and prognosis of fungal diseases.

Animal inoculation is sometimes helpful in establishing an etiology. Often clinical material will fail to grow on culture and stains may not show the organism, but experimental animals will become infected, and the etiology will be established. Animals are particularly useful in the survey of soils for the presence of pathogenic fungi. In present-day laboratory practice with the advent of newer cultural, serologic, and histologic techniques, animal inoculation is less often used and is reserved for particularly difficult cases.

CLINICAL MYCOLOGY

As previously outlined, the isolation and identification of fungal pathogens is a long, arduous, time-consuming procedure. There are a few simple diagnostic aids that may be done in the examining room or laboratory which will help to establish a fungal etiology for a disease process more rapidly.

For many years in dermatology the potassium hydroxide slide for examining skin scales has been used. This rapid procedure will demonstrate the presence of mycelial elements of dermatophytes and establish a diagnosis. It is also useful to visualize the short, hyphal strands and yeast forms of pityriasis versicolor and yeast and mycelial elements of cutaneous candidiasis. In addition, this procedure has many applications with other infections. Aspirates of the micropustules in a blastomycosis lesion, treated with KOH and examined, will show the large yeast cells and establish a diagnosis. Cultures may require four to six weeks or more to grow. Spherules in coccidiomycosis may also be demonstrated by this method. Concentrated or unconcentrated sputum, wound scraping, skin scraping, biopsy material, and specimens of all sorts can be examined by the KOH slide method, and often fungal elements can be seen. This procedure usually will not establish a specific etiology but only indicate the category of etiologic agent. In mycetoma, in which either a bacterial or mycotic etiology may be present, an examination of sinus tract exudate will usually reveal whether the infection is actinomycotic or eumycotic, and appropriate therapy may be instituted.

Another device that is useful in the diagnosis of cutaneous infections of fungal origin is the Wood's lamp. With a peak of 3650 Å, the lamp will demonstrate a characteristic fluorescence produced by some microbial agents. Classically, this procedure was used to demonstrate hair infection by *Microsporum* species in children. Surveys of large populations could be done rapidly and easily. The infected hairs show a bright green fluorescence. Today this is less useful because of the large increase in nonfluorescing tinea capitis due to *Trichophyton tonsurans*. The Wood's light will also show a gold fluorescence in areas of pityriasis versicolor; a bright green in toe webs, lesions, and wounds infected by *Pseudomonas;* and pink in erythrasma.

Specific diagnosis of fungal disease requires the isolation and identification of the agent. The previously noted procedures are only useful in indicating a possible etiology.

STANDARD REFERENCES

TEXTBOOKS ON MEDICAL MYCOLOGY

Conant, N. H., D. T. Smith, et al. 1971. *Manual of Clinical Mycology.* 3rd ed. Philadelphia, W. B. Saunders Company.

Emmons, C. W., C. H. Binford, J. P. Utz, and K. J. Kwon-Chung. 1977. *Medical Mycology.* 3rd ed. Philadelphia, Lea and Febiger.

Lacaz, Carlos da Silva, 1977. *Micologia Medica.* Fungos, Actinomicitose Algas de Interesse Médico Sao Paulo, Brazil, Sarvier.

LABORATORY MANUALS ON MEDICAL MYCOLOGY

Haley, L. D., and C. S. Callaway. 1978. Laboratory methods in Medical Mycology. 4th ed. U.S. Dept. Health, Education, and Welfare. HEW Publication No. (CDC) 78–8361. Washington, D.C., U.S. Govt. Printing Office.

Hazen, E. L., M. A. Gordon, et al. 1970. Laboratory Identification of Pathogenic Fungi, Simplified. 3rd ed. Springfield, Ill., Charles C Thomas.

Beneke, E. S., and A. Rogers. 1980. Medical Mycology Laboratory Manual. 4th ed. Minneapolis, Minn., Burgess Publishing Company.

Rebell, G., and D. Taplin. 1970. Dermatophytes, Their Recognition and Identification. 2nd ed. Coral Gables, Fla., University of Miami Press.

McGinnis, M. R. 1980. Laboratory Handbook of Medical Mycology. New York, Academic Press.

TEXTBOOKS ON GENERAL MYCOLOGY

Alexopoulos, C. J., and C. W. Mims. 1979. Introductory Mycology. 3rd ed. London, John Wiley and Sons.

Bessey, E. A. 1961. Morphology and Taxonomy of Fungi. New York, Hafner Publishing Company.

Burnett, J. J. 1968. Fundamentals of Mycology. New York, St. Martin's Press.

Ainsworth, G. C., and A. S. Sussman. 1966. The Fungi. Vols. 1–3, 4A, and 4B. New York, Academic Press.

DICTIONARIES

Ainsworth, G. C. 1971. Dictionary of the Fungi. Kew, Surrey, England, Commonwealth Mycological Institute.

Snell, W. H., and E. A. Dick. 1971. A glossary of Mycology. Cambridge, Harvard University Press.

REVIEWS AND ABSTRACTS OF CURRENT LITERATURE

Review of Medical and Veterinary Mycology. Kew, Surrey, England, Commonwealth Mycological Institute.

The Biology of Actinomycetes and Related Organisms. A quarterly review.

PATHOLOGY OF THE MYCOSES

Salfelder, K., and J. Schwarz. 1979. Atlas of Deep Mycoses. Philadelphia, W. B. Saunders Company.

Chandler, F. W., W. Kaplan, and L. Ajello. 1980. A color Atlas and Textbook of the Histopathology of Mycotic Diseases. London, Wolfe Medical Publishers.

1 Introduction to the Pathogenic Actinomycetes

The human pathogenic actinomycetes are so-called "higher" bacteria and are classified in the order Actinomycetales, allied to the coryneform group of bacteria (Part 17 of *Bergey's Manual*[11]), of the Kingdom Monera. This order includes some chronic, disease-producing organisms, such as the etiologic agents of tuberculosis and Hansen's disease (leprosy). By tradition, these two organisms have been studied along with other bacteria. The remaining pathogenic actinomycetes, however, were thought to be transitional forms between bacteria and fungi and were included in the sphere of medical mycology. The etiologic agents of lumpy jaw and actinomycotic mycetoma show some fungus-like characteristics, such as the branching of the organism in tissue, the formation of an extensive mycelial network that may occur in tissue or in culture, and the production of chronic disease states. However, cell wall analysis shows the presence of the characteristic bacterial muramic acid which, along with the lack of a membrane-bound nucleus, lack of mitochondria, typical bacterial size, and sensitivity to antibacterial antibiotics, defines these organisms as bacteria and not fungi. As far as their role as a phylogenetic "link" to the fungi is concerned, the presence in fungi of eukaryotic nuclei and mitochondria, and fungal conformity to Mendelian genetics make their derivation from prokaryotic bacteria, independent of other eukaryotic organisms, extremely improbable.

MORPHOLOGY

The actinomycetes grow in the form of fine, straight, or wavy filaments, 0.5 to 0.8 μ in diameter. These hyphae show both lateral and dichotomous branching and in some groups may grow out from the medium to form an aerial mycelium. On solid media, the filaments occur in tangled masses, while in liquid media and in tissue, there is a tendency for them to grow in clusters of radiating dendritic clumps, which are sometimes lobulated. There are several genera of particular medical interest: the anaerobic *Actinomyces*, *Bifidobacterium*, *Arachnia*, and *Rothia* and the aerobic forms *Nocardia*, *Actinomadura*, *Dermatophilus*, and *Streptomyces* (Table 1–1). Classification of species within the genera and even separation of the genera themselves are controversial.[10] There has been a constant flux of ideas and principles concerning the nomenclature of actinomycetes in the last few years, particularly with the application of numerical taxonomy. These changes are presented to an international committee on nomenclature that studies the proposals and publishes their findings in a newsletter. Alas!, no such orderly review of taxonomic changes exists for the field of the true fungi. The following characteristics are generally accepted by workers in the field of actinomycete taxonomy.

In the current nomenclature, nocardioform bacteria are defined as those that re-

Table 1-1A *Brief Key to the Actinomycetales of Medical and Veterinary Importance**

I. Mycelium not formed. Cell may be branched, rodlike, diphtheroid, or coccoid. No spores formed.

Family 1. *Actinomycetaceae*
Not acid-fast; usually anaerobic, facultatively anaerobic, or aerobic; usually without diaminopimelic acid (DAP). *Actinomyces, Rothia, Arachnia.*

Family 2. *Mycobacteriaceae*
Acid-fast at some stage of growth, cell wall type IV. *Mycobacterium.*

II. True mycelium produced.

Family 3. *Dermatophilaceae*
Mycelial filaments divide transversely and in at least two longitudinal planes to form masses of coccal motile cells. Aerial mycelium usually absent. Cell wall type III. *Dermatophilus.*

Family 5. *Nocardiaceae*
Mycelial elements commonly fragment to give coccoid or elongate spores, usually nonmotile. Aerial spores usually absent. Cell wall type IV. Sometimes acid-fast. *Nocardia, Rhodococcus.*

Family 6. *Thermomonosporaceae*
Mycelial elements fragment. Spore formed within a sheath on aerial or substrate mycelium. Cell wall type III. *Actinomadura.*

Family 7. *Streptomycetaceae*
Mycelial elements not commonly fragmenting. Aerial mycelium abundant and long spore chains formed (5 to 50 or more spores). Cell wall type I. *Streptomyces.*

Family 8. *Micromonosporaceae*
Mycelial elements not fragmenting. Spores formed singly in pairs or short chains on either (or both) submerged or aerial mycelium. Cell wall type II, III, or IV. *Micromonospora.*

---

*Families 4 and 9 do not so far contain organisms involved in medical or veterinary diseases. At the end of the description of the other families is a list of some of the genera included in that family that contain species of medical or veterinary interest. Family 9 Thermoactinomycetaceae contains species noted as allergens (see page 27).

Table 1-1B *The Pathogenic Actinomycetes, Related Organisms, and Their Diseases*

Disease	Organism	Geographic Distribution
Actinomycosis	*Actinomyces israelii* (man) *A. bovis* (cattle) *A. viscosis* *A. naeslundii* *Bifidobacterium eriksonii* *Arachnia propionica*	Endogenous in animals and man
Nocardiosis (pulmonary and systemic)	*Nocardia asteroides* *N. brasiliensis* *N. caviae*	Ubiquitous Mexico, South America, Africa, India, U.S. Ubiquitous
Mycetoma (actinomycotic)	*Actinomadura madurae* *A. pelletieri* *Nocardia caviae* *N. brasiliensis* *Streptomyces somaliensis*	Ubiquitous Africa, South America Ubiquitous Ubiquitous Africa, South America, United States, Arabia
Erythrasma	*Corynebacterium minutissimum*	Ubiquitous
Cracked heel	*Actinomyces keratolytica* (?)	India, United States
Trichomycosis axillaris	*Corynebacterium tenuis*	Ubiquitous
Epidemic eczema	*Dermatophilus congolensis*	Australia, Africa, United States

produce by fragmentation of any or all parts of their colonial structure into bacilli or coccoid elements.[19, 20] Included in this group are 7 families and 20 genera, such as *Actinomyces, Nocardia, Streptomyces,* and *Dermatophilus*—the commonly encountered actinomycotic agents of disease. The remaining 3 families and 10 genera of the order Actinomycetales contain organisms that form spores on a specific structure or part of the mycelium. These include *Actinoplanes* and *Ampullariella,* among others, few of which are of any medical interest.

Actinomyces includes organisms that are anaerobic or microaerophilic and nonacid-fast and in which the vegetative mycelium breaks up into bacillary or coccoid elements. The *Nocardia* are aerobic and sometimes partially acid-fast. They may fragment into bacillary or coccoid forms and produce chains of squared spores 1 to 2 μ long by simple fragmentation of hyphal branches. Segmentation of the filaments occurs in some species as early as 24 hours, while in others it is delayed three weeks or more; the segmented filaments fragment to form bacillary forms, 4 to 6 μ in length, which are morphologically indistinguishable from many other bacteria. In most smear preparations of the pathogenic forms, the filaments are broken up, and the appearance is that of ordinary bacilli. In stained smears of tissue or sputum, *Actinomyces* tend to be very small bacilli with one or two short bifurcations, while *Nocardia,* particularly *N. asteroides,* may show long sinuous threads with occasional branches, in addition to bacillary forms. Grains, or granules that are microcolonies, are commonly found in visceral actinomycosis, in visceral and subcutaneous infections of *N. brasiliensis,* and in subcutaneous infections of *Nocardia* species other than *N. asteroides. Nocardia asteroides* does not form grains in visceral infections and does so rarely, if at all, in subcutaneous disease.

The genus *Rhodococcus* (rhodochrous taxon)[18, 43] is intermediate between the genus *Nocardia* and the genus *Mycobacterium.* They are gram-positive, aerobic, partially acid-fast organisms and form yellow to orange to red colonies. These may be confused with *Nocardia* species, particularly *asteroides,* although the *Rhodococcus* are usually soft and mucoid. They are differentiated by the ability to degrade ethylene glycol in 7H-10 medium. They have been responsible for pulmonary and systemic infections in compromised hosts and are "borderline" pathogenic for experi-

mental animals.[21] *Dermatophilus congolensis* is regularly encountered as a pathogen of sheep and other animals. It produces muriform mycelium within which cells form a sporangium and later produce motile spores. Other genera such as *Øerskovia* and *Micropolyspora* are rarely encountered opportunists.[38]

In *Streptomyces* there is generally more aerial mycelium, no fragmentation of the mycelium into bacillary or coccoid forms, and no acid-fastness of mycelial elements. Chains of round-to-oval spores are produced consecutively within an aerial unit. These spores are partially acid-fast. The maturation of the spore-bearing filament is often associated with the formation of spirals; the coils of these spiral filaments may range in shape from being open and barely perceptible to being so compressed that adjacent turns are in contact. The spirals may be dextrose or sinistrose; both direction and tightness of coiling are constant within species. The spores are more resistant than the filaments and will survive at 60°C for as long as three hours, but they are less resistant than bacterial spores to heat and toxic agents. The tips of some filaments may become swollen and club-shaped.

All, or practically all, the actinomycetes are gram-positive, and most of the *Nocardia* are acid-fast. Grains will be stained by hematoxylin and eosin, but single filaments or bacilli will not. All morphologic forms are well demonstrated by methenamine silver stains, as are fungi. The periodic acid-Schiff stain (PAS) and its modifications (Gridley, Hotchkiss-McManus) stain fungi bright pink but do not stain bacteria. Thus, these stains can be utilized to distinguish between a fungal etiology or an actinomycete (bacterial) etiology in histopathologic preparations.

PHYSIOLOGY AND COMPOSITION

Recent work on the chemical composition of the actinomycetes has modified the previous concepts of classification of the aerobic actinomycetes. Mordarski[30, 31, 32] has found a particular lipid fraction in the *Nocardia* which she calls "lipid characteristic of *Nocardia*" (LCN-A). It is present in the *Nocardia* but not in the *Actinomadura, Streptomyces,* most *Mycobacterium,* or *Øerskovia* sp. It is also present in some *Corynebacterium* sp., some *Mycobacterium* sp., *Rhodococcus rhodochrous,*[22] and the genus *Bacterionema.*[30] Becker et al.[4] describe a simple paper chromatographic technique

Table 1–2   *Constituents of Cell Walls in Relation to Taxonomy**

Wall Type (W.T.)	Major Compounds in Wall	Whole Cell Type (C.T.)			Genus
		C.T.	Major Component		
I	LL-DAP† glycine				*Streptomyces*
II	*meso*-DAP glycine	D	xylose arabinose		*Micromonospora*
III	*meso*-DAP	B	madurose‡		*Actinomadura* *Dermatophilus* *Microbispora*
		C	galactose		*Nocardiopsis* *Thermoactinomyces* *Geodermatophilus*
IV	*meso*-DAP	A	arabinose galactose		*Nocardia* *Rhodococcus* *Corynebacterium* *Mycobacterium*
VI	Lysine, aspartic acid, galactose				*Øerskovia*

*From Berd, D. 1973. Am. Rev. Respir. Dis., *108*:909–112; Lechevalier, N. C. 1968. J. Lab. Clin. Med., *71*:934–944; Staneck, J. L., and G. D. Roberts, 1974. Appl. Microbiol., *28*:226–231.
   †DAP = diaminopimelic acid
   ‡3-0-methyl-D-galactose

for determining the cell wall composition of *Streptomyces* and *Nocardia* (Tables 1–2, 1–3). LL-diaminopimelic acid (DAP) and glycine are found in *Streptomyces; meso*-DAP, arabinose, and galactose are found in *Nocardia* and other genera (W.T. IV). The genera *Actinomadura,*[17] *Dermatophilus,* and *Microbispora* contain *meso*-DAP (W.T. III), with madurae as a constituent of the whole cell (C.T.B.) and so on as noted in Table 1–2.[6, 25, 28, 41] The polar lipid pattern for *Actinomadura* is distinct, as is the pattern for the genera *Streptomyces* and *Micropolyspora.*[36] The genus *Nocardia* also contains a distinct group

of lipid soluble, iron-binding compounds called nocobactins.[20] *Nocardiopsis dassonvillei* is wall type III, but without madurose, and contains galactose (cell type C). Because of this a new genus, *Nocardiopsis,*[29] has been proposed for it. The organism has been isolated from the soil and from animal infection. Goodfellow, applying the adansonian cluster technique to all the criteria used for identification, has delineated the several genera and their species.[20] The physiologic characteristics of the pathogenic *Nocardia* and *Streptomyces* species have been summarized by Pine.[35] The morphologic and physiologic constants of *Ac-*

Table 1–3   *Summary of Some Medically Important Aerobic Actinomyces**

Wall Type	Cell Type	Decomposition Media			Organism
		Casein	Tyrosine	Xanthine	
I		+	+	+	*Streptomyces*
		+	+	−	*S. somaliensis*
II	D	+	±	−	*Micromonospora*
III	B	+	±	−	*Actinomadura madurae* or *A. pelletieri*
III	C	+	+	+	*Nocardiopsis dassonvillei*
IV	A	−	±	−	*Rhodococcus* sp.
IV		−	−	−	*Nocardia asteroides*
IV		+	+	+	*N. brasiliensis*
IV		−	−	+	*N. caviae*

*Modified from Staneck, J. L., and G. D. Roberts. 1974. Appl. Microbiol., *28*:226–231.

tinomyces and related anaerobic forms are given by Berd[6] and Slack and Gerenscer.[40]

CULTIVATION

Both spores and mycelium of actinomycetes grow in subculture. On solid media, the growth of aerobic forms is dry, tough and leathery, sometimes wrinkled, and adherent to and piled above the medium; in many instances, it resembles the growth of *Mycobacterium*. *Nocardia* species, especially *N. asteroides*, grow well on the usual *Mycobacterium* media. In some cases, especially among the *Streptomyces*, the growth appears powdery or chalky, owing to the formation of aerial mycelia. In liquid media, growth occurs in the form of a dry, wrinkled surface film or, more often, as flakes or aggregates that adhere to the sides of the flask, especially at the surface, or sink to the bottom.[19, 21, 40]

Pigment formation, with colors ranging over the entire spectrum, is common among the actinomycetes. Differentiation is usually made between pigmentation of the vegetative mycelium and that of the spore-bearing aerial mycelium, as well as between pigment diffusing into the medium. Soluble purple and brown pigments are often observed on protein-containing media. These characteristics may be strain-variable, media-variable,

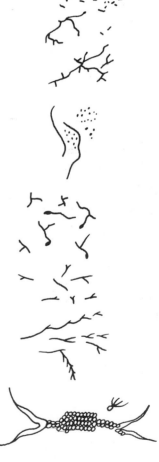

Actinomyces: form filaments but become diphtheroid in culture

Arachnia: Spider-web–like

Rothia: coccoid, diphtheroid, or filamentous or a mixture of these

Bacterionema: swollen, bacillus-like body

Bifidobacterium: diphtheroid and bifid forms

Mycobacterium: mostly rod shaped but may branch

Dermatophilus: broad filament that tapers at end; eight ranked motile spores

Geodermatophilus: tubular irregular elliptical motile and coccoid nonmotile spores are formed

Figure 1–1   Representative Genera of the Actinomycetales

Illustration continued on following page

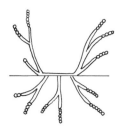

Micropolyspora: abundant aerial mycelium; short chains of unsheathed spores above and below surface basipetally produced.

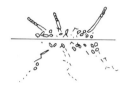

Nocardia: coccoid, diphtheroid fragments of filaments; aerial mycelium fragments to form unsheathed arthrospore-like units

Rhodococcus: diphtheroid, like frost; no right-angle branching or spore formation

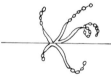

Saccharopolyspora: has aerial filaments that fragment into straight or spiral chains of unsheathed, spaced spores

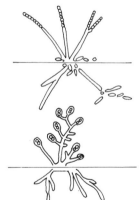

Actinomadura: short chains of spores within sheath; also fragments to form rods and coccoid cells

Microbispora: longitudinal pairs of spores in sheath on aerial filaments only

Figure 1-1 *Continued.*

Illustration continued on opposite page

and temperature-variable. The actinomycetes, especially the saprophytic forms, are physiologically active, utilizing a variety of nitrogen and carbon compounds, and many are actively proteolytic (Fig. 1–1). The important differences for species identification are discussed in later chapters. An earthy-to-musty odor, like that of freshly turned soil or of a damp basement, is produced by many species, especially of *Streptomyces*. The optimum temperature for growth is usually 20° to 30°C, though some of the pathogenic species grow best at 17°C, and thermophilic species, analogous to thermophilic bacteria, are known. Other forms, such as the *Rhodococcus rhodochrous*, grow at 10°C, a characteristic used in differentiating it from other species. The great majority of actinomycetes are aerobic and nonfastidious nutritionally, but a few are anaerobic or microaerophilic and are nutritionally exacting.

Differentiation of the actinomycetes is done in part on a morphologic basis and in part on a compositional and physiologic basis,

Nocardiopsis: long zigzag chains of spores spaced within a sheath

Streptomyces: extensive aerial mycelium; abundant spores in cobweb, flexuous, straight, or coiled branches

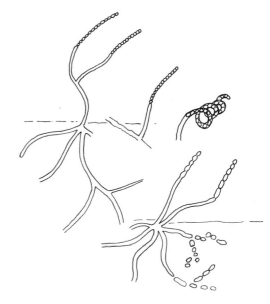

Nocardioides: fragments into coccoid and rodlike units; aerial mycelium produces short chains of irregular-sized spores

Micromonospora: no aerial mycelium; spores are subsurface singly or in grapelike clusters

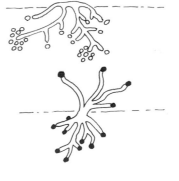

Thermoactinomyces: bacterial endospore produced singly in aerial and subsurface mycelium

Øerskovia: primary mycelium fragments into mono- and polytrichous motile spores

Figure 1–1 *Continued.*

the former including pigmentation and saccharolytic and proteolytic activity.

IMMUNOLOGY AND SEROLOGY

Immunologic investigations on actinomycetes have been numerous, confusing, controversial, and generally of limited value. These are reviewed in the book on the Actinomyces by Slack and Gerenscer.[40] Mostly this has been because of examination of only a few strains, misidentification, and lack of a standardized technique. Perhaps this situation has been best summarized by an early investigator, Bretey, who stated that *Nocardia* and *Streptothrix* are "bad antigens." Some of the more recent work has shown stable antigenic differences in these groups. González Ochoa and Vasquez-Hoyos[16] could serologically separate four groups of actinomycetes: (1) bovine and certain *Nocardia*, (2) the soma-liensis group, (3) the madurae group, and (4) the paraguayensis group. Genera could be delineated by a slide agglutination test of Schneidau and Shaefer,[39] separating *Nocardia* and *Streptomyces*. Kwapinski,[24] using complement fixation, has found patterns of antigens common to and specific for several genera, including *Mycobacterium, Streptomyces, Nocardia,* and *Actinomyces.* A complement-fixation test for the diagnosis of *N. asteroides* infection in cows has been devised.[34] Dyson and Slack[13] produced a specific skin test for *N. asteroides* infection in experimental animals. González Ochoa and Baranda,[15] with a carbohydrate fraction, developed a specific skin test for patients infected with *N. brasiliensis,* and Bojalil and Zamora[7] have a protein fraction specific for *N. asteroides* and *N. brasiliensis* infections.

Presumptive diagnosis of actinomycosis could not be reliably demonstrated by Georg et al.[14] using an immunodiffusion technique. Serologic procedures for antigenic grouping of strains of actinomyces have been investigated, and a polyvalent diagnostic reagent has been developed, using fluorescent antibody procedures.[10]

PATHOGENICITY

Several species of anaerobic actinomycetes are capable of producing disease in man. They include *Actinomyces israelii, A. naeslundii* (which may be a variant of *A. israelii*), *A. meyeri, A. viscosus, Arachnia propionica,* and *Bifidobacterium eriksonii.* Most are normal flora of the human buccal cavity and tonsillar crypts and are of limited invasive ability. They probably cannot elicit disease without the aid of trauma to the tissue and the presence of associated bacteria. *A. bovis* is found in lumpy jaw of cattle. Among the *Nocardia, N. brasiliensis* is probably the most virulent, while *N. asteroides* is considered an opportunist. *N. caviae, Actinomadura madurae, A. pelletieri,* and *S. somaliensis* are all soil saprophytes and are regularly associated with actinomycotic mycetoma of man and animals. These species and several others have been involved in pulmonary and systemic infections as well, especially in a setting of a compromised host. In such cases they are among the many microorganisms—bacteria, fungi, algae, and so forth—termed "opportunists." Infection by all acintomycetes initially elicits an acute in-

flammatory pyogenic response, followed by chronic inflammatory reaction and sometimes generalized spread. The degree and severity of disease depends on the physical status of the patient and the virulence of the organism.

TAXONOMY

An increasing number of agents and clinical types of infection that involve the Actinomycetales are now being recorded. Although most cases do not present as the classic "fungus-like" diseases of mycetoma, actinomycosis, or norcardiosis, the medical mycologist must be familiar with these as well as the usual clinical types of infection. In order to keep in context the familiar etiologic agents along with the related but rarely encountered species, a summary of the current taxonomy follows. This includes both groups of organisms and closely allied species.[19, 20] The present nomenclature is based on recent findings and summaries that have largely supplanted those listed in Section 17 of *Bergey's Manual*[11] and are contained in the *International Journal Systemic Bacteriology*, Vol. 30, 1980. The new taxonomy is based primarily on a combination of morphologic, chemical, and spore characteristics and covers pathogenic nocardioform actinomycetes and related groups. These and two coryneform actinomycetes tentatively called *Corynebacterium minutissimum* and *C. tenuis* are the agents of human and animal diseases covered in chapters 2, 3, 4, and 5 of this book (Table 1–4).

ORDER: ACTINOMYCETALES

Family I. Actinomycetaceae

The Actinomycetaceae are gram-positive, nonacid-fast, and catalase-positive or catalase-negative. They possess no mycelium. Filaments fragment into irregular rodlike or coccoid cells, with no spores; they exist as commensals in the oral and intestinal tracts of humans and animals.

Actinomyces Harz 1877 is usually catalase-negative. Initially the organisms form filamentous colonies, later becoming diphtheroid in continued culture. The products of glucose fermentation do not include propionic acid, and the cell wall contains neither

Table 1–4 *Order Actinomycetales: Nocardioform Actinomycetes and Related Taxa*

Family	Genus
1. *Actinomycetaceae*	*Actinomyces* Harz 1877 *Arachnia* Pine and Georg 1969 *Bacterionema* Gilmour, Howell, and Biddy 1961 *Bifidobacterium* Orla Jensen 1924 *Rothia* Georg and Brown 1967
2. *Mycobacteriaceae*	*Mycobacterium* Lehmann and Neumann 1896
3. *Dermatophilaceae*	*Dermatophilus* Van Saceghem 1915 *Geodermatophilus* Luedemann 1968
4. *Frankiaceae*	*Frankia* Brunchorst 1886
5. *Nocardiaceae*	*Micropolyspora* Lechevalier, Solotorvsky, and McDurmont 1961 *Nocardia* Trevisan 1889 *Rhodococcus* Zopf 1891 *Saccharopolyspora* Lacey and Goodfellow 1975
6. *Thermomonosporaceae*	*Actinomadura* Lechevalier and Lechevalier 1970 *Microbispora* Nonomura and O'Hara 1957 *Nocardiopsis* Meyer 1976
7. *Streptomycetaceae**	*Streptomyces* Waksman and Henrici 1943 *Intrasporangium* Kalakoutskii, Kirillova, and Krasilnikov 1967 *Nocardioides* Prauser 1976
8. *Micromonosporaceae**	*Micromonospora* Ørskov 1923
9. *Thermoactinomycetaceae**	*Thermoactinomyces* Tsiklinsky 1899
Not assigned	*Øerskovia* Prauser, Lechevalier, and Lechevalier 1970 *Promicromonospora* Krasilnikov, Kalakoutskii, and Kirillova 1961
Coryneform actinomycetes	
Corynebacteriaceae	*Corynebacterium* Lehmann and Neumann 1896

*Usually forms nonfragmenting mycelium

DAP nor arabinose. *A. bovis, A. israelii, A. meyeri, A. naeslundii, A. odontolyticus,* and *A. viscosus* have all been involved in classic actinomycosis or other types of infection in humans or animals (see Chapter 2).

Arachnia Pine and Georg 1969 is a facultative anaerobic and catalase-negative. Fermentation of glucose yields CO_2, acetate, and propionic acid. It contains LL-DAP, and the cell wall type I. The organism has been isolated from human lacrimal canaliculitis and is in the normal flora of humans but not animals. It has caused typical actinomycosis.

Bacterionema Gilmour, Howell, and Bibby 1961 is catalase-negative or -positive, aerobic or facultative anaerobic, and produces CO_2 and formic, acetic, propionic, and lactic acids. It is a long filamentous bacteria, characteristically ending in a swollen bacilluslike body. Its wall contains DAP and arabinose (type II A). It has been isolated from human saliva, plaque, and gingival and periodontal pockets

and also from monkeys. The species has not been associated with invasive disease.

Bifidobacterium Orla-Jensen 1924 includes diphtheroid cells, bifid forms, rods, and cocci. The organisms ferment glucose to acetic and lactic acid without CO_2 production. The wall contains neither DAP nor arabinose. *Bifidobacterium* is found in human vagina, mouth, and intestine, and also in sheep, guinea pigs, cattle, honey bees, and turkeys. *B. ericksonii* is regularly associated with actinomycosis and other types of human infections.

Rothia Georg and Brown 1967 is catalase-positive, aerobic, and may be all coccoid, diphtheroid, or filamentous or may be a mixture of types. The fermentation products include acetic, formic, and lactic acids, but not propionic acid. The cell wall contains neither DAP nor arabinose. *Rothia* is isolated from carious teeth, as oral flora, and from spinal fluid, blood, and abscesses; it has not been isolated from actinomycosis.

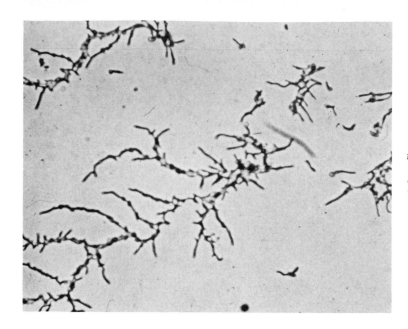

Figure 1–2 *Mycobacterium fortuitum.* Elongate filaments, little or no right angle branches as in *Nocardia* and other genera. ×1500. (Courtesy of W. Causey.)

Family 2. Mycobacteriaceae

The cells have no mycelium but form filaments. They fragment into coccoid, rodlike, or diphtheroid elements and are acidfast. The cell wall is type IV. The organisms contain mycolic acids (α branched β hydroxylated fatty acids). No spores or motile elements are present.

Mycobacterium Lehmann and Neumann 1896 is a catalase-positive, aerobic to microaerophilic, acid-fast actinomycete. It produces oxidative products and acids from sugars (Fig. 1–2). *Mycobacterium* includes obligate pathogens as well as soil and water saprophytes. The organism is most frequently rod shaped and may branch, but it does not form extensive mycelium. It is gram-positive and nonmotile, and it has no spores.

Family 3. Dermatophilaceae

The cells of this family possess branching filaments that divide longitudinally and transversely to form broad muriform septate mycelium. Within the septa, coccoid elements develop and may become motile.

Dermatophilus Van Saceghem 1915 is catalase-positive and is an obligate pathogen of animals. Its filamentous mycelium is broad and tapers at the end with the formation of right-angle lateral branches. The mycelium divides transversely and along horizontal and longitudinal planes, resulting in eight ranked encapsulated spores of equal size. These be-

come motile, with tufts of 5 to 7 flagella. The organism is weakly fermentative and the cell wall is type III B.

Geodermatophilus Luedemann 1968 is saprophytic in sea water and soil. Mycelium is not developed. Its colony consists of a tuber-shaped nonencapsulated, holocarpic thallus containing masses of cuboid cells 0.5 to 2.0 μm. This breaks up into coccoid or cuboid nonmotile cells and elliptical-to-lanceolate zoospores. The cell wall is type III C. Following germination, the cells form branched filaments that divide transversely and longitudinally.

Family 4. Frankiaceae

The organisms of *Frankiaceae* are obligate symbiotes in root nodules of nonleguminous, dicotyledonous plants belonging to 6 orders, 7 families, and 14 genera. Atmospheric nitrogen is fixed within the nodules. No spores are formed but "bacteroids" may develop. These endophytes have not been cultured *in vitro*. They may survive free in soil in the form of "bacterioids," however. Cell wall may be type II or type IV.

Frankia Brunchorst 1886 is an obligate symbiote within the root nodules of several orders, families, and genera as distinct as *Causerina*, *Alnus* (the alder tree), and *Myrica* (the wax myrtle tree). In the nodules, *Frankia* forms a mass within the center of the host cell. Near the periphery, spherical or club-shaped vesicles grow on radially arranged hyphae near the root cell wall. This association is actively involved in the fixation of nitrogen. If the host cell dies, vesicles are not formed and the mycelium fragments to form bacteroids, which are released into the soil; this is considered a resting stage.

Family 5. Nocardiaceae

In this family, mycelial filaments fragment to coccoid or elongate elements. Aerial spores are formed in *Nocardia* and *Micropolyspora,* and substrate spores in *Micropolyspora.* The cell wall is type IV, and all genera contain nocardiomycolic acids. Spores are not formed within a sheath.

Micropolyspora Lechevalier, Solotorvsky, and McDurmont 1961 forms substrate and aerial mycelium, which may bear short (1 to 20) chains of spores. The spores develop basipetally, terminally, or on lateral branches. They have thick walls, particularly at points adjacent to other spores. The wall is type IV A. *M. faeni* is abundant in hay and compost and is probably the chief allergen in farmer's lung syndrome and other such conditions.

Nocardia Trevisan 1889 is an aerobic, gram-positive, nonmotile, filamentous species of actinomycetes that may be acid-fast. The filaments fragment into coccoid and rodlike elements. Aerial mycelium are common, sometimes covering the colony in a white down, and may fragment to form chains of spores similar to arthroconidia in true fungi. Aesculin, allantoin, and urea are hydrolyzed, and nitrates are reduced; on high protein-rich media, a melanoid (tyrosine-based) pigment is formed. *Nocardia* are resistant to lysozyme. Acids are formed oxidatively from glucose, fructose, and glycerol and can utilize acetate, *n*-butyrate, *H*-malate, propionate, pyruvate, succinate, and paraffin as sole energy sources for growth. It is wall type IV A and contains LCN-A and nocardiomycolic acids. *N. caviae, N. brasiliensis,* and *N. asteroides* cause human and animal disease. All *Nocardia* are found in soil, fresh water, and marine environments. Some species are found metabolizing petroleum and have been recovered in test diggings half a mile or more deep in oil-bearing strata.

Rhodococcus Zopf 1891 consists of aerobic, gram-positive actinomycetes that are pleomorphic but form primary mycelia that fragment into irregular parts (Fig. 1–3). It is nonmotile and partially acid-fast and has no aerial mycelium. Colonies may be rough, smooth, or mucoid in consistency and yellow, orange, or reddish in color. They are sensitive to lysozyme. Its cell wall is type IV A and it contains LCN-A. Oxidative formation of acid from glucose, fructose, sucrose, mannitol, and sorbitol occurs and *n*-butyrate, *H*-malate, propionate, pyruvate, succinate, acetate, and many aliphatic and aromatic compounds are utilized as energy sources. Along with some *Nocardia,* they grow on petroleum slicks in water, tar, and asphalt and very deep in oil-bearing shale and strata. A few cases of human opportunistic infections have been recorded.[21]

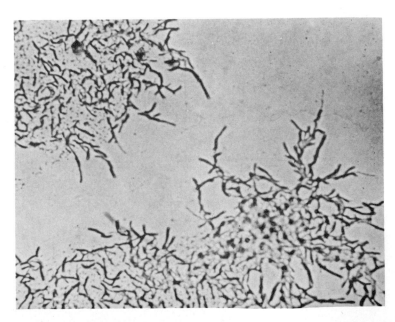

Figure 1–3 *Rhodococcus rhodochrous.* Irregular sinuous filaments, "like frost"; absence of right angle branching. ×1500. (Courtesy of W. Causey.)

Saccharopolyspora Lacey and Goodfellow 1975 is aerobic, gram-positive, and nonacid-fast and fragment into coccoid and rodlike units. Cell wall is type IV. Aerial mycelium segments into spores contained within a sheath in the form of long, straight chains or spirals. Spacing within the sheath is similar to that in *Nocardiopsis dassonvillei*. Free of the sheath, the spores of the type species *S. hirsuta* are hairy. They grow in temperatures ranging from 25 to 50°C, contain no LCN-A, and are susceptible to lysozyme and resistant to gentamycin. They are found in sugar cane bagasse and can contribute to extrinsic allergic alveolitis.[25]

Family 6. Thermomonosporaceae

This family represents fragmenting or nonfragmenting myceliate actinomycetes. Spores are formed within a thin fibrous sheath singly, in pairs, or in short chains on aerial or substrate mycelium or both. They are nonmotile and heat sensitive (in contrast to the *Thermoactinomycetaceae*). Cell wall is type III, and no LCN-A is present. Genera included are *Actinomadura*, *Nocardiopsis*, *Microbispora*, *Microtetraspora*, *Saccharomonospora*, and *Thermomonospora*.

Actinomadura Lechevalier and Lechevalier 1970 is an aerobic, gram-positive nonacid-fast species of actinomycetes forming branched subsurface mycelium and sparse to abundant aerial filaments. The filaments fragment to form spores within a sheath or lateral branches 5 to 15 in number. In some species these spores are formed as hooks or spirals and enclosed in gelatinous material, giving the appearance of a pseudosporangium. Cell wall is type III B, and no LCN-A is formed. They are common agents of mycetoma (Chapter 5).

Microbispora Nonomura and Ohara 1957 has spores that are formed in characteristic longitudinal pairs on aerial mycelium usually not on subsurface filaments. They are sessile or on short sporophores closely arranged on aerial filaments resembling a catkin (Fig. 1–4). Colonies are usually pink. On oatmeal and other solid media, crystals of iodinin are formed by some species. They are soil organisms, have wall type III B, and are thermotolerant. One species, *M. rosea*, has been isolated from pericarditis and pleuritis.[27]

Nocardiopsis Meyers 1976 is similar to *Actinomadura* but the cell wall is type III C (no madurose), and the zigzag spore chains within the sheath have spacings between the individual spores. It can be isolated from both soil and animal infections. It is a monotypic genus with *N. dassonvillei*, the only species yet described.[29]

Family 7. Streptomycetaceae

These are aerobic actinomycetes that form nonfragmenting or rarely fragmenting subsurface mycelium. They may produce spores, however, and may also develop into aerial mycelium bearing uniserial chains of spores

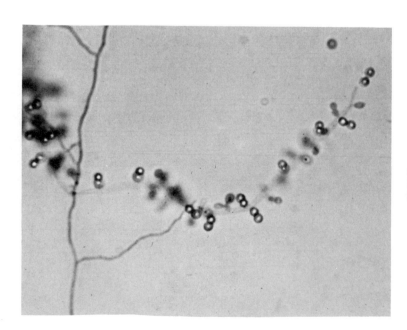

Figure 1–4 *Microbispora.* Paired spores within sheath appear regularly along fertile mycelium. ×19,000. (Courtesy of W. Causey.)

enclosed in a thin fibrinous sheath. Cell wall is type I (LL-DAP). The spore chains may be very long (50 or more spores in *Streptomyces* sp.) or short (2 to 5 spores in *Microellobosporia* sp.). There are many genera.

Streptomyces Waksman and Henrici 1943 is an aerobic actinomycetes with extensive, branching, substrate, and aerial filaments. Fragmentation of subsurface mycelium · is rare, as is subsurface spore production. Spores are formed on the abundant aerial mycelium by regular septation of the filament within a fibrous sheath. In some species, the sheath persists and coats each spore *(S. viridochromogenes)*; in others, the spores are loose within the sheath and, when freed, are free to any material *(S. venezuelae)*. The sheath deposits are responsible for the various architectural markings on the spores: warts, spindles, spines, undulations, waves, and so forth. This is an extremely common member of the soil flora. One, *S. somaliensis,* is an agent of actinomycotic mycetoma (see Chapter 5).

Intrasporangium Kalakoutskii, Kirillova, and Krassilnikov 1967 resembles *Streptomyces* but fragments easily and forms single spores within the myeclium that are similar to the endospores of bacteria.

Nocardioides Prauser 1976 is a grampositive, aerobic, nonacid-fast, mesophilic organism that fragments readily into rod- to coccuslike elements; it has wall type I.[37] It is susceptible to taxon-specific phages and is not affected by polyvalent *Nocardiae* or *Streptomyces* phage. *N. albus* is common in soil.

Family 8. Micromonosporaceae

In this family the filaments remain intact and have little or no aerial mycelium. Heat-sensitive spores are formed singly in the much branched substrate mycelium. The cell wall is type II.

Micromonospora Ørskov 1923 is an aerobic or anerobic species of Actinomycetes that usually lacks aerial mycelium. It forms orange colonies with black dots; by 14 days they are entirely black. Single spores are arranged in grapelike clusters in subsurface areas (cluster type sporophores) or dispersed throughout the mycelial network (open web types). They can be isolated from the alimentary tract of termites and soil. *M. gallica* was isolated once from a blood culture, and *M. caballi* from cutaneous actinomycosis of a horse.[3] *Micromonospora lustancia* was reported from mycetoma in Portugal and Italy.[3] This

organism was called *Actinomicromonospora* in that paper (*M. caballi* and *M. lusitancia* are tentative species).

Family 9. Thermoactinomycetaceae*

These are aerobic actinomycetes that form substrate and aerial mycelium, which carry heat-resistant endospores. These are produced endogenously within a mother cell. The cell wall is type III.

Thermoactinomyces Tsiklinsky 1899 is an actinomycetes that grows rapidly in temperatures ranging from 32°C to 65°C; it is resistant to 25 µg per ml novobiocin and forms single spores on substrate and aerial mycelium. The spores are similar in structure to bacterial endospores (a core surrounded by a cortex with a multilayered covering wall). They contain dipicolinic acid. Wall is type III C without arabinose, madurose, or xylose but with galactose. Thermoactinomyces is commonly found in soil, hay, humus composte, and heated environments.

A few species are of medical interest. *T. vulgaris* is an agent in extrinsic allergic alveolitis, farmer's lung, and in some forms of fog fever of cattle. It may also be involved in mushroom worker lung (Chapter 27). *T. sacchari* endospores are found in great abundance in self-heated, crushed sugar cane (bagasse) and are thought to be agents of the allergic respiratory disease, bagassosis. *T. dichotomica* may also be involved in allergic alveolitis.

Genera Not Assigned to Families

Oerskovia Prauser, Lechevalier, and Lechevalier 1970 is an aerobic, gram-positive actinomycete. It forms a primary mycelium that quickly fragments into motile rods. These rods are monotrichous when short and peritrichous when long. They germinate to form extensive mycelium. The colonies are yellow, without aerial mycelium, and nonacid-fast, and do not form spores. Its cell wall is type VI. A genus common in soil, *Ø. turbata* has been isolated from a case of endocarditis.[38]

Promicromonospora Krassilnikov, Kalakoutskii, and Kirillova 1961 is a mesophilic organism from garden soil. The substrate mycelium fragments into coccoid and diphtheroid ele-

---

*This is a tentative family.

ments similar to those of *Nocardia* (= *Proactinomyces*) and forms single spores on short sporophores or directly on the branched subsurface mycelium as in *Micromonospora*. The colonies are yellow, glabrous, or with a mat of asporogenous mycelium. No DAP is present, but lysine is. The wall type (possibly type VI) is similar to that of *Actinomyces israelii*.

Coryneform Actinomycetes

The coryneform group of organisms form uncertain relationships with *Nocardia* and *Mycobacterium* spp. The nature of these relationships is quite controversial, and no taxonomy including both has been agreed upon as yet. Suffice it to say that chemically and physiologically Section 1 (human strains) of the *Corynebacterium* genus (so designated in *Bergey's Manual*[11]) have in common many characteristics attributed to nocardioform actinomycetes. These include wall type IV A and possession of α branched β hydroxy acids (mycolic acids):

$$\underset{\substack{| \\ OH}}{R_1\text{—}\overset{\beta}{C}H}\text{—}\underset{\substack{| \\ R_2}}{\overset{\alpha}{C}H}\text{—COOH}$$

One mycolic acid from *C. diphtheriae* is a toxic glycolipid, a 6-6'-addition product of trehalose. It contains equimolar parts of corynemycolic acid ($C_{32}H_{64}O_3$) and corynemycolenic acid ($C_{32}H_{62}O_3$). This glycolipid is thus a lower mycolic analog of the cord factor of *M. tuberculosis*. Mycolic acids are not found in other *Corynebacterium* groups.

Corynebacterium Lehmann and Neumann 1896 is a gram-positive, nonmotile organism consisting of straight to slightly curved rods showing no branching. They stain irregularly, often have club-shaped swellings, and are not acid-fast. The two species of interest in our discussions are *C. minutissimum*, the etiologic agent of erythrasma, and *C. tenuis*, which is associated with trichomycosis axilaris. *C. minutissimum* appears to be a valid species and is allied to *C. xerosis*.

REFERENCES

1. Alashamaony, L., M. Goodfellow, and D. E. Minnikin. 1976. Free mycolic acid as a criteria in the classification of *Nocardia* and "rhodochrous" complex. J. Gen. Microbiol., *92*:188–189.

2. Alderson, G., and M. Goodfellow. 1976. Taxonomy of the genus *Actinomadura*. J. Appl. Bacteriol., Vol. 41.

3. Araviysky, A. N. 1968. *Micromonospora* in the discharge from giant ulcers in systemic mycosis. Mycopathologia, *35*:49–56.

4. Becker, B., M. P. Lechevalier, et al. 1964. Rapid differentiation between *Nocardia* and *Streptomyces* by paper chromatography of whole cell hydrolysates. Appl. Microbiol., *12*:421–423.

5. Berd, D. 1973. Laboratory identification of clinically important aerobic actinomycetes. Appl. Microbiol., *25*:665–681.

6. Berd, D. 1973. *Nocardia asteroides:* a taxonomic study with clinical correlations. Am. Rev. Respir. Dis., *108*:909–912.

7. Bojalil, L. F., and A. Zamora, 1963. Precipitin and skin test in the diagnosis of mycetoma due to *N. brasiliensis*. Proc. Soc. Exp. Biol. Med., *113*:40–43.

8. Bradley, S. G. 1975. Significance of nucleic acid hybridization to systematics of actinomycetes. Adv. Appl. Microbiol., *19*:59–70.

9. Bradley, S. G., and J. S. Bond. 1974. Taxonomic criteria for mycobacteria and nocardiae. Adv. Appl. Microbiol., *18*:131–190.

10. Brock, D. W., and L. K. Georg. 1969. Characterization of *Actinomyces israelii* serotypes 1 and 2. J. Bacteriol., *97*:589–593.

11. Buchanan, R. E., and N. E. Gibbons, eds. 1974. *Bergey's Manual of Determinative Microbiology*. 8th ed. Baltimore, Williams & Wilkins Company.

12. Committee of the Judicial Commission of the ICSB. 1980. First Draft. Approved lists of bacterial names. Int. J. Syst. Microbiol., *30*:225–420.

13. Dyson, J. E., and J. M. Slack. 1963. Improved antigens for skin testing in nocardiosis. I. Alcohol precipitates of culture supernates. Am. Rev. Resp. Dis., *88*:80–86.

14. Georg, L. K., G. W. Roberstad, et al. 1965. A new pathogenic anaerobic *Actinomyces* species. J. Infect. Dis., *115*:88–99.

15. González Ochoa, A., and F. Baranda 1953. Una prueba cutánea para el diagnóstico del micetoma actinomycósico por *Nocardia brasiliensis*. Rev. Inst. Salub. y Enferm. Trop., *13*:189–197.

16. González Ochoa A., and A Vasquez-Hoyos, 1953. Relaciones serologicas de los Principales actinomycetes pathogenicos. Rev. Inst. Salub. Enferm. Trop., *13*:177–187.

17. Goodfellow, M. 1971. Numerical taxonomy of the same nocardioform bacteria. J. Gen. Microbiol., *69*:33–80.

18. Goodfellow, M., and G. Alderson. 1977. The actinomycete-genus *Rhodococcus*: a home for the "rhodochrous" complex. J. Gen. Microbiol., *100*:99–122.

19. Goodfellow, M., G. H. Brownell, and J. A. Serrano. 1976. *The Biology of the Nocardiae*. London, Academic Press.

20. Goodfellow, M., D. E. Minnikin. 1977. Nocardioform bacteria. Annu. Rev. Microbiol., *31*:159–180.

21. Hasburchak, D. R., B. Jeffery et al. 1978. Infections caused by Rhodochrous. Am. J. Med., *65*:298–302.

22. Hecht, S. T., and W. A. Causey. 1976. Rapid method for the detection and identification of mycolic acids in aerobic actinomycetes and related bacteria. J. Clin. Microbiol., *4*:284–287.

23. Kurup, V. P., J. J. Barborial et al. 1975. *Thermoactinomyces candidus* a new species of thermophilic actinomyces. Int. J. Syst. Bacteriol., *25*:150–154.

24. Kwapinski, J. B. 1963. Antigenic structure of actinomycetales. VI. Serological realtionship between antigenic fractions of *Actinomyces* and *Nocardia*. J. Bacteriol., *86*:179–186.

25. Lacey, J., and M. Goodfellow. 1975. A novel actinomycete from sugar-cane bagasse. *Saccharopolyspora hirsuta* gen. et. sp. nov. J. Gen. Microbiol., *88*:75–85.

26. Lechevalier, N. C. 1968. Identification of aerobic actinomycetes of clinical importance. J. Lab. Clin. Med., *71*:934–944.

27. Louria, D. B., and R. E. Gordon. 1960. Pericarditis and pleuritis caused by a recently discovered microorganism, *Waksmania rosea*. Am. Rev. Respir. Dis., *81*:83–88.

28. Mariat, F., and H. Lechevalier. 1977. Actinomycetes aerobies pathogenes, generalities. Bacteriol. Med. *1*:566a–5662a.

29. Meyers, J. 1976. *Nocardiopsis*, a new genus of the order Actinomycetales. Int. J. Syst. Microbiol., *26*:487–493.

30. Minnikin, D. E., G. Pirouz, and M. Goodfellow. 1977. Polar lipids in the classification of some *Actinomadura* species. Int. J. Syst. Bacteriole., *27*:104–117.

31. Mordarska, H., and M. Mordarska. 1969. Comparative studies on the occurrence of lipid diaminopimelic acid and arabinose in *Nocardia* cells. Arch. Immunol. Ther. Exp. (Warsz), *17*:739–743.

32. Mordarski, M., K. P. Schaal et al. 1977. Interrelation of *Nocardia asteroides* and related taxa as indicated by deoxyribonucleic acid reassociation. Int. J. Syst. Bacteriol., *27*:66–70.

33. Nelson, E., and A. T. Henrici. 1922. Immunologic studies of actinomycetes with special reference to the acid-fast species. Proc. Soc. Exp. Biol. Med., *19*:351–352.

34. Pier, A. C., and J. B. Enright. 1962. *Nocardia asteroides* as a mammary pathogen of cattle. III. Immunologic reaction of infected animals. Am. J. Vet. Res., *23*:284–292.

35. Pine, L. 1977. The Actinomycetales. Parasitic or fermentative actinomycetes. *In* Laskin, A. I., and H. A. Lechevalier, eds. CRC Handbook of Microbiology, Vol. 1. Bacteria. 2nd ed. Cleveland, Ohio, USA, CRC Press, pp. 373–380.

36. Pirouz, G., M. Goodfellow, and D. E. Minnikin. 1975. Lipid composition of *Actinomadurae* and related bacteria. Proc. Soc. Gen. Microbiol., *3*:48.

37. Prauser, H. 1976. *Nocardioides:* A new genus of the order Actinomycetales. Int. J. Syst. Bacteriol., *28*:58–65.

38. Reller, L. B., G. L. Maddoux, et al. 1975. Bacterial endocarditis caused by *Oerskovia turbata*. ASM Abs. C136.

39. Schneidau, J. O., and M. F. Schaefer. 1960. Studies of *Nocardia* and the Actinomycetales. II. Antigenic relationships shown by slide agglutination. Am. Rev. Resp. Dis., *82*:64–76.

40. Slack, J. M., and M. A. Gerenscer. 1975. *Actinomyces*. Minneapolis, Burgess Publishing Co.

41. Staneck, J. L., and G. D. Roberts. 1974. Simplified approach to identification of aerobic actinomyces by thin layer chromatography. Appl. Microbiol., *28*:226–231.

42. Tsukamura, J. 1974. Differentiation of the *Mycobacterium' rhodochrous* group from Nocardiae by beta-galactosidase activity. J. Gen. Microbiol., *80*:553–555.

43. Tsukamura, J. 1974. A further numerical taxonomic study of the rhodochrous group. Jap. J. Microbiol., *18*:37–44.

2 Actinomycosis

DEFINITION

In its historical sense, the term actinomycosis defines a clinical entity consisting of a chronic suppurative and granulomatous disease characterized by peripheral spread and extension to contiguous tissue, rare hematogenous spread, and the formation of multiple draining sinus tracts. These sinuses drain from suppurative pyogenic lesions. The exudate contains firm, lobulated grains (sulfur granules) or microcolonies of the etiologic agent, which may reach 1 to 2 mm in diameter; it also consists of adherent cellular debris, associated microorganisms, and coccoid or bacillary forms of the etiologic agent. In addition to the disease described above, a wide variety of infections are evoked by the same anaerobic actinomycetes and will be included in this chapter. The pathogens are most commonly *Actinomyces israelii* in man and *A. bovis* in animals. Other less commonly implicated species include *A. naeslundii, A. viscosus. A. odontolyticus, A. meyeri, Bifidobacterium eriksonii,* and *Arachnia propionica.*

The usual disease affects the cervicofacial, thoracic, and abdominal regions in man and may result in severe disfigurement and disability. Localized lesions are amenable to therapy, but many cases of extensive involvement or hematogenous spread are fatal. Its ability to invade and destroy bone is notorious. Localized infections without granules and sinus tracts may be chronic and weakly invasive.[1, 42] Other such infections can be fulminant and rapidly fatal.[3, 76, 78]

Synonymy: Lumpy jaw, leptothricosis, streptotricosis.

HISTORY

Lumpy jaw, or actinomycosis, was once a fairly common disease of man and is still frequent in cattle. Today it is uncommon in general medical practice and is most often diagnosed in retrospect. This change is largely due to the widespread practice of giving a "shot" of antibiotic indiscriminately for almost any symptom. The etiologic agents, endogenous *Actinomyces* and related species, are quite sensitive to most antibacterial antibiotics, including penicillin and sulfas. Formerly, infection was often associated with tooth extraction or dental surgery, which provided traumatized tissue in which these organisms and associated (synergistic) microorganisms could grow. Prophylactic use of antibiotics following oral procedures has largely eliminated this hazard. Poor oral hygiene and carious teeth are still the most common predisposing factors, but new ones, such as the use of intrauterine devices (IUDs), are being added. An important aspect in the decline of infection is the high level of oral hygiene achieved in the "developed" nations of the world. That this is not the case in "emerging" nations is attested to by the continued numerous and increasing case reports of actinomycosis and related disease in these areas. Presently, the more common types of serious disease are those associated with abdominal injury or disease of the alimentary tract, and those in the thoracic area, the result of aspiration of buccal material, or extension from focal cervical lesions associated with poor oral hygiene and general health and the use of steroids.

Actinomycosis was undoubtedly observed early in the nineteenth century, as actinomycotic tumors were described erroneously in 1826 by Leblanc as osteosarcomas. It was first recognized as a specific parasitic disease in 1876 by Bollinger, who named the disease in cattle "lumpy jaw." In 1877, Harz, using material sent by Bollinger,[38] described the disease in cattle and called the etiologic agent *Actinomyces bovis* (ray fungus of cattle), not because of its cultural characteristics but because of its appearance in tissue. Harz's note and use of the name first appears in Bollinger's paper of 1877. A more complete, albeit rambling, account was published in 1879.[38] The syndrome of human actinomycosis had been recognized in 1854 by Graefe and in 1875 by Cohn,[19] who described what must have been authentic disease in the lachrymal canal. He called the agent *Streptothrix foersteri.* Bollinger's study material included other similar diseases of cattle, such as "woody tongue," which is caused by *Actinobacillus lignieresi.* Harz's use of the genus name *Actinomyces* was disputed for some years because it had been used previously. A yellow jelly fungus of the Basidiomycota had been named *Actinomyces horkelli* in 1827 by Meyen.[51] Cohn's name, *Streptothrix,* was also disputed, because it had been previously used in 1839 by Corda. The issue was resolved when it was noted that Meyen's jelly fungus had been described previously as *Tremella meteorica,* leaving the name *Actinomyces* available for the organisms of Bollinger and Cohn. Attempts to culture them were made on substrates such as plum distillate, bread, and cherry extract, but all failed. Aerobic *Streptomyces* contaminants often grew, however, though there were conflicting reports in the literature. The clubs on the periphery of the lumpy jaw granules were viewed as disklike lamella, and the name *Discomyces bovis* was proposed by Rivolta in 1878. He later retracted in favor of *Actinomyces.*

The first successful culture of the organism was probably that of Bujwid in 1889. He used Buchner's 10 per cent alkaline pyrogallic acid method, and since his material was of a human source it probably was the agent later called *A. israelii.* In 1878 Israel and Ponfick recognized in humans a disease similar to that in cattle; in 1891 Wolf and Israel[86] published an extensive paper on the disease and its etiology. The agent was recovered from cervicofacial and pulmonary cases in anaerobic but not aerobic culture. Growth was obtained in raw or soft boiled eggs. Experimental infections were produced in rabbits and guinea pigs, and the report was well illustrated and clearly described; their paper became a model for later medical writing.

In 1891 Bostroem[10] also published a paper on his researches. He described 11 bovine and 12 human cases in his well illustrated article. His aerobic cultures yielded a rust to brick red colony that, according to his excellent drawings, was surely a contaminating *Streptomyces.* Observing fodder material and grain awls in a cattle barn area, Bostroem postulated an "exogenous" source for the organism, although Israel had attempted and failed to find an external source. The exogenous theory for the etiologic agent gained popularity, though, and held sway for eighty years, until just recently. Both cattle and humans were depicted as chewing on straw or other plant material that caused abrasions to the mucosa and planted the organism in tissue. In 1910, however, Lord's[48] experiments *in vitro* demonstrated the presence of the organism in the normal mouth, in tonsillar crypts, and about carious teeth. These findings were confirmed by Emmons[25] and Rosebury,[70] and the modern concepts of the organism and disease were summarized by Erikson.[29] The disease is very easy to diagnose and, as observed by Cope,[21] "exists wherever there is a microscope and a laboratory." Before the advent of antibiotics, the disease was diagnosed very frequently, though accurate mortality rates are not known. Following the introduction in 1948 of penicillin therapy for the disease by Nichols and Herrell,[59] there was a sharp decrease in the number of deaths and severity of disease, especially of the cervicofacial type. Currently only the advanced cases of pulmonary and systemic disease come to medical attention; even these more advanced cases are amenable to long term antibiotic treatment.

ETIOLOGY, ECOLOGY, AND DISTRIBUTION

Presently, most workers agree that several commonly encountered species can lead to the disease syndrome "lumpy jaw," or actinomycosis. All are members of the family Actinomycetaceae, a group that consists of commensals found in the oral cavity and other mucosal area of various animal species. *A. bovis* is the usual cause in cattle, and *A.*

israelii the predominant organism in human infection, although it is occasionally found also in cattle. *A. naeslundii, A. viscosus, A. meyeri,* and *A. odontolyticus* as well as *A. israelii* are normal inhabitants of the human oral cavity and probably intestinal and vaginal areas also. All of the other species have been recorded in human disease. *A. odontolyticus* and *A. naeslundii* have not been recorded in animals; however, *A. israelii* and *A. viscosus* are found in animals and have been the etiologic agents of disease in cows, pigs, horses, dogs, cats, goats, and so forth. *A. viscosus,* first isolated from peridontal disease in guinea pigs, causes many different infections in a variety of animals. All animal isolates appear to be serotype 1, whereas the strains from human dental plaque are serotype 2.

In addition to the species of *Actinomyces,* other genera and species of Actinomycetaceae have been responsible for actinomycosis. *Arachnia propionica,* first isolated from human lacrimal canaliculitis, has been isolated from many typical cases of human actinomycosis. It is part of the oral flora of man but has not been isolated from human lacrimal ducts or the conjuctiva, nor has it been isolated from animals. *Bacterionema matruchotii* has been isolated from human gingiva, saliva, and peridontal pockets and from a monkey. It has not been recorded in other animals or in human disease, however. *Rothia dentocariosa,* also isolated from carious dentine, is part of normal human oral flora but has not yet been confirmed as an agent of human disease. The reports of its presence in blood, spinal fluid, and abscesses may represent contamination or transient hematogenous transport. Finally, the other genus of oral actinomycetes, *Bifidobacterium,* does contain an authenticated agent of actinomycosis, *B. eriksonii. Bifidobacterium* species have been found as normal flora in rats, bees, pigs, and turkeys. *B. eriksonii,* found so far only in humal oral areas, has been recovered from many cases of actinomycotic diseases of all types.

Possibly as a result of better diagnostic methods and acumen as well as better definition of the species, the primacy of *A. israelii* as the agent in human actinomycosis is being threatened. According to a recent survey published in the British Medical Journal,[2] *A. viscosus* accounted for most of the 43 infections in the year 1976. *A. israelii* and *Arachnia propionicus* were found in the others. *Actinomyces bovis,* for years considered the major agent of lumpy jaw, was found only once, in

a deep chronic sinus of the thigh in a dairy farmers wife. The lesion was a result of a cow bite.

Actinomycosis, even in closed lesions, appears to be a cooperative disease of *Actinomyces* and a mixed flora of other bacteria. Holm has referred to these as "associates" and claims that they exert a synergistic activity in the pathogenesis of actinomycosis. Laboratory investigation tends to confirm this. The injection of pure culture of *A. bovis* or *A. israelii* seldom produces lesions in animals without the aid of adjuvants, such as gastric mucin.[52] Fusiform bacilli, anaerobic streptococci, gram-negative bacilli, *Haemophilus aphrophilus,* or related organisms and *Actinobacillus actinomycetemcomitans,* named for its frequent association, are found together with *A. israelii* in cervicofacial disease. *Actinobacillus actinomycetemcomitans,* the streptococci, and *Haemophilus* are commonly found in the thoracic form of the disease, while various aerobic and anaerobic intestinal flora, including *Escherichia coli,* are found in abdominal or pelvic disease.

A. actinomycetemcomitans as well as other "associated bacteria" have been isolated from other types of human disease that bear no resemblance to actinomycosis. Many cases of endocarditis, one of thyroiditis[15] and one resembling Pott's disease[57] had *A. actinomycetemcomitans* as etiology.

Many unsuccessful attempts have been made to isolate the pathogenic *Actinomyces* from vegetation. The investigations of Lord, Emmons, Slack, and others lead to the conclusion that these organisms are endogenous in man and animals, leading a parasitic existence as nonpathogens in the mucous membranes of the oral cavity, in the caries and tartar of teeth, in tonsillar crypts, in the alimentary tract, and perhaps in the respiratory tract and vagina. Some authors indicate that granules without adherent neutrophils can be found in undiseased tonsils.

Several literature reviews prior to 1950 cite the case distribution as 50 to 60 per cent cervicofacial, 20 per cent abdominal, 15 per cent thoracic, and the remainder in other organs. Harvey et al.,[37] in a study of 37 cases, found 24 per cent cervicofacial, 13 per cent pulmonary, and 63 per cent abdominal disease, probably reflecting a decrease in focal oral infections associated with the use of antibiotics. However, Brown[13] in 1973 reviewed 118 cases and Weese and Smith[84] in 1975 compiled 57 cases in 36 years and did not find such a dramatic change in case type.

These discrepancies reflect the diagnostic acumen of the clinicians involved. In our experience the condition is regularly diagnosed, promptly treated, and seldom tabulated. Cases in which the diagnosis is not made result in serious disease and may present problems of clinical management.

CLINICAL DISEASE

Symptomatology

The clinical picture may vary with the location of the disease, and it is generally classified as cervicofacial, abdominal, or thoracic, depending on the site of primary infection. Rarely the disease may disseminate by hematogenous spread to other organs, or may be a primary infection in other areas, such as skin, cervix, or genitourinary tract.

Cervicofacial Actinomycosis. In the cervicofacial type, the organism enters through trauma to the mucous membrane of the mouth or pharynx, by way of carious teeth or through the tonsils. Salivary or lacrimal glands are sometimes involved.[42, 76] The common initial symptoms are pain and swelling along the alveolar ridge and associated areas. The regional lymph nodes are generally not enlarged at first. The swelling becomes firm; nodular or branny masses are described as "wooden" or "lumpy." The skin over the region may discolor (Fig. 2–1). The hard masses soften and become abscesses and, later, multiple granulomata. They break to the surface, forming multiple sinus tracts which discharge purulent material containing sulfur granules (Fig. 2–2). Pain is minimal, but trismus may develop, impairing mastica-

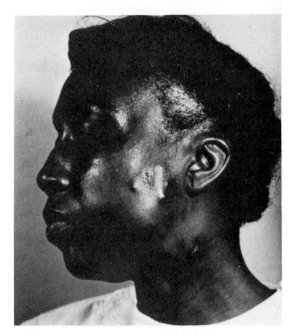

Figure 2–2 Actinomycosis. Multiple sinus tracts opening over ramus and mandible, "lumpy jaw."

tion; otherwise, there is little discomfort to the patient if the disease remains localized. Infections in the maxilla may extend to the cranial bones, giving rise to meningitis, or into the orbit of the eye and the middle ear. Mandibular disease invades the tongue (ligneous phlegmon) and sublingual salivary glands and, by extension, the neck and scapular and upper extremities. Direct extension into the lungs and pleural cavity may also occur. The sinus tracts may heal and reform in the same area later.

Early in the disease there is no bone involvement as shown by x-ray examination. Later, however, a periostitis develops, followed by destruction of bone cortex, by osteomyelitis, and occasionally by expansion of the cortex with cyst formation (Fig. 2–3). The marked ability of this disease to readily burrow through bony material is a point of contrast to some eumycotic infections.

The cervicofacial type of actinomycosis has the best prognosis. With surgical debridement and excision as an adjunct to proper antibiotic therapy, the disease may be cured without much difficulty, if it has remained localized.

Differential diagnosis includes other granulomatous lesions, especially tuberculosis. Also to be considered are the other mycotic diseases, such as blastomycosis and coccidioidomycosis, nocardiosis, glanders, gumma, tularemia, osteomyelitis, and neoplasm.

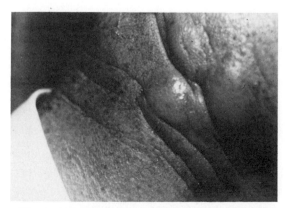

Figure 2–1 Actinomycosis. Erythematous swelling over jaw. This is an early lesion, and the nodule is hard and palpable. (Courtesy of D. Mincey.)

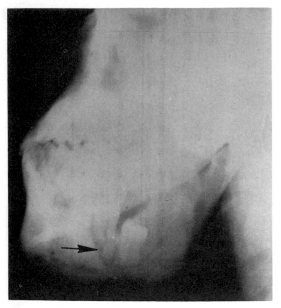

Figure 2–3 Actinomycosis. Destruction of bone around impacted tooth. (Courtesy of D. Mincey.)

Thoracic Actinomycosis. Thoracic actinomycosis may result from direct extension of the disease from the neck (Fig. 2–4), from extension of abdominal or hepatic infection through the diaphragm, or as a primary infection from aspiration of the organism through the mouth. The most common sites of infection are the hilar region and the basal parenchymal areas; this latter area is in contrast to tuberculosis, which is generally apical only. The initial symptoms are those of a subacute pulmonary infection, with a mild fever, cough, and production of purulent sputum without hemoptysis. As the disease progresses, small abscesses develop in the lung, and the sputum may become blood-streaked, suggesting lung destruction. In time, the infection spreads to include the pleura and thoracic wall and then penetrates to the surface to form typical discharging sinuses (Fig. 2–5,A). Increasing dyspnea, fever, night sweats, anemia, and general wasting are seen (Fig. 2–5, B).

The early physical signs of actinomycosis resemble those of tuberculosis. X-rays show massive smooth consolidations, usually bilateral, in the lower half of the lung field. Consolidations may project from the hilar area, suggesting neoplasm (Fig. 2–6).[47] Sometimes rarefactions are seen in these masses, indicating actinomycosis rather than tumor. Other types of presentation also occur, such

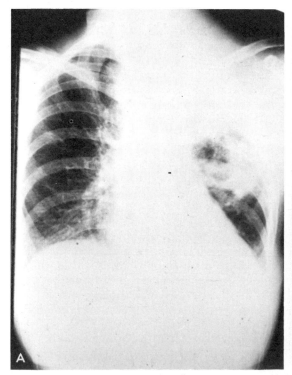

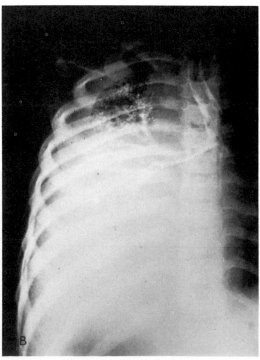

Figure 2–4. Actinomycosis. *A,* Pulmonary involvement resulting from extension of cervicofacial disease. There is diffuse, dense infiltrate of the upper lobe at the left lung and pleural thickening over the lateral upper lobe. Soft tissue swelling is seen in the left superclavicular region. There was a scrofula at this point. (Courtesy of J. Fennessey.) *B,* Basilar involvement of the right lung. This bronchogram demonstrates a large pleural effusion or empyema. There is incomplete filling of the lower lobe with no bronchiectasis. (Courtesy of J. Fennessey.)

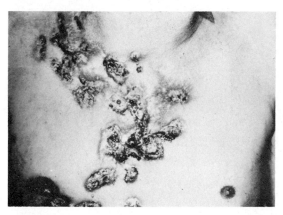

Figure 2–5 *A,* Actinomycosis. Multiple-discharging sinuses developing from hilar and cervical region. The disease is extended to the chest from the cervicofacial area (scrofuloderma type). Note the hyperplasia around the openings of the sinus tracts. *B,* Thoracic actinomycosis. Advanced disease in a cachectic patient. The lungs were seeded from aspirated material in the mouth and gingiva. The patient was an alcoholic with carious teeth.

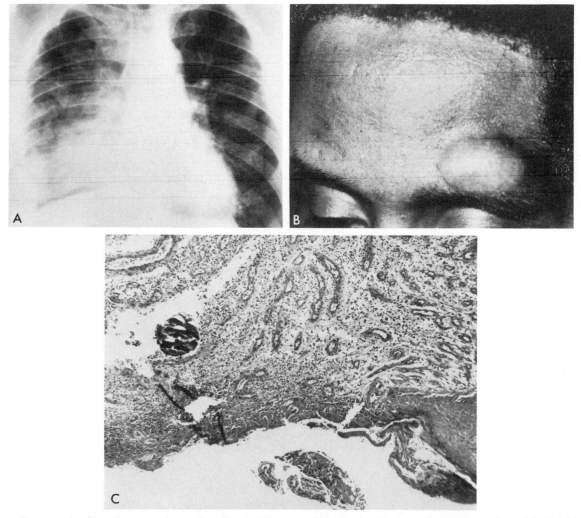

Figure 2–6 *A,* Actinomycosis. In this chest radiogram there is almost complete consolidation of the right lower lobe but the airways remain patent. There is a significant pleural reaction along the right lateral wall and a periosteal reaction on the upper surface of the eighth rib. *B,* Tender fluctuant nodule over left eye representing hematogenous dissemination from the primary pulmonary infection. *C,* Superficial ulceration from another site of focal dissemination. Acute and chronic inflammation is seen along with granulation tissue and sulfur granules in the dermis.[47]

as chronic bilateral patchy pneumonia, apical cavitary pneumonia simulating late-stage primary tuberculosis, and/or an enlarging mass infiltrating the fissures mimicking carcinoma.[4] Pleural effusion is frequent. Lesions usually are nonencapsulated, and adhesions are common. Ribs are involved and show destructive and proliferative changes, as previously described. Although dissemination is infrequently reported in other forms of actinomycosis, it is relatively common in primary thoracic disease. In some series dissemination has occurred in 50 per cent of cases.[81] Since the onset of thoracic disease is usually quite insidious, peripheral cutaneous foci resulting from hematogenous dissemination may be the first indication of disease.[47] Aside from cutaneous and subcutaneous abscesses, focal lesions of the bone also occur. In advanced cases, lesions are found in the central nervous system as a cerebral abscess and in other organs.

The disease must be differentiated from tuberculosis, the distribution of lesions being helpful, and from nocardiosis, lung abscesses, bronchiectasis, and tumor.

Abdominal Actinomycosis. This form of the disease may result from perforation of the intestinal wall by such objects as fish and chicken bones or from knife and gunshot wounds. Perforating ulcers may also initiate abdominal actinomycosis, but most frequently the primary source is the diseased appendix. Such organs as the fallopian tubes, gallbladder, and liver may appear to be the primary sites, but thorough examination and history usually reveal another primary source. Rare primary infections of each of these organs probably occurs.

The initial symptoms of abdominal actinomycosis are insidious and and related to the involved organ. In primary colon disease,[80] an indistinct irregular palpable mass may be present and must be differentiated from carcinoma. In this type of infection, extension to the liver is frequent, and jaundice may occur. Involvement of the gallbladder and urinary tract with accompanying dysfunction has been noted, but other causes of cystitis, pyelonephritis, and cholecystitis must be ruled out. Extension to the spinal column produces destructive bone lesions, with collapse and compression of the spinal cord and psoas abscess. After extensive spread to contiguous tissue, penetration of the abdominal wall and sinus formation may occur, with extrusion of pus and sulfur granules. Unless draining sinuses

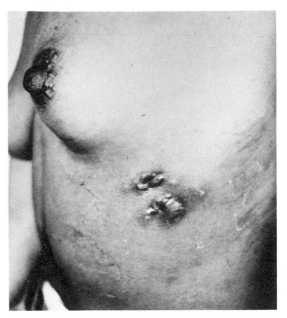

Figure 2–7 Actinomycosis on the skin of the chest wall. This was the result of a human bite.

develop, however, actinomycosis is rarely diagnosed before exploratory laparotomy.

X-rays of the abdomen (gallium scans are also useful[18]) show indistinct masses, enlargement of liver or spleen, or an abnormal process involving the vertebrae. The change in the latter is that of periostitis, with erosion of the articular facets, lamina, and transverse and spinous process. This is in contrast to tuberculous disease, which favors the vertebral bodies. Signs and symptoms include localized pain, with a palpable process, jaundice, weight loss, chills, spiking fever, vomiting, and nocturnal sweats. Differential diagnoses include neoplasms, amebiasis, liver abscess, psoas abscess of other origin, chronic appendicitis, salpingitis, and pyelonephritis.

Actinomycosis of Other Sites. Other areas of primary actinomycotic infection include the bladder, kidney, humerus, heart valve,[35] and CNS. Although hematogenous spread is rare, this must always be considered when lesions are found outside the usual areas. Direct inoculation into the skin, resulting in mycetoma, has been reported several times, and *A. israelii* is considered an etiologic agent of actinomycotic mycetoma. This disease and its pathology are discussed in a later chapter. Primary infections of the skin have been the result of human bites (Fig. 2–7), barbed wire, fist fights, hypodermic needles, insect stings, and so forth.[3] Primary infections of the bone, such as arthritis of the ankle, are recorded. Sometimes no primary site is discernible and it is assumed that infectious material was carried by the hematogenous route from mouth or colon to the site in the bone. Sherer and Dobbins have reviewed osseous actinomycosis.[73]

CNS involvement occurs in 3 to 11 per cent of cases in tabulated series.[21, 84] This is frequently the result of dissemination from a primary lung focus, although in a few cases no primary site was detected. The presenting symptoms are those associated with other bacterial brain abscesses: focal neurological signs, usually with the absence of fever or other symptoms of infection. The lesions may be single, multiple, or multiloculated. The enlarging lesion produces symptoms of increasing intracranial pressure. The foci are usually well encapsulated and only very rarely do they erode into a ventricle and result in meningitis. In very rare cases of basilar CNS disease, the infection was the result of direct extension to the calvarium or vertebral column from cervicofacial disease. Still more uncommon is dissemination from abdominal infection. In one case the sphenoid sinus appears to have been the primary site, and the patient presented with depression and headache.[63]

A new form of actinomycosis is being reported with some frequency. This is disease of the uterus, cervix, and vagina associated with the use of intrauterine devices. Abdominal pain with an enlarging mass is the usual presenting symptom. Masses may occur on the ovary or fallopian tube, in the abdomen, or in surrounding areas. Removal of the device and proper antibiotic treatment usually results in rapid resolution.[60, 72, 76, 83]

Lacrimal canaliculitis and conjunctivitis were described by Cohn in 1875.[19] *Actinomyces israelii*, *A. odontolyticus*, *Arachnia propionica*, and rarely *Nocardia asteroides*, *A. naeslundii*, or *Bifidobacterium eriksonii* have been isolated as agents.[13] Patients give a history of conjunctivitis that subsequently developed into a chronic intermittent creamy discharge from the corner of the eye. The disease is slowly developing and persists for months to two to three years or more. Probing of the canaliculi enables one to isolate one or more yellowish concretions, which can be removed by pressure or with a curette. They may adhere to the wall, however. The grains are soft and spongy to hard or granular. Infection with two or more agents is not uncommon.

Dental Plaque, Caries, and Periodontal Disease. Plaque initially results from the deposition of glycoproteins from saliva. These are then colonized by various bacteria, which may reach a population of 10^8 cells per gram wet weight. The composition of the population varies from tooth to tooth and from site to site on the same tooth. Plaque formation, both supragingival and subgingival, is the initial step in the development of caries and peridontal disease. Among the most important agents in the subsequent disease process are the anaerobic actinomycetes. Both *A. naeslundii* and *A. viscosus* produce extracellular polymers that enable them to adhere to the tooth surface or plaque or both. Following adherence, other bacteria aggregate with the *Actinomyces*. These include *Veillonella* and streptococci. The enamel surface beneath the plaque is subjected to the action of the metabolic products of the concentration of bacteria. These include organic acids such as lactic, acetic, formic, propionic, and succinic acid. As the pH is lowered to 5 or below, pits appear in the enamel. Proteolytic digestion of the dentine follows and the destructive caries process has been initiated. There are many other factors involved in the formation of caries, but the anaerobic actinomycetes appear to play a major role, along with *Streptococcus mutans* and perhaps other lactobacilli and filamentous bacteria.

Periodontal disease may also result from plaque. There are many contributing factors to periodontal disease, such as hypersensitivity to bacterial enzymes, endotoxins, and other metabolic products, but actinomycetes appear to be the predominant cause. Other bacteria preported to be involved in this disease include *Bacteroides*, *Bifidobacterium*, *Clostridium*, *Veillonella*, *Treponema*, and *Peptococcus* spp. The interaction of all these and their relation to periodontal disease are less well understood than the interactions leading to caries formation. The various anaerobic actinomycetes are universally present in such conditions and probably contribute to the pathology.

Actinomycosis presents such a variety of clinical signs and symptoms that it must be differentiated from a number of chronic infections and neoplastic diseases. In addition to those already mentioned, granuloma inguinale, staphylococcic actinophytosis (botryomycosis), sarcoidosis, carcinoma of the retroperitoneal tissues or iliac bones, typhoid fever, and the various deep mycotic infections should be ruled out.

Prognosis and Therapy. Prior to the use of antibiotics and sulfas, the prognosis of actinomycosis was poor. Therapy included surgical management of the involved area, along with irradiation, vaccination, and various types of nonspecific chemotherapy, such as thymol. The use of iodides probably resulted from their success in treatment of actinobacillus infection in cattle. However, they are of questionable value, their beneficial effect being derived from their direct action on the involved tissue rather than from their antimicrobial efficacy. The sulfonamides were introduced in 1938, and successful treatments with blood levels of 6 mg per 100 ml have been reported. Since the work of Nichols and Herrell,[59] penicillin has been considered the treatment of choice. Regimens vary from 1 to 6 million units to 10 to 20 million units IV per day, depending on the severity of disease. Therapy is administered for 30 to 45 days prior to surgical incision, drainage, or excision, or the lesions are allowed to heal by intention. Following surgical manipulation, treatment is continued for 12 to 18 months, with an oral dosage of 2 to 5 million units penicillin V per day. Two grams per day for a year of phenoxymethyl penicillin K is also recommended. Chlortetracycline, chloramphenicol, oxytetracycline, streptomycin, Terramycin, sulfadiazine, and other sulfonamides are useful as supplements to penicillin therapy. Successful treatment has been reported in cervicofacial actinomycosis using tetracycline alone administered in dosages of 250 mg, four times per day, for three weeks.[61] However, most authors agree that tetracycline is of limited value, especially in extensive disease. In cases in which penicillin allergy precludes its use, erythromycin and tetracycline have been found to be effective. *In vitro* sensitivity studies[41] indicate that all strains of *A. israelii* and *A. propionica* are sensitive to penicillin, erythromycin, tetracycline, clindamycin,[82] and lincomycin.[54] Owing to the adverse effects associated with clindamycin and lincomycin, the alternate to penicillin therapy is long-term erythromycin or tetracycline therapy. The organisms are resistant to sulfisomidine, metronidazole, and tinidazole. Lerner[46] investigated the effect of antibiotics on all species of *Actinomyces*. He found erythromycin and rifampin to be most active (MIC 0.008–0.25 μg per ml). Penicillin G, cephaloridine, minocycline, and clindamycin were also active (MIC 0.03–10 μg per ml). Cephalothin, ampicillin, lincomycin, tetracycline, doxycycline, and chloramphenicol also had MIC's within a therapeutic range.

The antimycotic agents — amphotericin B, nystatin, and griseofulvin — are not effective.

PATHOLOGY

Gross Pathology. Systemic actinomycosis often is not diagnosed until autopsy. Although the gross pathology may vary, depending on anatomic location, the usual picture is of a chronic, extending, suppurative, scarring, inflammatory process. Primary sites of infection are marked by a dense, cellular infiltrate, producing firm, nodular lesions of varying sizes which may be painfully swollen. This is followed by a softening of the tissues as granulomas form, and finally, suppuration within the granulomata to form multiple interconnecting abscesses and sinus tracts. This may be followed by cicatricial healing and spreading by burrowing, usually along fascial planes or muscle or tissue, leaving deep, communicating, scarred sinus tracts that may heal and later reopen. Debris and purulent material may be present within the lumen. Sinus tracts often burrow through bone, producing periosteal reaction and lytic lesions. The spinal column and ribs are more frequently involved than other osseous areas. The tendency of *Actinomyces* to involve bone is greater than that seen in *Nocardia asteroides* associated with nocardiosis but less than that of *N. brasiliensis* nocardiosis.

Autopsy. Cervicofacial actinomycosis is rarely encountered at autopsy, although extension from this site to the cervical spine and cranial cavity may result in fatal meningitis. An occasional complication is thrombosis of the jugular vein. Extension downward to the superior portion of the thoracic cage also occurs. Abdominal actinomycosis, on the other hand, is not uncommonly diagnosed at autopsy. Abdominal lesions are multiple abscesses, which most often are found in the abdominal wall. If the infection began in the large intestine or appendix, minute abscesses of the mucosa and submucosa or ulcerations may be found, along with rectal fistulae with multiple tracts. Extension from the retrocecal region along the iliopsoas muscle to the kidney may produce draining sinuses to the inguinal region, and may be accompanied by chronic active interstitial nephritis. The liver is frequently involved. Multilocular abscesses and pylephlebitis are seen. In cut section, foamy macrophages with large accumulations of lipids are seen at the edges of the abscesses and are responsible for the yellowish color observed on gross examination. Gray patches of scar tissue may also be present. Extension from the liver through the diaphragm to the pleural cavity is not uncommon. Involvement of the retroperitoneal tissues, psoas muscle, cecum, kidneys, spleen, fallopian tubes, and ovaries has been observed, with rare hematogenous spread to lungs, brain, meninges, or other tissues.

Primary actinomycosis of the lungs is manifested by multiple abscesses, bronchiectasis, pleural and pulmonary adhesions, and pleural thickening. Bronchial fistulae, emphysematous blebs in both lungs and pleura, dense fibrosis, empyema, and rare extension to the pericardium may occur. As discussed earlier, in about one quarter of reviewed cases, involvement of the bone has been described. Pathology of the scapula, humerus, lumbar vertebrae, ribs, sacrum, and long bones occurs. Occasional lesions involving arm, mastoid, orbit, axilla, and buttock have been described. The latter was a primary lesion, apparently the result of a bite (Fig. 2–7).

Histopathology. The histopathologic appearance of actinomycosis is similar in the various organs involved. The pathologic process consists of an acute pyogenic response, which evolves into a chronic granulomatous lesion, suppuration, and abscess formation. These may later heal, with fibrosis and scar formation. The characteristic granules, or drusen of German writers, have an adherent mass of polymorphonuclear neutrophils which are attached to the radially arranged eosinophilic "clubs" of the granule.

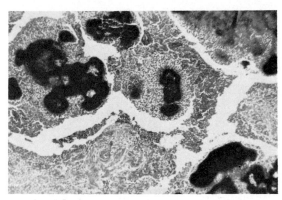

Figure 2–9 Actinomycosis. Multiple granules in sinus tracts. Hematoxylin and eosin stain. × 110.

This mass of leukocytes makes it easy to pick out granules in hematoxylin and eosin–stained preparations (Fig. 2–8). Enclosing the area is a band of celluar infiltrate, consisting of lymphocytes, plasma cells, epithelioid cells, and histiocytes, forming a huge abscess (Figs. 2–9 and 2–10). Peripherally, there is an area of granulation tissue containing lipoid-laden histiocytes (giving the area a yellowish color) and proliferating vascular structures. Giant cells are rare but may occur in the area. Since all lesions contain a mixed bacterial flora, some of the histopathologic changes may be due to associated bacteria. In many cases, abscesses are found to contain only such organisms and no granules of *Actinomyces israelii*. In these cases it is often necessary to search through many sections to find the typical gram-positive branching rods. Grains are usually found in the purulent areas of these sections.

The histologic picture of other bacterial and mycotic granulomata may resemble that

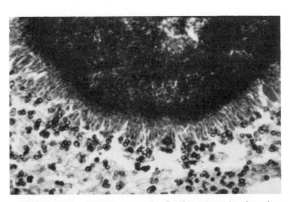

Figure 2–8 Actinomycosis. Sulfur granule showing eosinophilic "clubs" with attached neutrophils. Hematoxylin and eosin stain. × 400.

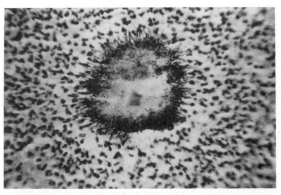

Figure 2–10 Actinomycosis. Gram stain of granule. The radiating rays (ray fungus) are gram-positive. × 400.

of actinomycosis. This is particularly true of staphylococcal actinophytosis (botryomycosis), which may present with a similar gross and microscopic appearance. The grains are of similar size and architecture, again with adherent neutrophils, but there are no eosinophilic "rays" radiating from the granule. Though the grain is basophilic on hematoxylin and eosin stain, it will be seen that there is central degeneration and a loss of basophilia. High-power examination shows that the organisms present in the granule are cocci. The histologic picture usually does not show the sharp demarcation between the leukocytic center and the peripheral zone. The lipid-laden histiocytes, a hallmark of actinomycosis, are missing.

The grains in actinomycosis, when examined microscopically, are seen to consist of dense rosettes of club-shaped filaments in radial arrangement. The individual rosettes vary from 30 to 400 μ, with the average between 100 and 300 μ. The large sulfur granules are visible to the naked eye and can often be seen on gauze dressings over an area of drainage. A single actinomyces filament has a diameter of 1 μ or less; thus it is easily differentiated from eumycotic filaments, which are from 3 to 4 μ. *Nocardia* filaments are also 1 μ in diameter. The rosette itself is made up of three kinds of structures: a central core of branching filaments, which are irregularly disposed but with a general radial arrangement; refringent, club-shaped bodies at the periphery, which are also radially arranged; and spherical coccuslike bodies interspersed throughout. The clubs may be plainly seen when the granules are crushed and examined in fresh preparation, or a stain such as eosin, which colors the sheaths of the clubs, may be used. The filaments are gram-positive; therefore, the Gram stain (Fig. 2–9) is useful for tissue sections in which grains are scarce and only unorganized filaments or coccoid forms are found. The Brown-Brenn or the MacCallum-Goodpasture modification of the Gram stain is useful. The methenamine silver stains are excellent because they stain all filaments regardless of age; however, they do not distinguish gram-positive actinomyces from gram-negative associates. In hematoxylin and eosin–stained sections, the core of the granules are basophilic with an eosinophilic periphery. Although quite useful when the grains are numerous, hematoxylin and eosin stain will not show single filaments. The organism is not acid-fast or

stained by special stains (PAS) for fungi. *Actinomyces* filaments are only occasionally seen not organized into grains. In such cases, they must be distinguished from *Nocardia asteroides*, which does not usually form grains. In general, *Nocardia* will show long, sinuous threads, in addition to short bacillary and coccoid elements, whereas *Actinomyces* consist of a fine meshwork of short, branching bacillary and coccoid forms.

The club-shaped bodies at the margin of the granules are conspicuous for their high refringency and generally structureless homogeneous appearance. They are pear-shaped swellings of the terminal ends of the filaments, which arise as distinct transformations of the filaments. In young colonies the hyaline substance that comprises the clubs is soft and water-soluble, but as the colony ages, the clubs attain a firmer consistency. Pine and Overman[65] have shown the clubs to have a different chemical composition from the cell walls and to be set down in layers that become firm with impregnation of $Ca_3(PO_4)_2$. Their formation appears to be associated with the resistance of tissues: when resistance to invasion is slight, they are absent; filaments alone are found. It is probable that the clubs are the result of antigen-antibody complexing. Clubs, as a rule, are more common in bovine than in human disease. They are infrequent or absent in grains of *A. naeslundii* and are not found in culture of *A. israelii* or *A. bovis*. Grains have not been reported in infection by *B. eriksonii*.

ANIMAL DISEASE

The majority of actinomycotic lesions in cattle are found in or about the head. In particular, the lower jaw is affected; hence the common name, "lumpy jaw," for this disease. Among dairy cattle, Guernseys more frequently show disease than other breeds.[5] The organism isolated is generally *A. bovis*, although *A. israelii* has been authenticated in two cases. In addition to the growth in the tongue and maxillary bone, actinomycotic lesions occur in the pharynx, lungs, skin, lymph glands, and subcutaneous tissue, especially of the head and neck, and occasionally in other organs, notably the liver. The growth of the parasite usually leads to the formation of a hard tumor, which gradually increases in size and burrows into the adjacent tissues, softening and disintegrating the bony structure of the head. At the same time, new tissue is forming so that great distortion often ensues. Extension of the disease takes place by gradual invasion of the contiguous tissues, metastases being uncommon. When death occurs, it is not as a rule due to any toxic effect but to the mechanical action of the tumor pressing upon or occluding the respiratory passages or interfering with the taking or mastication of food. Generalized actinomycosis in animals is rare. When it does occur, the bloodstream rather than the lymph seems to be the channel by which the disease is spread. Secondary abscesses are found mainly in the liver.

The characteristic yellow granules, or drusen, are found in the suppurating masses of the tumor. If the pus is shaken up with water in a test tube, the small granules become evident and sink to the bottom of the tube. These structures are so typical, their demonstration by microscopic examination in a case of obscure suppuration is sufficient to suggest a diagnosis. This must be substantiated by crushing the granules and staining for the typical forms. The new growth, granulation tissue, consists chiefly of epithelioid and spindle-shaped connective tissue cells; small giant cells are rarely present. To the naked eye, actinomycotic lesions in the lung and udder often resemble tuberculous nodules and, at times, have undoubtedly been mistaken for them. Microscopic examination suffices to establish their nature.

The incidence of actinomycosis in cattle is uncertain; in part, it is readily confused with actinobacillosis. A fair estimate of disease rate might be between 0.2 to 2 per cent in the United States.[5]

The "woody tongue" disease caused by the gram-negative *Actinobacillus lignieresii* is similar to actinomycosis in that the tongue, lymph nodes, and soft tissue are invaded, but there is little or no involvement of bone. Sinuses that contain a yellowish exudate and a few sulfur granules may form. Stains of the granules show them to be composed of numerous gram-negative rods instead of the gram-positive brunching rods of the anaerobic actinomycetes. *A. lignieresii* is gram-negative, nonmotile, nonencapsulated, catalase-positive, indole-positive nitrate-positive and produces acid from glucose, maltose, mannitol, and sucrose but not from xylose or inositol. Infections in man have been recorded.

In swine, actinomycosis most frequently involves udder, lungs, or internal organs rather than bone or jaw. *A. israelii*, *A. viscosus*, and *A. suis* have been isolated.[16] Erythromycin and sulphones have been used successfully[16] for treatment.

Canine actinomycosis has been reported on occasion. In the few cases where cultures were obtained *A. viscosus* was recovered,[22] soft tissue, jaw, internal organs, lung, skin, and flank have been involved. In the cat, pulmonary, epidural (*A. viscosus*), and tonsillar lesions have been recorded.[6, 49] Other cases have involved sheep, goats, horses, deer, moose, antelope, and mountain sheep. Few of these reports included isolation and culture, so the agent is unknown.

Experimental Animals

Attempts to infect experimental animals with *A. israelii* have in general been disappointing in that only a small proportion of inoculated animals develop disease, and lesions are limited and benign. Some success has been achieved by traumatizing the tissue first and including the associated bacterial flora with *A. israelii*. Experimental infection with pure cultures has been achieved in mice by Meyer[52] using hog gastric mucin to enhance the invasiveness of the organism. Others have had less success in rabbits, guinea pigs, and hamsters using repeated injections of the organism.[39] In experimental disease, the essential features of the natural infection are observed, including the formation of tuberclelike nodules and the development of structurally typical granules with clubs.[37]

IMMUNITY AND PATHOGENESIS

The fundamental mechanisms that allow the agents of actinomycosis to establish an infection are not known. The organisms are endogenous and live in equilibrium with the host under normal circumstance. Some underlying condition, injury, trauma, or some abrogation of defenses, be it only temporary, is needed before the disease can be elicited. Classically, cervicofacial actinomycosis is associated with caries, trauma to the jaw, and tooth extraction; thoracic with aspiration of the agent; and abdominal to trauma, surgery, or dissemination. Recently immunosuppression has been added to the list.[53] Puncture wounds resulting from bites (human), spines, fish bones in the intestine, hypodermic needles, swallowed pins, and so forth, have now been joined by those due to intrauterine devices (IUD),[67, 76] as resulting in localizing infections.

Whatever the predisposing event, the tissue is debilitated slightly and perhaps the redox potential is lowered. The agents proliferate, forming microcolonies within sinus tracts. The microcolonies appear to be fairly impervious to the cellular defenses. There are some laboratory data suggesting that *A. israelii* inhibits phytohemagglutination stimulation of lymphocytes from patients as well as normal controls.[53] As in true mycotic infections, cellular defenses are the chief mode of control and eradication. Apparently, any shift in balance either clears an early infection or permits the relentless chronic progressive nature of the disease, once established, to ensue. Very little is required in the initial infection to shift the equilibrium. Besides the presently available antibiotics, former methods such as heat, iodides, and thymol did result in clinical cures. Initial failure, however, allowed the organisms to elaborate their gamut of proteolytic enzymes and digest their way through the patient tissue, producing a chronic progressive disease.

SEROLOGY

Agglutinins, precipitins, and complement-fixing antibodies have been demonstrated in patients with actinomycosis, but there is little consistency in the reactions or correlation with the disease state. There is a notable lack of evidence that they are involved in combating or protecting the individual against disease. Skin test antigens elicit both immediate and delayed type reactions. Active defense against the organism probably occurs at a cellular level. It has been suggested that the "clubbing" of the organism seen at the periphery of the granule represents a response of the organism to the cellular defense of the host and is an antigen-antibody complex.

By use of the reciprocal agglutination absorption method, and subsequently fluorescent antibody, Slack and Gerencser[75] separated the various pathogenic actinomycetes into four groups: *A. naeslundii*, *A. bovis*, *B. eriksonii*, and *A. israelii*. Brock and Georg[11] have found two serotypes of the *A. israelii*. Serotype 1 represents 95 per cent of clinical isolants and differs from serotype 2 in that most stains show the classic molar tooth colony and have the ability to ferment arabinose. The majority of serotype 2 colonies are smooth and do not ferment the sugar. There is some evidence that *A. naeslundii* is a variant of *A. israelii*. A polyvalent fluorescent antibody is group-specific for the two serotypes of *A. israelii* as well as for *A. naeslundii* and is useful for identification of the organism and its demonstration in clinical material.[11] *A. viscosus* is now a frequent cause of actinomycosis. Human strains are all of serotype 2, whereas those found in peridontal disease of hamsters are serotype 1.

Serologic groups have been established within the genus *Actinomyces*. These are based on I.D., F.A., or C.W.A. (cell wall agglutination) tests. The fluorescent antibody test has been accepted most widely as a practical laboratory procedure. *A. bovis* is group B; *A. naeslundii*, group A; *A. israelii*, group D; *A. odontolyticus*, group E; and *A. viscosus*, group F. There are two serotypes within each group. By use of standard F.A. antisera, the identity of an unknown isolant can be confirmed.[34]

BIOLOGICAL STUDIES

Several interesting studies on the pathogenic mechanism of anaerobic actinomycetes have been reported recently. Using human polymorphonuclear neutrophiles, Engel et al.[26, 27] determined that *A. viscosus* had a direct chemotactic effect on such cells. They also have a cytomitogenic effect on serum and the ability to stimulate host immune cells to produce and release mediators of inflammation. Burckhardt et al.[14] found an extracellular glycan and cell supernate from *A. viscosus* that has mitogenic effects on B cells. The mitogens may plan a role in eliciting the plasma cell components found in lesions of actinomycosis.[26] The material responsible for cell mediated immune response in experimental infection was shown to be a soluble component of *A. viscosus*.[78]

The mechanisms of aggregation of the anaerobic actinomycetes and other bacteria have been investigated in relation to plaque formation.[50] Specific polysaccharides produced by *A. viscosus* are important in adhering to enamel, as are long surface appendages of *A. naeslundii*.[24]

Studies on the chemical composition of the anaerobic actinomycetales have been carried out. These reports indicate that in the genera *Actinomyces* and *Rothia*, neither LL or DL diaminopimelic acid (DAP) is present, LL-DAP is found in *Arachnia* and DL-DAP is present in *Bacterionema* species. Ornithine is absent in *Rothia, Arahnia, Bacterionema,* and *A. bovis* but present in the other *Actinomyces* species.[69] Lysine is present in the walls of all *Actinomyces* and *Rothia* species. Polar lipids have also been studied and differences between genera and species elucidated.[62]

LABORATORY DIAGNOSIS

Direct Examination

The isolation and identification of *Actinomyces* is difficult, but a tentative diagnosis can be made by finding the "sulfur granules" in purulent exudate. The pus may be spread out in a Petri dish and examined for the presence of granules. The granules vary in size from barely visible specks to 2.5 mm in diameter. They are yellowish-white to white, firm, and spherical or lobulated. They may be abundant in material from draining sinuses but are rare in sputum. A single granule may be placed in a drop of water, crushed under a microscope slide cover, and examined with the low-power objective. The granule will be seen as an opaque mass with a periphery of clear, gelatinous protrusions or club-shaped bodies (Fig. 2–11, *A*). Clubs are sometimes absent. A Gram stain of the material examined under oil immersion shows delicate, intertwined gram-positive filaments and coccoid and bacillary bodies. This is important in differentiating the granules of *Actinomyces* from those of other bacterial species. Staphylococcal granules are also gram-positive but are composed only of cocci, and gramnegative organisms, such as *Bacteroides* and *Proteus*, may also form loose pseudogranules. If granules are not present, a Gram stain of the material may reveal short-branching filaments, which may be *Actinomyces*. *Nocardia* can usually be ruled out by staining with the modified acid-fast stain.[7] They are acid-fast, as are the spores of *Streptomyces* species.

Culture Methods

Actinomyces may be cultured from purulent exudates, sputum, drainage material, and biopsy and autopsy material. If grains are present, they should be washed in several rinses of sterile saline to remove "associate" bacteria, then crushed in saline with a sterile glass rod against the side of a tube. The material can then be transferred to a brain-heart infu-

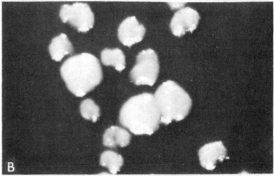
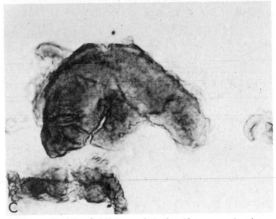

Figure 2–11 *A*, Wet mount of sulfur granule showing gelatinous club-shaped bodies. *B*, Molar tooth colonies of *Actinomyces israelii* growing on blood agar. One week, 37° C, anaerobic culture. *C*, Wet mount from an active lesion. The sulfur granule is hard and brittle and cracks in irregular planes. ×5.

sion blood agar plate, and a single loopful streaked onto several plates. Other primary isolation media have been recommended from time to time, but in the author's experience the previously mentioned procedure is sufficient in most cases. The plates are incubated under anaerobic conditions with CO_2 (a candle jar is sufficient for most isolants). A second set of plates is incubated aerobically to discern the oxygen requirements of the organism and for growth of associated bacteria.

The colonial morphology of *A. israelii, A. viscosus, A. bovis,* and *A. naeslundii* grown anaerobically on solid media is sufficiently distinctive that, with experience and the use of selected physiologic characteristics, the organisms can be differentiated from those of contaminating bacteria and from each other. At 48 hours, colonies of *A. israelii* are loose masses of branching filaments with a spider-like appearance or are white, rough, granular specks resembling sand grains. There are several variants in colonial morphology. These include: (1) very small colonies with only a few branching filaments, (2) larger colonies with many filaments, and (3) large colonies composed of very short filaments with angular branches, "frost," resembling *Rhodococcus. A. bovis* colonies are generally small, punctate, moist drops with entire edges, but when examined closely will be seen to have a smooth but grainy surface. *A. naeslundii* and *B. eriksonii* may be similar to either of the above at this stage, but typically *A. naeslundii* has a tangled dense center composed of a mass of diphtheroidal cells and rays of long, branched filaments projecting from the center of the colony. The colonies of both serotypes of *A. viscosus* resemble this, though the filamentous fringi is shorter in most human isolants.

After four to six days' incubation, the colonies are often less than 1 mm in diameter. At this time all organisms are usually opaque, dead white, or rarely have a slight gray or creamy tinge. *A. israelii* is generally a rough colony (R form), starting as a mass of branching filaments (spider) and developing into a lobulated, glistening "molar tooth" colony (Fig. 2–11,*B*). The rare S variant,or serotype

2 may be transparent and regular in form, thus resembling *A. bovis* (Fig. 2–12). *A. odontolyticus* may mimic any of these and is only separated by physiological characteristics. Mature colonies of *A. naeslundii* are usually smooth, circular, umbonate to convex, and entire. *A. viscosus* is circular or irregular, heaped (often with a central depression) and viscous in consistency. *A. odontolyticus* is deep red on blood agar and may be α or β hemolytic. The other species are not hemolytic.

Colonies are picked to freshly heated and cooled thioglycollate broth, heart infusion broth, or *Actinomyces* maintenance media.[7] In broth, *A. israelii* grows slowly, forming a hard, granular, fuzzy-edged colony. *A. bovis* generally grows as a smooth S form in which colonies are at first like dewdrops, then smooth, convex, and grainy with entire edges. The rare R variant may resemble *A. israelii*. In broth, *A. bovis* produces a soft, lobular colony that is easily dispersed. *A. naeslundii, A. viscosus,* and *B. eriksonii* resemble the latter. The *A. naeslundii* variant readily becomes microaerophilic or aerobic; *B. eriksonii* is strictly anaerobic. *A. bovis* and *A. israelii* may become microaerophilic. In physiologic tests, *A. israelii* usually ferments xylose and mannitol, reduces nitrate, and does not hydrolyze starch; *A. bovis* hydrolyzes starch, does not reduce nitrate, does not ferment mannitol and usually not xylose; *A. naeslundii* reduces nitrate but does not ferment xylose or mannitol. *A. viscosus* is catalase-positive, reduces nitrates, and ferments glucose and usually galactose and fructose but not arabinose, ribose, or xylose. *A. odontolyticus* is similar but is catalase-negative and ferments soluble starch. The anaerobic *Corynebacterium* organ-

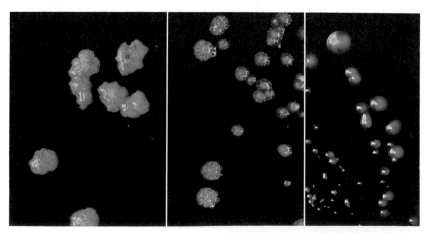

Figure 2–12 Colonies of *Actinomyces israelii* on brain-heart infusion agar after six days' incubation. *Left* and *center,* rough (R) "molar tooth" serotype 1. ×3. *Right,* colonies of the smooth (S) serotype 2. ×6. (From Rosebury, J., L. S. Epps, et al., 1944. J. Infect. Dis., 74:131–149.)

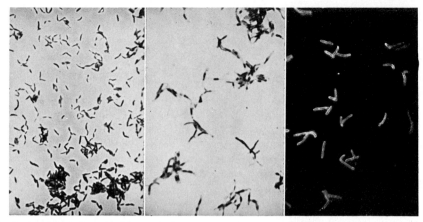

Figure 2–13 *Actinomyces israelii. Right,* darkfield. ×900. *Left and middle,* Gram stains of smooth and rough cultures, respectively. ×1200. (From Rosebury, J., L. S. Epps, et al. 1944. J. Infect. Dis. 74:131–149.)

isms are proteolytic, as determined by gelatin liquefaction, and are catalase-positive. As blood contains catalase, this test cannot be done on blood agars.

MICROBIOLOGY

Genus *Actinomyces* Harz 1877

Synonymy. *Streptothrix* Cohn 1875; *Cladothrix* Cohn 1875; *Discomyces* Rivolta 1878.

Actinomyces israelii (Kruse) Lachner-Sandoval 1898

Synonymy. *Streptothrix israeli* Kruse 1896; *Discomyces israeli* (Kruse) Gedoelst 1902; *Actinobacterium israeli* (Kruse) Sampietro 1908; *Nocardia israeli* (Kruse) Castellani and Chalmers 1913; *Cohnistreptothrix israeli* (Kruse) Pinoy 1913; *Oospora israeli* (Kruse) Sartory 1920; *Brevistreptothrix israeli* (Kruse) Lignières 1924. *Corynebacterium israeli* (Kruse) Haupt and Zeki 1933; *Proactinomyces israeli* (Kruse) Negroni 1934.

Colony Morphology. Brain-heart infusion blood agar plates. At 24 to 48 hours anaerobic incubation, serotype 1 *A. israelii* forms a minute, spiderlike colony with branching "mycelial" elements radiating from a central point. This is best observed by holding the plate at an angle to or above a light source. By ten days the colony is hard, lobulated, "molar tooth" gray-white, and glistening. The serotype 2 is small, shiny, entire, and "bacterial" at 24 to 48 hours and smooth convex entire and shiny at ten days.

Microscopic Morphology. Short rods with one or more branches and coccoid forms common in young cultures (Fig. 2–13). One μ diameters by 3 to 7 μ long. In old colonies and in tissue they may be long and extensive but less so than *Nocardia*. They are gram-positive and nonacid-fast.

Physiologic Characteristics.* Esculin hydrolysis +, catalase 0, indole 0, nitrate-variable, gelatin digestion 0, mannose A, lactose A, sucrose A, maltose A, raffinose A, xylose A, glycerol 0, mannitol variable, starch A, starch hydrolysis 0, microaerophilic.

Normal Habitat. Normal flora human oral, intestinal, and vaginal areas. Regularly associated with human actinomycosis. Also found as flora in many animals; has caused disease in cattle, swine, and so forth.

Actinomyces viscosus (Howell, Jordan, Georg, and Pine) Georg, Pine, and Gerencser 1969

Synonymy. *Odontomyces viscosis* Howell, Jordan, Georg, and Pine 1965.

Colony Morphology. BHI blood agar aerobic with CO_2. Dense center with a filamentous fringe, is opaque, cream to white, soft, mucoid to viscous in consistency. Hamster strains have an eccentric pit near apex. Human strains show various radial or concentric striations.

Microscopic Morphology. Coccoid, diphtheroid, and long sinuous branching filaments are found. At times each may predominate or there is a mixture of all.

Physiologic Characteristics. Esculin hydrolysis +, catalase + indole 0, nitrate 0, methyl red +, Voges-Proskauer +, starch hydrolysis 0, gelatin hydrolysis 0, arabinose 0, glucose A, glycerol A, inositol A, lactose A, maltose A, aerobic, facultative anaerobe.

---

*A = acid; 0 = not fermented or negative.

Normal Habitat. Oral cavity of man, rats, hamsters. Commonly involved in human and animal actinomycosis and other types of infection such as endocarditis.[35]

Actinomyces odontolyticus Batty 1958

Colony Morphology. BHI blood 48 hours usually smooth, and entire and may have a dense center. A few strains from spiderlike colonies similar to those of *A. israelii*. Some greening around the colony may occur, appearing as a type of α hemolysis. Mature colonies are red on blood agar.

Microscopic Morphology. Similar to *A. bovis*.

Normal Habitat. It is found in the oral cavity of man and has been isolated from systemic infection[55] and cervicofacial disease.[76]

Physiologic Characteristics. Esculin hydrolysis usually +, catalase 0, indole 0, nitrate +, methyl red +, Voges-Proskauer 0, starch hydrolysis 0, gelatin 0, arabinose ±, glucose A, glycerol A, inositol ±, lactose A, maltose A, mannitol 0, raffinose ±, salicin A, xylose ±, sucrose A, anaerobic, facultative aerobe.

Actinomyces bovis Harz 1877

Synonymy. *Discomyces bovis* (Harz) Rivolta 1878; *Sarcomyces bovis* (Harz) Rivolta 1879; *Bacterium actinocladothrix* Afanassieu 1888; *Nocardia actinomyces* Trevisan 1889; *Cladothrix bovis* Mace 1891; *Streptothrix actinomyces* Rossi-Doria 1892; *Oospora bovis* (Harz) Sauvageau and Radias 1892; *Actinocladothrix bovis* (Harz) Gasperinini 1892; *Actinomyces bovis sulfureus* (Harz) Gasperinini 1894; *Nocardia bovis* (Harz) Blanchard 1896; *Cladothrix actinomyces* (Harz) Mace 1897; *Streptothrix actinomycotica* (Harz) Foulerton 1899; *Streptothrix bovis communis* (Harz) Foulerton 1901; *Streptothrix bovis* (Harz) Chester 1901; *Sphaerotilus bovis* (Harz) Engler 1907; *Proactinomyces bovis* (Harz) Henrici 1939.

Colony Morphology. Brain-heart infusion blood agar plates. At 24 to 48 hours anaerobic incubation, the colony is small, smooth, glistening, entire, white, "bacteria"-like. At ten days it is smooth, flat-to-convex, opaque, white, grainy, similar to *A. israelii* serotype 2.

Microscopic Morphology. Short diphtheroid-like rods; coccoid bodies are found in young colonies. Short, branching rods occur in older colonies. Fragmentation to coccoid bodies occurs in tissue and culture. Gram-positive, nonacid-fast.

Physiologic Characteristics. Esculin hydrolysis +, catalase 0, indole 0, nitrate +, gelatin hydrolysis 0, mannose variable, lactose A, sucrose A, maltose A, raffinose 0, xylose 0, glycerol 0, mannitol 0, starch A, starch hydrolysis +, microaerophilic.

Normal Habitat. Normal flora of the oral cavity of cow and other animals. Not found in human flora and only rarely in human disease. The human infections were associated with cattle farms. Regularly causes actinomycosis in cattle and other animals.

Actinomyces naeslundii Thompson and Lovestedt 1951

Colony Morphology. Brain-heart infusion blood agar plates. After 24 to 48 hours anaerobic incubation, a small, smooth colony appears with mycelian radiation surrounding it. By ten days, the colony is convex-to-flat, entire edge, raised, rough, grainy, or similar to *A. bovis*.

Microscopic Morphology. Short rods, branching diphtheroid forms, and small coccuslike cells are seen in young colonies. Fragmentation to coccoid bodies occurs in time. Gram-positive, nonacid-fast.

Physiologic Characteristics. Esculin hydrolysis +, catalase 0, indole 0, nitrate 0, gelatin 0, mannose A, lactose A, sucrose A, maltose A, raffinose A, xylose 0, glycerol 0, mannitol 0, starch A, starch hydrolysis 0, microaerophilic.

Normal Habitat. It is part of the normal flora of human oral cavity. This species has caused many cases of human actinomycosis but has not been found in other animals.

Actinomyces meyeri is a recently described tentative species of *Actinomyces* (see Tables 2–1 and 2–2).

Bifidobacterium eriksonii* (Georg, Robertstad, Brinkman, and Hinklin) Georg 1974

Synonymy. *Actinomyces eriksonii* Georg, Robertstad, Brinkman, and Hinklin 1965.

Colony Morphology. Brain-heart infusion blood agar plates. At 24 to 48 hours anaerobic incubation, a dense mycelial colony appears, which is flat and granular with a conical center. By ten days, it is conical or convex, soft, entire, and white in color.

---

*Now considered in synonymy with *B. adolescentis* Reuter 1963.

Table 2–1 *Summary of Characteristic of the Etiologic Agents of Actinomycosis**

	A. bovis	A. israelii	A. viscosis†	A. naeslundii	Bifidobacterium eriksonii	Arachnia propionica
Aerobic	Variable	Variable	Variable	Variable	0	Variable
Anaerobic	+	+	±	+	+	+
Catalase	0	0	+	0	0	0
Litmus milk	Reduced	Reduced	Reduced	Reduced	Reduced with curd	Reduced
Propionic acid production	0	0	0	0	0	+
Fermentation:						
Glucose	A	A	A	A	A	A
Mannitol	0	A	0	0	A	A or variable
Mannose	0 or variable	A	A	A	A	A
Xylose	A or variable	A	0	0	A	0
Raffinose	—	A	A	A	A	A
Starch hydrolysis	+	0	0	0	+	0

*Data from Blair, J. E., et al. 1970. *Manual of Clinical Microbiology.* Bethesda, Md., American Society for Microbiology; Ajello, L., et al. 1963. *Laboratory Manual for Medical Mycology.* Public Health Service Publication No. 994. Washington, D.C., U.S. Government Printing Office.

†*A meyeri* is a new tentative species similar to *A. viscosis* and *A. bovis.* It is found as normal human mouth flora, has caused osteomyelitis and other infections, is distinguished from *A. bovis* by acid from ribose, and is catalase-negative.

Microscopic Morphology. Branching diphtheroid-like rods are seen in smears from young colonies. Fragmentation to coccoid bodies occurs in time. Gram-positive, nonacid-fast.

Physiologic Characteristics. Esculin hydrolysis +, catalase 0, nitrate 0, gelatin hydrolysis 0, glucose A, xylose A, mannitol A, raffinose A, starch hydrolysis +, mannose 0, strict anaerobe.

Rothia dentocariosa (Onishi) Georg and Brown 1967

Synonymy. *Actinomyces dentocariosa* Onishi 1949; *Nocardia dentocariosus* (Onishi) Roth 1957; *Nocardia salivae* Davis and Freer 1960.

Colony Morphology. Aerobic colonies on BHI are smooth, entire, and one mm at 24 hrs. Anaerobic colonies are small, highly filamentous and spiderlike, eventually become bread crumblike. No hemolysis on blood agar.

Microscopic Morphology. Coccoid, diphtheroid, or filamentous forms or a mixture. Young colonies are entirely coccoid; as they age they become filamentous.

Physiological Characteristics. Catalase +, indole 0, nitrate +, esculin hydrolysis +, gelatin hydrolysis 0, litmus milk 0, maltose A, glucose A, fructose A, arabinose 0, inositol 0, lactose 0, mannitol 0, mannose A, xylose 0, starch soluble 0, sucrose A. No DAP or arabinose in cell wall, but alanine, lysine, glutamic acid, and galactose present.

Table 2–2 *Colony and Microscopic Morphology of the Actinomyces*

	A. bovis	A. israelii	A. viscosis	Bifidobacterium eriksonii	A. naeslundii	Arachnia propionica
Microscopic morphology	Little branching, mostly diphtheroid, rarely filamentous	Rods, clubs, and branched forms common: rarely long and filamentous	Coccoid, diphtheroid, or filamentous in various combinations	Mostly diphtheroid, coccoid to awl shaped with bifurcate ends	Many branching rods and thick irregular forms, some diphtheroid	Short branching rods becoming, with age, long filamentous branched cells
Colony morphology	Shiny, grainy, entire-edged, convex colonies; some with scalloped edge	Very hard raised lobulated "molar tooth"; grow into agar and are removed whole; S or serotype 2 similar to A. bovis	Dense center mucoid or viscous consistency	Grainy lobulated edge convex colony	Similar to A. bovis	Filamentous or cobweb colony; flat to lumpy but less than A. israelii
Grains in man	+	+	+	−	Rare	+

Normal Habitat. It is part of the normal flora of the human mouth and throat. Though it was originally isolated from carious teeth, its relation to dental caries is unclear. So far it is not known to produce disease in humans or animals. It has been isolated from spinal fluid in blood culture, but clinical correlation was lacking.[9] It is as yet unknown for animals.

Arachnia propionica (Buchanan and Pine) Pine and Georg 1969

Synonymy. *Actinomyces propionicus* Buchanan and Pine 1962.

Colony Morphology. Brain-heart infusion blood agar plates. After 24 to 48 hours anaerobic incubation, a spiderlike colony similar to *A. israelii* appears. However, by ten days the colony becomes smooth, round, convex to flat, and appears more like *A. bovis*.

Microscopic Morphology. Typical *Actinomyces* morphology occurs in culture and tissue. Young colonies contain short-branching, diphtheroid-like forms, and fragmentation to coccoid bodies occurs in time.

Physiologic Characteristics. Esculin hydrolysis +, mannose A, mannitol A, raffinose A, xylose variable, starch hydrolysis 0, nitrate +, gelatin hydrolysis 0, microaerophilic. Cell wall contains LL-DAP (Type I).

Normal Habitat. It is part of the normal flora of the human mouth. It is regularly encountered as an agent of human actinomycosis[33] and lacrimal canaliculitis.[43]

Tables 2–1 and 2–2 summarize the preceding material.

REFERENCES

1. Adeniyi-Jones, C., J. A. Minielly, and W. R. Matthews, 1973. *Actinomyces viscosus* in a bronchial cyst. Am. J. Clin. Pathol., *60*:711–713.
2. Anon. 1977. Actinomycosis in 1976. Br. Med. J., *1*:1037.
3. Aulagi, A. 1972. Primary cutaneous actinomycosis. Br. Med. J., *2*:828–829.
4. Balikian, J. P., T. H. Cheng, et al. 1978. Pulmonary actinomycosis. A report of three cases. Radiology, *128*:613–616.
5. Becker, R. B., C. J. Wilcox, et al. 1964. Genetic aspects of Actinomycosis and Actinobacillosis in cattle. Ohio Res. Bull. No. 938, 24 pp.
6. Bestetti, G., V. Buhlman, et al. 1977. Paraplegia due to *Actinomyces viscosus* in a cat. Acta Neuropathol., *39*:231–235.
7. Blair, J. E., and E. H. Lennette, et al. 1970. Manual of Clinical Microbiology. Washington, D.C., American Society for Microbiology.
8. Blanksma, L. J., and J. Slipper. 1978. Actinomycotic dacryocystitis. Ophthalmologica, *176*:145–149.
9. Blevins, A., C. Semilic, et al. 1974. Common isolation of *Rothia dentocariosa* from clinical specimens studied in the microbiology laboratory. Abst. ASM, *73*:117.
10. Bostroem, E. 1891. Untersuchungen über die aktinomykose des menschen. Beitr. Pathol. Anat. Allg. Pathol., *9*:1–240.
11. Brock, D. W., and L. K. Georg. 1969. Characterization of *Actinomyces israelii* serotypes 1 and 2. J. Bacteriol., *97*:589–593.
12. Brock, D. W., L. K. Georg, J. M. Brown, M. D. Hicklin. 1973. Actinomycosis caused by *Arachnia propionica*. Am. J. Clin. Pathol., *59*:66–77.
13. Brown, J. R. 1973. Human Actinomycosis. A study of 181 subjects. Human Pathol., *4*:319–330.
14. Burckhardt, J. J., B. Guggenheim, and A. Heft. 1977. Are *Actinomyces viscosus* antigens B cell mitogens? J. Immunol., *118*:1460–1465.
15. Burgher, L. W., G. W. Loomis, and F. Ware. 1973. Systemic infection due to *Actinobacillus actinomycetemcomitans*. Am. J. Clin. Pathol., *60*:412–415.
16. Buschmann, G., and C. Duprée. 1976. Orale Behandlung der Gesäugeaktinomykose der Schweines. Dtsch. Tieraerztl. Wochenschr., *83*:14–17.
17. Causey, W. A. 1978. Actinomycosis. *In* Vinken, P. J., and G. W. Bruyn (eds.), Handbook of Clinical Neurology, Vol. 35, Infections of the Nervous System, pp. 383–394 Amsterdam, Elsevier-North Holland Pub. Co.
18. Chandarlapaty, S. K. C., M. Dusal, et al. 1975. Gallium accumulation in hepatic actinomycosis. Gastroenterology, *69*:752–755.
19. Cohn, F. 1875. Untersuchunger uber Bacterien II. Beitrage zur Biologie der Pflanzen, III:141–207.
20. Coleman, R. M., L. K. Georg, and A. Rozzell. 1969. *Actinomyces naeslundii* as an agent of human actinomycosis. Appl. Microbiol., *18*:420–426.
21. Cope, V. Z. 1939. Actinomycosis. London, Oxford University Press.
22. Davenport, A. A., G. R. Carter, and R. G. Schirmer. 1974. Canine actinomycosis due to *Actinomyces viscosus*: report of six cases. Vet. Med. Small Anim. Clin., *69*:1444–1447.
23. Ellen, R. P., and I. B. Balcerzak-Raczkowski. 1975. Differential medium for detecting dental plaque bacteria resembling *Actinomyces viscosus* and *Actinomyces naeslundii*. J. Clin. Microbiol., *2*:305–310.
24. Ellen, R. P., D. L. Walker, and K. H. Chan. 1978. Association of long surface appendages with adherence related functions of the Gram positive species *Actinomyces naeslundii*. J. Bacteriol., *134*:1171–1175.
25. Emmons, C. W. 1935. *Actinomyces* and Actinomycosis. Puerto Rico J. Public Health Trop. Med., *11*:63–76.
26. Engel, D., J. Clagett, et al. 1977. Mitogenic activity of *Actinomyces viscosus*. I. Effects on murine B and T lymphocytes and partial characterization. J. Immunol., *118*:1466–1471.
27. Engel, D., E. Epps, and J. Clagett. 1976. In vivo and in vitro studies on possible pathogenic mechanisms of *Actinomyces viscosus*. Infect. Immun., *14*:548–554.
28. Engel, D., H. E. Schroeder, and R. C. Page. 1978. Morphological features and functional properties of human fibroblasts exposed to *Actinomyces viscosus* substances. Infect. Immun., *19*:287–295.

29. Erikson, D. 1949. The morphology, cytology and taxonomy of the *Actinomyces* group. Medical Research Council (Great Britain) Special Report Series No. 240.

30. Flockton, H. I., and T. Cross. 1975. Variability in *Actinomyces vulgaris*. J. Appl. Bacteriol., *38*:309–313.

31. Georg, L. K., L. Pine, et al. 1969. *Actinomyces viscosus* comb. nov., A catalase-positive, facultative member of the genus *Actinomyces*. Int. J. Syst. Bacteriol., *19*:291–293.

32. Georg, L. K., G. W. Robertstad, et al. 1965. A new pathogenic anaerobic *Actinomyces* species. J. Infect. Dis., *115*:88–99.

33. Gerencser, M. A., and J. M. Slack. 1967. Isolation and characterization of *Actinomyces propionicus*. J. Bacteriol., *94*:109–115.

34. Gerencser, M. A., and J. M. Slack. 1976. Serological identification of *Actinomyces* using fluorescent antibody techniques. J. Dent. Res., *55*:184–191.

35. Gutschick, E. 1976. Endocarditis caused by *Actinomyces viscosus*. Scand. J. Infect. Dis., *8*:271–274.

36. Hammond, B. G., C. F. Steel, and K. S. Peindl. 1976. Antigens and surface components associated with virulence of *Actinomyces viscosus*. J. Dent. Res., *55*:19–25.

37. Harvey, J. C., J. R. Cantrell, et al. 1957. Actinomycosis: its recognition and treatment. Ann. Int. Med., *46*:868–885.

38. Harz, C. O. 1879. *Actinomyces bovis*, ein neuer Schimmel in den Geweben der Rindes, Jahresbericht der K. Central-Tierharztliche, Hochschule in Munchen fur, 1877–1878, *5*:125

39. Hazen, E. L., and G. N. Little. 1958. *Actinomyces bovis* and "anaerobic diphtheroids." J. Lab. Clin. Med., *51:968*–976.

40. Holmberg, K., and C. E. Nord. 1975. Numerical taxonomy and laboratory identification of Actinomyces and Arachnia and some related bacteria. J. Gen. Microbiol., *91*:17–44.

41. Holmberg, K., C. E. Nord, and K. Dornbusch. 1977. Antimicrobial in vitro susceptibility of *Actinomyces israelii* and *Arachnia propionica*. Scand. J. Infect. Dis., *9*:40–45.

42. Hopkins, R. 1973. Primary actinomycosis of the parotid gland. J. Oral Surg., *11*:131–138.

43. Jones, D. B., and N. M. Robinson. 1977. Anaerobic ocular infections. Trans. Am. Acad. Ophthalmol. Otolaryngol., *83*:309–331.

44. Lacey, J., and M. Goodfellow. 1975. A novel actinomycete from sugarcane bagasse: *Saccharopolyspora hirsuta* gen. et. sp. nov. J. Gen. Microbiol., *88*:75–85.

45. Lambert, F. W., J. M. Brown, and L. K. Georg. 1967. Identification of *Actinomyces israelii* and *Actinomyces naeslundii* by fluorescent-antibody and agar-gel diffusion techniques. J. Bacteriol., *94*:1287–1295.

46. Lerner, P. I. 1974. Susceptibility of pathogenic actinomycetes to antimicrobial compounds. Antimicrob. Agents Chemother., *5*:302–309.

47. Legum, L. L., K. E. Greer, and S. F. Glessner. 1978. Disseminated actinomycosis. South. Med. J., *71*:463–465.

48. Lord, F. T. 1910. Presence of actinomycosis in contents of carious teeth and tonsillar crypts of patients without actinomycosis. J.A.M.A., *55*:1261.

49. Lotspeich, M. 1974. Actinomycosis in a cat. Vet. Med. Small Anim. Clin., *69*:571–572.

50. McIntire, F. C., A. E. Vatter, et al. 1978. Mechanism of coaggregation between *Actinomyces viscosus* T14V and *Streptococcus sanguis* 34, Infect. Immun., *21*:978–988.

51. Meyen, F. J. F. 1827. *Actinomyces*. Strahlenpilz. Eine neue Pilz-Gattung Linnaea, *2*:433–444.

52. Meyer, E., and P. Verges. 1950. Mouse pathogenicity as a diagnostic aid in the identification of *Actinomyces bovis*. J. Lab. Clin. Med., *36*:667–674.

53. Miller, B. J., J. L. Wright, and B. P. Colquhoun. 1978. Some etiologic concepts of actinomycosis of the greater omentum. Surg. Gynecol. Obstet., *146*:412–414.

54. Mohr, J., E. R. Rhoades, and H. G. Muchmore. 1970. Actinomycosis treated with lincomycin. J.A.M.A., *212*:2260–2261.

55. Morris, J. F., and P. Kilbourn. 1974. Systemic actinomycosis caused by *Actinomyces odontolyticus*. Ann. Int. Med., *81*:700

56. Mousseau, P. A., and M. C. Mousseau-Brodu. 1973. L'actinomycose abdominale. J. Chir., *106*:569–588.

57. Muhle, I., J. Rau, and J. Ruskin. 1979. Vertebral osteomyelitis due to *Actinobacillus actinomycetemcomitans*. J. A. M. A., *241*:1824–1825.

58. Negroni, P. 1963. Datos estadisticos sobre 50 casos de actinomycosis y de su tratamiento vacunoterapica. Rev. Argent. Dermatol. *20*:458.

59. Nichols, D. R., and W. E. Herral. 1948. Penicillin in the treatment of actinomycosis. J. Lab. Clin. Med., *33*:521–525.

60. O'Brien, P. K. 1975. Abdominal and endometrial actinomycosis associated with an intrauterine device. Can. Med. Assoc. J., *112*:596–597.

61. O'Mahoney, J. B. 1966. The use of tetracycline in the treatment of actinomycosis. Br. Dent. J., *121*:23–25.

62. Pandhi, P. N., and B. F. Hammond. 1978. The polar lipids of *Actinomyces viscosus*. Arch. Oral Biol., *23*:17–21.

63. Per-Lee, J. H., A. A. Clairmont, J. C. Hoffman, et al. 1974. Actinomycosis masquerading as depression headache: case report — management review of sinus actinomycosis. Laryngoscope, *84*:1149–1158.

64. Pine, L., and Georg, L. K. 1969. Reclassification of *Actinomyces propionicus*. Int. J. Syst. Bacteriol., *19*:267–272.

65. Pine, L., and J. R. Overman. 1966. Differentiation of capsules and hyphae in clubs of bovine sulphur granules. Sabouraudia, *5*:141–143.

66. Prauser, H. 1976. *Nocardioides*, a new genus of the order, Actinomycetales. Int. J. Syst. Bacteriol., *26*:58–65.

67. Purdie, D. W., M. J. Carty, and R. I. F. McLeod. 1977. Tuboovarian actinomycosis and the IUCD. Br. Med. J., *2*:1392.

68. Radford, B. L., and W. J. Ryan. 1977. Isolation of *Actinomyces viscosus* from two patients with clinical infections. J. Clin. Pathol., *30*:518–520.

69. Reed, M. J., and R. T. Evans. 1974. Chemical structure of the cell wall peptidoglycan of *Actinomyces viscosus*. Arch. Oral Biol., *19*:1083–1086.

70. Rosebury, J., and L. S. Epps. 1944. A study of the isolation, cultivation and pathogenicity of *Actinomyces israelii* recovered from the human

mouth and from actinomycosis in man. J. Infect. Dis., *74*:131–149.

71. Scharfen, J. 1974. Untraditional glucose fermenting actinomycetes as human pathogens. I. *Actinomyces naeslundii* as a cause of abdominal actinomycosis. Zentralbl. Bakteriol., *232*:308–317.

72. Schiffer, M. A., A. Elguezabal, et al. 1975. Actinomycosis infections associated with intrauterine contraceptive devices. Obstet. Gynecol., *45*:67–72.

73. Sherer, P. B., and J. Dobbins. 1974. Actinomycotic arthritis: a case report. Med. Ann. D. C., *43*:66–68.

74. Slack, J. M., and M. A. Gerencser. 1966. Revision of serological group of *Actinomyces*. J. Bacteriol., *91*:2107.

75. Slack, J. M., and M. A. Gerencser. 1976. Proposal and description of ATCC 13683 and ATCC 12102 as neotype strains of *Actinomyces bovis* Harz 1877 and *Actinomyces israelii* (Kruse) Lachner-Sandoval 1898 respectively. Int. J. Syst. Bacteriol., *26*:85–87.

76. Stenhouse, D., D. G. McDonald. 1975. Cervicofacial and intraoral actinomycosis. Br. J. Oral Surg., *13*:172–182.

77. Suzur, F. 1974. Actinomycosis of the female genital tract. N.Y. State J. Med., *74*:408–411.

78. Thompson, L., and S. A. Lovestedt. 1951. An actinomyces-like organism obtained from the human mouth. Proc. Staff Meet. Mayo Clin., *26*:169–175.

79. Turner, D. W., B. S. Roberson, and R. W. Longton. 1976. Cell-mediated immune response to products of *Actinomyces viscosus* cultures. Infect. Immun., *14*:372–375.

80. Udagawa, S. M., B. A. Portin, and W. M. Bernhoft. 1974. Actinomycosis of colon and rectum: report of two cases. Dis. Colon Rectum, *17*:687–695.

81. Varkey, B., F. B. Landis, et al. 1974. Thoracic actinomycosis. Arch. Int. Med., *134*:689–693.

82. de Vries, J., and K. C., Bentley. 1974. Clindamycin in the treatment of cervicofacial actinomycosis. Int. J. Clin. Pharmacol., *9*:46–48.

83. Wagman, H. 1975. Genital actinomycosis. Proc. R. Soc. Med., *68*:228–229.

84. Weese, W. C., and I. M. Smith. 1975. A study of 57 cases of actinomycosis over a 36-year period. Arch. Int. Med., *135*:1562–1568.

85. Witwer, M. W., M. F. Farmer, et al. 1977. Extensive actinomycosis associated with an intrauterine contraceptive device, Am. J. Obstet. Gynecol., *128*:913–914.

86. Wolff, M., and J. Israel. 1891. Ueber reincultur des Actinomyces und eine eubertragborkeit auf thiere. Arch. Pathol. Anat. Physiol. Klin Med., *126*:11–59.

3 Nocardiosis

DEFINITION

Nocardiosis is an acute or chronic, suppurative (less commonly granulomatous) disease caused by the soil-inhabiting aerobic actinomycetes, *Nocardia asteroides, Nocardia brasiliensis*, and *Nocardia caviae*. There are three distinct clinical syndromes that may evolve: (1) primary cutaneous, (2) primary subcutaneous and (3) primary pulmonary and systemic. The subcutaneous disease conforms to the clinical entity known as actinomycotic mycetoma and is discussed in Chapter 5. Pulmonary, systemic disease and the rarely encountered primary cutaneous disease[69] are covered in this chapter. Primary pulmonary infection may be subclinical or pneumonic, chronic, or, rarely, acute, and may become systemic by hematogenous spread. The organism has a predilection for the central nervous system and, less commonly, other organs, such as the kidney. The patients affected are generally debilitated by other diseases or by medication, and *N. asteroides* and *N. caviae* are opportunists rather than primary pathogens. *N. brasiliensis* is more virulent and may be considered to be a primary pathogen. Primary cutaneous disease occurs in patients following trauma and contact with soil.[69] The infection may present as cellulitis, pustules, pyoderma, or a lymphocutaneous form mimicking sporotrichosis.[15] Sulfur granules are seen on occasion and the disease would probably evolve into a mycetoma. *N. asteroides* and *N. brasiliensis* are the etiologic agents. Sometimes cutaneous lesions appear to heal, but metastases appear later in the brain.

HISTORY

In 1888, Nocard[53] described an aerobic actinomycete causing *farcin du boeuf* in cattle in Guadeloupe. He called the organism *Streptothrix farcinica*. The disease was confused with farcy or glanders in Europe and the West Indies, and the early reports were so contradictory and equivocal it is difficult to equate the description with the specific disease, nocardiosis. Trevisan erected the genus *Nocardia* in honor of Nocard in 1889, with the type species *Nocardia farcinica*. The description was incomplete, and the cultures have been lost; for this reason, many authors contend the type species should be *Nocardia asteroides* (Eppinger) Blanchard. Eppinger[28] later (1891) described a pseudotuberculosis in a patient with miliary pulmonary involvement, bronchial node disease, and brain abscesses, which showed branching hyphae in pus. The name *Cladothrix asteroides* was given. The organism was renamed *Nocardia asteroides* by Blanchard in 1896.[18] Since then, the organism has been described in a number of medical and veterinary reports and called *Streptothrix carnea, S. freeri, S. asteroides, Actinomyces gypsoides*, and *Proactinomyces asteroides*. The early literature did not always differentiate the disease from actinomycosis; therefore, the incidence is difficult to assess. Henrici and Gardner in 1921 accepted 26 reported cases as valid. In 1943, Waksman and Henrici precisely differentiated the etiologic agent, *N. asteroides*, from other actinomycetes, and following this, in 1946 Kirby and McNaught[40] described 32 additional cases. Ballenger and Goldring[8] in 1957 accepted 95 cases as authentic, while Murray et al.[51a] in 1961 added more. Since that time, new aspects of nocardiosis have appeared in multiple case reports, particularly in relation to its association with other diseases and with the use of antileukemic drugs, cytotoxins, immune depressants, and corticosteroids.[81] The opportunistic nature of the infection is now emphasized[68] and there are approximately 1000 cases each year in the United States.[14, 30]

Although nocardiosis can occur in an apparently healthy individual, most cases are now seen in debilitated patients. This is not surprising in view of the enormous increase in opportunistic infections being recorded.[81] In renal transplant patients, 87 per cent of deaths are due to infections in some series[25a] The agents include gram-negative bacteria, *Candida, Aspergillus, Nocardia, Histoplasma,* and *Pneumocystis carinii.* In cancer patients, 14 per cent of all deaths are due to overwhelming fungal and nocardial infection. Formerly rare, nocardiosis is now a frequently encountered disease.

Recent studies, which include a more careful examination of the sputum and reevaluation of reported isolations, both in frank disease and as an incidental finding, indicate that the disease is more common than previously thought. Many of these infections may be of a minor, transient, benign, or chronic nature, without clinical pulmonary disease. Dissemination may occur in the absence of overt lung pathology, so that care must be taken when diagnosing disease in extrapulmonary areas as primary. There is also evidence that the organism may be a transient member of the normal flora of the tracheal and bronchial tree.[37]

The genus *Nocardia* formerly contained various species isolated from the clinical entity mycetoma. Most of these have been transferred to the genera *Streptomyces* and *Actinomadura. N. brasiliensis* is a common agent of mycetoma and can also cause systemic nocardiosis. *N. caviae* is rarely encountered in either clinical entity. *N. asteroides* accounts for most of the pulmonary and systemic cases of nocardiosis as well as about half of the primary cutaneous infections. The role of this organism in mycetoma is dubious; most reports of its isolation probably represent misidentified *N. caviae.* An organism designated *N. farcinica* (Kyoto-1) is closely related, if not identical, to *N. asteroides.* It accounts for a few cases of pulmonary infection.[36]

ETIOLOGY, ECOLOGY, AND DISTRIBUTION

Several species of the genus *Nocardia,* viz., *N. asteroides, N. brasiliensis, N. caviae,* and perhaps *N. farcinica,* are valid etiologic agents of the clinical disease nocardiosis in humans (Fig. 3–1). Recent studies by Gonzalez-Ochoa.[31, 32] emphasize that *N. brasiliensis* is more virulent than *N. asteroides* or *N. caviae:* the former is able to cause infection readily in experimental animals as well as systemic disease in normal patients, whereas the latter two species showed marked strain variation as far as experimental disease was concerned. It has been concluded that *N. asteroides* is rarely a primary pathogen but is rather an opportunist. Other investigators have reported varying results.[42]

Gordon and Hagen[33] first isolated *N. asteroides* from soil by the paraffin baiting technique. These findings have been confirmed by other investigators, and it appears that the organism has a worldwide distribution. *N. brasiliensis* and *N. caviae*[26] have also been recovered from soil. *N. asteroides* has been associated with disease in cattle, small animals,

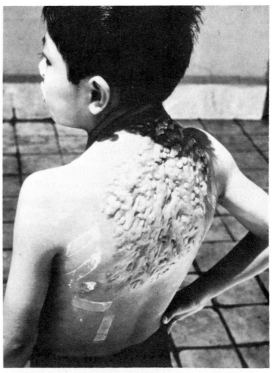

Figure 3–1 Nocardiosis due to *Nocardia brasiliensis.* Pulmonary infection with sinus formation and massive involvement of subcutaneous and cutaneous tissues. (Courtesy of H. Levine.)

and fish and has also been isolated from the normal skin.[76] Hosty,[37] while surveying the sputa from tuberculosis patients in an Alabama hospital, found *N. asteroides* in 175 of 85,000 specimens. The patients had no evidence of nocardial infection. Thus the *Nocardia* might be considered as a saprophyte or minor secondary invader in these cases. As these patients all had some lung disease, it is difficult to extrapolate to the normal population; moreover, survey studies have not been adequate. It is suggested, however, that *N. asteroides* may be at least a transient member of normal pulmonary flora. Such a carrier state was recently confirmed by Frazier et al.[30] Nine of 25 positive cultures for *N. asteroides* came from roentgenographically normal patients. Although preexisting lung disease may favor a frank infection by the organism, this is not a necessary prerequisite.

Nocardiosis has been reported in all parts of the world. No racial or occupational pattern is noted. Even though the organism is in the soil, rural incidence is no higher than urban. In reviewed cases, the male to female ratio is given as 2:1 or 3:1. Though infection has been reported in all age groups, the incidence is higher in the 30- to 50-year-old category. Nocardiosis in children has been reviewed by Idriss et al.[38]

About 1000 cases of nocardiosis occur in the United States each year.[14] *N. asteroides* accounts for about 90 per cent of cases, *N. brasiliensis* for nearly 7 per cent and *N. caviae* for less than 3 per cent. An organism presently defined as *N. farcinica* is recovered from a very few cases. In about 50 per cent of cases there is a known predisposing factor such as steroid therapy, cytotoxins, immunosuppression, leukemia, lymphoma, or neutropenia. Such infections are "opportunistic." In perhaps another 30 per cent a defect in cellular defenses is suspected or surmised. In at least 15 to 20 per cent of cases, however, no known or suspected defect can be found. These statistics relate to infections with *N. asteroides* and *N. caviae*. As stated earlier, *N. brasiliensis* is more virulent and may be considered essentially a primary pathogen. Thus in geographic areas where it is highly endemic, pulmonary nocardiosis is a primary bacterial pneumonia not requiring predisposing circumstances.

N. brasiliensis has been recovered from soil in several parts of North and South America, and a few reports indicate it to be of worldwide distribution. In the United States it is found sporadically in many states; it causes mycetoma and sporotrichoid lesions as well as pulmonary and systemic disease. In the central plateau of Mexico, in parts of Brazil, and in scattered areas in between, it is highly endemic. In these areas pulmonary and systemic infections account for a significant morbidity among the local population (Fig. 3–1).

CLINICAL DISEASE

Primary Pulmonary Disease. With rare exceptions, nocardiosis is a pulmonary disease of respiratory origin resulting from the inhalation of spores. Cerebral nocardiosis associated with pulmonary disease is frequently reported (about 27 per cent of pulmonary infection involves the central nervous system also), and there is conclusive evidence of hematogenous spread. Primary or secondary lesions in the gastrointestinal tract at a site of preexisting mucosal ulceration and, rarely, appendiceal involvement may result from ingestion of contaminated material or sputa from infected lungs. Nocardiosis has occurred at the site of a bruise where there was no rupture of the integument; therefore caution must be observed in assessing skin lesions as primary. However, authentic primary cutaneous nocardiosis has been described.[78]

In approximately 75 per cent of all reported cases, symptomatic pulmonary involvement has been noted. A pneumonia develops which may be lobular or lobar (Fig. 3–5D) and is often chronic. The lesion may be a solitary lung abscess, an acute necrotizing pneumonia, scattered infiltrates resembling miliary tuberculosis, a pulmonary mycetoma, or a progressive fibrosis and extension to the pleura and chest wall, with penetration simulating actinomycosis. It is evident, then, that with such a variety of presenting signs the clinical diagnosis is difficult and necessitates laboratory confirmation. The presenting symptoms are not distinctive. Fever is present, ranging from 37.2 to 41° C. Anorexia, malaise, weight loss, pleurisy, nocthidrosis, and, rarely, chills may be found. Cough is usually nonproductive and hacking at first but becomes mucopurulent and bloodstreaked. If cavitation occurs, massive hemoptysis may result. Consolidation of one or more lobes of the lung may occur, but caseation and granuloma formation are absent or rare. There is no favored lung site, as apical, hilar, and basilar regions are equally involved. The percussion note may be impaired, with diminished breath sounds over an affected area, and rales may be present. Mediastinal adenopathy may be so great as to cause a superior vena cava syndrome.

Secondary Sites. Though metastatic lesions may occur anywhere in the body, the brain is the most common secondary site of nocardiosis. Brain lesions may be multiple, but a large single abscess results by extension or coalescence (Fig. 3–2). Usually the meninges are not involved enough to cause diagnostic changes in the spinal fluid. The symptoms appear to be primary if lung involvement is minimal. The onset is sudden or gradual and the symptoms include, in descending frequency, headache, lethargy, convulsions, peripheral numbness, nuchal rigidity, psychic confusion, aphasia, tremor, and paresis. Rarely, involvement of the spinal column may lead to compression. Though several adjacent vertebrae may be involved, the disks tend to be spared. Mottled osteolytic lesions may occur in the vertebrae or cranial bones. Capsulization of nocardial abscesses is sometimes seen with vascularizations, which appear as "tumor blushes" on angiogram.

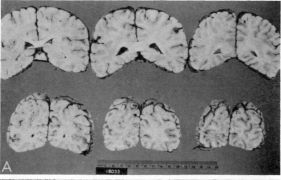

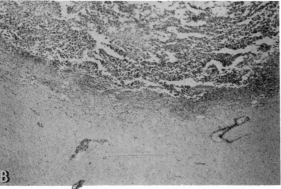

Figure 3–2 *A,* Nocardiosis. Abscess in brain from patient who had primary pulmonary disease. *B,* Histologic picture of brain lesion showing necrosis. ×100.

Recently, Beaman et al.[10] reviewed 243 cases of nocardiosis in the United States. Dissemination to the brain was found in 27 per cent, skin in 9 per cent, kidney in 8 per cent, pleura and chest wall in 8 per cent, eye in 3 per cent, liver in 3 per cent, and lymph nodes in 3 per cent. In 23 per cent the underlying condition was steroid therapy; 17 per cent, immunosuppression (leukemia, lymphoma, pancytopenia, and other deficiencies); and 12 per cent, pulmonary disease, together with a number of other conditions. In their cases, 49 per cent had no discernible underlying condition. Some cases of CNS disease occurred after inoculation of the organism into the skin through trauma.[61] Meningitis, though rare, has also been reported. In a case reviewed by Causey,[25] meningitis coexisted with pulmonary infection.

The kidney is probably the third most commonly involved organ in nocardiosis. Lesions extend from the cortex into the medulla (Fig. 3–5E). On occasion, the disease may metastasize to the pericardium, myocardium, spleen, liver, and adrenals. Lymphadenopathy occurs at various sites, but particularly in the cervical and axillary nodes. In contradistinction to actinomycosis, bone involvement is rare but reported. The eyes are rarely infected,[71] and symptoms include papilledema, blurring of visual fields, diplopia, and paralysis of the extrinsic muscle. Again, it is evident that these are vague findings that can be attributed to a number of other conditions. Such intraocular disease is the result of dissemination from a primary pulmonary focus. It is to be differentiated from primary keratitis following abrasion or injury.[64] *N. asteroides* is but one of dozens of organisms involved in such infections (Chapter 27). Canaliculitis of the lacrimals in the absence of other symptoms has been attributed to *N. asteroides* as well as to other actinomycetes.

X-ray Findings

A variety of manifestations can be seen in roentgenograms of pulmonary nocardiosis.[7, 63] These vary from a solitary nodule to lobar consolidation and single isolated or extensive bronchopneumonia with distant dissemination (Figs. 3–3, 3–4, and 3–5). Nodular, excavating and acute necrotizing patterns have also been described.[57] There is often marked pleural thickening (often 2 to 3 mm) and diffuse mottling. More often the abscesses in the lung or parenchyma, though

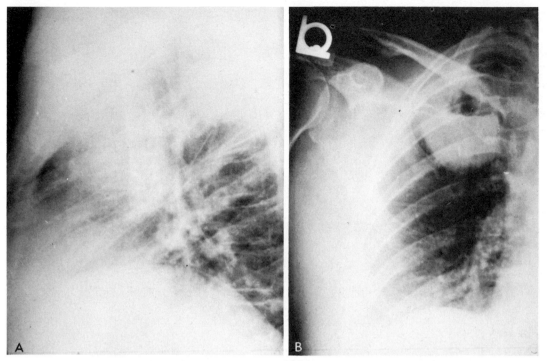

Figure 3–3   Nocardiosis. *A,* Pulmonary infection in a patient with lupus erythematosus on steroids. *B,* Large cavity in right upper lobe with an air fluid level. The cavity is well circumscribed in contrast to an acute bacterial disease which usually shows more of a reaction around lesions. (Courtesy of J. Fennessey.)

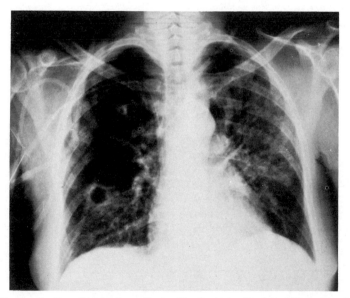

Figure 3–4   Nocardiosis. Cavitary lesion in patient with Crohn's disease. There is evident infiltrate, part of which is due to allergy to sulfas used in treatment. Infection resolved on trimethroprim-sulfa therapy.

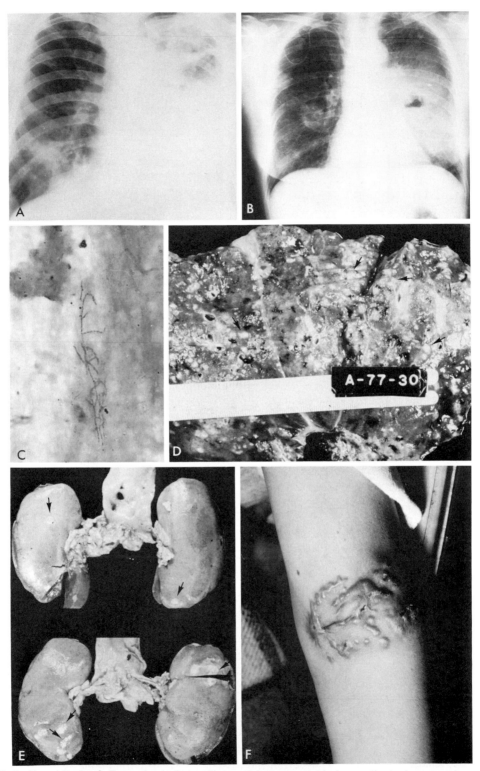

Figure 3–5 Nocardiosis. *A,* Extensive lesions of necrotizing pneumonia in a patient who was a heavy smoker but had no other underlying conditions. *B,* Nodular lesions in a woman receiving steroids. The large nodule on the left is excavating. *C,* Gram stain smear showing delicate beaded branching filament of *N. steroides.* ×1200. (Courtesy of V. Petrillo.[57]) *D,* Gross section of lung showing extensive consolidation. Arrows indicate nodular lesions. *E,* Hematogenous spread resulting in multiple abscesses, on the cortex (arrows) which then extend into the medulary section of kidney. (Courtesy of W. Causey.) *F,* Primary cutaneous infection due to *N. brasiliensis.* This serpinginous lesion developed after trauma and resulted in sinus tracts and grains as in mycetoma.

multiple, are too small to be visualized. As in other bacterial diseases, pleural effusion is common and a hallmark of pulmonary infection.

Differential Diagnosis

In pulmonary infection, nocardiosis mimics tuberculosis in all stages. Subsequent involvement of pleural and chest walls suggests actinomycosis. As there is no clinical syndrome that characterizes nocardiosis, other conditions must be considered. These include eumycotic infections of the lungs, bacterial brain abscess, carcinomatosis of lung and brain, sarcoma, or late syphilis. Systemic disease is often not diagnosed until surgical intervention or autopsy.

It is extremely important to differentiate nocardiosis from tuberculosis and actinomycosis, as each has a specific treatment.

Nocardiosis due to *N. asteroides* is a disease of the compromised patient. It is therefore a possible, often terminal complication in a number of disease states. Consequently, in patients suffering chronic progressive diseases or diseases causing impairment of normal immune mechanisms, the possibility of this or other facultative pathogens as infecting agents must be constantly considered. In general, both bacterial and eumycotic opportunistic infections are increasing in medical practice. In addition to immunosuppressive therapy, cytotoxins, and steroids, nocardiosis has been particularly associated with underlying diseases such as pulmonary alveolar proteinosis,[1] Cushing's syndrome, diabetes mellitus, and perhaps use of antibiotics.[68]

Though very rarely encountered, *N. caviae* is found in the same clinical settings as *N. asteroides.* It is not known whether the paucity of isolates reflects a lower virulence than *N. asteroides* or smaller populations in the natural reservoir. *N. caviae* infections are somewhat distinct in that grains similar to those seen in mycetoma and actinomycosis are often present. The range of clinical conditions reported[23, 25] include mycetoma, "mycotic" keratitis,[64] skin abscess, osteomyelitis, primary pulmonary infection, and fatal disseminated disease. Other conditions that were caused by *N. caviae* include pneumonia in a goat, mastitis in a cow, and infection of the hand of a baboon.

Rapidly developing systemic disease involving *N. brasiliensis* is occasionally reported.

Usually associated with classic mycetoma, this organism is regularly recovered from various areas outside its usual endemic zone as the cause of subcutaneous abscess with a fatal systemic course, skin abscess amenable to treatment, and rapidly fatal necrotizing lung infections.

Cutaneous Nocardiosis

A small number of cases of primary cutaneous disease involving species of the genus *Nocardia* appear sporadically in the literature. Satterwhite and Wallace[69] encountered seven patients in a 20-month period, indicating the disease is more common than previously suspected. All patients had a history of local trauma and contact with soil. The lesions included cellulitis, pustules, pyoderma, and a lymphocutaneous syndrome mimicking sporotrichosis.[15, 54] The last has been reported several times and usually involves *N. brasiliensis.* Some cases have even resolved with administration of iodides.[78] These infections, if in a normal host, are relatively benign. Often they resolve spontaneously or with a short course of sulfonamides. In a few cases, massive subcutaneous abscesses developed that disseminated to other organs. In some reports, minor diseases of the skin occurred before the appearance of cerebritis or infection in other anatomic areas.

The histopathology encountered in cutaneous infection is that of mixed pyogenic and granulomatous reaction. It is much more pyogenic than the histologic reaction associated with sporotrichosis. If allowed to progress, the infections caused by *N. brasiliensis* would probably become more granulomatous, demonstrate more granules, and develop into actinomycotic mycetoma.

PROGNOSIS AND THERAPY

In spite of the sensitivity of the infectious agent to several antibiotics and sulfas, the prognosis of nocardiosis is not good, especially if metastasis has occurred. Several factors contribute to this. Usually the patient is debilitated, his defenses compromised either by disease or by medication, so that he may be little more than culture media as viewed by the organism. Adequate and long-term therapy with appropriate drugs is necessary in nocardiosis. Before the advent of sulfonamide therapy, systemic disease was universally fatal. Uncomplicated pulmonary disease can be arrested, but systemic involvement presages difficulty. Central nervous system involvement is still about 87 per cent fatal. The overall mortality is currently estimated at 50 per cent.[14] This may be an exaggeration, as inapparent or mild infection often goes undiagnosed.

The treatment of choice in nocardiosis involves the sulfonamides. This has not changed since it was established in 1944, except that other drugs such as trimethoprim have been added to the regimen. Sulfonamide blood levels of 9 to 12 mg per 100 ml should be maintained and, in severe cases, 10 to 20 mg per 100 ml. This requires daily administration of 4 to 6 g of sulfadiazine or as high as 10 g.[3] Therapy should be continued for three to six months, or longer. Black and McNellis[17] recently studied the *in vitro* sensitivity of the sulfas and newer antimicrobial agents. None were superior to sulfadiazine and sulfamethoxazole, although fusidic acid was very effective. Clinically, the two sulfas are essentially equally effective. Presently the most widely accepted treatment regimen consists of a combination of sulfamethoxazole and trimethoprim. The dose is usually 800 to 1250 mg sulfa and 160 to 220 mg trimethoprim twice daily.[46] This combination has superseded sulfa alone because of cases in which CNS lesions developed during treatment. It was reported[3, 46] that sulfa levels in the cerebrospinal fluid will be one third of that of serum and that trimethoprim levels will be one half the serum level. This better penetration of the CNS protects the patient during the critical period of treatment. In *in vitro* studies using disk diffusion susceptibility and agar dilution techniques, Wallace et al.[79] tested 51 strains of *Nocardia* (48-*N. asteroides;* 2-*N. brasiliensis,* 1-*N. caviae*), and greater than 90 per cent of isolates were inhibited by amikacin and other aminoglycosides. Sulfasoxazole and trimethoprim-sulfamethoxazole appeared somewhat less effective, but only two thirds of the isolantes grew well enough to assay an effect. Those authors and others have emphasized the difficulties in antibiotic testing with *Nocardia* sp., and differences in susceptibility among strains and species have been noted by Lerner and Baum.[44]

In cases of serious sulfa sensitivity, other antibiotics have been used. These include minocycline, 200 to 600 mg per day; cycloserine, 250 mg every six hours; ampicillin, 1 to 6 g per day; and erythromycin, 1 to 3 g

per day.[35] Cycloserine has adverse reactions associated with its use, particularly CNS toxicity. Ampicillin in combination with erythromycin or sulfonamides was used for a time, but metastasis during therapy questions its protective efficacy. Minocycline may be the favored alternative to sulfa therapy.

Complications other than sensitivity during sulfa therapy have been reported. These include urinary concretions of sulfonamide crystals.[74] Treatment with fluids and alkali has resulted in improvement. Penicillin, chloramphenicol, and tetracycline, as well as other drugs, have been used in nocardiosis, but alone have failed to arrest the disease. Cephaloridine, 1 g every six hours, has been utilized successfully.

With adequate drug cover, surgical intervention is a valuable adjunct. Abscesses of the brain and lungs and empyema require drainage.[46] Excision or debridement of diseased tissue is also useful.

PATHOLOGY

Gross Pathology

Nocardiosis presents a similar picture in almost all involved organs. Multiple abscess formation with central necrosis, and little or no peripheral fibrosis are the usual findings, although some cerebral abscesses may be encapsulated. Only rarely is a tuberculoma or organized granuloma with giant cell infiltration seen. Caseation is infrequent. The picture, then, cannot be distinguished from infections caused by pyogenic bacteria. Grossly, the abscesses may appear homogeneous and shiny and may resemble lymphosarcoma. These lesions are generally soft and friable and on cut-section may contain granular, grayish-white material resembling liquefactive necrosis of tumor tissue (Fig. 3–5,*D*). There is less host response elicited in nocardiosis than in actinomycosis and, as a consequence, less fibrosis and scarring. The characteristic burrowing and sinus formation of the latter disease is generally absent in nocardiosis due to *N. asteroides,* as are true sulfur granules. Both, however, may be seen in infections due to *N. brasiliensis* and *N. caviae.*

The lung changes may be a combination of pneumonic consolidation, abscesses, diffuse miliary and nodular lesions, and cavities with adhesions. Pleural involvement gives rise to fibrinous pleurisy, empyema, multiple abscesses, and effusion with some fibrosis. Tracheobronchial adenopathy is found, but infarction and atelectasis are rare. Other organs present with suppurative abscesses of the general pyogenic cellular type.

Histopathology

Suppuration is the classic histologic picture of nocardiosis, regardless of the anatomic site of infection. The lesions are fibrogenic rather than granulomatous. There is a dense, polymorphonuclear leukocyte infiltrate within the nodules, as well as fibrin, lymphocytes, and plasma cells. Nodules may necrotize when they attain a large size, followed by cavitation. Fibrosis, Langhan's giant cells, and caseation necrosis rarely occur, thus allowing differentiation of the lesion from that generally seen in tuberculosis. Encapsulated lesions are sometimes found, especially in the brain.

The organism is easily missed in histologic examination, because it is not stained when the usual hematoxylin and eosin method is used. With the modified Gram's stain (Brown-Brenn, MacCallum-Goodpasture), delicate, multiple-branched filaments are seen, along with beaded chains of bacillary bodies. These may aggregate into a pseudogranule but are of loose consistency compared to the grains in actinomycosis. A modified acid-fast stain will demonstrate the organism, but its tendency to fragment into short, bacillary forms may cause it to be confused with the tubercle bacillus. The filaments are well demonstrated by the methenamine silver stains if staining time is extended. The special stains for fungi (PAS, Gridley) are not of use. Methods for staining *Nocardia* and *Actinomyces* have been reviewed by Robboy and Vickery.[67]

ANIMAL DISEASE

Natural infections by *N. asteroides* have been recorded in cats, cattle, dogs, horses, goats, pigs, rabbits, orangutan, monkeys, and fish. Considering that the organism is abundant in soil, it is not surprising that a variety of animals are infected. The original isolate from farcy of cattle was named *N. farcinica*, but it is probable that this organism was identical to the later described *N. asteroides*.[34] Infection in cattle can be epidemic. *N. asteroides* commonly infects the udder, causing mastitis; it may also be found in the jaw,

cervical lymph nodes, gastrointestinal tract, and other areas commonly invaded by *Actinomyces bovis*. It is also an important agent of bovine abortion. There is less bone involvement in nocardiosis, and granules are absent or rare. The disease entity farcy is characterized by tumefaction and granulomatous inflammation of the superficial lymph nodes, vessels, and subcutis, leading to suppuration. The organism is called *N. farcinica*.

In other animals epidemics among hogs and in trout hatcheries have also been recorded.[75] The dog is commonly infected, especially through cutaneous lesions. Similar types of pathology are seen in both animal and human infections. *N. asteroides* has also been isolated from several primates in captivity. There is no record of a particular underlying condition leading to infection, however. In monkeys and organgutans the pathology was similar to that seen in human pulmonary infections.[49] *N. caviae* is encountered in animal infections also. Bovine mastitis, pneumonia of a goat, and cutaneous disease in a baboon have been recorded. In some reports of mixed infection, *N. caviae* was accompanied by *Actinomadura madurae* or *N. autotraophica*.[24] Walton and Libke have reviewed nocardial infections in animals and their treatment.[80]

Another aerobic actinomycete, *Nocardiopsis dassonvillei*,[73] was isolated from cutaneous lesions similar to those in farcy and from granulomatous lesions of the intestine in cattle. Infection in man has thus far not been reported. The cell wall analysis and lipid content of this organism are similar to those of *Actinomadura madurae* and *A. pelletierii* and differ from other *Nocardia* and *Streptomyces* sufficiently that the organism is now classified in a new genus, *Nocardiopsis* (see Chapter 1).

EXPERIMENTAL INFECTION

Experimental infection in animals with *N. asteroides* has been investigated numerous times with varying results. Experimental disease does not appear to be of use in establishing identification of the species. Virulence of the organism in experimentally infected animals with varying degrees of immunosuppression has been investigated by Beaman and Maslan.[12] González-Ochoa[31, 32] studied the relative pathogenicity of *N. asteroides*, *N. brasiliensis*, and *N. caviae*. Grains were found in infections with the latter two species but not with *N. asteroides*. He concluded that the so-called *N. asteroides* isolated from mycetoma was in reality *N. caviae*. He emphasized the virulent nature of *N. brasiliensis* compared with *N. asteroides* and *N. caviae*. Other workers have reported regular and reproducible infections in mice with *N. asteroides*.[12, 29] In comparing *N. asteroides* and *N. brasiliensis*, Folb et al.[29] reported that suppurative abscesses were induced by *N. asteroides*, whereas granuloma resulted from infection with *N. brasiliensis*. They reported no difference in mortality between the two species.

The effect of immunosuppressive drugs on experimental infection has also been investigated by Beaman and Maslan.[12] Pulmonary infections in monkeys were produced by Mahajan et al.[47]

IMMUNITY

Natural resistance to infection by *N. asteroides* is probably high, as inhalation of the spores of this common soil organism is undoubtedly frequent. Natural barriers to infection such as the skin, mucociliary apparatus of the tracheobronchial tree, and phagocytosis and killing by wandering macrophages are effective in preventing infection in a normal, healthy individual. Transient colonization of the bronchi without infection has now been well documented. Recent work with experimental infections has documented the role of alveolar macrophages in clearing *N. asteroides* from lesions.[10] The possibility of hypersensitivity augmenting this effect has been debated for many years. Such a possibility was suggested by the work of Bojalil in developing a skin test antigen.[19] Many individuals living in areas highly endemic for *N. brasiliensis* reacted to antigen and have no evidence of disease. Early investigations indicated some role for humoral immunity as well. Drake and Henrici[27a] induced immunity of *N. asteroides* in rabbits and transferred this to guinea pigs by injections of serum from the immune donors. However, Krick and Remington,[41] using mice, were unable to transfer induced immunity either with serum or with lymphoid cells to nonimmune recipients. They did show that sublethal doses of the organism induce a population of activated peritoneal macrophages with increased microbicidal activity. Such populations were also produced by other, nonrelated, infections (e.g., intracellular parasites); the infected mice had a greater resistance to a primary nocardial infection. Similar results have been obtained using the RNA fraction from *Nocardia*.[77] In man, nei-

ther cellular nor humoral resistance factors have been documented. A degree of hypersensitivity to this constantly present organism is possible, however, as evidenced by low level reactivity in the general population to skin test antigens.

SEROLOGY

A skin test (asteroidin) described as not cross-reacting with tuberculin, was reported by Drake and Henrici, in 1943.[27a] Since then a number of skin test antigens have been produced,[60] but most have cross reactions with many diseases and reactions occur frequently within the normal healthy population. Skin tests for *N. brasiliensis* mycetoma were developed by Bojalil.[19] Pier and Enright[59] have developed a specific skin test for nocardiosis in cattle.

Vaccine prophylaxis was utilized without concrete evidence of efficacy before the advent of specific therapy. In experimental infections, Macotela-Ruiz and Mariat[48] could find no difference in susceptibility between animals sensitized with killed *Nocardia* cells and controls. Circulating antibodies have been reported in patients, but their role in defense has not been assessed.

Serologic procedures have been tried on sera from patients with nocardiosis, but no consistency of response has been found. Complement-fixing and agglutinating antibodies have been reported.[52] Recently, Pier and coworkers[60] developed a new complement-fixing antigen that has a sensitivity of 20 per cent or above, when tested in cases of human nocardiosis. Further clinical trials will indicate the usefulness of the procedure. The same authors[60] have reviewed serologic and immunologic testing used in animal infections. A fluorescent antibody technique for specific identification of *N. asteroides* has been attempted, but results were equivocal.[42, 43] At the present time, there is no reliable diagnostic or prognostic serologic procedure for the disease.

BIOLOGICAL STUDIES

Investigations concerning cell wall composition and associated carbohydrate components were reviewed in Chapter 1. The free mycolic acids of *N. brasiliensis*, *N. caviae*, and *N. asteroides* were examined by Minnikin et al.[50] The results indicate that these are valid species. These and other studies could not validate *N. farcinica* (ATCC 3318) as a separate species that would segregate from other reference strains of *N. asteroides*. In work by Riddell[66] using physiological and serologic criteria, most strains of *N. farcinica* appeared to be *Mycobacterium*, and the others *N. asteroides*. Taxonomic studies have also involved a cell component called nocobactins.[65] These are lipid-soluble, iron-binding compounds that appear to be species specific. Bacteriophage patterns can be used for typing of strains and may be useful in epidemiological studies.[62] Interesting work on L-forms of *Nocardia* has been reported by Bourgeois and Beaman.[22] In cell cultures of mouse peritoneal macrophage, wall-less intracellular spheroplasts of *N. asteroides* were persistent and multiplied. This has also been shown in rabbit alveolar macrophages.[10, 13] The ability to persist and reproduce in an altered state is found among many intracellular parasites for which cellular immunity is the primary mode of the host's defense. Beaman and Maslan[11] also found that variations in virulence may be due to related morphological changes during the growth cycle. Beadles et al.[9] could correlate variations in virulence and type of pathology induced with cell envelopes (Fig. 3–6).

LABORATORY DIAGNOSIS

Direct Examination

Sputum, pus, tissue material, and so forth can be examined for *N. asteroides*. The materials to be examined may be digested, then concentrated by centrifugation. *Nocardia* survives the processing of sputa for tubercle bacillas culture. In fact it grows out rapidly and luxuriantly on the usual tuberculosis media. Gram's stain of the material will show long, sinuous, branching, gram-positive filaments and fragmented bacillary bodies (Fig.

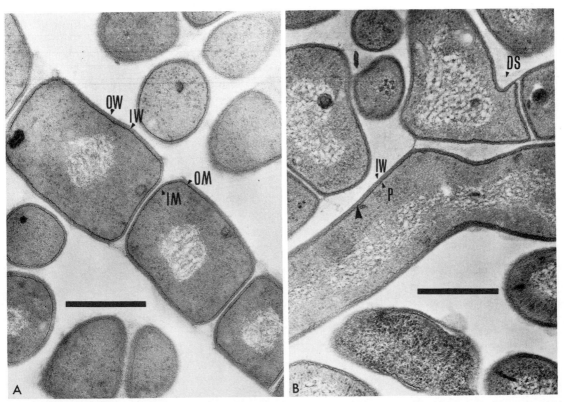

Figure 3–6 *A,* A filament of *N. asteroides* 14759 that has fragmented into sporelike short cells. There is a thin outer cell wall *(OW)* above a dark staining trilaminar, inner cell wall layer *(IW)*. The inner cytoplasmic membrane *(IM)* is also lighter staining than the outer membrane. The bar is approximately 0.5 μm. *B,* Filaments of *N. brasiliensis* T-11 with diffuse light staining substance *(DS)* apparent among the cells. There is a trilaminar inner wall *(IW)* similar to that seen in *N. asteroides*. There are many points of attachment *(p)* between the cell wall and the cytoplasmic membrane. (Courtesy of J. Land.)[9]

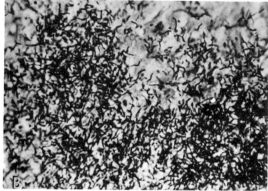

Figure 3–7 *A,* Gram stain of *Nocardia asteroides* in sputum. Note the long, sinuous, branching rods. × 1000. (From Rippon, J. W. *In* Burrows, W. 1979. *Textbook of Microbiology.* 20th ed. Philadelphia, W. B. Saunders Company, p. 727.) *B,* Actinomycosis. Grain stain of *Actinomyces israelii* in tissue debris. The bacilli are short, branching rods and tend to aggregate. × 1000.

3–7). The branching tends to be at long intervals and at right angles to the main axis of the mycelium. A modified acid-fast stain will show beaded or fragmented acid-fast bacillary forms. This distinguishes the organism from *Actinomyces,* but it may be confused with the tubercle bacillus. Both *N. asteroides* and *N. brasiliensis* are acid-fast; most other actinomycetes are not. Rarely, *N. asteroides* aggregates into a soft pseudogranule, whereas *N. brasiliensis* regularly forms true granules.

Culture Methods

Unlike *Actinomyces sp.,* *N. asteroides* is not fastidious and grows readily on ordinary laboratory media without antibiotics. It is aerobic but it grows out equally well under anaerobic conditions. Its optimal growth temperature is 37°C. Both *N. asteroides* and *N. brasiliensis* develop slowly on routine media.

By two to three weeks, they attain a diameter of 5 to 10 mm. The colonies are waxy, folded, and heaped at first. They may later develop areas of downy or tufted aerial mycelia. The whole surface may become dry and powdery. A musty, dirtlike odor is sometimes present. Though the classic colony of *N. asteroides* is glabrous, wrinkled, folded, and orange, the color range includes pink, white, buff, brown, lavender, and salmon (Fig. 3–8). The definitive work of Gordon and Mihm is recommended[34] as an aid in identification. The color range of colonies of *N. brasiliensis* is similar to that for colonies of *N. asteroides. N. caviae* produces a folded, heaped, convoluted colony on nutrient agar. The color is usually dull buff-pink.

Stained smears from growth on laboratory media will show coccoid and bacillary elements. The long, branched filaments are usually not seen from agar preparations but are easily obtained from growth in broth. Acid fastness is quite variable[67] from growth on agar but is enhanced if the organism is grown in litmus milk or on Löwenstein-Jensen media. Sporulation occurs by fragmentation into arthrospores. This can best be seen by the slide culture technique. Branching also may be observed, thus differentiating the organism from various mycobacteria and corynebacteria.

Colonies on Löwenstein-Jensen or similar isolation media for tubercle bacilli appear within one to two weeks. They are folded, moist, and glabrous and indistinguishable from so-called "atypical" mycobacteria. The

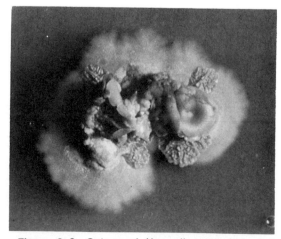

Figure 3–8 Colony of *Nocardia asteroides.* It is wrinkled, folded, glabrous, and bright orange. In one area, a white overgrowth is beginning where sporulation will occur.

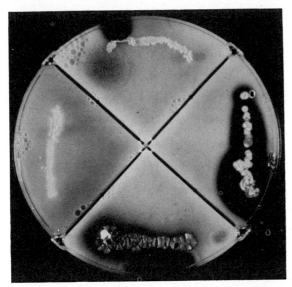

Figure 3–9 Example of physiological tests used in separation of *Nocardia* from other aerobic actinomycetes. This is a casein digestion plate. The right side has a positive control (*Streptomyces* sp.) and the left is a negative control (*N. asteroides* stock strain). The top is an unknown (*N. asteroides*) and the bottom is *Nocardiopsis dassonvillei.*

latter are strongly acid-fast and do not show branching.

The differentiation of *N. asteroides* from *N. brasiliensis* and *N. caviae*, together with other actinomycetes is detailed in the section on mycetoma. *N. asteroides* does not grow in gelatin test media, produce gelatinase, digest casein (Fig. 3–9) or tyrosine, or peptonize bromcresol purple milk. *N. brasiliensis* will have a positive reaction in all these previously mentioned tests. Both are urease-positive and partially acid-fast when the Kinyoun technique is modified by the use of 1 per cent H_2SO_4 to decolorize. *N. caviae* is separated from *N. asteroides* by its digestion of xanthine (Table 3–1).

MICROBIOLOGY

Genus *Nocardia* Trevisan 1889

Synonymy. *Proactinomyces* Jensen 1931; *Asteroides* Puroni and Leonardi 1935.

Nocardia asteroides (Eppinger 1891) Blanchard 1895

Synonymy. *Cladothrix asteroides* Eppinger 1891; *Streptothrix carnea* Rossi-Doria 1891; *Streptothrix eppingerii* Rossi-Doria, *Streptothrix asteroides* (Eppinger) Gasperini 1892; *Oospora asteroides* (Eppinger) Sauvageau and Radais 1892; *Actinomyces asteroides* (Eppinger) Gasperini 1894; *Actinomyces eppingeri* (Rossi-Doria) Berestnev 1897; *Discomyces asteroides* (Eppinger) Gedoelst 1902; *Streptothrix freeri* (Eppinger) Musgrave and Clegg 1907; *Actinomyces eppinger* (Rossi-Doria) Namyslowski 1912; *Actinomyces gypsoides* (Eppinger) Henrici and Gardner 1921; *Actinomyces asteroides* var. *serratus* Sartory, Meyer, and Meyer 1930; *Asteroides asteroides* (Eppinger) Puntoni and Leonardi 1935; *Proactinomyces asteroides* (Eppinger) Baldacci 1937; *Proactinomyces asteroides* var. *crateriforme* Baldacci 1938; *Proactinomyces asteroides* var. *decolor* Baldacci 1938.

Colony Morphology. Sabouraud's dextrose agar (SDA) 25°C or 37°C. The young colony is glabrous, folded, wrinkled, or granular, sometimes mycobacteria-like. The color is commonly orange but varies from yellow to pink to brown. In time, a fuzzy or chalky overgrowth of aerial mycelium occurs. A damp, soil-like odor may occur in time.

Microscopic Morphology. Delicate sinuous, irregularly branching mycelium is found in culture. The diameter averages 1 μ. In time, the mycelial strands fragment unevenly to form bacillary or coccoid bodies. The aerial mycelium also fragments, but

Table 3–1 *Separation of the Three Agents of Nocardiosis by Physiological Tests*

Species	Decomposition of			Acid-fast	·Lysozyme Resistance
	Casein	Tyrosine	Xanthine		
N. asteroides	−	−	−	+	+
N. brasiliensis	+	+	−	+	+
N. caviae	−	−	+	±	−
*N. Coeliaca	−	+	+	+	−
Streptomyces sp.	+	+	±	−	−

*Not pathogenic.

more regularly, and forms a series of squarish spores. Branching is less common and the mycelian strands are longer in tissue and culture as compared with *Actinomyces* species. It is gram-positive and irregularly acid-fast.

Physiologic Characteristics. Milk not coagulated, casein hydrolysis 0, tyrosine hydrolysis 0, xanthine hydrolysis 0, growth in 0.4 per cent gelatin 0, thermotolerant to 45°C for growth.[34] Flaky colonies on nutrient gelatin, no liquefaction. Not inhibited by 7 per cent salt, resistant to penicillin (5I.U.). *N. asteroides* as well as other *Nocardia* utilize a wide variety of carbon sources for energy, including paraffin, sebacic acid, acetate, butyrate-H-malate, jet fuel, asphalt, crude oil, and testosterone. Widely divergent physiological and virulence capacities are held by the organisms described as *N. asteroides* In the future, this heterogeneous group may be divided into several species. Reference Strain (R.S.) ATCC 19247.

Nocardia farcinica Trevasan 1889

Synonymy. *Streptothrix farcinica* (Trevisan Rossi-Doria 1891; *Bacillus farcinicus* (Trevisan) Gasperini 1892; *Actinomyces farcinicus* (Trevisan) Gasperini 1892; *Oospora farcinica* (Trevisan) Sauvageau and Radais 1892; *Actinomyces Lovis farcinicus* Gasperini 1894; *Streptothrix farcini bovis* Ritt 1899; *Bacterium nocardi* Migula 1900; *Streptothrix nocardii* (Migula) Foulerton and Jones 1901; *Cladothrix farcinica* (Trevisan) Macé 1901; *Bacillus nocardi* (Migula) Matzuschita 1902; *Discomyces farcinicus* (Trevisan) Krasil'nikov 1941.

This is the etiologic agent of farcy in cows in the Caribbean and Africa. The type culture ATCC 3318 has been found similar to *N. asteroides*.[16] However, strains isolated from disease in cows in Africa showed some consistent differences from *N. asteroides* as well as the type culture ATCC 3318. These include lack of benzamidase activity and lack of growth at 45°C. The colonies of *N. farcinica* are white, gray, or buff. The question of its taxonomy is still open, as is the isolation of *N. farcinica* from human disease.[36]

Nocardia brasiliensis (Lindenberg 1909) Castellani and Chalmers 1913

Synonymy. *Discomyces brasiliensis* Lindenberg 1909; *Streptothrix brasiliensis* (Lindenberg) Greco 1916; *Actinomyces mexicanus* (Lindenberg) Boyd and Crutchfield 1921; *Oospora*

brasiliensis (Lindenberg) Sartory 1920; *Actinomyces brasiliensis* (Lindenberg) Gomes 1923; *Nocardia pretoria* Pijper and Pullinger 1927; *Nocardia transvalensis* Pijper and Pullinger 1927; *Nocardia mexicanus* Ota 1928; *Actinomyces transvalensis* Nannizzi 1934; *Actinomyces pretoria* Nannizzi 1934; *Actinomyces violaceus* subsp. *brasiliensis* Krasil'nikov 1941; *Proactinomyces brasiliensis* (Lindenberg) Negroni 1954; *Proactinomyces mexicanus*, *Proactinomyces transvalensis*, and *Proactinomyces pretoria* Negroni 1954.

Colony Morphology. Optimum growth 30 to 37°C. The organism grows rapidly, producing a heaped, wrinkled folded colony. The color is yellow to yellow-white, tan, orange buff, orange or red-orange and closely resembles the colonies of *N. asteroides*. A white powdery overgrowth may occur and may indicate arthrospore production. Spore production is rare in *N. brasiliensis* as compared with *N. asteroides*. Up to 20 per cent of strains are nonacid-fast (partial), but all grow in lysozyme broth. In rare cases a brownish pigment diffuses into the media, particularly if it is rich in protein.

Microscopic Morphology. Branching filaments form extensive growth on most media. Diameter 1 μ. Fragmentation begins in the center of the colony at four days, producing irregular coccoid and rodlike cells. Growth occurs as long filaments in broth. Small beaded, acid-fast forms resemble a typical mycobacteria, but slide culture showing branching filaments of *N. brasiliensis* distinguish the two. *N. brasiliensis* grows as spherical colonies on the side of tubes containing 0.4 per cent gelatin; *N. asteroides* does not.

Physiological Characteristics. Decomposition of tyrosine, gelatin, and casein by *N. brasiliensis* distinguishes it from *N. asteroides*, and lack of xanthine decomposition distinguishes it from *N. caviae*. For energy, it utilizes a wide variety of carbon sources, such as paraffin acetate, butyrate, citrate, sebacic acid and propionate. Its growth is inhibited by 7 per cent salt, that of *N. asteroides* is not. Reference strain ATCC 19296.

Nocardia caviae (Erickson) Gordon and Mihm 1962

Synonymy. *Nocardia otitidis-cavarum* Snijders 1924; *Actinomyces caviae* Erickson 1935, *Nocardia caviae* (Erickson) Erickson 1935.

Colony Morphology. Growth is moderately rapid, and colonies resemble both *N.*

brasiliensis and *N. asteroides*. On special medium (Bennett's) a cream to peach color is produced. Tufts of short aerial mycelium are seen. Spores are uncommon. The tuft may be coiled.

Microscopic Morphology. By five days the extensively branched filamentous colony begins to fragment in the center. Bacillary and, later, coccoid elements are formed. It is often non–acid-fast unless grown on Löwenstein-Jensen, Dubos, or similar media or in litmus milk.

Physiological Characteristics. *N. caviae* decomposes xanthine and hydroxanthine but not casein, BCP milk, tyrosine, or gelatin. Acid is formed from inositol and mannitol, but not from arabinose, xylose, and galactose. *N. brasiliensis* ferments galactose. It survives 50°C for eight hours and grows sparingly at 40°C. Optimum temperature is 30°C. Reference strain ATCC 14629.

This organism can easily be confused with *N. coeliaca*, a non-pathogenic soil actinomycete. *N. coeliaca* is nitrate-negative and lysozyme-sensitive, decomposes tyrosine, and exhibits no growth at 45°C. Acid is formed from xylose and galactose. *N. caviae* has the opposite reaction in the aforementioned tests.

REFERENCES

1. Andriole, V. T., M. Ballas, and G. L. Wilson. 1964. The association of nocardiosis and pulmonary alveolar proteinosis. Ann. Intern. Med., *60*:266–275.
2. Anonymous. 1973. Nocardiosis, aspergillosis and reticulum cell sarcoma. Case Records of the MGH. New Engl. J. Med., *288*:1115–1121.
3. Aron, R., and W. Gordon. 1972. Pulmonary nocardiosis. Case report and evaluation of current therapy. South Afr. Med. J., *46*:29–32.
4. Arroyo, J. C., S. Nichols, and G. F. Carrol. 1977. Disseminated *Nocardia caviae* infection. Am. J. Med., *62*:409–412.
5. Bach, M. C., L. D. Sabath, and M. Finland. 1973. Susceptibility of *Nocardia asteroides* to 45 antimicrobial agents in vitro. Antimicrob. Agents Chemother., *3*:1–8.
6. Bach, M. G. 1975. The chemotherapy of infections due to *Nocardia*. Int. J. Clin. Pharmacol. Biopharm., *11*:283–285.
7. Balikian, J. P., P. G. Herman, and S. Kopit. 1978. Pulmonary nocardiosis. Radiology, 126:569–573.
8. Ballenger, C. N., Jr., and D. Goldring. 1957. Nocardiosis in childhood. J. Pediatr., *50*:145–169.
9. Beadles, T. A., G. A. Land, and D. J. Knezek. 1980. An ultrastructural comparison of cell envelopes of selected strains of *Nocardia asteroides* and *Nocardia brasiliensis*. Mycopathologia, *70*:25–32.
10. Beaman, B. L. 1977. *In vitro* response of rabbit alveolar macrophages to infection with *Nocardia asteroides*. Infect. Immun, *15*:925–937.
11. Beaman, B. L., and S. Maslan. 1978. Virulence of *Nocardia asteroides* during its growth cycle. Infect. Immun., *20*:290–295.
12. Beaman, B. L., and S. Maslan. 1977. Effect of cyclophosphamide on experimental *Nocardia asteroides* infection in mice. Infect. Immun., *16*:995–1004.
13. Beaman, B. L., and M. Smathers. 1976. Interaction of *Nocardia asteroides* with cultured rabbit alveolar macrophages. Infect. Immun., *13*:1126–1135.
14. Beaman, B. L., J. Burnside, et al. 1976. Occurrence of nocardial infections in the United States during a 24-month period (1972-74). J. Infect. Dis., *134*:286–289.
15. Belliveau, R. R., and F. Geiger. 1977. Lymphocutaneous *Nocardia brasiliensis* infection simulating sporotrichosis. West. J. Med., *127*:245–246.
16. Berd, D. 1973. *Nocardia asteroides*. A taxonomic study with clinical correlations. Am. Rev. Resp. Dis., *108*:909–917.
17. Black, W. A., and D. A. McNellis. 1970. Sensitivity of *Nocardia* to trimethoprim and sulphonamides *in vitro*. J. Clin. Pathol., *23*:423–426.
18. Blanchard, R. 1896. Parasites végétaux a l'exclusion des bactéries. *In* Bouchard (ed.) Traite de Pathologie Générale, Vol 2. Paris, G. Masson, pp. 811–926.
19. Bojalil, L. F., and A. Zamora. 1963. Precipitin and skin tests in the diagnosis of mycetoma due to *Nocardia brasiliensis*. Proc. Soc. Exp. Biol. Med., *113*:40–43.
20. Bollinger, O. 1877. Uber eine neue Pilzkrankeit beim Rind. Zbl. med. Wiss. *15*:481–485.
21. Boncyk, L. H., G. H. Millstein, and S. S. Kalter. 1976. Use of CO_2 for more rapid growth of the *Nocardia* species. J. Clin. Microbiol., *3*:463–464.
22. Bourgeois, L., and B. L. Beaman. 1974. Probable L-forms of *Nocardia asteroides* induced in cultured mouse peritoneal macrophages. Infect. Immun., *9*:576–590.
23. Causey, W. A., P. Arnell, et al. 1974. Systemic *Nocardia caviae*. Chest, *65*:360–362.
24. Causey, W. A., and B. Sieger, 1974. Systemic nocardiosis caused by *Nocardia brasiliensis*. Am. Rev. Respir. Dis., *109*:134–137.
25. Causey, W. A. 1974. *Nocardia caviae* — a report of 13 new isolations with clinical correlations. Appl. Microbiol., *28*:193–198.
25a. Cohen, M. L., E. B. Weiss, et al. 1971. Successful treatment of *Pneumocystic carinii* and *Nocardia asteroides* in renal transplant patients. Am. J. Med., *50*:269–276.
26. Conti-Diaz, I. S., L. Calegari, et al. 1973–74. Revision de los casos nacionales de infecciones por actinomicitos del genero *Nocardia*. Rev. Uruguaya Pat. Clin. Microbiol., *12*:25–33.
27. Dalovisio, J. R., and G. A. Pankey. 1978. *In vitro* susceptibility of *Nocardia asteroides* to amikacin. Antimicrob. Agent. Chemother. *13*:128.
27a. Drake, C. H., and A. T. Henrici. 1943. *Nocardia asteroides*, its pathogenicity and allergic properties. Am. Rev. Tuberc., *48*:184–198.
28. Eppinger, H. 1891. Uber eine neue pathogene *Cladothrix* und einedurch sie hervorgerufene pseudotuberculosis (Cladothrishica) Beitr. Pathol. Anat. Allg. Pathol., *9*:287–328.

29. Folb, P. I., R. Jaffe, et al. 1976. *Nocardia asteroides* and *Nocardia brasiliensis* infections in mice. Infect. Immun., *13*:1490–1496.

30. Frazier, A. R., E. C., Rosenow, et al. 1975. Nocardiosis: a review of 25 cases occurring during 24 months. Mayo Clin. Proc., *50*:657–663.

31. González-Ochoa, A. and A. Sandoval-Cuellar. 1976. Different degrees of morbidity in the white mouse induced *Nocardia brasiliensis, Nocardia asteroides* and *Nocardia caviae*. Sabouraudia, *14*:255–259.

32. González-Ochoa, A. 1973. Virulence of Nocardiae. Can. J. Microbiol., *19*:901–904.

33. Gordon, R. E., and W. A. Hagen. 1936. A study of some acid-fast actinomycetes from soil with special reference to pathogenicity to animals. J. Infect. Dis., *59*:200–206.

34. Gordon, R. E., and J. M. Mihm. 1962. The type species of the genus *Nocardia*. J. Gen. Microbiol., *27*:1–10.

35. Hoeprich. P. D., D. Brandt, et al. 1968. Nocardial brain abscess cured with cycloserine and sulfonamides. Am. J. Med. Sci., *255*:208–216.

36. Holm, P. 1975. Seven cases of human nocardiosis caused by *Nocardia farcinica*. Sabouraudia, *13*:161–169.

37. Hosty, T. S., C. McDurmont, L. K. Georg, et al. 1961. Prevalence of *Nocardia asteroides* in sputa examined by a tuberculosis diagnostic laboratory. J. Lab. Clin. Med., *99*:90–93.

38. Idriss, Z. H., R. J. Cunningham, et al. 1975. Nocardiosis in children. Report of 3 cases and review of the literature. Pediatrics, *55*:479–484.

39. Karassik, S. L., L. Subramanyam, et al. 1976. Disseminated *Nocardia brasiliensis*. Infection, *112*:370–372.

40. Kirby, W. M. M., and J. B. McNaught. 1946. Actinomycosis due to *Nocardia asteroides*. Arch. Intern. Med., *78*:578–591.

41. Krick, J. A., and J. S. Remington. 1975. Resistance to infection with *Nocardia asteroides*. J. Infect. Dis., *131*:665–672.

42. Kurup, P. V., H. S. Randhawa, R. A. Sandhu, and S. Abraham. 1970. Pathogenicity of *Nocardia caviae. N. asteroides*, and *N. brasiliensis*. Mycopathology, *40*:113–130.

43. Kurup, P. V., H. S. Randhawa, and N. P. Gupta. 1970. Nocardiosis: a review. Mycopathologia, *40*:194–219.

44. Lerner, P. I., and G. Baum. 1973. Antimicrobial susceptibility of Nocardia species. Antimicrob. Agents Chemother., *4*:85–93.

45. Lerner, P. I. 1974. Susceptibility of pathogenic actinomycetes to antimicrobial compounds. Antimicrob. Agents Chemother., *5*:302–309.

46. Maderazo, E. G., and R. Quintiliani. 1974. Treatment of nocardial infection with trimethoprim and sulfamethoxazole. Am. J. Med., *57*:671–674.

47. Mahajan, V. M., S. C. Padhy, et al.1977. Experimental pulmonary nocardiosis in monkeys. Sabouraudia, *15*:47–50.

48. Macotela-Ruiz, E., and F. Mariat. 1963. Mycetoma experimentale. Bull. Soc. Pathol. Exot., *56*:46–54.

49. McClure, H. M., J. Chang, et al. 1976. Pulmonary nocardiosis in an orangutan. Am. Vet. Med. Assoc., *169*:943–945.

50. Minnikin, D. E., P. V. Patel, et al. 1977. Polar lipid composition in the classification of *Nocardia* and related bacteria. Int. J. Syst. Bacteriol., *27*:104–117.

51. Mitchell, G., G. M. Wells, and J. S. Goodman. 1975. Sporotrichoid *Nocardia brasiliensis* infection. Response to potassium iodide. Am. Rev. Resp. Dis., *112*:721–723.

51a. Murray, J. F., S. M. Finegold, et al. 1961. The changing spectrum of nocardiosis. A review and presentation of nine cases. Am. Rev. Resp. Dis., *83*:315–330.

52. Musrafa, I. E. 1967. Studies of bovine farcy in the Sudan. I. Pathology of the disease. II. Mycology of the disease. J. Comp. Pathol. Ther., *77*:223–229, 231–236.

53. Nocard, E. 1888. Note sur la maladie des boeufs de la Guadaloupe connue sous le nom de farcin. Ann. Inst. Past., *2*:293–302.

54. Padron, S., et al. 1973. Lymphocutaneous *Nocardia asteroides* infection mimicking sporotrichosis. South. Med. J., *66*:609–612.

55. Palmer, D. L., R. L. Harvey, et al. 1974. Diagnostic and therapeutic considerations in *Nocardia asteroides* infections. Medicine, *53*:391–401.

56. Peterson, D. L., L. D. Hudson, et al. 1978. Disseminated *Nocardia caviae* with positive blood culture. Arch. Int. Med., *138*:1164–1165.

57. Petrillo, V. F., L. C. Severo, et al. 1981. Pulmonary Nocardiosis: Report of the first two Brazilian cases. Mycopathologia (in press).

58. Phillips, B. J., and W. Kaplan. 1976. Effect of cetylpyridinium chloride on pathogenic fungi and *Nocardia asteroides* in sputum. J. Clin. Microbiol., *3*:272–276.

59. Pier, A. C., and J. B. Enright. 1962. *Nocardia asteroides* as a mammary pathogen of cattle. III. Immunologic reaction of infected animals. Am. J. Vet. Res., *23*:284–292.

60. Pier, A. C., J. R. Thurston, et al. 1975. Serologic and immunologic tests for nocardiosis in animals. *In* Mycoses. Proceedings of 2nd Int. Conf. on the Mycoses. São Paulo, Pan Am. Health Org. (1975), pp. 162–167.

61. Poretz, D. M., M. N. Smith, et al. 1975. Intracranial suppuration secondary to trauma. Infection with *Nocardia asteroids*. J.A.M.A., *232*:730–731.

62. Pulverer, G. H. Schutt-Gerowitt, et al. 1975. Bacteriophages of *Nocardia asteroides*. Med. Microbiol. Immunol., *161*:113–122.

63. Rankin, R. S., and J. S. Westcott. 1973. Superior vena cava syndrome caused by *Nocardia* mediastinitis. Am. Rev. Resp. Dis., *108*:361–363.

64. Ralph, R. A., M. A. Lemp, and G. Liss. 1976. *Nocardia asteroides* keratitis: a report of a case. Brit. J. Ophthal., *60*;104–106.

65. Ratledge, C., and P. V. Patel. 1976. The isolation, properties and taxonomic relevance of lipid-soluble, iron-binding compounds (the nocobactins) from *Nocardia*. J. Gen. Microbiol., *93*:141–152.

66. Riddell, M., 1975. Taxonomic study of *Nocardia farcinica* using serological and physiological characters. Int. J. Syst. Bacteriol., *25*:124–132.

67. Robboy, S. J., and A. L. Vickery. 1970. Tinctorial and morphological properties distinguishing actinomycetes. New Engl. J. Med., *282*:593–596.

68. Saltzman, H. A., E. W. Chick, et al. 1962. Nocardiosis as a complication of other diseases. Lab. Invest., *11*:1110–1117.

69. Satterwhite, T. K., and R. J. Wallace. 1979. Primary cutaneous nocardiosis. J.A.M.A., *242*:333–336.

70. Shainhouse, J. Z., A. C. Pier, et al. 1978. Complement fixation antibody test for human nocardiosis. J. Clin. Microbiol., *8*:516–519.

71. Sher, N. A., C. W. Hill, et al. 1977. Bilateral intra-

ocular *Nocardia asteroides* infection. Arch. Ophthalmol., *95*:1415–1418.

72. Shome, S. K., and D. K. Sirkar. 1974. Efficiency of paraffin bait technique in the isolation of *Nocardia* from bronchopulmonary disorders. Mykosen, *17*:299–302.

73. Simonella, P., and N. R. Brizioli. 1968. Relazioni all tubesiolina nei bovinin portatori di *Nocardia dassonvillei.* Att: Soc. Ital. Sci. Vet., *22*:918–923.

74. Sooriyaaruchchi, G. S., and T. F. Hogen. 1978. Renal complications during sulfonamide therapy for systemic nocardiosis. Wisc. Med. J., *77*:515–518.

75. Snieszko, S. F., G. L. Bullock, et al. 1964. Nocardial infection in hatchery reared fingerling rainbow trout. J. Bacteriol., *88*:1809–1810.

76. Stropnik, Z. 1965. Isolation of *Nocardia asteroides* from human skin. Sabouraudia, *4*:41–44.

77. Sundararaj, T., and S. C. Agarwal. 1977. Cell mediated immunity to *Nocardia asteroides* induced by its ribonucleic acid protein fraction. Infect. Immun., *18*:253–256.

78. Vasarinish, P. 1968. Primary cutaneous nocardiosis. Arch. Dermatol., *98*:489–493.

79. Wallace, R. J., E. J. Septimus, et al. 1977. Disk diffusion susceptibility testing of *Nocardia* species. J. Infect. Dis., *135*:568–576.

80. Walton, A. M., and K. G. Libke. 1974. Nocardiosis in animals. Vet. Med. Small Anim. Clin., *69*:1105–1107.

81. Young, L. S., D. Armstrong, et al. 1971. *Nocardia asteroides* infection complicating neoplastic disease. Am. J. Med., *50*:356–367.

82. Zecler, E., Y. Gilboa, et al. 1977. Lymphocutaneous nocardiosis due to *Nocardia brasiliensis*. Arch. Dermatol., *113*:642–643.

4 Other Actinomycetous Infections

Included in this category are some minor infections caused by actinomycetes and diphtheroid organisms. Though some are common infections, they usually cause only slight discomfort to the patient. They are generally brought to the attention of the physician or dermatologist out of curiosity or concern by the patient about nodose axillary or pubic hair, discolored skin, discolored sweat, or plantar fissures. Rarely serious, they may be considered as nuisance curiosities.

Traditions die hard in science. Actinomy-cetes were once considered to be the link between bacteria and fungi. As previously discussed, it is now established that they are bacteria. Similarly, in the diseases included here, an actinomycete, generally *Nocardia*, was described as the etiologic agent for each one. More recent studies define the organisms as *Corynebacterium*, not actinomycetes. For convenience, however, they will be grouped together as actinomycetous infections.

Erythrasma

DEFINITION

Erythrasma (dhobie) is a chronic, mild, localized infection of the stratum corneum. It usually remains localized, most commonly involving the axilla or crural areas, but it may also involve the body folds and clefts and intertriginous areas. It is characterized by brownish-red punctate or palmate glistening lesions and dry, smooth, finely creased skin. The etiologic agent, *Corynebacterium minutissimum*, is apparently a common resident of the skin, particularly the toe webs. The condition is readily amenable to systemic antibacterial antibiotics, but recurrence is frequent.

The name dhobie refers to Indian servants of British army officers during the time India was a colony of Great Britain. They wrapped their loins in a tight cloth and often had a condition now called erythrasma of the groin. This is sometimes mistaken for "dhobie itch," which occurs in the same anatomic area but is caused by *Epidermophyton floccosum*, a dermatophyte.

HISTORY

This disease was first described by Buchardt in 1859,[7] and the term "erythrasma" was used by Bärensprung in 1862. Skin scales examined by these authors showed delicate filaments which were believed to be of fungal origin. They named the organism *Microsporum minutissimum*. The contagious nature of the disease was demonstrated by Köbner, who in 1884 was able to transmit the disease by rubbing scales from a patient on the skin of a pupil. Early workers did not accept the disease as a separate entity, and it was thought to be either a variant of pityriasis versicolor, a dermatophyte infection, or eczema marginatum. There was an early association of the disease with hyperhidrosis, and

Poehlman in 1928[25] stressed delicate skin, humidity, site, and secretions as factors predisposing to the disease. Though the disease is caused by a delicate gram-positive rod, it may be complicated by fungal or other bacterial infections, as emphasized by Gougerot in 1936.

ETIOLOGY, ECOLOGY, AND DISTRIBUTION

Though it has received various fungal and bacterial names, and as recently as 1958 was reported by Kalkoff to be gram-negative, the etiologic agent of erythrasma has been identified as gram-positive, filamentous, diphtheroid *Corynebacterium minutissimum*. The organism is included in the group of lipophilic diphtheroids and is related to *C. xerosis*. As a low-grade infection, erythrasma appears to be found throughout the world. The most common site of occurrence is the genital folds, where the mild form of the disease has been found in 4 to 20 per cent of populations examined. The patients are usually unaware of the infection. The incidence is higher among males in warm, humid climates. Another usually symptomless site is the toe web. Up to 25 per cent of students examined in the United States and England[33] demonstrated the presence of the organism. Again, most of the patients were without symptoms. However, some had fissuring and maceration along with other bacteria or a tinea pedis. Actual clinical disease is more common in adult males. This is the classic genitocrural type, and there may be a long history of fluctuating severity. A second, generalized type is predominantly present in middle-aged Negro women.

CLINICAL DISEASE

The lesions are punctate to palm-sized, well-circumscribed, maculopapular rashes. The color varies from pink to reddish to reddish-brown, depending on the age of the lesion. The involved area is glistening or greasy-looking and sometimes covered with small furfuraceous scales. The surface of older lesions may show fine creases. The scaly nature of the disease may require scratching to be revealed. The advancing border of the lesion is serpiginous and erythematous. There is no tendency for it to become elevated or vascularized, and lesions may remain symptomless, except for mild irritation and lichenification.

Erythrasma is most commonly found in areas of the groin and upper thigh in contact with the scrotum. Although both thighs may be involved, the left is more often infected because of more contact with the scrotum. Other common sites are the pubes, the axillae, the intergluteal folds, and under the breasts in women. Hair in the affected areas remains uninvolved. Occlusion and increased humidity tend to exacerbate mild disease.

In the toe web, there may be some scaling, fissuring, and maceration. In such cases, there is usually a mixed flora of other bacteria also present, including *Staphylococcus, Pseudomonas*, and *Proteus*. The disease may coexist with dermatophytosis. Infection is usually between the fourth and fifth, and third and fourth toes. Various studies[31, 33] have demonstrated that a mild subclinical form of the disease is very common in the population. Surveys are very easy to perform, taking advantage of the coral-red fluorescence of the lesion when seen under the Wood's lamp. By this method, as many as 25 per cent of the people surveyed were demonstrated to harbor organisms. They were usually symptom-free, however.

Much less commonly encountered than either of the preceding is the generalized form of erythrasma, which is found predominantly among middle-aged Negro women. There are well-defined, scaly, lamellar plaques on large areas of the trunk, proximal parts of the limbs, and the folds of the breasts (Fig. 4–1). Again, this condition is more common in hot, humid climates and is referred to as tropical erythrasma. Usually, there is a long history of chronicity, itching, and lichenification.

Chronic erythrasma that has lichenified may be mistaken for neurodermatitis or for tinea cruris. In contrast to dermatophytosis, however, in erythrasma there is little inflammation, no vesiculation, and no satellite lesions. In the obese patient there is replacement of the dry, scaly lesion with a shiny, smooth, red, wet intertrigo, which may confuse the picture. In the typical form of the disease in other areas, erythrasma is most commonly confused with pityriasis versicolor. The latter lacks the erythematous border and shows no tendency to localize in the body folds. A simple KOH mount readily differentiates the two. At some stages, tinea corporis, eczema, lichen simplex, contact dermatitis, and psoriasis may simulate erythrasma.

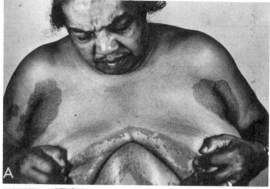

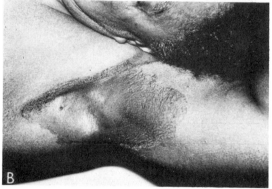

Figure 4-1   Erythrasma. *A,* Extensive lesions affecting the supramammary folds and the axilla. Lesions were also present on the trunk and legs. *B,* Well-defined, scaly, lamellar plaques in the axilla.

coccoid forms, rodlike organisms, and long filaments, 5 to 25 μ in length. These are seen with great difficulty by this procedure, and a much more effective method is to use transparent tape strippings from the area. The tape is pressed on the lesion and removed, carrying with it scales on the sticky side. If cellulose tape is used, a quick rinse with methylene blue can be used before the tape is pressed on a slide and examined by oil immersion. Vinyl tape allows a Gram or Giemsa stain to be done, revealing the organisms more clearly. The coccal forms are 1 μ in diameter, the bacilli 1 to 3 μ, and the filaments average 4 to 7 μ in length.

The Wood's light is of great diagnostic aid in erythrasma. A characteristic coral-red fluorescence can be seen in the active borders and in patches on old lesions (Fig. 4–2). In contrast, pityriasis versicolor will show some areas of a gold fluorescence. Ulcerated squamous cell carcinoma and carcinoma of the palate have shown some red fluorescence under the Wood's lamp, as have experimental carcinomas of rabbit skin.[33] The fluorescent substance is also produced by the etiologic agent of erythrasma in culture. It appears to be a porphyrin.

PROGNOSIS AND TREATMENT

Since the disease is of minor importance to the patient, many chronic cases are seen. The disease persists indefinitely except for occasional fluctuations in severity. In spite of local or systemic treatment, relapse is common. Local treatments include sodium hyposulfite (20 per cent) or 3 per cent sulfur ointment and the usual keratolytics, such as Whitfield's ointment. Systemic antibiotics, especially erythromycin, in doses of 1 g per day for five days, have evoked complete clinical cure.[31] Other effective antibiotics include tetracycline and Chloromycetin, but erythromycin appears to be the drug of choice. Penicillin and griseofulvin are without effect. Toe web infection is particularly resistant to treatment.

Experimental or natural disease in animals has not been reported.

LABORATORY DIAGNOSIS

Direct examination of scales from erythrasma in KOH mounts under oil will show small

CULTURE METHODS

The organism grows readily on a medium containing 20 per cent fetal calf serum and 2 per cent agar in tissue culture media No. 199. The colonies appear in two days, reaching a size of 2 to 3 mm. They are moist, translucent, convex, and nonhemolytic and, while young, will show the characteristic coral-red

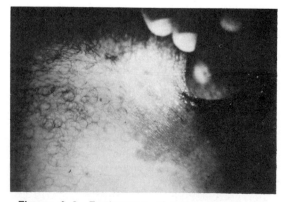

Figure 4-2   Erythrasma of groin. Wood's lamp demonstration of coral red fluorescence of lesions. (Courtesy of D. Taplin.)

fluorescence. Gram stains of colonies show typical diphtheroid dimensions and pleomorphism. Culture is not necessary for diagnosis.

MICROBIOLOGY

Corynebacterium minutissimum **(Burchardt) Sarkany, Taplin, and Blank 1961**

Synonymy. *Microsporon minutissimum* Burchardt 1859; *Microsporon gracilis* (Burchardt) Balzer 1883; *Sporotrichum minutissimum* (Burchardt) Saccardo 1886; *Microsporoides minutissimus* (Burchardt) Neveu-Lemaire 1906; *Discomyces minutissimus* (Burchardt) Verdun 1907; *Oospora minutissima* Ridet 1911; *Nocardia minutissima* (Burchardt) Verdun, 1912; *Actinomyces minutissimus* (Burchardt) Brumpt, 1927; *Leptothrix epidermidis.*

Trichomycosis Axillaris

DEFINITION AND ETIOLOGY

This is a superficial infection of the axillary or pubic hair, characterized by the formation of yellow (flava), red (rubra), or black (nigra) nodules around the hair shaft (Fig. 4–3). The organism involved has been isolated on occasion and named *Corynebacterium tenuis.* The studies of Crissey[9] indicate that the disease is prevalent in the temperate climate but is probably more widespread in the tropics where heat and moisture favor growth of the organism.

CLINICAL DISEASE

Patients with this condition are usually unaware of it, and the presenting complaint is of discolored, sweat-stained clothes. The concretion is most commonly yellow, least commonly black, and consists of firmly packed coccoid and bacillary forms in a mucinous mass.[21] The consistency is usually firm but may be soft and jellylike (zooglea). The affected hair appears lusterless and brittle and is easily broken. The infection does not extend to the hair base or root, thus differentiating it from ringworm. The flava variety resembles white piedra, but microscopic examination quickly differentiates them. Also included in the differential diagnosis are pediculosis, monilethrix (beaded hair), and trichorrhexis nodosa. Treatment includes depilation and daily application of formalin (2 per cent), bichloride of mercury (1 per cent), or sulfur (3 per cent). Recurrence is common.

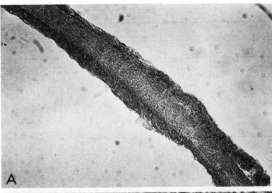
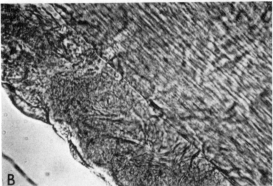

Figure 4–3 Trichomycosis axillaris. *A,* Concretion surrounding hair shaft. ×100. *B,* Higher magnification of concretion. There is no destruction of the hair itself. ×400.

LABORATORY DIAGNOSIS

KOH mounts of the hairs will show the nodules to be composed of delicate, short bacilli and coccoidlike diphtheroid organisms enmeshed in mucilaginous material.[21] The flava variety may show a distinct golden fluorescence. The organism has been cultured on enriched media. Work by McBride[18] has shown that the organisms readily colonize hair in broth cultures. They then concentrated colored salts, giving the nodules a red or black color. She suggests that there is only one species involved and that it has a unique ability to concentrate colored substances, apparently coming from the patient, giving rise to the variously colored nodules. Negroni has reported on the characteristics of an isolant from human infection. His organism demonstrated conspicuous branching and was acid-fast.[23] If other characteristics, such as LCN-A (lipid characteristic of *Nocardia*) determination and lysozyme resistance, are carried out, the etiologic agent may be redescribed as a *Nocardia*.

MICROBIOLOGY

Corynebacterium tenuis (Castellani) Crissey et al. 1952.

Synonymy. *Nocardia tenuis* Castellani 1912; *Discomyces tenuis* Castellani, 1912; *Cohnistreptothrix tenuis* Ota 1928; *Actinomyces tenuis* Dodge 1955.

Reference strain ATCC 15907 is labeled *Nocardia tenuis*, but a description is lacking.

Pitted Keratolysis

DEFINITION AND ETIOLOGY

Pitted keratolysis, cracked heel, or keratolysis plantare sulcatum is a superficial infection of the stratum corneum, characterized by circular areas of erosion on the plantar surfaces. The condition is much more prevalent than formerly recognized.[16, 36] Although the disease is usually asymptomatic, under conditions of heat and dampness it may lead to a severe and debilitating anhidrotic syndrome, with thinning and erosion of the skin. The disease is caused by an organism variously classified as *Actinomyces, Nocardia, Micromonospora, Streptomyces, Dermatophilus,* or *Corynebacterium*. The latter is probably the true agent, but it has not yet been published as such. The agent readily dissolves keratin and, in histologic section, can be found in the floor of the pits. It appears as coccoid bodies and filaments with diphtheroid morphology.

The disease was first described in India in 1910 by Castellani,[8] and was largely ignored. The studies of Zaias et al.[36] indicate it to be common in the normal population and especially in military personnel assigned to the tropics. Gill and Buckels described the condition in detail.[11] The symptomatic form was described by Lamberg.[16] He found that extreme humidity, intense heat, and poor sanitation exaggerated the disease and produced disability among military personnel, who sometimes required hospitalization. He suggested that prolonged wearing of boots and incubator-like conditions of tropical warfare were responsible for the exacerbation of disease.

CLINICAL FEATURES

The horny layer of the soles, toes, and heels shows numerous areas of erosion (Fig. 4–4). These are discrete, circular patches that later coalesce to form large areas of denuded stratum corneum. The edges may be discolored with a greenish or brownish outline. Fissures may appear within these areas and lead to secondary bacterial complications with an accompanying foul odor. Hyperhidrosis is noted. In the symptomatic form in which thinned, reddened areas are seen, anhidrosis occurs within the eroded lesions.[16] If the feet are soaked, the accompanying swelling of the horny layer clearly outlines the lesions.

Histologic examination shows a mild inflammatory reaction in the dermis without thickening, and an almost complete absence of the stratum corneum. On methenamine stain or modified Gram stain, filaments may be seen in the walls and floor of the pits (Fig. 4–4B). A similar disease, ulcus interdigitale, may occur on the toes.

PROGNOSIS AND THERAPY

The condition, when mild, resolves quickly with adequate personal hygiene and relief of the predisposing environment of heat and humidity. For symptomatic disease, effective therapy has been achieved with 20 to 40 per cent formalin in Aquaphor.[16] The normal sweating pattern returns within three days with bed rest, followed by relief of pain and clearing of the erythema.

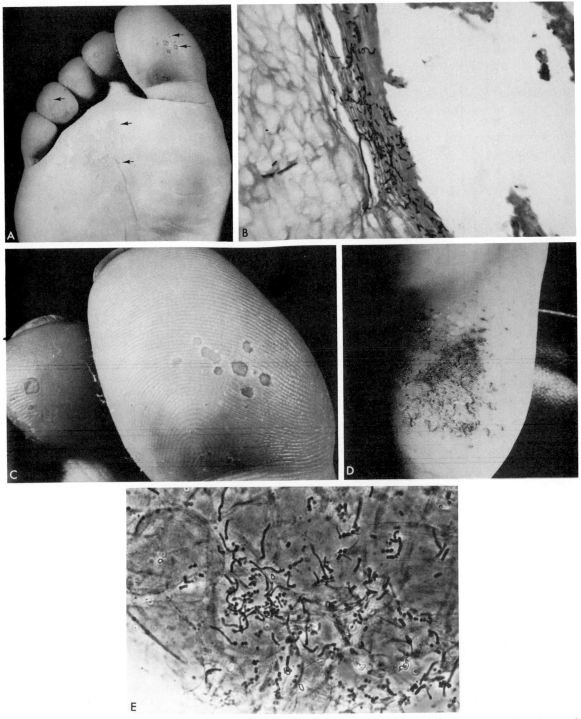

Figure 4–4   *A,* Pitted keratolysis. Plantar surface of toes and foot showing numerous areas of erosion. (arrows) (Courtesy of John Carmichael.) *B,* Pitted keratolysis. Methenamine silver strain of biopsy. Note the thin branching filaments and coccoid bodies at the base of the pit. ×440. *C,* Pitted keratolysis. Closeup of toe showing deep erosions. (Courtesy of John Carmichael.) *D,* Pitted keratolysis. Exensive involvement of heel area. The pits are soil stained. *E,* Dermatophilus-like organisms in scraping isolated from lesions of pitted keratolysis. GMS. ×1200. These organisms were also cultured. (Courtesy of John Carmichael.)

LABORATORY IDENTIFICATION

KOH mounts of infected scales will show filaments and coccoid forms resembling actinomycetes or diphtheroid organisms. Culture of the agent is difficult and, as yet, controversial. Taplin[36] has isolated a *Corynebacterium* sp. from the lesions, with which he has been able to reproduce the disease on a human volunteer, i.e., Mr. Taplin. However, a published account of the organism has not been made, so that the older designation of Acton and McGuire remains valid, though questionable.

Actinomyces keratolytica Acton and McGuire 1930[3]

This is a name without an organism, since the etiologic agent has not as yet been isolated. The name *Dermatophilus pedis* has also been proposed. Taxonomy must await the isolation of the responsible etiology.

Dermatophilosis

DEFINITION AND ETIOLOGY

Dermatophilosis (streptotrichosis, epidemic eczema, strawberry foot rot, contagious dermatitis) is a pustular, exudative dermatitis of animals and, rarely, man. The initial lesions (Fig. 4–5) are followed by the formation of scales and crusts that heal and fall off, leaving alopecia and scarred areas on the hide. In endemic regions, it may be highly contagious, and epidemics cause severe economic loss among domesticated animals (Fig. 4–6, *A–D*). The etiologic agent is an actinomycete of the Dermatophilaceae, *Dermatophilus congolensis*. The disease has been reported in cattle and sheep primarily, but it also occurs in goats, horses, antelopes, deer, zebras, swine, giraffes, foxes, Colombian ground squirrels, lizards,[5, 20] monkeys,[19] domestic cats (Fig. 4–7, *A, B*), and man.[14] Since the etiologic agent is a *Dermatophilus*, the name for the disease, dermatophilosis, is preferable to other suggested or historical designations.

A review of the disease in animals and humans was prepared by Stewart,[32] and Moreira et al.[22] have investigated the occurrence of the infection in South America.

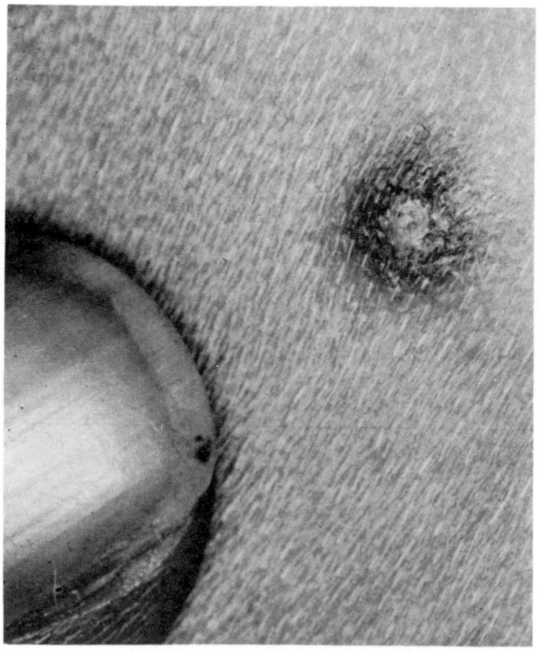

Figure 4–5. Dermatophilosis. Initial lesion on rabbit, the result of transmission by fly. (Courtesy of J. L. Richard, National Animal Disease Center, Ames, Iowa.)

HISTORY

This disease was first described by Van Saceghem in 1915[34] as a skin disease in cattle in the Belgian Congo. This work was largely ignored, and a similar disease in Australia received the name "lumpy wool" from Bull in 1929. The literature is confused by reports from various parts of the world, each using a different name, e.g., strawberry foot rot of sheep in Scotland (*Polysepta pedis*), reported by Thompson. The disease was first recognized in the United States in 1961.[10] Three separate outbreaks occurred within that year. It was found in calves from Texas, in horses from New England and New York, and in deer from New York. In the latter instance, the disease was transmitted to man by handling of the infected animals.[10, 15] Since then, it has appeared in Iowa, Kansas, and Georgia. It is now considered to be of worldwide distribution. Many names have appeared for the various isolates, but Gordon[13] concludes they are all the same species.

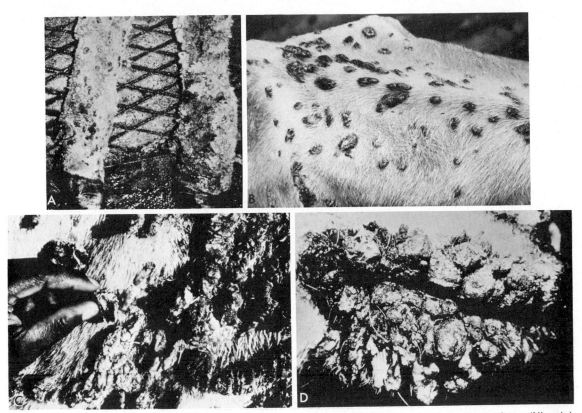

Figure 4–6 Dermatophilosis. *A,* Lesions on legs of cow in Nigeria. *B,* Crusts on back and flank of cow (Nigeria). *C,* Severe disease with pedunculated verrucous crusts on cow (Nigeria). *D,* Lesions around the mouth of a sheep. (Courtesy of J. L. Richard, National Animal Disease Center, Ames, Iowa.)

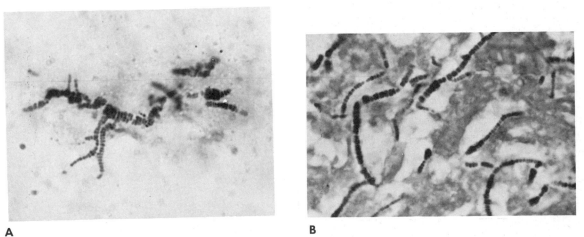

Figure 4–7 *A,* Dermatophilosis. Stained smear showing irregular, branched filaments. These are divided both longitudinally and transversely, forming packets of coccoid cells. *B,* Histopathologic section of a tumorlike lesion on tongue of cat. (Courtesy of J. L. Richard, National Animal Disease Center, Ames, Iowa.)

ECOLOGY

Roberts[27] believes the organism is a natural parasite of the epidermis of sheep. Though not recovered from soil, the organism can remain viable in dried crusts for long periods. It dies rapidly if exposed to moisture and probably does not occur in nature as a sabrophyte. It is known to be transmitted by direct contact under herd conditions, by flies, and probably by ectoparasites. Epidemics occur when the skin has been damaged by prolonged wetting. The macerated epidermis is less resistant to invasion, and the moisture facilitates release of the infective, motile zoospores and their penetration of susceptible epidermis.

CLINICAL DISEASE

The organism is a parasite of the epidermis and does not invade vascularized tissue. The initial symptom in animals is a pustular dermatitis, followed by formation of scabs and crusts. The hair is matted with exudate (lumpy wool). Healing occurs and the crusts fall off, leaving patches of alopecia. Lesions vary in size and may become confluent. There is some variation of the disease among horses, cattle, and sheep, but the general picture is similar. One review indicates that dermatophilosis occurred in less than 12 per cent of cattle during the rainy season but in only 4 per cent during the dry season.[6] In man, as reported by Dean,[10] four individuals contracted the disease by handling an infected deer. The onset was characterized by the appearance of multiple, nonpainful pustules on the hands within a week after exposure. The pustules were from 2 to 5 mm in size and contained a white to yellowish serous exudate. A small, shallow, reddish ulcer remained after expression of the fluid. This was followed by formation of a brownish scab that persisted for several days. The lesions healed spontaneously, leaving a purplish-red scar. There was no systemic illness.

Experimental disease has been produced in rabbits[14] and other animals and fowl.[1] Mice and guinea pigs appear to be resistant.

TREATMENT

In animals, various topical medicaments have been used with irregular results. These include dips of copper sulfate (1:500), 0.5 per cent zinc sulfate, and 1:5000 mercuric chloride or a dip containing arsenic. The organism is susceptible to the usual sulfonamides and antibiotics, but trials have been unrewarding. A long course of penicillin has been effective, but a single dose, regardless of amount, has been unsuccessful.

LABORATORY IDENTIFICATION

Scales and aspirates, stained by methylene blue or Giemsa stain, show branched filaments 2 to 5 μ in diameter, which are divided in longitudinal and transverse planes into packets of up to eight coccoid cells (Fig. 4–5). This picture is pathognomonic for the disease. Pier[24] has shown the immunofluorescent technique to be valuable for diagnosis. The organism is easily cultured from aspirates of unopened pustules at 37°C on blood agar without antibiotics.[2] If there is much contamination, the infected scabs can be rubbed on shaved rabbit skin and cultures made from the resulting pustules after four days. Within 24 hours, small (0.5 to 1.0 mm) colonies appear on agar, which are white to yellow, round or irregular, with a depressed periphery. After five days, a yellow-to-orange pigment appears (Fig. 4–8). The organism is gram-positive, with multiple branching filaments 0.5 to 1.5 μ in diameter. They show the characteristic transverse and longitudinal septation into packets of coccoid spores (a

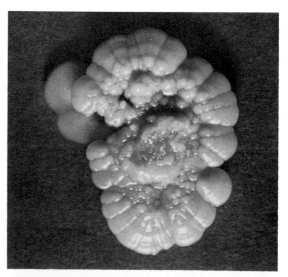

Figure 4–8 Dermatophilosis. Colony of *Dermatophilus congolensis* after five days growth on brain-heart infusion of glucose blood agar.

DERMATOPHILUS (LIFE CYCLE)

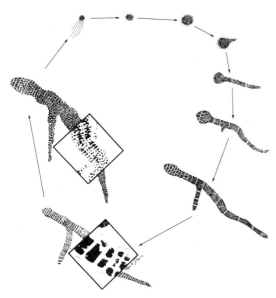

Figure 4–9 Dermatophilus life cycle. (Reprinted by permission from *Veterinary Bacteriology and Virology* by I. A. Merchant and R. A. Packer, 7th edition © 1967, by Iowa State University Press, Ames, Iowa.)

sporangium). Under proper conditions, these hatch and become flagellated and motile zoospores. They are not acid-fast (Fig. 4–9).[27]

The organism is catalase-positive and urease-positive and is proteolytic for gelatin and casein. It hydrolyzes starch but not xanthine or tyrosine. In enriched broth, it will produce acid but no gas from glucose and fructose and will not ferment sucrose, lactose, xylose, mannitol, dulcitol, salicin, and sorbitol.[13] The cell wall contains *meso*-diaminopimelic acid but no arabinose or galactose.

MICROBIOLOGY

Dermatophilus congolensis Van Saceghem 1915 emend, 1916[34]

Synonymy. *Dermatophylus congolensis* Van Saceghem 1915; *Actinomyces dermatonomus* Bull 1929; *Tetragenus congolensis* Van Saceghem 1934; *Actinomyces congolensis* (Van Saceghem) Hudson 1937; *Nocardia dermatonomus* (Bull) Henry 1952; *Streptothrix bovis* Snijders and Jensen 1955; *Polysepta pedis* Thomson and Bisset 1957; *Dermatophilus dermatonomus* (Bull) Austwick 1958; *Dermatophilus pedis* (Thomson and Bisset) Austwick 1958.

REFERENCES

1. Abu-Samra, M. T., and S. E. Imbabi. 1976. Experimental infection of domesticated animals and the fowl with *Dermatophilus congolensis.* J. Comp. Pathol., *86*:157–172.

2. Abu-Samra, M. T., and G. S. Walton. 1977. Modified techniques for the isolation of *Dermatophilus* spp. from infected material. Sabouraudia, *15*:23–27.

3. Acton, H. W., and C. McGuire. 1930. Keratolysis plantare sulcatum. India Med. Gaz., *65*:61–62.

4. Albrecht, R., S. Horowitz, et al. 1974. *Dermatophilus congolensis* chronic nodular disease in man. Pediatrics, *53*:907–912.

5. Anver, M. R., J. S. Park, and H. G. Rush. 1976. Dermatophilosis in the marble lizard (*Calotes mystaceus*) Lab. Anim. Sci., *26*:817–823.

6. Bidar, S. A. 1975. Seasonal prevalence of bovine dermatophilosis in northern Nigeria. Bulletin de l'Office International des Epizooties, *83*: 1131–1138.

7. Burchardt, M., and V. Barensprung. 1859. *In* Uhle and Wagoner. Patologie Général. Quoted in Balzer, F. 1883. De L'érythrasma (*Microsporum minutissimum*). Ann. Rev. Dermatol. Syphil. XI, *4*:681–688.

7a. Burchardt, M. 1859. Ueber eine Chloasma Vorkomende Pilzform. Medicinische Zeitung II (no 29), p 141.

8. Castellani, A. 1910. Keratoma plantare sulcatum. J. Ceylon Brit. Med. Ass., 7:10–11.

9. Crissey, J. T., G. C. Rebell, et al. 1952. Studies on the causative organism of trichomycosis. J. Invest. Dermatol., *19*:187–188.

10. Dean, D. J., M. A. Gordon, et al. 1961. Streptothricosis: A new zoonotic disease. N.Y. State J. Med., *61*:1283–1287.

11. Gill, K. A., and L. J. Buckels. 1968. Pitted keratolysis. Arch. Dermatol., *98*:7–11.

12. Gordon, H. H. 1975. Pitted keratolysis, forme fruste. A review and new therapies. Cutis, *15*: 54–58.

13. Gordon, M. A. 1964. The genus *Dermatophilus.* J. Bacteriol., *88*:509–522.

14. Gordon, M. A., and U. Perrin. 1971. Pathogenicity of *Dermatophilus* and *Geodermatophilus.* Infect. Immun., *4*:29–33.

15. Kaplan, W. 1966. Dermatophilosis—A recently recognized disease in the United States. Southwest. Vet., *20*:14–19.

16. Lamberg, S. 1969. Symptomatic pitted keratolysis. Arch. Dermatol., *100*:10–11.

17. Lloyd, D. H. 1976. Diagnostic features of *Dermatophilus* infection. Trans. R. Soc. Trop. Med. Hyg., *70*:285.

18. McBride, M., W. Duncan, et al. 1970. The effects of selenium and tellurium compounds on pigmentation of granules of trichomycosis axillaris. Int. J. Dermatol., *9*:226–231.

19. McClure, H. M., W. Kaplan, et al. 1971. Dermatophilosis in owl monkeys. Sabouraudia, *9*:185–190.

20. Montali, R. J., E. E. Smith, et al. 1975. Dermatophilosis in Australian bearded lizards. J. Am. Vet. Med. Assoc., *167*:553–555.

21. Montes, L. F., C. Vasquez, and M. S. Cataldi, 1963. Electron microscopic study of infected hair in trichomycosis axillaris. J. Invest. Dermatol., *40*:273–278.

22. Moreira, E. C., M. Barbosa, and Y. K. Moreira. 1974. Dermatophilosis in South America. Arqui de

Escola de Vet. de Univer. Fed. de Minas Garais, *26*:77–84.

23. Negroni, P. 1978. Estudios sobre la tricomicosis axilar. I. Caracteres microbiologicos del organismo cultirado. Rev. Argentina Micologia, *1*:12–16.

24. Pier, A. C., J. L. Richard, and E. F. Farrell. 1964. Fluorescent antibody and cultural techniques in cutaneous streptothricosis. Am. J. Vet. Res., *25*:1014–1020.

25. Poehlmann, A. 1928. Erythrasma. Handbuch Haut-Geschlechtskr (Jadassohn), *11*:711–723.

26. Roberts, D. S. 1961. The life cycle of *Dermatophilus dermatonomus*, the causal agent of ovine mycotic dermatitis. Aust. J. Exp. Biol. Med. Sci., *39*:463–476.

27. Roberts, D. S. 1965. Cutaneous actinomycosis due to the single species *Dermatophilus congolensis*. Nature, *206*:1068.

28. Ronchese, F., B. D. Walker, and M. R. Young. 1954. The reddish-orange fluorescence of necrotic cancerous surfaces under the Wood's light. Arch. Dermatol., *69*:31–42.

29. Rubel, L. R. 1972. Pitted keratolysis and *Dermatophilis congolensis*. Arch. Dermatol., *105*:584–586.

30. Salkin, I. F., M. A. Gordon, and W. B. Stone. 1975. Dual infection of a white tailed deer by *Dermatophilus congolensis* and *Alternaria alternata*. J. Am. Vet. Med. Assoc., *167*:571–573.

31. Sarkany, I., D. Taplin, and H. Blank. 1961. The etiology and treatment of erythrasma. J. Invest., *37*:283–290.

32. Stewart, G. H. 1977. Dermatophilosis: a skin disease of animals and man. Vet. Rec., *91*:537–544.

33. Temple, D. E., and C. R. Boardman. 1962. The incidence of erythrasma of the toewebs. Arch. Dermatol., *86*:518–519.

34. Van Saceghem, R. 1916. Etude complémentaire sur la dermatae des bovidis (impétigo contagieux). Bull. Soc. Pathol. Exot. Filiales, *10*:290–293.

35. Young, C. N. 1974. Pitted keratolysis—a preliminary report. Trans. St. John's Hosp. Dermatol. Soc., *60*:77–85.

36. Zaias, N., D. Taplin, and G. Rebell. 1965. Pitted keratolysis. Arch. Dermatol., *92*:151–154.

5 Mycetoma

DEFINITION

Mycetoma is a clinical syndrome of localized, indolent, deforming, swollen lesions and sinuses, involving cutaneous and subcutaneous tissues, fascia, and bone. It usually occurs on a foot or hand. The disease results from the traumatic implantation of soil organisms into the tissues. The lesions are composed of suppurating abscesses, granulomata, and draining sinuses, with the presence of "grains" which are characteristic granules of the etiologic agents. The triad of tumefaction, draining sinuses, and grains is used in a restrictive sense to define the term "mycetoma." The etiologic agents are a wide variety of bacteria (actinomycotic mycetoma) and fungi (eumycotic mycetoma) from plant debris and soil. The involved organisms may also cause other clinical diseases, such as actinomycosis, mycotic granuloma, and phaeohyphomycosis, but only when the above criteria are met is the diagnosis of mycetoma valid.

Synonymy. Madura foot, maduromycosis.

HISTORY

A condition of the foot in which fleshy tumors appeared and gradually increased in size was observed in ancient times among the peoples of the Indian subcontinent and is described in the Atharva-Veda under the name "Padavalmika" (foot anthill). It was differentiated from "Slipatham," or elephant foot (a filarial disease), and from several other conditions.[75]

Mycetoma as a medical entity was first reported (as Madura foot) by a Dr. Gill in a dispensary report of the Madras Medical Service of the British Army in India in 1842. He was stationed in the Madurai area. In his remarks, Dr. Gill notes that "when the leg has been amputated, the foot has been found to be one mass of disease of a fibrocartilaginous nature, with entire destruction of the joints, cartilages and ligaments; it has neither shape nor feature and is covered with large fungoid excescences discharging an offensive ichorous fluid."[27] This observation, made almost 150 years ago, is still an apt description of advanced cases of the disease. In another dispensary report the existence of the disease was confirmed by Colebrook in 1846. The term "mycetoma" was first used by Vandyke Carter in 1860[28] to designate all tumors produced by fungi. A detailed bibliography, descriptions, engravings depicting diseased appendages, and a detailed early history of disease is found in H. Vandyke Carter's book *On Mycetoma, or the Fungus Disease of India,* published in 1874.[27] He notes that the native name for the disease was Ghootloo Mahdee, descriptive of the egglike tumefactions. Further descriptions by Kanthack in 1892 and Vincent in 1894 delineated the disease.[45] They reported that there were several colors and types of grains and speculated that several "moulds" caused the disease. Kanthack named the agents *Oospora indica* var. *flava* for yellowish grains and *O. indica* var. *nigra* for the black.[45]

The first cases were from India, and the old term "Madura foot" is reflected in the orthographic derivatives: maduromycosis for the disease, *Madurella* and *Indiella* for fungal genera, and *Actinomadura madurae* as a specific etiologic agent. Brumpt in 1905[22] stressed that several fungi were capable of eliciting the same clinical disease, and Langeron[46] applied the term "mycetoma" to cases involving *Actinomyces* or *Nocardia* species and filamentous fungi. Brumpt erected the genus *Madurella* to include an organism described in 1902 by Laveran as *Streptothrix mycetomi* for the agent of black-grain mycetoma. He described *Indiella* sp. as the cause of white-grained disease. (What Brumpt called *Indiella* appears to be the actinomycete *Streptomyces somaliensis*.) These organisms were prevalent in India and Sudan. Black-grain disease was described in Europe and America[80] in the 1890's, but white-grain infection appeared to be more common. A rapidly growing hyphomycete was isolated from the first described case of white-grain disease. The patient was a butcher in Ibono, Sardinia. This case, reported by Tarozzi in 1909, was followed by a similar one with a similar isolant described by Radaeli[67] in 1911. (Details of these reports are found in Chapter 24.) The fungus was first called *Monosporium apiospermum* by Saccardo in 1911, but was later changed to *Scedosporium apiospermum* (discussed in Chapter 24). It is now known by its teleomorphic name *Pseudallescheria boydii.*

The division of the disease into two separate categories depending on the etiology was first suggested by Pinoy in 1913,[65] and was redefined by Chalmers and Archibald in 1916[29] and again by Lavalle in 1962.[47] They used the terms "actinomycosis" and "maduromycosis" to differentiate the two groups of causative agents.

The literature became confused with many ill-defined fungal and bacterial isolants, most of which have been reduced to synonymy with accepted species or discarded. As noted by Emmons,[34] "mycetoma" came to be used as a general term for the presence of any fungal organism in any tissue that was not a specifically defined mycosis. In present usage, the term is applied only when the clinical syndrome described above is found. The two categories are "actinomycotic mycetoma" (actinomycetoma), when the agent is an actinomycete, and "eumycotic mycetoma" (eumycetoma), when a true fungus is involved.[81]

ETIOLOGY, ECOLOGY, AND DISTRIBUTION

The single clinical entity mycetoma may be evoked by infection with a diverse group of bacteria and fungi (Table 5–1). About 50 per cent of cases are due to actinomycetes

Table 5–1 *Common Etiologic Agents of Mycetoma in Man and Animals*

Agent	Grain Color	Occurrence*
Actinomycotic Mycetoma (thin, branching, intertwined filaments)		
Actinomyces israelii	White to yellow	R
Nocardia asteroides	Grains rare-white	R
Nocardia brasiliensis	White	C
Nocardia caviae	White to yellow	R
Nocardia farcinica	White to yellow†	R
Actinomadura madurae	White to yellow or pink	C
Actinomadura pellitieri	Garnet red	O
Streptomyces somaliensis	Yellow to brown	O
Eumycotic Mycetoma (wide, branching, intertwined hyphae and cystlike cells)‖		
Pseudallescheria boydii	White	C
Madurella grisea	Black‡	R
Madurella mycetomatis	Black	C
Acremonium kiliense	White	R
Acremonium falciforme	White	E
Acremonium recifei	White	
Leptosphaeria senegalensis	Black	R
Leptosphaeria tompkinsii	Black	E
Exophiala jeanselmei	Black	E
Neotestudina rosatii	White	E
Pyrenochaeta romeroi	Black‡	R
Curvularia geniculata	Black§	E
Curvularia lunata	White	E
Cochliobolus spicifer	Black§	E
Fusarium sp.	White	E
Aspergillus nidulans	White	E
Botryomycosis (cocci or short rods)		
Actinobacillus lignieresii	Yellow§	
Staphylococcus epidermitis	White	
Staphylococcus sp.	White	
Streptococcus sp.	White	
Proteus sp.	White	
Escherichia coli	White	
Pseudomonas sp.	White	

*Key: *C* = common, *O* = occasional, *R* = rare, *E* = exceptional
†Some authors believe that this organism is within the species limits of *N. asteroides.*
‡It is believed that the two species are related, if not identical.
§From animals.
‖Some lesions induced by dermatophytes superficially resemble mycetoma; however, true grains and sinus tracts are lacking.

and 50 per cent of true fungi (eumycetes). With the exception of *Actinomyces israelii*, all are soil saprophytes or plant pathogens that reside on vegetable debris, thorns, sticks, and so forth, and gain entrance to the dermis through abrasion or implantation. In highly endemic regions, there is reason to believe that such factors as continued environmental exposure, inadequate nutrition and hygiene, and general health may play a role in determining the incidence of the disease. Though certain agents show some geographic and ecologic prevalence, most organisms implicated have been found in widely separated areas of the world, and it is probable that they are essentially of worldwide distribution. This again emphasizes occupational and health determinants as predisposing factors for infection.[52]

As in the case with *Sporothrix schenckii*, there is some evidence of inherent differences between soil isolants and their ability to evoke infection. Most strains of *Exophiala jeanselmei* isolated from soil and sewage were unable to grow above 30°C,[54] whereas human isolants readily grew at elevated temperature. Temperature tolerance would be a prime factor for determining potential pathogenicity. As is probable in other mycoses, man acts as a selective agent for strains that can adapt readily to the internal environment and defenses of living tissue.

In some parts of the world, mycetoma occurs commonly. In his classic paper, Abbott[1] noted the admission of 1231 patients with this disease to hospitals in Sudan during two and a half years. This did not include outpatient visits or those patients (particularly women) who for many reasons, including shyness, did not seek medical attention. In 1964 Lynch[48a] described several hundred additional cases and gave a conservative estimate of 300 to 400 new cases per year in that country alone. Sudan appears to have the highest number of cases per capita of any country in the world.[52] Interesting distribution patterns have been collated by Mariat et al.[54] for other regions of the world. *Madurella mycetomatis* and *Actinomadura madurae* are observed whenever the disease is recorded, but they are rarely reported in the United States. (The first record of disease in North America, a case of black-grain infection reported by Wright,[80] was probably *M. mycetomatis*.) Mycetoma due to *A. pelletieri* is very common in West Africa, *A. madurae* accounts for 70 per cent of cases in Tunisia, and *S. somaliensis* is prevalent in East Africa. In South America, *N. brasiliensis* and *M. grisea* account for almost all infections. Several French authors[32, 54] have also emphasized the high endemicity of mycetoma in the equatorial Trans-African Belt extending west from Senegal, Mauritania, Chad, and Nigeria through Algeria and Sudan to the Somali coast. The commonest agents in this region are *M. mycetomatis* (black grain), *Streptomyces somaliensis* (yellow grain), and *Actinomadura pelletieri* (red grain). A few cases caused by other species are also found. Some agents have been recovered from the thorns, soil and debris of the region.[54]

A second area of high endemicity is Mexi-co. Although the case rate does not approach that of the African states, mycetoma is a significant disease there. As noted by Lavalle,[47] there are interesting environmental parallels in the two regions. Both lie between the 14-degree and 33-degree north latitudes, are transected by the Tropic of Cancer, and have similar climatic conditions. The highly endemic regions of Mexico and Sudan have a rainy season extending from June to October, a dry, cool season from October to March, and hot and dry weather without rainfall from March to June. The amount of precipitation also coincides, varying from 50 to 500 mm. Again, the incidence of disease is highest in rural areas where exposure to the soil, occupation, health, and habits may be contributing factors. The predominant etiologic agents in Mexico differ from those in Sudan. *Nocardia brasiliensis* (86 per cent) and *Actinomadura madurae* (10 per cent) are most common in the former country. Lavalle[47] noted that 98 per cent of cases in Mexico are caused by actinomycetes; eumycotic infections are rarely encountered. The original region of high incidence, the Madura area of India, also shares many of these climatic conditions. In India about 65 per cent of cases are actinomycotic (*A. madurae, N. brasiliensis,* and a few *A. pelletieri*), with the remainder being almost all *M. mycetomatis.*

Other areas with significant incidence include the Mediterranean region, Greece, Italy, Rumania,[7, 8] Guatemala, the Caribbean Islands, Cambodia, Iran, and areas of South America. Cases are sporadically seen in Europe and the United States.[13, 37, 81] In the latter areas, *Pseudallescheria boydii* is the most common agent. Scattered case reports have come from all regions of the world, and the various etiologic agents that are very common in one region are rarely reported from other areas.[12]

In addition to the thirteen fungi and actinomycetes regularly isolated from clinical disease, rare cases have been substantiated with *Aspergillus* sp., *Fusarium* sp., and other soil forms[53, 54] as the etiologic agents. It appears that many fungi or actinomycetes, particularly strains able to grow at elevated temperatures and under the proper circumstances, may adapt to the tissue environment and evoke the clinical syndrome "mycetoma."

Most cases of mycetoma are seen in males from rural areas. Although the disease is found in all age groups, clinical attention is generally sought only after many years of development, the decades from 30 to 50 being the most frequently recorded for the disease. Actinomycotic mycetoma develops more rapidly than mycotic disease. Mariat et al.[54] found mycotic infections evolving 20 years or more in some patients, whereas severe clinical disease in actinomycotic infections were usually five years or less in total duration of development. All races are susceptible.

A clinical condition similar to mycetoma may be evoked by a number of other bacterial species. This condition, called "botryomycosis," is also characterized by tumefactions, sinus tracts, and grains. These grains are seen to be composed of masses of gram-positive and gram-negative cocci or short unbranched rods, whereas true actinomycotic mycetoma is due to branching gram-positive rods. Such lesions are often associated with children through a variety of underlying conditions as cystic fibrosis, or with adults through chronic cutaneous irritations and lesions, such as those found on the pinna of the ear or involving the ear canal, interdigital spaces, or lichenified lesions of the scalp. The etiologic agents are usually normal components of the flora of the skin or alimentary tract. This syndrome is discussed at the end of the chapter.

CLINICAL DISEASE

Symptomatology. Regardless of the etiologic agent, the clinical disease is essentially the same. Variations in the syndrome do occur and are related to the various etiologies, the anatomic site, the duration of the lesion, previous medical management, and the general health of the patient. There is some evidence that, after initial implantation, the organism may remain quiescent for some time, possibly requiring another insult to the area or time to adapt to the host. Almost all case histories record previous injury to the involved region, and noticeable symptoms occur after a lapse of several years. The primary lesion is a locally invasive, indolent, tumorlike process or a small, painless, subcutaneous swelling which slowly enlarges and softens to become phlegmonous (Fig. 5–1). It ruptures to the surface, forming sinus tracts, then burrows into the deeper tissues, producing swelling and distortion of the foot. The sole often has a convex rather than flat surface (Fig. 5–2). The foot becomes increasingly swollen and distorted, and may be painful. The surface is studded with numerous small

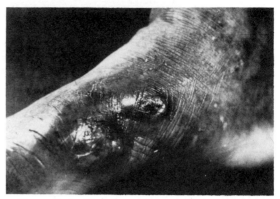

Figure 5–1 Mycotic mycetoma. Tumorlike process of the foot that has softened and ruptured. There are several draining sinus tracts. *Pseudallescheria boydii.* (Courtesy of S. McMillen.)

eminences, each containing the orifice of a sinus (Fig. 5–3). The organism invades subcutaneous connective tissue, bone (in some types), and ligaments, but tendons, muscles,

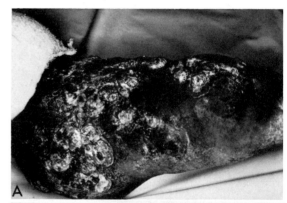

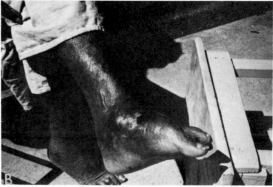

Figure 5–2 *A,* Actinomycotic mycetoma. Convex sole of foot in a patient with advanced disease. Tumorlike growths are present over the lower leg and foot. There is massive destruction of bone. The hyperplasia indicates an actinomycotic cause of disease. *Actinomadura madurae. B* Mycotic mycetoma. Swollen, distorted foot. In this case there is no invasion of muscle or bone and few draining sinus tracts. *Acremonium kiliense.*

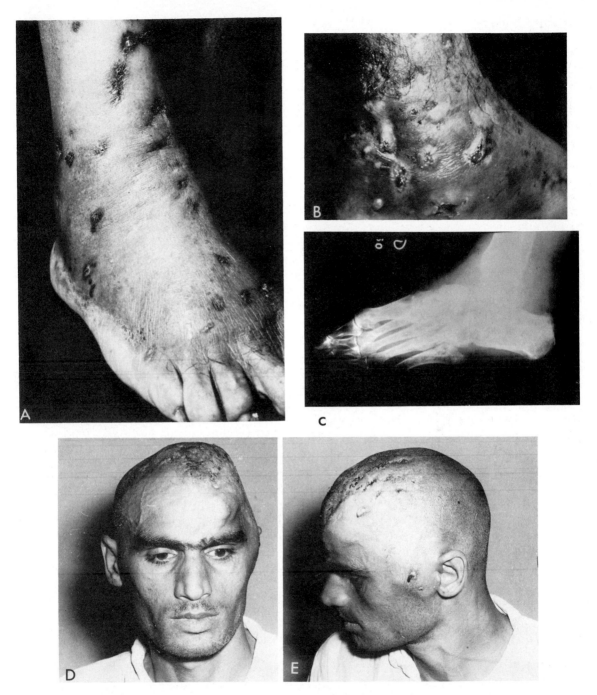

Figure 5–3 *A*, Mycotic mycetoma. Multiple sinus openings in an infection caused by *Madurella griesea*. The surfaces of the lesion are flat and somewhat discolored. The bone is not involved. *B*, Actinomycotic mycetoma. There is destruction of bone, distortion of the foot, and hyperplasia at the openings of the sinus tracts *Nocardia brasiliensis*. (Courtesy of S. McMillen.) *C*, Actinomycotic mycetoma. X-ray of foot showing destructive lesions of tarsals and metatarsals. (Courtesy of E. Macotela-Ruiz.) *D, E,* Mycetoma of the scalp and preauricular area caused by *Actinomadura pelletierii*. (Courtesy of T. R. Bedi.[11])

and nerve tissue are usually spared. Some actinomycetes, however, regularly penetrate muscle tissue. The burrowing follows the fascial planes as in actinomycosis. The abscesses suppurate and drain through multiple sinus tracts. The tracts may remain open for long periods of time, heal, and then reopen. The discharge is a serous exudate, or it may be serosanguinous, seropurulent, or purulent, the latter if there is secondary bacterial infection. Deep tissues are remarkably free of ancillary bacteria. The drainage, when expressed, contains numerous small particles or granules (grains) varying from $300\,\mu$ to 5 mm or more in size.

The grains are light-colored (white, yellow, cream), pink or red, or dark-colored (brown or black), depending on the etiologic agent (Fig. 5–4). Their presence serves to separate this disease from pseudomycetomatous diseases, such as yaws and sporotrichosis. Though the color and morphology of the grains are helpful in ascertaining the etiology, several species may produce similar granules, and it is unwise to ascribe a specific cause without further study. Examination of grains alone will permit the differentiation of actinomycotic mycetoma from eumycotic mycetoma and botryomycosis, and, in the hands of an experienced mycologist, a tentative diagnosis of the agent involved. Culture and laboratory identification, however, are always indicated. This is extremely important for the institution of effective therapy. As emphasized by Zaias et al.,[81] the triad of tumefaction, sinus formation, and grains is necessary for the diagnosis of mycetoma.

Following healing of the involved tissue, massive fibrosis occurs, which gives a tumorlike appearance and hard feel to the area.

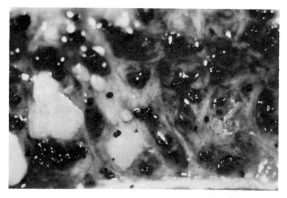

Figure 5–4 Mycotic mycetoma. Cut surface of tissue. Note large black grains surrounded by fibrotic tissue. *Madurella mycetomatis*. (Courtesy of S. McMillen.)

The disease usually remains localized, though it may slowly extend to contiguous tissue and produce severe disfigurement and disability. If the hand is involved, a similar clinical picture is seen, with swelling, sinus tracts, and slow, localized extension. In uncomplicated cases, there is little or no pain, fever, or lymphadenopathy, and review of systems is usually within normal limits.

Hematogenous spread is rare, but recorded. Intraosteal rather than periosteal lesions may indicate that the disease is becoming systemic. When the disease involves the buttocks, chest, or trunk (particularly with *N. brasiliensis*), there is a tendency for it to spread rapidly and widely. With most other agents, visceral involvement is rare, though the isolation of *Pseudallescheria boydii* with typical granules from lungs has been reported. This agent, along with the other agents, may be involved in meningitis, otitis, ocular infection, cystitis, septicemia, or other disease pictures not within the clinical definition of mycetoma.

In addition to the foot, other anatomic areas may be the site of primary or secondary infection. In eumycotic disease, dissemination of primary disease to other cutaneous areas is quite rare. If lesions occur on hands, arms, or neck, they are most probably the primary sites of invasions. In the Sudan and Somalia 80 per cent of cases involve the foot, 6 per cent the head, 6 to 10 per cent the leg, 2 to 3 per cent the buttock and perineum, and a few the abdominal wall, head, neck, knee, chest wall, or thigh. Actinomycotic agents, on the other hand, have a greater tendency to disseminate and evoke secondary lesions in other skin sites or internal organs. This is particularly true of *Nocardia brasiliensis*. In Mexico, where 92 per cent of the agents are actinomycotic and most of these are *N. brasiliensis*, only 35 per cent of infections involve the foot. Legs, arms, chest wall, and abdominal wall are the sites of lesions either primarily or by dissemination. Primary pulmonary nocardiosis is discussed in Chapter 3.

The x-ray picture in mycetoma will show necrosis, generalized osteoporosis, and some fusion of the smaller bones. These may be observed if the density of soft tissue is not too great. Areas of osteal hypertrophy and lysis may be present, as well as large, gouged-out lesions occupied by massive granules (Fig. 5–3, *C*).

Differential Diagnosis. Mycetoma, as defined, presents so typical a picture as to be readily diagnosed, especially in highly endemic areas. Actinomycosis is excluded from the mycetomas. It gives a similar picture of firm induration, swelling, and sinus tract formation with granules, but the onset, systemic distribution, endogenous nature, and overall clinical appearance are distinctive and form a well-defined entity. However, the etiologic agent, *Actinomyces israelii*, has been reported several times from the clinical syndrome of mycetoma. Systemic nocardiosis with its primary infection in the lungs and quite different clinical course again is a separate disease. Granules are lacking if the agent is *Nocardia asteroides*. *Nocardia brasiliensis* may be considered an agent of nocardiosis if the primary site of involvement is the chest and lungs. On occasion, dissemination to other areas can give rise to "mycetomas" in a variety of organ systems. Distinction of the two clinical conditions here seems futile.

Pulmonary "mycetoma" and aspergillomas are also excluded from the accepted definition of mycetoma. In these instances, such an organism, usually *Aspergillus* species, grows in ectatic bronchi or old tuberculous cavities, but lacks inherent qualities of extension or invasion into uninvolved tissue. The lesions may enlarge within the limits of debilitated tissue, and frank invasion of pulmonary parenchyma may ensue, but the pathologic picture is that of a mycotic granuloma rather than a mycetoma. Sinus tracts, tumefaction, and grains are lacking; instead, the organism occurs as "fungus ball."

Botryomycosis (staphylococcic actinophytosis) is a clinical entity that closely resembles mycetoma, with the production of swelling, grains, and draining sinuses (see section on botryomycosis). The agents involved include *Staphylococcus aureus*, *Pseudomonas aeruginosa*, *Proteus* sp., and possibly *Escherichia coli*. Examination of the grains quickly identifies this disease.

Several diseases may, at one stage or another, mimic mycetoma. Included are yaws, elephantiasis of the foot (no sinus tracts), cellulitis, and primary or secondary eumycotic infections, such as coccidioidomycosis, sporotrichosis, or ulcerated, hypertrophic chromoblastomycosis. Mycetoma in unusual anatomic sites may mimic progressive angioma of Darier, tuberculosis verrucosa, and acne conglobata.

Prognosis and Therapy. It is imperative that the general class of mycetoma be determined in order to initiate effective therapy. Actinomycotic disease is more amenable to treatment than are eumycotic mycetomas.

If the lesions remain localized, there is little impairment to general health, although extensive involvement may severely incapacitate the patient. Treatment is made particularly difficult by the inability of drugs to penetrate into cystic and fibrotic areas in sufficient concentration to inhibit the causal agent. Vigorous and prolonged treatment is necessary in order to ensure that an adequate dose is present in the involved tissue. A trial of specific medical treatment combined with conservative surgical management should be considered. Surgical intervention includes exploration and drainage of sinus tracts, debridement of diseased tissue, and removal of cysts from involved bones. Radical surgery and amputation should be considered only as a last resort.

Actinomycotic Mycetoma. Actinomycotic disease, even in advanced forms of the infection, is considered amenable to medical management. In a study of 144 patients with actinomycetoma, 63 per cent were cured, 22 per cent had great improvement, and 11 per cent had at least some degree of improvement. The most effective regimen involved either dapsone with streptomycin sulfate or sulfamethoxazole-trimethoprim with streptomycin. Sulfadoxine-pyrimethamine plus streptomycin or rifampin is also quite useful.[51] Dapsone is given as tablets (100 mg each) at the rate of 1.5 mg per kg, twice daily. Tablets of sulfamethoxazole-trimethoprim contain 100 mg of the former and 80 mg of the latter. The dose is 14 mg per kg twice daily, which amounts to 23 mg per kg of sulfamethoxazole and 4.6 mg per kg of trimethoprim daily. Higher doses sometimes result in leukopenia or hemoglobinemia. Streptomycin is given at the rate of 14 mg per kg per day, and rifampicin, 4.3 mg per kg twice daily. Tablets of 500 mg sulfadoxine and 25 mg pyrimethamine are given at the rate of 7.5 mg per kg twice weekly. Treatment requires 4 to 24 months, averaging 9 months. The superiority of these regimens is now quite well established. In some cases results are not always completely rewarding, however. Sulfadiazine and tetracycline over $2^{1}/_{2}$ years was used in an *A. madurae* infection with extensive involvement of bone.[43] Cure of a pedal mycetoma due to *A. madurae* was achieved using minocycline at 150 mg twice daily.[44] Various penicillins in combination with various other drugs were used in the past.[51] These modalities have been largely supplanted today.

Eumycotic Mycetoma. Mycetoma of fungal etiology is consistently resistant to chemo-therapy. The list of drugs tried without suc-cess is long and the results discouraging. The introduction of amphotericin B raised the hope that the disabling and disfiguring le-sions could be controlled by chemotherapeu-tic means. Blood levels of 1.5 to 2 μg per ml can be achieved safely in patients, but they are usually below the effective range for the agents involved. The *in vitro* sensitivity of *Madurella grisea* varies from 10 to 100 μg per ml, of *M. mycetomatis* from 0.62 to 25 μg per ml, and of *Leptosphaeria senegalensis* from 0.7 to 100 μg per ml. Some clinical success in mycetoma due to *M. mycetomatis* has been re-ported with combined therapy of amphoteri-cin B and surgical management. From the literature,[51] it appears that the drug is regu-larly used in the treatment of mycetoma produced by this agent and possibly by *M. grisea*, but is not very effective in cases caused by other agents. *M. grisea* has been treated successfully by Neuhauser[61] using 20 mg DDS per day. *In vitro* studies by Indira and Sirsi[42] showed partial inhibition of *M. grisea* by 10 μg per ml of amphotericin B and 20 μg per ml of nystatin. Hamycin and DDS were with-out effect. Aureofungin and pimaricin were quite inhibitory at 15 and 10 μg per ml, respectively. Clinical trials of these drugs have not been reported.

Nielson[59a] reports amphotericin B has little or no effect *in vitro* on *Pseudallescheria boydii.* Inhibited strains soon developed resistance (see Chapter 24 for *P. boydii*). One successful report of *in vitro* sensitivity and clinical cure of *P. boydii* mycetoma is that of Mathews,[54a] who used the compound D-25 (2,2'-dihydroxy-5,5'-dichlorodiphenyl sulfide). Treatment consisted of 0.5 to 1 g of the drug in oil injected into the affected area for intervals of six months. Nystatin has been reported as inhibitory, and was used success-fully in an ocular infection of *P. boydii*. Most agents of eumycotic mycetoma are inhibited by 28 μg per ml of nystatin *in vitro*, but there are few clinical trials reported. Mohr and Muchmore[55a] report a case of *P. boydii* myce-toma that would resolve when the patient was pregnant. Subsequently, the patient was given norethynodrel (5 to 20 mg per day), and a complete remission occurred. After extensive trials, griseofulvin was found to be of no value. Isolated limb perfusion tech-niques have been utilized to increase the drug level at the affected site. Its use has been of limited value and is appropriate only in spe-cialized cases.[81] Thiabendazole has been tried with some success. Clinical trials of 5-fluorocytosine are now being carried out in Mexico and Latin America.

The only mycotic agent of mycetoma that appears amenable to medical management is *P. boydii*. Miconazole was found to inhibit *P. boydii* at concentrations of 0.25 μg per ml. With a blood level of 1.0 μg per ml achievable following an intravenous dose of 9 mg per kg, the medical management of this type of the disease is promising. Clinical trials have not been completed (see Chapter 24). Mahgoub[53] tested other agents against *M. mycetomatis in vitro*. The minimum inhibitory concentration of clotrimazole varied from 1 to 200 μg per ml, with a cluster around 15 μg per ml. The MIC for griseofulvin in most strains was 200 μg per ml. In clinical trials neither drug was more curative than surgery alone.

The historic and perennial use of iodides cannot be ignored, as clinical remissions have occurred with prolonged therapy. As is the case with sporotrichosis, where the agent can grow in media containing 5 per cent KI, the effect is probably on the host tissue rather than on the organism. The drug acts by causing resolution of the granuloma formed around the parasite, thus exposing it to the action of the host's defense mechanisms.

PATHOLOGY

Gross Pathology. The gross pathology of mycetoma depends on the anatomic area affected by the disease. There is distortion and swelling of the involved region. Acute purulent abscesses develop, resolve, and re-develop, burrowing in several directions, un-dermining and invading the soft tissue and bone and erupting to the surface as draining fistulous tracts. As new areas become active, healing with extensive fibrosis occurs in older areas. If limbs are involved, the process may proceed throughout most of the lifetime of the patient without injury to the general health. Eventually there may be severe inca-pacitation of the involved appendage. The lesions develop where the organism is im-planted. In Africa and most other places, thorns or spines prick the legs or feet; thus the disease is generally limited to the lower limbs. In Mexico, many cases involve the up-per back, neck, and shoulder region, where the area is rubbed and abraded while carry-ing sugar cane, bundles of firewood, or sisal sacks; primary pulmonary disease is acquired by inhaling contaminated dust. Hands may be inoculated by working in the fields, clearing brush, or gathering reeds for making baskets. It can be seen that the mode of infection is the same as with chromoblastomycosis and sporotrichosis. In mycetoma, particularly actinomycotic mycetoma, if the head, chest, or neck regions are involved, complications may arise by extension of the destructive process into vital organs. Lymphatic and hematogenous spread are more common when these are the sites of primary involvement. The gross pathology is identical to that of actinomycosis.

Depending on the etiologic agent involved, there are some subtle differences in the gross appearance of the sinus openings, the tissue reaction, and the tissues invaded. These are discussed in the section on specific organisms. Secondary bacterial infection of the peripheral sinus tracts is common and may complicate the clinical and pathologic picture. Drainage from the sinuses is usually serous or serosanguinous, but may become purulent and foul-smelling because of secondary infections. There is often an oily or fatty content to the drainage related to the necrosis of subcutaneous adipose tissue. Grains may be found in expressed fluid, in purulent exudate, massed and clogging the sinuses internally, or lodged in cystic lesions of bones.

Histopathology. Histologically the lesions are similar regardless of the etiologic agent. The granules are seen in the center of an acute abscess, often coated with a crust of homogeneous material. This is surrounded by a large accumulation of neutrophils in all stages of degeneration. Immediately around the abscess there is an area of dense fibrosis and granulation tissue that is rich in capillaries, epithelioid cells, macrophages, and multinucleated giant cells. True grains in suppurating abscesses and typical clinical symptoms must be present for a diagnosis of mycetoma to be made. This disease is a serious condition and carries with it the associated difficulties in prognosis and therapy, often culminating in amputation. In other fungal diseases, accumulations of hyphae or yeast cells may be present which give the appearance of "pseudogranules." These may be accompanied by fungal elements in giant cells, but the lesions lack the suppuration found around the true granules of mycetoma. A more correct term for such lesions is "mycotic granuloma." These conditions are often more amenable to therapy and carry a better prognosis. Such diseases as coccidioidomycosis, blastomycosis, histoplasmosis, cryptococcosis, phaeohyphomycosis, sporotrichosis, and aspergillosis at some stage may simulate mycotic granuloma. Botryomycosis must also be excluded.

In tissue sections of biopsy or autopsy material from lesions of mycetoma, the grains

Table 5–2 *Actinomycotic Mycetoma Histology and Morphology of Colonies*

Species	Grain	Histology (H and E)	Colony and Microscopic Morphology
Nocardia asteroides	Rare; white, soft, irregular, 1 mm	Homogeneous loose clumps of filaments; rare clubs	Rapid growth at 37°C; glabrous, folded, heaped; orange-yellow, tan, etc.; short rods and cocci; rare branched filaments; acid-fast
Nocardia brasiliensis	White to yellow, lobed, soft, 1 mm	Same as above; clubs common	Rapid growth at 30°C; colony and microscopic morphology same as above; acid-fast
Nocardia caviae	Same as above	Same as above	Same as above; acid-fast
Actinomadura madurae	White, soft, oval to lobed, rarely pink, large, 5 mm	Center empty, amorphous; dense mantle peripherally basophilic wide pink border; loose fringe; clubs	Rapid growth at 37°C; cream-white, rarely clot-red; wrinkled, glabrous; delicate filaments, nonfragmenting, branched; non-acid-fast
Actinomadura pelletieri	Red, hard, small, oval to lobed, 1 mm	Round homogeneous dark staining; light peripheral band; hard — fractures easily; no clubs	Slow-growing (at 37°C); small, dry, glabrous; light to garnet red; delicate, nonfragmenting, branched filaments; non-acid-fast
Streptomyces somaliensis	Yellow, hard, round to oval, large, 2 mm	Variable size; amorphous center; light purple with pink patches; dark filaments at edge, entire; no clubs	Slow growth (30°C); cream to brown; wrinkled, glabrous; delicate, branched, nonfragmenting filaments; spores produced; non-acid-fast

Actinomyces israelii is also a cause of light-grained mycetoma. Consult Chapter 12 for identification.

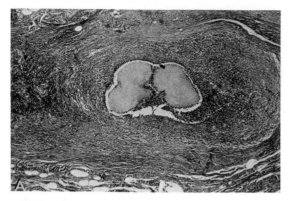

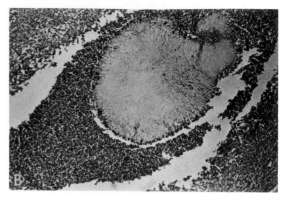

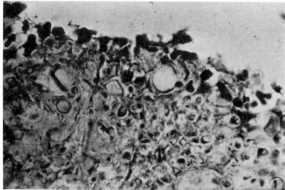

Figure 5–5 Mycotic mycetoma. *A,* Large grain within sinus tract. The periphery of the lesions shows extensive fibrosis followed by a little-organized granulomatous reaction. *Pseudoallescheria boydii.* × 100. *B,* Grain of *P. boydii.* Neutrophils are attached to the periphery of the grain. The grain itself has a slight basophilic outer edge and an eosinophilic center. × 400. (Courtesy of L. Ajello and L. Georg.) *C,* Edge of grain of *P. boydii.* Note distorted mycelium and large, cystlike vesicles. The grain is soft and ill-organized, and there is no cement between the mycelial strands.

appear with abscesses (Fig. 5–5 and Fig. 5–6). They are stained by hematoxylin and eosin. The center of actinomycotic grains is usually light-colored and unorganized. There is a strong basophilia surrounding the periphery of the grains and a wide-fringed eosinophilic border. In contrast, eumycotic grains vary and are discussed in the section on mycology. The usual picture of these, however, is broad, light pink mycelium in the periphery, surrounded by a basophilic area. The center may be strongly basophilic or unorganized. Black-grain mycetoma grains are, of course, brownish black in histologic section unstained or with hematoxylin-eosin staining (Fig. 5–6). The methenamine stains are useful for both types of agents. Thin, dendritic filaments are seen with the actinomycetes, and broad, hyphal units and grossly swollen cells are observed in eumycotic mycetoma. Subtle differences in grain characteristics are discussed later (Tables 5–2 and 5–5).

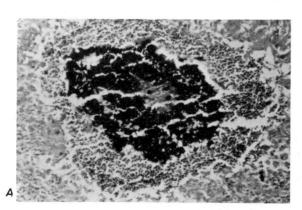

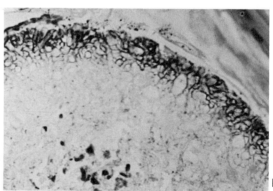

Figure 5–6 *A,* Mycotic mycetoma. Edge of black grain of *Exophiala jeanselmei.* The grain is soft and ill-organized. The fungus forms distorted mycelium with vesicles of cystlike structures. There is a covering of neutrophils attached to the grain, which appears free in the lumen of the sinus tract. The tract walls show histiocytes, giant cells, and a few lymphocytes. *B,* Black grain mycetoma caused by *Madurella griesea.* Hematoxylin and eoson stain. ×800. There is heavy pigmentation at periphery. Note swollen cystlike cells. (Courtesy of S. McMillen.)

In summary, it appears that eumycotic and actinomycotic mycetomas present similar pictures clinically and histologically, but with some minor differences. Both elicit a suppurative response initially. As the disease becomes chronic and established, the eumycotic infection is treated as a foreign body with granuloma formation, epithelioid hyperplasia, and giant cell formation. In actinocomycotic mycetoma, the acute suppurative pyogenic reaction persists, and the infection is treated in the same manner as a bacterial invasion. Other points of difference between the two classes of disease include the following: (1) actinomycotic agents produce more extensive and obliterative involvement of bone, with both lytic and hypertrophic changes, and there is late bone involvement with lytic but no hypertrophic effects in the course of disease with eumycotic agents; (2) the actinomycetes invade muscle more readily than eumycetes; and (3) the latter do not cause the cellular proliferation that leads to the raised border around the sinus openings on the skin surface.

ANIMAL DISEASE

Mycetoma is regularly reported in cattle under the name "farcy," and the ascribed agent is called *Nocardia farcinica*. There is much doubt about both the infectious nature of the disease and the validity of the agent. This disease is discussed in the chapter on nocardiosis. Natural infections in lower animals caused by the agents of actinomycotic mycetoma are rare. Ajello[2] described the disease caused by *N. brasiliensis* in a cat. A suppurating granuloma with granules developed following a bite on the hind leg which led to systemic involvement and death. The pathology was similar to that described for the human disease. Eumycotic mycetoma is also rare in lower animals. In 1967, Brodey[20] reviewed twelve cases involving dogs, cats, and horses, and reported a new case in a dog. The published reports all specified the finding of grains (usually dark) in suppurating granulomatous tissue. The disease was usually limited to the legs or feet, and the histology was that described in human disease. The agents when cultured, however, were quite different. Other fungi have also been implicated. Black grains caused by *Cochliobolus spicifer* in a cow,[63] *Curvularia geniculata* in a dog[20] and a horse,[15] and white grains by *Pseu-*

dallescheria boydii in various animals,[68, 69] have been reported. Some of these species have not been recovered in human cases.

Experimental infection in animals has been attempted many times with varying results. Most success has been achieved with the actinomycetes. Macotela-Ruiz and Mariat[50a] produced lesions with granules by injecting *N. brasiliensis* or *N. asteroides* into mice, hamsters, and guinea pigs. Some grains had club cells. They concluded that the hamster was the best animal for such studies. Using similar techniques, Rippon and Peck[70] were able to produce mycetoma in the legs of mice with *A. madurae* (Fig. 5–7). Their experiment was designed to show that the proteolytic enzyme, collagenase, was necessary for pathogenicity of this organism. Recent work by others has indicated differences in the degree of pathogenicity for three agents of mycetoma. When several strains of each organism are used, it has been found that *N. brasiliensis* is most virulent and always produces grains. Although *N. caviae* is less virulent, it still produces grains, but no grains has been found in *N. asteroides* infection, and this organism is significantly less pathogenic than the other agents. (see Chapter 3)

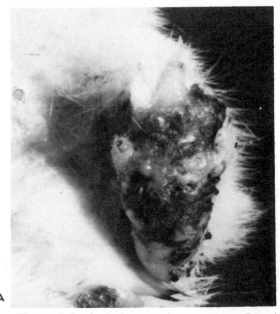

A

Figure 5–7 *A*, Actinomycotic mycetoma. Experimental disease produced in mice with *Actinomadura madurae*. Multiple white grains were extruded from the sinus tracts. There was complete destruction of the bones in the area.[70]

Illustration continued on following page

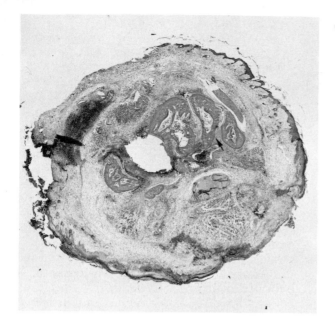

Figure 5–7 *Continued* *B*, Section through the limb of infected mouse. There is massive hyperplasia and fibrosis to make the appendage several times normal size. The muscle bundles (*arrow*) have been displaced and the bone obliterated. Hematoxylin and eosin stain. ×10. *C*, Elongate grain at *A. madurae* showing tufted outgrowth. The grain is the light area in the center and is completely surrounded by massed neutrophils (dark area) all contained within a sinus tract (*arrows*). *D*, A mouse with an experimental mycetoma of hind leg. The organism *A. madurae* has produced so much collagenase *in situ* that it caused the skin to peel off the body, the ears to wrinkle and flop, and the legs to splay.

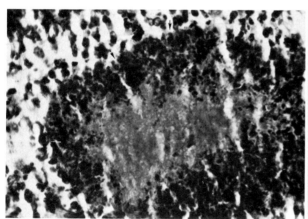

D

Such studies have been less successful with the eumycotic agents. Murray[57] reported infection in mice with *M. mycetomatis*; however, his results could not be reproduced by Indira.[42] In a series of papers, Avram[5, 6] has reported infection and grain production with *Acremonium kiliense (Cephalosporium falciforme)*, *P. boydii* and *M. mycetomatis*. It appears that true mycetoma can be produced in animals if correct procedures are used. Experimental infection does not carry any diagnostic value in the clinical laboratory, but it can be used for the evaluation of chemotherapeutic agents.

IMMUNITY

Though the etiologic agents of mycetoma are found throughout the world, the overall rarity of the disease attests to a high degree of natural resistance in the normal host, and a low grade of pathogenicity of the organisms. The disease has not been associated with any underlying debilities, as is the case with systemic nocardiosis. If any predisposing factors are to be considered, they would be general health and nutrition, repeated exposure or insult to the affected area, and personal hygiene.[53] The wearing of shoes is a preventative measure, although this increases the chances for ringworm of the foot, a much less hazardous choice.

SEROLOGY

At present, there are no established serologic procedures for the diagnosis or prognosis of mycetoma. As noted in the chapter on nocardiosis, González Ochoa and Baranda, studying the sera of patients with *N. brasilien-*

sis infection, found that about half had complement-fixing antibodies, a few had agglutinins, and none had precipitating antibody (see Chapter 3). There was no correlation between titer and duration, severity, or extent of the disease. Bojalil and Zamora[14a] prepared a purified polysaccharide from *N. brasiliensis* with which they were able to demonstrate precipitins in patients' sera, and a specific skin test reaction. In the latter instance, there were no cross-reactions with sera from patients infected by other *Nocardia* species, leprosy, or tubercle bacilli.

Murray and Mahgoub[56] prepared antigens from several eumycotic and actinomycotic agents of mycetoma using the gel diffusion technique. They found that sera from 69 patients and from mice experimentally infected with *M. mycetomatis* gave precipitin lines which were specific for the appropriate antigen. Control sera did not react, nor was there significant cross-reactivity with other fungal antigen. The gel diffusion technique is becoming a useful method in the diagnosis of several mycotic and actinomycotic infections and promises to be an easy and reliable diagnostic procedure in mycetoma infections as well. Since the etiologic agents of this disease are many and of diverse nature, a common or polyvalent antigen seems unfeasible; however, specific reactivity could be useful as an adjunct to cultural techniques when defining the etiology. The appearance of precipitin lines during active disease in histoplasmosis and aspergillosis and their disappearance after cure encourages the investigator that the same may be found for the serodiagnosis of mycetoma.

LABORATORY IDENTIFICATION

Direct Examination. Pus, exudate, or biopsy material from the patient may be examined for the presence of grains as well as the gauze bandage over the lesion. Though the grains vary in size from 0.5 to 2 mm or more, they are usually macroscopically visible, and a study of their morphology, texture, color, and shape may indicate with a fair degree of certainty the identity of the causative organism. As grains are quite similar, especially among the actinomycetes, culture is always necessary for definitive diagnosis. Examination of crushed material in potassium hydroxide preparations will allow delineation of actinomycotic and eumycotic mycetoma, as well as botryomycosis. Actinomycetes produce a grain composed of intertwined fila-

ments, 0.5 to 1 μ in diameter, as well as coccoid and bacillary elements. Eumycotic grains show intertwined, broad (2 to 5 μ) mycelial strands which may have large swollen cells (15 μ or more) at the periphery. A Gram stain will demonstrate the gram-positive filaments of the actinomyces, as well as the gram-positive cocci, or the gram-negative bacilli which produce botryomycosis. These all appear surrounded by a gram-negative matrix. Gram stain, however, is of limited value for observing true fungi. Zaias et al.[81] recommend Albert's stain for studying the fine morphology of fungal grains. Giemsa stain, as recommended by Lacaz, outlines in blue the crust or shell and the cementing substance of eumycotic granules. For preparation of stained smears, the granule should be crushed in its pus. *Nocardia asteroides, N. brasiliensis, N. caviae,* and *N. farcinica* are usually acid-fast; other species of the genus are not.

There are some minor differences in the morphology of the grains seen in direct preparation. These are illustrated in Tables 5–2 and 5–3.

Culture Methods. For culture, a deep biopsy is best, as it is usually free from contaminating bacteria and fungi. Otherwise the sample should be handled like the grains of actinomycosis. The granules are washed several times in sterile saline or Eugon broth, crushed with a sterile glass rod or in a Tenbroeck tissue homogenizer, and plated on appropriate media.

Actinomycetes are planted on blood agar or BHI agar. Several plates should be streaked and incubated in 37°C both anaerobically (for *A. israelii*) and aerobically. Although an anaerobic atmosphere including 5 to 10 per cent CO_2 is best for growing *Actinomyces* species, a candle jar will usually suffice. The *Nocardia* species will grow well on Sabouraud's yeast extract agar. The media should contain no antibiotics. Some actinomycetes grow more readily on Löwenstein-Jensen culture medium than on other media. After isolation of the colonies, specific identification can be made by physiologic characteristics.

Specimens containing eumycetes may be washed in antibiotic-containing saline and plated on Sabouraud's yeast extract (0.5 per cent) media without cycloheximide. Several plates should be planted, as the number of a single species present often indicates true etiology. Plates should be incubated at 25°C. As no true fungi are obligate anaerobes, anaerobic procedures are unnecessary. The

Table 5–3 *Actinomycotic Mycetoma: Morphology of Grains*

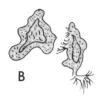

A, Nocardia asteroides: white, very soft, amorphous, more an intertwined mass of mycelium than a grain

B

B, Nocardia brasiliensis: white to yellow, soft to firm, large, lobed with outgrowths

C

C, Nocardia caviae: white, soft, amorphous, similar to *N. asteroides*

D

D, Actinomadura madurae: white, pink, rarely blood-clot red; soft to firm, often growing to very large size (5 mm or more with outgrowths); tends to be compact more than lobed

E

E, Actinomadura pelletierii: garnet red, hard, brittle, fan shaped

F

F, Streptomyces somaliensis: yellow, hard, round, edges entire; compact solid center shows ripples from microtome

growth of eumycotic agents is usually very slow, and plates should be kept for six to eight weeks. Identification depends on colony morphology, conidia types, and assimilation patterns detailed in the following section. Conidia production is often enhanced by growth on low-nutrient media, such as potato-carrot agar or cornmeal agar.

MICROBIOLOGY

The literature on mycetoma contains a long list of species as agents of the disease. Most of these are so poorly described that they are unidentifiable, or the apparent differences were not fully compared with valid species. Continued critical study by several authors, including Emmons, González Ochoa, Ajello, and Mackinnon, has caused this list to be reduced by reclassification and recognition of synonymous organisms. In this section, the more common and accepted species of the actinomycetes and eumycetes involved in human mycetoma will be described in detail. In addition the characteristic types of grains, colony morphology (Fig. 5–8), cultural characteristics, physiology, if relevant, and type of disease induced by each will be discussed. Preparation and use of special media and techniques are included in Part Five. The microbiology of *Nocardia asteroides* is discussed in Chapter 3, *Actinomyces israelii* in Chapter 2, and *Pseudallescheria boydii* in Chapter 24.

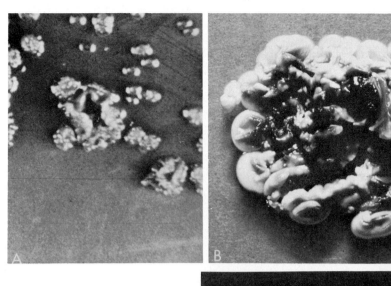

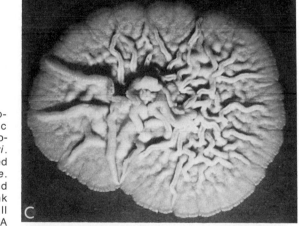

Figure 5–8 Colonial morphology of four common etiologic agents of actinomycotic mycetoma. *A, Actinomadura pelletieri.* This is a moist, red-pink, folded colony. *B, Actinomadura madurae.* This colony is moist, folded, and has areas of blood red-to-pink color. The usual isolant is dull white. *C, Nocardia brasiliensis.* A thin, glabrous, wrinkled, off-white colony. *D, Streptomyces somaliensis.* Creamy, folded colony with tufts of whitish grey mycelium.

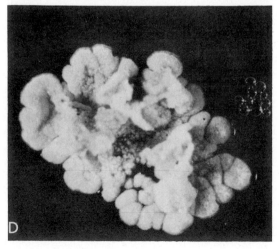

Actinomycetes

The identification of the aerobic actinomycetes depends on a combination of morphologic observations and physiologic reactions. In recent years considerable research and modifications of standard microbiological techniques have evolved and are evolving. Thus the techniques current one day may be changed the next. In this section the basic concepts of these techniques are reviewed; the specific details of the procedures will be given in the Part Five.

IDENTIFICATION OF AEROBIC ACTINOMYCETES

As discussed in Chapter 1, in modern systems of taxonomy, the considerations for the identification of aerobic actinomycetes are, firstly, the composition of the cell wall; secondly, the morphology of the colony and its filaments; and finally, physiologic reactions. It is seen that in dealing with such organisms a combination of techniques is utilized to identify these peculiar bacteria. Colony and microscopic morphology, the basis for delineation of true fungi, is used in order to determine the presence of aerial mycelium, its color, and its branching and spore formation, if any. Physiologic reactions such as sugar utilization, production of enzymes, and resistance to chemicals (a basic tool in bench bacteriology and zymology) are also required. Cell wall analysis is a unique requirement for this baffling group of organisms.

Cell Wall Composition. The final identification of an isolant termed an aerobic actinomycete depends to a large extent on the grouping determined by cell wall analysis. Many *Nocardia*, *Streptomyces*, and *Actinomadura* spp., in addition to *Micromonospora*, *Mycobacterium*, and *Rhodococcus* spp., are quite variable in their morphology and even their physiology. Thus, chemical composition is the first consideration in identification. Wall. type groups are given in Chapter 1, along with a summary of the families and genera of pathogenic and related species. The determination of composition is somewhat laborious but, once mastered, can become routine. Special equipment and expertise are required, and isolants needing this procedure are not regularly encountered. It is therefore recommended that cooperative regional laboratory centers servicing local hospitals and laboratories be set up or state or federal reference laboratories be ready to quickly carry out such tests.

Morphology. Growth of the organism in subculture is observed on a protein-containing media and a deficient media. Incubation is for 10 to 14 days at room temperature. On protein medium (nutrient agar such as Sabouraud's), color of aerial mycelium, morphology of vegetative mycelium, and production of diffusible or nondiffusible pigment are noted. Growth or lack of growth on deficient media (e.g., Czapeks-Dox agar, 7H10, 7H11) is useful, particularly in separating *Streptomyces* species. Odor of the colonies is also noted. Presence of aerial mycelium is best seen using a dissecting scope.

Slide cultures on both protein and deficient media are done in order to observe branching of filaments and spores and their arrangement. Care must be observed in interpretation, since *Mycobacterium* and *Corynebacterium* spp. can sometimes be "arranged" so as to appear branched. Slide cultures are incubated at 30 to 35°C for seven days. Examine under low power without removing cover slip or mounting. In this way the slides can be reincubated for late-appearing aerial growth and spores.

Physiologic Tests. Decomposition of amino acids and proteins along with utilization of sugars is an essential element in the identification of aerobic actinomycetes. A guide for the selection of media and details or procedures are given in the Part Five, and summaries of results are given in Tables 5–2 and 5–4. After selection of an isolated colony from the primary plating, a subculture is made on heart infusion agar (no added carbohydrate). After sufficient growth has occurred, a portion is streaked on one quadrant of a tyrosine plate, a casein plate, and a xanthine plate. A positive and negative control are also streaked (Fig. 3–9) on the same plates. Plates are incubated at room temperature for two weeks (casein), three weeks (xanthine), and four weeks (tyrosine). Sometimes results can be speeded by incubation at 35°C.

Lysozyme resistance is tested by inoculating some fragments in a tube containing glycerol and lysozyme. A control without lysozyme is also inoculated. This is the single most important test for separating *Nocardia* from other genera. Incubate one week at room temperature. Urea test broth (*not* agar) is used for aerobic actinomycetes. Fragments are inoculated into the tubes and incubated for four weeks.

There are several other useful growth tests. Inoculate *Nocardia* on Löwenstein-Jensen media, 7H10, 7H11, or into litmus milk to improve acid-fast reaction. *Mycobacterium fortuitum* grows on MacConkey agar without crystal violet, *Rhodococcus* grows sparsely, and aerobic actinomycetes not at all. Psychrophiles and psychrotolerant species are tested as follows: Two broth tubes with inoculum are incubated, one at 35°C and one in a 10°C water bath for seven days. No *Mycobacterium* spp.

Table 5–4   A, *Actinomycotic Mycetoma: Physiologic Reactions of Commonly Used Laboratory Tests*

| | Wall Type* | Decomposition of | | | | | Lysozyme Resistance | Acid-Fast | Acid from | | |
		Casein	Tyrosine	Xanthine	Hypoxanthine	Urea			Lactose	Xylose	Arabinose
N. asteroides	IV	−	−	−	−	+	R	+	−	−	−
N. brasiliensis	IV	−	+	−	+	+	R	+	−	−	−
N. caviae	IV	−	−	+	+	+	R	+	−	−	−
N. coeliaca	IV	−	+	+	+	+	S	+	+	+	+
A. madurae	III	+	+†	−	+	−	S	−	+	+	+
A. pelletieri	III	+	+	−	+	−	S	−	−	−	−
N. dassonvillei	III	+	+	+	+	+†	S	−	−	+	+
S. somaliensis	I	+	+	−	−	−	S	−	−	−	−
S. griseus	I	+	+	+	+	V	S	−	−	−	−
S. albus	I	+	+	+		+			−	−	−

*Wall types are defined in Chapter 1.
†Some exceptions. R = resistant, S = sensitive, V = variable

Table 5–4   B, *Complete List of Physiologic Reactions of Medically Important Aerobic Actinomycetes*

| Property | Species | | | | | | | | |
	N. asteroides	*N. caviae*	*N. brasiliensis*	*N. dassonvillei*	*A. madurae*	*A. pelletieri*	*S. somaliensis*	*S. albus*	*S. griseus*
Aerial hyphae	+	+	+	+	−	−	−	+	+
Acid-Fastness	+	+	+	−	−	−	−	−	−
Decomposition of									
Adenine	O	O	O	+	O	O	O	O	O
Casein	O	O	+	+	+	+	+	+	+
Hypoxanthine	O	+	+	+	+	+	O	+	+
Tyrosine	O	O	+	+	+	+	+	+	+
Urea	+	+	+	−	O	O	O	+	+
Xanthine	O	+	O	+	O	O	O	+	+
Resistance to Lysozyme	+	+	+	O	O	O	O	−	O
Acid from									
1(+)Arabinose	O	−	O	−	+	O	O	O	−
Cellobiose	O	O	O	+	+	O	O	+	+
i-Erythritol	O	O	O	O	O	O	O	+	O
Glycerol	+	+	+	+	+	O	O	+	+
d(+)lactose	O	O	O	O	−	O	O	+	+
d(+)maltose	O	O	O	+	−	O	−	+	+
d(−)mannitol	O	−	+	+	+	O	O	+	−
d(+)Melibiose	O	O	O	O	O	O	O	O	O
d-Sorbitol	O	O	O	O	O	O	O	O	O
d(+)Xylose	O	O	O	−	+	O	O	+	+
Nitrite from Nitrate	+	+	+	+	+	+	O	−	−

+ denotes 85% or more of strains tested
− denotes between 15 and 85% of strains
O denotes up to 15% of strains
Adapted from Mishra, S. K., R. E. Gordon, and D. A. Barnett. 1980. J. Clin. Microbiol., *6*:728–736.

Table continued on following page

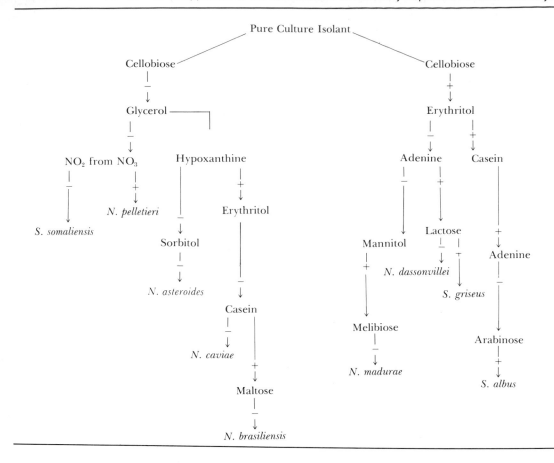

and very few *Nocardia* spp. will grow at this latter temperature, but *Rodococcus* spp. grow well.

Nocardia brasiliensis (Lindenberg) Castellani and Chalmers 1913 (for synonymy and details of culture see Chapter 3)

Clinical Disease. There is some cellular proliferation and local infiltration around the openings of the sinus tracts, often forming a raised border and giving the area a "bumpy" appearance. Punched-out lesions also may occur if there has been secondary bacterial infection. Primary cutaneous disease with serpiginous, raised, erythematous lesions (Fig. 3–5*F*) are also noted. Disseminated disease resembles disseminated actinomycosis or coccidioidomycosis, with subcutaneous abscesses, sinus tracts, and burrowing erosions. There is frequent and extensive involvement of bone. In conditions that mimic actinomycosis, the slight acid-fastness and aerobic growth of *N. brasiliensis* quickly differentiate the two. Optimum growth is 30° or 37°C.

Grains. The grains are usually less than 1 mm in size, white to yellowish in color, soft, and lobulated. "Clubs" may or may not be seen. They are scattered or may be clumped together in large masses. Dendritic acid-fast filaments are often seen to be fragmenting into bacillary bodies. Some differences in morphology of grains has been noted in different geographic locations.

Colonies. The organism grows rapidly, producing a heaped, wrinkled, and folded colony. The color is yellow to yellow-white, tan, orange, or red-orange and closely resembles the colony of *N. asteroides*. A white, powdery overgrowth may occur, indicative of fragmentation into arthrospores. Rarely, a brownish pigment diffuses into the media.

Stained smears show short, irregular bacilli and coccoid forms. Growth occurs as long filaments in broth cultures. Small, beaded, acid-fast forms may resemble atypical mycobacteria, but slide cultures showing branching filaments differentiate the two.

Culture. Decomposition of tyrosine, gelatin, and casein separate *N. brasiliensis* from *N. asteroides*. *N. brasiliensis* is acid-fast and does not attack xanthine or starch. These tests separate it from other *Nocardia* and *Streptomyces* species.

Nocardia caviae (Erickson) Gordon and Mihm 1962

Synonymy. *Nocardia otitidis cavarum* Snijders 1924; *Actinomyces caviae* Erickson 1935.

Although many references in the literature ascribe mycetoma to *N. asteroides*, it is probable that, in most cases, the causative agent is *N. caviae*. The two were distinguished by the critical examination of Gordon and Mihm in 1962.[38] *N. caviae* appears to be an infrequent cause of mycetoma in humans.

Clinical Disease. The few cases recorded resemble *N. brasiliensis* infection. *N. caviae* was first isolated from an ear infection of guinea pigs by Snijders in 1924.

Grains. Similar to *N. brasiliensis*.

Colonies. Growth is described as being moderately rapid, resembling that of both *N. asteroides* and *N. brasiliensis*. On special *Nocardia* media (Bennett's), a cream to peach color is formed. Tufts of short aerial mycelia are seen. The organism is acid-fast.

Culture. *N. caviae* decomposes both xanthine and hydroxanthine but not casein, BCP, milk, tyrosine, or gelatin. Acid is formed from inositol and mannitol, but not from arabinose and xylose. It survives at 50°C for eight hours and growth sparingly at 40°C. A more extensive characterization of *N. caviae* is given in Chapter 3.

Nocardia asteroides (Eppinger) Blanchard 1896 (Synonymy and details of culture are given in Chapter 3.)

Clinical Disease. In the authenticated cases of mycetoma involving *Nocardia asteroides* the lesions were not extensive and developed slowly. Balabanoff[9] discussed the possibility of primary lung infection with dissemination to the skin.

Grains. This fistulous discharge contains soft white intertwined hyphal masses, 20 to 200 μm in diameter, that resemble grains.

Actinomadura madurae (Vincent) Lechevalier 1970

Synonymy. *Streptothrix madurae* Vincent 1894; *Nocardia madurae* (Vincent) Blanchard 1896; *Actinomyces madurae* (Vincent) Lachner-Sandoval 1898; *Nocardia indica* Chalmers and Christopherson *Discomyces bahiensis* Piraja da Silva 1919; *Streptomyces madurae* (Vincent) Mackinnon and Artagaveytia-Allende 1956.

A. madurae is found wherever mycetoma exists. In most temperate and tropical areas it accounts for about 10 per cent of cases; it is rare in the United States and Canada, however. In Tunisia it accounts for 65 per cent of cases, and in India for 31 per cent of cases.

Clinical Disease. In infections produced by this organism, a wide raised border is often seen around the sinus openings because of infiltration and hyperplasia of epidermal cells. Bone involvement is notoriously common. Lytic and proliferative changes are seen, and osteophytes or spicules of bone may be found between cavities occupied by large grains. Proteolytic enzymes are elaborated *in vivo* and may affect the type of disease produced.[70] Muscle is readily invaded by the organism.

Grains. The granules produced by *A. madurae* are the largest seen in mycetoma. The average size is 1 to 5 mm, and 10 mm has been recorded. Usually they are white to yellowish, irregularly lobulated (mulberry), oval, serpiginous, or angular. The finding of pinkish to light red grains is recorded. The outside edge of the grain has a fringe of filaments which extend radially. These may be up to 50 μ in length. The granules have a shell of homogeneous eosinophilic material surrounding them but no cement. These are often seen as "clubs." The consistency is soft and the grain is easily crushed to a pasty mass.

Colonies. The usual colonial form is dirty white, glabrous or waxy, wrinkled, and folded, with a flat periphery. The colony is moderately fast growing and so membranous as to peel off the media when subcultures are attempted. A blood-clot red colonial form is not uncommon. The organism grows rapidly on Löwenstein-Jensen media and can be mistaken for atypical mycobacteria.

Stained smears show long, sinuous branched filaments that are usually less than 1 μ in diameter. Occasionally, chains of spherical arthrospores ranging in size from 0.5 to 1 μ in diameter are seen. The organism is gram-positive and nonacid-fast. It is nonfragmenting. Optimum growth occurs at 37°C.

Culture. *A. madurae* is both proteolytic and amylolytic, digesting casein, gelatin, tyrosine, BCP milk and starch. It produces acid from arabinose and xylose. It does not digest xanthine. Paraffin is not utilized. Nitrates are reduced to nitrites.

Actinomadura pelletieri (Laveran) Lechevalier 1970

Synonymy. *Micrococcus pelletieri* Laveran 1906; *Oospora pelletieri* (Laveran) Thiroux and Pelletieri 1912; *Nocardia pelletieri* (Laveran) Pinoy and Jouenne 1915; *Discomyces pelletieri* (Laveran) Neveu-hemaire 1921; *Actinomyces pelletieri* (Laveran) Brumpt 1927; *Nocardia africanus* (Laveran) Pijper and Pullinger 1927; *Aspergillus pelletieri* (Laveran) Smith 1928; *Nocardia genesii* (Laveran) Froes 1931;

Steptomyces pelletieri (Laveran) Mackinnon and Artagaveytia-Allende 1956.

This organism is rare in North and South America, accounting for only 2 per cent of cases. Only in Senegal (23 per cent) and Chad (25 per cent) is it regularly encountered. In India 6 per cent and in Somalia 3 per cent of cases are caused by this organism; the overall world figure is about 9 per cent.

Clinical Disease. Large, crateriform, lobulated nodules of the skin are seen around the openings of the draining sinuses in *A. pelletieri* infections.[11, 13] Marked epidermal hyperplasia and an exaggerated granulomatous inflammatory reaction cause this nodular appearance. Bone involvement is similar to that seen with *A. madurae*, though the granules are smaller, as are the lytic lesions and spicules. "Red-grain" mycetoma of the scalp has been said to be caused by this organism.[11]

Grains. The characteristic granule is small (300 to 500 μm) and garnet red. The morphology is oval to spherical, or irregular to lobulated. A regular curvilinear shape (fan shape) results from the splitting of the hard, friable grains. The edge is entire or slightly denticulate. It contains no cement, is very hard, and resists crushing. There is no filamentous fringe, as is found in *A. madurae* grains.

Colonies. Growth is very slow, and small, dry, granular or glabrous adherent colonies are produced. The color is at first pink to peach, developing into a deep cranberry red. Subcultures may in time become more mottled and peach-colored. White overgrowth may occur.

In slide preparations, delicate branched filaments are seen to be 0.5 to 1 μ in diameter. Conidia (arthrospores) are rare. This organism is not acid-fast and is gram-positive. Optimum growth occurs at 37°C.

Culture. *A. pelletieri* is proteolytic but not amylolytic. Casein, tyrosine, BCP milk, hypoxanthine, and gelatin are digested. Starch is usually not digested. Urea, potassium nitrate, and ammonium nitrate are not utilized. No acid is produced from arabinose, xylose, galactose, mannitol, or maltose. Acid is produced from trehalose. Paraffin is not utilized. Nitrates are reduced.

Streptomyces somaliensis (Brumpt) Waksman and Henrici 1948

Synonymy. *Indiella somaliensis* Brumpt 1906; *Nocardia somaliensis* (Brumpt) Chalmers and Christopherson 1916; *Actinomyces somaliensis* Brumpt 1927; *Streptothrix somaliensis* (Brumpt) Miescher 1917; *Discomyces somaliensis* (Brumpt) Neveu-hemaire 1921.

This organism is rarely documented in North or South America[59] but is common in Somalia, where it accounts for 50 per cent of cases. It accounts for 7 per cent of cases worldwide. It occurs variably in other African countries (Senegal 9 per cent, Chad 14 per cent), India (6 per cent), and Mexico (3 per cent).

Clinical Disease. In infections produced by this organism, the openings of the sinus tracts are not raised or indurated as in the infections produced by *Actinomadura* species. Bone is not as extensively involved, but muscle may be invaded.

Grains. Very hard, yellow to brown, large granules up to 2 mm are present. There is a firm, cementlike substance holding the filaments together. When crushed, grains break up into angular fragments. The shape is round to oval, and occasionally lobulate. The grains are rarely a pinkish color.

Colonies. On agar the colony develops rapidly in four to 11 days, and has a creamy, wrinkled, flaky growth. Growth is better on Löwenstein-Jensen media. Sectoring of the colony is common, giving areas of varying consistency and mottled coloring. Areas of brown or black may develop, and the underside often shows a yellow to ocher color. Tufts of brownish, aerial filaments develop, which in time may become gray. A diffusible brown pigment has been reported. The organism is nonfragmenting and may form arthroconidia in chains. Stained preparations show delicate branching filaments 0.5 to 1 μ in diameter. Optimum growth occurs at 30°C.

Culture. *S. somaliensis* is not proteolytic (gelatin). Casein and tyrosine are decomposed. The other tests (urea, hypoxanthine, xanthine, nitrate reduction, acid from sugars) are negative. About 4 per cent of strains are amylolytic (starch hydrolysis).

Streptomyces albus (Rossi Doria) Waksman and Henrici 1943

This organism has been observed in a relentless infection in Rumania[5] that required repeated amputations. This is the only confirmed report of an infection. Because it is a common contaminant, it is included in Table 5–4.

FUNGI

The eumycetes, of course, are true fungi and their identification is based on standard mycological techniques for the observation of colonial morphology and microscopic morphology. The colony obverse is examined for consistency, color, texture, rate of growth, and characteristic topography; the colony reverse is observed for diffusible and nondiffusible pigments, if any. The major determinant for identification, however, is the microscopic examination, which is best achieved using the slide culture technique (see Part Five). We utilize Sabouraud's agar on one side of the slide and Czapek-Dox or other nutritionally poor media on the other side. Incubation is at room temperature for two weeks or more. Slides can be checked regularly for conidia production. If conidia have not appeared, the slides are returned to the moist chamber for further incubation. Conidiation is sometimes induced by inoculation on nutritionally poor agars (pea, carrot-potato) or on plain water agar. As with other fungi, physiologic reactions are of little value for identification. A few tests that some authors feel are useful are listed in Table 5–5.

Table 5–5 *Mycotic Mycetoma*

Species	Grain	Histology (H and E)	Colony and Microscopic Morphology	St	Gel	G	Gal	L	M	S
				Physiologic Profile						
Pseudallescheria boydii	White, soft, oval to lobed, <2 mm	Hyaline hyphae, 5μ; huge swollen cells, <20 μ; no cement; red border; pink periphery	Rapid-growing; fluffy, mouse-fur gray; large 7-μ unicellular conidia on simple conidiophore; black cleistothecia; 30–37°C; also synnemata in *Graphium* state	+	+	+	V	0	0	V
Madurella grisea	Black, soft to firm, oval to lobed, <1 mm	Little dark cement in edge; polygonal cells in periphery; center hyaline mycelium	Very slow growing; leathery, tan-gray, later downy; sterile pycnidia; diffusible pigment; 30°C	+	−	+	+	0	+	+
Madurella mycetomatis	Black, firm to brittle, oval to lobed, <2 mm	1. Compact type with brown-staining cement 2. Vesicular type with brown cement only at edge; swollen cells, <15 μ; center hyaline mycelium	Very slow growing; downy, velvety, smooth or ridged; cream-apricot to ocher; diffusible brown pigment; black sclerotia, <2 mm; rare conidia, phialides; 37°C	+	±	+	+	+	+	0
Pyrenochaeta romeroi	Same	Same grains without vesicles	Hyalin and brown hyphae. Pycnioconidia in pycnidia	+	±	+	+	0	+	+
Acremonium kiliense	White, soft, irregular, <1.5 mm	No cement; hyaline hyphae, <4 μ; swollen cells, <12 μ	White glabrous colony, later downy; violet pigment diffusible; curved septate; conidia arranged as head on simple conidiophore; 30°C	0	±					
Exophiala jeanselmei	Black, soft, irregular to vermicular 0.2 to 0.3 mm	Helicoid to serpiginous; center often hollow; no cement; vesicular cells, <10 μ; brown hyphae	Slow-growing; leathery, black, moist, later velvety; reverse black; toruloid yeast cells, moniliform cells, long tubular phialides; 30°C	+	0	+	+	0	+	+
Leptosphaeria senegalensis	Black, soft, irregular, ~1 mm	Black hyphae; cement in periphery; center hyaline	Rapid-growing; downy gray; reverse black, rare rose pigment, diffusible; black ascostroma; <300 μ; septate ascospores, 25 × 10 μ			+	+	V	+	+
Neotestudina rosatii	Brownish white, soft, <1 mm	Polyhedral with filaments embedded in cement peripherally. Some vesicles in center	Slow growing, flat, folded, compact, light grayish brown. No conidia. In poor media black ascostroma form.							

*St, starch; Gel, gelatin; G, glucose: Gal, galactose; L, lactose; M, maltose; S, sucrose.

Pseudallescheria boydii (Negroni et Fischer) McGinnis, Padhye et Ajello 1981

Synonymy. *Allescheria boydii* Shear 1921; *Pseudoallescheria sheari* (as *shearii*) Negroni et Fisher 1944; *Petriellidium boydii* (Shear) Malloch 1970.

The teleomorph stage of this organism is being renamed *Pseudallescheria boydii* (Negroni et Fischer) McGinnis, Padhye et Ajello 1981 and a complete description will appear in Mycotaxon. The genus *Petriellidium* appears to be invalid.

Conidial state *Scedosporium apiospermum* (Saccardo) Castellani et Chalmers 1919

Synonymy. *Monosporium apiospermum* Saccardo 1911; *Monosporium sclerotiale* Pepere 1914; *Indiella americana* Delamare and Gatti 1929; *Arcomoniella lutei* Arêa Leão and Lobo 1940.

Clinical Disease. As with most cases of the eumycotic mycetomas, the swollen appendage is dotted with punctate, flat, fistulous openings and usually lacks the epidermal hyperplasia seen in many actinomycete infections. Bone involvement is also less marked. The osteomyelitis is primarily destructive without new bone formation. The organism is the commonest cause of mycetoma in Europe and the United States.[37, 81] It accounts for about 3 per cent of cases in the world. Its highest endemicity is in Rumania, where it causes 20 per cent of mycetomas recorded. It is encountered in Brazil[47] and sporadically in Africa and India. It has been isolated from soil and sewage[31] and has been involved in infection of the lung as well as other types of disease. The spectrum of other clinical conditions elicited by this organism is surveyed in Chapter 24.

Grains. The granules are large (up to 2 mm), white to yellowish, soft to firm, round to lobulated (Fig. 5–5, *A, B*). The hyphae are broad (up to 5 μ diameter), septate, and intertwined, and show numerous swollen cells (15 to 20 μ) at the periphery of the granule. These are sometimes referred to as chlamydospores but probably reflect a modification of the hyphae to the environment of the tissue (Fig. 5–5*C*). There is no evidence of cement between the hyphal strands.

Colonies. *P. boydii* grows rapidly on all laboratory media. There is an abundance of fluffy or tufted aerial mycelium, which is at first white but becomes brownish "house mouse" gray (Fig. 5–9). The reverse of the colony shows areas of gray to black pigmentation. Rare strains are ivory-colored and membranous; others are annular in growth, showing concentric rings.

The hyphae are hyaline and 1 to 3 μ in diameter. The large, pyriform annelloconidia are lemon-shaped, 4 to 9 × 6 to 10 μ, borne singly or in small groups on elongate, simple, or branched annellophores (Fig. 5–11), or

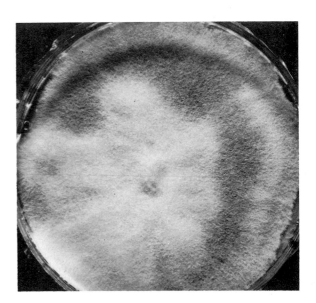

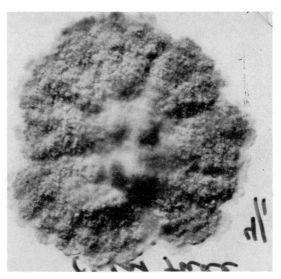

B

Figure 5–9 Mycotic mycetoma. *Pseuda llecheria boydii.* Soft, mouse-gray, furlike colony. Ten days' growth on Sabouraud's glucose agar. *B,* Colonial variant of *P. boydii* tufted gray-brown to tan colony with folding and wafts of white mycelium.

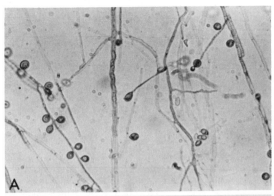

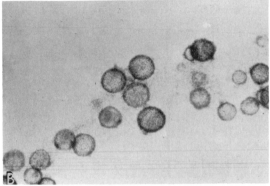

Figure 5-10 *A,* Culture mount in lactophenol cotton blue. Single conidia on short conidiophores from mycelium and on elongate condidiophores. ×400. (From Rippon, J. W. *In* Burrows, W. 1979: *Textbook of Microbiology,* 20th ed. Philadelphia, W. B. Saunders Company, p. 732.) *B,* Cleistothecia forming at edge of colony within agar. Ovoid to ellipsoide ascospores are the same size as conidia. ×50.

laterally on hyphae (Fig. 5–10,*A,* Fig. 5–11*B*). Tufts of conidiophores forming a synnemata are also found. Each annellophore ends in a single conidium (Fig. 5–12). The organism is homothallic, and some isolants produce ascocarps near the agar surface or in the mycelium near the edge of the colony (Fig. 5–10,*B*). Ascocarp formation is enhanced by placing strains on pea agar, potato-carrot agar, or plain water agar.

There is a conspicuous and persistent ascogonium. The cleistothecia measure 100 to 300 μ in diameter and have a yellow-brown to black wall composed of thick-walled polygonal cells. The fruiting body ruptures at maturity. The globose or subglobose asci are evanescent and not found in mature carps. Eight ascospores are produced in each ascus. They are 4 to 5 × 7 to 9 μ, faintly brown, and resemble the asexual annelloconidia. The latter have a truncated scar at the base of the cell where they were attached to the annellophores. Optimum temperature is 30° to 37°C.

Culture. *P. boydii* is both proteolytic and amylolytic. It assimilates urea, asparagine, potassium nitrate, and ammonium nitrate. Glucose is assimilated but not lactose or maltose.

Madurella mycetomatis (Laveran) Brumpt 1905 (as "*mycetomi*")

Synonymy. *Streptothrix mycetomii* Laveran 1902; *Glenospora khartoumensis* Chalmers et

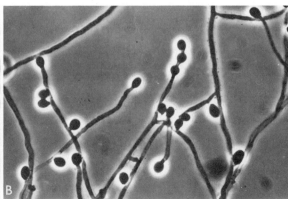

Figure 5-11 *P. boydii. A,* High-power view of annellide with an annellolconidia. *Arrow* indicates annellations. *P. boydii.* ×2300. (Courtesy of J. Carmichael.) *B,* Phase contrast showing single and short chains of annelloconidia. ×820. (Courtesy of J. Carmichael.)

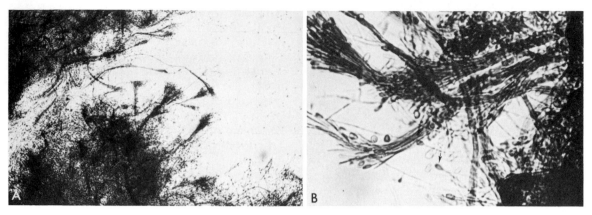

Figure 5–12 *A, P. boydii.* Elongate synnemata formation of the *Graphium* type conidiation. ×100. *B,* High-power view of single anneloconidia *(arrow)* and *Graphium* conidiation. (Courtesy of Janet Gallop.)

Archibald 1916; *Oospora tozeuri* Nicolle et Pinoy 1908; *Madurella tozeuri* Pinoy 1912; *Madurella tabarkae* Blac et Brun 1919; *Madurella americana* Gammel 1926; *Madurella ikedae* Gammel 1927; *Madurella lackawanna* Hanan et Zurett 1938; *Madurella virido brunnea* Redaelli et Ciferri 1942.

This organism is essentially never reported in North America or Central America, although the first case of mycetoma in the United States, reported by Wright in 1898,[80] may have been caused by *M. mycetomatis.* Approximately 13 per cent of cases due to this agent are found in South America, with the remainder in Africa and India. In the trans-African belt (particularly Somalia and Chad) it causes up to 40 to 50 per cent of cases and is second only to *N. brasiliensis* as the commonest agent worldwide. Of 63 cases of mycetoma in the United States, 4 were caused by *M. mycetomatis.*[24]

Clinical Disease. The clinical appearance is similar to that of *P. boydii* mycetoma. The infection remains localized and encapsulated at first. Later stages tend to respect anatomic barriers and spread along the fascial planes and the fibrous septa between the muscular bundles. In advanced stages, osteolysis is marked without accompanying osteogenesis.

Grains. *M. mycetomatis* grains are black, 0.5 to 1 mm in size, round or lobed. They may be aggregated to form a mass 2 to 4 mm in size. Grains are hard and brittle but can be crushed. They are composed of hyphae which are 2 to 5 μ in diameter, with terminal cells expanded to 12 to 15 μ in diameter and up to 30 μ. The hyphal fragments vary in content of pigment. Two types of grains are described. The first is a compact form filled with dark brown granular cement between the hyphal elements. In hematoxylin and eosin stains, it appears as a uniform rust brown color. A second, vesicular type of grain occurs in which the cement in the periphery is brown and filled with vesicles 6 to 14 μ in diameter. The center is light colored. This latter type resembles the grains of *M. grisea* (Table 5–6).

Colonies. The optimum temperature for growth is 37°C. The colony is at first leathery, folded, and heaped, white to yellow or ochraceous brown in color, and covered by a grayish down (Fig. 5–13, *A*). Later there is an overall growth of brownish aerial mycelia and diffusible pigment. The mycelia average 1 to 5 μ, with moniliform hyphae from 2 to 6 μ. Enlarged vesicle-like cells may attain a diameter of 25 μ. In old colonies (two months), black sclerotia, 1 mm in diameter, may be formed, composed of rounded, polygon-shaped mycelial elements. This is best observed on potato-carrot media. Old colonies take on a reddish-brown hue. Two types of conidiation have been described in growth on cornmeal agar, pyriform conidia (3 to 5 μ), with a truncated base borne on the tips of simple or branched conidiophores, and small, flask-shaped phialides producing rounded spores (3 μ in diameter) (Fig. 5–13*B*).

Culture. *M. mycetomatis* is amylolytic and weakly proteolytic. Glucose, galactose, and maltose are assimilated, but not sucrose. This latter reaction separates it from *M. grisea.*

Madurella grisea Mackinnon, Ferrada, et Montemayer 1949

This agent of mycetoma is found chiefly in the Western Hemisphere,[25, 37, 49, 61] although there are rare reports from other parts of the world.[52] Mackinnon et al.[49] described the organism as a new species in 1949 from cases of black grain originating in Argentina, Chile, Paraguay, and Venezuela. Some 40 cases are now recorded worldwide, four in the United States, about 30 in South America, and the others erratically distributed elsewhere.

Clinical Disease. The clinical picture is similar to that of *P. boydii* mycetoma.[54] An excellent description of the clinical disease is given by Butz and Ajello.[24]

Grains. The granules are black, round to lobed, and similar to those of the second type described for *M. mycetomatis*. The size is up to 1 mm; the consistency is soft at first, becoming hard and brittle. The brown cement is limited to the outer edges of the granule. The center appears hollow in sections and is composed of a loose, hyaline mycelial network. These cells are small (1 to 3 μ) and look like a chain of budding cells.

Colonies. The colonies are slow-growing, with an optimum temperature of 30°C. The appearance is dark, leathery, and folded, with a tan to gray fuzz developing over the surface. Colonies on Sabouraud's dextrose agar are usually sterile except for occasional chlamydospores (Fig. 5–14). On low-nutrient media, pycnidia with pycnidiospores may develop. There is some evidence that this species and *Pyrenochaeta romeroi* are related or identical.[58, 71]

Culture. Like *M. mycetomatis,* this species is amylolytic and weakly proteolytic. Glucose, galactose, sucrose, and maltose are assimilated, but not lactose. The lactose and sucrose reactions separate it from *M. mycetomatis.*

Exophiala jeanselmei (Langeron) McGinnis et Padhyė 1977

Synonymy. *Torula jeanselmei* Langeron 1928; *Pullularia jeanselmeii* Dodge 1935; *Phialophora jeanselmei* (Langeron) Emmons 1945; *Phialophora gougerotii* Borelli 1955.

This organism was first isolated in 1928 from a case of black-grained mycetoma of the foot by Jeanselme on the island of Martinique.[43] Since then it has occurred infrequently in cases of mycetoma in the United States. It was reclassified by Emmons as a species of *Phialophora* because of its mode of conidia

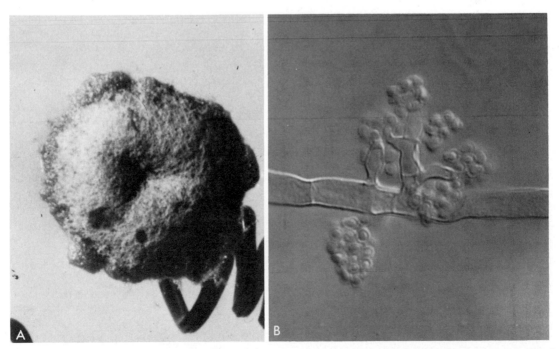

Figure 5–13 *Madurella mycetomatis. A,* Heaped, folded, dematiaecious colony on Sabouraud's agar. This is a recent isolation from a Chicago area patient. *B,* Culture mount shows flask-shaped phialides producing rounded conidia. ×1100. (Courtesy of M. McGinnis: Laboratory Handbook of Medical Mycology. New York: Academic Press, 1980.)

Table 5–6 *Mycotic Mycetoma: Morphology of Grains*

A

A, Madurella mycetomatis: black, hard, brittle, cemented together, consistency of coal

B

B, Madurella mycetomatis: vesicular, "open" grain, often with many large cyst-cells

C

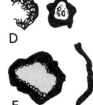

C, Leptosphaeria senegalensis: black, soft to medium, irregular, cemented

D

D, Madurella grisea or *Pyrenochaeta romeroi:* soft, irregular, center hyaline, periphery black

E

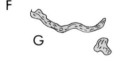

E, Exophiala jeanselmei: soft, open center, often filled with neutrophils, often wormlike and undulating, black

F

F, Petriellidium boydii: very large, soft, lobulated white grains

G

G, Acremonium sp.: white, soft, small, lobulated, vermiform

H

H, Neotestudina rosatii: white angulated mass breaking into sections

production.[34] Following extensive work on conidiogenesis, however, McGinnis and Padhye[55] have reclassified the fungus as *Exophiala jeanselmei.* It was originally designated as a species of *Torula* because of its "toruloid" or yeastlike phase seen in culture.[46] It is a common inhabitant of soil and has been isolated from wood pulp. It has been isolated from cases of mycetoma with soft, black, serpiginous granules, from subcutaneous abscesses that show brown vesicles and hyphal fragments, and from cases of phaeohyphomycosis with sclerotic round bodies in tissue and the hypertrophied verrucous skin lesions characteristic of that disease (Fig. 5–15). Isolants have also been recovered from onychomycosis and tinea pedis, but their significance in these conditions is as yet undetermined. The organism appears to be an opportunist with versatile adaptability.

Clinical Disease. As reported, it is similar to that of the other eumycotic mycetomas.

Grains. A vermiform or serpiginous undulating grain is seen that may be hollow in the center. It is black, of a soft consistency, and quite variable in size. It is composed mainly of swollen cells that are 5 to 10 μ in size. Brown hyphal elements are also seen.

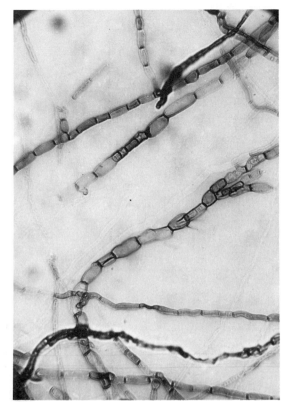

Figure 5–14 *Madurella grisea.* Mycelia of two widths and chains of arthroconidia characteristic of the species.

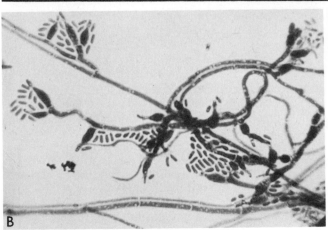

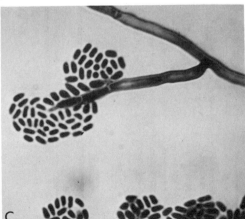

Figure 5–15 *A, Exophiala jeanselmei.* Black-gray velvety colony. *B,* Elongate and short, squat (lageniform) anellophores. *C,* Elongate annellophore that has a long, thin tapering end. (Courtesy of Libero Ajello.)

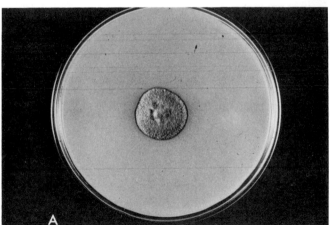

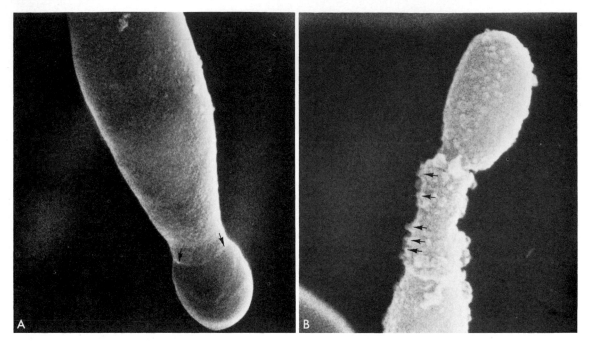

Figure 5-16   *E. jeanselmei.* Stages in conidiogenesis. *A,* First conidium has ruptured the apex of conidiogeneous cell (*arrows*). ×7920. *B,* Successive conidia have resulted in a series of annellations (*arrows*). ×14,000. (Courtesy of G. T. Cole.)

The granule is often helicoid, crescent-shaped, or stringing through a fistula, as a boat in a canal. It is quite distinct from other mycetoma grains.

Colonies.   To define the cultural characteristics of *E. jeanselmei* is perplexing. The initial slow-growing colony is mucoid or leathery, glabrous, and dark. Toruloid or yeast cells predominate. Conidiation at this point somewhat resembles *Aureobasidium pullulans.* This is the *Phaeococcus* state of the growth cycle. In time, a grayish-black, velvety, aerial mycelium develops with black pigment on the reverse side, giving a dematiaceous appearance. Moniliform hyphae are seen. In older colonies, tubelike annelides are formed with narrow, tapered tips.

The conidia are quite variable in size, with an average dimension of 5 to 6 μ × 1 to 2 μ. There is no attachment scar. They may be seen clustered about the tip of annelide. Optimum growth is at 30°C.

Culture.   *E. jeanselmei* is neither proteolytic nor amylolytic. The former characteristic distinguishes it from *Cladosporium* species that may have similar colonial characteristics. Glucose, galactose, maltose, and sucrose are assimilated. Lactose is not.

Acremonium kiliense Grütz 1925

Synonymy.   *Cephalosporium granulomatis* Weidman and Kligman 1945; *C. madurae* Padhye, Sukapure, et Thirumalachur 1962; *C. infestans* Gaind, Padhye, et Thirumalachur 1962.

Acremonium falciforme (Carrión) Gams 1971

Synonymy.   *Cephalosporium falciforme* Carrión 1951.

Acremonium recifei (Arêa, Leão, et Lobo) Gams 1971[36]

Synonymy.   *Cephalosporium recifei* Arêa, Leão, et Lobo 1934.

A. falciforme was first reported under the name *Cephalosporium falciforme* from Puerto Rico, where it was isolated from two cases of white-grained mycetoma.[26] Since then, several cases have been described in America (including three cases in the United States), Africa, and Rumania.[39] Isolated cases have occurred in Venezuela, Brazil, and Thailand. *A. kiliense* as *C. infestans*[35] and *C. madurae* came from black-grain disease in India and as *C. granulomatis*[37] from the United States. It has also been recovered as white-grain disease in India.[36] *A. recifei* has been isolated in France and Brazil. *Cephalosporium serrae*[3] from Venezuela is now included in the genus *Verticillium*.

Clinical Disease. In terms of grains and tissue reaction they are similar to that described for *P. boydii* mycetoma. The *grains* are whitish to yellow in color, soft, irregular, and up to 1.5 mm in diameter. Hyaline hyphae 3 to 5 μ in diameter are found, as well as numerous swollen cells, especially in the periphery of the grain. Some grains appear quite dark and have a gray to almost black color.

Colonies. The growth of all three species is relatively slow, and a tufted, down colony is produced that may be white, buff, pinkish, or lavender. A diffusible currant-red to violet pigment is seen on the reverse of the colony. Hyphae are hyaline and 2 to 4 μ in diameter. Typical *Acremonium* conidiophores and conidia are produced. The conidia are crescent- or cycle-shaped, single celled, multicellular, and 10×4 μ in diameter or nonseptate or straight, depending on the species. They are borne successively from the tips of conidio-

phores (phialides) and held there as a cephalic, "headlike" cluster by mucilaginous exudate (Fig. 5–17). Numerous chlamydoconidia of variable size are produced, and nodular bodies are regularly seen. Optimum temperature is 30°C.

A. falciforme has nonseptate to two-celled crescent conidia. *A. recifei* has nonseptate crescent conidia, and *A. kiliense* has straight conidia.

Leptosphaeria senegalensis Segretain, Baylet, Darasse, and Camain 1959

This recently described agent[73] of black-grained mycetoma has been isolated from ten cases in Senegal and Mauritania and appears to be a common incitent of the disease in that area. So far it is not recorded outside Africa, although a retrospective study of pathologic material indicates its possible presence in South India.[42]

Grains. The grains are soft, black, irregular, and about 1 mm in size (Fig. 5–18, *A*). Fragments and aggregates may coalesce to form large masses. These consist of a central

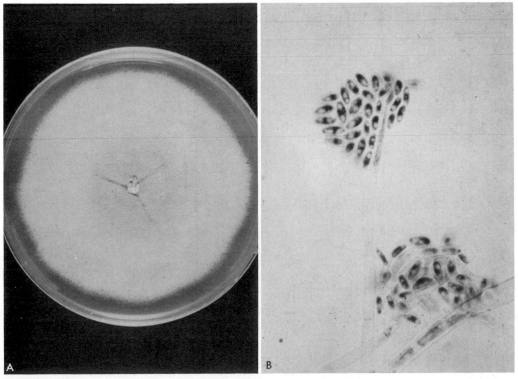

Figure 5–17 *Acremonium falciforme. A,* A two-week old colony on Sabouraud's agar. The colony is tufted downy off-white with a buff center. *B,* Microscopic view. ×100. Long, thin tapering phialides produce conidia. The conidia are one celled and vary from straight to curved. On the right are a few conidia with a pronounced (falc- or sickle-shaped) curvature. (Courtesy of M. Rinaldi.)

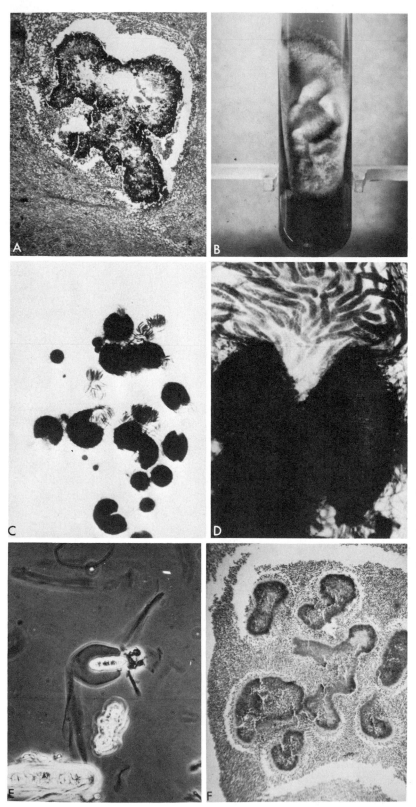

Figure 5–18 African and South American agents of mycotic mycetoma. (Courtesy of G. Segretain.) *A,* Soft black irregular grain of *Leptosphaeria senegalensis.* de Mann stain. ×20. *B,* Folded downy gray colony of *L. senegalensis* on Sabouraud's agar slant. *C,* Crushed ascostromata of *L. senegalensis.* The asci are very long and arranged parallel to each other. ×12. (From J. Baylet, R. Camain, and G. Segretain, 1959. Indentification des agents des Maduromycoses du Sénégal et de la Mauritanie. Description d'une espèce nouvelle. Bull. Soc. Path. Exot., *52:*448–477.) *D,* Crushed ascocarps of *L. senegalensis.* *E,* Elongate ascospore (encapsulated) of *L. senegalensis.* *F, Pyrenochaeta romeroi.* Soft, black grain. Hematoxylin and eosin stain ×10.

Illustration continued on opposite page

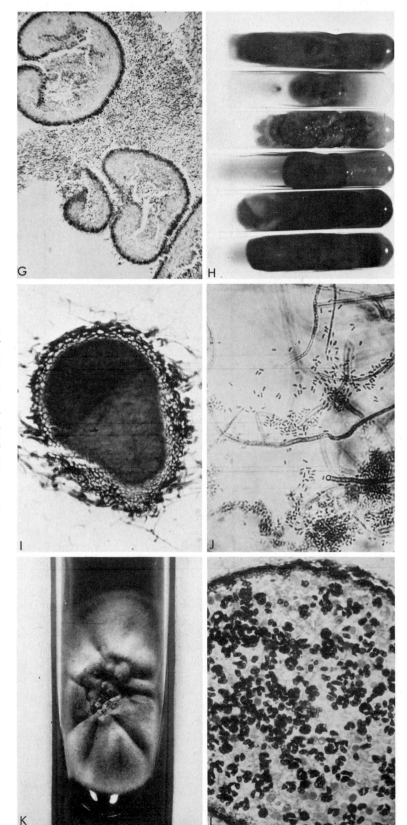

Figure 5–18 *Continued   G,* Detail of grain showing thin active edge, soft amorphous interior. Hematoxylin and eosin stain. ×45. *H,* Floccose, wooly, grayish black colonies of *P. romeroi. I,* Brownish ostiolate pycnidia of *P. romeroi* showing thick outer wall and setae. *J,* Crushed pycnidia showing elliptical phialoconidia. *K,* Folded black colony with radial grooves of *Neotestudina rosatii. L,* Section of ascocarp of *N. rosatii.* Two-celled ascospores are found in spherical asci.

Illustration continued on following page

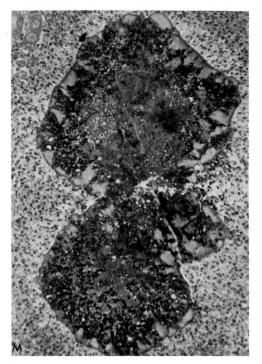

Figure 5–18 *Continued M.* Entire edged compact grain of *N. rosatii.* The peripheral cement is eosinophilic; the center basophilic.

core of hyaline hyphae, which are not imbedded in cement. There is a dense periphery of deeply pigmented hyphae interspersed with large vesicular cells in the cement substance. The latter cells also form a hyaline layer just underneath the dark outer periphery. The grains resemble those of *Exophiala jeanselmei, Pyrenochaeta romeroi,* and *Madurella grisea.* There is less lobulation, and the border is less even and more crenated than in the latter two organisms. Also, there are larger vesicles and more irregularly arranged hyphae in *L. senegalensis* grains.

Colonies. The organism grows rapidly, producing a downy grayish-brown colony (Fig. 5–18, *B*). The reverse is black, and a rose tint may develop in the agar. Ascostromata are usually not produced on Sabouraud's glucose agar. After several months on cornmeal agar, dark brown spherical locules are produced within an ascostroma, which are 100 to 300 μ in diameter. Asci and interspersed paraphyses are found within the locules. The asci are elongate, being 80 to 110 μ × 20 μ in size. The ascospores are eight in number, elongate to oval, 23 to 30 μ × 8 to 10 μ, five- to eight-celled, and encapsulated (Fig. 5–18,*C*). The walls are smooth and hyaline to light brown in color. A

species described as *Leptosphaeria tompkinsii* El-Ani has caused black-grain disease in Senegal[72] and in France from a native of Senegal.[64]

Pyrenochaeta romeroi Borelli 1959

This organism was isolated from black-grained mycetoma by Borelli in Venezuela. The granules are soft, black, and tubular in shape, and range from 0.5 to 1.5 mm in size (Fig. 5–18, *F, G*). There is a central network of hyphae and a thick band of polygonal inflated cells in the peripheral areas. The outer edge is composed of dark, swollen cells. The colony is floccose, wooly, and grayish-black, with a light-colored periphery (Fig. 5–18,*H*). The reverse is dematiaceous. The fungus grows rapidly at its optimum temperature of 30°C. Brownish-black osteolate pycnidia are produced, which are 40 to 100 by 50 to 130 μ in size (Fig. 5–18,*I*). They are covered with rigid or flexuose, rounded-end, dark setae. The pycnidiospores are elliptical, hyaline to dusky yellow in color, and average 1.5 to 2 × 1 μ or less (Fig. 5–18,*J*) in size. They are borne in chains on simple phialides within the pycnidia. The genus *Pyrenochaeta* occurs commonly in soil and decaying vegetation. Some morphologic studies indicate a close relationship or identity between *P. romeroi* and *M. grisea.*[4] However, other investigators describe serologic differences between the two.[58]

A second species of the genus was described by Borelli in 1976. *Pyrenochaeta mackinnonii* Borelli was isolated from one case of mycetoma.[16] Another pycnidia-producing species isolated from black-grain disease was also described in 1976 by Borelli.[17, 18] *Chaetosphaeronema larense* is a *nomen nudum,* however, since a proper description is lacking. This has been redescribed as *Pseudochaetosphaeronema larense* by Punithalingan (Nova Hedwigia, *31*:119–158, 1979).

Neotestudina rosatii Segretain et Destombes 1961

Synonymy. *Zopfia rosatii* Hawksworth et Booth 1974[40]; *Pseudophaeotrichum sudanense* Aue, Müller, et Stoll 1969.

This organism has been found in cases of white- to dark-grained mycetoma in the Trans-Africa Belt and Australia.[40, 48a] It is commonly isolated from soil in Somalia and has been recovered in India. The grains are light to brownish, soft, and 0.5 to 1 mm in size. In histologic preparations, peripheral cement is seen around convoluted hyphae. The center is basophilic in hematoxylin and eosin preparations (Fig. 5–18, M).

In culture, the organism grows slowly and produces a leathery, heaped, and folded colony with radial grooves and a flat periphery. The color is tan to brown, with a dark brown, dematiaceous reverse side. Spherical to oval nonosteolate ascostroma (about 350 μ) are produced on cornmeal or carrot-potato media after three weeks. They are superficial, scattered, black, and carbonaceous. The walls are composed of interwoven, dark-colored hyphae, and the interior is filled with colorless mycelium bearing asci. Pseudoparaphyses are rather sparse. They are hyaline and 1 to 2 μ in diameter and form a netlike mass. The central portion of the ascostroma deliquesces as ascogenous tissue and pseudo-paraphyses develop. Asci are crescent to spherical, 11 to 15 μ long, becoming 25 to 35 μ long when ascospores mature. They are bitunicate, with thick walls that deliquesce as

the eight ascospores mature. There are eight dark, curved, two-celled ascospores, which are 10 × 4 μ in diameter (Fig. 5–18, L). Optimum temperature for growth is 30°C.

Other species have been isolated from mycetoma of animals. *Curvularia geniculata* in dogs and horses,[15] and *Cochliobolus spiciferum* (as *Dreschslera speciferum*) in cows[63] are dematiaceous fungi producing a floccose gray-green to black colony. The reverse side is dark. *Curvularia* may produce radiating peripheral filaments and a pink-tinged mycelium in young colonies. The conidia are borne on septate, simple or branched, dark conidiophores. There are up to five cells per conidia, with the middle cell enlarged asymmetrically to produce the slight curve. *Dreschslera* conidia are produced on similar conidiophores (poroconidia), are thick-walled, and contain four or more cells formed by transverse septation. Grains are reported to be dark.[20] The disease has been described in cats, cows, dogs, and horses. Conidia are illustrated in the Part Five.

Botryomycosis

Botryomycosis or staphylococcic actinophytosis is a chronic, purulent, and granulomatous lesion of the dermis and subdermal tissue. Sulfur granules are produced. This disease mimics actinomycosis and mycetoma. It was originally described as being caused by fungi called "botryomycetes," because clumps of cocci were mistakenly observed as hyphae. The same organism was also wrongly associated with pyogenic granuloma. The disease occurs in animals as chronic, localized, or spreading swollen abscesses that rarely involve lymph nodes or visceral organs. It is a frequent complication of castration in pigs.

Human disease may develop anywhere on the skin, but particularly on the head, hands, pinna, and feet.[14, 79] The condition is usually localized but may spread to other organs, such as liver, lungs, kidney, heart, prostate, and lymph nodes. This is usually associated with debilitated patients. Several cases have been reported in children with cystic fibrosis and following surgery, skin abrasions, lacerations from auto accidents, and piercing of the ear. In a recent case reported by Bishop et al.,[14] no predisposing condition was found. The etiologic agent is usually *Staphylococcus aureus*, although there are cases recorded

caused by *Pseudomonas aeruginosa, Escherichia coli,* and *Bacteroides* and *Proteus* species. The general histologic picture is similar to actinomycosis or mycetoma. Neutrophils, lymphocytes, eosinophils, plasma cells, and fibroblasts, as well as histiocytes, spread extensively in cords and sheets. Scattered, foreign-body giant cells are seen. The granules occupy the suppurative center of granulomatous lesions, and are covered by neutrophils. The grains are soft, yellowish-white, lobulated, up to 1 mm in size, and composed of masses of cocci in an eosinophilic milieu (Fig. 5–19). This "cement" is PAS-positive. A "shell" similar to the clubbing of actinomycosis and mycetoma grains has also been described. In patients with this disease, an immune defect or an unusual tissue response for the handling of staphylococci has been postulated. Changes from the normal globulin patterns have not been observed, but the cases are generally associated with other idiosyncrasies of tissue defense and reaction. Little phagocytosis is apparent in the lesions. The organisms isolated are usually of a low order of virulence, but treatment is complicated by the inability of adequate concentrations of antibiotic to penetrate to the organ-

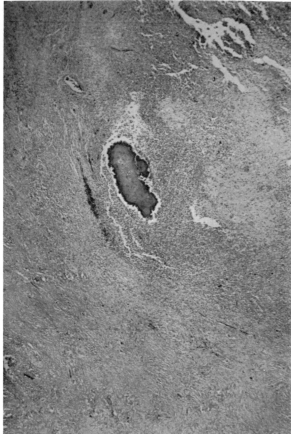

A

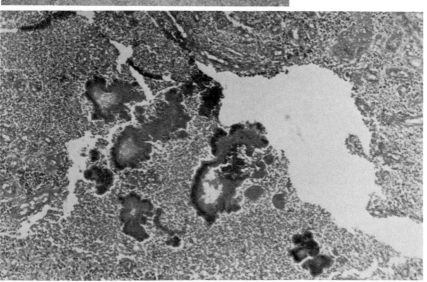

B

Figure 5–19 *A,* Botryomycosis. Small abscess in fibrous scar tissue in neck. Note granule composed of coccoid elements. (Courtesy of S. Thomsen.) *B,* Multiple soft grains in the same area demonstrating the varied morphology seen in tissue.

isms sequestered in the grains and granuloma. Treatment is determined by antibiotic sensitivity and necessitates a prolonged course. Surgical exploration and excision are sometimes helpful. The name "granular bacteriosis" has been suggested for this disease.

Mackinnon et al.[50] were able to reproduce the disease in experimental animals. They injected small amounts of *Pseudomonas aeruginosa* intratesticularly into guinea pigs. Granules up to 300 μ in diameter are produced, complete with eosinophilic "rays" and "clubs." In some animals, metastasis to the lung occurred.

REFERENCES

1. Abbott, P. 1956. Mycetoma in the Sudan. Trans. R. Soc. Trop. Med. Hyg., *50*:11–30.
2. Ajello, L., and W. W. Walker. 1961. Isolation of *Nocardia brasiliensis* from a cat. J. Am. Vet. Med. Assn., *136*:370–376.
3. Albornoz, M. B. 1974. *Cephalosporium serrae* agente etiologico de micetomas. Mycopathologia, *54*:485–498.
4. Alilou, M. 1977. *Madurella grisea* est-il synonyme de *Pyrenochaeta romeroi?* Bull. Soc. Fr. Mycol. Med., *6*:227–230.
5. Avram, A., and D. Hatmann. 1968. Quelques aspects concernant l'ostéite mycetomique. À propos d'une étude sur 23 cas roumains. Mykosen, *11*:711–717.
6. Avram, A. 1967. Grains expérimentaux maduromycosiques et actinomycosiques à *Cephalosporium falciforme, Monosporium apiospermum, Madurella mycetomi,* et *Nocardia asteroides.* Mycopathologia, *32*:319–336.
7. Avram, A. 1969. Micetomamele in Romania. Editura Acad. R. Soc. Romania, 00:000.
8. Avram, A., and G. Nicolau. 1969. Attempts to demonstrate complement-fixing antibodies in mycetoma. Mycopathologia, *39*:367–370.
9. Balabanoff, V. A. 1977. Madurafuas durch *Nocardia asteroides.* Castellania, *5*:91–94.
10. Baylet, J., R. Camain, and G. Segretain. 1959. Identification des Agentes des Maduromycoses du Senegal et de la Mauritanie. Description d'une espèce novelle. Bull. Soc. Pathol. Exot., *52*:448–477.
11. Bedi, T. R., S. Kaur, and B. Kumar. 1978. Red grain mycetoma of the scalp *(Actinomadura pelletieri).* A Case report from India. Mycopathologia, *63*:127–128.
12. Berd, D. 1973. *Nocardia brasiliensis* infection in the United States. Am. J. Clin. Pathol., *60*:254–258.
13. Bergeron, J., and J. F. Mullins, et al. 1969. Mycetoma caused by *Nocardia pelletiere* in the United States. Arch. Dermatol., *99*:564–566.
14. Bishop, G. F., K. E. Greer, and D. A. Horwitz. 1976. *Pseudomonas* botryomycosis. Arch. Dermatol., *112*:1568–1570.
14a. Bojalil, L. F., and A. Zamora. 1963. Precipitin and skin tests in the diagnosis of mycetoma due to *Nocardia brasiliensis.* Proc. Soc. Exp. Biol. Med., *113*:40–43.
15. Boomker, J., J. A. Coetzer, and D. B. Scott. 1977. Black grain mycetoma (maduromycosis) in horses. Onderstepoort J. Vet. Res., *44*:249–252.
16. Borelli, D. 1976. *Pyrenochaeta mackinnonii.* Nova species agente de micetoma. Castellania, *4*:227–234.
17. Borelli, D., and R. Zamora. 1973. *Chaetosphaeronema larense* nova species agente de micetoma — Presentacion del tipo, Boletin Mensul. Soc. Venez. Dermatol., *6*:17–18.
18. Borelli, D., R. Zamora, and G. Senabre. 1976. *Chaetosphaeronema larense* nova specie — agente de micetom, Gaceta Medica de Caracas, *84*:307–318.
19. Bourdin, M., and P. Destombes. 1975. Premier observation d'un mycétome a *Microsporum canis* chez un chat. Rec. Med. Vet., *151*:475–480.
20. Brodey, R. H. F., Schryver, et al. 1967. Mycetoma in a dog. J. Am. Vet. Med. Assn., *151*:442–451.
21. Brumpt, E. 1906. Les Mycetomes. Arch. Parasitol., *10*:489–572.
22. Brumpt, E. 1905. Les Mycétomes humaines. Compt. Rend. Soc. Biol., *58*:997–999.
23. Burgoon, C. F., F. Blank, et al. 1974. Mycetoma formation in *Trichophyton rubrum* infection. Br. J. Dermatol., *90*:155–162.
24. Butz, W. C., and L. Ajello. 1971. Black grain mycetoma. Arch. Dermatol., *104*:197–201.
25. Cardenas, J. V., V. Calle, et al. 1966. Micetomas. Antioquia Med., *16*:117–132.
26. Carrión, A. L. 1956. *Cephalosporium falciforme* n. sp. A new etiologic agent of maduromycosis. Mycologia, *43*:522–523.
27. Carter, H. Vandyke, 1874. On Mycetoma or The Fungus Disease of India. London, J. and A. Churchill.
28. Carter, H. Vandyke. 1859–1860. On Mycetoma. Trans. Med. Phys. Soc. of Bombay, Vols. VI & VII.
29. Chalmers, A. J., and R. G. Archibald. 1916–1917. A Sudanese Maduramycosis, Ann. Trop. Med., *10*:169–214.
30. Chausse, J. Davids, and L. Texic. 1968. Mycétome du pied autochtoe à *Pyrenchaeta romeroi.* Bull. Soc. Fr. Dermatol. Syphiligr., *75*:452–453.
31. Cooke, W. B., and Kabler, P. 1955. Isolation of potentially pathogenic fungi from polluted water and sewage. Pub. Health Rep., *70*:689–694.
32. Destombes, P., F. Mariat, L. Rosati, and G. Segretain. 1977. Le mycétomes en Somalie—conclusions d'une enquête menée de 1959 à 1964. Acta Trop., *34*:355–373.
33. Drouhet, E. 1955. The status of fungal diseases in France. In *Therapy of Fungus Diseases, an International Symposium.* Boston, Little, Brown & Co., pp. 43–53.
34. Emmons, C. W. 1945. *Phialophora jeanselmei* comb. n. from mycetoma of the hand. Arch. Pathol., *39*:364–368.
35. Gaind, M. L., A. A. Padhye, et al. 1962. Madura foot in India caused by *Cephalosporium infestans* n. sp. Sabouraudia, *1*:230–233.
36. Gams, W. 1971. Cephalosporium *Artige Schimmelpilze (Hyphomycetes).* Stuttgart, Gustaf Fischer Verlag.
37. Green, W. O., and T. E. Adams. 1964. Mycetoma in the United States. Am. J. Clin. Pathol., *42*:75–91.
38. Gordon, R. E., and J. M. Mihm. 1962. Identification of *Nocardia caviae* (Erikson) nov. comb. Ann. N.Y. Acad. Sci., *98*:628–636.
39. Halde, C., A. A. Padhye, et al. 1976. *Acremonium falciforme* as a cause of mycetoma in California. Sabouraudia, *14*:319–326.
40. Hawksworth, D. L., and C. Booth. 1974. A Revision of the genus *Zopfia* Rabenh, Mycological Paper No. 135, pp. 27–31. Kew Surrey, U.K., Commonwealth Mycological Institute.
41. Hubler, W. R., Jr., and W. R. Hubler. 1976. Actinomycotic mycetoma treated with minocyclinic: case report. Tex. Med., *72*:79–83.
42. Indira, P. U., and M. Sirsi. 1965. Studies on maduramycosis. Indian J. Med. Res., *56*:1265–1271.
43. Jeanselme, M. M., L. Huet, et al. 1928. Nouveau type de mycétoma à grains noire du a une

Torulla encore non decente. Soc. Fr. Dermatol. Syph., *35*:369–375.

44. Kamalam, A., and P. Subramaniyam. 1975. Restoration of bones in mycetoma. Arch. Dermatol., *111*:1178–1180.

45. Kanthack, A. A. 1893. Madura disease (mycetoma) and actinomyces. J. Pathol., *1*:140–162.

46. Langeron, M. 1928. Mycétoma à *Torulla jeanselmii* Langeron 1928. Nouveau type de mycétome à grains noire. Ann. Parasitol. Hum. Com., *6*:385–390.

47. Lavalle, P. 1962. Agents of mycetoma. *In* Dalldorf, G. (ed.). *Fungi and Fungous Diseases.* Springfield, IL, Charles C Thomas, pp. 50–65.

48. Leech, P. J., and B. G. Bedrock. 1970. An Australian case of maduromycosis. Aust. N.Z. J. Surg., *39*:293–295.

48a. Lynch, J. B. 1964. Mycetoma in the Sudan. Ann. R. Coll. Surg. Eng., *35*:319–340.

49. Mackinnon, J. E., L. V. Ferrada-Urzua, et al. 1949. *Madurella grisea* n. sp. A new species of fungus producing the black variety of maduromycosis in South America. Mycopathologia, *4*:384–392.

50. Mackinnon, J. E., I. A. Conti Diaz, et al. 1969. Experimental botryomycosis produced by *Pseudomonas aeruginosa.* J. Med. Microbiol., *3*:369–372.

50a. Macotela-Ruiz, E., and F. Mariat. 1963. Sur la production de mycétomes expérimentaux par *Nocardia brasiliensis* et *Nocardia asteroides.* Bull. Soc. Pathol. Exot., *56*:46–54.

51. Mahgoub, E. S. 1976. Medical management of mycetoma. Bull. W.H.O., *54*:303–310.

52. Mahgoub, E. S., and Ian G. Murray. 1973. *Mycetoma.* London, W. Heinemann Medical Books, Ltd., p. 10.

53. Mahgoub, E. S., S. A. Gumea, et al. 1977. Immunological status of mycetoma patients. Bull. Soc. Pathol. Exot. Filiales., *70*:48–54.

54. Mariat, F., P. Destomes, and G. Segretain. 1977. The mycetomas: clinical features, pathology, etiology and epidemiology. Contrib. Microbiol. Immunol., *4*:1–39.

54a. Mathews, R. S., C. E. Buckley III, et al. 1968. Local chemotherapy of deep seated fungus infections: Maduromycosis of the foot. Clin. Res., *16*:48.

55. McGinnis, M. R., and A. Padhye. 1977. *Exophiala jeanselmei* a new combination for *Phialophora jeanselmei.* Mycotaxon, *5*:341–352.

55a. Mohr, J. A., and H. G. Muchmore. 1968. Maduromycosis due to *Allescheria boydii.* J.A.M.A., *204*:335–336.

56. Murray, I. G., and E. S. Mahgoub. 1968. Further studies in the diagnosis of mycetoma by double diffusion in agar. Sabouraudia, *6*:106–110.

57. Murray, I. G., E. T. C. Spooner, et al. 1960. Experimental infection of mice with *Madurella mycetomi.* Trans. R. Soc. Trop. Med. Hyg., *54*:335–341.

58. Murray, I. G., and H. R. Buckley. 1969. Serological differences between *Pyrenochaeta romeroi* and *Madurella grisea.* Sabouraudia, *7*:62–63.

59. Negroni, R., L. Astarlaa, and J. L. Gonzalez (h). 1979. Micetoma podal por *Streptomyces somaliensis.* Presentation del primer caso argentino. Rev. Argent. Micol., *2*:20–25.

59a. Nielson, H. S., Jr. 1967. Effects of amphotericin B in vitro on perfect and imperfect strains of *Allescheria boydii.* Appl. Microbiol., *15*:86–91.

60. Nielson, H. S., N. F. Conant, T. Weinberg, and J. F. Reback. 1968. Report of a mycetoma due to *Phialophora jeanselmei* and undescribed characteristics of the fungus. Sabouraudia, *6*:330–333.

61. Neuhauser, I. 1955. Black grain maduromycosis caused by *Madurella grisea.* Arch. Dermatol., *72*:550–555.

62. Padhye, A. A., R. S. Sukapure, et al. 1962. *Cephalosporium madurae* n. sp. cause of madura foot in India. Mycopathologia, *16*:315–322.

63. Patton, C. J. 1977. *Helminthosporium speciferum* as the cause of dermal and nasal maduromycosis in a cow. Cornell Vet., *67*:236–244.

64. Piertrini, P., P. Lauret, et al. 1977. Deuxièmè cas de mycétome à *Leptosphaeria tompkinsii* El Ani observé à Rouen. Forme abcedée de la plante chez senegalais. Association d'un mycétome à grains blancs. Bull. Soc. Fr. Mycol. Med., *6*:85–88.

65. Pinoy, E. 1913. Actinomycoses and mycetomas. Paris Bull. Inst. Pasteur, *11*:929–938.

66. Pritchard, D., B. F. Chick, and M. D. Connole. 1977. Eumycotic mycetoma due to *Dreshlera rostrata* infection in a cow. Aust. Vet. J., *53*:241–244.

67. Radaeli, Fr. 1911. Micosi del piede da *Monosporium apiospermum.* Lo Sperimentale, *65*:f. 4.

68. Reid, M. M., I. W. Frock, et al. 1977. Successful treatment of a maduromycotic fungal infection of the equine uterus with amphotericin B. Vet. Med. Small Anim. Clin., *72*:1194–1196.

69. Reid, M. M., D. R. Jeffrey, et al. 1976. Å rare case of maduromycosis of the equine uterus. Vet. Med. Small Anim. Clin., *71*:947–949.

70. Rippon, J. W., and G. L. Peck. 1967. Experimental infection with *Streptomyces madurae* as a function of collagenase. J. Invest. Dermatol., *49*:371–378.

71. Segretain, G., and P. Destombes. 1969. Recherche sur les mycétomes à *Madurella grisea* et *Pyrenochaeta romeroi.* Sabouraudia, 7:51–61.

72. Segretain, G., M. Andre, et al. 1974. *Leptosphaeria tompkinsii* agent de mycétomes au Senegal. Bull. Soc. Fr. Mycol. Med., *3*:71–74.

73. Segretain, G., R. Baylet, et al. 1959. *Leptosphaeria senegalensis* n. sp. agent de mycétome à grains noirs. C. R. Acad. Sci. (Paris), *248*:3730–3732.

74. Segretain, G., and P. Destombes. 1961. Description d'un nouvelle agent de maduromycose. *Neotestudina rosatii* n. gen. n. sp. isole en Afrique. C. R. Acad. Sci. (Paris), *253*:2577–2579.

75. Sran, H. S., I. M. S. Narula, et al. 1972. History of mycetoma. Indiana J. of Hist. Med., *17*:1–17.

76. Verghese, A., and A. Klokke. 1966. Histologic diagnosis of species of fungus causing mycetoma. Indian J. Med. Res., *54*:524–530.

77. Waisman, M. 1962. Staphylococci actinophytosis (Botryomycosis): granular bacteriosis of the skin. Arch. Dermatol., *86*:525–529.

78. Weidman, F. D., and A. M. Kligman. 1954. A new species of *Cephalosporium* in madura foot. J. Bacteriol., *50*:491–495.

79. Winslow, D. J. 1959. Botryomycosis. Am. J. Pathol., *35*:153–167.

80. Wright, J. M. 1898. A case of mycetoma (Madura foot). J. Exp. Med., *3*:421–433.

81. Zaias, N., D. Taplin, and G. Rebell. 1969. Mycetoma. Arch. Dermatol., *99*:215–225.

The Pathogenic ——II
Fungi

6 Characteristics of Fungi

The fungi proper, or Eumycetes, are quite distinct from true bacteria in size, cellular structure, and chemical composition. Differences in cell wall composition and nuclear structure were discussed in the chapter on the actinomycetes. In this chapter some of the morphological and developmental aspects of fungi will be covered in more detail.

Fungi are eukaryocytic because of their nuclear structure and mode of genetic recombination and, based on ultrastructural studies, have been placed in a separate "kingdom" among all the various organisms constituting the biological world.

As a consequence of the general acceptance of Darwin's theory of evolution, published in 1869, biologists began to devise taxonomies for living organisms. In contrast to a classification system, which is a method of information retrieval, taxonomy has phylogenetic or evolutionary implications. Most 19th century biologists considered the fungi to be primitive plants and grouped them in the Thallophyta (as colorless algae) of the Kingdom Plantae. At that time only two kingdoms were recognized, plantae and animalia, encompassing all living things. The taxa were based, then as now, on similarities in reproductive structures. Some fungi were considered to be derived from green algae and others from red algae; however, we have examples of algae that have lost the ability to manufacture chlorophyll and have adopted a heterotrophic mode of existence. Owing to their compositional and physiologic characteristics, these organisms are not fungi but are achloric algae (see Chapter 26). Some early naturalists recognized this difficulty of classifying fungi as plants. A third kingdom of primitive organisms was proposed by Hogg[5] in 1860 and called the Protoctista. This group was to contain fungi and other simple organisms. In 1878 Haeckl[4] used the designation Protistenreich ("primitive kingdom") to include fungi, algae, and protozoa. These concepts were for the most part ignored, however, and for a hundred years the traditional plant-animal, two-kingdom view was almost universally accepted for all of biology.

Serious challenges to the prevailing ideas began in the 1950's. In mycology, G. W. Martin[8] reviewed the history of taxonomy and began to formulate a separate evolution for fungi. The noted biologist Copeland[3] revived the term protoctista in 1956, later calling it protista. He envisioned four kingdoms: plants, animals, protista, and bacteria. The protoctists included protozoa, algae (except blue-green algae), fungi, and slime molds. Blue-green algae were grouped with bacteria. Finally, in 1969, Whittaker published his five-kingdom modification of Copeland's Schema in *Science*.[12] This is the generally accepted system at present (Table 6–1). The Whittaker five kingdoms are (1) *Monera,* the prokaryotes (bacteria, actinomycetes, and blue-green algae), (2) *Protista,* which includes protozoa and other unicellular organisms, (3) *Fungi,* (4) *Plantae,* and (5) *Animalia.* Within the fungi there were eight phyla. Disagreements still abound, but most biologists accept Whittaker's classification or some modification of it. Presently the fungi are grouped into five phyla: Chytridiomycota, Zygomycota, Deuteromycota (fungi imperfecti), Ascomycota, and Basidiomycota (there are six phyla if lichens are included). Studies on the composition and biochemistry of fungi indicate that phylogenetically they have a greater

Table 6–1 *Five-Kingdom Classification of Living Things**

Kingdom	Characteristics	Phyla	Representative Organisms
I. Monera	Prokaryotic (anucleate), no nuclear membrane, no mitochondria, no mitotic apparatus, single circular chromosome; direct cell division, primarily by binary fission. Nutrition ingestive, absorptive, chemosynthetic, photoheterotrophic or photoautotrophic. Unicellular, filamentous or mycelial. If motile, flagella has one microtubule, no sterols in cell membrane. (DAP) lysine synthesis.	14	Bacteria, myxobacteria, actinomycetes, cyanobacteria (blue-green algae).
II. Protista	Eukaryotic (nucleate), nuclear membrane, more than one chromosome, heterotrophic or photoautotrophic nutrition, premitotic or mitotic division, diaminopimelic acid (DAP) lysine biosynthetic pathway, unicellular or multicellular. If motile, flagella or cilia composed of microtubules in the $9 + 2$ pattern. Flagella whiplash and/or tinsel type. Plastids and mitochondria.	30	Protozoans, mycetozoans (slime molds), brown algae, red algae, green algae, hyphochytrids, oomycetes.
III. Fungi	Absorptive nutrition, unicellular or mycelial. Cell walls with chitin-chitosan with B-glucan, mannan, α-glucan, chitin-mannan, or galactosamine-galactose polymers. If flagellate, flagellum posteriorly uniflagellate of the whiplash type. L-α-aminoapidic acid (AAA) lysine biosynthetic pathway.	6	Chytrids, zygomycetes, ascomycetes, basidiomycetes, deuteromycetes, lichens.
IV. Plantae	Photoautotrophic, highly differentiated, often with long diploid phase. DAP lysine pathway.	9	Liverworts, mosses, ferns, conifers, seed plants, etc.
V. Animalia	Heterotrophic, multicellular, diploid blastula.	32	Coelenterates, flatworms, mollusks, insects, reptiles, birds, mammals, etc.

*Modified from Whittaker, R. H. 1969. Science, *163*:150–160.

affinity to the euglenas (Euglenophyta) than to other protist groups.

The Fungi. As defined by Ainsworth[1] and modified by Ragan and Chapman[10] and McGinnis[9] fungi have the following diagnostic characteristics: *Nutrition:* heterotrophic and absorptive. *Thallus:* on or in the substrate; unicellular or filamentous (mycelial); aseptate, partially septate, or septate; typically nonmotile but some have a flagellated stage (zoospores). The flagellum is of the whiplash type and has the $9 + 2$ fibril construction. *Cell wall:* a well defined structure containing polysaccharides, sometimes polypeptides, and typically chitin. *Nuclear state:* eukaryotic, uninucleate, or multinucleate. The mycelium may be homo- or heterokaryotic, haploid, dikaryotic, or diploid. The latter condition is usually of limited duration but in rare instances persists throughout most of the life cycle (viz. the yeast *Saccharomyes cerevisiae).* *Life cycle:* simple to complex. *Sexuality:* asexual or sexual and homothallic or heterothallic. *Sporocarps:* simple, naked, and microscopic to complex within large carps and showing limited (pseudoparenchymatous) tissue differentiation. *Habitat:* on any organic material as saprobes, symbionts, parasites, or hyperparasites and of universal distribution in water and in soil.

In addition to similarities in structure, composition, and life cycle, the fungi share other basic characteristics (Table 6–1). These include a variety of biochemical mechanisms. One of the most interesting of these is the synthesis of the amino acid lysine. Most plants, protists, and bacteria, including actinomycetes, synthesize lysine by way of *meso* α, ϵ diaminopimelic acid (DAP). The oomycetes and slime molds also utilize this pathway. Fungi, however, synthesize the compound by a completely different pathway, which involves L α adipic acid (AAA), as do the Euglenophyceae. Several other biochemical and physiological attributes are also found uniquely among the fungi and affirm their separate development from protists and oomycetes. This topic is reviewed in Ragan and Chapman's book *Biochemical Phylogeny of the Protists.*[10]

Though fungi are essentially single-celled organisms, in some fungal species the cells may show various degrees of specialization. Examples of this are the pseudoparenchymatous tissue of the fruiting bodies of higher Ascomycetes and Basidiomycetes. The simplest morphologic form of the fungi is the unicellular, budding *yeast*. As complexity increases, elongation of the cell without separation of newly formed cells results in the threadlike *hypha*. An intertwined mass of hyphae is called a *mycelium*. The popular term mold (mould) also refers to the filamentous fungi. The terms *"hyphae"* and *"mycelia"* are used interchangeably. The mycelial mat is known as a *thallus*, but this term refers specifically to the colonial growth derived from a single spore or conidium. In the fungi discussed here, the thallus is a loose network of hyphae, also called a *colony*. Reproductive units are of several types. *Spores* are the units resulting from sexual mechanisms of reproduction or are produced asexually in saclike swellings called sporangia. In some groups spores are motile. *Conidia* are asexual, nonmotile, commonly deciduous propagules produced either exclusively (in Deuteromycota) or in addition to sexual or asexual spores (in Zygomycota, Basidiomycota, and Ascomycota). In some higher forms of fungi, the hyphae are cemented together to form large, structurally complex fruiting bodies such as mushrooms and puffballs, which may weigh more than 60 pounds. Even though they attain such great size, all fungi are still "primitive" organisms because their specialization into tissue is reversible. Any single cell separated from a 60-pound puffball is capable of regrowing the entire structure. This regenerative ability in general separates the protozoa, algae (protists), and fungi from the higher multicellular forms such as vascular plants and metazoan animals.

As to the origin of fungi, the debate concerning their evolution from other forms has been long and controversial. Modern scientific analysis has laid to rest the concept of a bacterial ancestor of fungi, in which the actinomycetes were the "link" between the two groups. The simplistic notion that fungi are algae without chlorophyll is also untenable at face value. There are several algae without chlorophyll; for example, *Prototheca* was probably derived from *Chlorella* but still retains many algaelike characteristics of composition, physiology, and reproduction which are quite distinct from fungi. Some primitive fungi are motile at one stage in their growth cycle as are algae and protozoa, but the chitinous cell wall of most fungi is uncommon in protists. Sessile protozoa have been pro-

posed as the immediate ancestors of the fungi. As noted previously, recent work has indicated that there are a variety of biochemical and physiologic characteristics in common between fungi and the euglenas. This group is now considered to be related to or possibly the ancestor of the fungi.[10]

Beginning with the primitive water molds, the chytrids, which at one stage in their life cycles are motile, we can see a close relationship within the remaining fungal divisions. The Zygomycota were probably the first terrestrial fungi and have no motile stage. Their sporangia and sporangiospores, however, are not very efficient aerodynamically, nor are they particularly resistant to adverse environmental conditions. They are replaced in the Ascomycota by sturdy conidia, and in Basidiomycota by basidiospores. With the development of successful methods for disseminating the species, fungi have utilized every possible organic substrate on the surface of the earth. The major phyla of the fungi as now recognized are given in Table 6–2. Fungi of medical interest are represented in almost all of these groups. The majority of medically important organisms, however, are found in Plectomycetes in the phylum Ascomycota. These include the etiologic agents of histoplasmosis, blastomycosis, and all the various dermatophyte infections. The remaining infectious fungi are scattered among the other taxonomic groups. Swamp cancer, a rare disease of horses, is believed to be caused by an Oomycete, a group formerly included in the fungi. The organism has a mycelium similar to that of fungi, but it differs in many other characteristics from true fungi. A discussion of the disease and the agent is found in Chapter 25.

Mycelium. Two main structural types of mycelium may be distinguished. In one of these, the cells making up the hyphae are not marked by cross walls or septa, thereby allowing protoplasm to flow (protoplasmic streaming) throughout the multinucleate structure. Within the tubelike hypha protoplasm freely travels back and forth. Such a structure is said to be nonseptate and is found in the Chytridiomycota and Zygomycota (Fig. 6–1). As growth modifies or environmental changes occur, septa are formed within the hyphal mass. Such septa are complete, thus separating the protoplasm from other sections of the colony. The colonies of *Mucor* or *Rhizopus* growing either in culture or in patient material are sparsely septate, the septa being

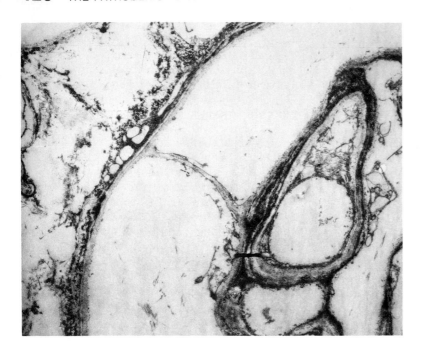

Figure 6–1. Septum of a Zygomycota. This is from a "fungus ball" that developed in a preformed cavity in the chest. The patient had had tuberculosis earlier in life. Shown here is one of few septa produced by the mucoraceous fungus. Serial sections of a septum revealed no pore. ×48,000. (Courtesy of Zelma Molnar.)

single, solid units (Fig. 6–1) and providing complete separation of the two parts of the mycelium.

In the two remaining groups of fungi, there are evident regular separations or septation.

The septation or cross walls of the Ascomycota and Basidiomycota are incomplete and have "holes" that allow cytoplasmic and sometimes nuclear flow between the cells. Thus the Ascomycota are septate but have an indeterminate number of nucleii within the cytoplasm (Fig. 6–2).

In the Basidiomycota, a special structure composed of the dolipore and parenthesome prohibits nuclear migration (Fig. 6–3). The dolipore is an expanded, liplike structure around either side of the opening of the septal wall. The parenthesome is a cap or cuplike structure a short distance from either side of the dolipore opening. Held in place by fibrils, it allows cytoplasmic, but not nuclear, migration and covers the dolipore lip when there is a lessening of the pressure on the other side of the septum. Each Basidiomycota cell contains two dissimilar haploid nuclei (dikaryon), one derived from one parent or mating type, and the other from the other parent. The mycelium of this latter group may also show a special bridge structure, a bumplike eminence connecting one cell to the next. This is called a clamp connection (Fig. 6–4). The clamp connection is a specialized hyphal bridge that allows the simultaneous mitosis of two nuclei to occur in

such a position that the dikaryons of compatible nuclei are duplicated in the proper relationship to each other. Although this is not the primary criterion for differentiation, the three main phyla of fungi may be distinguished in part by the characteristics of their mature nonsporulating hyphae. It is possible, therefore, to study electron micrographs of hypha in pathologic material or from cultures and to determine the group to which the organism belongs.

The mycelium is further delineated into three general types which differ in function. One of these, the vegetative mycelium, consists of masses of hyphae within the colony, adjacent to and growing into the substrate. The second is the submerged hyphae for nutrient absorption. They are concerned with the digestion and assimilation of food materials. Anastomoses of hyphal elements are commonly seen. The third type, the reproductive or fertile hyphae, usually extends into the air to form aerial mycelium, and gives rise to various types of conidia or other reproductive units. Fragments of any type of mycelium will grow and reproduce if transferred. The mode of propagule formation and the structure, size, shape, and morphology of the units produced and· unit-bearing elements are the characteristics by which fungi are differentiated, classified, and identified.[9]

In addition to reproducing through propagule formation and the vegetative growth of hyphae, many species of fungi reproduce by

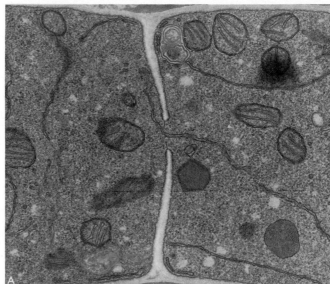

Figure 6–2. Septa of Ascomycota. *A,* in this thin section of the mycelium of *Chaetomium thermophile,* the perforate septum is clearly seen. Cytoplasm, mitochondria, and nuclei wander freely from cell to cell. × 51,800. (Courtesy of R. Garrison.) *B,* Multiple pored septum in *Geotrichum candidum (arrows).* ×87,800. (Courtesy of R. Garrison.)

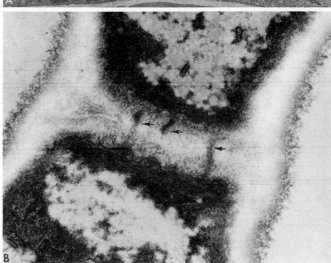

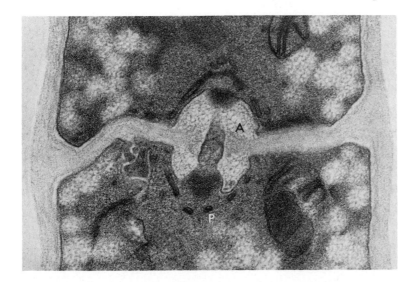

Figure 6–3. Septum of a Basidiomycota. In this section, the swollen lip of the dolipore is clearly evident (*A*) with the suspended parenthesome adjacent to it (*P*). ×75,600. (Courtesy of R. Garrison.)

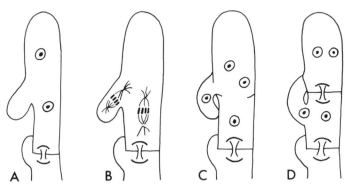

Figure 6–4. Clamp formation in Basidiomycota. *A,* Dikaryon cell. A backward protrusion develops from the growing end cell. *B,* One nucleus migrates into the neck of the protrusion and the other to the base of the cell. Both divide. *C,* Of the pair produced in the neck, one migrates to apex of cell, and the other migrates backward into growing clamp. Of the pair produced in the base of the cell, one migrates to apex, and one stays in the base. *D,* Septal rearrangement, nuclear migration, and formation of dolipore and parenthesome. Both the new cell and the cell from which it was formed have a pair of nucleii. These nucleii (dikaryon: n + n) represent the haploid nuclei of the original (parental) mating strains. The clamp is shut, and further nuclear migration is prevented.

the simple separation, from any part of the mycelium, of cells known as oidia (arthroconidia). These vegetative reproductive forms may give rise to a new mycelium or reproduce themselves by budding like the yeasts, depending on the available nutrients and the environment into which they are placed.

Yeasts. Defined morphologically, a yeast is a single-celled fungus that reproduces by simple budding (Fig. 6–5,*A*).[9] Under ordinary conditions, the life cycle of this fungus is spent as a single-celled, budding structure (blastoconidia, etc., formation), although it may elongate to form a pseudomycelium under the influence of certain environmental conditions and in some cases a true mycelium. In a stricter sense the true yeasts are Ascomycota, which, under proper conditions of nutrition, temperature, and sexual mating, produce ascospores. Beer yeast, *Saccharomyces cerevisiae,* is one of these. Certain of the imperfect fungi, such as *Candida* and *Cryptococcus,* retain a yeastlike morphology under most circumstances and are referred to as "asporogenous yeasts." The *Candida* fungi are probably derived from true yeasts and exhibit pseudomycelium formation under certain nutritional conditions. Cryptococci do not normally show any cellular elongation, though some mycelial variants having basidiomycetous hyphae suggest this genus is related to the Basidiomycota. In one species, *Cryptococcus neoformans,* a sexual stage was found and it is now included in Teliomycetes of the Basidiomycota. Yeasts are not primordial or primitive fungus cells, and it is

probable that they developed via evolutionary specialization to a "simpler" form. This is termed "secondary reduction" and is probably the same mechanism that accounts for the derivation of the "simple" mosses from the more complex ferns.

Many filamentous fungi grow and reproduce as yeastlike cells in certain environments. The best example of this is the parasitic yeast stage found in histoplasmosis, blastomycosis, and sporotrichosis. A suitable term for this transition is dimorphism. In the cases cited, the organisms are stimulated primarily by temperature and the tissue environment, and thus are thermal dimorphic organisms. *Coccidioides* also has a parasitic stage, which is not a budding yeast but a spherule with internal spore formation. It is induced by environmental or tissue factors and therefore can be considered dimorphic also. Many other pathogenic, filamentous fungi may be induced into a yeastlike stage by a particular emvironment. However, under normal conditions of growth, they are mycelial. Induced yeastlike growth may be concomitant with increased pathogenic potential in some species; in others, it is an artifact of growth conditions. In its simplest definition, a yeast is a fungus that is normally a single budding cell under optimal conditions of temperature and growth. If temperature or environmental factors or both are necessary for transition to a yeastlike phase, it is a mold that exhibits dimorphism. A parasitic stage refers to the morphology of the organism found in tissue, if it is different from its normal saprophytic growth. Some fungi that

Figure 6–5. Types of asexual reproduction in the fungi. *A,* Budding yeasts: left, *Saccharomyces cerevisiae,* which form holoblastic conidia that dehisce, (see Figure 6–9); right, *Schizosaccharomyces,* which splits in the middle by fission, the septa being formed by only the inner wall. Following separation, the yeast can expand before the outer wall is completed. *B, Rhizopus* sp., a Zygomycota with coencytic or sparsely septate (complete septum) hyphae, rootlike rhizoids, and a sporangium atop a sporangiophore. Spores are formed internally by a series of successive cleavage furrows. These furrows continue to form until an indeterminate number of single nucleate spores are formed. *C, Sporothrix.* Simple holoblastic conidia are budded from swollen tips of conidiogenous cells along hyphae or at the ends of conidiophores. The latter are proliferous and elongate sympodially to produce successive conidia. *D, Aspergillus* conidia are produced from phialides that radiate from a vesicle at the top of a conidiophore. *E, Penicillium.* Branched metulae bear phialides. They give rise to chains of conidia. *G, Coccidioides.* Thallic conidia are produced enteroarthrically and released by fracture of the wall of the degenerated adjacent cell. *H. Microsporum canis,* macroconidia and microconidia. *I, Trichophyton mentagrophytes:* microconidia *en grappes,* spiral hyphae, and some macroconidia. Hyphal appendages. *J,* Racquet mycelium. *K,* Nodular or knot body. *L,* Pectinate body. *M,* Chlamydoconidia characteristic of *Microsporum audouinii.*

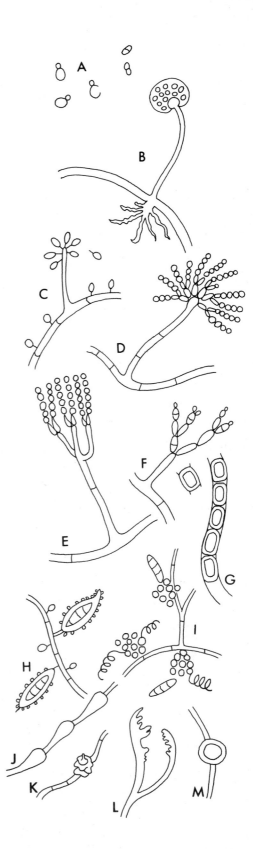

invade tissue (particularly opportunistic fungi) exhibit neither dimorphism nor a parasitic stage. *Aspergillus* and *Mucor* are mycelial in morphology, both as saprophytes and when causing disease in human tissue. Thermal dimorphism, nutritional dimorphism, and tissue dimorphism are discussed in detail later in this book.

CONIDIA FORMATION, SPORE FORMATION AND CLASSIFICATION

Two classes of reproductive units are distinguished in the fungi: (1) sexual spores, which are produced by the fusion of two nuclei that then generally undergo meiosis (sexual spores may or may not be derived from different mating strains); and (2) asexual propagules, which may be spores or conidia, depending on the mode of production, and which arise following mitosis of the parent nucleus. *Conidia* arise by *de novo* budding of a conidiogenous hypha or by differentiation of preformed hyphae. Asexual *spores* are commonly formed by consecutive cleavages of a sporangium. Asexual forms of reproduction represent the major method for the maintenance and dissemination of the species among many groups of fungi including most of the medically important species. Sexual methods of reproduction involve plasmogamy (cytoplasmic fusion of two cells), karyogamy (fusion of two nuclei), genetic recombination and meiosis. The resulting haploid spore is said to be a sexual spore.

If a sexual spore is produced only by fusion with a nucleus of one mating type to a nucleus of another mating type, the fungus is said to be *heterothallic*. If a nucleus from within the same thallus will serve for fusion with another nucleus of that thallus, the fungus is *homothallic*. Fungi in which a sexual spore is produced are known as "perfect fungi" and can be assigned to one of the four phyla: Chytridiomycota, Zygomycota, Ascomycota, and Basidiomycota (Table 6–2). Many perfect fungi produce numerous asexual units that serve for species dissemination and, on rare occasions only, reproduce sexually with genetic recombination and formation of sexual spores. In addition, many fungi have lost the ability to form sexual spores, or their proper mating type or inducing environment is not known. These latter fungi reproduce by asexual propagule-formation only and are called "Fungi Imperfecti" or Deuteromycota because no sexual spores

have as yet been found. Some other fungi produce no spores or conidia at all and are termed "mycelia sterilia."

For many years most of the pathogenic fungi were known in their asexual phase only, and were named and classed as Deuteromycota (Fungi Imperfecti).[9] Recent work has revealed that a sexual stage in many of these is produced as a result of contact with an opposite mating type. Since the type and form of sexual reproduction is the basis for biological taxonomy, the new sexual stage is given a descriptive name, allying it with related genera and species. Thus many pathogenic fungi now have two names, an imperfect or anamorph name (usually the first described) and a perfect or teleomorph name. Although it may be confusing at first, it is taxonomically legal. For example, the common systemic pathogen *Histoplasma capsulatum* (anamorph phase) is also known as *Ajellomyces capsulatus*, the descriptive name of its perfect phase (teleomorph phase). To add to the confusion, the dermatophyte called *Microsporum gypseum* is the imperfect phase of two species of perfect fungi, *Nannizzia gypsea* and *N. incurvata*. Although the asexual conidia of the two species are not discernibly different, the sexual fruiting bodies are. This emphasizes the concept that the term "Fungi Imperfecti" is a convenient descriptive filing cabinet (a classification system) for species waiting to be assigned to taxonomically meaningful categories, if their sexual stages can be discovered. The inherent pathogenicity of fungi is not related to their sexual or asexual stages. These stages merely describe methods of reproductive unit production, just as saprophytic phase or parasitic phase describes the morphologic response to the environment in which the fungus is growing.

There are three medically important phyla of fungi based on their methods of sexual reproduction: (1) Zygomycota, in which the fertilized cell becomes a zygote; (2) Ascomycota, where the sexual spores are contained in an ascus or sac; and (3) Basidiomycota, which produce sexual spores on the end of a basidium or club-shaped cell (Table 6–2).

Asexual Reproduction

In many fungi, the primary method of dissemination of the species is by asexual units. The units are spores or conidia, depending on their mode of production. Asexual spores produced in sporangia are com-

Table 6–2   *Phyla and Classes of Higher Fungi*

Phylum Chytridiomycota (gametes motile)
  Class Chytridiomycetes: Thallus holocarpic or eucarpic, penetrating substrate by means of rhizoids or mycelium or immersed in it; monocentric or polycentric. Zoospores motile by single whiplash flagellum, which contains a conspicuous globule; germination is monopolar.
  Class Harpochytridiomycetes: Thallus is uniaxial, eucarpic; no penetrating mycelium; epiphytic or epizooic. Basal portion attached by means of disklike holdfast on surface of substrate. Distal part sporogenous, lower part vegetative. After sporogenesis, lower part persists and regenerates sporogenous zone.
  Class Blastocladiomycetes: Thallus forms extensive mycelium into substrate and forms numerous asexual reproductive cells. These are thick-walled resting spores. Sex cells are motile, without globules, and germinate bipolarly.

Phylum Zygomycota (teleomorph state a zygote)
  Class Zygomycetes: Saprobic or parasitic or predaceous. If parasitic, mycelium is immersed in host.
    Order Mucorales: Saprobic, few parasitic; asexual reproduction by sporangia, sometimes conidia.
    Order Entomophthorales: Chiefly parasitic on lower animals, some saprobic; asexual reproduction by modified sporangia functioning as conidia or by a true conidium that is forcibly discharged.
    Order Zoopageles: As above; also predaceous, conidia not forcibly discharged.
  Class Trichomycetes: Commensals of arthropods; attached by holdfasts to digestive tract and not immersed in tissue.

Phylum Ascomycota (teleomorph state an ascus)
  Class Hemiascomycetes: Ascocarps and ascogenous hyphae lacking; thallus unicellular or short mycelial. Yeasts.
  Class Loculoascomycetes: Ascocarps and ascogenous hyphae present; asci bitunicate; ascocarp is an ascostroma
  Class Plectomycetes: Asci evanescent, scattered within an astomous ascocarp; the carp (body) is a gymnothecium or a cleistothecium; ascospores aseptate.
  Class Laboulbeniomycetes: Exoparasites on arthropods. Asci regularly arranged within an ascocarp (ostiolate perithecium); asci inoperculate, formed as a basal or peripheral layer of carp.
  Class Pyrenomycetes: Not exoparasites of arthropods. Asci formed in basal layer of perithecium, which is ostiolate (if not, asci not evanescent); asci operculate with an apical pore on slit.
  Class Discomycetes: Ascocarp an apothecium or modified apothecium, often macrocarpic (fleshy, large), epigean, or hypogean; asci operculate or inoperculate.

Phylum Basidiomycota (teleomorph state a basidium)
  Class Teliomycetes: Basidiocarps lacking or replaced by teliospores (encysted probasidia) grouped in sori or scattered within host tissue; as parasites on vascular plants, also as saprobes with yeastlike phase predominating.
  Class Hymenomycetes: Basidiocarp usually well developed but may be gymnocarpic (naked); basidia typically organized on hymenium (fertile layer lining the carp); basidiospores as ballistoconidia. Spores may be phragmobasidia (septate) or holobasidia (non-septate).
  Class Gastromycetes: Basidiocarps typically angiocarpic (fleshy, structured). Basidia not ballistoconidiogenous and spores are holobasidia.

Form-Phylum Deuteromycota (teleomorph state lacking)
  Form-Class Blastomycetes: Budding (blastoconidia) chief mode of life cycle; with or without pseudomycelium. True mycelium lacking or poorly developed. The asporongenous yeasts.
  Form-Class Hyphomycetes: Mycelium well developed; assimilative budding cells absent or occur under certain environmental conditions. Mycelium bears conidia directly or on special branches (conidiophores), which may be arranged together as synnemata but not acervuli or pycnidia. The family Moniliaceae include most fungi with colorless hyphae and the family Dematiaceae have detectable melanin pigment in hyphal walls.
  Form-Class Coelomycetes: Same as above but conidia-bearing structure as acervuli or pycnidia.

mon among the Zygomycota, although a few genera in this group also produce conidia such as chlamydoconidia. Conidia formation is very common among the Ascomycota, less so in the Basidiomycota. It is the only method of reproduction in the Deuteromycota. The latter division essentially represents Ascomycota that have lost the ability to reproduce sexually and produce only conidia. Deuteromycota (Fungi Imperfecti) are divided into three major classes. The first and most important in this context is the Blastomycetes, or asporogenous (nonsex-cell–producing) yeasts. These are differentiated primarily on the basis of physiological characteristics. The other two classes (Hyphomycetes and Coelo-

mycetes) produce hyphae and conidia and are differentiated on the basis of morphologic characteristics. Conidia are usually simple, unicellular bodies, but they may be multicellular and of various shapes, sizes, colors, morphology, and architecture. These characteristics plus their mode of formation, the structure of the conidia-bearing mycelium, and the arrangement of the conidia *in situ* are the bases used for the identification of species of fungi. In medical mycology these attributes are particularly important, as the perfect state (sexual fruiting body) is seldom found. A somewhat complex system of nomenclature for the different conidia types has been devised to aid in their differentiation. Familiarity with these terms is necessary for accurate identification of the isolated pathogen or accidental contaminant from pathologic material.[9]

Media. Since conidia and conidia production are very important for identification, some attention should be given to media on which the fungus is grown. The common isolation medium used in medical mycology is Sabouraud's agar, which contains peptone and sugar. For most organisms encountered, the type of growth produced on this medium is sufficient for identification, although it tends to promote mycelial growth and suppress conidiation. This is certainly not the best medium for the growth of most fungi, but it is the "traditional" one. Historically, the standard descriptions of the pathogenic fungi are based on the colonial morphology and conidia production when grown on Sabouraud's agar. Conidia are generally produced in greater abundance on "deficient" media, however. These are usually decoctions of some plant material. The most commonly used is cornmeal agar, a medium that is useful in inducing conidiation in most hyphomycetes. Other media include potato-dextrose, potato-carrot, tomato juice, lima bean, and so forth. Three media — Sabouraud's, cornmeal, and potato-dextrose — are sufficient stock for the medical mycology laboratory. In addition, Czapek-Dox (containing salts and glucose) is used in the identification of some contaminants, and blood agar is used for induction of dimorphism in systemic pathogens. Primary isolation media have previously been discussed

Conidia, Spores, and Hyphal Structures. Mycology has been termed an exercise in contemplative observation. One accumulates the characteristics of an isolant, its colonial color and morphology on the obverse, its pigmentation on the reverse, and its conidium or spore formation and hyphal structures on microscopic examination. Put together, these details lead to accurate identification, if they are accurately interpreted.

In the common Zygomycota, either from clinical material or as laboratory contaminants, the asexual spores are found within a sporangium. For example, in *Rhizopus oryzae,* the white cottony mold commonly isolated from cases of mucormycosis, soil, compost piles, and so forth, erect, unbranched hyphae (*sporangiophores)* arise from the coenocytic mycelium (Fig. 6–5,*B*). Near the apex of each strand a septum forms. The tip of the sporangiophore then swells into a globular *sporangium.* A small evagination of the supporting sporangiophore extends into the sporangium, acting as a sterile supporting structure, the *columella.* Within the sporangium, cleavage furrows form from the peripheral protoplasmic membrane, dividing the internal space into multinucleate protospores. More furrows form and result in the formation of numerous uninucleate *sporangiospores.* This process is called progressive cleavage formation. The internally produced asexual spores are released only by rupture or deliquescence of the sporangial wall.

In the higher molds, the asexual units are not enclosed in a sporangium. Instead they are produced free, either by segmentation, or by budding of the tips of hyphae or from the walls of the hyphae. Such propagules are given the general name *conidia,* and the hyphae that produce them are conidiogenous cells or conidiophores on conidiogenous (fertile) hyphae. Among such groups as the dermatophytes, in which different sizes of conidia are produced, the term *microconidia* is used to define conidia that are smaller and usually unicellular. If the conidia are larger they are *macroconidia* and may have more than one cell. Various conidia-bearing structures and modifications may be seen in the fungi. Together with conidial form, shape, septation, and color, these formed the basis of the traditional classification scheme. This was known as the saccardoan system after P.A. Saccardo.[11] In this system the numerous genera of imperfect Hyphomycetes (Fungi Imperfecti with mycelium) are grouped into sections depending on characteristics of the conidia. Thus the section *Didymosporae* contained organisms with ovoid conidia and one septum. The section containing pigmented conidia was *Pheodidymae.* Other sections included *Amerosporae, Phragmosporae, Dictyosporae, Scolecosporae,* and so forth. Unfortunately, organisms obviously related were placed in separate sections using this schema and, conversely, the imperfect stages of totally unrelated organisms were often lumped together.

Beginning in the 1950's, systematists such as S. J. Hughes in Canada, K. Tubaki in Japan, and G.L. Barron in the United States began to emphasize that the developmental stages of the conidium and the conidiogenous cell represented stable characteristics usable in the classification of fungi and that this was more of a "natural taxonomy." At present this approach, much modified and continually being modified, is the basis for nomenclature in the Hyphomycetes (Table 6–2). The adoption of this system has engendered much new terminology and the abandonment or modification of former definitions. A comprehensive treatment of medically important fungi based on conidiogenesis and current methods of taxonomy is given in the manual by McGinnis.[9] In the following section, an overview of current concepts of conidiogenesis in the Hyphomycetes will be presented.[2, 6, 7, 9]

There are two general methods of asexual conidiogenesis in the Hyphomycetes. These are called *blastic* and *thallic* conidia formation. *Blastic* or budded conidiation involves "blowing out" and *de novo* growth of a part of the hyphal element. *Thallic* conidiation involves the conversion of a part of the preformed hypha into a conidium. This may require some enlargement of the cell and secondary wall growth. Table 6–3 lists the types of conidia and their descriptive terms.

Thallic Development. Two processes of thallic conidiogenesis are distinguished: (1) thallic and (2) thallic-arthric conidiation. In *thallic conidiation,* a single, terminal, or intercalary conidium is formed (Fig. 6–6, *A*). This is seen in the dermatophyte genera *Microsporum, Trichophyton,* and *Epidermophyton.* The conidium originates as a short branch that may be unicellular or septate (macroconidium or microconidium). As the conidium matures, the distal wall expands and a septum is formed at or near the base, separating the conidium from the hyphae on which it was formed. Both of the wall layers of the mycelium are utilized in the formation of the conidium; thus, the full name of the process is holothallic conidiation. As development continues a second septum is formed just above the attachment to the hypha. The wall of the distal end of the conidium thickens, but the wall of the basal attachment cell remains thin (Fig. 6–6, *A*); in time, the cytoplasm within this cell disintegrates and the fully mature conidium is released by fracture or lysis of the thin basal cell wall.

Special terminology and abbreviations are used in referring to thallic conidia formation; these include

C.W.: cell wall

P.C.: the position of the conidia relative to the conidiogenous cell

O.A.: the order of arrangement of the conidium relative to other conidia

C.C.: the relation of the conidiogenous cell to the conidium that is next formed.

As noted, when the C.W. of the conidium is formed from both parts of the hyphal wall, it is *holothallic.* When the conidium is formed at the end of the hyphae branch, the P.C. is

Table 6–3   *Conidiogeny in the Hyphomycetes*

C.W.	P.C.	O.A.	C.C.	S.A.
Blastic Conidia				
holoblastic	terminal synchronous basipetal acropetal	solitary catenulate botryose	determinate proliferous sympodial percurrent basauxic retrogressive	porogenous
enteroblastic	terminal basipetal	solitary catenulate	determinate proliferous retrogressive	phialidic annellidic
Thallic Conidia				
holothallic	terminal intercalary	solitary	determinate	
holoarthric	terminal intercalary	catenulate	determinate proliferous	
enteroarthric	terminal intercalary	solitary catenulate	determinate proliferous	

KEY (see text for definitions of terms):

C.W. = conidial wall
P.C. = position of conidia in relation to conidiogenous cell or other conidia
O.A. = order of arrangement in relation to other conidia
C.C. = conidiogenous cell in relation to next conidia formed
S.A. = special attribute of conidiogenous cell, if any

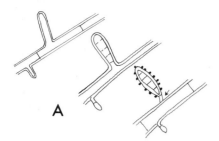

A, *Microsporum gypseum;* C.W. holothallic, P.C. terminal, C.C. determinate. In this case, the basal cell collapses as its protoplasm disappears. This aids in the release of conidia. Both macroconidia and microconidia are shown.

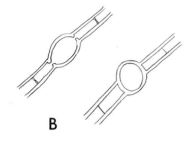

B, *Microsporum audouinii:* C.W. holothallic, P.C. intercalary or terminal, O.A. solitary, C.C. determinate chlamydoconidia.

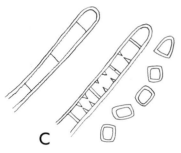

C, *Geotrichum candidum:* C.W. holoarthric, O.A. catenulate, P.C. terminal or intercalary, C.C. determinate. Units are arthroconidia.

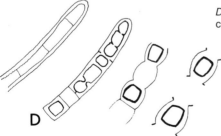

D, *Coccidioides immitis:* C.W. enteroarthric, P.C. terminal or intercalary, O.A. catenulate, C.C. determinate. Units are arthroconidia.

Figure 6–6. Thallic conidia. (From Hughes, S. J. 1958. Revisiones hyphomycetum aliquot cum appendice de nominibus rejiciendis. Can. J. Bot. *36*:727–836; Cole, G. J., and R. A. Samson 1978. Patterns of Development of Conidial Fungi. London, Pitman Publishing; and Kendrick, B. ed. 1972. Taxonomy of Fungi Imperfecto. Toronto, Univ. Toronto Press.)

terminal. If it is formed within the main hyphal trunk, the P.C. is *intercalary* (Fig. 6–6, *B*). In holothallic conidiation only one conidium is formed; thus, O.A. is *solitary.* The distal portion of the hyphal branch in holothallic conidiation is converted into one conidium and the lower part is a basal cell that disintegrates; since no more growth can occur, C.C. is said to be *determinate.*

In describing other methods of thallic conidia formation, several more terms can be introduced. If a chain of conidia is formed,

the O.A. is *catenulate.* If the conidiogenous cell grows to produce additional fertile areas, the C.C. is *proliferous.* Proliferous conidiogenous cells can be characterized as percurrent, sympodial, or basauxic, but these are encountered primarily among "Blastic" conidia-forming fungi. These terms will be further examined in a later section of this chapter.

A second type of thallic development involves conversion of determinant segments, terminal segments, or intercalary sections of hyphae into chains of deciduous conidia.

This is also called *thallic-arthric* development. In holoarthric development (Fig. 6–6,*C*), both walls of the fertile hyphae become the walls of the conidia. *Geotrichum candidum* is an example of this type of conidiation. In this species each arthroconidium contains one to four nuclei.

In the second type of *thallic-arthric* conidiation, the individual conidia differentiate as separate units within the outer wall of various sections of the preformed hyphae. The conidium itself is derived from the inner wall of the hyphae (*enteroarthric*). Between the developing conidia is preexisting hypha that degenerates. This degeneration aids in release of the endogenous conidia. When released, the conidium will still have attached to it fragments of the outer wall of the hypha in which it was formed as well as pieces of the wall of the degenerated adjacent cells. Such architectural markings are seen on the arthroconidia of *Coccidioides immitis* (Fig. 6–6, *D*).

Blastic Development. In this type of development there is a "blowing out" or budding or enlargement of the cytoplasmic material from the fertile hypha with *de novo* synthesis of new walls, cytoplasm, and structure. As in thallic development, two types of wall formation exist relative to the fertile hypha and the conidium that is produced from it. If both walls of the fertile hypha are used in the formation of the conidial wall (C.W.), it is *holoblastic*; if only the inner wall is utilized it is *enteroblastic*.

Many descriptive terms are used to define the conidium and conidiogenous cell; these may involve

1. Position of conidium (P.C.):
   a. *terminal*: on the end.
   b. *synchronous*: several developing at the same time from a *botryose* conidiogenous cell (Fig. 6–7, *D*).
   c. *basipetal*: the youngest cell is at the bottom of a chain, as in the phialidic *Aspergillus* or annellidic *Scopulariopsis* (there is usually no cytoplasmic connection between the succeeding cells in the chain) (Fig. 6–7, *I, J, K*).
   d. *acropetal*: the youngest conidium is at the top of the chain (Figs. 6–7, *C* and 6–8, *H*). There is usually a cytoplasmic connection from the conidiogenous hypha through the entire chain of conidia to the last conidium (e.g., *Cladosporium* spp., *Alternaria* spp., and *Septonema* spp.).

2. Order of arrangement (O.A.):
   a. *solitary*: one cell is formed at a time (Fig. 6–7, *A*).
   b. *botryose*: several cells develop from the same cell at the same time (Fig. 6–7, *B*).
   c. *catenulate*: the cells form a chain (Fig. 6–7, *I*).
3. Further development of conidiogenous cell (C.C.):
   a. *determinate*: cell ceases extension with beginning of formation of first conidium (Fig. 6–7, *A, C, H, I*).
   b. *proliferous*: cell has some type of continuous growth after the production of the first conidium; this extension takes one of three forms:
      i. *sympodial*: new growing point appears at side of or below first conidium (Fig. 6–7, *B, F* and 6–8, *F*).
      ii. *percurrent*: conidiogenous cell simply extends through the previous apex and new set of conidia is generated (Fig. 6–7, *E*).
      iii. *basauxic*: rapid growth of conidiogenous cell just under the newly formed conidium, causing conidium to be elevated (Fig. 6–7, *G*).
   c. *retrogressive*: after extension, mature conidiogenous cell produces a first conidium holoblastically; it then produces another conidium basipetally. A part of the outer wall of the conidiogenous cell is attached to the newly expanded enteroblastically produced conidium, however, and this slightly shortens the conidiogenous cell. Subsequent cell formation continues the process (Fig. 6–7, *K*), and the original conidiogenous cell disappears as fragments on each of the newly formed conidia.
   d. *stable*: determinate conidiogenous cells are not retrogressively converted into conidia.

Some special forms of conidiogenous cell development should be mentioned at this point. In some holoblastic conidia, a circumscissile rupture occurs on the wall of the conidiogenous cell. This is a pore, and the fungi are called porogenous (Fig. 6–8, *F*). A holoblastic conidium is budded out from this pore. (Indicated by arrows in Fig. 6–8, *F*).

In enteroblastic fungi two special forms of conidiogenesis are very commonly encountered: *phialide* formation (Fig. 6–7, *H*) and *annellide* formation (Fig. 6–7, *J*). In the

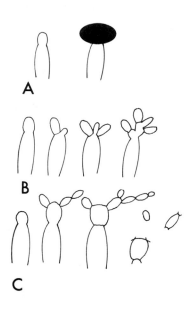

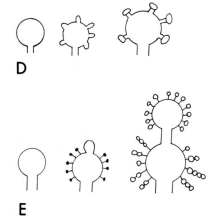

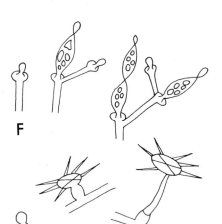

Figure 6–7. Blastic conidia.

Holoblastic

A, Nigrospora: P.C. terminal, O.A. solitary, C.C. determinate.

B, Tritiarchium: P.C. terminal, O.A. solitary, C.C. proliferous (sympodial).

C, Cladosporium: P.C. acropetal, O.A. catenulate, C.C. determinate.

D, Botrytis: P.C. synchronous, O.A. botryose (on an ampulla), C.C. determinate.

E, Gonatobotryum: P.C. synchronous for first conidia, acropetal for chains thereafter, O.A. botryose for first, then catenulate, C.C. proliferous (percurrent).

F, Alternaria: P. C. terminal, then acropetal; O.A. catenulate; C. C. proliferous (sympodial); S.A. porogenous.

G, Spegazzinia: P.C. terminal, O.A. solitary, C.C. proliferous (basauxic).

Illustration continued on opposite page

Figure 6–7 *Continued*

Enteroblastic

H. Phialophora: P.C. terminal, O.A. solitary, C.C. determinate, S.A. phialidic.

H

I, Aspergillus: P.C. synchronous, then basipetal; O.A. botryose, then catenulate; C.C. determinate; S.A. phialidic.

I

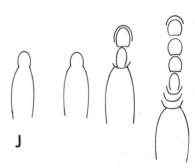

J, Scopulariopsis: P.C. terminal, then basipetal; O.A. solitary, then catenulate; C.C. proliferous (percurrent); S.A. annellidic.

J

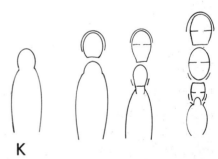

K, Cladobotryum: P.C. terminal, then basipetal; O.S. solitary, then catenulate; C.C. retrogressive.

K

former, a holoblastic conidium is formed at a fixed point on the conidiogenous cell. Its wall is formed from the inner wall of the conidiogenous cell, but as it emerges it carries with it part of the outer wall. Subsequent conidia are enteroblastic and formed basipetally and may be solitary or in balls, as in *Phialophora* (Fig. 6–7, *H*), or catenulate, as in *Aspergillus* (Fig. 6–7, *I*); these conidia consist entirely of material from the inner wall. The production of each new conidium in *Phia-*

lophora leaves a quantity of wall material on the opening of the phialide and forms a cup. This cup-shaped opening is called a collarette (Fig. 6–8, *C*). In *annellide* formation, the stable conidiogenous cell expands to bud out a holoblastic conidium. A septum at the base of this cell delimits the primary conidium. The conidiogenous cell then expands percurrently by elongating the inner wall, leaving a ridge or annellation at the site of the initial conidium (indicated by arrows in Fig.

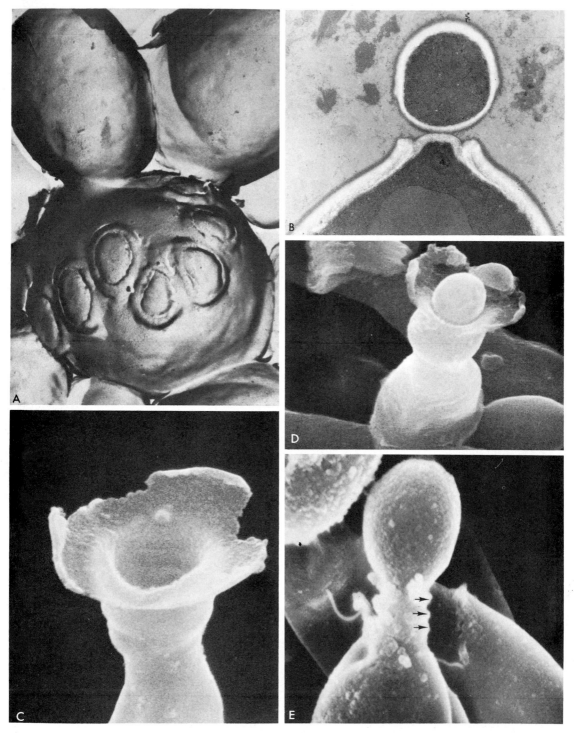

Figure 6–8. Aspects of fungal conidiogenesis. *A, Saccharomyces cerevisiae:* budding cells from a mother yeast cell. A series of bud scars is evident. Although a single yeast cell can form a number of bud cells, only a single daughter cell can arise from any given space on the parent cell; thus, the reproductive potential and probably the viability of a single cell are limited. *B, Candida albicans* bud dehiscing from a mother cell; morphogenesis of scar. (Courtesy of D. Ahearn.) *C,* Phialide. Opening of a phialide of *Phialophora verrucosa.* ×9450. (Courtesy of G. Cole.) *D,* Enteroblastically produced conidium from a phialide of *Phialophora verrucosa.* ×5200. (Courtesy of G. Cole.) *E,* Annellidic development of *Exophiala pisciphila.* Annellides (*arrows*) accompany percurrent proliferous growth of the conidiogenous cell with the production of each new conidium. ×14,400. (Courtesy of G. Cole.)

Illustration continued on opposite page

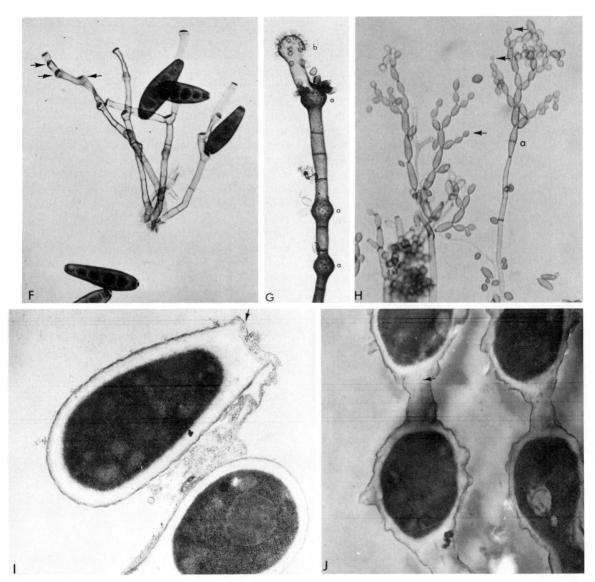

Figure 6–8 *Continued*

F, Sympodial porogenous conidiophore of *Drechslera.* ×1100. (Courtesy of J. Gallop.) *G, Gonatobotryum. a,* Scars of denticles and a few mature conidia arranged botryosely on conidiogenous cell (ampulla) and produced synchronously. Successive fertile areas have been formed by percurrent extension of the conidiogenous cell. *b,* Youngest fertile area of proliferous conidiogenous cell. New papillae are forming on the ampullae synchronously in a botryose arrangement. *H, Cladosporium,* The youngest conidium (*arrows*) is produced acropetalously (i.e., at the end) of a catenulately arranged (in chains) series of conidia. The conidiogenous cell (*a*) is determinate; that is, it neither extends after production of first conidia (proliferates) nor becomes smaller (retrogresses). ×450. (Courtesy of J. Gallop.) *I,* Microconidia of *Trichophyton mentagrophytes* peglike propagule that has dehisced from fertile mycelium by degeneration of connecting cell (*arrow*). ×25,000. *J,* Conidia of *Aspergillus fumigatus.* These conidia are produced enteroblastically from phialides but have a remnant of wall material connecting each conidium to the basipetally produced conidium beneath it. This connection breaks easily, releasing the individual conidia. In the center of the connecting material is a "strand" (*arrow*) of unknown origin and function. ×25,000. (Courtesy of J. Mullin and R. Harvey.)

6–8, *E*). The expanded conidiogenous cell now delimits a second conidium formed entirely from the inner wall. A second ridge is formed just above the first, and so on with the formation of each subsequent conidium.

Most conidia are released by one of three methods: (1) *fission*, in which a double septum is formed between the conidiogenous cell and the conidium (e.g., *Cladosporium*) or between adjacent conidia in a chain (e.g., *Geotrichum*); (2) *fracture*, the type of conidial release wherein the wall of the adjacent degenerated cell is ruptured, usually by mechanical stress, as in *Coccidioides*; and (3) *lysis*, the release of the conidium by dissolution of the wall of a basal attachment cell, such as the degenerated cell at the base of a macroconidia in *Trichophyton*. The dissolution may be autolytic or due to microorganisms in the vicinity. The fascinating study of conidiogenesis in Hyphomycetes has recently been examined in three highly recommended books by Cole and Samson,[2] Kendrick,[7] and McGinnis[9] and the paper by Hughes.[6]

Vegetative or Hyphal Structures. A number of structures are formed by the vegetative mycelium, which have no reproductive significance but are of considerable value in the differentiation of some clinically important fungi.

Spirals or coiled hyphae, similar to the coiled filaments of actinomycetes, are observed in a number of pathogenic fungi, especially dermatophytes (Fig. 6–5, *I*). They are bedspring-like, helical coils that are found at the ends of peridial hyphae surrounding an ascocarp, but these may be produced in the absence of ascocarps or peridial hyphae. They are very prominent in some strains of *Trichophyton mentagrophytes*.

Nodular organs are formed by some species, e.g., strains of *Microsporum canis* and *T. mentagrophytes* (Fig. 6–5, *K*). They are enlargements in the mycelium, which consist of closely twisted hyphae. Either side may branch and twine about the parent stem, or different hyphae may entwine together. The resulting structure has the appearance of a knot. This may represent an attempt to initiate a sexual fruiting body.

Racquet mycelium (mycelium *en raquette*) is a term applied to certain hyphae, usually larger than the others, that show a regular enlargement of one end of each segment, large and small ends being in opposition (Fig. 6–5, *J*). The appearance is that of a chain of tennis racquets. This structure is common in many fungi.

The term *pectinate body* is applied to unilateral, short, irregular projections or protuberances forming on one side of a hypha, which give it the appearance of a broken comb (Fig. 6–5, *L*). These are commonly seen in dermatophytes, especially *Microsporum audouinii*.

Favic chandelier (antler hyphae) is the name applied to the structure formed by numerous short, multiple branches appearing at the end of a hypha. These resemble a reindeer antler or a chandelier (Fig. 6–9, *A*). This structure is not common among the fungi, occurring primarily in *Trichophyton schoenleinii* and occasionally in *Trichophyton violaceum*.

Peridial hypha is a wide, indented, multiseptate hypha which may terminate in a spiral (Fig. 6–9, *F*). This structure is sporadically produced by some strains of fungi. It resembles the ornamentation found around the sexual fruiting body of some perfect dermatophytes or other Ascomycota, hence its name. It may be numerous in some strains of *Trichophyton mentagrophytes*.

Pycnidium is a mycelial structure resembling the fruiting body (perithecium) of some Ascomycota. However, the structure is filled with asexually produced conidia. The body is large, up to several millimeters in diameter, and may have a hard wall surrounded by peridial hyphae. Some strains of *T. mentagrophytes* produce numbers of pycnidia, especially when grown on soil-hair agar.

Sexual Reproduction

The sexual stage of fungi is not often observed in the diagnostic laboratory, but an acquaintance with the phenomenon is necessary for the serious student of medical mycology. The fungi of medical importance are assigned to three general divisions (phyla), depending on their mode of sexual spore formation. These are *Zygomycota, Ascomycota,* and the *Basidiomycota.*

Zygomycota. The Zygomycetes found commonly in the laboratory as contaminants or as agents of mucormycosis have a well-developed mycelial thallus. Fertile hyphae of the aerial mycelia produce numerous asexual spores in sporangia atop sporangiophores. There is an extensive vegetative mycelium. In the genus *Rhizopus*, there is also a multiple-branched, short, rootlike extension from the hyphae, called a rhizoid (Fig. 6–5, *B*), which penetrates the nutrient medium. The mycelia are coenocytic and thus multinucleate. The nuclei appear to divide directly by constriction without spindle formation or classic mitotic figures. This process is called *kary-choresis* and is common in the asexual growth of many fungi.

Sexual reproduction takes place by simple copulation of the tips of multinucleate hyphae. These tips consist of terminal swellings and arise as branches from the mycelial mat. The tips or gametangia are attracted to one another by sex hormones or fuse following chance contact. If compatible mating strains are necessary for initiation of the sexual cycle, the organism is heterothallic; if side branches from the same thallus fuse and form a *zygote*, the organism is homothallic. After contact, each of the hyphal tips swells, and a septal wall is formed separating the cytoplasm and nuclei in the swollen end from the rest of the hypha. The wall between the adjacent tips then dissolves, and there is mixing of the cytoplasm from the two mating strains followed by pairing of the nuclei. The new cell (zygote) which is the product of this fusion enlarges, and the walls become thick and pigmented (Fig. 6–9, *E*). When nuclear fusion takes place, a diploid nucleus is formed. In some species, only one nuclear pair survives to form spores; in other species several may survive. After a period of inactivity, the zygote cracks open, a sporangiophore emerges, and sporangium develops (Fig. 6–9, *E*). At any time during this process, the formation of haploid nuclei may occur by nuclear conjugation followed by reduction division. The timing of this and the mating type of the spores produced within the sporangia are dependent on the species. In some species, there are haploid spores from each of the parental mating strains; in some, only one sex

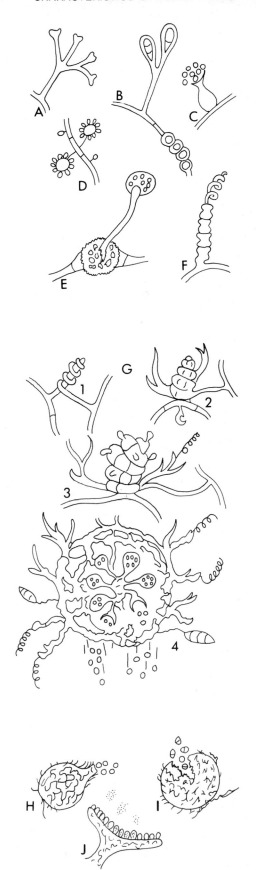

Figure 6–9. Asexual reproduction, sexual reproduction, and some hyphal appendages. *A,* Favic chandelier of *Trichophyton schoenleinii. B,* Spatulate (beaver tail) macroconidia and a series of chlamydoconidia in *Epidermophyton floccosum. C, Phialophora* phialoconidia from a phialide. *D, Histoplasma* verrucose macroconidia and microconidium. *E,* Zygote of Rhizopus. This is the sexual phase of the organism. The zygote has cracked and a sporangium has emerged. *F,* Peridial-like hyphae of dermatophytes. *G,* Sexual spore formation in the dermatophytes. (1) A rigid, erect antheridium is entwined by an archegonium. (2) Plasmogamy and (later) karyogamy occur between the cells of the archegonium and antheridium, forming ascogenous hyphae. (3) Papillae of ascus form initially, followed by croziers appearing from the ascogenous hyphae. The peridial hypha is now forming a carp. (4) The mature ascocarp containing asci and ascospores. The fruiting body produced by the dermatophytes is a gymnothecium. The ascospores can sift through the loose mycelial network of the gymnothecium. *H,* A perithecium. Mature ascospores are extruded through an ostiole. *I,* A cleistothecium. The ascospores are usually released only by dissolution of the ascocarp. *J,* An apothecium. The ascospores are usually released by being ejected into the air.

is produced, and in others both sexes and diploid spores are formed. The zygosporangium is similar to the asexual sporangium, and when it is mature, the walls rupture or lyse, releasing the spores. This pattern of sexual reproduction is common to the genera of medical importance: *Mucor, Absidia*, and *Rhizopus* spp. of the order Mucorales.

The agents of human subcutaneous disease entomophthoromycosis are in the genera *Conidiobolus* and *Basidiobolus* of the order Entomophthorales, the second major order of the class Zygomycetes. The mycelium is not as extensive as in the Mucorales, but sexual reproduction follows essentially the same pattern. Asexual reproduction differs from the preceding because conidia-like structures are produced on the ends of hyphal tips, which are forcibly ejected or "shot" into the surrounding area, seeking insect prey in the case of *Conidiobolus* or an insect as passage to frog dung in *Basidiobolus*.

Ascomycota. The patterns of sexual reproduction and the specialized hyphal mating branches and fruiting bodies produced are too varied among the ascomycetes to be encompassed in a general discussion. However, the end result of all these variations is the formation of the *ascus* or sac in which the sexual spores (*ascospores*) are produced. After nuclear conjugation and reduction division have occurred, the ascus usually contains eight haploid nuclei destined to be contained in spores (Fig. 6–10). The spores are produced by a process called free cell formation. Within the ascus, astral rays appear near one pole of the nucleus and radiate around and converge near the other pole, sectoring off a

mass of cytoplasm. Cell walls form along the path of the rays, and a spore is produced. The usual spore number is eight, four representing one mating type and four the other. If only four spores are produced, they are binucleate, containing two haploid nuclei of the same mating type. Homothallic sexual reproduction also occurs.

The simplest pattern of sexual reproduction is seen in the true yeasts. Two compatible mating types come in contact and fuse. The nucleus from one yeast cell enters the cytoplasm of the other. The recipient yeast then becomes an ascus, and the nuclei pair, conjugate and immediately undergo meiosis. Mitotic divisions then occur, and eight ascospores are formed.

In the group of fungi that produce mycelium, the process is more complex. The pattern found among the perfect dermatophytes, *Blastomyces*, and *Histoplasma*, will be outlined briefly (Fig. 6–9, *G*1–4). A special hyphal structure, the male unit (antheridium), is produced when it is in proximity to a compatible mating type. The antheridium is a rodlike structure around which the female structure (archegonium) wraps itself. Plasmogamy (fusion of cells by breakdown on the separating cell wall) occurs, and male nuclei migrate into the archegonium. Nuclear pairing occurs, and nuclei then swarm into the many hyphal branches that develop from the archegonium. These branches are called ascogenous hyphae, and a pair of nuclei migrate to their tips. The tip folds over to form a crook or crozier, and a septum is formed, enclosing the nuclear pair in this dikaryotic end cell (Fig. 6–10, *A*). Mitosis

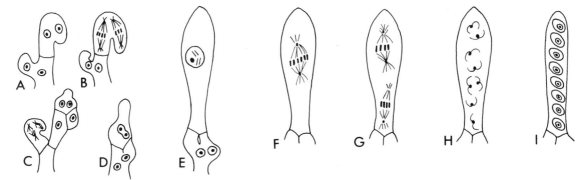

Figure 6–10. Ascospore formation. *A*, An ascogenous hypha containing both parental nuclei (dikaryon) forms a crozier. *B*, The two nuclei divide. One pair of nuclei wander to apex of cell. *C*, Rearrangement of the cell walls of the crozier segregates a pair of nuclei in the apical cell. This is called an *ascus mother cell*. *D*, Nuclear fusion occurs, and a diploid nucleus is produced. *E*, Diploid nucleus in an enlarged ascus mother cell, diplotene formation. *F*, Anaphase I. *G*, Anaphase II and reduction division. *H*, After anaphase III, ascospore delimitation begins. *I*, Mature ascus with eight ascospores.

occurs (conjugate nuclear division), forming four nuclei (Fig. 6–10, *B*). Septal formation now divides the crozier into three cells: the neck, having one nucleus; the bend, having two nuclei of opposite mating types; and the tip, containing one nucleus of the type opposite the one in the neck (Fig. 6–10, *C*). The nuclei in the bend cell conjugate to form a diploid nucleus. This cell, which is called an ascus mother cell, enlarges and becomes an ascus (Fig. 6–10, *D*, *E*). Reduction division and ascospore formation occur as previously described (Fig. 6–10, *F*, *G*, *H*, *I*). The tip cell and the neck cell fuse; their nuclei pair, and the fused cell enlarges to form a crozier, whereby the whole process is repeated again. While this is occurring adjacent mycelium may develop an extensive network around the asci-producing cells. These are called fruiting bodies or ascocarps. Some of the dermatophytes produce a carp consisting of a loose mesh of hyphae through which the ascospores can sift: a *gymnothecium* (Fig. 6–9, *G4*). If the carp is completely closed and ascospores are released only by disintegration of the carp walls it is called a *cleistothecium* (Fig. 6–9, *I*). This is the type seen in the perfect fungi having *Aspergillus* and *Penicillium* sp. as an anamorph stage. A *perithecium* is a carp with a pore or ostiole through which ascospores may be successively released as they mature and form asci in a hymenium or a basal cluster (Fig. 6–9, *H*). The common saprophytic fungus, *Neurospora crassa*, which has been used for many genetic studies, forms a perithecium. When the fruiting body is wide open and cup-shaped, the carp is called an *apothecium.* This group includes cup fungi *Peziza*, morels and truffles (Fig. 6–9, *J*).

In some fungi, the sexual spores are produced within locules set in a complex mass of mycelium called a stroma. This ascostroma is the communal structure supporting the individual locules. An ascostroma is found on the hair shaft in the disease called *piedra*, caused by *Piedria hortae.*

Basidiomycota. Until recently, few Basidiomycota were regularly encountered as agents of human infection. The perfect (sexual) phase of *Cryptococcus neoformans* has recently been found to produce basidiospores similar to that found in the Ustilaginales (an order containing the corn smut fungus *Ustilago*). The sexual phase of *Cryptococcus* was placed in the genus *Filobasidiella.* A few other Basidiomycota such as the mushrooms *Coprinus* and *Schizophyllum* have also been recovered from human infections. Such isolations are quite rare, however. An important category of human disease is the set of conditions known as mycotoxicosis (ingestion of toxic products) and mycetismus (ingestion of fungi that contain toxic products, i.e., mushroom poisoning). These often involve Basidiomycota species.

Spores of Basidiomycota usually germinate to form a mycelial mat (primarily mycelium) containing only haploid nuclei. Instead of two sexes, there are often four mating types. When compatible mating strains come in contact, the hyphal tips fuse, but nuclear fusion does not occur. This is called the *dikaryon* condition (n+n). Both nuclei divide simultaneously in a process called conjugate nuclear division. This is similar to the process that occurs in the crozier cell of the ascomycetes. When the nuclei are about to divide, a short branch from the side of the cell wall forms and arches back away from the hyphal tip (Fig. 6–4, *A*). This branch is called a *clamp connection.* One nucleus migrates into the neck of the connection and divides. Its spindle fibers point toward the hyphal tip (Fig. 6–4, *B*). One daughter nucleus remains in the clamp, while the other migrates to the hyphal tip (Fig. 6–4, *C*). Meanwhile the other parental nucleus divides in the center of the cell (Fig. 6–4, *B*, *C*). One daughter migrates to the tip; the other remains more or less in position. The clamp then completes its arch, coming in contact with the cell wall. It then fuses with it, and a septum is formed across the point of origin of the clamp (Fig. 6–4, *D*). Another septum forms through the hypha itself, separating the tip cell. The nucleus that was formerly in the clamp migrates back to the penultimate (original) cell containing the daughter nucleus of the other mating type. The end result is two cells, each of which has two nuclei representing the parental mating strains, and the dikaryon condition continues. This process is repeated each time a new cell is formed, with a clamp pointing back from the growing tip bridging each cell septum. The clamp connection, when present, is a reliable diagnostic characteristic of Basidiomycota. The dikaryon condition (secondary mycelium) may continue for long periods of time (years) before completion of the sexual cycle occurs. The paired nuclei fuse within a club-shaped terminal cell, the *basidium* (Fig. 6–11, *A*, *B*, *C*, *D*, *E*, *F*). Reduction division occurs, and four spores are produced. These spores are external to the basidium, attached to it by a short neck (*sterigma*) (Fig. 6–11, *G*). Commonly,

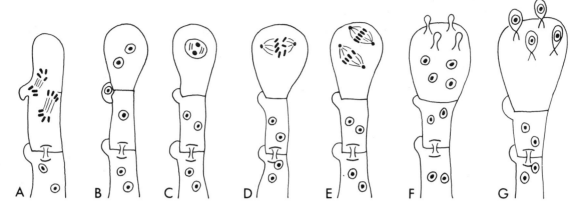

Figure 6–11. General pattern of basidiospore formation. *A,* After clamp connection formation and nuclear migration, a probasidium is formed that is still dikaryon (n + n). *B,* Dikaryon probasidium. *C,* Nuclear fusion to form diploid nucleus. *D* and *E,* Meiotic division to form four haploid nuclei. *F,* Young basidiospores developing on sterigmata. *G,* Four mature basidiospores on a basidium.

they are forcibly ejected by hydrostatic pressure. In the majority of the Basidiomycota, a large fruiting body, called a basidiocarp, may form by the massing together of hyphal elements (tertiary mycelium). In mushrooms, the basidia are produced along the gills on the underside of the cap. In the bracken fungi and boletes, they are produced in the lining of pores on the underside of the fruiting body or on the surface of the carp. In the corn smut and wheat rust, a typical carp is not formed.

THE FUNGI IN NATURE

The fungi, though simple organisms, are extremely adaptable to diverse environments. Their chief function in the interplay of biological forms is, to use the popular ecological jargon, to "recycle" organic debris. There is probably no organic substance in the biosphere that is free from attack by fungi. Wood, lignin, vegetation, chitin, keratin, fats, oils, phenols, asphalt, rubber tires, waxes, bones, and so forth, are degradable by fungi. They have been recovered from sulfuric acid-copper plating liquid at pH 0.05, and as fluffy fungus balls, from distilled water bottles, inorganic solutions, acids, and so forth, as "bottle imps" (usually *Paecilomyces lilacinus*), and they even attack pathology museum specimens contained in formaldehyde solution. The only requirements for growth are a little organic substrate and moisture. In the process of filling every ecological niche, some one hundred thousand species have evolved.

Most are saprobic, but many are plant parasites of great importance. Severe economic loss is caused annually by fungi which attack crops (wheat rust, corn smut, and so forth) and devastate forests (Dutch elm disease) and fruit trees (banana wilt). The history of mankind has been affected by fungi, as witness the potato famine which caused mass migration from Ireland to the new world, or the discovery of a fungal waste product, penicillin, ushering in the new age of chemotherapy and antibiotics. Fungi have adapted to exist as parasites on other fungi as well. Halophilic (salt-tolerant) fungi are known, appearing as white scum or "mycoderma" on brined foodstuffs. Thermophilic fungi are found in compost heaps, where the temperature may reach 45°C (113°F). Some of these temperature-tolerant organisms (as *Aspergillus fumigatus, Rhizopus oryzae,* and *Absidia corymbifera*) readily caused disease in debilitated patients. Cryophilic fungi live on snow banks in the Antarctic and on glaciers, digesting the meager organic material found in such environments. Fungal products and their metabolites are of great economic importance, for example, in the production of alcohol, citric acid, oxalic acid, the production or alteration of steroids, fermentation products, antibiotics, and even the flavoring of coffee and soy sauce.

Few of the fungi are significant parasites of animals. As discussed before, only a small number of fungi have the ability to invade and cause disease in man. Of these only the anthropophilic dermatophytes are obligate infectious agents of man. However, the remarkable adaptive ability of fungi is nowhere better exemplified than in modern medicine. In order to ameliorate other destructive processes or to treat or prolong life in disease states, therapeutics have often compromised natural defenses of the human host and allowed an ever growing list of saprophytic fungi to gain entrance and invade, usually as a terminal event in a disease. Leukemia, lymphoma, and organ transplant patients on cytotoxins and immunosuppressive agents, diabetics, or those on long-term steroid therapy are particularly vulnerable. Whereas medical mycology was formerly concerned with a few skin-infecting fungi and even fewer pulmonary and systemic infections, today it is concerned with the opportunistic infections of soil- and air-borne saprophytes adapting to and taking advantage of the presented organic substrate, the compromised patient. Glibly put, one could say the patient has become a culture medium, and these infections might be termed diseases of medical progress.

REFERENCES

1. Ainsworth, G. C., and A. S. Sussman (eds.). 1965. *The Fungi.* Vol. I. New York, Academic Press.
2. Cole, G. T., and R. A. Samson. 1978. *Patterns of Development of Conidial Fungi.* London, Pitman Publishing.
3. Copeland, H. F., 1956. *The Classification of Lower Organisma.* Palo Alto, Pacific Books.
4. Haeckl, E. 1878. *Generelle Morphologie der Organismen.* Berlin, Reimer.
5. Hogg, J. 1860. The Diversity of Life, Edinburgh, New Phil. J. N. S., *12*:216.
6. Hughes, S. J. 1958. Revisiones hyphomycetum aliquot cum appendice de nominibus rejiciendis. Can. J. Bot., *36*:727–836.
7. Kendrick, B. (ed.). 1971. *Taxonomy of fungi imperfecti.* Toronto, Univ. Toronto Press.
8. Martin, G. W. 1955. Are Fungi Plants? Mycologia, *47*:779–792.
9. McGinnis, M. R. 1980. *Laboratory Handbook of Medical Mycology.* New York, Academic Press.
10. Ragan, M. A., and D. J. Chapman. 1977. *A Biochemical Phylogeny of the Protists.* New York, Academic Press.
11. Saccardo, P. A. 1882. Sylloge Fungorum. 32 volumes. Pavia.
12. Whittaker, R. H. 1969. New Concepts of Kingdoms of Organisms. Science, *163*:150–160.

SUPERFICIAL MYCOSES

7 Superficial Infections

The superficial infections include diseases in which a cellular response of the host is generally lacking because the organisms are so remote from living tissue, or infection is so innocuous. There is essentially no pathology elicited by their presence, and patients may be unaware of their condition. These diseases are usually brought to the physician's attention because of their cosmetic effects on the patient.

Pityriasis Versicolor

DEFINITION

Pityriasis versicolor is a chronic, mild, usually asymptomatic infection of the stratum corneum. The lesions are characterized by a branny or furfuraceous consistency; they are discrete or concrescent and appear as discolored or depigmented areas of the skin. The affected areas are principally on the chest, abdomen, upper limbs, and back. The etiologic agents are the lipophilic yeasts *Malassezia furfur* and *M. ovalis*. Both are part of the normal flora of the human skin.

Synonymy. Tinea versicolor, dermatomycosis furfuracea, chromophytosis, tinea flava, liver spots.

HISTORY

This disease was first noted very early in the history of medical mycology by Eichstedt (1846).[24] It was fully described as a disease caused by a parasitic vegetable by Sluyter in his doctoral thesis in 1847.[55] Both investigators termed the disease pityriasis versicolor and both proposed a fungal etiology for the lesions. In 1853, Robin[54] named the agent *Microsporum furfur* because he thought it was similar to *Microsporum audouinii*, and he changed the disease epithet to tinea versicolor to relate it to the other ringworm (dermatophyte) infections. Malassez,[38] however, re-emphasized the yeastlike nature of the organisms in 1874 and considered them unlike any known mycelial fungi. Baillon recognized that the yeast of pityriasis was not related to the *Microsporum* species that were the agents of ringworm. He therefore erected a monotypic genus, *Malassezia,* in 1889 to encompass the etiologic agent as *Malassezia furfur.*[7] Numerous investigators claimed to have isolated the organism (among them Kobner in 1864, Schmitter in 1923, Macleod and Dowling in 1928, and Morris Moore in 1938), but it is doubtful that these investigators had the correct organism, because lipids were not incorporated into their isolation media. Oto and Huang in 1933 and Rhoda Benham in 1934 isolated *Pityrosporum ovale (M. ovalis)* and demonstrated it to be an obligate lipophile. *Malassezia furfur* was authentically isolated and characterized in 1951 by Morris Gordon,[27] who renamed it *Pity-*

rosporum orbiculare. The relation of *P. ovale* (*M. ovalis*) and *P. orbiculare* (*M. furfur*) to the disease as a clinical entity has been the subject of much controversy. It is now concluded that *M. furfur* is the correct name for the agent of pityriasis versicolor and that *P. orbiculare* is a synonym of it.[48] *M. ovalis* is also a common lipophilic yeast of the normal flora and at times occurs in great numbers. This organism may also be responsible for a similar clinical disease or it may contribute to pityriasis along with *M. furfur.* The preferred and historically accurate term for the disease is pityriasis versicolor. Tinea versicolor was based on the misconception that the disease was caused by a dermatophyte. The presently accepted designations for the organisms involved are *Malassezia furfur* and *M. ovalis.*

ETIOLOGY, ECOLOGY, AND DISTRIBUTION

The etiologic agent, *M. furfur,* has been shown to be a common endogenous saprophyte of the normal skin. Roberts[52, 53] has isolated it from a normal scalp in association with *M. ovalis;* from the chest in 92 per cent of people; and to a lesser extent from the back, trunk, limbs, and, occasionally, the face and other areas. In surveys by Noble and Midgley[45] involving nearly 600 children, direct correlation was demonstrated between age and carriage rate. Older and sexually mature children yielded more positive cultures for *M. furfur* and *M. ovalis* than did younger groups. Furthermore, carriage was more common in Caucasoids than in Negroids and in males than in females. It appears that the organisms are universally present as members of the normal flora of skin, and elicit the disease under special systemic or local conditions when an overgrowth of the organisms occurs.

The factors responsible for the overgrowth of the organisms in certain people resulting in the clinical condition of pityriasis versicolor are not yet known. A relation to squamous cell turnover has been suggested because of the occurrence of the disease in persons with increased endogenous or exogenously administered corticosteroid.[10, 33] Under such conditions, the turnover rate of the epithelium is slowed, and many patients develop pityriasis versicolor. The condition disappears when medications cease or endogenous levels are normalized. In the tropics, where a 50 per cent case rate in the population is

sometimes seen, this also could be attributed to a slowing of epithelial turnover. Other authors have suggested a genetic predisposition,[52] poor nutritional state, or the accumulation of extracellular glycogen in patients prone to the disease.[35] Poor health, chronic infections, excessive sweating, and at times, pregnancy have been regarded as predisposing factors. The low incidence of conjugal cases (7.5 per cent) also suggests individual predilection.[52] The sexes are about equally affected in adults, and the disease is of worldwide distribution and found in all races. It is quite common in temperate climates and is very prevalent in the tropics and subtropics. Case rates up to 50 per cent have been reported in Mexico, Samoa, Central and South America, India, parts of Africa, Cuba, the West Indies, and the Mediterranean region. The disease is seen in young adults and is usually established in the early twenties. It frequently becomes chronic, and there may be long periods when no lesions are seen. In temperate zones most cases are seen in the summer and autumn.[47] As these are lipophilic yeasts, there has been a suggestion that oil baths, skin lubricants, and the like may contribute to the condition.[4] In some climates, a severe form of the disease occurs in infants as a depigmenting diaper rash (achromia parasitica). *M. ovalis* is also capable of producing the same clinical condition, especially on the face, eyelids, and ears. Some investigators are of the opinion that there are only two species of the genus *Malassezia: M. furfur* and *M. pachydermitis.* They contend that *M. ovalis* is only a variety of *M. furfur.* More investigation should resolve this point.

CLINICAL DISEASE

Pityriasis Versicolor. The clinical lesions are very easy to diagnose, and the term versicolor is particularly suitable. The color varies according to the normal pigmentation of the patient, exposure of the area to sunlight, and the severity of the disease. Fawn, yellow-brown, or dark brown patches may be observed, or the lesion may consist of one continuous scaling sheet, involving the entire chest, trunk, or abdomen (Fig. 7–1). The lesions start as tiny, multiple, macular spots that soon scale and enlarge. They may coalesce to form gyrating areas of intermittent scaling of various shades and colors. Papular lesions are sometimes seen and are usually perifollicular (Fig. 7–2). There is seldom any irritation or inflammatory response, but in some patients a little reddening and some pruritus may occur. Many patients do not become aware of their condition until after exposure to sunlight. The chief complaint is a cosmetic one, in which areas of the skin fail to tan normally. The fungus or its products filter the rays of sunlight and interfere with

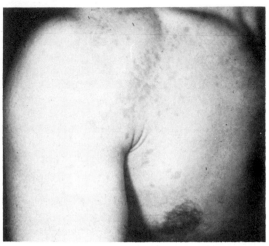

Figure 7–2. Pityriasis versicolor. The common form of the disease, represented by multiple punctate scaling macules with a slight fawn discoloration.

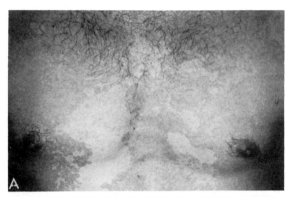

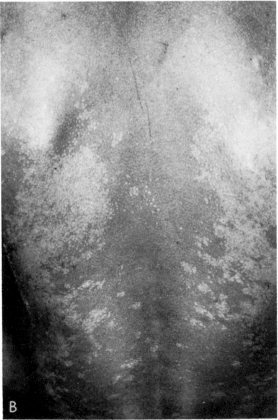

Figure 7–1. Pityriasis versicolor. *A,* Yellow-brown macules with scaling as a continuous sheath, involving the entire chest. Light areas of uninfected normal skin are in the center. *B,* Achromia parasitica. This patient had severe pityriasis as a child and now has a chronic case. In this form the light, unpigmented areas represent the formerly infected sites of the skin which are hypopigmented.

normal tanning, so that lesions are outlined as lighter areas of color than the surrounding skin. In chronic disease, hypopigmented areas (younger lesions) and hyperpigmented areas (older lesions) may coexist on the same patient.[4]

In dark-skinned infants, particularly in the tropics, a clinical variant is sometimes seen. The infection starts in the diaper-covered areas and spreads rapidly, causing marked depigmentation of the skin. This condition is referred to as pityriasis versicolor alba or achromia parasitica.

Folliculitis. More severe forms of disease are now attributed to *M. furfur.* It appears that in certain individuals *M. furfur* and probably *M. ovalis* may grow in large numbers in and about hair follicles and sebaceous glands. In histologic examination organisms are seen in the follicular ostia, the infundibular portion of the pilosebaceous canal, and often the surrounding dermis. The follicles are dilated, usually plugged, and contain keratinous debris. An inflammatory infiltrate is seen and the follicular epithelium may be ruptured. Neutrophils, lymphocytes, histiocytes, plasma cells, eosinophils, and foreign body giant cells in varying proportions constitute the cellular infiltrate. Clinically, the lesions are erythematous follicular papules or pustules 2 to 4 mm in size, distributed on the back and chest and occasionally on the shoulders, arms, neck, and flank.[49] This type of disease is found in young adults. A more severe form called acniform pityriasis folliculitis has also been described. In such patients abscesses containing the organism are found in the acroinfundibulum.[11]

Obstructive Dacryocystitis. *M. furfur* may colonize the lacrymal sac, causing enlargement and obstruction. In such cases, dacryoliths often develop, leading to inflammation and interference with normal tearing.

Differential Diagnosis. The diagnosis of pityriasis versicolor is quite easy to make by using a potassium hydroxide preparation of skin scrapings or by observing the yellow fluorescence of the affected area under the Wood's lamp. However, in some instances if organisms are sparse, microscopic observation of the fungus may be difficult. In addition, home treatment, especially application of ointments, may interfere with Wood's lamp examination. The infection may be confused with vitiligo or other pigmentary disorders, such as chloasma. In contrast to pityriasis versicolor, tinea circinata, seborrheic dermatitis, pityriasis rosea, erythrasma, secondary syphilis, and pinta are usually somewhat inflammatory. Pityriasis rosea is usually more acute, with rapid spreading after the appearance of the herald spot. Erythrasma is generally restricted to body folds and intertriginous areas. Demonstration of the scaly nature of pityriasis versicolor may require light scratching with a microscope slide, scalpel, or fingernail (the *coup d'ongle* of Besnier). The desquamative lamina are very characteristic of pityriasis versicolor. The squames can be loosened as a single unit encompassing the entire area of the lesion, if small. This can be done with stripping tape, and the squames subsequently stained by crystal violet, methylene blue, or iodine. In this manner the organisms may be readily seen. Most other dermatoses that simulate pityriasis versicolor have irregular or distorted scaling. If observation of the lesion is not hindered by obstructive medications, Wood's lamp examination will show a golden yellow fluorescence, usually in an area which is much greater than the apparent lesion.

Prognosis and Therapy. Numerous topical agents have been used for treating pityriasis versicolor with qualified success. With or without therapy, the lesions persist, spread, disappear, and reappear or become continuous and chronic. Keratolytic agents, such as Whitfield's ointment or salicylic acid, mild fungicides, such as sodium hyposulfite (20 per cent aqueous solution), or 2 per cent sulfur in ointment base will control and cure lesions. Recurrences are common. The treatment of choice appears to be a combination 25 per cent sodium thiosulfate (USP), and 1 per cent salicylic acid (USP) in a vehicle containing 10 per cent isopropyl alcohol (NF), propylene glycol (USP), menthol (USP), disodium edetate, colloidal alumina, and water. This preparation is applied to the affected areas as a thin film twice a day or more. The lesions clear in a week to several weeks. The patient usually remains free of recurrent lesions for a year or more. It is necessary to reassure the patient that the disease is not a serious mycosis and that recurrences are common and to be expected.

If patients object to the sulfur odor of the preceding preparation, miconazole (as micatin nitrate) is an excellent treatment.[46, 59] Selenium disulfide, 1 per cent, in a shampoo or water miscible base has been widely used, though some studies show disappointing results.[47] In particularly difficult cases (achromic or treatment resistant forms), retinoic acid (0.005 per cent as cream or lotion) has achieved success.[25] This drug has many side effects, however.

PATHOLOGY

The hypha and budding yeasts of *M. furfur* are generally restricted to the outermost layers of the stratum corneum (Fig. 7–3). They may be present in such large quantities that they appear as a continuous layer cemented together and replacing the outer cells of the corneum. Since there is little if any irritation, no pathology of the skin is seen in

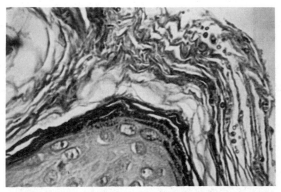

Figure 7–3. Pityriasis versicolor. Yeasts and mycelium in outer layers of stratum corneum. In this case, the lipophilic yeasts were most prevalent at the openings of sebaceous glands. Hematoxylin and eosin stain, ×400.

biopsy sections. The organisms are sometimes seen to mass near the hair follicle and extend into the follicular opening. Cells of the organism may be seen incidentally in biopsies from unrelated dermatoses and cause confusion in diagnosis. The yeast cells are up to 2 to 5 μ in diameter, and the hyphal fragments are 2 to 3 μ in width.

The yeast cells tend to be more delicate than *Candida,* although *M. furfur* cells up to 8 μ are seen. In biopsy, *M. furfur* and *M. ovalis* are seen to vary considerably in size. It is usually easy to distinguish these organisms from other fungi such as *Candida* or dermatophytes, which also may be seen in skin biopsies.

BIOLOGICAL STUDIES

A common observation in reviewing a series of cases of pityriasis versicolor is that both hypopigmented and hyperpigmented lesions occur, sometimes on the same patient. In studies by Nazzaro-Porro,[42, 43] some lipid fractions that were produced when *M. furfur* was metabolizing vaccenic and oleic acid were found to inhibit the dopa tyrosinase reaction. These fractions were C_9 to C_{11} dicarboxylic acids. Allen et al.[3] found an increased turnover rate of the stratum corneum (approximately 8 days versus 15 days for normal areas) in hyperpigmented lesions of pityriasis. Larger, singly distributed melanosomes were present in these areas, compared with smaller, multiple-packaged melanosomes in normal areas. They concluded that the brown color of lesions was due to an increase in melanosome size and a change in its distribution. Charles et al.[19] found abnormally small melanosomes in hypopigmented lesions. There is also some indication that *M. furfur* can activate the alternate complement pathway and induce psoriasis-like lesions.[54a]

The *in vitro* nutrition of *M. furfur* has been investigated. Nazzaro-Porro et al.[43] found fatty acids essential for growth. Most saturated or unsaturated fatty acids from the C_{12} to C_{24} series will support replication of the organism. Elongation of most strains to the mycelial form is accomplished on a medium containing salts, glucose, glycine, and Tween 80.[22] Ultrastructural studies on the organism in lesion material has been carried out by Pierard and Dock.[48] Experimental infections in guinea pigs using both *M. furfur* and *M. ovalis* were accomplished Drouhet et al.[23]

IMMUNITY AND SEROLOGY

It appears that normal skin with a normal epithelial turnover rate precludes a disease process by these members of the normal flora of the skin. If circumstances occur that result in the disease pityriasis versicolor, some hypersensitivity to fungal products does develop in most patients. Antibodies are detectable in chronic cases. The indirect immunofluorescent technique may be used to stain the organisms in skin scales and to verify identification of cultures of *M. furfur.*[36]

LABORATORY IDENTIFICATION

Direct examination of skin scales in KOH preparations shows the pathognomonic cluster of round, budding yeast cells (up to 8 μ in size, though averaging 4 μ) and the short,

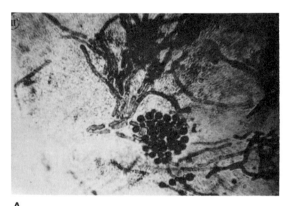

A

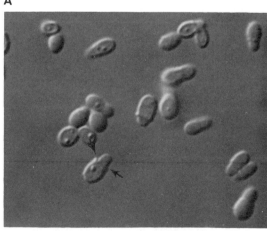

B

Figure 7–4. *A,* Pityriasis versicolor. Skin scales stained with methenamine silver. Short, branched mycelium and grouped, small yeast cells are diagnostic of this disease. × 400. *B,* Yeast cells of *Malassezia furfur.* From this picture it can be seen that the budding is phialidic. × 1100. (Courtesy of M. McGinnis.)

septate, occasionally branched, hyphal fragments (Fig. 7–4). Under rare circumstances, either the yeast form will predominate or the hyphal form will predominate. In culture the yeast cells are seen to be reduced phialides (Fig. 7–4, B).

Culture is not necessary for diagnosis, but the organism grows well at 37°C on malt agar or Sabouraud's agar containing streptomycin, penicillin, and Actidione and covered with a layer of olive oil. The colonies consist of yeastlike cells with only a rare elongation to form hyphal elements. Some strains are regularly more hyphal in nature, and there is speculation that these strains may be more likely to produce pityriasis versicolor in susceptible patients.[52]

MYCOLOGY

Malassezia furfur (Robin) Baillon 1889

Synonymy. *Microsporon furfur* Robin 1853; *Sporotrichum (Microsporon) furfur* Saccardo 1886; *Oidium (Microsporon) furfur* Zopf 1890; *Oidium subtile* Kotliar 1892; *Malassezia tropica* Castellani 1919; *Malassezia Macfadyeni* Castellani 1908; *Microsporum mansonii* Castellani 1908; *Cladosporium mansonii* Pinoy 1912; *Foxia mansonii* Castellani 1908; *Monilia furfur* Vaillemin 1931; *Pityrosporum orbiculare* Gordon 1951.

Two other species of this genus have been described. *M.* (or *P.*) *ovalis* and *M. pachydermatis*. *M. ovalis* is probably a synonym of *M. furfur*. *M. pachydermatis* is common on animal skin.

Tinea Nigra

DEFINITION

Tinea nigra is a superficial asymptomatic fungus infection of the stratum corneum characterized by brown to black nonscaly macules (Fig. 7–5). The palmar surfaces are most often affected, but lesions may occur on the plantar and other surfaces of the skin. The etiologic agent is *Exophiala werneckii* and perhaps other species of dematiaceous fungi.[12]

Synonymy. Tinea nigra palmaris, keratomycosis nigricans, cladosporiosis epidemica, pityriasis nigra, microsporosis nigra.

HISTORY

Tinea nigra was first described in 1898 as Caraté noir[40] by Montoya y Flores in Colombia. However, the description and illustrations of "Montoyella nigra" by Castellani in 1905 appear to be of pityriasis versicolor.[16] Castellani claimed that Manson first noted the disease in 1872 and in a later paper called the organism *Cladosporium mansonii*.[17] This name is now considered a synonym of *Malassezia furfur*, the etiologic agent of pityriasis versicolor.[37] The first authentic description of the disease was made by Alexandre Cerqueira (1891) in Bahia, Brazil. He gave it the name keratomycosis nigricans palmaris. However, it was not until 1916 that these findings, along with eight other cases, were published by his son, Castro Cerqueira-Pinto.[18] In 1921, Parreirus Horta discovered

the first case in Rio de Janeiro and subsequently isolated the fungus. The black dematiaceous fungus was called *Cladosporium werneckii*.[31] At present the accepted designation is *Exophiala wernickii* (Horta) v. Arx.[37]

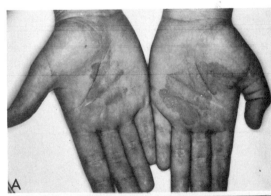

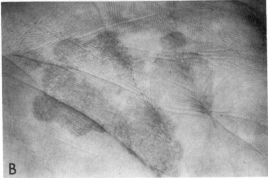

Figure 7–5. *A,* Tinea nigra. Brown-black, nonscaly macules. Bilateral disease. *B,* Enlarged area shows mottled olive-brown macules. The darkest areas are the advancing borders. There is no inflammatory reaction. (Courtesy of S. Lamberg.)

The work of Arêa Leão[5] and Aroeira Neves and Costa[6] in Brazil and Carrión[15] in Puerto Rico have clearly delineated the condition.

ETIOLOGY, ECOLOGY, AND DISTRIBUTION

The first name given by Castellani was *Cladosporium mansonii* (he termed *Montoyella nigra* a "temporary species"), and for some time this was the organism associated with infections in Asia. It appears that the single species, *Cladosporium wernickii* Parreiras Horta, 1921,[31] adequately and correctly defined the etiologic agent of the disease, and that *C. mansonii* is, in fact, a synonym of *Malessezia furfur*.[38a] The etiologic agent of tinea nigra is now placed in the genus *Exophiala* Carmichael. Organisms similar to *E. werneckii* are very abundant in soil, sewage, decaying vegetation, and humus. They also grow on wood and paint in humid environments and on shower curtains. The species is extremely variable and definitive classification is difficult. Some authors maintain that there are several species involved[8, 12]; others have suggested that the organism is identical to the black yeast, *Aureobasidium (Pullularia) pullulans*. Cooke[21] has found differences that are sufficient to separate *E. werneckii* from *A. pullulans,* however. It is possible that other species of dematiaceous fungi may induce tinea nigra, such as *Stenella araguata*. This organism was isolated from clinical disease by Borelli, who named it as a new species *Cladosporium castellani,*[12] but it appears to be a varient of *S. araguata*.*

The condition tinea nigra is generally considered to be a tropical disease and is common in Central and South America, Africa, and Asia. However, 15 cases were reported by Van Velsor[61] in North Carolina, and with increased awareness of the condition, new cases are being discovered more frequently in the United States and Europe.[58] About 75 cases acquired in North America have been recorded since 1950; the vast majority of infections, however, are traceable to visits in the American tropics and Caribbean islands. Most patients are less than 19 years old, but any age group may be involved, since the condition tends to be chronic. In most studies, affected females outnumber males three to one. No predisposing factors have been delineated, and impairment of the immune system does not appear to be relevant. Many patients were noted to be hyperhidrotic.

CLINICAL DISEASE

The lesions of tinea nigra are painless macules which are neither elevated nor scaly. They are sharply marginated and usually single. The condition often begins with the appearance of a light brown macule that spreads centrifugally and darkens. Rarely, eccentric areas of scaling are seen. The color is usually mottled, with deeper pigmentation seen at the advancing periphery. The most common site of infection is the palmar surface and the fingers, but infection of other areas has been reported, including plantar surfaces, neck, and thorax.

The primary importance of tinea nigra is that it is often misdiagnosed as other conditions, particularly malignant melanoma.[58] Such instances have resulted in surgical mutilation and debilitation of the patient, when a simple skin scraping and KOH mount examination would have revealed the true nature of the condition. As has been pointed out repeatedly, diseases which were once considered exotic will continually be appearing in local clinics as a result of jet age, worldwide travel. Not infrequently, vacationers bring home tinea nigra from the beaches of the Caribbean Islands, and the disease is no longer uncommon in the United States. The condition also simulates junctional nevus of the palm, contact dermatitis, pigmentation of Addison's disease, postinflammatory melanosis, melanosis from syphilis, pinta, and staining due to chemicals, dyes, and pigments.

Prognosis and Therapy. Tinea nigra responds readily to keratolytic agents. Daily applications of Whitfield's ointment will clear the condition readily. Tincture of iodine, 2 per cent salicylic acid, or 3 per cent sulfur is also effective. Griseofulvin appears to be ineffective. The application of 10 per cent thiabendazole suspension twice daily has been found effective,[14] as has miconazole nitrate in a cream base.

There is no tendency for this condition to recur except by reexposure to contaminated material.

PATHOLOGY

There is little or no reaction to the infection. Slight, abnormal thickening of the stra-

---

*Mycotaxon, 3:415–418, 1978.

tum corneum is sometimes seen with separation of the cells because of the large quantity of fungal elements present. Multiple-branched, brown hyphal filaments appear in masses in the upper layers of the stratum corneum, but the stratum lucidum is spared. Biopsy shows small areas of parakeratosis and a small amount of perivascular infiltrate.

The disease is not known in animals, but experimental infections in man and guinea pigs are easily induced by scarifying the skin, rubbing a culture of *E. werneckii* on the area, and then covering it with a bandage. The incubation period is 10 to 15 days.

No serologic or immunologic studies have been reported. There appears to be no association with other disease states and no genetic predisposition. The reports of several family members with the condition probably reflects common exposure.

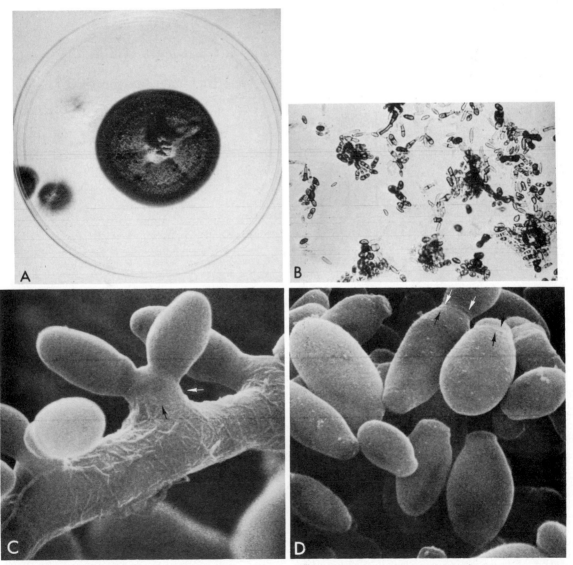

Figure 7–6. *A*, Colony of *Exophiala werneckii*. Colony after five weeks' growth on Sabouraud's glucose agar. Blackish gray mycelium covers an olive-black, moist colony. *B, Exophiala* typical two-celled yeast cells are producing annelloconidia. Some of the cells are deeply pigmented. ×400. *C*, Conidia of *Exophiala werneckii* are produced from intercalary annellides and from short lateral branches (*arrow*). ×6300. *D*, Seceded conidia are often yeastlike annellides and produce new conidia by polar budding. Such successive budding results in annellations (*arrows*). ×7500. (From Cole, G. 1978. Conidiogenesis in the Black Yeasts. *In* The Black and White Yeasts. Pan American Health Org. Sci. Pub. #356.)

LABORATORY IDENTIFICATION

For direct examination, epidermal scrapings are placed on a microscope slide with a drop of 10 per cent KOH and cleared by heating gently for a minute. Examination with the low power lens reveals brownish to olivaceous, multiple-branched, septate hyphae and budding cells. The hyphal elements are up to 5 μ in diameter. Terminal portions of the hyphae are usually hyaline. Older sections are twisted and tortuous, with numerous septations and thickening of the cell walls that become deeply pigmented. Chlamydoconidia, swollen cells, yeastlike cells, and fragmented hyphae are also seen. The mycelia differ from dermatophyte hyphae in that the latter are colorless (hyaline), usually are not so branched, and do not show the tapering contour of the terminal branches.

MYCOLOGY

Exophiala wernickii (Horta) von Arx 1970

Synonymy. *Cladosporium wernickii* Horta 1921; *Dematium werneckii* (Horta) Dodge 1935; *Pullularia werneckii* (Horta) de Vries 1952; *Cryptococcus metaniger* Castellani 1927; *Cladosporium metaniger* (Castellani) Ferrari 1932; *Pullularia fermentans* Wynne et Gott var. *castellanii* Wynne et Gott 1956; *Pullularia fermentans* Wynne et Gott var. *leaoi* Wynne et Gott 1956; *Careté noir* Montoya y Flores 1898.

Culture. In culture the organism grows slowly as a shiny, moist, adherent, yeastlike colony. Initially the colony is brownish, but it rapidly becomes olive to shiny greenish black

(Fig. 7–6). Three weeks may be necessary before growth is initiated. At this stage, microscopic examination shows the colonies to be composed primarily of two-celled, cylindrical to spindle-shaped yeastlike cells that taper toward the ends, which bear annellations. They are 3 to 10 μ in diameter. Some of the cells have a central cross wall. As the colony ages, one observes the development of an elongate mycelium, which is tortuous and has numerous septations. It becomes pigmented and develops a sleeve of conidia from undifferentiated, usually intercalary conidiogenous cells. The conidiogenous cells are annellides. Conidia are produced percurrently. At this stage the mycelium may be very thick (up to 7 μ), with thick-walled, squarish, pigmented cells in a chainlike arrangement (Fig. 7–6). The conidia are numerous and may develop from annellides anywhere on the mycelium. They vary from white elliptical cells (1 $\times$ 3 μ) to olive-colored, are one-celled and occasionally two-celled (2 to 4 $\times$ 20 μ). As the colony ages, a green-gray to gray-black fuzz develops over it. Sometimes a few chlamydoconidia with dark thick walls appear throughout the colony. Such conidia are singular and have one or more septations. By varying nutrition and environment, the organism will appear more mycelial or more unicellular.[32] Cysteine and CO_2 favor single cell growth, while O_2 and N_2 induce long mycelial strands.

Cladosporium mansonii Pinoy 1912 is now considered a synonym of *Malassezia furfur*. In a recent publication, Hermanides-Nijhof has put this organism in synonymy with *Sarcinomyces crustaceus* Lindner 1898, an organism on dried wood isolated from air.[29] The validity of this change has not been widely accepted.[37]

Piedra

DEFINITION

Piedra is a fungus infection of the hair shaft characterized by the presence of firm, irregular nodules. The nodules are composed of fungal elements cemented together anywhere along the hair shaft, and multiple infections of the same strand are common. Two varieties of piedra are recognized: white piedra, caused by *Trichosporon beigelii*, and black piedra, caused by *Piedraia hortae*.

Synonymy. Tinea nodosa, molestia de Beigel, trichomycosis.

HISTORY

The disease was first described as a fungus infection by Beigel in 1865 in his text[9] *The Human Hair: Its Growth and Structure*. He isolated the fungus (the "chignon fungus"), and Rabenhorst named it *Pleurococcus beigelii* in 1867.[50] Beigel was probably working with a contaminant fungus, for although his description of the infection in hair is accurate, the organism in his illustrations appears to be *Aspergillus*. The black variety of piedra was described by Malgoi-Hoes in 1901, and the two clinical types were fully differentiated by Horta in 1911.[30] He called the fungus he isolated *Trichosporon* sp. In 1913, Brumpt named the organism *Trichosporon hortai*. Fonseca and Arêa Leão, in 1928,[28] renamed the organism *Piedraia hortae* (as *hortai*) because a sexual stage was discovered and its relation to the Ascomycota became apparent.

ETIOLOGY, ECOLOGY, AND DISTRIBUTION

Piedraia hortae appears to be related to the Asterineae, a subfamily of loculoascomycetes that form hard masses on plants and trees in tropical climates.[37a] The disease is found commonly throughout the tropical areas of South America, the Far East,[1] and the Pacific Islands, but only sporadically in Africa and the rest of Asia. It is regularly found on monkeys and other primates in these regions. In a study of the pelts of primates in museum collections, Kaplan[34] and Ajello[2] found the infection to be very common. Sometimes the infection is encouraged as a mark of beauty by native peoples. They will purposely not oil their hair (usually done to discourage lice) and will sleep with their heads in a depression in the earth. Under these circumstances, multiple infections are seen, but the greater the number of nobs on the hair shafts, the more pulchritudinous the person is considered. Sometimes these nobs have a religious association.[41] Though scalp hair is most often affected, rare cases of infection on other body regions are reported.

White piedra is common on the temperate periphery of the tropical black piedra belt. The scalp is less frequently involved, and infection is most common in the bearded regions, the axilla, and the groin. The condition occurs sporadically in the United States and Europe, and more commonly in South America and the Orient. Infection in domestic animals, especially the horse, has been noted. *Trichosporon beigelii* and similar organisms are common in soil, air, sputum, and body surfaces.

The specific source of infection is unknown, but swimming in stagnant water has been suggested as a possibility, as the organism occurs in this environment in great numbers.

CLINICAL DISEASE

There is no discomfort or physical reaction on the part of the patient. The disease is only of cosmetic interest. Its chief importance is in differentiation from pediculosis. The nodules of black piedra are gritty, hard, brown to black encrustations that vary in size from microscopic to a few millimeters and in depth up to 150 μ. The infection starts under the cuticle of the hair shaft. As interpilar growth occurs and continues, the shaft may rupture and be weakened so that breakage may occur. The mass may enlarge and grow on the outside of the hair and completely envelop the hair shaft. In a mature nodule, the periphery is composed of aligned hyphal strands, whereas the cells in the central area

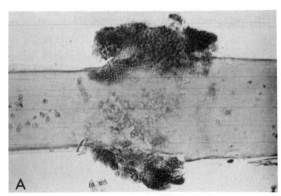

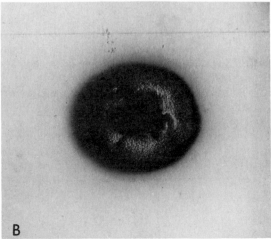

Figure 7–7. *A,* Piedra. Black nodules on hair shaft. Aligned hyphal strands on the periphery of the nodules and a stromalike center containing asci. *B, Piedraia hortae.* Dark, brown-black colony, which is elevated and cerebriform in the center and flat at the periphery. Note rusty-red, soluble pigment.

are cemented together to form a pseudo-parenchymatous mass which resembles organized tissue. Within this stroma are locules in which asci are produced (Fig. 7–7). In profile, the nodule appears like a mountain range in the center with a flat surrounding peripheral plane. Many cases have been reported in which white areas or sections that appeared to be amorphous debris were also seen on the hair shaft. The white areas usually represent colonization by *Trichosporon beigelii* in addition to *P. hortae,* so that mixed black and white piedra may occur. In the past some investigators considered *T. beigelii* to be the "imperfect" form of *P. hortae,* probably resulting from the observation of both organisms on the same hair shaft. In other cases the hair shaft may have areas colonized by the agents of trichomycosis axillaris (probably an actinomycete), as well as gram-positive cocci and general "zooglea." For these reasons the nodules seen in piedra are quite variable in appearance.

White piedra is characterized by the presence of a softer granule, which is white to light brown. The infection again appears to start beneath the cuticle, possibly following damage. The organism may grow inward and through the shaft to form nodular swellings spaced irregularly along the axis. The hair is weakened at these points and may break. Growth may occur primarily around the hair shaft, forming a soft sleeve of intertwined hyphae that fragment into yeastlike arthroconidia (Fig. 7–8). The regular cellular pattern of black piedra is not seen, nor are the nodules as adherent, for they may be stripped off the hair with ease. Several nodules may coalesce to form an extensive mass surrounding the hair shaft. It appears that infection begins soon after emergence of the hair from the follicle, the granule forming and hardening as the hair grows. The most common condition to be considered in the differential diagnosis is the presence of nits and lice. Microscopic examination will quickly rule out this condition. Piedra may simulate such developmental anomalies as monilethrix trichoptilosis and trichorrhexis nodosa. Trichomycosis axillaris may be distinguished by microscopic examination, since hyphal elements are 2 to 4 μ in piedra and 1 μ or less in trichomycosis axillaris. Piedra does not fluoresce under ultraviolet illumination as many trichomycosis axillaris infections do. The ovoid cells of black piedra may resemble the arthroconidia of dermatophytes. In piedra, the hair shaft is normal on either side of

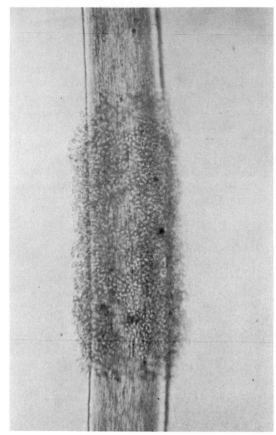

Figure 7–8. White piedra. Soft sleeve of intertwined hyphae and arthroconidia around the hair shaft.

the cell mass, whereas in tinea capitis the base of the hair shaft and the follicle are involved.

If therapy is desired, it may be achieved simply by shaving or cutting the infected hair. Topical fungicides, such as bichloride of mercury (1:2000), benzoic and salicylic acid combinations, 3 per cent sulfur ointment, and 2 per cent formalin, have been used.

Since, as is the case with the other superficial infections, there is no pathology or inflammatory response, no detectable serologic reaction has been described.

LABORATORY IDENTIFICATION

As previously described, direct microscopic examination of infected material in a 10 per cent potassium hydroxide preparation will differentiate not only the two types of piedra but also the simulating diseases. Black piedra is composed of a tightly packed stroma of regularly arranged, thick-walled, rhomboid cells (resembling arthroconidia) and dichotomously branched hyphae held together by a cementlike substance. The hyphae and cells are 4 to 8 μ in diameter, and there is even pigmentation in the walls. A sectioned nodule reveals the asci within locules imbedded in the cellular stroma. There are eight fusiform, curved ascospores (30 × 10 μ), the ends of which are prolonged into a spiral filament 10 μ long.

In contrast to black piedra, white piedra granules are softer, more easily detached from hair, and not as discrete. They often form a transparent, greenish, irregular sheath along the hair shaft. In addition, growth may be almost entirely intrapilar and cause only a raised cuticle. The mycelial elements are usually perpendicular to the surface of the hair and are not in an organized structure as in black piedra. The hyphae segment into oval to rectangular cells 2 to 4 μ in diameter, with occasional cells measuring up to 8 μ. Budding blastoconidia may also be seen, and bacteria may cohabit in the fungal mass to form a zooglea. No asci are seen.

Culture. Both organisms grow on ordinary laboratory media, but *Trichosporon beigelii* is inhibited by cycloheximide in mycologic selective media (Mycosel). *Piedraia hortae* grow very slowly at 25° C, developing into small, dark brown to black, conical, adherent colonies (Fig. 7–7). The center of the colony is elevated and cerebriform, and the periphery is flat. The colony may be glabrous when young but usually develops a short, greenish-

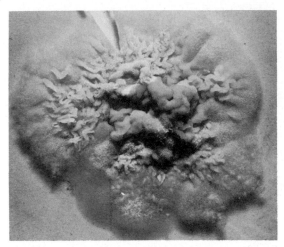

Figure 7–10. Thallus of *Trichosporon beigelii* of a buttercream color and consistency.

brown, aerial mycelium with time. A rusty red pigment may be seen in the media. Microscopic examination reveals thick-walled, closely septate hyphae, chlamydoconidia, and swollen, irregular cells. In the center of the colony, locules may be found in which asci develop in a manner similar to that seen on the nodules of the hair shaft.[56] A second species, *P. quintanilhae* Van Uden, Barros-Machado et Castelo Branco 1963,[60] has also been described. It has been recovered primarily from a variety of central African primates.[57] It differs from *P. hortae* in that it lacks terminal appendages on the ascospores.

Trichosporon beigelii grows rapidly on Sabouraud's agar, producing a cream-colored, yeastlike colony that develops radial furrows and irregular folds with age, and often separates the colony from the media (Fig. 7–9). The thallus somewhat resembles a buttercream cake frosting in color and consistency (Fig. 7–10). Old colonies may take on a gray cast. Microscopic examination shows hyaline, septate hyphae which fragment into oval or rectangular arthroconidia, 2 to 4 × 3 to 9 μ. Some blastoconidia are also seen. Electron micrographs of hyphal septa show dolipores and parenthesomes typical of the phylum Basidiomycota. It is probable that, if a perfect stage were found, it would be classified as a Basidiomycota. *T. beigelii* lacks the ability to ferment carbohydrates. It assimilates glucose, galactose, sucrose, maltose, and lactose, which distinguishes it from other members of the genus *Trichosporon*. Potassium nitrate is not utilized, and arbutin is split.

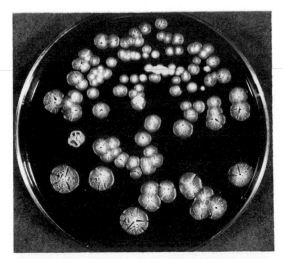

Figure 7–9. White piedra. *Trichosporon beigelii.* Soft, wrinkled, folded, friable, off-white colonies.

MYCOLOGY

Piedraia hortae (Brumpt) Fonseca and Arêa Leão 1928

Synonymy. *Trichosporon* sp. Horta 1911; *Trichosporon hortai* Brumpt 1913; *Trichosporon guayo* Delamare et Gatti 1928; *Piedraia sarmentoi* Pereira 1930; *Piedraia venezuelensis* Brumpt et Langeron 1934; *Piedraia surinamensis* Dodge 1935; *Piedraia javanica* Boedijn et Verbunt 1938.

Trichosporon beigelii (Küchenmeister and Rabenhorst) Vuillemin 1902

Synonymy. *Pleurococcus beigelii* Küchenmeister et Rabenhorst, 1867; *Sclerotium beigelianum* Hallier 1968; *Zoogloea beigelii* Eberth 1873; *Hyalococcus beigelii* Trevisan 1889; *Micrococcus beigelii* Migula 1900; *Trichosporum ovales* Paoli 1902; *Trichosporum glycophile* DuBois 1910; *Trichosporon ovoides* Behrend 1890; *Trichosporon giganteum* Unna 1895; *Trichosporon cerebriforme* (Kambayashi) Ota 1928; *Trichosporon granulosum* (Kambayashi) Ota 1928; *Trichosporon humakuaguensis* Mazza et Nino 1933; *Trichosporon minor* Arêa Leão 1940; *Piedraia colombiana* Dodge 1935; *Trichosporon cutaneum* Ota 1926.

REFERENCES

1. Adam, B. A., T. S. Soo Hoo, et al. 1977. Black piedra in west Malasia. Aust. J. Dermatol., *18*:45–47.
2. Ajello, L. 1964. Survey of tree shrew pelts for mycotic infections. Mycologia, *56*:455–458.
3. Allen, H. B., C. R. Charles, et al. 1976. Hyperpigmented tinea versicolor Arch. Dermatol., *112*:1110–1112.
4. Amma, S. M. 1978. Clinical and epidemiological studies on tinea versicolor in Kerala. Ind. J. Dermatol. Venerol. Leprol., *44*:345–351.
5. Arêa Leão, A. E., A. Curry, et al. 1945. Tinea nigra. Rev. Bras. Biol., 5:165–177.
6. Aroeira Neves, J., and O. G. Costa. 1947. Tinea Nigra. Arch. Dermatol., 55:67–84.
7. Baillon, H. 1889. Traite de Botanique Medica Cryptoganique Paris, Octave Doin Editeur, pp. 234–235.
8. Bastardo de Abornoz, M. C. 1966. Estudio micologica de las 4 primera cepas de *Aureobasidium wernecki* aislados en Venez. Med. Cutan., *4*:369–378.
9. Beigel, H. 1865. Cited by Langenon, M., in Darier, S., et al. 1936. Nouvelle Praqt. Dermatol. 2:377. Paris, Masson Cie.
10. Boardman, C. R., and Malkinson, F. D. 1962. Tinea versicolor in steroid-treated patients. Arch. Dermatol., 85:44–52.
11. Boganousky, A., and G. Lischka. 1977. *Pityrosporum orbiculare* bei akneiformen Eruptionen. Hautarzt, *28*:409–411.
12. Borelli, D., and C. Marcano. 1973. *Cladosporium castellanii* nova species agente de tinea nigra. Castellania, *1*:151–154.
13. Burchard, M., and V. Barensprung. 1859. *In* Uhle and Wagoner. Patologie Générale. Quoted in Balzer, F. 1883. De l'érythrasma. (*Microsporum minutissimum*). Ann. Rev. Dermatol. Syphiligr., II, *4*:681–688.
14. Carr, J. F., and C. W. Lewis. 1975. Tinea nigra palmaris. Treatment with thiabendazole topically. Arch. Dermatol., *111*:904–905.
15. Carrion, A. L. 1950. Yeast-like dematiaceous fungi infecting human skin. Arch. Dermatol., *61*:996.
16. Castellani, A. 1905. Tropical forms of pityriasis versicolor. Br. Med. J., *2*:1271–1272.
17. Castellani, A., and A. J. Chalmers. 1913. Tinea nigra. *In* Manual of Tropical Medicine. 2nd ed. London, Bailliere, Tindall and Cox.
18. Cerqueira-Pinto, A. G. C. 1916. Keratomycose nigricans palmar. Tése Fac. Med. Bahia, Brazil.
19. Charles, C. R., D. J. Sire, et al. 1973. Hypopigmentation in tinea versicolor: a histochemical and electromicroscopic study. Int. J. Dermatol., *12*:48–58.
20. Chong, K. C., B. A. Adam, et al. 1975. Morphology of *Piedraia hortae*. Sabouraudia, *9*:157–160.
21. Cooke, W. B. 1959. An ecological life history of *Aureobasidium pullulans* (de Bary). Arnaud. Mycopathologia, *12*:1–45.
22. Dorn, M., and K. Roehnert. 1977. Dimorphism of *Pityrosporum orbiculare* in defined culture medium. J. Invest. Dermatol., *69*:244–248.
23. Drouhet, E., D. Dompmartin, et al. 1977. Obtention chez le cobaye d'une dermatite a *Pityrosporum ovale* et *P. orbiculare*. Bull. Soc. Fr. Mycol. Med., *6*:167–171.
24. Eichstedt 1846. *In* Froriep's Notizen.
25. Fornasa, C. V., D. Simonetto, and C. de Rossi. 1978. Nuova terapia della pitiriasi veriscolor. G. Ital. Dermatol. Min. Dermatol., *113*:536–537.
26. Fonseca, O. da, and A. C. de Area Leao. 1928. Sobre os cogumelos da piedra Brasileira. Mem. Inst. Oswaldo Cruz Suppl. das Memorias no. 4 des de 1928.
27. Gordon, M. 1951. Lipophilic yeastlike organisms associated with tinea versicolor. J. Invest. Dermatol., *17*:267–272.
28. Gustafson, R. A., R. V. Hardcastle, et al. 1975. Budding in the dimorphic fungus *Cladosporium werneckii*. Mycologia, *67*:943–951.
29. de Hoog, G. S., and E. J. Hermanides-Nijhof. 1977. The Black Yeasts and Allied Hyphomycetes. Studies in Mycology No. 15. Centrael bureau voor Schimmel cultures. Baarn, Netherlands, pp. 141–177.
30. Horta, P. 1911. Sobre una nova forma de piedra. Mem. Inst. Oswaldo Cruz, *3*:87–88.
31. Horta, P. 1921. Sobre um caso de tinha preta e um novo cogumelo (*Cladosporium Wernecki*) Rev. Med. Cirug. Brazil., *29*:269–274.
32. Houston, M. R., K. H. Meyer, et al. 1969. Dimorphism in *Cladosporium werneckii*. Sabouraudia, *7*:195–198.
33. Jung, E. G., and Truniger, B. 1963. Tinea versicolor and Cushing-syndrome (Pityriasis versicolor and Cushing's syndrome). Dermatologica, *127*:18–22.

34. Kaplan, W. 1959. The occurrence of black piedra in primate pelts. Trop. Geog. Med., *11*:115–126.

35. Keddie, F. M. 1969. A novel reaction caused by tinea versicolor: Extracellular glycogen deposits. J. Invest. Dermatol., *53*:363–372.

36. Keddie, F., and Shadomy, S. 1963. Etiological significance of *Pityrosporum orbiculare* in tinea versicolor. Sabouraudia, *3*:21–25.

37. McGinnis, M. R. 1979. Taxonomy of *Exophiala werneckii* and its relation to *Microsporum mansonii*. Sabouraudia, *17*:145–154.

37a. Mackinnon, J. E., and G. B. Schouton. 1942. Investigaciones sobre las enfermedades de los cabellas de nominodas "piedra." Arch. Soc-Biol. Montevideo., *10*:227–266.

38. Malassez, L. 1874. Note sur le champignon da pityriasis simple. Ecole Prat. Hautes Etudes, Lab d' Histol. Coll. France Trav. *1*:170–183 pl. 9.

39. Menezes-Fonseca, O. J de, and S. A. Pecher. 1971. Tinea Nigra no Amazonas. Acta Amazonica, *1*:55–57.

40. Montoya y Flores, A. 1898. Recherches sur les Caratés de Colombie. These Med. Paris, *25*:48–49.

41. Moyer, D. G., and C. Keeler. 1964. Note on culture of black piedra for cosmetic reasons. Arch. Dermatol., *89*:436.

42. Nazzaro-Porro, M., S. Passi, et al. 1976. Growth requirements and lipid metabolism of *Pityrosporum orbiculare*. J. Invest. Dermatol., *66*:178–182.

43. Nazzaro-Porro, M., S. Passi, et al. 1977. Induction of hyphae in culture of *Pityrosporum* by cholesterol and cholesterol esters. J. Invest. Dermatol., *69*:531–534.

44. Nazzaro-Porro, M., and S. Passi. 1978. Identification of tyrosinase inhibitors in cultures of *Pityrosporum*. J. Invest. Dermatol., *71*:205–208.

45. Noble, W. C., and G. Midgley. 1978. Scalp carriage of *Pityrosporum* species: the effect of physiological maturity, sex and race. Sabouraudia, *16*:229–232.

46. Ortiz, Luis G., and C. M. Papa. 1978. Topical micanazole nitrate therapy in tinea pedis and tinea versicolor. Clin. Ther. *1*:444–450.

47. Parisis, N., J. Stratigos, et al. 1977. Pityriasis versicolor in Griechenland und ihre Prädispositions faktoren. Hautarzt, *28*:589–492.

48. Pierard, J., and P. Dock. 1972. The ultrastructure of tinea versicolor and *Malassezia furfur*. Int. J. Dermatol., *11*:116–124.

49. Potter, B. P., C. F. Burgoon, et al. 1973. Pityriasis folliculitis. Arch. Dermatol., *107*:388–391.

50. Rabenhorst, L. 1867. Zwei parasiten an den todten haaren der chignons. Hedwigia, *6*:49.

51. Reyes, O., and D. Borell. 1974. Caso de tina negra por cepa peculiar de *Cladosporium castellanii*. Dermatol. Venez., *13*:21–28.

52. Roberts, S. O. B. 1969. Pityriasis: a clinical and mycological investigation. Br. J. Dermatol., *81*:315–326.

53. Roberts, S. O. B. 1969. Pityrosporum orbiculare: incidence and distribution on clinically normal skin. Br. J. Dermatol., *81*:264–269.

54. Robin, C. 1853. Hist. Nat. Veg. Parasitol. P., 436–439.

54a. Rosenberg, E. W., P. Belew, et al. 1980. Effect of topical applications of heavy suspensions of killed *Malassezia ovales* on rabbit skin. Mycopathologia, *72*:147–154.

55. Sluyter, T. 1847 De vegetabilibus organismi animalis parasitis ac de novo epiphyto in pityriasi versicolore obvio. Diss. Inaug. Berlin.

56. Takashin, M., and R. Vanbreuseghem. 1971. Production of ascospores by *Piedraia hortai in vitro*. Mycologia, *63*:612–618.

57. Takashio, M., and C. De Vroey. 1975. Piédra noire chez des Chimpanzés du Zaire. Sahouraudia, *13*:58–62.

58. Vaffee, A. S. 1970. Tinea nigra palmaris resembling malignant melanoma. New Engl. J. Med., *283*:1112.

59. Van Cutsen, J., and A. Reyntjens. 1978. Miconazole treatment of pityriasis versicolor. A review. Mykosen, *21*:87–91.

60. Van Uden, N., A. De Barros-Machado, et al. 1963. On black piedra on central African mammals caused by the Ascomycete *Piedraia quintanilhae* nov. sp. Rev. Brasiliera Portugesa. Biol. Geral., *3*:271–276.

61. Van Velsor, H., and H. Singletary. 1964. Tinea nigra. Arch. Dermatol., *90*:59–61.

62. Von Arx, A. J. 1970. Genera of Fungi in Pure Culture, p 180.

63. Wolter, J. R. 1977. *Pityrosporum* species associated with dacryoliths in obstructive dacryocystitis. Am. J. Ophthal., *84*:806–809.

8 Dermatophytosis and Dermatomycosis

INTRODUCTION

The cutaneous infections of man include a wide variety of diseases in which the integument and its appendages, the hair and nails, are involved. Infection is generally restricted to the nonliving cornified layers, but a variety of pathologic changes occur in the host because of the presence of the infectious agent and its metabolic products. The majority of these infections are caused by a homogeneous group of keratinophilic fungi called the dermatophytes. In mycetoma it was seen that a great diversity of unrelated organisms cause essentially a single clinical entity. The dermatophytes, however, are a very similar and closely related group of fungi that cause a wide variety of clinical conditions. A single species may be involved in several disease types, each with its distinctive pathology. These fungi are among the commonest infectious agents of man, and no peoples or geographic areas are without "ringworm." An evolutionary development toward an accommodating host-parasite relationship can be seen among the dermatophytes which is absent among other fungal agents of human disease. This group of diseases is collectively referred to as dermatophytosis. Dermatophytosis is a clinical entity caused by members of the imperfect genera *Trichophyton*, *Microsporum*, and *Epidermophyton*.

In addition to the dermatophytic fungi, other molds are sometimes involved in cutaneous infections. These include a wide variety of soil-inhabiting yeasts and molds, and the disease produced by them is called dermatomycosis. Many reports in the literature erroneously cite contaminants and misidentified fungi as the etiologic agents in these infections. However, only a few of these cases are authentic, as dermatomycosis occurs much less frequently than dermatophytosis. Because the disease elicited by these organisms is similar to dermatophyte infections, all of them will be discussed in this chapter.

THE DERMATOPHYTOSES

Definition

In its restricted sense, dermatophytosis is a colonization by a dermatophytic fungus of the keratinized tissues — the nails, the hair, and the *stratum corneum* of the skin. The disease, if one results from such colonization, is a consequence of the host's reaction to the metabolic products of the fungus rather than to the invasion of living tissue by the organism. The severity of the disease depends on the strain or species of the dermatophyte and the sensitivity of the host to that particular fungus, along with idiosyncrasies of the individual host. Dermatophytosis is the preferred term, as it indicates a colonization or infection by a dermatophytic fungus. As previously indicated, dermatomycosis includes any fungus infection of the skin, such as secondary spread from a systemic mycosis, infection by *Candida* species, or colonization by a variety of soil-inhabiting organisms.

Dermatophytosis includes several distinct clinical entities, depending on the anatomic site and the etiologic agent involved. The pathology induced in the host initially is an eczemaform response, followed by allergic and inflammatory manifestations. The type and severity of these reactions are related to the immune state of the host, as well as to the strain and species of the organism causing the infection.

History

Because of its visibility, ringworm, like many other dermatologic conditions, has been noted and described from the earliest historical times. Growth of the fungus in skin and scalp is more or less equal in all directions, and the lesions produced tend to creep in a circular or ring form. For this reason, the Greeks named the disease herpes — a term which still persists, though modified, as *herpes tonsurans, herpes circinatus,* or *herpes desquamans.* The Romans associated the lesions with insects and named the disease tinea, meaning any small insect larva. This name is retained in the clinical terminology of the disease. Tinea also refers to a group of keratinophilic insects, the clothes moths. The English word, ringworm, then, is a combination of the meanings of the Greek and Latin terms.

One of the most disfiguring of the dermatophyte infections is the disease favus (L. honeycomb). The chronic inflammation, loss of hair, replacement of scalp by scar tissue, and formation of folded, crusted scutula have been recognized in central Europe and the Mediterranean area since classical times. In 1834, Remak examined material from favus and noted the presence of filaments resembling a mold.[215] He tried to reproduce the disease on his arm by rubbing the organism on his skin, but failed. Schoenlein[247] in 1839 described the filaments as being those of molds and concluded that favus was a disease caused by plants. The most remarkable work of that period was done by David Gruby.[97] In 1841, he published a paper describing the isolation of the fungus of favus (on potato slices) and the production of the disease by inoculation of this fungus onto normal skin. Thus he was the first to establish that a microorganism was responsible for human disease, and Koch's postulates for the criteria of the etiology of infection were fulfilled forty years before Koch formulated them. In addition, Gruby described the yeast, now called *Candida albicans,* from thrush, named the dermatophyte *Microsporum audouinii*[98] from tinea capitis, and recognized the endothrix form of trichophytosis. His work and that of Remak were generally ignored, possibly because of strong anti-Semitic feelings in medicine at that time. Schoenlein, because of such prejudice, was given most of the credit for the pioneering work on favus and the other dermatophytoses. In 1845 the fungus from favus was named *Oidium schoenleinii* by Lebert[148] and *Achorion schoenleinii* by Remak. Also in 1845, Malmsten erected the genus *Trichophyton* and described *T. tonsurans.*[156] *T. mentagrophytes* was defined in 1847 by Charles Robin. This author published, in 1853, a compilation of the early works on dermatophytoses in the book *Histoire Naturelle des Végétaux Parasites.* It was the first influential work to discuss topical therapy for dermatophyte infections and the importance of epilation in tinea capitis.

Early progress in the field became bogged down in numerous descriptions of fungi from skin infection. Small variations in clinical appearance or colony morphology seemed to warrant a new species. The literature was glutted with incomplete and inaccurate reports. By 1890 one of the great names in dermatology, Sabouraud, began to publish his systematic and scientific studies on the dermatophytoses.[241] He accumulated his work in the volume *Les Teignes,* published in 1910. This book is considered a classic in medical literature.[243] He included a classification system recognizing the three genera of dermatophytes, *Microsporum, Trichophyton,* and *Epidermophyton,*[242] along with the genus *Achorion,* which was based on clinical rather than botanical observation. His basic methodology and astute observations on treatment remained only slightly changed until the advent of griseofulvin. He again stressed epilation of hair, but by this time x-ray was practicable, and he wrote extensively on this subject. Many papers were published by him after *Les Teignes,* but these did little to clarify the taxonomy of dermatophytes. Some of his species have been reduced to synonymy, and the genus *Achorion* has been dropped. In general, as can be seen in Table 8–1, the major tenets of his classification still hold, a tribute to his intuitive insight.

In the succeeding years, the literature again became muddled and confused. Many clinicians described new species based on trivial differences in fungal colony color, morphology, conidial size or arrangement, as well as type, extent, topography, and inflammatory and therapeutic response of the disease. This was true for all other types of fungus infection as well. There are at least 172 synonyms for the yeast *Candida albicans.* A compendium of this early literature can be

Table 8–1. *The Classification of Sabouraud and its Present Equivalent*

Genus	Group	Type Species	Synonomy and Present Name
Trichophyton			
	Endothrix	*T. tonsurans*	
	Neoendothrix	*T. flavum*	*(T. tonsurans)*
	Ectothrix		
	Megaspores	*T. roseum*	*(T. megninii)*
	Faviformes	*T. ochraceum*	*(T. verrucosum)*
		T. violaceum	
	Microides		
	Gypseum	*T. mentagrophytes*	
	Niveum	*T. felineum*	*(T. mentagrophytes)*
Microsporum			
Neomicrosporum		*M. canis*	
Eumicrosporum		*M. audouinii*	
Achorion			*(Trichophyton)*
Neoachorion		*A. gallinae*	*(M. gallinae)*
Euachorion		*A. schoenleinii*	*(T. schoenleinii)*
Epidermophyton		*E. floccosum*	

found in the book *Medical Mycology*, published by C. W. Dodge in 1935.[57] This volume contains descriptions and references for thousands of species of fungi (including 118 dermatophytes) isolated from all types of human disease. He also discusses associations of certain anthropophylic fungi with particular groups or races of man, such as *Trichophyton violaceum* with Jewish and other Semitic peoples, and *Microsporum ferrugineum* with northern Chinese, Koreans, and Japanese. More recent studies tend to confirm his concepts on locally endemic dermatophytes associated with particular population groups. He also conjectures about many problems involved in immunology, pathology, and distribution of fungus diseases. In addition, many dermatologic conditions were described as having a fungal etiology that do not: namely, pemphigus, epithelioma, eczema, and so forth. Most of these reports resulted after isolation of some contaminant, or from incomplete, inaccurate, and careless observation. All of medicine suffered from such problems at that time; these have largely been eliminated today.

In the 1920's Hopkins and Benham began the scientific study of medical mycology. The laboratory at Columbia University was one of the first to systematically study fungi involved in disease. Rhoda Benham is considered the founder of modern medical mycology. In 1934, Chester Emmons redefined the dermatophytes according to the botanical rules of

nomenclature and taxonomy. He accepted the synonymy of the genus *Achorion* with the genus *Trichophyton*, as proposed by Langeron and Milochevitch in 1930[143, 144] and included all the then known dermatophytes in three genera. This was a landmark effort, as it emphasized the care and meticulous scientific observation necessary for accurate identifica-

Table 8–2. *The Classification of Dermatophyte Groups by Similarity of Colony Morphology As Proposed by Conant, 1954*[47]

Trichophyton Malmsten 1845
  I. Gypseum group
     1. *T. mentagrophytes*
  II. Rubrum group
     1. *T. rubrum*
  III. Crateriform group
     1. *T. tonsurans*
  IV. Faviform group
     1. *T. schoenleinii*
     2. *T. concentricum*
     3. *T. ferrugineum*
     4. *T. violaceum*
     5. *T. verrucosum*
  V. Rosaceum group
     1. *T. megninii*
     2. *T. gallinae**

Microsporum Gruby 1843
  M. audouinii
  M. canis
  M. gypseum

Epidermophyton Sabouraud 1907
  E. floccosum

*Now *M. gallinae.*

Table 8–3. *The Currently Recognized Dermatophytes*

Anamorph Genera and Species

Epidermophyton Sabouraud 1907
  *E. floccosum (Hartz 1870) Langeron et Milochevitch 1930
  E. stockdaleae Prochacki et Engelhard-Zasada 1974

Microsporum Gruby 1843
  M. amazonicum Moraes, Borelli, et Feo 1967
  *M. audouinii Gruby 1843
  M. boullardii Dominik et Majchrowicz 1965
  *†M. canis Bodin 1902
  M. cookei Ajello 1959
  †M. distortum DiMenna et Marples 1954
  †M. equinum (Delacroix et Bodin 1896) Gueguén 1904
  *M. ferrugineum Ota 1921
  *M. fulvum Uriburu 1909
  †M. gallinae (Mégnin 1881) Grigorakis 1929
  *M. gypseum (Bodin 1907) Guiart et Grigorakis 1928
  †M. nanum Fuentes 1956
  †M. persicolor (Sabouraud 1910) Guiart et Grigorakis 1928
  M. praecox Rivalier 1954
  M. racemosum Borelli 1965
  †M. rivariae Hubabek et Rush-Munro 1973
  M. vanbreuseghemii Georg, Ajello, Friedman, et Brinkman 1962

Trichophyton Malmsten 1845
  T. ajelloi (Vanbreuseghem, 1952) Ajello 1968

  T. concentricum Blanchard 1895
  †T. equinum (Matruchot et Dassonville 1898) Gedoelst 1902
  T. flavescens Padhye et Carmichael 1971
  T. georgiae Varsavsky et Ajello 1964
  T. gloriae Ajello 1967
  *T. gourvilii Catanei 1933
  T. longifusus (Florian et Galgoczy 1964) Ajello 1968
  *T. megninii Blanchard 1896
  *†T. mentagrophytes (Robin 1853) Blanchard 1896
    var. *mentagrophytes*
    var. *interdigitale*
    var. *erinacei*
    var. *quinckeanum*
  T. phaseoliforme Borelli et Feo 1966
  *T. rubrum (Castellani 1910) Sabouraud 1911
  *T. schoenleinii (Lebert 1845) Langeron et Milochevitch 1930
  †T. simii (Pinoy 1912) Stockdale, Mackenzie, et Austwick 1965
  *T. soudanense Joyeux 1912
  T. terrestre Durie and Frey 1957
  *T. tonsurans Malmsten 1845
  T. vanbreuseghemii Rioux, Tarry, et Tuminer 1964
  *T. verrucosum Bodin 1902
  *T. violaceum Bodin 1902
  *T. yaoundei Cochet et Doby Dubois 1957‡

*Commonly isolated from human infection.
†Commonly isolated from animal infection.
The remainder are soil keratinophilic fungi rarely if ever involved in disease. *T. thuringiense* Koch 1969 is now considered in synonymy with *T. terrestre* (fide Padhye), and *T. fluviomuniense* Miguens 1968 is a granular form of *T. rubrum* (fide Ajello), and *T. proliferans* English and Stockdale 1968 is in synonymy with *T. mentagrophytes* var. *erinacei*. Two more species, *T. fischeri* and *T. mariati*, have recently been isolated.
‡This species lacks a proper Latin diagnosis and is therefore a *nomen nudum.*

tion. Benham proposed a botanic grouping based on colony characteristics and, to a certain extent, clinical disease. In 1954, Norman Conant modified this and included it in his book, which has become a standard text in the field (Table 8–2). By means of nutritional studies, Lucille Georg[81] further clarified the identity of several organisms and established the use of physiologic characteristics and nutritional requirements in identification. She reduced the number of dermatophyte species to sixteen. With the description of new, valid species of keratinophilic soil saprophytes and skin pathogens, the number of recognized dermatophytes is now about 39. As reviewed by Ajello,[5] these include two species in the genus *Epidermophyton*, 16 in *Microsporum*, and 21 in *Trichophyton* (Table 8–3).

The next major development in the taxonomy of dermatophytes came in 1959. In this year Dawson and Gentles[53] described the perfect stage of a keratinophilic soil organism,

Trichophyton ajelloi as *Nannizzia ajelloi*. This led to the rapid discovery of the ascomycetous form of many dermatophytes (Table 8–4). Although previously in 1927 Nannizzi had described the perfect stage of *M. gypseum*,[177, 178] but his work was severely attacked by Langeron and Milochevitch in 1930[144] and subsequently ignored by the scientific world. In 1960, however, Donald Griffin, using hair placed on soil, reisolated the perfect stage of *M. gypseum* in Australia, thus vindicating Nannizzi's original observation. Phyllis Stockdale in 1961 independently isolated and described another of the perfect states of *M. gypseum* as *Nannizzia incurvata*.[261] In 1963 she described the second perfect state, *N. gypsea*. This fungus has an anamorph (imperfect) state identical to *N. incurvata* and is the one recovered by Griffin. Stockdale thus demonstrated that an imperfect taxon (species) such as *M. gypseum* may be the anamorph of more than one teleomorph (sexual) species.

Table 8–4. *The Ascigerous Genera and Species of Dermatophytes*

Teleamorph State	Anamorph State
Arthroderma	*Trichophyton*
A. benhamiae Ajello et Cheng 1967	T. mentagrophytes
A. ciferrii Varsavsky et Ajello 1964	T. georgiae
A. flavescens Padhye et Carmichael 1971	T. flavescens
A. gertleri Bohme 1967	T. vanbreuseghemii
A. gloriae Ajello 1967	T. gloriae
A. insingulare Padhye et Carmichael 1972	T. terrestre
A. lenticularum Pore, Tsao et Plunkett 1965	T. terrestre
A. quadrifidum Dawson et Gentles 1961	T. terrestre
A. simii Stockdale, Mackenzie, et Austwick 1965	T. simii
A. uncinatum Dawson et Gentles 1959	T. ajelloi
A. vanbreuseghemii Takashio 1973	T. mentagrophytes
Nannizzia	*Microsporum*
N. borellii Moraes, Padhye, et Ajello 1975	M. amazonicum
N. cajetani Ajello 1961	M. cookeii
N. fulva Stockdale 1963	M. fulvum
N. grubia Georg, Ajello, Friedman, et Brinkman 1967	M. vanbreuseghemii
N. gypsea Stockdale 1963	M. gypseum
N. incurvata Stockdale 1961	M. gypseum
N. obtusa Dawson et Gentles 1961	M. nanum
N. otae Hasegawa et Usui 1974	M. canis
N. persicolor Stockdale 1967	M. persicolor
N. racemosa Rush-Munro, Smith, et Borelli 1970	M. racemosum

Since then, many of the imperfect or anamorphic "species" have been found to have one or more ascigerous states. This is particularly true of the soil and animal origin dermatophytes.

At present, the use of physiologic characteristics, mating types, and critical antigenic analysis, in addition to the classic methods of descriptive morphology, has placed the taxonomy of dermatophytes and other pathogenic fungi on a firm scientific basis.

Another epochal contribution by Gentles was to revolutionize the therapy of ringworm infections. In 1958 he reported that the oral administration of griseofulvin cured experimental dermatophytosis in a guinea pig.[75] Independently, Martin[159a] reported similar results. This was the first major change in the treatment of ringworm since Sabouraud plucked the heads of children in Paris. Williams in 1958 described the first patient to be cured by griseofulvin. The patient, a child, had tinea capitis due to *M. audouinii*. Previously a patient with a life-threatening *T. rubrum* infection had responded dramatically to the drug. This report and the subsequent work of Blank and Roth drew worldwide attention to the efficacy of griseofulvin in human disease. Blank and coworkers then established dosage and treatment schedules, which led to general acceptance of the drug as the treatment of choice in almost all forms of dermatophyte infection.

Clinical trials were carried out in many parts of the world and included the whole gamut of dermatophytic diseases. These trials, conducted by Riehl, Pardo-Castello,[198] Degos, Esteves, Neves,[181] and others, have been reviewed by Blank and are the standards for therapy.[34] Recently several new topical drugs have been introduced. Of these, tolnaftate (Tinactin) has gained wide popularity but is of limited value in some types of infection. More recently a few of the imidazoles—clotrimazole, ketoconazole, miconazole, and econazole—have been used as topical agents in the treatment of certain dermatophyte infections. These are broad spectrum agents that have some effect against yeasts and bacteria as well as the dermatophyte fungi. This increases the chance of cure in dermatophytosis complex infections when other organisms have colonized the primary or dermatophytosis simplex lesion. Another useful agent, haloprogin, is also broad spectrum; however, this agent as well as some of the others is associated with skin sensitization and drug induced eruptions. Treatment failures and relapses occur with all presently available drugs. The need for better therapeutic agents is apparent.

Dermatophytes remain a major public health problem. The commonness of ringworm infections and the occasional disfiguring severity of the disease have prompted continuing interest and investigation into their etiology, distribution patterns, improved therapy, and mechanisms of pathogenicity.

Etiology, Ecology, and Distribution

The group of organisms known as the keratinolytic fungi (families Gymnoascaceae and Onygenaceae of the Ascomycota) are homogeneous not only in appearance but also in physiology, taxonomy, antigenicity, growth requirements, and limits of infectivity and disease. The ability of these microorganisms to invade and parasitize the cornified tissues is closely associated with, and dependent upon, the utilization of keratin. Keratin is a highly insoluble scleroprotein, and its use as a substrate is rare in nature. Certain insects, including the clothes moth (*Tinea*), the carpet beetles *(Dermestes)*, the biting lice (Mallophaga), and possibly some streptomycetes, and some chytrids[130] also utilize keratin.

It had long been recognized that the dermatophytes were very closely related to one another. The asexual stages of all the dermatophyte species were placed in three genera. When the sexual stages were discovered in some dermatophytes, it was again found that only two very similar genera of the ascomycetes could encompass all species. The two perfect genera corresponded closely to the imperfect genera, i.e., all *Microsporum* species with perfect stages belong to the genus *Nannizia;* all *Trichophyton* species to the genus *Arthroderma*. Antigenically and physiologically, the dermatophytes are also closely related. It has been almost impossible to differentiate genera by an antigenic mosaic. Separation of species also has not been achieved. No serologic tests are available that delineate more than a "group" reaction which includes all dermatophytes. In addition, very few physiologic and nutritional differences between species and genera have been described. The few differences found have been useful in separating similar species (Table 8–11).

Etiology. At present, there are 39 species of dermatophytes recognized as acceptable[4] (Table 8–3). Many of these have been found only in soil and as yet have not been reported as causing human or animal disease. The teleomorphic (perfect or sexual) stage of 21 dermatophytes has been described. The taxonomy of these fungi as reviewed by Ajello[5] appears to be biologically correct and is the one generally accepted by mycologists.

A recent revision by Vanbreuseghem was deemed inaccurate because he lumped together obviously unrelated species, e.g., some *Trichophyton* and *Microsporum* species, in a genus, *Sabouraudites.* An attempt by Benedek to define the "ancestral forms" of dermatophytes apparently was based on unorthodox views of genetics and contaminated cultures.[26]

As far as human disease is concerned, most infections throughout the world are caused by 11 species. Within the continental United States, the list is pared to six. Worldwide travel, however, has increased the host exposure to formerly geographically limited species. Many infections produced by exotic organisms or the rare infections produced by soil keratinophiles present great problems in diagnosis and identification.[223, 227] The soil contains many keratinophilic fungi closely related to the dermatophyte genera. Most of these organisms are rarely, if ever, isolated from human infection. However, they may be transient inhabitants of the skin and particularly of animal fur. These genera include *Ctenomyces, Gymnoascus, Chrysosporium, Malbranchea, Aphanoascus,* and so forth. Sometimes they are isolated from the human skin, but there is question of their significance. *Chrysosporium* species are frequently cultured from feet. Whether they represent transients or are involved in disease is difficult to determine. However, the clinician and mycologist must be aware of the potential for infection by these soil forms. On occasion, reports can be verified of their evoking clinical disease, and was found recently with *Aphanoascus fulvescens.*[226] The isolant was morphologically quite unlike the usual dermatophyte and might have been considered a contaminant. This report emphasizes the difficulties that arise in identification of species other than those routinely isolated.

Taxonomy. For many years, the similarity of dermatophytes to soil-inhabiting Ascomycota had been recognized. Matruchot and Dassonville (1899)[162] suggested that the asexual dermatophytes were related to the Gymnoascaceae. *T. mentagrophytes* was even transferred to the teleomorphic genus, *Ctenomyces,* by Langeron in 1930.[144, 219] Pycnidia were formed which resembled the cleistothecia of Ascomycota, but only asexual conidia were found. Nannizzi probably was the first to find a sexual fruiting body in a dermatophyte. He published a paper in 1927 describing ascocarps in a culture of *M. gypseum.*[178] He renamed the organism *Gymnoascus gypseus.* This work was not accepted because he used unsterilized soil in his cultures, and his findings were thought to be due to a contaminant. With the work of Dawson and Gentles,[53] it was established that the teleomorphs of the dermatophytes were Ascomycota in the family of soil keratinophiles, the Gymnoascaceae. So far, all species described are heterothallic except *T. georgiae*. Interestingly, the teleomorphic states of *Blastomyces dermatitidis* and *Histoplasma capsulatum* are also in the same family. It is known that "inducer substrates," such as soil, oatmeal, keratin, iron filings, beetle wings, tomato juice, and chicken feathers, are sometimes necessary to evoke a fruiting body, even when mated pairs are used. By now many dermatophyte species are known to have a teleomorphic or sexual state (Table 8–4), and the idiosyncrasies of the proper romantic environment to encourage mating are known also. In general, as the dermatophyte becomes more anthropophilic, it gradually loses not only the ability to produce asexual conidia but also the ability to form a sexual state. Thus, such species as *M. audouinii*, *T. rubrum*, and *T. schoenleinii* have so far defied efforts to induce a sexual state. Animal isolants of *T. mentagrophytes* mate easily and produce fertile gymnothecia of *Arthroderma benhamiae* containing asci with ascospores. In man these strains induce a severe inflammatory disease. In contrast, *T. mentagrophytes* var. *interdigitale*, usually isolated from non-inflammatory lesions of humans, rarely mates with the testor strains of *Arthroderma benhamiae* and is considered an anthropophile.

Epidemiology. The major importance of knowing the teleomorphic stage of dermatophytes is for purposes of identification. New isolants, or isolants that vary in normal colonial morphology, can be mated with testor strains. By this method several previously described species of dermatophytes were found to be variants of recognized species. Another interesting use for the mating phenomenon is in epidemiologic studies. It was found that most isolants of *T. mentagrophytes* from troops in Vietnam were one mating strain of *A. benhamiae*, whereas in the United States another mating strain predominates. This provided evidence that the infections were acquired in Vietnam and not carried from the United States with clinical disease exacerbated because of the wet tropical environment. Certain enzymatic differences have been discovered between the mating types of the same species. In experimental dermatophytic infections, some of these have been related to differences in

pathogenicity, inflammatory response, and duration of infection.[220, 224] These and other aspects of the organisms are reviewed in the section on biological studies.

Ecology. By placing hair on soil (hair baiting), Vanbreuseghem[273] and later workers found a rich flora of keratinophilic fungi in this milieu. The significance of this discovery is only now being appreciated. Apparently the soil abounds with many closely related species of keratinophiles. Most of these are seldom if ever involved in dermatophytoses, but they have the potential for it. One can see a natural evolution from keratin-utilizing soil saprophytes (geophilic species) to association with and finally invasion of cornified substrate in living animals (zoophilic species) and man (anthropophilic species) (Table 8–5). The first animals to act as hosts were probably the soil-inhabiting rodents. Almost all rodents harbor a population of keratinophilic fungi in their dander or in their burrows.[69] In accepting a primarily parasitic existence, many changes have occurred in the fungus. Conidia production, which is very abundant in soil organisms, is gradually diminished. The strict anthropophiles, such as *M. audouinii*, *T. rubrum*, *T. schoenleinii*, and *T. violaceum*, produce very few conidia in culture. Some hair-invading species have evolved a "parasitic" arthroconidium which is produced only during active infection but will remain viable and infectious on fomites for several years. In general, the greater the tendency toward intermittent, chronic, or continuous infection, the lesser the inflammatory

Table 8–5. *Ecology of Human Dermatophyte Species*

Anthropophilic	Zoophilic	Geophilic
Cosmopolitan Species		
E. floccosum	*M. canis*	*M. gypseum*
M. audouinii	*M. gallinae*	*M. fulvum*
T. mentagrophytes	*T. mentagrophytes*	*T. ajelloi*
var. *interdigitale*	var. *mentagrophytes*	*T. terrestre*
T. rubrum	*T. verrucosum*	
T. schoenleinii	*T. equinum*	
T. tonsurans	*M. nanum*	
T. violaceum		
Rare and Geographically Limited Species		
M. ferrugineum	*M. distortum*	*M. racemosum*
T. concentricum	*T. mentagrophytes* var.	
T. gourvilii	*erinacei*	
T. megninii	*T. mentagrophytes* var.	
T. soudanense	*quinckeanum*	
T. yaoundei	*T. simii*	
	M. persicolor	

response evoked by the organism. The dermatophytes appear to reach a type of equilibrium with the host in matters of growth and lack of irritation. Examples of these are Adamson's fringe in tinea capitis and occult tinea pedis in which the dermatophyte (strains of *T. mentagrophytes*) is essentially a part of the "normal flora" of the foot. Along with diminished conidia production, these organisms are usually physiologically less active, particularly in regard to production of proteolytic enzymes when they are growing on their preferred or particular host. This equilibrium may become rather host specific. Zoophilic strains of *T. mentagrophytes*, which cause a minor or subclinical infection in animals, will evoke a severe inflammatory response in humans. The mild human infection is caused by the downy or var. *interdigitale* form of *T. mentagrophytes*. Both varieties are able to produce various proteolytic and other types of enzymes in *in vitro* culture. Yet these toxic and allergenic materials are not usually produced when the organism is colonizing the host with which it has evolved an association. If it elaborated these materials an inflammatory reaction would occur speeding up epidermal shedding and aborting the infection. It has been suggested that there is some inhibitor in the keratin of the preferred host that "turns off" the production of inflammatory agents by the colonizing fungus. When the fungus has colonized a different host, however, the inhibitor is lacking; proteolytic enzymes, allergens, and so forth are produced and a severe inflammatory disease usually results.

In surveying the infectious dermatophytes and the various species of host animals that are colonized, one discerns varying degrees of adaptation and host specificity. *M. audouinii*, an agent of "gray patch" tinea capitis, usually produces little inflammation in most north European children and American children of north European origin.[33, 201] If other children are infected, the resulting disease may be more severe. This species is truly anthropophilic, because it is unable to colonize animals other than man. This is also true of some of the other anthropophilic species, such as *T. rubrum*, *T. schoenleinii*, *T. tonsurans*, and *T. violaceum*. Some dermatophytes show a preference for particular races or population groups of a species. *M. ferrugineum* is associated with northern Chinese, Koreans, and Japanese. When infection occurs in black children, an inflammatory disease often results. *T. concentricum* is associated with particular

peoples of Indonesia and the South Pacific along with some Indians in Central and South America. Cohabiting members of other races remain free of infection even after years of intimate contact. Numbers of other examples of this type of specialization also exist.

In the listing in Table 8–5, it can be seen that there are many species of dermatophytes adapted to colonizing the human host (anthropophilic species). Few other animals have so many dependent dermatophyte species. Since infection is a chance occurrence, perhaps this is because our relative hairlessness affords more area of *stratum corneum* available for colonization. This has been cited as the reason for the commoner occurrence of tinea capitis in short-haired boys than in long-haired girls. Another factor contributing to frequency of dermatophytosis in humans is the close association of humans to other humans. A successful dermatophyte must be transferred readily to a new source of keratin. Thus we have family-centered diseases such as the favus of *T. schoenleinii*, dormitory- and barracks-associated epidemics of tineas due to *Epidermophyton floccosum*, and so on. In the animal world, dermatophytosis is more common in social than in solitary species. Cattle have a variety of associated dermatophytes, as do pigs, horses, and dogs (Table 8–9); all are herd or pack animals. Some dermatophyte species such as *M. canis* and *T. mentagrophytes* have strains associated with a large number of animal hosts (Table 8–10). These strains are probably in the process of evolving into separate species that will be specific for a particular host. Both *M. canis* and *T. mentagrophytes* are known to be complexes, each of which represents several teleomorph (sexual) species.

The major importance of defining isolants and species as geophilic, zoophilic, or anthropophilic is to determine the source of infection. Tinea capitis due to *M. audouinii* almost invariably results from contact with other infected children and may mean a school-centered epidemic. Thus, it becomes a problem in public health and disease control. On the other hand, an infection with *M. canis* usually presages a family-centered epidemic and indicts the family or neighbor's cat.

Distribution. Certain of the dermatophytes are geographically restricted and endemic only in particular parts of the world[203] (Table 8–7). *M. ferrugineum* is found in Japan and adjacent areas (Fig. 8–1); *T. concentricum* in the South Seas and small areas of northern

NIIIIII *Trichophyton simii*

////// *Trichophyton megninii*

||||||| *Trichophyton concentricum*

≡≡≡ *Microsporum ferrugineum*

Figure 8–1. Distribution of *Trichophyton simii, Trichophyton megninii, Trichophyton concentricum,* and *Microsporum ferrugineum.*

South and Central America (Fig. 8–1); *T. yaoundei, T. gourvillii,* and *T. soudanense* in central and west Africa (Fig. 8–3). Other species may be of sporadic but worldwide distribution. Dermatophytes that are endemic within a population are carried by that population to new places. Troop movements, migration of labor, emigration, social habits, and rapid, worldwide travel have all contributed to the changing distribution of ringworm. For example, until recently *T. tonsurans* was rarely isolated in the United States. With the immigration of peoples from Mexico, Puerto Rico, and other Latin American

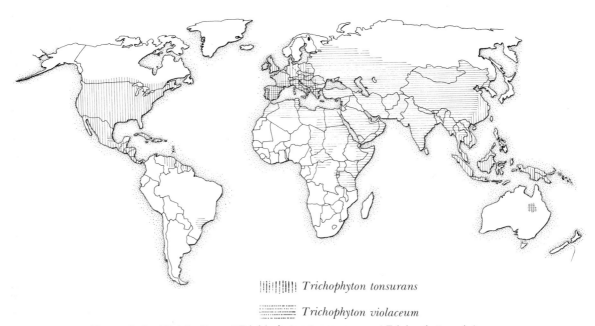

||||||| *Trichophyton tonsurans*

≡≡≡ *Trichophyton violaceum*

Figure 8–2. Distribution of *Trichophyton tonsurans* and *Trichophyton violaceum.*

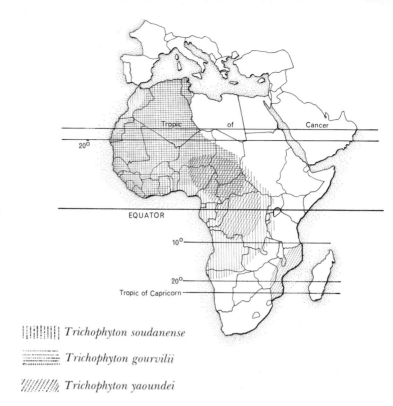

Figure 8–3. Distribution of *Trichophyton soudanense, Trichophyton gourvilii,* and *Trichophyton yaoundei.*

||:||:||| *Trichophyton soudanense*

≡≡≡ *Trichophyton gourvilii*

///////, *Trichophyton yaoundei*

countries, this organism, which is endemic in those regions, has become quite common in several cities (Fig. 8–2). In New Orleans, Charleston, New York, and Chicago, it is tending to supplant *M. audouinii* as the most frequent cause of tinea capitis[206] (Fig. 8–4). Sometimes a new population will show a quite different response to a dermatophyte strain that is endemic in another region. Taplin[35] observed that South Vietnamese troops rarely had the severe disabling form of tinea pedis caused by the endemic strain of *T. mentagrophytes.* American troops, however, in the same region and under the same swamp conditions, developed severe inflammatory lesions which sometimes covered the entire lower half of the body. The strain was quite distinct morphologically and was apparently carried by the local water rat. ARVN (Army of the Republic of Vietnam) troops were troubled by *T. rubrum* tinea corporis, which was uncommon among U.S. troops.

Other dermatophytes have a sporadic distribution. *T. schoenleinii* is rare in the United States except in small endemic foci, such as towns in Appalachia. In these areas all residents are descended from immigrants who lived in the same central European village where the disease was endemic (Fig. 8–4). The presence of the African species *T. soudanense* in Brazil and the United States may

be a legacy of the infamous slave trade that flourished between the fifteenth and nineteenth centuries.[227] The patients involved had not been outside their respective countries. With jet age travel, *T. simii* can be picked up from monkeys in India (Fig. 8–1), and diagnosed in the midwestern United States.[223] *T. soudanense* acquired in Africa[274] has been identified in Germany, Belgium, and England.[44, 48, 232] These citations emphasize that the diagnosis and identification of dermatophytes will be more difficult in the future as endemic species are transported to other areas of the world. The geographic distribution of anthropophilic, zoophilic, and geophilic dermatophytes has been reviewed by Philpot.[203]

Though they are not debilitating or fatal, dermatophytoses are among the most prevalent of human infectious diseases. Tinea pedis affects almost everyone who wears shoes throughout the world. The principal etiologic agents are *T. mentagrophytes* and *T. rubrum.* In contrast, the most frequent causes of the other dermatophytoses (tinea capitis, tinea corporis, and tinea unguium) vary from area to area. Ajello[3a] has tabulated these for tinea capitis (Table 8–7). Millions of adults and children in the United States suffer from one or more types of dermatophyte infection. It is estimated that, in the United States

Table 8–6. *Dermatophyte Infections—Clinical Diseases and Common Etiologies*

Disease	Dermatophyte Involved	Disease	Dermatophyte Involved
Tinea capitis	*Microsporum*, any species	Tinea imbricata	*T. concentricum*
	Trichophyton, any species except	Tinea cruris	*E. floccosum*
	T. concentricum		*T. rubrum*
Tinea favosa	*T. schoenleinii*		*T. mentagrophytes* (*Candida albicans*)*
	T. violaceum (rare)	Tinea pedis	*T. rubrum*
	M. gypseum (rare)		*T. mentagrophytes*
Tinea barbae	*T. mentagrophytes*		*E. floccosum* (*Candida albicans*)*
	T. rubrum	Tinea manuum	*T. rubrum*
	T. violaceum		*E. floccosum*
	T. verrucosum		*T. mentagrophytes*
	T. megninii	Tinea unguium	*T. rubrum*
	M. canis		*T. mentagrophytes*
Tinea corporis	*T. rubrum*		(rare: *T. violaceum*, *T. schoenleinii*,
	T. mentagrophytes		*T. tonsurans*)
	M. audouinii		*(C. albicans* and other fungi are
	M. canis, almost any dermatophyte,		involved in a similar clinical
	and *Candida albicans**		disease, onychomycosis)
Tinea of animals (hair, skin, claws, feathers)	*M. canis*	Tina favosa (favus of animals)	*T. equinum*
	M. nanum		*M. gallinae*
	T. mentagrophytes		
	T. verrucosum		

Candida albicans, an opportunistic yeast of the normal human flora, can elicit infections that mimic true dermatophyte disease, particularly in the clinical types marked with an asterisk.

alone, the public expenditure for medication for ringworm infections is about $40,000,000 per year.[3a]

Source of Infection

The infections transmitted from animal to man (zoophilic) and from soil to man (geophilic) have already been discussed. The methods of transmission of the anthropophilic species remain a complex and controversial problem. *T. schoenleinii* and *T. violaceum* are endemic in certain middle European and Mediterranean regions. They are probably transmitted by direct personal contact, particularly in families. *T. concentricum*[160, 257] is passed from mother to child soon after birth. Infection of entire populations of children in schools and institutions during epidemics of *M. audouinii* has been recorded. Transmission is probably indirect, occurring via fallen

||||||||| *Microsporum audouinii*

||||||||| *Trichophyton schoenleinii*

Figure 8–4. Distribution of *Microsporum audouinii* and *Trichophyton schoenleinii*.

Table 8–7. *Prevalent Agents of Tinea Capitis in the World**

North America	Europe
United States	Denmark, England, Finland, France
M. audouinii	*M. canis, M. audouinii*
T. tonsurans	
M. canis	
Mexico	Greenland
T. tonsurans	*T. schoenleinii*
Canada	Italy, Portugal, Spain, USSR, Yugoslavia
M. canis	*T. violaceum*
T. tonsurans	
Caribbean Area	Africa
Cuba, Puerto Rico	Algeria, Egypt, Libya, Tunisia
M. canis	*T. violaceum*
Dominican Republic	Angola, Congo, Mozambique, Zaire
M. audouinii	*M. ferrugineum*
	T. yaoundei, T. Gourvilii,
	T. soudanense
	Morocco, Spanish Sahara
South America	*T. schoenleinii*
Argentina, Chile, Uruguay	Cape Verde Islands
M. canis	*M. canis*
Brazil	Nigeria
T. violaceum	*M. audouinii*
Peru, Venezuela	
T. tonsurans	Asia
	China, India, Israel
Australasia	*T. violaceum*
Australia and New Zealand	Japan
M. canis	*M. ferrugineum*
	Iran, Turkey
	T. schoenleinii

*Modified from Ajello, L. 1960. Geographic distribution and prevalence of the dermatophytes. Ann. N.Y. Acad. Sci., *89*:30–38.

hairs and loosened, infected squames rather than by direct contact. The organism has been isolated from combs, brushes, the backs of theater seats, and caps. *T. tonsurans* and *T. violaceum* have been isolated from all of the above and from bed linens. *E. floccosum* has been recovered from towels, undergarments, and jock straps. *T. mentagrophytes* is regularly isolated from locker room floors trod over by "athletes' feet." It is apparent that essentially all persons come into contact with dermatophytes during their lives, yet only a small percentage manifest clinical symptoms of disease.

In light of the prevalence of these organisms, Sulzberger and Baer[18] concluded that dermatophytosis is not a contagious disease. They felt that there must be other factors important for manifestation of disease, since we continually wade through a sea of dermatophytes and only a few develop clinical symptoms. Occult tinea pedis is very common, but the difficulties of establishing experimental disease in human volunteers indicates a high degree of natural resistance against a significant disease process. It appears that trauma, occlusion, and maceration are necessary for symptomatic infection.

Marching, temperature stresses, moisture, and sweaty socks have been cited as necessary factors for exacerbation of previous subclinical disease. An opposing view is that contact with infectious material is the sole factor necessary for disease.[61, 63] Gentles[77] and Georg[79a] concluded on epidemiologic evidence that tinea pedis is a highly contagious disease. It appears then that there is no simple answer to the mode of spread of anthropophilic dermatophyte infections.

Strain and species differences may account for some of the problem. Infections produced by the granular form of *T. mentagrophytes* are characterized by inflammation and rapid resolution of infection.[224] *T. rubrum* infections often result in chronic disease, characterized by long periods of quiescence and intermittent recrudescence. Once the disease is established in them, individuals may be designated "*T. rubrum* people," for there is a lifelong association with this fungus. The factors involved in this association have been investigated, but no clear-cut determinants have been described. Epithelial turnover has been suggested, as is the case in pityriasis versicolor. Genetic predisposition is possible, but begs the question without defining factors. Hypersensitivity or "immunity," if such exists, is very transient following infection.[93] Reactivity to the trichophytin skin test appears to be of some significance in diagnosis, prognosis, or resistance to infection.[121, 218, 282] Recent work has indicated that the composition and amount of amino acids found in the sweat of infected patients differs from those found in normal subjects.[208] The authors suggest that this is a factor predisposing to chronic infection. In other studies, certain amino acids, particularly those involved in the urea cycle, were shown to stimulate germination of *M. audouinii* in infected hairs. In contrast to this, other amino acids were inhibitory to germination.[170, 225] Stimulation or inhibition of germination by other amino acids has been found in a variety of dermatophyte species.[107]

Many other aspects of dermatophyte infections remain undefined. The waxing and waning of epidemics caused by endemic species is known to occur, but is unexplained. *M. audouinii* will infect almost all of the population of a given area, then disappear for long periods of time. The incidence and the severity of *T. violaceum* infection in Yugoslavia and certain other countries varies markedly over a period of years. Simple explanations, such as the presence of a susceptible population, or the appearance of a more virulent strain, do not seem to account for all the vagaries encountered. Seasonal variation in incidence also occurs, and records from clinics and diagnostic laboratories indicate some periodicity. In the midwest, *Candida* infections are most prevalent in the late summer and early fall. *M. audouinii* tinea capitis used to increase with the opening of the school year. Tinea corporis is common in the midwinter "pale" season, whereas tinea pedis is exacerbated by the sweaty conditions of a humid summer. Kaplan and Ivens examined seasonal incidence of ringworm in dogs and cats.[129] They found that there was a seasonal variation, but the pattern appeared to be different for each dermatophyte.

Pathogenicity and Pathogenesis

Of the many fungi that produce disease in man, only the dermatophytes show evolution toward a parasitic mode of existence. The discovery of the many soil keratinophiles and their relation to known disease-producing dermatophytes has been mentioned previously. When scrutinized closely, the biologic distance between a "soil dermatophyte" like *Trichophyton ajelloi* and an obligate "skin dermatophyte" like *T. rubrum* does not appear very great. These similarities have led to speculation about the phylogenetic history of skin dermatophytes. As modified from various sources, these hypotheses include (1) evolution in the soil of a specialized group of fungi with keratinolytic ability; (2) association with furred animals and ability to produce transient infections, e.g., *T. ajelloi* and *M. gypseum;* (3) adaptation to growth in a living keratinizing zone; (4) development of accommodation and equilibrium to the host, e.g., little irritation to the specific host as exemplified by Adamson's fringe in some forms of tinea capitis and occult or symptomless dermatophytic infections (colonizations), a type of tinea pedis; (5) development of specialized methods of reproduction for successful dissemination of the parasite from host to host, e.g., arthroconidia in ectothrix and endothrix hair infections and in infected squames of the glabrous skin; and (6) adaptation to specific animal hosts, increasing duration of survival, dissemination, and chronicity of infection, e.g., *T. rubrum* in man, *T. mentagrophytes* var. *erinacei* in hedgehogs, *T. verrucosum* in cattle, *M. canis* in cats, *M. nanum* in pigs, and *T. mentagrophytes* var. *quinckeanum* in rodents. As the molecular structure of keratin varies from species to species, it has been suggested that different keratinases have been evolved with relative specificity for a particular host.[94, 292] This may also explain in part the racial preferences of some anthropophilic dermatophytes and the preference for particular breeds of an animal species by some zoophilic dermatophytes.

There are several dermatophytes that run the gamut of this active adaptation (evolution). Animal reservoirs of the granular form of *T. mentagrophytes* (var. *mentagrophytes*) have long been recognized. Infection in man by such strains evokes a primary irritant reaction, followed by a severe inflammatory response and rapid termination of infection. This fungus readily invades hair and produces numerous "saprophytic" conidia in culture. *T. mentagrophytes* var. *interdigitale* elicits little inflammatory response, produces chronic infections, does not readily infect experimental animals, and in culture produces few "saprophytic" conidia. Yet rapid serial transmission of these strains in guinea pigs will transform the var. *interdigitale* into the granular var. *mentagrophytes* type with a concomitant increase in severity of infection and in spore production.[79] There is some evidence that this occurs in human epidemics also.[245] In contrast to the severe infection described by Sabouraud and produced by the present endogenous strains in Africa, *M. audouinii* has become a fairly benign and widely dispersed anthropophilic species. Most *T. rubrum* infections are chronic and not severe in the normal host, and the isolants produce a few spores in culture. However, some rare strains that have been recovered in Miami, Vietnam, Tunisia, and Nigeria sporulate heavily and are associated with a more inflammatory type of infection.

Another aspect of pathogenicity is also demonstrated by *T. mentagrophytes* and by *M. canis*. Organisms given these names have been isolated from a wide variety of animal hosts (Table 8–10) and man. What is called *T. mentagrophytes* is the imperfect (anamorph) stage of at least two or three sexual species (teleomorph stages); thus we speak of a *T. mentagrophytes* complex. What appears to be happening is that the "species" is in active evolution for infection on a particular host or a range of hosts and will in time segregate into several separate species. This also seems to be happening in *M. canis* as well. One sexual stage (there are probably more) of *M. canis* is known. When testing clinical isolates[287] of *M. canis* for mating compatibility, however, it was found that almost all were of the minus (−) mating type. Perhaps in the evolution toward successful parasitism, a selective advantage occurred in this mating type, with the result that the other mating type is dying out. This process may already have occurred in the anthropophilic species *T. tonsurans*, *T. rubrum*, and *M. audouinii*.

The dermatophytes show a high degree of specificity as to the tissues attacked. While these fungi are well adapted to parasitize the horny layer of the epidermis and the hair and nails, they appear unable to invade and infect other organs of the body in normal patients. Intravenous injection of *Microsporum* conidia or hyphal suspensions of virulent *T. mentagrophytes* does not produce an infection of the internal organs; rather, the fungi become localized in the skin and produce infection only in areas previously damaged by scarification.[37] Dermatophyte growth is quite sensitive to temperature. Normal body temperature (37°C) inhibits growth of most strains and species. Modestly elevated temperatures (41°C), as shown by Lorincz and Sun, kill the organisms and cure experimental infection in animals.[153] However, it has been shown that dermatophytes can be "trained" to grow at elevated temperatures and even assume a yeastlike phase similar to that of several deep infecting fungi.[228] In this transient condition, the organisms can invade deep tissues of experimental animals. As shown by Lorincz et al.[152] and Roth et al.,[233] fresh serum inhibits the growth of dermatophytes. Experimental and clinical evidence of this has been reported. For example, when serum is washed out of tissue cultures of skin, *T. rubrum* invades the living cellular areas, and extensive invasion of epithelium and dermis by this same fungus was seen in a patient with a low level of serum antidermatophyte activity.[233] Disseminating invasive granulomatous disease owing to *M. audouinii* has also been reported in an immunoincompetent patient[8] in whom a specific factor for lymphocyte blastogenesis was lacking (Fig. 8–19,*C*).

Pathogenesis. The natural history of dermatophyte infection is the same initially in all types of disease. Colonization begins in the horny layer of the skin, and the ultimate outcome of disease depends on host, strain, species variation, and anatomic site. On the glabrous skin, the infection spreads centrifugally, showing the classic "ringworm" pattern. The host reaction may be limited to patchy scaling, or proceed to a toxic eczemaform eruption. Later, an inflammatory reaction may occur. In most all cases, apparent resolution occurs, and clinical symptoms disappear. However, organisms may persist for years, and the host becomes a normal carrier. Stress or trauma may exacerbate clinical disease. The natural evolution of dermatophytosis as exemplified by tinea capitis has been delineated by Kligman.[136] He divides the infection into several stages:[137] a period of incubation, enlargement, and spread; a refractory period; and a stage of involution. These stages were studied in detail using human volunteers.* Tinea capitis was produced in them using hair infected with *M. audouinii* as inoculum. The following sequence of events was found to occur. In the incubation period (three to four months), hyphae grew in the stratum corneum of the scalp and the follicular orifices. The mycelium entered the follicle and grew downward into it along the surface of the hair shaft. During the period of spread, the fungus grew radially from the point of origin in

---

*A review of experimental human infections has been published by Knight.[140]

the scalp *stratum corneum*. It invaded new follicles as they were encountered. The hyphae penetrated to the upper limits of the zone of active keratinization without invading it. The downward growth at this stage was equal to the upward growth of the forming hair. A ring of extending hyphae was formed just above the keratogenous zone, which is known as Adamson's fringe. As the growing hair carried hyphae to the surface, the specialized parasitic conidia (arthroconidia) were produced. This sequence occurred only in growing (anagen) hairs. In telogen, or in epilated hairs, the hair is attacked as it would be by a saprophytic keratinophile, with production of penetrating organs and fronds, and macro- and microconidia. If hair growth is interrupted by normal catagen or x-ray treatment, the delicate equilibrium is upset. Growth of the fungus is impeded, the hair is lost, and so is the parasite. The refractory period of the infection is characterized by disappearance of hyphae in the scalp, and no new lesions develop. Large quantities of arthroconidia are formed in the infected hairs. Gradually a period of involution occurs, and the hair returns to normal. Infected hairs are shed, and the infection resolved. Reinfection and exacerbation may occur but are rare. Susceptibility to experimental infection appears to be independent of previous history of dermatophytosis or trichophytin skin test reactivity, but in natural infection distinct patterns are seen. In tinea capitis owing to *M. audouinii*,[72] a single course of infection is the rule. This is usually benign, but inflammatory disease (kerion) may occur in hypersensitive patients.[209] Delayed hypersensitivity appears to be associated with immunity or refractivity to infection in tinea pedis as well.[121] Lack of skin test reactivity or correlation of specific blockage of reactivity with chronic infections has been described in some studies. This emphasizes the importance of cell-mediated immunity in dermatophyte infections.[93, 121, 282] However, susceptibility and immunity to natural infection is a particularly controversial subject[93] and will be reviewed in the section on immunity.

The preceding discussion underlines the delicate host-parasite balance that exists in a dermatophyte infection. In patients treated with oral griseofulvin, this host-parasite balance is abrogated. The drug is laid down in the keratinizing tissues and the fungus ceases to penetrate and is shed with the keratinized debris. On the other hand, if the serum inhibitory factor or factors are absent or diminished, the fungus is no longer impeded in its invasion. As first noted by Majocchi in 1883, granulomatous lesions may develop as the fungus invades the dermis and subcutaneous tissues. Although the Majocchi granuloma, as defined by Pillsbury, refers to small granulomata in hair follicles following shaving of the legs (apparently traumatizing the tissue and implanting the fungus), the original description probably referred to disease in patients with an underlying disease. These patients have a lessened ability to contain infection, and widespread granulomatous lesions may develop. In almost all cases the organism is *T. rubrum*. Such lesions have been associated with abnormalities of carbohydrate metabolism, lymphomas, Cushing's syndrome, and impaired immune response (delayed hypersensitivity). Fetal systemic invasion by dermatophytes has been recorded in Japan, Russia, Rumania, and Portugal.[14, 184] In these cases the fungus involved was *T. violaceum*.

Clinical Disease

The many species of dermatophytes elicit a number of well-defined clinical syndromes. The same species may be involved in quite different symptomatic disease, depending on the anatomic site. Because these syndromes form distinct entities, they will be discussed separately along with the organism commonly associated with them.

The clinical conditions are (1) tinea capitis (ringworm of the scalp); (2) tinea favosa (favus due to *T. schoenleinii*); (3) tinea corporis (ringworm of the glabrous skin); (4) tinea imbricata (ringworm due to *T. concentricum*); (5) tinea cruris (ringworm of the groin); (6) tinea unguium and onychomycosis (ringworm of the nail); (7) tinea pedis (ringworm of the feet); (8) tinea barbae (ringworm of the beard); and (9) tinea manuum (ringworm of the hand).

With the possible exception of white-spot tinea unguium, all infections begin in the horny layer of the epidermis. Those which infect the hair follicles, hair, and nails soon invade these structures, frequently producing little more than a transient scaling of the epidermis. Infections which are limited to the epidermis may affect the drier parts of the skin, including the palmar and plantar surfaces, or the moister regions, such as the inguinocrural folds and the interdigital spaces. Although a wide variety of clinical types may be observed, the differences are more apparent than real. The histopathologic changes are fundamentally the same in all types of disease. The entire gamut of disease entities produced by dermatophyte infection can be mimicked by *Candida albicans*. The most comprehensive review of the various forms of clinical dermatophytoses is given in the text edited by Baker.[21]

Tinea Capitis

DEFINITION

Tinea capitis is a dermatophyte infection of the scalp, eyebrows, and eyelashes caused by species of the genera *Microsporum* and *Trichophyton*. The disease varies from a benign scaly noninflamed subclinical colonization to an inflammatory disease characterized by the production of a scaly erythematous lesion and by alopecia that may become severely inflamed with the formation of deep, ulcerative kerion eruptions. This often results in keloid formation and scarring, with permanent alopecia. The type of disease elicited is dependent on the interaction of the host and the etiologic agent.[44]

Synonymy. Ringworm of the scalp and hair, tinea tonsurans, herpes tonsurans.

FUNGI INVOLVED

The commoner types of ringworm are classed according to the site of formation of their arthroconidia (Table 8–8). *Ectothrix* infection is defined as fragmentation of the mycelium into conidia around the hair shaft or just beneath the cuticle of the hair with destruction of the cuticle (Fig. 8–5). In *Endothrix* infections arthroconidia formation occurs by fragmentation of hyphae within the hair shaft without destruction of the cuticle (Fig. 8–6). Small-spored (conidia) ectothrix organisms include *Microsporum audouinii, M. canis*, and *M. ferrugineum*. Masses of conidia (1 to 3 μ) are produced in a mosaic pattern; all are Wood's lamp–positive. *M. gypseum* and *M. fulvum* produce few conidia (Fig. 8–5,*B*) and are Wood's lamp–negative.[68] *T. mentagrophytes* forms small conidia (3 to 4 μ) in chains on the surface of the hair. *T. verrucosum* forms large conidia in chains (8 to 12 μ). *T. megninii* also forms medium sized conidia in chains. The principal organisms producing endothrix infections are *T. tonsurans, T. violaceum*, and the African species *T. yaoundei, T. soudanense*, and *T. gourvillii*. *T. rubrum* is not infrequently involved in tinea capitis, but rarely invades hair. Both endothrix and ectothrix conidiation have been described for this species as well as endoectothrix involvement in villous hair. Fig. 8–6*B*

SYMPTOMATOLOGY

Anthropophilic Tinea Capitis

Ectothrix. "Gray-patch ringworm" ectothrix or prepubertal tinea capitis is a very common disease of children that often reaches epidemic proportions. The classic causative agents are *M. audouinii* in Europe and America and *M. ferrugineum* in Asia. Infection begins as a small erythematous papule around a hair shaft. Within a few days it pales and becomes scaly, and the hair assumes a characteristic grayish, discolored, lusterless appearance. The hair becomes weakened and may break off a few millimeters above the scalp. The lesion spreads, with the formation of more papules in a characteristic ring form, and may coalesce with other infected areas (Fig. 8–7). All hairs within the area are infected. Itching may become intense, and some alopecia may be seen in the infected areas. There may be little inflammatory reaction in *M. audouinii* infections, although on occasion ulcerations and kerions may form (Fig. 8–9*C*). Usually the infection is quite benign and the disease resolves spontaneously. There is some indication that *M. audouinii* infection is less inflammatory in white children than in black children. Similarly, *M. ferrugineum* is less inflammatory in certain oriental groups than in other races. Ectothrix infections can also be elicited by zoophilic and geophilic species as well. These are usually quite inflammatory, leading to abortion of the infection and rapid resolution of the disease. The gray patch seen clinically does not delineate the extent of the infection of the scalp. A Wood's lamp examination shows the characteristic brilliant green fluorescence of infected hairs to extend well beyond the symptomatic areas. Occult infections are also commonly discovered by this examination. Infection rate for boys is up to five

Table 8–8. *The Common Dermatophytes and Their Diseases*

			Species	*Disease in Man*	*Geographical Distribution*
Invading the Hair and Hair Follicles	*Small Conidia Varieties*	*Ectothrix Type*	*Microsporum audounii**	Prepubertal ringworm of the scalp; suppuration rare; child to child	Commonest in Europe, producing about 90 per cent of infections; in U.S. 50 per cent, becoming rare
			Microsporum canis*	Prepubertal ringworm of scalp and glabrous skin; suppuration not infrequent; kerion occasional; from pets	Uncommon in Europe, except England and Scandinavia; responsible for about half the infections in U.S.
			Microsporum gypseum	Ringworm of the scalp and glabrous skin; suppuration and kerion common; from soil	Relatively rare in U.S.; common in South America
			Microsporum fulvum	Ringworm similar to that of *M. gypseum*	Same as above
			Microsporum ferrugineum*	Similar to *M. audouinii*	Africa, India, China, Japan
	Large Conidia Varieties	*Endothrix Type*	*Trichophyton tonsurans*	Black-dot ringworm of the scalp; smooth skin; sycosis; tinea unguium; suppuration common; the hair follicles are atrophied	Common in Europe, Russia, Near East, Mexico, Puerto Rico, and South America, but uncommon in U.S. until recently
			Trichophyton violaceum	Black-dot endothrix in both scalp and smooth skin; onychomycosis; suppuration is the rule and kerion frequent	Common in southern Europe, the Balkans, and the Far East; rare in the U.S.
			Trichophyton soudanense *Trichophyton gourvilii* *Trichophyton yaoundei*	Inflammatory, scarring ringworm of scalp	Central and West Africa
		Ectothrix Type	*Trichophyton mentagrophytes var interdigitale*	Commonest cause of intertriginous dermatophytosis of the foot ("athlete's foot"); ringworm of the smooth skin; suppurative folliculitis in scalp and beard	Ubiquitous
			Trichophyton verrucosum	Ringworm of the scalp and smooth skin; suppurative folliculitis in scalp and beard; from cattle	Ubiquitous
			Trichophyton megninii	Sycosis is the most common lesion; infection of smooth skin and nails	Sporadic distribution; Portugal, Sardinia
	No. Conidia in Hair		*Trichophyton schoenleinii**	Favus in both scalp and smooth skin; scutulum and kerion	Europe, Near East, Mediterranean region; rare in U.S.
	Rare in hair		*Trichophyton rubrum*	Psoriasis-like lesions of smooth skin; tinea unguium; mild suppurative folliculitis in beard; rare invasion of scalp hair endo- and ectothrix described; endoectothrix in villous hair	Ubiquitous
Not Invading the Hair and Hair Follicles			*Epidermophyton floccosum*	Cause of classic eczema marginatum of crural region; causes minority of cases of intertriginous dermatophytosis of foot; not known to infect hair and hair follicles	Ubiquitous, but more common in tropics
			Trichophyton concentricum	Cause of tinea imbricata; infection of hair and nails uncertain	Common in South Pacific islands, Far East, India, Ceylon; reported in west coast of Central America and northwest coast of South America

*Infected hairs show fluorescence by Wood's lamp.

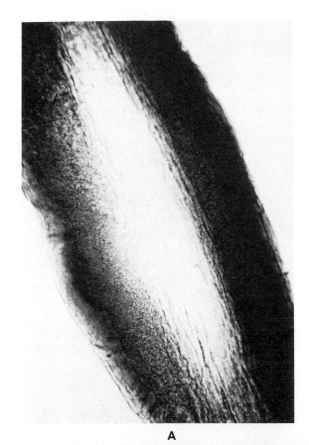

A

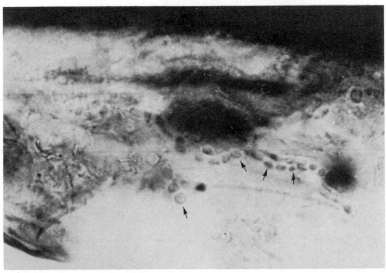

B

Figure 8–5. *A,* Ectothrix infection of hair. Arthroconidia form a sheath around the hair shaft. *Microsporum audouinii.* ×500. *B,* Ectothrix infection due to *Microsporum gypseum.*[68] Only a few scattered arthroconidia are seen around the hair shaft (arrows). (Courtesy of I. Alteras.)

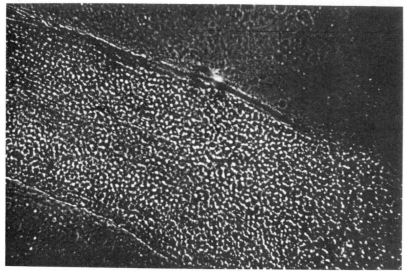

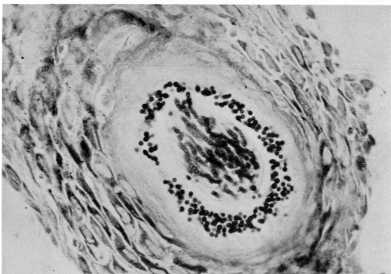

Figure 8–6. *A*, Endothrix infection of hair. Arthroconidia form within the hair shaft. *Trichophyton violaceum*. ×700. *B*, Ectoendothrix arthroconidiation. Chains of conidia are found in the medulla of the hair shaft as well as around the cortex. This is villous hair infected by *T. rubrum*. Such hairs act as reservoirs for the fungus and lead to recurrent tinea corporis. The production of such arthroconidia occurs only in particular hosts—"*T. rubrum* people"—and appears to be influenced by the amino acid content of perspiration. (Courtesy of A. Eng.)

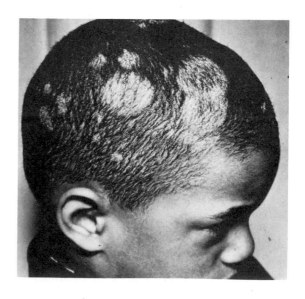

Figure 8–7. "Grey-patch ringworm" of *M. audouinii*, an anthropophilic species. (Courtesy of S. Lamberg.)

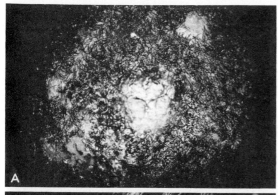

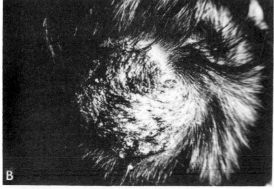

Figure 8–8. Tinea capitis. *A*, Boggy area of inflammatory infiltrate called a *kerion. Microsporum canis. B*, Large kerion consisting of crusts, matted hair, exudate, and scalp debris. *Microsporum canis.*

times that for girls. However, the reverse is true after puberty.[33, 201]

Uneventful spontaneous cure usually occurs in *Microsporum* infections. This may coincide with the onset of puberty. Rothman[235, 236] considered that the change in composition of sebum with an increase in fungistatic fatty acids might account for this resolution of infection. Even-numbered fatty acids of medium length possess the greatest fungistatic activity. However, some workers have felt that other factors are important. Kligman[137, 138] concluded that infection simply undergoes spontaneous resolution and that this may occur before, during, or after puberty.

Endothrix. All endothrix producing agents are anthropophilic. Similar to ectothrix dermatophytes, they are associated with particular populations or races and are geographically limited (Fig. 8–2 and 8–3). *T. tonsurans* is endemic in the western Mediterranean areas; it arrived in the Western Hemisphere with migration and colonization by peoples from this region. *T. violaceum* is common in the eastern Mediterranean areas[155]

and most of Asia. There are three endothrix agents limited to Africa (Fig. 8–3). In the usual hosts, these various agents tend to cause a relatively benign albeit chronic disease; however, there appears to be a greater tendency for severe infections than is seen in anthropophilic ectothrix disease.

"Black-dot ringworm" endothrix infection produced by *T. tonsurans* and *T. violaceum*, differs from *Microsporum* infection in a number of respects. While initial infection of the hair follicle and hair follows essentially the same course, the lesions are small and often contain only two or three affected hairs in a given area. If the lesion is large, all the hairs within the area are not affected. The lesions are multiple, numerous, and scattered over the scalp (Fig. 8–9,*A*). Chronic infections commonly occur, but the inflammatory reaction is minimal. In other patients extensive involvement can be seen and the bald patches tend to lose the circumscribed outline of the lesions, and a diffuse alopecia is found. The infected hairs break off sharply at the follicular orifice, leaving a conidia-filled stub or "black dot" (Fig. 8–9,*B*). The stub sometimes curls as it grows and may be subsurface. In these cases it is necessary to excise the stub with a scalpel in order to culture it. The lesion plaques tend to be angulated, often forming polygons. Sometimes there is moderate inflammation with delayed scarring, which gives an appearance similar to lupus erythematosus. In *Trichophyton* infections, a more severe type of inflammatory reaction is more common than in microsporosis. Kerion formation followed by scarring and permanent alopecia are regularly seen. The var. *sulfureum* of *T. tonsurans* has been associated with the production of multiple kerions and erythema nodosum of the scalp.[71, 276] Multiple infections often occur which involve other dermatophytes. In some cases, three species of dermatophytes, or a *Candida* sp. and *Staphylococcus* sp., have been isolated from lesions[276] (Fig. 8–9). There has been an increasing incidence of *T. tonsurans* in the United States.[206] Hairs infected with endothrix *Trichophyton* do not fluoresce. As a result, in mass examinations for tinea capitis conducted by public health officers, a significant percentage of infections are not detectable. For this reason, dermatophyte test media should be used along with Wood's lamp examination in such surveys.[270] Unlike gray-patch ringworm, endothrix infections tend to become chronic and continue into adult life. This is especially true when *T. violaceum* is the etiologic agent.

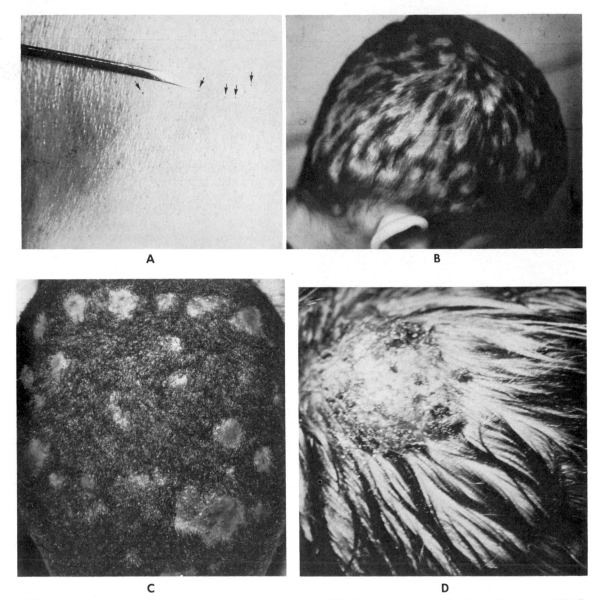

Figure 8–9. *A,* Endothrix tinea capitis due to *T. tonsurans.* This benign infection is in a Latin American child. *B,* Infection by the same species in a northern European child. The reaction to the infection is more inflammatory. (Courtesy of S. McMillen.) *C,* Tinea capitis. Multiple kerions and patchy alopecia in scalp infected by *T. tonsurans* var. *sulfureum, M. audouinii, Candida albicans,* and *Staphylococcus.* (From Varadi, D. P., and J. W. Rippon, 1967. Arch. Dermatol., *95*:299–301.) *D, Microsporum gypseum.* Kerion formation on a child's head infected by the geophilic species. (From Feuerman, E. I., Alteras, et al. 1975. Mycopathologia *55*:13–17.)

Zoophilic and Geophilic Tinea Capitis

Ectothrix. As stated previously, colonization of man by zoophilic or geophilic dermatophytes usually results in an inflammatory disease. "Gray patch" tinea capitis is regularly produced by *Microsporum canis.* This infection is Wood's lamp–positive and its course is similar to that caused by *M. audouinii.* Kerion, keloid, and severe inflammatory disease are much more frequent, however. The geophilic species *M. gypseum* and *M. fulvum* are unusual agents of tinea capitis. The infection is Wood's lamp–negative and the infection is quite severe, with keloid formation being common.

Ectothrix infection with animal origin *Trichophyton* spp. (*T. mentagrophytes* var. *mentagrophytes* and *T. verrucosum*) is distinguished clinically by a more marked inflammatory reaction than are other forms of tinea capitis.

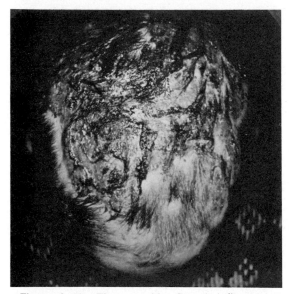

Figure 8–10. Tinea capitis. Severe inflammatory reaction and suppurative folliculitis produced by infection with *T. mentagrophytes*. This zoophilic strain was mating type a (*Arthroderma benhamiae* mating type a) and produced large quantities of elastase in culture and experimental infections.[220] The infection was acquired from a cow.

They often produce suppurative folliculitis (Fig. 8–10). These infections are rare and are seen mostly in rural areas. The organisms are usually acquired from animals, cattle being the commonest source. Kerion formation is the rule in such infections. There is a massive acute inflammatory infiltrate in the skin of the affected areas and deeper tissue. In contiguous areas, a suppurative folliculitis occurs, with indolent cutaneous and subcutaneous infiltration (kerion celsi) (Fig. 8–11). These are boggy on palpation, and pus oozes from the follicles. Several areas of the scalp may be involved simultaneously, and the size of the lesions is variable. The swelling is painful and the hairs easily dislodged. It is important to note that the pus is not due to secondary bacterial invasion but is attributable to the fungus alone, and therefore surgical intervention is not required. On occasion, dull-red sharply defined plaques studded with pustules may be seen, a condition referred to as agminate folliculitis. Resolution of the infection is accompanied by scarring and patchy permanent alopecia. Individuals with dark, thick hair tend to have more severe inflammatory reactions than those with light, thin hair.

The dermatophytid ("id") reaction may occur in tinea capitis, but less commonly than in tinea pedis. The "id" reaction is an allergic manifestation of infection at a distal site, and the lesions are devoid of organisms. A series of grouped vesicles (pompholyx type) that are tense, itchy, and sometimes painful are found anywhere on the body. However, lesions on the trunk are common in tinea capitis, whereas fingers are more often involved when the primary lesion is a tinea pedis. The lesions may evolve into a scaly eczematoid reaction or a follicular papulovesicular response which covers wide areas of the body and causes great distress and discomfort to the patient.

A "blue-dot" tinea capitis has also been observed.[147] The infecting organism was a *Penicillium* species. *Aureobasidium pullulans* was also isolated in the same case. The involved hairs showed mycelium and conidia around the shaft and in the follicles. A small, crusted area developed around each of the broken hairs, and mechanical removal of the hairs was followed by redevelopment of the lesions. The disease resolved spontaneously after two and one half years. A series of five such cases has been reported.

It was observed by Margarot and Deveze[158]

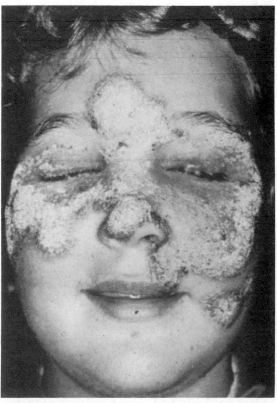

Figure 8–11. Extensive cutaneous and subcutaneous involvement in an infection by *T. verrucosum*. The infection was acquired from a cow. (Courtesy of D. Windhorst.)

in 1925 that infected hairs and some fungus cultures fluoresce in ultraviolet light. This "black light" is commonly known as the Wood's lamp. The light is filtered through a Wood's nickel oxide glass (barium silicate with NiO), which passes only the longer ultraviolet rays (peak at 3650 Å). Hair infected with *M. canis, M. audouinii*, and *M. ferrugineum* fluoresces a bright green color. Hair infected with *T. schoenleinii* may show a dull green color. *T. verrucosum* exhibits a green fluorescence in infected cow hairs, but infected human hairs do not fluoresce. Hedgehog hairs infected with *T. mentagrophytes* var. *erinacei* and all *T. simii* infections also fluoresce a bright green.[223] *M. distortum, M. nanum, M. vanbreuseghemii*, and *M. gypseum* have been reported on occasion to produce infections that fluoresce.

The fluorescent substance appears to be produced by the fungus only in actively growing infected hairs. Saprophytic growth of the organisms on epilated hairs does not produce the active principle. Comparative studies of the water-soluble fluorescent substances show the spectroscopic pattern to be similar in hairs infected by *M. canis, M. audouinii*, and *T. schoenleinii*. From infrared spectrophotometric studies, Wolf[290] concluded that these substances were pteridines (pyrimidine-4′,5′:2,3-pyrazine). The exact structure has not as yet been worked out. Infected hairs remain fluorescent for many years even after the conidia have died.

DIFFERENTIAL DIAGNOSIS

The diagnosis of tinea capitis, especially in children, is suggested by erythematous, scaling patches of alopecia. The presence of brittle, lusterless hair or boggy, infiltrative, ulcerative lesions may also be a clue. Seborrheic dermatitis, psoriasis, lupus erythematosus, alopecia areata, pseudopelade, impetigo, trichotillomania, pyoderma, folliculitis decalcans, and secondary syphilis are considered in the differential diagnosis. Examination of potassium hydroxide mounts usually determines the proper diagnosis if it is a tinea.

In seborrheic dermatitis, the hair involvement is diffuse rather than patchy, the hairs are not broken and the scalp is red, scaly, and itchy. This and other chronic scaling diseases, such as psoriasis, may cause accumulation of scales in matted masses on the scalp. This condition is called "pityriasis amiantacea" (asbestos pityriasis).[25] Scales are more promi-

nent in psoriasis, but again hairs are not broken. Impetigo is difficult to differentiate from inflammatory ringworm, but the pain is usually less severe in the latter. Alopecia areata may have an erythematous border in the early stages of the disease, but reversion to normal skin color occurs. Also in this condition there is a lack of scaling, and the hairs on the border do not break off; however, they pull out easily. These hairs are of the "exclamation point" morphology.

The vesicles of the "id" reaction must be differentiated from toxic reactions of many etiologies, including pompholyx, dyshidrosis, and other causes of vesicles and subcorneal bullae.

HISTOPATHOLOGY

There is no specific histologic picture for dermatophyte infections, and hyphae must be shown with special stains. Fungi are seen sparsely in the *stratum corneum*, penetrating in between and through the squames[108] (Fig.

Figure 8–12. *A*, Dermatophyte infection. The hyphal strands penetrate through the stratum corneum but do not invade the living cells of the epidermis. Periodic acid–Schiff stain. ×400. (Courtesy of M. Sulzberger.) *B*, Mycelium growing down a hair shaft to the bulb. ×440.

8–12). Hyphae extend down into the hair shaft and penetrate into the hair lying parallel to it. Hyphal tips grow downward within the hairshaft to the edge of the living keratinizing cells and form "Adamson's fringe." The overall histologic picture in tinea capitis is that of a subacute or chronic dermatitis. In the mildest form, there is intercellular edema of the rete. Parakeratosis, slight vasodilation, and a perivascular infiltrate in the upper dermis are also seen. An exaggerated allergic reaction leads to folliculitis and perifolliculitis, with formation of kerion celsi. This consists initially of an acute leukocytic reaction in the deep dermis and even subcutaneous tissues. The cellular infiltrate gradually becomes that of the chronic inflammatory type. The infiltrate is massive and obliterates much of the normal architecture of the tissue. Generalized follicular keratosis (lichen spinulosus) may be seen as transient eruptions in rapidly developing inflammatory ringworm.

PROGNOSIS AND THERAPY

Most mild ringworm infections of the microsporosis (gray-patch) type resolve uneventfully with time, usually in early adolescence. The more inflammatory the reaction, the earlier the termination of disease. This is particularly true of animal-acquired ringworm (*M. canis, T. verrucosum,* and *T. mentagrophytes*). Thus most ectothrix infections involute during the normal course of disease and without treatment. However, the patients spread the organisms to others during the infection period. Endothrix infections, on the other hand, tend to become chronic and last far into adult life. *T. violaceum* causes a particularly persistent infection in which patients become vectors for spreading the disease within family groups and the community.[234]

Patients should be actively treated to terminate such infections and prevent their spread.

Topical treatment of tinea capitis appears to be without benefit. The infection is unaffected by local applications of fungistatic preparations, keratolytic agents, and antifungal drugs, such as griseofulvin and tolnaftate (Tinactin). However, the addition of topical fungistatic agents (benzoic acid, magenta paint, and so forth) to a more effective treatment regimen such as systemic administration of griseofulvin is an important adjunct to therapy. By this means conidia are killed and infected debris is removed, thereby preventing spread of infection.

Griseofulvin is the most effective drug presently available for treating tinea capitis. Dosage schedules vary, but the usual standard treatment (micronized griseofulvin) is 500 mg per day for adults and 250 mg for children in four divided doses. Coarseparticle griseofulvin is given in a dosage of 1 g per day for adults and 500 to 750 mg per day for children. Patients are encouraged to take the drug after a fatty meal to increase absorption from the alimentary tract. After three days, vigorous daily scrubs of the scalp are begun to eliminate infectious debris. Treatment may require several months. Contraindications to this therapy include patients with porphyria or hypersensitivity to griseofulvin.

Griseofulvin has all but eliminated the necessity for x-ray epilation. Though once very widely used[6] and until recently called upon in particularly stubborn cases,[250] x-ray is now contraindicated in tinea capitis. Several recent studies involving many patients indicated a probable association between use of x-ray epilation and a variety of sequelae, including neoplasm.[171, 185] Since present day treatment regimens involving drugs, manual epilation, and scrubs almost always prove efficacious, radiation therapy is unwarranted.

Tinea Favosa

DEFINITION

Favus is a clinical entity characterized by the occurrence of dense masses of mycelium and epithelial debris which form yellowish, cup-shaped crusts called scutula. The scutulum develops in a hair follicle, with the hair shaft in the center of the raised lesion. Removal of these crusts reveals an oozing, moist red base. After a period of years, atrophy of the skin occurs, leaving a cicatricial alopecia and scarring. Scutula may be formed on the scalp or the glabrous skin. The Latin name favus refers to the similarity of appearance of scutula and honeycombs.

Synonymy. Favus, honeycomb ringworm, erbgrind (Ger.), teigne faveuse (Fr.).

ORGANISMS INVOLVED

The vast majority of cases of favus are caused by *Trichophyton schoenleinii*. The peculiar and characteristic clinical disease evoked by this organism has led mycologists like Remak (1845)[215] and Vanbreuseghem (1962) to place the organism into a separate genus, *Achorion*. However, extensive mycologic investigation of cultural characteristics does not warrant this transfer. Other organisms are capable of producing the same clinical entity. *T. violaceum*, on occasion, may evoke a similar disease, and *Microsporum gypseum* has also been isolated from a few cases. Similar diseases are also produced in animals. *M. gallinae* produces favus of gallinaceous birds, *T. equinum* favus of horses, and *T. mentagrophytes* of the *quinkeanum* variety and *M. persicolor* both produce "mouse favus." *T. schoenleinii* is highly endemic in central and southern Europe, the Middle East, Iran, Kashmir, and Greenland (Fig. 8–4). In North America there are a few small endemic foci, for example, a mountain region in Kentucky, the Gaspé coast of Quebec,[32] a rural area near Montreal, Quebec,[120] and a village near Chichicastenango in Guatemala. The disease is extremely common among the Bantu in South Africa, where it is called "witkop." An endemic focus also exists around São Paulo in Brazil. Formerly, patients with favus were not permitted entry into the United States.

SYMPTOMATOLOGY

Three grades of severity are defined when the infection involves the scalp. The mildest form consists of redness of the scalp in a general follicular distribution and some matting of hair, but without hair loss. In the second grade of severity, the patients show formation of scutula, loss of hair, more redness, and more widespread involvement. Extensive loss of hair (usually over one third of the scalp or more), atrophy of the skin in the bald areas, healing and scarring in the central areas, and formation of new scutula and crusts at the periphery constitute grade three of the disease.

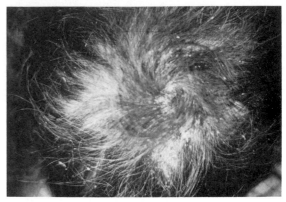

Figure 8–13. Tinea favosa. Seborrheic stage of disease showing matted hair and lesion with erythematous base. The infected hair is grey, whereas the normal hair is pigmented.

The infection begins as small, yellowish-red, subcuticular puncta. There is an erythematous reaction on the scalp and a variable degree of seborrhea and flaking. At this stage the puncta may develop into the cup-shaped yellowish crusts or the lesions may resemble seborrheic eczema. In the latter type of disease, there is extensive scaling, some matting of hair, and an erythematous base to the lesions (Fig. 8–13). The condition closely resembles seborrheic dermatitis and "tinea amiantacea" and must be differentiated from them. Tinea favosa and tinea capitis have a patchy distribution, whereas seborrheic dermatitis is more generalized and usually affects other parts of the body as well. Soon after infection, the hair becomes lusterless and grey as mycelium penetrates the shaft. On microscopic examination mycelium, "air bubbles," and fat droplets are seen intrapilarly. Arthroconidia are very rarely formed and are irregular in shape. As the lesion progresses, the hair is shed and the follicle atrophies (Fig. 8–14,*A,B*). Typically the cup-shaped crusts develop and regress as the lesion advances peripherally. Over a period of years an insidious atrophy of the skin develops as central clearing occurs. Unlike tinea capitis, there is no tendency for the infection to involute at puberty. The infection may last the lifetime of the patient. The extent and appearance of the lesions depend to a large degree on the general hygiene practiced by the patient. Matted hair, debris, and scutula (sometimes called "godets") along with serous exudate, secondary bacterial involvement, pus, and general filth may be present, and the scalp has an unpleasant "cheesy" or "mousy" odor. Removal of scu-

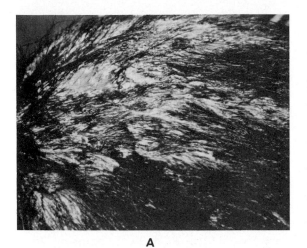
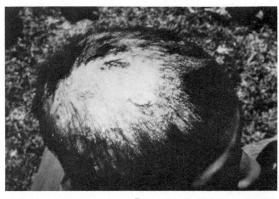

A

B

Figure 8–14. *A*, Tinea favosa. Middle stage of the disease. These are large crusts and scutula covering most of the top of the scalp. *B*, Tinea favosa. Advanced disease involving the entire head. There are areas of atrophied skin and alopecia, in addition to some active lesions of infection. From a small endemic focus in Guatemala. (Courtesy of R. Mayorga.)

tula and debris aids in the treatment of the disease.

Tinea favosa also presents on the glabrous skin. In addition to vesicular, papular, or papulosquamous lesions, typical scutula are formed. The cup-shaped crusts may be very numerous and resemble a range of volcanoes. The skin may atrophy in these areas, a condition not seen in ordinary tinea corporis. *Trichophyton schoenleinii* may also invade the nails. The disease produced in nails is indistinguishable from other forms of tinea unguium.

HISTOPATHOLOGY

Mycelium is present in the horny layer of the scalp, within and around the hairs, and in the scutulum. The scutula consist of intertwined mycelial masses, scales, sebum, and other debris cemented together to form a cup-shaped crust. The periphery is composed of well-preserved hyphae, while in the center, dead and degenerating mycelium and granular debris are found. The scutulum rests on an atrophic epidermis. The dermis shows a mild to moderately severe inflammatory reaction with a round cell infiltrate. The horny layer of the skin often extends over the edge of the scutulum. The etiologic agent, *T. schoenleinii*, elaborates many proteolytic enzymes, including a collagenase and an elastase, which may account for some of the bizarre pathology seen in this disease.[221]

DIFFERENTIAL DIAGNOSIS

When typical, yellowish, cup-shaped scutula are present, along with a mousy or musklike odor of the scalp and a dull green fluorescence of the hairs under Wood's lamp, the diagnosis is readily apparent. The presence of these symptoms appears to vary geographically. In Iran, less than half of the patients examined showed scutula, whereas they were present in a much higher percentage of people surveyed in Quebec.[32] Marked scalp atrophy is also seen in pseudopelade, lupus erythematosus, and lupus vulgaris. Staphylococcic pyodermas and seborrheic eczema also simulate some of the clinical features of tinea favosa.

PROGNOSIS AND TREATMENT

The infection rarely involutes spontaneously, though the crusts, inflammatory reaction, and debris may gradually subside. The thin, atrophied skin of the scalp is denuded of most hair follicles and sebaceous glands, which are replaced by scar tissue. A cicatricial alopecia may cover essentially the entire scalp. However, within the area, incongruously, a few tufts of normal-appearing hair may remain. Mycelium is usually seen in these hairs and is still sparsely present in the scalp as well. In treated cases, the ultimate prognosis for the scalp is favorable. The extent of sequelae depends on the stage of the disease when arrested. It is important to realize that this is a family-centered infection, in contrast to most cases of tinea capitis which are transmitted by group exposure. Several generations in the same family may be infected and show various stages of the disease.

T. schoenleinii has approximately the same sensitivity to griseofulvin as the other dermatophytes. Resolution of infections has been accomplished with long-term use of the drug.

The dosage schedule is the same as for tinea capitis. Cleaning of the debris, removal of crusts, and general improvement of scalp hygiene aid in clinical management. All family members should be treated simultaneously. Transmission probably requires long-term association and exposure, so that casual contacts are usually not infected. X-ray epilation was the standard treatment[250] before griseofulvin but is no longer used.

Tinea Corporis

DEFINITION

Tinea corporis is a dermatophyte infection of the glabrous skin most commonly caused by species of the genera *Trichophyton* and *Microsporum*. The infection is generally restricted to the *stratum corneum* of the epidermis. The clinical symptoms are a result of the fungal metabolites acting as toxins and allergens. Lesions vary from simple scaling, scaling with erythema and vesicles to deep granulomata. Villous hair in the involved area may be invaded, and the follicle often acts as a reservoir for recrudescence of the disease (Fig. 8–6,*B*).

Tinea corporis is a universal affliction of man. The disease is found in all areas of the earth, from the arctic to the equator to the antarctic. However, it is generally more common in the tropics than in temperate climates.

Synonymy. Ringworm of the body, tinea circinata, tinea glabrosa, scherende Flechte (Ger.), herpès circiné trichophytique (Fr.).

ORGANISMS INVOLVED

All species of dermatophytes are able to produce lesions of the glabrous skin, even though some species are more commonly associated with other types of infections. In addition, many species of soil keratinophilic fungi are capable of evoking clinical ringworm. In contrast to tinea capitis, there is little tendency for geographic dominance of a particular species in tinea corporis. Probably the most universally encountered species is *Trichophyton rubrum*, followed in frequency by *T. mentagrophytes*. In areas where there is a heavy infection rate of tinea capitis caused by an endemic species of fungus, there will be a predominance of that species causing tinea corporis. Thus, *T. tonsurans* causing tinea capitis in children is a common agent of tinea corporis in adults who handle children (nurses, mothers, and so forth), and an outbreak of tinea capitis in children owing to *Microsporum canis* and *M. audouinii* will manifest itself in an associated adult population as tinea corporis. Epidemics of tinea in animals are also reflected by epidemics of dermatophytosis of animal origin in humans.[16]

SYMPTOMATOLOGY

Natural infection begins with the deposition of infected scales containing hyphae, or arthroconidia on the skin of a susceptible person. Infection may be transmitted by direct contact with an infected individual or animal; by fomites, such as clothing, furniture, and so forth; or by spread from existing, sometimes subclinical, lesions elsewhere (e.g., occult tinea pedis).

Invasion of the horny layer of the skin occurs at the site of inoculation. This is followed by centrifugal spread from the initial site. An incubation period of one to three weeks elapses before clinical signs are evident. The formation of the characteristic rings of inflammatory reaction is more evident in infections of the smooth skin than in hirsute areas. This annular appearance results from the elimination of the fungus and its irritating products from the center of the lesion as the margin spreads peripherally. A second centrifugal spread of the fungus may occur from the original site, with the formation of concentric rings (Fig. 8–15). Hair follicles act as reservoirs of infection, as is commonly seen with *T. rubrum* infection (Fig. 8–6,*B*).[56] Some lesions lack any tendency toward spontaneous healing and remain scaly (e.g., *T. tonsurans*) or vesicular (e.g., *T. menta-*

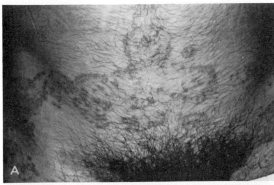

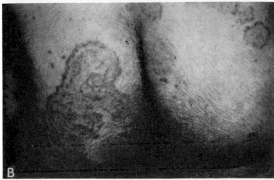

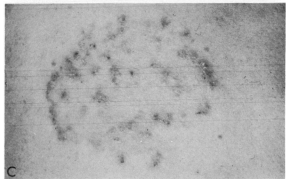

Figure 8–15. Tinea corporis. *A,* Annular appearance of lesions on trunk and (*B*) on gluteal area. (Courtesy of S. Lamberg.) *C,* Annular lesion on the thigh. Such concentric rings are the most commonly encountered acute form of tinea corporis caused by *T. rubrum.*

grophytes). As with other types of ringworm, the more inflammatory the reaction, the shorter the duration of the infection.

Two types of lesions are most commonly encountered. One is dry and scaly *annulare* (annular patches), and the other *vesiculare* ("iris" form). The first begins as a small spreading, elevated area of inflammation. The margin remains red and sometimes slightly swollen, while the central area becomes covered with small scales (Fig. 8–15). Spontaneous healing occurs in the center as the circinate margin advances. Lesions resolve after a few months or may become

chronic and last for the lifetime of the individual. Organisms commonly producing this type of lesion include *T. rubrum* and *Epidermophyton floccosum*. *T. rubrum* is most often associated with the concentric ring forms of tinea corporis (Fig. 8–15,*A,B,C*), whereas a maculopapular rash with a serpiginous border and a hyperemic central area ("eczema marginatum") is more often seen in infections caused by *E. floccosum*. *T. rubrum* infects any area of the body, whereas *E. floccosum* is generally restricted to the crural areas and the feet.

In the second type of lesion, vesicles appear irregularly or immediately behind the advancing hyperemic and elevated margin. A crust is formed, then healing follows in the center of the lesion to leave a more or less pigmented area (Fig. 8–16). The lesions may become pustular when the fungus invades the hair follicles. For any dermatophyte infection, the severity of response is proportional to the involvement of the hair. In both types of lesions, the fungi are most numerous at the margin of the advancing lesion; therefore scales and vesicles from this region are best suited for study and isolation of the etiologic agent. Lesions usually resolve in a few weeks or months. Chronic infections are uncommon. The usual fungi involved are *T. mentagrophytes* and *T. verrucosum*.

Variations of the above two types of clinical lesions have been described. Several confluent annular patches may converge to form a polycyclic lesion. Extensive hyperkeratosis on a red base is called the *psoriasiform* lesion (Fig. 8–17,*A*). The *plaque type* lesion is a stationary, well-defined area with minimal scaling and redness (Fig. 8–18,*A*). It is most often caused by *T. rubrum*, which commonly

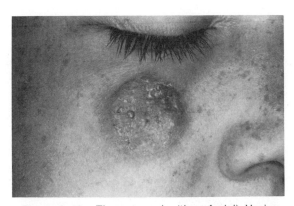

Figure 8–16. Tinea corporis. (tinea faciei). Vesicular type lesion. Vesicles and crusts are seen over entire area. *T. verrucosum.* The infection was acquired from a cow.

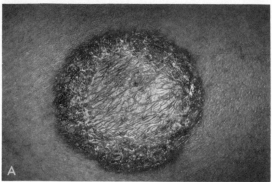

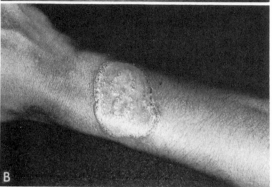

Figure 8–17. Psoriasiform lesion. *A*, Silvery scales covered this infiltrated maculopapular area. Some vesicles are evident in several places on the surface. The periphery was quite erythematous and inflamed, but the overlying epidermis of the center was of a purplish red hue and had a consistency like that of cigarette paper, with many wrinkles and folds. *T. rubrum. B*, Plaquelike lesion. The outer ring is infiltrated and has a red rolled border with a few vesicles, as does the inner ring. The remaining surface of the lesion is smooth, firm, infiltrated, and plaquelike.

becomes chronic, and at times is barely visible. In Figure 8–18,*B* such a lesion has responded to topical therapy; however, the lesion may reappear sometime later from a reservoir of the fungus in the villous hair (Fig. 8–6,*B*). Subsequent lesions are usually of the same morphology as well as in the same location. Pruritus is common in all types of infection. Furthermore, any of these types of lesions may appear to resolve spontaneously or may respond to treatment but reappear months or years later. This is particularly true of the psoriasiform and plaquelike forms of the disease.

Other, more severe types of lesions are sometimes seen. These include *granulomatous lesions, verrucous lesions*, and *tinea profunda*. Small, deep granulomata may be produced around hair follicles. Microscopically these are seen to be perifollicular granulomas caused by fragments of infected hair penetrating the follicle wall (Fig. 8–19,*A*). The lesions are most commonly seen on the legs of women after shaving of leg hairs.[289] A more serious condition with a similar histologic picture is Majocchi's granuloma. Involvement is more extensive, and the granulomas are large and sometimes vegetating. The organism producing both types of infection is *T. rubrum*. Patients with Majocchi's granuloma usually have some underlying disease. Blank and Smith[34] described a patient with an impaired cellular defense mechanism whose infection was accompanied by numerous subcutaneous nodules and abscesses. Biopsy

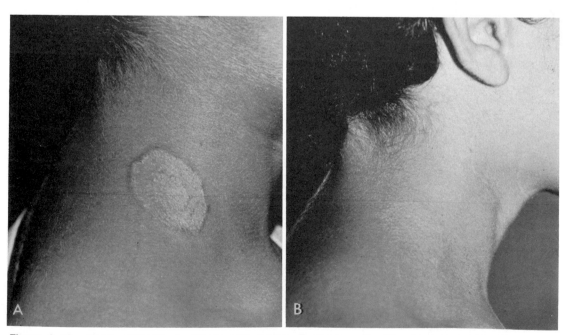

Figure 8–18. Tinea corporis. *A*, Plaque-type lesion. *T. rubrum. B*, After treatment with thiabendazole. (Courtesy of F. Battistini.)

182

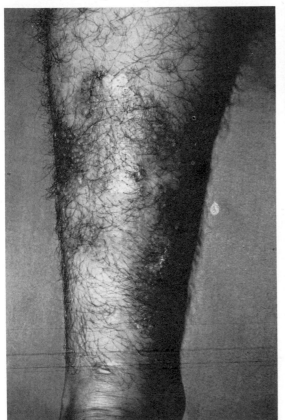

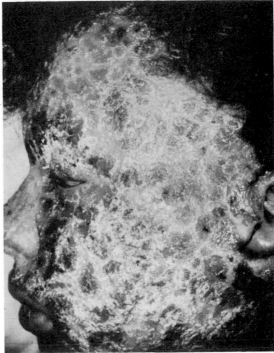

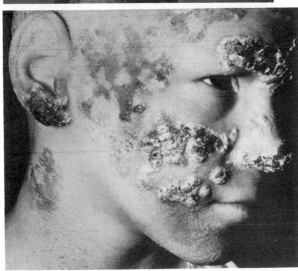

Figure 8–19. *A*, Tinea corporis. Granulomatous reactions around hair follicles, resembling Majocchi's granuloma. There is a serous exudate from some follicles. *T. mentagrophytes*. This infection was the result of wearing boots without socks. (note the lesion stops along a line corresponding to the boot top). The lesions are infiltrated and of a deep purplish hue with an overcast of gray. The overlying epidermis is taut, smooth, and glistening. *B*, Tinea profunda. In this verrucous form of tinea corporis, vegetating lesions were formed on all body sites. A defect in cellular immunity was presumed but not detected. *T. rubrum*. (Courtesy of I. Neuhauser.) *C*, Granulomatous disease associated with systemic spread in a patient with immune deficiency. *M. audouinii*. (From Allen, D. E. R. Snyderman, et al. 1967. Am. J. Med. *63*:991–1000.)

demonstrated invasion of living tissue by mycelium, and the organism was identified as *T. rubrum*. A similar type of disease has been attributed to *M. audouinii* by Allen et al.[8] In this case, the patient had anergy and depressed cellular immunity associated with a lack of a plasma factor required for lymphocyte blastogenesis. A generalized systemic infection developed and was successfully treated with amphotericin B and plasma infusion (Fig. 8–19,*B*). Another rare form of dermatophyte infection is referred to as "pseudomycetomas." In these cases, soft lobulated masses of mycelium are found deep in tissue.[5a] This type of lesion is discussed in the section on tinea pedis.

A verrucous type of ringworm of the glabrous skin has been ascribed to *E. floccosum*.

The face, forehead, ear, and buttocks were covered with verrucous nodules, and the trunk and legs had large, scaly plaques. In some personally observed cases, vegetating masses were seen on the hands and wrists. No invasion of living tissue was observed in biopsy sections of lesions, and no underlying disease was discovered. The lesion resembled those seen in *Candida* granuloma. In other cases, such as trichophytosis totalis (Fig. 8–19,C) (due to *T. rubrum*), agammaglobulinemia was noted. Patients with this disease who have extensive or unusually severe infection generally respond to griseofulvin treatment but universally relapse when the drug is discontinued.

Tinea profunda represents an exaggerated inflammatory response on the glabrous skin and is the equivalent of a kerion of the scalp. Such a reaction can be triggered by the use of topical or systemic steroids.[183]

HISTOPATHOLOGY

A toxic reaction in the epidermis is the first tissue response to the presence of the fungus in the *stratum corneum*. This may subside, become chronic, or develop into an allergic reaction. It presents as a typical eczemaform pattern, although later the histologic picture is essentially that of a nonspecific, subacute chronic dermatitis. Special fungus stains demonstrate mycelium in the horny layer of the skin. If the biopsy is from a dry scaly area, hyperkeratosis and parakeratosis are present. Some acanthosis may occur, which, in combination with papillary edema and a perivascular round cell infiltrate, leads to flattening of the rete. The appearance of tinea corporis, then, is initially one of simple inflammation, and no characteristic features are present. The histologic picture of vesicular lesions is similar, but, in addition, there are subcorneal and intraepidermal vesicles and a more pronounced cellular infiltrate. Intracellular edema affects all parts of the epidermis. Acanthosis is also more marked, and there is often a leukocytic invasion of the epidermis.

In the nodular granulomatous perifolliculitis caused by *T. rubrum*, conidia and mycelium are seen in the hair follicle and in the inflammatory infiltrate of the dermis. Irregularly shaped, conidialike structures have been described which may attain 6 μ in diameter. It is the presence of the conidia and mycelium in tissue after the rupture of the follicle wall which stimulates the foreign body reaction and granuloma formation. Lymphocytes, histiocytes, epithelioid cells, and some foreign body giant cells are seen.[168]

DIFFERENTIAL DIAGNOSIS

Though many diseases may mimic tinea corporis at some stage, the diagnosis is usually straightforward. Tinea corporis is characterized by the appearance on the glabrous skin of an annular papulosquamous lesion, which is either simply scaly or has a vesiculopustular component at the periphery and is crusted in the center. However, atypical infections are numerous, and the possibility should always be entertained that any red scaly rash on the body may be of fungal etiology. Psoriasis, pityriasis rosea, nummular eczema, granuloma annulare, annular secondary syphilis, lichen planus, seborrheic dermatitis, contact dermatitis, fixed drug eruption, pityriasis versicolor, dermatocandidiasis, and erythema annulare are some of the diagnoses to be considered. Nummular eczema is commonly confused with tinea corporis, but the papulovesicular plaques tend to be more symmetrical. The symmetrical pattern of seborrheic dermatitis also distinguishes it from fungal infection, and there usually is involvement of the scalp and intertriginous areas as well. Although the "herald patch" of pityriasis rosea is indistinguishable from tinea corporis, the restriction of the former to the trunk and the symmetrical distribution of lesions delineate this disease. Tinea corporis is readily and rapidly diagnosed by the potassium hydroxide mount (Fig. 8–41). Examination of scales will show septate hyphae and squared or rounded, irregularly arranged arthroconidia. Cultures should always be taken. In our experience, the number of culture-positive cases when the KOH was negative ranges from 5 to 15 per cent.

PROGNOSIS AND THERAPY

In normal patients tinea corporis resolves spontaneously after a few months. There is less tendency toward chronicity than in tinea pedis and tinea cruris. Treatment aids in resolution of lesions and effects a clinical cure. Reinfection of the same area may occur within a few weeks to months, if the patient is again exposed to infectious material. In some patients lesions of tinea corporis reappear at regular intervals. This is particularly true of infections caused by *T. rubrum*. This species often forms ectoendoconidia in villous hair (Fig. 8–6,B). Such hair acts as a reservoir for the fungus and leads to recurrence of the disease.

Uncomplicated tinea corporis of the annular, plaque, or vesicular type can be treated topically. There are several drugs presently available that are useful in treatment regimens. These include tolnaftate, miconazole, clotrimazole, haloprogin, the triclosans, and econazole.[237, 259] These are quite effective, either as lotions, solutions, or creams for lesions of limited size in accessible areas. The imidazole and haloprogin preparations have the advantage of being broad spectrum antibiotics and effective against *Candida* spp. and in some cases bacteria. Logamel, a new substituted triclosan is also broad spectrum and has about the same effectiveness as the imidazoles.[237] Tinea corporis due to *T. rubrum, T. mentagrophytes, M. canis, M. audouinii*, and generally *T. tonsurans* are amenable to treatment by these drugs. More vigorous therapy is sometimes required for infections by *T. verrucosum* and *T. violaceum*. Cleansing of the area to remove scales is a useful adjunct to therapy. Older modes of treatment include 5 per cent ammoniated mercury ointment, Pragmatar ointment, Aquafor containing 3 per cent sulfur and 3 per cent salicylic acid, tincture of iodine, and 1 per cent sodium omadine.[255] These will also effect clearing of the infection in about two weeks, in uncomplicated cases. Widespread tinea corporis and the more severe types of lesions (granulomatous, verrucous, and tinea profunda) may require systemic griseofulvin therapy.

The treatment schedule is usually 1 g administered daily in divided doses or 500 mg of the micronized preparation. Itching, erythema, and scaling diminish by one week or less, owing to the anti-inflammatory properties of griseofulvin. Although griseofulvin applied topically has not been effective in treating infections, some evidence indicates that when incorporated in DMSO to increase its penetration, it is clinically useful.[67] The course of therapy is continued for six weeks to two months. When tinea pedis and especially tinea unguium are also present, a longer course of treatment is required. Excellent results have been obtained using thiabendazole (10 per cent in Aquaphor cream). Battistini* obtained nearly 100 per cent cure in 250 cases.

The response to treatment is quite variable regardless of the drug used. Clinical evidence indicates increasing resistance of chronic infections to treatment with griseofulvin, and resistant mutants have been produced in the laboratory,[149] as well as having been isolated from cases of treatment failure.[15] There are some possible side effects in any of the therapeutic agents now utilized. Though never associated in human therapy, griseofulvin has been implicated as inducing tumors in rodents,[240] and a few cases of porphyria have occurred in patients during treatment. Other side effects are very rarely encountered. Haloprogin is associated with skin sensitivity with some frequency, and allergy to miconazole has been reported.[244]

---

*Personal communication 1970.

Tinea Imbricata

DEFINITION

Tinea imbricata is a geographically restricted form of tinea corporis caused by *Trichophyton concentricum*. It is characterized by polycyclic, concentrically arranged rings of papulosquamous patches of scales scattered over and often covering most of the body.

Synonymy. Tokelau, Burmese, Chinese, Indian ringworm, Lofa tokelau, tinea circinata tropical, Gogo.

DISTRIBUTION

The disease occurs in the Pacific islands of Oceania, Southeast Asia, and Central and South America. The distribution of the disease may be of anthropological value. It has been suggested that the infection was introduced to the western coasts of South and Central America by pre-Columbian voyagers from Polynesia.[28] This contradicts Thor Heyerdal's theory of population migration, as presumed in Kon Tiki. Tinea imbricata occurs sporadically among the Indians of Mexico,[28, 277] Guatemala, Panama, and Brazil (Fig. 8–1). The organism appears to be population-group specific, as it is found only in Indonesians and Polynesians and their probable descendants in the Americas. Peoples of other races usually do not acquire the infection even after years of intimate association with infected individuals.

SYMPTOMATOLOGY

The disease begins with the appearance of brownish maculopapules which gradually increase in size. The central portions of the lesions become detached, and fissures develop toward the periphery. Around this border, a brownish zone appears. The margins are elevated. By passing a finger over the lesion from the center point to the active edge, a slight resistance can be felt. Layers of the stratum corneum become detached, with the free edges facing the center (Fig. 8–20). This process continues until numerous concentrically arranged, imbricated rings are formed from the initial central lesion. The process by which annular rings are formed in tokelau is quite different from that of annular tinea corporis. Scaling may be profuse and itching severe at first. There is little or no erythema. When the disease is chronic, there is little irritation or discomfort to the patient (Fig. 8–21). The polymorphic, polycyclic patches are considered marks of beauty by some of the populace in Thailand and other countries. The face and scalp may be in-

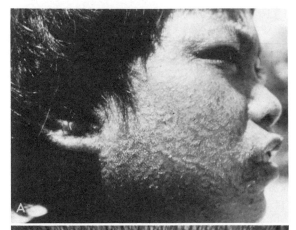

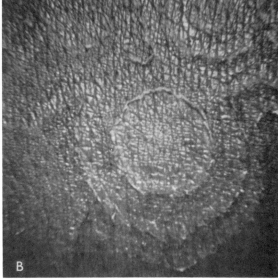

Figure 8–21. Tinea imbricata. *A,* Chronic involvement of face. *B,* Close-up of lesion. Note free edge of scale is toward center. (Courtesy of C. Halde.)

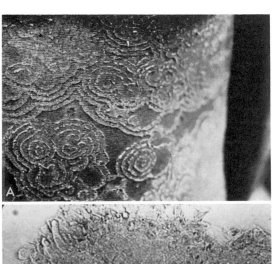

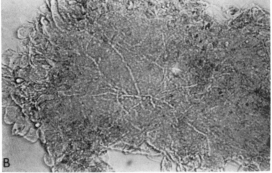

Figure 8–20. Tinea imbricata. *A,* Concentric rings forming over chest. Free edge of scales face center of lesion. *B,* Potassium hydroxide mount of infected skin scale. (Courtesy of C. Halde.)

volved, but the hair is spared. If the infection involves the skin over the antecubital fossa, transverse parallel ridges are formed.[160] Nail infection is not uncommon. The etiologic agent. *T. concentricum*, may be accompanied in these lesions by other dermatophytes, e.g., *T. rubrum* and *E. floccosum.* In these cases, the symptoms are usually more severe.

The disease is more common in rural than urban areas. It is assumed that the organism is transmitted by direct, intimate contact, as from a mother to her baby. Thus, the first lesions are found on the face where the cheek of the child touches the breast of the mother. Once established, the disease becomes chronic and is of lifelong duration. There is no sex or age predilection, and a high percentage of an entire community is often affected. In some studies, it was not uncommon to find some members of a family with long-established disease, while other members were without infection. This emphasizes the low infectivity of *T. concentricum*. In a survey of dermatomycoses of some islands in the Pacific Ocean, tinea imbricata was found only among Polynesians and not among Caucasians. Some of the latter had lived for generations in the islands and had had direct contact with the Polynesians.[160, 257]

DIFFERENTIAL DIAGNOSIS

The appearance of the polycyclic, polycentric rings with no evidence of erythema is so characteristic that there is little confusion in diagnosis. If the morphology is obscured, the scaling may resemble a form of ichthyosis.

PROGNOSIS AND THERAPY

The disease is chronic and extremely resistant to treatment. It responds readily to griseofulvin, but the required course of therapy is long, and relapse is common. The new lesions are multifocal and inflammatory.[160] Relapse may be accompanied by a rather severe vesicular reaction, with an erythematous base. Itching and irritation are sometimes severe, and the patient is worse off than before therapy was instituted. Complete cures have been recorded with a total dose of 24 g of griseofulvin given over 18 days.[42]

Tinea Cruris

DEFINITION

Tinea cruris is a dermatophyte infection of the groin, perineum, and perianal region, which is acute or chronic and generally severely pruritic. The lesion is characteristically sharply demarcated, with a raised, erythematous margin and thin, dry epidermal scaling. The disease is found in all parts of the world but is more prevalent in the tropics. It tends to occur when conditions of high humidity lead to maceration of the crural region. A similar condition may involve the axilla or other intertriginous areas.

The disease is most common in men (Fig. 8–22) and rarely involves women, except when it is transmitted by intimate contact or fomite (Fig. 8–23). It may reach epidemic proportions in athletic teams, troops, ship crews, and inmates of institutions. In such cases, it is probably most commonly transmitted by towels, linens, and clothing. *E. floccosum*, an etiologic agent of this condition, has been isolated from blankets and sheets.[24]

Synonymy. Ringworm of the groin, Dhobie itch, eczema marginatum, jock itch, gym itch.

ORGANISMS INVOLVED

Infection of the groin usually accompanies dermatophyte disease of the feet, so that the flora involved is commonly the same. In surveys conducted in England, Northern Ireland, Portugal, and Denmark, up to one half of the cases were caused by *Epidermophyton floccosum*. In the United States, *T. rubrum* appears to be the predominant species responsible for tinea cruris. A similar, common disease is caused by *Candida albicans*. *Trichophyton mentagrophytes* is associated with the more inflammatory pustular type of tinea cruris. An unusual and interesting case was a report of *M. gallinae* as an agent of tinea cruris.[87]

SYMPTOMATOLOGY

The disease begins as a small, round, swollen area of inflammation. At first, this spreads to produce a circinate lesion, which later, because of differences in rate of spread, becomes serpiginous (Fig. 8–22,*A*). In lesions caused by *E. floccosum*, there is a well-marginated, raised border (eczema marginatum) studded with numerous small vesicles or vesiculopustules filled with a serous exudate. Clipping the top of the vesicles for examination aids in the demonstration of the fungus. The central portion is brownish to red in color and covered with thin, branny, furfuraceous scales. The lesions are commonly bilateral, but not necessarily symmetrical. In most instances, the infection begins on the thigh where it is in contact with the scrotum, and spreads rapidly. There is evidence that the disease begins with dermatophytic colonization of the scrotum, although the infection is not clinically evident. The infection is transferred to the inner aspect of the thigh (usually the left) and subsequently develops into a pruritic, erythematous, symptomatic rash.[145] It may sometimes involve the penis, although this is rare (Fig. 8–22,B). The disease usually involves the inner thighs and spreads downward farther on the left, because of the lower extension of the scrotum on that side. Sometimes the gluteal and pubic regions are involved. If *E. floccosum* is the etiologic agent, infection rarely extends farther. In *T. rubrum* infections, the lesions frequently extend over the body, particularly to the waist, buttocks, and thighs (Fig. 8–15). Infections with *T. mentagrophytes* may rapidly involve the chest, back, legs, and feet and cause a severe, incapacitating inflammatory disease.

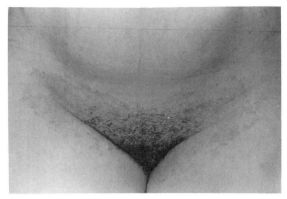

Figure 8–23. Tinea cruris. Extension of lesion from crural area. *T. rubrum*. (Courtesy of F. Battistini.)

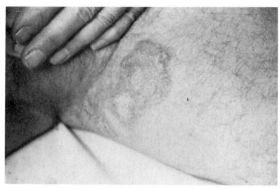

A

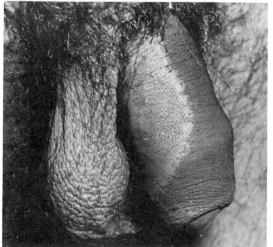

B

Figure 8–22. *A*, Tinea cruris. Serpiginous lesion of left thigh in contact with scrotum. *E. floccosum*. This type of lesion is also called eczema marginatum. There is some infiltration of this maculopapular rash, the color is slightly reddish, and the surface is branny and furfuraceous. *B*, Tinea cruris involving the shaft of the penis. This is a rarely encountered form of the disease and is most often associated with *T. rubrum*. (Courtesy of B. Kumar.)

In acute infections, erythema is present, and there is intense itching. Older lesions are often lichenified, leathery, and plaquelike. Rarely, lesions appear as solitary vesiculopustules without a marginated border and have little tendency to spread. If there is no secondary bacterial involvement, the lesions are dry and branny. Rarely they may be soggy, whitish, and macerated, more closely resembling a *Candida albicans* infection. In contrast to dermatocandidiasis, the scrotum and penis are usually symptomatically spared in tinea cruris (Fig. 8–22,B). Experimental inoculations to the scrotum do result in evident clinical disease, but it is of shorter duration than that of the thigh. It appears, however, that the scrotum may act as a reservoir for recurrent infections.[145]

PREDISPOSING FACTORS

Concentration of individuals in intimate surroundings, such as barracks, dormitories, and locker rooms, leads to rapid spread of the infection among isolated populations. This type of epidemic usually involves *Epidermophyton floccosum*. This anthropophilic organism is not associated with chronic human infections, but it does produce resistant arthroconidia in skin scales that remain viable for long periods of time. Thus, rugs, shower stalls, locker room floors, and furniture may harbor infectious squames for years after use by an infected person. Both tinea cruris and tinea pedis may be acquired in this way. *Trichophyton rubrum* does not appear to remain viable in shed scales for an appreciable time; however, it is associated with long chronic infections in which infectious squames are continually being dropped. *T. mentagrophytes* infections of animal origin are severe and of short duration. They can be acquired from fomites harboring animal (particularly rodent) dander. These factors must be taken into consideration when dealing with the public health aspects of tinea cruris or any dermatophytosis.

Perspiration, humidity, irritation from clothes, and other factors that cause maceration of the crural skin increase the susceptibility to dermatophyte infection. Such diseases as diabetes, neurodermatitis, leukorrhea, and friction from skin folds in obese persons frequently are cited as predisposing factors.

DIFFERENTIAL DIAGNOSIS

The typical appearance of eczema marginatum, with its raised border and actively advancing periphery, is diagnostic of tinea cruris. If the lesions are weeping and satellites occur beyond the main lesion, the etiologic agent is probably *Candida albicans*. This is especially true in women, in whom the infection is often associated with vaginal candidiasis. Candidal infection is much more common in diabetics than dermatophytosis. Other conditions that may simulate tinea cruris include erythrasma, seborrheic dermatitis, psoriasis, lichen planus chronicus, and contact dermatitis. Seborrheic dermatitis is usually more symmetrical in distribution and rarely restricted to the groin. Psoriasis is more scaly and has a more erythematous base than tinea cruris. In the obese, bacterial infection or simple intertrigo may be common. Lichen planus, contact dermatitis, and eczema are unlikely to be symmetrical and may closely simulate tinea cruris. Use of the Wood's lamp to demonstrate coral red fluorescence is helpful in ruling out erythrasma. In all cases, diagnosis is more frequently made by examination of a potassium hydroxide mount. Cultures are necessary in suspected lesions, even when the KOH is negative.

PROGNOSIS AND THERAPY

With adequate local or systemic therapy and sterilization of clothing, linens, and so forth to prevent reinfection, the prognosis is good unless the etiologic agent is *T. rubrum*. In the latter case a chronic disease may ensue, involving the body, feet, hands, and nails. In all cases of tinea cruris, special attention should be paid to the increased probability of the presence of a clinically evident or occult tinea pedis.[231] Recurrence in uncomplicated cases is prevented by elimination of predisposing factors. Tight-fitting underwear, athletic supporters, sweating, macerating conditions, and obesity are all potential inciters of exacerbation.

Topical treatment with tolnaftate (Tinactin) is usually successful and has few side effects. The topical agents listed in the section on tinea corporis are also useful in tinea cruris. Because of greater sensitivity of the skin in the crural area, other forms of topical therapy often result in a weeping, irritating dermatitis, which is frequently complicated by secondary bacterial infection. Systemic griseofulvin (500 mg per day) is most useful, particularly in *T. rubrum* infections. Symptoms are relieved within three days, and lesions involute within four to six weeks. The course of treatment should be continued until all clinical, microscopic, and cultural signs of the disease disappear. The rapidity with which symptomatic relief is obtained following administration of griseofulvin suggests there is an effect other than an initial direct action on the fungus. Deposition of griseofulvin in the horny layer with its subsequent inhibition of growth of the dermatophyte would require a longer time lapse to produce relief of symptoms. It has been suggested that, in addition to its influence on fungal metabolism, the initial effect of the drug is one of detoxification or anti-inflammatory action. Another topical agent, thiabendazole (10 per cent cream), is also very effective, as is haloprogin and the imidazoles. The latter are also effective in dermatocandidiasis. There appears to be a greater incidence of skin sensitization associated with the use of haloprogin than with tolnaftate or the imidazoles. Ganor et al.[74] found 2 per cent miconazole cream quite effective in conditions caused by *Trichophyton rubrum*, less so with other etiologic agents.

Tinea Unguium

DEFINITION

Tinea unguium is an invasion of the nail plates by a dermatophyte. For consistency of terminology, this disease is differentiated from onychomycosis, which is an infection of the nails caused by nondermatophytic fungi and yeasts.[24] The disease, tinea unguium, is of at least two types: (1) leukonychia mycotica (superficial white onychomycosis), in which invasion is restricted to patches or pits on the surface of the nail; and (2) invasive, subungual dermatophytosis (ringworm of the nail), in which the lateral or distal edges of the nail are first involved, followed by establishment of the infection beneath the nail plate.

Synonymy. Ringworm of the nail, dermatophytic onychomycosis.

INTRODUCTION

Persistent subungual dermatophytosis was noted by Mahon in the 1860's.[198, 294, 295] The infection was contracted in a fingernail used for epilation of hairs from patients with favus. This classic form of tinea unguium was considered rare until recently. Current surveys indicate the case rate is increasing, and up to 30 per cent of patients with fungus disease of the skin also have tinea unguium.[296] Twenty per cent of all nail disturbances[198] are due to fungi. Some of the apparent increase is probably attributable to better diagnostic methods. The disease and its mycology have recently been reviewed by Zaias.[296]

In the past, leukonychia mycotica was difficult to distinguish from other forms of leukonychia caused by systemic disorders or local insults. The disease was described in 1921 by Ravaut and Rabeau and its association with *T. mentagrophytes* established.[211] These investigators were unable to see mycelium in the nail scrapings, but fungi were isolated in cultures. Jessner[118] described several more cases and named the disease leukonychia trichophytica. The term leukonychia mycotica was first used in 1926 by Rost. A comprehensive review of the clinical and mycological aspects of this disease, under the name superficial white onychomycosis, is given in Zaias.[296] Leukonychia mycotica is the preferred term, as it avoids confusion with other conditions and their etiologic agents. The disease is restricted to toenails.

ORGANISMS INVOLVED

Almost all species of dermatophytes have been isolated from ringworm of the nail. The etiologic agents are usually those which are common or endemic in the population. Since invasive ringworm of the nails is generally associated with infection of other cutaneous areas, the causative agents are often the same. Tinea unguium of the fingernail is most commonly due to *T. rubrum*; however, the toenail may be infected by a variety of organisms. Nail involvement associated with tinea corporis and tinea pedis occurs most commonly with *T. rubrum*, *T. mentagrophytes*, and *E. floccosum*; with tinea capitis and tinea favosa, *T. tonsurans*, *T. violaceum*, *T. megninii*, and *T. schoenleinii*; and with tinea imbricata, *T. concentricum*. Rarely encountered species include *M. gypseum*, *M. canis*, *M. audouinii*, *T. soudanense* and *T. gourvillii*.

Leukonychia mycotica is frequently an isolated lesion which is not associated with other forms of dermatophytosis. The most commonly isolated etiologic agent is *T. mentagrophytes*. Both the granular and interdigitale types of this fungus are encountered. Rarely involved organisms include *Acremonium roseogriseum*, members of the *Fusarium oxysporum* group, and *Aspergillus terreus*.

SYMPTOMATOLOGY

Invasive or subungual ringworm of the nail usually begins at the lateral or distal edge of the plate. A minor paronychia usually precedes the infection and may become chronic or resolve. Paronychial inflammation results in production of a pitted or grooved surface on the nail. The initial symptom of nail involvement is a small, well-outlined, yellow or whitish spot which spreads to the base of the nail, or may remain stationary for years (Fig. 8–24). In established infections, the nail plate is brittle, friable, and thickened and may crack because of the piling up of subungual debris. Its color is often brown or black (Fig. 8–25). The accumulation of subungual keratin and debris is considered to be the characteristic feature of tinea unguium.[260] Under normal conditions, the nail bed does not contribute to the keratinization of the nail plate, nor does it form keratin under it. External stimulation or irritation due to the presence of the fungus in the nail bed evokes

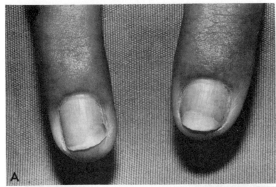

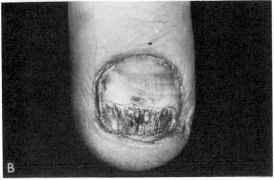

Figure 8–24. *A,* Tinea unguium. Invasive type. Initial infection at distal edge of nail plate. (Courtesy of F. Battistini.) *B,* Advanced disease showing grooved dark brown coloration.

the production of soft, friable keratin. This loosens the nail, and as keratin accumulates, distortion and apparent thickening of the plate occur. This is in distinct contrast to candidal onychomycosis, in which accumulated debris is absent and the nail is usually not thickened. The cheesy mass of epidermal detritus and keratin provides a fertile milieu for the rapid growth of fungi. Direct invasion of the hard nail plate may occur from below. Irritation of the nail bed stimulates more keratinization, and the nail becomes grossly

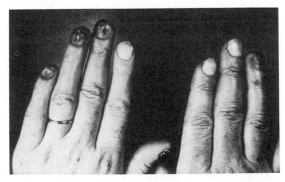

Figure 8–25. Tinea unguium. Advanced disease involving several nails. The infection is chronic and of many years' duration, but a few nails remain uninvolved. *T. rubrum.*

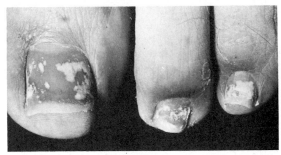

Figure 8–26. Tinea unguium. Leukonychia mycotica. Multiple white, irregular lesions on surface of nail. *T. mentagrophytes.* (Courtesy of N. Zaias.)

distorted. A diverse flora consisting of other fungi and bacteria is found as a secondary invader in the debris. The nail matrix is spared. In some cases, minimal architectural changes and discoloration occur, and in rare instances no visible abnormalities are apparent. Another complication that is more common in *T. rubrum* infection is the cracking and separation of the distal part of the plate, leaving a thin furrowed base with ragged edges (Fig. 8–24,*B*). This occurs more commonly in fingernail infections. Involvement of the whole nail may lead to destruction of the entire appendage.

Leukonychia mycotica (superficial white spot tinea unguium) begins as an opaque, circumscribed area on the surface of the nail plate. These lesions are usually punctate at first, irregular in outline, and may be numerous or solitary (Fig. 8–26). They begin at any place on the nail surface — in the center of the nail, near the lunula, in the free edge, or in the lateral folds. An established infection sometimes spreads to involve the entire surface or may remain restricted. The surface of the nail is soft and crumbly. The infection is otherwise asymptomatic and, because of its separation from living tissues, does not elicit an inflammatory response. The infection may be chronic and last for many years.

HISTOPATHOLOGY

In the subungual invasive type of tinea unguium, fungi are readily discernible in PAS-stained sections of the infected nail. The hyphal filaments and arthroconidia are aligned horizontally between the lamellae of the nail and are generally limited to the lowermost portion of the plate. Onycholysis is not a feature of dermatophytic infection; rather, the lamellae are mechanically separated by the fungi growing between them.[210] Distribution and concentration of the fungus in the nail are quite variable and lead to difficulty when the diagnosis is made by potassium hydroxide mounts or histopathologic sections. There is little or no inflammatory response in the underlying tissues.

In contrast to subungual tinea unguium, the mycelial elements in leukonychia mycotica are restricted to the uppermost portion of the nail plate. Fungal invasion rarely involves the deeper layers. Abundant hyphae are seen. As described by Zaias,[296] these are larger and broader than those seen in subungual involvement. They appear to be similar to the "penetrating organs," "eroding fronds," and "carpal bodies" that have been described in soil organisms and dermatophytes during their saprophytic utilization of keratin.[62] Aggregates or masses of distorted hyphae and irregularly shaped arthroconidia are often present in sections. In contrast to the more "parasitic" picture seen in subungual tinea unguium, leukonychia mycotica represents an essentially "saprophytic" condition.

DIFFERENTIAL DIAGNOSIS

Subungual tinea unguium is often difficult to diagnose because of the scarcity of fungi and their location in the lowermost sections of the nail plate. Abnormalities of the nail which simulate the condition are congenital, the result of systemic disease, or due to external causes. The congenital conditions include nonmycotic leukonychia, clubbing, Beau's lines (transverse grooves or lines), and pachyonychia congenita. Among the external causes are contact irritants, trauma, onychomycosis due to filamentous fungi and yeasts, onychogryposis, onychophagy, onychotillomania, viral and bacterial infections, neoplasms, ingrowing toenails, subungual exostoses, and fibromas of tuberous sclerosis. Many skin diseases affecting the dorsal skin of the fingers or toes may cause dystrophic nails: eczema, lichen planus, bacterial paronychia, Darier's disease, scleroderma, syringomyelia, Raynaud's disease, hyperthyroidism, keratoderma blennorrhagica, keratoderma palmaris, acrodermatitis perstans, exfoliative dermatitis, and idiopathic onycholysis. Subungual conditions that rarely mimic tinea unguium are hemorrhages due to bacterial endocarditis or trichinosis, the periungual bulla of pemphigus, and argyria. It is apparent that the differential diagnosis covers a wide range of dermatologic and systemic diseases. The condition most closely simulating tinea unguium is psoriasis of the nails. It is extremely difficult to differentiate the two without evidence of psoriasis on other areas of the body. Distorted, deformed, thickened, discolored nails with an accumulation of debris beneath them, particularly with ragged and furrowed edges, strongly suggest tinea unguium. *Candida* onychomycosis lacks gross distortion and accumulated detritus. Most of the other diseases listed commonly involve several nails and are symmetric in distribution. Inexplicably, tinea unguium may involve a single nail and have an asymmetric distribution. A chronic infection of many years' duration may involve one nail, while another in close proximity remains normal. Leukonychia mycotica is mimicked by leukonychia of other etiology, particularly trauma. Mycologic confirmation is the final proof of diagnosis in all cases.

PROGNOSIS AND TREATMENT

Tinea unguium is the form of dermatophytosis that is most resistant to treatment. It rarely, if ever, resolves spontaneously, and recurrence of the disease in clinically cured nails is common. Topical treatments alone have had a very poor record of cure. Formerly, evulsion of the nail followed by application of fungistatic agents was the only procedure which met with any success. Complete ablation has sometimes been used,[89] but this is an extreme and unwarranted procedure. The condition is chronic, lifelong, resistant to treatment, and may occur in infants[119] as well as adults.

Systemic griseofulvin therapy has resulted in complete remission of the disease in some patients. The course of therapy is long (a year or more), and good results are not assured. In one study 80 per cent of patients with fingernail infections were cured after ten months on a dose of 1 g per day, compared with 12 per cent of patients with toenail infections.[116] In another study 8 of 14 patients still had infected toenails after 15 months of griseofulvin therapy, whereas almost total cure of fingernail infections was achieved.[260] Filing down of the nail to paper-thin consistency, followed by soaking in potassium permanganate (1:4000) or painting with phenol, 10 per cent salicylic acid, 1 per cent iodine, or chrysarobin (20 per cent in chloroform), is a useful adjunct to systemic griseofulvin therapy.[116] Some success has been noted with thiabendazole. Glutaraldehyde, 25 per cent diluted 1:1 with phosphate buffer, has effected cure in several cases but was ineffective in others. In an extensive series, Achten et al.[2] used long-term therapy with 2 per cent miconazole in alcohol on a variety of forms of onychomycosis. Their rate of clinical cure was equal to or better than that cited for systemic griseofulvin.

Onychomycosis

DEFINITION AND ORGANISMS INVOLVED

As a general term, onychomycosis includes any infection of the nail produced by a fungus. Since infections caused by dermatophyte fungi have a characteristic evolution, pathogenesis, and therapy, they are considered separately and termed tinea unguium. The remaining infections are caused by a heterogeneous group of filamentous fungi and yeasts. Most of these have been found in dystrophic nails, but their significance in either initiation or aggravation of the condition is usually questionable. Primary invasion of the nail plate by *C. albicans*, *Geotrichum candidum* and *Scopulariopsis brevicaulis* is well established.[38] In a recent review, Zaias[296] lists the following fungi as confirmed etiologic agents of onychomycosis: *Aspergillus candidus*, *A. flavus*, *A. fumigatus*, *A. sydowi*, *A. terreus*, *A. ustus*, *A. versicolor*, *Acremonium* sp., and *Fusarium oxysporum*. Other species sometimes encountered include *Arthroderma tuberculatum* and *Phyllosticta sydowi*. In our clinic, the presence of *Cheatomium globosum* has been confirmed in three cases. Isolation of many other species has been reported, and in several of them, filaments were seen by direct microscopy.[294, 295, 296] Aside from *C. albicans*, the most frequently implicated yeasts are *C. parapsilosis* and *Trichosporon beigelii*[216] (Fig. 8–27,*B*). Dystrophic nails harbor a large flora of saprophytes and secondarily invading fungi, yeasts, and bacteria. The black yeastlike organism *Hendersonula toruloidea* has been involved in infections of the nails as well as the feet.[76, 159]

HISTOPATHOLOGY

Active invasion of the nail plate occurs with *Scopulariopsis brevicaulis*. The characteristic conidia can be seen within the body of the nail. Candidiasis of the nails is usually associated with chronic paronychia, which results in distortion of the nail architecture, and a chronic inflammatory response (Fig. 8–27). The yeasts are seen within the nail as well as adnexal tissue. The other fungi are usually located in grooves and cavities of the nail where there is an accumulation of debris. Active invasion of the nail plate is less frequent and of sporadic distribution. In sec-

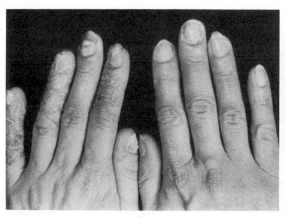
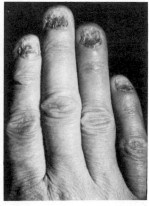

A **B**

Figure 8–27. *A*, Onychomycosis. *Candida albicans*. This is a patient with chronic mucocutaneous candidiasis associated with a thymoma. The architecture of the nail is distorted, and there is a chronic paronychia. *B*, Severe infection and destruction of the nails caused by *Trichosporon beigelii*. (Courtesy of A. Restrpo-Moreno.)

tions mycelial filaments are seen which resemble those found in tinea unguium.

PROGNOSIS AND TREATMENT

Most cases of onychomycosis occur in abnormal nails. The disease often resolves when the antecedent condition is corrected. In chronic or genetic diseases of the nail in which therapy is not possible, the fungal infection may become chronic. Essentially all of the organisms causing onychomycosis are resistant to griseofulvin. *Candida* infections and their associated paronychia are effectively treated by nystatin ointment (100,000 U per g). Aqueous nystatin is more effective, but the drug is unstable in this milieu. Topical amphotericin B, gentian violet (0.5 per cent), resorcin (10 per cent solution in 70 per cent alcohol), and iodine (1 per cent in chloroform) have been used to treat yeast and filamentous fungal infections of the nail. Some success has been found using thiabendazole (10 per cent in cream base) under occlusive dressing. Glutaraldehyde may prove the treatment of choice after several series have been studied.

Tinea Barbae

DEFINITION

Tinea barbae is a dermatophyte infection of the bearded areas of the face and neck and therefore is restricted to adult males. Lesions are of two types: a mild superficial type that resembles tinea corporis, and a type in which there is a severe, deep, pustular folliculitis (Fig. 8–28).

Synonymy. Tinea sycosis, barbers' itch, trichophytie sycosique [Fr.], parasitare Bartifinne [Ger.], ringworm of the beard.

ORGANISMS INVOLVED

Tinea barbae infections are more common in rural areas, and the organisms are usually

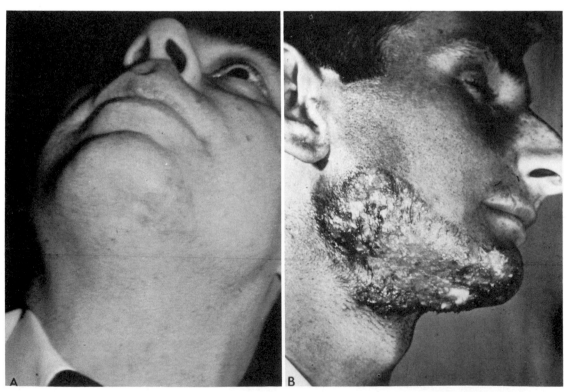

Figure 8–28. Tinea barbae. *A,* Mild superficial type due to *T. mentagrophytes* (JWR). Other isolants of this species may produce a more severe inflammatory disease. *B,* Deep, pustular folliculitis in a patient with *T. verrucosum* infection.

acquired from animals; therefore they are generally zoophilic dermatophytes. As noted previously, the severity of infection caused by zoophilic dermatophytes is often greater than that produced by anthropophilic fungi. In addition, the severity of the host reaction is also much greater when hair is involved. The combination of these two factors may explain the extremely severe reactions seen in some patients with tinea barbae. The most common organisms involved are *T. mentagrophytes* and *T. verrucosum*, both of which may be acquired from cows. *T. mentagrophytes* is also acquired from horses and dogs. *M. canis* is an uncommon cause of tinea barbae. In areas where *T. schoenleinii* and *T. violaceum* are endemic, they are frequently involved in this disease. Though the latter are anthropophilic fungi, they evoke a severe infection, probably because of hair and follicular involvement. *T. rubrum* is an infrequent cause of tinea barbae and may represent infection acquired from other parts of the body or transmitted as "barbers' itch" from unsanitary barbering practices. A geographically restricted species, *T. megninii*, is not infrequently isolated from barber-transmitted infections in its endemic areas. This organism, not prevalent in any country, is found in Portugal, Sardinia, Sicily, Africa (as *T. kuryangei*)[275] and rarely in other parts of Europe.

Once a frequently encountered infection, tinea barbae now is uncommon. Most infections were acquired in barber shops when men frequently had a "shave and haircut" and the barber used the same razor on succeeding customers. The introduction of disinfecting dips for barbering tools and the common use of "at home" safety razors has largely eliminated this disease. At present most infections are acquired from animals.

SYMPTOMATOLOGY

The superficial type of tinea barbae resembles the lesions of tinea corporis. There is central scaling and a vesiculopustular border. The host reaction is less severe, though alopecia may develop in the center of the lesion. Hair and follicle involvement is less pronounced than in the deep type of infection. *T. rubrum* is generally the causative agent.

The deep or pustular type of tinea barbae is characterized by the presence of deep, follicular pustules that may result in the formation of the nodular, kerion-like lesions seen in tinea capitis (Fig. 8–28). These pustular lesions are initially truly mycotic, and the pus is full of fungal arthroconidia. The reaction may be so severe that most of the hair is shed, leading to spontaneous resolution of the disease. Permanent alopecia and scarring are common. The lesions are boggy and edematous. The hairs, when epilated, are seen to have a pussy, whitish mass involving the root and surrounding tissue. Draining sinuses develop and undermine the surrounding tissue. Slight pressure evokes extrusion of purulent material. The lesions are usually solitary and most frequently are found on the maxillary regions. Occasionally the whole bearded area is involved, and extensive reddish-purple verrucose indurations are formed. Enlarged regional lymph nodes, mild pyrexia, and general malaise may accompany severe infections, especially those caused by *T. verrucosum*.[103] The upper lip usually is spared in tinea barbae, in contrast to the bacterial infection sycosis vulgaris.

HISTOPATHOLOGY

The cellular reaction to tinea barbae is similar to that produced in the more severe types of tinea capitis. Organisms may be seen in the hair shaft and the follicle, and large numbers of arthrospores are present both on the shaft and free in the cellular debris. Sometimes organisms are absent and only an acute pyogenic infiltrate is seen. In chronic and resolving lesions, a chronic inflammatory infiltrate with giant cells may be present.

DIFFERENTIAL DIAGNOSIS

A history of contact with animals together with the presence of the severe, inflammatory pustular lesions evoked by *T. verrucosum* or *T. mentagrophytes* var. *mentagrophytes* suggests a diagnosis of tinea barbae. The follicular pustules, brittle, lusterless, easily epilated hair, and the presence of actively spreading peripheral borders compose a classic picture of the disease. If the causative agent is *M. canis*, fluorescence of hairs under a Wood's lamp will be seen. The *Trichophyton* species do not fluoresce. Potassium hydroxide slide mounts readily show the presence of fungal elements and differentiate this disease from sycosis vulgaris. The milder forms of ringworm are less painful and tender than pyodermas caused by staphylococci. Infection by dermatophytes may also involve the eyebrow, but the conjunctiva is spared. Eyebrow infections without other involvement have been noted, particularly in children, and *M. canis* is frequently the etiologic agent (Fig. 8–29). Other conditions that may mimic tinea barbae are contact dermatitis, iododerma, bromoderma, cystic acne, actinomycosis, and pustular syphilids.

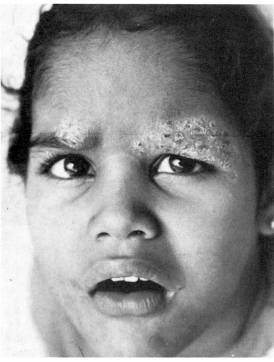

Figure 8–29. Tinea of the eyebrow caused by *M. canis.* (Courtesy of F. Battistini.)

PROGNOSIS AND TREATMENT

Since the majority of cases of tinea barbae are the inflammatory type, spontaneous resolution usually occurs. Duration of the infection varies with the organism involved. Since *T. verrucosum* and *T. mentagrophytes* var. *mentagrophytes* are the most virulent organisms, infections produced by them generally resolve in one to three weeks. *M. canis* infection lasts from two to four weeks. Chronic infections lasting more than two months are not uncommon when *T. rubrum* or *T. violaceum* is the etiologic agent.

Griseofulvin may be of some value in treatment of tinea barbae, particularly the chronic type. Rapid disappearance of the general malaise, pain, and discomfort, together with failure to develop satellite lesions and more rapid resolution of the disease, has been reported after treatment of the severe *T. verrucosum* infections.[103] The dose of griseofulvin is 500 mg micronized daily divided in two parts. Therapy should be continued for two to three weeks following disappearance of symptoms.

Formerly, manual or x-ray epilation together with compression using permanganate soaks (1:4000) or Vleminckx's solution (1:33) was employed. None of these regimens are presently indicated, especially not x-ray epilation. Ammoniated mercury (5 per cent), Quinolor, Desenex, Sopronol, or Asterol was sometimes applied to the lesion. Some of the above may still be useful in resistant cases as an adjunct to griseofulvin therapy. Clipping and shaving of the bearded areas are recommended, along with warm compresses and debridement of diseased tissue.

Tinea Manuum

DEFINITION

Most dermatophyte infections of the hand, particularly of the dorsal aspect, are similar to tinea corporis. Tinea manuum refers to those infections in which the interdigital areas and the palmar surfaces are involved and show characteristic pathologic features. The disease was first described by Fox in 1870[70] and by Pellizzari[200] in 1888. The subject was reviewed by Mitchell in 1951.[169, 188] Along with tinea pedis, tinea manuum is one of the commoner types of chronic dermatophytosis

in the adult. It has been postulated that this is related to the lack of sebaceous glands and their fungistatic lipids in these two areas.[235]

ORGANISMS INVOLVED

Although almost all dermatophytes are potential invaders of the hand, the majority of infections are caused by *T. rubrum*, *T. mentagrophytes*, and *E. floccosum*. Tinea manuum is almost always associated with tinea pedis, so that the flora of the latter is usually the etiologic agent of the hand infection.

SYMPTOMATOLOGY

The clinical symptoms displayed by tinea manuum vary considerably, even though the etiologic agent is most often one species, *T. rubrum*. Five clinical forms have been described. Diffuse hyperkeratosis of the palms and fingers is the most common (Fig. 8–30). This condition is usually unilateral. The second type is crescentic exfoliating skin involvement similar to that seen in tinea pedis. Vesicular, circumscribed patches constitute the third type and are most frequently caused by *T. mentagrophytes*. Discrete, red papular and follicular patches are the fourth, and erythematous scaly sheets on the dorsum the fifth type of infection. The latter two types

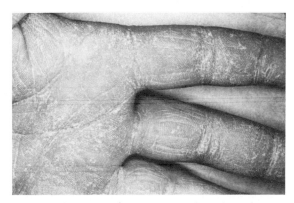

Figure 8–30. Tinea manuum. Diffuse hyperkeratosis of palm and finger. *T. rubrum*. In this chronic infection, only the palmar surface was involved; the dorsum was spared. There were no other lesions on the body.

are most commonly caused by *T. rubrum*. No particular predisposing factors are noted except anatomic deformity and occupational compression of the interdigital spaces. The latter leads to maceration and produces a predisposing condition similar to that affecting the feet. *T. mentagrophytes* var. *interdigitale* is a common invader in such cases. *E. floccosum* has been isolated from rare cases of tinea manuum in which verrucous vegetating processes were seen.

DIFFERENTIAL DIAGNOSIS

The "id" (dermatophytid) reaction occurs on the hand and is nonspecific in appearance. It results from dermatophyte infection elsewhere on the body, usually the feet. It may resemble dyshidrotic pompholyx or be desquamative and thus resemble tinea manuum. These lesions resolve when the infection at the primary site is cleared. Other conditions that mimic tinea manuum are psoriasis, contact dermatitis, neurodermatitis, chronic pyoderma, secondary syphilis, and dermatocandidiasis of the hand. Tinea manuum is usually unilateral.

PROGNOSIS AND TREATMENT

They are the same as for tinea corporis and tinea pedis.

Tinea Pedis

DEFINITION

Tinea pedis is a dermatophyte infection of the feet involving particularly the toe webs and soles. The lesions are of several types, varying from mild, chronic, and scaling to an acute, exfoliative, pustular, and bullous disease.

Synonymy. Athlete's foot, ringworm of the foot.

INTRODUCTION

Ringworm of the foot is by far the most common fungus disease of man and is among the most prevalent of all infectious diseases. It is said to be a penalty of civilization and the wearing of shoes. In Western Samoa the disease was found among Europeans who

wore shoes but not in barefoot natives.[160] The moisture and warmth of the toe clefts that are induced by shoes and socks provide a humid tropical environment that encourages the growth of the fungi. These effects are more pronounced in the space between the fourth and fifth toes, and it is this site that is most frequently involved. Based primarily on historical accounts of western medicine and to a certain extent on descriptions in ancient Greek, Roman, and Indian literature, tinea pedis is a very new disease. In none of these texts is a condition described that is identifiable as tinea pedis. Historically, the most prominent and commonest form of dermatophytosis until the late 19th century was tinea capitis. Since that time its incidence has waned, whereas tinea pedis, particularly the chronic forms, has increased.

Though tinea pedis is a very common disease, it was not recognized until relatively recently. Tilbury Fox[70] reported on tinea manuum in 1870, but the first recognized case of tinea pedis appears to be that noted by Pellizzari in 1888.[188, 200] Sabouraud in Paris and Whitfield in Britain wrote on the clinical and mycologic aspects of the disease. Whitfield reported the first British case in 1908. Systematic studies of the mycology and the histologic and clinical appearances of tinea pedis and tinea manuum were published in 1914 by Kaufman-Wolf.[132]

The increase in incidence of tinea pedis beginning in the late 19th century may correlate with the introduction or generalized distribution of *Trichophyton rubrum* into Europe and America effected by worldwide travel. Some authors speculate that the original endemic area of this species was southeast Asia.[3a] The native populations in this geographic area carry *T. rubrum* as a chronic or occult tinea corporis. Tinea pedis is rare in this population, because the people generally do not wear shoes.

Tinea pedis is now of worldwide occurrence and is distributed equally among the sexes. In contrast to many other forms of dermatophyte infection, tinea pedis is generally a disease occurring in adult life. Although the infection has been described in children only a few months old,[119] the incidence increases with age. It is probable that infection is related to repeated exposure to dermatophytes, so that people using common bathing facilities such as shower stalls in gymnasiums, barracks, and so forth are more prone to acquire infections earlier in life.[10] Repeated exposure, macerating conditions due to ill-fitting shoes and "sweaty socks,"[77] and possibly genetic factors are suggested as the most likely predisposing conditions for the disease. Strauss and Kligman[265] felt that the periodic recurrences of tinea pedis were due to a constant supply of fungi existing in occult lesions. English, on the other hand, feels that these recurrences represent reinfection.[61]

As mentioned in the previous section tinea manuum and tinea pedis are the commonest forms of chronic dermatophytosis. Two possibilities have been raised in relation to this. Both tinea pedis and tinea manuum occur in anatomic areas lacking sebaceous glands with their fungistatic lipids,[235] and both are associated most commonly with infection by *Trichophyton rubrum*.

Tinea pedis is one of the most perplexing of infectious diseases. Though estimates of the infection rate for the population range between 30 and 70 per cent, the majority of these are occult or subclinical cases. The reasons why some people contract the disease and others with the same exposure do not are unknown. Deliberate attempts to induce infection both in the feet of normal patients and in those with a past history of infection resulted only in production of acute inflammatory lesions. These resolved spontaneously or with the aid of topical fungistatic drugs. Repeated attempts to produce tinea pedis in volunteers by immersing their feet in water laden with fungi did not result in a single clinical case.[19] This experiment led to the postulation that infection was universal, and that tinea pedis was not a contagious disease but rather was dependent on host and environmental factors for expression of the clinical symptoms. Other workers were able to isolate dermatophytes from shower stalls, shoes, floors, and so forth and correlate exposure with infection and species concerned.[10, 24, 63] In more recent experimental work, Taplin was able to reproduce the severe inflammatory type of tinea pedis.[35] He inoculated socks with strains of *T. mentagrophytes* of animal origin. The socks were then worn under occlusive boots, and a severe, persistent tinea pedis was produced that occasionally spread up the leg to involve the groin and buttocks.

There are several perplexing aspects to the incidence, epidemiology, and distribution of tinea pedis. As was noted in the section on tinea cruris, many everyday activities may expose the individual to infected squames containing viable dermatophytes. Besides locker room floors,[77] shower stalls, and public corridors, rugs, furniture, drapes, and linens may also act as fomites harboring the infected squames. Thus, hotel rooms, offices, neighbors' houses, libraries, and other public facilities must be considered in determining the occurrence of an isolated case or an epidemic outbreak. There is also some variation in the duration of viability of the species involved and the susceptibility of the host to the infectious agent. As noted previously, *T. rubrum* does not appear to remain viable long in infected squames that have been shed, but it induces a chronic infection so that the environment is continually being showered with live mycelial units. *E. floccosum* is not associated with chronic infections but forms resistant arthroconidia in skin scales that remain viable for long periods of time. Furthermore, this organism is more infective than most dermatophytes. Thus, an individual refractive to infection by *T. rubrum* even after continual exposure in a home environment may pick up an infection caused by *E. floccosum* from the rug in a hotel room where infected squames has been deposited long before. Infection caused by *T. mentagrophytes* often indicates rodents in the area, and the rare case due to *M. canis* points to a colonized cat.

ORGANISMS INVOLVED

At present three dermatophytes, *T. mentagrophytes*, *T. rubrum*, and *E. floccosum*, cause the majority of cases of tinea pedis. The relative incidence of these varies considerably in different reports.[146] An interesting example of this was provided by English and Gibson,[64a] who surveyed the feet of all children in one school. The rate of occurrence of *T. mentagrophytes* was eight times that of other dermatophytes. Yet, in clinical records both in England and the United States, *T. rubrum* was the predominant species. They postulated that *T. mentagrophytes* was more common in occult or inapparent infections, whereas *T. rubrum* caused a chronic and disfiguring type of disease for which patients sought medical advice. *E. floccosum* is a rather constant third. It is particularly prevalent in summer months and may account for 20 per cent of cases in the United States. In other areas (Asia), this species accounts for 44 per cent of cases.[245] In regions where *T. violaceum* is endemic, this organism is also involved in tinea pedis.[146] *T. megninii*, *M. persicolor*,[193] and *M. canis* are responsible for a small percentage of cases.[181] *T. rubrum* was probably endemic in Japan, China, and the Far East until its recent spread to the rest of Asia, Europe, America, and Australia.[201] This worldwide distribution has occurred in the past forty years, and is postulated to have resulted from the vast migration of peoples and troops during the World Wars. *T. rubrum* is now the most cosmopolitan of dermatophytes.

SYMPTOMATOLOGY

Tinea pedis occurs in such a variety of clinical types that a satisfactory classification is difficult to devise. The four most commonly seen forms are (1) chronic intertriginous, (2) chronic papulosquamous hyperkeratotic, (3) vesicular or subacute, and (4) acute ulcerative vesiculopustular.

1. The intertriginous form of tinea pedis is the most common type. It appears as a chronic dermatitis, with peeling, maceration, and fissuring of the skin.[77,111] The areas between the fourth and fifth toes and the third and fourth toes are most often involved. The webs and subdigital and interdigital surfaces are the favored sites. The area is covered with dead, white, macerated epidermis and debris, and there is often a foul odor present. Beneath the debris, the epidermis is erythematous and weeping. The denuded epidermis also harbors the fungus. In successive exacerbations the infection may spread to adjacent areas of the feet, to include the sole, heel arch, and dorsal surface. This condition is very persistent and is associated with hyperhidrosis. The infection may be intensified by hot, humid, summer weather and become severely pruritic. When *E. floccosum* is the fungus involved, marked scaling of the toe and sole, accompanied by numerous punctate satellite lesions, is seen. This may develop into brownish macules (Fig. 8–31). The intertriginous form of tinea pedis is usually amenable to topical treatment.

Uncomplicated intertriginous tinea pedis is referred to as *dermatophytosis simplex* by Leyden and Kligman.[150a] In this form the dermatophyte alone is responsible for the clinical condition and the infection will respond to an antimycotic agent. Sometimes, after the integrity of the stratum corneum has been abrogated by the dermatophyte, many bacteria including lipophilic diphtheroids will greatly increase in number and contribute to the pathology of the condition. In such cases of *dermatophytosis complex*, an antibacterial as well as antimycotic agent may be necessary for effective therapy.

2. The chronic, papulosquamous, hyperkeratotic type of tinea pedis is very persistent and difficult to treat. It is characterized by the presence of areas of pink skin covered by fine silvery white scales (Fig. 8–32). It is commonly bilateral (Fig. 8–33). Though usually patchy in distribution, the lesions may involve the whole foot, in which case the disease is termed "moccasin foot." The etiologic agent is usually *T. rubrum* or *T. mentagrophytes* var. *interdigitale*.

3. The vesicular form of tinea pedis is most often caused by *T. mentagrophytes*. The lesion is characterized by the appearance of vesicles, vesiculopustules, and sometimes bullae. The involved area may extend from the intertriginous areas to include the dorsal surface of the foot, the instep, and less frequent-

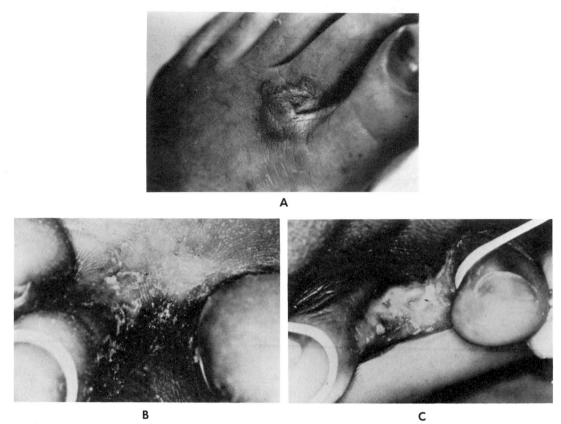

A

B   **C**

Figure 8–31. *A*, Tinea pedis. Intertriginous form between first and second toes. In this lesion, there is an inflammatory response. The agent was *T. mentagrophytes* of rodent origin. *B*, Dermatophytosis simplex. An infection involving only a dermatopyte (anthropophilic *T. mentagrophytes*). There is no erythema and there are few scales. *C*, Dermatophytosis complex. The barrier of the stratum corneum was broken by a dermatophyte, and superinfection by bacteria has occurred. The lesions are pruritic and painful. (Courtesy of A. Kligman.)[150a]

ly the heel and anterior areas (Fig. 8–34). The eruptions vary in size up to 7 to 9 mm and are isolated or occur in patches. The vesicles are tense and contain a clear, serous exudate. After rupturing, they dry to leave a ragged collarette. The fungus is best demonstrated by clipping off the top of a vesicle or bulla and using this for direct microscopic observation or culture. The fungus is located on the inner top of the vesicle roof. The acute form of the disease frequently resolves spontaneously but often recurs under hot, humid,

Figure 8–32. Tinea pedis. Chronic papulosquamous hyperkeratotic type. Note areas of pink skin covered with scales. *T. rubrum.*

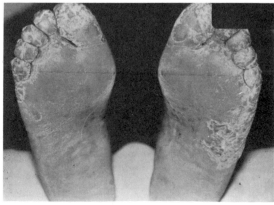

Figure 8–33. Tinea pedis. Chronic papulosquamous type involving both feet. *T. rubrum.*

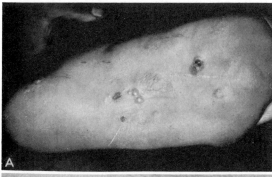

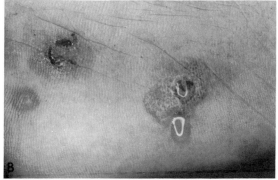

Figure 8–34. Tinea pedis. *A,* Vesicular form involving intertrigo and instep. *B,* Close-up of vesicle showing serous exudate. *T. mentagrophytes.* (Courtesy of S. Lamborg.)

macerating conditions. In such cases, the disease is often quite inflammatory and may be incapacitating. This form is most often responsible for production of the "id" reaction on other areas of the body (Fig. 8–35). A cellulitis, lymphangitis, and lymphadenitis are occasionally seen, and the disease often resembles erysipelas.

4. In the acute ulcerative form of tinea pedis, the rapid spread of an eczematoid vesiculopustular process is seen. This form of the disease is complicated by secondary bacterial infection. The vesicle fluid is purulent, and ulceration of the epidermis occurs. In rare instances, this process is so fulminating that vast areas or even the entire surface of the sole is shed. In addition, there is pronounced cellulitis, lymphadenitis, lymphangitis, and pyrexia. The "id" reaction is common and may be widespread. Overtreatment with topical preparations often exaggerates this condition (dermatosis medicamentosa).

HISTOPATHOLOGY

During the acute stage of tinea pedis, intracellular edema and spongiosis with a leukocytic infiltrate are seen in the epidermis. The

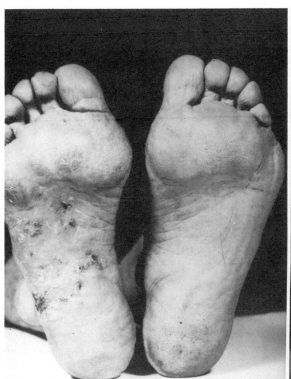
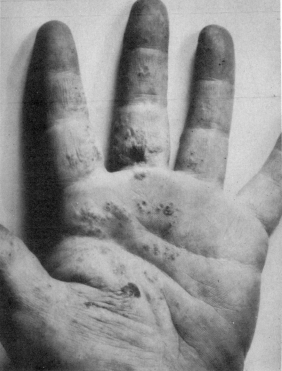

Figure 8–35. Tinea pedis. Vesicular type and associated "id" reaction on hand.

vesicles are in the upper portion of the epidermis and lie just beneath the stratum corneum. Parakeratosis is present. In chronic lesions, hyperkeratosis, acanthosis, and a chronic inflammatory infiltrate are found.

DIFFERENTIAL DIAGNOSIS

Branny, furfuraceous, scaly patches or groups of vesicles over the soles, together with fissures and a macerated epidermis, are highly suggestive of tinea pedis. Direct microscopy, with a potassium hydroxide mount and culture on selective media, confirms up to 95 per cent of cases. In some instances, several cultures are necessary because of the heavy overgrowth of bacteria and the presence of saprophytic fungi. Multiple or mixed infection with one or more species of dermatophyte and concomitant infection with *Candida albicans* is not uncommon. This may complicate therapy, but most topical drugs for dermatophytes are also antibacterial and anti-yeast. Systemic griseofulvin, however, is not effective against bacteria or *Candida* species.

Erythrasma of the toe cleft is practically impossible to differentiate from intertriginous tinea pedis. Use of the Wood's lamp to show the coral red fluorescence characteristic of erythrasma can be used to exclude this disease. Bacterial infections caused by *Pseudomonas, Micrococcus,* and *Acinetobacter* species also mimic tinea pedis. The clinical condition known as athlete's foot may be caused by many organisms of which fungi are but one. It is not unusual in the course of development of severe chronic disease that the fungal component will die out and be supplanted by various bacterial species. This requires a change in therapeutic approach.

The differential diagnosis of tinea pedis includes contact dermatitis, pustular psoriasis, idiopathic hyperkeratosis, dyshidrosis, acrodermatitis perstans, dermatitis repens, erysipelas, pyodermas, candidiasis, secondary syphilis, arsenical keratosis, and fixed drug eruptions.

On occasion, especially in chronic lesions subjected to repeated injury, the dermatophytes may be seen in histologic sections in the form of soft lobulated grains or granules. These somewhat resemble the grains found in mycetoma and have been termed "pseudomycetoma." Such lesions can be seen in tinea corporis and tinea capitis as well. They lack the characteristics of a true mycetoma and represent only a clinical variant of disease.[5a]

PROGNOSIS AND THERAPY

Uncomplicated tinea pedis is generally amenable to topical or systemic therapy unless the causative agent is *T. rubrum*. Tolnaftate is widely used and seems to effect a clinical cure in most forms of the disease. It is well tolerated by most patients and has been advocated as a prophylactic agent for people prone to infections or exposed to infectious material.[40] Other effective treatments include the imidazoles, haloprogin, and a variety of topical agents. The importance of the organism of erythrasma and other bacteria in exaggerating the disease should be recognized. Good hygiene and control of secondary bacterial invaders often lead to remission of symptoms.

Chronic infection involving *T. rubrum* usually necessitates griseofulvin therapy. Systemic griseofulvin therapy (500 mg micronized per day) usually requires two to six weeks before symptomatic improvement occurs, and up to six months or more is needed in resistant cases. Even on such a regimen, some treatment failures are noted.[15] Vascular differences and leaching of the drug from toe webs have been suggested as responsible for the lack of effect of griseofulvin in some cases; in others, organisms resistant to high concentrations of the drug have been isolated.* Bullae and vesicles usually clear after a short time, but there may be an explosive vesicular reaction in some patients. New sterile blisters may occur on the sole and a severe id on the hands. A mechanism similar to the Herxheimer reaction, which occurs in treated syphilis, has been postulated.

In resistant forms of tinea pedis, combined topical therapy with tolnaftate or other drugs and oral griseofulvin has been used with success. The older topical agents, such as undecylenic acid, salicylic acid, benzoic acid, and sulfur ointment, are not considered to be of great value in treatment. Undecylenic acid continues to be popular, though rigorous clinical studies fail to confirm it has any effect more significant than simple good foot hygiene.[251] Some studies[11] have shown a degree of efficacy for this preparation, but sensitivity of organisms to the compound is quite variable. In general practice, continued use of the agent is often necessary to control symptoms. These patients and those that are completely unresponsive to undecylenic acid usually respond dramatically to tolnaftate or other topical agents. Thymol, phenol-camphor, and

---

*Personal communication, W. Artis, 1980.

strong salicylic acid are irritating and sensitizing. Chronic infections with minor flaking and scaling are not uncommon when *T. rubrum* is the etiologic agent, and these persist despite treatment. Occult or subclinical infections caused by the "downy" or var. *interdigitale* form of *T. mentagrophytes* are also quite persistent, and exacerbations frequently occur. Haloprogin is about as efficacious as tolnaftate and also clears *Candida* infections. Skin sensitization appears to occur more commonly with the former preparation, however. The imidazoles are also efficacious.

ANIMAL DISEASE

Natural Infections

Dermatophyte infections of wild and domestic animals have been recognized for many years. It has been pointed out repeatedly that animals act as a reservoir for human dermatophytosis.[16, 80, 157, 189] Ringworm disease in domestic animals constitutes a constant source of infection for persons in contact with them. Thus zoophilic dermatophyte infections are particularly common in rural areas. Fungi from domestic animals, such as dogs and cats, may initiate an epidemic among children. In addition, wild animals also harbor ringworm and may be an indirect source of human infections, since the infected hairs shed from these animals may contaminate dwelling places and working areas.[16, 157] Sometimes contamination from rodent carriers leads to outbreaks of severe dermatophytosis in a human population.[35]

Though long recognized, animal ringworm has only recently been studied in detail. The specific pathologic picture of the infected animal, the dermatophytes involved, and the frequency of transmission of such infections to humans have been reviewed by Georg,[80] Otčenásek[189] and Mantovani.[157] The organisms involved are either species-specific dermatophytes, such as *M. equinum* and *T. equinum* in horses and *M. nanum* in pigs, or universal infecting agents of man and animals, such as *T. mentagrophytes* and *M. canis*. The former species are very rarely isolated from human infection, whereas the latter are commonly involved in human disease. Animals also may serve as a vector for animal dermatophytoses and cause epidemics among other animal species. Rodents and cats are probably the chief disseminators of general dermatophytosis. Cases are also on record of transmission from man to animals.[127] Some of the well-recognized dermatophytoses of animals are tabulated in Tables 8–9 and 8–10. Complete lists of animal species and reports of disease are given in Ainsworth,[3] Georg,[80] Menges,[165, 167] and Kaplan.[129]

Ringworm of Cats and Dogs

A more apt name for *Microsporum canis* would be "*M. felis*," as cats are the major reservoir for this dermatophyte. The disease in cats is so minimal in the majority of cases that it goes unnoticed. Such infected animals spread the disease to children in the family and neighborhood, as well as to dogs and other cats, and leave a trail of infected debris in surrounding areas.[16, 157, 252] The resulting public health problems are obvious.

Clinical Symptoms. Ringworm of the cat caused by *M. canis* is most often subclinical or inconspicuous. In many cases, attention is first drawn to the animal after the development of ringworm by human contacts. The head is the most common site of infection, with areas of hair loss around the nose, eyes and ears. In clinically apparent lesions, a mild, noninflammatory scaling and patches of alopecia and broken hairs are found. Rarely there is an inflammatory reaction and crust formation. In young animals, lesions are more clearly defined. Discrete circinate lesions are found which show hair loss, scaling, and a vesicular border. In sickly kittens, the infection often extends to the whole body, forming weeping, crusted lesions, and the disease resembles a severe mange infestation. Favic scutula are sometimes found when the disease is caused by *T. mentagrophytes*. The latter organism evokes a more pronounced inflammatory response. Infection between the paws is common, but whiskers and nails may also be involved.

The disease in dogs is more obvious than in cats. Circular lesions up to 2.5 cm in diameter may appear on any part of the body (Fig. 8–36). Transmission from man to animals is recorded, but this is very rare. A boxer contracted a tinea corporis infection caused by *T. rubrum* from his owner, who had the habit of rubbing the dog with his bare feet. The owner had hyperkeratotic tinea pedis.[127] Infections caused by *M. canis* will fluoresce with the Wood's lamp. The characteristic greenish color of the arthroconidia sheath around infected hairs is easily seen. Nonfluorescent hairs infected by *T. mentagrophytes* will show

Table 8–9. *Domestic and Wild Animals and the Dermatophytes Causing Natural Infection*[80, 189]

Animal	Species Recovered and Frequency*		Animal	Species Recovered and Frequency*	
Cat	M. canis	U	Pigs	M. canis	R
	M. distortum	R		M. nanum	U[174]
	M. gallinae	R		T. mentagrophytes	F
	M. gypseum	F		T. verrucosum	R
	T. mentagrophytes	F			
	T. schoenleinii	R	Horses	T. equinum	U
	T. verrucosum	R		M. distortum	R
	T. violaceum	R		M. gypseum	F
	M. vanbreuseghemii[175]	R		T. mentagrophytes	F
Dog	M. audouinii	R		T. verrucosum	F
	M. canis	U		M. equinum	R
	M. cookeii	R			
	M. distortum[230]	R	Rodents	T. mentagrophytes var.	U (hedgehogs)
	M. gypseum	F	(domes-	erinacei	
	M. persicolor	R	tic and	T. mentagrophytes	U
	M. vanbreuseghemii	R	wild)	M. canis	F
	T. equinum	R		M. gallinae	R
	T. megninii	R		M. gypseum	F
	T. mentagrophytes	F		M. persicolor	U (voles)
	T. rubrum	R		M. vanbreuseghemii	R
	T. simii	R			
	T. verrucosum	R	Monkeys	M. audouinii	R
	T. violaceum	R		M. canis	U
	E. floccosum	R		M. cookeii	R
				M. distortum	R
Cattle	T. mentagrophytes	F		M. gypseum	F
	T. verrucosum	U		T. mentagrophytes	U
	M. canis	R		T. rubrum	R
	M. gypseum	R		T. simii	U
Sheep	M. canis	R	Fowl	M. gallinae	U
	T. mentagrophytes	R		T. simii	F
	T. verrucosum	R			

*U = usual; F = frequent; R = rare.

Table 8–10. *A Survey of Hosts of* Microsporum canis *and* Trichophyton mentagrophytes*

Microsporum canis		Trichophyton mentagrophytes	
man	leopard	man	kangaroo
	lion		mouse
cat	lynx	buffalo	mule
cattle	mink	cat	muskrat
chimpanzee	orang-utan	cattle	nutria
chinchilla	polecat	chinchilla	opossum
dog	rabbit	coypu	polecat
donkey	rhesus monkey	dog	porcupine
fox	sheep	donkey	rabbit
gibbon	swine	fox	raccoon
goat	tiger	goat	rat
gorilla	weasel	guinea pig	sheep
guinea pig		hamster	squirrel
horse		hare	swine
hyena		hedgehog	
jaguar		horse	chicken

*From Otčenásek, M. 1978. Mycopathologica, 65:67–72.

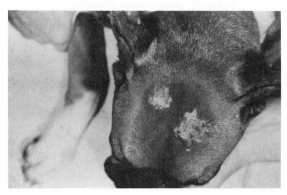

Figure 8–36. Tinea of dog. Circinate lesion on head showing alopecia. *M. canis.* (Courtesy of S. McMillen.)

arthroconidia in chains on direct examination.[79, 80] Infections caused by *M. gypseum* are clearly circular, with alopecia, erythema, scaling, and a peripheral yellowish-white crust.[166]

Topical treatments are the same as those listed for human infections but are of questionable value. *M. canis* has a tendency to become chronic, though disease caused by other species may resolve spontaneously with good nutrition and hygiene. Rapid resolution of symptoms follows treatment with oral griseofulvin. The dosage recommended is 60 mg for adult cats and 40 mg for month-old kittens. The drug is given daily until gross symptoms subside (usually two weeks). After this it is given two times a week. In a clinical trial 10 of 18 cats were cured in three weeks; others required up to five weeks.[125] The response of dogs to oral griseofulvin is even more rapid. Some seasonal variation in the incidence of ringworm has been found.[129] Individual patterns were found for the several species.

Ringworm of Horses

Clinical Symptoms. The most common clinical picture of dermatophytosis of the horse is that of dry, raised scaling lesions on any part of the animal (Fig. 8–37). Areas that are macerated or rubbed frequently, such as the saddle and girth area, and the hind quarters are the most common sites of infection. Colts and yearlings are most susceptible. The initial lesion is a swelling that can be felt through the hair. The lesions often become small, inflamed ulcers with pussy exudate (girth itch). The hairs appear to be glued together, and the entire mass may be removed as a unit. This condition is called favus of horses. Lesions enlarge with the loss

of peripheral hairs, and a chronic infection is established. As lesions heal, crusts fall off, leaving large bald areas and a "moth-eaten" appearance. Infected debris is found on brushes, saddle gear, and buildings. *T. equinum* is the organism responsible in the great majority of cases, but epizootics of this disease caused by *M. gypseum* are recorded.[128] *M. equinum* is a rare etiologic agent. None of the common horse-infecting dermatophytes are fluorescent, with the exception of *M. equinum.* This species fluoresces with the same color and intensity as infections caused by *M. canis.* Infections caused by *T. equinum* resolve rapidly with incorporation of griseofulvin as a feed additive.[217] Such feed additives appear to be the most practical method for treating dermatophytosis in horses.

Ringworm of Cattle

Estimates of the prevalence of ringworm in cattle average 20 per cent, and the vast majority of infections are due to *T. verrucosum.*[189] It is generally considered that calves and yearlings are more susceptible than older animals, for figures of up to 40 per cent of the total number of infected have been given for this age group. Although there is a consensus that the infection rate is higher in winter than summer, some authors feel that crowded conditions with increased contact between animals and the presence of infected debris in buildings account for both the higher incidence in calves and the greater infection rate in winter. In support of this, McPherson[164] and others[124] found that *T. verrucosum* in infected squames and hair on building woodwork remained viable and infective for from 15 months to 4½ years. This has recent-

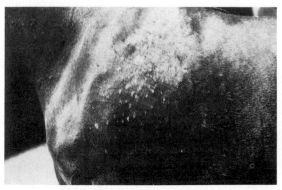

Figure 8–37. Tinea of horse. Multiple, dry, scaly, raised lesion. *T. equinum.*

ly been confirmed by Ray and Misra.[212] Therefore, ample opportunity would be provided for infection of new stock brought into the premises. In addition, this is also a ready source of infection for humans, since children have contracted *T. verrucosum* infection from debris on clothing worn by others in a dairy barn.[212] *T. verrucosum* arthroconidia located in a 1.5-mm thickness of skin were able to withstand ultraviolet light equivalent to 437 hours of midday, midsummer sunshine at medial latitudes. Infections may resolve spontaneously, with a subsequent degree of resistance to reinfection evoked. Adult cows with no past history of disease were as susceptible to experimental infection as were calves.[249, 258] Kane and Smitka[124] have devised a selective medium for the early detection of *T. verrucosum*.

Clinical Symptoms. Ringworm in cattle begins as scattered, discrete, circinate lesions, with slight skin scaling and hair loss. The symptoms may remain stationary and a chronic infection may be established, or, more commonly, the disease develops acutely into large, circumscribed plaques up to four inches in diameter, particularly when the flanks or back are involved (Fig. 8–38). These are thickened and covered with a grayish-white crust. Frequently the crusts are large, knob-like bumps that stud large areas of the infected animal. These crusts, which at first are white to gray, become brownish and asbestos-like. They are firmly attached to the animal, and when they are removed, a weeping, bleeding erythematous base is seen. The lesions are often severely inflamed and pruritic. Ulcerations and secondary bacterial infections are frequent. Spontaneous healing follows this stage of the disease, and the lesions become dry, scaly patches, with alopecia and scar formation. In other anatomic areas such as ears (Fig. 8–38,*A*) or tail, the lesions are often benign and subclinical and become chronic but usually resolve in time.

There is at present no satisfactory topical treatment for ringworm of cattle. Systemic griseofulvin has been used successfully,[248] but is prohibitively expensive for treating an entire herd. Good hygiene and sanitation are important for controlling the disease. A fungicidal spray, Captan, in a concentration of 0.45 to 0.5 lb per 20 gallons water, used at a rate of 1.0 to 1.5 gallons per animal, has been recommended,[20] as has natamycin (pimaricin).[258] This sterilizes infected material and reduces the reservoir of infection. Defungit (bensuldazic acid) used as a wash in a concentration of 0.5 to 1.0 per cent was therapeutically effective under field conditions.[135] As in human dermatophytosis, thiabendazole has been used with some success in cattle ringworm.[180] The mycelium of *Penicillium griseofulvum* has been included in the feed of cattle. This prevented infection and cured those already infected;[59, 109] however, the use of griseofulvin additives in feed has been discouraged because of possible accumulation in meat and dairy products intended for human consumption. Vaccines are now being tested.[12]

Ringworm of Pigs

Reports of dermatophyte infection in pigs were rare until 1964. It was believed that these animals were seldom, if ever, subject to dermatophyte disease. A few scattered reports of infection by *T. mentagrophytes* were listed,[84] and Dawson and Gentles mentioned the occurrence of *M. nanum* in pigs in Kenya. Subsequently Ginther et al. in a series of papers described the infection in the United States as being very common in swine, affecting up to 27 per cent of a single herd.[85, 86] Apparently the disease is so benign it has gone unobserved for years, but recently it has been shown to have a worldwide distribution.[189] Although disease has been found in all breeds, Yorkshires are the most frequently involved. Infection by *M. nanum* in swine farmers and children in rural areas has been recorded.[73, 174]

Clinical Symptoms. The lesions caused by *M. nanum* in pigs are mild and slightly inflamed at first. They then expand centrifugally and may extend to involve large areas of the body (Fig. 8–38,*C*). The initial reaction subsides rapidly, and only inconspicuous scaling and brownish discoloration remain. Lesions behind the ear are most common in chronic infections. No alopecia or systemic disturbances are reported, and the disease becomes chronic and subclinical. Once it is established there is little tendency for spontaneous cure. Infection is extremely rare in piglets or young swine. Occasional transmission to children (as tinea capitis) has been recorded. Such a case in Cuba represented the first isolation of the organism.[73] As infection appears to cause no distress to animals nor is it economically important, treatment of the disease has not been established. In *T. mentagrophytes* infection of swine, a more severe, inflammatory response and pruritus were reported.[84]

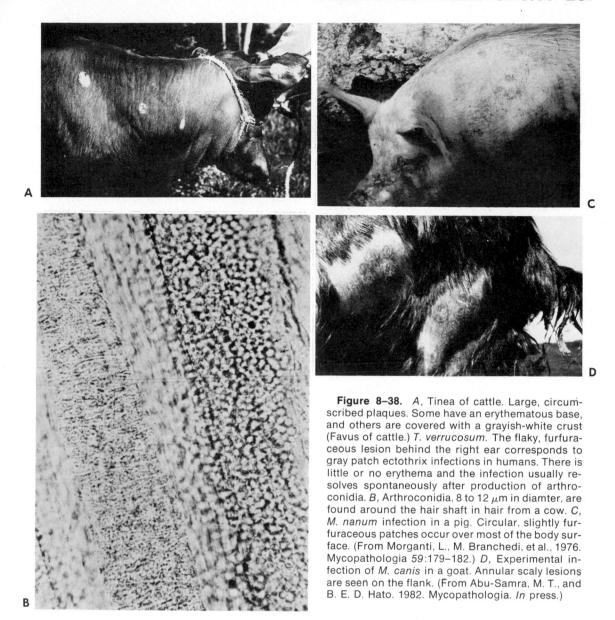

Figure 8–38. *A,* Tinea of cattle. Large, circumscribed plaques. Some have an erythematous base, and others are covered with a grayish-white crust (Favus of cattle.) *T. verrucosum.* The flaky, furfuraceous lesion behind the right ear corresponds to gray patch ectothrix infections in humans. There is little or no erythema and the infection usually resolves spontaneously after production of arthroconidia. *B,* Arthroconidia, 8 to 12 μm in diamter, are found around the hair shaft in hair from a cow. *C, M. nanum* infection in a pig. Circular, slightly furfuraceous patches occur over most of the body surface. (From Morganti, L., M. Branchedi, et al., 1976. Mycopathologia 59:179–182.) *D,* Experimental infection of *M. canis* in a goat. Annular scaly lesions are seen on the flank. (From Abu-Samra, M. T., and B. E. D. Hato. 1982. Mycopathologia. *In* press.)

Fowl Favus

Ringworm (favus) of children due to *M. gallinae* was first described in 1881 by Megnin[164a] and was extensively studied by Sabouraud.[242] In the United States, the first report in fowl was by Beach and Halpin in 1918.[23a] They called the disease "white comb," and it has appeared sporadically since. Its reported occurrence is world-wide but it is considered to be a rare infection except in Brazil, where it appears to be frequently encountered. The etiologic agent, *M. gallinae,* is one of the few dermatophytes that infect birds. Disease is most common in gallinaceous birds, such as chickens, turkeys, and grouse. Infection of man, dogs, and children associated with transmission from fowl has been recorded.[87] *T. simii* also infects chickens.[271]

Clinical Disease. The disease is characterized by a white, moldy, patchy overgrowth on the comb and wattle. Thick white crusts often develop, and in severe cases the infection becomes generalized to involve the base of feathers. The classic favic scutula commonly described in earlier reports is rarely seen today. Infection of chickens with *T. simii* evokes a more inflammatory reaction, with focal necrosis and crusting.

Ringworm of Monkeys

Dermatophyte disease of monkeys in their native habitat appears to be uncommon. Except for *T. simii* and *M. canis*, few fungal species have been isolated from natural infections. In captivity, however, monkeys acquire infection from a number of sources, such as handlers, rodents, domestic animals, and soil. The list of frequently involved species includes *M. gypseum*, *M. canis*, and *T. mentagrophytes*. All of these evoke a rather severe inflammatory disease, with crusting and hair loss. In contrast, *T. simii* produces silvery scaling lesions with little or no erythema and minimal hair loss. The face is the favored location for infection. *M. gallinae* has also been recovered from infection in a monkey.[90]

Other Animals

Many other animal species have been reported to have ringworm. Natural infection in sheep, goats, rabbits,[102] chinchillas (fur slipping), rats, mice, muskrats, foxes, lions, tigers, guinea pigs, and so forth is recorded[189] (Table 8–10). The majority of these infections are caused by *T. mentagrophytes* and some by *M. gypseum*. Naturally occurring tinea pedis and tinea unguium of guinea pigs due to *T. mentagrophytes* have been reported.[239] Mouse favus, which is characterized by the presence of numerous white crusted lesions and scutula on the head and body, was ascribed to *T. quinckeanum*. Mating studies, however, have shown that this is not a separate species but a variant of *T. mentagrophytes*. Essentially all rodents carry *T. mentagrophytes* as normal flora, with little or no symptoms of disease. A number of other dermatophytes and keratinophilic fungi are associated with rodent fur—*T. mentagrophytes* var. *erinacei*[17] in hedgehogs, *M. persicolor*[186] in voles, *Arthroderma curreyi* in wild rabbits, *T. phaseoliforme*, *M. amazonicum*, and *T. terrestre* in mice, and *Chrysosporium* species in many rodents. Griseofulvin has been used in the treatment of natural infections in rabbits and other rodents.[102]

Experimental Infections

Experimental infection of animals, especially guinea pigs and rabbits, is commonly used to study the pathogenicity, immunity, treatment, and prophylaxis of dermatophyte infection (Fig. 8–39, 8–40). Disease is most easily established with zoophilic[49, 223] or geophilic species.[75, 224, 239] The area to be infected is plucked, and the prepared site is scarified and inoculated with a suspension of microconidia. A most useful device is the Sterneedle multiple puncture gun, which is dipped into the conidia and applied to the plucked area. Occlusive dressing is sometimes used. Within one week lesions develop. They are usually inflammatory, and crusting is frequently seen. Resolution of the lesions occurs within three to four weeks, although rarely a chronic, sparsely scaly infection is established which lasts for months or years.[214] Infection involves the stratum corneum of the skin, and hair is frequently invaded with the production of arthroconidia. Application of a neck collar prevents licking and chewing of lesions.[126] Experimental infection using anthropophilic species is more difficult to produce. Successful inoculations with *T. rubrum* in rabbits was recorded by Reiss.[214] Castration of the experimental animals augmented the disease, and infections lasting up to nine months were noted. Variations in susceptibility to dermatophyte infection of different laboratory strains of mice and gnotobiotic mice have been examined, but no consistent pattern has been found.[246]

Literature concerning the treatment of experimental infections in animals is sparse. *T. mentagrophytes* of rabbits has been successfully treated with oral griseofulvin. The drug was included in their food at a rate of 0.375 g griseofulvin per pound of feed.

Immunology of Dermatophytosis

No subject in the field of medical mycology has evoked more controversy than "immunity and resistance" in dermatophyte infections. The voluminous literature begins in the early nineteen hundreds and continues to accumulate unabated to the present day.[93, 150] A review of the work on experimental, acquired resistance, natural resistance, hypersensitivity, immunization, and antibody response leads to the following summation. There is no single, clear-cut mechanism which will explain all aspects of susceptibility and immunity to dermatophyte infection. The differences in response to infection, susceptibility to reinfection, and even acquisition of the infection varies considerably with each species of fungus and to a great degree with the individual host involved. Very few derma-

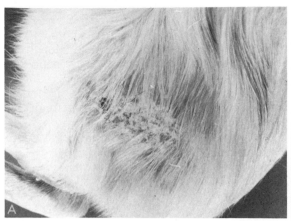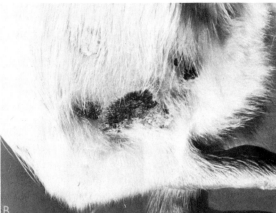

Figure 8–39. Experimental infection in guinea pigs. *A*, Lesion produced by *Arthroderma benhamiae* mating type A. The infection was a chronic, scaling lesion. *B*, Crusted lesion produced by *A. benhamiae* mating type a.[224]

tophyte species evoke the same response in all patients or in experimental animal and human infections.[140, 150, 254] In a given population with equal exposure, only a certain percentage will manifest clinical symptoms of disease. A given individual may have a short, protracted course of infection or no infection with one dermatophyte species, but will de-

Figure 8–40. *T. simii* infection. *A*, Lesion on arm of human subject who caught it from monkey. *B*, Experimental lesion in guinea pig.

velop a mild chronic infection of lifelong duration with another species. The relevant aspects of natural resistance[7] to dermatophytosis are discussed in the sections on clinical disease. They have been extensively reviewed by Lepper[150] and Grappel.[93]

The skin is the primary barrier or defense of the body against invasion by microorganisms from the external environment. In dermatophyte infection, this defense is not abrogated. A peculiar relationship exists between the dermatophyte and its host that is unparalleled by other microbial agents of disease. The organism is truly a dermato- or ectophyte, as it does not invade living cells, and its nutritional demands involve no depletion of metabolizable substances from the host. In this sense it is not a parasite. The immune and inflammatory response evoked in the living tissue beneath the site of infection or distal to it are incidental to the disease. They reflect the lack of or low degree of adaptation between the fungus and the host. Attempts to correlate circulating antibody response or specific immune mechanisms to resolution of infection and resistance to reinfection have been unrewarding.

A degree of acquired resistance to the disease has been observed in patients and in experimental infection in animals. Clinical records indicate that, in a large series of children treated for tinea capitis, none returned with a second infection.[72] Similar findings were seen in agricultural workers infected with *T. verrucosum*.[150] These observations are interpreted to mean that there was increased resistance to reinfection as a result of the initial infection. On the other hand, multiple episodes of tinea pedis occur, and reinfection (exacerbation?) is common in patients with this disease. Many of these patients have had a previous history of tinea capitis or tinea corporis. Repeated tinea corporis infection is also common, and each episode differs little from the initial infection.[56] Laboratory animals generally acquire gradually increasing resistance to reinfection with either the same or different species of dermatophytes. In guinea pigs, this resistance is transitory and has been correlated with cutaneous hypersensitivity.[55, 288] Both responses are maximal two to three weeks following resolution of primary lesions. Rabbits, mice, and rats appear to develop neither resistance to reinfection nor hypersensitivity.[138]

Hypersensitivity can be demonstrated by skin tests using trichophytin.[94, 121, 208] Intradermal injections of this substance elicit either a delayed (tuberculin) or immediate (urticarial) response. The latter may be passively transferred and is associated with a reaginic antibody tentatively identified as an IgE. Purification and chemical analysis of trichophytin shows it to be a galactomannan peptide.[93, 150, 190] Degradation studies indicate that the immediate reaction is associated with the carbohydrate fraction and delayed reactivity with the peptide moiety. Patients with chronic, widespread infections caused by *T. rubrum* often show only the urticarial reaction when challenged with trichophytin. The same reaction can be elicited in atopic patients without fungous disease. Preliminary impressions from several studies indicate that such patients and those with chronic mucocutaneous candidiasis are hyporeactive in their cell-mediated immune systems (CMI). Patients with either acute or chronic infection have an activated CMI, as demonstrated by leukocyte adherence inhibition test. In chronic disease, however, there is a specific serum factor that blocks this reaction.[282]

This depression is antigen-specific in localized infections but nonspecific in generalized disease. The delayed hypersensitivity reaction is found in either experimental or natural infection by *T. mentagrophytes* and several other dermatophytes. It is associated with the degree of inflammatory response evoked by the primary infection. In children, severe inflammatory disease and sometimes kerion formation is associated with hypersensitivity (by trichophytin test), whereas noninflammatory disease was found in nonreactive individuals. The etiologic agent in this series was *T. tonsurans*.[209] Patients with chronic *T. rubrum* infections often have a negative delayed reac-

tion to trichophytin. Many of these patients show an immediate type reaction, however, and a few have circulating antibodies.[95] Patients with clinically evident infection by *T. mentagrophytes* almost always have a positive reaction to trichophytin.

In guinea pig infection, delayed hypersensitivity correlates with resistance. Attempts to conclusively demonstrate this in human disease have as yet been unsuccessful, although the work of Jones et al.[121] is suggestive. They concluded that delayed sensitivity is a correlate of immunity, whereas its absence or its coexistence with immediate reactivity is associated with susceptibility to chronic or recurrent tinea infections. Patients with atopy are particularly prone to chronic infections.

Trichophytin, as presently prepared, is not species-specific and its properties and composition are essentially the same regardless of the species of dermatophyte used in its preparation.[190] The relative antigenicity of different preparations is influenced primarily by the media in which the fungus is grown. Trichophytin-like substances have been isolated from species of *Aspergillus* and *Penicillium*, as well as dermatophytes. Positive skin tests to trichophytin have been elicited in patients with penicillin hypersensitivity[281] and in those with cutaneous tuberculosis. Other studies have detected a high percentage of positive reactors among the normal populace, which was free of fungous infection. This may reflect occult colonization, however. The demonstration of delayed or immediate-type hypersensitivity to intradermal injection of trichophytin appears to be of limited diagnostic and prognostic value. The development of species-specific antigens would greatly aid in the understanding of dermatophyte disease.

Dermatophytid or "id" reactions are secondary eruptions occurring in sensitized patients as a result of circulation of allergenic products from a primary site of infection. The "id" reaction is frequently found in patients with an absence of delayed reaction to trichophytin. The morphology and site of the "id" lesions vary. The condition, as first described by Jadassohn, resembled lichen scrofulosorum. It is commonly associated with tinea capitis in children. Small, grouped, or diffusely scattered follicular lesions are found on the body. They are symmetrical and central in distribution, but may extend to involve the limbs and face. Horny spines are sometimes observed on top of the involved follicles. Lesions on the fingers are frequently found in patients with tinea pedis. Generally they are papular, but they may be vesicular (pompholyx type), bullous, or rarely pustular. They appear on the sides of the fingers and wrist, or are grouped on all parts of the body. The dermatophytid reaction responds to desensitization and resolves spontaneously with the elimination of the primary disease. The condition is sometimes exacerbated or exaggerated during treatment with systemic griseofulvin or injection of trichophytin.

The serological aspects of dermatophytoses in man and animals have also been reviewed by Lepper and Grappel.[93, 94, 150] Circulating antibodies to natural or experimental dermatophytosis have been demonstrated by a number of techniques.[95] Their significance remains to be established, but they appear most commonly in patients with chronic infections.[122] IgE antibodies have also been detected in some cases. Gotz et al.[92] feel that they actually account for some clinical manifestations of infection and contribute to the resulting pathology. Precipitating or complement-fixing antibodies and indirect hemagglutination titers were found in rabbits and cattle that were naturally or experimentally infected by several dermatophytes, but no direct relationship to resistance has been found.[150] In man, circulating antibodies that react to antigens from dermatophytes have been detected in patients with and without disease. The C-reactive protein in the β-globulin fraction of human serum reacts with C-reactive substances found in glycopeptide fractions of E. floccosum, Aspergillus fumigatus, or the galactomannan peptide from T. mentagrophytes.[151] Inoculation of living or killed dermatophytes or their extracts results in the formation of circulating antibodies. Just as in other serologic tests, the antisera produced reacts with many species of dermatophytes and other fungi.[9, 93]

Humoral substances other than antibodies have also been described as having an effect on dermatophytes or their products. Normal human or animal serum contains substances identifed as α_2 macroglobulins, which inhibit proteolytic enzymes of dermatophytes. Perhaps such factors from serum, and possibly from keratin as well, act as blocking agents and are the basis of host specificity in anthropophilic and zoophilic species. Several "serum factors" that inhibit growth of dermatophytes have been described.[152, 233] In inflammatory lesions percutaneous leakage of such factors bathing the dermatophyte in situ may aid in resolution of the infection.

Immunization by injection of live or killed fungi, their extracts, or their metabolic products has been attempted many times. In guinea pigs, the injection of acetone-dried powdered mycelium of T. mentagrophytes imparted both an immediate and a delayed-type hypersensitivity.[9] Upon challenge with a homologous organism, an increased inflammatory response was noted, and the disease was attenuated. However, resistance was transitory, and complete susceptibility returned after a few months. In the same animals, an immunologically active ribonucleic acid or oligoribonucleotide derived from T. mentagrophytes had little skin reactivity, but it did evoke antibodies.[114] Ribonucleic acid fractions have been used successfully for immunization against tuberculosis or other diseases that evoke a strong hypersensitivity reaction. Topical application of T. mentagrophytes extracts to the feet of a group of infection-free persons conferred some resistance when they were subsequently challenged by infection with this organism.[114]

At present, there is no evident correlation between resistance to reinfection and antibody levels as measured by standard in vitro techniques. Although a transient resistance to reinfection has been demonstrated, there appears to be little direct relationship between the degree and type of cutaneous hypersensitivity and lasting immunity to dermatophyte disease. The available data necessitate acceptance of the premise that immunity can exist in degrees.[114]

BIOLOGICAL STUDIES

Because of the frequency of the infections they elicit and their cosmopolitan distribution, the dermatophytes have been the subject of much basic research since their discovery in the late 19th century. These studies have included mechanisms of pathogenicity, comparative physiology, analysis of composition, basic genetics, and genetics as related to infecion. A cursory overview is presented in this section.

Physiology of Infection. Severity of infection produced by the dermatophytes varies from host to host and from species to species. Such observations have been noted since direct correlation of the fungus and the disease produced were made beginning in the 1850's. Differences in the individual host and in racial groups were discussed in the sections on epidemiology and immunity. Investigations have also been carried out correlating enzymatic capacity and the type of disease evoked. Extracellular enzymes produced by dermatophytes were described in a number of papers,[41, 43, 51, 131, 220, 221, 224] and studies designed to show an association between these and the capacity to infect were also carried out. Inflammatory disease in humans[220] and in animals[224] was demonstrated to be a function of the production of proteolytic enzymes, particularly elastase and collagenase, by the fungus during infection. In some instances this was also shown to be associated with the mating type of the fungus, as in *Nannizzia fulva*.[224] In most species, however, it was found to be independent of the sexual type. In genetic studies on *Arthroderma benhamiae* by Cheung and Maniotis,[43] the locus for mating type segregated independently of both the colonial morphology locus and a locus for determining elastin hydrolysing capacity. The colonial morphology locus (two alleles) governed the granular or downy colonial phenotype and is closely linked to the locus for extracellular elastolytic ability. High production of enzyme is found in strains of the granular morphology. Georg[79] demonstrated that the downy form of *T. mentagrophytes* produced less severe lesions in guinea pigs than the granular form. The high production of elastin hydrolysing capacity in this dermatophyte may be the result of the quantity of enzyme produced or a defective regulatory locus for the protease.[43] Regulation of protease activity by a host factor has been invoked as the basis of host specificity of dermatophyte infection, but experimental evidence for this is so far lacking.

Genetics. The genetics of pigment production in culture was studied and reviewed by Ghani et al.,[83] and the production of various mutants by Howard and Dabrowa.[112] These latter investigators were able to produce temperature-sensitive mutants, some of which were "pleomorphic" and some of which were not. They also produced slow-growing mutants that apparently had defects in mitochondrial ribosome assembly and deficiencies in cytochromes. Some of these slow-growing mutants showed changes in ergosterol content. Lower ergosterol content correlated with increased resistance to amphotericin B just as with other fungi.[91] The genetics of "pleomorphic growth" among dermatophytes is complex and appears to involve one or more gene mutations. An analysis of pleomorphism in *Microsporum gypseum* was carried out by Weitzman.[283] The general field of the genetics of pathogenic fungi was reviewed by Kwon Chung.[142]

The genetic analysis of mating behavior in dermatophytes revealed that several imperfect (anamorph) "species" are in reality complexes of perfect (teleomorph) species. The first of these was the *Microsporum gypseum* complex noted by Stockdale.[261] Since then, complexes have been found in *Trichophyton mentagrophytes* with two perfect stages so far (*Arthroderma benhamiae*[5] and *Arthroderma vanbreuseghemii*,[267]), three at least in *T. terrestre*,[195, 196] and perhaps more than one in *M. canis*.[286] In an analysis of *M. canis* from clinical material, Weitzman and Padhye[287] found most isolates to be of the (−) mating strain. This may reflect a selective preference held by one mating type over the other as far as invasive ability or pathogenicity is concerned. Such a disproportionate distribution of one mating type has also been shown in clinical vs. soil isolates of *Histoplasma capsulatum*. In studies of *T. mentagrophytes* var. *erinacei* associated with the hedgehog, Takashio[268] found all isolates from animals to be of the (+) mating type. Minus mating types could be produced *in vitro* by mating in the laboratory, but these apparently die out in nature. Thus var. *erinacei* may be evolving into a host specific species with concomitant loss of one mating type. By using the *T. simii* sex stimulation test, all strains of *T. rubrum* are of the minus mating type,[110] *Microsporum audouinii* of the (+) and *T. tonsurans* also of the (−) mating type. The opposite mating types of these anthropophilic species may have been lost during evolution.

Physiology and Growth. The germination of dermatophyte microconidia and arthroconidia as affected by the environment was studied by several investigators. Arthroconidia of *M. audouinii* on hair developed germ tubes most rapidly in the presence of components of the urea cycle.[225] This was also found in studies on the arthroconidia of *T. rubrum* by Miyazi and Nishimura.[170] Hashimoto et al.[106] found L-leucine induced germination of the microconidia of *T. mentagrophytes* more than other amino acids. Germination was unaffected by carbohydrates, salts, vitamins, or other compounds in this species. Formation of the germ tube was accompanied by loss of resistance to heat and stain, reduction of dry weight and specific gravity, and development of active glucose utilization.

Exogenous factors also affect production of extracellular enzymes. Meevootisom and Niederpruem[182] found that exogenous glucose suppressed elastase as well as keratinase and protease production in all dermatophytes tested. Various other carbohydrates and amino acids (individually) suppressed general protease production, but ammonium phosphate was not inhibitory. Substances such as ferric ion and mercuric ion stimulated elastase but inhibited the keratinases. The keratinases[269] produced by dermatophytes were described by Yu et al.,[292, 293] and their importance in infection and immunity to disease was reviewed by Grappel.[94]

Chemical Composition. The chemical composition of dermatophytes, particularly in relation to immunity and pathogenicity, has also been studied by many investigators. The relation of chemical composition and acquired immunity was reviewed by Grappel,[93, 94] and the association of lipids with pathogenesis by Vincent.[278] By analysis, the chemical composition of dermatophytes is, in general, similar to other ascogenous fungi.[279] There are some differences, however. The changes in phospholipid composition during growth have been studied in *M. gypseum* by Khuller et al.[134] and in *T. rubrum* by Das and Banerjee.[51] Carotene pigment formation during arthroconidia formation[106] has also been studied. The production of these pigments is stimulated by elevated temperature and may be involved in resistance to external factors by arthroconidia. As a cautionary note on evaluating the results obtained in research using dermatophytes, Jesenska[117] found wide variation in growth rate, conidiation, sensitivity to drugs, and pathogenicity in strains obtained from a single isolant of *T. mentagrophytes.*

Ecology of Growth. Investigations on the interactions of dermatophytes with other microorganisms are just beginning. Dermatophytes are known to produce a variety of antibiotic substances, including penicillin.[281, 291] In addition, a number of lipids found in dermatophytes possess antimycotic activity.[278] The possible role of dermatophyte antimicrobial substances on the bacterial population during infection has been noted by Bibel et al.[29, 30] There was a decrease in bacterial flora of infected skin as demonstrated by quantitative culture methods[29] and by electron micrography, and a dissolution of bacterial cells adjacent to dermatophyte hyphae.[30]

Conidia Production. Arthroconidia are normally produced by dermatophytes only under circumstances of active parasitization. Agents of tinea capitis such as *Microsporum audouinii* produce arthroconidia only when invading actively growing hair and agents of tinea cruris and tinea pedis (such as *Epidermophyton floccosum*) form them in the stratum corneum of infected individuals. They represent the mechanism for transferring the infection from one individual to another. The factors affecting their formation are being studied by Hashimoto et al. They found temperature, atmospheric composition, and media components important in inducing arthroconidia formation of an isolant of *Trichophyton mentagrophytes.*[60] The strain they used was isolated from a tinea capitis acquired from contact with cattle. So far these investigators have not been able to induce arthroconidia formation artificially in anthropophilic dermatophytes that is comparable to that produced during natural infection.

The morphology of growth and conidia production is also affected by temperature. All strains of *Trichophyton verrucosum* grow as chains of budding chlamydoconidia at 37°C (Fig. 8–55,*B*). Other dermatophytes, such as *T. rubrum* and *M. audouinii,* can be induced to form this "yeastlike" growth under the influence of elevated temperature and reducing agents.[228] In this form they produce systemic disease in experimental infections.

LABORATORY IDENTIFICATION

Direct Examination. The diagnosis of dermatophyte infection is most easily confirmed by direct microscopic examination of skin scrapings in a potassium hydroxide slide mount. This is prepared by placing the material to be examined (skin scales, nail scrapings, hair stubs, and so forth) on a microscope slide and adding a drop of 10 per cent KOH, which is mixed well with the specimen and the pieces separated in as thin a layer as possible. A cover slip is placed over the KOH-specimen mixture, and the slide is gently heated. Boiling is avoided, as this precipitates KOH crystals. The slide is allowed to cool and "ripen" a few minutes before examination. The KOH "clears" the specimen by digesting proteinaceous debris and bleaching pigments and loosening the sclerotic material without damaging the fungus. The hyphae of the fungus are unaffected by this treatment and stand out as highly refractile, long, undulating, branched, septate threads (Fig. 8–41). These threads are seen to course in, around, and through epidermal scales. In a well-cleared mount, one is able to discern nuclei, organelles, and fat droplets within the mycelium. Mature hyphae will show numerous

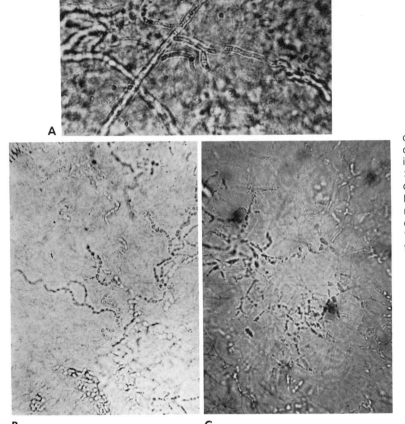

Figure 8–41. *A*, Potassium hydroxide mount of skin from tinea corporis. Note refractile, branching, septate hyphae. *T. rubrum* ×440. *B*, Skin scales from tinea cruris caused by *E. floccosum*. Note the thick-walled arthroconidia. These are more resistant to drying and heat than mycelium. *C*, "Mosaic fungus." Cholesterol crystals among the edge of squames.

septations. In time, these may fragment into rounded or barrel-shaped arthroconidia. This is particularly true in *Epidermophyton floccosum, T. verrucosum*, and other species that remain viable in the environment for long periods of time. Examination of hair stubbles often requires more time for clearing, especially if it is dark pigmented hair. The base of the hair shaft and follicular debris are the areas where fungi are most likely to be seen. The arthoconidia are outside the hair shaft in chains in a mosaic pattern, or intrapilar, depending on the species involved and whether it is endothrix or ectothrix. Nail scrapings are the most difficult to examine. Fungi are seldom in abundance, and a day or two may be required for good KOH digestion. For all types of specimens, hyphae must be differentiated from other artifacts.[253a] These include fibers of cotton, wool, and synthetic materials, starch grains, fat droplets, vegetable detritus, and "mosaic fungus" (Fig. 8–41,*C*). This last artifact causes the most difficulty for beginners. "Mosaic" is a network

of material, including cholesterol crystals, which is deposited around the periphery of keratinized epidermal cells. It can be seen to follow the outline of the cell but not to go through it. This observation, together with the abrupt changes in width, the lack of internal organelles, and the flat crystalline structures with reentrant angles, differentiates it from true hyphae. A host of other debris is seen in nail scrapings. Mycelial elements of *Candida* species in skin and nails are less refractive when seen in KOH mounts and show pseudomycelium, hyphae, and budding yeast cells. *Malasezzia furfur*, the etiologic agent of pityriasis versicolor, has a distinct morphology. Incidental saprophytic fungi and pollen grains may also be found. In experienced hands, the KOH mount is one of the most useful procedures in medical mycology. Besides its application in diagnosis of dermatophyte infections, it can be used in the examination of other specimens, such as sputum, pus, exudates, biopsy material, urine, and fecal material. In such cases, and for the beginner examining skin scrapings, it is helpful to add a stain to the KOH. The most convenient procedure of those recommended is the addition of Parker Superchrome Blueblack ink or some comparable substance to the KOH solution. The proportions depend on the preference and aesthetic taste of the user. The ink selectively colors the hyphae, making them more pronounced.

The selection of the specimen to be examined determines the success of the procedure to a large extent. In tinea pedis, the outer debris and macerated tissue should be removed from the sole and interdigital spaces. Cleaning the site with an alcohol sponge is helpful. (Do not use cotton swabs, as this introduces cotton fiber artifacts into the KOH mount.) As in all fungal lesions, the organisms are most prevalent in the edge or active site; therefore the epidermal scales should be removed from the periphery of the lesion. They may be placed directly on a slide for KOH mounting or into a sterile petri dish or sterile envelope for transport to the examining laboratory. In such containers, the organisms in skin scales remain viable for some time. An air-tight, damp container would promote the growth of saprophytic fungi. Scalpel, scissors, or preferably a double-edged Foman knife are used to obtain scrapings. If bullae or vesicles are present, these are clipped, examined, and cultured. Any antifungal or other topical medicaments must be removed before culturing material from lesions. This often involves vigorous washing. Even if the KOH preparation is negative, culture is warranted.

The procedure used for tinea corporis lesions is basically the same. Scrape the lesion from the center to the outside edge in order to loosen scales. In some infections it is often necessary to scrape until a serous exudate is formed, particularly when the etiologic agent is *T. rubrum*. In tinea capitis and tinea barbae, infected hairs are selected from lesions. In microsporosis this is aided by plucking hairs that fluoresce when a Wood's lamp is held above the head. In trichophytosis the abnormal or distorted hairs near the border of the lesions are picked. In "black-dot" ringworm it is often necessary to use a scalpel to excise the twisted deformed root that may be subsurface. Scutula and crusts from tinea favosa, animal ringworm, and inflammatory ringworm are usually highly contaminated. They should be macerated, washed in alcohol or antibiotic solution, and small, separated sections examined and cultured. Hair and crusts

from inflammatory lesions, particularly from the central areas, are often devoid of fungi. Removing the crust and examining the outer edge of the moist red base is usually successful.

Success in finding fungi in tinea unguium and onychomycosis is by far the most difficult to obtain. In white-spot infection the superficial layers are scraped off, cultured, and examined. In invasive tinea unguium most of the upper nail is scraped off and discarded. Only the deepest layers harbor fungi. Sometimes one can remove the debris from under the distal end of the toenail and then vigorously scrape the undersurface of the juncture of the nail with the nail bed. If oozing and slight bleeding occur, the chances of success are greater.

Culture Methods. When the index of suspicion is high, all specimens should be cultured, even if the KOH preparation is negative. Because ringworm can appear quite variable and culture is such an easy procedure, it is advised as a routine part of dermatologic examination. For mycologic examination, cultures are planted by furrowing the specimen into the media with the scraping knife. As much material as the surface of the agar will reasonably accommodate should be included. A wide agar slant or specimen jar is recommended. Attention to these details increases the chance of success.

The standard media for primary isolation of dermatophytes is Sabouraud's agar, containing cycloheximide and an antibacterial antibiotic. The cycloheximide (Actidione) in a concentration of 0.1 to 0.4 mg per ml suppresses the growth of most saprophytic fungi without deterring the growth of dermatophytes. The various antibacterial antibiotics used include chloramphenicol (0.05 mg per ml) or Aureomycin (0.1 mg per ml); both are satisfactory. Growth is relatively slow; usually ten days to three weeks are required at the optimum temperatures of 25° C. *T. verrucosum* and rare strains of *T. tonsurans* grow better at 37° C. When growth becomes evident on the primary isolation media, mycelial strands are transferred to a slide culture preparation (see Section Five). On slide culture, two media are advised: cornmeal with 1 per cent glucose to stimulate pigment production of *T. rubrum*, and Sabouraud's agar without antibiotics for visualization of the normal morphology of spores, conidia arrangement, and mycelial appendages (Fig. 8–42). The slide culture allows observation of the culture while it is growing. When conidia

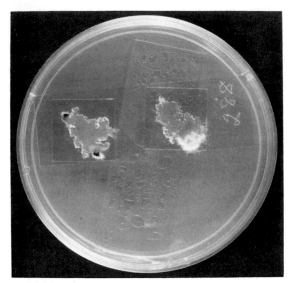

Figure 8–42. Slide culture. The right-hand side is Sabouraud's glucose agar and the left is cornmeal with 1 per cent glucose. The latter stimulates pigment production in *T. rubrum*. The former is for growth of the dermatophyte in its typically described morphology.

are evident, a lacto-phenol cotton blue mounting can be made for accurate observation. Sometimes vitamin-enriched, casein digest media enhance the growth and production of macro- and microconidia. This is particularly true of the faviform *Trichophyton* species. L. K. Georg in a series of papers delineated nutritional requirements for some dermatophytes.[81] These are very useful in identifying isolates of species closely resembling one another (Table 8–11). The requirement for thiamine can be fulfilled by the pyrimidine moiety in most species. A variety of other nutritional tests have been described. Philpot[202] has reviewed these procedures and listed those helpful in distinguishing morphologically similar isolants.

Some other techniques are useful in sepa-

Table 8–11. *Nutritional Requirements or Growth Enhancement Factors of Dermatophytes*[*202]

Species	Requirement
T. equinum	Nicotinic acid†
T. megninii	L-histidine
T. tonsurans	Thiamine
T. verrucosum	Inositol and thiamine, 80% of isolants; thiamine only, 20%
T. violaceum	Thiamine

*The tubes inoculated are ammonium nitrate or casein, enriched with the supplement. Most all dermatophytes grow better in vitamin-enriched medias.

†Some nicotinic acid–independent strains have been isolated in New Zealand and Australia.

rating closely related species. The presence of perforating organs in *in vitro* hair cultures (Fig. 8–51) separates *T. mentagrophytes* from *T. rubrum*.[4] *T. rubrum* produces a red pigment on potato-dextrose slants or cornmeal agar with 1 per cent glucose. *T. tonsurans* and most isolants of *T. mentagrophytes* do not.

For the small office without access to mycologic consultation or in large public health field studies, the routine isolation and identification of dermatophytes is difficult. Recently two new primary medias have been developed which are very helpful in primary isolation of dermatophytes. They are both based on pH change caused by the proteolytic activity of dermatophytes, which is lacking in most saprophytic fungi.[88] The first of these to gain popularity was the ink-blue agar of Baxter.[23] The agar is blue, but a colorless area is seen around a growing dermatophyte colony. Some of the drawbacks of this medium are that the mycelium is stained by the dye, thereby masking its natural color, and a number of contaminating bacteria cause color change. A far more satisfactory formulation was devised by Taplin et al.[270] The indicator is phenol red (pH 6.8 yellow, pH 8.4 red), and gentamycin, chlortetracycline, and cycloheximide are included to inhibit growth of bacteria and saprophytic fungi. The medium called DTM is commercially available. It has proved very useful in large field surveys because the color change is readily evident and fairly specific for dermatophyte fungi. Some workers feel that the recovery rate is lower on DTM than on Sabouraud's agar. This was not borne out by Sinski et al.[253] They found no difference in ability to support growth among DTM, Sabouraud's media, and certain batches of Youssef medium. Other batches of Youssef media and Littman oxgall agar were inferior in supporting growth or inhibiting contaminants. The red color in DTM does mask somewhat the colonial morphology and pigmentation. In all cases, the identification of organisms should be confirmed by the use of the slide culture technique in the hands of an experienced mycologist. In most cases, however, the presence or absence of a dermatophyte is sufficient knowledge to indicate therapy.

MYCOLOGY

As noted earlier, the number of species of dermatophytes is large; when they are added to the numerous keratinophilic fungi, they amount to a formidable group to describe and identify. Repeated exposure is required in order to become familiar and competent in identifying species and knowing their variability. In the present section some of the clinically important species of dermatophytes will be described, along with their ascigerous state when present. Because so many of these species have had numerous citations in the literature under other names, the synonymy is given for each valid species. The major differential characteristics are given in outline form as obverse and reverse colony morphology, microscopic morphology, and special characteristics if present. The most useful and complete manual covering the identification of dermatophytes and related soil forms is that published by Rebell and Taplin.[213]

Microsporum Gruby 1843

This genus is characterized by the presence of fusiform, obovate to spindle-shaped macroconidia. The walls are thick (to 4 μ in *M. canis*) and pitted, asperulate, echinulate, or spiny. This latter architectural feature may be restricted to the distal end and appears only on a few mature conidia. The macroconidia range in size from 7 to 20 × 30 to 160 μ, with from 1 to 15 septations. The form and size of these conidia are important in identification. The microconidia, on the other hand, are sessile or stalked, clavate, and 2.5 to 3.5 × 4 to 7μ in size. They are usually not helpful in identification, as they are similar to those found in several other genera, and vary little from species to species within the genus *Microsporum*. When found, the teleomorph stage of the members of the anamorph genus *Microsporum* is in the genus *Nannizzia* of the family Gymnoascaceae in the order Eurotiales of the phylum Ascomycota.

Nannizzia Stockdale 1961

The chief distinguishing features of this genus are the peridial hyphae found around the sexual fruiting body, the gymnothecia. These hyphae are generally verticillately branched, have asperulate to echinulate hyphal cells with one or more constrictions, and terminate in smooth-walled, blunt-ended cells, spikes, or spirals (Fig. 8–50a). The individual cells of the peridial hyphae lack the knobby protuberances found in the genus *Arthroderma*, the teleomorph stage associated with the genus *Trichophyton*.

Microsporum audouinii Gruby 1843

Synonymy. *Trichophyton decalvans* Malmsten 1848; *Microsporum villosum* Minne 1907; *Microsporum umbonatum* Sabouraud 1907; *Microsporum velvetieum* Sabouraud 1907; *Microsporum tardum* Sabouraud 1910; *Microsporum tomentosum* Sabouraud 1910; *Microsporum depauperatum* Guéguen 1911; *Microsporum rivalieri* Vanbreuseghem 1963; *Closteroaluriosporia audouinii* (Gruby) Grigorakis 1925; *Martensella microspora* Vuillemin 1895; *Sabouraudites audouinii* (Gruby) Ota et Langeron 1923; *Sabouraudites langeronii* Vanbreuseghem 1950.

The correct ending of this genus name should conform to the Greek "-on," as the root word is Greek in origin and would agree with *Trichophyton* and *Epidermophyton*. David Gruby, when naming this genus, made an error in syntax which unfortunately has continued. The isolation of this organism has been decreasing in the last 40 years throughout its endemic regions.[201] The reasons for this decreased incidence are unknown.

Colony Obverse. (Fig. 8–43,*A*) On SGA* the growth is slow, forming a flat, spreading, dense colony with a silky, furry, matted consistency and radiating edges. The color is white or gray to tan or rust-buff. Rarely seen colonial forms include a rust-brown or rose-tan color with radial grooves, as in var. *langeronii*. The var. *rivalieri* has a grayish-white folded thallus with a satiny sheen similar to *T. schoenleinii* (Fig. 8–44,*A*). This latter variety is common in Africa and found in Florida.

Reverse. Salmon, rust, or peach color pigment is characteristically produced by *M. audouinii*. If produced in quantity, it stains the mycelium. Many isolants, however, have little or no pigment.

Microscopic Morphology. Generally few conidia are seen (Fig. 8–45,*A*). The usual microscopic examination shows thick-walled terminal or intercalary chlamydoconidia that are fairly characteristic and permit identification of the culture. Racquet hyphae, pectinate bodies, and, rarely, some irregular microconidia may be present. The macroconidia, if seen, are irregular in shape, elongate, thick-walled, and echinulate (3 × 4 μ).

---

*Sabouraud's glucose agar is the standard medium on which are based the descriptions of growth characteristics and colony morphology of medically important fungi. Unless otherwise noted, the summary of characteristics is growth on SGA at 25° C for two weeks.

Figure 8–43. *Microsporum.* Typical colonial morphology of commonly isolated species. *A, M. audouinii.* Silky, flat growth with pleomorphic tufts. The reverse is a salmon color. *B, M. canis.* Floccose growth. The reverse is chrome yellow. *C, M. gypseum.* Powdery beige-colored surface. The reverse is variable in color. *D, M. ferrugineum* showing the folded, wrinkled, rust-colored colony. The reverse is variable in color.

Differential Characteristics. On polished rice, growth of *M. audouinii* is almost imperceptible in contrast to the luxuriant, pigmented growth of *M. canis.* Growth on yeast extract media or soil-hair agar stimulates the production of macroconidia. Intermediate forms between *M. canis* and *M. audouinii* are not uncommon, and their occurrence suggests a close relationship, if not recent evolutionary derivation, between the two. No perfect stage for these species has yet been found. All strains of *M. audouinii* tested by the *T. simii* sex stimulation procedure show a (+) mating type. *M. audouinii* is anthropophilic, small-spored ectothrix, and fluorescent in infected hairs. This species probably evolved in Africa, and the var. *rivalieri* may represent the ancestral form. Since only one mating type appears to exist now, it might have had a selective advantage over the other during its adaptation to an anthropophilic organism.

Microsporum canis Bodin 1902

Synonymy. *Microsporum felineum* (Bodin) Mewborn 1902; *Microsporum lanosum* (Bodin) Sabouraud 1907; *Microsporum caninum* Sa-

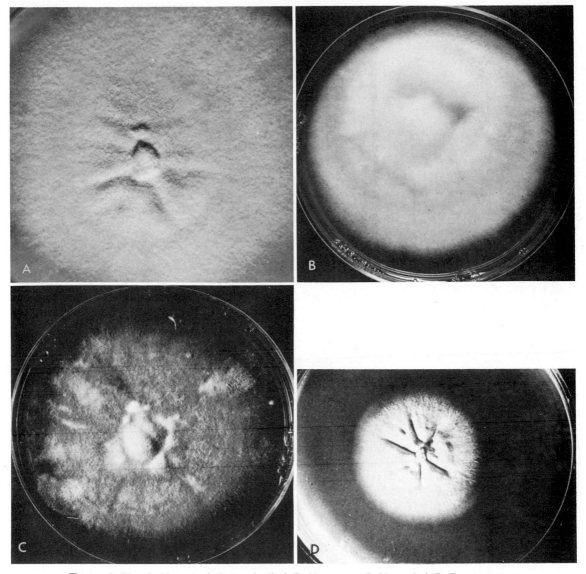

Figure 8–44. *A, M. audouinii* var. *rivalieri. B, M. nanum. C, M. cookei. D, T. verrucosum.*

bouraud 1908; *Microsporum stillianus* Benedek 1937; *Microsporum peudolanosum* Conant 1937; *Microsporum simiae* Conant 1937; *Microsporum obesum* Conant 1937; *Closterosporia lanosa* Grigorakis 1925; *Closterosporia felinea* Grigorakis 1925; *Closterosporia felinea* Grigorakis 1925.

Teleomorph State. *Nannizzia otae* Hasegawa et Usui 1975.[104, 105]

Colony Obverse. (Fig. 8–43,*B*). The growth is rapid, producing a woolly or cottony, white to yellowish, flat to sparsely grooved colony with radiating edges. It rapidly becomes pleomorphic. Dysgonic isolants are glabrous, heaped, and deeply pigmented.

Reverse. The underside of the colony is characteristically a deep chrome yellow. This is best viewed in young growth. On potato-glucose medium, a sparse mycelium and an abundant lemon pigment are seen.

Microscopic Morphology. The large macroconidia (8 to 20 × 40 to 150 μ) are produced in abundance (Fig. 8–45,*B*). They have thick walls (2 μ), up to 15 septa, and are spindle-shaped and echinulate, or pitted. Curved or hooked ends are seen. The dysgonic variant produces wide hyphae arranged as fascicles and clublike nobs and elongated conidia with few septa,[64] and var. *obesum* produces fat, three-celled conidia. The var. *obesum* is associated with monkeys and

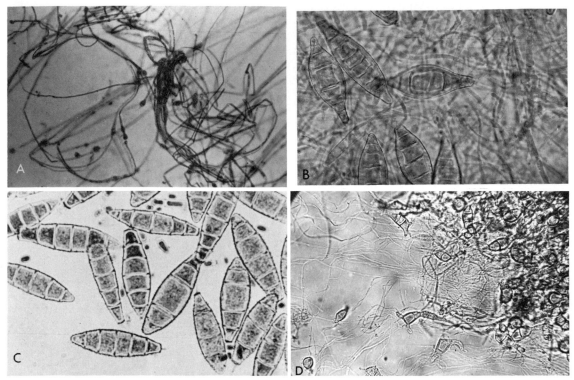

Figure 8–45. *A, M. audouinii.* Microscopic view showing hyphae and a few chlamydoconidia ×440. *B, M. canis.* Macroconidia. (Courtesy of M. T. Abu-Samra.) *C, M. gypseum,* Macroconidia. *D, M. distortum.* Macroconidia.

apes[139] and may be a separate species. The var. *equinum,* associated with horses, is now regarded as a separate species properly referred to as *M. equinum.* The microconidia are slender and clavate, similar to many other species. Racquet hyphae, pectinate bodies, nodular bodies, and chlamydoconidia are seen.

Differential Characteristics. The growth is abundant on polished rice grains. *M. canis* is zoophilic and native to cats, dogs, and probably horses, apes, and monkeys (Table 8–10). Though not considered geophilic, it has been isolated from soil in Hawaii and Rumania. *M. canis* is small-conidiated, ectothrix, and fluoresces in infected hairs. Colony variants of this organism are common.

N. otae. The perfect state of *M. canis* was first described in Japan by Hasegawa and Usui.[105] In recent studies by Weitzman and Padhye[287] almost all clinical isolants are the (−) mating type and *M. canis* may be a complex of species. Gymnothecia are globose, 280 to 700 μ in diameter. At first they are white and later become pale buff. The peridium consists of a network of hyaline, septated, echinulate hyphae showing a pronounced constriction at the junction of each cell. The hyphae branch dichotomously or rarely verticillately. The distal branches curve over the gymnothecium. They terminate in three types of appendages: smooth-walled septate hyphae 150 μ in length, spiral hyphae, or a macroconidium. Asci are thin-walled, subglobose and evanescent, and 5 to 7 μ in size, with 8 ascospores. The ascospores are lenticular, smooth and hyaline and 2.5 to 3.8 × 2 to 2.5 μ in size.

Microsporum ferrugineum Ota 1921

Synonymy. *Microsporum japonicum* Dohi et Kambayashi 1921; *Microsporum aureum* Takeya 1925; *Microsporum orientale* Carol 1928; *Trichophyton ferrugineum* (Ota) Langeron et Milochevitch 1930.

Colony Obverse (Fig. 8–43,*D*). The growth is slow, forming a heaped, folded, glabrous, reddish-yellow to orange-yellow thallus with a waxy surface. This is the usual strain found in the Far East. A fine, velvety, white overgrowth may occur. Variants without pigment resemble *T. verrucosum* and are common in the Balkans.

Reverse. No characteristic pigment is seen.

Microscopic Morphology. Distorted mycelium without conidia is the usual microscopic picture. Faviform, abnormal, mycelial elements and coarse hyphae with prominent crosswalls (bamboo hyphae) are seen.

Differential Characteristics. On deficient media (dilute SGA) and potato-glucose with charcoal, macroconidia of the *Microsporum* type are occasionally seen. On Löwenstein-Jensen medium, the color is light yellow compared to the dark red-brown of *T. soudanense.* It is anthropophilic, small conidiated, ectothrix, and fluorescent in infected hairs.

Microsporum gypseum (Bodin) Guiart and Grigorakis 1928

Synonymy. *Achorion gypseum* Bodin 1907; *Microsporum flavescens* Horta 1911; *Microsporum scorteum* Priestley 1914; *Microsporum xanthodes* Fischer 1918.

Teleomorph States. *Nannizzia gypsea* (Nannizzi) Stockdale 1963; *Nannizzia incurvata* Stockdale 1961.

Colony Obverse. (Fig. 8–43,*C*). Colonies grow rapidly, producing a flat, spreading, powdery surface that is rich cinnamon-buff to brown, occasionally with overtones of violet. The powder consists of masses of macroconidia. The edges of the colony are entire to scalloped or ragged. Diffuse pleomorphism rapidly develops.

Reverse. A variety of pigments or none are produced.

Microscopic Morphology. Macroconidia are produced in great abundance. They are thin-walled, 8 to 16 × 20 to 60 μ, roughened, and have 4 to 6 septa (Fig. 8–45,*C*). The usual variety of other conidia including microconidia, is seen. Hair penetration organs are produced.

M. gypseum is geophilic and abundant in soil throughout the world. It is ectothrix but produces few arthroconidia (Fig. 8–5,*B*). It is usually associated with an inflammatory disease and may cause a kerion formation.[68] Fluorescence is absent or dull in infected hairs. The complex is made up of two species whose teleomorph states are *Nannizzia incurvata* and *Nannizzia gypsea.* The anamorph stage (*M. gypseum*) of these two species is essentially the same and can only be identified by mating with appropriate tester strains. However, *N. gypsea* generally produces a more spreading and a coarser granular colony; the macroconidia are slightly wider, and the surface color is brighter and sometimes a redder color than *N. incurvata.* *N. incurvata* is pale buff and finely granular and occasionally has a reddish to yellow reverse. Differences in the ornamentation of the macroconidia have been seen by the scanning electron microscope.[280]

N. gypsea. The gymnothecia are globose, beige to buff, and 300 to 800 μ in diameter. The peridial hyphae are asperulate, with septa constricted 1 to 3 times. They are verticillately branched, and these branches curve away from the main axis (Fig. 8–50,*D*). This is in contrast to those of *N. incurvata,* which curve toward the main axis. The branches terminate in elongate, smooth-walled, tapering spikes up to 250 μ in length or, rarely, end in tightly coiled spiral hyphae. Sometimes ellipsoid macroconidia are found at the ends of peridial hyphae. Mutants with smooth-walled macroconidia have been described. The asci are evanescent, thin-walled, 5 to 7 μ in diameter, and eight-spored. Ascospores are smooth, ovoid, 2 × 4 μ in diameter, and appear yellow in mass.

N. incurvata. The gymnothecia, asci, and ascospores are similar to those of *N. gypsea.* The peridial hyphae again are asperulate, septate, with 1 to 3 constrictions and verticillately branched. The branches curve inward toward the main axis and away from the gymnothecial body (Fig. 8–50,*C*). They end as blunt tips, spikes, or spiral hyphae.

Microsporum fulvum Uriburu 1909

Teleomorph State. *Nannizzia fulva* Stockdale 1963.

Until its delineation by Stockdale in 1963, this fungus was treated as a variety of *Microsporum gypseum.* A teleomorph stage was discovered, and it is now considered to be a separate species.

Colony Obverse. Although it resembles the *M. gypseum* complex, the colony is more floccose and tawny buff in color. The periphery is often white.

Reverse. A dark red undersurface is occasionally seen; otherwise it is colorless to yellow brown.

Microscopic Morphology. The macroconidia of *M. fulvum* are more clavate, cylindrical, or bullet-shaped and are less often in large clusters than those of *M. gypseum.* Microconidia 2 to 3.5 × 3 to 8 μ are produced, but these are indistinguishable from those of other species. Numerous spiral hyphae, which are often branched, are seen. Hair penetration organs are produced.

This is another geophilic species of worldwide distribution. It is a sparsely conidiated ectothrix and nonfluorescent in infected hairs. Its teleomorph stage is *N. fulva.*

N. fulva. The gymnothecia are larger (500 to 1300 μ) than *N. gypsea* but otherwise indistinguishable. The asci, ascospores, and peridial hyphae are also similar. Identification depends on mating with tester strains. The sexes are distinguishable by the production of elastase.[224] This enzyme is associated with the (+) mating type and is absent in the minus type. *M. fulvum* is a less pathogenic species than *M. gypseum;* however, its (+) mating type is somewhat more virulent in experimental infection than the (−) type.[224] Others have had inconsistent results in determining elastase production.[284]

Microsporum nanum Fuentes 1956

Teleomorph State. *Nannizzia obtusa* Dawson et Gentles 1961.

Colony Obverse. (Fig. 8–44,*B*). A rapidly growing, thin, spreading, powdery colony is produced that resembles a variant of *M. gypseum.* It is white to yellow, changing to pinkish-buff in color and has a fringed edge.

Reverse. A red brown pigment is frequently seen.

Microscopic Morphology. The characteristic ovoid to pear-shaped macroconidia are produced in great numbers. They are one- to three-celled, 12 to 18 × 4 to 8 μ in diameter, with thin, verrucous walls. Microconidia are clavate. They occur rarely on SGA but are numerous on soil-hair agar and distinguish this species from *Chrysosporium* sp. The latter have large, ovoid microconidia, with a wide attachment base that can be confused with the macroconidia of *M. nanum.* Hair penetration organs are produced by *M. nanum.*

This is a geophilic and zoophilic species associated with pigs, although the first isolant was from a boy's scalp.[73] It has been isolated from soil in pig yards. In man a tinea corporis contracted from pigs has been seen.[133,174] The organism is a sparsely conidiated ectothrix and nonfluorescent in infected hair. The teleomorph stage is *N. obtusa.*

N. obtusa. Gymnothecia are globose, 250 to 450 μ in diameter, and pale buff or yellowish. Peridial hyphae are pale yellow, hyaline, dichotomously branched at an obtuse angle from the main axis, and rarely verticillately branched. The cells of the peridial hyphae are thick-walled, echinulate, cylindrical, and 5 by 13 μ in size, with one to two slight constrictions in the center. The peridial hyphae may end in a long, smooth, tapering spike or tightly coiled spiral hyphae. Asci are subglobose, 5.5 to 5 to 6 μ in size, and evanescent. Eight oblate, smooth-walled, yellowish ascospores are formed. They are 2.7 to 3.2 by 1.2 to 2 μ in size, and yellow in mass. The walls may be somewhat roughened.

Microsporum distortum di Menna et Marples 1954

This species is a rare cause of tinea capitis in New Zealand and Australia and of infection in lower animals such as dogs. Infection in the U.S. has been traced on occasion to contact with monkeys from South America.[230] This species is similar to and possibly is a variant of *M. canis.* It is small-conidiated, ectothrix, and fluorescent in infected hairs.

Colony. In general the colonial growth is similar to *M. canis,* but usually has less pigmentation on the underside. The colony is flat, velvety to fuzzy, and white to tan in color.

Microscopic Morphology. The macroconidia are similar to those of *M. canis* but grossly bent and distorted. Sessile clavate microconidia are produced. They are usually more abundant than isolants of *M. canis* (Fig. 8–45,*D*).

Differential Characteristics. *M. distortum* like *M. canis* grows abundantly on rice grains; only the morphology of the macroconidia differentiates the two species.

Microsporum gallinae (Mégnin) Grigorakis 1929

Synonymy. *Epidermophyton gallinae* Mégnin 1881; *Achorion gallinae* (Mégnin) Sabouraud 1910; *Trichophyton gallinae* (Mégnin) Silva et Benham 1952.

Colony Obverse. This species grows rapidly, producing a conical, slightly folded, downy to satiny white colony with an entire to slightly scalloped edge. The mycelium may be stained pink (Fig. 8–46,*A*).

Reverse. Within a few weeks a diffusible pigment is formed. Initially it is yellow, but becomes a bright Montmorency sour cherry red or strawberry red (Fig. 8–46,*B*).

Microscopic Morphology. The recent finding of echinulations on the macroconidia has prompted the restoration of this species to the genus *Microsporum.*[90] The macroconidia are 6 to 8 × 15 to 50 μ in size and are often elongate with a blunt tip (spatulate or slipper-shaped). There are two to ten cells,

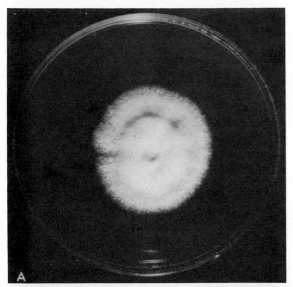

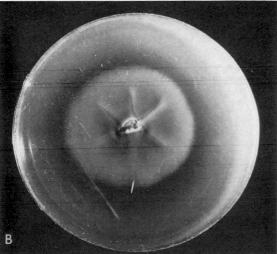

Figure 8–46. *M. gallinae A*, Obverse. *B*, Reverse, showing diffusible cherry-red pigment.

and the walls are usually smooth or sometimes echinulate at the tip. The conidia are often curved, with a flared distal end resembling a threatening cobra. They are often attached to dentate, pectinate, or leaflike hyphae. Clavate and pyriform microconidia are found (Fig. 8–55,*D*). Thiamine or yeast extract added to the media increases conidiation. This is a zoophilic species associated with fowl and is rarely involved in human infection.[87, 272] It is ectothrix and nonfluorescent in infected hairs.

Special Characteristics. *M. gallinae* is distinguished from *T. megninii* by good growth on ammonium nitrate media. The latter species requires L-histidine.

Microsporum persicolor (Sabouraud) Guiart et Grigorakis 1928

Synonymy. *Trichophyton persicolor* Sabouraud 1910.

Teleomorph State. *Nannizzia persicolor* Stockdale 1967.[262]

For many years there was doubt as to the validity of this species, and most mycologists regarded it as a variety of *T. mentagrophytes*. The discovery of a perfect stage in the genus *Nannizzia* has verified it as a distinct species. This species does not appear to invade hair.

Colony Obverse. A rapidly growing, flat to gently folded, fluffy, yellowish buff to pale pink thallus is produced (Fig. 8–47,*B*).

Reverse. Pigmentation of the underside is variable. The color ranges from peach or rose to deep shades of ochre. On sugar-free media, sectors of rose to red to deep wine tints are seen. This distinguishes it from *T. mentagrophytes*, which does not produce color on these media. This red color is also produced on pablum or rice grain media.

Microscopic. Microconidia are usually abundant. They are clavate or fusiform to globose, and arranged in clusters resembling the microscopic morphology of *T. mentagrophytes*. Spiral hyphae are common (Fig. 8–47,*C*). The macroconidia are sparsely produced. They are elongate, fusiform to clavate, and usually six-celled. The walls are thin and smooth with some echinulations at the tip (Fig. 8–47,*D*). Hair penetration organs are produced.

M. persicolor grows better than *T. mentagrophytes* on ammonium nitrate media, and the predominance of stalked, elongate, clavate microconidia in the former is also a distinguishing feature. This species is zoophilic and is a rare pathogen for man, but it is frequently found in bank voles and field voles. Human infections are noted for the severity of the disease evoked.[193] The spectrum of human involvement includes tinea capitis (Fig. 8–47,*A*), tinea corporis, and tinea pedis. In most cases reported there was some contact with rodents and the organism seems to be of worldwide distribution.[66, 186] It has been found in soil. The teleomorph stage is *N. persicolor*.[262]

N. persicolor. Globose gymnothecia are formed that vary from 250 to 900 μ in diameter and are the usual buff color. The peridial hyphae are branched, asperate to echinulate, and hyaline to pale beige in color. The branches curve outward on the gymnothecia and often end in spirals, which may

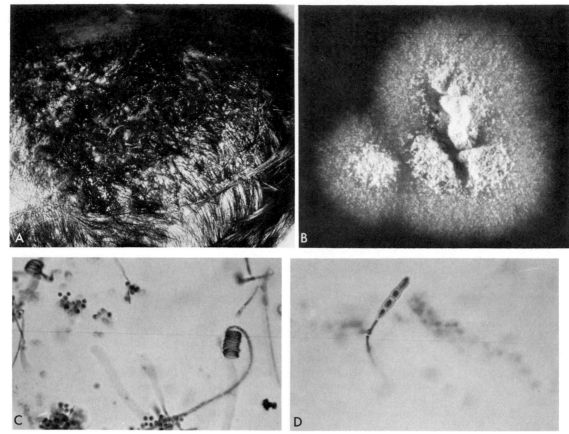

Figure 8–47. *Microsporum persicolor. A*, Tinea capitis showing erythematous, vesiculopustular lesions with the formation of kerion, areas of necrosis, periorbital swelling, fever, and regional lymphadenitis. The patient, an eight-year-old girl, had pet mice. (Courtesy of T. Davis.) *B*, Colony of isolant is folded peach- buff with pale granular surface and a vinaceous reverse. *C*, Microconidia and numerous spiral hyphae. *D*, Macroconidium is thin walled with six septa. This was one of several cases from the Chicago area.

be dichotomously branched. The cells of the peridial hyphae are symmetrical and have a single central constriction. Asci are evanescent, ovoid, and 4.3 to 6 × 5 to 7 μ in size. Eight ascospores are produced. They are hyaline, lenticular, smooth-walled, yellow in mass, and 2.3 to 3.2 × 1.5 to 2.1 μ in size.

Microsporum cookei Ajello 1959

Teleomorph State. *Nannizzia cajetani* Ajello 1961.

This is a geophilic species reported in rodents, dogs, and rarely in man. It is not known to invade hair.

Colony Morphology. The spreading, rapidly growing colony is coarsely granular and usually a deep pink (Fig. 8–44,*C*). The morphology is similar to strains of *M. gypseum*. The color varies from greenish buff to brown, and a deep vinaceous red is usually found on the reverse.

Microscopic Morphology. The macroconidia are thick-walled, echinulate, 30 to 50 × 10 to 15 μ in size, resembling *M. gypseum* except in wall width. The microconidia are obovoid and abundant. Hair perforating organs are formed.

N. cajetani. Gymnothecia are 368 to 686 μ in size, globose, and pale buff to yellow. Peridial hyphae are septate; the cells are elongate, echinulate, and minimally constricted. The ends taper to a long, smooth-walled spike up to 480 μ in length or to tight spiral hyphae. Asci are globose to ovate, 6 to 9 μ in diameter, and contain eight ascospores. These are lenticular, golden in mass, smooth-walled, and 3 to 3.6 × 1.8 μ in size.

Microsporum vanbreuseghemii Georg, Ajello, Friedman et Brinkman 1962

Teleomorph State. *Nannizzia grubyia* Georg, Ajello, Friedman, et Brinkman 1962.[82]

This is a geophilic species rarely involved in

ringworm of man, dog, squirrel, and cat.[175] It is ectothrix and nonfluorescent in infected hair.

Colony Morphology. A fast-growing, flat, spreading colony is produced that is cottony and cream-yellow to lavender-pink. The reverse is colorless to yellow. The fungus rapidly becomes pleomorphic.

Microscopic Morphology. Macroconidia are abundant, 59 to 62 × 11 μ in size, echinulate, thick-walled (2 to 3 μ), and cylindro-fusiform in shape, with seven to ten cells. They are similar to those of *T. ajelloi* except that the latter has smooth-walled spores and a purple pigment. *M. praecox* and *M. racemosum* have fewer cells in the macroconidia than *M. vanbreuseghemii*. The microconidia are pyriform to obovate. This species readily infects guinea pigs and produces hair perforation organs. The teleomorph stage is *N. grubyia*.

N. grubyia. The gymnothecia are globose, white to buff in color, and 150 to 600 μ in diameter. Peridial hyphae are hyaline, uncinately or dichotomously branched, curving away from the main axis. The tips end in long, smooth, blunt points or curved, clawlike, phalangiform cells. Spiral hyphae and macroconidia are also found at the tips. Asci are evanescent, globose, 4.8 × 6 μ in diameter, containing eight lenticular, smooth-walled ascospores which are 2.4 × 3 μ in size and pale yellow in mass.

Microsporum racemosum Borelli 1965

Teleomorph State. *Nannizzia racemosa* Rush-Monro, Smith, et Borelli 1970.

This is a geophilic species isolated initially from soil in South America and then Romania. It has been isolated from two cases of tinea corporis in the Chicago area[52, 222] (Fig. 8–48) and thus is cosmopolitan in distribution.

The colony is flat, rapidly spreading, granular, and white with a powdery cream-colored central area (Fig. 8–48, *B*). The reverse is wine red. On microscopic examination the microconidia are noted for their arrangement in racemes (Fig. 8–48,*C*). The macroconidia are tapered and sparsely echinulate, have from five to ten cells, and may have a rat-tail terminal filament.

Microsporum praecox Rivalier 1953

This is a rare cause of pustular ringworm isolated from a case of pustular disease on the wrist in France[229] and recently in the United States.[285] The colony is flat, granular, and buff or powdery with much soft folding (Fig. 8–49, *A*). The macroconidia are thin-walled and similar to those of *M. gypseum* but more elongate (Fig. 8–49, *B*). In the variant described by Weitzman and McMillen[265] distorted branching conidia were noted (Fig. 8–49,*C*).

Microsporum amazonicum Moraes, Borelli and Feo 1967

Teleomorph State. *Nannizzia borellii* Moraes, Padhye, and Ajello 1975.

This organism has been obtained several times as normal flora of the spiny rat (*Proechimys guyannensis*).[172, 173] The colony is flat, granular, and "medlar fruit color." The macroconidia are ellipsoid, thin-walled, and rough and have one to three septa. The gymnothecia of the teleomorph state are similar to those of other *Nannizzia* species.

Microsporum boullardii Dominik et Majchrowicz 1965[58]

This was isolated from soil collected in Conakry, Guinea, in Africa. The bullet-shaped macroconidia are similar to those of *M. fulvum* but are longer, with more septa. They are less than 60 μ, more cylindrical or cigar-shaped in contrast to the conidia of *M. praecox* and *M. racemosum*. In other respects the culture resembles *M. fulvum*, but it does not produce ascospores when mated with testers of *N. fulvum*.

Microsporum equinum (Delacroix et Bodin) Gueguén 1904

This is a rare cause of infection among horses. The colony and conidia resemble *M. canis*, but the macroconidia have only one to four cells and are poorly developed.[100]

Trichophyton Malmsten 1845

The macroconidia characteristic of this genus are elongate, clavate to fusiform, generally thin-walled (up to 2 μ), smooth, and have zero to ten septa. The size range is 4 to 8 × 8 to 50 μ. As with other dermatophytes, the microconidia are usually uninucleate, and the cells of the macroconidia multinucleate. Macroconidia are few or absent in many species. Microconidia are globose in shape, 2.5 to 4μ

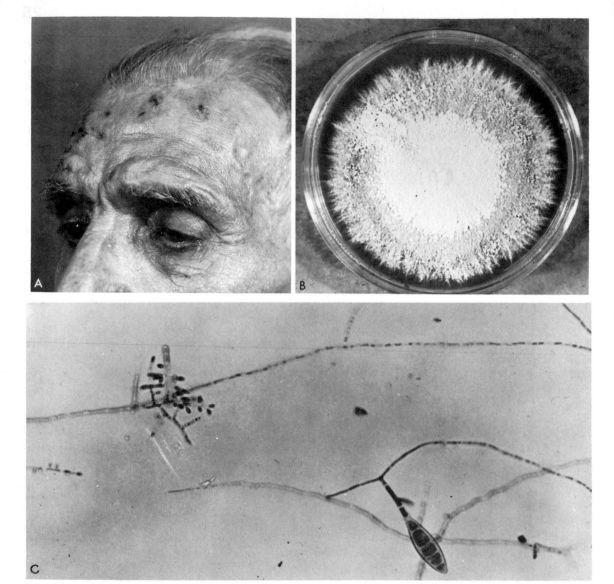

Figure 8–48. *Microsporum racemosum. A,* Excoriated macules on forehead. *B,* Flat granular colony. *C,* Microconidia arranged in racemes with tapered macroconidium. (Courtesy of V. Daum.)[52]

in diameter, or clavate, pyriform, and 2 to 3 × 2 to 4μ in diameter. The teleomorph stage, when found, is in the genus *Arthroderma.*

Arthroderma Berkeley 1860

The peridial hyphae around the gymnothecia of *Arthroderma* are usually dichotomously branched, with asperulate hyphal cells that are noticeably constricted at the septa. The individual cells have a central constriction, and at either end have a prominent knobby protuberance (Fig. 8–50, *B*). They have an overall dumbbell-shaped appearance. This genus is quite similar to *Nannizzia,* and few

distinguishing characteristics are constant.[163, 194]

Trichophyton mentagrophytes (Robin) Blanchard 1896

Synonymy. *Microsporon mentagrophytes* Robin 1853; *Achorion quinckeanum* Blanchard 1896; *Trichophyton felineum* Blanchard 1896; *Trichophyton gypseum* Bodin 1902; *Trichophyton granulosum* Sabouraud 1909; *Trichophyton radiolatum* Sabouraud 1910; *Trichophyton laticolor* Sabouraud 1910; *Trichophyton denticulatum* Sabouraud 1910; *Trichophyton farinulentum* Sabouraud 1910; *Trichophyton asteroides* Sabouraud 1910; *Trichophyton inter-*

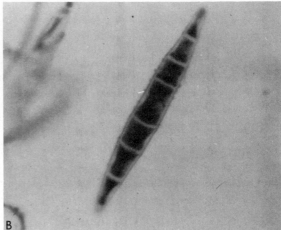

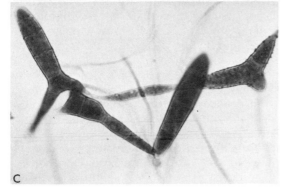

Figure 8–49. *Microsporum praecox. A*, Softly folded granular colony that is buff in color. *B*, Fusiform macroconidium with thin walls and echinulations. *C*, Bizarre verrucose branching macroconidia from a case of tinea corporis in an adult female described by Weitzman and McMillen. (Courtesy of S. McMillen.[285])

Figure 8–50. Peridial hyphae from the gymnothecia of *Nannizzia* and *Arthroderma* spp. *A, Nannizzia.* Peridial hyphae are asperulate, more or less symmetrically constricted at the juncture of the cells with no constriction in the center of the cell (*arrow*). Also seen are an echinulate macroconidium, asci, and ascospores. *B, Arthroderma.* The hyphae are also asperulate, dichotomously branched, and the cells are constricted at the septa; there is also a constriction in the center of each cell (*arrow). C, Nannizzia incurvata.* Note the branches of the peridial hyphae curve toward the central axis (*arrow*): adaxial. *D, Nannizzia gypsea.* The branches of the peridial hyphae curve away from (*arrow*) the central axis: abaxial.

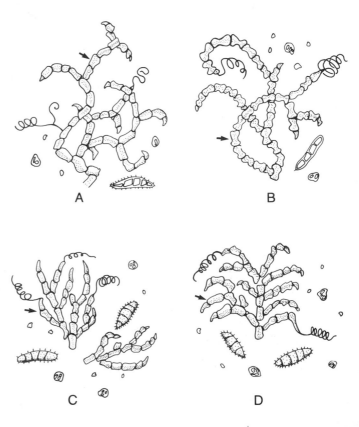

digitale Priestley 1917; *Trichophyton "C"* Hodges 1921; *Trichophyton Kaufmann-Wolf* Ota 1922; *Trichophyton pedis* Ota 1922; *Trichophyton quinckeanum* (Zopf) MacLeod et Muende 1943.

Teleomorph States. *Arthroderma benhamiae* Ajello et Cheng 1967 and *Arthroderma vanbreuseghemii* Takashio 1973.

T. mentagrophytes is the commonest dermatophyte of man and animals. Its morphology is so variable that it has been given a long list of binomials based on colony color, morphology, conidiation, and host association. The characteristics are so inconstant, and so many intermediates are seen that such names are not useful or practical. Almost all these variants mate with the tester strains of *A. benhamiae* or *A. vanbreuseghemii*, so that a single anamorph species is recognized. The dropping of most varietal designations is encouraged. *T. mentagrophytes* is universal, and anthropophilic and zoophilic forms are found. If recovered from soil it is in association with animal dander such as that found in rodent burrows. It is not considered geophilic. It is usually small-conidiated, ectothrix, and nonfluorescent in infected hair. Some strains (var. *quinckeanum* type) may produce endothrix infections and show a dull fluorescence. All strains produce hair perforating organs (Fig. 8–51). Guinea pigs are easy to infect with granular strains (var. *mentagrophytes*), but less so with the downy (var. *interdigitale*) type.[79]

The var. *quinckeanum* is associated with favus of mice.[31] Most of the strains mate with the A(+) or a(−) tester strains of *A. benhamiae*; however, Weitzman and Padhye[286] have shown that 19 isolants also mate with *A. simii* 678 A,[223] but not with other *A. simii* testers. The var. *erinacei*[161, 256] is usually isolated from hedgehogs,[65] though occasionally from human infections.[238] All strains from nature are A(+) mating type and urease-negative; however, urease-positive organisms and a(−) mating types can be produced in the laboratory.[268] The var. *erinacei* mate with *A. benhamiae* but not *A. vanbreuseghemii*. The var. *interdigitale* is an anthropophilic fungus isolated from mild or occult tinea pedis. Takashio[266, 267] has found that none of 26 isolants mated with *A. benhamiae*, *A. vanbreuseghemii*, or *A. simii* and considers that *T. interdigitale* should be retained as a distinct species.

Colony Observe (Fig. 8–53,*C*). The anthropophilic form grows as a flat, downy thallus with white edges and a cream-tinted central area. On potato-glucose agar, a stellate colony with sparse mycelium and numerous conidia is seen. Zoophilic isolates produce a flat, rapidly growing, granular colony that is cream, yellowish, buff to tan, or reddish-brown in color. Mycelium is usually sparse, and the powdery appearance is due to quantities of microconidia. The edges are often raylike. Some strains, particularly those from Southeast Asia, show a powdery, lavender-tinged surface. Numerous variations of colony morphology occur.

Reverse. Pigmentation is variable; colorless, yellow-brown, reddish-brown, brown, and a deep wine red resembling *T. rubrum* are seen. In differentiating pigmented colonies from *T. rubrum,* one should note that the latter usually produces a red pigment on potato-glucose and cornmeal 1 per cent glucose agar, while *T. mentagrophytes* does not. Almost all strains of *T. mentagrophytes* are urease-positive except the var. *erinacei,* which is negative. *T. rubrum* is urease-negative.

Microscopic Morphology (Fig. 8–54,*B*). Probably the most consistent feature of *T. mentagrophytes* is production of globose microconidia in grapelike clusters *(en grappe)*. These are most abundant in zoophilic-granular strains, and less so in downy strains. In the latter instance the conidia are more clavate-shaped and resemble those of *T. rubrum*. Macroconidia are thin-walled, smooth, and variable in shape. Their size ranges from 4 to 8 × 20 to 50 μ and they have three to five cells. They are generally cigar-shaped with a narrow attachment base, and here, in contrast to *T. rubrum,* the end cell of these conidia may have a terminal filament or "rat tail" appendage. The typical picture of *T. mentagrophytes* seen under the microscope is massed microconidia, some macroconidia, and several spiral hyphal cells, all in clusters on the vegetative hyphae. Structures resembling peridial hyphae, antlerlike hyphae, arthroconidia, nodular bodies, racquet mycelium, and chlamydoconidia are also seen. The variety designated as *quinckeanum* (mouse favus) has a gently folded thallus and lateral clavate microconidia. In some strains, isolated from tinea pedis and having deep yellow to orange pigmentation, nodular bodies are numerous and microconidia few in number. Hair perforating organs (Fig. 8–51) and lack of pigment on cornmeal glucose and potato-glucose agar separate *T. mentagrophytes* from *T. rubrum.*

The teleomorph states are *Arthroderma benhamiae* and *A. vanbreuseghemii*. Other teleomorph states may also exist, as some strains

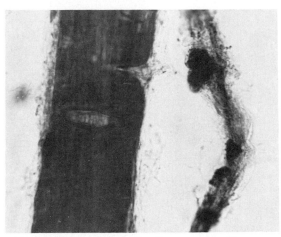

Figure 8–51. Hair perforating organs. These are wedge-shaped indentations produced by hyphae of some dermatophytes while growing saprophytically on hair. ×440.

from human occult infections and isolants from rodent fur do not mate with either set of tester strains.

A. benhamiae. The gymnothecia are globose, white to pale buff, and 250 to 500 μ in diameter (Fig. 8–52). The peridial hyphae are dichotomously branched, curving out from the main axis (abaxial), and the cells are asperulate and dumbbell-shaped. The cell number in each branch is usually four, in contrast to the two to three cells in terminal branches of *A. simii.* The end cells of *A. benhamiae* end in tapering, smooth-walled spikes or spirals. Asci are evanescent globose to ovate, 4.2 to 7.2 × 3.6 to 6.0 μ in size, and eight-spored. The ascospores are smooth, lenticular, and 1.2 to 1.8 × 1.2 to 2.8 μ in size.

A. vanbreuseghemii. The cells of the peridial hyphae are asymmetrically constricted at the septal junctures in *A. simii* and *A. benhamiae,* and the terminal branches are composed of one to three cells in *A. simii* and three or more cells in *A. benhamiae.* In contrast, the septal constrictions of the peridial

Figure 8–52. Gymnothecia of *A. benhamiae (an-amorph* is *T. mentagrophytes).* They are growing on hair in soil. (Courtesy of G. Rebell.)

hyphae in *A. vanbreuseghemii* are quite symmetrical and the terminal branches are composed of one to three cells. The hyphae also tend to twist in relation to the main axis, whereas in the other two species branches tend to curve away from the main axis.

Trichophyton rubrum (Castellani) Sabouraud 1911

Synonymy. *Trichophyton purpureum* Bang 1910; *Trichophyton rubidum* Priestley 1917; *Trichophyton "A"* Hodges 1921; *Trichophyton "B"* Hodges 1921; *Trichophyton marginatum* Muijs 1921; *Trichophyton plurizoniforme* MacCarthy 1925; *Trichophyton lanoroseum* MacCarthy 1925; *Trichophyton coccineum* Katoh 1925; *Trichophyton spadix* Katoh 1925; *Trichophyton multicolor* Magalhaes et Neues 1927; *Trichophyton kagawaense* Fujii 1931; *Trichophyton rodhainii* Vanbreuseghem 1949; *Epidermophyton rubrum* Castellani 1910; *Epidermophyton perneti* Castellani 1910; *Epidermophyton salmoneum* Froilano de Mello 1921; *Trichophyton fluviomuniense* Miguens 1968.

T. rubrum is anthropophilic and has recently become the most common and widely distributed dermatophyte of man. It is very rarely isolated from animals and never from soil. It is extremely variable in its morphology and lacks a teleomorph stage for positive identification. Scalp infections are uncommon, and hair is rarely invaded. In scalp infections involving hair, ecto- and endothrix conidia that are nonfluorescent have been described.

In cases of tinea corporis, villous hair is often invaded where arthroconidia are produced (Fig. 8–6). This serves as a reservoir for repeated exacerbations of disease. *T. rubrum* is also frequently associated with chronic tinea unguium and tinea pedis. Certain individuals having once become parasitized by *T. rubrum* appear to remain carriers for life.

All strains of *T. rubrum* tested by the *Arthroderma simii* sex stimulation test have been of the (−) mating type.[110] The opposite mating type may have been lost during the evolution to an anthropophilic fungus dependent on man for dissemination of the species. The (−) mating type possibly possessed a selective advantage over the (+) mating type in this regard.

Colony Obverse (Fig. 8–53,*A*). The typical thallus is slow-growing, downy white, generally devoid of conidia, and pigmented on the reverse. This is the type commonly isolated from chronic tinea pedis and chronic tinea corporis. Other isolants are less cottony, less

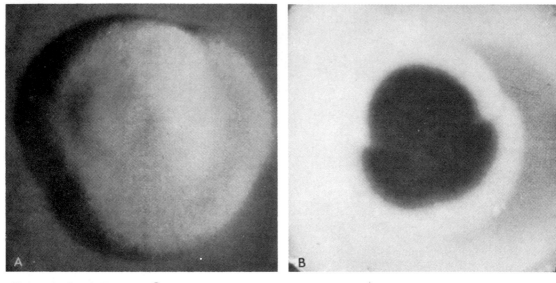

Figure 8–53. *A, T. rubrum* showing fluffy white mycelium. *B, T. rubrum* showing reverse demonstration of wine-red pigment.

Illustration continued on opposite page.

pigmented, and produce numerous macro-conidia. The coniediating strains usually come from inflammatory tinea corporis, tinea capitis, and granulomatous lesions. Occasional isolants are heaped up, folded, glabrous, intensely pigmented, devoid of conidia, and resemble *T. violaceum*. This morphology is seen in the var. *rodhainii* type. Folded colonies of this strain with diminished pigmentation resemble *T. tonsurans*.

Reverse (Fig. 8–53,*B*). The typical pigment is an intense, nondiffusing port wine or venous blood red. If produced in abundance, it not infrequently stains the mycelium on the obverse. Pigment is slow in developing, and is usually first noted on the edge of the colony in agar slants where the media is dried. The color may be yellow initially, developing through a melanoid-green, and finally becoming red. Several pigments are formed by the organism, differing in quantities at different times.[297] A black melanin-like diffusible pigment is sometimes produced. This is the var. *nigricans* and is also called *T. olexae*. The mating type and other characteristics of these strains are the same as those of other isolants of *T. rubrum*.[110] Some strains, particularly those from patients on griseofulvin therapy, fail to show pigment. Growth on potato-glucose and cornmeal glucose agar enhances pigmentation and separates *T. rubrum* from other red pigmented species.

Microscopic Morphology. The common isolants of *T. rubrum* produce few conidia. There are clavate "tear drop" microconidia (2 to 3×3 to 5 μ) produced lateral to the hyphae. Clusters of conidia in an arborescent or "pine tree" arrangement are seen. Macroconidia are absent or rare except in granular coniediating strains. When present these conidia are long, narrow, fusiform (pencil-shaped), usually without a stipe, multicelled, and often develop in groups directly from hyphae. Peridial hyphal-like structures, chlamydoconidia, pectinate hyphae, and many aberrant structures are also seen (Fig. 8–54,*A*).

Differential Characteristics. Pigment production on special media, negative urease test, and lack of *in vitro* hair penetrating organs differentiates *T. rubrum* from *T. mentagrophytes*. *T. violaceum* grows poorly without thiamine, and *T. megninii* requires L-histidine. Neither the amino acid nor the vitamin is required by *T. rubrum*.

Trichophyton tonsurans Malmsten 1845

Synonymy. *Trichophyton epilans* Boucher et Mégnin 1887; *Trichophyton sabouraudi* Blanchard 1896; *Trichophyton crateriforme* Sabouraud 1902; *Trichophyton flavum* Bodin 1902; *Trichophyton acuminatum* Bodin 1902; *Trichophyton effractum* Sabouraud 1910; *Trichophyton fumatum* Sabouraud 1910; *Trichophyton umbilicalum* Sabouraud 1910; *Trichophyton regulare* Sabouraud 1910; *Trichophyton exsiccatum* Sabouraud 1910; *Trichophyton polygonum* Sabouraud 1910; *Trichophyton plicatile* Sabouraud 1910; *Trichophyton pilosum* Sabouraud 1910; *Trichophyton sulfureum* Sa-

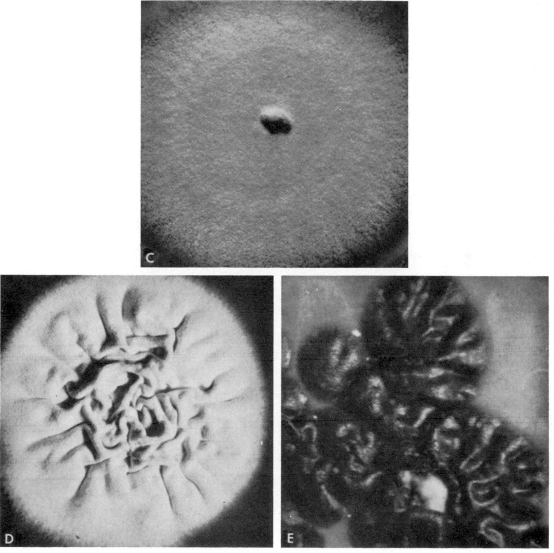

Figure 8–53 *Continued. C, T. mentagrophytes.* Flat granular colony, cream-colored in central area. *D, T. tonsurans.* Cerebriform colony with suedelike surface. The central convolutions often crack. *E, T. violaceum.* Glabrous, slightly folded, violet-colored surface. There are white pleomorphic patches.

bouraud 1910; *Trichophyton cerebriforme* Sabouraud 1910; *Trichophyton ochropyrraceum* Muijs apud Papengaaji 1924.

This is another species that varies particularly in color, texture, and morphology of the thallus. It is anthropophilic, endothrix, and nonfluorescent in infected hair. The young colonies are sometimes dully fluorescent. *T. tonsurans* was probably originally endemic in central and southern Europe and became prevalent in South America. Recently it has spread to the United States and other countries through the migration of infected peoples.[206] Infection in a horse and a dog is reported. All strains tested by the *A. simii* sex

stimulation test were of the (−) mating response.[110]

Colony Obverse (Fig. 8–53). There are four common colonial forms: crateriform, cerebriform, plicatile, and flat. All originally had different specific names. The most frequently isolated strains show a flat growth initially, which is powdery and yellow-tinged. This colony develops into a folded thallus (plicate), with a grayish to buff, suedelike surface. Pink tints are sometimes seen, especially if the reverse is heavily pigmented. The var. *sulfureum* type is less powdery, with a deep yellow to chartreuse, suedelike colony resembling *E. floccosum.* This strain is associated

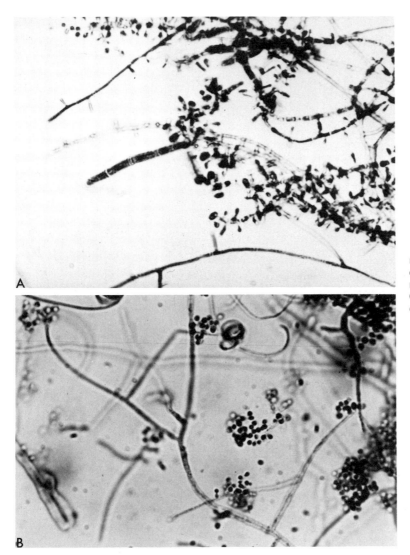

Figure 8–54. *A, T. rubrum* in rare granular form showing macroconidia and "tear drop" microconidia. ×440. *B, T. mentagrophytes.* Globose microconidia in grapelike clusters and spiral hyphae. ×440.

with particularly severe inflammatory disease that often results in lesion formation.[276] A particular colony type may be locally endemic. Pale and white strains are also seen.

Reverse. The pigment on the under surface is commonly yellow-brown to reddish-ochre or deep mahogany red. It sometimes diffuses into the media.

Microscopic Morphology (Fig. 8–55, *A*). The characteristic microconidia are variable in size and shape but usually abundant. Their size is 2 to 5 × 3 to 7 μ. They are clavate, tear-shaped, often borne on elongate stipes (match stick conidia), and formed in clusters on multiple branched, thickened terminal hyphae. Several conidia may be expanded and balloon-shaped. Filiform or claw-shaped conidia arranged laterally on irregular hyphae resemble a centipede. Macroconidia

occur less frequently, are irregular in shape, and are somewhat thick-walled. Chlamydoconidia, racquet mycelium, and irregular arthroconidia-like structures are found.

Differential Characteristics. Growth is minimal or absent in thiamine-less media, a characteristic differentiating this species from *T. rubrum* and *T. mentagrophytes.* L-Histidine is required for the growth of *T. megninii* but not of *T. tonsurans.*

Trichophyton violaceum Sabouraud apud Bodin 1902

Synonymy. *Trichophyton glabrum* Sabouraud 1910; *Achorion violaceum* Bloch 1911; *Favotrichophyton violaceum* Dodge 1935.

An anthropophilic species endemic in

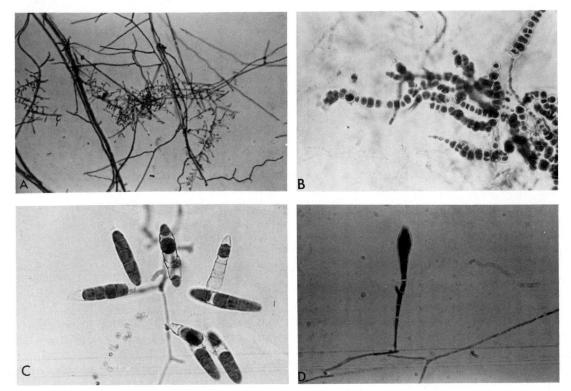

Figure 8–55. *A, T. tonsurans.* Microscopic view showing elongate "match stick" microconidia in groups and at right angles to mycelium. ×150. *B. T. verrucosum.* Chains of chlamydoconidia produced at 37° C. *C, T. simii.* Macroconidia with endochlamydoconidia. *D, M. gallinae.* Spatulate macroconidia. ×440.

South America, Mexico, Europe, Asia, and Africa, it is rare in the United States. It is an endothrix organism and nonfluorescent in infected hair. Animal infection is rare, but cases of disease in calf, dog, cat, horse, mouse, and pigeon have been reported.

Colony Obverse (Fig. 8–53, *D*). *T. violaceum* is very slow growing, producing a conical or verrucous (faviform) thallus that is heaped up, folded, glabrous or waxy, and deep violet in color. The organism rapidly becomes pleomorphic, sectoring as pale to white mycelia that quickly overgrow the original colony. Old stock cultures become flat, white, and fluffy. Occasional isolants are nonpigmented *(T. glabrum).*

Reverse. The purple pigment that stains the mycelium is also found on the underside.

Microscopic Morphology. Distorted hyphae and the lack of conidia are typical of strains grown on the usual media. The hyphae contain cytoplasmic granules. On media enriched with thiamine, a few clavate microconidia are produced, and rarely an elongate macroconidia is seen.

Differential Characteristics. The partial requirement for thiamine separates this organism from *T. gourvillii. T. rubrum,* and other species that may produce purple pigmented colonies.

Trichophyton verrucosum Bodin 1902

Synonymy. *Trichophyton album* Sabouraud 1908; *Trichophyton ochraceum* Sabouraud 1908; *Trichophyton discoides* Sabouraud 1910; *Trichophyton faviforme* Sabouraud 1892; *Favotrichophyton verrucosum* Neveu-Lemaire 1921.

T. verrucosum is principally associated with cattle and is of worldwide distribution. In cattle, epidemics of the disease are called "barn itch." It is a large-conidia ectothrix, and fluorescence of hair in infected cattle has been reported. Infected human hair is not fluorescent. It is associated with extremely severe infections of the scalp, especially in children,[264] and of the beard as tinea barbae.

Colony Obverse. The colony is very slow growing, producing a knoblike or slightly folded, heaped, glabrous, gray-white colony (Fig. 8–44, *D*). This morphology is referred to as the var. *album* type. Two other color variants are described; var. *ochraceum* has a flat, yellow, glabrous colony, and var. *discoides* has a flat, tomentose, gray-white thallus. On blood agar or media enriched with thiamine and inositol, growth is more rapid and spreading. A medium for early detection of *T. verrucosum,* dependent on its hydrolysis of casein, has been devised.[124] Growth is enhanced by a temperature of 37°C. No characteristic pigment is produced on the reverse.

Microscopic Morphology. On unenriched media, distorted hyphae are seen, with some suggestion of antlerlike branching similar to that of *T. schoenleinii.* No conidia are seen. On enriched media, clavate microconidia and rarely elongate fusiform or the rather characteristic rat tail macroconidia are produced. These have from three to five cells and sometimes are shaped like a string bean. At 37°C the fungus grows as a chain of chlamydoconidia (Fig. 8–55,*B*). All of 35 strains tested produced these chlamydoconidia at 37°C in a report by Kane and Smitka.[124] This was the most reliable test for identification of the fungus.

Differential Characteristics. All strains require thiamine, and most require inositol. All produce chlamydoconidia at 37°C.

Trichophyton schoenleinii (Lebert) Langeron and Milochevitch 1930

Synonymy. *Oidium* (Lebert) *schoenleini* Lebert 1845; *Achorium schoenleini* Remak 1845; *Favuspiltz B* Quincke 1886; *Oospora porriginis* Saccardo 1886; *Schoenleinium achorian* Johan Olsen 1897; *Grubyella* (Lebert) *schoenleini* Ota and Langeron 1923; *Arthrosporia* (Lebert) *schoenleinii* Grigorakis 1925.

The classic cause of favus in man is *T. schoenleinii.* It is endemic throughout Eurasia and Africa but occurs erratically in the Western hemisphere. There are endemic foci in Kentucky, the Gaspé Peninsula, Quebec, Guatemala, and São Paulo. This fungus is rare in animals, but infection of dog, cat, hedgehog, cow, horse, mouse, rabbit, and guinea pig is reported.

Colony Morphology. A slow-growing, glabrous (waxy), or suedelike, off-white colony is produced. There are gentle folds (faviform), but the colony may become very distorted and convoluted, rising off the agar surface or into it, causing cracking and splitting. The growth of some strains is mainly subsurface. In time, old laboratory strains become flat and downy. Some initial isolants are yeastlike in consistency and morphology and grow well at 37°C (Fig. 8–56,*A*).

Microscopic Morphology. Conidia are not seen. The hyphae form antlerlike or chandelier-like structures (Fig. 8–57). The branches often have a swollen (nail-head) tip. These are quite characteristic, though a few other species may occasionally form them also. A few distorted clavate microconidia are found in some isolants, particularly when the organism is grown in rice grains. Macroconidia are essentially never observed, even on enriched media.

Differential Characteristics. This fungus grows well at 37°C and does not require thiamine or other special nutrients.

Trichophyton concentricum Blanchard 1896

Synonymy. *Trichophyton mansonii* Castellani 1905; *Trichophyton castellanii* (Perry 1907) Castellani 1908; *Endodermophyton concentricum* Castellani 1910; *Endodermophyton indicum* Castellani 1911; *Endodermophyton mansoni* Castellani 1919; *Endodermophyton tropicale* Castellani 1919; *Endodermophyton roquettei* Fonseca 1925; *Endodermophyton castellanii* Perry 1910; *Lepidophyton* sp. Tribondeau 1899; *Aspergillus lepidophyton* Pinoy 1903; *Aspergillus tokelau* Wehmer 1903.

The fungus causes tinea imbricata and is sporadically distributed in South Asia, the Pacific, Mexico,[277] and Central and South America. It does not invade hair, and no animal infections have been reported.

Colony Morphology. The thallus is raised, deeply folded, and convoluted. It is glabrous and white at first, becoming cream, amber, brown, or coral red with time. A fuzzy or velvety growth of aerial mycelium may occur. The colony is slow-growing and has a diameter of 5 to 20 mm after ten days (Fig. 8–56,*B*).

Microscopic Morphology. Distorted, convoluted hyphae without conidia are seen. The branching mycelium resembles the antlers of *T. schoenleinii,* but the nail head ends are lacking. On bean pod media, lateral irregular macro- and microconidia have been reported to develop. In about 50 per cent of isolants, growth is stimulated by thiamine.

Trichophyton simii (Pinoy) Stockdale, Mackenzie, and Austwick 1965

Synonymy. *Epidermophyton simii* Pinoy 1912; *Pinoyella simii* (Pinoy) Castellani and Chalmers 1919; *Trichophyton mentagrophytes* Emmons 1940.

Teleomorph State. *Arthroderma simii* Stockdale, Mackenzie, and Austwick 1965.

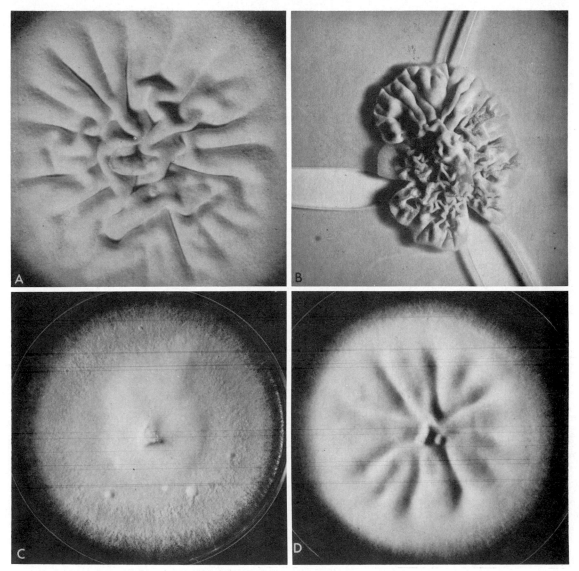

Figure 8–56. *A, T. schoenleinii. B, T. concentricum. C, T. simii. D, T. megninii.*

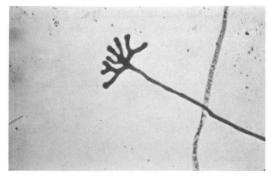

Figure 8–57. "Favic chandelier" of *Trichophyton schoenleinii*. Lactophenol cotton blue. ×400. (From Rippon, J. W. *In* Burrows, W. 1973. *Textbook of Microbiology.* 20th ed. Philadelphia, W. B. Saunders Company, p. 707.)

This organism is a frequent cause of ringworm in monkeys and chickens in endemic areas. It is uncommon in man but has been isolated regularly from soil and small mammals. Travelers are not infrequently infected. It causes an ecto-endothrix type of hair invasion. Infected guinea pig hair fluoresces a bright blue-green.[223]

Colony Obverse. The fungus grows rapidly, producing a flat, granular colony with a central umbo. Buff is the usual color, but it varies from cream to white (Fig. 8–56,*C*).

Reverse. In time, the colony may have a pigmented undersurface. The color is usually vinaceous (red-brown), but may range from yellow to madder rose. On malt extract agar, quantities of red-brown pigment diffuse into the media. On glucose peptone agar, the colony is yellow, and the reverse straw to salmon.

Microscopic Morphology. Macroconidia are usually produced in great abundance. They are thin-walled, smooth and clavate, cylindriform or fusiform in shape, and have four to ten septa. Conidia often occur in clusters. The cells of the macroconidia frequently enlarge, become thick-walled, and are termed endochlamydoconidia. The cells between the enlarged chlamydoconidia are often empty and rupture, causing fragmentation of the macroconidia. The resultant free chlamydospores are convex and have a collar left from the adherent portion of the broken cell. Clavate, elongate, pyriform, or peg-shaped microconidia are produced laterally on hyphae (Fig. 8–55,*C*).

A. simii. The gymnothecia are buff, globose, and from 200 to 650 μ in diameter. The morphology is somewhat similar to that of *Arthroderma uncinatum*. Peridial hyphae are hyaline, asperulate, yellow, and dichotomously branched. There are up to three cells per branch, ending in a spike or spiral hypha. Asci are evanescent, subglobose, 5 to 7.7 μ in diameter, with eight lenticular, smooth-walled ascospores. The latter are yellow in mass and 1.7 to 2.1 $\mu \times 2.9 \times 3.3$ μ in size.

The *Arthroderma simii* sexual stimulation test[263] can be utilized to determine the mating types of some other species. In the test, a known mating type of *A. simii* is placed on the plate with a strain of another species. If the mating types are opposite, at the zone of contact dense growth and gymnothecial initials (infertile) are formed. If the mating types are the same, the colonies remain separated by a clear zone. This test has been used to indicate the mating type of species in which the perfect stage has not yet been described, e.g., *T. rubrum, T. tonsurans,*[110] and *M. audouinii.*

Trichophyton equinum (Matruchot et Dassonville) Gedoelst 1902

Synonymy. *Trichophyton* sp. Matruchot and Dassonville 1898; *Megatrichophyton equinum* Dodge 1935; *Ectotrichophyton equinum* Dodge 1935.

This species is a cause of infection frequently found in horses and rarely in man. It is of worldwide distribution. All strains require nicotinic acid for growth except those from Australia and New Zealand, which are autotrophic (var. *autotrophicum*).

Colony Morphology. The thallus is fluffy, resembling the var. *interdigitale* type of *T. mentagrophytes*. Growth is rapid and flat, but the colony may develop gentle folds. The color is cream-white to yellow. The reverse is bright yellow, changing to dark pink or brown. This pigment may diffuse in the media.

Microscopic Morphology. Thin, elongate to pyriform, stalked microconidia are formed laterally along the hyphae. Rarely they are clustered. Macroconidia are similar to those of *T. mentagrophytes* and are rarely seen in culture. They are fusiform or clavate.

Differential Characteristics. Nicotinic acid (niacin) is required for growth of *T. equinum* except for the strains isolated in New Zealand and Australia. This species is reported to grow on horse hair and not on human hair, but this may not be a reliable characteristic.

Trichophyton megninii Blanchard apud Bouchard 1896

Synonymy. *Trichophyton roseum* Sabouraud apud Bodin 1902; *Trichophyton rosaceum* Sabouraud 1909; *Trichophyton vinosum* Sabouraud 1910; *Ectotrichophyton megninii* Castellani and Chalmers 1919; *Megatrichophyton megninii* Neveu-Lemaire 1921; *Aleuriosporia rosacea* Grigorakis 1925; *Megatrichophyton roseum* (Bodin) Dodge 1935; *Trichophyton kuryangei* Vanbreuseghem et Rosenthal 1961; *Trichophyton à culture rose* Sabouraud 1893.

This is a rare species found in Europe and Africa, particularly in Sardinia and Portugal, and recently in Ontario.[123] Most often this organism causes tinea barbae and is ectothrix, with conidia in chains. Animal infection is unknown. An endemic focus has recently been reported from Burundi.[275]

Colony Obverse. The thallus grows moderately rapidly to form a flat or gently folded center with radial furrows. The surface is felt or suedelike, white in color, but often stained pink (Fig. 8–56,*D*).

Reverse. A deep red pigment is elaborated on the underside of the colony. It is less intense than that of *T. rubrum*. The color is more of a bordeaux red than the port wine red color of the latter species.

Differential Characteristics. The requirement for L-histidine separates *T. megninii* from other red pigmented species. Conidia are similar to those of *T. rubrum*, but the urease reaction of *T. megninii* is positive[123] and that of *T. rubrum* is negative.

Trichophyton soudanense Joyeux 1912

Synonymy. *T. sudanense* Bruhns et Alexander 1928.

This is an endothrix species which is endemic in Central and West Africa (Fig. 8–3), with occasional isolations reported in Europe, Brazil, and the United States. It is not known to infect animals.[48, 227, 274]

The thallus is slow-growing, flat to folded in the center of a tough, leathery consistency, and varying from yellow to apricot in color. The colony has a fringed or raylike radiating edge (Fig. 8–58,*A*). The reverse is a deep

yellow. There is also a violet variant. Round chlamydoconidia, short, segmented hyphae (arthroconidia), and reflexive or right-angle branching mycelium that are bunched into a bushlike bundle are seen on microscopic examination. Pyriform microconidia are formed laterally on the mycelium. On Löwenstein-Jensen media, the colony is dark compared to *M. ferrugineum*, which is a light yellow.

Trichophyton gourvilii Cantanei 1933

Synonymy. *Trichophyton german II* Gregorio 1931; *Favotrichophyton gourvili* (Catanei) Dodge 1935.

This is an endothrix species endemic in

Figure 8–58. *A, T. soudanense. B, T. gouvilii. C, T. yaoundei. D, T. ajelloi.*

West Africa (Fig. 8–3). The thallus is folded, heaped, and waxy but becomes flat and velvety on transfer (Fig. 8–58,*B*). The color is lavender pink to deep garnet red. Typical *Trichophyton*-type micro- and macroconidia develop. Its lack of requirement for histidine and thiamine distinguish it from similar species.

Trichophyton yaoundei Cochet and Doby Dubois 1957

This is an endothrix species endemic in equatorial Africa, particularly common in Cameroun, Zaire, and Mozambique (Fig. 8–3). It is not known to infect animals. The thallus appears glabrous and moist, and is slow-growing, heaped, folded, and white-cream in color (Fig. 8–58,*C*). In time, the colony becomes deep tan to chocolate brown. Some strains grow almost entirely submerged in the medium. Pyriform microconidia are formed. The *Trichophyton*-type macroconidia are rarely seen. Branched, antlerlike mycelial appendages are formed, as well as chlamydospores. It has no specific nutritional requirements. The original report of this species lacks a Latin description. It is a distinct species but a *nomen nudum*.

Trichophyton ajelloi (Vanbreuseghem) Ajello 1967

Synonymy. *Keratinomyces ajelloi* Vanbreuseghem 1952.

Teleomorph State. *Arthroderma uncinatum* Dawson et Gentles 1961.

This is a very common soil keratinophilic fungus. It has been isolated rarely from tinea corporis of man,[205] and from infections of cattle, dogs, horses and squirrels. In man, the lesions often resemble psoriasis. It is readily isolated from soil by hair baiting. The colony grows rapidly, producing a flat, thin powdery to downy, cream, yellow, tan, or orange thallus (Fig. 8–58,*D*). The reverse is colorless, red, or a deep blue-black. The thick-walled, smooth macroconidia are numerous. They are fusiform to cylindrical in shape, 20 to 65 × 5 to 10 μ in size, and contain 5 to 12 cells. Pyriform, sessile microconidia are also abundant. Hair perforating organs are produced. Guinea pigs and mice can be infected.

A. uncinatum. Gymnothecia are globose, white to buff, and 300 to 900 μ in diameter. The peridial hyphae are hyaline, uncinately branched away from the main axis, and have knobby, asperulate, centrally-restricted, peridial cells. The tips are blunt to clawlike or have spirals arising either at the end or later-

ally on the body of the hyphae. Asci and spores are similar to those of *A. benhamiae*.

Other Trichophyton species. There are several other *Trichophyton* species and related keratinophils which are abundant in soil (Fig.' 8–59). These are rarely involved in human or animal disease. *T. terrestre* Durie and Frey 1957 represents a complex of species with at least three known teleomorph stages: *Arthroderma quadrifidum* Dawson and Gentles 1961; *A. lenticularum* Pore, Taso, et Plunkett 1965; and *A. insingulare* Pedhye et Carmichael 1972.[195,196] *T. georgiae* Varsavsky et Ajello 1964 has the teleomorph stage, *Arthroderma ciferrii* Varsavsky et Ajello 1964. *T. gloriae* Ajello 1967 is the anamorph form of *Arthroderma gloriae* Ajello 1967 and was isolated from soil in the southwestern United States. It is homothallic. *T. vanbreuseghemii* Rioux, Jarry, et Juminer 1964 was isolated from soil in Tunisia. Its teleomorph stage is *Arthroderma gertleri* Bohme 1967; it has now been found to be of worldwide distribution in soils.

A. tuberculatum Kuehn 1960, *A. cuniculi* Dawson 1963, *A. curreyi* Dawson 1963, and *A. flavescens* Padhye et Carmichael 1971 (*T. flavescens*), among others, are common soil keratinophiles often associated with animal dander, rabbit and rodent lairs, bird nests and so forth. Most have a *Chrysosporium* conidial state or may resemble *Trichophyton terrestre*. Some are homothallic. Such species are not infrequently isolated from feet, scalp, other areas of skin, and even sputum. Their resemblance to dermatophytes often leads to difficulty of interpretation. In general, few if any authentic cases of infection can be ascribed to them. The medical mycologist must be familiar with these and other soil keratinophiles, because they frequently cause confusion in the diagnostic laboratory.

Other new species of dermatophytes have recently been described. More investigation will be required in order to determine their validity as distinct species. Ozegovic[191] has isolated a new dermatophyte from human infection and has named it *Trichophyton teheraniensis*, and Krivanec et al.[141] have recovered an organism from badger burrows called *Arthroderma melis*. More such reports occur each year. Most of these are variants of known species and do not represent a new organism.

Epidermophyton Sabouraud 1907

This is a genus of two species characterized by an expanded clavate macroconidia that has a rounded or blunt terminus and **smooth**

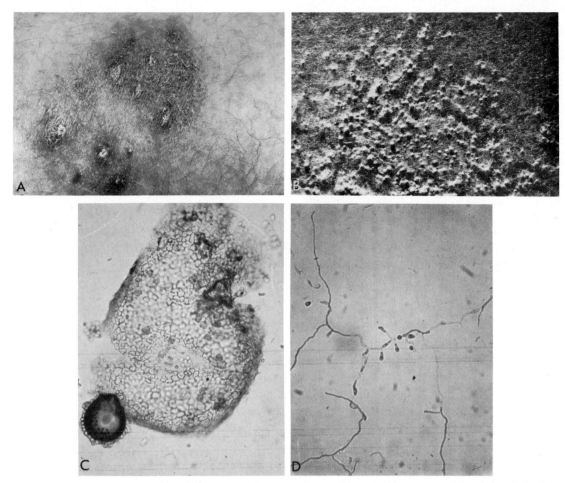

Figure 8–59. Soil keratinophilic fungi that rarely cause disease in man. *Aphanoascus fulvescens. A,* Lesion on inner thigh. *B,* Surface of colony covered with numerous brown-black cleistothecia. *C,* Crushed cleistothecium to show thick-walled rhomboid cells of the pseudoparenchymous wall. Note groups of mature ascospores. *D,* Microconidia of the Chrysosporium-type and intercalary chlamydoconidia.[226]

walls. Microconidia are absent. Hair is not known to be invaded.

The original description of this genus is inadequate and is therefore a *nomen nudum.* Ajello, Rogers, and McGinnis have recently published a paper in Taxon to remedy this and request conservation of the name *Epidermophyton.*[163a]

Epidermophyton floccosum (Harz) Langeron and Milochevitch 1930

Synonymy. *Trichothecium floccosum* Harz 1870; *Trichophyton interdigitale* Sabouraud 1905; *Trichophyton cruris* Castellani 1908; *Epidermophyton inguinale* Sabouraud 1910; *Epidermophyton cruris* Castellani and Chalmers 1910; *Epidermophyton plicarum* Nicolau 1913; *Epidermophyton elypeisoforme* MacCarthy 1925; *Acrothecium floccosum* Harz 1871; *Blastotrichum floccosum* Berlese and Boglino 1886; *Clasterosporia inguinalis* Grigorakis 1925; *Fusoma cruris* Vuillemin 1929.

This is an anthropophilic species. Chronic infections are not common, so that rapid transmission within population groups is necessary for maintenance of the species. It is worldwide in distribution, but is found particularly in the tropics and subtropics. It probably originated in India and Southeast Asia (Dhobie itch) and was subsequently distributed by the colonialists. Animal disease is unknown and the organism is not recovered from soil.

Colony Morphology (Fig. 8–60). The growth of the thallus is slow and it is frequently grainy, lumpy, and sparse on initial isolation. When developed, the colony is gently folded, fuzzy or suedelike in texture, and is characteristically olive, khaki, or

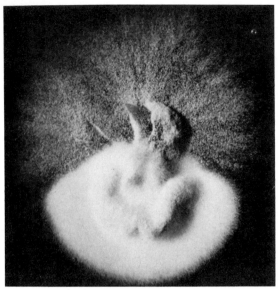

Figure 8–60. *E. floccosum* showing gently folding, suedelike texture. The color is chartreuse. A white pleomorphic overgrowth has begun to cover the surface.

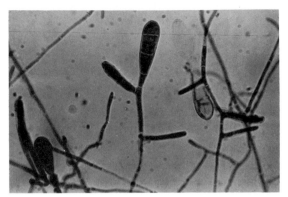

Figure 8–61. *E. floccosum.* Beaver tail–shaped macroconidia. (From Rippon, J. W. *In* Burrows, W. 1973. *Textbook of Microbiology.* 20th ed. Philadelphia, W. B. Saunders Company, p. 709.)

chartreuse in color. Yellow, yellow-brown, and white variants are reported. The underside is colorless to deep yellow-brown.

Microscopic Morphology. The characteristic macroconidia are usually abundant. They are 7 to 12 × 20 to 40 μ in size, beaver tail-shaped or clavate, and have smooth thin walls (Fig. 8–61). These conidia often occur in groups. Rarely, pits and knobs have been seen on them. Microconidia are absent. Chlamydoconidia are frequently very abundant, particularly as the colony ages (Fig. 8–62). They are rounded and thick-walled. Macro-conidia sometimes transform into chlamydoconidia. Racquet mycelium, spirals, and nodular bodies may be found. No specific growth requirements have been reported. Natural infection in animals has not been described, and experimental infection in laboratory animals is rarely successful.

Epidermophyton stockdaleae Prochacki and Engelhardt-Zasada 1974

This species of the genus *Epidermophyton* was isolated by a modified hair-baiting technique from a pig yard in Poland.[207] It resembles the anthropophilic *E. floccosum* in colony morphology, lack of microconidia, and club-shaped macroconidia. The latter are large 20 to 60 × 4 to 13 μ with up to nine septa. No other isolants have yet been reported.

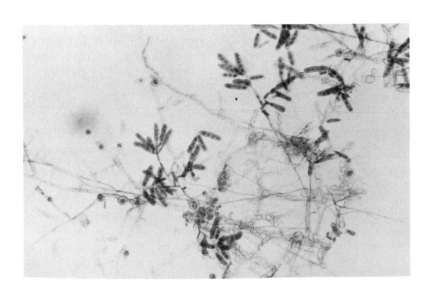

Figure 8–62. *E. floccosum.* Variation in shape of macroconidia and abundant chlamydoconidia. No microconidia are produced.

DERMATOMYCOSIS

From time to time a variety of species of soil fungi have been isolated from human disease. Most of these are contaminants, and an etiologic role as an infectious agent is generally lacking. In the past few years a few soil fungi have now been unequivocally associated with infection of human skin. Probably the first of such reports on a valid skin infection by a soil organism was that of Gentles and Evans.[76] They described *Hendersonula toruloidea* Nattrass 1933, a dematiaceous pycnidia forming hyphomycete from infection of the foot and nail. Since then, several reports of skin and nail infections associated with this organism have been published.[159] In one from Algeria, a verrucous disease was produced. This patient had an immune defect, however, so this should be considered an opportunistic infection.

Scytalidium hyalinum Campbell and Mulder 1977[39] was isolated from eight skin and nail infections. The patients were from Jamaica or West Africa. In a recent report by Peiris et al.,[199] all patients were from the West Indies.

The disease seems to be chronic, with mild scaling of the skin and minimal distortion of the nail. *Scytalidium* is a conidial anamorph of *Hendersonula* and has pale brown one- or two-celled arthroconidia in culture.

Some yeastlike fungi have also been involved in dermatophyte-like disease. Almost all of these are due to *Candida* species or other normal flora. A few cases have represented infection by continued contact with a fungus from an external source. These include the "bakers' dermatitis," colonization and infection by bakers' yeast *Saccharomyces cerevisiae, Trichosporon capitatum* from printer's fluid,[99] and others.

Hyphomycetes have also been recovered from infections of the skin in otherwise normal individuals. Collins and Rinaldi[46] described pustular erythematous lesions associated with *Fusarium moniliforme (Gibberelia fujijuroi)*. This strain produced lesions in the skin of mice similar to that seen on the patient. *Scopulariopsis brevicaulis* has been reported several times as causing similar disease.[38] These and other rare fungal diseases are discussed in Chapter 27.

REFERENCES

1. Abu-Samra, M. T., and B. E. D. Hato, 1982. Experimental infection of goats and guinea pigs with *Microsporum canis* and trials with canesten cream and neguvon solution. Mycopathologica, in press.
2. Achten, G., H. Degreef, and P. Dockx. 1977. Treatment of onychomycosis with a solution of miconazole 2% in alcohol. Mykosen, *20*:251–256.
3. Ainsworth, G. C., and P. K. C. Austwick. 1959. Fungal Diseases of Animals. Commonwealth Agriculture Bureau, Farnham Royal, Bucks, England.
3a. Ajello, L. 1960. Geographic distribution and prevalence of the dermatophytes. Ann. N.Y. Acad. Sci., *89*:20–38.
4. Ajello, L., and L. K. Georg. 1957. *In vitro* hair cultures for differentiating between atypical isolates of *T. mentagrophytes* and *T. rubrum.* Mycopathologia, 8:3–17.
5. Ajello, L. 1978. Present day knowledge of imperfect *Epidermophyton, Microsporum* and *Trichophyton* species. Hautarzt, *29*:6–9.
5a. Ajello, L., W. Kaplan, et al. 1980. Dermatophyte mycetoma: fact or fiction. Proc. V. Int. Conf. Mycoses. P.A.H.O. Sci. Pub. 396 pp. 135–140.
6. Alberts, R. E., and A. R. Omram. 1966. Follow-up study of patients treated by x-ray epilation for tinea capitis. Am. J. Public Health, *56*:2114–2120.
7. Allen, A. M., J. H. Reinhardt, et al. 1973. Griseoful-vin in the prevention of experimental human dermatophytosis. Arch. Dermatol., *108*:233–236.
8. Allen, D. E., R. Snyderman, et al. 1977. Generalized *Microsporum audouinii* infection and depressed cellular immunity associated with a missing plasma factor required for lymphocyte blastogenesis. Am. J. Med., *63*:991–1000.
9. Alteras, I. 1969. The therapeutic effect of vaccines extracted from keratinophilic soil fungi. Mycopathologia, *38*:145–150.
10. Alteras, I., I. Cojocary, et al. 1967. Occupational mycotic infections in swimming pools and public baths. Dermatol-Vener., *12*:409–414.
11. Amsel, L. P., L. Cravitz, et al. 1979. Comparison of *in vitro* activity of undecylenic acid and tolnaftate against athlete foot fungi. J. Pharm. Sci., *68*:384–385.
12. Anon. 1975. Specific prophylactic vaccine against bovine trichophytosis. Moscow USSR Vneshtovquizdat. No. 21M 313 27.
13. de Area Leão, A. E., and M. Goto. 1950. O Tokelau entre os Indios do Brazil. Hospital, Rio de Janeiro, *37*:225–240.
14. Araüysky, A. N. 1964. Rare mycological findings in pathological material. Mycopathologia, *22*:185–200.
15. Artis, W. M., B. M. Odle, et al. 1981. Therapeutic failure of griseofulvin in *Trichophyton rubrum* infections correlates with *in vitro* resistance of the infecting fungus. Arch. Dermatol., *117*:16–19.

16. Badillet, G. 1977. Population Parisenne et derma- tophytes transmis par les animaux. Bull. Soc. Fr. Mycol. Med., 6:109–114.

17. Badillet, G., P. Pietrini, and M. Martinez. 1978. A propos de trois nouveaux isolements de *Tricho- phyton erinacei*. Bull. Soc. Fr. Mycol. Med., 7:157–160.

18. Baer, R. L., and S. Rosenthal. 1966. The biology of fungus infections of the feet. J.A.M.A., 197:1017–1020.

19. Baer, R., S. A. Rosenthal, et al. 1956. Experimental investigations on mechanisms producing acute dermatophytosis of feet. J.A.M.A., 160:184– 190.

20. Baker, N. F., and D. W. Davis. 1954. Spray applica- tion for treatment of ringworm in cattle. Vet. Med., 49:275–276.

21. Baker, R. D. 1971. (ed.). The Pathologic Anatomy of Mycoses. Dritter band, Funfter teil, Berlin, Springer Verlag, Chapter VII, pp. 211–377.

22. Barker, S. A., and M. D. Trotter. 1960. Isolation of purified trichophytin. Nature, 188:232–233.

23. Baxter, M. 1965. The use of ink-blue in the identi- fication of dermatophytes. J. Invest. Derma- tol., 44:23–25.

23a. Beach, B. A., and J. G. Halpin. 1918. Observations on an outbreak of favus. J. Agri. Res. 15:415– 420.

24. Beare, J. M., J. C. Gentler, et al. 1972. Chapter 25, Rook, A, et al. (eds.), Textbook of Dermatolo- gy, Philadelphia, F. A. Davis.

25. Becker, S. W., and K. B. Muir. 1929. Tinea amian- tacea. Arch. Dermatol., 20:45–53.

26. Benedek, T. 1967. On the ancestral form of true organ of fructification in the saprophytic state of the dermatophytes of the faviform group: *Favomicrosporon pinetti* n. sp. and its perfect form: *Anixiopsis stercoraris* (Hansen) Hansen 1897 Mycopathologica, 31:81–143.

27. Benito, J. Velasco, A. Martin-Pascual, and A. Gar- cia Perez. 1979. Epidemiological study of der- matophytoses in Salamanca (Spain). Sabourau- dia, 17:113–123.

28. Biase, W. D. 1974. Parasitismo e migra des humans pré-historicas das Américas. Rev. Brasilieira Med., 32:156–159.

29. Bibel, D. J., and J. R. Hebrun. 1975. Effect of experimental dermatophyte infection on cuta- neous flora. J. Invest. Dermatol. 64:119–123.

30. Bibel, D. J. and R. J. Smiljanic. 1979. Interactions of *Trichophyton mentagrophytes* and micrococci on skin culture. J. Invest. Dermatol., 72:133– 137.

31. Blank, F. 1957. Favus of mice. Can. J. Microbiol., 3:885–896.

32. Blank, F. 1962. Human favus in Quebec. Derma- tologica, 125:369–381.

33. Blank, F., S. J. Mann, and P. A. Peale. 1974. Distribution of dermatophytosis according to age, ethnic group, or sex. Sabouraudia, 12:352–361.

34. Blank, H. (ed.). 1960. Griseofulvin and derma- tomycoses. An International Symposium spon- sored by the University of Miami. Arch. Der- matol., 81:649–882.

35. Blank, H., D. Taplin, et al. 1969. Cutaneous *Tricho- phyton mentagrophytes* infections in Viet Nam. Arch. Dermatol., 99:135–144.

36. Borelli, D. 1965. *Microsporum racemosum* nova spe- cies. Acta. Med. Venez., 12:148–151.

37. Brocq-Rousseau, D., D. A. Urbain, et al. 1926. Sur l'electivite cutanee des teignes animales quel

que soit leur voie d'introduction dans l'organ- isme. C. R. Soc. Biol., 95:966–967.

38. Buchvald, J. 1977. Finding of *Scopulariopsis brevi- caulis* (Sacc) Baines 1907, in pathologically changed nails and skin. Bratisl. Lek. Listy., 67:3–8.

39. Campbell, C. K., and J. L. Mulder. 1977. Skin and nail infection by *Scytalidium hyalinum* sp. nov. Sabouraudia, 15:161–166.

40. Charney, P., V. M. Torres, et al. 1973. Tolnaftate as a prophylactic agent for tinea pedis. Int. J. Dermatol., 12;179–185.

41. Chattaway, F. W., D. A. Ellis, et al. 1963. Peptidases of dermatophytes. J. Invest. Dermatol., 41:31– 37.

42. Chermisrivathana, S., and P. Boonsri. 1961. A case of tinea imbricata (Hanuman ringworm) treat- ed with fulvicin. Aust. J. Dermatol., 6:63–66.

43. Cheung, S. C., and J. Maniotis. 1973. A genetic study of an extracellular elastin hydrolysing protease in the ringworm fungus *Arthroderma benhamiae*. J. Gen. Microbiol., 74:299–304.

44. Clayton, Y. M., and G. Midgley. 1977. Tinea capitis in school children in London. Hautarzt, 28:32– 34.

45. Coisco, W. A. C. 1967. Algunas aspectos clinicas y micologicas de la tinas del cuero cabellundo en la Republica Dominicana. Rev. Dom. Derma- tol. 1:4–12.

46. Collins, M. S., and M. G. Rinaldi. 1977. Cutaneous infection in man caused by *Fusarium monili- forme*. Sabouraudia, 15:151–160.

47. Conant, N. F., P. T. Smith, et al. 1954. *Manual of Clinical Mycology*, 2nd ed. Philadelphia, W. B. Saunders Company.

48. Cos, D. R., and F. Blank. 1977. Tinea capitis due to *Trichophyton soudanense*. Arch. Dermatol., 113:1600.

49. Cox, W. A., and J. A. Moore. 1968. Experimental *Trichophyton verrucosum* infections in laboratory animals. J. Comp. Pathol., 78:35–41.

50. Cruickshank, C. N. D., M. D. Trotter, et al. 1960. Studies on trichophytin sensitivity. J. Invest. Dermatol., 35:219–233.

51. Das, S. K., and A. B. Banerjee. 1977. Phospholipid turnover in *Trichophyton rubrum*. Sabouraudia, 15:99–102.

52. Daum, V., and D. J. McCloud. 1976. *Microsporum racemosum:* first isolation in the United States. Mycopathologia, 59:183–185.

53. Dawson, C. O., and J. C. Gentles. 1959. The perfect stage of *Keratinomyces ajelloi*. Nature (Lond.), 183:1345–1346.

54. Day, W. C., P. Toncic, et al. 1968. Isolation and properties of an extracellular protease of *Tri- chophyton granulosum*. Biochem. Biophys. Acta 167:597–606.

55. Delameter, E. D. 1941. Experimental studies with dermatophytes. III. Development and dura- tion of immunity and hypersensitivity in guin- ea pigs. J. Invest. Dermatol., 4:143–158.

56. Desai, S. C., M. L. A. Bhatikas, et al. 1963. Biology of *Trichophyton rubrum* infections. Indian J. Med. Res., 51:233–243.

57. Dodge, C. W. 1935. *Medical Mycology*. St. Louis, C. V. Mosby.

58. Dominik, T., and I. Majchrowicz. 1965. Sec- ond contribution to the knowledge of keratino- lytic and keratinophilic soil fungi in the re- gion of Szczecin. Ecol. Polska. Ser. A., 13: 415–447.

59. Edgeson, F. A. 1970. Mass treatment of ringworm

of cattle with griseofulvin mycelium. Vet. Rec., 86:58–59.

60. Emyantoff, R. G., and T. Hashimoto. 1979. The effect of temperature, incubation atmosphere, and medium composition on arthrospore formation in the fungus *Trichophyton mentagrophytes*. Can. J. Microbiol., 25:362–366.

61. English, M. P. 1957. *Trichophyton rubrum* infection in families. Br. Med. J., 1:744–746.

62. English, M. P. 1962. The saprophytic growth of keratinophilic fungi on keratin. Sabouraudia, 2:115–120.

63. English, M. P. 1969. Tinea pedis as a public health problem. Br. J. Dermatol., 81:705–707.

64. English, M. P. 1978. The dysgonic strain of *Microsporum canis*. Mycopathologia, 64:73–81.

64a. English, M. P., and M. D. Gibson. 1959. Studies in epidermiology of tinea pedis. I and II. Br. Med. J. 1:1442–1445, 1446–1448.

65. English, M. P., C. D. Evans, et al. 1962. Hedgehog ringworm. Br. Med. J., 1:149–151.

66. English, M. P., L. Kapica, et al. 1978. On the occurrence of, *Microsporum persicolor* in Montreal, Canada. Mycopathologia, 64:35–37.

67. Epstein, W. L., V. P. Shah, et al. 1975. Topically applied griseofulvin in prevention and treatment of *Trichophyton mentagrophytes*. Arch. Dermatol., 111:1293–1297.

68. Feuerman, E. J., I. Alteras et al. 1976. Kerion-like tinea capitis and barbae caused by *Microsporum gypseum* in Israel. Mycopathologia, 58:165–168.

69. Feuerman, E. J., I. Alteras. 1975. Saprophytic occurrence of *Trichophyton mentagrophytes* and *Microsporum gypseum* in coats of healthy laboratory animals. Mycopathologia, 55:13–17.

70. Fox, T. 1870. Tinea circinata of the hand. Br. Med. J., July 30. p. 116.

71. Franks, A. G., E. M. Rosenbaum, et al. 1952. *T. sulfureum* causing erythema nodosum and multiple kerion formation. Arch. Dermatol., 65:95–96.

72. Friedman, L., and V. J. Derbes. 1960. The question of immunity in ringworm infection. Ann. N.Y. Acad. Sci., 89:178–183.

73. Fuentes, C. A. 1956. A new species of *Microsporum*. Mycologia, 48:613–614.

74. Ganor, S., S. Radai, et al. 1977. Tinea inguinalis treated with miconazole cream. A double-blind study. Israel J. Med. Sci., 13:587–589.

75. Gentles, J. C. 1958. Experimental ringworm in guinea pigs; oral treatment with griseofulvin. Nature (Lond.), 182:476–477.

76. Gentles, J. C., and E. G. V. Evans, 1970. Infection of the feet and nails with *Hendersonula toruloidea*. Sabouraudia, 8:72–75.

77. Gentles, J. C., and J. C. Holmes. 1957. Foot ringworm in coal miners. Br. J. Int. Med., 14:22–29.

78. Georg, L. K. 1952. *Trichophyton tonsurans* ringworm — A new public health problem. Public Health Reports, 67:53–56.

79. Georg, L. K. 1954. The relationship between the downy and granular forms of *Trichophyton mentagrophytes*. J. Invest. Dermatol., 83:123–141.

79a. Georg, L. K. 1960. Epidemiology of dermatophytes. Ann. N.Y. Acad. Sci., 89:69–77.

80. Georg, L. K. 1960. Animal ringworm in Public Health. U.S. Dept. HEW. Public Health Series publication #727.

81. Georg, L. K., and L. B. Camp. 1957. Routine nutritional tests for the identification of dermatophytes. J. Bacteriol., 74:477–490.

82. Georg, L. K., L. Friedman, and S. A. Brinkman. 1962. A new species of *Microsporum* pathogenic to man and animals. Sabouraudia, 1:189–196.

83. Ghani, H. M., J. H. Lanchaster, et al. 1974. Genetic analysis of pigmentation in *Arthroderma benhamiae*. J. Gen. Microbiol., 84:205–208.

84. Ginther, O. J., L. Ajello, et al. 1964. First American isolation of *Trichophyton mentagrophytes* in swine. Vet. Med. Small Anim. Clin., 59:1038–1042.

85. Ginther, O. J., and L. Ajello. 1965. Prevalence of *Microsporum nanum* infection in swine. J. Am. Vet. Med. Assn., 146:361–365.

86. Ginther, O. J., G. R. Bubash, et al. 1964. *Microsporum nanum* infection in swine. Vet. Med. Sm. Anim. Clin., 59:79–84.

87. Gip, L. 1964. Isolation of *Tricophyton gallinae* from 2 patients with tinea cruris. Acta Derm. Venereol., 44:251–254.

88. Goldfarb, N. J., and F. Hermann. 1956. A study of pH changes by molds in culture media. J. Invest. Dermatol., 27:193–201.

89. Goldstein, N., and G. Woodward. 1969. Surgery for tinea pedis. Syndactylization or amputation of toes for chronic severe fungus infections. Arch. Dermatol., 99:701–704.

90. Gordon, M. A., and G. N. Little. 1968. *Trichophyton (Microsporum) gallinae* ringworm in a monkey. Sabouraudia, 6:207–212.

91. Gottlieb, D., H. E. Carter, et al. 1958. Protection of fungi against polyene inhibition by sterols. Science, 128:361.

92. Gotz, H. B., Deichmann, and E. M. Zabel. 1978. On the question of immunoglobulin formation A, G, M, and E in Dermatomycoses. Mykosen, 21:266–269.

93. Grappel, S. F., C. T. Bishop, and F. Blank. 1974. Immunology of dermatophytes and dermatophytosis. Bacteriol. Rev., 38:222–250.

94. Grappel, S. F., and F. Blank. 1972. Role of Keratinases in dermatophytosis. I. Immune responses of guinea pigs infected with *Trichophyton mentagrophytes* and guinea pigs immunized with keratinases. Dermatologica, 145:245–255.

95. Grappel, S. F., F. Blank, et al. 1972. Circulating antibodies in dermatophytosis. Dermatologica, 144:1–11.

96. Gray, H. R., J. E. Dalton, et al. 1960. *Trichophyton tonsurans* infection of the scalp in central Indiana. J. Indiana State Med. Assoc., 53:75–80.

97. Gruby, D. 1841. Sur les mycodermes que constituent la teigne faveuse. C. R. Acad. Sci., 13:309–312.

98. Gruby, D. 1843. Recherches sur la natur, la siege et le developpement du porrigo de calvans ou phytoalopecie. C.R. Acad. Sci., 17:301–303.

99. Grunder, K. 1976. *Trichosporon capitatum* als Erreger von Dermatomykosen. Hautarzt, 27:422–425.

100. Gueguén, F. 1904. Le champignons parasites de l'homme dt des généralités, classification, biologie, technique, clefs analytiques, synonymie, diagnoses animaux, histoire parasitologique, bibliographie. Paris, xvii, 299, pp, 12 pls.

101. Hackett, L. P., L. J. Dusci. 1978. Determination of griseofulvin in human serum using high performance liquid chromatography. J. Chromatogr., 155:206–208.

102. Hagen, K. W. 1969. Ringworm in domestic rabbits:

oral treatment with griseofulvin. Lab. Anim. Care, *19*:635–638.

103. Hall, F. R. 1966. Ringworm contracted from cattle in western New York state. Arch. Dermatol., *94*:35–37.

104. Hasegawa, A., and K. Usui. 1974. The perfect state of *M. canis.* Jap. J. Vet. Sci., *36*:447–449.

105. Hasegawa, A., and K. Usui. 1975. *Nannizia otae* sp. nov. The perfect state of *Microsporum canis* Bodin. Jap. J. Med. Mycol., *16*:148–153.

106. Hashimoto, T., J. H. Pollack, et al. 1978. Carotenogenesis associated with arthrosporulation of *Trichophyton mentagrophytes.* J. Bacteriol., *136*: 1120–1126.

107. Hashimoto, T., C.D.R. Wu, and H. J. Blumenthal. 1972. Characterization of L-leucine induced germination of *Trichophyton mentagrophytes.* J. Bacteriol., *112*:967–976.

108. Hetherington, G., R. C. Freeman, et al. 1969. Intracellular location of hyphae in experimental dermatomycosis. Experientia, *25*:889–890.

109. Hiddleston, W. A. 1970. Antifungal activity of *Penicillium griseofulvin* mycellium. Vet. Rec., *86*:75–76.

110. Hironaga, M., S. Watanabe. 1977. Studies on the genera *Arthroderma-Trichophyton,* Jap. J. Med. Mycol., *18*:161–168.

111. Holmes, J. G. 1958. Tinea pedis in miners. In Riddell, R. W., and G. T. Stewart (eds.), *Fungous Diseases and Their Treatment.* London, Butterworth and Co.

112. Howard, O. H., and N. Dabrowa. 1979. Mutants of *Arthroderma benhamiae.* Sabouraudia, *17*:35–50.

113. Hubalek, Z., and F. M. Rush-Munro. 1973. A dermatophyte from birds. *Microsporum ripariae* sp. nov. Sabouraudia, *11*:287–292.

114. Huppert, M., and E. L. Keeney. 1959. Immunization against superficial fungous infection. J. Dermatol., *52*:15–19.

115. Ito, Y. 1965. On the immunologically active substances of the dermatophytes. J. Invest. Dermatol., *45*:285–294.

116. Jeremiasse, H. P. 1960. Treatment of nail infections with griseofulvin combined with abrasion. Trans. St. John Hosp. Dermatol. Soc., *45*:92–93.

117. Jesenska, Z. 1976. Heterogenicity of dermatophyte population. Dermatol. Monatsschr., *162*:190–191.

118. Jessner, M. 1922. Uber eine neue Form von Nagelmykosem leukonychia trichophytica. Arch. Dermatol. Syph. (Berlin), *141*:1–8.

119. Jewell, E. W. 1970. *Trichophyton rubrum* onychomycosis in a four month old infant. Cutis, October, p. 1121.

120. Joly, J., G. Delage, et al. 1978. Favus. Twenty indigenous cases in province of Quebec. Arch. Dermatol., *114*:1647–1648.

121. Jones, H. E., J. H. Reinhardt, et al. 1973. A clinical, mycological and immunological survey for dermatophytosis. Arch. Dermatol., *108*:61–65.

122. Kaaman, T. 1978. The clinical significance of cutaneous reactions to trichophytin in dermatophytosis. Acta Dermatol. Venereol., *58*:139–143.

123. Kane, J., and J. B. Fisher. 1975. Occurrence of *Trichophyton megninii* in Ontario. Identification with a simple cultural procedure. J. Clin. Microbiol., *2*:111–114.

124. Kane, J., and C. Smitka. 1978. Early detection and identification of *Trichophyton verrucosum.* J. Clin. Microbiol., *8*:740–747.

125. Kaplan, W., and L. Ajello. 1960. Therapy of spontaneous ringworm in cats with orally administered griseofulvin. Arch. Dermatol., *81*:714–723.

126. Kaplan, W., and L. K. Georg. 1957. A device to aid in the development of mycotic and other skin infections in laboratory animals. Mycologia, *49*:604–605.

127. Kaplan, W., and R. H. Gump. 1958. Ringworm in a dog caused by *Trichophyton rubrum.* Vet. Med., *53*:139–142.

128. Kaplan, W., J. L. Hopping, et al. 1957. Ringworm in horses caused by *Microsporum gypseum.* J. Am. Vet. Med. Assoc., *131*:329–332.

129. Kaplan, W., and M. S. Ivens. 1961. Observations on the seasonal variations in incidence of ringworm in dogs and cats in the United States. Sabouraudia, *1*:91–102.

130. Karling, J. S. 1946. Keratinophilic chytrids. Am. J. Bot., *33*:751–757.

131. Kashkin, A. P., and Y. N. Voevodin. 1976. Proteolytic enzymes of *Trichophyton mentagrophytes* (Robin) Blanchard. Mikol. Fitopatol., *10*:179–185.

132. Kaufman-Wolf, M. 1914. Uber Pilzerkrankungen der Hande und Fusse. Dermat. A. (Dermatologica), *21*:385–396.

133. Keely, R., and S. Searls. 1977. *Microsporum nanum* infection in Victoria. Aust. J. Dermatol., *18*:137–138.

134. Khuller, G. K., J. N. Verma, et al. 1978. Changes in the phospholipid composition of *Microsporum gypseum* during growth. Indian J. Med. Res., *68*:234–236.

135. Klatt, P. 1969. Treatment of bovine ringworm with two new antimycotics. Blue Book Vet. Prof., *16*:23–26.

136. Kligman, A. M. 1952. The pathogenesis of tinea capitis due to *Microsporum audouinii* and *Microsporum canis.* J. Invest. Dermat., *18*:231–246.

137. Kligman, A. M. 1955. Tinea capitis due to *M. audouinii* and *M. canis.* Arch. Dermatol., *71*:313–348.

138. Kligman, A. M. 1956. Pathophysiology of ringworm infections in animals with skin cycles. J. Invest. Dermatol., *27*:171–185.

139. Klokke, A. H., and G. A. DeVries, 1963. Tinea capitis in Chimpanzees caused by *Microsporum canis* Bodin 1902 resembling *M. obesum* Conant 1937. Sabouraudia, *2*:238–270.

140. Knight, A. G. 1972. Experimental human dermatophytosis. J. Invest. Dermatol., *59*:354–358.

141. Křivanec, K., V. Janeckova, and M. Otčenašek. 1977. *Arthroderma melis* spec. nov. — a new dermatophyte species isolated from badger burrows in Czechoslovakia. Cesk. Mykol., ·*31*:91–99.

142. Kwon-Chung, K. J. 1974. Genetics of fungi pathogenic for man. *CRC Critical Reviews in Microbiology.* Cleveland, Ohio, Chemical Rubber Co., pp. 115–133.

143. Langeron, M., and S. Milochevitch. 1930. Morphologie des dermatophytes sur les milieux naturels et milieux a base de polysaccharides. Ann. Parasitol. Hum. Comp., *8*:422–436.

144. Langeron, M., and S. Milochevitch. 1930. Morphologie des dermatophytes. Ann. Parasitol. Hum. Comp., *8*:465–508.

145. La Touche, C. J. 1967. Scrotal dermatophytosis. Br. J. Dermatol., *79*:339–344.

146. Lavalle, P. 1966. Tinea pedis in Mexico. Dermatologia (Mex.), *10*:313–329.

147. Leavell, V. W., E. Tucker, et al. 1966. Blue dot infection of the scalp in two brothers. J. Ky. Med. Assoc., *64*:1107–1110.

148. Lebert, H. 1845. Physiologie pathologique on recherches cliniques, experimentales et microscopiques sur l'inflammation, la tuberculisation, les tumeurs, la formation du cal etc. 2 vol. pp. 490 in section Mémoire sub la teigne *2*:477–497 pl. 223.

149. Lenhart, K. 1970. Griseofulvin resistant mutants in dermatophytes. Mykosen, *13*:139–144.

150. Lepper, A. W. D. 1969. Immunological aspects of dermatomycoses in animals and man. Rev. Med. Vet. Mycol., *6*:435–442.

150a. Leyden, J. J., and A. M. Kligman. 1978. Interdigital athlete's foot. The interaction of dermatophytes and resident bacteria. Arch. Dermatol. *114*:1466–1472.

151. Longbottom, J. L., and J. Pepys. 1964. Pulmonary aspergillosis. Diagnostic and immunological significance of antigens and C-reactive substance in *Aspergillus fumigatus*. J. Pathol. Bacteriol., *88*:141–151.

152. Lorincz, A. L., J. O. Priestly, et al. 1958. Evidence for a humoral mechanism which prevents growth of dermatophytes. J. Invest. Dermatol., *31*:15–17.

153. Lorincz, A. L., and S. H. Sun. 1963. Dermatophyte viability at modestly raised temperatures. Arch. Dermatol., *88*:393–402.

154. MacLennan, R. 1960. A trial of griseofulvin in tinea imbricata. Tr. St. John's Hosp. Dermatol. Soc., *45*:99–100.

155. Malhorta, V. K., M. P. Gary, et al. 1979. A study of tinea capitis in Libya (Benghazi). Sabouraudia, *17*:181–184.

156. Malmsten, P. H. 1845. *Trichophyton tonsurans,* harskarande Mogel Bidrag till utredande af de sjukdomar som valla harets affal. Stockholm, 1. p. (Published in translation by F. C. H. Creptin as *"Trichophyton tonsurans* der haarscheerende Schimmel, Ein Beitrag zur Auseinandersetzung der Krankheiten welche das Abfallen des Haares be wirken." Arch. Anat. Physiol. Wiss. Med. (J. Muller) 1848 1–19 1 pl.

157. Mantovani, A., and L. Morganti 1977. Dermatophytozoononoses in Italy. Vet. Science Comm., *1*:171–177.

158. Margarot, J., and P. Devèze. 1929. Aspect de quelques dermatosia on lumière ultraparavidette: note préliminaire. Bull. Soc. Sci. Med. Biol. Montpellier., *6*:375–378.

158a. Margarot, J., and P. Devèze. 1929. La Lumière Wood en dermatologie. Ann. Dermatol. Syphilogr. VI. *10*:581–608.

159. Mariat, F., B. Liautaud, et al. 1978. *Hendersonula toruloidea,* agent d'une dermatite verruqueuse mycosique observée en Algéria. Sabouraudia, *16*:133–140.

159a. Martin, A. R. 1958. The systematic treatment of dermatophytoses. Vet. Rec., *70*:1232.

160. Marples, M. J. 1960. Microbiological studies in Western Samoa II. The isolation of yeastlike organisms from the mouth with a note of some dermatophytes isolated from Tinea. Tr. Roy. Soc. Trop. Med. Hyg., *54*:166–170.

161. Marples, M. J., and J. M. B. Smith. 1960. The hedgehog as a source of human ringworm. Nature, *188*:867–868.

162. Matruchot, L., and C. Dassonville. 1899. Sur le champignon d'herpes (*Trichophyton*) et les formes voisines, et sur la classification des Ascomycetes. Bull. Soc. Mycol. Fr., *15*:240–253.

163. Matsumoto, T. 1971. On the family Gymnoascaceae: Especially as the perfect states of the dermatophytes. Jap. J. Dermatol., *81*:25–32.

163a. McGinnis, M. R., L. Ajello, and D. P. Rogers. 1981. Proposal to conserve *Epidermophyton* Saboraud. 1907. Taxon *30*:351–353.

164. McPherson, E. A. 1957. The influence of physical factors on dermatomycosis in domestic animals. Vet. Rec., *69*:1010–1013.

164a. Mégnin, P. 1881. Nouvelle maladie parasitaire de la peau d'un coq. C.R. Soc. Biol., *33*:404–406.

165. Menges, R. W., and L. K. Georg. 1957. Survey of animal ringworm in the United States. Public Health Rep., *72*:503–509.

166. Menges, R. W., and L. K. Georg. 1957. Canine ringworm caused by *M. gypseum.* Cornell Vet., *47*:90–100.

167. Menges, R. W., G. J. Lane, et al. 1957. Ringworm in wild animals in southwestern Georgia. Am. J. Vet. Res., *18*:672–677.

168. Mikhail, G. R. 1970. *Trichophyton rubrum* granuloma. Int. J. Dermat., *9*:41–46.

169. Mitchell, J. H. 1951. Ringworm of hands and feet. J.A.M.A., *146*:541–546.

170. Miyazi, M., and K. Nishimura. 1971. Studies on Arthrospores of *Trichophyton rubrum.* Jap. J. Med. Mycol., *12*:18–23.

171. Modan, D., B. Baidatz, et al. 1974. Radiation-induced head and neck tumors. Lancet, *1*:277–279.

172. Moraes, M., A. Padhye, et al. 1975. The perfect state of *Microsporum amazonicum.* Mycologia, *67*:1109–1113.

173. Moraes, M., D. Borelli, et al. 1967. *Microsporum amazonicum* nova species. Med. Cutan., *11*:281–286.

174. Morganti, L., M. Branchedi, et al. 1976. First European isolation of swine infection by *Microsporum nanum.* Mycopathologia, *59*:179–182.

175. Morganti, L., A. A. Padhye, et al. 1975. Recovery of *Nannizzia grubyia* from a stray Italian cat (*Felis catus*). Mycologia, *67*:434–436.

176. Mullins, J. F., L. J. Willes, et al. 1966. *Microsporum nanum.* Arch. Dermatol., *94*:300–303.

177. Nannizzi, A. 1926. Ricerche sui rapporti morfologici e biologici tra gymnoascaeeae e dermatomiceti. Ann. Mycol., *24*:85–129.

178. Nannizzi, A. 1927. Ricerche sull' origine saprofitica dei funghi delle tigne. Il *Gymnoascus gypseum* sp. n. forma ascofora del *Sabouraudites (Achorion) gypseum* (Bodin) Ota et Langeron. Atti R. Accad. Fisiocrit. Siena X, *2*:89–97, 4 figs.

179. Neisser, A. 1902. Plato's Versuche uber die Herstellung and Verwundung von Trichophytin. Arch. Dermatol. Syph., *60*:63–76.

180. Neuman, M., and N. Platzner. 1968. Treatment of bovine ringworm with thiabendazole. Refuah. Vet., *25*:40–46.

181. Neves, H. 1960. Mycological study of 519 cases of ringworm infections in Portugal. Mycopathologia, *13*:121–132.

182. Niederpruem, D. J., and V. Meevootisom. 1979.

Control of exocellular proteases in dermatophytes and especially *Trichophyton rubrum.* Sabouraudia, *17*:91–106.

183. Nishiyama, C., T. Isikawa, et al. 1977. Six cases of tinea profunda probably induced by use of steroid ointments. Jap. J. Med. Mycology *18*:22–28.

184. Oliveira, H., R. Trincao, et al. 1960. Tricofitose cutanea generalizada com infeccao systemica por *Trichophyton violacium.* J. Med. (Porta), *41*:629–642.

185. Omran, A. R., R. E. Shore, et al. 1978. Follow-up study of patients treated by x-ray epilation for tinea capitis: psychiatric and psychometric evaluation. Am. J. Public Health, *68*:561–567.

186. Onsberg, P. 1978. Human infections with *Microsporum persicolor* in Denmark. Br. J. Dermatol., *99*:531–536.

187. Onsberg, P. 1979. Gymnoascaceae and Onygenaceae as contaminants of skin, hair and nails. Mykosen, *22*:325–327.

188. Ormsby, O. S., and J. M. Mitchell. 1916. Ringworm of hands and feet. J.A.M.A., *67*:711–716.

189. Otčenásek, M. 1978. Ecology of the Dermatophytes. Mycopathologica, *65*:67–72.

190. Ottaviano, P. J., H. E. Jones, et al. 1974. Trichophytin extraction. Biological Comparison of Trichophytin extracted from *Trichophyton mentagrophytes* grown in a complex medium and a defined medium. Appl. Microbiol., *28*:271–275.

191. Očegovie, L. 1973. Eine neue vom Menschen isolierte Trichophyton-Art *(Trichophyton teheraniensis).* Dermatol. Monatsschr., *159*:463–468.

192. Padhye, A. A., and L. Ajello. 1977. The taxonomic status of the hedgehog fungus *Trichophyton erinacei.* Sabouraudia, *15*:103–114.

193. Padhye, A. A., F. Blank, et al. 1973. *Microsporum persicolor* infection in the United States. Arch. Dermatol., *108*:561–562.

194. Padhye, A. A., and J. W. Carmichael. 1971. The Genus *Arthroderma* Berkeley. Canad. J. Bot., *49*:1525–1540.

195. Padhye, A. A., and J. W. Carmichael. 1972. *Arthroderma insingulare* sp. nov. another gymnoascaceous state of the *Trichophyton terrestre* complex. Sabouraudia, *10*:47–51.

196. Padhye, A.A., and J. W. Carmichael. 1973. Mating reactions in the *Trichophyton terrestre* complex. Sabouraudia, *11*:64–69.

197. Pankova, Ya. V. 1969. On the clinical picture of rubrophytosis in Itsenko-Cushings disease. Vestn. Dermatol. Venereol., *43*:63–69.

198. Pardo-Castello, V., and O. A. Pardo. 1960. Disease of the Nail. 3rd ed. Springfield, Ill. Charles C Thomas.

199. Peiris, S., M. K. Moore, and R. H. Marten. 1979. *Scytalidium hyalinum* infection of skin and nail. Br. J. Dermatol., *106*:579–584.

200. Pellizzari, C. 1888. Recherche sur *Trychophyton tonsurans.* Giornale Italiano della malattie veneree, *29*:8–40.

201. Philpot, C. 1977. Some aspects of the epidemiology of tinea. Mycopathologia, *62*:3–13.

202. Philpot, C. M. 1977. The use of nutritional tests for the differentiation of dermatophytes. Sabouraudia, *15*:141–150.

203. Philpot, C. M. 1978. Geographical distribution of dermatophytes: a review. J. Hyg., *80*:301–313.

204. Post, K., and J. R. Saunders, 1979. Topical treatment of experimental ringworm in guinea pigs

205. Presbury, D. G., C. Young. 1978. *Trichophyton ajello* isolated from a child. Sabouraudia, *16*:233–235.

206. Prevost, E. 1979. Nonfluorescent tinea capitis in Charleston, S. C. A diagnostic problem. J.A.M.A., *242*:1765–1767.

207. Prochacki, H., and C. Engelhardt-Zasada. 1974. *Epidermophyton stockdaleae* sp. nov. Mycopathologia, *54*:341–345.

208. Pushkarenko, V. I., G. D. Pushkarenko, et al. 1969. Osobennosti aminokistotnogo sostava pota so stop u litz s klinicheski vyrazhennoi epidermofitiei. Vestn. Dermatol. Venerol., *49*:47–48.

209. Rasmussen, J. E., and A. R. Ahmed. 1978. Trichophytin reactions in children with tinea capitis. Arch. Dermatol., *114*:371–372.

210. Raubitschek, F., and R. Maoz. 1957. Invasion of nails *in vitro* by certain dermatophytes. J. Invest. Dermatol., *28*:261–268.

211. Ravaut, P., and H. Rabeau. 1921. Sur une forme speciale de trichophytic ungueale. Ann. Dermatol. Syphil. (Paris), *2*:363–365.

212. Ray, S. K., S. K. Misra. 1976. A note on the viability of arthrospores of *Trichophyton verrucosum.* Ind. Vet. J., *53*:74–75.

213. Rebell, G., and D. Taplin. 1970. *Dermatophytes: Their Recognition and Identification.* 2nd ed. Coral Gables, Fla. U. of Miami Press.

214. Reiss, F. 1944. Successful inoculations of animals with *Trichophyton purpureum.* Arch. Dermatol., *49*:242–248.

215. Remak, R. 1840. Zur Kenntnis von der planzlichen natur der parrigo lupinasa. Med. Zeitung, *9*:73–74.

216. Restrepo, A., and L. de Uribe. 1976. Isolation of fungi belonging to the genera *Geotrichum* and *Trichosporum* from human dermal lesions. Mycopathologia, *59*:3–9.

217. Reuss, U. 1978. Ein beitrag zur Behandlung der Trichophytie bei pferden. D. T. W., *85*:231–232.

218. Reyes, A. C., and L. Friedman. 1966. Concerning the specificity of dermatophyte reacting antibody in human and experimental animal sera. J. Invest. Dermatol., *47*:27–34.

219. Rioux, J. A., D. M. Jarry, et al. 1966. *Ctenomyces, Arthroderma,* ou *Trichophyton?* Fin d'une controverse et nouvelle acception due terme de dermatophyte. Ann. Parasit. Hum. Comp., *41*:523–534.

220. Rippon, J. W. 1967. Elastase: Production by ringworm fungi. Science, *157*:947.

221. Rippon, J. W. 1968. Collagenase from *Trichophyton schoenleinii.* J. Bacteriol., *95*:43–46.

222. Rippon, J. W., and T. W. Andrews. 1978. Case report *Microsporum racemosum.* Second clinical isolation from the United States and Chicago area. Mycopathologia, *64*:187–190.

223. Rippon, J. W., A. Eng, and F. Malkinson. 1968. *Trichophyton simii* infection in the United States. Arch. Dermatol., *98*:615–619.

224. Rippon, J. W., and D. Garber. 1969. Dermatophyte infection as a function of mating type and associated enzymes. J. Invest. Dermatol., *53*:445–448.

225. Rippon, J. W., and L. J. Lebeau. 1965. Germination of *Microsporum audoninii* from infected hairs. Mycopathologia, *26*:273–288.

226. Rippon, J. W., F. C. Lee, et al. 1970. Dermatophyte

infection caused by *Aphanoascus fulsvecens.* Arch. Dermat., *102*:552–555.

227. Rippon, J. W., and M. Medenica. 1964. Isolation of *Trichophyton soudanense* in the United States. Sabouraudia, *3*:301–304.

228. Rippon, J. W., and G. H. Scherr. 1959. Induced dimorphism in dermatophytes. Mycologia, *51*:902–914.

229. Rivalier, E. 1953. Description de *Sabouraudites praecox* nova species suivie de remarques sur le genera Sabouraudites. Ann. Inst. Pasteur, *86*:276–284.

230. Robertstad, G. W., B. Bennett, and L. Ajello. 1974. Isolation of *Microsporum distortum* from a dog in south western United States. J. Cut. Pathol., *1*:117–119.

231. Rosenthal, S. A., R. L. Baer, et al. 1956. Studies on the dissemination of fungi from the feet of subjects with and without fungus disease of the feet. J. Invest. Dermatol., *26*:41–51.

232. Rosman, N. 1977. Imported superficial fungal infections in Denmark. Curr. Therap. Res., *22*:100–103.

233. Roth, F. J., C. C. Boyd, et al. 1959. An evaluation of the fungistatic activity of serum. J. Invest. Dermatol., *32*:549–556.

234. Rothman, S., G. Knox, et al. 1957. Tinea pedis as a source of infection in the family. Arch. Dermatol., *75*:270–271.

235. Rothman, S., and A. L. Lorincz. 1963. Defence mechanisms of the skin. Ann. Rev. Med., *14*:215–242.

236. Rothman, S., A. Smiljanic, et al. 1947. The spontaneous cure of tinea capitis in puberty. J. Invest. Dermatol., *8*:81–98.

237. Roubicek, M., and A. Krebs. 1977. Logamel a new broad spectrum antimicrobial corticoid combination for topical use. Results of a double blind comparative trial in patients with superficial dermatomycosis. Praxis, *66*:585–588.

238. Rush-Monro, F. M. 1978. *Trichophyton erinacei.* Rev. Med., *19*:639–646.

239. Rush-Munro-F. M., A. J. Woodgyer, and M. R. Hayter. 1977. Ringworm in guinea pigs. Mykosen, *20*:292–296.

240. Rustia, M., and P. Shubik. 1978. Thyroid tumours in rats and hepatomas in mice after griseofulvin treatment. Br. J. Cancer, *38*:237–249.

241. Sabouraud, R. 1894. *Les Trichophytons Humaines.* Paris, Rueff et Cie.

242. Sabouraud, R. 1910. Les maladies du cuir chevelu. In *Les Maladies Cryptogamiques,* Vol. 3. 420–446. In *Les Teignes,* Paris, Masson et Cie.

243. Sabouraud, R. 1910. *Les Teignes.,* Paris, Masson et Cie.

244. Samsoen, M., and G. Jelen. 1977. Allergy to daktarin gel. Contact Dermatitis, *3*:351–352.

245. Sanderson, P. H., and J. C. Sloper. 1953. Skin disease in the British Army in S. E. Asia. III. Relationship between mycotic infections of body and feet. Br. J. Dermatol., *65*:362–372.

246. Schmitt, J. A., and R. G. Miller. 1967. Variation in susceptibility to experimental dermatomycosis in genetic strains of mice. Mycopathologia, *32*:306–312.

247. Schoenlein, J. L. 1839. Zur pathologie der impetigenes. Archiv. fur Anat. Phys. Wiss. Med. p. 82.

248. Schulz, J. A., and R. Lippmann. 1968. Eradication of bovine ringworm by gricinvet. Mh. Veterinaermed., *23*:531–534.

249. Sellers, K. C., W. B. V. Sinclair, et al. 1956. Preliminary observations on natural and experimental ringworm in cattle. Vet. Rec., *68*:729–732.

250. Shanks, S. C. 1967. Vale epilatio. X-ray epilation at Goldie Leigh Hospital Woolwich (1922–58). Br. J. Dermatol., *79*:237–238.

251. Shapiro. A. L., and S. Rothman. 1945. Undecylenic acid in the treatment of dermatomycosis. Arch. Dermatol. Syph., *52*:166–171.

252. Simic, L., and S. Perisic. 1969. Microsporosis caused by *M. canis* in humans and a dog transmitted by an imported cat. Mykosen, *12*:699–703.

253. Sinski, J. T., L. M. Kelley, et al. 1977. Dermatophyte isolation media. Quantitative appraisal using skin scales infected with *Trichophyton mentagrophytes* and *Trichophyton rubrum*, J. Clin. Microbiol., *5*:34–38.

253a. Sinski, J. 1974. Dermatophytes in human skin, hair, nails. Springfield, Ill., Charles C Thomas.

254. Sloper, J. C. 1955. A study of experimental human infection due to *T. rubrum, T. mentagrophytes* and *E. floccosum.* J. Invest. Dermatol., *25*:21–28.

255. Smith, E. B., R. T. Jessen, and J. A. Ulrich. 1978. Sodium omadine lotion in tinea pedis. Curr. Therap. Res., *23*:433–435.

256. Smith, J. M. B., and M. J. Marples. 1963. *Trichophyton mentagrophytes* var. *erinaiei.* Sabouraudia, *3*:1–10.

257. Smith, J. M. B., and M. J. Marples. 1964. Ringworm in the Solomon Islands. Trans. Roy. Soc. Trop. Med. Hyg., *58*:63–67.

258. Spanoghe, L., and E. P. Oldenkamp. 1977. Mycological and clinical observations on ringworm in cattle after treatment with natamycin. Vet. Rec., *101*:135–136.

259. Stahl, D., and P. Onsberg. 1978. Local treatment of dermatomycosis. Mykosen, *21*:48–52.

260. Stevenson, C. J., and N. Djavahiszwili. 1961. Chronic ringworm of the nails; long term treatment with griseofulvin. Lancet, *1*:373–374.

261. Stockdale, P. M. 1963. The *Microsporum gypseum* complex (*Nannizzia incurvata* Stockd., *N. gypsea* (Nann.) comb. nov. *N fulva* sp. nov.). Sabouraudia, *3*:114–126.

262. Stockdale, P. M. 1967. *Nannizzia persicolor* sp. nov., the perfect state of *Trichophyton persicolor* Sabouraud. Sabouraudia, *5*:355–359.

263. Stockdale, P. M., D. W. R. Mackenzie, and P. K. C. Austwick. 1965. *Arthroderma simii* sp. nov. The perfect state of *Trichophyton simii* (Pinoy) comb. nov. Sabouraudia, *4*:112–113.

264. Stocker, W. W., A. J. Richsmeier, et al. 1977. Kerion caused by *Trichophyton verrucosum.* Pediatrics, *59*:912–915.

265. Strauss, J. S., and A. M. Kligman. 1957. An experimental study of tinea pedis and onychomycosis of the foot. Arch. Dermatol., *76*:70–79.

266. Takashio, M. 1973. Etude des phénomènes de reproduction liés au viellissement et au rajeunissement de cultures de champignons. pp 536–549. Ann. Soc. Belg. Med. Trop., *53*:429–580.

267. Takashio, M. 1973. Une nouvelle forme sexuée du complexe *Trichophyton mentagraphytes, Arthroderma vanbreuseghemii* sp. nov. Ann. Parasitol. Hum. Comp., *48*:713–732.

268. Takashio, M. 1975. Single ascospore strains from the mating between *Trichophyton mentagrophytes*

varerinacei and *Arthroderma benhamiae*. Trans. Br. Mycol. Soc., *65*:67–75.

269. Takiuchi, I., and D. Higuchi. 1977. Isolation purification and biochemical properties of keratinase elaborated by *Microsporum gypseum*. Jap. J. Dermatol., *87*:305–309.

270. Taplin, D., N. Zaias, G. Rebell, et al. 1969. Isolation and recognition of dermatophytes on a new media. Arch. Dermatol., *99*:203–209.

271. Tewari, R. P. 1969. *Trichophyton simii* infections of chickens, dogs, and man in India. Mycopathologia, *39*:293–298.

272. Torres, G., and L. K. Georg. 1956. A human case of *Trichophyton gallinae* infection. Arch. Dermatol., *74*:191–195.

273. Vanbreusegham, R. 1949. La culture des dermatophytes in vitro sur des cheveux isoles. Ann. Parasit., *24*:559–573.

274. Vanbreusegham, R. 1968. *Trichophyton soudanense* in and outside Africa. Br. J. Dermatol., *80*:140–148.

275. Vanbreusegham, R., and S. A. Rosenthal. 1961. *Trichophyton kuryangei* n. sp. nouveau dermatophyte Africaine. Ann. Parasitol. Hum. Comp., *36*:797–803.

276. Varadi, D. P., and J. W. Rippon. 1967. Scalp infection of triple etiology. Arch. Dermatol., *95*:229–301.

277. Velasco-Castrejón, O., and A. Gonzalez-Ochoa. 1975. La *Tinea imbricata* en la sierra de Puebla, México. Rev. Invest. Salud. Publica, *35*:109–116.

278. Vincent, J. 1977. The importance of fatty acids in pathogenesis of dermatophytosis. Current Therapeutic Res., *22*:83–91.

279. Vincent, K. 1978. Dermatophyte lipids. Prog. Chem. Fats Other Lipids., *16*:171–177.

280. Visset, M. 1972. Les formes conidiennes du complexe *Microsporum gypseum* deservers en microscapu electronique a balayarge. Sabauraudia, *10*:191–192.

281. Wallerstrom, A. 1967. Production of antibiotics by *Epidermophyton floccosum*. *1*. The antibiotic spectrum of crude filtrates. Acta Pathol. Microbiol. Scand., *71*:287–295.

282. Walters, A. J., J. E. D. Chick, et al. 1974. Cell-mediated immunity and serum blocking factors in patients with chronic dermatophyte infections. Int. Arch. Allergy Appl. Immunol., *46*:849–857.

283. Weitzman, I. 1964. Variation in *Microsporum gypseum*. I. A genetic study of pleomorphism. Sabouraudia, *3*:195–204.

284. Weitzman, I., M. A. Gordon, et al. 1971. Determination of the perfect state, mating complex and elastase of *Microsporum gypseum* complex. J Invest. Dermatol., *57*:278–282.

285. Weitzman, I., and S. McMillen. 1980. Isolation in the United States of a culture resembling *M. praecox*. Mycopathologica, *70*:181–186.

286. Weitzman, I., and A. Padhye. 1976. Is *Arthroderma simii* the perfect state of *Trichophyton quinckeanum*. Sabouraudia, *14*:65–74.

287. Weitzman, I., and A. Padhye. 1978. Mating behavior of *Nannizzia otae (Microsporum canis)*. Mycopathologia, *64*:17–22.

288. Wenk, P. 1962. Causes of spontaneous recovery from trichophytosis in guinea pigs. Z. Tropenmed. Parasit., *13*:201–215.

289. Wilson, J. W., D. A. Plunkett, et al. 1954. Nodular granulomatous perifolliculitis due to *Trichophyton rubrum*. Arch. Dermatol., *69*:258–277.

290. Wolf, F. T., L. A. Jones, et al. 1958. Fluorescent pigment of *Microsporum*. Nature (London), *182*:475.

291. Youssef, N., C. E. Wyborn, et al. 1978. Antibiotic production by dermatophyte fungi. J. Gen. Microbiol., *105*:105–111.

292. Yu, R. J., S. A. Harmon, et al. 1969. Hair digestion by a keratinase of *Trichophyton mentagrophytes*. J. Invest. Dermatol., *53*:166–171.

293. Yu, R. J., S. R. Harmon, et al. 1971. Two cell bound keratinases of *Trichophyton mentagrophytes*. J. Invest. Dermatol., *56*:27–32.

294. Zaias, N. 1966. Superficial white onychomycosis. Sabouraudia, *5*:99–103.

295. Zaias, N. 1969. Fungi in toe nails. J. Invest. Dermatol., *53*:140–142.

296. Zaias, N. 1972. Onychomycosis. Arch. Dermatol., *105*:263–274.

297. Zussman, R. A., I. Lyon, et al. 1960. Melanoid pigment production in a strain of *T. rubrum*. J. Bacteriol., *80*:708–713.

THE SUBCUTANEOUS MYCOSES

9 Chromoblastomycosis and Related Dermal Infections Caused by Dematiaceous Fungi

INTRODUCTION TO THE SUBCUTANEOUS MYCOSES

The subcutaneous mycoses include a heterogeneous group of infections characterized by the development of a lesion at the site of inoculation. Unlike the systemic mycoses, whose primary mode of entry is usually pulmonary, these infections are the result of traumatic implantation of the fungus into the skin. In general, the ensuing disease remains localized to this area or slowly spreads to surrounding tissue, a picture similar to that seen with the mycetomas. In some diseases, slow extension via lymphatic channels is a frequent occurrence (sporotrichosis); in others, hematogenous and lymphatic dissemination is rarely recorded (chromoblastomycosis).

The type of disease evoked is an interesting interplay between host reactions and defenses, and the relative virulence of the infecting agent. The species concerned are common soil saprophytes whose ability to adapt to the tissue environment and elicit disease is extremely variable. The agent of sporotrichosis is a thermodimorphic fungus similar to *Histoplasma* and *Blastomyces*, but it is of relatively limited virulence. Most soil-isolated strains of this organism are essentially nonpathogenic. The agents of chromoblastomycosis are of even less pathogenic potential. The degree and type of dimorphism exhibited by these organisms depend on the relative resistance of the host and on the tissue in which they are growing. Some

species adapt relatively easily and are the planate-dividing, yeastlike bodies of classic verrucous dermatitis. In debilitated patients, however, the same organisms are mycelial in morphology. In contrast, other species are rarely found in "chromoblastomycosis" and are restricted to opportunistic infections of the severely compromised host.

The entity termed "entomophthoromycosis" is even more perplexing. Again, the disease is subcutaneous, with little tendency to spread. Furthermore, it is completely different from the disease caused by other members of the Zygomycota, i.e., mucormycosis. Essentially nothing is known about its frequency, its mode of transmission, or the relative pathogenicity of the infectious agent. It does not exhibit dimorphism in tissue and is grouped here only because it is "subcutaneous." Lobo's disease (keloid blastomycosis) and rhinosporidiosis are also included in this very heterogeneous group of diseases. The etiologic agents of these two diseases have not as yet been cultivated *in vitro* and even less is known about their life history and epidemiology.

CHROMOBLASTOMYCOSIS

Definition

The recently used term "chromomycosis" included a group of clinical entities caused by a variety of dematiaceous (pigmented) fungi. The most common form of this disease was known as verrucous dermatitis or chromoblastomycosis. Chromoblastomycosis was originally used when this clinical entity was first differentiated in the 1910's and 1920's. Later, other types of infections began to be recorded in which dematiaceous fungi were also involved; however, the organisms appeared as hyphal elements instead of the "sclerotic cells" or "planate-dividing yeasts." For this reason "blasto" was dropped from the name of the disease, and an all-inclusive term, "chromomycosis," was coined. In time, many more types of clinical diseases were noted in which dematiaceous fungi were found. This made the term chromomycosis useless, since it included too many varieties of infections, pathologies, and etiologic agents. Presently there is no consensus among experts in the field concerning the nomenclature of the diseases. For the purposes of this text, the author has decided to utilize the following terminology. The older term, *chromoblastomycosis*, is reinstituted to encompass a specific clinical entity (verrucous dermatitis). *Phaeomycotic cyst* will be used for the diseases called phaeosporotrichosis and cystic chromomycosis. *Cutaneous phaeohyphomycosis* is the term recommended for the occasional infection of the skin by a dematiaceous fungus. (This could also be included in "dermatomycosis" of the previous chapter.) Following the suggestion of Ajello,[1] the other clinical entities involving dematiaceous fungi, e.g., cerebral and systemic infections, are included in the term phaeohyphomycosis and discussed in Chapter 26. In this chapter chromoblastomycosis and its related dermal entities, phaeomycotic cyst and phaeomycotic colonization (cutaneous phaeohyphomycosis), will be discussed.

In chromoblastomycosis, as in other subcutaneous mycoses, the fungi gain entrance through the skin by traumatic implantation. The lesions develop at the site of inoculation and are most commonly limited to the cutaneous and subcutaneous tissue. The response is one of hyperplasia, characterized by the formation of verrucoid, warty, cutaneous nodules, which may be raised 1 to 3 cm above the skin surface. These roughened, irregular, pedunculated vegetations often resemble the florets of a cauliflower (Figs. 9–1, 9–3). The fungi are most often seen in tissue as planate-dividing, yeastlike bodies (sclerotic cells) (Figs. 9–4, 9–5). The disease is usually confined to the lower legs and feet, but involvement of the hands, buttocks, ears, chest, abdomen, and other surfaces has been recorded. Some of these represent lymphatic or hematogenous spread. The clinical entity chromoblastomycosis is caused by a very limited series of soil-inhabiting dematiaceous fungi.

The remainder of the diseases included in the chromoblastomycosis group are localized infections in which the same species or a variety of other species are involved. The tissue reactions evoked differ from those of verrucous dermatitis and are quite variable. The morphology of the organisms is also variable. They appear much less as planate-dividing sclerotic cells and are more often seen as distorted, septate, hyphal strands and a variety of swollen, rounded forms. These infections often occur in debilitated patients, whereas "chromoblastomycosis" occurs in normal hosts in whom infection is a function of continued trauma and exposure.

Synonymy. Chromomycosis, verrucous dermatitis, phaeosporotrichosis, cladosporiosis, dermatitide verrucosa chromoparasitaria, infections by dematiaceous fungi.

HISTORY

Although chromoblastomycosis is essentially a tropical and subtropical disease, the first authentic case was reported from Boston. Lane[54] and Medlar[67] in 1915 described a patient from New England with verrucous lesions on the foot. The fungus isolated was named *Phialophora verrucosa* by Thaxter, but the physician describing the case gave essentially no credit to this mycologist, and so the specific epithet is usually ascribed to Medlar. Prior to this, in 1914, Rudolph[80] had published a report of a case of chromoblastomycosis from Brazil, but there was no description of the fungal etiology. Pedroso and Gomes reported on four cases of the disease in 1920.[76] One of the patients had been under observation since 1911, and thus is considered to be the first case recognized (Fig. 9–1). The fungus isolated from this patient was not similar to the one of Thaxter and was named *Hormodendrum pedrosoi* by Brumpt in 1922.[15] In 1924 a few more cases were reported from Brazil by Terra, Torres, Fonseca, and Areã Leão. Carini[21] described two additional cases and coined the term "chromoblastomycosis" for the disease. He noted that the fungus isolated and that the morphology of the sclerotic cells in tissue were similar to that seen previously in tuberculoid lesions of kidney and lung in a Brazilian big frog (*Leptodactylus*

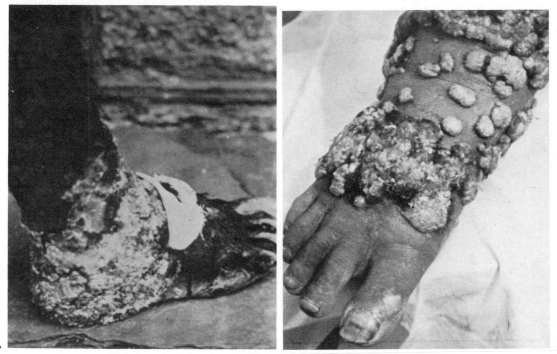

Figure 9–1. Chromoblastomycosis. *A*, First case of the disease as described by Pedroso and Gomes.[76] (From Pedroso, A., and J. M. Gomes. 1920. Ann. Paulistas Med. Cir., *9*:53.) *B*, Typical cauliflower-like lesions on the foot of a patient in Costa Rica. Both legs were involved. (Courtesy of D. Moore and Col. H. Hunter.)

pentadactylus). Carini was thus the first to describe the disease in animals. The first case outside of the Americas was reported by Montpellier and Catanei[69] in 1927. Their patient was from North Africa (Algeria), and they named the isolant *Hormodendrum algeriensis*. This is now considered in synonymy with *Fonsecaea pedrosoi*. By 1929 cases from Poland and Sumatra had been described.

It was not until 1933 that a second case, reported by Wilson et al.,[100] was found in the United States. This report and another by Martin, Baker, and Conant[62] in 1936 represented the only three cases known in the United States for a long time. In the past few years, however, the disease has been recognized more frequently.[2, 97]

Other etiologic agents of chromoblastomycosis have been reported over the years. In 1936 Carrión noted an agent differing from *Hormodendrum (Fonsecaea) pedrosoi*[22] and called it *H. compactum*. At the same time, a detailed mycological investigation of the various fungi isolated from chromoblastomycosis was made by Negroni.[71] He noted that the fungi from South America had two common types of conidiation: *Hormodendrum (Cladosporum)* and *Acrotheca*. He proposed a new genus — *Fonsecaea* — to include these fungi. From the first

Japanese case of chromoblastomycosis-like disease Kano[49] isolated yet another agent. He called it *Hormiscium dermatitidis*, a fungus more often encountered in systemic opportunistic infections and cystic disease than in true chromoblastomycosis. This species is now called *Wangiella dermatitidis*. Yet a different isolant from cases in South Africa and Australia was described by Simpson in 1946[86]; he called it *F. pedrosoi* var. *cladosporium*. While studying this isolant, Carrión found a similarity to isolants by O'Daly from cases in Venezuela. Trejos renamed the organism *Cladosporium carrionii*,[92] and this is now found to be the commonest agent of the disease encountered in Australia and Venezuela.[56]

In 1907, Beurmann and Gougerot[10] reported on a clinical variant of sporotrichosis appearing as an intramuscular abscess. The isolated organism was a dematiaceous fungus and was named *Sporotrichum gougerotii* by Matruchot in 1910[63] but was transferred to the genus *Phialophora* by Borelli in 1955. The organism has recently been reduced to synonymy with *Exophiala jeanselmei*.[66] About 50 cases of infection by this fungus have now been recorded. Mariat et al.[61] proposed the name phaeosporotrichosis for this type of disease. However, it is usually self-limiting and cystlike in character and does not resemble sporotrichosis. A more suitable descriptive name would be phaeomycotic cyst. Most of the isolants have been *E. jeanselmei*, an organism of low virulence. The majority of patients were somewhat debilitated, so that this could be considered an opportunistic infection.

The final category is that of phaeomycotic colonization. In this disease a wide variety of dematiaceous fungi are found colonizing preexisting lesions. They may then contribute to or exaggerate the pathology of the original lesion. This type of infection was recently reviewed by Lazo.[55]

Reports implicating commoner agents of chromoblastomycosis and some rarely isolated species are appearing more frequently in the literature. Some of these describe the disease in partially debilitated or compromised patients. These cases, therefore, represent opportunistic infections. The tissue reaction, as would be expected, is not the same as the one seen in chromoblastomycosis (verrucous dermatitis), nor is the morphology of the fungus the same. These etiologic agents are all soil-inhabiting dematiaceous fungi. They are light-brown, pigmented organisms of varied morphology in tissue and are included in the category of invasive phaeohyphomycosis along with cerebral and systemic infections associated with dematiaceous fungi (see Chapter 26).

ETIOLOGY, ECOLOGY, AND DISTRIBUTION

Etiology. The etiologic agents of chromoblastomycosis are soil-inhabiting fungi of the form-family Dematiaceae of the Hyphomycetes. Their saprophytic occurrence is well documented, and they are among the commoner fungi found in decaying vegetation, rotting wood, and forest litter. The mycelium, conidia, and sclerotic cells of the organisms are pigmented in shades of light brown, yellow-brown, and brown-black. The sclerotic cells and yeastlike cells are found in chromoblastomycosis, and the variously arranged and distorted hyphal elements occur in other forms of phaeohyphomycosis. The several species involved appear to be closely related and are very difficult to distinguish one from the other. Probably no aspect of taxonomy in medical mycology has evoked so much debate and controversy as the classification of the etiologic agents of these diseases. The organisms elaborate a wide variety of conidial types, depending on the strain, the substrate used, and the various physical conditions under which they are grown. The first described agent, *Phialophora verrucosa*, is fairly easily identified and remains the type species of the genus. The second agent, described as *Hormodendrum pedrosoi*, has been placed at times in several genera. The various other organisms isolated have been placed in the genera *Phialophora, Hormodendrum, Hormiscum, Fonsecaea, Rhinocladiella*, and *Torula*. There is as yet no agreement on their taxonomy. The final determination of good species will depend on finding either a stable, distinguishing characteristic or, hopefully, the sexual stage of the agents, a method that has become very important in the taxonomy of dermatophytes. For convenience in this chapter, the author has adopted the arbitrary taxonomy most frequently used in recent publications. The various arguments concerning classification will be discussed in the section on mycology. The principal etiologic agents of chromoblastomycosis are *Fonsecaea pedrosoi, F. compacta, Phialophora verrucosa*, and *Cladosporium carrionii*. A few cases of chromoblastomycosis have also been caused by *Cladophialophora ajelloi*[13a] and *Rhinocladiella aquaspersa*.[12] These six fungi are the only recognized agents of chromoblastomycosis. The commonest agent by far is *F. pedrosoi*. (N.B. Current taxonomic studies indicate that *F. compacta* may be a dysplastic variant of *F. pedrosoi*[13a] and *C. carrionii* should be included in the new genus *Cladophialophora*.)

The disease reported by Kano[49] as being caused by *Hormiscium dermatitidis* (now in the genus *Wangiella*) appears not to have been chromoblastomycosis, and this fungus has not been isolated from a true case since. *Wangiella dermatitidis* is found regularly as an agent of phaeomycotic cyst[41] and phaeohyphomycosis. Phaeomycotic cyst is almost always associated with *Exophiala jeanselmei*. This organism is also frequently recovered from black-grained mycetoma and occasionally from systemic phaeohyphomycosis.[31] Other agents of phaeomycotic cyst are *Phialophora verrucosa*,[47a] *P. richardsiae*,[91] and *E. spinifera*.[72] So far some 28 species have been recovered from invasive phaeohyphomycosis. They are discussed in Chapter 26.

Phaeomycotic colonization has also been attributed to a number of dematiaceous fungi, including genera such as *Alternaria, Cladosporium, Aureobasidium*, and *Curvularia*.[55]

In most cases such colonization of cracked, fissured skin, preexisting ulcers, and lichenified or eczematized areas is purely saprophytic; however, the fungi may contribute to the pathology, and the colonization may be the nidus for eventual entry and invasion of the skin and other organs.

Ecology. All of these various agents have been recovered from decaying vegetation or soil habitats. Conant, in 1937, demonstrated that *P. verrucosa* was identical to *Cadophora americana*, a cause of "bluing" of lumber. *E. jeanselmei* is more frequently associated with mycetoma than phaeomycotic cyst, and has been isolated from wood pulp in Sweden and Italy under the name *Cadophora lignicola.* Several studies by Putkonen[77, 88] and others have implicated the Finnish sauna as the point source of numerous infections with *F. pedrosoi. P. verrucosa* has also been found in such environments. *C. carrionii* has been isolated from decaying wood.

Distribution. Though the first published case of chromoblastomycosis came from Massachusetts, the disease is infrequently reported in temperate climates or among a shoe-wearing population. The organisms abound in soils in all parts of the world, but because of repeated exposure, particularly among poor rural peoples without shoes, the disease is more frequent in tropical and subtropical climates. Puncture wounds are the main mode of entrance for the agents of chromoblastomycosis. The verrucous dermatitis type of disease is always associated with a history of such a trauma or repeated trauma. In some systemic infections, particularly those involving the brain, the primary site may be pulmonary. In opportunistic infections, the organisms gain entry by various means. These types of infections are discussed in Chapter 26.

More cases of the verrucous type of disease are reported among males than females, but this is probably because males have greater opportunity for soil contact and predisposition to injury while working. Patients with the disease most frequently fall within the 30- to 50-year-old range. The disease develops slowly and may be of prolonged duration. Paradoxically, the disease is rarely found in children exposed to the same environmental conditions as adults, especially in temperate climates. It has been suggested that implanted organisms may remain quiescent for long periods of time. It is possible that repeated trauma and tissue injury to the area are required before the organisms are able to

incite a disease process. Good hygiene and adequate nutrition may help the individual abort a potential infection. A tabulation of reported cases of verrucous dermatitis is noted in Al-Doory.[2]

Though verrucous chromoblastomycosis is found throughout the world, most cases come from the American tropics and subtropics.[57, 94, 97, 103] It is very common in Mexico, particularly in the states of Veracruz and Tabasco. When it is looked for, numerous cases are found throughout Central America. In Cuba and the Dominican Republic, there is also a high frequency of disease among the rural population. In all these areas, the most commonly isolated agent is *F. pedrosoi*, and feet and legs are the most frequent sites of infection. By contrast, in Venezuela the body sites most frequently involved are the shoulders, chest, and trunk. The agent involved is primarily *C. carrionii*, particularly in the arid Lara and Falcon states.[20] However, in the more humid areas, where climatic conditions are similar to the other countries cited, the primary agent is *F. pedrosoi*. A few cases were reported in Puerto Rico where early studies on the disease were carried out by Carrión.[22, 23] In Colombia[94] and Ecuador, 75 per cent or more of cases involved *F. pedrosoi*, the rest being caused primarily by *P. verrucosa.* Several hundred cases have been reviewed by Lacaz[53] in Brazil. Various clinical presentations were noted, and most cases were caused by *F. pedrosoi*. A few scattered reports come from Peru, Argentina, Martinique,[18] Russia,[60] Finland,[77, 88] Czechoslovakia,[98] Rumania, Japan, Algeria, Italy, and Australia, among others. In Australia[56] and South Africa[86] the commonest agent is *C. carrionii.* Local outbreaks have been recorded in the United States.[46] A review of the epidemiology, geographic distribution, and etiologic agents was published by Brygoo and Destombes.[17] Only about four cases in the world literature are ascribed to *F. compacta:* two in Puerto Rico, one in Tennessee, and one in Martinique.

Cystic disease and phaeomycotic colonization have been reported from many areas of the world. No pattern of age or sex distribution has been identified, but it appears these are probably opportunistic infections associated with slight debilitation of normal cell defense responses.[52]

Chromoblastomycosis

This was the first form of disease associated with dematiaceous fungi to be recognized, and it is still the most frequently encountered. The lesion appears at the site of some trauma or puncture wound; however, the trauma may have occurred so long before that a history may be lacking. Initially the lesion is a small, raised, erythematoid, nonpruriginous papule.[79a] Rarely there is some pruritus.[22] The original papules or pustules are violaceous in color and histologically consist of an effusive round cell infiltrate. Often the lesions are scaly, and organisms can be seen as distorted hyphal elements in skin scrapings (Fig. 9–2,A). Such lesions could be considered phaeomycotic colonization. In time (often months or years), a new crop of lesions appears in the same area or adjacent areas. The latter follow the distribution of local lymphatic channels. Frequently these lesions become raised to 1 to 3 mm above the skin surface and are hypertrophic, with a scaly, dull, red to greyish surface (Fig. 9–2,

B). Sometimes there is peripheral spread (Fig. 9–2, C), with healing in the center, as in cutaneous blastomycosis;[53] usually, however, the lesions tend to enlarge and become grouped. After many years they may become elevated to 1 to 3 cm, pedunculated and verrucous, and resemble florets of the cauliflower (Fig. 9–3). The surface is studded by round ulcerations that are 1 mm in diameter or covered by hemopurulent material. Observation of these "black dots" (Fig. 9–3, C), or "cayenne pepper," and the associated histopathology have led to the designation of chromoblastomycosis as a "minimycetoma" by Zaias.[103]

It is in these mature lesions of chromoblastomycosis that the planate-dividing, yeastlike bodies (sclerotic cells) are found (Figs. 9–4 and 9–5). These are referred to as "copper pennies" or "Medlar bodies." The second

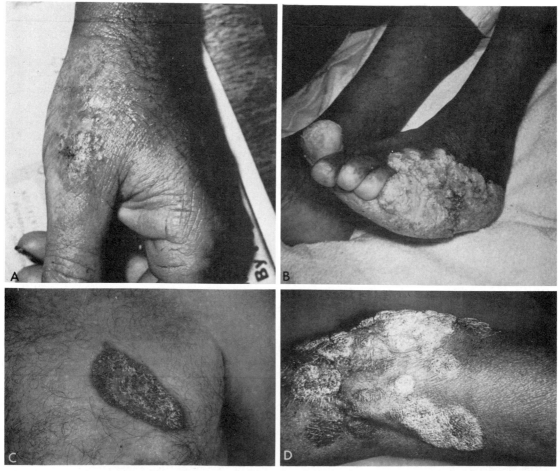

Figure 9–2. Chromoblastomycosis. *A*, Erythematous, violaceous pustules on the hand of a lumber worker. Skin scraping showed distorted hyphal elements. *Phialophora verrucosa* was cultured. This is an example of cutaneous colonization that was penetrating deeper into the dermis. *B*, Chronic verrucous chromoblastomycosis of foot due to *Fonsecaea pedrosoi*. *C*. Clinical variant seen on chest. The center is a healing atrophic scar, and an active raised border is present. In this case colonization had extended from the dermis into the pannicle. *D*, Old chronic disease with some scarring and keloid formation. Many lesions were inactive.

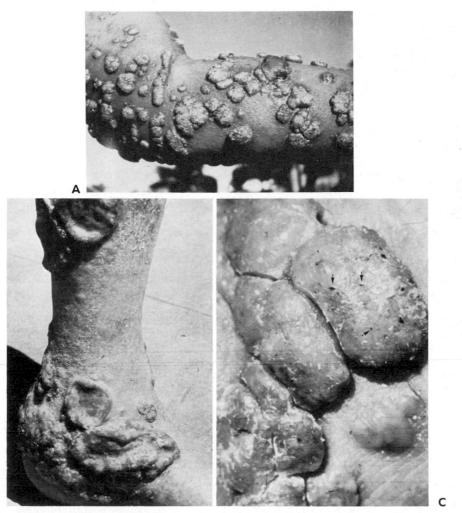

Figure 9–3. *A*, Chromoblastomycosis. Pedunculated verrucous lesions resembling florets of cauliflower. (Courtesy of F. Pifano.) *B*, Broad extensive lesions of 20 years' duration. *C*, Close-up of lesions showing "black dot" at the apices of microabscesses (arrows). (Courtesy of L. Ajello.)

most commonly encountered type is the annular, flattened, papular type with a raised active border. The center, through healing, has become cicatricial (Fig. 9–2,*D*). If the infection involves the chest wall, a raised lesion with a spreading erythematous border (Fig. 9–2,*C*) is noted. Scrapings and biopsies from such lesions show transition forms, from elongated hyphae, hyphae with round cells, and distorted elements to sclerotic cells, the latter in the microabscesses deep within the epidermis. Abrasions leading to infections may also present with this clinical and histologic picture in other areas, such as the arm, wrist (Fig. 9–2,*A*), and trunk. Lesions on areas that receive continuous trauma, such as the plantar surface, may be ulcerative and craterlike in appearance (Fig. 9–8). In material from such lesions various morphological forms of the fungus are seen.

Many other clinical variants occur. Lesions often are traumatized and ulcerate and become secondarily infected with bacteria. These may have a purulent exudate rather than the normal, dry, crusted appearance and have a putrescent odor. Spreading lesions with atrophic, scarred centers and raised borders are not infrequently seen (Fig. 9–2,*B, D*).[53] These must be differentiated from blastomycosis. Sometimes in healing lesions, there is extensive keloid formation resulting in a fibroma-like appearance.

In general, the disease remains localized to the immediate area of the initial infection. In old cases, lesions in all stages of development may be seen. There is no apparent discomfort to the patient. Secondary infection sometimes leads to considerable lymph stasis, which may result in elephantiasis. Since chro-

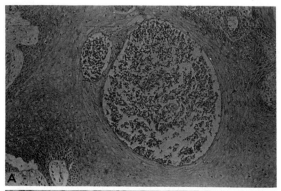

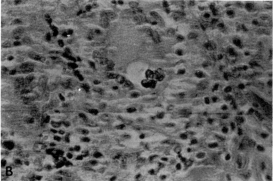

Figure 9–4. Chromoblastomycosis. *A*, Massive pseudoepitheliomatous hyperplasia with small abscesses. Hematoxylin and eosin stain. ×100. *B*, High-power view of abscess showing brown sclerotic bodies of fungus within giant cell. Hematoxylin and eosin stain. ×400.

moblastomycosis itself is usually not debilitating, this complaint may be the initial reason for the patient to seek medical attention. There is no invasion of the bone or muscle, or fistula formation as is commonly seen in mycetoma. In rare cases there is hematogenous spread to uninvolved areas of the body, and verrucous lesions may appear on other appendages as well as on the chest, abdomen, or trunk.[4] About a dozen cases with dissemination to the brain are now recorded.[34] *F. pedrosoi*, *W. dermatitidis*, and *P. verrucosa* have all been isolated from brain abscesses. In this tissue they appear as long-branching, septate, brown hyphal strands indistinguishable from those seen in phaeohyphomycosis of the brain caused by *C. bantianum*. The latter organism produces disease in the absence of preexisting skin lesions. In experimental studies, all species, but especially *C. bantianum* and *W. dermatitidis*, are neurotropic.[3, 48]

Differential Diagnosis. The early skin lesions of verrucous chromoblastomycosis that show centrifugal spread resemble those of blastomycosis. However, in chromoblastomycosis the lesions usually lack the

sharply raised border and the multiple tiny pustules which are present in cutaneous blastomycosis. In addition to blastomycosis, the differential diagnosis must include yaws, tertiary syphilis, tuberculosis verrucosa cutis, mycetoma, leishmaniasis, candidiasis, and sporotrichosis. Chromoblastomycosis has also been confused with lupus vulgaris, lupus erythematosus, and leprosy. Lymph stasis in advanced disease along with multiple flat lesions gives the appearance of tropical "mossy foot." In all cases, demonstration of the fungus by direct examination of a potassium hydroxide mount and by culture establishes the diagnosis.

Phaeomycotic Cyst

This group includes reports of about fifty cases of infection usually resulting from puncture wound in which there were single or multiple deep cysts containing masses of brown pigmented fungi of varied morphology.[52, 61, 102] The lesions are subcutaneous or intramuscular and tend to be stationary. They may be misdiagnosed as any one of a

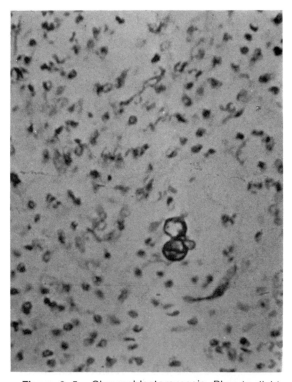

Figure 9–5. Chromoblastomycosis. Planate-dividing, rounded, sclerotic body of fungus in verrucoid chromoblastomycosis. ×400. *Phialophora verrucosa*. (From Rippon, J. W. *In* Freeman, R. 1979. Textbook of Microbiology. *20th ed. Philadelphia, W. B. Saunders Company, p. 754.*)

variety of other diseases.[89] From these, the most commonly isolated agent is *E. jeanselmei*, but a few cases are caused by a variety of other species, such as *P. richardsiae*, *E. spinifera*, *P. verrucosa*,[47a] and *W. dermatitidis*. The clinical designation phaeosporotrichosis[32] has been proposed for this group, but the disease does not resemble sporotrichosis, and the etiologic agents are not related. The organisms involved are probably less virulent than the usual agents of chromoblastomycosis and only rarely are able to evoke a disease process.

The clinical picture most commonly recorded is an abscess which is cutaneous, subcutaneous, or intramuscular (Fig. 9–6, *A, B*). This may evolve into a hard, cystlike process of several centimeters in diameter, covered by a raised, thickened epidermis (Fig. 9–6, *A*). Occasionally the cyst may ulcerate and ex-

trude pus containing brown pigmented hyphae and sclerotic cells (Fig. 9–12, *J*). This is especially true when *E. jeanselmei* or *W. dermatitidis* is the etiologic agent. In most cases the regional lymph nodes are not involved, and dissemination is very rare. Aspiration of the lesions sometimes results in chronic fistula formation. In one case involving *W. dermatitidis*, the subcutaneous mass extended over most of the thigh, although it is usually contained in a fibrous cyst (Fig. 9–7). The patient was a diabetic, and the lesion started at the site of an insulin injection. Several such cases of phaeomycotic cyst and cystlike infections have been seen, usually in debilitated patients. In the past they were referred to as chromomycosis, cystic chromomycosis, and phaeosporotrichosis. Such lesions may develop further if the patients become more debilitated. Dis-

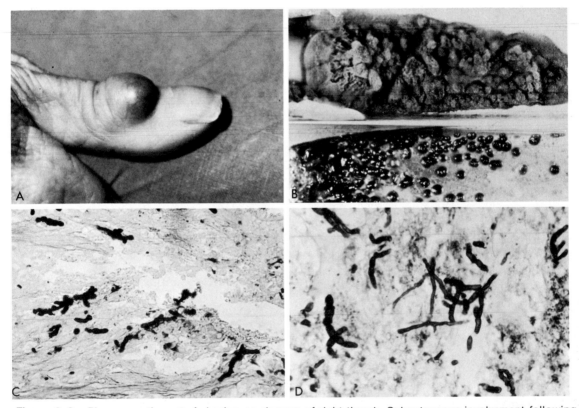

Figure 9–6. Phaeomycotic cyst. *A*, Lesion on dorsum of right thumb. Subcutaneous involvement following nonpenetrating injury. The lesion is fluctuant, tender, blue-gray, with no connection to surface. *B*, Culture from lesion identified as *Wangiella dermatitidis*. The lower tube demonstrates the initial moist "black yeast" or *Phaeococcomyces* state at 12 days. The upper tube shows the dark hyphal (dematiaceous) overgrowth at four weeks. (Courtesy of K. Greer. From Greer, K., G. P. Gross, et al. 1980. Cystic chromomycosis (chromoblastomycosis) due to *Fonsecaea dermatitidis*.[41]) *C*, Distorted mycelium and rounded cells from a 2-cm-wide, deep subcutaneous cyst in a patient with chronic leukemia. There was no sinus track to the surface. Patient treated with surgical excision and 5 fluorocytosine. *W. dermatitidis*. (Courtesy of A. Fathizadeh.) GMS. ×450. *D*, *Phialophora parasitica* from a deep dermal abscess. The patient had a renal transplant and was on steroids. (From Ajello, L., L. K. Georg, et al. 1974. Mycologia *66*:490–498.)

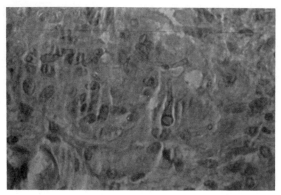

Figure 9–7. Phaeomycotic cyst. Distorted hyphae and round bodies in a phaeomycotic cyst. The subcutaneous mass extended over the thigh of the patient, who was a diabetic. The lesion started at the site of insulin injections. (Courtesy of G. Hambrick.)

seminated systemic infections may begin as cystic disease[31] or as cutaneous colonization. In this form, the disease — now called phaeohyphomycosis — is often fulminant, with a fatal outcome. Severe debilitation predisposes to this clinical manifestation. With the increasing use of macrodisruptive

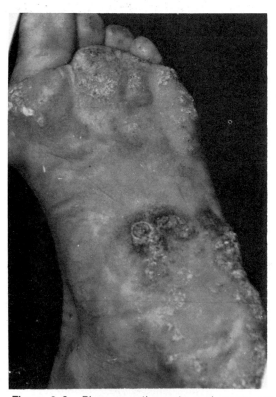

Figure 9–8. Phaeomycotic cysts and cutaneous colonization developing into chromoblastomycosis at the site of repeated injury and contact with soil. Chronic verrucous lesions and a purulent lesion resulting from bacterial superinfection. (Courtesy of F. Battistini.)

medications (long-term steroids, immuno-suppressives, cytotoxins), more nonspecific or opportunistic phaeohyphomycoses will be observed.

Many transition forms among these clinical diseases exist, adding confusion to the already confused literature. In patients of reasonable health but with repeated injury and soil exposure, several types of dematiaceous fungal colonization and infection may coexist (Fig. 9–8).

Phaeomycotic Cutaneous Colonization

Colonization of skin areas by dematiaceous fungi is a regularly encountered observation (Fig. 9–9). These may be in cracked, fissured areas of the sole of the foot, ulcerations of various causes, and lichenified or eczematized skin. Often, repeated scratching of an area will lead to transient colonization by fungi. In most cases such colonizations are completely benign; however, it has been suspected that they may contribute to the pathology of the lesion, if only as a primary irritant. Lazo[55] and others reported on such colonization by dematiaceous fungi, sometimes in combination with yeasts and other fungi. Application of topical antifungal agents and institution of good hygiene lead to clinical improvement and disappearance of the lesions. The commonly reported agents are *Alternaria, Curvularia, Cladosporium,* and *Aureobasidium*.

Under certain circumstances and when particular fungal species are involved, phaeomycotic colonization may be the beginning stages of chromoblastomycosis or systemic or cerebral phaeohyphomycosis. This must be kept in mind before concluding that such a colonization is completely benign and without consequence.

The differential diagnosis of phaeomycotic cyst includes the gumma of tertiary syphilis, sebaceous cyst, foreign body granuloma, tendon sheath ganglion, and aberrant sporotrichosis. Demonstration of pigmented fungi in material from lesions and their isolation in culture establish the diagnosis.

PROGNOSIS AND THERAPY

Chromoblastomycosis usually remains localized and will not debilitate the patients if left untreated. Secondary infection resulting in lymph stasis and elephantiasis may inter-

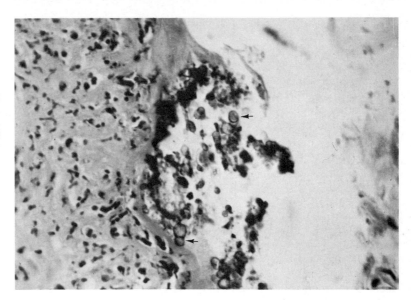

Figure 9–9. Phaeomycotic cutaneous colonization. This was an incidental finding on several eczematous lesions on the foot. Other lesions were not colonized. In the lower portion of the cyst some round, thick-walled brown cells can be seen (arrows). *Aureobasidium pullulans.* In similar lesions from other patients *Cladosporium* sp., *Alternaria* sp., etc. are frequently cultured. Conidia are often seen in histopathologic sections.

fere with normal activity. There is also a potential hazard of systemic spread, especially to the brain, depending on the species involved.[4]

In the early stages of the disease, the most reliable therapy is surgical excision or electrodesiccation, or cryosurgery.[77a] Most cases seen in clinical practice are well advanced and require medical management. The older modes of therapy included administration of iodides (1 to 9 g per os per day for several years), iontophoresis with copper sulfate,[62] and vitamin D injections, among others. There is a report of successful therapy using surgery and compresses soaked in the fungicide sodium dimethyl dithiocarbamate. Two newer approaches to treatment with drug therapy have received considerable attention. The first is the use of amphotericin B, either intralesionally[99] or topically as a lotion containing 30 mg per ml t.i.d.[47] Intravenous administration has also been used with success. A second promising drug for treatment is thiabendazole. This has been used in a series of studies in Central and South America.[7, 8] In the first report, 2 g per day of the drug was given orally, supplemented by local application of additional drug in DMSO. In a second report, the oral dose was not well tolerated, and only local applications in water-soluble base (Velvachol) were used.

At present the treatment of choice for chromoblastomycosis and other infections caused by dematiaceous fungi is 5-fluorocytosine.[59, 73, 96] In a series of 12 cases, Lopes found improvement or cure in almost all patients receiving 4 doses per day of 100 mg per kg.[58] Gonzales Ochoa[39] has successfully used this drug to treat two advanced cases of the disease. The regimen was 5 g (500 mg tablets) per day for 220 days in one case and 88 days in the other. A dose of up to 10 g per day is now recommended. Morison et al.[70] were successful in treating extensive disease caused by *F. pedrosoi* with a schedule of 100 mg per kg for two weeks, 200 mg per kg for two weeks, and 150 mg per kg for five months. Treatment failure has been noted, especially in long-standing disease.[59] These responded to thiabendazole.

Treatment of cystic chromomycosis has usually consisted of surgical removal. Sometimes contamination of the surrounding tissue during the procedure results in recurrence and spread of the infection. Excision and debridement followed by oral 5-fluorocytosine therapy may be recommended in cystic disease or in cases of significant colonization. Local thermal therapy in the form of a battery driven "pocket warmer" was useful in one case.[101]

The prognosis and seriousness of colonization or infection by dematiaceous fungi depend largely on the fungus species and the general health of the patient. For this reason culture and identification of the fungus are very important. Several species, notably *Cladosporium bantianum, W. dermatitidis,* and, to a lesser extent, *P. verrucosa* and *F. pedrosoi,* have been found in cerebral phaeohyphomycosis. The latter two species have also occasionally been responsible for widely disseminated disease in normal hosts. All these species plus a few others have caused systemic infections in compromised patients. On the other hand, secondary colonization of cutaneous ulcers or fissures by relatively nonvirulent *Alternaria* or *Cladosporium* species in an otherwise normal host is usually an incidental finding. Other significant infections caused by dematiaceous fungi are reviewed in Chapter 26.

PATHOLOGY

Chromoblastomycosis is generally confined to the body surfaces. The lesions range from erythematous macules to warty, hyperkeratotic, pedunculated papules. Not infrequently the surgical pathologist discovers "Medlar bodies" or "copper pennies" in a specimen marked epidermoid tumor or carcinoma. In histologic section, the nonulcerated lesion that is not secondarily infected shows a very high degree of pseudoepitheliomatous hyperplasia. It is not as extreme as in cutaneous blastomycosis or coccidioidal granuloma, but it is considerable and could be termed extreme acanthosis (Figs. 9–4, 9–5). This extreme acanthosis appears to be a cellular response to the fungi contained in the microabscesses and has been called transepithelial elimination.[6] Miliary abscesses and poorly defined granulomatous nodules are seen in this disease, as in the latter two diseases. Changes consistent with chronic fibrosing inflammation are seen between the nodules. Numerous neutrophils cluster in the centers of the granulomata. Foreign body giant cells are present, as well as lymphocytes, plasma cells, and other cells of the chronic inflammatory reaction. The histologic picture is that of a foreign body granuloma interspersed with microabscesses. The fungus is visible in unstained as well as stained sections. The chestnut or light yellow-brown bodies appear as rounded yeasts (5 to 15 mm in diameter), with thick, planate, septal walls, which are often grouped or in chains (Fig. 9–5). These are the sclerotic cells of the fungus.[85] Sometimes they lie within giant cells. In early lesions in which the organism is found mostly in the epidermis and stratum corneum, hyphal strands are seen more commonly than sclerotic cells. This is also observed when the organism metastasizes to other body sites. In Giemsa-stained sections, the fungus has a greenish-brown color. Since the organism is readily seen in hematoxylin and eosin–stained material, special fungus stains are not needed.

Ulceration and secondary bacterial invasion alter the histologic picture of chromoblastomycosis. Instead of the chronic inflammatory and granulomatous picture, an acute or chronic pyogenic reaction is seen.

Phaeomycotic cyst usually presents as a hard, defined, subcutaneous mass of several centimeters diameter. There may be marked acanthosis above the lesion. The lesion itself is a cyst limited by a thick wall of hyaline connective tissue. It is lined with granulomatous inflammatory tissue, with a center of necrotic debris. In the necrotic center and the inner walls of the cyst, abundant brown pigmented fungi are seen. Their morphology is quite variable. Hyphal elements ranging from 3 to 8 μ are common, with a few distorted bodies of up to 20 μ in diameter. Small, yeastlike rounded bodies and moniliform hyphae of gradually decreasing diameter may be present. Sometimes only compact masses of hyphae similar to those seen in phaeohyphomycosis of the brain are found.

Some clinical and histologic variants occur. Fistulous tracts lined with loose hyphal strands that resembled mycetoma have been reported. Such lesions lacked true grains,[51, 89] and *E. jeanselmei* was recovered. This organism is frequently seen in true black-grained mycetoma, however, and the aforementioned cases represent yet another clinical variation: a pseudomycetoma.

ANIMAL DISEASE

Natural infection of animals with the organisms that cause chromoblastomycosis occurs but is rarely reported. A few cases have been found in dogs, cats, horses, and frogs. In a case involving a horse, the lesions were described as similar to those seen in human verrucous chromoblastomycosis. Biopsy revealed brown-pigmented, hyphal strands and yeastlike bodies. These were contained in granulomatous lesions consisting of chronic and acute inflammatory cells.[86] The organism was identified as a *Hormodendrum* species. In an interesting study of toads (*Bufo* sp.) in Colombia, Velasquez and Restrepo[95] reported that natural infection occurred in 12 of 66 specimens examined. Granulomatous lesions were located in the liver, lungs, and kidney. Numerous sclerotic cells similar to those seen in human verrucous chromoblastomycosis were noted. The organisms isolated were *F.*

pedrosoi (three strains), *E. jeanselmei* (three strains), and *C. carrionii* (seven strains).[95] Other reports of the organisms in toads[37] and frogs[21] have been recorded, as well as epidemics involving both toads and frogs[9, 81] (Fig. 9–10, *A, B*). One of the latter occurred in Lansing, Michigan, and involved *F. pedrosoi*. Schmidt and Hartfiel[82] have published a review of the disease in amphibians.

Several cases of infection in cats have also been reported. Lesions were described as an ulcerated mass[44] or as a granulomatous dermatitis.[43] *E. jeanselmei* was most frequently recovered. *W. dermatitidis* has been isolated from asymptomatic bats in Brazil.[78]

Experimental infection of animals has regularly produced the verrucous type of chromoblastomycosis (Fig. 9–10, *C, D*). Intracutaneous injection results in an ulcerating local lesion that heals slowly[13] but sometimes produces chronic verrucous lesions resembling human chromoblastomycosis. Systemic infection involving most internal organs can be produced by intravenous inoculation of the fungus, particularly *P. verrucosa*. Brain abscesses have developed following injection with *C. bantianum, F. pedrosoi,* and *F. compacta. F. dermatitidis* is also neurotropic.[48] *C. carrionii* is not neurotropic, and this characteristic is used to distinguish it from *C. bantianum*.

BIOLOGICAL STUDIES

Few biological studies on the dematiaceous fungi have been forthcoming. In early studies, deSilva-Hutner[85] induced sclerotic cell formation in a synthetic "lymph." The comparative taxonomy of the etiologic agents has been reviewed by McGinnis,[66] and scanning electron microscope studies on conidiogenesis were carried out by Cole.[24]

In experiments on thermotolerance, Padhye et al.[75] found all strains of *W. dermatitidis* able to grow at 40°C but no higher. Roberts and Szaniszlo[79] concluded that multicellularity was the result of the inhibition of budding and separation, thus leading to sclerotic bodies consisting of several cells. Their work was based on temperature-sensitive mutants. Few experimental human infections have been recorded[12]; however, in a laboratory accident, infection by *C. carrionii* resulted in an erythematous papulopustule.[19] Curettage of the lesion was followed by electrodesiccation, and a complete cure was achieved.

IMMUNOLOGY AND SEROLOGY

Serologic evaluation of patients with chromoblastomycosis does not yet appear to be a practical adjunct to diagnosis. In one series, precipitating antibodies were found in 12 of 13 patients with the disease. These appeared to be specific for the causative agent. By cross-precipitation, it was shown that *F. pedrosoi, F. compacta,* and *P. verrucosa* were closely related and distinct from *C. carrionii* and *E. jeanselmei*.[16] Conflicting results were found in another study.[28] Precipitins begin to disappear following successful treatment. Iwatsu et al.[47b] isolated a skin test–reactive substance that caused a reaction in patients infected with *F. pedrosoi*. Although direct microscopic demonstration of the fungus is sufficient for diagnosis, the rise and fall of precipitin titers would be of prognostic importance following the institution of therapy.

Several studies have been made of the antigenic relationships between the various etiologic agents of chromoblastomycosis. So far, no consistent patterns have been found that firmly delineate the species involved. In an early study Conant[27] demonstrated by the complement fixation test that several strains of *P. verrucosa* were antigenically similar. Using a fluorescent antibody technique, Gordon and Al-Doory[40] were able to distinguish *C. carrionii* from *C. bantianum,* but there were many cross-reactions between conjugates prepared from the pathogenic species, *F. pedrosoi, F. compacta, W. dermatitidis,* and soil saprophytes (*Cladosporium* sp.). Cooper and Schneidau[28] demonstrated antigenic relationships between several genera using immunodiffusion and immunoelectrophoresis techniques. They found common antigens among *Cladosporium, Phialophora,* and *Fonsecaea,* but concluded that *Cladosporium* and *Phialophora* are more closely related to one another than to *Fonsecaea*.

Natural resistance to infection is high in normal individuals, and disease occurs only under circumstances of repeated exposure and trauma. Hypersensitivity has been demonstrated in patients with verrucous chromoblastomycosis. It diminishes after clinical cure. The relation of this to resistance to reinfection is not known.

LABORATORY IDENTIFICATION

Direct Examination. Laboratory diagnosis of chromoblastomycosis by direct exami-

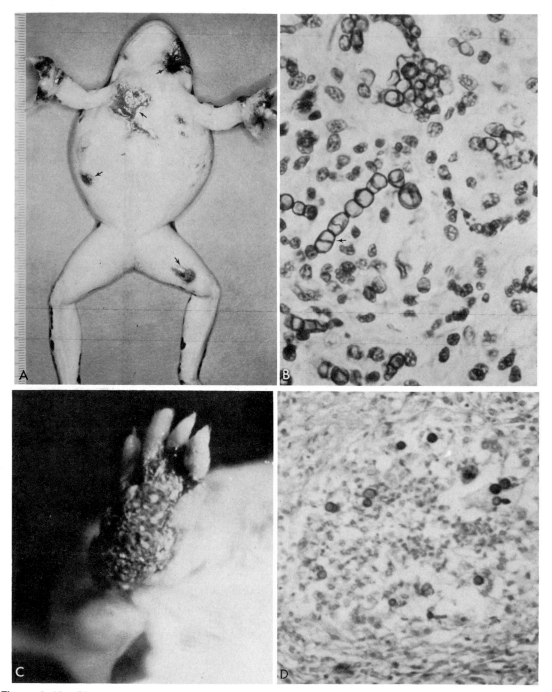

Figure 9–10. Phaeomycotic disease in frogs. *A*, Cutaneous lesions of a frog (*Rana pipiens*). The organism was *Fonsecaea pedrosoi*. *B*, Sclerotic cells in the kidney of a frog. (Courtesy of E. S. Beneke.) *C*, Experimental chromoblastomycosis in mice mimicking human infection. The organism was *Fonsecaea pedrosoi*. *D*, Section of lesion with demonstration of planate-dividing sclerotic cells. (Courtesy of A. González-Ochoa.)

nation of suspicious material is relatively easy, but it must always be confirmed by culture. This is important in prognostic evaluation, as some species disseminate to the brain and other organs. Skin scrapings, crusts, aspirated debris, and biopsy and excised material can be examined in a potassium hydroxide mount. Brown-pigmented, branching, hyphal strands (2 to 6 μ wide) are easily seen in skin scrapings, crusts, and aspirates. In pus from cysts, very distorted hyphae (3 to 8 μ wide) and pleomorphic brown bodies (up to

20 μ in diameter), which sometimes appear to bud, may be found. In granulation tissue obtained by curettage, excision, or biopsy from verrucous chromoblastomycosis, the sclerotic bodies predominate. They are thick-walled, brown-pigmented, exhibit planate division, and may be grouped in a chainlike formation. The size varies from 4 to 12 μ. Material from mycotic lesions of the brain can be similarly examined. Hyphal forms predominate in such cases.

Culture Methods. Since the agents of chromoblastomycosis are not inhibited by cycloheximide (Actidione) or chloramphenicol, selective media using these antibiotics may be used. Cultures should be kept at 25°C for at least six weeks. In contrast to the saprophytic species of the soil, most of the commonly isolated agents of chromoblastomycosis grow slowly. The specific identification of the organism is taxonomically taxing. It depends on conidia types, percentage of conidia types present, fine details of conidia production, and perhaps the position of the planets. Part of the problem is that there are still no clearly delineated taxa for these organisms. Three general types of conidiation are found in this group (Figs. 9–11 and 9–12):

1. *Phialophora* type. In this type there is a distinct conidiogenous cell called a phialde, which occurs terminally or along the mycelium (Fig. 9–11, *B*). This structure is generally flask-shaped and has a rounded, oval, or elongate base, a constricted neck, and an opening that may have a flaring collarette or lip. The conidia are formed at the end of the flask (semiendogenous conidiation) and are extruded through the neck. They may accumulate around the neck area, giving a picture of "flowers in a vase" (Fig. 9–12,*D,E,F*). The conidia are oval, smooth-walled, and hyaline, and have no attachment scars.

2. *Rhinocladiella (Acrotheca)* type. The conidiophores are simple and sometimes not differentiated from the vegetative hyphae. Oval conidia are produced irregularly on the top and along the sides of the conidiophore (acropleurogenous conidial arrangement) (Fig. 9–11, *C*). They are usually single and do not bud. However, occasionally chains formed by budding and transition to the *Cladosporium* type conidiation are seen. When the conidia are detached, small bud scars can be found on the conidiophore, and there is a single scar on the conidium at the attachment site. Hyphae, conidia, and conidiophores are a pale greenish brown.

3. *Cladosporium (Hormodendrum)* type. In this type of conidiation there is a simple stalk that serves as a conidiophore (Fig. 9–11, *A* and *A'*). It is usually slightly enlarged at the distal end. Two or more conidia are formed at the tip. These in turn bud and form secondary conidia at their distal poles (acropetalous conidiation). Conidiation continues, with the formation of long chains. The youngest conidium is the one distal to the conidiophore. Detached conidia show a thick-

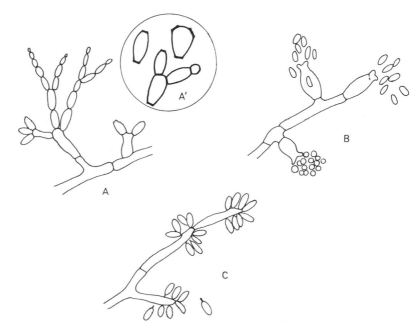

Figure 9–11. *Chromoblastomycosis.* Conidiation types. *A, Cladosporium* type conidiation consists of a simple conidiophore that buds to form conidia. In *A'*, note the thickened scars or disjunctors. A "shield" cell is shown that has three disjunctors. *B, Phialophora* type conidiation. Vase-shaped phialides give rise to conidia with no attachment scars. *C, Rhinocladiella* type conidiation. The conidia are borne at the ends and sides of conidiophores and have one scar of attachment.

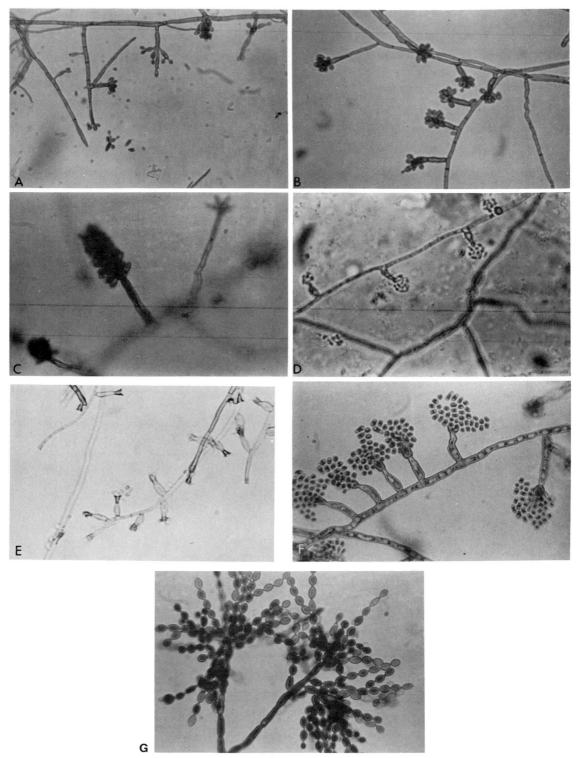

Figure 9–12. Chromoblastomycosis. Conidiation types. *A, F. pedrosoi. Cladosporium* type and *Rhinocladiella* type conidiation. *B, F. pedrosoi.* Several young conidiophores, showing *Rhinocladiella* conidiation. ×400. *C*, Older conidiophores, with conidia coming down sides of conidiophore. ×400. *D, F. pedrosoi.* Series of phialides showing "flowers in a vase" conidiation of the *Phialophora* type. *E, P. verrucosa.* Shows various sizes and morphology of phialides. *F, P. verrucosa.* Accumulation of conidia around flaring lip of phialide. ×400. *G, Cladosporium carrionii.* Long, flexuose chains of elliptical conidia. ×400.

Illustration continued on opposite page

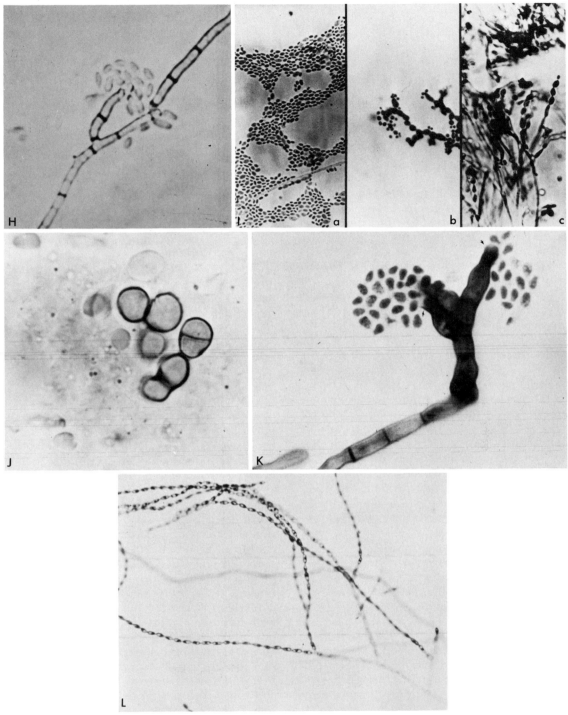

Figure 9–12 *Continued.* *H, Wangiella dermatitidis* phialide with a mass of conidia. This culture had numerous phialides of this morphology and "moniliform" hyphae, as in Figure 9–12, *I.* (Courtesy of L. Ajello.) ×600. *I,* Polymorphism in *Wangiella dermatitidis. a,* "black yeast" or *Phaeococcomyces* state. On initial isolation, *W. dermatitidis* is a glistening black or dark brown colony resembling the initial growth of *E. jeanselmei* and *E. spinifera. b,* Phialides produced after reversion to mycelial growth. *c,* Chains of acropetally arranged conidia. (Courtesy of G. Cole.) *J, W. dermatitidis* planate-dividing sclerotic bodies in a case of cystic phaeohyphomycosis.[41] (From Greer, K. E., G. P. Gross et al. 1980. Arch: Dermatol. *115*:1433–1434. Cystic chromomycosis (chromoblastomycosis) due to *Fonsecaea dermatitidis.*) *K,* Phialides of *F. pedrosoi* (arrows). *L,* Long chains of elliptical conidia in culture of *Cladosporium bantianum.*

ening or scar, called a disjunctor, where they were connected to the other conidia. All of the conidia within the chain will have two or three (if it formed a branch) disjunctor scars, except the terminal conidium, which will have one. A conidium having three disjunctors is described as "shield-shaped." It is thus possible, by examining the individual conidium carefully, to distinguish the type of conidiation that was involved. Hyphae, conidia, and conidiophores are usually dark oliveaceous brown. *P. verrucosa*, *P. richardsiae*, and *P. parasitica*[1] conidiate almost exclusively by the *Phialophora* type of conidiation. Some strains, however, have been said to show *Cladosporium* and *Rhinocladiella* type conidiation in addition to phialides.

W. dermatitidis, *E. jeanselmei*, and *E. spinifera* grow initially as elliptical aseptate yeasts. Mycelium gradually forms as the colony ages. *F. pedrosoi* (Fig. 9–12) and *F. compacta* produce primarily *Cladosporium*-type conidiation, some *Rhinocladiella*, and, rarely, *Phialophora* type conidiation. The type and degree of conidiation are influenced by the media on which the organisms grow. In "synthetic lymph," even the sclerotic cells found in lesions can be produced *in vitro*.[85] *C. bantianum* and *C. carrionii* (Fig. 9–12) conidiate exclusively by the *Cladosporium* type of conidiation. (*C. carrionii* has been reported to produce a few phialides on Borellis's lactrimal medium.[12]) They can be differentiated by their thermotolerance and by the neurotropism exhibited by *C. bantianum* in experimentally infected mice.

Although many studies have been made of this carbon and nitrogen utilization, physiologic differentiation of the species is not yet well standardized. Fuentes[38] has shown that the pathogenic species do not liquefy gelatin, coagulate milk, or digest starch. These are useful in distinguishing the etiologic agents of chromoblastomycosis from soil saprophytes and air contaminants. By use of biochemical, immunological, and morphological techniques, Padhye[74] and McGinnis[66] were able to demonstrate that *E. jeanselmei* and *P. gougerotii* were the same species.

MYCOLOGY

The taxonomy of the agents of chromoblastomycosis has been one of the most confusing in the field of medical mycology. At the present writing this problem is still unsettled, so that choosing the terminology to be used is arbitrary at best. Some of the current schools of thought will be presented and then discussed. The nomenclature of these organisms as reviewed by McGinnis[66, 66a] will be used in this text.

The type species of the genus *Phialophora* is *P. verrucosa*. Since its conidiation is exclusively of the *Phialaphora* type, there appears to be no difficulty with its nomenclature.[25] This is also true of species that demonstrate only *Cladosporium* type conidiation, e.g., *C. bantianum*. The other agents of chromoblastomycosis have usually been described as belonging to the genus *Hormodendrum*. This genus, however, has been reduced to synonymy with the genus *Cladosporium*. Negroni in 1936 described a new genus, *Fonsecaea*,[71] which, as redefined by Carrión, included *F. pedrosoi*, *F. compacta*, and *F. dermatitidis*.[23] The latter organism has now been placed in a new genus *Wangiella*.[65] The erection of the genus *Fonsecaea* was necessary because the species concerned produce various types of conidia by various methods, a property which excludes them from the genus *Cladosporium* (*Hormodendrum*), where they were originally placed. Emmons[35] proposed emending the definition of the genus *Phialophora* so that it could include the organisms now constituting the genus *Fonsecaea*. However, since these fungi only rarely produce phialides, this transfer seems unwarranted. Schol-Schwarz[83] has stated that all the *Fonsecaea* species and a few of the *Phialophora* species should be placed in the genus *Rhinocladiella*, but there are many objections to this suggestion.[48] As it is now defined, this latter genus could not accommodate the species of *Fonsecaea* again because of the variety of conidia produced. It is obvious that the organisms now included in *Fonsecaea* are closely related. Therefore, in order not to destroy already well-defined taxonomic groups by emendation, the designation *Fonsecaea* has been retained for use in this text.[66a] It is also possible that these particular species are intermediate forms between the two genera *Phialophora* and *Cladosporium;* thus this designation *Fonsecaea* may represent a natural phylogenetic classification. Indeed, there is serologic evidence for this progression.[28] Until such time as the perfect states of these fungi are discovered or consistent distinguishing characteristics are found, the arguments are moot.

Detailed descriptions of the etiologic agents of chromoblastomycosis are given below and summarized in Table 9–1. *E. jeanselmei* is described in the chapter on mycetoma, a disease with which it is more frequently associated.

Table 9–1 *Characteristics of the Agents of Chromoblastomycosis and Phaeomycotic Cyst*

Species	*Dermo-tropic*	*Neuro-tropic*	*Growth at 37°C*	*Gelatin*	*Primary Culture — Yeast*	*Growth Rate*	*Conidia Formation*
Fonsecaea pedrosoi	+	±	+	−	−	Slow	*Cladosporium* primarily in short chains or, less commonly, from tips and sides of conidiophore; rarely from phialides. *Rhinocladiella* type conidiation often predominates.
F. compacta	+	−	+	−	−	Slow	Same as above, but *Cladosporium* conidial heads reduced and compact
Wangiella dermatitidis	+	+	+	−	+	Slow	Yeast forms early; abstrictions from tip and sides of conidiophores; semiendogenously from elongate phialides, also from peg-like phialides
Phialophora verrucosa	+	±	+	−	−	Slow	*Phialophora* type from flaring cups on flask-shaped phialides; other conidiation types rare or absent
P. richardsiae	+	−	+		−	Rapid	Phialoconidia semiendogenously; long phialides with saucer-shaped lips or collarettes
Exophiala jeanselmei	+	−	±	±	+	Slow	Yeast at first; abstrictions from sides and tips; semiendogenously from elongate annellides
E. spinifera	+	−	±		+	Slow	Abstrictions from tips, sides, and along mycelium; conidiophores spikelike, not flared; bears conidia semiendogenously
Cladosporium carrionii	+	−	±	−	−	Slow	Long chains of conidia from branched conidiophore only; acropetalous; phialides on some media
C. bantianum	±	+	+	−	−	Slow	Similar to above, differentiation by thermotolerance (43°C), neurotropism, irregular spore size, and sparsely branching conidial chains of 35 or more elliptical conidia
C. species	−	−	−	+	−	Rapid	Same as *C. carrionii;* usually not thermotolerant or pathogenic for animals

Fonsecaea pedrosoi (Brumpt) Negroni 1936

Synonymy. *Hormodendrum pedrosoi* Brumpt 1922; *Acrotheca pedrosoi* Fonseca et Leão 1923; *Trichosporium pedrosoi* Langeron 1929; *Gomphinaria pedrosoi* Dodge 1935; *Hormodendroides pedrosoi* Moore et Almeida 1936; *Phialophora pedrosoi* (Carrion) Redelli et Ciferri 1942; *Harmodendrum algeriensis* Montepellier and Catanei 1927; *Trichosporium pedrosianum* Ota 1928; *Hormodendrum rossicum* Merlin 1930; *Botyroides monophora* Moore et Almeida 1936; *Phialoconidiophora guggenheimia* Moore et Almeida 1936; *Hormodendrum japonicum* Takahashi 1937.

Colony Morphology. In culture *F. pedrosoi* grows very slowly, producing a black-brown, grey-black, olive-grey, or black colony. The texture is velvety to fluffy, and the surface varies from flat to heaped and folded. Radiations and zonations occur in some strains. The colony characteristics may vary considerably between isolants, and the organism is indistinguishable macroscopically from *P. verrucosa.* (Fig. 9–14, *A*).

Microscopic Morphology. All three types of conidiation exist in this species, the proportion varying with the strain and the media used for growth (Figs. 9–11 and 9–12, *A–D*). The *Cladosporium* type sometimes predominates, with frequent admixture of the *Rhinocladiella* type and an occasional phialide (Fig. 9–12, *K*).

In the initiation of conidiogeny often the first conidia are produced holoblastically from an elongate conidiogenous cell. The conidia arise from short denticles (Fig. 9–13). The conidiogenous cell then swells, extends, and produces more conidia on denticles (sympodially). The primary conidia are elongate, cylindrical to obovoid (Fig. 9–12, *C*). At this stage the conidiation appears to be primarily of the *Rhinocladiella* type. Secondary conidia may arise from the apex of the primary conidia, however (Fig. 9–13, *D*, 1, 2). Successive budding forms short *Cladosporium*-like chains. The species is quite polymorphic, and variations in these patterns are to be expected. Few conidia are produced on Sabouraud's glucose agar, but on deficient media,

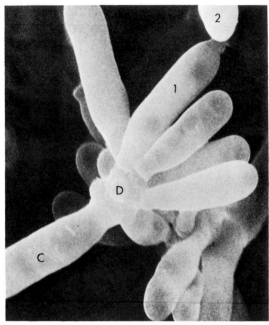

Figure 9–13. Conidiogenesis in *Fonsecaea pedro-soi*.[24] The conidiogenous cell (*C*) has given rise to conidia on denticles (*D*). The primary conidia (1) then buds to form secondary conidia (2). This conidiation is similar to that observed in *Rhinocladiella* species except that the fertile area is usually confined to the upper part of the conidiogenous cell. This type of conidiation is usually the dominant type in most isolants of *F. pedrosoi*. Scanning electron micrograph. (Courtesy of G. T. Cole.) ×7650.

such as cornmeal agar, conidiation is enhanced. As distinct from species of *Cladosporium*, in *F. pedrosoi* conidia are borne in short rather than long chains from the distal end of the conidiophore. The conidiophores are of varying lengths. The conidia are elliptical, cylindric to obovoid, and 1.5 to 3 μ in width by 3 to 6 μ in diameter. They exhibit two dark, thick scars (disjunctors) where they were attached in chains. "Shield cells" with three scars may be found among conidia on the terminal end of conidiophores if there were two fertile poles on the cell. The last cell of the chain has only one scar. *Rhinocladiella*-type conidiation is formed from bare hyphae or from knobby, denticulate conidiophores. The conidia are arranged acropleurogenously (at the tip and along the side of the conidiophore) and are attached by a short apiculus. The denticle remains on the conidia when it is detached. Conidia are usually found on the upper end of the conidiogenous cell. This end becomes swollen and extends sympodially (Fig. 9–13). The *Phialophora*-like conidia are produced from phialides similar to those of

P. verrucosa, but they may be scant or lacking. Sclerotic bodies and cleistothecia-like bodies have also been described. They are hard and round and have spinelike mycelia arising from the peripheral wall.

Fonsecaea compacta (Carrión) Carrión 1940

Synonymy. *Hormodendrum compactum* Carrión 1935; *Phialoconidiophora compactum* (Carrión) Moore et Almeida 1936; *Phialophora compactum* (Carrión) Redelli et Ciferri 1942.

Colony Morphology. This very slow growing fungus produces a folded, heaped, brittle colony, which is dark olive-black and develops a brownish-black fuzz with age. It is this fuzz that contains most of the conidia. The colony is essentially indistinguishable from that of *F. pedrosoi*.

Microscopic Morphology. The conidiophores are similar to those of *F. pedrosoi* except for the more compact form of the conidial heads. The conidia are subspherical to ovoid, $1.5 \times 2 \times 3$ μ in size, whereas the spores of *F. pedrosoi* are elliptical and larger. The other types of conidiation are the same as seen in *F. pedrosoi* and *F. compacta* and may only represent a variant of the latter species.[66a] Hundreds of isolants of *F. pedrosoi* are known, whereas only a few of *F. compacta* have been found (Puerto Rico and Tennessee). There is a recent report of its isolation in Martinique.[18] Some authors consider this species to represent a dysplastic strain of *F. pedrosoi*.[13a]

Wangiella dermatitidis (Kano 1934) McGinnis 1977

Synonymy. *Hormiscum dermatitidis* Kano 1934[49] and 1937[50]; *Hormodendrum dermatitidis* (Kano) Conant 1953; *Phialophora dermatitidis* (Kano) Emmons 1965; *Exophiala dermatitidis* (Kano) de Hoog 1977.[30]

This species is sufficiently different in its conidiation from other fungi[65] that McGinnis has placed it in a separate genus. It has not been isolated from chromoblastomycosis but has been recovered from endocarditis,[36] phaeohyphomycosis,[49, 50] phaeomycotic cyst,[41, 45, 47a] brain disease,[84] opportunistic infections,[93] from a frog,[37] and from bat intestinal contents.[78] It has been isolated from several natural sources such as palm tree, plant debris and wasp nests.[27a]

Colony Morphology. There are two colony types. The colony which grows on initial isolation is moist, glistening, yeastlike in appearance, and olive to black in color (Fig. 9–6,*B*). At this stage it is identical to *E. jeanselmei*. This is the "*Phaeococcomyces*" or "black yeast" stage of growth. After four to five weeks, tufts of submerged or twisted ropy strands of mycelium develop around the colony edge. These give rise to olive-gray aerial mycelia (Fig. 9–6, *B*). In time, patches of the same type of mycelium appear over the surface of the colony (Fig. 9–14, *C*), with occasional areas of white hyphal growth.

Microscopic Morphology. The initial moist colony is composed of budding yeast-like cells similar to those of *Phaeococcomyces* (Fig. 9–12, *I*). When mycelia appear, phialides are produced which are similar to those seen in *P. verrucosa*. However, the lips of the phialides are usually inconspicuous (Fig. 9–12, *H*). The phialides may also be short and round (Fig. 9–15, *C*), and conidia may bud secondarily as yeastlike cells (Fig. 9–15, *A*). Abstriction scars from multiple budding cells can be seen (Fig. 9–15, *E*). Moniliform hypha is also found.[24] Conidiophores with pleurogenously arranged conidia similar to those found in *F. pedrosoi* are also seen (Fig. 9–15, *B*). The degree of polymorphism in this species is influenced by pH and carbon source.[65] Conidiation of all types is usually sparse. *W. dermatitidis* has been isolated from many cases of cystic and disseminated phaeohyphomycosis and is similar to the *Torula bergerii* reported from Canada. In cases of cystic disease (Fig. 9–12, *J*) sclerotic cells have been found in tissue, but only hyphal strands have been seen in infections of the brain and mucosa, and in opportunistic infections.

Phialophora verrucosa Medlar 1915

Synonymy. *Cadophora americana* Nannfeldt 1927; *Phialophora macrospora* Moore et Almeida 1936; *Fonsecaea pedrosoi* var. *phialophorica* Carrión 1940.

Colony Morphology. This species grows slowly as a dark olive-gray to black colony, which is initially dome-shaped and later becomes flattened.[25] Some strains are heaped and folded or have radial grooves (Fig. 9–14,*D*). A gray aerial mycelium eventually covers the colony, which is compact, tough, and leathery. On cornmeal agar, growth is scant but conidiation is enhanced.

Microscopic Morphology. Well-formed, flask-shaped or vase-shaped phialides are formed along the vegetative hypha. These are 3 to 4 μ wide and 4 to 7 μ in length and have a pronounced, flaring collarette. Within their neck, small, elliptical, thin-walled conidia, 1×3 to 2×4 μ in size, are formed semiendogenously. The conidia are produced in succession but are not attached in chains. They are surrounded by adhesive material, so they may accumulate at the lips of the phialide, forming a large ball. As each conidium is released from the phialide, it leaves a residue of wall material around the lip or collarette. In this way, the size of the lip is increased as each succeeding conidium is released, so that mature phialides have deeply pigmented, wide, flaring lips. Very rarely one can find examples of other types of conidiation in mature cultures. Conidia may be arranged at the tip (acrogenous) or at the tip and along the sides (acropleurogenous) of a conidiophore. Sometimes these conidia may bud to produce secondary and tertiary conidia, thus forming the arborescent heads characteristic of the *Cladosporium* type of conidiation. Cleistothecia-like bodies have been found in some strains, but they did not contain ascospores. The perfect stage remains unknown.

In studies by Zweibel and Wang,[104] most of the isolants from wood or natural sources had very deep collarettes, whereas those from human infections were shallow. There were also differences in the electrophoretic patterns of culture filtrate proteins. All isolants were pathogenic in mice, however. *P. verrucosa* is a very common isolant from chromoblastomycosis and also from phaeomycotic cyst, cerebral phaeohyphomycosis, and opportunistic infection. In cases of chromoblastomycosis it exists as a planate-dividing sclerotic cell; in other types of clinical infection it is found as distorted mycelial units and rounded forms.

Phialophora richardsiae (Melin et Nannfeldt) Conant 1937

Synonymy. *Cadophora richardsiae* Melin et Nannfeldt 1934; *Cadophora brunnescens* Davidson 1935; *Phialophora brunnescens* (Davidson) Conant 1937; *Phialophora calciformis* Smith 1962.

This organism was originally described as *Cadophora richardsiae* by Melin and Nannfeldt when isolated from rotting wood pulp. This species is a common lignicolous decay fungus.[27] It has been isolated only a few times from human disease.[91] The lesion was a subcutaneous, well-encapsulated cyst, with the

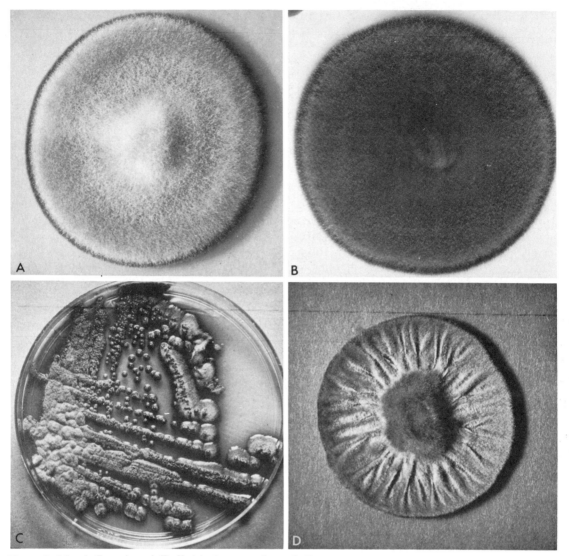

Figure 9–14.   *A, F. pedrosoi.* Fluffy black-gray colony. *B, F. compacta.* Fuzzy black colony. *C, W. dermatitidis.* Two colony types: glistening yeastlike colonies and transition to fuzzy mycelial form. On the left hand side some white cottony tufts are appearing. *D, P. verrucosa.* Olive-gray colony with radial grooves.

usual histologic morphology of a foreign body granuloma. The fungus was found in the central necrotic area and consisted of dark hyphal strands and a few budding cells.

Colony Morphology. The organism grows rapidly in culture, producing a woolly or tufted brownish-gray colony with concentric zonation. Conidiation is almost exclusively of the *Phialophora* type.

Microscopic Morphology. A simple tapering branch (1 to 2 μ at base, 2 to 10 μ in length) from the vegetative mycelia produces semiendogenously elliptical, hyaline conidia (1 to 2 × 2 to 4 μ). The tip of the phialide is simple, narrow, and barely discernible. As the phialide matures, the flaring saucer-shaped collar is produced, which is typical of

the genus. Also as the phialide matures, the conidia become thicker-walled, pigmented, and more spherical.

Exophiala spinifera (Nielsen et Conant) McGinnis 1977

Synonymy. *Phialophora spinifera* Nielsen et Conant 1968; *Rhinocladiella spinifera* (Nielsen et Conant) de Hoog 1977.

This species has been recovered once from a nasal lesion in an elderly woman. The lesion was a granuloma, in which the fungus was found to consist of spherical budding cells.[72]

Colony Morphology. The organism grows slowly, producing a moist, olive to black, yeastlike colony. It may remain yeast-

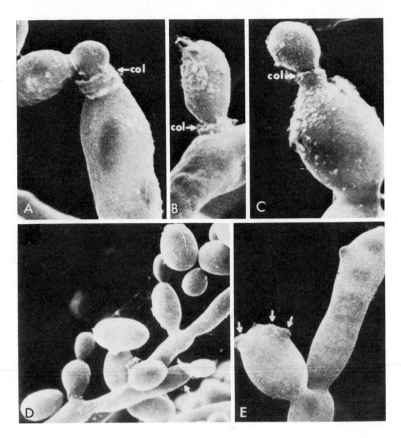

Figure 9–15. Polymorphous conidiogenesis in *Wangiella dermatitidis.* *A,* Cylindric phialide with short collarette is giving rise to a conidium. ×12,150. *B,* Peglike collarette on the side of a mycelium is giving rise to a conidium. ×12,150. *C,* Rounded, globose phialide with a conidium. ×11,200. *D,* Yeastlike conidiogenesis along a hyphal strand (pleurogenous) and near a phialide. *E,* Abstriction scars (*arrows*) at the points of budding from a fertile cell. ×4100. (From Cole, G. T. 1978. Conidiogenesis in the black yeasts. Proc. IV Int. Conf. Mycoses. P. A. H. O. Sci. Pub. 356, p. 69.[24])

like almost indefinitely on Sabouraud's glucose agar or develop short, naplike, mycelial elements over the surface of the colony. The growth is mycelial, and conidiation is enhanced when the organism is grown on cornmeal agar.

Microscopic Morphology. Yeasts of the *Phaeococcomyces* type are found in the initial colony. Heavy conidiation occurs on the hyphal strands. Masses of conidia are borne on short protuberances (apiculae) from the mycelium and from developing spinelike conidiophores. The latter are simple or branched, appear rigid, and are produced at right angles to the hyphae. The tip is smooth or extended and not flared. The conidia are 2.5 × 3.5 μ in size. Short, flask-shaped annellophores with flaring apical collars are also found. The conidia from these are 1.0 × 1.5 μ in size. Moniliform hyphae are also present.

Cladosporium carrionii Trejos 1954[92]

Synonymy. *Fonsecaea pedrosoi* var. *cladosporium* Simson 1946;[87] *Fonsecaea cladosporium* Cowell 1952.

This organism is a regular isolant from verrucous chromoblastomycosis in Austral-

ia.[56, 92] It is also commonly found in Venezuela and South Africa.[87] Isolants from phaeomycotic cysts and opportunistic infections have also been reported.

Colony Morphology. The organism grows slowly, producing a small, smooth or folded, dark olive-black, compact colony (3 to 4 cm after one month). The edges are entire and are marginated by black submerged hyphae.

Microscopic Morphology. Only the *Cladosporium* type of conidiation is regularly found. Elongate conidiophores produce long, flexuose, branching chains of conidia, 1.5 to 3.0 by 2.0 to 7.5 μ in size (Fig. 9–12,*A*). The conidiophore is usually quite distinct (macronematous) from the vegetative hyphae. The organism differs from most saprophytic *Cladosporium* species in its inability to digest gelatin. The morphologically similar species, *Cladosporium bantianum,* is differentiated by (1) neurotropism in experimental animal infections, (2) growth at 43°C, (3) less regular conidia size, and production of sparsely branching long chains of 35 or more elliptical conidia, and (4) more rapid growth. *C. carrionii* is (1) dermo- but not neurotropic, (2) limited in growth to 35 to 36°C, (3) regular in conidia size, and (4) very slow growing.

On lactrimel medium[12] some strains of *C. carrionii* produce phialides. This may necessitate renaming the organism as a member of the genus *Cladophialophora*.[13a]

Cladosporium bantianum (Saccardo) Borelli 1960

Synonymy. *Torula bantiana* Saccardo 1912; *Cladosporium trichoides* Emmons 1952.

This organism is the most commonly isolated species from phaeohyphomycosis of the brain. It has also recently been isolated from a case of chromoblastomycosis involving the knee and lower limb[73] and a cutaneous infection.[42] In the former case,[73] the isolant was identified as *Cladophialophora ajelloi*[13a] since the fungus produces phialides as well as *Cladosporium*-type conidiation. Whether or not it is identical to the *Torula bantiana* of Saccardo is still a matter for debate. As pointed out by Emmons,[35] the latter species is described as having conidia in unbranched chains which are 8 to 11 × 5 μ in size. Most recent isolants have conidia 4 to 7 × 2 to 3 μ in size which are usually in branched chains. However, in Saccardo's illustration the chains are branched. Some conidia up to 15 and 20 μ in length have been found. The conidiophore is not distinct from (micronematous) the vegetative hyphae. The descriptions of the lesions produced by *T. bantiana* and *C. bantianum* in the brain are similar. The latter organism is neurotropic in experimental animal infections. Intravenous inoculation of 10^5 cells regularly produces cerebral lesions in mice. The first isolant of the organism was from a cerebral infection described by Banti.[5]

Colony Morphology. The organism grows moderately fast and produces a spreading, slightly folded, olive-gray to olive-brown colony (Fig. 9–16).

Microscopic Morphology. Elongate, brown, septate conidiophores are produced which are indistinct from the vegetative hyphae of other species and thus are different from conidiophores produced by other species of *Cladosporium*. Conidia are produced in long, branched, flexuose chains (Fig. 9–12,*L*). Most conidia are 2 to 3 × 4 to 7 μ in size, but some are 3 × 15 to 20 μ (Fig. 9–16).

Chmelia slovaca Svoboda 1966[90] was isolated from a verrucoid lesion of the external ear, in which thick-walled, sclerotic cells were found. The fungus eventually grew as a feltlike colony, though it was quite yeasty at first. The cells were budding and yeastlike and were

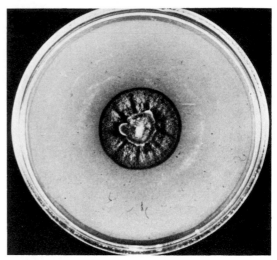

Figure 9–16. *Cladosporium bantianum.* Folded, olive-gray compact colony. Tufts of white mycelial overgrowth are beginning to form. (Courtesy of L. Ajello.)

similar to the "*Pullularia*" stage of other species. Chlamydoconidia that were septate in many planes, blastoconidia, and tangled, coiled hyphae were described. The author claimed to have reproduced the disease in animals. Work by Borelli and Marcano did not confirm this, and they questioned whether this species was actually capable of causing disease.[14]

Phialophora parasitica Ajello, Georg, Steigbigel et Wang 1974

This organism was isolated from a case of cystic disease in a renal transplant patient on steroids. In tissue, the fungus was in the form of short hyphal units[1] (Fig. 9–6,*D*). The colony is flat, effuse, moist, and gray olivaceous with a gray olivaceous reverse and grows to a diameter of 5 cm by 21 days on potato glucose agar. The phialides are macronematous simple, smooth, and hyaline to pale brown. They occur as intercalary or terminal conidiogenous cells and are 3 × 12 μ, tapering at the tip. The collarette is small. The conidia are 1.3 × 3.4 μ, cylindric to allantoid. Yeast of the *Phaeococcomyces* type and secondary conidia are also found.

Acrotheca aquaspersa Borelli 1972[12] was isolated from a case of chromoblastomycosis from Mexico. It produces planate dividing sclerotic cells in tissue in humans and mice. The colony is flat and dematiaceous. *Acrotheca*-type conidia production is the predominant form of conidiation. The conidiophores are long and branched and conidiation is acropleurogenous. The conidia disperse in water. Phialides and chains of conidia are also found. De Hoog[30] in 1977 reclassified this isolant as a strain of *Ramichloridium cerophilum*. This fungus has recently been reclassified, however. For a moment, it was *Rhinocladiella cerophilum*[66]; it is now *R. aquaspersa*. Next week, who knows?!

Phialophora hoffmanii (Beyma) Schol-Schwarz 1970[83]

Synonymy. *Marginomyces hoffmanii* Beyma 1939; *Phialophora luteo-viridis* (Beyma) Schol-Schwarz 1970; *P. aurantiaca* Beyma 1940; *P. mutabilis* (Beyma) Schol-Schwarz 1970; *Marginomyces mutabilis* Beyma 1944–1945.

This organism has been isolated from subcutaneous cysts. The colony is yeastlike at first, but later is woolly.[11] Coloration is ocherous, amber, house mouse gray to pale olive gray, with an amber to olive gray reverse. Conidiophores are micronematous to macronematous, simple, and smooth and end in a small collarette. Small phialides are 1×3 μ, large 3×21 μ, and are nonseptate. Conidia are allantoid 1.3×4.2 μ. Spirals, yeastlike cells, chlamydoconidia, and secondary conidia are common. Isolants have also been recovered from butter and a burned child[66] and from at least two cases of endocarditis. (See Chapter 26.)

Phialophora repens (Davidson) Conant 1937[27]

Synonymy. *Cadophora repens* Davidson 1935.[29]

This organism is regularly isolated as an agent of blue staining of lumber.[29] It has also been isolated from a subcutaneous cyst.[68] The patient had lepromatous leprosy and was on steroids. The lesion was granulomatous and on the scalp. The hyphae were often nonpigmented and massed into soft masses similar to grains of mycetoma. No sinus tracts were found. The masses were 20 to 110 μ in diameter. Some free hyphal strands 3 μ in diameter were seen among the epithelioid cells. *P. repens* is rapidly growing, producing a flat, cottony effuse colony that is light to dark brown, olivaceous to isbelline. Tufts of white aerial hyphae are found and the reverse is olive. Phialophores are single or in tufts, semi- to macronematous, with a short truncated collarette. Conidia are hyaline, curved, or cylindric, 1.3 to 2.8 $\times$ 2.5 to 8 μ, accumulating at the phialide tip. Coiled hyphae but no yeasts are found.

REFERENCES

1. Ajello, L., L. K. Georg, R. T. Steigbigel, and C. J. K. Wang. 1974. A case of phaeohyphomycosis caused by a new species of *Phialophora*. Mycologia, *66*:490–498.
2. Al-Doory, Y. 1972. *Chromomycosis*. Missoula, MT, Mountain Press Publishing Company.
3. Aravysk, R. A., and V. B. Aronson. 1968. Comparative histopathology of chromomycosis and cladosporiosis in experimental infection. Mycopathologia, *36*:322–340.
4. Azulay, R. D., and J. Serruya. 1967. Hematogenous dissemination in chromoblastomycosis. Report of a generalized case. Arch. Dermatol., *95*:57–60.
5. Banti, G. 1911. Sopra un caso di oidiomicosi cerebrale. Dell Accademia Medico Fisica Fiorentina. p. 49.
6. Batres, E., J. E. Wolf, et al. 1978. Transepithelial elimination of cutaneous chromomycosis. Arch. Dermatol., *114*:1231–1232.
7. Battistini, F., and R. N. Sierra. 1968. Tratamiento de un caso de cromomicosis con aplicaciones locales de thiabendazole. Derm. Venez., *7*:3–10.
8. Bayles, M. A. H. 1971. Chromomycosis. Treatment with thiabendazole. Arch. Dermatol., *104*:476–485.
9. Beneke, E. S. 1978. Dematiaceous fungi in laboratory-housed frogs. Proc. IV Int. Conf. Mycoses. P.A.H.O. Sci. Pub. *356*:101–108.
10. Beurmann, L de, and H. Gougerot. 1907. Association morbides des sporotrichoses. Bull. Me. Soc. Med. Hop. Paris, *24*:(3 ser.).
11. Beyma, F. H. 1939. Beschreibung einiger neuer pilzarten aus dem "Centraalbureau voor Schimmel cultures" Baar. (Netherlands) V. Mitteilung. Zentralbl. Bakteriol. Parasit. Kde (Abt. II), *99*:381–394.
12. Borelli, D. 1972. *Acrotheca aquaspersa* nova species: agente de cromomicosis. Acta Cient. Venez., *23*:193–196.
13. Borelli, D. 1972. A method for producing chromomycosis in mice. Trans. R. Soc. Trop. Med. Hyg., *66*:793–794.
13a. Borelli, D. 1978. Causal agents of chromoblastomycosis (chromomycetes). Proc. V Int. Conf. Mycoses. P.A.H.O. Sci. Pub., *396*:334–335.
14. Borelli, D., and C. Marcano. 1969. Observaciones sobre *Chmelia slovaca*. Derm. Venez., *8*:740–747.
15. Brumpt, E. 1922. Precis de Parasit. 3e Ed. Paris, Masson et Cie, p. 1105.
16. Buckley, H., and I. G. Murray 1966. Precipitating antibodies in chromomycosis. Sabouraudia, *5*:78–80.
17. Brygoo, E. R., and P. Destomes. 1975. Geographic-

al distribution of chromoblastomycosis and its various pathogenic agents. Bull. Soc. Fr. Mycol. Med., *4*:181–183.

18. Cales, D., and R. Helenon. 1976. La chromomycose a la Martinique. Afr. Med., *15*:457–466.

19. Callaway, J. L., D. L. Laymon, and N. Conant. 1970. Chromoblastomycosis — laboratory infection with *Cladosporium carrionii*. Cutis, *6*:60–62.

20. Campins, H., and M. Scharyj. 1953. Cromoblastomicosis; Comentarios sobre 34 cases, con estudio clinico, histologico and micologico. Gac. Med. (Caracas), *61*:127–151.

21. Carini, A. 1910. Sur une moisissure qui cause une maladie spontannée du *Heptodactylus pentaduetylus*. Ann. Inst. Pasteur, *24*:157–172.

22. Carrión, A. L. 1936. Chromoblastomycosis. A new clinical type caused by *Hormodendrum compactum*. Puerto Rico J. Public Health Trop. Med., *11*:663–682.

23. Carrión, A. L., and M. Silva-Hunter. 1971. Taxonomic criteria for fungi of chromoblastomycosis with reference to *Fonsecaea pedrosoi*. Int. J. Dermatol., *10*:34–43.

24. Cole, G. T. 1978. Conidiogenesis in the black yeasts. Proc. IV Int. Conf. Mycoses. P.A.H.O. Sci. Pub. *356*:66–78.

25. Cole, G. T., and B. Kendrick. 1973. Taxonomic study of *Phialophora*. Mycologia, *65*:661–668.

26. Conant, N. F. 1937. The occurrence of a human pathogenic fungus as a saprophyte in nature. Mycologia, *29*:597–598.

27. Conant, N. F., and D. S. Martin. 1937. The morphologic and serologic relationships of various fungi causing dermatitis verrucosa (chromoblastomycosis). Am. J. Trop. Med., *17*:553–578.

27a. Conti Diaz, I., et al. 1978. Isolation and identification of black yeasts from the external environment in Uruguay. Proc. IV Int. Conf. Mycoses P.A.H.O. Sci. Pub., *356*:109–114.

28. Cooper, B. H., and J. D. Schneidau. 1970. A serological comparison of *Phialophora verrucosa, Hormodendrum pedrosoi* and *Cladosporium carrionii* using immunodiffusion and immunoelectrophoresis. Sabouraudia, *8*:217–226.

29. Davidson, R. W. 1935. Fungi causing stain in logs and lumber in southern states including five new species. J. Agric. Res., *50*:789–807.

30. de Hoog, G. S., and E. J. Hermanides-Nijhof. 1977. The black yeasts and allied hyphomycetes, p. 118. In Studies in Mycology No. 15. Baarn, The Netherlands, Centraal bureau voor Schimmel cultures.

31. Di Salvo, A. F., and W. H. Chew. 1968. *Phialophora gougerotii:* An opportunistic fungus in a patient treated with steroids. Sabouraudia, *6*:241–245.

32. Discamps, G., and J. C. Doury. 1976. La phaeosporotrichose. Med. Trop., *36*:475–476.

33. Duque, H. O. 1961. Cromoblastomicosis. Revision general y estudio de la Enfermedad en Colombia. Antioquia Med., *11*:499–521.

34. Duque, H. O. 1961. Meningo-encephalitis and brain abscess caused by *Cladosporium* and *Fonsecaea*. Am. J. Clin. Pathol., *36*:505–517.

35. Emmons, C., C. H. Binford, J. P. Utz, and K. J. Kwon-Chung. 1977. Medical Mycology, 3rd ed. Philadelphia, Lea and Febiger, p. 481.

36. Engleman, R. M., R. M. Chase, et al. 1970. Mycotic infections on prosthetic and homograph heart valves: report of the first case of endocarditis caused by *Hormodendrum dermatitidis*. Ann. Surg., *173*:455–461.

37. Frank, W., and U. Roester. 1970. Amphibien als Träger von *Hormiscium (Hormodendrum) dermatitidis* Kano 1937, einem Erreger der Chromoblastomykose (Chromomykose) des Menschen. Z. Tropenmed. Parasit. *21*:93.

38. Fuentes, C. A., and Z. E. Bosch. 1960. Biochemical differentiation of the etiologic agents at chromoblastomycosis from non-pathogenic *Cladosporium* species. J. Invest. Dermatol., *34*:419–422.

39. Gonzalez-Ochoa, A. 1970. Curation de la criptococosis y de la Chromomicosis con 5-fluorcitosina. Rev. Invest. Salud. Publica., *30*:63–76.

40. Gordon, M. A., and Y. Al-Doory. 1965. Application of fluorescent antibody procedures to the study of pathogenic dematiaceous fungi. J. Bacteriol., *89*:551–556.

41. Greer, K. E., and G. P. Gross, et al. 1981. Cystic chromomycosis (chromoblastomycosis) due to *Fonsecaea dermatitidis*. Arch. Dermatol., *115*:1433–1434.

42. Gugnani, H. C., A. Suseelan, et al. 1977. Cutaneous cladosporiosis due to *Cladosporium trichoides*. J. Trop. Med. Hyg., *80*:177–178.

43. Haschek, W. M., and O. B. Kasale. 1977. A case of cutaneous feline phaeohyphomycosis caused by *Phialophora gougerotii*. Cornell Vet., *67*:467–471.

44. Hill, J. R., G. Migaki, and R. D. Phemister. 1978. Phaeomycotic granuloma in a cat. Vet. Pathol., *15*:559–561.

45. Hohl, P., A. E. Prevost, and H. P. Holly. 1982. Ineffectiveness of amphotericin B in a human case of phaeohyphomycosis due to *Wangiella dermatitidis*. Mycopathologia, in press.

46. Howes, J. K., C. B. Kennedy, et al. 1954. Chromoblastomycosis. Report of nine cases from a single area in Louisiana. Arch. Dermatol., *68*:83–90.

47. Hughes, W. T. 1967. Chromoblastomycosis: Successful treatment with topical amphotericin B. J. Pediatr., *71*:351–356.

47a. Iwatsu, T., and M. Miyaji. 1978. Subcutaneous cyst caused by *Phialophora verrucosa*. Mycopathologia, *64*:165–168.

47b. Iwatsu, T., M. Miyaji, et al. 1979. Skin test active substances prepared from culture filtrate of *Fonsecaea pedrosoii*. Mycopathologia, *67*:101–105.

48. Jotisankasa, V., H. S. Nielsen, et al. 1970. *Phialophora dermatitidis:* its morphology and biology. Sabouraudia, *8*:98–107.

49. Kano, K. A. 1934. A new pathogenic *Hormiscium* (Kunze) causing chromoblastomycosis. Aichi Igakkai Zasshi, *41*:1657–1673 (in Japanese).

50. Kano, K. 1937. Über die chromoblastomykose durch einen noch nicht als pathogen beschrieben Pilz: *Hormiscum dermatitidis* n. sp. Arch. Dermatol. Syph. Orig., *176*:282–294.

51. Kempson, R. H., and W. H. Steinberg. 1963. Chronic subcutaneous abscesses caused by pigmented fungi. A lesion distinguishable from cutaneous chromoblastomycosis. Am. J. Clin. Pathol., *39*:598–606.

52. Katrajaras, R., and S. Chongsathien. 1979. Subcu-

taneous chromomycosis abscesses caused by *Phialophora gougerotii*. Int. J. Dermatol., *18*:150–154.

53. Lacaz, C. S., J. M. Cruz Hurtado, et al. 1966. Dermatite verrucosa cromoparasitiaria. O. Hosp., *70*:9–17.

54. Lane, C. G. 1915. A cutaneous disease caused by a new fungus *Phialophora verrucosa*. J. Cutan. Dis., *33*:840–846.

55. Lazo, R. F. 1978. Superficial mycoses induced by black yeasts. Proc. IV Int. Conf. Mycoses. P.A.H.O. Sci. Pub., *356*:25–33.

56. Leslie, D. F., and G. L. Beardmore. 1979. Chromoblastomycosis in Queensland: a retrospective study of 13 cases at the Royal Brisbane Hospital. Aust. J. Dermatol., *20*:23–30.

57. Londero, A. T., and C. D. Ramos. 1976. Chromomycosis: a clinical and mycological study of thirty five cases observed in the hinterland of Rio Grande do Sul, Brazil. Am. J. Trop. Med. Hyg., *25*:132–135.

58. Lopes, C. F., E. O. Cissulpino, et al. 1971. Treatment of chromomycosis with 5-fluorocytosine. Int. J. Dermatol., *10*:172–191.

59. Lopes, C. F., R. J. Alvarenga, et al. 1978. Six years experience in treatment of chromomycosis with 5-fluorocytosine. Int. J. Dermatol., *17*:414–418.

60. Malkina, A. Y., and N. N. Darchenkova. 1977. Distribution of chromomycosis in the world. Vestn. Dermatol. Venerol., *1*:41–45.

61. Mariat, F., G. Segretain, et al. 1967. Kyste sous-cutane mycosique (phaeosporotrichose) a *Phialophora gougerotti*. Sabouraudia, *5*:209–219.

62. Martin, D. S., R. D. Baker, and N. F. Conant. 1936. A case of verrucous dermatitis caused by *Hormodendrum pedrosoi*. Am. J. Trop. Med., *16*:59.3.

63. Matruchot, O. O. 1910. Sur un nouveau group de champignons pathogenes agents des sporotrichoses. Comp. Rem. Acad. Sci. (Paris), *150*:543–545.

64. Mauceri, A. A., S. I. Cullen, et al. 1974. Flucytosine. An effective oral treatment for chromomycosis. Arch. Dermatol., *109*:873–875.

65. McGinnis, M. R. 1977. *Wangiella* a new genus to accommodate *Hormiscium dermatitidis*. Mycotaxon, *5*:353–363.

66. McGinnis, M. R. 1978. Human pathogenic species of *Exophiala, Phialophora* and *Wangiella*. Proc. IV Int. Conf. Mycoses P.A.H.O. Sci. Pub., *356*:37–59.

66a. McGinnis, M. R., and W. A. Schell. 1980. The genus *Fonsecaea* and its relationship to the genera *Cladosporium, Phialophora, Ramichloridium*, and *Rhinocladiella*. Proc. Vth Int. Conf. Mycoses P.A.H.O. Sci. Pub., *396*:215–234.

67. Medlar, E. M. 1915. A cutaneous infection caused by a new fungus *Phialophora verrucosa* with a study of the fungus. J. Med. Res., *32*:507–522.

68. Meyers, W. M., J. R. Dooley, and K. J. Kwon Chung. 1975. Mycotic granuloma caused by *Phialophora repens*. Am. J. Clin. Pathol. *64*:545–551.

69. Montpellier, J., and A. Catanei. 1927. Mycose humaine due à un champignon du genre *Hormodendrum, H. algeriensis* n. sp. Ann. Dermatol. Sypl., *8*:626–635.

70. Morison, W. L., B. Connor, and Y. Clayton. 1974. Successful treatment of chromoblastomycosis with 5-fluorocytosine. Br. J. Dermatol., *90*:445–449.

71. Negroni, P. 1936. Estudio del primer caso argentino de cromomicosis, *Fonsecaea* (Neg.) *pedrosoi* (Brump) 1921. Dept. Nat. de Higiene. Rev. Inst. Bacteriol., *7*:419–426.

72. Nielsen, H. S., and N. F. Conant. 1968. A new human pathogenic *Phialophora*. Sabouraudia, *6*:228–231.

73. Nsanzumuhire, H., D. Vollum, and A. A. Potera. 1974. Chromomycosis due to *Cladosporium trichoides* treated with 5-fluorocytosine. Am. J. Clin. Pathol., *61*:257–263.

74. Padhye, A. A. 1978. Comparative study of *Phialophora jeanselmei* and *P. gougerotii* by morphological, biochemical and immunological methods. Proc. IV Int. Conf. Mycoses P.A.H.O. Sci. Pub., *356*:60–65.

75. Padhye, A. A., M. R. McGinnis, and L. Ajello. 1978. Thermotolerance of *Wangiella dermatitidis*. J. Clin. Microbiol., *8*:424–426.

76. Pedroso, A., and J. M. Gomes. 1920. Four cases of dermatitis verrucosa produced by *Phialophora verrucosa*. Ann. Paulistas Med. Cir., *9*:53.

77. Putkonen, T. 1961. Die chromoykose in Finnland. Der mogliche Anteil der Finnishen Sauna an ihrer Verbreitung. Hautarzt, *17*:507–509.

77a. Ramirez, M. 1973. Treatment of chromomycosis with liquid nitrogen. Int. J. Dermatol., *12*:250–254.

78. Reiss, N. R., and W. Y. Mok. 1979. *Wangiella dermatitidis* isolated from bats in Manaus, Brazil. Sabouraudia, *17*:213–218.

79. Roberts, R. L., and P. J. Szaniszlo. 1978. Temperature-sensitive multicellular mutants of *Wangiella dermatitidis*. J. Bacteriol., *135*:622–632.

79a. Roux, J., M. Fissarou, et al. 1967. Uncas de chromoblastomycose au debut. Bull. Soc. Fr. Syphil. *74*:655–656.

80. Rudolph, M. 1914. Ueber die brasilianisele figueira. Arch. für Schiffs. Tropeh-Hyg., *18*:498.

81. Rush, H. G., M. R. Anver, and E. S. Beneke. 1974. Systemic chromomycosis in *Rana pipiens*. Lab. Animal Sci., *24*:646–655.

82. Schmidt, R. E., and D. A. Hartfiel. 1977. Chromomycosis in amphibians — review and a case report. J. Zoo Animal Med., *8*:26–28.

83. Scholl-Schwarz, M. B. 1968. *Rhinocladiella*, its synonym *Fonsecaea* and its relation to *Phialophora*. Antonie Van Leeuwenhoek J. Microbiol. Serol., *34*:119–152.

84. Shimazone, Y., I. Kiminori, et al. 1963. Cerebral chromomycosis. *Fonsecaea dermatitidis*. Folia Psychiatr. Neurol. Jpn., *17*:80–96.

85. Silva, M. 1960. Growth characteristics of the fungi of chromoblastomycosis. Ann. N.Y. Acad. Sci., *89*:17–29.

86. Simpson, J. G. 1966. A case of chromoblastomycosis in a horse. Vet. Med. Small Anim. Clin., *61*:1207–1209.

87. Simson, F. W. 1946. Chromoblastomycosis. Some observations on the type of the disease in South Africa. Mycologia, *38*:432–449.

88. Sonck, C. E. 1975. Chromomycosis in Finland. Dermatologia, *19*:189–193.

89. Suescun, T., J. D. Jaramillo, and S. Prada. 1977. Abscesos subcutaneos por hongos dermatiaceos. Mycopathologia, *62*:97–102.

90. Svobodova, Y. 1966. *Chmelia slovaca* sp. novo. A

new dematiaceous fungus pathogenic for man and animals. Biologia (Bratislava), *21*:81–88.

91. Swartz, I. S., and C. W. Emmons. 1968. Subcutaneous cystic granuloma caused by a fungus of wood pulp *(Phialophora richardsiae)*. Am. J. Clin. Pathol., *49*:500–505.

92. Trejos, A. 1954. *Cladosporium carrionii* n. sp. Rev. Biol. Trop., *2*:75–112.

93. Tsai, C. Y., U. C. Lii, et al. 1966. Systemic chromoblastomycosis due to *Hormodendrum dermatitidis* (Kano) Conant. Am. J. Clin. Pathol., *46*:103–114.

94. Velasquez, J., A. Restrepo, and G. Calle. 1976. Cromomicosis-experiencia de doce anos. Acta Med. Colombiana, *1*:165–171.

95. Velasquez, L. F., and A. Restrepo. 1975. Chromomycosis in the toad *(Bufo marinus)* and a comparison of the etiologic agent with fungi causing human chromomycosis. Sabouraudia, *13*: 1–9.

96. Vitto, J., D. J. Santa Cruz, et al. 1979. Chromomycosis. Successful treatment with 5-fluorocytosine. J. Cutan. Pathol., *6*:77–84.

97. Vollum, D. I. 1977. Chromomycosis: a review. Br. J. Dermatol., *96*:454–458.

98. Vortel, V. and Z. Kraus. 1975. Another case of chromomycosis in the territory of Czechoslovakia identified histologically. Česká Dermatol., *50*:325–326.

99. Whiting, D. A. 1967. Treatment of chromoblastomycosis with high local concentrations of amphotericin B. Br. J. Dermatol., *79*:345–351.

100. Wilson, S. J., S. Hulsey, et al. 1933. Chromoblastomycosis in Texas. Arch. Dermatol., *27*:107–122.

101. Yanase, K., and M. Yamada. 1978. Pocket warmer therapy of chromomycosis. Arch. Dermatol., *114*:1095.

102. Young, J. M., and E. Ulrich. 1953. Sporotrichosis produced by *Sporotrichum gougerotii*. Arch. Dermatol., *67*:44–53.

103. Zaias, N. 1978. Chromoblastomycosis — a superficial minimycetoma. Proc. IV Int. Conf. Mycoses P.A.H.O. Sci. Pub., *356*:17–19.

104. Zweibel, S. M., and C. J. K. Wang. 1978. Reexamination of *Phialophora verrucosa*. Proc. IV Int. Conf. Mycoses. P.A.H.O. Sci. Pub., *356*:91–100.

10 Sporotrichosis

DEFINITION

Sporotrichosis is most commonly a chronic infection characterized by nodular lesions of the cutaneous or subcutaneous tissues and adjacent lymphatics that suppurate, ulcerate, and drain. The etiologic agent, *Sporothrix schenckii,* gains entrance by traumatic implantation into the skin or, very rarely, by inhalation into the lungs. Secondary spread from subcutaneous lesions to articular surfaces, bone, and muscle is not infrequent, and the infection may also occasionally involve the central nervous system, lungs, or genitourinary tract.

HISTORY

The first case that unquestionably presented with the clinical picture of sporotrichosis was recorded by Schenck in 1898[82] from the Johns Hopkins Hospital in Baltimore. This report described the fungus isolated from the lesion as resembling a "sporotricha," a name given it by the consultant mycologist, E. F. Smith. Previously, Link in 1809 and Lutz in 1889 had described cases that in all probability were sporotrichosis, but no fungus was isolated. The second recorded case, also from the United States (Chicago), was reported by Hektoen and Perkins[36] as a lesion that developed on the finger of a boy following a blow by a hammer. The lesion apparently healed spontaneously; however, 65 years later this patient still had a positive sporotrichin skin test but had a negative serum agglutination titer.[66] A fungus was isolated from the lesion and referred to as *Sporothrix schenckii,* but in the title of the paper only. A very cursory but recognizable description of the organism was given in the text. Later, in 1903, the disease was described in France by de Beurmann and Ramond.[25] The isolant from their cases was termed *Sporotrichum beurmanni* by Matruchot and Ramond in 1905. The isolant from Schenck's original case had lost its pigment by that time, and when it was examined by the French authors was considered to be a species distinct from the organism recovered by de Beurmann and Matruchot. However, Matruchot in 1910 redescribed Schenck's organism as *Sporotrichum schencki.*[63] Curiously, he dropped the final "i" of the specific name, and until recently this spelling was commonly used. Between 1906 and 1911 de Beurmann and Gougerot identified at least ten more cases of sporotrichosis and tabulated some two hundred additional cases by 1912. This formed the basis of their excellent review of the disease,[24] which is still considered to be a classic in the field of medical literature. In addition to the cutaneous form of disease, they included for the first time descriptions of disseminated, pulmonary, osseous, and mucosal involvement.

In 1907 the first case of natural infection in animals was recorded by Lutz and Splendore in rats in Brazil.[55] Two years later a similar organism isolated from a horse in Madagascar was named *S. equi* by Carougeau.[15] Since that time, reports of the disease in man and animals have been published from all parts of the world but with intriguing and enigmatic shifts in incidence and geographic distribution. Though several hundred cases had been reported by 1915, after the first two decades of this century, sporotrichosis became quite rare in France and in Europe in general. The disease is still uncommon in those areas.

Cawley[15a] could tabulate only 200 cases in the United States before 1932. Since that time there has been a distinct rise in the number of infections recognized and increasingly frequent reports of "opportunistic" or disseminated disease.

Epidemics of sporotrichosis have been reported from time to time. One of the most famous of these was in South Africa. In a space of two years, almost 3000 cases occurred involving miners who brushed against timbers of a mine shaft on which the fungus was growing.[27, 86] The epidemic was terminated by treating the timbers with a fungicide.

Strains of fungi isolated from human infections in France and America were studied by Davis in 1921.[21] He concluded that they were identical and called the etiologic agent *Sporotrichum schencki*. However, Carmichael in 1962[14a] pointed out the differences in conidiation between the genera *Sporothrix* and *Sporotrichum* and determined that the correct binomial for the organism was *Sporothrix schenckii*.

ETIOLOGY, ECOLOGY, AND DISTRIBUTION

Etiology. All forms of sporotrichosis in man are caused by the single species *Sporothrix schenckii*. The etiologic agent of the disease in horses, which was first described by Carougeau as *S. equi*, has also been shown to be identical to *S. schenckii*. Sporotrichosis in other animals is also caused by this species.

With few exceptions, the fungus gains entry into the body through some trauma to the skin. The most common histories obtained are scratches from thorns or splinters, cuts from sedge barbs, or handling of reeds, sphagnum moss,[75] or grasses. Sometimes brushing against infected tree bark or timber will result in disease. These modes of transmission emphasize the saprophytic association of the organism with plant life. Sporotrichosis has also occurred following parrot bites, dog bites, insect stings,[62] injury by metal particles, handling fish,[64] hammer blows,[36] and other traumas.[91] In these instances, infection probably followed contamination of the wound with soil. The organism has been isolated from fleas, ants, and horse hair, which possibly represent vectors for the conidia. In a few rare cases, human infection has occurred after contact with infected animals in experimental laboratory studies. The disease has also been transmitted via contaminated dressings from suppurating lesions, and there is at least one case in which the disease was transmitted by direct contact from the cheek of a mother to the cheek of a child.[87]

Ecology. *S. schenckii* is an organism commonly found on decaying vegetation, and it has been isolated many times from soil.[38, 57, 65] Its distribution in the soil and the conditions that enhance its occurrence have been the subject of much speculation. In contrast to *Coccidioides immitis*, there is as yet no clear-cut explanation for the elusive and transient distribution of *Sporothrix schenckii* in soil. After the famous epidemic involving gold miners in South Africa, Findley[29] made a careful study of the conditions that favored growth of the organism in the mines. He found that the fungus grew well on untreated mine poles at a temperature of 26 to 27°C and at a relative humidity of 92 to 100 per cent. In the original report of the epidemic,[86] it has been noted that in mines where the timbers were infected with various lignicolous Basidiomycetes, such as *Poria* sp., the timbers did not harbor *S. schenckii*, and the workers in those mines were free of infection. Findley also determined that, in areas outside the mines, the infection was most prevalent in the temperate highland plateau. Here the average relative humidity is 65 per cent, and the rainfall is 25 to 30 inches per year. Mackinnon has studied the epidemiology of the disease in an endemic area in Uruguay.[57] He found that there was a seasonal distribution of cases, with increased frequency during the hot, rainy autumn season (March to July). The conditions favoring optimal growth of the fungus were a humidity of 90 per cent and a temperature above 15°C. Yet in a similar study by González-Ochoa[34, 35] in Mexico, the greatest frequency of infection coincided with the dry and cooler parts of the year. The most highly endemic region of that country is the temperate plateau, where rainfall is sparse throughout the year. The seasonal distribution of cases was winter, 51.4 per cent; fall, 23 per cent; summer, 17 per cent; and spring, 9 per cent. This is in contrast to the situation in Uruguay. The geopathology of sporotrichosis was reviewed by Gonçalves.[33]

Distribution. Sporotrichosis is the most common subcutaneous and deep mycosis in Mexico. It outnumbers cases of mycetoma, which is also highly endemic there. In some villages in Jalisco and Michoacan, sporotrichosis is so common that lesions can be regularly spotted on inhabitants walking through the market place. Yet the area is quite dry and almost arid in contrast to other highly endemic areas of Brazil, Uruguay, and South Africa which are moist and humid. Most of the patients in Mexico give a history of working with grass — either gathering it or using it as packing material or for making baskets.[35, 95] In the majority of cases in Uruguay, the infection was associated with hunting armadillos. The fungus was found not in the animals but in their nests and burrows.[57] Straw was reported to be the common source of a family epidemic in Brazil.[85] In an endemic area near Lake Ayarza in Guatemala, only men who handled fish acquired the infection.[64] An epidemic occurred among brickyard workers who, after the drinking of much beer, began tossing bricks at each other, causing many skin abrasions. *S. schenckii* was isolated from the packing straw used for the bricks and from the many sporotrichotic lesions of the workers.

In the United States, France, Canada,[62] and other temperate countries, infection in soil is usually associated with gardening and contact with gardening soil.[73] It is frequently associated with sphagnum moss,[75] and the disease is considered an occupational hazard of greenhouse workers and rose growers. In contrast to the situation in other endemic areas, the organism appears to thrive in moist, rich loam and humus. In a fascinating example of epidemiologic detective work, Mariat[58a] traced the source of an infection acquired from a house plant to the potting soil of oak and beech litter that had been gathered fifty miles from Paris.

Although members of the genus *Sporothrix* are known to cause disease in plants, e.g., carnation bud rot, *S. schenckii* has not yet been conclusively proved to be a plant pathogen. It does appear able to grow on injured or debilitated plant tissue. This was shown experimentally by Benham,[10] who injected conidia into carnation buds, which subsequently became diseased. The fungus appears to be fairly resistant to drying but is sensitive to direct exposure to sunlight[57] and severe winter weather.[65] It will grow on meat products in cold storage vaults, however.[2] This led to speculation that "hot dogs" could be a potential source of food-borne sporotrichosis.[6] Several animal studies indicated that this was a very remote possibility,[20, 23] although packers and handlers of sausages, salami, and other types of meat products may be at some risk.

Cases of sporotrichosis have been found in all age groups. Primary infections in children as young as ten days (following a rat bite) and in adults in their seventh and eighth decades have been recorded. The apparent differences in distribution of the disease among the sexes is probably related to occupation and exposure. Though the ratio in many case studies is 3:1,[7] male to female, others have noted the reverse distribution. In some reports from Brazil, 60 to 70 per cent of the cases occurred in females. In Mexico, González-Ochoa[34] found the ratio to be about 1:1. The men acquired the disease on the legs from thorns and sedges or on the fingers and wrists from gathering grass, the women on the fingers from cultivating decorative house plants and making baskets, and the children on the face from scratches of branches. In this study, over 60 per cent of cases were found in persons less than 30 years of age. Primary lesions were most often seen in the young, whereas chronic disease was found with greater frequency in older patients. Since the number of cases appears to decrease markedly in this same population after the age of 50, it appears that many infections heal spontaneously.

Although the etiologic agent, *S. schenckii*, is found in soils all over the world, incidence of the disease varies considerably as does the type of disease contracted. In the early 1900's, sporotrichosis was very common in France, and many cases of extracutaneous dissemination were recorded. Since that time it has become rare in that country and in Europe in general. At the present time the greatest number of cases come from Mexico, Central America, Colombia,[96] and Brazil, with a scattered distribution throughout the rest of the world. De Beurmann and Gougerot[24] in 1912 were the first to suggest that sporotrichosis could be considered an opportunistic infection. Most of their patients had some underlying disease, and essentially all of the cases of extracutaneous dissemination occurred in those with severe debilitating conditions or malnutrition. This is still a common finding, in addition to the association of systemic disease with alcoholism.[70, 81]

Mariat,[58a] citing the rural conditions in France during 1900 to 1914, has suggested that malnutrition is a significant factor in infection. He noted some experimental evidence that mice on protein-deficient diets were more susceptible to severe disease. Dietary deficiency could play a major role in the infection in rural areas of Mexico, Brazil, and Central America, where the case rates are so high.

To a certain extent all fungus infections are opportunistic, and normal, healthy, well-nourished adults seldom develop disease unless they receive an overwhelming inoculum. In the epidemiologically unique case of the miners in South Africa, the patients were in good health, and among the 3000 cases no disseminated disease was found. Frequent injury from brushing the mine poles and continual exposure to large concentrations of conidia were suggested as the prime factors leading to infection in this instance. In contrast, exposure to small amounts of conidia in an endemic area may gradually confer immunity. González-Ochoa[34, 35] has shown that a group of grass handlers who had been on the job for more than ten years had no clinical disease, but 100 per cent of them had a positive sporotrichin skin test. When disease does occur in this population, it is the fixed cutaneous type, and many cases appear to heal spontaneously.

Not only has the epidemiology of sporotrichosis changed over the years, but it appears that the pathology has also. In most of the early reports from France and the United States, numerous cigar-shaped bodies were found in lesions. More recently, lesions from cases in the United States and South Africa are described as containing very few organisms. The so-called asteroid body, which is very commonly seen in South Africa[53] and Japan, is very rarely found in cases from the United States, Mexico, or Central America.[33, 53, 54] The asteroid body is an eosinophilic mass surrounding the organism in tissue and represents an antigen-antibody complex. It was first seen in Brazil in 1907[55] but was not seen again in that country until 1964. It has again become a frequent finding there.

It is apparent that there are many facets of the epidemiology and pathogenicity of sporotrichosis that are not yet clearly defined. The organism is only weakly pathogenic, and it is known that the inherent virulence of strains isolated from soil is quite variable.[38] However, this factor alone fails to explain the great variability in incidence, distribution, and severity of sporotrichosis that has been seen since its discovery in 1898.

CLINICAL DISEASE

Within a few years of the discovery of sporotrichosis, a wide spectrum of disease types was noted.[23, 24] Most cases were the so-called "gummatous" type, presenting as large ulcerations involving skin, subcutaneous tissue, and local lymph channels.[23] However, disseminated disease as well as mucous membrane involvement and the primary pulmonary form were soon described. The type of disease and its pathology depend on the site of inoculation of the organism and the response of the host to it. For purposes of discussion, the clinical types are divided into five categories: lymphocutaneous, fixed cutaneous, mucocutaneous, extracutaneous and disseminated, and primary pulmonary. The first four result from traumatic implantation and the last from inhalation of conidia.

Lymphocutaneous Sporotrichosis

Lymphocutaneous sporotrichosis comprises up to 75 per cent of all cases in most literature surveys.[28, 64, 96] The fungus gains entrance by traumatic implantation, and the first signs of infection may appear as soon as five days later. The average incubation time is three weeks, but in some patients the initial injury may have taken place six months or more before symptoms occur. The first sign of disease is the appearance of a small, hard, movable, nontender and nonattached subcutaneous nodule. This develops into a bubo, which attaches itself to the overlying skin (Fig. 10–1). The lesion becomes discolored, varying from pink to purple and sometimes black. This is accompanied by local or regional adenopathy. At this stage the disease may be mistaken for cutaneous anthrax. The lesion then penetrates the skin and becomes necrotic, thus forming the so-called sporotrichotic chancre. Not infrequently, the primary lesion begins as a small ulcer instead of a nodule at the site of the injury. This usually indicates location of the organisms in the upper dermis or epidermis rather than in subcutaneous tissue. The initial lesion remains for several weeks or months and tends to heal with scarring as new buboes develop in other areas. Some lesions may remain active for years. Failure to treat either type of lesion or treatment with topical agents alone results in a chronic course of infection. Some cases of spontaneous resolution are known, however.[34]

In chronic sporotrichosis, the lymphatics that drain the area of the initial lesion are involved. Within a few days or weeks of the appearance of the primary lesion, multiple subcutaneous nodules develop along the local lymphatic channels (Fig. 10–1,A). These nodules, which are similar to the primary lesion,

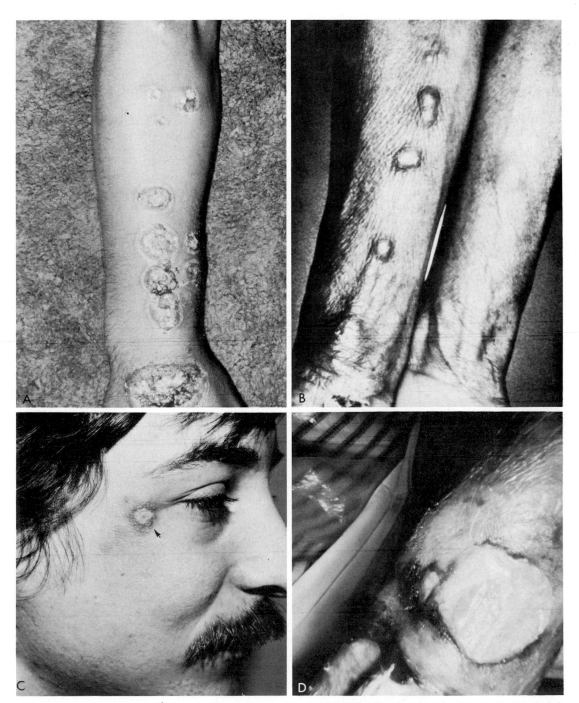

Figure 10–1. *A,* Lymphocutaneous sporotrichosis. The initial lesion is discolored, ulcerated, and draining. Secondary lesions are elevated, but only two have ulcerated. Note the top lesion, recently developed, is a deep subcutaneous nodule. (Courtesy of A. Restrepo.) *B,* Lymphocutaneous disease with only the initial lesion breaking down to ulcerate. The arm on the right side of the picture shows a positive sporotrichin skin test. (Courtesy of A. Restrepo.) *C,* Pyoderma-like lesion (arrow) in a student greenhouse worker resulting from a cut caused by a cycad.[73a] (Courtesy of M. Pepper.) *D,* Gummatous sporotrichosis presenting as pyoderma gangrenosa. Multiple clean-cut ulcers exposing underlying fascia and tendons. (From Stroud, J. D. 1968. Arch. Dermatol. 97:667–670.)

are movable at first but later become attached to the overlying skin (Fig. 10–3,*B*). They become discolored and suppurate, and their connecting lymphatics become hard and cordlike. The clinical picture of an ulcer on the finger or wrist with an associated chain of swollen lymph nodes extending up the arm is so pathognomonic as to be an "over the telephone" diagnosis (Fig. 10–1,*A,B*). A thin, seropurulent discharge may drain from the first few lesions; however, as larger and more distant nodes become involved, there is less tendency to necrosis and drainage. The secondary lesions are more gummatous and tend to persist for months or years. Untreated sporotrichosis usually becomes chronic, but some cases heal spontaneously.

In a few instances, "gummatous" sporotrichosis may develop[15a, 89, 94] (Figs. 10–1,*D* and 10–2). Such cases are characterized by the relentless development of more and more clear-cut ulcerations on wrists, legs, or forearms. These lesions are deep and expose underlying fascia, bone, and tendons (Fig. 10–1,*D*). The presentation is similar to pyo-derma gangrenosa. An unusual form of hypersensitivity has been suggested for this exaggerated response and may be similar to gumma formation in late syphilis and the "Lucio phenomenon" in leprosy.[89]

Fixed Cutaneous Sporotrichosis

In highly endemic areas, a significant portion of the population becomes sensitized (without infection) to *S. schenckii*. This is demonstrated by reaction to the sporotrichin skin test.[34, 35] Primary infection in such people is quite commonly restricted to the site of inoculation and is called the fixed cutaneous type of sporotrichosis. The commonest sites of infection are the face, neck, and trunk (Fig. 10–3,*A*, 10–4). In an endemic area of Australia, fixed cutaneous lesions occurred most often on the legs, probably the results of scratches from plants.[7]

The lesions manifest themselves as ulcerative (Fig. 10–4), verrucous, acneform, infiltrated, or erythematoid plaques, or as scaly, patchy, macular or papular rashes that do not involve local lymphatics and remain "fixed." Small satellite lesions are common (Fig. 10–3,*A*, 10–4). This type of sporotrichosis may account for 40 to 60 per cent of cases in some geographic areas. In a survey of 150 cases near Medellin, Colombia, Velasquez et al.[96] reported 40 per cent were of the lymphocutaneous type, 45 per cent of the fixed cutaneous type and 15 per cent polymorphic type. In most large series of cases, however, the fixed cutaneous type of infection usually does not exceed 20 to 25 per cent. Infections in this category very seldom evolve into systemic disease. Since the lesions are so variable in appearance and often become crusted or weeping, they must be differentiated from many cutaneous diseases. Some of these are, in order of commonness, verrucous tuberculosis, papular necrotic tuberculosis, crustaceous syphilids, chromoblastomycosis,[96] and cutaneous leishmaniasis. The fixed cutaneous lesions on occasion heal spontaneously but otherwise are resistant to local therapy. They may remain active for years, alternately healing and reppearing at the same location, often with a change in morphology.

Mucocutaneous Sporotrichosis

Infection of the mucosa without infection in other areas was described in early reports of sporotrichosis,[24] but this type of disease

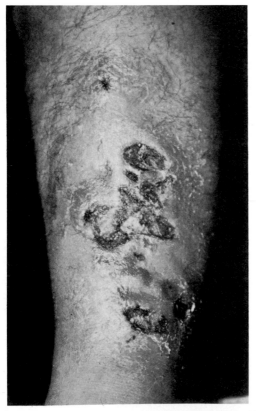

Figure 10–2. *"Gummatous"* sporotrichosis. Necrotic, ulcerating lesion of the leg. The lesion was diagnosed initially as stasis ulcer. (Courtesy of S. MacMillen.)

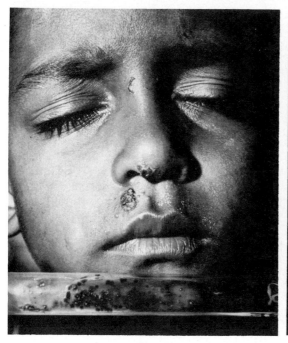 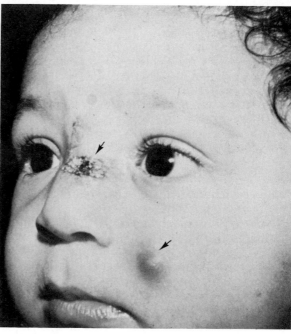

A   B

Figure 10–3. *A*, Fixed cutaneous sporotrichosis. Maculopapular rashes, satellite lesions and the culture of the organism. (Courtesy of F. Battistini.) *B*, Lymphocutaneous sporotrichosis with primary lesion on the nose (arrow). A secondary raised erythematous lesion is forming on the cheek (arrow). Both this lesion and those on the boy (*A*) are the result of scratches with pineapple leaves.

remains relatively rare. Involvement of the mucocutaneous areas secondary to dissemination is not infrequent, however. In either case the morphology of the lesions is reported to be similar. In the mouth, pharynx, or nose the lesion is erythematous, ulcerative, and suppurative at first and eventually becomes granulomatous, vegetative, or papillomatous. The infection is usually accompanied by pain (which is unusual in cutaneous lesions), swelling, and inflammation of the involved areas. Regional lymph nodes become

hard and enlarged. Chronicity is not common, and the lesions usually heal with nondeforming scar formation. Even after healing, lesions may still harbor a small flora of *S. schenckii.* Mucous membrane sporotrichosis may resemble aphthous ulcers, oral lichen planus, or secondary cutaneous leishmaniasis.

Extracutaneous and Disseminated Sporotrichosis

This is a convenient but somewhat artificial grouping of disease types. Although it is rare, disease of the skeletal system is the most common type of infection next to the cutaneous forms of sporotrichosis. In almost all cases there was a history of the presence of cutaneous lesions prior to osseous involvement,[70, 99] but in a few instances, direct inoculation into the knee has been recorded.[101] Rarely, dissemination involving bone, joints, and other systems may be associated with cortisone therapy.[58, 70] It is also possible that some cases represent spread from inapparent primary pulmonary infections.[14, 81]

Bone, Periosteum, and Synovium. In a review of 30 cases of extracutaneous sporo-

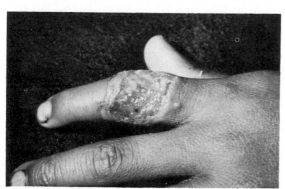

Figure 10–4. Fixed cutaneous sporotrichosis. Verrucous, ulcerative lesion of finger with a satellite lesion. (Courtesy of F. Pifano.)

trichosis, 80 per cent had lesions involving bony tissue. All 30 had had cutaneous or subcutaneous lesions also.[14, 101] The disease is described as a destructive arthritis, with osteolytic lesions, tenosynovitis, and periosteitis. Lesions of the metacarpals and phalanges were most frequently noted, although the tibia is also commonly involved. Pathologic fracture at this latter site was noted very early in the French literature.[23, 24] In descending frequency, other sites of bony involvement include carpal, metatarsals, tarsals, radius-ulna, femur, and ribs.

Sporotrichotic arthritis also occurs in the absence of clinically apparent cutaneous or pulmonary disease. This adds to the difficulty of diagnosis. In a recently reported review of seven cases, Crout et al.[19] found the average time from onset of joint symptoms to the establishment of diagnosis to be 25 months. It must be emphasized that in evaluations of serologic tests, the immunodiffusion and slide agglutination procedures are almost 100 per cent positive in articular disease. Thus, these can be used in screening patients with arthritis of obscure etiology.

In articular sporotrichosis symptoms of pain, swelling, and chronic progressive limitation of motion are usually recorded.[99, 101] Pain is not always a constant feature.[58] A viscous, serosanguineous joint effusion that has an elevated protein and cell count is usually present. Biopsy of synovial tissue reveals a granulomatous inflammation. Successful isolation of the fungus usually results from direct inoculation of such fluid onto culture media. In general the organisms are more numerous in cases of extracutaneous sporotrichosis than in cutaneous disease. The studies of Crout et al.[19] indicated that tissue from open synovial biopsy was superior to synovial fluid for recovery of the organism. Concomitant synovial fluid and synovial tissue culture was superior to either alone. Amphotericin B and surgical debridement were successful in treatment of the infection, but many patients require arthrodesis.

Eye and Adnexae. The eye and adnexae have been involved in about 50 cases of sporotrichosis. The lids, conjunctiva, and lacrimal apparatus are involved in about two thirds of cases. In about 70 per cent, there were no other sites of infection. These therefore represent primary infections due to deposition of the fungus, probably on organic debris, into the conjunctival and lacrimal area or following surgical manipulation.[44] Preauricular adenopathy is reported to be

absent[4] in some cases, and in only 5 of 48 cases were adjacent nodes enlarged.[101] The lesions are ulcerative and gummatous and run a course similar to primary cutaneous infection. Not infrequently such lesions extend to involve the orbit.

Sinusitis. On rare occasions the nose and sinuses may represent the portal of entry of *S. schenckii*. In a report by Agger et al.,[1] a 27-year-old male diabetic had right-sided ocular pain and necrotizing ethmoid sinusitis, a syndrome resembling rhinocerebral mucormycosis. Enucleation of the eye and amphotericin B were required to arrest the infection.

Systemic Disease. Disseminated sporotrichosis with involvement of organ systems other than the skeletal is quite rare. Sometimes there is some underlying disease, such as diabetes, sarcoidosis, a history of long-term cortisone therapy,[58] alcoholism,[81] or as an opportunistic infection in patients with neoplasias.[80] Dissemination associated with neoplasias is very rare however. In this form of sporotrichosis, the organisms are usually more numerous than in the cutaneous forms of the disease and are easily demonstrated in histologic sections by special stains (Fig. 10–9). Whereas lymphatic spread is usually local and develops slowly, hematogenous dissemination may result in the sudden widespread involvement of many organ systems with multiple lesions. Most of these occur in cutaneous, osseous, or muscle tissue, but there are records of polynephritis, orchitis, epididymitis, and mastitis. Involvement of liver, spleen, pancreas, brain,[81] thyroid, and myocardium is very rare but has been reported.[81, 101] Dissemination of the disease is frequently associated with a fever of 39°C or higher, anorexia, weight loss, and stiffness of joints.

Meningitis. Infection of the central nervous system is extremely rare.[31, 45, 84] In the three or four recorded cases, the primary site of infection was obscure in at least two, but the others had previous lesions in cutaneous tissue. Barrier breaks are suspected in the other cases. In a report by Schoemaker,[84] disease occurred in a patient two years following myelography, culminating in neurologic decline and death. Another similar case was treated successfully with amphotericin B.[45] In the recent case reported by Freeman and Ziegler,[31] the source of infection was obscure, but the disease responded to amphotericin B. The presenting symptoms in all cases included dizziness, headache, confusion, and weight loss. The spinal fluid protein was elevated to about 400 mg per 100 ml, and a pleocytosis of 200 to 400 cells per ml (predominantly lymphocytes) was present, with other parameters within normal limits. *S. schenckii* was recovered from spinal fluid by culture. In the fatal cases, granulomatous microabscesses were scattered throughout the cerebral cortex.[84]

Pulmonary Sporotrichosis

In disseminated sporotrichosis, the lung is very rarely involved when the primary infection is at some other body site. Until recently only about 30 cases of pulmonary sporotrichosis had been recorded. It now appears that most of these were primary disease, with infection resulting from the inhalation of conidia, a situation analogous to primary pulmonary histoplasmosis or coccidioidomycosis.[40, 67, 69, 76, 78, 103] In large urban hospitals, pulmonary sporotrichosis is not infrequent and is now considered one of the major mycoses of the lung. However, it is seldom diagnosed, or, if diagnosed at all, it is retrospectively determined by serologic evaluation using immunodiffusion. It appears to be particularly prevalent among chronic alcoholics.

The disease manifests itself as two general types, both of which closely resemble forms of tuberculosis. The first and most frequently reported type is chronic cavitary disease[40, 103] (Fig. 10–5). The general course of this form is similar to that of other pulmonary mycoses. The infection begins as an acute pneumonitis or bronchitis that is sometimes minimal and may go unnoticed or be accompanied by fever, cough, and malaise. Apical areas of the lung appear to be the favored sites of infection, thus leading to an initial diagnosis of tuberculosis. The disease becomes a chronic pneumonitis, with nodular masses and development of thin-walled cavities with fibrosis and pleural effusions. If left untreated, these infections may remain stationary, but usually the disease progresses and extends. Cavitation sometimes becomes massive, with caseation necrosis often leading to a fatal outcome. Recovery has been effected with iodides, amphotericin B, or surgical excision.[40] Spontaneous recovery is unknown. In several of the reported cases of chronic cavitary disease the proper diagnosis was not made until many years after the onset of symptoms.[103] In all such cases in which it was carried out, the immunodiffusion or some other serologic

procedure was positive. Therefore, these tests can be used as a screening procedure when the etiology of a chronic pulmonary disease is not apparent.

A second type of pulmonary infection has been described which involves the lymph nodes primarily. The disease is acute and rapidly progressive,[76] but resolution of the lesions and recovery are frequent (Fig. 10–6). In this type of infection, the tracheobronchial lymph nodes are involved, and the parenchyma of the lung per se is largely spared. Hilar lymphadenopathy may be of such magnitude as to cause bronchial obstruction. The disease of the lymph nodes may remain stationary for long periods of time, and spontaneous resolution is not infrequent. The syndrome of massive adenopathy is usually diagnosed as primary tuberculosis, but this can be ruled out by a negative, second-strength PPD or 1:100 O.T. In both forms of pulmonary sporotrichosis, the diseases to be considered next in the differential diagnoses are histoplasmosis, coccidioidomycosis, and sarcoidosis. The fungal infections may be eliminated by negative skin and complement fixation tests, and sarcoidosis by a negative Kveim test. Cultural and serologic confirmations are necessary to establish a diagnosis of sporotrichosis. Cultures are best done from bronchial washings or biopsy material, since sputum cultures are usually overgrown by *Candida,* and *S. schenckii* has been recovered on occasion as transient normal flora. Use of Smith medium inhibits the growth of *Candida* and allows the development of *S. schenckii* from sputum. On primary isolation from sputum, *S. schenckii* is sometimes colorless and yeastlike and resembles *Geotrichum.* For this reason it is often discarded as a contaminant.

The sporotrichin skin test using whole yeast cell antigen is usually strongly positive in cases of pulmonary sporotrichosis. After fading, this reaction sometimes reappears during therapy because of release antigen. González-Ochoa has described a polysaccharide skin test that is more specific[34, 35] than the commercially available sporotrichin test. Complement fixation titers, immunodiffusion lines, and agglutination titers appear in the serum of infected patients and can be used to follow the progress of therapy. The latter test appears to be the most accurate for this purpose,[42, 78, 100] but immunodiffusion is quite specific and easily performed. Precipitin lines disappear with recovery from disease.

X-ray Findings. The two major forms of pulmonary sporotrichosis are reflected in the

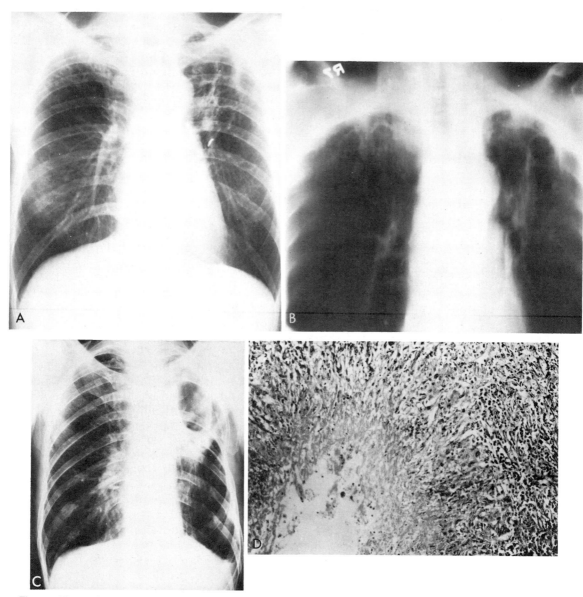

Figure 10–5. *A*, Chronic pulmonary sporotrichosis. There is a marked loss of volume in the left lobe, diffuse interstitial infiltrate, and many cystic areas in the apex. Similar changes are present in the upper right lobe, but they are less severe. (Courtesy of H. Grieble.) *B*, Tomogram showing multiple, thin-walled cavities in the apices. *C*, Large cavitary sporotrichosis of left lung. The patient was diagnosed as having tuberculosis, even though culture and smear were negative. During the 9½-year course, the upper left lobe was removed and cavities developed in the lower lobe. The disease was arrested by potassium iodide after the proper diagnosis was made. (From Rippon, J., and L. Adler. 1979. Clin. Microbiol. Newsletter, Vol. 1, No. 15, pp. 5–6.) *D*, Fibrocaseous lesion in the left upper lobe following resection.

x-ray findings. The first is an acute or chronic pneumonitis, with widespread areas of miliary-type infiltration suggestive of tuberculosis, segmental involvement, or localized expanding areas of opacity simulating tumor. In any of these, extensive cavitation may develop with time (Fig. 10–5). The second acute form involves the tracheobronchial lymph nodes. The lesions are most frequently in the hilar areas, often with mediastinal widening (Fig. 10–6). These may remain unchanged for long periods of time, even following resolution of symptoms. Mottling throughout several fields and atelectasis are described in some cases, which represents a mixed type of nodal and parenchymal involvement.[69, 76]

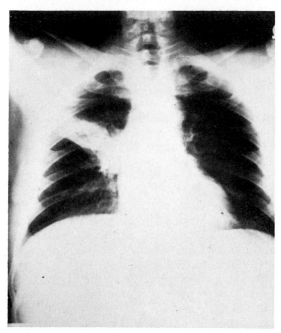

Figure 10–6. Acute pulmonary sporotrichosis. There is dense pulmonary consolidation in the middle right lung, extending from the hilum to the pleural surface. There is pronounced hilar adenopathy and some mediastinal adenopathy. The infection resolved without treatment, leaving only enlarged tracheobronchial nodes. (Courtesy of S. Kabins.)

DIFFERENTIAL DIAGNOSIS

Lymphocutaneous sporotrichosis is so constant in its clinical picture and in the evolution of the disease that diagnosis can be made with considerable confidence on first examination. Tularemia, anthrax, and some other bacterial infections may simulate part of the clinical picture, but these diseases are usually more acute. Sporotrichosis should be considered in any case in which multiple polymorphous eruptions or ulcers occur, particularly if they do not resolve with usual topical therapy. Other mycoses to be considered are chromoblastomycosis, blastomycosis, mycetoma, paracoccidioidomycosis, and granulomatous trichophytosis. Recently a sporotrichoid-like lesion was described, with *Pseudallescheria boydii* as the agent (see Chapter 24). Gummatous syphilis, pyogenic lesions, cutaneous and lymphatic tuberculosis, staphylococcal lymphangitis, glanders, and drug eruptions such as bromoderma also must be considered. Histologic evidence of the organism is often lacking in primary sporotrichosis, and culture is the definitive procedure for diagnosis. Serologic tests may aid diagnosis, but they are usually negative in the cutaneous forms of the disease. In disseminated sporotrichosis, organisms are more numerous and usually evident in tissue sections and biopsy. Cultural confirmation is always necessary, however. Pulmonary disease is almost always diagnosed initially as tuberculosis, but this is true of all fungus infections of the lungs. Other mycoses, sarcoidosis, and tumor must also be considered.

PROGNOSIS AND THERAPY

Lymphocutaneous, fixed cutaneous, and mucocutaneous sporotrichoses are chronic infections that may alternately develop and regress for years in untreated patients. There is little distress or impairment of activity to the patient, and spontaneous cure occurs and is probably not infrequent, particularly in the fixed type of disease. On the other hand, the prognosis in disseminated sporotrichosis, as with all disseminated fungus disease, is grave. Organisms are numerous, and host defenses are inadequate. Spontaneous cure is unknown. Pulmonary disease limited to the hilar nodes may remain quiescent and often resolves. In cases in which there is more parenchymal involvement or in which chronic pneumonitis has developed, progression is common, and the disease may be fatal if not treated.

Following the suggestion of Sabouraud, de Beurmann[24] was the first to employ iodides in the chemotherapy of sporotrichosis. Remarkable success was encountered, and this therapy remains the treatment of choice today.[7, 9, 75] Treatment schedules vary somewhat, but all utilize rapidly increasing doses given daily. The drug is administered orally in milk as a saturated solution of potassium iodide. The initial dose is 5 drops (1 ml) three times a day. This is increased by 3 to 5 drops each dose each day (minimum increase 1.5 ml per day) as tolerated, until 30 to 40 or more drops each dose are given (4 to 6 ml per dose, 12 to 18 ml per day). If signs of intoxication occur, such as indigestion, rash, lacrimation, cardiac problems, or swelling of salivary glands, the schedule can be tapered or the drug can be given intravenously. Treatment is continued for at least four weeks after clinical symptoms have resolved. Open lesions can be treated topically with a solution of 2 per cent KI containing 0.2 per cent iodine. Treatment failures are known, and they usually do not respond to retreatment with iodides but require other therapeutic modalities.[8] The mode of action of potassium iodide is unknown. The fungus will grow *in vitro* on media containing a concentration of the drug up to about 10 per cent. Potassium iodide is known to cause resolution of granulomas and other abscesses. It has been suggested that this is because of enhancement of proteolysis and clearing of debris by proteolytic enzymes.[49] A direct effect on the fungus by iodine cannot be discounted.[97]

In very early infections and in uncomplicated cases in which lesions are few, the application of heat is often sufficient to bring about resolution of the disease.[64] Immersion in hot water or battery-operated "pocket warmers" have been utilized.

Amphotericin B is the most effective drug used for the treatment of relapsed lymphocutaneous sporotrichosis and pulmonary and disseminated disease.[8, 19, 43, 70, 76] The two latter forms of infection may respond to iodides, and this should always be tried first. The usual dosage schedule of amphotericin B for mycotic infections is appropriate here. Infusion of the drug into the affected area has not been remarkably successful.

Relapse of the disease has occurred in a few patients treated with both iodides and amphotericin B.[8] This is particularly true in articular involvement and systemic disease. Dihydroxystilbamidine was successfully used in one personally observed case, and others have reported similar results.[101] Dosage schedules vary from 50 to 225 mg per day (2 mg per kg) dissolved in 5 per cent glucose and given slowly by intravenous infusion. The total levels necessary for treatment with this drug are not as high in sporotrichosis as in blastomycosis. The course of therapy consists of ten days of the treatment followed by ten days of rest. Three such treatment regimens are usually sufficient but may be altered depending on clinical response.

In one report, griseofulvin (250 mg four times per day for three months) has been used successfully in fixed sporotrichosis.[48] Many treatment failures have occurred with this drug, and it has not gained wide popularity.[43] Some investigators are using 5-fluorocytosine in a daily dose of 100 mg per kg, and it has been of value when other modalities have failed. Beardmore[8] reported on a recalcitrant sporotrichosis of 11 years that had failed to respond to oral KI, griseofulvin, micronazole, curettage and cauterization, and intralesional injections of amphotericin B. It was finally controlled, if not wholly, at least in part, by 5-fluorocytosine.

Ulcerated cutaneous lesions and lesions of the mucosa are often secondarily infected by bacteria. Antibacterial antibiotics are a useful adjunct in such cases.

PATHOLOGY

Gross Pathology. Because there have been so few cases of systemic sporotrichosis studied at autopsy, patterns of gross changes seen in this disease have not been described. Most cases have revealed osteomyelitis and destructive lesions of the digits and joints, widespread ulcerations of the skin, some enlargement of the spleen and liver, and a few purulent lesions of the kidney, pancreas, and other internal organs.[53, 54] In the rare cases of cerebral involvement, numerous granulomatous microabscesses have been reported scattered throughout the cerebral cortex. In other cases, granulomatous basilar meningitis was present.[84] In only two reports was gastrointestinal involvement noted,[53] and both patients had primary infections at another anatomic site. Thus, the possibility of primary infection of the gastrointestinal tract is remote.[12]

Sufficient numbers of cases of pulmonary sporotrichosis have been studied to establish a discernible pattern of anatomic change. Once they are established in the lung, parenchymal lesions seldom resolve; they either remain fixed or expand with time, causing more and more distress to the patient. Resection of the lung has been successful in treating some of these cases.[40, 76, 103] Upon removal, the lesions are described as gray-white plaques, necrotic granulomata, thin-walled cavities within the lung tissue, or bronchiectatic cavities. The lesions are usually found in the upper lobes of the lung, particularly in the apical areas, although involvement of the middle and lower lobes is recorded. A thin-walled cavity is highly suggestive of fungal etiology, especially when tuberculosis has been ruled out. The lesions average 3 cm in diameter and are solitary or few in number. Rarely, close examination will reveal several small developing nodules peripheral to the main lesion. On cut section, the cavities are filled with a gray-white exudate and have a ragged necrotic lining where the organisms are found. In the bronchiectatic cavity described in one report,[76] there was a lining of necrotic material, and the space was filled with a red liquid.

Histopathology. One of the most difficult aspects of obtaining a diagnosis of sporotrichosis is the dearth of organisms present in biopsy material. Many cases of sporotrichosis are missed initially as the lesions are surgically removed because they are diagnosed as epitheliomas. Even histologically the two appear similar. It is extremely important, therefore, that cultures of biopsy material be performed whenever the diagnosis of sporotrichosis is being considered. In the typical clinical cases, culture and response to iodide treatment corroborate the correct diagnosis. The organisms are elusive in histologic sections, even when diligently looked for. In cases in which the index of suspicion is not high, they are usually not seen. Fetter[28] has modified standard fungal staining procedures to include treatment of the smear with 1:1000 malt diastase. The procedure dissolves all nonfungal polysaccharide, so that when sections are subsequently stained by PAS or methenamine silver, the organisms are much easier to find.

The histologic features of primary cutaneous sporotrichosis are a combination of granulomatous and pyogenic reactions. Lurie in 1963[53] classified the various granulomatous patterns seen as (1) sporotrichotic, (2) tuberculoid, and (3) foreign body. The reaction seen depends somewhat on the site of the lesion. The basic lesion of the sporotrichotic granulomatous reaction consists of masses of epithelioid histiocytes, which have some tendency to form concentric zones. The central area of the lesion consists of neutrophils or necrotic material surrounded by an infiltrate of neutrophils and some plasma cells and lymphocytes. In the tuberculoid reaction, this area sometimes merges into a zone of epithelioid cells mixed with fibroblasts, lymphocytes, and Langhans' giant cells (Fig. 10–7 and Fig. 10–8). Other lesions, however, may

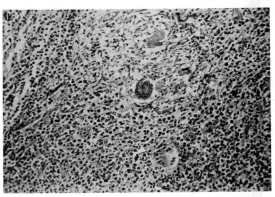

Figure 10–8. Sporotrichosis. High-power view of giant cells, fibroblasts, and lymphocytes. Hematoxylin and eosin stain. (Courtesy of L. Ajello.) ×440.

have a prominent outer layer of plasma cells, which may suggest the presence of a syphilid. In some cases, the histologic picture of a foreign body granuloma without any evidence of a pyogenic component is observed. Sometimes microabscesses without an epithelioid histiocytic component are found (Fig. 10–9). This is particularly notable in disseminated disease.

In chronic sporotrichotic lesions, the pseudoepitheliomatous hyperplasia may be so extensive as to suggest a neoplasm. The rete ridges are elongated and broadened and extend into the corium, which is usually severely inflamed with some round cell infiltration into the rete itself. This combination of pseudoepitheliomatous hyperplasia and a mixed pyogenic-granulomatous-cellular reaction is highly suggestive of secondary cutaneous blastomycosis and coccidioidomycosis. In histologic sections demonstrating the above picture, sporotrichosis should be suspected if there is no evidence of other fungal etiology.

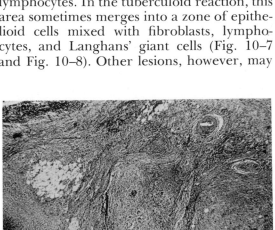

Figure 10–7. Sporotrichosis. Tuberculoid granuloma type lesion. Hematoxylin and eosin stain. (Courtesy of L. Ajello.) ×100.

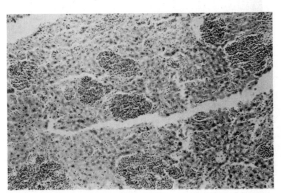

Figure 10–9. Sporotrichosis. Systemic disease. Numerous small microabscesses in a liver. Hematoxylin and eosin stain. ×110.

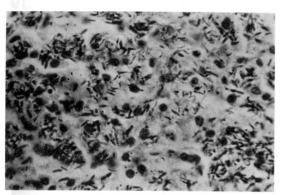

Figure 10–10. Sporotrichosis. Material from experimental orchitis of guinea pig. Gram stain, showing elongate, irregularly staining yeasts. ×400.

Confirmation of the diagnosis often requires obtaining additional material from the patient to be used for culture and obtaining a serum sample for serologic procedures.

As stated before, demonstration of the fungus in tissue section is very difficult. Application of the diastase method[28] or fluorescent antibody techniques[17, 41] (Fig. 10–11) aids in detection of the fungi. Even then the organisms are not numerous, and many sections may have to be examined. When they are found, the fungi are seen to be rounded and to multiply budding, yeastlike cells, 3 to 5 μ in diameter. Gram stains of material from lesions show the organisms to be Gram-positive and irregularly stained (Fig. 10–10). In specimens in which the organisms are few in number, the cells tend to be of smaller size, and budding is infrequent. In lesions in which the fungi are more numerous, they appear cigar-shaped, 3 to 5 μ or wider in diameter, with up to three buds. These cells are most frequently seen in secondary foci of infection. Some lesions contain fungal cells with very irregular morphology, and the presence of mycelial elements in tissue has been reported on occasion.[56] Cigar-shaped cells and large spherical bodies up to 8 μ in diameter are very plentiful in experimentally infected mice (Fig. 10–11). There is no capsule around the yeast cells of *S. schenckii*; however, artifacts caused by cytoplasm shrinking away from the wall or the wall splitting and shrinking during fixation of tissue have given rise to erroneous reports of encapsulated organisms (Fig. 10–12,*B*).[54]

The asteroid body, when seen, is considered characteristic of sporotrichosis, but it has been found in cases of other mycoses as well. Typically it consists of a rounded or oval, basophilic, yeastlike body 3 to 5 μ in diameter. Radiating from this yeast cell are rays of eosinophilic substance up to 10 μ in thickness (Fig. 10–12,*A*). The asteroid phenomenon was first described in 1907 by Splendore,[55] who coined its name. Some authors considered it to be a characteristic of certain strains which was related to their virulence, but this was later disproved. Only ten cases with asteroid bodies were recorded between 1908 and 1940. It is now regularly seen in cases reported from Japan, South America, and South Africa.[13] Careful examination of serial tissue sections is often necessary to detect it, which probably accounts for the paucity of reports of its occurrence. In meningeal involvement, such eosinophilic masses may surround several yeast cells, suggesting that they are encapsulated.[54] Such cells are often seen in cerebral and meningitic sporotrichosis (Fig. 10–13,*B*). Although there had been much speculation as to the nature of the eosinophilic substance forming the rays of the asteroid body, it has now been established that it is an antigen-antibody complex.[53, 54] Hoeppli observed a similar phenomenon around the ova of schistosomes in tissue, and this has been used as the basis for the development of a specific serologic test for schistosomiasis.[72] Similar eosinophilic material is regularly found around the grains in actinomycosis and surrounding the mycelia in entomophthoromycosis as well as in rare cases of candidiasis, blastomycosis, paracoccidioidomycosis, mucormycosis, aspergillosis, and coccidioidomycosis. Moreover, in these diseases the phenomenon is probably the result of an immune reaction.

In disseminated sporotrichosis, ulceration of cutaneous nodules occurs infrequently, and the overlying dermis and epidermis show little pathologic change. The nodule itself is

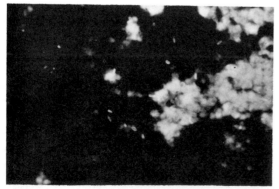

Figure 10–11. Sporotrichosis. Impression smear from liver of experimentally infected mouse stained by fluorescent antibody technique. Small yeast cells are visible. (Courtesy W. Kaplan.)

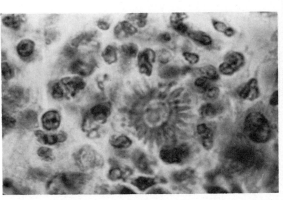 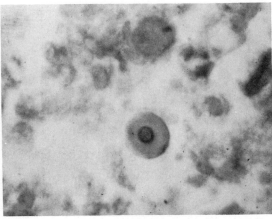

A   B

Figure 10–12. *A*, Sporotrichosis. Asteroid body in testicular tissue of a hamster. Hematoxylin and eosin stain. ×1000. (Courtesy of F. Mariat.) *B*, Yeast cell with a halo resembling a capsule. These cells are frequently seen in cerebral or meningitic sporotrichosis and can be confused with *Cryptococcus. (*Courtesy of J. Schwarz.)

not unlike the granulomata of primary lesions and consists of an outer zone of round cells and plasma cells, an inner layer of histiocytes and giant cells, with a central area of necrosis and neutrophilic infiltrate. The presence of asteroid bodies is reported to be more common in these secondary lesions than in the primary ones.[53, 54] In systemic disease of debilitated patients, the histologic picture is similar to that seen in other opportunistic fungous infections. Small microabscesses, some with necrotic centers, are seen in various organs (Fig. 10–9). Rarely, hyphal elements can be seen in tissue.[66]

ANIMAL DISEASE

Natural infection among animals is common, and sporotrichosis is considered a frequent and serious disease of horses.[22, 30] In addition, the disease has been recorded in dogs,[26, 52] cats,[32, 102] boars, rats, mules, foxes, camels, dolphins,[68] and fowl.[102] Historically, infection in rats in Brazil and in mules in Madagascar was described soon after the first human cases were recorded.[15, 55]

In horses the pathology is similar to that seen in human disease. After introduction of the fungus, subcutaneous nodules form, followed by involvement of the local lymphatic channels and subsequent ulceration and discharge of pus. The fixed cutaneous type of disease has also been described in horses. Treatment with intravenous sodium iodide (2 doses of 2 oz each on alternate days combined with 1 oz of organic iodine in feed) has cured the infection.[30] Treatment failures following iodide therapy have been recorded. In a case

reported by Davis,[22] a regimen of 10 g of KI per day orally and 6 g of NaI intravenously arrested but did not cure the infection. However, griseofulvin, 10 g per day for two weeks followed by 5 g per day for 46 days, led to clinical cure. Sporotrichosis in horses must be differentiated from equine epizootic lymphangitis, which is caused by *Histoplasma farciminosum.*

Lymphocutaneous sporotrichosis occurs in dogs also, but there is a greater tendency for the disease to disseminate in this animal than in other species. Most infections are fatal in spite of treatment. In a few cases of cutaneous disease, treatment with KI has effected cure.[52] Sodium iodide, 20 per cent given at 1 ml per 4.5 body weight, three times daily was successful in three cases of cutaneous disease in dogs as reported by Scott et al.[83]

Sporotrichosis as a canine otitis externa has also been recorded.[26] The disease persisted for over five years and was described as consisting of multiple cutaneous nonsuppurative nodules in the auricle. The lesions resolved with oral KI and topical griseofulvin. The latter probably had no effect on the course of the infection. A granulomatous lymphadenitis that was severe and necrotizing was described in a Pacific white-sided dolphin (*Lagenorhynchus obliquidens*) held in captivity for three years. *S. schenckii* was identified by histologic and fluorescent antibody techniques as the etiologic agent.[68]

Disseminated disease in a cat was described in a report by Garrison et al.[32] The initial lesion was a pyogranulomatous ulceration of the right forepaw. At autopsy the fungus was found in the axillary lymph nodes, the lungs, and the liver. Garrison made a fine structure

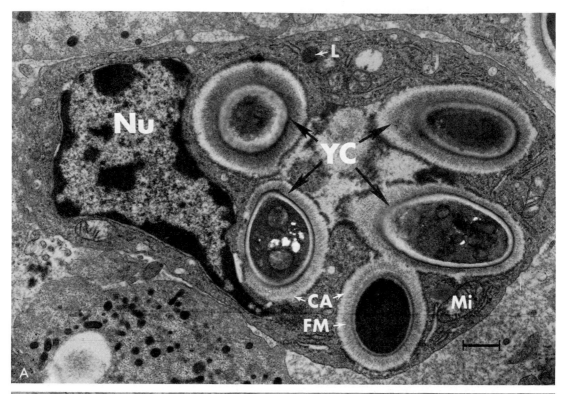

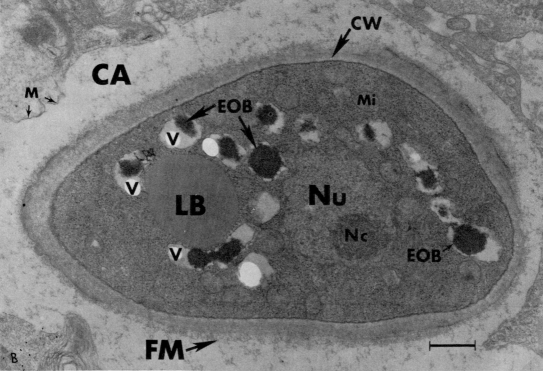

Figure 10–13. Electron photomicrographs of *S. schenckii* in infected tissue. *A*, Longitudinal thin section of a macrophage containing five yeastlike cells (*YC*) of *S. schenckii* in the plane of section. The macrophage cytoplasm contains a single nucleus (*Nu*), scattered mitochondria (*Mi*), and a lysosome (*L*). Note the clear outer area (*CA*) and microfibrillar material (*FM*) at the outermost portions of the fungal cell walls. Bar equals 1.0 μm. *B*, Longitudinal thin section of a free ovate yeastlike cell of *S. schenckii*. Note the microfibrillar material (*FM*) along the outer surface of the cell wall (*CW*), the electron transparent area (*CA*), remnants of the limited membrane (*M*), nucleus (*Nu*), with nucleolus (*Nc*), mitochondria (*Mi*), lipid body (*LB*), and electron opaque bodies (*EOB*) contained within vacuoles (*V*). Bar equals 0.5 μm.

Illustration continued on opposite page

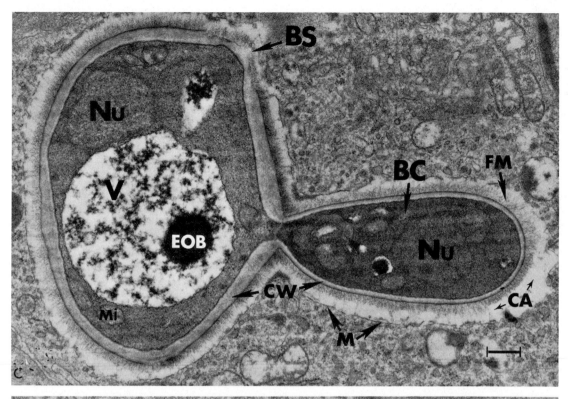

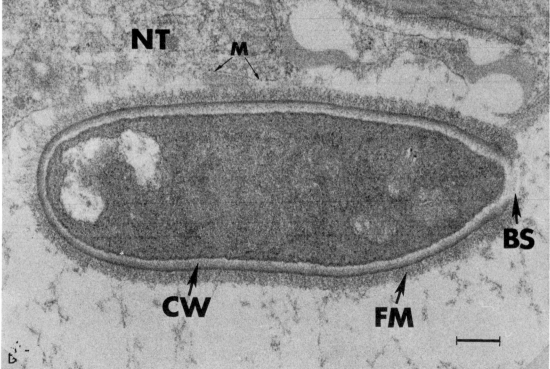

Figure 10–13 *Continued   C,* Longitudinal thin section of a budding yeastlike cell *(BC)* of *S. schenckii.* Note the limiting membrane *(M),* the clear area around the cell *(CA),* the microfibrillar material *(FM)* at the cell wall *(CW),* lipid body *(LB),* the nucleus *(Nu),* the mitochondria *(Mi),* and the large vacuole *(V)* containing electron opaque material and body *(EOB).* Bar equals 0.5 μm. *D,* Longitudinal thin section of a bacillary form of yeastlike *S. schenckii.* Note the necrotic host tissue *(NT),* remnants of the limiting membrane *(M),* the polar bud scar *(BS),* and microfibrillar material *(FM)* along the cell wall *(CW).* Bar equals 0.25 μm. (Courtesy of R. Garrison.)

study of the interaction of the host phagocytic cells and the yeasts of *Sporothrix schenckii*. Both free and phagocytized yeast cells were seen. The free cells were within an extracellular electron transparant vacuolar area probably of host origin (Fig. 10–13). A vacuolar area containing yeast was seen within macrophages as well.

Experimental infections are easily produced in rats, mice, and hamsters. Mice are particularly susceptible to disease. Scrapings of the mycelial mat or suspensions of yeast may be injected intraperitoneally into mice or intratesticularly into mice or guinea pigs. Within ten days a severe peritonitis or orchitis develops. Gram stain of pus from such lesions demonstrates numerous irregularly staining Gram-positive cigar-shaped yeast cells (Fig. 10–10). In testicular lesions of guinea pigs asteroid bodies are sometimes seen (Fig. 10–12,*A*). The disease is progressive in mice, in which dissemination with destructive lesions of bones and other internal organs leads to death, usually by three weeks.[93] Most isolants from human disease regularly kill mice if given in a dose of 10^5 yeast cells. However, Howard and Orr[38] have demonstrated that few soil isolates are capable of growth at 37°C or are pathogenic for mice. This indicates that man acts as a selective agent for those strains that are able to evoke disease. This is probably true for many other "pathogenic" fungi. Serial animal passage has been shown to increase virulence. By this method, Mariat[59, 60, 73] has adapted strains of *Ceratocystis stenoceras* to produce a disease similar to sporotrichosis. He believes that *S. schenckii* and *C. stenoceras* may be identical. Charoenuit and Taylor[16] were able to produce a self-limited lymphatic infection or a disseminated fatal disease in hamsters by means of varying the size of the inoculum. They also demonstrated the immunizing efficacy of the ribosomal fraction of trypsinized cells.

BIOLOGICAL STUDIES

Sporothrix schenckii is one of the five species of thermal dimorphic fungi of medical interest. Many studies have been made of the morphologic changes that occur under the influence of temperature. These studies are considered in the sections on colony morphology and mycology. In addition to morphologic changes there are also changes in physiology and cellular composition that ac-

company dimorphism. In a series of investigations, Lloyd, Travassos, and Mendonça-Previato et al.[50, 51, 92] have shown the cell wall composition to differ in the yeast, mycelial, and conidial structures of the organism. Monorhamnosyl-rhamnomannans were found in the yeast cell wall, in the yeast primorida on the hyphae, and in conidia. Rhamnose is uncommon among fungal species. Dirhamnosyl-rhamnomannans were found in hyphal walls but not in other cell structures. Sera from human patients react primarily with the monorhamnomannans, as would be expected, but there are some antibodies to the dirhamnomannanosides that may indicate some formation of mycelium *in vivo*.[92] The morphologic, physiologic, and compositional changes that occur during dimorphism are reviewed in Rippon.[77]

The relative virulence of isolants from cases of human infection compared with that of isolants from nature was first systematically studied by Howard and Orr[38] in 1963. Investigators consistently show a lesser pathogenic potential in isolants from nature. A difference in lipids relevant to this difference in virulence was suggested by the work of Stretton and Dart.[88] The major lipid component in strains from human disease was a C_{18} diene, whereas saprophytic isolants had a lesser amount of this component and more of other compounds, particularly C_{16} saturates. Virulence related to lipid composition has been suggested in *Blastomyces dermatitidis* and *Coccidioides immitis* as well. Kwon-Chung[46] compared isolants from lymphocutaneous and fixed cutaneous disease. The former grew readily at 37°C and 25°C, whereas isolants from the latter type of infection failed to grow above 35°C. This was also reflected in the type of disease produced in experimentally infected mice.

There have been several attempts to produce disease by the gastrointestinal route. De Beurmann and Gougerot[23, 24] fed newborn guinea pigs large numbers of yeast cells in milk. A few infections resulted. Davis[20] again tried this method and could only obtain infection in rats given very large amounts of inoculum. These studies have not been repeated utilizing modern methods of graded inoculum. These studies have now been repeated utilizing modern methods of graded inoculum, several host parameters, and so forth. The results were negative.*

---

*Kennedy, M. J., P. S. Bajwa, and P. A. Voltz. 1982. Gastrointestinal inoculation of *Sporothrix schenckii* in mice. Mycopathologia. In press.

IMMUNOLOGY

Natural resistance to sporotrichosis is quite high. Since the organism is universally present in soil, factors other than exposure must be important in areas where the disease is prevalent. Mariat[58a] considers nutritional state as well as repeated exposure to be of great importance in the majority of patients who contract the disease. There is some experimental evidence in support of this. Most cases of lymphatic disease are contracted by the malnourished rural poor who are in constant contact with the soil and hence the fungus. Pulmonary disease, however, is most frequently found in chronic alcoholics. Kedes[43] has coined the "alcoholic rose gardener syndrome" for patients who have constant exposure to soil and are imbibers.

In highly endemic areas, hypersensitivity develops in persons with no clinical signs of infection. In studies of grass handlers by González-Ochoa,[34, 35] the development of positive skin tests correlated with the length of time spent at this occupation. After ten years all workers had positive skin tests. In many individuals this hypersensitivity appears either to prevent infection or to modify the course of disease. The fixed cutaneous type of sporotrichosis is most common in this population. A similar survey was conducted in areas where clinical sporotrichosis had never been reported; there were essentially no reactors to the skin test.[39] This indicates that this procedure is a valuable epidemiologic tool for assessing the prevalence of the disease or exposure to the organism.

The cell mediated immunity (CMI) in patients with cutaneous disease differs from the CMI in patients with systemic infections. Plouffe et al.[74] demonstrated that patients with infection restricted to the skin had normal CMI when measured by response to antigen by the conventional assays of lymphocytic transformation. These were abnormal or lacking in patients with protracted systemic illness. Use of transfer factor in such patients showed improvement in parameters of CMI and control of their disease.

SEROLOGY

Strains of *S. schenckii* isolated from human and animal disease and from soil are all antigenically similar by cross-agglutination tests.[5] Antigens prepared from these strains are used for skin testing, complement fixa-

tion tests, and immunodiffusion tests. There appear to be few cross-reactions with other fungal diseases.

The standard sporotrichin skin test consists of the intradermal injection of a 1:1000 dilution of heat-killed, packed yeast cells. A reaction of 5 mm of induration after 24 hours constitutes a positive test. As noted above, this test is useful for determining exposure to the fungus. A test using a polysaccharide derived from a crude antigen preparation has been developed by González-Ochoa and Figueroa. This test appears to be more specific than the standard skin test for diagnosis of actual infection.[35, 36, 58a] However, some authors have found both the crude antigen and the polysaccharide derivative equally sensitive in the diagnosis of the disease.[79]

Extensive reviews of the serologic procedures used in the diagnosis and prognostic evaluation of sporotrichosis have been published by Karlin and Nielsen[42] and Blumer et al.[11] Karlin and Nielsen[42] concluded that the yeast cell agglutination test and the latex particle agglutination test were the most sensitive serologic tests for all forms of the disease. The yeast agglutination test developed by the authors utilized lyophilized yeast cells. The material was suspended in pH 7 buffer to an O.D. of 0.12 at 420 mμ, which is equivalent to a final concentration of 0.1 mg per ml. Yeast cells that had been centrifuged for 20 minutes at 2500 rpm and then diluted 1:2400 (v/v) could also be used as antigen. In the performance of the test, serial dilutions of the patient's sera were made in phosphate buffer (pH 7) to a final volume of 0.5 ml. To each tube was added 0.5 ml of yeast antigen, followed by incubation at 37°C for one hour, then storage at 4°C overnight. After this, the tubes were centrifuged and gently agitated. The last tube containing well-formed yeast aggregates was read as the end point. A titer of 1:40 or below was considered questionable by the authors, but a higher titer was felt to be strong evidence for the presence of infection. Similar results were found by Welsh and Dolan.[100] Comparison studies were done on the same sera, utilizing the lyophilized yeast-antigen suspension in complement fixation tests and a broth culture of yeast as antigen for latex agglutination and immunodiffusion tests. They found that the yeast agglutination and latex particle agglutination tests were positive in all cases, but the immunodiffusion test was positive in 80 per cent and the complement fixation only in 39 per cent. The agglutination and immunodiffusion tests were recommended for diagnosis and prognosis of sporotrichosis. Agglutinin titers fell and precipitins disappeared with resolution of the disease.

The immunodiffusion (ID) test is the easiest serologic procedure to perform and, along with the latex agglutination, the most specific. It is almost always positive in any extracutaneous form of sporotrichosis. The value of ID as a screen in pulmonary, articular, osseous, and disseminated disease is well established.[78, 103] In a comparison of the various serologic tests for all forms of sporotrichosis, Blumer et al.[11] concluded that the latex slide agglutination test was superior to all others. It was positive in 94 per cent of 80 cases of sporotrichosis of various types. Furthermore, this test along with the immunodiffusion was the most specific, in that it did not cross-react with sera from other mycotic, bacterial, and parasitic infections. The indirect fluorescent antibody (IFA) had significant numbers of false positive reactions; a few false positives also occurred in the complement fixation (CF) and tube agglutination (TA) tests. The ID was positive in all extracutaneous disease but in only 50 per cent of cases of localized cutaneous infection. The relation of the bands seen in ID test to various enzymes extracted from the fungus was studied by Walbaum et al.[98] They were able to identify 30 bands and characterize each as to its associated fungal enzyme.

LABORATORY IDENTIFICATION

Direct Examination. In the cutaneous forms of sporotrichosis, there are so few organisms present in pus, exudates, biopsy material, and aspirates that, in general, direct examination of such material is unrewarding. An exception to this is the fluorescent antibody staining technique published by Kaplan and coworkers,[41] and recently by Chuang et al.[17] Utilizing this procedure, material from lesions, histologic slide preparations, and mycelia or conidia from cultures can be specifically stained.

Culture Methods. *S. schenckii* grows well on almost all culture media. Aspirates from cutaneous nodules, pus, exudate, and material from curettage or swabbings from open lesions can be planted on Sabouraud's glucose agar or blood agar and incubated at 25 or 27°C. Surgical incision and biopsy are contraindicated as methods for obtaining material for culture, since these procedures may result

in spread of the disease. The fungus is resistant to cycloheximide; therefore, Mycosel or Mycobiotic agar also can be used for culture. The percentage of cultures that become positive in cases of the disease is very high, and this is the most reliable method for diagnosis. Growth usually occurs within three to five days. Cultures should be held for at least four weeks before being discarded as negative.

The colony morphology of the initial growth of *S. schenckii* is quite variable. In cultures of material from cutaneous lesions and articular aspirates, a black to brown fungal colony appears, which is shiny at first and later becomes fuzzy. Young colonies of many of these isolants and also those from cases of pulmonary sporotrichosis are white, glabrous, and yeastlike at all temperatures (Figs. 10–14 and 10–15,*A*). In time, the colony becomes wrinkled and membranous, later developing areas of discoloration that normally become black and mycelial (Fig. 10–15,*B*). Coloration is inconstant, and strains vary considerably. Even the same strain will vary from transfer to transfer. Some colonies remain white and glabrous, which may cause them to be misread as *Geotrichum* species and discarded as contaminants (Fig. 10–15,*A*). Special media, such as cornmeal, malt, or Czapek's agar, are useful in inducing mycelium formation, conidiation, and pigmentation.

Demonstration of dimorphism is important for specific identification of *S. schenckii*. To induce mycelial to yeast transformation, the fungus is inoculated on moist blood agar tubes and incubated at 37°C. Conversion to the yeast form sometimes is restricted to the outer edge of the colony. Animal inoculation

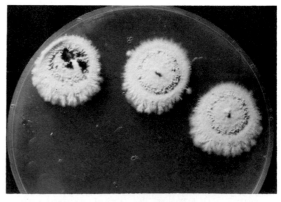

Figure 10–14. White colonies of the initial isolant of *S. schenckii* from a case of articular disease. Subcultures of the organism became black.

is also used for identification, especially of isolants that convert to the yeast phase with difficulty *in vitro*. There are no physiologic tests of importance for identification. *S. schenckii*, as well as many other fungal species, has a requirement for thiamine.

MYCOLOGY

Sporothrix schenckii Hektoen et Perkins 1900

Synonymy. *Sporotrichum* sp. Smith 1898; *Sporotrichum schencki* Matruchot 1910; *Sporotrichum beurmanni* Matruchot and Ramond 1905; *Sporotrichum asteroides* Splendore 1909; *Sporotrichum equi* Carougeau 1909; *Sporotrichum jeanselmei* Brumpt and Langeron 1910; *Sporotrichum councilmani* Wobach 1917; *Sporotrichum grigsby* Dodge 1935; *Sporotrichum fonsecai* Filho 1930; *Sporotrichum cracoviense* Lipinski 1928; *Rhinocladium schencki* Verdun and Mandoul 1924; *Rhinotricum schenckii* Ota 1928.

The first report of *Sporothrix schenckii* infection noted that the organism was dimorphic.[82] Though it was not seen in lesion material, the parasite grew as a mycelial fungus in culture but formed cigar-shaped yeasts when it was infected into animals.[82] Shortly thereafter, Lutz and Splendore[55] demonstrated temperature-dependent dimorphism *in vitro*. *S. schenckii* grows as a budding yeast-like organism at 37°C and as a conidia-producing hyphal fungus at 25°C. As indicated previously, strains isolated in nature vary considerably in their ability to grow at 37°C.[38] Infection may be a selective process of those strains capable of adaptation to higher temperature and growth within animal tissue. Thermal dimorphism and infection as a result of this selective process are also exhibited by *Histoplasma capsulatum*, *Blastomyces dermatitidis*, and *Paracoccidioides brasiliensis*.

Early investigators concluded that the growth form seen in tissue and *in vitro* at 37°C was an extension of the conidiation process. Acropetalous budding of conidia borne on mycelia had been observed *in vitro*, and it was thought that the conidia in tissue formed buds and thus became blastoconidia. However, in a detailed investigation of the process, Howard[37] demonstrated that M-Y transformation in tissue culture involves production of yeast cells directly from the mycelium itself. Conidia were seen to germinate into

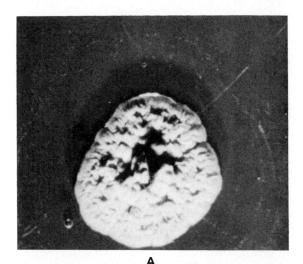

A

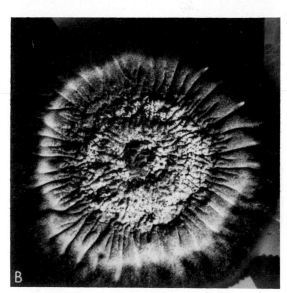

Figure 10–15. *A, S. schenckii.* White colony from pulmonary infection. This strain remained white on subculture and resembles the colonial morphology of *Geotrichum* sp. *B, S. schenckii.* Folded black-brown colony with radial grooves.

short mycelial units, which in turn gave rise to yeast cells. Two morphologic transformations of the mycelium were found to occur. The first involved formation of club-shaped structures at the hyphal tips or on lateral branches, which then gave rise to budding units, the blastoconidia (yeasts). The second method was the formation of "oidia" within the mycelium, followed by fragmentation. These free oidia subsequently bud and are then blastoconidia also. The cycle for this type of yeast cell formation was summarized as mycelium → oidia → blastoconidia. Previous work by the author had shown a similar mechanism in the transformation of *H. capsulatum* and *B. dermatitidis* to yeast-form growth. The budding from club-shaped structures, however, was not found in the latter two species. The yeast to mycelium transformation is effected by elongation of the parent cell, a rearrangement of the cytoplasmic membrane system, followed by septal formation.[47] The elongation continues and branching occurs. It appears that all dimorphic fungi convert from the yeast to the mycelial form of growth in a similar manner.[77]

Colony Morphology. SGA, 25°C. The initial colony of *S. schenckii* isolated from clinical material is moist, glabrous, and yeastlike but becomes tough, wrinkled, and folded in time. The color is usually dirty white at first, although in some strains it is yellow, brown, or quite black (Fig. 10–15,*B*). Pigmentation is extremely variable. Uusually the whitish colony develops an overgrowth of fuzzy mycelium and turns darker in sectors. Old laboratory strains often lose their pigmentation altogether and are a dirty white color. Primary isolants vary considerably in their colony morphology (Fig. 10–16) and may be creamy white, cottony white, cream-brick pink, various shades of brown, and intense black. Some white colonies may not show darkness for many months or never; however, most become dark-streaked in time.

At 37°C on SGA, BHI blood, or other media containing high concentrations of glucose, the organisms grow in the yeast phase. The colony morphology varies from pasty,

Figure 10–16. Colonial forms of *Sporothrix schenckii*. *A*, Top view of glabrous, wrinkled, crusted colony. The dark area is dirty brown-black; the periphery is off-white to cream. This is the "muddy coconut macaroon" colony from an infection associated with peritoneal dialysis. *B*, Side view resembling a volcano on an island. *C*, Primary isolation of black wrinkled and folded mycelial colonies (25°C) from a cavitary pulmonary disease. *D*, Transfer to blood agar at 37°C; glistening white yeastlike colonies.[103]

white, and yeastlike to a greyish yellow, sometimes glabrous, bacteria-like colony (Fig. 10–16,*D*).

Microscopic Morphology. SGA, 25°C. Thin septate, branching hyphae, which generally do not exceed 1 or 2 μ in diameter, are formed at room temperature. Not infrequently they will appear as twisted ropes containing several mycelial strands (Fig. 10–17). At first, conidiation is from long, slender, tapering conidiophores rising at right angles from the hyphae. The conidiophores are erect or recumbent, 1 to 2 μ in diameter at the base, narrowing to 0.5 to 1 μ at the tip (Fig. 10–17). The length is quite variable. The apex of the conidiophore may expand to form a denticulate vesicle, and the phore elongates sympodially to form another apex. Simple, ovate hyaline conidia, 2 to 3 × 3 to 6 μ are formed at first on the apex. Their arrangement suggests a palm tree or a flower head. Sometimes the conidia are angular or obovate. With age, conidiation increases, so

that conidia are formed along the sides of the conidiophore and eventually along the undifferentiated hyphae as well. Dense sleeves of conidia are seen in old cultures. Free conidia have a frilly annular notch where they were attached to the conidiophore. Some conidia may bud once or twice to form acropetalous clusters similar to the *Cladosporium* type of conidia formation seen in *Fonsecaea pedrosoi* or related dematiaceous species. A second conidial form produced by some strains consists of brown, thick-walled triangular one-celled macroconidia[61] (Fig. 10–17,*C*). These pigmented conidia may be more resistant to untoward environmental factors than are the thin-walled hyaline conidia. Malt extract rather than Sabouraud's glucose agar favors macroconidia formation. Some strains of *S. schenckii* are morphologically similar to *Ceratocystis,* and it has been suggested that it may be a conidial form of *Ceratocystis* species.[59, 60] Mariat has produced a sporotrichosis-like disease in mice with isolants of *Ceratocystis steno-*

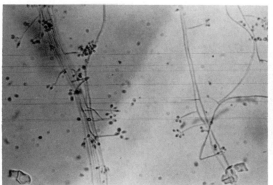

A

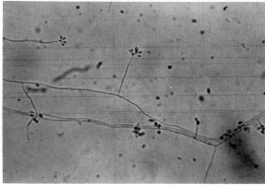

B

C

Figure 10–17. *S. schenckii. A,* Twisted mycelial strands bearing delicate conidiophores and conidia. 25°C. *B,* Tapering, elongate conidiophores showing expanded denticulate vesicle at apex, bearing conidia sympodially. 25°C. × 100. *C,* Obovate, ovoid, and triangular "macroconidia."[61] These conidia have thick walls and are deeply pigmented. They appear to be more resistant to drying and ultraviolet light than hyaline conidia. They are also more virulent for mice. × 2060. (Courtesy of R. Garrison.)

ceras or *C. minor.*[60] The organism recovered from tissue, however, was no longer able to form perfect (sexual) stage spores. Other authors have claimed the perfect stage to be *Dolichoascus schenckii.*[90]

At 37°C in the yeast form of growth, the cells are spherical or ovate blastoconidia. The size is variable, but averages 1 to 3 × 3 to 10 μ. Several buds may appear on the yeastlike cell and resemble the budding conidia found in culture at 25°C.

"Fungi are a mutable and treacherous tribe" is an old adage describing the only constant in the study of mycology: variation. Two variants of *S. schenckii* have been described recently which illustrate this point. Ajello and Kaplan[3] have reported on an isolant that regularly forms sclerotic bodies *in vitro. In vivo* this organism forms thin-walled globose cells (1.6 to 4.8 × 1.6 to 6.9 μ), which reproduce by budding, and thick-walled globose cells (11.7 to 18.7 × 14 to 23.4 μ), which reproduce by septation or by single or multiple budding. The name proposed by them for this isolant is *Sporothrix schenckii* var. *luriei.* The second variant, reported by Brand and Van Niekerk,[13] was isolated from 25 cases of sporotrichosis in South Africa. These isolants did not produce any pigment in culture. In experimental infections, the authors report that only a monocytic response was found in animals, whereas typical cultures induce an acute pyogenic and granulomatous response. The colonial features of these isolants were more typical of *Ceratocystis* species than of the usual strains of *S. schenckii.*

Another species, *S. cyanescens* de Hoog et de Vries, has been reported as occurring in a human infection (Mycol. Study No. 7., p. 21, CBS publications).

REFERENCES

1. Agger, W. A., R. H. Caplan, and D. G. Maki. 1978. Ocular sporotrichosis mimicking mucormycosis in a diabetic. Ann. Ophthalmol., *10*:767–771.

2. Ahearn, D. G., and W. Kaplan. 1969. Occurrence of *Sporotrichum schenckii* on cold stored meat products. Am. J. Epidemiol., *89*:116–124.

3. Ajello, L., and W. Kaplan, 1969. A new variant of *Sporothrix schenckii.* Mycosen, *12*:633–644.

4. Alvarez, G. E., and A. Lopez-Villegas. 1966. Primary ocular sporotrichosis. Am. J. Ophthalmol., *62*:150–151.

5. Andrieu, S., J. Biquet, and S. Massamba. 1971. Étude immunologique comparée de *Sporothrix schenckii* et des souches saprophytes voisines. Sabouraudia, *9*:206–209.

6. Anon. 1972. Hot dogs: a potential source of pathogens. J.A.M.A., *222*:633.

7. Auld, J. C., and G. L. Beardmore. 1979. Sporotrichosis in Queensland: a review of 137 cases at the Royal Brisbane Hospital. Aust. J. Dermatol., *20*:14–22.

8. Beardmore, G. L. 1979. Recalcitrant sporotrichosis: a report of a patient treated with various therapies including oral miconazole and 5-fluorocytosine. Australas. J. Dermatol., *20*:10–13.

9. Becker, F. T., and H. R. Young. 1970. Sporotrichosis. A report of 21 cases. Minn. Med., *53*:851–853.

10. Benham, R., and B. Kesten. 1932. Sporotrichosis: its transmission to plants and animals. J. Infect. Dis., *50*:437–438.

11. Blumer, S. O., L. Kaufman, et al. 1973. Comparative evaluation of five serological methods for diagnosis of sporotrichosis. Appl. Microbiol., *26*:4–8.

12. Boggs, T. R., and H. Fried. 1925. *Sporothrix* infection of large intestine and fingernails. Bull. Johns Hopkins Hosp., *37*:164–169.

13. Brand, F. A., and V. Van Niekerk. 1968. An atypical strain of *Sporothrix* from South Africa. J. Pathol. Bacteriol., *96*:39–44.

14. Brook, C. J., K. P. Ravikrishnan, and J. G. Weg. 1977. Pulmonary and articular sporotrichosis. Am. Rev. Respir. Dis., *116*:141–143.

14a. Carmichael, J. W. 1962. *Chrysosporium* and some other aleurosporic hyphomycetes. Can. J. Bot., *40*:1137–1173.

15. Carougeau, P. 1909. Sur une nouvelle mycose souscutanée de équides. J. Med. Vet. Zootech., *60*:8–22, 75–90, 148–153.

15a. Cawley, E. P. 1949. Sporotrichosis: a protean disease with a report of disseminated subcutaneous gummatous case. Ann. Int. Med., *30*1287–1292.

16. Charoenuit, Y., and R. L. Taylor. 1979. Experimental sporotrichosis in Syrian hamsters. Infect. Immun., *23*:366–372.

17. Chuang, T. Y., J. S. Deng, et al. 1975. Rapid diagnosis of sporotrichosis by immunofluorescent methods. Chin. J. Microbiol., *8*:259–261.

18. Conti Diaz. I. A., and E. Civila. 1969. Exposure of mice to inhalation of pigmented conidia of *Sporothrix schenckii.* Mycopathologia, *38*:1–6.

19. Crout, J. E., N. S. Brewer, et al. 1977. Sporotrichosis arthritis. Ann. Int. Med., *86*:294–297.

20. Davis, D. J. 1919. The permeability of the gastrointestinal wall to infection with *Sporothrix schenckii.* J. Infect. Dis., *16*:688–693.

21. Davis, D. J. 1921. The identity of American and French sporotrichosis. U. Wis. Studies Sci., *2*:104–130.

22. Davis, H. H., and W. E. Worthington. 1964. Equine sporotrichosis. J. Am. Vet. Assoc., *145*:692–693.

23. de Beurmann, L., and H. Gougerot. 1907. Sporotrichoses tuberculoides. Ann. Derm. Syphiligr. IV, *8*:497–544, 603–635, 655–679.

24. de Beurmann, L., and H. Gougerot. 1912. Les Sporotrichoses. Paris, Felix Alcan.

25. de Beurmann, L., and L. Ramond. 1903. Abcès

sous-cutanes multiples d'origine mycosique. Ann. Dermatol. Syphiligr IV, *4*:678–685.

26. Dion, W. M., and G. Speckman. 1978. Canine otitis externa caused by the fungus *Sporothrix schenckii*. Can. Vet. J., *19*:44–45.

27. Du Toit, C. J. 1942. Sporotrichosis on the Witwatersrand. Proc. Mine Med. Offrs. Assoc., *22*:111–127.

28. Fetter, B. F., and J. P. Tindall. 1964. Cutaneous sporotrichosis. Clinical study of nine cases utilizing an improved technique for demonstration of organism. Arch. Pathol., *78*:613–617.

29. Findley, G. H. 1970. The epidemiology of sporotrichosis in the Transvaal. Sabouraudia, *7*:231–236.

30. Fishburn, F., and D. C. Kelley. 1967. Sporotrichosis in a horse. J. Am. Vet. Med. Assoc., *151*:45–46.

31. Freeman, J. W., and D. K. Zigler. 1977. Chronic meningitis caused by *Sporotrichum schenckii*. Neurology, *27*:989–992.

32. Garrison, R. G., K. S. Boyde, et al. 1979. Spontaneous feline sporotrichosis: a fine structural study. Mycopathologia, *69*:57–62.

33. Gonçalves, A. P. 1973. Geopathology of sporotrichosis. Int. J. Dermatol., *12*:115–118.

34. González-Ochoa, A. 1965. Contribuciones recientes al conociemiento de la esporotrichosis. Gac. Med. Mex., *95*:463–474.

35. González-Ochoa, A., E. Ricoy, et al. 1970. Valoracion comparatira de los antigenos polisacarido y cellular de *Sporothrix schenckii*. Rev. Invest. Salud. Publica, *30*:303–315.

36. Hektoen, L., and C. F. Perkins, 1900. Refractory subcutaneous abscesses caused by *Sporothrix schenckii*. J. Exp. Med., *5*:77–89.

37. Howard, D. H. 1961. Dimorphism in *Sporotrichum schenckii*. J. Bacteriol., *81*:464–469.

38. Howard, D. H., and G. F. Orr. 1963. Comparison of strains of *Sporotrichum schenckii*, isolated from nature. J. Bacteriol., *85*:816–821.

39. Ingrish, F. M., and J. D. Schneidau. 1967. Cutaneous hypersensitivity to sporotrichin in Maricopa County, Arizona. J. Invest. Dermatol., *49*:146–149.

40. Jung, J. Y., C. H. Almond, et al. 1979. Role of surgery in the management of pulmonary sporotrichosis. J. Thor. Cardiovasc. Surg., *77*:234–239.

41. Kaplan, W., and M. S. Ivens. 1961. Fluorescent antibody staining of *Sporotrichum schenckii* in cultures and clinical material. J. Invest. Dermatol., *35*:151–159.

42. Karlin, J. V., and H. S. Nielsen. 1970. Serologic aspects of sporotrichosis. J. Infect. Dis., *121*:316–327.

43. Kedes, L. H., J. Siemski, et al. 1964. The syndrome of the alcoholic rose gardener: sporotrichosis of radial tendon sheath. Report of a case with Amphotericin B. Ann. Intern. Med., *61*:1139–1141.

44. Kim, K. H., U. S. Jeon, and S. B. Suh. 1975. A case of sporotrichosis developed after double eyelid operation. Korean J. Dermatol., *13*:193–197.

45. Klein, R. C., M. S. Ivens, et al. 1966. Meningitis due to *Sporotrichum schenckii*. Arch. Int. Med., *118*:145–149.

46. Kwon-Chung, K. J. 1979. Comparison of isolates of *Sporothrix schenckii* obtained from fixed cutaneous lesions with isolates from other types of lesions. J. Infect. Dis., *139*:424–431.

47. Lane, J. W., and R. G. Garrison. 1970. Electron microscopy of the yeast of mycelial phase conversion of *Sporothrix schenckii*. Can. J. Microbiol., *16*:747–749.

48. Leavell, U. W., E. Tucker, et al. 1965. A case of sporotrichosis cleared when given griseofulvin. J. Ky. Med. Assoc., *63*:415–416.

49. Lieberman, J., and N. B. Kurnick. 1963. Induction of proteolysis within purulent sputum by iodides. Clin. Res., *11*:81.

50. Lloyd, K. O., and L. R. Travassos. 1975. Immunological studies on L-rhamno-D-mannans of *Sporothrix schenckii* and related fungi by use of rabbit and human antisera. Carbohydr. Res., *40*:89–97.

51. Lloyd, K. O., L. Mendonça-Previato, and L. R. Travassos. 1978. Distribution of antigenic polysaccharides in different cell types of *Sporothrix schenckii* as studied by immunofluorescent staining with rabbit antisera. Exp. Mycol., *2*:130–137.

52. Londero, A. T., R. M. DeCastro, et al. 1964. Two cases of sporotrichosis in dogs in Brazil. Sabouraudia, *3*:273–274.

53. Lurie, H. I. 1963. Histopathology of sporotrichosis. Arch. Pathol., *75*:421–437.

54. Lurie, H. I., and W. J. S. Still. 1969. The capsule of *Sporotrichum schenckii* and the evolution of asteroid body. Sabouraudia, *7*:64–70.

55. Lutz, A., and A. Splendore. 1907. Über eine bei Menschen und Ratten beobachtete Mykose. Ein Beitrag zur Kenntnis der sogenannten Sporotrichosen. Zentralbl. Bakteriol., I, *45*: 631; *46* (1908): 21 and 97.

56. Maberry, J. D., J. F. Mullins, et al. 1966. Sporotrichosis with demonstration of hyphae in human tissue. Arch. Dermatol., *93*:65–67.

57. Mackinnon, J. E., I. H. Conti-Diaz, et al. 1969. Isolation of *Sporothrix schenckii* from nature. Sabouraudia, *7*:38–45.

58. Manhart, J. W., J. A. Wilson, et al. 1970. Articular and cutaneous sporotrichosis. J.A.M.A., *214*:365–367.

58a. Mariat, F. 1968. The epidemiology of sporotrichosis. In Wolstenholme, G. E. W., ed., *Systemic Mycoses.* London. A. Churchill., pp. 144–159.

59. Mariat, F. 1971. Adaptation de Ceratocystis à la vie parasitaire chez l'animal — étude de l'aquisition d'un pou voir pathogene comparable à celui de *Sporothrix schenckii*. Sabouraudia, *9*:191–205.

60. Mariat, F. 1977. Taxonomic problems related to the fungal complex *Sporothrix schenckii/Ceratocystis* spp. Recent Advances in Medical and Veterinary Mycology. K. Iwata, ed., U. Tokyo Press, 265–270.

61. Mariat, F., R. Garrison, et al. 1978. Premières observations sur les macrospores de *Sporothrix schenckii*. C. R. Acad. Sc. Paris 286 Série D1429-32.

62. Mathieu-Serra, A., and J. P. Collins. 1978. A propos d'un cas de sporotrichose. Union Med. Can., *107*:802–804.

63. Matruchot, L. 1910. Les champignons pathogenes, agents des sporotrichoses, C.R. Acad. Sci., *150*:543–545.

64. Mayorga, R., A. Caceres, et al. 1978. Étude d'une zone d'endemie sporotrichosique au Guatemala. Sabouraudia, *16*:185–198.

65. McDonough, E. S., A. L. Lewis, et al. 1970. *Sporothrix (Sporotrichum) schenckii* in a nursery barn containing sphagnum. Public Health Rep., *85*:579–586.

66. McFarland, R. B. 1966. Sporotrichosis revisited. 65 year follow up of the second reported case. Ann. Intern. Med., *65*:363–366.

67. Michelson, E. 1977. Primary pulmonary sporotrichosis. Ann. Thorac. Surg., *24*:83–86.

68. Migaki, G., R. L. Font, et al. 1978. Sporotrichosis in a Pacific white-sided dolphin (*Lagenorhynchus obliquidens*). Ann. J. Vet. Res., *39*:1916–1919.

69. Mohr, J. A., C. D. Patterson, et al. 1972. Primary pulmonary sporotrichosis. Am. Rev. Resp. Dis., *106*:260–264.

70. Molstud, B., and R. Strom. 1978. Multiarticular sporotrichosis. J.A.M.A., *240*:556–557.

71. Nicot, J., and F. Mariat. 1973. Caractères morphologiques et position systématique de *Sporothrix schenckii*, agent de la sporotrichose humaine. Mycopathol. Mycol., *49*:53–65.

72. Oliver-González, J. 1954. Anti-egg precipitins in serum of humans infected with Schistosoma mansonii. J. Infect. Dis., *95*:86–91.

73. Park, C. H., C. L. Greer, et al. 1972. Cutaneous sporotrichosis: Recent appearance in northern Virginia. Am. J. Clin. Pathol., *57*:23–26.

73a. Pepper, M. C., and J. W. Rippon. 1980. Sporotrichosis presenting as facial cellulitis. J.A.M.A., *243*:2327–2328.

74. Plouffe, J. F., J. Silva, et al. 1979. Cell-mediated immune response in sporotrichosis. J. Infect. Dis., *139*:152–157.

75. Powell, K. E., A. Taylor, et al. 1978. Cutaneous sporotrichosis in forestry workers. Epidemic due to contaminated sphagnum moss., *240*:232–235.

76. Ridgeway, N. A., F. C. Whitcomb, et al. 1962. Primary pulmonary sporotrichosis. Report of two cases. Am. J. Med., *32*:153–160.

77. Rippon, J. W. 1980. Dimorphism in Pathogenic Fungi. CRC Critical Reviews in Microbiology, *8*:49–97.

78. Rippon, J., and L. Adler. 1979. Chronic Pulmonary Sporotrichosis. The importance of appropriate fungal serology. Clin. Microbiol. Newsletter Vol. 1, No. 15 pp. 5–6.

79. Rocha Posada, H. 1968. Prueba cutanea con esporotricina. Mycopathologia. *36*:42–54.

80. Rosen, P. O. 1976. Opportunistic fungal infections in patients with neoplastic diseases. Pathol. Ann. (1976) *11*:255–315.

81. Satterwhite, T. K., and W. V. Kagler. 1978. Disseminated sporotrichosis. J.A.M.A., *240*:771–772.

82. Schenck, B. R. 1898. On refractory subcutaneous abscesses caused by a fungus possibly related to sporothrichia. Bull. Johns Hopkins Hosp., *9*:286–290.

83. Scott, D. W., J. Bentinck-Smith, and G. F. Haggerty. 1974. Sporotrichosis in three dogs. Cornell Vet., *64*:416–426.

84. Shoemaker, E. R., H. D. Bennett, et al. 1975. Leptomeningitis due to *Sporotrichum schenckii*. Arch. Pathol., *64*:222–227.

85. Silva, Y. P., and N. A. Guimaraes. 1964. Esporotrichose familiar epidemica. O Hospital (Rio de J.), *66*:573–579.

86. Simson, F. W. 1947. Sporotrichosis infection in mines of the Witwatersrand. A symposium. Proc. Transv. Mine Med. Officers Assoc.,

87. Smith, L. M. 1945. Sporotrichosis: Report of four clinically atypical cases. South. Med. J., *38*:505–515.

88. Stretton, R. J., and R. K. Dart. 1976. Long chain fatty acids of *Sporothrix (Sporotrichum) schenckii*. J. Clin. Microbiol., *3*:635–636.

89. Stroud, J. D. 1968. Sporotrichosis presenting as pyoderma gangrenosum. Arch. Dermatol., *97*:667–670.

90. Thibaut, M. 1972. La forme parfaite du *Sporotrichum schenckii* (Hektoen et Perkins 1900): *Dolichoascus schencki*: Thibaut et Ansel 1970 Nov. gen. Ann. Parsitol. Hum. Comp., *47*: 431–441.

91. Thompson, D. W., and W. Kaplan. 1977. Laboratory acquired sporotrichosis. Sabouraudia, *15*:167–170.

92. Travassos, L. R., and S. Mendonca-Previato. 1978. Synthesis of monorhamnosyl L-rhamno-D-mannans by conidia of *Sporothrix schenckii*. Infect. Immun., *19*:1–4.

93. Tsubura, E., and J. Swartz. 1960. Treatment of experimental sporotrichosis in mice with griseofulvin and amphotericin B. Antibiot. Chemother. (Basel), *10*:753–757.

94. Urabe, H., and T. Nagushima. 1970. Gummatous sporotrichosis. Int. J. Dermatol., *4*:301–303.

95. Velasco Castegon O., and A. González-Ochoa. 1971. Esporotricosis en individuos con esporotricino reaccion positiva previa. Rev. Invest. Salubr. Publica., *31*:53–55.

96. Valasquez, J. P., A. Restrepo, and G. Calle. 1976. Experiencia de 12 anos con la esporotricosis polimorfismo clinico de la entidad. Antioquia Medica, *26*:153–169.

97. Wada, R. 1968. Studies on mode of action of potassium iodide upon sporotrichosis. Mycopathologia, *34*:97–107.

98. Walbaum, S., T. Duriez, et al. 1978. Study on an extract of *Sporothrix schenckii* (yeast form). Electrophoretic and immunoelectrophoretic analysis; characterization of enzyme activities. Mycopathologia, *63*:105–111.

99. Weitzner, R., E. Mak, and Y. Hertratanakul. 1977. Articular sporotrichosis. Ann. Int. Med., *87*:382.

100. Welsh, M. S., and C. T. Dolan. 1973. Sporothrix whole yeast agglutination test. Ann. Clin. Pathol., *59*:82–85.

101. Wilson, D. E., J. J. Mann, et al. 1967. Clinical features of extracutaneous sporotrichosis. Medicine (Baltimore), *46*:265–280.

102. Werner, R. E., B. G. Levine, et al. 1971. Sporotrichosis in a cat. J. Am. Vet. Med. Assoc., *159*:407–412.

103. Zvetina, J. R., J. W. Rippon, and V. Daum. 1978. Chronic pulmonary sporotrichosis. Mycopathologia, *64*:53–57.

11 Entomophthoromycosis

DEFINITION

Entomophthoromycosis is a chronic inflammatory or granulomatous disease which is generally restricted to the subcutaneous tissue or the nasal submucosa. It is caused by species of the order Entomophthorales[22, 35, 52a] of the Zygomycetes and includes two clinically and mycologically distinct diseases. The first, entomophthoromycosis conidiobolae, involves the nasal submucosa and is characterized by the presence of polyps or extensive palpable subcutaneous masses that generally remain restricted to that area. Clinical variants, including pulmonary and systemic infections, have also been described. The etiologic agent, *Conidiobolus coronatus,* is a soil organism that is pathogenic for termites, other insects, and spiders. *Conidiobolus incongruus* has been documented in one case of pericardial and pulmonary disease.[20]

The second disease, referred to as entomophthoromycosis basidiobolae, is characterized by massive, palpable, indurated, nonulcerating, subcutaneous masses on the limbs, trunk, chest, back or buttocks. The etiologic agent, *Basidiobolus ranarum* (and possibly other species), is a soil fungus found in leaf detritus and the intestinal tract of amphibians and reptiles. Both diseases are chronic and develop slowly. Although they are seldom life-threatening, they are very disfiguring. In contrast to mucormycosis, an infection caused by species of the Mucorales which are also in the class Zygomycetes, the entomophthoromycoses have not yet been associated with underlying predisposing factors and therefore do not appear to be opportunistic infections. A major difference between mucormycosis and entomophthoromycosis is in the pathology evoked by infection. Mucormycosis is characterized by invasion and occlusion of blood vessels resulting in embolic phenomena, ischemia, and necrosis. In entomophthoromycosis, vascular invasion by the hyphae does not occur. Other major differences exist between the two disease entities involving prognosis, clinical development, treatment, and pathology. Thus the separation of the diseases entomophthoromycosis and mucormycosis is based on clinical, pathological, and etiological differences. To include these two disease entities in the category zygomycosis is unwarranted. The disease mucormycosis is considered in another chapter.

Synonymy. Subcutaneous phycomycosis, rhinophycomycosis, phycomycosis entomophthorae, rhinoentomophthoromycosis.

Entomophthoromycosis Conidiobolae

Infection by *Conidiobolus coronatus* has been reported in man, horses,[24, 31] a dolphin,[51] and a chimpanzee.[48] In most cases, the primary infection appears to have originated in the nasal mucosa, and disease is usually restricted to the local subcutaneous tissue. Rarely it may spread to involve the paranasal sinuses, pharynx, facial muscles, and subcutaneous fat. The fungus, which was first isolated in 1897, is a common soil saprophyte found in leaf

303

detritus and is known to be a pathogen of several arthropods, including spiders and insects. The first report of infection in higher animals was made by Emmons and Bridges,[24] who described nasal polyps in horses from Texas. These animals were abnormal in no other way, and it was assumed that the infection was the result of traumatic implantation from vegetable material or from infected insects. The first human case with substantiating mycologic evidence was reported by Bras et al. in 1965.[8] The patient, who was from the Grand Cayman Island, had an extensive swelling of the nasal area. On direct examination of scrapings and biopsy material, broad, thin-walled, sparsely septate hyphae were seen, and *C. coronatus* was isolated in culture. A subsequent report published many years later noted that the infection was chronic and persistent.[30] Similar clinical conditions had been noted previously by Ash and Raum in 1956,[2] by Blache et al.[7] from a patient in the Cameroons in 1961, and by several other workers from Africa and Puerto Rico.[11, 25] These were only clinical and histologic reports, however, and the identity of the etiologic agent was unknown. In a review, Fromentin and Ravisse[25] tabulated over 55 cases reported up to 1977. Of these, 44 were in Africa (Nigeria 20, Cameroons 10, Ivory Coast 4, Zaire 4, Central African Empire 2, Kenya 2, Senegal 1, Malagasy 1); 5 cases were in India; and 6 were in the Americas (Colombia 1,[46] Brazil 2, Grand Cayman 1, Puerto Rico 1, and the United States 1).

ETIOLOGY, ECOLOGY, AND DISTRIBUTION

The only etiologic agent that has been isolated from the culturally confirmed cases of entomophthoromycosis conidiobolae is *Conidiobolus coronatus,* with the exception of one case in which *C. incongruus* was reported.[20] In essentially all the cases in which *C. coranatus* was the etiologic agent the infection began in the nasal septum and secondarily involved the sinuses, the pharynx, and subcutaneous tissue of the central region of the face. *Basidiobolus ranarum* has also been isolated from the same clinical condition. Almost all patients[11,45] have come from the tropical rain forests of Africa, particularly Nigeria. The other recorded cases have been scattered throughout the world, mainly in tropical or subtropical climates. The one infection so far recorded as being caused by *C. incongruus*

occurred in the United States. The infection involved the pericardium and lungs of a 15-month-old boy.[36] These reported cases are too few in number to allow speculation as to the existence or distribution of the virulent form of the organism, since *C. coronatus* is found worldwide and in great abundance in warmer climates in moist, decaying leaves. *C. incongruus* was first recovered from plant detritus in the eastern United States in 1960[17] and has been recovered from natural sources only once since.[3]

As is so commonly seen in other systemic and subcutaneous mycoses, the incidence of disease is higher among males than females. Over 80 per cent of the reported cases occurred in males. It is doubtful that this can be attributed to exposure alone. Unlike entomophthoromycosis basidiobolae, in which most patients are children, entomophthoromycosis conidiobolae is found mainly in adults. Clark[11] notes that the youngest patient was an 8-year-old girl and the oldest a 60-year-old man, but the majority were in the second to fourth decades of life. When they have been recorded, almost all cases of disease have occurred in agricultural workers whose exposure to soil organisms is greater than that in other occupations.

CLINICAL DISEASE

Symptomatology. The usual course described in case reports is that of a nasal swelling beginning in the inferior turbinates, which slowly grows or sometimes rapidly extends to include the submucosa, sutures, ostia, foramina, and paranasal sinuses. Disease is usually bilateral but may be unilateral. The expanding mass causes disfigurement of the adjacent tissue but is usually painless (Fig. 11–1). It is palpable and not attached to the overlying skin but rather anchored to the structures beneath it. The overlying epidermis does not ulcerate, though it may become acanthotic and erythematous. The mass may be uneven and bumpy (Fig. 11–2), and clog the passage of the nares by pushing the turbinates against the septum. The accompanying edema may extend to include the cheeks, forehead, and lips. The eyelids may be swollen shut, resulting in leukoma of the eye. The x-ray picture shows an opaque antrum, obliteration of the nasal air space, and mucosal thickening. The blood count is not elevated, there is no fever, and the patient is otherwise normal.

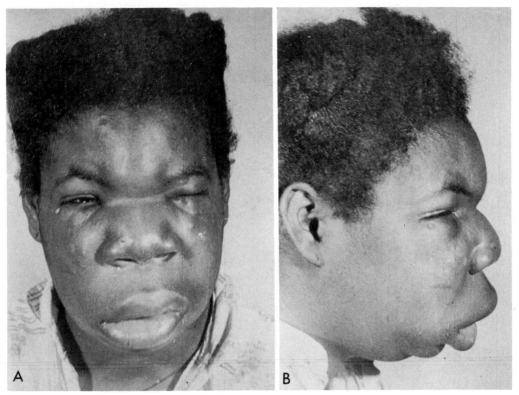

Figure 11-1. Entomophthoromycosis conidiobolae. Infection by *C. coronatus. A,* Bilateral disortion of the subcutaneous tissue of the nasal region. *B,* Profile showing distortion. (From Restrepo, A., D. L. Greer, et al. 1967. Am. J. Trop. Med. Hyg., *16*:35.)

Differential Diagnosis. The clinical symptoms of entomophthoromycosis may simulate rhinocerebral-mucormycosis, which also begins in the nasal mucosa. The latter disease is associated with debilitated patients and runs a rapid, often fatal, course. Entomophthoromycosis simulates other conditions of the nasal area, such as pyogenic abscess, neoplasia, pyomyositis, tuberculosis, dracunculiasis, and onchocerciasis. The correct diagnosis is easily obtained by examination of direct smear (Fig. 11-4), histologic examination (Fig. 11-3) and culture of affected tissue.[57]

A disease which may closely simulate entomophthoromycosis conidiobolae was first described by Miloshev et al. in 1966.[44] A series of cases of this disease, paranasal gran-

Figure 11-2. *A,* Entomophthoromycosis conidiobolae. *A,* Later stage in same patient. The lesions have regressed in some areas and become nodular in others. *B,* This is a recurrence of an infection treated with various antifungal agents and surgery.[30] (Courtesy J. K. Herstoff.)

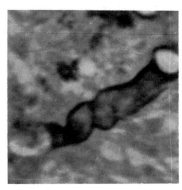

Figure 11–3. Entomophthoromycosis conidiobolae. Hyphae in tissue. (Courtesy of D. Greer.)

uloma caused by *Aspergillus flavus,* was reported by Mahgoub.[40] The condition is chronic and presents as a unilateral, nodular, painless mass which causes proptosis, destruction of the ethmoid air cells, and opacities in the maxillary antrum seen on x-ray. Nasal obstruction usually does not occur.

PROGNOSIS AND TREATMENT

Unlike rhinocerebral-mucormycosis, entomophthoromycosis conidiobolae is relatively benign and sometimes clears spontaneously. Most patients, however, require treatment with potassium iodide.[25] The suggested dose is 30 mg per kg body weight given in increments to attain 2 to 3 g per day for an adult. Treatment is continued for three months or more. It is important that the infection be adequately treated, since recurrent infections are more resistant to iodides. Steroids, trimethoprim, and sulfamethoxazole are sometimes included in the regimen. The lesions generally respond rapidly to this therapy, but a few have required amphotericin B treatment in addition.[24] Recalcitrant and advanced disease also requires amphotericin B. A case reported by Restrepo et al.[46, 47] did not respond to either potassium iodide or 2 g of amphotericin B. The patient reported by Herstoff et al.[30] was treated with amphotericin B as well as KI, 5FC, sulfas, trimethoprim, rifampin, mycophenolic acid, and surgery, all to no avail. This patient became discouraged with his medical management and left the treatment center.

PATHOLOGY

Gross Pathology. When they are removed surgically, the submucosal masses are firm, whitish in color, and have linear strands of fibrous tissue coursing through them. Focal areas of yellowish material that simulate caseous necrosis are seen. Fluid expressed from these areas contains abundant hyphal elements.

Histopathology. The histologic features of entomophthoromycosis conidiobolae and entomophthoromycosis basidiobolae are identical and contrast sharply to that of mucormycosis. The two major points of difference between entomophthoromycosis and mucormycosis are (1) the eosinophilic sheath (Splendore-Hoeppli phenomenon) found around the hyphae in the former disease, and (2) the lack of vascular invasion by the Entomophthorales so characteristic of the pathology found in mucormycosis. In addition, the hyphal elements of fungi causing mucormycosis are sparsely septate in tissue, whereas frequent septation is seen in entomophthoromycosis.

In all "Zygomycete" diseases, hyphal elements are readily stained in routine hematoxylin-eosin preparations.[57] In contrast, with special fungus stains, such as Gridley and PAS, the hyphal elements are not well stained; therefore, these procedures are not of sufficient help to warrant their use. Even the very sensitive GMS stain may only stain the fungi weakly. The hyphae of *C. coronatus* in tissue may lie singly or in clusters (Fig. 11–3). Their mean diameter is 8 μ, with variations between 4 μ and 10 μ and sometimes up to 22 μ. They are regularly septate but have broader hyphal units and are fewer in number than fungal elements seen in aspergillosis. The walls are thin but easily defined and surrounded by bright, radiating, granular, eosinophilic material which may be between 2 μ and 6 μ in thickness. The eosinophilic material has a fringelike or satellite arrangement and is similar to that seen in some cases of sporotrichosis (asteroid bodies), schistosomiasis, coccidioidomycosis, blastomycosis, paracoccidioidomycosis, and, rarely, in candidiasis. The hyphae do not have a predilection for any particular site in infected tissues. They do not infiltrate the walls of blood vessels, are not seen in the lumen of vessels, and, in contrast to mucormycosis, there is no vascular thrombosis. Panarteritis and endarteritis are sometimes observed, however.

The cellular reaction evoked by the infection is an acute or chronic inflammatory reaction or a combination of the two. The chronic reactions tend to be of the granulomatous type. The organisms are not always found in the midst of the exudate but are seen near it. The acute reaction consists of masses of eosinophils, lymphocytes, and variable numbers of plasma cells. Neutrophils and fibroblasts are less commonly found than in chronic lesions. Vascular proliferation occurs.[57] The eosinophils may be so numerous as to form an eosinophilic abscess, as seen in some infections caused by helminths, particularly microfilariae. For this reason, cases of entomophthoromycosis were once regarded as worm infestations.

The chronic inflammatory reaction is characterized by granulomatous infiltrates. The chief cellular components are foreign body giant cells containing phagocytized hyphal elements, histiocytes, lymphocytes, and some eosinophils and plasma cells. Peripherally there is fibroblastic activity, and large collagen cords are sometimes seen. There is no caseation or coagulation necrosis. Hypha viewed in cross section may have the appearance of empty coccidioidal spherules. In hematoxylin-stained sections these elements may closely resemble old, heavily mineralized capillaries with which they sometimes have been confused.

The eosinophilic precipitate surrounding the fungal elements in tissue is probably due to an antigen-antibody reaction. Histochemical tests demonstrate the presence of phospholipids, neutral lipids, PAS-positive material, and globulins. The precipitate has a yellow autofluorescence which is useful for observing the organisms when they are sparse in tissue.

Natural animal disease has been reported in the horse.[31] The lesions were described as nasal polyps and large granulomata of the nasal cavity. Disease in chimpanzees and dolphins also was reported.[48, 51] Attempts at experimental infection of several animals have so far been unsuccessful.[13] Lopez-Martinez et al.[38] tried a variety of routes of inoculation in hamsters, mice, and guinea pigs. Infections were transient, resolved spontaneously, and did not compare with the human entomophthoromycosis.

IMMUNOLOGY AND SEROLOGY

Natural resistance to infection is high, as the etiologic agent is found worldwide, but the disease is rare. There is no satisfactory serologic procedure for diagnosis, so the true extent of subclinical disease or of cases of spontaneous remission is unknown.

LABORATORY DIAGNOSIS

Direct Examination. To provide material for direct examination, soft intact vesicles are punctured or scrapings of the affected nasal mucosa are made. In case of infection, a potassium hydroxide mount reveals broad hyphae with occasional septations (Fig. 11–4). The hyphal walls are doubly refractile, and there is some branching. Granular inclusions are readily seen.

Culture Methods. Pathologic material for culture can be gently broken apart in a solution containing 500 mg of streptomycin and 1 megaunit of penicillin. It is then plated on Sabouraud's agar with or without antibacterial antibiotics and incubated at 25° or 37° C. Mycosel and other media containing cycloheximide cannot be used. Growth can be observed at 48 hours.[46]

To isolate the fungus from soil or leaf detritus, the material is shaken in sterile saline and allowed to stand. The supernatant is

Figure 11–4. Entomophthoromycosis conidiobolae. Direct examination of scraping from nasal mucosa. (From Restrepo, A., D. L. Greer, et al. 1967. Am. J. Trop. Med. Hyg., *16*:34–39.)

decanted and pipetted onto sterile Whatman No. 1 filter paper, which is then inserted in the cover of a Petri dish and placed over a plate containing Sabouraud's agar. The Petri dish is inverted and incubated at 25° C. Leaves can be pressed directly onto moistened filter paper inside the lid of such a plate. Conidia are propelled from the lid to the agar, and visible colonies may be observed in 24 to 72 hours.[12]

MYCOLOGY

Conidiobolus coronatus (Costantin) Batko 1964

Synonymy. *Boudierella coronata* Costantin 1897; *Delacroixia coronata* (Costantin) Saccardo and Sydow 1899; *Entomophthora coronata* (Costantin) Kevorkian 1937; *C. coronatus* (Costantin) Srinivasan and Thirumalachar 1967; *C. villosus* Martin 1925.

Colony Morphology. The colony grows rapidly and is glabrous and adherent at first. Furrows and folding occur, particularly when the organism is grown at 37° C. In time, the colony becomes covered with short, white, aerial mycelia and conidiophores (Fig. 11–5). The sides of the lid of the dish or tube soon become covered with conidia which are forcibly discharged by the conidiophores. The color of the colony becomes tannish to light brown with age.

Microscopic Morphology. Sporangiola are formed, but these are single celled and appear as large conidia (25 to 45 μ in diameter). They are produced on the end of short, erect, unbranched sporangiophores. The

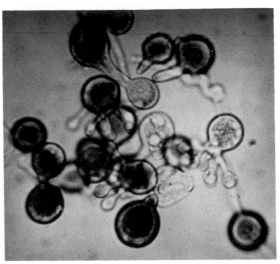

Figure 11–6. Corona of papillae on spores of *C. coronatus.* In the lower center of the illustration is a newly discharged primary spore that has a broadly rounded apex at the top. This spore is typical of the genus *Conidiobolus.* Other spores have produced the papillae that will form secondary spores. The villose spore characteristic of the species *C. coronatus* is not shown. (Courtesy of D. Greer.)

spores are ejected and travel distances up to 30 mm. If they land on nutrient medium, they will germinate and produce one or more hyphal tubes. The spores differ from those of *Basidiobolus* sp. because they have prominent papillae on the wall that may give rise to secondary spores. Several papillae and spores may be produced, giving the original spore a corona of secondary spores (Fig. 11–6). A spore may also produce multiple short, hairlike appendages called villae. This characterizes the species *C. coronatus.* Spores falling on glass will produce a short sporangiophore and eject another spore. This spore will in turn germinate to produce another spore. This process repeats itself until stored nutrients are exhausted. Many chlamydoconidia submerged in the medium are also present in some colonies, but zygospores are rare.[35]

C. incongruus Drechsler 1960 is similar to *C. coronatus,* since it also produces multiplicative spores with papillae, but the villose corona characteristic of *C. coronatus* is not produced.

ENTOMOPHTHOROMYCOSIS BASIDIOBOLAE

The first report of this disease in which there was mycologic confirmation came from Indonesia. In 1956, Lei-Kian-Joe et al. de-

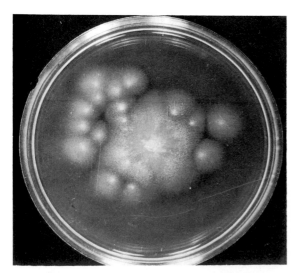

Figure 11–5. Culture of *C. coronatus.*

scribed three cases in which there were extensive palpable masses in the subcutaneous tissue and muscle fascia.[33] The disease spread widely to involve the neck, arms, and upper chest, but eventually the lesions healed spontaneously. The etiologic agent was identified as *Basidiobolus ranarum,* but has since been reclassified as *B. haptosporus.* The taxonomy of the etiologic agent is at present unclear. Previously the disease had been reported to occur in a horse[56] and possibly a man.[10] After the first report by Joe, several more cases were found in Indonesia.[37, 54] These were more severe and did not heal spontaneously. Two additional patients with the disease were encountered in England;[52] one of these had lived in Indonesia. Since then many cases have been reported from Uganda[11] and India[34, 37] as well. The mycosis is now known to occur in many tropical African countries, Southeast Asia, and recently Brazil.[6] In the last 20 to 25 years, about 150 cases of the disease have been recorded. Almost all were in children and they had large subcutaneous masses as the primary symptom of the disease. The first case of infection diagnosed in the United States, however, involved the maxillary sinus and palate of a 49-year-old hypoglycemic, asplenic man.[18]

The disease is primarily one of the subcutaneous tissue, but on occasion deeper structures have been invaded. Two deaths have been noted; in one patient the large intestine and pelvis were involved, and in the other the neck.[11, 21, 25]

The disease syndrome became well known under the name subcutaneous phycomycosis. Recent changes in nomenclature, however, preclude use of the term phycomycosis. Several infections have involved the deeper tissue, lungs, and alimentary tract, invalidating the use of the designation subcutaneous. Therefore the name *Entomophthoromycosis basidibolae,* first suggested by C. W. Emmons, is recommended.

ETIOLOGY, ECOLOGY, AND DISTRIBUTION

The original isolant from the case of Lei-Kian-Joe[33] was identified by Emmons as *B. ranarum* because some rough-walled zygospores were seen. The classification of the genus is tenuous and based on the morphology of the zygospore coat and thermotolerance of the fungus. Greer and Friedman[28] found only smooth-walled zygospores in human isolants and called the fungus *B. meristosporus,*

differentiating it from *B. haptosporus* only by the thermotolerance of the former. Drechsler,[14, 15, 16] who created the species *B. meristosporus,* was unsure of the correct epithet. Srinivasan and Thirumalachar concluded that the organisms were *B. haptosporus.* Since there is great variability in thermotolerance of isolants from soil or from human cases, Clark and Greer had concluded that the name *B. haptosporus* is correct and that *B. meristosporus* should be considered in synonymy with it. Recently King et al.[35] examined numerous isolants of *Basidiobolus* sp. from saprophytic and parasitic habitats. They concluded that the characteristics used to separate the species were so variable that distinction between the strains was impossible. Based on their results, McGinnis[43] suggests that such isolants be considered *B. ranarum* until a detailed taxonomic revision is published.

B. ranarum is a ubiquitous species that occurs in decaying vegetation and the gastrointestinal tract of many reptiles and amphibians. It does not appear to cause any disease in these animals. It has not yet been isolated from insects, but spores sometimes become attached to insects, and they may serve as vectors for dissemination. The fungus does not tolerate cold and dies quickly if kept in a cold environment. The organism has been isolated from tropical and subtropical environments throughout the world.[11, 12]

Entomophthoromycosis basidiobolae is primarily a disease of children. Of the 78 cases tabulated by Clark,[11] 46 per cent of the patients were 10 years old or younger, and 82 per cent were under 20 years of age. The male-female ratio varied between 3 to 1 and 6 to 1, which tends to demonstrate a predominance of disease in male children. However, too few cases have been recorded to delineate patterns of sex distribution. Of the cases tabulated by Clark, the largest number were reported from Uganda (35), Nigeria (17), and Indonesia (9). Other countries in which cases were reported include India, Burma, Senegal, Kenya, Ghana, Iraq, Sudan, 4 so far from Brazil,[6] and 1[18] (possibly 2[50]) from the United States. Tio[54] has since reported 12 more cases from Indonesia, and the worldwide total is now well over 150 as recorded by Fromentin and Ravisse.[25]

CLINICAL DISEASE

Symptomatology. The infection begins as a subcutaneous nodule that gradually increases in size. The portal of entry of the fungus is unknown, but in a few cases mosquito or other insect bites were noted before the onset of symptoms. This may indicate that there is an arthropod vector. Introduction of the agent by an intestinal route has also been suggested. The subcutaneous swelling has a firm consistency and is well circumscribed and painless, though there may be some pruritus. The mass is palpable and attached to the overlying skin but not the underlying muscle fascia. It is primarily contained within the pannicle. When fingers are placed under the edge of the mass, it is found to be freely moveable. The skin tends to be atrophic and discolored or hyperpigmented, but it does not ulcerate. The mass continues to grow and sometimes involves the whole shoulder, arm and upper body, face and neck, or the entire leg and buttocks (Fig. 11–7). In a few cases, involvement of underlying organs, such as liver, intestine, and muscle, has been reported.[6, 21] There is evidence suggestive of pulmonary infection in a case reported by Bittencourt et al.[6] and maxillary sinus and palate involvement in the adult case of Dworzack et al.[18] The identity of the organism in the latter case is still in question. There are no reports of involvement of regional lymph nodes or hematogenous or lymphatic dissemination. A few cases regress spontaneously, but most require iodide[34] or

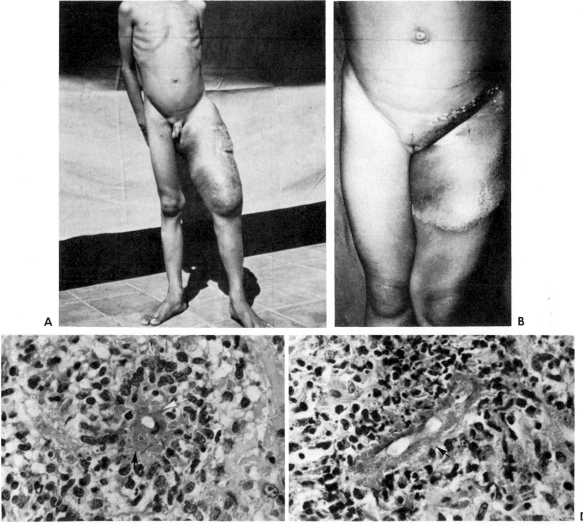

Figure 11–7. Entomophthoromycosis basidiobolae. *A,* Subcutaneous lesion involving entire thigh and buttock of Indonesian boy. There was secondary bacterial infection following a biopsy. (Courtesy of C. Halde.) *B,* Indurated plaque involving left thigh, groin, and part of vulva in 6-year-old girl from St. Terezinha, Brazil. *C,* Cross section and *D,* oblique section of biopsy. The eosinophilic (Splendore-Hoeppli) sleeve is evident around the hyphae (arrows). (Courtesy A. T. Londero.) (Bittercourt, A. L., A.T. Londero et al. 1979. Mycopathologia, *68*:101–104.)

amphotericin B treatment.[18] In the cases recorded by Tio[53] there was a marked leukocytosis (up to 29,000) and eosinophilia (up to 30 per cent); no other abnormalities or predisposing factors were noted. Tio[54] noted that *in vitro* the fungus was unaffected by incorporation in culture media of eight times the concentration of KI found in blood following the usual pharmacologic dose.

Differential Diagnosis. The clinical history along with the histologic picture is so characteristic that the diagnosis is readily made. Cultural confirmation is always necessary. There are a variety of diseases that simulate this condition. These include neoplasias, pyogenic abscess, elephantiasis, and helminth infections. Another disease common in Uganda that simulates entomophthoromycosis basidiobolae is Buruli ulcer. This is caused by *Mycobacterium ulcerans*, which also appears to be a lipophilic organism. The disease it produces also spreads slowly and involves extensive areas of subcutaneous tissue.

Histologically, the rare infections caused by *Mortierella* sp. resemble those of entomophthoromycosis. There is a marked eosinophilia, and sparse, poorly staining hyphae are present. An eosinophilic sheath surrounding the hyphae may be seen.

PROGNOSIS AND THERAPY

The prognosis of entomophthoromycosis basidiobolae is generally good. Among the more than one hundred and fifty cases reported, only a few deaths have been directly attributable to the disease.[21] The gross disfigurement produced resolved, with little residual disease. Although a few cases regress spontaneously, most require treatment. Therapeutic trials with nystatin and griseofulvin have been unsuccessful, and the organism is insensitive to the drugs *in vitro*.[53, 54] Iodide therapy (potassium iodide, 30 mg per kg body weight per day for a month or more) has been remarkably successful.[34] Amphotericin B has been used in a few patients who had taken iodides for up to nine months with little or no effect. The response to this drug was disappointing in most instances.

In cases in which iodide therapy has been effective, the healing process was accompanied by diminution of the eosinophilic sheath around the hyphae and gradual disappearance of hyphae. The tuberculoid granuloma remains for some time, and necrotic tissue is replaced by fibrosis.

PATHOLOGY

Gross Pathology. In biopsy specimens the overlying skin is seen to be atrophic but rarely ulcerated. The subcutaneous mass itself is firm and thickened with fibrous tissue containing occasional yellowish foci of necrosis. Masses of mycelium in a suppurative exudate may be expressed from these areas.

Histopathology. The histologic appearance of entomophthoromycosis basidiobolae is identical to that of entomophthoromycosis conidiobolae. The hyphal elements (10 to 40 μ) may be sparse in the granuloma, but the large eosinophilic sheath and its autofluorescence are helpful in locating the organism (Fig. 11–7 *C, D*). The hyphae stain poorly with all stains, and it is only the eosinophilic sheath that indicates their presence. In both diseases, the hyphae show varying numbers of septations. *Mortierella* sp. in tissue also have an eosinophilic sheath, and collection of eosinophils is found in the surrounding tissue.[52]

ANIMAL DISEASE

Natural infection in animals has been recorded only in horses,[56] and only once, in 1925. The organism seems to be a natural inhabitant of the intestinal tract of reptiles but is not associated with any clinical disease. Attempts at experimental infection of several animals have not been successful.[28, 54] Eades and Corbel[19] produced progressive disease when conidia were injected intracerebrally in mice.

BIOLOGICAL STUDIES

As with other fungi that are occasional pathogens of man, only those that are thermotolerant (growing at or near 37° C) are able to incite infection. Greer and Friedman[28] found strains of *B. haptosporus* isolated from human infection able to grow at 37° C, whereas other isolants from saprophytic sources and *B. ranarum* are unable to grow at this temperature. *B. ranarum* also produces an odor of hexachlorobenzene. Reinvestigation of these findings has indicated that great variability occurs even in subcultures of the same isolant.[35] Tyrrel[55] also found the lipid content of several "species" to be the same. He concludes that there are no characteristics allowing separation of the various isolants into species based on morphological or physiological attributes at this time. The name *B.*

ranarum has priority over other specific epithets.

C. coronatus was investigated by Bievre and Mariat.[4, 5] The outer cell wall layer consisted of polyosides, glucomannans, and glucans. The major central layer contained mannans and glucans, with another layer containing glucans and chitin. This is similar to other Zygomycetes. *C. coronatus* isolated from human infection produced more elastase, esterase, and collagenase than did saprophytic isolants in studies by Fromentin et al.[26] Several aspects of the fine structure of *C. coronatus* cells as well as electron cytochemical localization studies of certain hydrolytic enzymes were reported by Garrison et al.[27] A summary of composition, multiplication, and basic biology of *Conidiobolus* and *Basidiobolus* was compiled by Fromentin and Ravisse.[25]

IMMUNOLOGY AND SEROLOGY

The rarity of infection in spite of the probable common exposure of man to this fungus in tropical and subtropical environments indicates a high natural resistance to this disease. No satisfactory serologic procedures are available as yet to allow investigation of the extent of subclinical or spontaneously resolved infections. Greer and Friedman[28] found antigenic differences between *B. meristosporus (B. haptosporus)* and *B. ranarum.* Their studies indicated that there was a common antigenic identity among several isolants of *B. haptosporus* from human disease, although there is the possibility that more than one species exists. Chemical analysis of the polysaccharides of the two species of *Basidiobulus* did not reveal significant differences, however.[29]

LABORATORY IDENTIFICATION

Direct Examination. Examination of biopsy material macerated in a potassium hydroxide mount demonstrates the same general findings noted for entomophthoromycosis conidiobolae. Broad, septate, branching hyphal elements are seen. There are numerous inclusions within the cell and doubly refractile cell walls.

Culture Methods. Biopsy material may be plated on Sabouraud's agar or any medium not containing Actidione. In the experience of Tio,[53, 54] growth is more easily ob-

tained from large lumps of tissue than from thin slices or macerated pieces. The plates are incubated at 25° to 30° C. Growth is observable in 48 to 72 hours, and colonies attain a diameter of 7 cm in four days. Isolation of the organism from leaf detritus and the intestinal contents of reptiles is accomplished in the same manner described for *C. coronatus.*[12]

MYCOLOGY

Basidiobolus ranarum Ęidem 1886

Synonymy. *B. meristosporus* Drechsler 1955;[14-17] *B. heterosporus* Srinivasan et Thirumalachar 1965;[49] *B. haptosporus* var. *minor* Srinivasan et Thirumalachar 1965.[49] *B. haptosporus* Drechsler 1947.[17]

Colony Morphology. Almost all strains grow rapidly at 30° C. Ability to grow at 37° C is variable, and the same isolant will differ in its response from one plating to another.[11] The colony is flat, folded, furrowed, grayish, and waxy in consistency. Large vegetative hyphae (8 to 20 μ) are formed, which become increasingly septate as production of spores begins.

Microscopic Morphology. After seven to ten days, the colony becomes overgrown with mycelium as masses of zygospores, chlamydoconidia, and sporangiola are formed. The zygospores are from 20 to 50 μ in diameter and have a smooth to undulated thick wall (Fig. 11–8). Most of the isolants from human infection have produced zygotes with smooth walls. They were therefore identified as *B. haptosporus* or *B. meristosporus.* In examining isolants from a variety of sources, King et al.[35] found this characteristic as well as the production of streptomyces-like odor to be inconsistent and variable. McGinnis[43] suggests that these species, together with *B. lacertae, B. magnus, B. philippinensis,* and *B. myxophilus,* be considered in synonymy with *B. ranarum.* The only other recognized species of *Basidiobolus* is *B. microsporus* Benjamin 1962.

The zygospores of both *Basidiobolus* species have a prominent "beak" attached to one side, representing the remnants of the copulatory tubes (Fig. 11–8). Sporangiophores are produced, each of which bears a unicellular sporangium. The apical portion of the sporangiophore at first enlarges. This then becomes the vesicle immediately below the sporangium (sporangiola). The spore is blown out of the top of this swelling, carrying with it the top of the vesicle as a residual attachment. The subsporangial swelling or

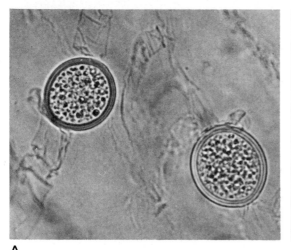
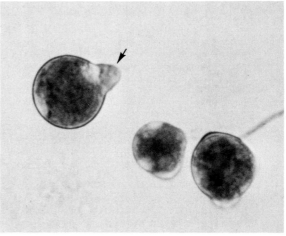

A   B

Figure 11–8. Zygospores of *B. ranarum*. Note "beak" (B, *arrow*). (Courtesy of D. Greer and A. Rogers.)

vesicle then emits a stream of fluid that propels the spore, a characteristic of the genus *Basidiobolus*.[32] Secondary or replicative spores can be produced from discharged primary conidium.

REFERENCES

1. Ader, P., and J. K. Dodd. 1979. Mucormycosis and entomophthoromycosis. A bibliography. Mycopathologia, *68*:67–100.
2. Ash, J. E., and M. Raum. 1956. *An Atlas of Otolaryngic Pathology.* Washington, D.C., Armed Forces Institute of Pathology, p. 179.
3. Batko, A. 1964. Notes on entomophthoraceous fungi in Poland. Entomophaga, Mem. Horts. Ser. *2*:129–131.
4. Bievre, C. de, and F. Mariat. 1978. Extraction et purification des polyosides de *Conidiobolus coronatus*. Bull. Soc. Fr. Mycol. Med., 7:265–269.
5. Bievre, C. de, H. Fromentin, and F. Mariat. 1978. Étude de la composition en sucres de souches de *conidiobolus* saphrophytes ou pathogène pour l'homme et l'animal. Mycopathologia, *63*:167–172.
6. Bittercourt, A. L., A. T. Londero, et al. 1979. Occurrence of subcutaneous zygomycosis caused by *Basidiobolus haptosporus* in Brazil. Mycopathologia, *68*:101–104.
7. Blache, R., P. Destombes, et al. 1961. New subcutaneous mycoses in Southern Cameroons. Bull. Soc. Pathol. Exot., *54*:56–63.
8. Bras, G., C. C. Gordon, et al. 1965. A case of phycomycosis observed in Jamaica; infection with *Entomophthora coronata*. Am. J. Trop. Med. Hyg., *14*:141–145.
9. Burkett, D. P., A. M. Wilson, et al. 1964. Subcutaneous phycomycosis. Br. Med. J. *1*:1669–1672.
10. Casagrandi, C. 1931. Sur la presence de Basidiobolus dans l'homme. Boll. Soc. Internaz. Microbiol. Sez. Ital., *9*:63–64 and Sulla presenza di Basidioboli nell' uomo. Riv. Biol., *13*:1–8, 1931.
11. Clark, B. M. 1968. The epidemiology of phycomyco-

sis. *In* Wolstenholme, G. E., and R. Porter (eds.), *Symposium on Systemic Mycoses.* Boston, Little, Brown and Company, pp. 179–192.
12. Coreman-Pelsener, J. 1973. Isolation of *Basidiobolus meristosporus* from natural sources. Mycopathologia, *49*:173–176.
13. Della Torre, B., and L. Mosca. 1965. Experimental phycomycosis in rodents. Mycopathologia, *26*:417–452.
14. Drechsler, C. 1947. A *Basidiobolus* producing elongated secondary conidia with adhesive beaks. Bull. Torrey Bot. Club., *74*:403–413.
15. Drechsler, C. 1955. A southern *Basidiobolus* forming many sporangia from elongated adhesive conidia. J. Wash. Acad. Sci., *45*:49–56.
16. Drechsler, C. 1958. Formation of sporangia from conidia and hyphal segments in an Indonesian *Basidiobolus*. Am. J. Bot., *45*:632–638.
17. Drechsler, C. 1960. Two new species of *Conidiobolus* found in plant detritus. Am. J. Bot., *47*:368–377.
18. Dworzack, D. L., A. S. Pollack, et al. 1978. Zygomycosis of the maxillary sinus and palate caused by *Basidiobolus haptosporus*. Arch. Intern. Med., *138*:1274–1276.
19. Eades, S. M., and M. J. Corbel. 1979. Experimental cerebral lesions produced by inoculation with Basidiobolus strains. Mycopathologica, *67*:187–197.
20. Eckert, H. L., G. H. Khoury, et al. 1972. Deep *Entomophthora* phycomycotic infection reported for the first time in the United States. Chest, *61*:392–394.
21. Edington, G. M. 1964. Phycomycosis in Ibadan, West Nigeria. Trans. R. Soc. Trop. Med. Hyg., *58*:242–245.
22. Eidam, Eduard. 1886. *Basidiobolus*, eine neue Gattung der Entomophthoraceen. Cohn's Beitr. Biol. Pflanzen, *4*:181–251.
23. Emmons, C. W., Lie-Kian-Joe, et al. 1957. *Basidiobolus* and *Cercospora* from human infections. Mycologia, *49*:1–10.
24. Emmons, C. W., and C. H. Bridges. 1961. *Entomophthora coronata,* the etiological agent of a phycomycosis of horses. Mycologia, *53*:307–312.
25. Fromentin, H., and P. Ravisse. 1977. Les ento-

mophthoromycosis tropicales. Acta Trop., *34*:375–394.

26. Fromentin, H., H. Hurion, and F. Mariat. 1978. Collagenase, esterase, et elastase do souches pathogene et saprophyte de *Entomophthora coronata*: cinetique de production. Ann. Microbiol. (Inst. Pasteur) *129A*:425–431.

27. Garrison, R. G., F. Mariat, et al. 1975. Ultrastructural and electron cytochemical studies of *Entomophthora coronata*. Ann. Microbiol. (Inst. Pasteur), *126B*:149–173.

28. Greer, D. L., and L. Friedman. 1966. Studies on the genus *Basidiobolus* with reclassification of the species pathogenic for man. Sabouraudia, *4*:231–241.

29. Greer, D. L., and E. Barbosa. 1967. Some chemical constituents of cell-bound and extracellular polysaccharides of *Basidiobolus ranarum,* isolated from nature and *B. meristosporus,* isolated from subcutaneous phycomycosis. Sabouraudia, *5*:329–334.

30. Herstoff, J. K., H. Bogaars, and C. J. McDonald. 1978. Rhinophycomycosis entomophthorae. Arch. Dermatol., *114*:1674–1678.

31. Hutchins, D. R., and K. G. Johnson. 1972. Phycomycosis in a horse. Aust. Vet. J., *28*:237–254.

32. Ingold, C. T. 1934. The spore discharge mechanism in *Basidiobolus ranarum.* New Phytologist, *33*:273–277.

33. Joe, L. K., T. E. Njo-Injo, et al. 1956. *Basidiobolus ranarum* as a cause of subcutaneous phycomycosis in Indonesia. Arch. Dermatol., *74*:378–383.

34. Kamalam, A., and A. S. Thambiah. 1979. Entomophthoromycosis basidiobolae — successfully treated with Kl. Mykosen, *22*:82–84.

35. King, D. S. 1979. Systematics of fungi causing entomophthoromycosis. Mycologia, *71*:731–745.

36. King, D. S., and S. C. Jong. 1976. Identity of the etiologic agent of the first deep entomophthoraceous agent in the United States. Mycologia, *68*:181–183.

37. Koshi, G., T. Kurien, et al. 1972. Subcutaneous phycomycosis caused by *Basidiobolus.* A report of 3 cases. Sabouraudia, *10*:237–243.

38. Lopez-Martinez, R., C. Toriello, et al. 1978. Study of the pathogenicity of *Conidiobolus coronatus* in experimental animals. Mycopathologia, *66*:59–65.

39. MacLeod, D. M., and E. Müller-Kögler. 1973. Entomogenous fungi. *Entomophthora* species with pear-shaped to almost spherical conidia (Entomophthorales: Entomophthoraceae). Mycologia, *65*:823–893.

40. Mahgoub, E. S. 1971. Aspergillosis in Sudan. Compte Rendus Ve Congres de la Société Inter. de Mycol. Hum. et Anim. p. 173–174.

41. Martinson, F. D. 1971. Chronic phycomycosis of the upper respiratory tract. Am. J. Trop. Med. Hyg., *20*:449–455.

42. Martinson, F. D., and B. M. Clark. 1967. Rhinophycomycosis entomophthoral in Nigeria. Am. J. Trop. Med. Hyg., *16*:40–47.

43. McGinnis, M. R. 1980. Recent taxonomic developments and changes in medical mycology. Annu. Rev. Microbiol. *34*:109–135.

44. Miloshev, K., S. Mahgoub, et al. 1966. Aspergilloma of paranasal sinuses and orbit in Northern Sudanese. Lancet, *1*:746–747.

45. Renoirte, R., J. Vandepitte, et al. 1965. Phycomycóse nasofaciale (rhinophycomycose) due to an *Entomophthora coronata.* Bull. Soc. Path. Exot., *58*:847–862.

46. Restrepo, A., D. L. Greer, et al. 1967. Subcutaneous phycomycosis: Report of the first case observed in Columbia, South America. Am. J. Trop. Med. Hyg., *16*:34–39.

47. Restrepo, M., L. F. Morales, et al. 1973. Rinoficomicosis par *Entomophthora coronata* en equinas. Antioquia Medica, *23*:13–25.

48. Roy, A. O., and H. M. Cameron. 1972. Rhinophycomycosis entomophthorae occurring in a chimpanzee in the wild of East Africa. Am. J. Trop. Med. Hyg., *21*:234–237.

49. Srinivasan, M. D., and M. J. Thirumalachar. 1967. Studies on *Basidiobolus* species from India with discussion on some of the characters used in the speciation of the genus. Mycopathologia (Den Haag), *33*:56–64.

50. Straatsma, B. R., L. E. Zimmerman, and J. D. Grass. 1962. Phycomycosis: a clinico pathologic study of fifty-one cases. Lab. Invest., *11*:963–985.

51. Sweeney, J. C., G. Migaki. 1976. Systemic mycosis in marine animals. J. Am. Vet. Med. Assoc., *69*:946–948.

52. Symmers, W. S. 1960. Mucormycotic granuloma possibly due to *Basidiobolus ranarum.* Br. Med. J., *5182*:1331–1333.

52a. Thaxter, R. 1888. The Entomophthoreae of the United States. Mem. Boston Soc. Nat. Hist., *4*:133–201.

53. Tio, T. H., M. Djojopranoto, et al. 1966. Subcutaneous phycomycosis. Arch. Dermatol., *93*:550–553.

54. Tio, T. H., and G. A. Vries. 1977. Subcutaneous zygomycosis. South African Med. J., *52*:77–78.

55. Tyrrel, D. 1967. The fatty acid composition of 17 entomophthora isolates. Can. J. Microbiol., *13*:755–760.

56. Van Overeem, C. 1925. Beitrage zue Pilzflora von Niederlandisch Indiana 10. Über ein merkwuerdiges Vorkommen von Basidiobolus ranarum. Eidana. Bull. Jardin Bot. Buitzern Borg. Ser. III, *7*:423–431.

57. Williams, A. O. 1969. Pathology of phycomycosis due to *Entomophthora* and *Basidiobolus* species. Arch. Pathol., *87*:13–20

12 Lobomycosis

DEFINITION

Lobomycosis is a chronic, localized, subepidermal infection characterized by the presence of keloidal, verrucoid, nodular lesions or sometimes by vegetating crusty plaques and tumors. The lesions contain masses of the spheroidal, yeastlike organism tentatively referred to as *Loboa loboi*. There is no systemic spread. The disease has been found in man and dolphins.

Synonymy. Keloid blastomycosis, Lobo's disease.

HISTORY

The disease was first described in 1931 by Jorge Lobo.[20] The patient, who was an Indian from the Amazon valley, had cutaneous keloidal lesions without lymphangitis or systemic symptoms. A fungus isolated from the patient appears to have been lost. The isolant, which is listed as J. B. 525 in the Instituto Oswald Cruz, is purported to have been isolated from the first case of the disease. It is now regarded as a mislabeled strain of *Paracoccidioides brasiliensis*.[8, 18] Lobo considered the disease a mild form of paracoccidioidomycosis, but subsequent clinical, histologic, and mycologic investigation has delineated it as a separate disease.[19] The two infections may coexist in the same patient, however.[15] Since the time of its original description, lobomycosis has been reported sporadically throughout the American tropics in man and recently in dolphins from the waters of Florida and the Caribbean (Fig. 12–1). A great deal of literature has been devoted to the nomenclature of the organism,[5, 16] but the controversy

will not be resolved until the etiologic agent is cultured. The disease was called "keloid blastomycosis," but lesions other than the keloidal type have been found in patients.[1, 14, 31] Although "Lobo's disease" has been proposed as the name for the infection, the term "lobomycosis" is preferred, as it more clearly defines this entity. The final designation of the disease awaits successful isolation and study of the etiologic agent.

ETIOLOGY, ECOLOGY, AND DISTRIBUTION

The etiologic agent of lobomycosis has not been isolated in culture. Until it is, the final designation of its true taxonomic relationships cannot be made, and discussion of its nomenclature is a moot point. Since some term must be used for the organism, the one proposed by Ciferri,[9] *Loboa loboi*, is retained, as it has become widely adopted and implies no relationship to other organisms. Most of the other names proposed have been based on studies of contaminants.

The ecology of the organism also remains a mystery. Until recently, the disease was known only in humans. There seemed to be a common denominator of habitation in tropical rain forests and bush country, particularly the Mato Grosso in Brazil and areas of Surinam, where most of the infections have occurred; the geographic and climatic "reservárea" of Borelli.[6] The discovery of the disease in dolphins in a marine environment makes previous assumptions as to the saprophytic nature of the fungus difficult to maintain.[7, 10] One possibility that might account for both its inability to be cultured *in*

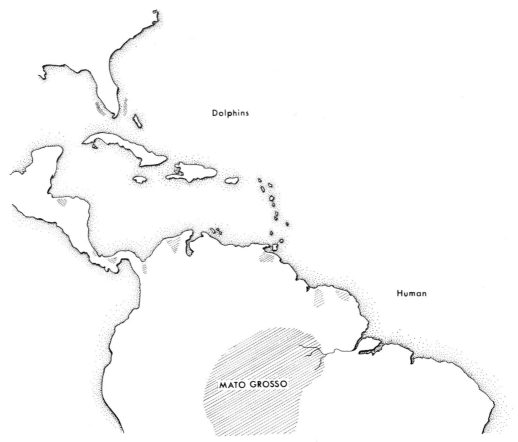

Figure 12–1. Distribution of lobomycosis based on case reports of the disease in man and dolphins.

vitro and its clinical restriction to the cooler parts of the body is that the organism is an obligate parasite of some lower animal form.

Most of the human cases of lobomycosis have been in patients living in the Amazon valley of Brazil. Lacaz[16] tabulated 69 cases from this region up to 1970. The Caiabi Indians have a legend that traces the introduction of the malady into their community by diseased children captured from the Ipeni tribe.[2] The second largest series of cases is reported from Surinam. In this study, Wiersema[31] recorded 13 cases in Negroes, most of whom lived along the Saramaccaner River in the interior of the country. Scattered reports also place the disease in Colombia (with 8 cases),[14] Venezuela,[3] French Guiana,[11, 25, 29] Panama, and Costa Rica. A report of the disease in Honduras was not confirmed. Over 100 cases have now been tabulated.[1, 14] In almost all of these the patients were male and agricultural workers or rural inhabitants. The age at acquisition is difficult to assess, because the infection has a protracted and benign course. In a case described by Jaramillo et al.,[14] the disease started *de novo* in a very young female, was excised, and reappeared many years later. A report from Panama indicated that the lesion developed following a bee sting that had occurred 20 years previously.[30] Wiersma[31] stresses previous trauma as a site of infection, but in most clinical histories the event is so minimal that it is not remembered.

CLINICAL DISEASE

In cases in which it has been possible to determine, the initial infection occurs at the site of some trauma to the skin.[31] The lesions begin as small, hard nodules resembling keloids that are sharply defined, freely moveable, and have a smooth surface. The color of the nodules may be slightly brownish with telangiectasia. The surrounding cutaneous area is normal, and erythema is absent. The lesions are painless or slightly pruritic. The infection is restricted to the subepidermal area and does not penetrate into the subcutaneous tissue or spread internally. The disease may be transferred to other areas of the skin by subsequent abrasions and autoinoculation,[31] in which case groups of nodules may be found on several different areas of the body (Fig. 12–2). The observation that groups of lesions occur on different regions of the body has led to speculation that hematogenous or lymphatic spread of disease may take place. There is no evidence to support this. Regional lymph nodes are only rarely involved, and reports of this probably represent mechanical dispersion of fungi following injury. The organism is so large that penetration through uninjured skin seems improbable. Lesions spread slowly in the dermis and continue to develop over a period of many years. Older lesions become verrucoid and may ulcerate. The areas of involvement may be quite extensive, since spread by autoinoculation occurs over the years.

Other clinical forms of the disease have been reported[1, 14] in addition to the keloidal type previously described. These include the infiltrative type, usually associated with the initial stages of infection; the gummatous type, in which the organisms are very deep in the tissue, the ulcerated type, in which the epidermis is lost and secondary bacterial infections occur; and the verrucous type sometimes observed in old, chronic lesions (Figs. 12–2 to 12–6).

Systemic involvement has not been recorded, but in 8 of 99 cases reviewed by Azulay[1] organisms were found in adjacent lymph nodes.

There appears to be no tendency for lesions to heal spontaneously. In patients with a long-term history of disease, mature verrucoid lesions are found in some areas, while young hard nodules are found in others. The favored body sites are exposed areas, such as ears, legs, arms, and face.

The differential diagnosis depends on the age of the lesions. Young, sharply defined nodules resemble keloids, fibromas, or Darier-Ferrand's dermatofibrosarcoma protuberans. The older verrucoid lesions are similar to those of chromoblastomycosis, mycetoma, chronic pyogenic vegetations, or granulomata.[19, 31] Since lesions are not macroscopically distinctive, histologic examination is necessary for definitive diagnosis.

Lobomycosis causes little discomfort to the patient, and there are no generalized symp-

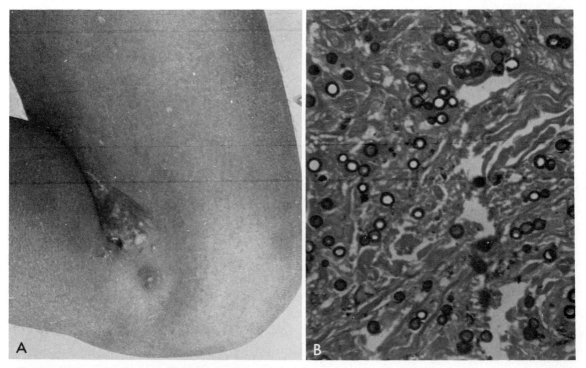

Figure 12–2. Lobomycosis. *A,* Old subepidermal nodules and recent satellite lesions. The latter are probably the result of autoinoculation by scratching. (Courtesy of F. Battistini.) *B,* Biopsy of lesion, demonstrating numerous yeast cells. GMS. ×400.

Figure 12–3. Lobomycosis. Nodular lesion on ear of 30 years' duration. (Courtesy of F. Battistini.)

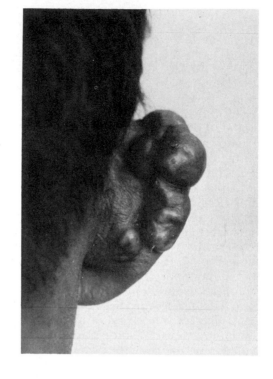

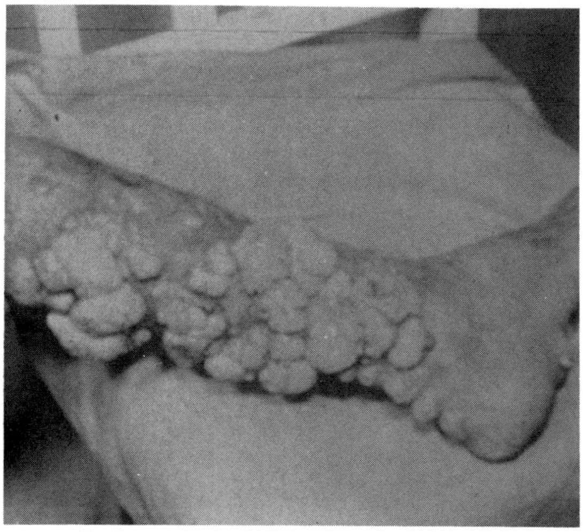

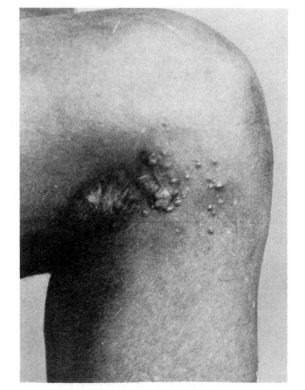

Figure 12–4. Lobomycosis. Extensive verrucoid lesions on legs.

Figure 12–5. Lobomycosis. Old nodular lesions and new satellite lesions representing autoinoculation into new areas. This is the same patient as in Figure 13–2, *A* after two more years. Note new crops of nodules.

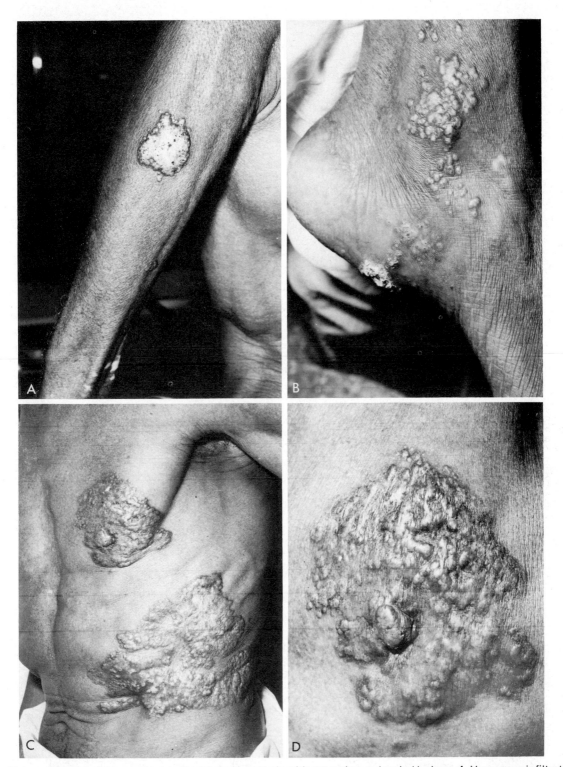

Figure 12–6. Lobomycosis. A patient from Venezuela with extensive and varied lesions. *A,* Verrucous infiltrate of long duration that ulcerated in some places. *B,* Lesions on the ankle and instep of the sole. Some were pruritic and spread by autoinoculation. Constant trauma to the lower lesions resulted in chronic verrucoid morphology resembling chromoblastomycosis or pedal sporotrichosis. *C,* Extensive lesions on the chest, some of which have healed and formed cicatrizations. *D,* Close-up of lesions on the chest. The nipple has become verrucoid and pedunculated. (Courtesy of K. Salfelder.)

toms. In some cases, there is a history of the presence of lesions for 30 or 40 years. There is no evidence of predisposing factors, and patients appear otherwise normal.

The most successful treatment for the condition is wide surgical excision of the affected area. Care should be taken to prevent contamination of surgical wounds, as relapse is not uncommon. The sulfa drugs, especially sulfadimethoxine, have been used for therapy with varying results, as has clofazimine.[30]

PATHOLOGY

The typical nodules consist of subepidermal granulomata that usually spare the overlying skin and do not penetrate into deep subcutaneous tissue. The epidermis is atrophic and shows neither the pseudoepitheliomatous hyperplasia nor the intraepidermal abscesses that are common in blastomycosis and coccidioidomycosis. The histology of the subepidermal granuloma is variable. In the

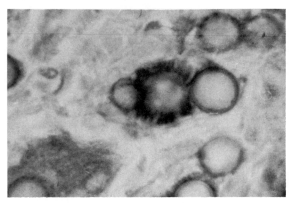

Figure 12–8. Lobomycosis. Radiating, eosinophilic, "asteroid"-like bodies around a yeast cell.

usual case, there is extensive hyaline fibrosis interspersed with masses of histiocytes and giant cells. The proportion of giant cells present increases with the age of the lesion. There is no evidence of necrosis or suppuration. The fungi are very numerous and hyaline in the lesions and are mostly located within giant cells and macrophages. Giant cells with a diameter of 40 to 80 μ may contain up to ten yeast cells. The yeast cells are uniform in size (about 10 μ), and generally form a single chain. They are connected one to the other by short, bridgelike structures (Fig. 12–7).[19, 31] Chains of 20 or more yeast cells with several branching points are sometimes found outside the macrophages. The wall of the fungus cell is 1 μ in thickness. In hematoxylin and eosin-stained preparations, the walls are poorly colored, and the cytoplasm is usually shrunken into the center. Several dotlike nuclei may be seen. Radiating, eosinophilic, "asteroid" bodies are sometimes found (Fig. 12–8).[21] These probably represent the Splendose-Hoeplii phenomenon, which occurs in other fungal and parasitic diseases. In GMS-stained material, the organism appears a very intense black color. The cell wall can be observed to narrow toward the juncture of individual cells and to be continuous from one cell to the next. Bud scars can also be found where disjunction of cells has occurred. In the Gridley stain, the yeast cells have a yellowish-brown cytoplasm, dark brown nuclei, and a pink cell wall. PAS-positive material can be seen radiating from or forming an envelope around the outside of the cell wall. Distinct, rough, spikelike formations are sometimes noted, particularly on the wall of the central cell. This material is stainable by the acid mucopoly-

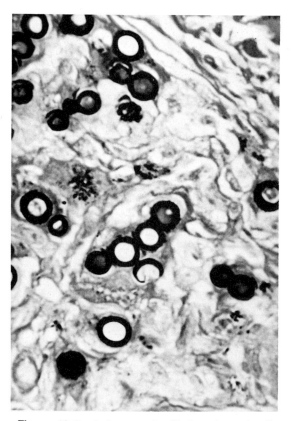

Figure 12–7. Lobomycosis. Chains of yeast cells. Note the branch and the connecting bridgelike structure. Some of the yeasts are empty of cellular contents. GMS. ×900.

saccharide and alcian blue methods but not by GMS. Some of the yeast cells may have a bud, and a few show branching (Fig. 12–7). The cell wall is microporous, allowing macrophages to introduce their pseudopodia into the interior of the cell. Macrophages are often seen filled with the shells of dismembered cell walls.[7]

The numerous small buds, which are characteristic of *Paracoccidioides,* are not observed. There are no capsules present, nor is there great variation in size of yeasts as seen in cryptococcosis. In several cases examined by Wiersema,[31] there was little collagen associated with the stroma of the keloid, but a fine reticulin meshwork between the giant cells was observed by silver stain. He also observed hourglass–shaped yeast cells with a large connecting isthmus and a suggestion of pseudo-hyphae in some sections.

In older lesions, ulceration and some pyogenic infiltrate is seen. The ulceration appears to be mechanical rather than related to the fungus. The corneal layer in such lesions is often markedly thickened, and crusts are formed. Parakeratosis and acanthosis are present, along with empty yeast cells and giant cells, which often contain much PAS-positive debris. The histopathologic appearance of lobomycosis is so characteristic that a diagnosis is readily made.

ANIMAL DISEASE

Natural infection of animals has only recently been discovered. Lobomycosis had been known only in humans until the diagnosis of the disease was made in bottle-nosed dolphins *(Tursiops truncatus).*[7, 22] The first case occurred in a female dolphin found in the Gulf Intercoastal Waterway near Sarasota, Florida, an area quite removed from any known human cases. The migration range of this species of dolphin has not been fully determined, but it apparently does not include South or Central America. Other observers on both Florida coasts have concluded that this infection occurs with some frequency among several species of dolphins. Recently deVries and Laarman[10] diagnosed the infection in a dolphin *(Sotalia guianensis)* in an estuary of the Surinam river. This places the animal in the same locale in which human infection has occurred. Both species of dolphin in which infection has been noted spend time in estuaries, rivers, or brackish water inlets near the sea coast.

Grossly, the lesions on the dolphin resemble those of old lesions found in human cases of disease (Fig. 12–9). They are verrucoid, crusty plaques with ulcerations and occur on "air-exposed" surfaces, i.e., on the dorsal fin, the tail flukes, and the top of the head and caudal peduncle. Histologically, the granulomata consist of histiocytes and giant cells. Yeasts are very numerous and uniform in size. They occur in long chains of many yeast cells and are found within giant cells. Many of the severely infected animals were malnourished, cachectic, or moribund. Systemic infection has not been documented, however.

Most reports of attempts to transfer lobomycosis to experimental animals have been negative,[7] but a few successful inoculations have been performed. Diaz[12] produced lesions in the cheek pouches of hamsters, but Wiersema[31] failed in attempts to infect mice intraperitoneally, intratesticularly, or intracutaneously. However, he was able to produce chronic disease with a typical histologic appearance in hamsters. In these experiments, biopsy material from human lesions was macerated, and yeast cells in chains were injected into the foot pads of the animals. Infection took a long time to develop, but after 8 months typical chronic nodular lesions were present. Sampaio was partially successful in infecting turtles and tortoises[26] and recently described experimental disease in armadillos.[27] The lesions developed more rapidly than in natural human infection, and liquifactive necrosis occurred in addition to the usual pathology observed in the disease in man. The idea of using armadillos had been suggested previously by Jaramillo et al.[14] It is emphasized that lesions in experimental as well as in natural infection require much time to develop. Recently Ajello has succeeded in transferring the infection through four generations of mice using as inoculum site the hind footpad. These observations will be published shortly.

IMMUNOLOGY AND SEROLOGY

There have been too few cases of lobomycosis studied to detect any patterns of immunity or factors predisposing to disease. In the series of cases reported, the patients appeared to be otherwise normal and healthy. No detectable antibodies have been noted, and as yet there are no serologic procedures for diagnosis. Lacaz et al.[17] have established

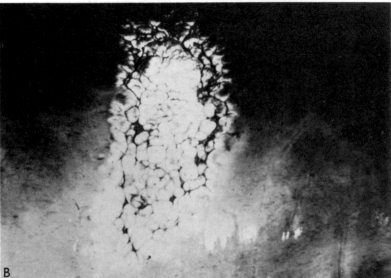

Figure 12–9. Lobomycosis. Lesion on the skin of a bottle-nosed dolphin. (From Migaki, G., M. G. Valerio, et al. 1971. J. Am. Vet. Med. Assoc., *159*:578–582.)

immunologically that paracoccidioidomycosis and lobomycosis are distinct diseases. They also discovered that the first isolant reputed to come from a case of lobomycosis was in reality a culture of *Paracoccidioides brasiliensis*. In an attempt to demonstrate common antigens between *Loboa loboi* and other fungi, Silva et al.[28] labeled with fluorescein sera from patients with lobomycosis. The sera reacted with cells of *P. brasiliensis, Candida albicans* serotypes A and B, and yeast-form *Sporothrix schenckii*. The sera did not react on biopsy tissue from a case of lobomycosis; therefore, it is doubtful that specific antibod-ies for the fungus of lobomycosis were present. Skin test antigens (lobins) have been described, but their specificity and usefulness have not been substantiated.

LABORATORY IDENTIFICATION

Direct Examination. Material obtained by curettage, surgical excision, or biopsy can be macerated and a potassium hydroxide mount made. Chains of uniform yeast cells will be abundant in case of disease. The average cell size is 9 to 10 μ, with a few ranging from 7 to 12 μ.

The etiologic agent of lobomycosis has not as yet been successfully or repeatedly grown in culture. In all instances the organisms isolated have later been found to be contaminants.[8, 18]

MYCOLOGY

Loboa loboi (Fonseca Filho and Arêa Leao) Ciferri, Azevedo, Campos, and Siquerira Carneiro 1956

Synonymy. *Glenosporella loboi* Fonseca Filho and Arêa Leão 1940; *Glenosporopsis amazonica* Fonseca Filho 1943; *Paracoccidioides loboi* Almeida and Lacaz 1949; *Blastomyces loboi* Langeron and Vanbreuseghem 1952.

An isolant that was said to be obtained from the first case of lobomycosis was described by Fonseca and Leão as *Glenosporella loboi* in 1940. The fungus was used to infect experimental animals. On the basis of the pathology produced, the name *Paracoccidioides loboi* was given to this organism by Almeida and Lacaz in 1949. It has since been shown to be a strain of *P. brasiliensis*.[8, 18] Fonseca in 1943 isolated another fungus from a patient with lobomycosis and described it as *Glenosporopsis amazonica*. This isolant and one supplied by Borelli have since been identified by Raper[23] as *Aspergillus penicilloides,* an osmophilic *Aspergillus* of the *A. restrictus* group. Another strain of fungus isolated from a patient has now been identified as the pedunculated yeast, *Sterigmatomyces halophilus* Fell 1966.[18] All isolants with the exception of the *P. brasiliensis* have been nonpathogenic for laboratory animals and embryonated eggs.

The etiologic agent of lobomycosis remains to be cultivated *in vitro*. Its *in vivo* morphology is the same in the two mammalian species in which the disease has been recorded — man and dolphin. The organism is spherical, elliptical or lemon-shaped, has a uniform diameter of 9 to 10 μ and a doubly refractile wall, and is multinucleate. The yeasts reproduce by budding at the terminus of the cell. The buds usually remain attached by means of a tubular bridge, and chains of up to 20 yeasts are seen. More than one bud may occur, leading to branched chains of cells.

As some designation of the etiologic agent is necessary, the author follows the suggestion of Ciferri et al.[9] in referring to the organism as *Loboa loboi*. This implies no relationship to other organisms and awaits *in vitro* isolation and study for true identity of the agent.

References

1. Azulay, R. D., J. A. Carneiro, et al. 1976. Keloid blastomycosis (Lobo's disease) with lymphatic involvement: a case report. Int. J. Dermatol., *15*:40–42.
2. Baruzzi, R. G., R. M. Castro, et al. 1967. Ocorrencia de blastomicose queloideana entre indios Caiabi. Rev. Inst. Med. Trop. São Paulo, *9*:135–142.
3. Battistini, F., S. G. Jover, et al. 1966. Dos casos de Blastomicosis Queloideana o Enfirmedad de Jorge Lobo. Rev. Dermatol. Venez. *5*:30–36.
4. Borelli, D. 1962. Lobomicosis experimental. Dermatol. Venez., *3*:72–82.
5. Borelli, D. 1968. Lobomicosis; Nomenclatura de su agente. Med. Cutan., *3*:151–156.
6. Borelli, D. 1969. La reservárea de la lobomicosis. Comentarios a un trabajo del doctor Carlos Pena sobre dos casos colombianos. Mycopathologia, *37*:145–149.
7. Caldwell, D. K., M. C. Caldwell, et al. 1975. Lobomycosis as a disease of the Atlantic bottle nosed dolphin (*Tursiops trucatus* Montagu 1821). Am. J. Trop. Med. Hyg., *24*:105–114.
8. Carneiro, L. S. 1952. Contribuiqao ao Estudo Microbiologica do Agente. Etiologico da Doenga de Jorge Lobo. Tese. Imprensa Industrial. Recife., Pernambunco, Brasil.
9. Ciferri, R., P. C. Azevedo, et al. 1956. Taxonomy of Jorge Lobo's disease fungus. Univ. Recife Inst. Micology Publ. No., *53*:1–21.
10. DeVries, G. A., and J. J. Laarman. 1973. A case of Lobo's disease in the dolphin *Sotalia guianensis*. Aquatic Mammals, *1*:26–33.
11. Destombes, P., and P. Ravisse. 1964. Étude histologique de une cas guyanais de blastomicose cheloidienne (maladie d.J. Lobo). Bull. Soc. Pathol. Exot., *57*:1018–1024.
12. Diaz, L. B., M. M. Sampaio, et al. 1970. Jorge Lobo's disease. Observations on its epidemiology and some unusual morphological forms of the fungus. Rev. Inst. Med. Trop. São Paulo, *12*:8–15.
13. Guamaraes, F. N. 1964. Inoculacoes em Hamsters da blastomicose sul americana (doenca da Lutz, da blastomicose queloidiforme (doenca de Lobo) e da blastomicose dos-Indios do Tapajosxing. Hospital (Rio), *66*:581–593.
14. Jaramillo, D., A. Cortés, et al. 1976. Lobomycosis. Report of the eighth Colombian case and review of the literature. J. Cutan. Pathol., *3*:180–189.
15. Lacaz, C., R. G. Ferrí, et al. 1967. Blastomicose queloideana associada a blastomicose sulamericana. Registro de um caso. Hospital (Rio), *71*:7–11.
16. Lacaz, C. S., and M. Rose. 1969. Bibliografia sobre blastomicose salamericana (doença de Lutz) e blastomicose queloidiforme (duença de Lobo) (1909-1968). São Paulo, Instituto de Medicina Tropical.
17. Lacaz, C. S., R. G. Ferrí, et al. 1962. Aspectos immunoquimicos na Blastomicose sul Ameri-

cana e Blastomicose Queloidiana. Rev. Med. Circ. Farm., *298*:63–74.

18. Lacaz, C. S., and O. J. de M. Fonseca. 1971. Estudo de culturas isoladas do blastomicose queloidiforma (doença de Jorge Lobo). De nomincao ao seu agente etiologica. Rev. Inst. Med. Trop. São Paulo, *13*:225–251.

19. Leite, J. M. 1954. Doença de Jorge Lobo. Contribuqao a seu estudo antatomopatologico. Tese. Oficinus Graficos da Revista Veterinairia. Belém. Pará, Brasil.

20. Lobo, J. 1931. Um caso de blastomicose produzida por uma especie nova, encontrada em Recifie. Rev. Med. Pernambuco *1*:763–765.

21. Michahanay, J. 1963. Corpos asteroides ora blastomicose de Jorge Lobo. A proposito deum novo caso. Rev. Inst. Med. Trop. São Paulo, *5*:33–36.

22. Migaki, G., M. G. Valerio, et al. 1971. Lobo's disease in an Atlantic bottlenosed dolphin. J. Am. Vet. Med. Assoc., *159*:578–582.

23. Raper, K., and D. Fennell. 1965. *The Genus Aspergillus*. Baltimore, The Williams and Wilkins Co., p. 234.

24. Robledo, V. M. 1965. Enfermedad de Jorge Lobo (Blastomicosis queloidiana). Presentaction de un neuvo caso Colombiano. Mycopathologia, *25*: 373–380.

25. Roche, J. C., L. Monod. 1976. Apropos du 10e cas de lobomycose observé en Guyane Francaise. Bull. Soc. Pathol. Exot., *69*:540–546.

26. Sampaio, M. M., L. Dias, et al. 1971. Bizarre forms of the aetiological agent in experimental Jorge Lobo's disease in tortoises. Rev. Inst. Med. Trop. São Paulo, *13*:191–193.

27. Sampaio, M. M., and L. Braga-dias. 1977. The armadillo *Euphractus sexcinctus* as a suitable animal for experimental studies of Jorge Lobo's disease. Rev. Inst. Med. Trop. São Paulo, *19*:215–220.

28. Silva, M. E., W. Kaplan, et al. 1968. Antigenic relationship between *Paracoccidioides loboi* and other pathogenic fungi determined by immunofluorescence. Mycopathologica, *36*:98–105.

29. Silverie, C. R., P. Ravisse, et al. 1963. La blastomycose cheloidienne ou maladie de Jorge Lobo en Guyane francaise. Bull. Soc. Pathol. Exot., *56*:29–35.

30. Tapia, A., A. Torres-Calcindo, et al. 1978. Keloidal blastomycosis (Lobo's disease) in Panama. Int. J. Dermatol., *17*:572–574.

31. Wiersma, J. P., and P. L. A. Niemel. 1965. Lobo's disease in Surinam patients. Trop. Geogr. Med., *17*:89–111.

13 Rhinosporidiosis

DEFINITION

Rhinosporidiosis is an infection of the mucocutaneous tissue caused by *Rhinosporidium seeberi*, an as yet unisolated and unclassified fungus. It is a chronic granulomatous disease characterized by the production of large polyps, tumors, papillomas, or wartlike lesions that are hyperplastic, highly vascularized, friable, and sessile or pedunculated. The nose is most commonly affected, with the conjunctiva the second most frequently involved site. Areas of infection rarely involved include the anus, penis, vagina, ears, pharynx, and larynx. The name of this clinical condition connotes an infection by the fungus *Rhinosporidium* and does not exclude sites of primary disease other than the nasal region. The infection has also been described in a variety of wild and domestic animals.

HISTORY

Interestingly, both this disease and one caused by a morphologically similar etiologic agent were discovered and described by essentially the same group of investigators in a geographic area where these particular infections are rarely encountered. Coccidioidomycosis was discovered by Posadas in 1892, and rhinosporidiosis was reported by Seeber in 1900. Both men were students of Professor R. Wernicke in Buenos Aires at the same time. These diseases were at first thought to be protozoan in origin, and both are quite uncommon in the country of their discovery.

Guillermo Seeber[33] published the first case report of rhinosporidiosis in a 19-year-old agricultural worker. The patient had a large nasal polyp that impeded breathing. He noted in his thesis that Malbran, who was also from Argentina, had seen a similar case in 1892. Independently O'Kinealy in 1903 reported a case from India under the name "localized psorospermosis," which he had first seen in 1894. Ellet in the United States also described the disease in the nose of a patient first seen in 1897. The patient was a native-born farmer in Tennessee. A full report of the case was published in 1907 by Wright.[36] In 1900 in the Programa de Zooligia Medica, Wernicke described the organism from Seeber's case as a *Coccidium,* but this paper appears to have been lost. Seeber considered it to be a *Coccidioides* related to the agent of Posadas' disease. Minchin and Fanthum in 1905[23] detailed the appearance of the organism and concluded it was a Sporozoa related to Neosporidia and Haplosporidia. Unaware of the Argentinean publication, they named it *Rhinosporidium kinealyi.* Many more reports of the disease appeared after the first published cases. Ashworth[4] made a very detailed analysis of the organism and its development in tissue and, concluding that it was a fungus, gave it the name *Rhinosporidium seeberi.* He also published the first report of the disease in Scotland. The patient was an Indian medical student with nasal polyps. An autochthonous case in a European was described by Denti in 1925.[11] The infection involved the palpebral conjunctiva in a 56-year-old male from Lombardy in Italy. Karunaratne in 1964[17] published a detailed account of the disease in man and reviewed the literature.

325

ETIOLOGY, ECOLOGY, AND DISTRIBUTION

The etiologic agent is a fungus called *Rhinosporidium seeberi.* Throughout the years many attempts to culture the organism[14] have failed, and the taxonomic affinities of this organism are still lacking. Nevertheless, based on his studies, Ashworth[4] concluded it to be a lower aquatic fungus. He determined that the nutritive reserve of cells in the trophic stage is fatty material, the multiple nuclear division takes place simultaneously prior to formation of spores, and that the spherule walls contain cellulose or a similar substance. These are all characteristics of the Olpidiaceae of the Chytridiales, a group of water molds in which Ashworth provisionally placed the organism. Rao[31] has demonstrated chitin rather than cellulose in the spherule walls.

A saprobic existence for the organism has never been established, and inference from case histories is always tenuous. If the organism is indeed a *Chytridium,* an aqueous natural habitat is probable. Many case studies have noted an association of disease in the patient with frequent bathing or working in stagnant fresh water. Mandlick found a 20 per cent infection rate in a group of workers engaged in removing sand from a river bottom. The patients' co-workers on shore did not have any disease.[22] Similarities of the organism to *Ichthyosporidium,* a fungus infecting salmon and trout, have been noted, and a parasitic existence of *Rhinosporidium* in an aquatic form of life conjectured.[17] In arid countries most infections are ocular, and dust is postulated to be a vector.[18]

About 2000 cases of disease had been recorded up to 1964 when Karunaratne reviewed the world literature.[17] Of these, 88 per cent were from India and Ceylon. In one series, Allen and Dave reported seeing 60 cases within an 18-month period in India.[3] In these areas the infection is so common that many cases are not reported. The areas with the greatest endemicity are Kerala and Tamil Nadu states. After India, most published cases are reported from South America, particularly Brazil and Argentina. Forty-one cases of both animal and human infection were reported by Niño.[25, 26] All occurred in a highly endemic focus in the Villa Angela region in the province of Chaco in Argentina. Thirty of the infections were in horses and six in humans. All involved the nose or adjacent areas. Some 15 reports have come from Brazil[5] and 11 from Mexico,[12, 29] with a scattering throughout Colombia, Venezuela, and other countries. In an area of southern Brazil at the latitude corresponding to that of the Chaco of Argentina, another highly endemic region has been delineated. Londero et al.[21] recorded 13 horses, 4 mules, and a cow with the disease.

About 40 reports, including one of the first, have been published from the United States. In about half these, the nasal area was involved; most of the remainder involved conjunctival disease. The disease has been recorded from almost all parts of the world, including 51 cases in Uganda[28] and a few in Madagascar,[6] Ghana,[9] Iran, Russia[13, 16] Southeast Asia, the Near East, and Europe.[2, 11] Many of the European cases were in individuals who were from India or had visited there for extended periods.

Though the age of patients varies from 3 years to 90 some years, most patients are between 20 and 40 years of age when diagnosed.[19, 20] Since the lesions do not cause any undue discomfort to the patient and the infection is very slow in developing, it is often difficult to ascribe age at onset. One of the youngest cases on record was in a Texas girl, age 3, who had severe nosebleed as a presenting symptom.[27] In both human and equine cases of the disease, males account for 70 to 80 per cent of cases,[17, 25] but this varies with age, site of infection, and geographic location. In prepubertal cases, the infection appears to occur about equally between the sexes. Eye infections seem to be more common in women, but no clear pattern is evident. When location is recorded, most patients are from rural environments, and in many there is an association with work or play in fresh water. Several report infection at the site of a previous injury.[17] Ocular infections are more frequent in arid areas and appear to occur following dust storms and injury to the eye.[18]

CLINICAL DISEASE

The name "rhinosporidiosis" implies a sporozoan infection of the nose, which is the commonest site of infection. Of the 2000 cases tabulated, about 70 per cent involved the nasal area, 15 per cent the eye, and 8 per cent some other mucosal area or, very rarely, a cutaneous site.[17] The development of lesions and their gross appearance varies somewhat, depending on the anatomic site involved.

Nasal Disease. The exact mode of infection is unknown, but some initial trauma may be a predisposing factor. Karunaratne[17] conjectures that the Muslim custom of mechanically cleansing the nose before entering a mosque may predispose to infection. Once disease is established in the nose, trauma to adjacent sites followed by autoinoculation may serve to spread the disease.

The commonest sites for initial infection are the mucous membrane of the septum, interior turbinate, and nasal floor.[32] Other areas, such as middle turbinate, middle meatus, and nasal roof, are less frequently involved. The first symptom noted by patients is the feeling of the presence of a foreign body in the nose. The infection may be accompanied by mild to intense pruritus and coryza. The lesion, which at first is sessile, develops into a pedunculated polyp (Fig. 13–1). In time the growth obstructs the air passage, and it is this symptom for which attention is sought. The nasal discharge is mucus tinged with blood and contains spores and sporangia. Mild bleeding is frequent, as the lesions are quite friable, but epistaxis may be severe. As the lesion grows it may take on a bizarre shape and appearance. Fully developed tumors are usually polypoid, globoid, and pedunculated (Fig. 13–2). The lesions may be part pedunculate and part sessile or wholly sessile. Papillary projections and lobules are sometimes seen, giving the lesion a raspberry, strawberry, or, when very corrugated, a cauliflower appearance. The color is bright pink initially, becoming deep red. The red color that develops is due partly to profuse vascularization and partly to free blood and old hemorrhagic material. The lesion is often spleenlike in appearance. On close examination, the polyps are mottled owing to the presence of numerous whitish, macroscopically visible spherules. Fibrous cords are often numerous, and the whole lesion, when cleansed of blood, may have a grayish cast. Fully mature polyps hang down through the nasal meatus and may extend beyond the lip. If they are located high up on the turbinates or nasal septum, the lesions may hang into and beyond the nasopharynx, interfering with respiration and feeding (Fig. 13–3). These lesions contain much mucus, sometimes mucinous cysts, have bulbous ends, and a general appearance of a ripe fig. They may weigh 20 g or more. Lesions in the nasopharynx alone without nasal disease are uncommon, but have been reported in about

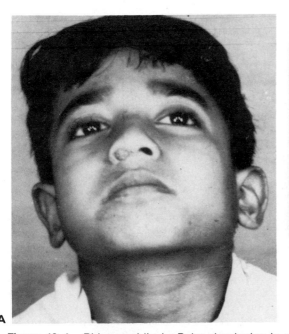

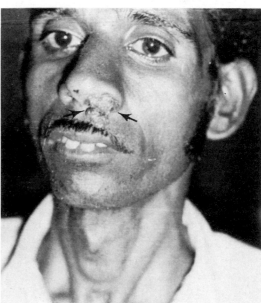

A   B

Figure 13–1. Rhinosporidiosis. Polyp developing in nasal opening. (Courtesy of S. Banerjee.) *B,* Recurrent rhinosporidiosis. This patient has had polyps removed surgically 49 times.[8] (Courtesy of V. Chitravel.)

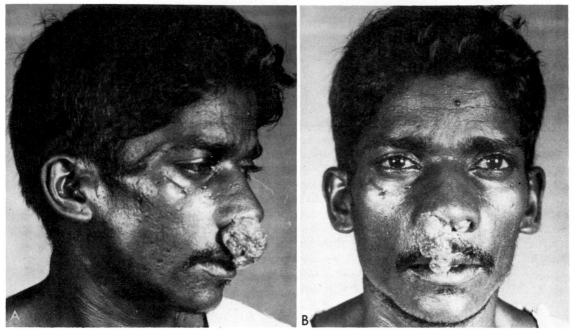

Figure 13–2. Rhinosporidiosis. *A* and *B,* Fully developed polyploid pedunculated tumor. (Courtesy of C. Satyanarayana.)

30 patients. The polyps are similar to those of nasal origin and occasionally lead to obstruction, dyspnea, and dysphagia.

Ocular Disease. The eye and adnexae account for about 15 per cent of cases in

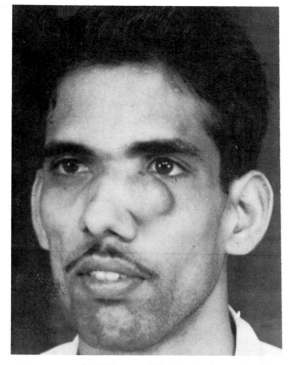

Figure 13–3. Rhinosporidiosis. Lesions developing far up in the turbinates. (Courtesy of S. Banerjee.)

India and other moist tropical environments. In dry, dusty areas, as in Transvaal, Iran, and arid areas of the Indian subcontinent, almost all cases are ocular. Kaye,[18] commenting on the disease in South Africa, reported that infection occurred after dust storms. He theorized that dust not only caused eye injury but also transmitted the fungus, thereby leading to infection. About half the published cases of rhinosporidiosis in the United States are nasal, and most of the remaining are ocular.[27, 30] The majority of these reports come from Texas, with the others scattered throughout the remaining states.

In almost 90 per cent of ocular infections, the palpebral conjunctiva is involved (Fig. 13–4). In the remaining cases, disease was of the bulb, limbus, caruncle, or canthi. Though bilateral involvement[24] and multiple lesions are not infrequent in nasal disease, most infections of the eye are unilateral, and the lesions single.[27] The growths are sessile or stalked, the attachment being to the upper or lower fornix or tarsal conjunctiva. The lesions are often small and flat, accommodating themselves to lie between the lid and eyeball. They may cause no discomfort to the patient, who may not be aware of their presence. They are freely moveable, granular, pink to red in appearance, and with careful examination can be seen to contain whitish spherules. The patient usually does not present with any symptoms other than a growth in the eye.

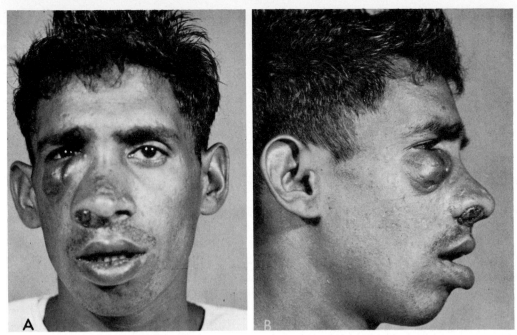

Figure 13–4. Rhinosporidiosis. *A*, Involvement of palpebral conjunctiva and nose. *B*, Profile of same patient. (Courtesy of P. Kulkavni.)

When lesions become large, other symptoms may occur, however. These include excessive tearing, redness of the eye, discharge, photophobia, conjunctival infection, and eversion of the lid. Some lesions are so deep blue-red and spleenlike in appearance that they suggest hemangioma, and often they are diagnosed clinically as such.

Cutaneous Disease. Lesions on the skin are infrequent, and most are associated with adjacent mucocutaneous disease. On rare occasions, skin sites at a distance from mucosa may be infected from scratching and autoinoculation.[3] Skin lesions alone, i.e., not associated with disease elsewhere, are very rare but have been reported. These include a single lesion on the scalp and one on the abdomen, and there has been one report of multiple skin lesions over several areas of the body.[17] Skin lesions begin as tiny papules that become wartlike growths. Their surface is crenated. Since they are friable, they are often ulcerated and secondarily infected with bacteria. Cutaneous lesions rarely become pedunculated. The growths that occur in the rare cases of hematogenous dissemination are described as firm, hard, subcutaneous nodules that may remain unattached to the overlying skin or invade through it. Cutaneous lesions are painless and cause no discomfort to the patient unless they are in areas where they are continually traumatized, such as the sole of the foot.

Other Areas of Involvement. Lesions occurring in other mucosal areas are infrequently encountered but reported. These include the larynx, hard palate (Fig. 13–5), epiglottis (Fig. 13–6), vagina, vulva, uvula,

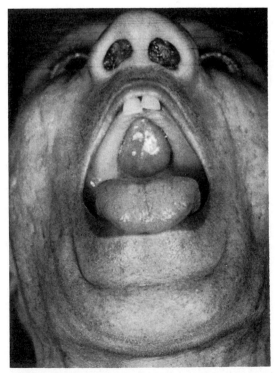

Figure 13–5. Rhinosporidiosis. Lesion on the hard palate. (Courtesy of C. Satyanarayana.)

329

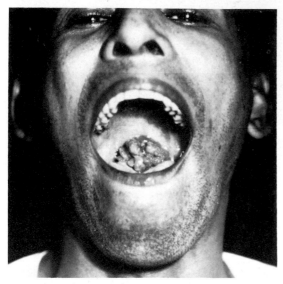

Figure 13–6. Rhinosporidiosis. Lesions involving epiglottis. (Courtesy of C. Satyanarayana.)

and anus. Vaginal and anal lesions were described as resembling condylomata, rectal polyps, or hemorrhoids. The urethra has been involved in a few cases in males. These lesions were red, pedunculated, knobby masses extending beyond the meatus of the penis. The Muslim habit of removing the last drops of urine following micturition by rubbing the meatus of the penis with a stone is cited as causing repeated trauma to the area which may predispose to infection.[31] Disease of the parotid gland, trachea, and bronchus[34] have also been reported. In the case of bronchial disease, complete obstruction of air passages occurred with a fatal outcome.

In the commonly encountered cases of rhinosporidiosis, the organisms are limited to the polyps; regional lymph nodes and adjacent tissue are not involved. There are a few reports of disseminated rhinosporidiosis, however. Agrawal described a case in which spherules were found infecting the spleen, liver, lung, and other viscera as well as being found in laked blood and urine.[1] In the one case of cerebral infection, lesions were found in the brain, the adjacent blood vessels, and the nose without involvement of other areas.[2] A few cases in the external ear have been recorded.[17] The lesions had the appearance of ordinary aural polyps. Only when they enlarged and caused obstruction and pressure were symptoms present. Invasion and destruction of the bone has also been recorded.[7] Lesions occurred in osseous tissue underlying nasal, pharyngeal, or digital sites of infection.

DIFFERENTIAL DIAGNOSIS

The lesions of rhinosporidiosis are most often red, friable, sessile growths or polyps of the mucosal surfaces. These may resemble mucoceles, hemangiomas, condylomata, or neoplasms. Cryptococcus is the only other fungus known to elicit polypoid tumors. Direct examination of biopsy material allows the diseases to be easily differentiated.

PROGNOSIS AND THERAPY

Rhinosporidiosis is a chronic disease that usually has a long history without pain, discomfort, or debility to the patient. Disease of 30 to 40 years' duration has been noted in some reports.[17] These infections usually consist of single small lesions that cause difficulty only when they are large enough to obstruct a passage or cause pressure on vascular or neural bundles. In the rare cases of systemic disease, dissemination appears to have occurred early in the course of the infection.

Recurrence of infection is a characteristic of rhinosporidiosis, and many patients have had several surgical procedures for removal of growths. In an early case report in the United States, Wright in 1907 described a young Tennessee farmer who had had nasal polyps removed three times.[36] Ten operations in 15 years have been noted in some records, and in a recent case series reported by Chitravel et al.[8] disease had recurred 49, 21, and 23 times (Fig. 13–1, *B*). No predisposing susceptibility or immunity to reinfection has been demonstrated in rhinosporidiosis.

Treatment involves surgical removal of the affected tissue. This is accomplished by use of a hot or cold snare to avoid spreading the infection to adjacent tissue.[8, 32] Recurrent disease following electrodessication in a 13-year-old girl was reported by Portilla-Aguilar et al.[29] Antrostomy, resection of the tumor, and electrocoagulation of the zone of implantation led to cure. Local bacterial infection and some fatal septicemias have occurred following unskilled surgery. Copious bleeding is also a complication.[8] Local injection of amphotericin B may be used as an adjunct to surgery to prevent reinfection and spread;

however, no studies have demonstrated inhibition of growth of the organism by this agent. Other drugs such as dapsone have been without efficacy.

PATHOLOGY

Gross Pathology. Very few cases of generalized and fatal rhinosporidiosis are known. The best described case is the one reported by Agrawal.[1] The patient, a 30-year-old Hindu, had had an infection of the eyelid, skin, and palate for more than one year. At autopsy, white, firm nodules were found in the lung and on the pleura, vocal cords, and epiglottis. Granulomata were also found in striated muscle and skin. Spherules were found in the sinusoids of the liver and in the spleen and kidney, but these had mostly degenerated.

Gross examination of the specimen from a usual case of rhinosporidiosis reveals it to have the appearance of an ordinary nasal polyp. In contrast to the loose, edematous, myxomatous stroma of the latter, polyps of rhinosporidiosis are rather dense, and mucinous cysts are usually absent. Opaque grayish-white granular material is apparent; this represents the mature sporangia. Cut sections show that such sporangia are of varying sizes, the more mature being closer to the epithelial surface. Some of them are collapsed and assume a semilunar shape. Polyps from conjunctival infections are flattened, soft, reddish-pink to dark red, and less lobulated. Minute opaque spherules are also readily visible.

Histopathology. The layers of transitional epithelium are often invaginated to become flask-shaped and may form pseudocysts. Such areas contain spores, pus, and mucous material. The epithelium is generally hyperplastic, though it may be quite thinned in some areas. Mature sporangia often lie just beneath the thinned areas. Spores accompanied by neutrophils are sometimes found in the epithelium. The major portion of the growth consists of very vascular, fibromyxomatous connective tissue in which the parasites are found in varying stages of development. The cellular infiltrate consists of plasma cells, lymphocytes, histiocytes, and neutrophils. Occasionally eosinophils are present in large numbers, but this finding is not so constant as in ordinary nasal polyps. The cellular areas may contain mainly lymphocytes or plasma cells. Giant cells are not uncommon, especially in older lesions. Microabscesses are frequent, and there is often evidence of chronic trauma and hemorrhage. As is the case in coccidioidomycosis, freshly liberated spores from a spherule incite a polymorphonuclear response. The eosinophilic material characteristic of asteroid body formation in many fungal diseases has not been noted.

The life cycle of the organism in tissue was outlined in detail by Ashworth.[4] The infecting spore appears to be able to penetrate into the mucosal epithelium and to begin maturation in the subepithelial tissue. This spore, called the "trophic stage" by Ashworth, is 6 to 10 μ in size, has a chitinous wall, a clear protoplasm, and a vesicular nucleus with a nucleolus. He also noted the presence of a karysome-like body. As the spherule grows to a diameter of 10 to 12 μ, globular and granular material appears in the cytoplasm. This material is fatty in nature and is stained by the usual fat stains. At 50 μ the first nuclear division occurs (Fig. 13–7). Ashworth determined that there were four chromosomes. Succeeding nuclear divisions take place synchronously. At 100 μ in size, a layer of cellulose-like (probably an α glucan) material is laid down on the inner surface of the wall of the spherule. The cellulose material is about 3 μ thick, and at maturity the wall itself sometimes exceeds 5μ. The cellulose is thin in the region where a pore later develops. The pore allows the escape of mature spores. When 2000 nuclei are present, the cytoplasm is seen to condense around them. An annulus of uncondensed material, 12 μ thick, surrounds the inner wall. More nuclear divisions occur until about 16,000 young spores are present. The spherule is about 150 μ in diam-

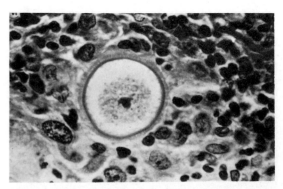

Figure 13–7. Rhinosporidiosis. Spherule just before nuclear division. Note the globular and granular material in the cytoplasm and the prominent nucleolus. Hematoxylin and eosin stain. ×400.

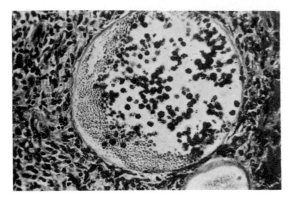

Figure 13–8. Rhinosporidiosis. Spherule with maturing endospores. Many prospores fail to develop. Hematoxylin and eosin stain. ×400.

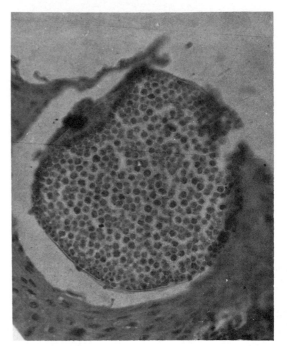

Figure 13–10. Rhinosporidiosis. Mature spherule releasing spore through pore. The spherule has broken through the lining of the epithelium. [From Satyanarayana, C. 1966. Chapter 13 in *Clinical Surgery. Rhinosporidiosis.* C. Rob and R. Smith (eds.), London, Butterworth and Co.]

eter at this point. It enlarges to 250 to 350 μ, the wall becomes thinner, and the annulus begins to disappear. The mature spores migrate toward the periphery near the pore. Approximately one third of the spores fail to mature (Fig. 13–8). Rupture of the pore occurs, possibly from internal pressure, and the spores are released into the surrounding connective tissue. The spores at maturity are 7 to 9 μ. They contain a nucleus, a basophilic karyosome-like body, and a globular cytoplasm containing eosinophilic material. An influx of foreign body giant cells is incited by rupture of the sporangia. These may be seen to invade the empty spherule together with the neutrophils. Mature sporangia usually rupture through the epithelial layer, and nasal exudate may contain numerous spores (Fig. 13–9 and 13–11). Karunaratne[17] presents evidence that freed spores lodge in the epithelium, become surrounded by tissue, and begin the maturation cycle over again. Controversy exists as to the fate of the spores

released into tissue, as they stain differently from the so-called trophic state. The ultrastructure of the maturing spherule has been examined by Vanbreuseghem.[35]

Rhinosporidium seeberi is easily observed in the usual hematoxylin and eosin-stained sections. Maturing sporangia and their spores are stained by the Gridley, PAS, and GMS stains. The early trophic stage, however, does not take the Gridley or PAS stain but is readily colored by the GMS method. The

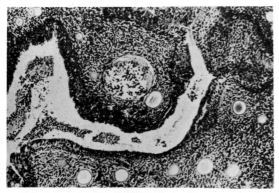

Figure 13–9. Rhinosporidiosis. Numerous spherules of various sizes in pedunculated lesion of nose. Hematoxylin and eosin stain. ×100.

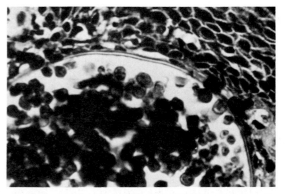

Figure 13–11. Rhinosporidiosis. Mature spherule showing thick wall and endospores near surface epithelium. Nuclei can be seen within the endospores.

inner cellulose-like material of the sporangium and the outer layer of newly released spores take the Mayer's mucicarmine stain. With the exception of the capsule of *Cryptococcus,* this is the only fungus which is colored by this method, but confusion of the two is not a problem. Small empty spherules may resemble those seen in coccidioidomycosis.

ANIMAL DISEASE

Although a prolonged and chronic disease in experimental animals has not yet been produced, natural infection in animals occurs frequently. About 100 cases in horses have been recorded, 90 per cent of which were in males. The histopathology was identical to that seen in human disease. Two dozen cases in cattle, 12 in mules, and one in a dog are also on record. Almost all of these were in male animals. The lesions in animals were all in the nose except for one in the larynx. No eye infections have been reported. Many attempts to transfer the disease to experimental animals have been made. In a few instances, granulomata have been produced, but sustained and progressive disease with propagation of the parasite has not occurred.[14] Natural infection was also reported in the wood duck (*Aix sponsa*) by Davidson and Nettles.[10] A multilobed granular polyp was seen in the nasal passage of an otherwise healthy bird. The authors consider this further evidence of the aquatic habitat of the organism and speculate that migratory birds could carry the fungus over great distances.

IMMUNOLOGY AND SEROLOGY

Essentially nothing is known about immunology or resistance to the disease. The patients who have been studied have not had any debilitating diseases, and infection does not appear to be opportunistic in that sense. No serologic studies are available.

LABORATORY IDENTIFICATION

Direct Examination. Examination of the lesions often reveals macroscopically visible subsurface sporangia. Dissected or excised tissue or nasal discharge can be slightly macerated and examined in a potassium hydroxide preparation. Mature sporangia up to 350 μ in diameter and spores 7 to 9 μ can be seen.

Only spores are seen in nasal discharge, but a few sporangia may be present.

Culture Methods. *Rhinosporidium seeberi* has defied laboratory cultivation. Diagnosis of the disease can be made from histopathologic section or by direct examination.

MYCOLOGY

Rhinosporidium seeberi (Wernicke) Seeber 1912

Synonymy. *Coccidium seeberi* Wernicke 1900; *Coccidium seeberia*-Wernicke apud Belou 1903[5a]; *Coccidioides seeberi* Wernicke 1907[4]; *Rhinosporidium kinealyi* Minchin et Fanthum 1905; *Rhinosporidium equi* Zschokke 1913; *Rhinosporidium ayyari* Allen et Dave 1936.

The early investigators considered the organism to be a Sporozoa. When its fungal nature was demonstrated by Ashworth,[4] it was considered to be a Phycomycete related to the Chytridiales. Later, C. W. Dodge classified it among the Endomycetales as an ascomycete, as he considered that the internal spores produced were ascospores. Since a saprobic existence of a life cycle outside man has not been found and its cultivation *in vitro* has not yet been substantiated, it is not possible currently to delineate the true taxonomic relationships of the organism. Most mycologists feel it is a *Chytridium* related to the Olpidiaceae or, as suggested by C. W. Emmons, a *Synchytrium.* The latter group of fungi are parasites of plants, producing galls, or "plant polyps," on the host, and sporulation is very similar to that observed in *Rhinosporidium.*

REFERENCES

1. Agrawal, S., K. D., Sharma, et al. 1959. Generalized rhinosporidiosis with visceral involvement. Report of a case. Arch. Dermatol., *80:*22–26.
2. Alessandrini, O. O. 1926. Cited in Ruiz, F. R., and T. Ocana. 1930. Nueva observacione sobre *Rhinosporidium seeberi.* Rev. Med. Latino Am., *16:*24–30.
3. Allen, F. R. W. K., and M. Dave. 1936. Treatment of rhinosporidiomycosis in man based in sixty cases. Indian Med. Gaz., *71:*376–395.
4. Ashworth, J. H. 1923. On *Rhinosporidium seeberi* (Wernicke 1903) with special reference to its sporulation and affinities. Trans. R. Soc., Edin., *53:*301–342.
5a. Belou, P. 1903. Tratado de Parasitologia Animal. 1st ed. Buenos Aires.

6. Brygoo, E. R., C. Bermond, et al. 1959. First Madagascan case of rhinosporidiosis. Bull. Soc. Pathol. Exot., *52*:137–140.

7. Chatterjee, P. K. 1977. Recurrent multiple rhinosporidiosis with osteolytic lesions in hand and feet. J. Laryngol. Otola, *91*:729–734.

8. Chitravel, V., B. M. Sundaram, et al. 1981. Recurrent rhinosporidiosis in man: case reports. Mycopathologia, *73*:79–82.

9. Christian, E. C., and J. Kovi. 1966. Three cases of rhinosporidiosis in Ghana. Ghana Med. J., *5*:63–64.

10. Davidson, W. R., and V. F. Nettles. 1977. Rhinosporidiosis in a wood duck. J. Am. Vet. Med. Assoc. *171*:989–900.

11. Denti, A. V. 1925. Un caso di sporidiosi congiuntivale in Lombardia. Boll. d'oculisti cutica *2*:71–87.

12. Gonzalez-Mendoza, A., and B. Austria. 1975. Rinosporidiosis en Mexico. Revision de la litaratura nacional y comentarios epidiomiologicos a proposito de la observacion de dos nuevos casos. Bol. Soc. Mex. Microbiol., *9*:149–153.

13. Gremeshkevich, G. K., M. N. Malets, et al. 1975. Rinosporidoz slizistoi obolochki vek. Vestnik Oftal'mologii No. 5, pp. 89–90.

14. Grover, S. 1970. *Rhinosporidium seeberi*: A preliminary study of the morphology and life cycle. Sabouraudia, *7*:249–251.

15. Kameswaran, S. 1966. Surgery in rhinosporidiosis. Experience with 293 cases. Int. Surg., *46*:602–605.

16. Karpova, M. F. 1964. On the morphology of rhinosporidiosis. Mycopathologia, *23*:281–286.

17. Karunaratne, W. A. E. 1964. Rhinosporidiosis in Man, London, The Athlone Press.

18. Kaye, H. 1938. A case of rhinosporidiosis on the eye. Br. J. Ophthalmol., *22*:447–455.

19. Kutty, M. K., T. Sreedharan, et al. 1963. Some observations on rhinosporidiosis. Am. J. Med. Sci., *246*:695–701.

20. Kutty, M. K., and P. N. Unni. 1969. Rhinosporidiosis of the urethra. A case report. Trop. Geogr. Med., *21*:338–340.

21. Londero, A. T., M. N. Santos, et al. 1977. Animal rhinosporidiosis in Brazil. Report of three additional cases. Mycopathologia, *60*:171–173.

22. Mandlick, G. S. 1937. A record of rhinosporidial polypi with some observations on the mode of infection. Indian Med. Gaz., *72*:143–147.

23. Minchin, E. A., and H. B. Fantham. 1905. *Rhinosporidium kinealyi* n.g.n. sp. A new sporozoon from the mucous membrane of the septum nasi of man. Quart. J. Microbiol. Sci., *49*:521–532.

24. Neumayr, T. G. 1964. Bilateral rhinosporidiosis of the conjunctiva. Arch. Ophthalmol., *71*:379–381.

25. Niño, F. L., and R. S. Freire. 1964. Exitenciade un foco endemico de rinosporidiosis en las provincia del Chaco. V. Estudio de neuvas observaciones y consideraciones finales. Mycopathologia, *24*:92–102.

26. Niño, F. L., and R. S. Freire. 1966. Existencia de un foco endemico de rinosporidiosis en la provincia del Chaco. Neuvas observaciones de rhinosporidiosis equina. Caracteres ecologicos de la region de Villa Angela. Rev. Med. Vet. Buenos Aires, *47*:421–437.

27. Norman, W. B. 1960. Rhinosporidiosis in Texas. Arch. Otolaryngol., *72*:361–363.

28. Owor, R., and W. M. Wamukota. 1978. Rhinosporidiosis in Ugana: a review of 51 cases. East Afr. Med. J., *55*:582–586.

29. Protilla Aguilar, J., M. Reynosos-Garcia, et al. 1977. Rhinosporidiósis nasal. Gac. Med. Mex. *113*:363–365.

30. Peters, H. J., and C. G. DeBelly, 1969. Conjunctival polyp caused by *Rhinosporidium seeberi*. Report of a case. Am. J. Clin. Pathol., *51*:256–259.

31. Rao, S. N. 1966. *Rhinosporidium seeberi*: A histochemical study. Indian J. Exp. Biol., *4*:10–14.

32. Satyanarayana, C. 1966. Chapter 13 in *Clinical Surgery. Rhinosporidiosis.* C. Rob and R. Smith (eds.)., London, Butterworth and Co.

33. Seeber, G. R. 1900. Un neuvo esporozuario parasito del hombre. Dos casos encontrades en polipos nasales. Tesis Univ. Nat. de Buenos Aires.

34. Thomas, T., N. Gopinath, et al. 1956. Rhinosporidiosis of the bronchus. Br. J. Surg., *44*:316–319.

35. Vanbreusegeham, R. 1973. Ultrastructure of *Rhinosporidium seeberi*. Int. J. Dermatol., *12*:20–28.

36. Wright, J. 1907. A nasal sporozoon (Rhinosporidium kinealyi). N.Y. Med. J., *86*:1149–1153.

THE SYSTEMIC MYCOSES

14 The True Pathogenic Fungus Infections and The Opportunistic Fungus Infections

INTRODUCTION

The systemic diseases caused by fungi fall into two very distinct categories. These categories are delineated by the interaction of two factors: inherent virulence of the fungus and constitutional adequacy of the host (Table 14–1). The first category includes infections caused by the true pathogenic fungi: *Histoplasma, Coccidioides, Blastomyces,* and *Paracoccidioides.* The second group of infections are termed "opportunistic" because the organisms involved are of inherently low virulence, and disease production depends on diminished host resistance to infection. The common etiologic agents of opportunistic infections are *Aspergillus, Candida, Rhizopus,* and *Cryptococcus.*

I. TRUE PATHOGENIC FUNGUS INFECTIONS

The two categories of disease are distinct in essentially all aspects of host-parasite interaction. The true pathogenic fungi are those species that have the ability to elicit a disease process in the normal human host when the

Table 14–1 *The Systemic Mycoses*

	True Pathogenic Fungus Infections	*Opportunistic Fungus Infections*
Diseases	Histoplasmosis Blastomycosis* Paracoccidioidomycosis Coccidioidomycosis	Aspergillosis Candidiasis Mucormycosis Cryptococcosis*
Host	Normal	Abrogated
Portal of entry	Primary infection pulmonary	Various
Prognosis	99% of cases resolve spontaneously	Recovery depends on severity of impairment of host defenses
Immunity	Resolution imparts strong specific immunity	No specific resistance to reinfection
Host response	Tuberculoid granuloma; also mixed pyogenic	Depends on degree of impairment — necrosis to pyogenic to granulomatous
Morphology in tissue	All agents show dimorphism to a tissue form	No change in morphology†
Distribution	Geographically restricted	Ubiquitous

*These diseases have significant exceptions to the usual patterns.
†*Candida* sp. is found as mixed yeasts and mycelial elements in tissue.

inoculum is of sufficient size. As previously discussed, pathogenicity in fungi is an accidental phenomenon and is not essential to the survival or dissemination of the species involved. In recent years it has been determined that the vast majority of such infections, usually more than 90 per cent, are either completely asymptomatic or of very short duration and quickly resolved. Resolution of the infection is accompanied by a strong specific resistance to reinfection that is of long duration. In the few individuals who have residual or chronic infection, the usual cellular response is a granulomatous process resembling that seen in tuberculosis.

Another characteristic of the true pathogenic fungi is that they have a very restricted geographic distribution. It is also necessary for the patient to encounter the fungus when it is conidiating. Thus infection by *Coccidioides* requires a person to be present in the small ecologic areas favored by the fungus and to be there at the time of year when the fungus is growing and fruiting. The same is also true for *Histoplasma, Blastomyces,* and *Paracoccidioides.* The major endemic regions of these organisms are in the Americas and, with the exception of *Paracoccidioides,* primarily in the United States.

Sex, age, and race are important factors in the statistics of pathogenic fungus infections. Adult males constitute the majority of individuals with serious disease. It also appears that dark-skinned individuals have a greater risk of developing the severe disseminated forms of the disease than do Caucasians. Rural dwellers and agricultural workers are particularly susceptible to these infections because of their constant exposure to the soil-inhabiting fungi.

The pathogenic fungi exhibit a morphologic transition from the mycelial or saprophytic form to the parasitic form found in infected tissue (Table 14–2). In most of these organisms the change is a conversion to a budding, yeastlike form. This change is governed principally by the temperature of incubation. Thermal dimorphism is exhibited by all of the major organisms that produce systemic infections and also by one agent that produces subcutaneous infections (sporotrichosis). This morphologic change in the dimorphic fungi is a response or adaptation for survival by the organism finding itself in an unfavorable environment. Of the thousands of species of fungi only these few species exhibit this morphologic change under these particular environmental circumstances. What selective advantage this has for the organism, if any, has yet to be determined. Man-to-man transmission of these agents is unknown or extremely rare, so that dissemination of the species does not occur by this route; disease production appears to be an accidental phenomenon. The "parasitic" or tissue phase of all these fungi can be induced *in vitro.* This morphologic change to the tissue phase is also accompanied by profound changes in the metabolism, physiology, metabolic rate, cell wall composition, RNA poly-

Table 14–2   *The Dimorphic Pathogenic Fungi*

Disease and Etiologic Agent	*Saprophytic Phase (25°C)*	*Parasitic Phase (37°C)*
	Systemic Thermal Dimorphic Fungi	
Blastomycosis *Blastomyces dermatitidis*	Septate mycelium; conidia are pyriform, globose, or double. Colonies are white or beige, fluffy or glabrous.	Budding yeast with broad based bud, 8 to 20 μ.
Histoplasmosis *Histoplasma capsulatum*	Septate mycelium, microconidia, tuberculate macroconidia. Colonies are white or buff, fluffy.	Small, single budded yeasts, 1 to 5μ; 5 to 12μ in var. *duboisii.*
Coccidioidomycosis *Coccidioides immitis*	Septate mycelium, fragment to arthroconidia. Colonies are buff or white, fluffy or "moth-eaten."	Spherules 10 to 80μ. Endospores produced.
Paracoccidioidomycosis *Paracoccidioides brasiliensis*	Similar to *Blastomyces dermatitidis.*	Large, multiple budding yeasts 20 to 60μ.
	Subcutaneous Thermal Dimorphic Fungus	
Sporotrichosis *Sporothrix schenckii*	Septate, delicate mycelium; conidia on denticles from delicate conidiophores. Colonies are verrucous, black, white, or grey; glabrous or fluffy.	Fusiform, oval budding yeasts, 5 to 8μ.

merases, regulators such as "histins," lipid content, ubiquinone content, and so forth. Though temperature is the single most important factor affecting this transformation, other components also may exert an influence. These include oxidation-reduction potential, availability of sulfhydryl groups, CO_2 tension, and others. The relative importance of these factors as well as the degree and kinds of compositional changes varies with the species involved. It can be seen that dimorphism in pathogenic fungi is a very complicated process, and this subject has been the object of a great amount of investigation. Salient features of the changes as now understood are summarized in the Biological Studies section of each of the diseases.*

There are a number of other similarities among the true pathogenic fungus infections in regard to clinical course and manifestations, general pathology, prognosis, and response to treatment. In addition, the diagnostic laboratory tests and methods of isolation and identification of the causal agents are also similar. These are discussed in detail in the individual chapters concerning each disease. In the present section some general patterns of clinical types and pathologic responses are summarized in order to compare and contrast the diseases involved.

In regard to clinical manifestations and types of pathology observed, histoplasmosis is similar to coccidioidomycosis and blastomycosis is similar to paracoccidiodomycosis, as can be seen in Table 14–3. A comparison of the pathologies of the systemic mycoses to other mycoses is found in the introductory chapter of this book.

II. THE OPPORTUNISTIC FUNGUS INFECTIONS

The second group of systemic fungus infections are those caused by opportunistic fungi. Although objections have been raised to the use of the term "opportunistic," it seems quite appropriate. The organisms involved have a very low inherent virulence, and the patient's defenses must be abrogated before infection is established. Formerly, these diseases occurred quite rarely, but in recent years they have become much more common and of great medical significance. The rise in the incidence of these opportunistic infections has paralleled the use of antibiotics, cytotoxins, immuno-suppressives, steroids, and other macro-disruptive procedures that result in lowered resistance of the host. In contrast to the picture seen in infections caused by pathogenic fungi, the usual cellular response to opportunistic fungi is a suppurative necrotic process which is, at most, a poorly organized, granulomatous reaction. If the patient survives his debilitating disease or the medical procedure, he is usually able to contain the fungal infection by granuloma formation, fibrosis, and calcification.

The opportunistic fungi also differ from the true pathogens in geographic distribution. A patient with lymphoma would have to travel to the San Joaquin Valley to contract coccidioidomycosis, but *Candida* and *Aspergillus* are always present in his environment. *Candida* is found in small numbers in the normal gut, but it proliferates in great numbers if the equilibrium of the bacterial flora is upset or if it gains entrance into the body through a "barrier break." The conidia of *Aspergillus* are continually present in the air in all parts of the world. Unlike disease produced by the pathogenic fungi, establishment of opportunistic infections depends primarily upon altered host resistance rather than inoculum size. These infections are usually insidious and may not be diagnosed until autopsy. *Candida* and *Aspergillus* account for the vast majority of opportunistic fungus infections in all types of debilitated patients. Mucormycosis occurs less commonly but is dramatic in its presentation and often is rapidly fatal.

Opportunistic infections differ from those produced by pathogenic fungi in several other ways (Table 14–1). Recovery from the former does not establish a specific immunity, and reinfection may occur if general resistance is lowered again. There are no differences in susceptibility ascribable to age, sex, or race. The population that is afflicted with these infections is determined only by the type and severity of the underlying disease — two factors which decide the final outcome of the infection, almost independently of therapy. Lastly there is a striking difference between the tissue forms of the two groups of systemic fungus diseases. Whereas a dimorphism is exhibited by the etiologic agents of the pathogenic fungus infections, no such transition is exhibited by the organisms in opportunistic fungus diseases. *Aspergillus* growing on an agar plate, in decaying leaves,

*For a more detailed analysis of the subject of dimorphism, see Reference 3.

Table 14–3 Clinical and Pathological Patterns of the True Pathogenic Systemic Mycoses

I. Primary Infection	Histoplasmosis	Coccidioidomycosis	Paracoccidioidomycosis	Blastomycosis
Pulmonary	Commonly resolves spontaneously	Commonly resolves spontaneously	Resolved subclinical disease occurs	Resolved subclinical disease rare(?)
Disseminated infection	Commonly resolves spontaneously	Commonly resolves spontaneously	Acute phase undetermined	Acute phase undetermined

Usual Histopathology

Histoplasmosis: Intracellular yeasts in histiocytes; tuberculoid granuloma; Langhans' giant cells; caseation necrosis. Histology is the most similar to TB of all fungous diseases; calcification common.

Coccidioidomycosis: Tuberculoid granuloma with patches of pyogenic reaction; caseation rare; calcification common.

Paracoccidioidomycosis: Tuberculoid granuloma and mixed pyogenic reaction; calcification rare; yeasts occur singly and intracellularly in giant cells.

Blastomycosis: Tuberculoid granuloma and mixed pyogenic reaction; calcification rare; yeasts occur singly and intracellularly in giant cells.

II. Secondary Disease	Histoplasmosis	Coccidioidomycosis	Paracoccidioidomycosis	Blastomycosis
Chronic Pulmonary	Commonly as "coin" lesion; histoplasmoma; "bubble," "popcorn," or "shotgun" calcifications	Common; as "coin" lesions; coccidioidoma; and "bubble" calcifications	50% of cases present as single solid expanding mass	50% of cases present as single solid expanding mass
Chronic Disseminated	Not common; involves reticuloendothelial system, liver, spleen, adrenals, marrow, intestine; mucocutaneous, osseous, and rarely meningeal	Common; cutaneous, osseous, meningeal; later subcutaneous, mucocutaneous, and generalized	50% of cases present as mucocutaneous disease; rare or later subcutaneous, adrenal, "actinic," lymph nodes, osseous, and generalized	50% of cases present as cutaneous disease; osseous, genitourinary, adrenal also common; later subcutaneous, meningeal, and generalized

or in the lungs of a leukemic is mycelial in form. *Mucor* produces hyphal strands in moist bread or in the brain of a diabetic. The morphologic transitions exhibited by *Candida* are determined by nutrition rather than temperature or the *in vivo* environment. Although the presence of mycelial strands indicates colonization and invasion, neither yeast nor mycelium can be considered the "parasitic" form. The opportunistic fungus diseases include aspergillosis, mucormycosis, cryptococcosis, and candidiasis. In addition to these, there are a number of rare infections caused by a variety of soil fungi and fungi that constitute our normal cutaneous yeast flora.

Although no dimorphism is exhibited by the agents of opportunistic fungal infections, there are certain similarities among these agents. Of all the fungal species that may infect a debilitated patient, four species account for the vast majority of such infections: *Candida albicans, Aspergillus fumigatus, Cryptococcus neoformans,* and *Rhizopus oryzae.* All of these species are able to grow or have optimum growth at body temperature. They are all also tolerant of the tissue environment, its reduced oxidation-reduction potential, and its particular CO_2 and O_2 tensions; they are, however, all subject to destruction by normal cellular defenses of the host, just as are other species (Figs. 14–1 through 14–4). Inadequate cellular defenses, changes in normal flora, or physiologic changes (e.g., ketone bodies in an acidotic diabetic) allow growth, colonization, and possibly invasion of tissue structure by these species. The interactions of these fungi and the cellular defenses of the host are very complex and are at present the subject of intense research activity. A detailed analysis, review of research, and summarization of these host-parasite interactions has been compiled by R. Diamond.[1, 2] An outline of the known pathogenetic factors for the species of fungi involved is given in the Biological Studies sections of the individual diseases.

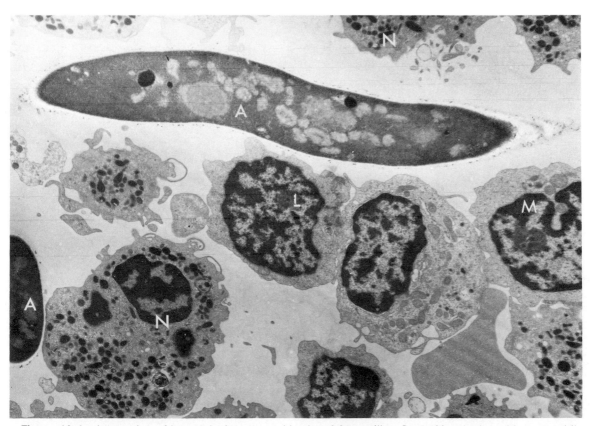

Figure 14–1. Interaction of human leukocytes and hypha of *Aspergillus.* Start of incubation with neutrophils close but not in contact with hypha. *A,* hypha of *Aspergillus; N,* neutrophils; *L,* lymphocyte; *M,* monocyte.

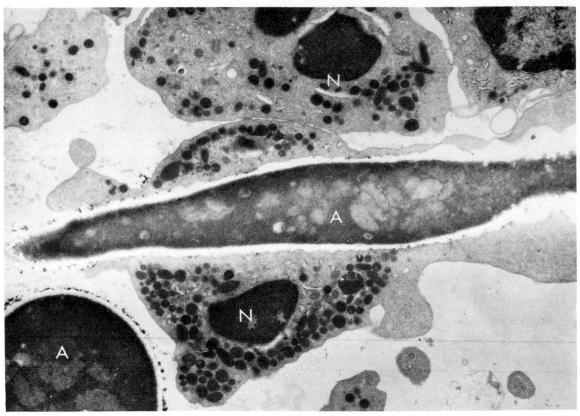

Figure 14–2. Hypha of *Aspergillus* surrounded by neutrophil after thirty minutes' incubation. *A*, hypha of *Aspergillus; N*, neutrophils; *L*, lymphocyte; *M*, monocyte.

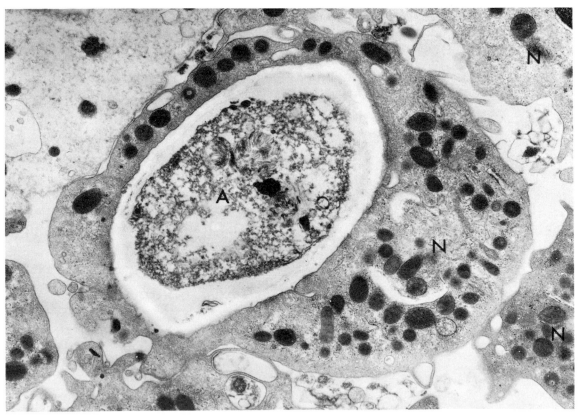

Figure 14–3. Cross section of hypha after two hours' incubation. The hypha is surrounded by neutrophil. Damage to the hypha is indicated by loss of internal structures. *A, Aspergillus; N*, neutrophil; *L*, lymphocyte; *M*, monocyte.

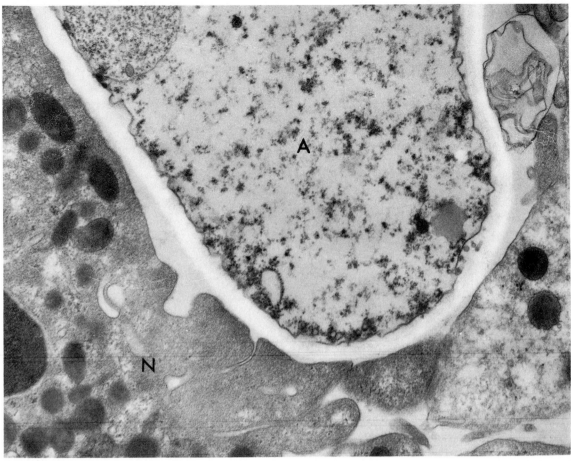

Figure 14–4. Hypha of *Aspergillus* surrounded by neutrophil. The internal structure has become amorphous as damage progresses. *A*, hypha of *Aspergillus; N*, neutrophil; *L*, lymphocyte; *M*, monocyte. (From Diamond, R. D., R. Krzesicki, et al. 1978. Am. J. Path. *91*:313–328.) (Courtesy of R. Diamond.)

References

1. Diamond, R. D., R. Krzesicki, et al. 1978. Damage to hyphal forms of fungi by human leukocytes *in vitro*. A possible host defense mechanism in aspergillosis and mucormycosis. Am. J. Pathol., *91*:313–328.
2. Diamond, R. D., R. Krzesicki, et al. 1978. Damage of pseudohyphal forms of *Candida albicans* by neutrophils in the absence of serum in vitro. J. Clin. Invest., *61*:349–359. For mechanisms of attachment of neutrophils to *Candida albicans* pseudohyphae in absence of serum and of subsequent damage to pseudohyphae by microbicidal processes of neutrophils *in vitro*, see *ibid.*, pp. 360–369.
3. Rippon, J. W. 1980. Dimorphism in pathogenic fungi. CRC Crit. Rev. Microbiol. *8*:49–79.

15 Histoplasmosis

DEFINITION

Histoplasmosis is a very common granulomatous disease of worldwide distribution caused by the dimorphic fungus *Histoplasma capsulatum*. Infection is initiated after inhalation of conidia of the organism and results in a variety of clinical manifestations. Approximately 95 per cent of cases are inapparent, subclinical, or completely benign. These are diagnosed only by the x-ray finding of residual areas of pulmonary calcification and a positive histoplasmin skin test. The remaining patients may have a chronic progressive lung disease, a chronic cutaneous or systemic disease, or an acute fulminating, rapidly fatal, systemic infection. The latter form is particularly common in children. The etiologic agent has been found in practically all habitable areas of the earth in which it has been sought. Its growth is particularly associated with the presence of guano and debris of birds and bats. The fungus can survive and be transmitted from one location to another in the dermal appendages of both and also in the intestinal contents of bats, but wind is probably the most important agent of dissemination. The infection is also very common in wild and domestic animals in endemic areas. A clinically distinct form of disease, histoplasmosis duboisii, common in Africa, is caused by *H. capsulatum* var. *duboisii* (see Chapter 16). Both forms exist in Africa, and the var. *duboisii* may represent the primitive progenitor of *H. capsulatum*. The fungus is a heterothallic ascomycete whose teleomorph state is *Ajellomyces capsulatus,* which is classified in the family Gymnoascaceae, order Eurotiales of the Ascomycota.

Synonymy. Darling's disease, reticuloendotheliosis, reticuloendothelial cytomycosis, cave disease, Ohio Valley disease, Tingo Maria fever.

HISTORY

The most important event in the history of histoplasmosis occurred in 1944 when Amos Christie discovered that Darling's disease, which was first described in 1905, was not a rare medical curiosity but a very common, widely distributed, pulmonary infection. Since that time mycologists, epidemiologists, pathologists, radiologists, and clinicians have been uncovering the infection wherever they look for it. It now appears that histoplasmosis is one of the more common infectious diseases in the world. Its popularity is ever increasing, as witnessed by the publication of at least five books, four national symposia, and the voluminous case reports, research communications, reviews, and reviews of reviews on this subject that continue to appear. Even so, many aspects of this disease and its agent remain enigmatic.

In 1905 the newly appointed pathologist of the Ancon Canal Zone Hospital, Samuel Taylor Darling, did an autopsy on a Negro from Martinique. The patient had died of an overwhelming infection, with gross lesions that resembled tuberculosis. However, on microscopic examination of the lesions, round, intracellular bodies were found. The 33-year-old Darling was influenced by the recent findings of Donovan and Leishman concerning protozoan disease and by the prediction of Manson and Ross that kala-azar could exist in America. Darling believed the organisms he found in his patient to be protozoan

342

because of their size and staining characteristics. They were found in histiocytes, resembled plasmodia, and seemed to have a capsule, so he coined the name *Histoplasma capsulatum* for them. During this time many thousands of workers had been employed to build the Panama canal and many were dying of various tropical diseases, so that within a few years Darling had performed some 33,000 autopsies. His first case of histoplasmosis was found on December 7, 1905,[36] and he discovered two more in January and August of the following year. The second case was also in a Negro from Martinique, and the third was in a Chinese who had resided for 15 years in Panama. After studying these cases he realized that this disease was similar to kala-azar in that it also induced splenomegaly and intracellular organisms were present in lesions. He decided it was a new disease, however, since the organism lacked a blepharoplast and had the aforementioned capsule. His description and illustrations were so complete that there can be no question as to the identity of the organism. Richard P. Strong had published a description of a similar disease a few months earlier.[138] He had been working on various tropical ulcers in the Philippines and described lesions in which intracellular organisms were found on histologic examination. Although it is probable that the disease he studied was histoplasmosis, his description and drawings were less complete than those of Darling.

In 1913, Henrique da Rocha-Lima, a Brazilian studying in Hamburg, compared the microscopic sections from Darling's case with epizootic lymphangitis of horses.[37] The latter disease was known to be caused by a fungus, and on the basis of histologic similarity, he concluded histoplasmosis also was a mycotic infection. Twenty years elapsed from the first cases of Darling until the next report of the disease. In 1926 Riley and Watson[115] described the infection in a woman resident of Minnesota. As a result, they noted that histoplasmosis was therefore not restricted to the tropics and might be the cause of obscure cases of splenomegaly with emaciation, anemia, and pyrexia. In the same year another case was reported by Phelps and Mallory.[103] Their patient had been a banana plantation worker in Honduras.

The next significant discovery in the history of histoplasmosis occurred at Vanderbilt University. In 1929, Dr. Catherine Dodd diagnosed the first case antemortem.[43] The patient, a child, had parasites observable in blood smears subjected to supravital staining. Material from the child was cultured on a variety of media by William de Monbreun.[39] To his surprise, a mold grew out of the pathologic material. Thereafter, in a series of carefully executed studies he established the dimorphic nature of this fungus, the cultural characteristics of the mycelial and yeast stage, and the production of disease in experimental animals. He thus fulfilled Koch's postulates. His paper describing these findings was presented in November, 1933, at a meeting of the American Society of Tropical Medicine. As is the case with so many discoveries, there was a prior claim to the cultivation of *Histoplasma capsulatum*. A few months before de Monbreun's presentation, Hansmann and Schenken read a paper at a meeting of the American Association of Pathologists and Bacteriologists describing the growth of the organism and called it a *Sepedonium* species.[67] Though their work was not as detailed as that of de Monbreun, all three share credit for establishing that *H. capsulatum* is a dimorphic fungus and not a protozoan.

It had been noted by many radiologists and pathologists that there were many apparently healthy people with small calcifications in their lungs who were tuberculin-negative. Review of these cases showed that a very high percentage of involved people lived in the Mississippi-Ohio Valley regions of the midwestern United States. Amos Christie had had experience in California, where the mild form of coccidioidomycosis had recently been delineated. He was newly appointed to the faculty at Vanderbilt and became puzzled about the lung calcifications in tuberculin-negative people. Influenced by his California experience, he began skin testing a number of such children with an extract of the mycelial phase of the fungus, a histoplasmin. He correlated skin test reactivity with the presence of pulmonary calcifications, thus proving, in 1944, that a mild form of the disease exists.[27] A much larger skin testing series was carried out by Carroll Palmer in 1945,[99, 100] which firmly established the occurrence of the subclinical form of disease and its great prevalence. Interestingly, in the same year Parsons and Zarafonetis[101] published a paper reviewing all known cases of histoplasmosis from the year 1905 to 1945. The total number was 71, and the authors concluded that the disease was widespread, but rare and always fatal.

Great interest was aroused by the work of

Christie and Palmer, and the first seminar on histoplasmosis was held at the National Institutes of Health on September 13, 1948.[109] The convener was Norman Conant, and papers by Emmons, Campbell, Loosli, Salvin, and others confirmed the efficacy of the complement fixation test and the skin test in diagnosis of the disease, and the occurrence of the disease in a wide variety of wild animals. The stage was now set for rapid development in the investigation of all phases of the disease. Negroni in 1940[96, 97] had studied the first case in South America, thus extending the known range of the disease as far south as Argentina. He also induced the yeast form of growth *in vitro*. The organism was suspected to occur in the soil, and in 1948, Emmons[50] isolated it from rat burrows. Zeidberg et al.[147] also recovered the organism from soil and noted the very important association of fungus with bird dung. Skin testing with histoplasmin was now being done everywhere. In 1945 Furcolow established a center in Kansas City to study the prevalence of the disease in this highly endemic area. Here he isolated the organism from the air, water supply, and practically every chicken house and bird roost in Missouri.[52] The first "epidemic" of histoplasmosis occurred in 1947 in Camp Gruber, Oklahoma. Since then, famous epidemics have occurred in Milan, Michigan; Mason City, Iowa; Mexico, Missouri; Dalton, Georgia; Indianapolis, Indiana[47]; Rogers City, Michigan[57]; and many other places. Ironically one such epidemic (Delaware, Ohio, 1970) was associated with Earth Day ecologic activities.

Demonstration of the organism in tissue was still difficult. Using a variant of the McManus periodic acid-Schiff stain, Hotchkiss was able to stain the polysaccharide in the yeast cell walls. Kligman et al.[84] in 1951 established the usefulness of this stain for fungi in tissue. Pathologists such as Puckett were finding that many "tuberculomas" were in fact "histoplasmomas," when they used this new method of staining.[127] The work of Schwarz, Straub, and Servianski in 1955[127, 128, 137] established the clinical and pathologic characteristics of the initial mild infection. They found that following inhalation of conidia a Ghon complex was formed in time, along with pulmonary calcifications and, more importantly, small foci of splenic calcifications. This indicated that there was rapid hematogenous spread of the organisms early in the infection, probably within macrophages. The entire disease process could still occur without symptoms notable to the patient.

By now, the whole world was becoming aware of the disease, and histoplasmosis had become famous. Furcolow et al. in 1952 organized the first national conference, and in 1962 a second.[4] Almost everywhere investigators looked they could find the organism in pathologic material, isolate it from soil, or find reactors to the histoplasmin skin test. Darling's rare medical curiosity of 1905 had become a very common infectious disease. In the United States alone, it is estimated that 40,000,000 people have had the disease, and there are 200,000 new infections every year.[1, 52] An excellent review of the biology of histoplasmosis is found in Domer and Moser,[46a] and Goodwin et al. have summarized the state of understanding of the various clinical manifestations.[61, 62]

Recently another group at Vanderbilt University has delineated a variety of clinical manifestations of histoplasmosis heretofore unsuspected. In addition to actual infection, some of these manifest as exaggerated cellular response to the continued inhalation of the conidia in highly endemic areas.[61, 62] This appears to be the basis of chronic cavitary disease, histoplasmomas, and a group of conditions characterized by excessive fibrotic response and collagenosis.

ETIOLOGY, ECOLOGY, AND DISTRIBUTION

Hotspur:　　　　　　　　Nay, I will, that's flat.
He said he would not ransom Mortimer,
Forbade my tongue to speak of Mortimer.
But I will find him when he lies asleep,
And in his ear I'll hollo "Mortimer."
Nay, I'll have a starling shall be taught to speak
Nothing but "Mortimer," and give it him,
To keep his anger still in motion.

—Wm Shakespeare, 1 Henry IV, i, 3, 232–239

The etiologic agent of histoplasmosis is the dimorphic fungus *Histoplasma capsulatum*. In culture at temperatures below 35° C and on natural substrates it grows as a white to brownish mycelial fungus. The organism elaborates characteristic echinulate, oval, or pyriform macroconidia (8 to 16 μ in diameter) and small (2 to 5 μ) microconidia. When inhaled into the alveolar spaces, it is primarily the latter that sprout and then transform into small budding yeasts that are 2 to 5 μ in diameter. The yeast cells are found within cells of the reticuloendothelial system. In culture at a temperature of 37° C the organism also grows in the yeastlike form. A clinically distinct form of the disease, histoplasmosis duboisii, is caused by *H. capsulatum* var. *duboisii*. Whereas the larger yeast cells of the latter are found mainly in giant cells, the small yeast forms of *H. capsulatum* are generally within histiocytes.

The ecology of *H. capsulatum* has been the subject of numerous investigations. It is firmly established that the organism grows in soil with high nitrogen content, generally associated with the guano of birds and bats.[6, 40, 41, 57, 145, 147] The first isolation of the organism from a natural environment was from soil near a chicken house,[50, 147] and since that time it has been recovered on numerous occasions from bat caves, bird roosts, chicken houses, silos inhabited by pigeons, and other such environments. In avian habitats, the organism seems to grow preferentially where the guano is rotting and mixed with soil rather than in nests or fresh deposits. In the laboratory, it has been shown to grow on shed feathers, and when the organism is injected into birds, it can be recovered from feathers *in situ*. Even in highly endemic areas, the presence of the organism is often restricted to small areas where birds congregate in large numbers. Not all guano appears to serve equally well as a substrate. The most highly endemic areas of the United States (Missouri, Kentucky, Tennessee, Southern Illinois, Indiana and Ohio) are also the areas with the greatest concentration of starlings. Their habit of congregating in great numbers results in formation of large deposits of guano. It is possible that the Ohio-Mississippi Valley is heavily infested with *H. capsulatum* as a result of the importation of these birds from Europe. Negroni[97] suggests that comparable areas of South America escape heavy infestation because the blight of the starling has not yet been introduced. In that continent the main breeding grounds for histoplasmosis are the chicken habitats and bat caves.

An interesting sidelight to the epidemiology of histoplasmosis concerns the importation and establishment of the starling *(Sturnus vulgaris)* into the United States.[11, 31] A group of Swiss immigrants living in New York City were devoted both to birds and to the arts, especially Shakespeare. Among this group was a Eugene Schieffelin. He undertook to import into the United States the birds mentioned in Shakespeare plays and poems. The starling is mentioned in Henry the Fourth, Part One. Mr. Schieffelin wrote to a Mr. C. F. Pfluger of Portland, Oregon, that forty pairs of starling were liberated in Central Park in New York in 1890 and forty more in 1891 (Fig. 15–1,*B*). Importation had been tried several times previously, but this one appears to have established the bird in America.[81, 104] By 1895 it was common in the vicinity of New York City and Long Island and thence it spread westward and southward. It may have been aided in this spread by the demise of the passenger pigeons, the last specimen of this species having died in the Cincinnati zoo in 1914. Formerly it had flocked by the thousands in the Midwest, a region or niche now occupied by the starling. *Histoplasma* probably existed in these areas before the arrival of the starling, but its present great abundance may be due to exploitation of the guano of the starling-associated habitat.

Though the organism grows in great abundance only in relatively restricted environmental conditions, *H. capsulatum* is found throughout the world. In two of the most detailed studies of climatic conditions favoring growth, Furcolow[52] and Carmona[23] found a mean temperature of 68 to 90° F (22 to 29° C), an annual precipitation of 35 to 50 inches (~ 1000 mm), and a relative humidity of 67 to 87 per cent or more during the growing season to be related to the greatest presence of positive skin tests and soil isolations. These conditions appear to be the most favorable for proliferation of the organism in what may be termed an "open" environment. This would include many areas in tropical, subtropical, and temperate regions of the world where there is adequate moisture. When skin testing has been done in such areas, high rates of reactivity have been found. Inexplicably some areas, such as most of Europe, seem to have escaped heavy infestation.

Outside of areas with appropriate environmental conditions, there also occur scattered areas with high endemicity. These are usually associated with existence of caves inhabited by bats or birds and may be considered "closed" environments. Though the surrounding countryside may be too dry or cold for sustained proliferation of the organism, the protected and relatively stable conditions inside the caves allow such proliferation. This probably accounts for the so-called "cave fever" in areas where otherwise there is a low incidence of the disease.[6] The chief vector of dissemination in the open environment is the wind. Though birds and bats may contribute to the dissemination of viable conidia in this type of environment, in the closed areas they most certainly do. Several investigators[41, 85] have shown that bats may contaminate the environment with organisms present in their intestinal contents. They often have yeast-containing ulcers within their intestinal tract. As these animals migrate to new caves, they

seed them with the fungus. Caves which are otherwise suitable for growth but are too wet remain free of infestation.[23] So far, attempts to isolate organisms from the cloaca of birds have been unsuccessful, and it appears that birds do not become infected with the fungus.

Many other factors have been postulated as influencing the epidemiology and distribution of histoplasmosis. These include wind direction, flooding of riverbanks, the presence of limestone in soil, red-yellow podzolic soil, and so forth.[52] Unfortunately it appears that *H. capsulatum* will not fit as neatly into a particular environmental niche as does *Coc-*

A

Figure 15–1. *A,* The incidence and prevalence of histoplasmosis in North and South America based on skin testing surveys. The darker shaded areas indicate zones of very high endemicity.

Illustration continued on opposite page

Figure 15–1 *Continued. B,* Spread of the European Starling *(Sturnus vulgaris)* from its importation in 1896 to the midwest in 1926. By 1936 vast flocks of the birds inhabited the Ohio-Mississippi valley, concentrating in roosts where guano accumulated. The bird reached California, becoming common thereafter. It has colonized 3 million square miles during the first 50 years after introduction. (From Cook, N. T. 1928. The spread of the European starling in North America. U.S. Dept. Agriculture Circular No. 40.) *C,* Starling roost near Fort Campbell, Kentucky. The silhouettes of many birds can be seen in the trees. (Courtesy of G. Land.) *D,* Warning sign in a woods heavily contaminated with starling guano and *H. capsulatum.* The guano was 5 to 6 inches thick in places, with *Histoplasma* growing in essentially pure culture. Arrow indicates starling atop warning sign. Vicinity of Fort Campbell, Kentucky. (Courtesy of C. Smith.)

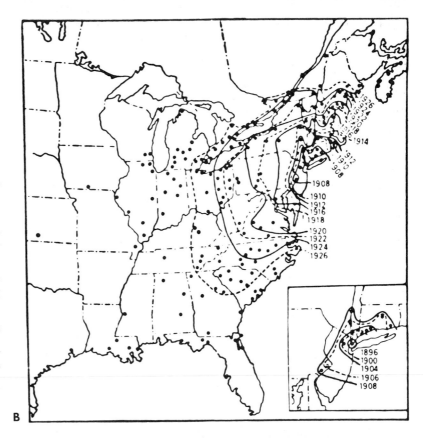

B

C

D

cidioides immitis into its Lower Sonoran Life Zone.

A skin test for histoplasmosis was first used in 1941 by Van Pernis. The histoplasmin that is presently used was standardized by Emmons a few years later. It is the filtrate of an asparagine glucose broth in which the organism (mycelial phase) has been growing for 2 to 4 months at 25° C. The medium is the same "nonallergic" broth used for making old tuberculin (OT). Histoplasmin is injected intradermally as a 1:100 or 1:1000 dilution. An area of induration of at least 5 mm after 48 hours is considered a positive test. Cross-reactions occur in patients with blastomycosis and, to a lesser extent, coccidioidomycosis and possibly other fungal diseases. Since the early 1950's hundreds of thousands of people have been skin tested in all parts of the world. From such studies it appears that the largest area with high prevalence (80 to 90 per cent) is the middle section of the North American continent (Fig. 15–1). However, scattered, often isolated, areas with high prevalence of the disease exist in all parts of the world. These include southern Mexico, northern Panama,[88] Honduras, Guatemala, Nicaragua, Venezuela, Colombia, Peru, Brazil,[92] Surinam, Jamaica,[26] Puerto Rico, Belize,[111] Alaska,[30] Burma,[112] Indonesia, Philippines, Turkey, Israel,[5] Italy,[93, 132a] Switzerland,[17] Australia, Asia,[112] and others.

In a series of studies conducted in the highly endemic area of Kansas City, Furcolow[53] found that, by age 20, between 80 and 90 per cent of the population had a positive histoplasmin skin test. The same is true in the Cincinnati–southern Ohio region, southern Illinois, central Missouri, and areas of Kentucky, Tennessee, and Arkansas. Focal areas of high endemicity also occur in Michigan, Minnesota, Georgia, and Louisiana.[136] At present the histoplasmin skin test merely indicates that one has probably lived in the central United States at one time, and the test itself has essentially no diagnostic value.

Epidemics of acute respiratory histoplasmosis are frequently recorded. These occur when there is exposure to an aerosol containing numerous spores. The circumstances surrounding such exposures are varied and fascinating. Cases include a family that cleaned a silo in Indiana; a Boy Scout troop that cleared a park in Mexico, Missouri; children in a school in Milan, Michigan where the air conditioner intake was located over a pile of starling droppings, workers who developed "carcinoma" of the lung while cleaning herring gull detritus in a port facility in Rogers City, Michigan[57]; and workers cleaning out a bamboo field in Louisiana,[136] tearing down an amusement park in Indianapolis,[47] cutting up a fallen tree in Tennessee,[142] and cleaning blackbird roosts in South Carolina.[40] As pointed out by DiSalvo,[40] these are usually "microfocus" epidemics: exposure of a small number of people to a large number of conidia in a small area and usually all at the same time. These are often anthropurgic, as they are the result of some human activity. Very heavy exposure may cause death, whereas moderate exposures lead to infections of varying severity which usually resolve spontaneously.[40, 61] About a hundred well-documented "epidemics" have thus far been recorded. These are particularly interesting studies for epidemiologists and public health workers.[40] Among the excellent studies is one by Dodge et al.[44] demonstrating the association of a bird roosting site with infection of school children. It serves as a model of epidemiologic detective work.

All ages and sexes are subject to primary pulmonary infection, but the sex distribution and form of disease change with increasing age of the patient. Certain select children acquire a fulminating, rapidly fatal disease in which there is massive proliferation of histiocytes and the *Histoplasma* organism within them. Both sexes are equally affected by all forms of the disease until the age of puberty. Then, chronic progressive disease is at least three times more frequent in adult males than in females. There is no documented racial difference in susceptibility. Point exposure is much more important than occupation. A rapidly progressive opportunistic infection occurs in some patients with the lymphoma-leukemia-Hodgkin's group of diseases or those on steroid therapy or other immunosuppressive agents. *Histoplasma* is now considered a regularly encountered opportunist in such circumstances.[38, 42, 79] This organism appears to be involved in opportunistic infections more often than the other "true" pathogenic fungi. The commonness of the infection and the persistence of subclinical disease in people exposed to the organism many years previously are some of the reasons that account for this.

CLINICAL DISEASE

Histoplasma capsulatum is a pathogenic fungus. Inhalation of a sufficient quantity of conidia will cause an infection in the lungs of the normal healthy patient. This is followed by a rapid, transient, hematogenous and systemic spread of the organism. In the vast majority of patients, the infection is aborted, leaving only residual calcifications in the lung and sometimes the spleen. Such a course constitutes the benign form of histoplasmosis. Resolution of the disease confers a certain degree of immunity to reinfection and, in addition, varying grades of hypersensitivity to the antigenic components of the organism. As a consequence, massive reinfection may result in a fatal acute allergic reaction in highly sensitized lungs.

In a few patients the disease becomes chronic and progressive and follows a course similar to that of chronic tuberculosis. Such infections are included in a second category of disease, termed opportunistic. Other types in this category are the mysterious histiocytosis accompanying the fulminant disease of childhood; the uncommon systemic disease associated with the aforementioned lymphomatous diseases and immunosuppression; the slowly progressive disease of adults, wherein an immune defect is usually undetectable, and the chronic pulmonary disease associated with structural defects of the lung. A third category of clinical manifestations is associated with aberrant fibrosis and hypersensitivity. The gradually enlarging histoplasmoma,[60] superior vena cava syndrome, and bronchocentric granulomatosis[78] with mucoid impaction and hypersensitivity are all tentatively included in this category.

As the preceding summary indicates, histoplasmosis manifests itself in a bewildering array of clinical types and variations. It has been called the syphilis of the fungus world. Because there are now so many known clinical forms of histoplasmosis, new and revised classification schemes of these clinical diseases appear regularly. The one adopted for this text utilizes the schema formulated by Goodwin et al. and Baum et al. in their several publications.[60,61,62,128]

Symptomatology. With the exception of rare, accidental, cutaneous inoculation, infection by *H. capsulatum* occurs by way of the lungs. The resulting disease may be divided into three major categories: benign infection, opportunistic infection, or aberrant fibrosis and hypersensitivity. Each of these has acute and chronic subcategories. Presumed *Histoplasma* choroidities (uveitis) is discussed in Chapter 25 with mycotic infections of the eye.

Clinical Forms of Histoplasmosis

I. Benign infection.
  A. Usual dose: Endemic histoplasmosis
    1. Endemic subclinical disease: "Skin converters"
    2. Endemic symptomatic disease: Summer sickness in children and "fungus flu" in adults
    3. Primary cutaneous disease
  B. Heavy dose: Epidemic histoplasmosis
    1. Acute pulmonary histoplasmosis
      a. Primary infection (symptoms in 10 to 18 days)
      b. Reinfection (symptoms in 3 to 7 days)
    2. Acute disseminated histoplasmosis
II. Opportunistic infection
  A. Disseminating histoplasmosis (with time course of disease)
    1. Fulminant disease of children (2 to 10 weeks)
    2. Moderately chronic disease of adults (several months)
    3. Mildly chronic disease of adults (10 to 20 years)
    4. Fulminant disease of adults associated with immunosuppression and/or lymphomatous diseases (rapid but variable)
  B. Chronic pulmonary histoplasmosis
    1. Colonization of preexisting or acquired structural anomalies
    2. Smoker's emphysema and "migrating flashes" on x-ray examination
    3. Cavitary histoplasmosis
III. Aberrant fibrosis and hypersensitivity disease
  A. Histoplasmoma
  B. Mediastinal fibrosis
    1. Superior vena cava syndrome
    2. Bronchocentric granulomatosis

Benign Infection

Usual Dose: Endemic Histoplasmosis

Endemic Subclinical Disease. From the studies of Christie, Peterson, Palmer, Furcolow, and others, it is estimated that at least 95 per cent of all primary cases of histoplasmosis are not referable to specific symptomatology. Multiple calcifications may be seen in the lungs of patients living in highly endemic areas, and such patients cannot give a history of relevant symptoms. The only other indications that infection has occurred are a positive skin test and a few calcifications in other organs such as the spleen. The latter attests to the wide dissemination of the organism, even in subclinical disease.

Endemic Symptomatic Disease

Mild Disease. Mild disease may present with a flulike syndrome that includes nonproductive cough, pleuritic pain, shortness of breath, and hoarseness. Histoplasmosis is a common cause of "summer fever" in children.[91] There is a sudden onset of a febrile illness, hilar and mediastinal enlargement, and clusters of parenchymal infiltrates (Fig. 15-2, B). The enlarged nodes compress and irritate the bronchi, leading to a "brassy," nonproductive cough and, occasionally, pericarditis.[83] Symptoms resolve in a few days to three weeks. Such symptomatic infections are usually recognized in infants and young children, and supportive therapy is usually sufficient. In a small percentage of patients, however, owing to age-related or genetic deficiencies of the immune system, this self-limiting disease may destablize into a serious systemic infection with hepatosplenomegaly and generalized organ involvement leading to a fatal outcome if specific treatment is not instituted.

Moderately Severe Disease. In moderately severe disease, the preceding signs are exaggerated and also include fever, night sweats, weight loss, some cyanosis, and occasional hemoptysis. In this latter group of patients, the organism can sometimes by cultured from sputum, and when fever is present, it has been recovered from bone marrow obtained by sternal puncture. This can be observed in inexperienced adults who have recently arrived in an endemic area, particularly if the course is complicated by some other respiratory illness. The syndrome is termed "fungus flu" of adults. Again, a few patients develop disseminated progressive disease, though it is usually self-limited (See Section II. A. 2.), which follows. Death has resulted from disseminated intravascular coagulation in such patients.[141]

X-ray Findings. The roentgenographic picture of all these types of primary disease is similar. There is the appearance of multiple lesions scattered in all lung fields. Initially they appear as disseminated infiltrates or discrete nodular foci of activity (Fig. 15-2, A). Hilar lymphoadenopathy and a Ghon-like complex are almost always present. This may lead to an initial diagnosis of tuberculosis, or an expanding lesion may appear to be a lymphoblastoma or other neoplasm. At all stages of the disease the appearance is essentially identical to tuberculosis and can only be differentiated from it by proper serologic and cultural procedures.[131a]

The radiologic findings of primary acute disease include (in order of commonness) nodules, infiltrates, adenopathy (particularly asymptomatic cases), cavitating lesions, and rarely pleural effusion.[30a] No chest findings at all can be seen in both asymptomatic and symptomatic cases, however. In children, adenopathy is more pronounced and cavities are rare but do occur.[91] The lesions resolve slowly. Intense inflammatory reaction infiltrates the parenchymal lesions, regional lymph nodes, and any metastatic foci. This manifestation of hypersensitivity results in caseous necrosis and fibrotic encapsulations of lesions (Fig. 15-2). A few undergo complete resolution or may leave small areas of fibrosis; other lesions heal by fibrosis, and characteristically these calcify (Fig. 15-2, B, C). The time required for this to occur varies from a few months in children to as long as five years in adults. As pointed out by Schwartz and Straub[137] and Goodwin et al.,[61, 62] the resolved primary complex of histoplasmosis has a consistent x-ray pattern. The lymph nodes are large and the tubercles small, and there are halos around the primary calcifying nodules. They are usually numerous, round, oval, and of uniform size (Fig. 15-2, D). The healing lesions of tuberculosis tend to be smaller and more irregular in shape. The rule is that a parenchymal calcification of more than 4 mm or a hilar or mediastinal calcification of more than 1 cm is generally histoplasmosis. Though "buck shot" calcifications are seen in both histoplasmosis and tuberculosis, "pop corn" type (Fig. 15-3), "mulberry" type, and other large, grouped calcifications in paratracheal nodes are generally associated with histoplasmosis. In histoplasmosis, calcified lesions can also be seen in the spleen and occasionally in the liver.

There are other residual manifestations of primary histoplasmosis. One very common one is frequently picked up on routine x-ray (Fig. 15-2, E). This is the presence of a solitary lesion that has not yet calcified. Such

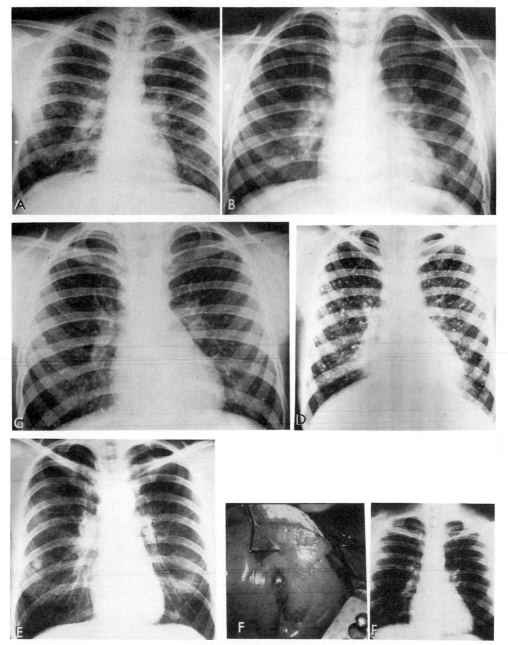

Figure 15–2. Radiologic aspects of histoplasmosis. *A,* Diffuse interstitial and nodular infiltrates are seen in both lungs and are associated with hilar and mediastinal involvement. This is symptomatic primary histoplasmosis, or "fungus flu." *B,* Multiple miliary nodules in both lungs of a child associated with hilar and mediastinal adenopathy, summer sickness of children. *C,* Two years later there are multiple calcified densities throughout both lung fields. *D,* Epidemic primary disease. Numerous small, calcified areas in lung parenchyma. Bilateral involvement. (Courtesy of J. Fennessy.) *E,* A well-circumscribed peripheral nodule in right middle lobe associated with enlarged hilar nodes; a residual coin lesion. *F,* Well-circumscribed peripheral nodule in right lower lobe and appearance of lesion at surgery. (Courtesy of F. Kittle.)

"coin lesions" are difficult to distinguish from various neoplasms and are often surgically removed to establish a diagnosis (Fig. 15–2, *F*).

Another manifestation of histoplasmosis that mimics tuberculosis is the histoplasmo-

ma. These are found in the lungs of adults and are usually 2 to 3 cm in size. The central area is necrotic, and there is a fibrotic capsule. Calcification proceeds in the central portion, but the lesion appears to enlarge with alternating concentric rings of fibrosis and calcifi-

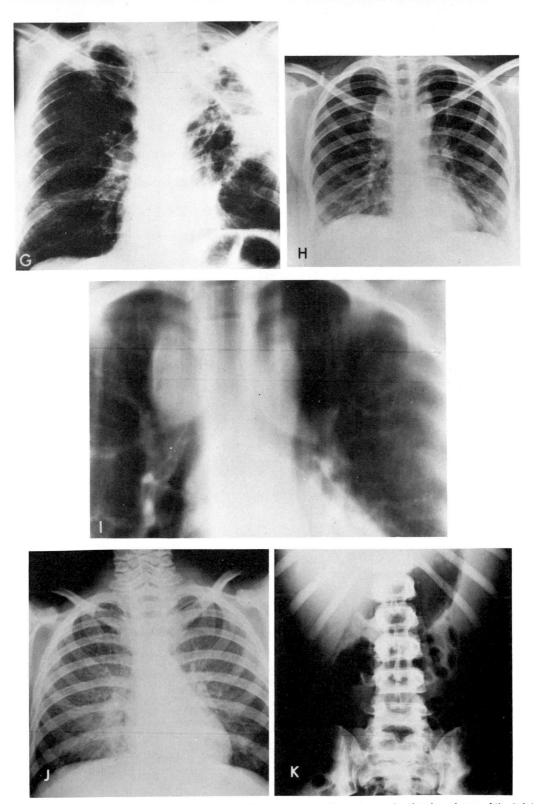

Figure 15–2 *Continued.* *G,* Progressive cavitary histoplasmosis. There is a reduction in volume of the left lung, with multiple cavities in upper lobe. A very large lateral cavity has an air-fluid level. There is also marked pleural thickening. *H,* Superior mediastinal syndrome. There is a large right-sided mediastinal mass in the region of the tracheobronchial nodes which is smooth in outline and shows no calcification. *I,* A tomogram confirms the location in tracheobronchial node and shows displacement of upper lobe bronchus downward. *J,* Fatal disseminated histoplasmosis in a 10-year-old child with leukemia. There are diffuse interstitial and pulmonary inflltrates of both lungs associated with hilar adenopathy and hepatosplenomegaly. *K,* Plane film of abdomen showing splenomegaly in same case.

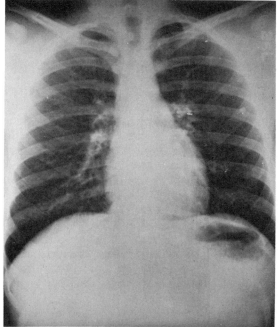

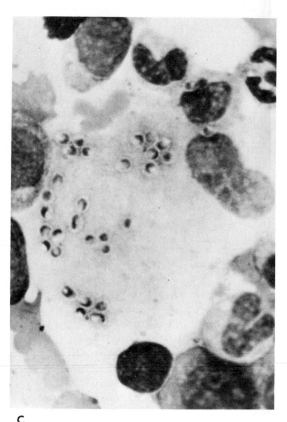

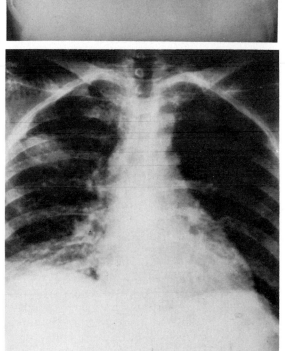

Figure 15–3. *A,* Resolved histoplasmosis. Left hilar lymph nodes contain dense and irregular "popcorn"-type calcifications. There is also a small calcified lesion in the periphery of the left lung. (Courtesy of H. Grieble.) *B,* Progressive disseminated histoplasmosis of the moderately chronic type of the adult. The histoplasmosis occurred after a viral pulmonary episode. Patient has symptoms of adrenal insufficiency, weight loss, night sweats, fever, adenopathy (cervical, axillary, and inguinal), is cachectic, and has hepatosplenomegaly. *C,* Sternal puncture of the same patient. Twenty-five to 30 yeast cells, some budding, are within a macrophage. Giemsa. ×1200.

cation. Again, they may have the appearance of neoplasms. Histoplasmomas are discussed in more detail in part III.

Calcification is even more common in histoplasmosis than in tuberculosis and is also frequently encountered in coccidioidomycosis. It is unusual in cryptococcosis and blastomycosis.

Primary Cutaneous Histoplasmosis. Primary cutaneous histoplasmosis is a very rare but intermittently reported entity.[139, 140] It is usually the result of inadvertent injection of contaminated material, as in a laboratory accident or during an autopsy. An indolent chancriform ulcer develops with regional adenitis. The lesions involute and heal spontaneously after a few months. Very rarely do such lesions occur in immunosuppressed patients. In this type of patient the course can be progressive and become serious.[32, 59]

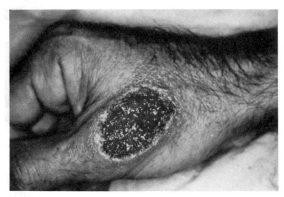

Figure 15–4. Primary cutaneous histoplasmosis. Well-demarcated ulcer with raised border. The lesion was the result of the inoculation of conidia by a thorn. The hand is that of the late explorer Francis Brenton. (Courtesy of S. McMillen.)

In a personally observed case of primary cutaneous histoplasmosis, the infection was the result of implantation by a thorn while the patient was collecting bat guano from a hollow tree in Letitia, Colombia (Fig. 15–4). *H. capsulatum* was isolated from the guano and the lesion. The patient's hand healed without complication.

Heavy Dose: Epidemic Histoplasmosis

The second type of benign primary histoplasmosis is the epidemic form. This has been described under a variety of names as cave fever, acute miliary pneumonitis, primary atypical pneumonia, and many more. The syndrome occurs when individuals are exposed to large quantities of conidia. The symptoms, severity, time course, and outcome of such an encounter depend on the previous experience of the patient. If this is a primary infection, symptoms do not occur for 10 to 18 days, during which time there is multiplication of the organism. At onset there is fever, malaise, chills, and myalgias: a flulike illness. Patchy pneumonitis that is up to 5 mm or more in diameter is seen on x-ray examination; hilar adenopathy is present, and occasionally there is pleural involvement. The latter may last two to three months. With the development of hypersensitivity, *erythema nodosum* and *erythema multiforme* may appear. The pulmonary lesions usually clear rapidly, with calcifications (Fig. 15–2,*D*), and only very rarely developed into progressive disease.

In both primary and reinfection epidemic histoplasmosis there is dissemination of the organisms within macrophages to other organ systems. The resulting lesions almost always heal with the same time course as the pulmonary lesions. In some cases, however, the fungi remain viable for long periods of time and one of the various forms of chronic disseminated disease may develop.

If the patient has encountered the organism before, a syndrome called reinfection histoplasmosis occurs. Symptoms develop in three to seven days. The organism does not appear to multiply to any significant degree in reinfection disease. An influenza-like disease also occurs but with fine interstitial nodular infiltrates (1 to 2 mm in size), leading to granuloma formation rather than to an exudative pneumonitis. There is no pleural effusion, hilar adenopathy, or calcification of resolved lesions. This syndrome also may develop into disseminated disease, although very infrequently.

Opportunistic Infection

As indicated in the preceding sections, the encounter of the normal human host with this organism of low virulence usually results in a benign, self-limited infection that is most often without clinical symptoms. In a few select patients, however, the encounter results in the establishment of one of several serious diseases that may ultimately be fatal. The several types of serious disease occur in patients with some abnormality or predisposing factor and are therefore opportunistic infections. The two general categories of opportunistic infection — disseminated histoplasmosis and chronic cavitary disease — have different and unrelated predisposing conditions, but the organism is able to exploit both sets of abnormalities. In disseminated disease, there is some defect of cellular immunity; in chronic cavitary disease, a structural or anatomic defect predisposes to colonization by the fungus.

Disseminated Histoplasmosis

Disseminated histoplasmosis is a rare opportunistic infection that develops in individuals with some degree of cellular immune defect, often undefined. The disease may occur in the very young (about one third of cases); in patients with lymphatic or hematopoietic malignancies; iatrogenically in patients on steroids or immunosuppressive therapy for some other condition; and idiopathically, when the defect is unknown, currently undefined, or transient. Passing viral infections have been suggested as a cause of the last. The case rate of disseminated disease is estimated to be 1:100,000 to 1:500,000 of persons infected per year. Presently such patients are classified in four categories: (1) fulminant disease of children, (2) moderately chronic disease of adults, (3) mildly chronic disease of adults, and (4) fulminant infection of adults. As pointed out by Goodwin et al.,[61] disseminated histoplasmosis is a disease of the mononuclear phagocytic system, and the severity and manifestations of this disease relate to the degree or permissiveness of parasitization of the macrophage. This is referred to as the MDPP (macrophage degree of permissiveness of parasitization) and often correlates to the severity of the infection and the prognosis of the disease.

Fulminant Disease of Children. This is the most rapidly developing and most often fatal form of disseminated histoplasmosis. The macrophages are in greater-than-normal number and are packed with yeast cells (Fig. 15–10). These yeast-laden cells clog capillaries, leading to stasis and circulatory collapse. Interestingly, this was the first premortem form of histoplasmosis recognized.[43] Over 80 per cent of such patients are infants of less than one year of age. There is high fever, malaise, anorexia, weight loss, gross hepatosplenomegaly (6 cm or more below the costal margin), interstitial pneumonia, and the hematologic disorders of anemia, leukopenia, and thrombocytopenia. The entire course terminates fatally in 2 to 10 weeks if untreated. Amphotericin B has been used successfully in some cases. Organisms can be cultured and seen in sputum and blood smears and from sternal biopsies. In sternal or nodal biopsy, the numbers and size of the parasitized histiocytes can be appreciated. They often cause obliteration of normal architecture. It has been suggested that such patients have a reticuloendothelial defect, but there is as yet no evidence to substantiate this view.

Moderately Chronic Disease of Adults. Progressive histoplasmosis of the adult was the form of the disease first described by Darling and the only form known for many years. Pulmonary symptoms may or may not be prominent (Fig. 15–3, B,C), but hepatosplenomegaly, a characteristic of leishmanial disease as well, is quite apparent. Numerous organisms are seen within the macrophages and extracellularly (Fig. 15–3,C and 15–7,B). Other significant symptoms are low-grade intermittent fever, weight loss, and weakness. There is a slow, unrelenting downhill course that terminates fatally in 6 to 12 months if untreated. Infants and older children sometimes manifest this syndrome.

A characteristic of this infection type is focal destructive organ disease. Thus, patients will often have associated symptoms of Addison's disease, if the adrenals are involved, in addition to meningitis, intestinal ulceration, or endocarditis. Signs and symptoms referable to any or all the previously mentioned organ systems are regularly encountered. There is also evidence of "migrating foci," wherein one organ is involved and heals and the disease occurs in other organs. A few patients demonstrate other symptoms as well. These include anemia, leukopenia and thrombocytopenia, and occasionally oropharyngeal ulcerations, cutaneous and osseous lesions, or polyarthritis.[29, 114]

Mildly Chronic Disease of Adults. This type of disease occurs almost exclusively in adults and is the commonest form of disseminated histoplasmosis of the adult. The hallmark of this type of infection is the oropharyngeal ulcer, which occurs in 75 per cent or more of patients. The course is slow and protracted, with symptoms of low-grade intermittent fever, weight loss, and fatigue. Recurrent episodes have been documented for 10 to 20 years before a diagnosis was made. Chronic disease and migrating foci of infection have been noted in such infections, including occasional instances of meningitis, cerebritis, Addison's disease,[34] endocarditis,[16] and intestinal ulcerations.[98] Hepatomegaly occurs in 50 per cent and splenomegaly in 30 per cent of patients, and cutaneous lesions may appear in late stage disease. Mucosal ulcers are the most prominent finding, however. Painful swellings on the buccal mucosa; hoarseness; enlargement of the vocal chords; denuded (Fig. 15–5) and nonulcerated nodules of the tongue and gingiva; and ulcers with thickened rolled borders are some of the morphologies observed. Most often the diagnosis is not made until a biopsy is taken to rule out malignancy. Several cases have occurred with lesions on the penis as the presenting symptom (Fig. 15–6). The complement fixation test is positive in about two thirds of cases and the skin test in one third, so that neither are very useful in the diagnosis of this form of histoplasmosis. Biopsies with appropriate stains and culture are the more useful examinations.

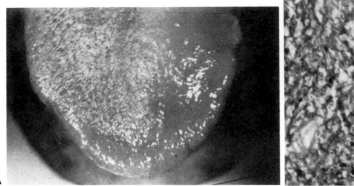
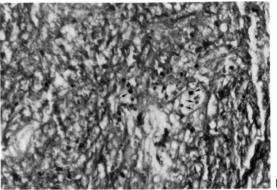

A   B

Figure 15–5. *A,* Mildly chronic type of disseminated histoplasmosis. Denuded erythematous lesion on tongue. *B,* Biopsy of the patient shows few scattered extracellular yeasts (arrows) with little evidence of budding. GMS. ×450.

In the fulminant and moderately chronic disease of children, the sex ratio is about 1:1. In the mild chronic disease, however, there is a 3:1 to 12:1 male dominance. The underlying defect in these patients has thus far defied definition. Parasitization of the macrophage cells is considerably less than that seen in the fulminant disease of children or the moderately chronic disease of adults. It is considerably more, however, than that which occurs in the benign acute infections of all age groups (Fig. 15–7,*A*). Whether this is a meaningful index has yet to be determined. This form of disease often presents as single organ involvement as endocarditis,[58] meningitis,[107] subcutaneous nodules,[82] or cutaneous lesions.[96]

Fulminant Histoplasmosis of the Adult. This is the rapidly developing, usually fatal form in the immunosuppressed patient. The conditions predisposing to this type of disease are the various lymphomatous and hematopoietic malignancies and their treatments, drug-induced (usually steroid) immunosuppression during the course of

organ transplant, and treatment for hepatitis,[103] sarcoid. lupus erythematosus, rheumatoid arthritis, pancytopenia, and various other conditions. There are no constitutional symptoms that characterize this disease, and the only constant findings are fever and yeast cells in bone marrow examinations and sometimes in peripheral blood smears.[42] The fungus-laden macrophages are huge, similar in appearance to those seen in fulminant disease of childhood (Fig. 15–10), and found at autopsy in many organ systems. There is an absence of discrete granulomas, and minimal nonspecific tissue reaction is seen in histologic examination of biopsy or autopsy material. Pulmonary symptoms are noncontributory. They range from normal to diffuse miliary infiltrates (Fig. 15–2,*J*), lobar infiltrates, or hilar adenopathy. At autopsy the lung is almost always involved. There is speculation about the origin of the infection. In endemic areas recent infection from exogenous sources is probable. There is, however, convincing evidence for the possible reactivation of an old, long quiescent infection in some cases, particularly when the patient had left the endemic area many years previously. Owing to the difficulties of diagnosis, most cases of this type of histoplasmosis are fatal. In a few cases in which examination of sternal puncture was done early in the course of the disease, amphotericin B therapy was instituted and was curative. Serologic tests are generally not useful.

Chronic Pulmonary Histoplasmosis

Colonization of Structural Anomalies. Chronic pulmonary histoplasmosis is a condition that usually develops in a setting of

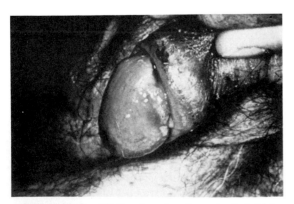

Figure 15–6. Mildly chronic disseminated histoplasmosis. Granulomatous lesion of penis.

chronic obstructive lung disease (COPD) with centrilobular and bullous emphysema. The patient is most commonly a male in the third to fourth decade living in an endemic area and regularly encountering conidia of the organism. Often the emphysema is exacerbated by, or the result of, smoking. Preexisting anatomic defects may also serve as the site for colonization as well as induced or developmental anomalies The natural history of the disease is not clearly defined, but in its more common form it appears to be a self-limiting process that resolves spontaneously in a few weeks. Recurrent episodes are noted in only 20 per cent of patients.[61] The severity of these recurrences, particularly those resulting in progressive cavitary lesions, is thought to parallel the course of the underlying pulmonary process. Evidence is accumulating that the fatal progressive cavitary form of histoplasmosis is a rare occurrence and that the contributory role of the fungus to the developing pathology of the disease is minimal or at least obscure. Death is due to deterioration of pulmonary functions; histoplasmosis is an incidental event.

Smoker's Emphysema and "Migrating Flashes." There are at least two types of preexisting lesions colonized by the fungus, and these result in distinctly different types of disease process. The first is infection of small emphysematous air spaces that develop into interstitial pneumonitis. The cellular response is mononuclear, and there is little replication of the organism. The resulting local lesion heals by encapsulation and disappears, or the antigen-rich effusion spills into the local bronchial tree, causing a transient segmental pneumonitis. Such lesions are most common in the apical-posterior areas of the lung. The process clears in a few days, leaving an infarct-like necrotic area within the area of pneumonitis. Other areas may become involved subsequently, leading to the appearance of "migrating flashes" on x-ray examination. The necrotic area becomes dense in two to four weeks, eventually undergoing contraction and disappearing completely during a two- to four-month course. Some fibrotic streaks may remain. This is the benign self-limited form of chronic pulmonary histoplasmosis.

Cavitary Histoplasmosis. The second type of chronic disease involves infection of large bullous spaces. These infected cavities tend to remain infected and lead to chronic progressive disease. Organisms are found in the necrotic lining, are small in number, and show little evidence of active replication. In the particularly hypersensitive host, the wall of the cavity continues to necrose and enlarge. These are described as "marching cavities" on x-ray examination. In general, such cavities are adjacent to pneumotic lesions of the type described previously and are located in the extreme apices of the lung (Fig. 15–2, G). During the development of lesions, thickening of the walls can cause obliteration of adjacent cavities by effusion and obstruction. Other lesions remain as thin-walled cavities. These are rarely chronically infected, but if the wall has a thickness greater than or equal to 2 mm, persistent colonization is probable. Walls with a thickness of 4 mm may develop in highly sensitive patients. These cavities tend to enlarge by continual necrosis and reach huge proportions. Although thin-walled cavities seldom give rise to complications and do not require treatment, those with thick walls erode surrounding structures and require amphotericin B therapy. Surgical management must sometimes be considered.[51, 74, 108]

The signs and symptoms associated with chronic pulmonary histoplasmosis are common to, but less pronounced than, those of cavity tuberculosis. Mild intermittent fever, cough, sputum production, malaise, fatigue, and weight loss can occur. At least one third of patients complain of persistent deep chest pain. In one fifth of patients no symptoms are noted and disease is discovered on routine x-ray examination. In chronic cavitary disease, cough with sputum production is more pronounced. Bouts of hemoptysis occur when lesions erode into vessels. Death in these patients seldom results from the infection but is caused by advancing pulmonary insufficiency (Fig. 15–2, G). The complement fixation test is positive in only 35 per cent of cases. In general, serologic procedures are unrewarding. Cultural recovery of the organism is often difficult. In cases of thick-walled cavitary disease, up to 70 per cent of sputum cultures are positive, but in cases of thin-walled disease and healing pneumatic episodes, the recovery rate is essentially nil.

There is a third category of chronic progressive pulmonary histoplasmosis, in which the contribution of preexisting lung disease is difficult to assess. The lesions are again apical, cavitary, and progressively destructive. The course varies in duration but is relentlessly downhill unless treated with amphotericin B, sometimes combined with surgery.[51] Such patients may have a cellular defect, and their disease represents a pulmonary manifestation of moderately chronic disseminated histoplasmosis of the adult.

Aberrant Fibrosis and Hypersensitivity Disease

In the vast majority of patients who become infected with *Histoplasma capsulatum*, the disease is self-limited and the pneumatic lesions heal with encapsulation, contraction, and calcification. The primary pneumatic lesion is, histologically, a tuberculoid-type granuloma consisting of a concentrated accumulation of histiocytes (now transformed to epithelioid cells), fibroblasts, lymphocytes, and a few plasma cells and multinucleated giant cells. There is caseous necrosis in the center of the lesion, and in this milieu the fungus remains inactive but viable for years or decades, even though encapsulation and calcification have occurred. Fungi can be recovered from calcified remnants of parenchymal nodules and hilar and mediastinal nodes (Ghon complex), just as the tubercle bacillus can persist in such residua of primary tuberculosis.

In a few patients, the sequence of events just outlined is altered owing to some as yet undefined aberration of normal granuloma formation and fibrosis. In its simplest form the granuloma remains as a thin-walled cavity that is referred to as a coin lesion. This may alternatively excavate and fill over a period of years or remain completely dormant (Fig. 15–2, *F*). In other patients, more serious developments occur, resulting in histoplasmomas, mediastinal fibromas and granulomata, bronchopulmonary[75] and cutaneous fistulas, and bronchocentric granulomatosis.[78]

Histoplasmoma

Histoplasmoma, like tuberculoma and coccidioidoma, is an enlarging fibrous mass that develops around a healed primary focus of infection in the lung. During its active phase the original lesion is 2 to 4 mm in size; it subsequently heals by fibrous encapsulation and calcification. The wall is about 1 mm thick in the histoplasmoma; fibrosis continues, however, and concentric layers of collagen and calcification are added. The growth rate is usually 1 to 2 mm annually, reaching 3 to 4 cm over a 10- to 20-year period. Of itself the lesion is of no consequence, but its growth may erode into vital structures. In periods of activity, a reaction can be seen around the nodule on x-ray examination. This reaction consists histologically of clumps of fibroblasts, lymphocytes, and mononuclear cells. The continued fibrosis is probably a response to antigenic material issuing from the center of the lesion. After a few years the central collagen layers become disrupted and concentric calcifications occur. This gives a radiologic appearance of a central target calcification (the primary lesion), with surrounding rings of calcium deposits. This picture helps to differentiate it from a neoplasm, the most important differential diagnosis. Histoplasmomas often are seen to enlarge rapidly on serial radiographs, causing concern to the clinician. These lesions most commonly develop in the peripheral parenchyma and do no significant damage to important structures. They sometimes develop close to significant sites, and some complications may arise. Their counterpart lesions in the mediastinum frequently lead to life-threatening sequelae, however.

In time, most histoplasmomas cease growth and completely calcify, and the patient is left with a multiringed marble ball in his lung (Fig. 15–8, *B*). If they do not cease growing they may require a surgical procedure, particularly if adjacent important structures are threatened.[108]

Mediastinal Fibrosis and Granulomatosis

Mediastinal granuloma formation represents coalescence and excessive fibrosis in which the initial lesion is in one or several nodes in the mediastinum. Several nodes may become matted together, break down, and be encapsulated to form a mass up to 10 cm in diameter. The wall is usually 4 to 5 mm thick but may continue growing to become 9 mm in thickness. Since these lesions are most often in the right peritracheal areas, such a large mass impinges on vital adjacent structures, including the superior vena cava, the pulmonary artery, the pulmonary veins, and the bronchi. This leads to significant complications, such as the superior vena cava syndrome (Figs. 15–2, *H, I* and 15–11); otherwise, the mass is harmless. Very rarely such continued growth and necrosis leads to fistula formation. Such fistulas may erode into the bronchi or even through the chest wall to the skin.[75]

Mediastinal fibrosis is the counterpart of peripheral histoplasmoma formation. The involved primary sites are hilar or mediastinal lymph nodes. Again, the patient is particularly hypersensitive to the antigenic components of the fungus and exaggerated fibrosis or collagenosis occurs. The core is a small caseous node that has layers of fibrous walls developing around it. Since the mediastinum space is small and contains numerous important structures, the enlarging mass leads to a number of complications. Stenosis of the right stem bronchus and superior vena cava occurs in the right paratracheal area; compaction of the right and left main stem bronchi develop if nodes of the subcarina are involved. Hilar node enlargement manifests as main vessel and bronchial stenosis. Surgical removal of such masses is frequently necessary to relieve the resulting stenoses and to differentiate such masses from neoplasms. Amphotericin B or other antifungal therapy is of no benefit.

There are several other entities that involve aberrant tissue response to the presence of *Histoplasma* or other fungi. One that has been described recently is bronchocentric granulomatosis.[78] In the evolution of this process, a series of granuloma develop in response to a fungus. This results in stenosis of bronchi and mucoid impaction. On histologic examination, a few fragments of fungi can be seen within the granulomas.

In addition to the clinical syndromes described in the previous sections, histoplasmosis can occur as the obscure cause of single organ system disease or can mimic a totally unrelated infectious or noninfectious condition. Most often this is a manifestation of the mildly chronic recurrent form; however, in some cases the genesis of the disease defies explanation. Such manifestations include pericarditis,[28, 63, 83, 146] peritonitis,[114] polyarthritis,[29] exfoliative erythroderma,[122] granulomatous cutaneous disease,[139] and others.

DIFFERENTIAL DIAGNOSIS

As stated previously, at all stages of its pathogenesis, histoplasmosis mimics tuberculosis. The differentiation of these diseases is always difficult. Only culture and adequate serologic evidence provide the correct diagnosis. Histoplasmosis may coexist with any number of granulomatous diseases of the lung, including tuberculosis, sarcoidosis, actinomycosis, or other mycotic infections.[120]

Primary histoplasmosis in its acute stage closely resembles the acute infection produced by other mycoses, viral and bacterial pneumonias, lipoid pneumonia, and Hamman-Rich syndrome (diffuse interstitial pulmonary fibrosis).

Acute disseminated histoplasmosis with its accompanying hepatosplenomegaly, lymphopathy, anemia, and leukopenia resembles the acute stage of the visceral leishmaniasis as well as many of the lymphomatous diseases. In most forms of histoplasmosis, thick blood smears or material from sternal punctures often reveal the organism more readily than culture or direct examination of sputum (Fig. 15–7). The blood smears can be made from the buffy coat of centrifuged citrated blood or from the bottom of the tube, where sedimented, heavily infested cells are found. Other diseases to be considered in the differential diagnosis of histoplasmosis are infectious mononucleosis, brucellosis, malaria, and Gaucher's disease. When cutaneous or mucocutaneous lesions are present, they may suggest neoplasias, sporotrichosis, syphilis, toxoplasmosis, bacterial cellulitis, tuberculosis cutis, or other systemic mycotic infection.[97, 139]

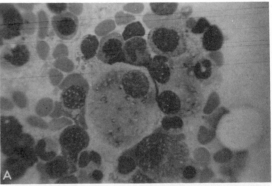
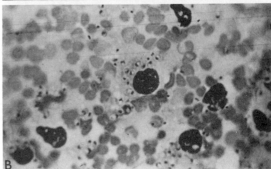

Figure 15–7. Histoplasmosis. *A*, Sternal puncture showing intracellular yeast cells. Benign form of acute primary disease. Some yeasts are killed and fragmented. Wright's stain. ×400. *B*, Final stage of moderately chronic disseminated disease. Numerous organisms are seen in macrophages and extracellularly. Budding of the yeast cells is evident. GMS. ×400.

The histoplasmin skin test is of no diagnostic value, as it only connotes past or present exposure to the organism. Serial complement fixation tests, especially when there is a rise in titer, are very significant. There are exceptions, however, in which disseminating disease is accompanied by a drop in complement fixation (CF) titers.[51] It must be remembered that the histoplasmin skin test may cause a one- to two-tube false rise in the complement fixation test for histoplasmosis or coccidioidomycosis.

PROGNOSIS AND THERAPY

In essentially all cases of histoplasmosis, when the patient does not have an underlying disease or the inoculum is not overwhelming, the disease resolves uneventfully. Even in moderately severe primary infections, bed rest and supportive measures are sufficient to effect resolution of the disease. However, disseminated disease, chronic cavitary, mucocutaneous, or systemic infections require antimycotic therapy.

The treatment to be used is the same as that for all the systemic mycoses: amphotericin B.[61, 62, 121] The dosage schedule is also generally the same, as are its side effects and measures to ameliorate these side effects (see Chapter 18). The total course of therapy is 0.5 g in severe primary disease, and 1 to 2 g in chronic disease. For the treatment of disseminated histoplasmosis, Drutz et al.[48] and Goodwin et al.[61, 62] used a dosage of 0.6 mg per kg per day for six weeks in adults and more when tolerated in infants. A blood level of 1.56 μg per ml for ten weeks was maintained in adults. If blood levels cannot be followed, the patient is administered 30 to 50 mg per kg daily or three times a week to a total dose of 2.0 gm. Clinical cure is achieved in almost all patients who can tolerate the drug and complete the course of therapy. In contrast with coccidioidomycosis and blastomycosis, recovery in histoplasmosis is rapid and relapse is essentially unknown. In the rare case of cerebral involvement, slow (over 6 hours) intrathecal administration of 0.2 to 0.3 mg of amphotericin B administered by means of an Ommaya reservoir has been recommended.[61, 62]

Endocarditis is another form of histoplasmosis in which the usual course of 1 to 2 g amphotericin B is generally not curative. In a review of twenty-eight cases of the disease by Bradsher et al.,[16] few patients responded to even several repeated courses. In their case, combined amphotericin and surgery were effective. The same experience was reported by Gaynes et al.[58] in a patient with an infected prosthetic valve.

Ethyl vanillate, MDR-12, nystatin, 2-hydroxystilbamidine, sulfadiazine, 5-fluorocytosine, and antibacterial or antitubercular antibiotics are ineffective in histoplasmosis. Sulfadiazine shows some effect in the treatment of disease in experimental animals, and before amphotericin B was available appears to have been efficacious in a few human cases.[61, 62] The imidazoles (miconazole and ketoconazole) are still being evaluated clinically. So far a large number of relapses have occurred with either therapy. Ketoconazole has the advantage of oral administration and can be considered if use of amphotericin B is precluded.

After medical management has brought about clinical arrest or the disease spontaneously resolves, surgical excision of large cavities or granulomatous masses may be considered.[74, 108] Prophylactic use of amphotericin B is advisable to preclude reactivation or dissemination during the surgical procedure.

PATHOLOGY

Gross Pathology. The gross pathology induced by acute primary disease is essentially unknown, as this form of infection is a mild, uncomplicated disorder, and few patients come to autopsy. To extrapolate from the animal experiments of Proknow,[110] following infection there is an initial alveolitis, and macrophages engulf the organisms. Subsequently there is an invasion by neutrophils and lymphocytes, and a progressive inflammatory reaction of the usual pyogenic type occurs. The macrophages probably carry the organisms to other body parts very early in the course of infection. The pyogenic response in the lung is followed by the formation of epithelioid cell tubercles indistinguishable from those of tuberculosis. In the few fatal cases of acute histoplasmosis due to overwhelming exposure to conidia, there was massive pleural and alveolar effusion, causing death by asphyxiation within a few days. When the course was prolonged, numerous yellowish nodular masses (forming tubercles) and confluent nodular masses were found in the lungs, along with great enlargement of the hilar and mediastinal lymph nodes. In less severe disease, where there was a need to

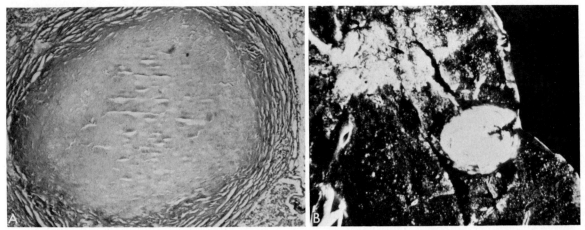

Figure 15–8. *A,* Resolved acute histoplasmosis. Fibrocaseous nodule in the lung. Encapsulation is proceeding around the periphery, and caseous necrosis has occurred in the center. The few organisms present in the central area may remain viable for an extended period of time. Hematoxylin and eosin stain. ×50. (Courtesy of W. Kaplan.) *B,* Histoplasmoma. The core is composed of a healed primary nidus of infection which resolved and was calcified. Over the years, layers of fibrous material were laid down around it and calcified; thus, a large mass composed of concentric rings of calcified debris was formed. (Courtesy of W. Causey.)

remove a nodule for diagnosis, the lesion was an epithelioid tubercle with Langhans' giant cells that would have been called, if caseating, "histologically compatible with tuberculosis, bacilli not seen" or, if not caseating, sarcoid. Healed lesions from subclinical or mild disease completely calcify after several years (Fig. 15–8). They are small, 15 mm or less, and may be scattered throughout all lung fields. Similar nodules are present in the spleen (Figs. 15–9, *A, B*) and less frequently in the liver. The organisms in tubercles are moribund but often still viable. Culture is very difficult, often requiring hypertonic "L" form or protoplast media; otherwise, they

are demonstrable only by the Grocott-Gomori methenamine stain or other special fungal stains.

In chronic progressive cavitary disease, the lesions are identical to those of cavitary tuberculosis. Caseation, necrosis, and layers of fibrosis and calcification are present. The cavity may be 8 or more cm in diameter, and a few have been noted to contain *Aspergillus* fungus balls. *H. capsulatum* yeast cells can be demonstrated in the walls, and the hyphae of the *Aspergillus* are in the lumen.[120, 129]

The pulmonary "coin" lesion, which is usually solitary, may vary from 0.5 to several cm in size (Fig. 15–2, *F*). It is usually located

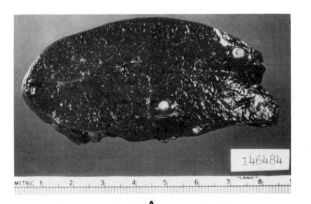

A

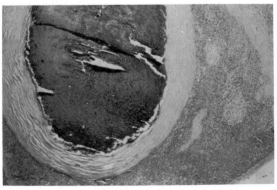

B

Figure 15–9. *A,* Histoplasmosis. Spleen from a patient who had primary histoplasmosis many years previously. Well-circumscribed, partially calcified, light yellow, raised nodules are present in cut surface of spleen. They are from 0.2 to 0.4 cm in diameter. The lesions are typical of healed fibrocalcific granulomata of old resolved histoplasmosis. (Courtesy of S. Thomsen.) *B,* Histoplasmosis. Fibrocaseous lesion of spleen completely calcified. This represents a residual lesion from an infection acquired many years previously. The lesion was inactive and the organism dead. Yeasts were visible on GMS stain. Hematoxylin and eosin stain. ×100. (Courtesy of S. Thomsen.)

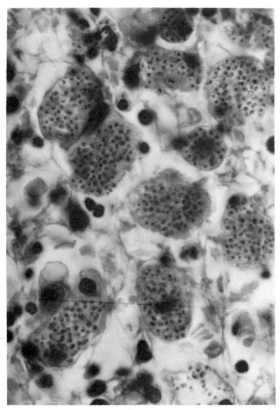

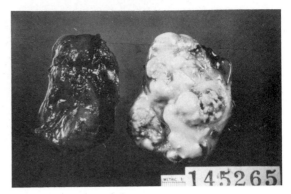

Figure 15–11. Histoplasmosis. Superior mediastinal syndrome. The mass removed was well encapsulated and nodular. On cut surface it could be seen that the lymph nodes were nearly all destroyed by necrotizing and caseous granulomata. The necrotic center contained a thick, creamy, light-green fluid (see Fig. 15–2,*H,I*). (Courtesy of S. Thomsen.)

Figure 15–10. Histiocytes containing numerous yeast cells. This was from a case of fulminant disease of childhood. Hematoxylin and eosin stain. ×4000. (Courtesy of E. Humphreys.)

there is evident caseous necrosis, and such patients may have signs of Addison's disease.[34, 61, 62]

In mildly chronic disease, all of the above mentioned pathology is present but exaggerated.[8, 62] In addition, ulcerating lesions may be found in the intestine, larynx, pharynx, genitals, tongue, meninges, and endocardium (Fig. 15–13, *B,C*). In the latter instance, extensive vegetations resembling those of *Candida* endocarditis are present. Recurrent ulcerations of the intestine may provoke a malabsorption syndrome,[98] and severe pericarditis can cause valvular stenosis.[28] Any of these symptoms may present as an isolated event. A complete analysis of gross and histopathologic appearance of organ systems in all forms of histoplasmosis has been compiled by Schwarz,[129] and Or-

in one of the lower lobes and generally attached to the pleura. There is a fibrous thickening of the overlying pleura. The lesion itself has a fibrous capsule with some calcific material and a caseous center. The organisms present are frequently moribund but viable and are in the central caseous part. The histoplasmoma has essentially the same characteristics except that it is larger, has concentric rings of fibrosis and calcification, and may be seen to expand with time (Fig. 15–8, *B*).[60]

In the rapidly progressive fatal form of histoplasmosis, there is, as first noted by Darling in 1905, gross enlargement of the liver and spleen. This may be the only abnormality apparent at autopsy. Histologically, phagocytized yeast cells are seen in the areas rich in reticuloendothelial cells, such as liver, spleen, bone marrow (Fig. 15–10), lymph nodes (Fig. 15–12), and sometimes the adrenals. If the disease had a more protracted course (moderately chronic form), tubercles are present in essentially all organs, lymph nodes, and tissue. In certain organs such as the adrenals,

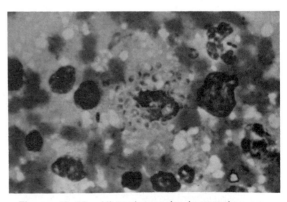

Figure 15–12. Histoplasmosis. Impression smear of liver from a case of fulminant disseminated disease of the adult. Intracellular organisms are seen, and their internal morphology is easily discerned. Wright's stain. ×400.

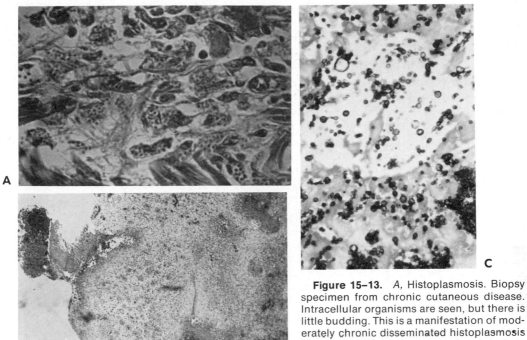

A

B

C

Figure 15–13. *A,* Histoplasmosis. Biopsy specimen from chronic cutaneous disease. Intracellular organisms are seen, but there is little budding. This is a manifestation of moderately chronic disseminated histoplasmosis of the adult. The appearance of such lesions usually heralds persistent fungemia and a rapid deterioration of the patient. Hematoxylin and eosin stain. ×440. (From Studdard, J. W., F. Sneed et al. 1976. Ann. Rev. Resp. Dis. *113*:689–693.) *B,* Histoplasma endocarditis vegetations from an infected prosthesis. Hematoxylin and eosin stain. ×100. (From Gaynes, R. P., P. Gardner, and W. Causey. 1982. Prosthetic valve endocarditis caused by *Histoplasma capsulatum.* In Press.) *C,* Tissue from valvular lesion stained by GMS. *Histoplasma* yeast cells are usually uniform in size and morphology but are sometimes aberrant in form, as seen here. ×440.

chard et al.[98] have reviewed the literature on gastrointestinal histoplasmosis.

Histopathology. The histologic appearance of the lesions depends on the recentness of the infection and severity and form of the disease. In the acute, rapidly fatal form (histiocytomycetic disease), numerous yeast cells are seen only within the histiocytes. Neutrophils, plasma cells, and lymphocytes are not abundant and do not contain yeast cells. The yeast cells within the histiocytes are budding and of uniform size (about 3 μ) and are visible in Gram, Giemsa, Wright's, or hematoxylin-eosin stained smears. They resemble *Leishmania donovani* but lack the rod-like parabasal body (kinetoplast) (Figs. 15–7 and 15–12). The internal morphology is best seen with Giemsa stains. *Histoplasma* yeast cells are stained by any of the special fungus stains, whereas *Leishmania* is not. *Toxoplasma* also must be ruled out in such cases. It is smaller and not found within histiocytes and does not take the special fungus stains. *Candida glabrata* also may be intracellular. Only culture or the use of the immuno-

fluorescent technique on tissue sections will differentiate them.

In the less fulminating forms of the disease, epithelioid granulomas are formed that contain plasma cells, lymphocytes, macrophages, neutrophils, and giant cells. The organisms are seen in any of the phagocytic cells, but in smaller numbers than in fulminant disease. Budding is less prominent, and the uniformity of size and the staining characteristics are the basis for presumptive diagnosis. This is also true in biopsy material from cutaneous and mucocutaneous lesions (Fig. 15–13,*A*). The stain of choice in such cases is the methenamine silver stain. In chronic cavitary disease, the organisms are even fewer in number and very rarely are seen to bud.

In old lesions, coin lesions, histoplasmomas, and calcified nodules, the organisms are usually moribund but viable and very few in number. They are difficult to find in histologic sections. In the infarct-like nodules of chronic pulmonary histoplasmosis, a few cells are seen in the center of the process. In chronic cavitary disease, cells are found in the necrotic lining rather than in the deeper tissue substance. In some areas they may be abundant but are very irregular in size and morphology. This is particularly true in very old chronic cavities. Organisms up to 20 μ have been found (Fig. 15–13, *C*). In many lesions of chronic or resolved pulmonary disease, the yeast cells are very rare, and few if any will show budding. In such cases, cultures are usually negative, and a diagnosis must be made on histologic evidence alone after extensive search of many sections. The yeast cells of *H. capsulatum* must be differentiated from newly released spores and young spherules of *Coccidioides immitis*, from *Cryptococcus neoformans*, and from the small cell form of *Blastomyces dermatitidis*. There is usually enough capsular material around even dead cryptococci to stain with Mayer's mucicarmine stain. The so-called "capsule" of *H. capsulatum* represents unstained cell wall, and it does not take the mucicarmine stain. If the lesion in question is caused by *C. immitis*, spherules or parts of spherules are found after intensive search. Even the small forms of *B. dermatitidis* show a broad-based bud and can be seen to be multinucleate in a hematoxylin and eosin stain. *H. capsulatum* is uninucleate and has a narrow-necked bud. Small hyphae have been reported to be present in a case of *Histoplasma* endocarditis, but such a finding is rare. In such a case, differentiation of the disease from candidiasis is necessary. In lesions in which differentiation of *Histoplasma* from other etiologic agents cannot be made, specific fluorescent antibody staining can be attempted.[76]

ANIMAL DISEASE

Natural disease in wild and domestic animals is very common in endemic areas. Dogs seem to be particularly susceptible,[18, 135] and the infection has been reported in cats,[102] swine, cattle, and horses.[64] The symptoms range from pulmonary and cutaneous lesions to ocular involvement,[102] nodules on the tongue,[18] mycotic abortion,[64] and generalized dissemination to other organ systems.[135, 137] Among wild animals, histoplasmosis has been reported in numerous species, ranging from the house mouse to the Kodiak bear (the latter housed in an Ohio zoo). Indeed the only recorded case of histoplasmosis in Switzerland is that of a badger.[17] This animal was found along the roadside, and the fungus was demonstrated in its submandibular lymph nodes. The only animal species considered to be an important vector for dissemination of the fungus is the bat.[41]

The type and severity of infection in animals and the pathology elicited are as protean as those found in human cases. They vary from small calcified nodules found incidental to other diseases to an acute, disseminating, rapidly fatal form. There is no substantiated evidence that natural infection occurs in avian species. Menges, in Chapter 20 of the symposium on Histoplasmosis, has reviewed the animal species in which infection has been reported and the varieties of pathology encountered.[4] Epizootic lymphangitis caused by *H. farciminosum* is discussed in Chapter 27.

Experimental Infection. The mouse is the laboratory animal most susceptible to experimental infection. Rowley[118] demonstrated that from 1 to 10 yeast cells will infect a mouse, and Ajello and Runyon[3] found that one macroconidium was sufficient to evoke disease. For this reason the mouse can be used as an adjunct to other procedures for isolation of the yeast from patient material, for determining the presence of the organism in soil samples, and for testing the efficacy of chemotherapeutic agents in experimental infections.

The procedures for use of the mouse as an aid to isolation of the fungus are well standardized. Material obtained at biopsy can be macerated and injected intraperitoneally. Sputum, however, must be mixed with antibiotics before injection. Usually one ml of a mixture of penicillin (7 mg per ml) and streptomycin (10 mg per ml) is added to 5 ml of sputum. Since sublethal doses of *H. capsulatum* are known to immunize mice and abort infections, mice should be autopsied from two to four weeks after inoculation. The spleen and liver are removed and minced with a scissors and smeared on Sabouraud's or blood agar plates. These are sealed with tape and incubated at room temperature for at least four weeks. The initial growth may be glabrous but is usually a white fuzz. For identification, a portion of the colony is transferred to blood agar slants and incubated at 37° C to demonstrate dimorphism. The mycelial growth occurring at 25° C is examined for the presence of tuberculate macroconidia. Care should be exercised in handling the mycelial phase.

The procedure for isolation of the organism from soil is similarly executed.[89] A saline suspension of the soil is mixed with antibiotics in the same concentrations as described for sputum. The material is then injected intraperitoneally. The animals are usually held for one month before autopsy. Mice may also be infected by housing them with contaminated soil.

For the purpose of testing therapeutic agents, it is necessary to assay the virulence of the particular strain of *Histoplasma* to be used. Though most strains will kill mice when a dose of 10^6 or 10^7 cells is injected intraperitoneally, some strains are less virulent.[35] In antibiotic testing it is preferable to determine the LD_{50} and inoculate this dose by the intravenous route. The tail vein is suitable for this procedure, and with experience few technical problems are encountered. Accurate counts of the number of viable units injected can be determined by the Janus green method of Berliner and Reca.[14]

Dogs are quite susceptible to experimental histoplasmosis. The early sequence of events following infection has been studied by administering aerosols of conidia to dogs.[110] Other laboratory animals, such as rabbits, hamsters, and rats, vary considerably in their susceptibility to infection. Experimental disease has also been produced in monkeys.[77]

Experimental disease has been produced in poikilothermic animals also. In one experiment lizards and frogs were injected with either the yeast or mycelial form of the fungus. Regardless of which form was injected, lesions in animals incubated at 25° C contained mycelium, and yeast cells were found in those incubated at 37° C.[126]

Pigeons and chickens have been injected with *H. capsulatum*. Infection is produced, but it is of short duration. The body temperature of birds is thought to be too high for a progressive disease to develop.[128]

BIOLOGICAL STUDIES

One of the most intriguing aspects of its biology is the dimorphism exhibited by *H. capsulatum*. As with the other pathogenic fungi and in contrast to opportunistic agents of fungus disease, a radical morphologic and physiologic change accompanies *in vivo* existence. In *H. capsulatum*, the form found in its natural saprophytic habitat is a mycelium producing numerous conidia, but when this fungus is inhaled by susceptible animal species, it grows as a small, delicate, budding yeast. The factors that govern this dimor-

phism have been the subject of much investigation.[117]

Dimorphism or polymorphism occurs in many biological groups. The cestodes and trematodes show great variation in morphology of larval and adult stages. This is also true of the glochidium stage when compared with the adult clam, or the tiny gametophyte stage relative to the large sporophyte plant of ferns. In almost all cases these morphologic changes are part of maturation or developmental stages in a life cycle. Such developmental dimorphism is found among some fungi (e.g., smuts and rusts), but it is not common in the fungal kingdom. Dimorphism and polymorphism associated with nutritional or environmental factors are frequently encountered among fungal species, however. Dutch elm disease (*Ceratocystis ulmi*), an ambrosia fungus of beetles, exists in yeast or mycelial form, depending on phosphate concentration and CO_2 contents. CO_2 and glucose concentration in fluid media influences the growth of *Mucor* sp. to form mycelium chlamydoconidium or yeast, a phenomenon first noted by Pasteur. The presence or absence of fermentable carbohydrates is the most important determinant of yeast or mycelium growth ratios in *Candida albicans* and other members of the genus. In contrast, dimorphism primarily as a response to environmental temperature is found almost exclusively among the true pathogenic fungi. As noted in Chapter 14, there are five species of fungi that demonstrate this thermal dimorphism in association with invasion of and proliferation in animal tissue. In two species, *Blastomyces dermatitidis* and *Paracoccidioides braziliensis*, temperature alone will suffice for conversion of mycelium to yeast phase. *Sporothrix schenckii* is aided in conversion by reducing agents, but this requirement is not absolute, and *Coccidioides* assumes its spherule form most readily in liquid media with high CO_2 tension and a temperature of 40° C. The exception of the group is *Histoplasma capsulatum*. This organism has requirements in addition to temperature for its conversion to the yeast stage.

Initially it was thought that substances from serum or tissue within the host were necessary for conversion to the yeast. *In vitro* conversion to the yeast form of growth could be accomplished on blood agar but not on nutrient agar incubated at 37° C, even though nutrient agar incubated at this temperature was known to be sufficient for the growth of the yeast form of *B. dermatitidis*.

Since *H. capsulatum* grew as a mycelium at 25° C on blood agar, it appeared that a factor in addition to temperature affected the mycelial to yeast conversion. Investigations by Pine,[105, 106] Scherr,[125] and others established that it was free -SH groups provided by the blood in blood agar which effected the transformation. They demonstrated that, in a salt medium containing glucose, mycelial to yeast conversion could be accomplished if cysteine was included to provide the free -SH groups. It was later shown that this conversion could be effected by electrolytically lowering the oxidation-reduction potential of liquid media to that of living tissue.[116] Cysteine is the most effective -SH containing compound for converting and maintaining the yeast form of growth. In a study of a series of low molecular weight, sulfur-containing compounds, Garrison et al.[54] have shown that cysteine has the greatest stimulatory effect on yeast respiration, and that it stimulated growth of the yeast form to a greater degree than the mycelial form. The reduced environment appears to enhance cell wall production as a consequence of its stimulation of energy metabolism.

The early studies cited previously were concerned primarily with external factors that influenced conversions to the yeast phase in *H. capsulatum*. Recently, work has turned to the internal events involved in this morphologic change. Nickerson[117] described a protein disulfide reductase that softened an area in the cell wall in *Candida* and allowed bud formation. This has been noted to occur in *Histoplasma* as well. To keep this enzyme active, a continuous source of -SH groups is necessary, and a role for a sulfide reductase was demonstrated. Many internal regulators are involved in this complex process, and their roles are only now being elucidated.[15, 117, 134] The requirement for lowered oxidation-reduction potential is concerned with a redox-sensitive component of terminal respiration, ubiquinone. This in turn is a rate-limiting factor in activating the sulfite reductase system. In addition, protein regulators of the processes are involved, including "histins" and other component parts, such as several RNA polymerases and cyclic AMP.[15] At this point it appears that an elevated temperature is required to initiate the enzymatic mechanisms for conversion to the yeast phase (the duration of the exposure to that temperature may be quite short); thereafter, a lowered oxidation-reduction potential, however derived, is sufficient to maintain yeast phase growth. Complete understanding of all the interelated processes requires further investigation.

It has now been established that temperature is the single most important factor regulating the mycelial to yeast conversion of *H. capsulatum*, and therefore it is a thermal dimorphic fungus. Pine[105] and Scherr[125] demonstrated this temperature dependence *in vitro*, and Howard[70] found that phagocytized intracellular yeast cells in tissue culture replicated as yeasts when maintained at 30° C. If the phagocytized yeasts were incubated at 25° C, however, they sprouted to form mycelium. A similar temperature dependence of *in vivo* growth form has been demonstrated in poikilothermic animals.[126] The morphogenesis of the yeast form and its growth in tissue culture have been investigated and reviewed by Howard.[68-73] Nutritional and physiologic aspects of both forms have been reviewed by Bauman[9] and Rippon.[117]

Domer et al. have studied the differences in cell wall composition accompanying morphogenesis.[45, 46] They found quantitatively more chitin and less mannose and amino acids in the walls of the yeast form than in the mycelial form. The composition of the mycelial form cell wall closely resembled that of the yeast *Saccharomyces cerevisiae*. They found that there are at least two chemotypes of the cell walls of the yeast form. Chemotype II was similar in its monosaccharide and chitin content to the yeast form of *B. dermatitidis*.[45]

There were also differences between the α glucan (usually associated with yeast cell walls) and β glucan (mycelial wall) composition of different strains of the fungus. The complexity of cell wall composition has been recently confirmed by San Blas et al.[123] These chemical differences in cell wall composition are undoubtedly reflected in the different antigenic patterns of the yeast form of *H. capsulatum*.

The course of experimental histoplasmosis has been examined by a number of investigators. In a recent series, Artz and Bullock[7] observed histologic, serologic, and immunologic parameters during disseminated disease. Infection was produced by IV injection of 5×10^3 cells in mice, and the disease resolved over an eight-week period. CF titers to the yeast phase (y) antigen appeared at week 1, peaked at 3, and declined thereafter. M phase CF titers came later (week 3), and peaked at 18. Granulomata in the spleen occurred by week 1 and resolved by week 3. Thymocytes depleted in early disease returned to normal by week 8. The spleen produced an immunosuppressor substance that depressed delayed-type hypersensitivity, impaired blastogenic transformation by splenocytes *in vitro,* depressed cytotoxic activity of spleen cells, and depressed humoral antibody production to various antigens. With resolution of the infection, the dominant suppressor activity returned to helper activity, and delayed type hypersensitivity, cytotoxic capacity, and plaque-forming ability of the splenocytes also returned to normal levels. Hyperactive suppressor T cell activity also has been noted in human infection. In this instance it was compartmentalized and restricted to the cerebrospinal fluid in a case of meningitis;[33] serum and CNS serologies were negative.

Other biological studies have included the standardization of growth curves for production of antigens. Reca and Campbell[113] found that different conditions of growth will markedly affect the antigens produced by the organism. Protoplasts of the yeast phase can be produced by a number of agents.[13] The presence of "L" forms in infected tissue has been postulated but not established. "L" forms have been found in other fungus infections. Electron microscopic studies of the organism have also been carried out[55] (Fig. 15–17).

In the study of epidemics, it is necessary to recover the organism from the point source of infection. This is easily done by the mouse inoculation procedure of Zeidberg and Ajello,[147] but this technique may require several weeks for recovery of the organism and results. Gaur and Lichwardt have developed a rapid preliminary visual screen,[56, 57] which they applied to a study of a point source epidemic involving herring gull detritis in port facilities in Rogers City, Michigan.

IMMUNOLOGY AND SEROLOGY

The presence of a high natural resistance to infection by *H. capsulatum* is attested to by the fact that in the vast majority of infected persons there are few, if any, symptoms, and the disease resolves rapidly. In cases of overwhelming exposure to fungus, disease is established in all patients and also occurs in a few patients who have had only minimal exposure. Subtle, as yet undefined, immune differences or slight, even transient, debilitation may account for these latter cases. As

noted previously, *H. capsulatum* is infrequently an opportunistic fungus invader in the usual sense, which tends to make the problem of the patient population that develops serious disease more enigmatic. It now appears that recovery from mild infection renders the patient firmly resistant to reinfection. Circulating antibodies are considered of little consequence in the immunity established following resolution of infection. In experiments using tissue culture it was shown that the phagocytosis and intracellular growth of *H. capsulatum* yeast cells was not affected by the presence of serum containing antibodies to the organism.[69] However, Howard et al.[72, 73] have demonstrated that mononuclear phagocytes from immunized animals restricted the intracellular growth of the yeast cells as compared with phagocytes from nonimmune animals. It appears, then, that immunity to the infection is concerned with cellular rather than humoral defenses. Further, it appears that the permissiveness of the macrophages as far as intracellular replication of the yeast phase is concerned may be the first determinant in the outcome of infection by *Histoplasma capsulatum.* The factors involved in this permissiveness will be a fascinating area of research in the future.

Serologic Tests

The Skin Test. As has been noted previously, past or present exposure to the fungus usually confers a long lasting reactivity to the histoplasmin skin test. Thus this test is of little diagnostic or prognostic value except in cases of disseminated disease, in which absence of reactivity denotes anergy,[53, 62] or in very early infections, in which the test may also be negative. This procedure also may evoke a false rise in the complement fixation titer for histoplasmosis and also for coccidioidomycosis. The skin test usually becomes positive within two weeks after exposure and remains so for long periods of time. With no new exposure to the fungus it is felt that sensitivity wanes in most patients by ten or more years. People continually exposed to the microconidia or those housing quiescent but still viable yeasts maintain reactivity.

The Complement Fixation Test. Serial samples of serum used in the complement fixation test are of great value in establishing the diagnosis as well as delineating the prognosis of the disease. This was first demonstrated by Salvin in experimentally infected animals.[21] Complement fixing antibodies may appear as early as two weeks following infection, and almost all patients will have a demonstrable titer by four weeks. Following resolution of the disease, the titer gradually falls and usually disappears by the ninth month. There are exceptions, however, in which dissemination has occurred while titers dropped.[51] Reaction to yeast phase antigen generally remains longer than to that prepared from the mycelial phase. In some recorded cases, complement fixing titers after clinical recovery have remained positive for up to four years when the histoplasmin (mycelial) antigen was used, and up to nine years with the yeast cell antigen. However, a high titer (1:32 or over) that remains at that level or rises is usually considered to be indicative of active or progressive disease.

Titers are quite variable in disseminated disease. Only two thirds of patients with moderately or mildly chronic disseminated infection have definable patterns of complement fixation titers. The immunodiffusion test is generally positive, however. The greatest difficulty in interpreting tests concerns the migrating focal infections and single organ involvement found in mildly chronic disease. In some cases wherein endocarditis was the only known site of infection, titers were extraordinarily high,[58] whereas in silent meningitis and cerebritis all serum serologies and skin test activity were negative but CF of spinal fluid was positive.[107] The serologic pattern and its interpretation is not as well established in histoplasmosis as it is in coccidioidomycosis.

There are many pitfalls in the use of complement fixation tests for diagnostic and prognostic evaluation. The careful analysis of Campbell[20-22] has shown that high antibody titers to histoplasmin may be present in cases of cryptococcosis, blastomycosis, and more rarely coccidioidomycosis. Cross-reactions occur in the sera from experimentally infected animals also.[119] In addition, it has been noted that administration of the histoplasmin skin test may cause a one- to two-fold increase in complement fixation titers.[21, 22] Other workers, however, could not corroborate this.[124] Campbell has also demonstrated that the reactivity of the antigen varies from batch to batch with age of the cells, with composition of the media, and with other as yet unknown factors such as, perhaps, the phase of the moon.[21, 49] This is not surprising, considering the multiple antigenic sites present on the organism. Efforts to fractionate and isolate a specific antigenic component to be used in the complement fixation test have been reviewed by Larsh and Bartels.[90] Of the many fractions isolated, the most promising from the point of view of specificity is a polysaccharide derived from histoplasmin. Bauman and Smith[10] found the CF test with histoplasmin (M phase) to be positive in 73 per cent of proven cases and the CF with yeast antigen in 94 per cent. There were more false positives when the yeast phase antigen was used, however. The immunodiffusion (ID) test using histoplasmin was positive in 90 per cent of cases in the same series. At present, therefore, the complement fixation test is useful but cannot be said to give unequivocal evidence for the presence or absence of disease.

The immunodiffusion test using concentrated histoplasmin is a useful adjunct to the complement fixation test. It appears to become positive by the third or fourth week following infection. As first described by Heiner,[80, 90] the antigen and patient's sera were allowed to diffuse toward each other from wells in an Ouchterlony plate. He noted that two lines were of significance. An m line was present in persons who had recovered from the disease or in cases of early infection. An h line closer to the well containing the serum corresponded with the presence of active infection (Fig. 15–14). It disappears with resolution of disease but may be present for up to two years following clinical cure. A c band, which is sometimes found, appeared to represent a common antigen between *H. capsulatum* and *B. dermatitidis*. Other bands designated x and z have appeared in some patients' sera, but their significance is unknown. Further evaluation of what appeared to be an extremely important diagnostic tool has shown variations and cross-reactions in this test also,[90] but it remains one of the most important serologic procedures for diagnosis and prognostic evaluation of histoplasmosis.

The latex agglutination test, in which serum is mixed with latex particles coated with histoplasmin, has been extensively investigated by Hill and Campbell.[66] It may be positive before the complement fixation test and is also useful in cases in which the patient's serum is anticomplementary.

The use of a battery of serologic tests tends to rule out both false-positive, false-negative, and cross-reactions. Their interpretation is always difficult, and the only unequivocal evidence for the diagnosis of histoplasmosis is

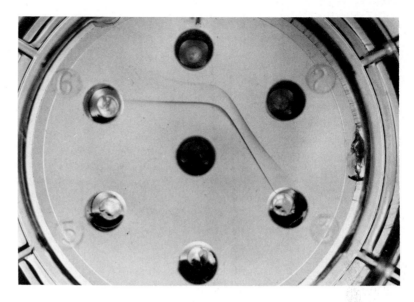

Figure 15–14. Histoplasmosis. Immunodiffusion plate with a known positive serum in one outer well and an unknown one in an adjacent well. The antigen is in the center well. Note the lines of identity in the two tests. The h line is the finer line near the serum-containing well. The m line is near the antigen-containing well.

isolation of the fungus *H. capsulatum.* Only when a single specific antigen has been isolated, purified, and standardized will the serologic testing in this and other fungus diseases be of absolute diagnostic value. It is possible, however, that developing a test for the presence of antigens derived from the organism that may be present in the patient's blood would be a more useful procedure than trying to evaluate the presence of the various antibodies. This is already a practical procedure in the serodiagnosis of cryptococcosis.

Cell-fixed antibodies have been demonstrated in experimental disease.[70] In this study, spleen cells from infected animals are mixed with *Histoplasma* yeast cells, which stick to the sensitized cells, forming rosettes.

Skin testing for histoplasmosis has been discussed at length in other sections of this chapter. The recommended procedure is to inject 0.1 mg of a 1:100 dilution of a standardized histoplasmin intracutaneously. The test is read after 48 and 72 hours, though it may not become maximal for five days. An area of induration of 5 mm or more is considered positive. Care should be exercised in administration of the test. Hypersensitive individuals may develop a severe reaction, with fever and nausea with necrosis at the area where the test was administered. A positive test indicates past or present exposure to the organism. A negative test in a patient with known histoplasmosis indicates anergy and suggests a grave prognosis. A tine test that allows easier application of antigen has recently been developed.[25] Since skin testing

has little diagnostic value and may interfere with the complement fixation tests for histoplasmosis and coccidioidomycosis, the general use of this procedure is to be discouraged.

Several studies have been made of the appearance of particular species of globulins in patients infected with *H. capsulatum.* Results initially indicated that, in new infections, IgM globulins were the first to increase in quantity. This was followed by increases in IgG and IgA. In chronic disease, the IgM and IgG levels were normal, but there was an absolute increase in IgA.[143, 144] More recent and extensive work reviewed by Larsh and Bartels[90] has shown that there is wide variation among patients in the types and quantities of the immunoglobulins in any stage of the disease.

Antigenic analysis of *H. capsulatum* has revealed that there are several serotypes. Kaufman and Blumer[80] have shown that there are four antigenic components and several antigenic patterns. The 1,4 type is identical serologically to *H. capsulatum* var. *duboisii,* indicating that the latter organism is probably a variety of *H. capsulatum* and not a separate species. Pine and Boone[106a] have demonstrated that some differences in serologic types are reflected by differences in cell wall composition. Serology using known antihistoplasma antisera can be used to confirm the identity of an isolated fungus.[133]

A fluorescent antibody specific for the yeast cells of *H. capsulatum* has been developed by Kaufman and Kaplan.[76] The usefulness of this procedure in making or confirming a diagnosis in the absence of cultures has been reviewed by Kaplan.[76]

Summary of Practical Serologic Testing in Histoplasmosis

Screening Tests. The latex agglutination becomes positive early in the disease. Immunodiffusion can differentiate active from inactive disease. It is positive about the same time as the latex agglutination (two to five weeks after development of symptoms). The fluorescent antibody test is also very useful when available. The CF test can also be used in screening.

Confirmatory Tests and Tests for Prognostic Evaluation. The CF test is positive later in disease (six weeks or more) than most other tests. A titer of 1:32 is considered significant, but some patients with active disease have titers of only 1:8 or 1:16. Serial titers should be done. A rise in titer indicates dissemination. At this writing it appears that the combination of the CF test and ID are the most important serologic procedures for diagnosis and evaluation of prognosis. In the study by Bauman and Smith,[10] no cases were missed when both were used.

LABORATORY DIAGNOSIS

The diagnosis of histoplasmosis is made by the isolation and identification of *Histoplasma capsulatum*. Serologic techniques may be useful but cannot be considered absolute. Demonstration of the organism in stained sputum and preparations of biopsy or necropsy material, especially if confirmed by immunofluorescence, can be reliable affirmation of the diagnosis in the absence of culture. But since various "bodies" seen in stained material may resemble *Histoplasma* yeast cells, the isolation of the organism in culture remains the absolute criterion for diagnosis.

Direct Examination. Detection of the organism in sputum by direct examination is difficult. The KOH procedure, which is useful in other fungus diseases, is usually negative in histoplasmosis. Sputum is a more useful specimen when cultured. Sputum, the buffy coat of centrifugal blood specimens, the sediment of such specimens, biopsy material, and sternal puncture material are best examined following staining. The material is spread on a slide and fixed for ten minutes

with methyl alcohol. The fixed material is then stained either by the Wright or Giemsa method. From personal experience, it is felt that the Giemsa method is superior. Overstaining tends to emphasize the yeast cells. The organism is seen within macrophages and is 2 to 4 μ in diameter. It is usually ovoid, with the bud at the smaller end. The large-form cells of the var. *duboisii* are described in Chapter 16. There is a halo of unstained material around the cell, representing the cell wall. Within the cell there is a large vacuole, and a crescent-shaped, red-staining mass of protoplasm is usually present at the larger terminus of the cell. The buds are delicate and usually detach during the staining procedure. Though they are predominantly in macrophages or monocytes, yeast cells may appear extracellular because of destruction of the phagocytic cell. Many other yeasts, foreign bodies, artifacts, and parasites may simulate the appearance of *H. capsulatum* yeast cells.

Culture. In cases of primary pulmonary histoplasmosis, the most valuable specimen for isolation of the organism is sputum. The first morning sputum is collected after the patient has rinsed out his mouth. Needless to say, the organism is found in sputum; saliva will not be useful for culture. Gastric lavage done early in the morning before feeding the patient may also be rewarding. Sputum is planted directly on blood agar and Sabouraud's agar plates and incubated at 25°C. Areas of sputum that are purulent or blood-streaked are selected for culture. Gastric lavage is centrifuged, and the sediment is planted on he same media as sputum. From personal experience and that communicated by Larsh, blood agar without antibiotics is the medium of choice. With diligent and frequent examination of such plates, the tiny colonies of *H. capsulatum* can be subcultured, even when there is abundant growth of contaminating bacteria and *C. albicans*. The medium recently developed by Smith and Goodman,[132] however, is recommended as the best means of recovering the organism from heavily contaminated material or sputum containing numbers of *Candida albicans* yeast cells. It is a yeast extract agar with ammonium hydroxide. The latter kills or inhibits many bacteria, yeasts, and saprophytic fungi, and it neutralizes acid produced by the yeasts. This extract has been found useful for the isolation of all pathogenic fungi, including *Pseudallescheria boydii* from sputum, and should be used for routine sputum cultures for fungi.

Other pathologic material, such as biopsy specimens, sternal punctures, aspirations, and so forth, are also minced and spread over blood agar and Sabouraud's agar plates. Plates should be sealed with tape, or stacks of them enclosed in plastic bags to prevent drying out. All plates should be kept for from six to twelve weeks. Any colony showing hyphal growth is subcultured for further identification. The initial colonies on blood agar are usually glabrous, cerebriform, and pinkish to reddish brown. In time they become mycelial and white to brownish. The colony is indistinguishable from that of *B. dermatitidis* and many other fungi. Demonstration of the characteristic tuberculate macroconidia and conversion to the yeast phase are necessary for identification. The latter is accomplished by growth on blood agar slants at 37°C. The exoantigen procedure can also be used.

MYCOLOGY

Histoplasma capsulatum Darling 1906

Synonymy. *Cryptococcus capsulatus* (Darling) Castellani et Chalmers 1919; *Torulopsis capsulatus* (Darling) Almeida 1933; *Posadasia capsulata* (Darling) Moore 1934; *Pasadasia pyriforme* Moore 1934; *Histoplasma pyriforme* Dodge 1935.

Teleomorph State. *Ajellomyces capsulatus* (Kwon-Chung) McGinnis et Katz 1979.[94]

Synonymy. *Emmonsiella capsulata* Kwon-Chung 1972.

Colony Morphology. SGA, 25°C. The colony that develops on Sabouraud's agar is initially white or buff-brown. These two morphologic types have been termed A (albino) and B (brown) by Berliner[12] (Fig. 15–15). Both types may be isolated from the same patient, and with subculture the brown type shows sectoring and eventual conversion to the albino type. As reported by Berliner, these types remain constant when the organism is kept in the yeast form. The brown type has more numerous tuberculate macroconidia than the white.

Microscopic Morphology. SGA, 25°C. The characteristic tuberculate macroconidia are seen on examination of hyphae (Fig. 15–16, *A*). They are borne on narrow tubular conidiophores, 8 to 14 μ in diameter, and are round or pyriform (Fig. 15–17, *A*). The tubercles or fingerlike projections are quite variable in size and morphology, and in subculture only about 30 per cent of the macroconidia show tubercles (Fig. 15–18). With special stains they appear to be mucopolysaccharide in nature, and they do not contain cytoplasm. The saprophytic fungus *Sepedonium* also produces tuberculate macroconidia but no microconidia (see sec-

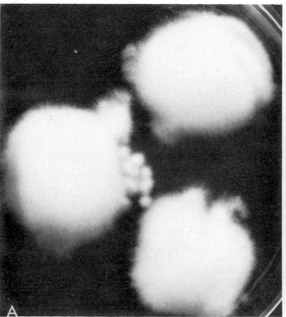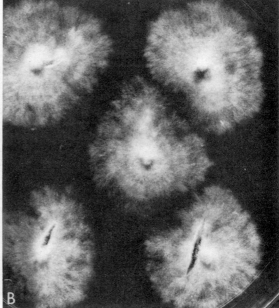

Figure 15–15. Histoplasmosis. *A,* Type A (albino) colony form. *B,* Type B (brown) colony form of *H. capsulatum.* There is sectoring of B to A. (Courtesy of M. Berliner.)

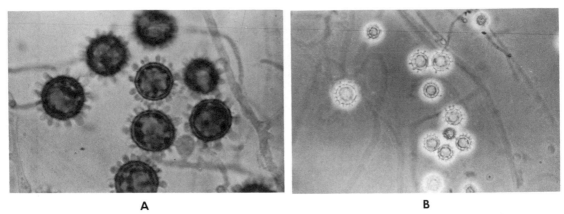

A   B

Figure 15–16. *A,* Histoplasmosis. Characteristic macroconidia of *H. capsulatum. B, Chrysosporium* state of *Renispora flavissima.* This organism is almost identical in culture characteristic and conidial morphology to *Histoplasma capsulatum.* Slide culture 12 days, ×830. (Sigler, L., P. K. Gaur, et al. 1979. *Renispora flavissima,* a new gymnoascaceous fungus with tuberculate chrysosporium conidia. Mycotaxon *10*:133–141.)[131]

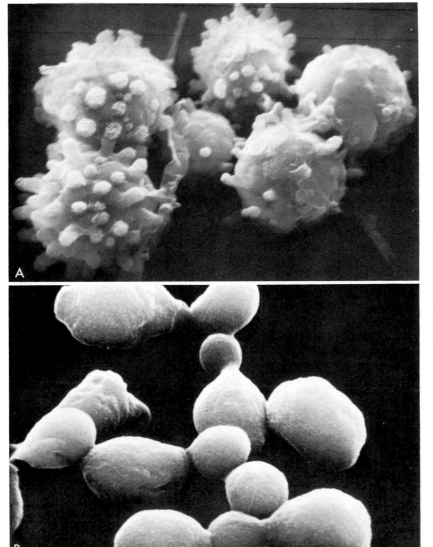

Figure 15–17. Scanning electron micrographs of the conidial (*A*) and yeast (*B*) phases of *Histoplasma capsulatum.* The tuberculations (echinulations) are well demonstrated in *A.* In *B,* the narrow neck of the budding yeast of *H. capsulatum* is apparent. (Courtesy of R. Garrison.)

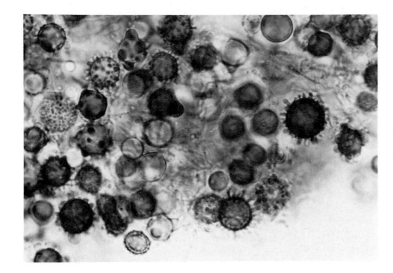

Figure 15–18. Histoplasmosis. Variation in size and morphology of macroconidia. (From Berliner, M. 1968. Sabouraudia, *6*:111–118.)[12]

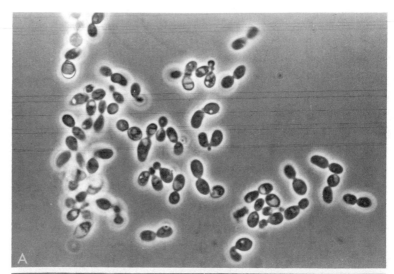

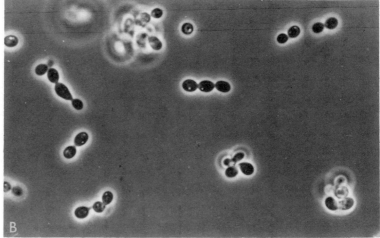

Figure 15–19. Histoplasmosis. Microscopic morphology of yeast form cells from culture. ×800. (Courtesy of M. Berliner.)

tion on contaminants), nor is it dimorphic or pathogenic for animals. The latter organism is found in soil and parasitizes fleshy fungi. As recently pointed out by Sigler and Carmichael, there are many fungi with *Chrysosporium*-type conidia that may resemble the anamorph state of *Histoplasma capsulatum*. One recently described by them is the *Chrysosporium* state of *Renispora flavissima* (Fig. 15–16, *B*). Microconidia are also produced and are particularly abundant in fresh isolants of *H. capsulatum*. These conidia are borne at the tips of short, narrow phores and at right angles to the vegetative hyphae. They are spherical, 2 to 4 μ in diameter, and occasionally there is a secondary lateral conidium that sometimes appears to bud, giving the unit a dumbbell shape. The walls are smooth but often somewhat roughened. After many subcultures, the white mycelial growth produces few conidia of either type.

Colony Morphology and Microscopic Morphology. Blood agar, 37°C. The yeast form of *H. capsulatum* does not develop directly from either type of conidium when transferred to an incubation temperature of 37°C but arises from within the mycelium itself.[68] Conidia first germinate and then convert to the yeast phase. The colony produced is at first glabrous and tenacious. Examination shows it to be a mixture of pseudohyphae, hyphae in the process of conversion, and freely budding yeast cells. On subsequent transfer, the colony becomes white, smooth, and yeastlike. The cells are now oval, budding yeasts. The average size is 2 to 3 × 3 to 4 μ (Fig. 15–19). Under certain conditions, such as growth in tissue explants, a size of 20 μ may be attained.[129] The buds occur at the narrow end of the oval yeast cell. They have a narrow neck (0.2 to 0.3 μ) (Fig. 15–17, *B*), and the attachment is often drawn out as a narrow thread. This characteristic is useful in differentiating it from small forms of *B. dermatitidis*, which have a broad-based bud. Giemsa stained cells show that a *H. capsulatum* yeast cell contains a single nucleus with a densely staining nucleolus. The yeast cells of *B. dermatitidis* contain from six to eight nuclei. As reported by Howard,[70, 71] the generation time

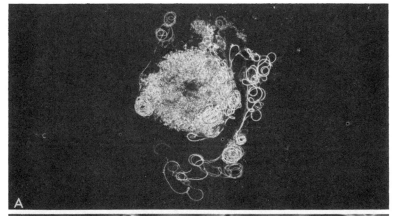
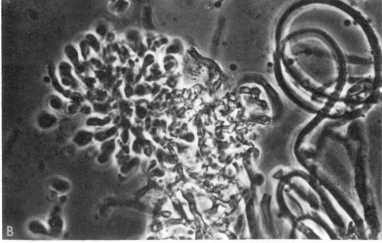

Figure 15–20. *Ajellomyces capsulatus.* Coiled hyphae of mated strains. (Courtesy of M. Berliner.)

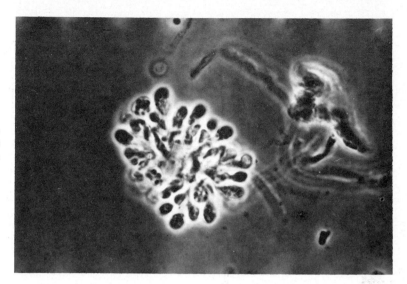

Figure 15–21. *Ajellomyces capsulatus.* Mature gymnothecium showing asci and ascospores. (Courtesy of M. Berliner.)

within mouse histiocytes is 11 hours. Comparable times *in vitro* are evident by the growth curve studies of Reca and Campbell.[113]

Ajellomyces capsulatus (Kwon-Chung) McGinnis et Katz 1979

The teleomorph state of *H. capsulatum* was first reported to be *Gymnoascus demonbreunii*, a homothallic fungus, by Ajello and Cheng.[2] It appears that this was a soil keratinophilic contaminant. The work of Kwon-Chung[86, 87] conclusively demonstrated that *H. capsulatum* has a teleomorph state. She named it *Emmonsiella capsulata*. Recently McGinnis and Katz[94] have restudied the organism and concluded it could be placed in the genus *Ajellomyces* along with *A. dermatitidis*, the teleomorph state of *Blastomyces dermatitidis*. *A. capsulatus* is heterothallic and morphologically consistent with fungi classed in the family Gymnoascaceae of the Ascomycota. In mated pairs, tightly coiled hyphae radiate from a common source at the base of a naked young ascocarp (Fig. 15–20). In appearance *A. capsulatus* is similar to the teleomorph stage of *B. dermatitidis (Ajellomyces dermatitidis)* but differs from it in that the highly branched hyphae arising from such coils are irregularly curved and are not constricted at the cross walls.[87] The mature gymnothecia are globose, have a buffy pigment, and range in size from 80 to 250 μ in diameter (Fig. 15–21). The asci are pyriform. The ascospores are globose, eight in number, hyaline, smooth, and 1.5 μ in diameter. The mating type distribution from soil isolants is approximately 1:1 for the (+) and (−) mating types. Isolants from clinical disease, however, are predominantly of the − mating type.[87] This may indicate a difference in virulence between the mating types of *A. capsulatus.* Gymnothecia are also produced when tester strains of *A. capsulatus* are paired with strains of *H. capsulatum* var. *duboisii*. Aberrant ascospores are produced, and their viability has at this writing not been ascertained.

REFERENCES

1. Ajello, L. 1970. *The Medical Mycological Iceberg.* Proc. Int. Sympos. Mycoses, Sci. Pub. PAHO No. *205*:3–10.
2. Ajello, L., and S. L. Cheng. 1967. Sexual reproduction in *Histoplasma capsulatum.* Mycologia, *59*:689–697.
3. Ajello, L., and L. C. Runyon. 1953. Infection of mice with single spores of *Histoplasma capsulatum.* J. Bacteriol., *66*:34–40.
4. Ajello, L., E. W. Chick, et al. 1971. *Histoplasmosis.* Proceedings of Second National Conference. Springfield, Ill., Charles C Thomas.
5. Ajello, L., E. S. Kuttin, et al. 1977. Occurrence of *Histoplasma capsulatum* Darling 1906 in Israel with a review of the current status of histoplasmosis in the middle east. Am. J. Trop. Med. Hyg., *26*:140–147.
6. Ajello, L., P. E. C. Manson-Bahr, et al. 1960. Amboni caves, Tanganyika. A new endemic area for *Histoplasma capsulatum.* Am. J. Trop. Med. Hyg., *9*:633–637.
7. Artz, R. P., and W. E. Bullock. 1979. Immunoregulatory responses in experimental disseminated histoplasmosis depression of T-cell-dependent and T-effectory responses by activation of splenic suppressor cells. Infect. Immunol., *23*:893–902.
8. Baker, R. D. 1964. Histoplasmosis in routine autopsies. Am. J. Clin. Pathol., *41*:457–470.
9. Bauman, D. S. 1971. Physiology of *Histoplasma capsulatum. In Histoplasmosis.* Proceedings of Second National Conference. Ajello, L., E. W.

Chick, et al. (eds.), Springfield, Ill., Charles C Thomas, Chapter 12.

10. Bauman, D. S., and C. D. Smith. 1975. Comparison of Immunodiffusion and Complement fixation tests in the diagnosis of histoplasmosis. J. Clin. Microbiol., 2:77–80.

11. Bent, A. C. 1950. Life histories of North American wagtails, shrikes, verios and their allies. U.S. Nat. Hist. Bull. 197.

12. Berliner, M. 1968. Primary subcultures of *Histoplasma capsulatum*. 1. Macro- and micromorphology of the mycelial phase. Sabouraudia, 6:111–118.

13. Berliner, M. D., and M. E. Reca. 1966. Vital staining of *Histoplasma capsulatum* with Janus Green B. Sabouraudia, 5:26–29.

14. Berliner, M. D., and M. Reca. 1969. Protoplasts of systemic dimorphic fungal pathogens. Mycopathologia, 37:81–85.

15. Boguslawski, G., and D. A. Stegler. 1979. Aspects of the physiology of *Histoplasma capsulatum* (a review). Mycopathologia, 67:17–24.

16. Bradsher, R. W., C. G. Wickre. 1980. *Histoplasma* endocarditis, cured by amphotericin B combined with surgery. Chest, 78:791–795.

17. Burgisser, von H., R. Fankhauset, et al. 1961. Mykose bei einem Dachs in der Schweiz histologisch Histoplasmose. Pathol. Microbiol., 24:794–802.

18. Burk, R. L., and B. D. Jones. 1978. Disseminated histoplasmosis with osseous involvement in a dog. J. Am. Vet. Med. Assoc., 172:1416–1417.

19. Campbell, C. C. 1965. Problems associated with antigenic analysis of *Histoplasma capsulatum* and other mycotic agents. Am. Rev. Respir. Dis., 92:113–118.

20. Campbell, C. C. 1967. Serology in the respiratory mycoses. Sabouraudia, 5:240–259.

21. Campbell, C. C. 1971. History of the development of serologic tests for histoplasmosis. *In Histoplasmosis*. Proceedings of Second National Conference. Ajello, L., E. W. Chick, et al. (eds.), Springfield, Ill., Charles C Thomas, Chapter 42.

22. Campbell, C. C., G. B. Hill, et al. 1962. *Histoplasma capsulatum* isolated from a feather pillow associated with histoplasmosis in an infant. Science, 136:1050–1051.

23. Carmona, F. J., 1971. Analisis estadistico y ecologio-epidemiologico de la sensibilidad a la histoplasmina en Colombia, 1950–1968. Antioquia Med., 21:109–154.

24. Chandler, J. W., T. K. Smith, et al. 1969. Immunology of the mycoses. II. Characterization of immunoglobulin and antibody responses in histoplasmosis. J. Infect. Dis., 119:247–254.

25. Chase, H. V., P. J. Kadull, et al. 1968. Comparison of histoplasmin tine test with histoplasmin Mantoux. Am. Rev. Respir. Dis., 9:1059–1059.

26. Chen, W. N., and O. B. James. 1972. Histoplasmosis in Jamaica. West Indian Med. J., 21:220–224.

27. Christie, A., and J. C. Peterson. 1945. Pulmonary calcification in negative reactors to tuberculin. Am. J. Public Health, 35:1131–1147.

28. Cintron, G. B., J. A. Snow, et al. 1977. Pericarditis mimicking tricuspid valvular disease. Chest, 71:770–772.

29. Class, R. V., and F. S. Cascio. 1972. Histoplasmosis presenting as acute polyarthritis. N. Engl. J. Med., 287:1133–1134.

30. Comstock, G. W. 1959. Histoplasmin sensitivity in Alaskan natives. Am. Rev. Respir. Dis., 79:542.

30a. Connell, S. V., and J. R. Muhm. 1976. Radiologic manifestations of pulmonary histoplasmosis: A ten year review. Radiology, 121:281–285.

31. Cooke, N. T. 1928. The spread of the European starling in North America. U.S. Dept. Agriculture Circular No. 40.

32. Cott, G. R., T. W. Smith, et al. 1979. Primary cutaneous histoplasmosis in immunosuppressed patient. J.A.M.A., 242:456–457.

33. Couch, J. R., N. I. Abdou, et al. 1978. Histoplasma meningitis with hyperactive T cells in cerebrospinal fluid. Neurology, 28:119–123.

34. Crispell, K. R., W. Parson, et al. 1956. Addison's disease associated with histoplasmosis. Report of four cases and review of the literature. Am. J. Med., 20:23–29.

35. Daniels, L. S., M. D. Berliner, et al. 1968. Varying virulence in rabbits infected with different filamentous types of *Histoplasma capsulatum*. J. Bacteriol., 96:1535–1539.

36. Darling, S. T. A. 1906. A protozoan general infection producing pseudotubercles in the lungs and focal necrosis in the liver, spleen and lymph nodes. J.A.M.A., 46:1283–1285.

37. da Rocha-Lima, H. 1912–1913. Beitrag zur kenntnis der Blastomykoses Lymphangitis epizootica und Histoplasmosia. Zentralbl. Bakteriol., 67:233–249.

38. Davies, S. F., M. Khan, et al. 1978. Disseminated histoplasmosis in immunologically suppressed patients occurrence in nonendemic area. Am. J. Med., 64:94–100.

39. de Monbreun, W. A. 1934. The cultivation and cultural characteristics of Darling's *Histoplasma capsulatum*. Am. J. Trop. Med., 14:93–135.

40. DiSalvo, A. F., and W. M. Johnson. 1979. Histoplasmosis in South Carolina: support for the microfocus concept. Am. J. Epidemiol., 109:480–492.

41. DiSalvo, A. F., L. Ajello, et al. 1969. Isolation of *Histoplasma capsulatum* from Arizona bats. Am. J. Epidemiol., 89:606–614.

42. Dismukes, W. E., S. A. Royal, et al. 1978. Disseminated histoplasmosis in corticosteroid treated patients: Report of five cases. J.A.M.A., 240:1495–1498.

43. Dodd, K., and E. H. Tomkins. 1934. A case of histoplasmosis of Darling in an infant. Am. J. Trop. Med., 14:127–137.

44. Dodge, H. J., L. Ajello, et al. 1965. The association of a bird roosting with infection of school children by *Histoplasma capsulatum*. Am. J. Public Health, 55:1203–1211.

45. Domer, J. E. 1971. Monosaccharide and chitin content of cell wall of *Histoplasma capsulatum* and *Blastomyces dermatitidis*. J. Bacteriol., 107:870–877.

46. Domer, J. E., J. G. Hamilton, et al. 1967. Comparative study of the cell walls of the yeastlike and mycelial phase of *Histoplasma capsulatum*. J. Bacteriol., 94:466–474.

46a. Domer, J. E., and S. A. Moser. 1980. Histoplasmosis. A review. Rev. Med. Vet. Mycol., 15:159–182.

47. Drutz, D. J. 1979. Urban coccidioidomycosis and histoplasmosis. Sacramento and Indianapolis. New Engl. J. Med., *301*:381–382.

48. Drutz, D. J., A. Spikard, et al. 1968. Treatment of disseminated mycotic infections. Am. J. Med., *45*:405.

49. Ehrhard, H. B., and L. Pine. 1972. Factors influencing the production of H and M antigens by *Histoplasma capsulatum:* Effect of physical factors and composition of medium. Appl. Microbiol., *23*:250–261.

50. Emmons, C. W. 1949. Isolation of *Histoplasma capsulatum* from soil. Public Health Rep., *64*:892–896.

51. Fouts, J. B., R. E. Brashear, et al. 1976. Chronic pulmonary histoplasmosis with declining complement fixation titers and persistence of positive sputum culture. Am. Rev. Respir. Dis., *113*:677–682.

52. Furcolow, M. L. 1958. Recent studies on the epidemiology of histoplasmosis. Ann. N.Y. Acad. Sci., *72*:127–164.

53. Furcolow, M. L. 1963. Tests of immunity in histoplasmosis. New Engl. J. Med., *268*:357–361.

54. Garrison, R. G. 1970. The uptake of low molecular weight sulfur containing compounds by *Histoplasma capsulatum* and related dimorphic fungi. Mycopathologia, *40*:171–180.

55. Garrison, R. G., and J. W. Lane. 1973. Scanning-beam electron microscopy of the conidia of brown and albino filamentous varieties of *Histoplasma capsulatum*. Mycopathologia, *49*:185–191.

56. Gaur, P. K., and R. W. Lickwardt. 1980. Mating types and convertibility among soil isolates of *Histoplasma capsulatum*. Mycologia, *72*:404–405.

57. Gaur, P. K., and R. W. Lickwardt. 1980. Preliminary visual screening of soil samples for the presumptive presence of *Histoplasma capsulatum*. Mycologia, *72*:259–269.

58. Gaynes, R. P., P. Gardner, and W. Causey. 1981. Prosthetic valve endocarditis caused by *Histoplasma capsulatum*. Arch. Intern. Med., *141*:1533–1537.

59. Giessel, M., and J. M. Rau. 1980. Primary cutaneous histoplasmosis; a new presentation. Cutis, February, 1980, pp. 152–154.

60. Goodwin, R. A., and J. D. Snell. 1969. The enlarging histoplasmona. Am. Rev. Respir. Dis., *100*:1–12.

61. Goodwin, R. A., and R. M. Des Prez. 1978. Histoplasmosis. Am. Rev. Respir. Dis., *117*:929–956.

62. Goodwin, R. A., J. L. Shapiro, et al. 1980. Disseminated histoplasmosis: clinical and pathologic correlations. Medicine, *59*:1–31.

63. Gregoriades, D. G., H. L. Landeluttig, et al. 1961. Pericarditis with massive effusion due to histoplasmosis. J.A.M.A., *178*:331–334.

64. Hall, A. D. 1979. An equine abortion due to histoplasmosis. Vet. Med. Small Anim. Clin., *74*:200–201.

65. Heiner, D. C. 1958. Diagnosis of histoplasmosis using precipitin reactions in agar gel. Pediatrics, *22*:616–629.

66. Hill, G. B., and C. C. Campbell. 1962. Commercially available histoplasmin sensitized latex particles in an agglutination test for histoplasmosis. Mycopathologia, *18*:169–179.

67. Hansmann, G. H., and J. R. Schenken. 1934. A unique infection in man caused by a new yeast-like organism, a pathogenic member of the genus. *Sepedonium*. Am. J. Pathol., *10*:731–738.

68. Howard, D. H. 1962. The morphogenesis of the parasitic forms of dimorphic fungi. Mycopathologia, *18*:127–139.

69. Howard, D. H. 1965. Intracellular growth of *Histoplasma capsulatum*. J. Bacteriol., *89*:518–523.

70. Howard, D. H. 1967. Effect of temperature of intracellular growth of *Histoplasma capsulatum*. J. Bacteriol., *93*:438–444.

71. Howard, D. H. 1973. Fate of *Histoplasma capsulatum* in guinea pig polymorphonuclear leukocytes. Infect. Immun., *8*:419–421.

72. Howard, D. H., V. Otto, et al. 1971. Lymphocyte mediated cellular immunity in histoplasmosis. Infect. Immun., *4*:605–610.

73. Howard, D. H., and V. Otto. 1977. Experiments on lymphocyte-mediated cellular immunity in murine histoplasmosis. Infect. Immun., *16*:226–331.

74. Hughes, F. A., C. E. Eastbridge, et al. 1971. Surgical treatment of pulmonary histoplasmosis. *In* Ajello, L., E. W. Chick, et al. (eds.) *Histoplasmosis*. Proceedings of Second National Conference. Springfield, Ill., Charles C Thomas, Chapter 54.

75. Johns, L. E., R. G. Garrison, et al. 1973. Bronchopleurocutaneous fistula due to infection with *Histoplasma capsulatum*. Chest, *63*:638–641.

76. Kaplan, W. 1971. Application of the fluorescent antibody technique to the diagnosis and study of histoplasmosis. *In* Ajello, L., E. W. Chick, et al. (eds.), *Histoplasmosis*. Proceedings of Second National Conference. Springfield, Ill., Charles C Thomas, Chapter 41.

77. Kaplan, W., L. Kaufman, et al. 1972. Pathogenesis and immunological aspects of experimental histoplasmosis in cynomolgus monkeys *(Macaca fascicularis)*. Infect. Immun., *5*:847–853.

78. Katzenstein, A. L., and A. A. Liebow. 1974. Bronchocentric granulomatosis, mucoid impaction and hypersensitivity reaction to fungi. Am. Rev. Respir. Dis., *111*:497–537.

79. Kauffman, C. A., K. S. Israel, et al. 1978. Histoplasmosis in immunosuppressed patients. Am. J. Med., *64*:923–932.

80. Kaufman, L., and S. Blumer. 1966. Occurrence of serotypes among *Histoplasma capsulatum* strains. J. Bacteriol., *91*:1434–1439.

81. Kessel, B. 1953. Distribution and migration of the European starling in North America. Condor, *55*:49–67.

82. King, R. W., S. Kraikitpanitch, et al. 1977. Subcutaneous nodules caused by *Histoplasma capsulatum*. Ann. Intern. Med., *86*:586–587.

83. Kirchner, S. G., R. M. Heller, et al. 1978. The radiologic features of *Histoplasma* pericarditis. Pediatr. Radiol., 7:7–9.

84. Kligman, A. M. 1951. The Hotchkiss-McManus stain for the histopathologic diagnosis of fungus disease. Am. J. Clin. Pathol., *21*:86–91.

85. Klite, P. D., and F. H. Dierchel. 1965. *Histoplasma capsulatum* in fecal contents and organs of bats in the Canal Zone. Am. J. Trop. Med. Hyg., *14*:433–439.

86. Kwon-Chung, K. J. 1972. Sexual stage of *Histoplasma capsulatum*. Science, *175*:326.

87. Kwon-Chung, K. J. 1973. Studies on *Emmonsiella capsulata* I Heterothallism and development of the ascocarp. Mycologia, *65*:109–121.

88. Larrabee, W. F., L. Ajello, et al. 1978. An epidemic of histoplasmosis on the isthmus of Panama. J. Trop. Med. Hyg., *27*:281–285.

89. Larsh, H. W., A. Hinton, et al. 1953. Efficacy of the flotation method in isolation of *Histoplasma capsulatum* from soil. J. Lab. Clin. Med., *41*:478–485.

90. Larsh, H. W., and P. A. Bartels. 1970. Serology of histoplasmosis. Mycopathologia, *41*:115–131.

91. Little, J. A. 1960. Benign primary pulmonary histoplasmosis: a common cause of unexplained fever in children. South Med. J., *53*:1238–1240.

92. Londero, A. T., and C. D. Ramos. 1978. The status of histoplasmosis in Brazil. Mycopathologia, *64*:153–156.

93. Mantovani, A., A. Mazzoni, et al. 1968. Histoplasmosis in Italy. 1, Isolation of *Histoplasma capsulatum* from dogs in the province of Bologna. Sabouraudia, *6*:163–164.

94. McGinnis, M. R., and B. Katz. 1979. *Ajellomyces* and its synonym *Emmonsiella*. Mycotaxon, *8*:157–164.

95. Nicholas, W. M., J. A. Weir, et al. 1961. Serologic effects of histoplasmin skin testing. Am. Rev. Respir. Dis., *83*:276–279.

96. Negroni, P. 1940. Estudio micológico del primer caso Argentino de histoplasmosis. Rev. Inst. Bacteriol. Malbran., *9*:239–294.

97. Negroni, P. 1965. *Histoplasmosis*. Springfield, Ill., Charles C Thomas.

98. Orchard, J. L., F. Laparello, et al. 1979. Malabsorption syndrome occurring in the course of disseminated histoplasmosis: case report and review of gastrointestinal histoplasmosis. Am. J. Med., *66*:331–336.

99. Palmer, C. E. 1945. Nontuberculous pulmonary calcification and sensitivity to histoplasmin. Public Health Rep., *60*:513–520.

100. Palmer, C. E. 1946. Geographic differences in sensitivity to histoplasmin among student nurses. Public Health Rep., *61*:475–487.

101. Parsons, R. J., and C. Zarafonetis. 1945. Histoplasmosis in man. A report of 7 cases and a review of 71 cases. Arch. Intern. Med., *75*:1–23.

102. Peiffer, R. L., and P. V. Belkin. 1979. Ocular manifestations of disseminated histoplasmosis in a cat. Feline Practice, *9*:24–29.

103. Phelps, B. M., and F. B. Mallory. 1926. Toxic cirrhosis and primary liver cell carcinoma complicated by histoplasmosis of the lung. Fifteenth Ann. Rep. Med. Depart. United Fruit Co. 1926, 115–123.

104. Phillips, J. C. 1928. Wild Birds introduced or transplanted. Tech. Bull. No. 61 USDA p. 54–55.

105. Pine, L. 1960. Morphological and physiological characteristics of *Histoplasma capsulatum*. In Sweany, H. D. (ed.), *Histoplasmosis*. Springfield, Ill., Charles C Thomas, pp. 40–75.

106. Pine, L. 1970. Growth of *Histoplasma capsulatum*. VI. Maintenance of the mycelial phase. Appl. Microbiol., *19*:413–420.

106a. Pine, L., and C. J. Boone. 1968. Cell wall composition and serologic reactivity of *Histoplasma* serotypes and related species. J. Bacteriol., *96*:789–798.

107. Plouffe, J. F., and R. J. Fass. 1980. Histoplasma meningitis: Diagnostic value of cerebrospinal fluid serology. Ann. Intern. Med., *92*:189–191.

108. Polk, J. W. 1971. Surgery for cavitary histoplasmosis. In Ajello, L., E. W. Chick, et al. (eds.), *Histoplasmosis*. Proceedings of Second National Conference. Springfield, Ill., Charles C Thomas, Chapter 55.

109. Proceedings of the Seminar on Histoplasmosis. September 13, 1948. Bethesda, Md., National Institutes of Health.

110. Procknow, J. J., M. I. Page, et al. 1960. Early pathogenesis of experimental histoplasmosis. Arch. Pathol., *69*:413–426.

111. Quiñones, F., J. P. Koplan, et al. 1978. Histoplasmosis in Belize, Central America. Am. J. Trop. Med. Hyg., *27*:558–561.

112. Randhawa, H. S. 1970. Occurrence of histoplasmosis in Asia. Mycopathologia, *41*:75–89.

113. Reca, M. E., and C. C. Campbell. 1967. Growth curves with yeast phase of *Histoplasma capsulatum*. Sabouraudia, *5*:267–273.

114. Reddy, P. A., P. A. Brasher, et al. 1970. Peritonitis due to histoplasmosis. Ann. Intern. Med., *72*:79–81.

115. Riley, W. A., and C. J. Watson. 1926. Histoplasmosis of Darling: with report of a case originating in Minnesota. Am. J. Trop. Med., *6*:271–282.

116. Rippon, J. W. 1968. Monitored environment system to control growth, morphology, and metabolic rates in fungi by oxidation-reduction potential. Appl. Microbiol., *16*:114–121.

117. Rippon, J. W. 1980. Dimorphism in Pathogenic Fungi. CRC Critical Reviews in Microbiology, *8*:49–79.

118. Rowley, D. A., and M. Huber. 1955. Pathogenesis of experimental histoplasmosis in mice. I. Measurement of infecting dosages of the yeast phase of *Histoplasma capsulatum*. J. Infect. Dis., *96*:174–183.

119. Salfelder, K., and J. Schwartz. 1964. Cross reaction to *Histoplasma capsulatum* in mice. Sabouraudia, *3*:164–166.

120. Salfelder, K., et al. 1973. Multiple deep fungus infections. Curr. Top. Pathol., *57*:123–177.

121. Saliba, N. 1971. Amphotericin B, basic techniques and dosage for histoplasmosis. In Ajello, L., E. W. Chick et al. (eds.), *Histoplasmosis*. Proceedings of Second National Conference. Springfield, Ill., Charles C Thomas, Chapter 48.

122. Samovitz, M., and T. Dillon. 1970. Disseminated histoplasmosis presenting as exfoliative erythroderma. Arch. Dermatol., *101*:216–219.

123. San Blas, G., D. Oraz, et al. 1978. *Histoplasma capsulatum:* chemical variability of the yeast cell wall. Sabouraudia, *16*:279–284.

124. Saslaw, S., R. L. Perkins, et al. 1967. Histoplasmosis. To skin test or not to skin test. Proc. Soc. Exp. Biol. Med., *125*:1274–1277.

125. Scherr, G. H. 1957. Studies on the dimorphism of *Histoplasma capsulatum*. Exp. Cell Res., *12*:92–107.

126. Scherr, G. H., and J. W. Rippon. 1959. Experimental histoplasmosis in cold blooded animals. Mycopathologia, *11*:241–249.

127. Schwarz, J., F. N. Silverman, et al. 1955. The relation of splenic calcification to histoplasmosis. New Engl. J. Med., *252*:887–891.

128. Schwarz, J., G. L. Baum, et al. 1957. Successful infection of pigeons and chickens with *Histoplasma capsulatum*. Mycopathologia, *8*:189–193.

129. Schwarz, J. 1971. Histoplasmosis. *In* Baker, R. D. (ed.), *Handbuch der Speziellen Pathologischen Anatomie und Histologie Dritter Band Funfter Teil.* Berlin, Springer-Verlag.

130. Serosi, G. A., J. D. Parker, et al. 1971. Histoplasmosis outbreaks, their patterns. *In* Ajello, L., E. W. Chick, et al. (eds.), *Histoplasmosis.* Proceedings of Second National Conference. Springfield, Ill., Charles C Thomas, Chapter 16.

131. Sigler, L., P. K. Gaur, et al. 1979. *Renispora flavissima,* a new gymnoascaceous fungus with tuberculate chrysosporium conidia. Mycotaxon, *10*:133–141.

131a. Silverman, F. N. 1960. Roentgenographic aspects of histoplasmosis. *In* Sweeny, H. C. (ed.), *Histoplasmosis.* Springfield, Ill., Charles C Thomas, pp. 337–381.

132. Smith, C. D. and N. L. Goodman, 1975. Improved culture method for the isolation of *Histoplasma capsulatum* and *Blastomyces dermatitidis* from contaminated specimens. Am. J. Clin. Path., *63*:276–280.

132a. Sotgin, G., A. Montovani, et al. 1970. Histomosis in Europe. Mycopathologia, *41*:53–74.

133. Standard, P. G., and L. Kaufman. 1976. Specific immunological test for the rapid identification of members of the genus Histoplasma. J. Clin. Microbiol., *3*:191–199.

134. Stegler, D. A., and G. Boguslawski. 1979. Cysteine biosynthesis in a fungus histoplasma capsulatum. Sabouraudia, *17*:23–24.

135. Stickle, J. E., and T. N. Hribernik. 1978. Clinicopathological observations in disseminated histoplasmosis in dogs. J. Am. Anim. Hosp. Assoc., *14*:105–110.

136. Storch, G., J. C. Burford, et al. 1980. Acute histoplasmosis. Description of an outbreak in Northern Louisiana. Chest, *77*:38–42.

137. Straub, M., and J. Schwarz. 1960. General pathology of human and canine histoplasmosis. Am. Rev. Respir. Dis., *82*:528—541.

138. Strong, R. P. 1906. Study of some tropical ulcerations of skin with particular reference to their etiology. Philippine J. Sci., *7*:91–116.

139. Studdard, J. W., F. Sneed, et al. 1976. Cutaneous histoplasmosis. Ann. Rev. Respir. Dis., *113*:689–693.

140. Tesh, R. B., and J. O. Schneidan. 1966. Primary cutaneous histoplasmosis. New Engl. J. Med., *275*:597.

141. Velasco Castrejon, O., and A. Gonzalez Ochoa. 1974. Sindrome de coagulasion diseminada en histoplasmosis primaria. Rev. Invest. Salud. Publica., *34*:121–124.

142. Ward, J. I., M. Weeks, et al. 1979. Acute histoplasmosis: clinical epidemiological and serologic findings of an outbreak associated with exposure to a fallen tree. Am. J. Med., *66*:587–595.

143. Walter, J. E. 1969. The significance of antibodies in chronic histoplasmosis by immunoelectrophoretic and complement fixation test. Am. Rev. Respir. Dis., *99*:50–58.

144. Walter, J. E., and G. B. Price. 1968. Chemical, serologic and dermal hypersensitivity activities of two fractions of histoplasmin. Am. Rev. Respir. Dis., *98*:474–479.

145. Wilcox, K. R. 1958. The Walworth, Wisconsin, epidemic of histoplasmosis. Ann. Intern. Med., *49*:388–418.

146. Young, E. J., B. Vainrub, et al. 1978. Pericarditis due to histoplasmosis. J.A.M.A., *240*:1750–1751.

147. Zeidberg, L. D., L. Ajello, et al. 1952. Isolation of *Histoplasma capsulatum* from soil. Am. J. Public Health, *42*:930–935.

16 Histoplasmosis Duboisii

DEFINITION

Histoplasmosis duboisii, or African histoplasmosis, is a clinically distinct form of histoplasmosis. It is characterized by the presence of granulomatous and suppurative lesions, primarily of the cutaneous, subcutaneous, and osseous tissues. There usually is little evident involvement of the lung. The histologic picture is that of masses of large yeast forms within numerous giant cells. These points differ from the *capsulatum* type histoplasmosis, in which the lung is always involved and the yeast forms are small and intracellular in histiocytes. The etiologic agent has been called *Histoplasma duboisii*, although whether it is a separate species or variety of *H. capsulatum* remains to be determined.

It is unwise and often misleading to give a geographic designation to a disease. Therefore the term "histoplasmosis duboisii" is preferable to the name "African histoplasmosis," which is sometimes used. In fact, the disease has recently been recorded in Japan.[35]

HISTORY

In 1943, J. T. Duncan first observed a large yeast form of *Histoplasma* in the biopsy of an annular, papulocircinate skin lesion. The patient was an English mining engineer who worked in Ghana.[15, 16] Duncan found no cultural differences between the organism isolated and *H. capsulatum* but noted that the large yeast cells in tissue resembled *Blastomyces*. He did not choose to give the organism a special designation and did not publish his results until 1947.[14] In 1945, Catanei and Kervan described a large-celled type of histoplasmosis in a patient from the Sudan.[8] In retrospective studies it may be that histoplasmosis duboisii was the disease reported by Blanchard and Lefrou in 1922 as a saccharomycete infection in a Congolese patient. It also appears to be the "levures" found in a Senegalese by Brumpt in 1936.[9, 33] In 1952, Dubois isolated a fungus from a patient with cutaneous lesions. Also in 1952, Vanbreuseghem studied this fungus from material sent him by Dubois and designated it a new species *Histoplasma duboisii*.[14] However, in 1957, after study of several strains, Drouet named the etiologic agent *Histoplasma capsulatum* var. *duboisii*.[13]

ETIOLOGY, ECOLOGY, AND DISTRIBUTION

Whether the etiologic agent is a stable variant of *H. capsulatum* or a separate species has been the subject of much discussion and debate. Now that the perfect stage of *H. capsulatum* has been found (*Ajellomyces capsulatus*), and since it is heterothallic, it will be possible to mate the *duboisii* strains and resolve the issue. Taschdjian[32] has already shown that hyphal fusion of the two strains occurs, which may indicate they are the same species or closely related. In addition, preliminary studies by Kwon-Chung indicate that matings between the two are fertile.

Vanbreuseghem has summarized the differences between the *duboisii* type and the *capsulatum* type of histoplasmosis.[27, 33] He stressed that the characteristic giant forms, increased cell wall thickness, and lipoidal bodies of the yeast phase of var. *duboisii* were distinctly different from those of var. *capsulatum.* If experimental animals such as guinea pigs and hamsters are given large inocula of either type of yeast, they rapidly die, and small-form yeasts are seen in histologic sections. If small doses are used, by four to eight months only giant cells are found in animals inoculated with the *duboisii* strain, whereas small yeasts are present if var. *capsulatum* is injected.

In 1967 Al-Doory and Kalter[3] reported the isolation of a *duboisii* type *Histoplasma* from pooled soil samples obtained from a village near Darajani in Kenya, but the validity of their isolant is questioned by other investigators. So far this is the only record of the recovery of the organism from soil. Some human infections have been associated with chicken runs and bat-infested caves. These are the same ecologic environments associated with the *capsulatum* type histoplasmosis. In one case, a schoolgirl contracted the disease after sweeping out a schoolroom contaminated with bat guano.[5] A lesion developed at the site of an abrasion. This may be a case of either direct inoculation or hematogenous localization following inhalation of conidia.

So far almost all cases of var. *duboisii* (about a hundred) have been found in Africa be-

tween the two major deserts: the Sahara to the north and the Kalahari to the south. As pointed out by Vanbreuseghem,[33] the area is encompassed by 15 degrees latitude north and 10 degrees latitude south of the equator, and from the Senegal on the Atlantic coast to Uganda in the east (Fig. 16–1). This is almost identical to the endemic area of African trypanosomiasis, and is a region with a high average rainfall, high humidity, and little variation in diurnal temperature. A single possible case of the *duboisii* type has been reported from Japan.[35]

In the reviews by Cockshott and Lucas[9, 10] and Vanbreuseghem,[33] the age range of patients was quite variable. Patients were found to be from 2 years to 70 years of age, but there was a cluster of cases in the second decade. The male-female ratio was 2:1 or more. All races were affected, and, as in paracoccidioidomycosis, the possibility that there are many subclinical cases of disease exists. The endemic areas of histoplasmosis delineated by the few skin test surveys made in Africa with histoplasmin (*capsulatum*) do not necessarily correspond to areas in which *duboisii* type histoplasmosis has been found. Clinically identifiable *capsulatum* type histoplasmosis is not uncommon in many regions of Africa. Histoplasmin prepared from *duboisii* strains is sometimes negative in patients with the *duboisii* type infection, as is histoplas-

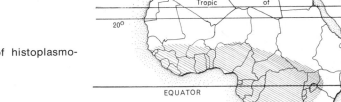

Figure 16–1. Incidence of histoplasmosis duboisii in Africa.

min prepared from *capsulatum* strains. Early infection or anergy cannot always be ruled out in such cases.

CLINICAL DISEASE

Histoplasmosis duboisii is still virtually an undefined disease. Cockshott and Lucas have tabulated about one hundred cases and these have presented a somewhat similar disease pattern.[9, 10] The cases fall roughly into three categories: (1) solitary lesions of the skin, subcutaneous tissue, or bone; (2) disseminated or multiple lesions of the above tissues and other organs; and (3) disease confined primarily to the lung.[12, 18] The questions of portal of entry and the presence of subclinical disease have been much discussed. If the pattern known for the other systemic mycoses holds true with this disease, the primary infection is pulmonary. This is the portal of entry in coccidioidomycosis, blastomycosis, paracoccidioidomycosis, and *capsulatum* type histoplasmosis, even though very rare primary cutaneous disease is known for all these entities. *Duboisii* yeast cells have been recovered several times from sputum and in one case in which there was no clinical or radiologic suggestion of a lung lesion. However, in the several necropsy examinations performed on infected patients, pathologists have been unable to discover any pulmonary focus of infection. It has been demonstrated in animal studies that the *duboisii* organism is not highly virulent. As is now presumed in paracoccidioidomycosis, the *duboisii* organism may be inhaled and then transported hematogenously to a favorable site for proliferation.

The clinical categories delineated by Cockshott and Lucas and others are presented here.

1. Pulmonary disease. The acute form of *duboisii* type histoplasmosis involving the lung and corresponding to *capsulatum* type disease has not yet been recognized. There are, however, a few reports of chronic progressive and cavitary disease in which the var. *duboisii* was involved. Derrien[12] noted multiple expanding cavities in a European living in Chad, and Gentilini[18] found hilar enlargement with no other pathology in a missionary who had worked in Africa. Pulmonary involvement has also been described in disseminated fatal infections.

2. Localized type. In these cases there is a single lesion which may be represented by a circumscribed skin lesion[17] (as in Duncan's first case), a subcutaneous granuloma,[17] or a single lesion of bone.[23] There are no signs of systemic disturbance, such as fever, anemia, or weight loss. The disease is chronic, and the lesions go through phases of quiescence and recrudescence. Often they heal spontaneously.

3. Disseminated type. As the term indicates, multiple lesions occur that may involve the skin, lymph nodes (Fig. 16–2, *A, B*), bones, intestine,[1] and abdominal viscera. Sys-

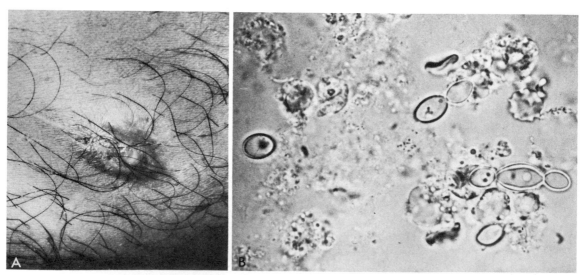

Figure 16–2. *A,* Draining sinus from enlarged matted area on left femoral lymph node. *B,* Direct mount from aspirate of cervical lymph node. Note large narrow-neck yeast cells. ×400. (From Nethercott, J. R., R. K. Schachter, et al. 1978. Arch. Dermatol., *114*:595–598.[23])

temic *duboisii* histoplasmosis often runs a rapidly progressive and fatal course, particularly in patients with hepatic and splenic involvement. The lesions occur randomly throughout the body. Of particular interest is the involvement of bone. In one series, skeletal lesions demonstrated by x-ray were present in two thirds of all cases.[10, 11] They are often multiple and resemble the cranial defects seen in multiple myeloma or those seen in the phalanges and carpals in sarcoidosis. Lymphadenopathy may be very prominent, as is seen in paracoccidioidomycosis. This may be confined to one area, such as the groin, or to several areas, or it may be generalized.

As would be expected in a systemic disease, fever, weight loss, and other signs of debilitation are present. Anemia may be quite severe, as there is often localization of the organisms in the marrow and loss of blood from skin ulcers. *Capsulatum* histoplasmosis involves the reticuloendothelial system and can be seen within histiocytes in material obtained by sternal puncture and elsewhere. The organism rarely deposits in bone. The *duboisii* type also involves the reticuloendothelial system but is found within giant cells and causes a destructive granulomatous process in the bones. Some writers have commented on the presence in infected patients of dysproteinemia, a low albumin/globulin ratio. However, in contrast to Europeans, this low a/g ratio is normal for Nigerians and other Africans. The white cell count is usually normal.

Cutaneous lesions are papular,[17] nodular, ulcerative, circinate, eczematoid, or psoriasiform. The verrucoid or chancriform lesion characteristic of primary inoculation mycosis is seldom described. In disseminated disease, numerous skin granulomata form and are seen to resolve as new ones evolve. In one case observed, the time from the first appearance of the lesions to their resolution was from two to four months. The lesions appeared first as flat papules that grew into dome-shaped nodules, which were paler than the dark Negro skin. The lesions were usually sessile, but a few became pedunculated. As they enlarged, pits were formed, and the central area became denuded, forming an ulcer with a rolled border. Some small lesions healed without ulceration, leaving a depigmented or hyperpigmented macule. Spontaneous healing in the center with annular expansion of the active border may mimic lesions seen in blastomycosis.

Subcutaneous lesions may arise from foci in the superficial flat bones, ribs, or skull, or they may develop without underlying bony involvement. They may be abscesses that suppurate and burrow to the skin or become freely moveable granulomata. The acute phase is one of swelling, tenderness, and pain, followed by evolution into a soft, fluctuant, cold abscess. This is similar to the development of subcutaneous lesions in disseminated blastomycosis.

Mucosal involvement occurs in a few cases. Lesions at this site are described as dome-shaped papules about 1 cm in diameter, which are found on the glossal or buccal mucosa.

The intestine has only rarely been involved. One patient died as the result of a perforation of an ileocecal granuloma.[9] However, intestinal involvement may be the only pathology observed in a few cases. Adekunle et al.[1] reported stricture of the jejunum, with tubercles scattered over the surface of the bowel and adjacent mesentery in a 36-year-old woman. The patient was cured with amphotericin B therapy.

Hepatic and splenic involvement has also been described in a few instances. Most postmortem cases have shown infiltration of the liver and spleen, with typical granulation tissue containing numerous giant cells and large yeasts.

Osseous involvement is a very common feature of disseminated *duboisii* histoplasmosis.[9, 10, 21] The marrow elements of the skeletal system are involved with particular frequency. Granulomatous lesions develop which destroy bone trabeculae and expand to erode cortical bone. If the periosteum is involved, it is at first lifted off the surface, and there is production of new extracortical bone. The granulomata often are not contained within the bony process but extend to adjacent soft tissue. In a survey of 56 cases, the skull was involved in 12, the jaw in 4, the scapula in 3, the clavicle in 3, the sternum in 2, the spinal column in 5, the arm in 3, the forearm in 10, the hand in 3, the thigh in 8, the leg in 6, and the foot in 4. Multiple lesions of the skull are the rule, and the ribs also may have several foci. Involvement of the spine may result in paraplegia. A "spinal block" occurred in one patient following destruction of the pedicle of the seventh dorsal vertebra with extension to the nerve trunk. The peculiar predilection of *duboisii* histoplasmosis for the bone is a characteristic of the disease.

Summary of the Differences Between *capsulatum* Histoplasmosis and *duboisii* Histoplasmosis. Histoplasmosis of the *capsulatum* type has two stages: (1) the primary, often subclinical, type is pulmonary and usually heals, leaving calcified opacities in the lung; (2) the second stage, in the rare patient who has generalized disease, is systemic spread to involve (in addition to the lungs) the liver, spleen, lymph nodes, and oro- and naso-pharynx. Bone involvement is very rare. The organism is most commonly seen within his-tiocytes.

The primary portal of entry for histo-plasmosis duboisii is not known. Although it is presumed to be the lung, the organ-ism usually does not remain there long enough to cause discernible pathology. The first clinical signs of disease are often single granulomata involving skin, lymph nodes, or bone. The disseminated disease involves mul-tiple lesions in bone, skin, lymph nodes, liver, and spleen. The bone, especially the marrow elements, is most commonly involved. The organisms are found within giant cells. The course of both forms of disseminated his-toplasmosis, if untreated, is progressive and fatal.

DIFFERENTIAL DIAGNOSIS

The single lesion of histoplasmosis duboisii resembles many other dermatologic condi-tions. These include eczema, lichen planus, psoriasis, the tineas, annular tuberculosis, Majocchi erythema annulare centrifugum, molluscum contagiosum, rodent ulcer, and warts. In disseminated disease, a tentative diagnosis is made by the appearance of multi-ple abscesses in bone, subcutaneous tissue, and pleomorphic skin lesions. Blastomycosis also exists in the endemic region and may present with a similar clinical picture. The yeast cells of both appear similar on direct examination of lesion material.

PROGNOSIS AND THERAPY

The isolated lesion of *duboisii* histoplasmo-sis may heal spontaneously but is amenable to surgical removal. Disseminated disease is quite grave and is generally fatal when liver and spleen are involved. Amphotericin B has been used successfully for treatment in many disseminated cases. The recommended dose is 0.25 mg to 1 mg per kg diluted in 5 per cent glucose. This is given as a slow infusion over six hours. Alternate-day therapy can be em-ployed. A total of 1 to 3 g is necessary for a full course of treatment. The side effects are frequent, and consist of fever, rigors, and headache, as well as nausea and vomiting.

Salicylates, barbiturates, and chlorpromazine are usually sufficient to control the side ef-fects. Cockshott and Lucas[9, 10] recommend administration of 200 mg sodium amytal and 100 mg chlorpromazine by mouth to adults one-half hour before infusion of amphoteri-cin B. This may be repeated after four hours. Dramatic results are seen within one week of therapy. A full curative course may require from 6 to 22 weeks. Amphotericin B has been used in combination with trimethoprim-sulfa.[17]

PATHOLOGY

Gross Pathology. In the necropsy studies that have been done on patients who have died of histoplasmosis duboisii, granuloma-tous and suppurative lesions have been found. Both may occur in the same individu-al. The lesions involve the cutaneous and subcutaneous tissues, bones, joints, and ab-dominal viscera. Parenchymal lung lesions are rare, and the alimentary tract is not often involved. The central nervous system is not involved, except when a vertebral lesion causes compression of the cord. Lesions in the skull do not penetrate the dura mater.

Histopathology. The histologic picture is quite characteristic. In contrast to the gran-uloma of *capsulatum* type histoplasmosis, in which giant cells are infrequent, the *duboisii* type lesion consists largely of aggregates of giant cells. These may be very large (up to 200 μ or more) and contain numerous ovoid, double contoured, walled yeast cells, with a diameter of 12 μ to 15 μ. The yeasts are sometimes present in chains of four or five cells (Fig. 16–3). They resemble the yeast cells seen in blastomycosis but lack the broad-based bud. Other types of inflammatory cells, especially neutrophils, may also be present. In caseating lesions, extracellular degenerat-ing yeasts may be found. These vary in size from very small bodies to large spheroidal empty cells that may be crenated and crescent-shaped. The fungus is visible with the usual hematoxylin-eosin stain but is better demonstrated by the methenamine silver, Gridley, or periodic acid-Schiff stains (Hotchkiss-McManus). Organisms may be rare in healing fibrotic lesions.

ANIMAL INFECTION

Natural infection in animals has been re-corded in baboons of the species *Papio papio*

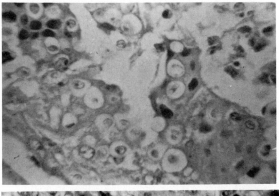

A

Figure 16–3. *A,* Histoplasmosis duboisii. Large yeast cells. The cytoplasm has shrunk away from the wall. Nuclei are visible. ×450. *B,* Midpower view of skin nodule from gluteal area of baboon in Lansing, Michigan, primate center. Giant cells are packed with large yeast cells. ×400. *C,* Low power view of same section stained with GMS.

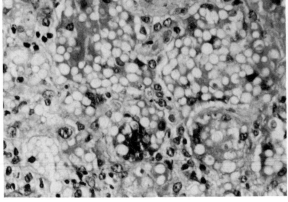

B

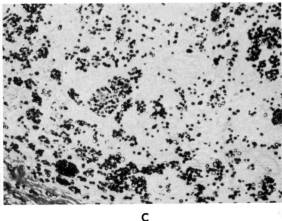

C

and *Papio cyanocephalus.*[22, 34] Both had granulomatous lesions of the skin (Fig. 16–3, *B, C*), subcutaneous tissue, and lymph nodes, and in one invasion of the bone occurred. In neither were the lungs or internal viscera involved, or at least disease was not detected. The granulomatous reaction with many giant cells observed in histopathologic sections was similar to that seen in human cases. In one case the animal had been in captivity for five years in England,[34] and in the report by Mariat and Segretain[22] the five animals with the disease had been at the Institut Pasteur in Paris for 18 months to two years. This indicates the long duration of the infection before clinical symptoms are observable. These animals had been captured in Gambia and Guinea. E. S. Beneke has found an infected baboon in the primate center at Lansing, Michigan. The lesions were in the gluteal area (Fig. 16–3, *B,C*).

Experimental infection in animals has been studied by a number of authors,[24, 33] who have shown the *duboisii* organism to be of relatively low virulence. Whereas *H. capsulatum* uniformly produces fatal infection in hamsters, this was rarely achieved by inoculation of the *duboisii* strains. Infection which spontaneously resolved was produced in guinea pigs, rabbits, and pigeons. When the fungus is first injected, small-cell yeasts (*capsulatum* type) are formed which are 2 to 5 μ. These are gradually replaced by the large cells of the *duboisii* type. Once the transition has occurred, no small cells are seen. Small coccoid bodies (0.5 to 1.5 μ) representing degenerating yeasts are visible in Kupffer cells. The initial cellular reaction is histiocytic but is replaced by giant cells, particularly in infections of hamsters and mice. Some of the giant cells are between 220 and 500 μ and contain huge numbers of organisms.

Biological Studies. Many physiologic studies have been carried out in an attempt to distinguish the *capsulatum* and *duboisii* forms of *Histoplasma.*[2, 6, 25, 26] Montemartini and Ciferri reported that histidine or ammonium sulfate could serve as a sole nitrogen source for *H. capsulatum,* but not for the *duboisii* strain. Urease production by *H. capsulatum* and lack of it by *duboisii* was reported by Coremans.[11] Rosenthal and Sokolsky[28] found tyrosine but not gelatin to be hydrolyzed by *H. capsulatum,* while the *duboisii* strains hydrolyzed gelatin but not tyrosine. Berliner[6] also found *H. capsulatum* was unable to hydrolyze gelatin, while *duboisii* strains were able to liquefy it within 24 to 96 hours. Blumer and Kaufman[7, 19, 20] reevaluated the work of Coremans, Rosenthal and Sokolsky, and

Berliner. They could not verify the findings of these investigators and concluded that strain variations were too great to allow differentiation of *duboisii* and *capsulatum* biochemically.

Transition of the two types to the yeastlike phase *in vitro* was extensively studied by Pine et al.[25] They found that both types of *Histoplasma* initially grow as small yeasts. However, the *duboisii* strains gradually transform into large, thick-walled yeasts, whereas the *capsulatum* strains remain typically small (2 to 5 μ). Once the *duboisii* strains transform to the larger size, their daughter cells are always of the large *duboisii* cell type. Both the *capsulatum* type and *duboisii* type cells may produce spheroplasts, which are degenerating forms, often devoid of cytoplasm, and subject to crenation. Such spheroplasts have been found in pus and necrotic tissues from patients with both types of infection. San Blas and Moreno[29] studied the biochemical composition of the cell wall of the var. *duboisii*. They found a variety of glucans and mannans but could not conclude whether the organism was a variety of *H. capsulatum* or a separate species.

IMMUNOLOGY AND SEROLOGY

Inherent immunity or immunity resulting from resolution of clinical or subclinical infections has not been established in histoplasmosis duboisii. Skin testing using histoplasmin *(capsulatum)* has been done extensively in Africa, and there are high reaction rates in some areas.[4, 30] These rates often do not correspond with the known cases of histoplasmosis duboisii. In the few clinical cases in which both *duboisii* histoplasmin and *capsulatum* histoplasmin have been used, the results were equivocal.[10, 15, 26]

Fluorescent antibody studies have shown a close relationship between the two types of *Histoplasma*. Originally it was thought that the two could be separated by "specific" antisera. However, Kaufman and Blumer[19, 20] through reciprocal cross-staining and adsorption procedures defined five serotypes of *H. capsulatum:* 1, 2; 1, 4; 1, 2, 3; 1, 2, 4; and 1, 2, 3, 4. This led to a reinvestigation of the *duboisii* strains. All *duboisii* strains were antigenically indistinguishable from the 1, 4 serotypes of *H. capsulatum.*

Ajello,[2] in reviewing the genus *Histoplasma*, was in agreement with Duncan,[15] who in 1958 said, "The species *H. duboisii* rests on a very

slender basis, merely its morphology under certain conditions of parasitic life and to some extent an analogous tendency to develop large cells in yeast phase in culture in response to unfavorable environmental factors, which seem to be more marked in this species than in *H. capsulatum*." Both authors concluded that the identity of the organism as a type or as a separate species should be preserved. The clinial disease produced by *duboisii* strains is directly different from that produced by *H. capsulatum*. Since the sexual mating types of *Histoplasma capsulatum* have been discovered, the matter of species separation will be settled presently.

LABORATORY IDENTIFICATION

Direct Examination. The organisms are usually numerous in pus from skin lesions, abscesses, draining sinuses from bone lesions, or biopsy material. In potassium hydroxide mounts, the yeasts appear as large (12 to 15 μ in diameter), thick-walled yeasts. They may have fat droplets within the cells, and the presence of broad-based buds has been reported in a few instances. The latter is a characteristic of the yeast cells of *Blastomyces dermatitidis,* which have been reported to cause disease in several parts of Africa also. It is therefore necessary to isolate the organism in culture for correct diagnosis. The *duboisii* organism can be seen occasionally in sputum when no clinical disease is evident.

Culture Methods. Material for culture can be placed on Mycosel agar or other antibiotic-containing media and incubated at 25°C. It grows slowly, producing a flat, brown, glabrous colony or a fluffy white to beige colony. Growth sometimes is not initiated for four to six weeks. In a manner similar to *H. capsulatum*, the *duboisii* strains grow on almost all laboratory media and can be converted to the yeast phase by inoculation on blood agar or KY agar[25] and then incubated at 37°C.

MYCOLOGY

Histoplasma capsulatum var. *duboisii* Drouhet 1957

Synonymy. *Histoplasma duboisii* Vanbreuseghem 1952.[14]

Colony Morphology. The colony morphology of cultures grown at 25°C is identical

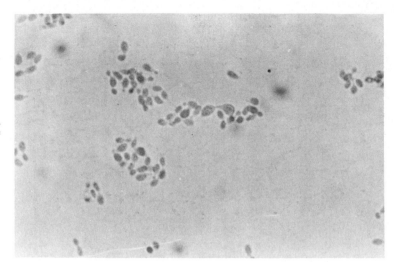

Figure 16–4. Histoplasmosis duboisii. Variation in morphology of cells grown at 37° C on blood agar.

to that of *H. capsulatum.* On microscopic examination, microconidia and tuberculate macroconidia are produced that are identical to those of normal strains of *H. capsulatum.*

Morphology at 37°C. The organism converts easily to the yeast phase. Initially on blood agar the small yeast forms are seen, but in time large, thick-walled cells (12 to 15 μ) are produced (Fig. 16–4). Great morphologic variation is encountered when KY agar is used. Once they are converted, the *duboisii* cells reproduce by the production of large *duboisii* yeast cells, and small *capsulatum* type cells are no longer seen.

REFERENCES

1. Adekunle, O. O., P. Sudhakaran, et al. 1978. African histoplasmosis of the jejunum. Report of a case. J. Trop. Med. Hyg., *8*:88–90.
2. Ajello, L. 1968. Comparative mycology and immunology of members of the genus *Histoplasma.* Mykosen, *11*:507–514.
3. Al-Doory, Y., and S. S. Kalter. 1967. The isolation of *Histoplasma duboisii* and keratinophilic fungi from soils of East Africa. Mycopathologia, *31*:289–295.
4. Ball, J. D., and P. R. Evans. 1954. Histoplasmin sensitivity in Uganda. Br. Med. J., *2*:848–849.
5. Basset, A., M. Basset, et al. 1963. Formes cutanées de l'histoplasmose Afrique. Bull. Soc. Fr. Dermatol. Syphiligr., *70*:61–64.
6. Berliner, M. D. 1967. Gelatin hydrolysis for identification of the filamentous phases of *Histoplasma, Blastomyces,* and *Chrysosporium.* Sabouraudia, *5*:274–277.
7. Blumer, S., and L. Kaufman. 1968. Variations in enzymatic activities among isolates of *Histoplasma capsulatum* and *Histoplasma duboisii.* Sabouraudia, *6*:203–206.
8. Catanei, A., and P. Kervran. 1945. Nouvelle mycose humaine observée au Soudan francais. Arch. Inst. Pasteur Alger., *23*:169–172.
9. Cockshott, W. P., and A. O. Lucas. 1964. Histoplasmosis duboisii. Quart J. Med., *33*:223–238.
10. Cockshott, W. P., and A. O. Lucas. 1964. Radiological findings in *Histoplasma duboisii* infections. Br. J. Radiol., *37*:653–660.
11. Coremans, J. 1963. Un test biochemique de differentiation de *Histoplasma duboisii* Vanbreuseghem 1952 avec *Histoplasma capsulatum* Darling 1906. C. R. Soc. Biol. (Paris), *157*:1130–1132.
12. Derrien, J. P., J. Vedy, et al. 1978. Histoplasmose pulmonaire africaine à *Histoplasma duboisii.* Premier cas tchadien. Bull. Soc. Med. Afr. Noire Lang Fr., *23*:210–213.
13. Drouhet, E. 1957. Quelques aspects biologiques et mycologiques de l'histoplasmose. Pathol. Biol. (Paris), *33*:439–461.
14. Dubois, A., P. G. Janssens, et al. 1952. Un cas d'histoplasmose africaine avec une note mycologique sur *Histoplasma duboisii* n. sp. par R. Vanbreuseghem. Ann. Soc. Belge Med. Trop., *32*:569–584.
15. Duncan, J. T. 1947. A unique form of Histoplasma. Trans. R. Soc. Trop. Med. Hyg., *40*:364–365.
16. Duncan, J. T. 1958. Tropical African histoplasmosis. Trans. R. Soc. Trop. Med. Hyg., *52*:468–474.
17. Egre, J. U., and H. C. Gugnani. 1978. African histoplasmosis in eastern Nigeria: report of two culturally proven cases treated with septrin and amphotericin B. J. Trop. Med. Hyg., *81*:225–229.
18. Gentilini, M., M. Desporte, et al. 1977. Histoplasmose pulmonaire africaine à *Histoplasma duboisii.* Ann. Med. Interne., *128*:451–455.
19. Kaufman, L., and S. Blumer. 1966. Occurrence of serotypes among *Histoplasma capsulatum* strains. J. Bacteriol., *91*:1434–1439.
20. Kaufman, L., and S. Blumer. 1968. Development and use of a polyvalent conjugate to differentiate *Histoplasma capsulatum* and *Histoplasma duboisii* from other pathogens. J. Bacteriol., *95*:1243–1246.
21. Lucas, A. O. 1967. The clinical features of some of the deep mycoses in West Africa. *In* Wolstenholme, G. E. W. and R. Porter (eds.), *Systemic Mycoses.* CIBA Foundation Symposium. Boston, Little, Brown and Company, pp. 96–112.
22. Mariat, F., and G. Segretain. 1956. Étude mycologi-

que d'une histoplasmose spontanée de singe africain *(Cynocephaus babuin)*. Ann. Int. Pasteur (Paris), *91*:874—891.

23. Nethercott, J. R., R. K. Schachter, et al. 1978. Histoplasmosis due to *Histoplasma capsulatum* var. *duboisii* in a Canadian immigrant. Arch. Dermatol., *114*:595–598.

24. Okudaira, M., and J. Swartz. 1961. Infection with *Histoplasma duboisii* in different experimental animals. Mycologia, *53*:53–63.

25. Pine, L., E. Drouhet, et al. 1964. A comparative morphological study of the yeast phases of *Histoplasma capsulatum* and *Histoplasma* duboisii. Sabouraudia, *3*:211–224.

26. Pine, L., L. Kaufman, et al. 1965. Comparative fluorescent antibody staining of *Histoplasma capsulatum* and *Histoplasma duboisii* with a specific anti-yeast phase *H. capsulatum* conjugate. Mycopathologia, *24*:315–326.

27. Renoirte, R., J. L. Michaux, et al. 1967. Nouveaux cas d'histoplasmose africaine et de cryptococcose observes en Republique Democratique du Congo. Bull. Acad. R. Med. Belg., *7*:465–527.

28. Rosenthal, S. A., and H. Sokolsky. 1965. Enzymatic studies with pathogenic fungi. Int. J. Dermatol., *4*:72–79.

29. San Blas, G., and N. Moreno. 1977. Estudio Bioquemico de la pared celular de la fase micelial de *Histoplasma duboisii*. Acta Cient. Venez., *28*:333–334.

30. Stott, H. 1954. Histoplasmin sensitivity and pulmonary calcifications in Kenya. Br. Med. J., *1*:22–25.

31. Swartz, J. 1953. Giant forms of *Histoplasma capsulatum* in tissue explants. Am. J. Clin. Pathol., *23*:898–903.

32. Taschdjian, C. L. 1952. Hyphal fusion studies on *Histoplasma capsulatum* and *Histoplasma duboisii*. Mykosen, *2*:1–6.

33. Vanbreuseghem, R. 1964. L'histoplasmose africaine un histoplasmose causes pour *Histoplasma duboisii* Vanbreuseghem 1952. Bull. Acad. R. Med. Belg., *4*:543–585.

34. Walker, J., and E. T. C. Spooner. 1960. Natural infection of the African baboon *Papio papio* with the large cell form of *Histoplasma*. J. Pathol. Bacteriol., *80*:436–438.

35. Yamato, H., H. Hitomi, et al. 1957. A case of histoplasmosis. Acta Med. Okayama, *11*:347–364.

17 Coccidioidomycosis

DEFINITION

Coccidioidomycosis is a benign, inapparent, or mildly severe upper respiratory infection which usually resolves rapidly. Rarely, the disease is an acute or chronic, severe disseminating, fatal mycosis. Recovery from the mild forms of the disease usually results in lifelong immunity to reinfection. If infection is established, the disease may progress as a chronic pulmonary condition or as a systemic disease involving the meninges, bones, joints, and subcutaneous and cutaneous tissues. Such involvement is characterized by the formation of burrowing abscesses. The initial tissue response and that found in rapidly disseminating disease is suppuration. However, in established chronic and slowly advancing infection, a granulomatous reaction is found, with some areas showing a mixed-type cellular infiltrate. The presence of *Coccidioides immitis,* the etiologic agent of coccidioidomycosis, is associated with a hot, semi-arid environment. It is probably the most virulent of the fungal pathogens. The highly endemic areas of disease include the southwestern United States and northern Mexico, with endemic foci in Central America, Venezuela, Colombia, Paraguay, and Argentina. The disease has been reported to occur in Russia, central Asia, Nigeria, and Pakistan, but these reports lack varification or have been shown to be in error.[127]

Synonymy. Posadas' disease, coccidioidal granuloma, Valley fever, desert rheumatism, Valley bumps, California disease.

HISTORY

Coccidioidomycosis was the first of the severe fatal mycoses in which an inapparent or mild form of disease was found to occur commonly in inhabitants of its endemic region. The pathogenesis of the common, mild disease rather than the rare, severe infection was delineated by Dickson and Gifford[29] in the late 1930's, just as it was in the 1940's for histoplasmosis by Christie and Palmer. This pattern of disease is now being discovered for other systemic mycoses. Coccidioidomycosis is probably the most studied and best understood of the systemic human mycotic infections, although many unknown aspects of the disease remain to be investigated.

As pointed out by Fiese,[38] human coccidioidomycosis is probably a relatively new disease. The endemic areas where it is found were very sparsely populated until the advent of European explorers and the subsequent settling of agricultural and ranching populations in these regions. The indigenous populations that existed there, such as the Yokuts of the San Joaquin valley, were soon eliminated by exposure to the European diseases of influenza, cholera, syphilis, and smallpox. We do not know if coccidioidomycosis existed among them.

As was the case for several of the mycoses, the discoverers of coccidioidomycosis initially described the organism as a protozoan. The patient in whom the disease was first found, Domingo Escurra, was a soldier from the Argentine pampas. He had had recurrent tumors of the skin for four years before entering the University hospital in Buenos Aires in 1891. There, his disease was studied by Alejandro Posadas,[100, 101] a student of the famous pathologist Robert Wernicke.[138] The patient lived another seven years, during which time the investigators noted the progress and pathologic development of the disease. They considered the disease to be a neoplasia, a form of mycosis fungoides (tumors resembling a mushroom), but recognized the presence within the lesions of an as yet undescribed parasite. This organism was likened to the protozoa of the order Coccidia. In 1892 Posadas[100] published the preliminary reports of the case in Argentina, and Wernicke described the same patient[138] in a paper printed in Germany. Posadas was able to reproduce the disease in various animals by inoculating them with lesion material from the patient. The disease was considered to be rare in Argentina, and the second case from the country was not reported until 35 years later.[87] When the known cases of coccidioidomycosis in Argentina were reviewed by Negroni in 1967,[89] only 27 had been reported. Thus the clinically evident form of coccidioidomycosis is rare in the nation where it was first discovered, as is another mycosis, rhinosporidiosis, which was first discovered by the same investigators, Wernicke and Posadas. An interesting sidelight concerns the mortal remains of the first patient with coccidioidomycosis. In 1948 Dr. Flavio Niño found an unidentified head resembling that of the patient described by Posadas in the anatomy museum of the medical school. Further study confirmed this was so, and therefore the specimens from the first case of coccidioidomycosis were rediscovered after being lost for half a century. This head and other appendages of the patient are now a featured exhibit of the medical school museum (Fig. 17–1).

At almost the same time that Posadas was describing the disease in Argentina, Emmet Rixford was studying the first case in California. The patient, Joas Furtado Silverra, a Portuguese from the Azores, had emigrated to the San Joaquin Valley in 1886. He soon developed recurrent nodular cutaneous lesions similar to those observed in Posadas' patient. This particular type of presentation is now known to be fairly rare when compared with other forms of the disease (see Fig. 17–7, *B*). Rixford and Gilchrist[107] studied the parasite, which they observed in the lesion material. At the suggestion of C. W.

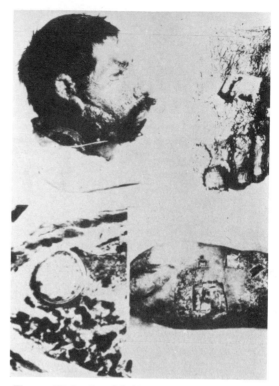

Figure 17–1 Coccidioidomycosis. Museum specimens of the first case of Posadas and Wernicke, showing the recurrent verrucous granulomata of skin and a tissue section containing a spherule. These specimens were recovered by Dr. Flavio Niño. The patient, Domingo Escurra, was a soldier on the pampas of Argentina.

Stiles, they named the organism *Coccidioides* (Coccidia-like) *immitis* (*im* = not, *mitis* = mild) and described it as a protozoan of the class Sporozoa. The organism appeared to them to be related to the members of the order Coccidia, which includes the causative agent of coccidiosis of chickens. Bacteriologic studies were performed on biopsy material from the patient, but the agar plates were noted to be overgrown by a "mould" and discarded as "contaminated." A second fatal case of disease was found by them in 1894, again in a Portuguese from the Azores. In 1896, Rixford and Gilchrist published a detailed account of these two cases.[107] They recognized the similarity of the organism seen in their patients to that found in Wernicke's case, but considered the parasite to be the etiologic agent of a distinct and new infectious disease which was not related in any way to neoplasia.

The true nature of the etiologic agent of coccidioidomycosis was elucidated by Ophüls.[92, 93, 94] In 1900, the third American

As yet no
organism e
C. immitis t
Most fungal
it to be r
arthroconidi
Although th
ed from soi
occurrence

The most
cidioidomyc
States and
delineated
firmed by
ecologic ch
Sonoran Li
ized by the
tridentata (
opuntias ar
(Fig. 17–3
there are
mice), Dij
(ground s
fox, the
and a few
are preda
and Speoty
ditions of
than comp
inches per
one seasor
the summe
38° F. The
of the so
appears t
of the w
Swatek,[128]
region th
to a few

Figure
cosis. Typi
the fungus
like area i
in Guatem
rounded b
gle, this h
suitable f
the fungu
this region
test reacti
few miles
(Courtesy

case was discovered, again in a Portuguese from the Azores. Bacteriologic investigation revealed growth only of a mold that was considered, at least at first, to be a contaminant. The regularity of its isolation, however, eventually led to the association of the mold with the disease. Thus Ophüls and Moffitt in 1900[94] described the etiologic agent as being a fungus. In 1905 Ophüls published his works on the life cycle of *C. immitis*.[92]

In 1915, ten years after the work of Ophüls, Ernest Dickson reviewed the 40 known cases of the disease and stressed the importance of its occurrence in the southern California region. The experience of Ophüls and others indicated that the lung was the portal of entry for the fungus, which later disseminated to the skin and other organs. The discovery of healed pulmonary lesions in patients that did not have disseminated disease began to suggest that there was a milder form of the infection. Subsequently, a laboratory accident dramatically proved this point. A student named Chope, in Ophüls' laboratory, inadvertently inhaled a quantity of conidia of *C. immitis*. When he become ill, he was declared to have coccidioidal granuloma, a universally fatal infection.[50, 54] Though it was severe at first, the disease resolved instead of progressing toward a fatal outcome. After recovery, the student decided to leave California and moved to Baltimore. In 1932 Stewart and Meyer[125] isolated *C. immitis* from soil in the San Joaquin Valley near a site where four Filipinos had contracted their severe or fatal infections. Thus, they established a soil reservoir for the organism in the area. The stage was now set for the next important discovery in the history of the disease.

Dickson continued his work at Stanford University, compiling statistics on the incidence and pathology of coccidioidomycosis. At the same time Myrnie A. Gifford, who practiced medicine in Bakersfield, located in the San Joaquin Valley, had examined a patient with fever, pleurisy and pneumonia who later developed erythema nodosum. This condition was referred to as Valley bumps and Valley fever. A few months later she was able to recover *C. immitis* from the sputum of another patient with the same syndrome. In 1936 she noted that a bout of the bumps (erythema nodosum) had preceded development of coccidioidal granuloma in three of 15 patients. The association of bumps and infection was emphasized by Dickson in 1937 in a publication that named

the disease "coccidioidomycosis" and delineated a primary and secondary phase.[29, 30] By 1938, the cooperative studies of Dickson and Gifford had resulted in establishing that Valley fever and its synonyms were in fact mild forms of coccidioidomycosis and that the disease was much more common than previously suspected.[29, 30]

With a grant from the Rosenberg Foundation, Dickson, Gifford, and a new member of the team, C. E. Smith, began an extensive study of coccidioidomycosis in the San Joaquin Valley. Smith criss-crossed the desert in an old Ford named the "Flying Chlamydospore." He gathered data concerning the incidence, severity, and epidemiology of the disease. He developed a precipitin test and standardized the coccidioidin skin test.[122] With such data he and his associates were able to define the various forms that the initial phase of the disease may take, establish the duration of the incubation period, and study epidemics resulting from simultaneous exposure of many people to areas heavily contaminated with fungal conidia.

In the 1940's airfields were built in endemic areas in the Valley.[121] This brought about the exposure of hundreds of nonimmune men to the conidia-containing dust of the area, and large numbers of infections resulted. Smith formulated dust control methods, such as oiling roads, planting grass, and using swimming pools rather than dusty athletic fields for recreation. Such methods reduced the infection rate by 65 per cent.[120, 121] Meanwhile the disease was discovered to be a problem in other areas of the Southwest. Its endemicity was established in Arizona, Nevada, New Mexico, Texas, Utah, and northern Mexico.[48] The great risk of disease in nonimmune individuals forced the closing of a prisoner of war camp in Florence, Arizona. Recently, new areas of endemicity have been established in other countries. In 1948, Campins et al.[15] diagnosed the first case in Venezuela, Subsequently, cases and endemic foci of disease have been found in Honduras in 1950, in Guatemala in 1960,[86] in Colombia in 1967,[110] and in El Salvador.[86]

Since the discovery of the benign form of coccidioidomycosis, the literature produced on this subject has been voluminous. Studies and reviews of the clinical and pathologic forms of the disease and its epidemiology, ecology, serology, and therapy, have made it one of the most "famous" of the mycotic infections. An excellent review of the subject was published by Marshall Fiese in 1958.[38]

Coccid
three
in 197
of the
bibliog
that b
1981,
updat
1958.

**ETIOL
DISTR**

Etic
etiolo
strain:
fairly
physic
Hupp

However, survival of the organism in soil with normal bacterial and fungal flora is greatly decreased. It appears that *C. immitis* is ill fitted to survive in competition with other soil microorganisms and indeed is inhibited by some. Egeberg et al.[33, 34] found that, in the soils particularly favored by *C. immitis,* the most important inhibitory organisms were *Bacillus subtilis* and *Penicillium janthinellum.* These species proliferate during the rainy season, but as the temperature increases and evaporation increases salinity of the soil, their growth is inhibited. *B. subtilis* is sensitive to high salt concentration, and *P. janthinellum* to a temperature of 100° F. *C. immitis* is very tolerant of a wide range of salt concentrations and almost uniquely tolerant to boron-containing salts.[123] Growth also occurs up to a temperature of 130° F. Both spherules and arthroconidia surive for long periods under adverse conditions,[41, 42] although the arthroconidia survive extreme conditions much longer. It appears, then, that the factors which delineate the ecologic areas inhabited by *C. immitis* are dependent upon the tolerance of the fungus to adverse soil composition and high temperatures, which are inhibitory to competing organisms. It must be noted that the organism is absent in many areas of the Sonoran Life Zone and exists in considerable numbers in jungle-like environments of Colima, Michoacán, and Guerrero states in Mexico and the cool, moist, sandy soil of Pacific Beach section of San Diego. In addition, infection has been noted in the free-ranging, wholly aquatic mammal, the California sea otter *(Enhydra lustris),*[21] which would remain far removed from desertlike areas.

C. immitis presumably is disseminated primarily by dust aerosols that are prevalent in early summer and continue until the first rains of winter. The conidia may be distributed naturally by wind storms, by manmade disruptions during construction work, and by farming, or in "digs" for archeological or biological specimens. The latter have been the origin of many small "epidemics."[137] Even the digging by desert rodents will form infectious aerosols. Swatek et al.[129] studied translocation of the organism which was not associated with rodent burrows. They found the fungus in the upper two inches of virgin desert soil and concluded that the wash of a water-soil slurry during the rainy season would also spread the organism to new areas.

Endemic Areas

United States. Sensitivity to the intradermal injection of coccidioidin has been used to delineate the endemic areas of the disease. The most comprehensive survey and review of skin test sensitivity in the United States is that of Edwards and Palmer.[32] The most highly endemic area in the world is Kern County, California. In some areas within this region, the positive skin test rate is 90 per cent or more, the average being 50 to 70 per cent. In the San Joaquin Valley, the prevalence of positive skin tests gradually tapers off north of this area. The northernmost foci of infestation so far recorded are Red Bluff and Chico, California.[137] Outside of California, very high rates of skin test sensitivity (50 per cent or more) are found in Maricopa (Phoenix) and Pima (Tucson) counties in Arizona and in several west and southwest counties of Texas near the Mexican border. Rates of reactivity vary considerably over the remaining areas of the southwest. In regions where histoplasmosis is endemic, there is a low incidence of positive reactors, which reflects cross-reactions between the antigens of coccidioidin and histoplasmin rather than true presence of coccidioidomycosis. It is estimated that 100,000 cases of coccidioidomycosis[31] occur per year in the United States, with 70 deaths annually.[31] In endemic areas it is as common as chicken pox. The disease is a problem, not because of the mortality it causes, but because of the morbidity associated with it.

Mexico. There are three endemic areas of northern Mexico which are ecologically comparable to the San Joaquin Valley.[48] The first is termed the "northern zone," which is a continuation of endemic areas in the southwestern United States. Skin reactivity rates of 50 per cent or more are found in northern Baja California, the Sonora, and the Chihuahua states. The rate is less in the other northern states of Coahuila, Nuevo Leon, and Tamaulipas. The second zone of endemicity is the Pacific littoral zone. This encompasses the Sonora and Sinoloa states west of the Sierra Madre Occidental Mountains. The reactivity rate here is over 50 per cent, but it diminishes in the coastal states to the south. The northern zone descends toward the central states of Mexico but is separated from them by the Sierra Madre Oriental Mountains. This area constitutes the central zone of endemicity. Coccidioidin reaction

Fig
incide

rates are lower than in the other regions but may reach 50 per cent in some scattered areas. These endemic areas correspond roughly to the limits of the Lower Sonoran Life Zone in Mexico. In addition, however, there are two highly endemic areas that are in tropical regions. The first, in the states of Colima and Michoacan, has a rate of reaction to coccidioidin of up to 30 per cent. The annual rainfall in this area is 717 mm. A second area in the state of Guerrero has a reactivity rate of 10 per cent and a rainfall of 1309 mm. Endemic foci are also found in other countries of Central and South America that are not defined ecologically as Lower Sonoral Life Zones. It appears, therefore, that *C. immitis* may survive under a variety of ecologic conditions.

Central and South America. The highest endemic area of coccidioidomycosis outside the southwestern United States — northern Mexico region appears to be in Venezuela. In the dry, arid, northwestern states of Falcon, Lara, and Zulia, skin test sensitivities of 50 per cent or more have been found. The average temperature varies from 24 to 27° C, with a rainfall of 600 mm or less.[15]

In Central America, two endemic areas have so far been discovered. One is the Montagua Valley in Guatemala (Fig. 17–3), and the other is the Comayagua Valley of Honduras. Skin reaction rates up to 30 per cent have been found there. These regions are semi-arid, with xerophilic vegetation and a rainfall of 30 inches (800 mm). This is higher by far than the Sonoran Life Zone but similar to the endemic region in central Paraguay.

The other endemic regions for the disease are found in Paraguay, Argentina, and Colombia. The area in Paraguay includes the north and the western open plains of the Gran Chaco. The climate is hot and dry, with an average rainfall of up to 20 to 28 inches per year in certain parts. Among the Indians of the western Chaco, skin test sensitivities of 43 per cent have been found.[3] In Argentina the endemic areas are found in the central arid areas of the Patagonia from the 40th parallel north to the Salta-Hondo-Dulce Rivers on the 27th parallel.[89] The highest skin test reactivity (20 per cent) was found at the northern end of the zone near the Rio Hondo. Cases also have been noted in Bolivia.[3, 11, 129]

A recently discovered endemic focus in Colombia is in an area adjacent to Venezuela. A part of the region is in the departments of Guajira and Magdalena, where there is rainfall of 500 mm per year, and the remainder is in the state of Cesar that has 1000 to 2000 mm of precipitation yearly.[110] It appears that an important factor in all these areas where rainfall is higher than normal for the Sonoran Life Zone is that the rains occur all in one short season and are followed by hot, dry, dusty weather.

Cases have been reported outside the American endemic zones. Most of these have been traced to either acquisition of the infection while visiting an endemic area or transmission by fomite.[2, 3] A recent case occurring in Australia was acquired after a trip to the arid interior of that country, but so far *C. immitis* has not been found to occur there.[131] Reports of the disease in Russia have been published for some time. Stephanishtcheva et al.[124] record a total of 35 cases up to 1971. The disease called "coccidioidosis sui generis" is now known to be caused by organisms other than *C. immitis*.[127] Coccidioidomycosis has also been recorded in buffalo in West Pakistan.[64] Corroboration of this report is also lacking. There is also a recent report of canine disease in Nigeria[90]; again, this lacks confirmation.

Fomites account for a number of cases of the disease occurring far from the endemic areas. Instances of Europeans being infected from straw used in packing dishes made in Arizona or sedges used in stuffing for furniture from the southwestern United States have been noted.[114] Recently a fish handler was infected by material in a crate of fish from an endemic area,[134] and a cotton gin operator in Mississippi was infected by conidia on California-grown cotton.[45]

Distribution. There are considerable differences in the severity of coccidioidomycosis which are related to sex and race. Skin test surveys show there is no difference between the sexes in acquisition of primary infections, and, until the age of puberty, the dissemination rates are about equal. In adult males, however, the risk of dissemination is 265 per 100,000, compared with 74 per 100,000 for females.[38] Following initial infection, erythema nodosum, a manifestation usually associated with high resistance to development of severe disease, is found in 25 per cent of women but only in 5 per cent of men. This difference in susceptibility of adult males holds true for all pathogenic fungus infections but not for opportunistic infections. The exception is in pregnant women, in whom the dissemination rate about equals or exceeds that of men.[119, 133] The explanation for these observations obviously involves differences in hormones and hormonal balance, but this subject has not yet been adequately investigated.

An essentially unique feature of coccidioidomycosis is the purported increased susceptibility of persons with pigmented skin. Most all the cases of disseminated disease in the early literature were among Portuguese and Filipino farm laborers. In the many studies that have been made since, differences in rates of occurrence of disseminated disease have consistently been noted. Equalizing as much as possible for occupational exposure and socioeconomic conditions, Filipinos and Negroes run the highest risk of dissemination, with the rate considerably less for Indians and Indian mixtures (Mexicans). Caucasians have the lowest incidence of dissemination. A little more than 1 per cent of all primary infections develop into serious disease, and of these, dissemination is ten times more likely to occur in Negroes than Caucasians. These studies were done primarily by Smith et al. in the 1940's.[120, 122] In reviewing the subject in 1978, Drutz and Catanzaro[31] also felt that the combined experience indicated a greater incidence of disseminated disease in pigmented population groups, although they pointed out that many studies were poorly controlled. Some recent reviews indicate that when properly matched, the rate for fatal disease is no greater in non-whites than in whites.[66] At the present time there is no consensus on this point.[58, 116] New studies indicate an association of severe disease with group B blood type and HLA-A9; both parameters are more common in persons of black and Filipino origin than in those of Caucasian background.

Primary infection with *C. immitis* is dependent on exposure to the fungus and can occur at any age. Agricultural workers and construction crews are particularly prone to infection because of exposure to conidia-bearing dust. Considering the number of "epidemics" recorded among them, archeology students could be considered involved in a "high-risk" occupation.

Coccidioides immitis is a pathogenic fungus and is probably the most virulent of all of the etiologic agents of the human mycoses. Inhalation of a few conidia will produce infection in the normal human host, and a sufficient quantity of conidia will produce an overwhelming disease. In the great majority of patients, primary disease is followed by recovery (without treatment) and development of a strong specific immunity to reinfection. The few patients who develop serious disease probably have some as yet undefined difference in their defense mechanisms from those that recover. These differences may be genetic or transient, but they influence the ultimate outcome of the host-parasite interaction. The same is true to a certain extent in all of the pathogenic fungus infections. These differences in host susceptibility are the areas most in need of imaginative investigation at present. In a few rare instances, coccidioidomycosis may be an opportunistic infection in patients with debilitating disease.[108] This again is related to exposure. A leukemic patient would have to travel to an endemic area to acquire a *Coccidioides* infection, whereas *Candida* and *Aspergillus* are always with him. It must be remembered also that an old, residual, coccidioidal granuloma may reactivate years later if the patient becomes debilitated (see Figure 17–5, *C*). In such cases the infection could be considered opportunistic.

CLINICAL DISEASE

Infection by *Coccidioides* results after inhalation of dust containing arthroconidia. Initial infection is followed by the production of two vastly different clinical diseases. The first is a completely inapparent or moderately severe disease followed by complete resolution of the infection and establishment of strong immunity to reinfection. The second is the rare form in which the infection becomes established and is followed by a chronic progressive disease or an acute, rapidly fatal dissemination. Such diversity in the disease syndrome following inhalation of infectious units is also seen in histoplasmosis, paracoccidioidomycosis, and other fungus infections. The extensive studies by Smith, Fiese, and others on coccidioidomycosis have shown that in about 60 per cent of patients primary infections are asymptomatic, and 40 per cent have mild to severe acute pulmonary disease. However, only in about 0.5 per cent (one or two per thousand) is a serious disease established. This is known as secondary coccidioidomycosis. The clinical classification presented here reflects these two major forms of the disease.

Clinical Types of Coccidioidomycosis

I. Primary coccidioidomycosis
   A. Pulmonary
      1. Asymptomatic

2. Symptomatic
B. Cutaneous (rare)
II. Secondary coccidioidomycosis
A. Pulmonary
1. Benign chronic
2. Progressive
B. Single or multisystem dissemination
1. Meningeal
2. Chronic cutaneous
3. Generalized

PRIMARY COCCIDIOIDOMYCOSIS

Pulmonary

Asymptomatic Pulmonary Disease. This classification includes the majority of patients who become infected. There are no symptoms or at least none significant enough to have been remembered by the patient. Such patients are the "skin test convertors" who become reactive to intradermal injections of coccidioidin in the absence of demonstrable disease. There is no residual scar or lesion in the lungs, and it is only the conversion to coccidioidin sensitivity that indicates an infection has occurred. Such a clinical course does not always indicate a particularly strong resistance to the infection, however. In some patients with extrapulmonary involvement (e.g., meningitis), no demonstrable pulmonary lesions could be found, and there was no history of symptoms coincident with exposure.[17]

The asymptomatic group of patients who have no residual pathology merges gradually into the groups with some indication of past infection in addition to a positive skin test. In these there may be a history of a mild, undiagnosed, flu-like disease. On roentgenographic examination such patients usually show small, healed areas of fibrosis and, commonly, calcifications representing old hilar, parenchymal, or pleuritic lesions.

Symptomatic Pulmonary Disease. Symptoms vary considerably in this group of patients. Following exposure and an average incubation period of 10 to 16 days, they become mildly ill with a "cold" or may have severe respiratory disease. The extremes of the incubation period are seven days to four weeks, and the duration of the primary disease is from a few days to several weeks. The incubation period and duration and severity of symptoms are usually a function of the magnitude of the exposure. Inhalation of a few conidia is associated with an illness of

short duration. Sometimes no respiratory disease is noted, and symptoms are confined to allergic manifestations, such as toxic erythemas (Fig. 17–4), arthralgias, episcleritis, and so forth.

Patients have, in varying degrees, one or more of the following symptoms and signs.

1. *Fever.* Most patients are febrile for some period during their illness. There is nothing characteristic about its pattern. The fever usually rises to a peak with diurnal variation and then diminishes. It may last a few hours to a few months; may reach 40.5° C, and can be accompanied by night sweats. Persistence or recrudescence of fever following initial recovery often heralds dissemination. Spread of disease may occur in the absence of and without increase of fever, however.

2. *Pain.* Chest pain is very common and is usually the first sign of the disease. It occurs in 70 to 90 per cent of symptomatic patients. Although it may manifest only as a dull discomfiture, pleuritic pain may be so severe and sudden in onset as to emulate rib fracture, myocardial infarction, or cholecystitis. Pain is often accompanied by friction rub

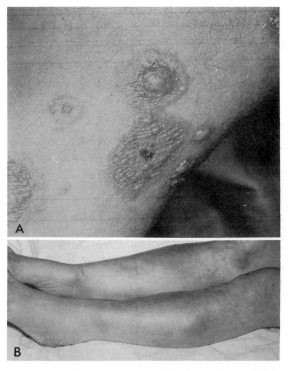

Figure 17–4 Allergic manifestations of coccidioidomycosis. *A,* Erythema multiforme on extensor surface of arm. Intact bullous eruption on erythematous base. Target or "bulls-eye" lesion. *B,* Erythema nodosum. Erythematous tender nodules with a bluish cast on the anterior aspect of lower extremities.

and, sometimes, demonstrable pleural fluid levels. Interference with respiration often occurs as the pleuritic pain is made worse by coughing or deep inspiration. Interference with swallowing may result from substernal pain caused by enlargement of the mediastinal nodes.

3. *Respiratory embarrassment.* This symptom is uncommon but may result from the presence of pleural effusion or pneumonic spread of the disease. Although quite rare, shortness of breath may also be due to spontaneous pneumothorax.

4. *Cough.* Cough is less common and less severe in coccidioidomycosis than in other pulmonary infections. It is often nonproductive, but with severe pulmonary involvement sputum may be white or purulent and sometimes blood-streaked. In such cases, the organisms can often be seen on direct examination of a potassium hydroxide mount of the sputum, but their absence does not rule out the diagnosis.

5. *Anorexia.* This symptom frequently occurs even in mild disease. The duration is usually short, but some patients lose 20 to 30 pounds within a three-week period. In disseminated disease, anorexia may persist and lead to profound cachexia.

Other symptoms are present in varying degrees. Generalized aching, malaise, myalgia, and backaches are often present. In mild cases malaise and lassitude may be the only complaints. Headache may be quite severe in acute, uncomplicated disease. It may subside in a week or persist for a much longer period of time. Spinal fluid is normal at this stage. Meningitis is an infrequent sequela of primary disease and is accompanied by abnormal spinal fluid findings.

X-ray. The roentgenographic findings of symptomatic disease vary from a normal chest picture to extensive pulmonary infiltration, lymphadenopathy, or massive pleural effusion. In the mildest infections, only superficial inflammation of the bronchial mucosa or the alveolar spaces is present, and there is nothing detectable in the thoracic shadow. However, about 80 per cent of patients show some pulmonary changes on x-ray. Infection may involve any area of the lung, including the apices, but involvement of the lower lobes or the bases of the upper and middle lobes is more common. Thus primary coccidioidomycosis more closely resembles primary atypical rather than bacterial pneumonia.

There is no single x-ray pattern noted, and individual roentgenograms often defy classification. Probably the most common finding is that of a pneumonic infiltrate. There are soft, hazy, uniform lines which radiate from the hila toward the base or periphery. Although it may persist for months, this type of infiltrate usually resolves about a week to ten days after cessation of clinical symptoms. In other cases only a soft, fuzzy, hilar thickening is found which usually resolves within two weeks. Pleural involvement is frequently (< 70 per cent) manifested as pleuritic chest pain, less frequently (< 20 per cent) as blunting of the costophrenic margin, and occasionally evident as discernible quantities of pleural fluid (< 6 per cent).[8] Pleural effusion is often (> 50 per cent) found in patients manifesting erythema modosum.[81]

The best characterized lesions are well circumscribed nodules found in the lung parenchyma. These are most often basilar in the middle or lower lobes. They usually have a diameter of 2 to 3 cm and are single or, rarely, multiple. Such lesions resemble the nodules of primary or metastatic tuberculosis. They usually resolve, leaving little residue, or become thin-walled cavities. Such cavities fibrose and later calcify. Large inactive cavities may persist for years.

An infrequently seen pattern is that of hilar and mediastinal lymphadenopathy. Although hilar nodes are often (< 20 per cent) involved, mediastinal nodes appear to be secondarily invaded. Such a picture is often found in patients who later have disseminated disease. When hilar nodes alone are involved, they regress in a few days to a week, but mediastinal nodes may take months to heal, even if dissemination does not occur. Enlarged nodes most often accompany parenchymal infiltrates in acutely ill patients. Rarely they are the only sign present and can be confused with neoplasias, such as Hodgkin's disease. Another pattern observed in about one-fifth of patients consists of small pleural effusions. They may suffice only to obscure the costophrenic sulcus and usually resolve rapidly.

Primary infection that is becoming rapidly progressive and disseminated is indicated by progression of the above findings on serial x-rays. Pneumonia and pleural effusion become extensive, and there is widening of the mediastinum. Fuzzy miliary foci that become confluent are seen, and consolidation of one or more lobes usually indicates an imminent fatal terminus.

The other laboratory findings in primary coccidioidomycosis include an elevated leuko-

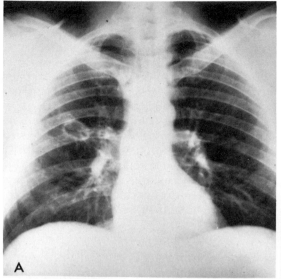

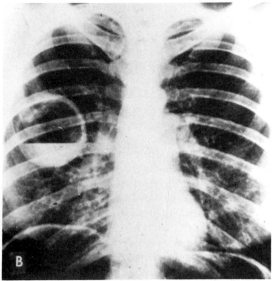

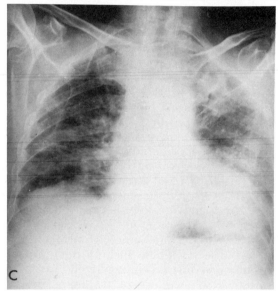

Figure 17–5 Radiologic aspects of secondary coccidioidomycosis. *A,* "Coin lesion." In the superior segment of the right lower lobe, there is a thin-walled, sharply defined cavity. This lesion has been inactive for years, and essentially no reaction is seen around it in this x-ray. (Courtesy of F. Kittle.) *B,* Chronic coccidioidoma. There is a large, sharply defined cavity which has an air-fluid level in the middle right lung, lower lobe. *C,* Disseminated disease from an old cavity lesion. The lungs are underinflated. Both lung fields show diffuse interstitial infiltrates, particularly evident on the right. In addition, there is a thin-walled cavity on the left apically located that is surrounded by intense tissue reaction. In the left base there is a confluence of shadows consistent with alveolar disease. The picture on the left is consistent with a diagnosis of pneumonia, whereas that on the right would probably be interpreted as hematogenous spread. (Courtesy of H. Grieble.)

cyte count, often with an eosinophilia; an elevated sedimentation rate; and conversion of the coccidioidin skin test to positive. The latter occurs from 2 to 21 days following initial symptoms. Humoral antibodies develop later. In very mild infections precipitins and complement fixing antibodies do not appear. In more severe primary infections they are present for varying lengths of time but disappear with resolution of the disease. Persistence of a complement fixation titer is associated with establishment of serious disease and probable dissemination. Another test that can gauge progress of the disease is the erythrocyte sedimentation rate. It is markedly elevated, even in mild or moderate primary disease. It gradually falls with recov-

ery. Persistence of, or a steady increase in, an elevated sedimentation rate denotes development of progressive disease and possible dissemination.

Allergic Manifestations of Primary Disease. Toxic erythema, erythema nodosum, and erythema multiforme are common cutaneous symptoms of early primary coccidioidomycosis. They may occur in the absence of other clinical signs. The appearance of toxic erythema has no prognostic significance, but development of erythema nodosum or multiforme usually heralds development of strong resistance to the infection. Toxic erythema frequently appears within a week of the infection (before coccidioidin skin test sensitivity) and disappears rapidly. There is a fine, macular, diffuse, erythematous rash involving both the extremities and the trunk and may include an oral exanthem. It is most common in children and young adults but is found in 10 per cent of older adults.[53] The origin of the erythema is unknown. Though toxic erythemas are found in many acute febrile illnesses, in the endemic areas they are almost pathognomonic for coccidioidomycosis.

In about 5 to 10 per cent of cases some other form of allergic manifestation occurs. An erythema nodosum or multiforme-like toxic eruption (Fig. 17–4) is found in 25 per cent of adult white females and 4 per cent of white males. It is much less common in dark-pigmented people. Other manifestations include arthralgias, arthritis ("desert rheumatism"), and episcleritis. These, along with erythema nodosum ("desert bumps") and erythema multiforme ("valley fever"), constitute the common benign but symptomatic primary coccidioidomycoses of the highly endemic regions.

Development of any form of "desert bumps" or "desert rheumatism" usually denotes strong resistance to the infection. Very rarely, however, the disease may progress in patients with such symptoms, most often to pulmonary cavitation or meningitis. In the latter case, the patients are usually skin test negative.

The lesions of erythema nodosum are most often restricted to the lower extremities. Usually they are found on the anterior tibial areas, are more numerous about the knees, and may extend to the thighs. Typically they appear as a crop of bright red, tender, itching, or painful raised nodules. They are firm and elastic and deeply imbedded in the skin (Fig. 17–4, B). They range in size from a few millimeters to several centimeters. The acute lesions regress within a few days. Although the color is initially red to purple, after resolution, the areas may show some degree of brownish postinflammatory hyperpigmentation, which lasts for several weeks or months.

The erythema multiforme–like toxic eruption develops on the upper half of the body. The favored sites are the neck, face, collar, upper back, thorax, and occasionally the arms. They appear as reddish to purple nodules, papules, macules, or vesicles and fade in time to a violaceous or brownish hue (Fig. 17–4, A). Sometimes they are accompanied by erythema nodosum on the legs. One crop of lesions is the usual pattern, but several crops may appear. Reappearance of such lesions often follows fatigue from physical exertion.

"Desert rheumatism" is the name for the arthritis that sometimes accompanies primary disease. The symptoms vary from vague arthralgias to periarticular swelling, stiffness, erythema, and heat. Joint effusion is not found. Other less common allergic phenomena involve the eye. These include episcleritis, phlyctenular conjunctivitis, and keratitis, all of which resolve usually within a few days. When they occur they are often overshadowed by the more spectacular skin rashes and joint pains that may be present in the same patients.

Primary Cutaneous Coccidioidomycosis

This is the rarest form of infection, and it has been adequately documented only in about three cases. The first case was reported in 1927 and involved inoculation of conidia by a prick of a cactus needle, as described by Guy and Jacob.[50] Introduction of spherules into the abraded skin of his finger by a mortician embalming a patient who had died of the disease was the second.[141] The third case involved a laboratory accident in which a graduate student introduced conidia into her thumb.[132] In all three cases a chancre-like lesion developed, there was regional adenitis, and the lesions healed uneventfully within a few weeks. Wilson et al.[141] have set up rigid criteria for documentation of primary cutaneous disease. These include no history of pulmonary disease immediately preceding onset, clear evidence of traumatic inoculation, regional adenitis in nodes that drain the area, and a chancriform lesion. The lesion is similar to the primary lesion seen in sporotrichosis. It is a painless, firmly indurated nodule or nodular plaque with central ulceration. Reported cases of primary cutaneous disease, in which the lesions were described as being multiple or verrucous or were abscessed, torpid, cutaneous ulcers, in reality represent secondary lesions following an inapparent primary pulmonary infection. Primary cutaneous disease has been recently reviewed by González-Ochoa and Velasco-Castrejón.[49, 134] Descriptions of primary infection occurring through the oral and nasal mucosa have not been adequately documented. A case of primary endophthalmitis has been reported, however.[51]

SECONDARY COCCIDIOIDOMYCOSIS[31, 56]

Pulmonary

Benign Chronic Pulmonary Disease. The symptoms of primary pulmonary coccidioidomycosis ordinarily subside by the end of the second or third week of clinically apparent disease. In patients that retain clinical symptoms or roentgenographic abnormalities beyond the sixth to eighth week, some manifestation of secondary or persistent coccidioidomycosis is going to develop. There are two major categories: (1) chronic benign pulmonary disease with associated nodular and cavitary lesions and (2) progressive pulmonary disease resulting from persistent or progressive pneumonia or miliary coccidioidomycosis. Resolution of any one of these forms is accompanied by fibrosis, bronchiectasis, and calcification.

Nodules. Nodule formation is a common sequela of coccidioidal pneumonia. The lesion is spherical, is dense, usually in the mid-lung fields (within 5 cm of the hila), and develops over a period from 10 days to several months. These lesions may remain stable for long periods of time. The organism in this and in cavitary disease remains viable. In one recorded case the organism was recovered in culture 15 years after the infection.[23] Thus, these lesions may be the source for later opportunistic dissemination.

Cavitary Disease. In about 2 to 8 per cent of symptomatic infections, residual chronic cavitation occurs. It is considered benign because dissemination is a feature of the primary infection and rarely develops from chronic cavitary disease. As a complication of primary disease, cavitation occurs more frequently in Caucasians, a group that is much less prone to disseminated disease. Development of cavitations is not necessarily related to the severity of the primary infection and may occur as early as the tenth day of infection.

The commonest presentation of this form of disease is a single, chronic, thin-walled cavity which often has a fluid level and little surrounding tissue reaction. Early in the course of the infection it can be seen as a central clearing in an area of dense infiltrate or within a nodule. The cavity may have a thick, "shaggy" wall and enlarge by internal dissolution until a "thin"-walled lesion is formed. These walls, which separate the cavity from the normal lung, are actually quite thick. Over a period of time they may vary in size owing to air trapping, fluid collection, and connection with patent bronchi. In time the cavity stabilizes. It is usually symptomless and discovered only on routine x-ray (Fig. 17–5, *A*). If symptoms occur, the most common indication of the presence of the lesion is hemoptysis, which is also the most frequent and serious complication. In a few patients, cough, production of sputum, and malaise are also present. The distribution of the lesions in the lung fields is the same as that of the primary pulmonary disease from which it develops, and multiple cavities can be found.

Hemoptysis is always a danger in the cavitary form of the disease. In addition, if the lesions are peripheral they may rupture, causing pleural effusion, empyema, spontaneous pneumothorax, or bronchopleural fistula. Resolution of these cavities is often resistant to chemotherapy, and they must be removed surgically. On surgical removal, the cavity is found to have a tough fibrotic wall, with a granular interior containing spherules and sometimes hyphae of *C. immitis*. Enlarging cavities are usually the result of bacterial superinfection. In about 50 per cent of cases, spontaneous closure occurs, with a mean time of two years; others resolve in as long as four years. In cavities that do not close, complications such as infection of the cavity by fungi or bacteria, pyopneumothorax, and pulmonary hemorrhage may occur at any time. The infectious complication is the most common. Secondary infection with bacteria often results in rapid change in size and shape of the cavity in addition to causing fever and malaise. *Aspergillus* fungus balls occur in these and cavities of any origin. Since the bronchi are usually not patent, the sputum culture is negative, but diagnosis can be made on x-ray examination and by a means of the characteristic immunodiffusion test reaction (see Chapter 22). Fungus balls consisting of mycelial units of *C. immitis* have been recorded.[37] They are identical in presentation to aspergilloma, but sputum culture is usually positive for *C. immitis* and, of course, the serologic evaluation is different. These cases should not be referred to as mycetomas, because that clinical entity involves tumefaction, sinus tracts, and grains. True pulmonary mycetoma occurs in some fungal infections, viz., *Pseudallescheria boydii*.

Though they are typically thin-walled and cystlike, there are many variations of coccidioidal cavities. The commonest is one with a thick wall, presenting as a moderately dense pulmonary shadow. Such lesions are difficult to differentiate from tuberculous cavities or bacterial lung abscesses. The size of all types of cavities varies from a few millimeters to several centimeters.

Coccidioidoma. Another form of chronic residual disease is the coccidioidoma, which is comparable in morphology and development to the histoplasmoma or tuberculoma. The coccidioidoma in some patients forms from a resolved or arrested pneumonitis or granuloma. It may result from an exaggerated hypersensitivity in the same manner as the immunopathologic mechanism proposed for the genesis of the histoplasmoma. Although the coccidioidomas does not represent a hazard to the patient, since reactivation of such lesions is very rare, it does represent a diagnostic difficulty because of its resemblance to neoplasias. Their size varies from a few millimeters to 4 or 5 centimeters (Fig. 17–5,*B*). The outline is often lobulated or irregular, reflecting the coalescence of several sites of pneumonitis. Most frequently they are single, but multiple lesions may be seen. Although they are usually stable for long periods of time, in some excavation occurs followed by refilling. This is most often associated with superinfection by bacteria. Some calcification of the lesion is common. On surgical removal, this type of lesion is seen to have a thick, fibrous wall with a center that is soft, yellow-gray or yellowish, necrotic, and caseous. On histologic examination, spherules and sometimes hyphae are seen. The spherules are often atypical and may be wrinkled and distorted empty ghosts. A few viable organisms are usually present. Satellite tubercles from the main cavity are common. The complement fixation titers in both chronic cavity disease and coccidioidoma are low or absent.

Less common complications of primary disease include bronchiectasis, pulmonary fibrosis, empyema, pneumothorax, hydropneumothorax, and chronic pericarditis. Since primary disease is frequently endobronchial, chronic or slowly progressive bronchiectasis may result. There is chronic productive cough, hemoptysis, fever, and frequent bacterial or viral superinfection. The complement fixation titers are usually higher in such patients than in those with cavitary disease or coccidioidoma. Persistent pulmonary fibrosis of itself is not a serious complication, but on x-ray examination it causes diagnostic prob-lems. Empyema is an infrequent but more serious complication which results from rupture of a chronic cavity or follows the pleural effusion of the primary disease. Together with pneumothorax and hydropneumothorax, it is associated with a brief spread of the infection to other pleural areas (Fig. 17–5,*C*). Occasionally more serious consequences develop, such as multiple sinuses draining through the chest wall (Fig. 17–6). Chronic empyema can be a very debilitating condition and requires chemotherapy. In contrast, pericarditis is a benign residual complication of primary disease. Mechanical construction of the heart, however, may require surgical correction.

Progressive Pulmonary Disease.[27, 31] Although residual pulmonary disease remains stable, though chronic, it may also become progressive. This sometimes begins immediately after the primary disease has begun to resolve or after long periods of time following stabilization. Moderate debilitation, malnutrition, old age, and chronic pulmonary disease are factors that predispose to reactivation and progression of persistent coccidioidomycosis. This progressive disease may take several forms. One is a simple, locally extending lesion. Others include persistent and progressive pneumonia, miliary disease, enlarging or multiplying cavities and nodules, abscessing nodules, extending infiltrates, or lobar consolidation. Such activation and progression are associated with increasing symptoms of severe chest disease, and the patient should be followed by serial x-ray examinations.

Progressive chest disease, particularly following previously stabilized and dormant disease, is a very serious condition (Fig. 17–5,*C*). The infection may stabilize again, but there is usually extrapulmonary dissemination or a relentless pulmonary course with a fatal outcome. It is therefore necessary to initiate specific chemotherapy in such circumstances.[111] In about 84 per cent of cases of fatal progressive pulmonary coccidioidomycosis the patients were immunocompromised, and in 50 per cent the patients died of pulmonary disease with no evidence of dissemination to other organ systems.

Two forms of progressive disease are of particular interest. The first is progressive coccidioidal pneumonia. This form is characterized by very slow development (up to 15 years), with symptoms of chronic cough, anorexia, fever, hemoptysis, dyspnea, and chest

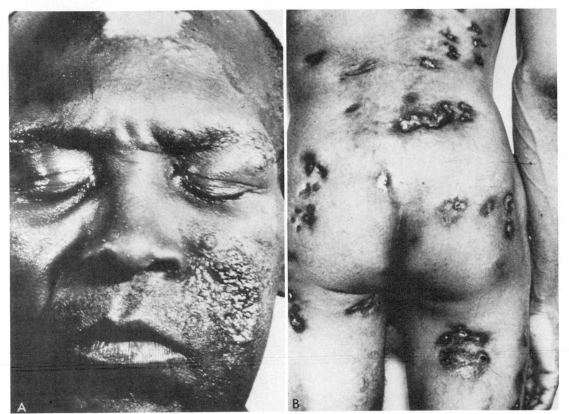

Figure 17–6 Coccidioidomycosis. *A,* Chronic lesions of the face. Active lesions are seen on the cheek. An atrophic, depigmented scar representing a healed lesion is on the forehead. *B,* Chronic granulomatous lesion and suppurating lesions in an advanced case of disseminated coccidioidomycosis.

pain. On x-ray, biapical disease is evident with multiple cavities, retractions, and fibrotic scars. Thus, it is similar to chronic cavitary tuberculosis and chronic cavitary histoplasmosis. The complement fixation test is usually positive and sputum culture is usually positive, but the skin test is almost always negative. Some authors consider this as coccidioidal colonization of old tuberculous cavities or emphysematous blebs in a manner similar to that occurring in histoplasmosis. As yet, evidence for this is lacking.

The second form of progressive disease is miliary coccidioidomycosis. This occurs in about 4 per cent of severe symptomatic infections and usually occurs very early in the course of disease. It presents as rapidly progressive pulmonary failure and, unless treated immediately, will have a fatal outcome. Such infections often are associated with disseminated disease, especially meningeal. Thorough examination of organ systems, especially spinal tap, is therefore warranted, even if the pulmonary disease is under control.

Single or Multisystem Dissemination[27, 31, 142, 144]

Dissemination in coccidioidomycosis is dependent on several factors. Overwhelming exposure to the fungus may result in an almost immediate dissemination and a severe, rapidly fatal disease. In Negroes and Filipinos dissemination usually follows closely the onset of the primary infection. It is the general rule that if dissemination is going to occur, it will do so within the first few weeks of primary disease. The signs that herald dissemination are a persistent elevated sedimentation rate and complement fixation titer, along with persistent and increasing fever and malaise. In patients with these signs dissemination is often rapidly fatal. It results in acute meningitis, involvement of multiple organ systems, and frequently cutaneous and subcutaneous abscesses.

Other patients may contain their infections to a greater extent and have apparent remissions. They appear more resistant to the infection and generally have a marked eosinophilia along with their pneumonitis and pleural effusion, but their defenses are in some way less than adequate to contain the disease. Symptoms such as anorexia and fever reappear, and a rise in the complement fixation titer at this point indicates imminent progress and dissemination of the infection. The disease that results in such patients frequently has a more protracted course than that in patients with a rapidly fatal infection. Lung lesions and the involvement in other organs go through periods of quiescence and recrudescence.

A third category of patients includes those in whom dissemination occurs late in the course of the disease (Fig. 17–5,C). This form of infection is particularly prevalent in white males. Late dissemination probably represents the emergence of active infection from inapparent disease. The course may be quite insidious, as is found in cases of chronic meningitis which may become symptomatic long after the primary infection has abated.

In patients receiving steroid therapy, inactive lesions of coccidioidomycosis may be reactivated, and dissemination may occur.[108] In such cases, the organism is considered to be an opportunist. Opportunistic dissemination also may occur in some cases of leukemia, lymphoma, or other neoplasias.

In coccidioidomycosis, as in other systemic infections, women only rarely manifest disseminated disease. The exception to this is when pregnancy and coccidioidomycosis coincide.[133] The rule is that the later in the pregnancy infection occurs, the greater tendency there is to dissemination. Transplacental infection occurs and, unless treated, is fatal to both mother and child. Many cases of successful treatment of pregnant women with amphotericin B are now on record, however. Instances of congenital disease are rare.[115]

Meningitis. Involvement of the meninges may be acute, subacute, or chronic. It is most often acute and rapidly fatal in Negroes and Filipinos. The disease in Caucasians, however, is more frequently a subacute or long, chronic process. Meningitis is found in about 25 per cent of cases of secondary coccidioidomycosis. After tuberculosis and cryptococcosis, it is the third most common form of chronic granulomatous meningitis. The disease is uniformly fatal if not treated but may have a protracted course with periods of remission.

The acute form of coccidioidal meningitis develops as a sequela of primary pulmonary disease. There is no remission of symptoms associated with the primary infection, and fever, lassitude, anorexia, and weight loss continue. Neurologic signs may be minimal at first. The spinal fluid is turbid and yellow, and there is a marked pleocytosis. The complement fixing titers of serum and spinal fluid are often quite high.

Chronic meningitis may be very insidious in its course. In a series of cases reported by Caudill et al.,[17] over half the patients had had no previous history of coccidioidal disease. In all patients, the skin test was negative, but significant complement fixing titers were present in serum and spinal fluid. The initial spinal fluid examination was grossly abnormal, and a pleocytosis, elevated protein, and decreased glucose were found. The authors conclude that early examination of spinal fluid is necessary to exclude this and other treatable granulomatous meningitides in patients presenting vague, protracted, neurologic signs and psychiatric syndromes. In other series, the complement fixing antibody was present in the spinal fluid in only 75 per cent of cases. Unfortunately, the correct diagnosis in some patients is not made until autopsy.[16] Complications of meningitis, usually types of obstructive hydrocephalus, are frequent. Chemotherapy is required, and intrathecal combined with intravenous amphotericin B has been used successfully for treatment.[17, 145]

Chronic Cutaneous Disease. This was the form of coccidioidomycosis seen in the first case of Posadas and those of Rixford. Both patients had long courses of disease with recurrent lesions until they died of multisystem involvement, 11 and 8 years after the appearance of the first cutaneous lesions. This form of disseminated disease is seen most often in Negroes and other dark-skinned individuals and is quite rare in Caucasians. Lesions often appear first on the nasolabial folds, face, scalp, or neck. At first the lesions are epidermal thickenings without inflammation. Gradually they become larger and wartlike to form the characteristic type lesions termed "verrucous granuloma," which are similar to those of chronic cutaneous blastomycosis. The lesions may grow and spread and resemble fungating epitheliomas. Other lesions remain small and eventually heal, leaving atrophic scars (Fig. 17–6). As disseminated disease progresses to involve other tissues, skin lesions begin to appear as indolent ulcers (Fig. 17–7). These are frequently seen over joints and represent sinus

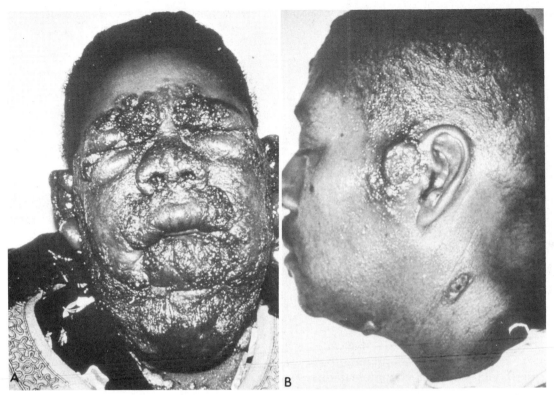

Figure 17–7 Chronic cutaneous coccidioidomycosis. *A,* Granulomatous lesions on face, neck, and chin. *B,* Advanced chronic cutaneous disease. This is the rare nodular form and similar to the morphology present in the first cases of Posadas and Wernicke and of Rixford. (Courtesy of M. Gifford.)

tracts from subcutaneous and osseous foci of infection (Figs. 17–6, *B* and 17–8).

Generalized Disease. Dissemination often is followed by the establishment of miliary lesions in many organ systems. Commonly involved are lymph nodes, which sometimes heal by fibrosis. Splenic lesions are granulomatous, tumorlike, or suppurative, depending on which stage the disease was in when the organ became involved. Osteomyelitis is found in about half the cases of disseminated infection, but it can commonly present as single organ system disease.

The bones most commonly involved are the ribs, the vertebrae, and those of all the extremities (Fig. 17–9). In long bones, lesions are commonly found in the distal portions and within the prominences. This is similar to the distribution of lesions found in blastomycosis. Lesions in the flat bones are usually centrally located. The reaction is most often osteolytic, but in chronic and solitary lesions osteoblastic activity is found. Sometimes chronic osseous disease is the only extrapulmonary lesion noted. Bone scans are often useful in picking up occult disease.[7] Joints are

the sites of lesions, with accompanying arthritis, in about a third of disseminated cases.

Involvement of the vertebrae also occurs. The accompanying spondylitis is characterized not only by the relative sparing of the disk but also by involvement of all other parts, including arches and processes. Vertebral collapse may ensue, but there is a notable lack of gibbus deformity that usually accom-

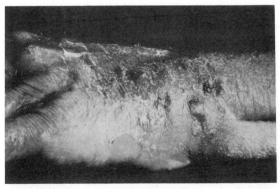

Figure 17–8 Advanced disseminated coccidioidomycosis. The nodular lesions are now mixed with suppurative draining lesions. Subcutaneous and osseous lesions were present in the arm and wrist.

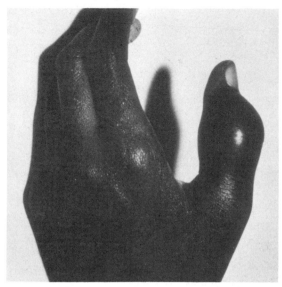

Figure 17–9 Coccidioidomycosis. Dactylitis. The lesion is swollen and hot but has not eroded through the skin.

panies acute anterior angulation. These manifestations contrast coccidioidal disease with tuberculous spondylitis but not with actinomycotic involvement. Both actinomycotic and coccidioidal disease also show a predominance of sclerotic rather than lytic lesions and the development of paraspinous abscesses.[24] Furthermore, both may show a periostitis, with erosion of the articular facets, lamina, and processes, again in contrast to tubercular disease. Coccidioidomycosis presenting as Pott's disease is recorded, however.[139] Skull lesions are characteristically circular or irregular and destructive, involving both tables or the outer table only. They are indistinguishable from those of blastomycosis. Other mycotic lesions of the cranium are exceedingly rare.

Involvement of other organs varies considerably in cases of disseminated coccidioidomycosis.[31, 56] The skeletal muscles are usually spared, but a psoas abscess may be found, again similar to involvement seen in actinomycosis. The adrenals are rarely foci of infection. Thus the complication of Addison's disease is much less frequently encountered than in blastomycosis and histoplasmosis. Pericardial granulomata are present in about one fourth of cases. Involvement of heart muscle is extremely rare. In contrast to the findings in histoplasmosis and paracoccidioidomycosis, the gastrointestinal tract is usually spared. Miliary granulomata or abscesses

may be found in the renal parenchyma, but the pelvis is generally without lesions. Coccidioiduria in which the only other manifestation was a stable cavity in the lung has been recorded.[98] The male genital system is very rarely involved, but when it is, the prostate is the usual site of infection. Involvement of this organ in this disease is much less frequent than in systemic blastomycosis. Lesions in the female genitalia are also uncommon. In a recent report by Hart et al.[52] there was some evidence for a primary uterine infection with a possible fomite as vector of the organism. Involvement of the larynx in the absence of other disease is recorded for both infant and adult.[136]

DIFFERENTIAL DIAGNOSIS

In the endemic areas, coccidioidomycosis should be considered in the differential diagnosis of any nonspecific illness. This is also true with persons who have visited these areas at any time prior to onset of symptoms, particularly when the signs and symptoms are vague and nonspecific.

Primary disease is often confused with other acute pulmonary infections, such as influenza, primary atypical pneumonia, bronchitis, bronchial pneumonia, or simply a "cold." Since this form of the disease most often resolves uneventfully and without treatment, the patient is none the worse. Secondary coccidioidomycosis, however, requires therapy so that it is imperative to make a specific diagnosis. The disease must be differentiated from tuberculosis, neoplasias, other mycotic infections, syphilis, tularemia, glanders (melioidosis), and osteomyelitis of bacterial origin.

PROGNOSIS AND THERAPY

Recovery from primary disease occurs in almost all symptomatic and asymptomatic cases. Therefore primary disease is treated with bed rest and restriction of activity. Exceptions to this are manifest in one of the three major categories that predispose to a serious clinical course: race, pregnancy, and immunosuppression. In addition to bed rest, steroids are sometimes used to control severe allergic manifestations of uncomplicated primary infection, but patients so treated should be closely watched, as steroids may promote dissemination. Surgical intervention should be considered in benign residual disease of the lung because of the danger of recurrent hemoptysis. Cavities sometimes close spontaneously, however. All other forms of secondary coccidioidomycosis have a grave prognosis unless specific chemotherapy is instituted.

The drug of choice in coccidioidomycosis is amphotericin B. The late William Winn had extensive experience in treating disseminated disease, and he formulated the generally accepted dosage schedule.[31, 142] The regimen is essentially the same as that used for other systemic mycoses: 0.6 mg per kg of body weight per day for the adult and a little more for children. The total dose is 500 to 2500 mg, but persistent disease may require more. Patients are started on a 5.0 mg (test dose) schedule initially; this is gradually increased as tolerated until 40 to 60 mg are being administered each day. As the disease becomes stabilized, therapy can be lowered to 25 to 50 mg per day. Alternate day therapy may be required (50 to 70 mg per day). Twenty-five to 35 mg of drug generally produces a serum concentration of 1.55 μg per ml. This is predictable, as is the sensitivity of *C. immitis* to amphotericin B. It is therefore unnecessary to expose the technical staff to the dangers inherent in a sensitivity test on the organism. The drug is diluted in 500 ml of 5 per cent dextrose solution. It must be made up fresh, as the drug is unstable in solution over long periods of time. Storing the solution in brown bottles or covering the bottle is not necessary, since the drug is stable exposed to light for the six-hour infusion time.[9] Additives are used to control side effects when necessary. These include 10 to 30 mg of diphenhydramine hydrochloride (Benadryl) or 10 mg of chloroprophenpyridamine maleate for nausea. Phlebitis can be controlled by the addition of 10 to 30 mg of heparin. The solution is administered intravenously as a slow infusion over 4 to 6 hours. A 22- to 24-gauge needle is recommended. An *in vitro* method for monitoring therapy has been described by Borchardt et al.[10]

Renal toxicity is the most significant side effect of drug therapy. It is mandatory to follow the blood urea nitrogen level and the creatinine level during therapy. The latter is now considered the most sensitive index of renal damage. Most nephrotoxicity is reversible, but irreversible changes occur when more than an average of 3 g of amphotericin B is given. Pretreatment evaluation of kidney function will often forewarn the possibility of difficulties and can be used as a guide to modify the dosage regimen.

Meningitis usually requires the use of intrathecal, in addition to intravenous, amphotericin B therapy, because the drug does not pass the blood-brain barrier in sufficient quantity to attain therapeutic concentrations in the CNS. The treatment regimen may involve the use of the Ommaya reservoir, and often the Pudez ventriculoatrial shunt when hydrocephalus is present. Details of such treatments are given by Drutz,[31] Winn,[143] and Zealear and Winn.[149] Many complications arise from the treatment for meningeal disease, including obstruction of flow of CSF (resulting in noncommunicating hydrocephalus), ventricular spread of the infection, and lack of penetration of IV-administered amphotericin B. Intrathecal drug use is accompanied by severe inflammatory reaction or even acute toxic delirium,[146] which complicates evaluation of therapy. Complement fixation (CF) antibody titer is probably the best parameter to follow for such evaluation. At the present time meningeal disease can be arrested but seldom cured. Thus, intermittent suppressive therapy may be required for many years or for life.

Hundreds of other drugs have been tried for the treatment of coccidioidal disease. (A partial list of these is given in Fiese.[38]) Few of these, however, have been found to be efficacious. Those most recently evaluated are miconazole, ketoconazole, and econazole. Miconazole IV has been associated with a significant relapse rate and adverse reactions such as nausea, phlebitis and pruritis (the latter was probably caused by the vehicle). It is presently being evaluated in combination with amphotericin B for meningeal disease.[25] At this writing, the drug appears to have limited value in the treatment of coccidioidomycosis.

Ketoconazole has been used in the treatment of progressive cutaneous and subcutaneous disease. In preliminary trials using a dosage of 200 mg per day per os, the agent arrested the course of the infection. Treatment schedules are long — a year or more — but symptoms of the disease were noted to have improved by three months. The drug has the distinct advantage of being administered orally. It has an MIC of 0.77 μg per ml, similar to that of miconazole, and interferes with the ergosterol metabolism of the organism. So far, side effects have been limited to slight nausea and pruritus in a few patients. The relapse rate is quite high, however, and wide application of this drug is unlikely. This agent is also being tested in other mycotic infections, as is the structurally related compound econazole. Evaluation of the latter drug is pending. It is hoped that for coccidioidomycosis as well as for the other systemic mycoses, a better and less toxic drug than amphotericin B will be found.

PATHOLOGY[56]

Gross Pathology. Death in the primary stage of coccidioidomycosis is very infrequent, so that the early gross and histopathologic changes that occur have not been well described. From the few natural cases that have been studied and from data obtained in experimental disease in animals, the following sequence has been delineated. Conidia that land on the bronchi evoke an endobronchial reaction, such as a bronchitis or a bronchopneumonia. Conidia reaching the alveolar space induce an alveolitis, with an initial cellular invasion of macrophages. This is rapidly followed by an acute pyogenic reaction. The course varies considerably at this point. Necrosis, caseation, and granuloma formation may follow in any combination and in any degree. Thus the lesions seen in gross specimens show localized or patchy pneumonitis, a caseous pneumonitis, or involvement of whole lobes with consolidation and necrotic cavitation (Fig. 17–10). In the milder forms of disease, microabscesses and tubercles are found, and in severe infections numerous necrotic granulomatous nodules are present. Frequently, small areas of necrosis lead to cavity formation. The peribronchial and peritracheal lymph nodes are enlarged, as they become infected very early in acute primary disease. With resolution of primary infection, nonnecrotic granulomatous areas hyalinize and appear later as small bands of scar tissue or as firm nodules several centimeters in diameter. Cavities are formed in areas of necrosis. These become surrounded by a granulomatous reaction, thus forming thick-walled, well-organized nodules or cavities with central necrosis or thin-walled cavities with central caseation. The former may develop into coccidioidomas, and the latter are seen as the diagnostically troublesome, thin-walled "coin lesions" (Fig. 17–5,*A*). Such lesions most often are solitary, but may be multiple in either one lobe or several lobes.

Figure 17–10 Coccidioidomycosis. Cut section of lung. There is focal consolidation in upper and lower lobes with generalized consolidation and focal abscess formation in the middle lobe. Fibrosis is evident in hilar lymph nodes.

They very closely resemble the circumscribed nodules known as histoplasmomas and tuberculomas. In endemic areas such lesions can be given a presumptive diagnosis of coccidioidal granuloma; in nonendemic areas the true etiology usually must await surgical removal and pathologic examination.

The gross pathology of progressive and disseminated disease has been the subject of a number of reviews.[56] The types of lesions and their distribution are essentially identical to those of blastomycosis. In both diseases pulmonary involvement may be minimal or extensive lung disease may be present, but invariably there is extensive cutaneous and subcutaneous involvement. In addition, both diseases frequently cause lesions in bone with the same favored sites. It is easy to appreciate why these two diseases were often confused in their early history. The pulmonary findings are similar to those seen in resolved primary or residual disease. In disseminated disease, however, miliary abscesses and nodules caused by hematogenous spread of the organism may be found. It must be remembered that, although the lung is the portal of entry for the organism, many cases of fatal dissemination have been recorded for which no pulmonary pathology was found.

One of the commonest causes of death in coccidioidomycosis is meningitis. Of 181 such cases tabulated by Huntington,[56] leptomeningitis was present in 100. The characteristic lesion is a diffuse granulomatous meningitis that envelops the brain. Such lesions often cause an obstructive hydrocephalus. The brain matter itself is very rarely the site of disease. In the few cases in which infection of the brain occurred, the lesions varied from small, soft granulomas to large abscesses with ragged borders.[38]

Based on autopsy studies, statistically the spleen and liver are second only to the lungs as sites of disease. Both organs are usually infected late in the course of the disease, so there is no clinical evidence of involvement. Often spherules and endospores are seen with little or no cellular reaction around them, indicating an anergic patient. Other lesions vary from acute pyogenic tumors to well-formed granulomas. Sites of disease in other organs in descending frequency are kidney, skin, bone, and adrenals.

Clinically important disease is produced in the lung, meninges, skin, and bone (in that order) during the course of the infection. In these areas well-formed granulomas as well as areas of necrosis and abscess formation are to be expected. Other organs are involved late in the disease during the terminal "fungemia," and such lesions are usually nothing more than small areas of necrosis and abscess formation.

Histopathology. The cellular reaction evoked in infection by *C. immitis* is of three basic types. These are pyogenic (purulent), granulomatous, and mixed. The pyogenic reaction occurs around the initial inflecting conidia and also within granulomata at the time of the rupture of the spherule and release of endospores (Fig. 17–11). This is thought to be the result of the chemotactic response of neutrophils elicited by the substances released during spherule rupture, al-though no differential chemotactic response for inflammatory cells has ever been demonstrated. Granuloma formation occurs around the developing spherule (Fig. 17–12,*A*). Spherules are frequently found within histiocytes during the early development of the granuloma, and in later stages within foreign body giant cells. The granulomatous response demonstrates the sequential invasion of lymphocytes, plasma cells, monocytes, histiocytes, epithelioid cells, and giant cells. After organization, these lesions exhibit fibrosis, caseation, and calcification (Fig. 17–12,*B*). Calcification in coccidioidomycosis occurs as frequently as, or to a greater extent than, that in histoplasmosis. In lesions in which the organisms are growing and reproducing, a mixed-type cellular reaction is found. This mixing of the granulomatous response associated with spherules and the pyogenic response evoked by release of endospores was described in detail by Ophüls[93] in defining the coccidioidal granuloma. The intermittent pyogenic reaction occasioned by the release of endospores from spherules is presumed to be responsible for this histologic picture. Thus, the histology associated with coccidioidomycosis differs from tuberculosis and histoplasmosis and more closely resembles that of melioidosis and blastomycosis. In old, inactive lesions in which the organisms are dead or moribund, one is more likely to see a "pure" granuloma. As the resistance of the patient diminishes in disseminated disease, purulent lesions become more prominent.

In analyzing the histologic response seen in coccidioidomycosis, Drutz and Catanzaro[31] noted an immunopathologic spectrum among patients. This ranged from a predom-

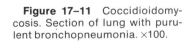

Figure 17–11 Coccidioidomycosis. Section of lung with purulent bronchopneumonia. ×100.

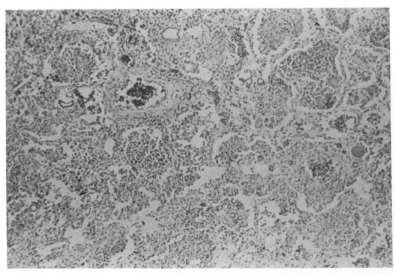

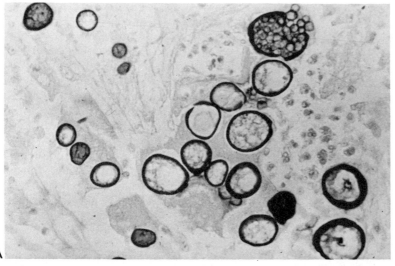

Figure 17–12 *A,* Coccidioido-mycosis. Gridley stain of section from lymph node showing various stages in the development of spherules. ×400. (From Rippon, J. W. *In* Freeman, R. A. 1979. Burrow's Textbook of Microbiology. 21st ed. Philadelphia, W. B. Saunders Company, p. 764.) *B,* Caseous necrosis in a pulmonary granuloma. Later organization will lead to capsulization, fibrosis, and calcification. These lesions are similar to those seen in histoplasmosis and tuberculosis. (Courtesy of W. Kaplan.)

inate formation of caseous granulomata in the highly "immune" patient to prominent suppuration in the "non-immune" patient, in whom serious or fatal disease was more likely. There were also other associated manifestations with these two patient types. The "immune" patient had maximal delayed type hypersensitivity (skin test reaction) and absent or minimal antibody response. In addition, few organisms were seen in tissue, and there was evidence of the destruction of fungal units. An intact cutaneous response to "recall" skin tests such as mumps, trichophytin, streptokinase-streptodornase (SK/SD), and so forth, was also present in these patients. In the patient likely to develop serious disease or in whom dissemination had already occurred, there is anergy to coccidioidin or spherulin skin testing, persistent or rising titers of hu-

moral antibodies, lack of or diminished "recall" skin testing, loss of sensitization ability to dinitrochlorobenzene, and suppurative abscesses in osseous, muscular, and cutaneous lesions with numerous organisms observable in tissue. Without therapy, these patients would succumb to generalized dissemination. Presently available therapy is not curative in such cases and thus treatment must be pursued for many years.

The character of the cutaneous lesions of chronic coccidioidomycosis is particularly noteworthy. The pseudoepitheliomatous hyperplasia may be as pronounced as that seen in cutaneous blastomycosis (Fig. 17–13). More than once skin biopsies from patients with this form of disease have been read as neoplasia. One can understand why Posadas thought his case was a type of mycosis fun-

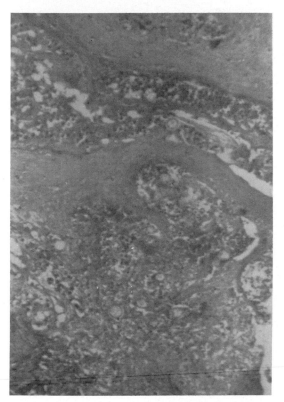

Figure 17-13 Coccidioidomycosis. Pseudoepitheliomatous hyperplasia of skin. Note the spherules. Hematoxylin and eosin stain. ×100.

goides. The spherules of *C. immitis* are much more easily seen than the yeast cells of *Blastomyces*, so that misdiagnosis is less frequent in coccidioidomycosis than in blastomycosis. As the patient's defenses diminish in progressive disseminated disease, suppurative necrotic lesions of the cutaneous and subcutaneous tissue are found.

Cutaneous lesions not associated with the presence of organisms are found in primary disease. These include erythema nodosum and multiforme-like eruptions. These are toxic erythemas and consist of panniculitis and angiitis.

There are also other histologic findings not associated with the presence of organisms. One is the diffuse myocarditis that is not infrequently found in disseminated disease. The reaction is a nonspecific inflammatory response, and its etiology is not known. Another is the renal lesion associated with amphotericin B toxicity.[105] At first the lesions are mainly confined to the tubules and are reversible with cessation of chemotherapy. Later, severe glomerular necrosis occurs and is usually considered irreversible. The drug remains in the body at low levels following cessation of therapy. Thus the glomerular necrosis may occur after the end of the treatment regimen.

Coccidioides immitis is easily seen in routine hematoxylin and eosin stained tissue sections. Following inhalation, the arthroconidium develops into a spherule, within a few hours or days. Arthroconidia are barrel-shaped and average 2.5 to 4 by 3 to 6 μ in size. They become more rounded as they transform into spherules (Fig. 17–12). At maturity, these are 30 to 60 μ in diameter (Figs. 17–14 and 17–15). The wall is thick (to 2 μ) and quite prominent. The cytoplasm is eosinophilic and contains many nuclei. As the spherules near maturity, endospore production begins by a process called "progressive cleavage." Fur-

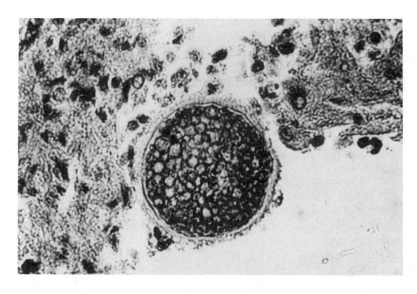

Figure 17-14 Coccidioidomycosis. Mature spherule with endospores. Hematoxylin and eosin stain. ×500.

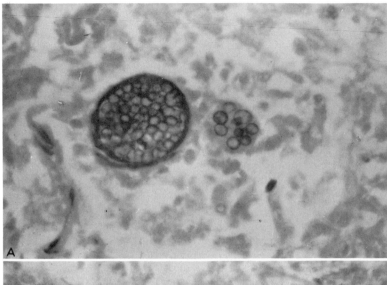

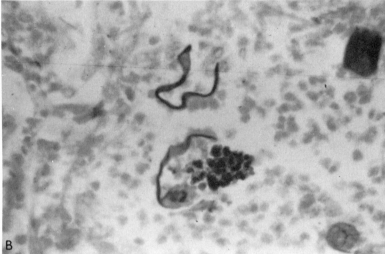

Figure 17–15. Coccidioidomycosis. *A,* Spherule with endospores and released endospores. Note that the endospores somewhat resemble budding yeast cells of *Blastomyces* or other fungi. *B,* Endospores and cell wall of empty spherule. Gridley stain. ×450. (Courtesy of P. Graff.)

rows form and divide the protoplasm into multinucleate masses called "protospores." Often there is a central vacuole in the spherule. Secondary cleavage lines are formed which divide the contents of the spherule into uninucleate endospores that are 2 to 5 μ in diameter. At maturation, the spherule wall breaks and the endospores are released. It is at this time that neutrophils can be seen invading the spherule and surrounding the newly released spores. It is probable that this is the time when the patient's defenses are most effective in killing these newly released spores and eliminating the infection. The surviving spores gradually evolve into spherules, and the process is repeated. In some tissue sections, spherules may be seen to have eosinophilic radiations emanating from the wall (Fig. 17–16). This is probably an example of the Splendore-

Hoeppli phenomenon, the eosinophilic material representing antigen-antibody complexes.

In sections from old inactive lesions, the spherules may appear very distorted, broken, crescent-shaped, and empty (Fig. 17–15, *B*). Endospores which approximate the size of neutrophils are often difficult to distinguish from artifacts and leukocytic debris in hematoxylin and eosin stains. The special stains for fungi, such as Gridley, methenamine silver, and Hotchkiss-McManus (PAS), are useful in such cases. Since the organisms are usually dead, or the specimen was immersed in formaldehyde before mycotic disease was suspected, cultures cannot be performed, and it is often difficult to make a specific diagnosis on histopathologic grounds alone. Small, empty spherules may approximate the size of the yeast cells of *B. dermatitidis*, and persistent

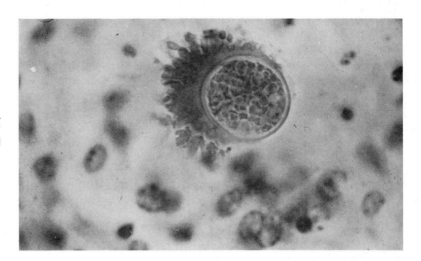

Figure 17-16 Coccidioidomycosis. Spherule with "asteroid" formation. Hematoxylin and eosin stain. ×450.

endospores may be confused with the cells of *Cryptococcus neoformans, Histoplasma capsulatum,* and *Paracoccidioides brasiliensis.* The appearance of all these organisms in old lesions is atypical. Cryptococcosis can be diagnosed histopathologically by the use of Mayer's mucicarmine stain. The close approximation of two small cells of *Coccidioides* may be misinterpreted as budding, so that this cannot be relied upon for discerning the correct etiology (Fig. 17–15). Search of many fields is often necessary to find a spherule with some endospores. Use of specific fluorescent antibody is an aid in these cases.[67]

In old cavitary lesions of the lung and in some granulomata, hyphae of *C. immitis* are found (Fig. 17–17). In one series the incidence was 73 per cent in cavities and 30 per cent in granulomatous lesions.[103] Hyphae have also been described in meningeal lesions. The hyphae observed do not contain arthroconidia; therefore, unless spherules are present, a specific etiology cannot be assigned. With careful search, however, a few endospore-containing spherules can be seen. It must be remembered that hyphae of *Aspergillus* may also be found in old cavities of coccidioidal granulomata. Fungus balls formed from the hyphae of *C. immitis* are reported.[37]

ANIMAL DISEASE

Natural Infection. Natural infection among animals is very common in the zones which are highly endemic for coccidioidomycosis. Emmons[35] in 1942 was the first to note coccidioidal lesions in the lungs and lymph nodes of native rodents in the southwestern

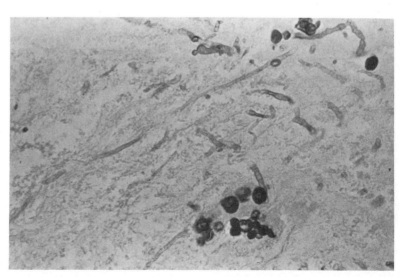

Figure 17-17 Coccidioidomycosis. Spherules and mycelium in fibrocaseous nodule. ×400.

deserts of the United States. The species included deer mice, pocket mice, kangaroo rats, and ground squirrels. It has been postulated that such infected animals may act as vectors of dissemination. This is as yet an unsettled question. The disease in such animals seems to be minimal, and predators feeding on them do not usually acquire the infection.[12] However, it was found that the burial of carcasses of animals that had died of experimental disease rendered the surrounding soil areas positive for the growth of *C. immitis* for many years.[84] There are very few species of native mammals that occupy the endemic zones of the *C. immitis,* and none of these seem to be susceptible to the serious forms of coccidioidomycosis.

The first report of natural infection in domestic animals was that of Giltner[46] in 1918. He described lesions in the bronchial and mediastinal lymph nodes of cattle. He reproduced the disease in swine by injection of lesion material. Since then, coccidioidomycosis has been reported in horses,[26, 75] cattle, swine,[134] llamas, burros, rabbits, dogs,[147] sheep, sea otters,[21] cats,[147] tigers,[55] and primates. The disease in most of these animals is minimal; the disseminated form is regularly found only in primates and dogs.

Several investigators have determined the susceptibility of monkeys to the disease and its transmission to cage mates.[69] As few as ten arthroconidia were able to cause infection in this animal. When infected monkeys were housed with noninfected monkeys, transmission of the disease occurred in a few cases.

The first case of disseminated disease in dogs was noted in 1940 by Farness. This occurred in his Great Dane, Sarah.[38] Since that time severe disease in dogs has been reported frequently. Maddy[82] reviewed several hundred canine cases and concluded that dissemination in such breeds as the boxer and Doberman pinscher is even more common than in man. The systemic disease is quite similar to that seen in human cases,[90] except that osseous lesions are proliferative rather than lytic. Benign disease akin to Valley fever has also been noted in some dogs. Infection in the cat has rarely been reported, and this species appears to be quite resistant to infection compared with the dog.[147]

From the first report in 1918 until the present, there have been many records of the disease in cattle. Essentially all of these have been noted in abattoirs where a few lesions are found in the lymph nodes or lungs. Disseminated disease is extremely rare, so that the presence of such lesions in lymph nodes does not require condemnation of the carcass. Maddy has skin tested some 12,000 cattle from different areas of Arizona. He found the reaction rate varied from 0 to 84 per cent. The overall average was 24.6 per cent.[85] In endemic areas, horses and sheep also have high rates of skin test sensitivity. However, disseminated disease is also very rare in these animals, though it does occur.

Experimental Infection. Experimental disease can be produced in most laboratory animals. The intravenous inoculation of ten arthroconidia or less evokes a chronic infection in mice. At a dosage of 100 to 500 arthroconidia, a rapidly fatal disease is produced. The virulence of different strains is variable, however.[41] Experimental infection in bats has been produced by Krutzsch and Watson.[70] They speculate on the importance of natural infections in bats as a vector in disseminating the organism to new environments. Inoculation of conidia into poikilothermic animals does not elicit disease.

BIOLOGICAL STUDIES

Because of the great importance of coccidioidomycosis in its endemic areas, many investigations of the biology of *C. immitis* have been carried out. These have been concerned with practical aspects, such as vaccines, soil fungicides, and so forth, and many basic studies on the pathogenesis, metabolism, physiology, and development of the fungus have also been done.

No serious work on a vaccine has been done with any of the other mycoses. Because it is known that strong immunity follows resolution of primary infection and because of the significant susceptibility of certain racial groups and occupations in endemic areas, much effort has gone into the development of a vaccine. Both living and nonliving vaccines have been tried. Attenuated spherules have been produced that markedly increase the resistance of mice to challenge infections.[76-79] Such attenuation was found to be an unstable characteristic, however, and thus the use of such vaccines was deemed dangerous. Scalarone and Levine[112] found that mice were more resistant to challenge infections following intravenous or intramuscular injections of formalin-killed spherules. A stronger immunity was acquired using the intramuscular route as compared to the intravenous route. The intravenous route induced higher precipitin titers but weaker delayed hypersensitivity. This route of injection also did not induce the leukocytolytic and macrophage aggregation response of white blood cells in the presence of *Coccidioides* antigen. These responses were conferred by intramuscular injection of spherule vaccine. The use of these vaccines in humans has conferred skin test reactivity on some,[76] but in other subjects it was lacking or at most weak.[97] Controlled clinical trials of the vaccine have not been carried out. Other types of vaccines prepared from arthroconidia and cellular components have also been investigated.[19] At this writing, a safe, effective vaccine does not appear to be forthcoming.

The activation of cellular defenses by the *Coccidioides* organism has been the subject of much investigation. Both mycelial and spherule extracts of the organism have been shown to attract leukocytes. This attraction is dose-related[44] and complement-mediated. The chemotactic substances from the mycelial and spherule phases differ from one another, as shown in chemical analysis described by Galgiani et al.[44]

Another practical aspect of coccidioidomycosis that has been investigated is the possibility of soil disinfection. This would be particularly appropriate in highly endemic areas where construction or agricultural work is contemplated. So far, one partially effective fungicide has been found. Used as a soil spray, 1-chloro-2-nitropropane leaves an area noninfectious for 24 hours after use.[34] Results of field tests have not as yet been reported.

Studies on the virulence of *C. immitis* have been carried out by several investigators. Friedman et al.[41] found considerable variation among strains of *C. immitis* in their ability to cause disease in mice. Gale et al.[43] have determined that the dissemination rate was higher in pigmented than in albino mice. In addition, they demonstrated the toxicity of culture filtrates of the organism. Mice surviving toxic challenge were more resistant to subsequent infection when challenged with an infection by injection of conidia.

The effect of amphotericin B on *C. immitis* has also been investigated. The growth inhibiting property is associated with injury to the cell membrane, particularly at loci containing sterols. Exposure to the drug results in a lesion which causes loss of intracellular metabolites, particularly potassium, inorganic phosphate, and sugars. However, even after treatment of endospores with 4 μg per ml of the antibiotic for five days, capacity to grow was restored by subsequent treatment with cysteine or glutathione.[118] The organisms thus treated do not become spherules but rather grow as mycelial units. The effect is thought to be owing to the lowering of the oxidation-reduction potential of the growth medium, which now permits growth of antibiotic-treated cells, but only in the mycelial form.

The phenomenon of dimorphism in *C. immitis* has been investigated by a number of workers. Similar to the other thermal dimorphic fungi, such as *Histoplasma capsulatum* or *Blastomyces dermatitidis*, morphogenesis in *C. immitis* is governed by temperature. Temperature and other factors, including CO_2 tension, are also important. Increased CO_2 and growth in a liquid medium are sufficient to produce "culture spherules" (Fig. 17–23). Such "culture spherules" may not be altogether comparable to "tissue spherules," however.[14] Even with all the environmental conditions favorable for spherule conversion the yield may not be great at 37°C. At 40°C, however, all new growth is as spherules. On almost all liquid and solid media, direct conversion of arthroconidia is obtained. The sequence of changes from arthroconidia to spherules during morphogenesis has been studied by Sun et al.[126] using electron micrographs.

All the enzymes of the Embden-Meyerhof-Parnas pathway and the pentose shunt have been found in *C. immitis*. The enzymes of the Krebs cycle are present except for α-ketoglutaric dehydrogenase. However, the finding of isocitric lyase suggests the operation of the glyoxylate shunt[80] in the organism. This metabolic pathway is common among many fungi. In *C. immitis*, the enzyme levels were always higher in the mycelial as compared to the "culture spherule" stage.

The cell walls of all growth forms of *C. immitis* contain an abundant chitinous "core." In addition, there are β-glucans and some other polysaccharides. The inner wall of the spherule is lined with phospholipids, the composition of which changes during morphogenesis from mycelium to arthroconidia to spherule.[140] Among the interesting components of the cell is the rare sugar 3-O-methylmannose,[99] known previously only in three species of bacteria. The compositional and morphological changes during conversion of arthroconidia to spherules are reviewed by Rippon.[106]

IMMUNOLOGY AND SEROLOGY

Immunity. The innate human immunity to infection by *C. immitis* is extremely high. At the time of initial inhalation of conidia, this is a general resistance or nonspecific immunity to the infection. As discussed in previous sections, in fully 60 per cent of people infected, the entire disease process and its resolution is completely asymptomatic. Most of the remaining 40 per cent have a mild febrile to moderately severe respiratory disease. Following recovery from the infection, the patient is left with a strong specific immunity to reinfection. This immunity can be gauged to a certain extent by responsiveness to the coccidioidin skin test. It is also reflected by the allergic manifestations present during the initial infection. If dissemination is going to occur, it does so at an early stage of primary infection. Recovery from primary infection essentially rules out dissemination, even in those with residual cavitary disease. Only in patients with persistent adenopathy is systemic disease probable, even when the infection in the lung has resolved.[65] In patients demonstrating toxic erythemas, dissemination is even less likely. As has been noted, these various responses are related to sex and race. Thus the white female is most likely to demonstrate allergic manifestations and least likely to have disseminated disease. On the other hand, dark-skinned men very rarely have allergic lesions, frequently show persistent adenopathy,[65] and most often have disseminated disease. There are, of course, exceptions to this general pattern. In a few documented cases, dissemination has occurred in the presence of erythema nodosum. This may have been related to the infectious dose. However, cases of so-called "late" dissemination are more likely to be "late" clinical appearance of what was an occult, quiescent, disseminated infection.

Following inhalation of the arthroconidia and initiation of the infectious process, an alveolitis with a pyogenic cellular response ensues. As in other types of infectious diseases, macrophages phagocytize, degrade, and process the invading microorganism. The derived antigenic material is appropriately displayed so as to sensitize lymphocytes. The sensitization of the T lymphocytes is manifested by the development of delayed hypersensitivity to antigens of the organism and "superactivated" macrophages with enhanced microbial killing ability. Sensitization of B lymphocytes results in the production of the usual sequences of IgM followed by IgG antibodies (some IgD and IgE have also been detected). This overall marshaling of the various components of the immune response is influenced by populations of T lymphocytes with "suppressor" or "helper" function. Any variance or defect in this system, whether transient or permanent, can result in persistence of the infection, and the degree of the variance can affect the severity and ultimate outcome of the disease. In coccidioidomycosis, the appearance of delayed type hypersensitivity is indicated by reactivity to the coccidioidin or spherulin skin test. The tube precipitin (TP) test measures the production of IgM antibodies, and the various IgG antibodies are measured by the complement fixation (CF) and immunodiffusion (ID) tests.

Serology. Our knowledge of the various serologic and immunologic patterns and their relation to clinical disease is greater in coccidioidomycosis than in any other of the mycoses. Beginning with Smith,[122] who correlated the results of 39,500 sera with the various clinical forms of the infection, and adding the work of Huppert,[57] Pappagianis,[97] Levine,[77] and others, we now have valuable serologic aids for the diagnostic and prognostic evaluation of this disease. These include the well-standardized skin test reagent, coccidioidin, the complement fixation test, the precipitin tests, and the more recently developed latex agglutination, immunodiffusion, and fluorescent antibody tests.

Coccidioidin Skin Test. The standard procedures for the manufacture and use of coccidioidin were developed by C. E. Smith and associates.[122] These were modified from those used in the production of the tuberculin skin test reagent. The coccidioidin skin test consists of the intradermal injection of 0.1 ml of a 1:100 dilution of standardized mycelial coccidioidin. The reaction is read at 24 and 48 hours. The induration must be 5 mm or more to be considered positive. There may be an immediate response, but this is ignored. Uusually there is erythema accompanying the induration, which may consist only of a faint blush. The following precautions and sources of error are to be noted. The needle used should be a disposable one or one that has not previously been used for injection of biological materials. Tuberculin is not inactivated by autoclaving and may evoke a false-positive reaction. Subcutaneous injection of the material will also lead to false results. The maximum reaction is reached at an average of 36 hours. Reading too early or too late may lead to false interpretations. A few patients demonstrate reactivity to other fungal skin test antigens, though usually to a lesser degree than to coccidioidin. Patients with any allergic manifestations may have a severe reaction to the standard dose of coc-cidioidin, and dilutions should be used. Le-vine et al.[78] have prepared a skin test antigen from the spherules of *C. immitis*. It appears to be much more sensitive than standard coc-cidioidin and therefore may be more useful in diagnosis of the disease and in epidemio-logic studies.

Erythema nodosum, erythema multiforme, and desert rheumatism are accompanied by a strong and often severe reaction to the intra-cutaneous injection of coccidioidin. The se-quelae include headache, fever, worsening of the toxic lesions, swelling, and possible necro-sis at the test site. For this reason, in patients manifesting any allergic reactions, a 1:1000 or 1:10,000 dilution of coccidioidin is used. Allergy to coccidioidin can be passively trans-ferred by the intradermal injection of leuko-cytes from skin test–positive patients into the skin of coccidioidin test–negative patients.[104] This cannot be accomplished by injecting serum. Repeated skin testing does not appear to induce a positive skin response or a detect-able antibody titer. In contrast to histoplas-min, the use of the coccidioidin skin test reagent does not cause a false rise in the complement fixation titer. However, patients originally negative to dilute coccidioidin (1:100) may be positive following injection of more concentrated material.[104]

Patients with primary coccidioidomycosis develop skin test sensitivity quite rapidly. The reaction is positive in 87 per cent of cases during the first week and in 99 per cent after the second week of clinical symptoms. The latex agglutination test is sometimes positive before this. The extremes recorded for devel-opment of coccidioidin sensitivity are 2 to 21 days after onset of symptoms. A negative skin test may be the result of testing too early in the course of the disease or an indication of impending dissemination. The latter is par-ticularly true when the test was previously positive and then converts to negative. Such anergy is associated with a grave prognosis. The skin test is also often negative in chronic progressive pulmonary disease; in the persist-ent, very slowly developing meningeal dis-ease; in chronic single and multiple organ disease; and in fulminant terminal illness. Therefore, a negative skin test should never be used as "proof" that a patient does not have coccidioidomycosis. In patients with negative skin test in the face of known infec-tion, other "recall" skin tests are negative, and the ability to be sensitized by DNCB (dini-trochlorobenzene) is also lost. This is similar to the picture encountered in leprosy or miliary tuberculosis. Our present under-standing of the immunopathology of the dis-ease suggests that this may result from block-ing factors in serum that inhibit or depress transformation of lymphocyte enhancement of "suppressor" function of T cells or inter-ruption of the circulation of T lymphocytes (sequestering).[96]

According to the method of Smith, coc-cidioidin is produced by growing the fungus in a modification of the asparagine, salts, glucose medium used for preparation of PPD tuberculin. Ten strains of *C. immitis* are used. These were isolated from patients with vari-ous clinical forms of the disease and from various geographic areas. No significant dif-ferences in the coccidioidin produced by the various strains have been observed, although fine analysis has shown some antigenic differ-ences between some isolants.[57] The fungus is grown for 6 to 10 weeks in static broth culture, after which aqueous Merthiolate is added, and the culture filtrate is separated out by passage through a millipore or Berke-feld filter. The material is tested for sterility, potency, and specificity. Many lots must be discarded because of the lack of specificity and potency. Coccidioidin is also used in the complement fixation and precipitin tests. Batches are standardized for each test sepa-rately, and a single batch may be excellent for one test and unusable for the others. The material appears to be stable indefinitely. The complement fixing antigens are destroyed by autoclaving, but those responsible for the skin test and the precipitin test are not.

A tine test has been developed using coc-cidioidin.[135] In comparison to the standard intracutaneous injection of 0.1 ml of a 1:100 dilution, it was found to have certain advan-tages for large scale surveys but lacked the flexibility of differing dosages needed in clin-ical evaluation.

Tube Precipitin Test (TP). In essentially all symptomatic and asymptomatic cases of primary coccidioidomycosis, the coccidioidin skin test will be positive and is usually the first serologic procedure to become so. Other se-rologic procedures, such as the tube precipi-tin and complement fixation tests, are posi-tive for varying lengths of time in most patients that have symptomatic disease. Pre-cipitins usually appear within the first week of clinical illness and persist for only a few days or weeks, even if dissemination occurs. Within the first few weeks of clinical disease, 90 per cent of symptomatic patients will have a positive tube precipitin test, but by the fourth month only 10 per cent do. Eventually patients with all forms of the disease become negative by this test. Therefore, a positive tube precipitin test indicates early active disease. The precipitin test is highly specific for coccidioidomycosis, and cross-reactions are very rare.

The test is carried out by the usual method for detecting precipitating antibodies. Into each of four tubes is placed 0.2 ml of undiluted serum. This is followed by the addition of 0.2 ml of undiluted coccidioidin in the first tube, 0.2 ml of a 1:10 dilution in the second, and 0.2 ml of a 1:40 dilution in the third. A control is run using plain asparagine broth. The solutions are mixed and read daily for five days. A positive result is indicated by the appearance of a flocculus or precipitate at the bottom of the tube.

Complement Fixation Test (CF). Complement fixing antibodies are the last to appear, but they have the greatest prognostic significance. Sometimes their presence is not detected until three months after the appearance of clinical symptoms. They, like the precipitins, appear in 90 per cent of symptomatic patients, but in only 10 per cent of asymptomatic cases. Though they usually disappear by six to eight months, they may persist at low titers for years in patients with resolved, uncomplicated disease. In disseminated infection, they rise to very high titers and persist through the fatal terminus of generalized disease. For this reason a persistent or gradually rising CF titer combined with clinical symptoms indicates present or imminent dissemination.

The CF test is based on the quantitative Kolmer complement fixation technique for syphilis. Serial dilutions of sera are made to 1:512. Constant amounts of the other reagents are added. After proper incubation the test is read by noting the presence or absence of hemolysis. Details are given in Section Five. A single positive complement fixation test may indicate either early active primary disease, a residual disease, such as coccidioidoma, or progressive disease. Therefore, serial samples are necessary for proper evaluation.

The CF test for coccidioidomycosis is quite specific, and cross-reactions with other mycotic infections are rare. In cases of acute or disseminated histoplasmosis, however, the test may be weakly positive. Though the coccidioidin CF test is not usually positive in blastomycosis, the blastomycin CF test may be positive in dilutions of serum of 1:128 or higher in cases of coccidioidomycosis. This antigen (blastomycin) is of absolutely no value in any serologic procedure for any disease.

Immunodiffusion and Latex Particle Agglutination. The tube precipitin (TP) and complement fixation tests (CF) have proved to be very useful in the diagnosis and prognosis of coccidioidomycosis. Unfortunately, as routine screening procedures or in small diagnostic laboratories they have many drawbacks. Chief among these are time, the necessity for special equipment and technical skill, and expense. For this reason two other procedures have largely replaced the standard methods for initial screening. These are the immunodiffusion test (ID) and the latex particle agglutination test (LPA). Both have been developed and tested in large measure by Huppert and associates.[57, 96] As a single test, the LPA test was positive in 70.6 per cent of cases of disease, compared to 13.5 per cent for the tube precipitin (TP) test. The former had a 3 per cent false-positive rate. Again, as single tests, the CF and immunodiffusion (ID) tests were of equal sensitivity, being positive in about 80 per cent of cases. However, as paired procedures, the LPA and ID detected 93 per cent of cases, compared to 84 per cent for TP with CF. Thus as an initial diagnostic screen, the combination of the two tests (LPA and ID) is superior on several counts. If these two are positive, a CF should be done for confirmation and setting of a base line for later serial examinations.

Using culture filtrates, toluene extracts of mycelium, and other procedures, Huppert et al.[57-62] have found several antigenic types and combinations in different strains of *C. immitis.* This raises the following interesting possibility. What we think of as the limited distribution of the fungus is in large part the result of skin testing with coccidioidin prepared from strains found in essentially only one geographic area. The finding by Huppert et al. of strains with no antigens in common had suggested that the organism and infection could exist in other parts of the world. Further evaluation has not indicated this to be the case.[96] Whenever possible, skin testing for sensitivity survey should be done with organisms indigenous to the area or with a polyvalent antigen derived from a variety of isolants. At this writing four important antigens have been described from strains of *C. immitis.* They are detectable as independent lines on a double diffusion ID plate. Different strains may contain all, none, or any combina-

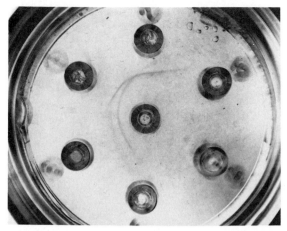

Figure 17–18 Coccidioidomycosis. Serology. Immunodiffusion plate. The center well contains antigen, and three wells contain sera from patients with different types of the disease. The wells with two precipitin lines are from cases of active disease. The serum in the well showing one line is from a patient with a "coin lesion."

tion of these antigens. The antigen 1 of Huppert et al.[57, 61] is involved in the tube precipitin test and the latex particle test. It is prepared from toluene lysates of mycelium. Many strains lack this antigen. Antigen 2 is the specific antigen involved in the complement fixation (CF) and immunodiffusion (ID) tests and is the standard antigen for these tests. Two other antigens have also been noted in ID tests by these investigators. The identity and significance of these other lines have not been determined.

The ID test becomes positive at about the same time as the CF test (Fig. 17–18). As far as has been determined, the duration of antibodies detectable by the two tests is roughly parallel. The ID test for coccidioidomycosis is performed, as are other double diffusion tests. Serum from the patient is placed into one well in an agar plate, and antigen is placed in an adjacent well. They are allowed to diffuse toward each other. The plates are read at 14 hours and each day for five days or more. Lines appear which represent specific antigen-antibody combinations. (See Section Five under Aspergillus antigen testing for use of EDTA to avert false positives. Quantitation of the test can be done using serum dilutions. The prognostic significance of the appearance and disappearance of particular lines has yet to be determined. Multiple lines are usually seen in active disease, whereas a single line is frequently associated with stable chronic infection. Spherules and arthroconidia also contain a

number of different antigens,[73, 74] But their diagnostic significance has not yet been delineated.

The latex particle agglutination test[62] is performed by placing two drops of inactivated serum and one drop of sensitized latex particles in a small tube. This is rotated at 180 rpm for four minutes. The tubes are read for the presence of agglutinated particles. The latex particles are sensitized by being coated with coccidioidin antigen.

Specific fluorescent antibody techniques have been developed for detection of spherules in tissue and arthrospores in soil. In addition, FA inhibition tests are sometimes positive in sera that is anticomplementary or that is negative by the tube precipitin test.[67]

The detection of antigen in serum would probably be the most specific diagnostic and prognostic test for the disease. So far the development of this procedure has not been completed.

Summary of Practical Serologic Procedures. For initial diagnosis and screening of sera, the LPA and ID are positive in 90 per cent of cases with clinical symptoms. The complement fixation (CF) and tube precipitin (TP) tests are used for diagnosis as well as for confirmation. Serial studies of the CF titer are used in prognostic evaluation. The skin test is widely used for epidemiologic studies; a single test is of value in diagnosis only in patients with a history of a previous negative reaction. It should not be used routinely. A negative skin test in a patient that previously had a positive test indicates anergy associated with advanced disseminated disease. The tube precipitin test, when positive, indicates early active disease.

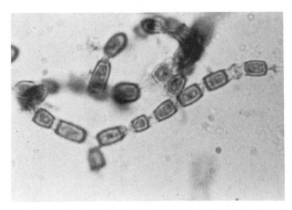

Figure 17–19 Coccidioidomycosis. Arthroconidia alternating with empty cells in mycelial form of *C. immitis.* ×450.

LABORATORY IDENTIFICATION

Direct Examination. Because of the hazards involved in the handling of cultures of *C. immitis,* direct examination of clinical materials and histopathologic studies are of great importance in the diagnosis of this disease. Mature spherules of the organism are large and, when present, are easily seen on direct microscopic examination. Young spherules and newly released endospores are difficult to differentiate from host cells and other artifacts, however. Sputum, pus, gastric washings, and exudates from cutaneous lesions can be examined using a 10 per cent potassium hydroxide mount and subdued lighting. The spherules containing endospores vary in size from 10 to 80 μ, with an average diameter of 30 to 60 μ (in experimentally infected mice, they may reach 200 μ). The wall of the spherule is doubly refractile and up to 2 μ in thickness (Fig. 17–20). The endospores have an average diameter of 2 to 5 μ. Their size and numbers are quite variable. Immature spherules and endospores are also variable in size, and they resemble nonbudding yeast cells of *Blastomyces dermatitidis;* therefore, a careful search should be made for spherules containing endospores. Once they are seen, this is sufficient corroborative evidence for the tentative diagnosis of coccidioidomycosis. In sputum or other samples held for periods of time, the spherules may form sprout mycelium and germ tubes. Numerous artifacts may be present in sputum or other material that resemble the young spherules of *C. immitis.* Iodine or Sudan IV stain can be used to differentiate starch and fat globules from the real organism.

Culture Methods. In cases in which the organism is not demonstrable in direct examination, or to confirm the identity of the organism, culture of the material may be performed. The fungus grows on almost all routine laboratory media with or without antibiotics. Because of the hazards inherent in culture work, the use of Petri dishes is not advised without special precautions. Material to be examined can be spread on slants or bottles of Sabouraud's agar and incubated at room temperature. Growth usually occurs by the third to fourth day, and conidiation by the tenth to 14th day. If subcultures are to be done, they should be made in a gloved isolation hood. When the culture is ready for examination, formaldehyde may be poured over the slants. Teased mounts of the mycelium are made with lactophenol cotton blue. Identification depends on finding the thin-walled rectangular, ellipsoidal, or barrel-shaped arthroconidia that are 2.5 to 4 by 3 to 6 μ in size. Confirmation of identification, however, requires the production of spherules in experimentally infected animals or in liquid culture at 40°C. Many soil keratinophils that may be incidentally isolated from sputum resemble *C. immitis.* Demonstration of thermal dimorphism is the only absolute criterion.

To obviate the dangers in handling live cultures of *C. immitis,* Kaufman and Standard[68] have developed an exoantigen test. In this procedure, 8 ml of a 1:5000 solution of Merthiolate in phosphate buffered saline is placed on a slant of the fungal growth. It is allowed to react for 24 hours at room temperature. A 5-ml sample is concentrated 25 times by an Amicon concentrator. A portion of this

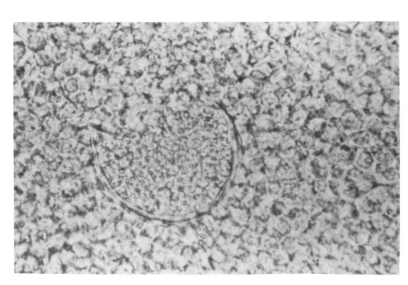

Figure 17–20 Coccidioidomycosis. Spherule present in wet mount of pus. ×450.

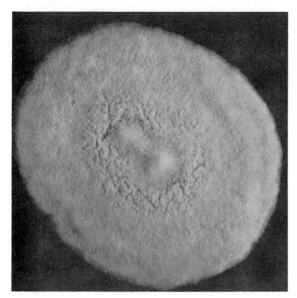

Figure 17–21 Coccidioidomycosis. Typical colonial morphology of *C. immitis*. The cracked, powdery appearance of the center indicates fragmentation to form arthroconidia.

sample is placed in an immunodiffusion plate against a known positive antiserum. A precipitin band indicates that the culture is most probably *C. immitis*. This is a very useful procedure. Sigler and Carmichael[117] have indicated that many arthroconidia producing soil organisms have similar or identical morphological characteristics to *C. immitis,* but Huppert et al.[63] have demonstrated the specificity of the exoantigen test for *C. immitis* in such "look-alike" cases.

MYCOLOGY

Coccidioides immitis in Rixford et Gilchrist 1896[107]

Synonymy. *Posadasia esferiformis* Cantón 1898; *Blastomycoides immitis* Castellani 1928; *Pseudococcidioides mazzai* da Fonseca 1928; *Geotrichum immite* Agostini 1932; *Coccidioides esferiformis* Moore 1932; *Glenospora metaeuropea* Castellani 1933; *Glenospora louisianoideum* Castellani 1933; *Trichosporon proteolyticum* Negroni et de Villafane Lastra 1938.

C. immitis is a dimorphic fungus. It grows as a mold with a septate mycelium in soil and culture media, and as an endosporulating spherule in animal tissue and under certain *in vitro* growth conditions at elevated temperatures. Unlike most of the other pathogenic fungi, the thermal dimorphism exhibited by *C. immitis* results in an endosporulating spherule rather than a budding yeast.

Colony Morphology. SGA, 25°C. Growth of *C. immitis* is apparent within three to four days after inoculation. At first, it is moist, glabrous, and grayish, but the colony rapidly develops abundant, floccose, aerial mycelium that soon covers the slant (Fig. 17–21). The mycelium is initially white, but usually becomes tan to brown with age.

As demonstrated by Emmons[36] and Huppert et al.,[59] there is considerable variation in the colonial morphology of *C. immitis* (Fig. 17–22), and Sigler and Carmichael[117] have shown (Fig. 17–24) that numerous soil fungi have similar morphology at 25°C. (Only *C. immitis* converts to the endosporulating spherule at 37 to 40°C in culture or in experimentally infected animals.) Some of the strains of *C. immitis* studied by Huppert et al.[59] were so different in their morphology that they would not have been correctly identified by routine laboratory procedures. A few strains produced colonies that had no aerial mycelium, others had radial grooves, and in others the color varied from gray to

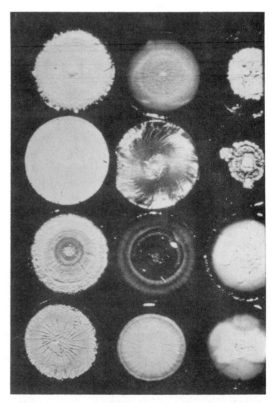

Figure 17–22 Coccidioidomycosis. Variation in morphology, texture, and color in isolants of *Coccidioides immitis*. (From Huppert, M., S. H. Sun, et al. Copyright 1967. Natural variability in *Coccidioides immitis,* In *Coccidioidomycosis.* Proceedings of Second Coccidioidomycosis Symposium. Ajello, L. (ed.), Tucson, University of Arizona Press, p. 323, by permission.]

pale lavender, pink, buff, cinnamon, yellow, and brown. Some strains produced a diffusible brown pigment. As stated previously, however, all of the isolants, regardless of their morphologic variation, were pathogenic for mice and produced spherules in tissue or at elevated temperature in culture.

Microscopic Morphology. Conidiation begins within a few days after the initiation of growth. The conidia usually appear first on side branches of the vegetative hyphae. The hyphae themselves are thin and septate, but the side branches are almost twice as thick and have numerous septations. Thick-walled arthroconidia are then produced, alternating with thin-walled empty cells (disjunctors). The arthroconidia are barrel-shaped 2.5 to 4 by 3 to 6 μ in size, and are released by fragmentation of the mycelium. They retain portions of the walls of the disjunctor cells as ornaments on the walls of either end. This characteristic is often helpful in distinguishing the species. The cytology and nuclear cycle of arthroconidia formation have been carefully detailed by Kwon-Chung.[71] As the culture ages, the vegetative hyphae also fragment into arthroconidia. No other type of conidia formation is seen in culture of this fungus, but racquet mycelium may be found.

Huppert et al. also found a wide variation in conidiation. This variation occurred within the same strain or was constant within a strain. A few straight or flexuose cells ranged in size from 1.5 to 2.3 by 1.5 to 15 μ. More commonly, the size range was 3 to 4.5 by 3 to 12 μ. Some conidia were thin-walled and elongated, and some had rounded ends. In some, conidia-bearing hyphae occurred in clusters and branched at acute angles from vegetative hyphae, and in others, the conidiating hyphae tended to lie in parallel rows.

Dimorphism with the production of spherules is produced *in vitro* using particular media and increased CO_2 tension (see Section Five) with incubation at 37 to 40° C (Fig. 17–23). Spherules are most easily produced in liquid modified Converse medium (MCM) with CO_2 added at 37° C, or better yet at 40° C. Some spherule formation can be seen in agar slants 40° C using a variety of media, however. The MCMA of Brosbe[13] is probably the best medium for initiation and propagation of the spherule stage of *C. immitis*. Details of the development of the spherules in tissue are discussed in the section on histopathology.

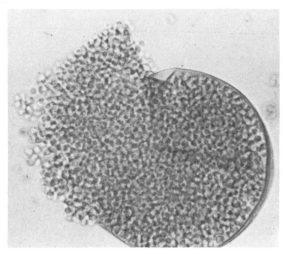

Figure 17–23 Coccidioidomycosis. "Culture spherule." ×400. (Courtesy of H. Levine.)

The taxonomic relationship of *C. immitis* to other fungi is one of the most baffling in medical mycology. The organism was at first thought to be a protozoan. When the fungal nature of the organism was discovered, its spherules were considered to be asci, and it was classed as an Ascomycete.[92] However, the number of endospores produced was irregular, and they were formed by progressive cleavage — characteristics not usually observed in fungi of the class Ascomycetes.

In the 1920's Castellani grouped *C. immitis* with the Hyphomycetes or Fungi Imperfecti, and many mycologists agreed with that classification at the time. Endosporulation by progressive cleavage is a characteristic of the zygomycetous fungi in their formation of sporangia and sporangiospores. Thus in 1926 Ciferri and Redaelli suggested that *C. immitis* was a Zygomycete, but the problem of its taxonomy remains unsettled. The soil or saprophytic form of the organism, with its septate mycelium and production of alternately spaced arthroconidia, is similar to those of several other soil fungi,[36, 117] including haploid Basidiomycetes and Hyphomycetes. However, it most closely resembles the imperfect stage of some of the Gymnoascaceae, which include members of the genus *Auxarthron* and the group designated "Gymnoascus imperfecti."

C. immitis has been reported to be keratinophilic,[89] and in soil it is frequently found in association with other keratinophilic species of the genera *Arthroderma*, *Trichophyton*, and *Microsporum*.[95] None of these organisms has as yet been shown to form an endosporulating spherule. The only other patho-

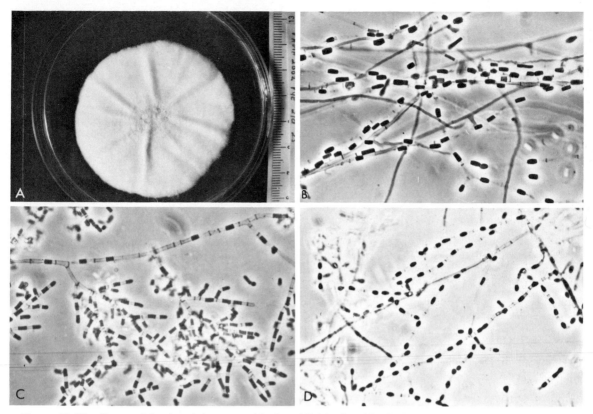

Figure 17–24. Some of the fungi that resemble *Coccidioides immitis. A,* Colony of the Malbranchea state of *Uncinocarpus reesii,* 21 days on PYE at 25°C. *B,* Arthroconidia of the **Malbranchea** state of *Uncinocarpus reesii.* ×830. *C, Malbranchea dendritica.* ×830. *D, Malbranchea gypsea.* ×830. (Courtesy of L. Sigler and J. W. Carmichael.)[117]

genic fungus demonstrating this characteristic in tissue is *Rhinosporidium seeberi,* but the two organisms do not appear to be closely related. The Zygomycetes in general do not have septate mycelium, and do not form arthroconidia. Thus, we have a fungus (*C. immitis*) that has characteristics of a Zygomycete in its parasitic stage and is gymnoascaceous in its saprophytic stage. The problem will be fully settled only when the perfect state of *C. immitis* is found. If, in fact, *C. immitis* were found to be in the Gymnoascaceae, then essentially all the true pathogenic fungi, both dermatophytes and systemic infecting organisms, would be classed in that small family of the Ascomycota.

Recently, Sigler and Carmichael[117] addressed the problem of the taxonomic relationships of *C. immitis.* They feel that the organism is closely related to other arthroconidia-producing fungi referable to the *Malbranchea* genus. This attitude is being adopted by many mycologists at the present time. Changing the name of the organism now appears unwarranted, however. The authors also emphasize that many species of soil keratinophils that produce arthroconidia are indistinguishable morphologically at 25° C from *C. immitis* (Fig. 17–24).

REFERENCES

1. Ajello, L., K. Maddy, et al. 1965. Recovery of *Coccidioides immitis* from the air. Sabouraudia, *4*:92–95.
2. Ajello, L. (ed.) 1967. *Coccidioidomycosis.* Proceedings of Second Coccidioidomycosis Symposium. Tucson, University of Arizona Press.
3. Ajello, L. 1971. Coccidioidomycosis and histoplasmosis. A review of their epidemiology and geographic distribution. Mycopathologia, *45*:221–230.
4. Ajello, L. (ed.) 1977. *Coccidioidomycosis. Current clinical and diagnostic status.* New York, Stratton Intercont. Med. Book Corp.
5. Al-Doory, Y. 1972. A bibliography of coccidioidomycosis. Mycopathologia, *46*:113–188.
6. Al-Doory, Y. 1975. *The Epidermiology of Human Mycotic Infestions.* Springfield, Ill, Charles C Thomas.
7. Armbruster, T. G., T. G. Goergen, et al. 1977. Utility of bone scanning in disseminated coccidiodomycosis case report. J. Nucl. Med., *18*:450–454.
8. Birsner, J. W. 1954. The roentgenographic aspects of five hundred cases of pulmonary coccidioid-

omycosis. Am. J. Roentgenol. Radium Ther. Nucl. Med., *72*:536–573.

9. Block, E. R., and J. E. Bennett. 1973. Stability of amphotericin B in infusion bottles. Antimicrob. Agents Chemother., *4*:648–649.

10. Borchardt, K., K. Litwack, et al. 1973. *In vitro* monitoring of amphotericin B therapy in disseminated coccidioidomycosis. Arch. Dermatol., *108*:119–120.

11. Borelli, D. 1970. Prevalence of systemic mycoses in Latin America in International Symposium on Mycoses. Scientific Publication 205. Pan American Health Organization. Pan American Sanitary Bureau, Regional Office of the World Health Organization, Washington, D.C., p. 28.

12. Borelli, D., and C. Marcano. 1972. Transmission experimental de micosis profundas par "predacion." Med Cutan., VI 247–252.

13. Brosbe, E. A. 1967. Use of refined agar for the *in vitro* propagation of spherule phase of *Coccidioides immitis.* J. Bacteriol. 93:497–498.

14. Burke, R. C. 1951. *In vitro* cultivation of the parasitic phase of *Coccidioides immitis.* Proc. Soc. Exp. Biol. Med., 76:332–335.

15. Campins, H. 1967. Coccidioidomycosis in Venezuela. *In* Ajello, L. (ed.) 1967. *Coccidioidomycosis.* Tucson, University of Arizona Press.

16. Case Records of the Massachusetts General Hospital. 1971. Coccidioidal meningitis. New Engl. J. Med., *11*:621–630.

17. Caudill, R. G. 1970. Coccidioidal mengingitis. Am. J. Med., *49*:360–365.

18. Collins, M. 1975. Inhibition of growth of *Coccidioides immitis* on Sabouraud medium containing polymyxin B. J. Clin. Microbiol. *1*:335–336.

19. Converse, J. 1965. The effect of nonviable and viable vaccines in experimental coccidioidomycosis. J. Bacteriol., *74*:106–107.

20. Converse, J. L., and R. E. Reed. 1966. Experimental epidemiology of coccidioidomycosis. Bacteriol. Rev., *30*:679–694.

21. Cornell, L. H., K. G. Osborn, et al. 1979. Coccidioidomycosis in a California sea otter *(Enhydra lutris).* J. Wildl. Dis., *15*:373–378.

22. Cox, R. 1979. Cross reactivity between antigens of *Coccidioides immitis, Histoplasma capsulatum* and *Blastomyces dermatitidis* in lymphocyte transformation assays. Infect. Immun., *25*:932–938.

23. Cox, A. J., and C. E. Smith, 1939. Arrested pulmonary coccidioidomycosis granuloma. Arch. Pathol., *27*:717–734.

24. Dalink, M. F., and W. H. Greendyke, 1971. The spinal manifestations of coccidioidomycosis. J. Can. Assoc. Radiol. 22:93.

25. Davis, S. J., and W. H. Donovan. 1979. Combined intravenous miconazole and intrathecal amphotericin B for treatment of disseminated coccidioidomycosis. Chest, 76:235–236.

26. DeMartini, J. C., and W. E. Riddle. 1969. Disseminated coccidioidomycosis in two horses and a pony. J. Am. Vet. Med. Assoc., *155*:149–156.

27. Deppisch, L. M., and E. M. Donowho. 1972. Pulmonary coccidioidomycosis. Am. J. Clin. Pathol., *58*:489–500.

28. Dereskiski, S. C., D. Pappagianis, et al. 1979. Association of ABO blood group and outcome of coccidioidal infection. Sabouraudia, *17*:261–264.

29. Dickson, E. C. 1937. "Valley fever" of the San Joaquin Valley and fungus Coccidioides. California West. Med., *47*:151–155.

30. Dickson, E. C., and M. A. Gifford, 1938. Coccidioides infection (Coccidioidomycosis): the primary type of infection. Arch. Intern. Med., *62*:853–871.

31. Drutz, D. J., and A. Catanzaro. 1978. Coccidioidomycosis: state of the art. Am. Rev. Respir. Dis. *117*:559–585 and 727–771.

32. Edwards, L. B., and C. Palmer. 1957. Prevalence of sensitivity to coccidioidin, with special reference to specific and nonspecific reactions to coccidioidin and histoplasmin. Dis. Chest., *31*:35–60.

33. Egeberg, R. O., A. E. Elcovin, et al. 1964. Effect of salinity and temperature on *Coccidioides immitis* and three antagonistic soil saprophytes. J. Bacteriol., *88*:473–476.

34. Elconin, A. F., M. G. Egeberg, et al. 1967. A fungicide effective against *Coccidioides immitis* in the soil. *In* Ajello, L. (ed.) 1967. *Coccidioidomycosis.* Tucson, University of Arizona Press. pp. 319–322.

35. Emmons, C. W. 1942. Isolation of *Coccidioides* from soil and rodents. Public Health Rep., *57*:109–111.

36. Emmons, C. W. 1967. Fungi which resemble *Coccidioides immitis. In* Ajello, L. (ed.) 1967. *Coccidioidomycosis.* Tucson, University of Arizona Press. pp. 333–338.

37. Fee, H. J., J. M. McAvoy, et al. 1977. Unusual manifestation of *Coccidioides immitis* infection. J. Thorac. Cardiovasc. Surg. 74:548–550.

38. Fiese, M. J. 1958. *Coccidioidomycosis.* Springfield, Ill., Charles C Thomas, p. 182.

39. Flynn, N. M., P. D. Hoeprich, et al. 1979. An unusual outbreak of wind borne coccidioidomycosis. New Engl. J. Med., *301*:358–361.

40. Fosberg, R. G., B. F. Baisch, et al. 1969. Limited pulmonary reaction for coccidioidomycosis. Ann. Thorac. Surg., *7*:420–427.

41. Friedman, L., C. E. Smith, et al. 1962. Studies of the survival characteristics of the parasitic phase of *Coccidioides immitis* with comments on contagion. Ann. Rev. Respir. Dis., *85*:224–231.

42. Friedman, L., W. G. Roessler, et al. 1956. The virulence and infectivity of twenty-seven strains of *Coccidioides immitis.* Am. J. Hyg., *64*:198–210.

43. Gale, D., E. A. Lockhart, et al. 1967. Studies of *Coccidioides immitis.* 1. Virulence factors of *C. immitis.* Sabouraudia, *6*:29–36.

44. Galgiani, J. N., R. A., Isenberg, et al. 1978. Chemotaxis activity of extracts from the mycelial and spherule phase of *Coccidioides immitis* for human polymorphonuclear-leukocytes. Infect. Immun., *21*:862–865.

45. Gehlbach, S. H., J. D. Hamilton, et al. 1973. *Coccidioidomycosis.* An occupational disease in cotton mill workers. Arch. Intern. Med., *131*:254.

46. Giltner, L. T. 1918. Occurrence of coccidioidal granuloma (oidiomycosis) in cattle. J. Agric. Res., *14*:533–542.

47. Goldstein, E. 1978. Miliary and disseminated coccidioidomycosis. Ann. Intern. Med., *89*:365–366.

48. Gonzalez-Ochoa, A. 1967. Coccidioidomycosis in Mexico. *In* Ajello, L. (ed.) 1967. *Coccidioidomy-*

cosis. Tucson, University of Arizona Press. pp. 293–300.

49. González-Ochoa, A., and O. Velasco-Castrejón. 1976. Inoculación primaria cutánea accidental de coccidioidomicosis a partir de un caso clínico Rev. Invest. Salud. Publica (Mexico), *36*:227–234.

50. Guy, W. H., and F. M. Jacob. 1927. Granuloma coccidioides. Arch. Dermatol., *16*:308–311.

51. Hagele, A. J., D. J. Evans, et al. 1967. Primary endophthalmic coccidioidomycosis: Report of a case of exogenous, primary coccidioidomycosis of the eye diagnosed prior to enucleation. *In* Ajello, L. (ed.). 1967. *Coccidioidomycosis.* Tucson, University of Arizona Press. pp. 37–40.

52. Hart, W. R., R. P. Prins, et al. 1976. Isolated coccidioidomycosis of the uterus. Hum. Pathol. 7: 235–239.

53. Harvey, W. C., and W. H. Greendyke. 1970. Skin lesions in acute coccidioidomycosis. Am. Fam. Physician *2*:81–85.

54. Hektoen, L. 1907. Systemic blastomycosis and coccidioidal granuloma. J.A.M.A., *49*:1071–1077.

55. Henrickson, R. V., and E. L. Biberson. 1971. Coccidioidomycosis accompanying hepatic disease in two Bengal tigers. J. Am. Vet. Med. Assoc., *161*:674–677.

56. Huntington, R. W. 1971. Coccidioidomycosis. *In* R. D. Baker (ed.), *Handbuch der speziellen pathologischen anatomie and histologie. Dritter Band/Funfter teil.* Berlin, Springer-Verlag, pp. 147–210.

57. Huppert, M. 1970. Serology of coccidioidomycosis. Mycopathologia, *41*:107–113.

58. Huppert, M. 1978. Racisim in coccidioidomycosis? *118*:797–798.

59. Huppert, M., S. H. Sun, et al. 1967. Natural variation in *Coccidioides immitis. In* Ajello, L. (ed.) 1967. *Coccidioidomycosis.* Tucson, University of Arizona Press, pp. 323–330.

60. Huppert, M., J. W. Bailey, et al. 1967. Immunodiffusion as a substitute for complement fixation and tube precipitin tests in coccidioidomycosis. *In* Ajello, L. (ed.) 1967. *Coccidioidomycosis.* Tucson, University of Arizona Press, pp. 221–226.

61. Huppert, M., K. R. Krasnow, et al. 1977. Comparison of coccidioidin and spherulin in complement fixation tests for coccidioidomycosis. J. Clin. Microbiol., *6*:33–41.

62. Huppert, M., E. T. Peterson, et al. 1968. Evaluation of a latex particle agglutination test for coccidioidomycosis. Am. J. Clin. Pathol., *49*:96–102.

63. Huppert, M., C. Sun, et al. 1978. Specificity of exoantigens for identifying cultures of *Coccidioides immitis.* J. Clin. Microbiol., *8*:346–348.

64. Illahi, A., H. Afzal, et al. 1966. Coccidioidomycosis in buffaloes in West Pakistan. Pakist. J. Anim. Sci., *3*:14–20.

65. Jenkins, D. W. 1977. Persistant adenopathy in coccidioidomycosis: an indication for therapy. South Med. J., *70*:531–532.

66. Johnson, W. M. 1967. Coccidioidomycosis mortality in Arizona. *In* Ajello, L. (ed.) 1967. *Coccidioidomycosis.* Tucson, University of Arizona Press, pp. 33–44.

67. Kaplan, W. 1967. Application of the fluorescent antibody technique to the diagnosis and study of coccidioidomycosis. *In* Ajello, L. (ed.) 1967.

Coccidioidomycosis. Tucson, University of Arizona Press, pp. 227–231.

68. Kaufman, L., and P. Standard. 1978. Improved version of the exoantigen test for identification of *Coccidioides immitis* and *Histoplasma capsulatum* cultures. J. Clin. Microbiol. *8*:42–45.

69. Kruse, R. H., T. D. Green, et al. 1967. Infection of control monkeys with *Coccidioides immitis* by caging with inoculated monkey. *In* Ajello, L. (ed.) 1967. Coccidioidomycosis. Tucson, University of Arizona Press, pp. 387–396.

70. Krutzsch, P. H., and R. H. Watson. 1978. Isolation of *Coccidioides immitis* from bat guano and preliminary findings on laboratory infectivity of bats with *Coccidioides immitis.* Life Sci., *22*:679–684.

71. Kwon-Chung, K. J. 1969. *Coccidioides immitis:* Cytological study on the formation of the arthrospores. Can. J. Genet. Cytol., *9*:43–53.

72. Lacy, G. H., and F. E. Swatek. 1974. Soil ecology of *Coccidioides immitis* at Amerindian middens in California. Appl. Microbiol. *27*:379–388.

73. Landay, M. E., R. W. Wheat, et al. 1967. Serological comparison of the three morphological phases of *Coccidioides immitis* by agar gel diffusion method. J. Bacteriol., *93*:1–6.

74. Landay, M. E. 1973. Spherules in the serology of Coccidioides immitis. II. Complement fixation tests with human sera. Mycopathologia, *49*:45–52.

75. Langham, R. F., E. S. Beneke, et al., 1977. Abortion in a mare due to *Coccidioides immitis.* Am. Vet. Med. Assoc., *170*:178–180.

76. Levine, H. B., and C. E. Smith, 1967. The reactions of eight volunteers injected with *Coccidioides immitis* spherule vaccine; first human trials. *In* Ajello, L. (ed.) 1967. *Coccidioidomycosis.* Tucson, University of Arizona Press, pp. 197–200.

77. Levine, H. B., and G. M. Scalarone. 1971. Deficient resistance to *C. immitis* following intravenous vaccination. III, Humoral and cellular respones to intravenous and intramuscular doses. Sabouraudia, *9*:97–108.

78. Levine, H. B., A. González Ochoa, et al. 1973. Dermal sensitivity to *Coccidioides immitis:* A comparison of response elicited by spherulin and coccidioidin. Am. Rev. Respir. Dis., *107*:379–386.

79. Levine, H. B. 1970. Development of vaccines for coccidioidomycosis. Mycopathologia, *41*:177–185.

80. Lones, C. 1967. Studies of the intermediary metabolism of *Coccidioides immitis. In* Ajello, L (ed.) 1967. *Coccidioidomycosis.* Tucson, University of Arizona Press, pp. 349–353.

81. Lonky, S. A., A. Canazaro. 1976. Acute coccidioidal pleural effusion. Am. Rev. Respir. Dis., *114*:681–688.

82. Maddy, K. 1958. Disseminated coccidioidomycosis of the dog. J. Am. Vet. Med. Assoc., *132*:483–489.

83. Maddy, K. T. 1965. Observations on *Coccidioides immitis* found growing naturally in soil. Arizona Med., *22*:281–288.

84. Maddy, K. T., and H. G. Crecelius. 1967. Establishment of *Coccidioides immitis* in negative soils following burial of infected animals and animal tissues. *In* Ajello, L. (ed.) 1967. *Coccidioidomycosis.* Tucson, University of Arizona Press, pp. 309–312.

85. Maddy, K. T., H. G. Crecelius, et al. 1960. Distribution of *Coccidioides immitis* determined by testing cattle. Public Health Rep., *75*:955–962.

86. Mayorga, R. 1967. Coccidioidomycosis in Central America. *In* Ajello, L. (ed.) 1967. *Coccidioidomycosis.* Tucson, University of Arizona Press, pp. 287–292.

87. Mazza, S., and S. Parodi. 1929. Micosis chaquena producida por el *Pseudococcidioides mazzai.* Prensa Med. Argent., *16*:268–272.

88. Meyer, R. D., F. R. Sattler, et al. 1978. Miconazole for treatment of disseminated coccidioidomycosis. Unfavorable experience. Chest, *73*:825–831.

89. Negroni, P. 1967. Coccidioidomycosis in Argentina. *In* Ajello, L. (ed.) 1967. *Coccidioidomycosis.* Tucson, University of Arizona Press, pp. 273–278.

90. Oduye, O. O. 1972. Canine disseminated coccidioidomycosis in Nigeria. Bull. Epiz. Dis. Afr., *20*:113–120.

91. Omieczynski, D. J., and F. E. Swatek. 1967. The comparison of two methods for direct isolation of *Coccidioides immitis* from soil using three different media. *In* Ajello, L. (ed.) 1967. *Coccidioidomycosis.* Tucson, University of Arizona Press, pp. 265–272.

92. Ophüls, W. 1905. Further observations on a pathogenic mould formerly described as a protozoan. *(Coccidioides immitis, Coccidioides pyogenes).* J. Exp. Med., *6*:443–485.

93. Ophüls, W. 1905. Coccidioidal granuloma. J.A.M.A., *45*:1291.

94. Ophüls, W., and H. C. Moffitt. 1900. A new pathogenic mould (formerly described as a protozoan: *Coccidioides immitis).* Preliminary report. Phila. Med. J., *5*:1471–1472.

95. Orr, C. F. 1968. Some fungi isolated with *Coccidioides immitis* from soils of endemic areas in California. Bull. Torrey Bot. Club. *95*:424–431.

96. Pappagianis, D. 1978. Coccidioidomycosis. *In* Samter, M. (ed.), *Immunological Diseases.* Boston, Little, Brown and Company, p. 652.

97. Pappagianis, D., C. E. Smith, et al. 1967. Serologic status after positive coccidioidin skin reactions. Am. Rev. Respir. Dis., *96*:520–523.

98. Peterson, E. A., B. A. Friedman, et al. 1976. Coccidioiduria: clinical significance. Ann. Intern. Med., *85*:34–38.

99. Porter, J. F., E. R. Scheer, et al. 1971. Characterization of 3–O methylmannose from *Coccidioides immitis.* Infect. Immun., *4*:660–661.

100. Posadas, A. 1892. Un neuvo caso de micosis fungoidea con psorospermias. Circulo Med. Argent., *5*:585–597.

101. Posadas, A. 1894. Contribution to the study of the etiology of tumors: generalized infectious psorospermosis (in Spanish) unpublished thesis, Buenos Aires.

102. Posadas, A. 1900. Psorospermiose infectante generalise. Rev. Chir. (Paris), *21*:277–282.

103. Puckett, T. F. 1954. Hyphae of *Coccidioides immitis* in tissues of the human host. Am. Rev. Tuberc., *70*:320–327.

104. Rapaport, F. T., H. S. Lawrence, et al. 1960. Transfer of delayed hypersensitivity to coccidioidin in man. J. Immunol., *84*:358–367.

105. Rhodes, E. R., J. J., McPhaul, et al. 1967. The interpretation of renal changes associated with administration of amphotericin B.: Impor-

tance of pre-treatment studies. *In* Ajello, L. (ed.) 1967. *Coccidioidomycosis.* Tucson, University of Arizona Press, pp. 137–140.

106. Rippon, J. W. 1980. Dimorphisim in Pathogenic Fungi. CRC Crit. Rev. Microbiol., *8*:49–97.

107. Rixford, E., and T. C. Gilchrist, 1896. Two cases of protozoan (coccidioidal) infection of the skin and other organs. Johns Hopkins Hosp. Reg., *1*:209–269.

108. Roberts, P. L., J. H. Knepshield, et al. 1968. Coccidioides as an opportunist. Arch. Intern. Med., *121*:568–570.

109. Roberts, P. L., J. H. Knepshield, et al. 1968. Coccidioides as an opportunist. Arch. Intern. Med., *121*:568–570.

109. Roberts, J. J. M. Counts, et al. 1970. Production *in vitro* of *Coccidioides immitis* as a diagnostic aid. Am. Rev. Respir. Dis., *102*:811–813.

110. Robledo, V. M., A. M. Restropo, et al. 1968. Encuesta epidemiologica sobre coccidioidomicosis en algunas zonas aridas de Colombia. Antioquia Med., *18*:503–522.

111. Salkin, D. 1967. Clinical examples of reinfection coccidioidomycosis. Am. Rev. Respir. Dis., *95*:603–611.

112. Scalarone, G. M., and H. B. Levine. 1969. Attributes of deficient immunity in mice receiving *Coccidioides immitis* spherule vaccine by the intravenous route. Sabouraudia, *7*:169–177.

113. Schmidt, R. J., and D. H. Howard. 1969. Possibility of *C. immitis* infection in museum personnel. Public Health Rep., *83*:882–888.

114. Schwarz, J., and C. A. Kantman. 1977. Occupational hazards from deep mycoses. Arch. Dermatol., *113*:1270–1275.

115. Shafai, T. 1978. Neonatal coccidioidomycosis in premature twins. Am. J. Dis. Child, *132*:634.

116. Sievers, M. L. 1979. Coccidioidomycosis and race. Am. Rev. Respir. Dis., *119*:839.

117. Sigler, L., and J. W. Carmichael. 1976. Taxonomy of *Malbranchea* and some other hyphomycetes with arthroconidia. Mycotaxon, *4*:349–488.

118. Sippel, J. E., and H. B. Levine. 1969. Annulment of amphotericin B inhibition of *Coccidioides immitis* endospores. Effects on growth respiration and morphogenesis. Sabouraudia, *7*:159–168.

119. Smale, L. E., and K. G. Waechter. 1970. Disseminated coccidioidomycosis in pregnancy. Am. J. Obstet. Gynecol., *107*:356–361.

120. Smith, C. E. 1940. Epidemiology of acute coccidioidomycosis with erythema nodosum. Am. J. Public Health, *30*:600–611.

121. Smith, C. E., R. R. Beard, et al. 1946. Effect of season and dust control on coccidioidomycosis. J.A.M.A., *132*:833–838.

122. Smith, C. E., M. T. Saito, et al. 1956. Pattern of 39,500 serologic tests in coccidioidomycosis. J.A.M.A., *160*:546–552.

123. Sorenson, R. H. 1967. Survival characteristics of diphasic *Coccidioides immitis* exposed to the rigors of a simulated natural environment. *In* Ajello, L. (ed.) 1967. *Coccidioidomycosis.* Tucson, University of Arizona Press, pp. 313–318.

124. Stephanishtcheva, Z. C., A. M. Arievitch, et al. 1972. Investigation of deep mycoses in the USSR. Int. J. Dermatol, *11*:181–183.

124a. Stevens, D. 1981. Coccidioidomycosis. A Text. New York, Plenum.

125. Stewart, R. A., and K. F. Meyer, 1932. Isolation of *Coccidioides immitis* (Stiles) from soil. Proc. Soc. Exp. Biol. Med., *29*:937–938.

126. Sun, S., S. S. Sekhon, et al. 1979. Electron microscopic studies of saprobic and parasitic forms of Coccidioides immitis. Sabouraudia, *17*:265–273.

127. Suteeva, T. G., O. B. Minsker, et al. 1978. Results of a 10-year study of a disease called Coccidioidomycosis. *In*: The Epidemiology of Nererologii No. 124–130.

128. Swatek, F. E. 1970. Ecology of Coccidioides immitis. Mycopathologia, *40*:3–12.

129. Swatek, F. E. 1975. The epidemiology of coccidioidomycosis. *In*: The Epidemiology of Human Mycotic Infections. Y. Al-Doory (ed.) Springfield, Ill. Charles C Thomas, p. 74.

130. Swatek, F. E., D. T. Omiecznski, et al. 1967. *Coccidioides immitis* in California. *In* Ajello, L. (ed.) 1967. *Coccidioidomycosis*. Tucson, University of Arizona Press, pp. 255–264.

131. Symmers, W. S. C. 1971. An Australian case of coccidioidomycosis. Pathology, *3*:1–8.

132. Trimble, J. R., and J. Doucette. 1956. Primary cutaneous coccidioidomycosis. Report of a case of laboratory infection. Arch. Dermatol., *74*:405–410.

133. Van bergen, W. S., F. J. Fleury, et al. 1976. Fatal maternal disseminated coccidioidomycosis in a non-endemic area. Am. J. Obstet. Gynecol., *124*:661–663.

134. Velasco-Castrejon, O., and A. González-Ochoa. 1975. Coccidioidomicosis adquirida en la ciudad de Mexico. Un caso atipico desde el punto de vista epidemiologico. Rev. Invest. Salud. Publica (Mexico) *35*:97–102.

135. Wallraff, E. B., and P. R. O'Bar. 1969. Evaluation of a coccidioidin tine test as a skin testing technique. Am. Rev. Respir. Dis., *99*:943–945.

136. Ward, P. H., G. Berci, et al. 1977. Coccidioidomycosis of the larynx in infants and adults. Ann. Otol. Rhinol. Laryngol., *86*:655–660.

137. Werner, S. B., D. Papagianis, et al. 1972. An epidemic of coccidioidomycosis among archeology students in northern California. New Engl. J. Med., *286*:507–512.

138. Wernicke, R. 1892. Ueber einen Protozoenbefund bei mycosis fungoides. Zentralbl. Bakteriol., *12*:859–861.

139. Wesselius, L. J., R. J. Brooks, et al. 1977. Vertebral coccidioidomycosis presenting as Pott's disease. South Am. Med., Assoc., *238*:1397–1398.

140. Wheat, R., and E. Scheer. 1977. Cell walls of *Coccidioides immitis:* neutral sugars of aqueous alkaline extract polymers. Infect. Immun., *15*:341–347.

141. Wilson, J. W., C. E. Smith, et al. 1953. Primary cutaneous coccidioidomycosis: the criteria for diagnosis. California Med., *79*:233–239.

142. Winn, W. A. 1962. The diagnosis and treatment of coccidioidomycosis. Arizona Med., *19*:211–217.

143. Winn, W. A. 1964. The treatment of coccidioidal meningitis. The use of amphotericin B in a group of 25 patients. California Med., *101*:78–89.

144. Winn, W. A. 1967. A working classification of coccidioidomycosis and its application to therapy. *In* Ajello, L. (ed.) 1967. *Coccidioidomycosis*. Tucson, University of Arizona Press, pp. 3–10.

145. Winn, W. A. 1967. Coccidioidal meningitis, a follow-up report. *In* Ajello, L. (ed.) 1967. *Coccidioidomycosis*. Tucson, University of Arizona Press, pp. 55–62.

146. Winn, R. E., J. H. Bower, et al. 1979. Acute toxic delerium. Neurotoxicity of intrathecal administration of amphoterion B. Arch. Intern. Med., *139*:706–707.

147. Wolf, A. M. 1979. Primary cutaneous coccidioidomycosis in a dog and a cat. J. Am. Vet. Med. Assoc., *174*:504–506.

148. Wright, E. T., and L. H. Winer. 1971. The natural history of experimental *Coccidioides immitis* infection. Int. J. Dermatol., *10*:17–23.

149. Zealear, D. S., and W. Winn. 1967. The neurosurgical approach in the treatment of coccidioidal meningitis: report of ten cases. *In* Ajello, L. (ed.) 1967. *Coccidioidomycosis*. Tucson, University of Arizona Press, pp. 43–54.

18 Blastomycosis

DEFINITION

Blastomycosis is a chronic granulomatous and suppurative disease having a primary pulmonary stage that is frequently followed by dissemination to other body sites, chiefly the skin and bone. The primary infection in the lung is often inapparent. The name blastomycosis can refer to any infection caused by a yeastlike organism, so that reference must be made in terms of specific etiology. The causative agent is the dimorphic fungus *Blastomyces dermatitidis,* which has been assumed to be a soil saprophyte in nature, although its ecologic niche has not as yet been delineated. The disease is most prevalent in men in the fourth through sixth decades of life. Formerly thought to be restricted to the North American Continent, the disease has been described from divergent parts of Africa, Asia, and Europe. Unlike the other common systemic mycoses, it does not appear to have a common mild subclinical form, and the frequency of the self-limited, spontaneously resolving infection is as yet unknown.[39, 108, 118, 119, 141]

Synonymy.   North American blastomycosis, Gilchrist's disease, Chicago disease.

HISTORY

In 1894 T. Caspar Gilchrist described a new type of skin disease caused by a yeast (blastomycetic dermatitis) before the American Dermatologic Association.[45a] The patient's disease had been diagnosed as scrofuloderma of the hand. The attending physician (Dr. Duhring) had sent a biopsy specimen to Gilchrist for pathologic diagnosis.

Gilchrist could find no tubercle bacilli but noted "curious bodies" distributed throughout all sections. He at first thought they were protozoan parasites but noted that they appeared to be budding and hence were yeastlike. The presentation of this case in May of 1894 preceded that of Busse's case of "*Saccharomyces hominis*" (cryptococcosis) by six months. At the time of Gilchrist's first publication, he thought his organism and that of Busse were identical.[45a, 46] A yeast (blastomycete) had been recovered from culture of Busse's patient, but further study of Gilchrist's blastomycete was prevented because the lesion from which it had come was completely excised before he could take material for culture.

A second case was published by Gilchrist and Stokes in July, 1896.[47] A 33-year-old male was referred to Gilchrist by a Dr. Halsted with the diagnosis of lupus vulgaris. The disease started behind the ear and gradually extended to involve much of the face. Over the next 11½ years, lesions appeared on the hand, scrotum, thigh, and back of the neck, in that order. Gilchrist noted that many lesions spontaneously healed, leaving scar tissue, and that active lesions had an elevated border with minute pustules in which the organism could be demonstrated in a potassium hydroxide mount. He also noted that the local lymph nodes were not involved. With our present understanding of the disease, we would assume that Gilchrist's patient had chronic cutaneous blastomycosis secondary to pulmonary infection rather than a primary cutaneous infection. Gilchrist was able to culture the organism from this case and noted it grew at first as a soft mass that produced "prickles" and eventually a fluffy mycelium. He thus concluded it was different from

428

Busse's organism, which remained a yeast in all subcultures. He and Stokes named the fungus *Blastomyces dermatitidis* in 1898.[48] The patient was lost to follow-up, as he had to return home to a sick wife, but Gilchrist declared the prognosis "good" because his previous lesions had healed. Gilchrist also produced the disease in lungs and other organs in experimentally infected animals.

In the next few years many other cases of blastomycosis were recognized, especially in the Chicago area; thus the disease became known as the "Chicago disease." H. T. Ricketts of Chicago in 1901 described fifteen cases of cutaneous disease[110] and redescribed the fungus as *Oidium dermatitidis*. Other clinicians of the Chicago area were the first to use x-ray for treatment (Montgomery, in 1902), defined more clearly the epidemiology and pathology of the disease (Hektoen, 1907),[62] and were the first to use potassium iodide in therapy (Montgomery, 1908). In 1908, Montgomery and Ormsby reviewed the cases that had been reported in the Chicago area as well as in the world literature,[93] including the cases of Busse and Curtis, which were later realized to be cryptococcosis. In their own cases, however, they did describe the systemic nature of the blastomycosis and its essential pathology.

For many years Gilchrist's disease was confused with cryptococcosis (under the name torulosis), coccidioidomycosis, and paracoccidioidomycosis.[138] These diseases were not fully delineated until the early 1930's. Almeida defined paracoccidioidomycosis, and Benham[8] and Conant[21] resolved the mycologic problems of the other two. A detailed study of the basic mycology of *B. dermatitidis* had previously been published by Hamberger in 1907.[55] He emphasized the temperature-dependent diphasic nature of this fungus in contrast with yeasts that remained as yeasts both in tissue and *in vitro* at room temperature.

Until 1950 it was assumed that there were two distinct forms of blastomycosis — systemic and cutaneous. The route of infection was thought to be through the skin in the cutaneous form, and by way of the lungs for the visceral type. Careful analysis of pathologic material from both types of the disease led Schwarz and Baum in 1951[123] to suggest that all cases of blastomycosis begin as primary pulmonary infections. Subsequent studies have tended to confirm this.[13, 125] The confusion about what constituted a primary cutaneous lesion in this and other systemic mycoses was finally clarified by Wilson et al.[148, 149]

They used primary cutaneous sporotrichosis as the model infection. The lesion produced in this disease is chancriform, and there is local lymphatic involvement. When cases of cutaneous blastomycosis in the literature, including those of Gilchrist, were carefully reviewed, the lesions were noted to be of the chronic granulomatous type without apparent involvement of the local lymphatics and, therefore, secondary manifestations of primary pulmonary disease. As is the case with the other systemic mycoses, essentially all infections are acquired by inhalation of the conidia. Primary cutaneous disease does occur but is extremely rare.[148] Also very rare are the reported cases of sexual transmission of the disease from prostatic fluid containing the organism[23, 34, 36] or other person-to-person contact.

Designating a disease by its geographic distribution is an unfortunate and often confusing practice. African histoplasmosis has been found in Japan; maduromycosis (after Madura, India) and European blastomycosis (cryptococcosis) are worldwide in distribution. Gilchrist's disease was called North American blastomycosis for many years because all cases had been found in the North American continent. In 1952 Broc and Haddad[11] described a case which was similar to Gilchrist's disease in Tunisia. The fungus isolated was named "*Scopulariopsis americana*," which is now considered a synonym of *B. dermatitidis*. Since that time autochthonous cases have been found and confirmed in South Africa,[131] Uganda, Nigeria,[4] Zaire, Morocco, Tanzania, and the Malagasy Republic. It is apparent that the disease is widely distributed in Africa. By 1980 some 60 cases had been reported in 15 African countries.[4, 131] Cases acquired by travel to endemic regions have appeared in reports for many years.[20] Infections transmitted by fomites have been recorded in Switzerland[146] and England.[31] More recently, autochthonous cases have appeared in Poland,[72] Lebanon,[60] Saudi Arabia, and Israel.[73] Some reports of infection in South America also appear to be valid,[2] thus indicating a worldwide distribution of the etiologic agent.

Until relatively recently, treatment and management of blastomycosis was very difficult. Chronic pulmonary and disseminated disease was uniformly fatal, regardless of therapeutic modality. The slowly evolving chronic cutaneous form of the disease sometimes responded to such treatment as iodides and x-ray therapy, but relapse of the disease was the rule. Iodides and x-ray along with vaccines, propamidine, and undecyclenic acid offered little hope for cure and were beset with many side effects. Schoenbach in 1951[122] successfully treated the disease with stilbamidine, but the peripheral neuropathy associated with the drug restricted its wide usage. A more easily tolerated derivative of the drug, 2-hydroxystilbamidine, was introduced by Snapper and McVay in 1953.[136] It became the standard of therapy and remains so in disease of children[19] and in limited pulmonary infections of adults. In extensive disease, however, the relapse rate was very high. In 1957 amphotericin B was introduced by Harrell and Curtis[56] and became the standard of therapy for essentially all forms of the disease, replacing 2-hydroxystilbamidine.

ETIOLOGY, ECOLOGY, AND DISTRIBUTION

Etiology. The only etiologic agent of blastomycosis is *Blastomyces dermatitidis*. McDonough and Lewis described the teleomorph stage in 1967[88] as *Ajellomyces dermatitidis*. It has since been shown that most isolants from human and animal disease from various geographic locations in the United States and Canada mate with the testor strains. It therefore seems reasonable to conclude that a single species is involved in this disease in North America. Isolants from Africa form gymnothecia but few fertile ascospores,[87] so that the possibility of another species in Africa and Asia still exists. The choice of the generic name *Blastomyces* for the imperfect stage was unfortunate, as this term had been used previously in mycology and thus was illegal under the International Botanical Code. Since it is so widely accepted now, it would serve no purpose to change it, as the organism does not fit well into any of the other described genera of Hyphomycetes.

Ecology. The natural habitat of *Blastomyces dermatitidis* remains an enigma. Since it now appears that essentially all infections are acquired by inhalation of conidia into the lungs and follow the pattern of disease established for the other systemic mycoses, the organism should be a saprophyte in soil, producing mycelium and air-borne conidia. All attempts to isolate the organism from soil in endemic areas have failed, with only a few exceptions.[28] Denton and DiSalvo[25] recovered the organism from 10 of 356 soil samples obtained near Augusta, Georgia. These were collected on only two days — March 20, 1962, and February 7, 1963. The areas yielding positive samples included chicken houses, cattle loading ramps, an abandoned kitchen, a rabbit pen, and a mule stall. Later that year (December 10, 1963) two more positive soil samples were collected from the abandoned kitchen.[27] To date this kitchen is the only "hot spot" where the fungus has been isolated more than once (Fig. 18–1).

McDonough[87] has studied the fate of the organism in soil. He found that yeast cells rapidly lyse when placed in soil,[90, 91] and mycelium soon disappears under these same conditions. Conidia placed in soil survive for

Figure 18–1 The only site of recovery of *Blastomyces dermatitidis* on more than one occasion. *A,* Abandoned kitchen near Augusta, Georgia. Three positive samples were obtained on two samplings. *B,* Shelf in the kitchen. The fungus was isolated from decaying debris collected near the boiler lid. (From Denton, J. F., and A. F. DiSalvo. 1964. Am. J. Trop. Med. Hyg., 13:716–722. Courtesy of A. DiSalvo.)

a few weeks only. These same conditions also eradicate *Coccidioides immitis*. *B. dermatitidis* may not be able to compete with the normal flora of soil and perhaps survives only in a very restricted ecologic environment.

Judging from epidemiological data, it appears that the organism grows on decaying organic material[120, 146] and the conidia are distributed when the soil is disturbed.[70, 118, 141] It is reported as being recovered in its yeast stage from pigeon manure used as fertilizer.[120] In that instance, the fertilizer was the source of the organism in a case of acute progressive blastomycosis in a horticulturist. This is the only record of isolation from the environment in association with human disease. The ecological niche inhabited by the fungus remains a mystery.

Several reviews of case studies[13] have noted that primary infection appears to occur during the cooler, wetter months of the year. This correlates with the reports of recovery of the organism from a saprophytic environment in December, February, and March.[25-28] It is possible that the organism is dormant most of the time, flourishing only rarely under particular climatic and environmental conditions of the colder seasons. It has been reported to grow on tree bark[27] and decorticated wood.[30] A very early and as yet unconfirmed report of the isolation of *B. dermatitidis* from decaying wood was made in 1914 by Stober.[139] The available evidence indicates that this isolation is definitely possible, since the fungus is lysed in ordinary soil but flourishes in the laboratory on woody plant material.[30]

Distribution. The geographic distribution of blastomycosis is also problematic. Unlike histoplasmosis and coccidioidomycosis, which occur within well-established geographic areas, delineation of the endemicity of blastomycosis has been hampered by the lack of two important epidemiologic tools. There is no acceptable and useful skin test available to detect subclinical and resolved infection in the population, and the organism cannot be regularly recovered from a saprophytic environment. Our knowledge of the endemic range of blastomycosis is based, therefore, only on reports of clinically apparent human and animal disease.[2]

The overall range of autochthonous cases in the American continent includes the middle western states, the southeastern states (excluding Florida), and the Appalachian states (Fig. 18-2). In general, this is the drainage pattern for the Mississippi and Ohio River basins and, to a certain extent, for the Missouri River.[39, 40] From Minnesota, the endemic zone extends into the southern section of Manitoba and southwest Ontario in Canada.[50, 69, 127] In western New York and eastern Ontario, another endemic zone originates, which then follows the St. Lawrence River through Quebec. The highest number of Canadian cases are recorded from the latter province.[50] Sekhon[127] has reviewed the Canadian reports of the disease and tabulated 120 cases. He also described the first three infections in Alberta. New England is spared the infection, as are the plains states, the two tiers of mountain states, and the west coast. All cases of blastomycosis in California have been traced to other sections of the country.[137] In analyzing 1476 human cases and 384 canine cases, Furcolow et al.[40] found Kentucky had the highest number of human infections. The highest incidence was largely to the south of the Ohio River and to the east of the Mississippi (excepting Arkansas and Louisiana). Other areas of disease concentration are North Carolina, except for the coastal areas, and the western shore of Lake Michigan, particularly the Chicago and Milwaukee regions. The prevalence of canine disease showed a distribution similar to human cases. Infection in dogs often is a harbinger of the outbreak of human case clusters.[121] Most infected humans and dogs came from rural environments, although epidemics have occurred in urban areas as well.[70] A notable number of human infections occurred in agricultural workers or hunters.

Epidemics of blastomycosis have been recorded in the past and are still regularly reported.[118, 119, 141] The first was that of Smith et al. in 1955.[135] In this report several people in the same community developed pleural symptoms at the same time during the winter season. No particular point source could be detected. In a report by Harris et al. in 1957[57] an Arkansas farmer and many of his hunting hounds developed blastomycosis at about the same time following a hunting expedition in the local forest. Tosh et al.[141] and Serosi et al.[118, 119] reported on a common source outbreak of the disease that occurred in 21 persons. They present evidence that some of the patients had mild infections that resolved spontaneously. Similar epidemics have been recorded in a Chicago suburb[70] and in Eagle River, Wisconsin, as well as several other geographic areas. Since resolved infections leave no residua such as calcified lesions in the lung, skin test sensitivi-

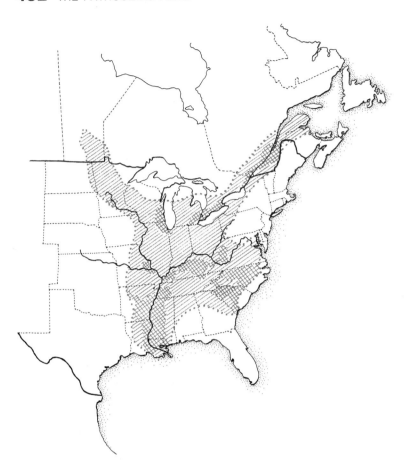

Figure 18–2 Blastomycosis. Incidence and prevalence of blastomycosis in North America. The dotted line indicates the known endemic region. The hatched areas are those with the highest incidence.

ty, or positive serology, the extent or frequency of these point source exposures cannot be accurately tabulated. Their discovery at all depends on astute clinicians and laboratory personnel.

The discovery of blastomycosis in Africa has been too recent and the cases too few to generalize on the epidemiology of the disease there. The most interesting observation has been the great diversity of countries reporting cases over a wide geographic range. The first case, found in Tunisia by Broc and Haddad,[11] was reported in detail by Vermeil.[145] A subsequent case has been found from the same region,[103] and the identity of the culture confirmed. Since then, cases have come from the Congo,[43] South Africa,[131] Nigeria,[4] Angola,[9] Uganda,[35] Zimbabwe, Morocco, Tanzania, and the Malagasy Republic.[16] Segretain[126] maintains that the African form is characterized by fewer ulcerated cutaneous lesions and more gummas or subcutaneous abscesses. As noted in earlier sections, blastomycosis has now been described in several Middle Eastern countries[60, 73] and Poland.[72] The latter case was of a 51-year-old farmer who had never been abroad. He was success-fully treated with amphotericin B. Martinez-Baez et al.[86] recorded the disease in Mexico, and it may also occur in Central America.[86] A report of the disease in South America by Polo et al.[104] is questionable, though a few other case reports may be valid.[2]

Blastomycosis has been acquired by handling fomites. There is a case recorded in a tobacco worker in Switzerland[146] and in a packing material handler in England.[31]

The sex and age distribution of blastomycosis is generally the same as that of other systemic nonopportunistic fungus infections. Of the 1114 cases for which data were available, 89 per cent were in men.[40] Other studies have found variation in sex distribution.[151] The infection was three times as great among Negroes as Caucasians, and the former tended to be younger. Though primary infections have been recorded in children as young as three months and in a patient in the ninth decade, the peak incidence is in middle-aged adult males.[97] In over 60 per cent of cases, the patients were between 30 and 60 years of age. Disease in women is usually occult or slowly progressive. Rapid dissemination can occur during pregnancy or with other hormonal imbalances.[94] About forty cases in children were tabulated in a review by Yogev and Davis[152] and Laskey and Serosi.[80]

As has been established by analysis of numerous case reports and autopsies, blastomycosis is acquired by the inhalation of fungal conidia. Very rarely and under extraordinary circumstances, human-to-human transmission has apparently occurred. Some cases of conjugal transmission of yeast cells in semen have been noted.[34, 36] Transmission from human to human has also been considered in a case that involved two workers who both acquired blastomycosis.[106a] It was surmised that there was aerosol transmission of yeasts from one patient to the other. A more acceptable explanation, however, is that there was a common source of conidia from the saprophytic mycelial phase in the environment of the workers.

CLINICAL DISEASE

The clinical forms of blastomycosis as well as its possible saprobic occurrence and epidemiology are enigmatic. Beginning with Gilchrist's case in 1896 and for many years thereafter, it appeared that there were two distinct forms of the disease dependent on site of inoculation. Pulmonary disease was associated with disseminated blastomycosis and usually had a chronic, relentless, and eventually terminal outcome. The cutaneous form, which was not associated with apparent pulmonary involvement, waxed and waned over a period of many years. Since the work of Schwartz and Baum in 1951[123] and publication of the description of primary cutaneous disease by Wilson in 1955,[149] it has been thought that essentially all infections begin in the lung.[13, 125] The difference between the two groups of patients is apparently a reflection of the relative ability of the host to handle the disease.

If we extrapolate from the known clinical picture of coccidioidomycosis and histoplasmosis, then large numbers of self-resolving subclinical pulmonary infections of blastomycosis should occur. At present we do not know if this entity exists. Although apparent spontaneous recovery has been noted several times,[108, 119] most of the evidence tends to indicate that once an infection is established the disease progresses either as an occult insidious process or as a chronic expanding and eventually systemic infection.

The clinical disease can be divided into five categories: (1) primary pulmonary disease, which may have severe presenting symptoms or be an inapparent infection that resolves spontaneously or disseminates to another form; (2) chronic cutaneous disease, which may have occult osseous lesions as well; (3) single organ system involvement, which may remain occult for many years; (4) generalized systemic disease, involving multiple organ systems and usually running a rapid course; and, finally, (5) inoculation blastomycosis, a self-limited primary infection.

Primary Pulmonary Blastomycosis

The inhalation of the fungal conidia initiates an alveolitis with invasion by macrophages. This is followed by an inflammatory reaction consisting of exudation of cells (predominantly polymorphonuclear leukocytes) and, later, granuloma formation. In this stage or at any stage in any organ, the process may be suppurative or mixed suppurative and granulomatous. *Blastomyces* yeast cells are found within pulmonary macrophages, which may then disseminate to other organ systems. The primary lesions often resemble those of tuberculosis or histoplasmosis, characterized by a Ghon complex, with parenchymal infiltration, lymphangitis, and lymphadenitis. Three general categories of clinical sequence evolve from the primary pulmonary infections: (1) resolution of lung disease with appearance in other organ systems, (2) resolution of lung disease with no evident remaining foci, and (3) severe or progressive pulmonary involvement. The initial symptom is usually a mild progressive respiratory infection, with a dry cough, some pleuritic pain, hoarseness, and low-grade fever. Patients in the first category are seen to undergo primary organization and healing, leaving a small fibrotic scar. There is little evidence of the caseation seen in tuberculosis or histoplasmosis, and frequently such lesions heal by fibrosis and absorption, leaving no residual evidence of infection. Spread by macrophages and deposition of the organism at distant sites may have already occurred, however. This type of patient develops the chronic cutaneous, occult osseous, or other type of organ system disease. Calcification of primary lesions is infrequently observed; thus, there is no residual evidence of primary infection in the lungs.

The second category of patients includes those in whom the lesions resolve with no residua or remaining foci. The acute self-resolving primary pulmonary disease was described in detail by Sarosi et al.[118, 119] The clinical symptoms were of two types. The first type, similar to symptoms of acute histoplasmosis, consisted of fever, productive cough, arthralgia, and myalgia. The second type consisted of mild to severe pleuritic pain. These cases represented a point source of infection, and so far all have resolved without treatment. Several such "point-source" epidemics have now been recorded, and authenticated cases of pulmonary colonization followed by re-

solution of the infection have been described. The frequency of such resolved primary pulmonary disease has yet to be assessed.

If primary pulmonary infection does not resolve, the third type of disease, called severe progressive blastomycosis, can develop. The infection evolves into an acute lobar pneumonia, acute bronchopneumonia with rapid hematogenous dissemination, or a more chronic type of infection, such as a suppurating pyogenic process or an expand-

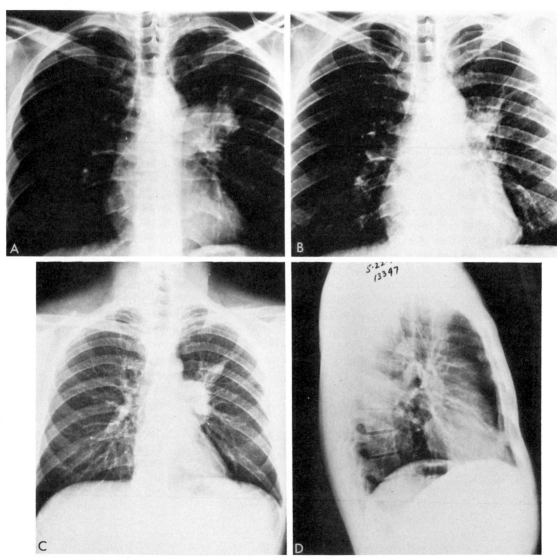

Figure 18–3 Radiologic aspects of blastomycosis. *A,* Large, dense, irregular infiltrates are found in the middle lobe abutting the hilum. Satellite densities are scattered to the main bronchi. On bronchogram the bronchi were found to be patent. (Courtesy of F. Kittle.) *B,* Same patient after two months of amphotericin B therapy. This patient was an example of the post-primary disease developing into chronic progressive phase illness. There is a marked decrease in the infiltrate. *C,* Irregular lobulated 4 × 6 cm mass in the superior segment of the lower lobe. There is some atelectasis. *D,* On lateral film the lesion is seen to be located in the posterior portion. The diagnosis of carcinoma was made and a lobectomy performed (see Fig. 18–13). This patient could remember no clinical illness prior to the slow onset of low grade symptoms and represents the indeterminable phase of illness. (Courtesy of F. Kittle.)

ing "crab claw" shadow granuloma. The patient's symptoms reflect the increasing severity of the disease. Sputum production increases and is now purulent and blood-tinged; the temperature is elevated; and dyspnea and weight loss occur, along with night sweats and increasing weakness. The pleura may be involved, but pleural effusion in blastomycosis[64] is less common than in actinomycosis. The physical signs of the disease at this stage are dullness to percussion and altered breath sounds with transitory and variable rales. Sometimes a discharging sinus or subcutaneous abscess develops over the thorax.

The lungs may also become involved by endogenous reactivation of disease in other organ systems. As in reactivated tuberculosis, there is minimal perihilar activity but rapid miliary spread.[78, 112]

X-ray Examination. The radiologic appearance of pulmonary blastomycosis is quite varied, depending on the stage of infection and type of disease produced. As delineated by Pfister,[101] Cush,[24] and Laskey,[79] a variety of patterns may be observed and none is specific for blastomycosis. In the acute primary disease a diffuse infiltrative pattern or bilateral lower lobe densities may be observed. These are common in the symptomatic as well as the asymptomatic patients. The latter, though untreated, demonstrate healing or radiographically quiescent residua and, in time, may have completely normal-appearing x-rays of the chest.[141]

A second type of picture is that of subacute or post-primary progression. The initial lesions heal, but there is development of the disease elsewhere in the pulmonary parenchyma. In time, these patients and those with no recall of symptoms form the third, or indeterminate, category. They now show a widening of hilar shadows or unilateral disease resembling tuberculosis or neoplasm (Fig. 18–3). The latter is particularly important, as blastomycosis has been misdiagnosed as carcinoma more often than any other fungus infection.[65, 117] The condition of these patients may evolve into the fourth radiographic category, the chronic pulmonary phase of illness (Fig. 8–3,*A,B*). Such individuals have low-grade clinical illness of more than a month's duration and show scarring as well as active disease on chest film. Disease at this stage may be chronic or slowly progressive. If the disease remains untreated, there is occasionally a sudden burst of activity with rapid progression of the infectious process. The outcome is usually fatal, even with proper medical management. The fifth and final pattern is seen in this active miliary form of disease.[52] Spread may occur from an unresolved pulmonary lesion, or a miliary picture is produced from hematogenous seeding originating from an endogenous source in some other organ system[112] (Fig. 18–4).

As delineated by Pfister et al., a picture suggestive of blastomycosis includes dense, fibrotic pleura without either extensive calcification or an accumulation of effusion fluid. The disease may occur anywhere in the lungs, but the posterior segments of the upper lobes are the most common sites of the infiltrative granulomatous type.[101] In most series, the apex is usually spared, but middle lobe and basilar involvement are frequent. Although they are less common than in histoplasmosis or coccidioidomycosis, in some series cavities varying in size from 1 to 8 cm occurred in 25 per cent of cases.[13, 79] As noted previously, hematogenous spread may produce a picture of miliary disease. If the disease has progressed this far, resolution of the pulmonary lesion does not occur, and the infection progresses to death if untreated.

Chronic Cutaneous and Osseous Disease

Cutaneous blastomycosis is the most common form of extrapulmonary disease. Thus

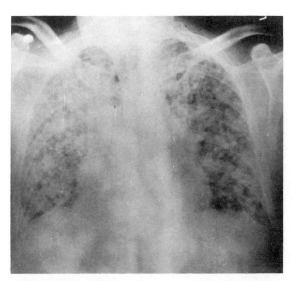

Figure 18–4 Miliary blastomycosis. Hematogenous seeding from reactivation of a cerebellar focus of infection. (From Rippon, J. W., and J. R. Zvetina. 1977. Mycopathologia *60:*121–125.)

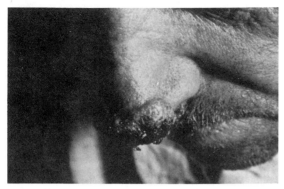

Figure 18–5 Blastomycosis on nose. This is a papulopustular lesion with some scarring in the center. Note small pustules on the outer edge of the lesion. The organism is easily demonstrated from aspirates of this material.

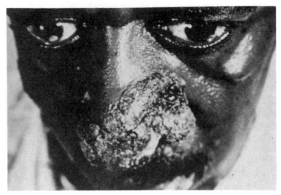

Figure 18–6 Blastomycosis. Evolution of the lesion into an ulcerated granuloma with a serpiginous advancing border. The central area is covered with crusts.

skin lesions are often the most frequent presenting symptom of the disease. For this reason many, if not most, cases of this type of disease are first seen in dermatology clinics. In most series, about 50 per cent of patients have pulmonary disease and the other 50 per cent have cutaneous alone or cutaneous disease with some involvement of other organ systems. Up to 80 per cent of all patients have some skin lesions, and often the organism is first demonstrated in aspirated material from such sites.[149, 151] The first signs to appear are subcutaneous nodules or papulopustular lesions that ulcerate. The initial lesions occur singly or in groups and are most commonly found on exposed peripheral areas, such as the face, hand, wrist, and lower legs, or on mucocutaneous areas, such as the larynx,[102, 107] tongue,[131] or mouth,[98] (Fig. 18–5). In time, as the disease progresses, lesions also occur on the trunk and other unexposed areas. Within weeks or months the lesions evolve into ulcerated verrucous granulomas with serpiginous, advancing borders which are raised 1 to 3 mm and have a sharp sloping edge (Fig. 18–6). The central area is covered with crusts and characteristically contains "black dots," representing degenerating papillary vessels.[82] This violaceous, discolored, crusty, verrucous lesion has often been misdiagnosed as basal cell carcinoma[20] (Fig. 18–7); however, in blastomycosis small microabscesses occur at the periphery. Aspirated material from these can be examined in a potassium hydroxide mount for demonstration of the yeast cells of the organism. The ulcers heal from the center by fibrosis and cicatrization. Biopsy from the center of the lesion is usually devoid of organisms, and

only fibrosis and scar formation are seen. The yeasts are most numerous at the active edge of the lesion; therefore, biopsy and culture should be taken from this site. Lymphadenitis and lymphadenopathy are usually not present in secondary cutaneous blastomycosis. Over a period of years these lesions become deforming, thin, atrophic scars, which may cover large areas of the face, neck, or other areas (Fig. 18–8).

Subcutaneous Fulminant Disease. Cutaneous lesions associated with underlying bone involvement or generalized systemic disease may first appear as sinus tracts exuding purulent material containing numerous organisms. Such lesions usually lack the verrucous advancing margin seen in the usual chronic cutaneous disease. They begin as deep, subcutaneous cold abscesses (Fig. 18–9) that eventually burrow to the surface as large pustular or bullous lesions with purulent exudate. Numerous yeast cells can be seen in aspirated material. This phase, if untreated,

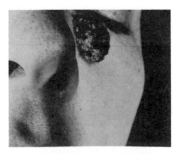

Figure 18–7 Blastomycosis. Dark, crusty, verrucous lesion resembling basal cell carcinoma. There were no other lesions on the patient, and the lung fields were clear. Aspiration of a small microabscess revealed the yeast cells of *B. dermatitidis*.

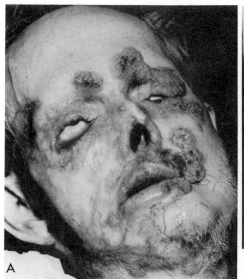
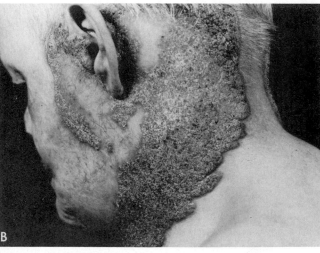

Figure 18–8 Blastomycosis. *A,* Secondary cutaneous lesions in a case of long duration. *B,* The fungus is found in the advancing border within the verrucous vegetations. The central areas clear and heal, with formation of scar tissue. The marked telangiectasia resulted from the use of x-rays as treatment for the disease. (Courtesy of A. Lorincz.)

is a harbinger of rapid dissemination to all organ systems and a fulminant fatal terminus of the process.

Osseous Blastomycosis. In the several reviews of case reports of blastomycosis, from 25 to 50 per cent of patients have had disease of the bone.[44] Not infrequently, the only presenting symptom is an occult osteolytic lesion or monoarticular arthritis.[116] The most frequent sites of bone infection are the vertebrae, ribs, skull, long bones, and short bones. The long bones were found to be most commonly involved.[44] The manifestations of osseous disease are protean, and there is no distinctive radiologic picture. The lesions

generally consist of a focal or diffuse suppurative osteomyelitis at the epiphyseal end of long bones. Osteolytic and osteoblastic processes may be seen by x-ray examination. The borders of the osteolytic lesion are sharp and well defined. There is a sharp juncture between the area of involvement and normal bone (Fig. 18–10,*B*). In contrast, tuberculosis of the bone is notable for the rough irregular borders, usually with some evidence of demineralization of the adjacent bony areas. Blastomycotic lesions often resemble infarctions, particularly in long bones. The subarticular and subepiphyseal regions are the favored sites. These may erode into the cap-

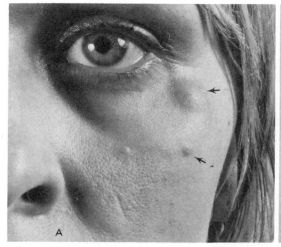
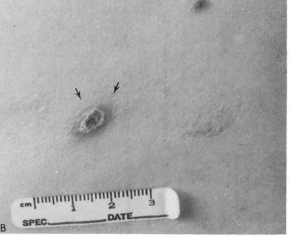

Figure 18–9 *A,* Deep subcutaneous abscess (arrows) of disseminating blastomycosis in a pregnant woman. The lesions would burrow to and erode through the skin. *B,* Presenting as pustules one cm or more in diameter.

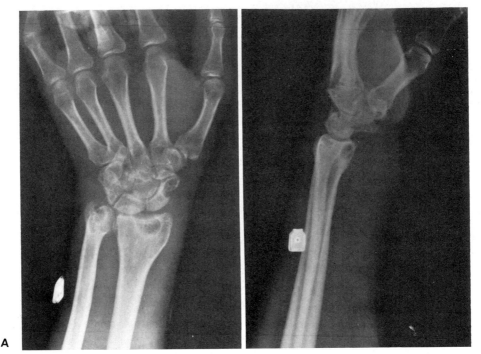

A   B

Figure 18–10   *A,* Left wrist. Destructive areas are seen in the carpus, the radial metaphysis, and the distal ulna. The destructive focus in the radius connects to the articular surface, and there is a slight widening of radiocarpal articular space. The foci have sclerotic margins. This was a case of monoarticular involvement diagnosed as rheumatoid arthritis. No organisms were seen in biopsy, but it was culture-positive. The lung fields were clear. *B,* Lateral view shows marked soft tissue swelling and sharp borders of the lesions.

sule to produce septic arthritis.[116] The tissue reaction resembles that of granuloma formation of tuberculosis more than the cystlike lesions of coccidioidomycosis. The proliferative response in blastomycosis is less prominent than in actinomycosis.

Involvement of the vertebrae is similar to that seen in tuberculosis. In both diseases, granulomatous lesions destroy the disk spaces, erode the vertebrae anteriorly, and produce paraspinal masses. The anterior longitudinal ligaments are dissected from beneath, but the spinal segments are spared. Compression of the spinal cord may result. Paravertebral or psoas abscesses may occur as a result of extension of the bony lesion.

Isolated lesions can occur on any bony site. The skull is not infrequently listed (Fig. 18–11,*A,B*). Blastomycosis of the fronto-ethmoid complex in a 5-year-old boy was reported by Devgan et al.[29] This could possibly represent a primary portal of entry. A solitary lesion of the sacroiliac joint, personally seen, had the appearance of osseous tuberculosis. (Fig. 18–11,*C*). The patient had no other signs or symptoms except soft tissue swelling over the involved area. The lesion cleared with amphotericin B therapy.

Blastomycosis of the joints often presents as a hot, swollen, septic arthritis, which is sometimes accompanied by a chronic exudative sinus tract. The periarticular tissues may be destroyed along with the synovial membrane and articular ligaments, with resultant subluxation. Since bony involvement is so frequent in blastomycosis and so often occult, complete roentgenographic examination of the entire skeleton should be made in all cases of the disease. In low-grade chronic blastomycosis, a single lesion is usually found, and the number of organisms present is small. In generalized systemic disease, however, numerous organisms are present, and several bones are involved.

Systemic Blastomycosis

Patients with extensive and unresolving pulmonary involvement almost always develop generalized systemic disease. These patients appear to be less able to contain the infection than those with low-grade chronic disease. Whereas few organisms are seen in biopsies from the latter, lesion material in generalized systemic disease contains nu-

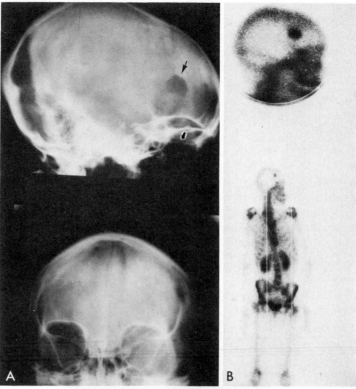

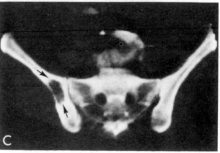

Figure 18–11 *A,* Isolated lesion of blastomycosis involving the skull *(arrow).* The lung fields were clear and there was no evidence of foci in other organ systems. Skull film. *B,* Bone scan of patient. *C,* Isolated lesion in another patient *(arrows).* No other evidence of disease except this lytic process in the sacroiliac joint. There was some soft tissue swelling.

merous fungi. Showers of organisms from the pulmonary foci serve to seed other organ systems. Disseminated infections may manifest as cutaneous and osseous lesions that develop more rapidly and have less tendency to resolve than chronic disease. Usually multiple organ systems are involved, and the patient may have a rapidly deteriorating course.

There are always exceptions to any clinical rule. Disseminated blastomycosis may develop in patients with negative chest films. This occurs when there is endogenous reactivation of some other site, as reported by Laskey and Sarosi.[78] In another case the patient presented with labyrinthitis and an intracranial tumor. He went on to develop miliary foci in the lung and died. At autopsy the seeding lesion appeared to have been in the cerebellum (Fig. 18–14,*A*).[112]

Disseminated blastomycosis is almost always a disease of men in the middle decades. Rapidly spreading systemic disease may also develop in women during particular stages of life, such as the prepubertal or postmenopausal period, or during alterations in the hormonal balance. At particular risk are women in pregnancy (Fig. 18–9). Such patients usually respond well to amphotericin B therapy, and no harm befalls the fetus.[94]

Urogenital. The third most commonly involved site of extrapulmonary blastomycosis is the urogenital system. In published reports, the incidence varies from 5 to 22 per cent. Epididymitis appears to be the most frequently encountered condition, and it tends to be recurrent. The epididymis is swollen and tender, and testicular involvement is not infrequent. Scrotal ulcers and a draining sinus from an orchitis are sometimes seen.[47] In our experience, the prostate is also commonly involved in disseminated blastomycosis. The prostate is enlarged, boggy, and slightly tender. Organisms can be demonstrated in material obtained after prostatic massage. The disappearance of the fungus from this material may be used as an index of therapeutic progress. Involvement of the female reproductive system is quite rare, as is disease of the kidney. Conjugal transmission of the infection has been reported several times.[23, 34, 36] As in paracoccidioidomycosis and histoplasmosis, Addison's disease may result from blastomycosis of the adrenals.[32, 59] Chandler,[18] reporting a recent case, advised that patients with blastomycosis should have their adrenal function evaluated. In the case he reported, treatment with amphotericin B precipitated an addisonian crisis. In a case reported by Arora et al.,[3] hyperprolactinemia, galactorrhea, and amenorrhea occurred during chest wall involvement by blastomycosis. These endocrine abnormalities disappeared after treatment with amphotericin B.

Central Nervous System. The central nervous system is involved by hematogenous spread from other foci. This is not a common complication, being found in only 3 to 10 per cent of infections.[125] Sometimes exclusive neurological involvement is seen and presents difficulties for correct diagnosis.[49] The most frequent neurologic symptoms are headache, convulsions, confusion, coma, paraparesis, hemiparesis, and aphasia. Other patients can present with labyrinthitis, intracranial tumor, or personality changes.[112] The cerebrospinal fluid protein and pressure are elevated. The leukocyte count is variable, and the sugar is usually normal or low. Generally the organism is not recovered in culture. Of the nine patients recorded by Buechner and Clawson,[12] there was a 67 per cent mortality rate. Leers et al.[83] described a cerebellar lesion that was not associated with meningitis or a detectable primary site of infection. A cerebellar lesion was also the reactivating focus in a case of fatal disseminated disease.[112]

Granulomata of the liver and spleen are infrequently seen. In contrast to histoplasmosis and paracoccidioidomycosis, gastrointestinal involvement is quite rare. Endophthalmitis has been recorded in a patient who had no other evidence of disease.[37] As would be expected, occult foci can be reactivated in the immunosuppressed patient. Opportunistic blastomycosis occurs,[60, 96] but not with near the frequency of opportunistic histoplasmosis or coccidioidomycosis.

Inoculation Blastomycosis

This is a very rare entity, and most cases are the result of laboratory accidents[81] or accidental implantation during autopsy examination or while embalming patients who have died of blastomycosis.[149] The skin of the fingers or the hands is usually involved. The lesions that develop are of the chancriform type. The formation of an indurated ulcer is accompanied by lymphangitis and regional adenopathy. These are mild forms of the disease and heal spontaneously without peripheral extension or dissemination. Larsh and Schwarz[81] reviewed the subject in 1977 and discussed eight documented cases.

Blastomycosis of Children

Although chronic progressive or cutaneous disease is seldom reported in individuals less than 15 years of age, acute pulmonary blastomycosis is being described with some frequency in this age group. Usually such cases are encountered in a point-source epidemic[119] and the infections resolve without treatment. Without the association of adult human or animal cases, such infections in children would probably go undiagnosed.

Severe acute pulmonary blastomycosis requiring treatment also occurs in endemic areas. Powell and Schuit[106] discussed fourteen children, ages 2½ to 15 years, in whom the illness was remarkably similar. All had disease limited to the lungs with fever, malaise, and respiratory symptoms. Following treatment with amphotericin B or dihydroxystilbamidine, thirteen of the fourteen had persistently abnormal roentgenograms. In a report by Chesney et al.[19] one patient of four years had *B. dermatitidis* isolated from sputum, the other of three months of age was culture-positive from nasopharyngeal aspirate. The family dog had died of blastomycosis prior to the onset of the patient's disease. The authors conclude that the treatment of choice for children is dihydroxystilbamidine rather than amphotericin B. This is because of the severe side effects of the latter drug in children.

In recent reviews of blastomycosis in children by Yogev and Davis[152] and Laskey and Sarosi,[80] the authors note that the full spectrum of adult disease can be manifested in younger individuals as well. They conclude, however, that secondary blastomycosis is rare in the younger age groups and that most infections are of the acute primary pulmonary type.

DIFFERENTIAL DIAGNOSIS

Blastomycosis must be differentiated from any chronic granulomatous or suppurative pulmonary disease. The list includes histoplasmosis, because of the overlap of endemic areas, tuberculosis, silicosis, sarcoid, and to a lesser extent actinomycosis, nocardiosis, and other bacterial diseases. High on the list of differentials is pulmonary neoplasms. Cutaneous lesions resemble scrofuloderma, lupus vulgaris, epitheliomas, bromoderma, iododerma, nodular syphilids, granuloma inguinale, swimming pool granuloma, and similar diseases.

There does not seem to be an association of blastomycosis with underlying or debilitating disease. In the many reviews of complications that occur in patients with neoplasias and leukemias, of those receiving steroids, no increase in incidence of blastomycosis was found. It is possible that subtle immunologic differences are present in the patient who develops the disease. This is particularly true in patients with the systemic form rather than the chronic cutaneous form of the disease. There is as yet no evidence to confirm this hypothesis. Blastomycosis may coexist with tuberculosis, histoplasmosis, and bronchogenic carcinoma, as well as other diseases with severe pulmonary involvement.

PROGNOSIS AND THERAPY

Prior to the advent of effective chemotherapy, the diagnosis of blastomycosis usually meant a fatal outcome. In cases of systemic disease the mortality rate was 92 per cent. Patients with chronic cutaneous disease usually had a more protracted but also fatal course. There are reports of spontaneous resolution of clinically apparent disease,[119, 141] but it must be remembered that infections may remain quiescent for years before recurring.[76, 80]

The treatment of choice in all forms of blastomycosis is amphotericin B; however, dihydroxystilbamidine, which is much less toxic, can be used in some circumstances. The organism is quite sensitive to amphotericin B (0.03 to 1.0 μg per ml *in vitro*), and relapses occur only in cases of inadequate treatment. In the reviews by Abernathy and Jansen in 1960,[1a] Parker et al. in 1969,[99] Witorsch and Utz in 1965,[151] Lockwood et al. in 1969,[85] and Klapman in 1970,[71] clinical cure was achieved when a total of 2 or more grams of the drug were used. There was a significant relapse rate if less than 1.5 g were used. This therapeutic regimen is the one commonly used for the other systemic mycoses. The initial dose is 10 to 20 mg in 300 to 500 ml of 5 per cent glucose given as a slow infusion over a three- to six-hour period, depending on the patient's tolerance. The dose is increased by 10 to 20 mg daily. Initially alternate day therapy may be necessary if side effects are severe. Diphenhydramine (10 to 30 mg) is used to control the nausea; 100 units of heparin may be added to avert phlebitis, and aspirin is used for pain and headache. The dosage is increased until the standard amount of 0.6 mg per kg body weight per day is achieved, and blood levels of 0.5 to 3.5 μg per ml are obtained. The blood:spinal fluid concentration ratios are 30:1 to 50:1. Intrathecal injections may be necessary in cases of meningeal involvement.[97] A 5-ml quantity of spinal fluid is withdrawn, and 20 mg of hydrocortisone is mixed with it. This is slowly instilled intraspinally. After a few minutes, another 5 ml of spinal fluid is withdrawn, mixed with a solution containing 0.5 mg of amphotericin B, and reinstilled. This procedure may be repeated two or three times weekly until a total dose of 15 mg has been given. This is considered adequate for therapy of meningitis. A treatment schedule for systemic disease in children has been worked out by Turner and Wadlington,[143] and Chesney et al.[19] Because of the side effects associated with amphotericin B therapy, the latter authors recommend the use of dihydroxystilbamidine in children when the disease is restricted to the lungs. Oral amphotericin B has been used, but results have been variable and serum levels unpredictable.

The aromatic diamidine, hydroxystilbamidine, had been used with some success in the treatment of blastomycosis. The drug is administered as a daily dose of 225 mg in 500 ml of saline. Total dose of 8 g is recommended. When stilbamidine was first used in the 1950's there were serious side effects associated with it, including peripheral neuropathy, and changes were seen in liver function tests. With the use of the dihydroxy derivative of the drug, first published by Snapper and McVay,[136] these problems have largely been obviated. In several series that compared the efficacy of dihydroxystilbamidine to amphotericin B,[14, 71, 85] the following conclusions were presented: dihydroxystilbamidine, when used in all forms of blastomycosis, has a relapse rate of up to 30 per cent; however, in patients with limited disease (i.e., few lesions in the lung and/or skin) and in patients with preexisting kidney impairment, serious consideration should be given to dihydroxystilbamidine. It also can be tried in amphotericin B–resistant cases or cases of relapse. The drug is not recommended for disseminated disease, central nervous system involvement, or osseous lesions.

Older modes of therapy, such as iodides, x-ray, and vaccines, have been replaced by modern therapeutic procedures. Allergic reactions may occur during therapy, but these can usually be controlled with the judicious use of steroids and do not require desensitizing procedures.

Several of the imidazoles (miconazole, econazole, and ketoconazole) have had trials in experimental animal infections or in human disease. Though some success has been reported from time to time with miconazole, there is a high relapse rate. Significant side effects are minimal, however, and Bleeker and Haanen[9] reported the cure of chronic disseminated blastomycosis (chest wall and skin of leg) with miconazole. It appears that in cases of very limited disease miconazole may be useful, although extensive disease is refractory to this therapy. As yet there have been too few trials with the other imidazoles to warrant judgment. Surgical procedures are of value when large abscesses require drainage. Thoracic surgery to remove large pulmonary lesions or cavities should not be performed until the exudative phase is controlled by medical means. Busey has reviewed the long-term status of several patients treated surgically compared with those who had medical management alone.[13] Although some patients with minimal involvement were cured by resection, the authors concluded that amphotericin B should be used to treat all cases before surgical procedures are instituted.

PATHOLOGY[20, 125]

Gross Pathology. The sharp differentiation between the two categories of blastomycosis is again evident in the gross and histologic pathology of the disease. In patients in whom there is minimal or no apparent pulmonary disease, the cutaneous, laryngeal, and mucocutaneous lesions are so verrucous and hyperplastic that they lead to a preliminary diagnosis of carcinoma. Even the cut surface of these lesions and the carcinoma-like granulomata removed from lungs give no indication of their true etiology. Accurate diagnosis depends on careful histologic examination and culture of such material. At autopsy the healed primary lesion of patients with generalized blastomycosis often appears only as a focus of pulmonary scarring or minor pleural fibrosis. In contrast to histoplasmosis, calcified fibrocaseous nodules are not a usual finding.

Whereas granuloma formation is characteristic of prolonged chronic disease, disseminated blastomycosis tends to be suppurative or mixed suppurative and granulomatous. The pulmonary foci may be scattered, small, multiple nodules (Fig. 18–12) or large, caseous nodules, abscesses, and enlarging gran-

Figure 18–12 Cut surface of the lung from a case of endogenous miliary dissemination. There are numerous small multiple yellow necrotic nodules *(arrows).*[112] See Figure 18–17.

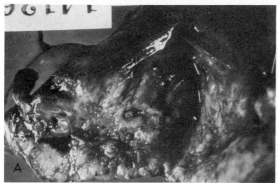

Figure 18-13 Blastomycosis. *A,* Cut surface of lung. The involved area is firm and granular. The texture and infiltrative pattern of the peripheral portions of lesions are indistinguishable from carcinoma of the lung. The patient had been diagnosed by x-ray as having carcinoma. See Figs. 8-3, *C, D, F;* 8-15, *B.* The reaction was predominantly of the suppurative necrotizing granulomatous type. *B,* Same lung after perfusion. (Courtesy of S Thomsen.)

ulomata (Fig. 18-13). Caseation occurs less commonly than in tuberculosis or histoplasmosis. Although cavitation is infrequent, it may occur at single or multiple sites. Pleurisy and pleural fibrosis are also encountered, but pleural effusion is rare.[64] Acute disease is characterized by pneumonic-type consolidations without effusions. Pleural reactions occur in late stage or chronic illness. If the disease was very active, invasion of other systems in the chest area may have occurred. Pericarditis and endocarditis, which are not usually associated with fungus infections, have been reported.[20] Often this is from an extension of adjacent lung lesions,[105] although localization by hematogenous spread has been recorded. Congestive heart failure resulting from pericardial adhesions and the presence of massive amounts of pericardial pus have also been noted.

The disease may extend from the pleural cavity into the vertebrae and ribs and through the chest wall, thus simulating actinomycosis. As in the latter disease, draining sinus tracts may burrow through from any osseous or subcutaneous abscess to the skin. The pus is characteristically pink-tinged because of the extravasated red blood cells. It usually contains numerous organisms which are easily demonstrated in potassium hydroxide mounts. Psoas abscesses similar to those seen in tuberculosis are also found. Lymph nodes become masses of abscesses and granulomas, with some areas of necrosis and some of fibrosis.

Abscesses of the brain and meninges are found which also mimic tuberculosis. Blastomycosis is an uncommon cause of granulomatous meningitis, but it is fourth on the list headed by tuberculosis, cryptococcosis, and coccidioidomycosis. Blastomycotic meningitis may occur without evident foci elsewhere in the body. Meningitic disease occurs as exudative, purulent, fibropurulent, or granulomatous and is either focal or diffuse. Microscopically, the lesions are histiocytic or predominately polymorphonuclear in reaction. Lesions in the cerebrum, cerebellum, or spinal cord are often hard and tumorlike or appear as necrotic abscesses (Fig. 18-14).

Prostatic involvement is common in systemic disease, whereas the uterus and tubes are not involved. Spleen, liver, and kidneys are usually spared, although minute abscesses are found in the late stages of extensive systemic disease.

Histopathology. The tissue response in blastomycosis varies from that of epithelioid granulomata to chronic suppuration, necrosis, and fibrosis or a combination of all of these. Again the reaction varies according to the type of disease in the patient, e.g., epitheliomatous hyperplasia, and few organisms are seen in the chronic type of infection, whereas suppuration, necrosis, and numerous organisms are more commonly encountered in the generalized systemic form.

The histologic diagnosis of blastomycosis depends on the demonstration of the yeasts of *Blastomyces dermatitidis*. The organisms are generally of uniform size, varying from 8 to 15 μ in diameter, depending on the age of the lesions (Figs. 18-18, 18-21, 18-22). Some organisms may be as large as 20 to 30 μ in diameter. The wall is thick and rigid and is termed "double-contoured," or doubly refractile. The cytoplasm usually shrinks away from the wall, leaving a space, but the cytoplasm stains more prominently than the wall in hematoxylin-eosin preparations. In well-fixed, well-stained sections, the multinucleate nature of the cytoplasm is demonstrat-

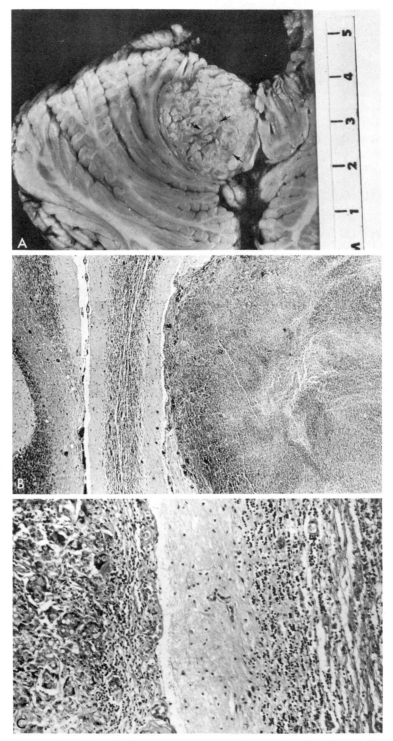

Figure 18–14 *A,* Cut section of cerebellum. A 1.5-cm lesion *(arrows)* in the inferomedial aspect of the right cerebellum. The lesion was necrotic and purulent. *B,* Edge of cerebellar lesion with central suppuration and granulomatous margin. Hematoxylin and eosin stain. ×40. *C,* Section of cerebellar cortex with granulomatous margin. The Langhans' giant cell contains the yeasts of *B. dermatitidis.*[112]

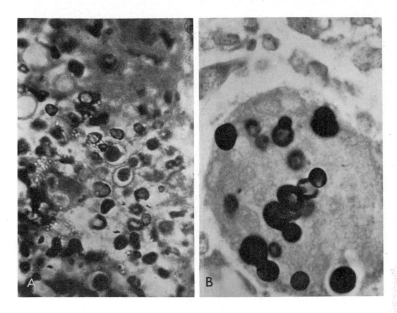

Figure 18–15 Blastomycosis. *A,* Budding yeast cell. The cytoplasm has shrunk from the side of the colorless wall but can be seen to be multinucleate. Hematoxylin and eosin stain. ×440. *B,* Numerous yeast cells, some of which are budding, in a giant cell. Note the broad-based buds. Methenamine silver stain. (GMS) ×440.

ed. The most characteristic feature of *B. dermatitidis* is the broad-based bud. An unequivocal histologic diagnosis cannot be made unless this form is seen (Fig. 18–15). Although small forms no larger than *Histoplasma* yeast cells have been described,[5, 123, 124, 125] these have broad-based buds, in contrast to *H. capsulatum*, which has a thin-necked bud on the yeast cell. Stainability of the organisms is quite variable also. Even in sections treated with special fungus stains, the yeast cells may be difficult to see. In hematoxylin and eosin preparations, the wall is colorless, and the organism may appear only as an outline in a giant cell (Fig. 18–16). The GMS or Gridley stains demarcate the wall, but its outline may be irregularly colored, especially if dead or dying organisms such as those found in chronic disease are present. In material from disseminated cases, viable organisms which were well fixed and stained are easily seen. Hyphae are very rarely present in tissue.

The finding of cells without buds necessitates the differentiation of *B. dermatitidis* from *Coccidioides immitis*, *Cryptococcus neoformans*, *Paracoccidioides brasiliensis*, and *Histoplasma capsulatum* var. *duboisii*. The yeasts of *B. dermatitidis* closely resemble those of *H. capsulatum* var. *duboisii*, and both are endemic in some of the same geographic regions. However, the latter organism lacks the broad-based bud. *H. capsulatum* is uninucleate and has a narrow-necked bud. The yeasts of *B. dermatitidis*, including the small forms, usually have several nuclei in addition to their broad-based buds. It must be remembered that both species have been isolated in culture from the

same lung.[10] Groups of young spherules of *C. immitis* are especially difficult to differentiate from *B. dermatitidis*. Only with the finding of a mature spherule with endospores can one be sure of the diagnosis. The capsule of *C. neoformans* stains a brilliant pink color with Meyer's mucicarmine stain, even in old lesions in which little capsular material is present. The yeast of *B. dermatitidis* is stained only faintly by this procedure. Needless to say, in all cases isolation by culture and identification of the etiologic agent is the only unequivocal method for diagnosis of the disease.

In primary pulmonary infection, the initial cellular reaction is inflammatory. Numerous

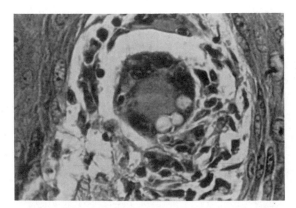

Figure 18–16 Blastomycosis. In this hematoxylin and eosin preparation, the yeast cell is outlined in a giant cell. This was from a case of cutaneous blastomycosis in which there was extensive pseudoepitheliomatous hyperplasia. Interspersed in the epidermis were a few microabscesses containing neutrophils, lymphocytes, and an occasional giant cell. Yeast cells were very rarely encountered.

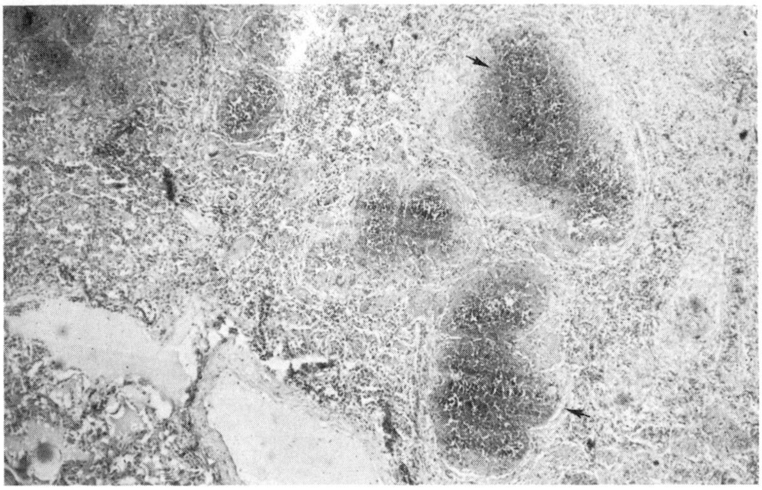

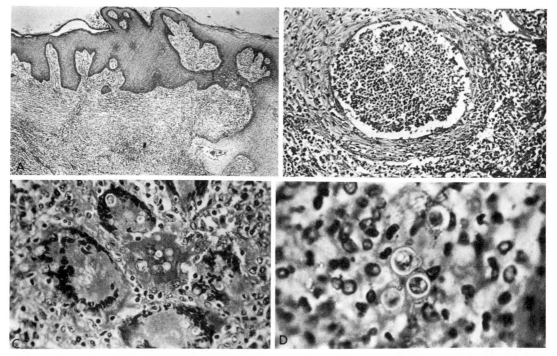

Figure 18–17 Miliary blastomycosis. Lower power view of histologic section of lung. The numerous necrotizing granulomatous lesions have almost completely replaced the pulmonary architecture (*arrows*). Hematoxylin and eosin stain. ×40.[112] (From Rippon, J. W., and J. R. Zvetina. 1977. Mycopathologia *60*:121–125.)

polymorphonuclear neutrophils are seen. In time, a chronic granulomatous reaction with focal suppurative areas is found — the picture of an epithelioid cell granuloma (Fig. 18–17). Multiple, small abscesses are present within the granuloma. These contain leukocytes, debris, and usually a few giant cells. It is within these abscesses that the organisms are found. Not infrequently they are seen as

shadows outlined within the giant cells (Fig. 18–16). The hyperplasia may be so great as to suggest neoplasia, and the microabscess is often the only key to the true etiology.

In extensive disseminated blastomycosis, organization into well-formed granulomata is less pronounced, and a suppurating necrotic abscess is the histologic picture seen. As the patient's defenses diminish, the organism

Figure 18–18 Chronic cutaneous blastomycosis. *A*, Extensive acanthosis and pseudoepitheliomatous hyperplasia. The downward growth has resulted in isolated islands in the dermis. A few microabscesses are seen in the upper epidermis. There is an infiltrate in the dermis of lymphocytes and plasma cells with a few neutrophils. ×100. *B*, A microabscess lying within the epidermis. In the neutrophilic debris were a few yeast cells. ×400. *C*, Giant cells from an organized granuloma of the lung, showing intracellular yeast cells. ×1000. *D*, Extracellular organisms that resemble the young spherules of *C. immitis*. ×1000.

proliferates in great numbers, and at autopsy masses of yeasts may be found in all organ systems.

The histologic picture of chronic cutaneous disease is quite different from that of disseminated infection. The organisms are carried to the papillary dermis, probably by macrophages. They then incite an acanthosis that is more extensive in this than in any other fungal disease. This pseudoepitheliomatous hyperplasia has often been diagnosed as squamous cell carcinoma (Figs. 18–18,*A,B*; 18–19). However, the epithelial cells are well differentiated when the acanthosis is due to blastomycosis. If examined carefully, microabscesses that contain neutrophils, some lymphocytes, and an occasional giant cell (Fig. 18–18) can be seen. The fungus that is generally present in very small numbers in this form of blastomycosis is found within these abscesses. Diligent search of serial sections may be required to find the yeasts. Though the lesions begin in the dermis, prolific downgrowth of the rete ridges often results in the finding of microabscesses in the epidermal areas. Beneath the extremely irregular border of the epidermis, a band of infiltrate is seen. The cells include neutrophils, lymphocytes, and a few giant cells. Many dilated blood vessels are present, but fibrosis is not extensive, and connective tissue stroma is minimal. In healing lesions, the fibrosis is more apparent but not dense or firm.

ANIMAL DISEASE

Natural disease among dogs is a very common finding in endemic areas. Furcolow et al.[40] tabulated 384 canine cases, but this probably represents only a minimal indication of the true incidence. The disease in dogs was generally found to occur in the endemic areas known from records of human cases, being particularly prevalent in Arkansas, Mississippi, Wisconsin, Minnesota, Illinois and Kentucky. The association of canine disease with outbreaks of human infections is so close that, as mentioned earlier, blastomycosis of the dog is considered a harbinger of human illness.[121]

The disease in dogs exhibits clinical symptomatology and pathology similar to human blastomycosis. Selby[128] has emphasized the occurrence of cutaneous ulcerative lesions, while others have found massive pulmonary involvement and osseous disease. In the case

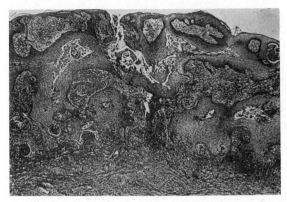

Figure 18–19 Chronic cutaneous blastomycosis. This case was diagnosed as basal cell carcinoma because of the very extensive cellular proliferation. On histologic examination, the cells were seen to be well differentiated, and a few yeast cells were found in the microabscesses. ×100.

described by Shull et al.[130] symptoms included hematuria, dyspnea, and anorexia. Yeast cells of *B. dermatitidis* were seen in the urine sediment. At autopsy the fungus was found in lymph nodes, the lungs, and the kidneys. This differs from the usual human pathology, since the kidney is rarely involved, even in late stage dissemination. Urogenital disease in dogs has been reported with some frequency.[17, 130, 150] As noted in earlier sections, urogenital involvement is common in human infections.

An essentially unique feature of canine disease is ocular blastomycosis. Although dissemination to the eye is rarely reported in human disease, it appears to be a common finding in canine studies.[17, 100, 132, 150] The symptoms usually include blepharospasm, diffuse edema in the cornea, and increased intraocular pressure. Yeast cells are found in the choroid and often in the vitreous and aqueous. Sometimes eye findings are the only focus of disease, but the eye does not appear to be a primary portal of entry. More often, ocular lesions are seen in association with pulmonary and cutaneous involvement. In one study still in progress, 25 per cent of forty cases of canine blastomycosis in the Chicago area had involvement of the eye.

Treatment with amphotericin B has been successful.[15] A total dose in the range of 3.75 mg per kg is recommended. The efficacy of serologic procedures in the diagnosis of canine disease has been reviewed by Turner et al.[142] Precipitins and complement-fixing antibodies were not dependable indicators of disease. Culture of lesion material on laboratory media was more efficient than mouse inoculation for recovery of the organism.

Disease in the horse has been recorded several times, particularly in Kentucky. Cutaneous lesions in the lower legs are the most common presenting signs, with lung involvement almost always present. An interesting case in another animal was the occurrence of blastomycosis in a northern sea lion (*Eumetopias jubata*).[147] This 8-year-old female had been housed in the Chicago Zoo for six years. She died of massive infection of the lung and meningitis from the "Chicago disease." The disease has also been recorded in the cat[61, 129] but is rare. The cat appears to be resistant to blastomycosis even in highly endemic areas. This is also true of the other systemic mycoses.

Experimental disease can regularly be produced in laboratory animals.[26] In a study of comparative susceptibilities of such animals, Conti Diaz et al.[22] inoculated guinea pigs and hamsters intratesticularly and mice intravenously with 9400, 940, and 94 mycelial fragments. The organism was recovered in all three species at all doses. Soil flotation followed by injection of the supernatant intravenously into mice was used successfully by Denton and DiSalvo[26] to isolate the organism from soil. It is possible that the intratesticular inoculation of soil samples may be a better method than intravenous inoculation, as this organ is cooler than other parts of the body and most conducive to the growth of the fungus.

A mouse model of pulmonary blastomycosis has been developed by Harvey et al.[58] Using intranasal inoculation, they could titer the virulence and repeatedly predict the outcome based on body weight and age; they thereby found a wide range of virulence among various isolants of the organism.

Sputum and contaminated clinical material mixed with an antibacterial antibiotic can be injected intraperitoneally into mice as an adjunct to cultural isolation. Mice are killed after three to four weeks and examined for the presence of pus and abscesses in the omentum and peritoneal cavity. Such lesions are usually near the pancreas, and the omentum may be adherent to other internal organs. Intravenous inoculation may result in pulmonary disease.

Landay et al.[75] have compared the relative susceptibility of the two sexes of hamster to experimental infection. As has been observed in other mycoses, males were more susceptible than females. Experimental cutaneous disease was studied by Salfelder.[115] Subcutaneous injection of *B. dermatitidis* yeast cells into animals produced an active skin lesion that did not ulcerate. Regional lymphadenitis occurred, but this was sometimes difficult to discern. The lesions healed spontaneously without hematogenous dissemination. The relative pathogenicity of the yeast and mycelial forms of the organism have been studied by Guidry and Bujard,[53] using the chorioallantoic membrane of embryonated chick eggs. At 37°C the yeast cells rapidly penetrated the mesoderm and proliferated. At the same temperature, mycelial elements did not invade until they converted into the yeast phase. At 31°C mycelial elements did not convert and were confined to the ectoderm.

Experimental disease in dogs has been studied by Ebert et al.[33] They found that injection of 10^5 cells was sufficient to produce disease. The clinical course was more severe than in experimental canine histoplasmosis. Amphotericin B therapy was found to prevent the death of some of the animals, but not all survivors were cured. All control animals died of disseminated disease. The natural occurrence in bats of several mycoses (histoplasmosis and sporotrichosis) has been recorded, and the bat has been postulated to be a natural vector of the etiologic agents of these diseases. Tesh and Schneidau[140] produced experimental blastomycosis in bats (*Tadarida brasiliensis*) by inoculating them with 5.5×10^5 fungal cells. All animals developed severe disease, and most of them died. *B. dermatitidis* was isolated in fecal cultures obtained during the course of the disease.

BIOLOGICAL STUDIES

The morphologically distinct saprophytic and parasitic forms of *B. dermatitidis* have interested investigators for many years. It was noted early that physiologic changes occur concomitantly with morphologic changes. Levine and Ordal[84] demonstrated a severalfold increase in metabolic rate of the yeast form compared to the mycelial form of the fungus. Later work has shown that temperature is the only factor controlling dimorphism in the organism. There are some differences in enzyme production between the mycelial and yeast forms,[7, 111] for in general more proteolytic enzymes are elaborated in the former than in the latter form. There are also some differences in enzyme production associated with the two sexual mating types.[111] Physiologic differences between the growth phases have also been noted. Bawdon and Garrison[6] found that both phases took up leucine and cystine by permease systems when the amino acids were in low concentrations. At higher concentration, uptake was by simple diffusion. Determination of the Michaelis-Menton constants for the permeases indicated that both leucine and cystine were incorporated eleven times more rapidly in the mycelial phase than in the yeast phase. Physiologic and compositional comparisons of *Blastomyces dermatitidis* to *Histoplasma capsulatum* are discussed in Chapter 15.

As is the case with other dimorphic fungi, more chitin is present in the yeast form than in the mycelial. The yeast cell wall also contains about 95 per cent α-glucan, whereas the cell wall of the mycelial form contains only 60 per cent; the remaing 40 per cent is β-glucan.[38] Fundamental differences in membrane organization and morphology have been discerned in ultrastructural studies of the two morphologic forms by Garrison et al.[41] Conversion of the yeast to mycelium is accompanied by membrane reorganization and the appearance of woronin bodies where mycelial septa are going to form. Garrison[42] also studied the conversion of the conidium to the yeast phase. Under conditions favoring conversion the conidia readily germinate, forming short "germ tubes." These subsequently enlarge to form yeast mother cells (YMC), the inner wall of the germ tube developing into the wall of the yeast mother cell. The yeast mother cells then begin to produce yeast cells. In contrast, mycelial units under conditions favoring conversion demonstrate profound degenerative changes with only a few yeast cells developing. The conidium is thus the primary infective unit of this fungus.

The yeast form in nature usually exists only in infected animals. Saliva containing yeast cells may contaminate the soil, but the cells quickly lyse.[90] McDonough has shown that both live and dead yeast cells lyse when placed in soil; therefore *B. dermatitidis* probably exists in nature only as a mycelial saprophyte. The mycelium has also been shown to lyse and die in soil. The fungus grows on woody material,[30] however, and this may be its niche in nature. In between "blooms" of activity it probably survives only as a conidium. Conversion of the conidium and the production of disease may represent a case of accidental pathogenicity, which therefore

represents a dead end as far as dissemination of the species is concerned. Under conditions of high temperature, however, in a milieu rich in nitrogen and partially decayed plant material, the organism has been found in its yeast phase. It was discovered by Sarosi and Serstock[120] in a bag of pigeon manure being used for fertilizer and was the point source for human infection.

Studies of the chemical composition of the yeast of *B. dermatitidis* have shown that it differs from most other pathogenic fungi in having a very high lipid content. In addition, both the yeast and mycelial forms produce an unusual metabolic product, ethylene.[95] This gas is also elaborated by ripening fruit and incites the degeneration of chlorophyll. So far the gas has been found only in a few other fungi, including *Penicillium digitatum* and the etiologic agents of histoplasmosis, blastomycosis, and paracoccidioidomycosis. A role in pathogenesis has been postulated, but no experimental evidence has been established. Factors affecting the conversion of the mycelial phase to the yeast phase and physiologic and compositional differences between Y and M cells of *Blastomyces dermatitidis* are reviewed by Rippon[112a] and Gilardi.[45]

Several interesting studies have been done on the interaction of components of *B. dermatitidis* with human leukocytes. Culture filtrates of the fungus contain levels of granulocyte chemotactic activity.[133] This activity was higher than that observed for filtrates from *Histoplasma* or *Cryptococcus neoformans*; however, levels of intracellular killing were much lower for *B. dermatitidis* than for the other fungi. Repine et al.[109] describe a factor in the serum of patients with untreated blastomycosis that inhibits neutrophil locomotion. This factor was not present in patients with histoplasmosis, coccidioidomycosis, cryptococcosis, sporotrichosis, or treated blastomycosis. A specific lymphocyte-transforming principle from *B. dermatitidis* has been isolated by Hall et al.[54] Many other investigations delineating the aspects of cellular defense mechanisms and infection by *Blastomyces dermatitidis* are now under way.

IMMUNOLOGY AND SEROLOGY

The immunology and serology of blastomycosis is the least well understood of any of the systemic mycoses. No specific skin test is available, so that the existence of resolved subclinical infections which could lead to spe-

mating types of *Ajellomyces dermatitidis*. Mycopathologia, *60*:65–72.

112. Rippon, J. W., J. R. Zvetina, et al. 1977. Miliary blastomycosis with cerebral involvement. Mycopathologia, *60*:121–125.

112a. Rippon, J. W. 1980. Dimorphism in pathogenic fungi. CRC Critical Rev. Microbiol. *8*:49–97.

113. Rose, H. D., and B. Varkey, 1978. Miconazole treatment of relapsed pulmonary blastomycosis. Am. Rev. Respir. Dis., *118*:403–408.

114. Ross, M. D., and F. Goldring. 1966. North American blastomycosis in Rhodesia. Cent. Afr. J. Med., *12*:207–211.

115. Salfelder, K. 1965. Experimental cutaneous North American blastomycosis in hamster. J. Invest. Dermatol., *45*:409–418.

116. Sanders, L. L. 1967. Blastomycosis arthritis. Arthritis Rheum., *10*:91–98.

117. Sanders, J. S., G. A. Sarosi, et al. 1977. Exfoliative cytology in the rapid diagnosis of pulmonary blastomycosis. Chest, *72*:193–196.

118. Sarosi, G. A., and K. J. Hammerman. 1974. Clinical features of acute pulmonary blastomycosis. N. Engl. J. Med., *290*:540–542.

119. Sarosi, G. A., and R. A. King. 1976. Follow-up of an epidemic of blastomycosis. Am. Rev. Respir. Dis., *116*:785–788.

120. Sarosi, G. A., and D. S. Serstock. 1976. Isolation of *Blastomyces dermatitidis* from pigeon manure. Am. Rev. Respir. Dis., *114*:1179–1193.

121. Sarosi, G. A., M. R., Eckman, et al. 1979. Canine blastomycosis as a harbinger of human disease. Ann. Intern Med., *91*:733–735.

122. Schoenbach, E. B., J. M. Miller, et al. 1951. Systemic blastomycosis treated with stilbamidine. J.A.M.A., *146*:1317–1318.

123. Schwarz, J., and G. L. Baum. 1951. Blastomycosis. Am. J. Clin. Pathol., *21*:999–1029.

124. Schwarz, J., and G. L. Baum. 1953. North American blastomycosis. Geographic distribution pathology and pathogenesis. Docum. Med. Geogr. Trop. (Amst.), 5:29–41.

125. Schwarz, J., and K. Salfelder. 1977. Blastomycosis: a review of 152 cases. Curr. Top. Pathol., *65*:165–200.

126. Segretain, G. 1968. La blastomycose in *Blastomyces dermatitidis*. Son existence en Afrique. Maroc Med., *48*:20–27.

127. Sekhon, A. S., M. S. Bogorus, et al. 1979. Blastomycosis: Report of three cases from Alberta and a review of Canadian cases. Mycopathologia, *68*:53–63.

128. Selby, L. A., R. T. Habermann, et al. 1964. Clinical observations on canine blastomycosis. Vet. Med., *59*:1221–1228.

129. Sheldon, W. G. 1966. Pulmonary blastomycosis in a cat. Lab. Anim. Care, *16*:280–285.

130. Shull, R. M., D. W. Hayden, et al. 1977. Urogenital blastomycosis in a dog. J. Am. Vet. Med. Assoc., *171*:730–735.

131. Simon, G. B., S. D. Bersonet, et al. 1977. Blastomycosis of the tongue. South Afr. Med. J., *52*:82–83.

132. Simon, J., and L. C. Helper, 1970. Ocular disease associated with blastomycosis in dogs. J. Am. Vet. Med. Assoc., *157*:922–925.

133. Sixbey, J. W., B. T. Fields, et al. 1979. Interactions between human granulocytes and *Blastomyces dermatitidis*. Infect. Immun., *23*:41–44.

134. Smith, C. D. 1964. Evidence of the presence in yeast extract of substances which stimulate the growth of *Histoplasma capsulatum* and *Blastomyces dermatitidis* similar to that found in starling manure extracts. Mycopathologia, *22*:99–105.

135. Smith, J. G., Jr., J. S. Harris, et al. 1955. An epidemic of North American blastomycosis. J.A.M.A., *158*:641–646.

136. Snapper, I., and L. V. McVay. 1953. The treatment of North American blastomycosis with 2-hydroxystilbamidine. Am. J. Med., *15*:602–623.

137. Sorenson, R. H., and D. E. Casad. 1969. Use of case survey technique to detect origin of Blastomyces infection. Public Health Rep., *84*:514–520.

138. Stein, R. O. 1914. Die Gilchristche Krankheit (Blastomycosis Americana) und ihre Beziehung zu den in Europa beobachteten Hefeinfektionen. Arch. Dermatol. Syph. (Berlin), *120*:889–924.

139. Stober, A. M. 1914. Systemic blastomycosis. Arch. Intern. Med., *13*:509–556.

140. Tesh, R. B., and J. D. Schneidau, Jr. 1967. Experimental infection of bats *(Tadarida brasiliensis)* with *Blastomyces dermatitidis*. J. Infect. Dis., *11*:188–192.

141. Tosh, F. E., K. J. Hammerman, et al. 1974. A common source epidemic of North American blastomycosis. Am. Rev. Respir. Dis., *109*:525–529.

142. Turner, C., C. D. Smith, et al. 1972. The efficiency of serologic and cultural methods in the detection of infection with *Histoplasma* and *Blastomyces* in mongrel dogs. Sabouraudia, *10*:1–10.

143. Turner, D. J., and W. B. Wadlington. 1969. Blastomycosis in childhood: Treatment with amphotericin B and a review of the literature. J. Pediatr., *75*:708–715.

144. Utz, J. P., H. J. Shadomy, et al. 1968. Clinical and laboratory studies of a new micronized preparation of hamycin in systemic mycoses in man. Antimicrob. Agents Chemother., *7*:113–117.

145. Vermeil, C., A. Gordeetf, et al. 1954. Sur un case Tunisien de mycose generalise mortelle. Ann. Inst. Pasteur (Paris), *86*:636–646.

146. Wegmann, T. 1952. Blastomykose und andere Pilzerkrankungen der lunge. Dtsch. Arch. Klin. Med., *199*:192–205.

147. Williamson, W. M., L. S. Lombard, et al. 1959. North American blastomycosis in a northern sea lion. J. Am. Vet Med. Assoc., *139*:513–515.

148. Wilson, J. W., E. P. Cawley, et al. 1955. Primary cutaneous North American blastomycosis. Arch. Dermatol, *71*:39–45.

149. Wilson, J. W. 1963. Cutaneous (chancriform) syndrome in deep mycoses. Arch. dermatol., *87*:81–85.

150. Wilson, R. W., A. A. van Dreumel, et al. 1973. Urogenital and Ocular lesions in canine blastomycosis. Vet. Pathol., *10*:1–11.

151. Witorsch. P., and J. P. Utz. 1965. North American Blastomycosis: A study of 40 patients. Medicine (Baltimore), *47*:169–200.

152. Yogev, R., and A. T. Davis, 1979. Blastomycosis in children: a review of the literature. Mycopathologia, *68*:139–143.

19 Paracoccidioidomycosis

DEFINITION

Paracoccidioidomycosis is a chronic granulomatous disease that characteristically produces a primary pulmonary infection, often inapparent, and then disseminates to form ulcerative granulomata of the buccal, nasal, and occasionally the gastrointestinal mucosa. The lymph nodes are very commonly involved, and sometimes there is extension to cutaneous tissue or systemic involvement of multiple organ systems. The disease in its inception and development is similar to blastomycosis and coccidioidomycosis. The infection is geographically restricted to areas of South and Central America. The only etiologic agent is *Paracoccidioides brasiliensis*. The name of the disease, paracoccidioidomycosis, is particularly unwieldy. Because of its historical development, this term means a fungus disease, like a fungus disease, like a protozoan disease.

Synonymy. South American blastomycosis, Brazilian blastomycosis, Lutz-Splendore-Almeida's disease, paracoccidioidal granuloma.

HISTORY

Adolfo Lutz in 1908[63] described two patients who had extensive granulomatous lesions in the nasopharyngeal region and intense cervical adenopathy. In material from the lesions, he observed a fungus which reproduced in tissue by multiple budding but otherwise resembled the parasitic form of *Coccidioides immitis*. The fungus was also grown on various media at room temperature in a filamentous form and thus was noted as being dimorphic. Lutz coined the term "hyphoblastomycosis" for infections produced by dimorphic fungi and named this newly discovered disease "pseudococcidioidal granuloma." He did not name the organism, but he did stress the differences in the tissue forms in this disease compared with those of the recently described coccidioidomycosis and of Gilchrist's disease (North American blastomycosis). The Italian microbiologists Carini and Splendore published several studies of the organism between 1908 and 1912.[24, 112] Splendore classified it as an ascomycetous yeast in the genus *Zymonema*, a group in which he also included *Candida albicans* and all dimorphic fungi.[112] In many reports that followed, the disease was confused with coccidioidomycosis and other mycoses. In coccidioidomycosis, the fungus forms endospores in tissue, and the endospores are released by fracture of the spherule wall. It was thought that in the reproduction by *P. brasiliensis* the endospores passed through pores in the cell wall to the outside and remained attached to the surface of the parent cell. This process was termed "cryptosporulation," based on misinterpretation of the budding process. The excellent work of F. Almeida published between 1927 and 1929 finally clarified the subject and separated the two diseases.[5] P. Negroni studied the organism and in 1921 demonstrated dimorphism *in vitro*.[78] Conidiation of the mycelial phase was confirmed in 1951 by Neves and Bogliolo.[84] After reviewing the characteristics of the fungus *in vitro* and *in vivo*, Almeida and Lacaz in 1928 suggested the genus name *Paracoccidioides*, and Almeida in 1930 designated the fungus *Paracoccidioides brasiliensis*.[5] The similarity of conidiation of the mycelial form of this or-

459

ganism and that of *Blastomyces dermatitidis* led Conant and Howell to rename it *Blastomyces brasiliensis* in 1942.[26] Later studies have shown that the two fungi are distinct, and the original designation of Almeida and Lacaz is retained.

In Lutz's two cases and in the many cases reviewed and tabulated by Splendore,[112] the most evident lesions were the mucocutaneous ones about the mouth. The presence of lesions in the alimentary tract and apparent systemic dissemination from this area led to the long-held conclusion that the disease began by either the inoculation of the fungus into the oral mucosa or the ingestion of vegetable material containing the organism, followed by infection of the gastrointestinal tract. In 1936, Motta and Pupo[74] defined four broad categories that may be encountered in cases of paracoccidioidomycosis: tegumentary (mucocutaneous), lymphatic, visceral, and mixed. The tegumentary was considered the most common, and involvement of the lungs appeared to be rare. Only recently, because of the careful investigations of several authors, has it become apparent that the primary infection is pulmonary. These studies have been aided greatly by skin test surveys. Just as histoplasmosis was considered to be a rare disease until 1946, since only a few cases of the advanced fatal form were known, paracoccidioidomycosis was also thought to be rare. It now appears that subclinical and resolved infections are common in the endemic areas.

In 1914, Stein in Austria developed a skin test for blastomycosis. Patients with the Brazilian disease paracoccidioidomycosis did not react consistently with this test, indicating that the diseases were indeed separate entities. A primitive complement fixation (CF) test and a precipitin test were described by Moses in 1916.[73] He noted that eight of ten patients with the disease had demonstrable CF titers. Fonseca and Area Leão in 1927[35] were the first to devise a specific skin test for patients with paracoccidioidomycosis. They had misidentified their culture as *Coccidioides immitis*, however. This demonstrates the confusion that still existed at that time regarding these diseases. Almeida and Lacaz[7] developed a polyvalent skin test antigen from the Sabouraud's broth filtrate of 19 different cultures of *P. brasiliensis*. They called this preparation "paracoccidioidin." A specific polysaccharide developed by Fava-Netto in 1955[32] has been used more recently. This has shown that subclinical infection is quite frequent and that the disease

is much more common than formerly appreciated.

Therapy for the disease had been ineffective prior to 1940. Until then the usual iodides, vaccines,[6] x-rays, and other modalities used in other systemic mycoses had been used in this infection. Ribeiro in 1940[105] demonstrated the efficacy of sulfonamides, particularly sulfapyridine, in the treatment of paracoccidioidomycosis. This became the standard of treatment and is still used in the milder forms of infection. Some eighteen years later, Lacaz and Sampaio[57] introduced therapy with amphotericin B. This drug is effective in all forms of the infection. More recently, the imidazoles miconazole[113, 114] and ketoconazole[103] have gained favor in the therapy of the disease. These drugs appear to be quite efficacious and are less toxic than is amphotericin B.

There have been a few isolations of the organism from soil[3, 25, 79, 80]; however, the saprophytic habitat of this organism has not been defined. Knowledge is also incomplete concerning the extent and frequency of subclinical infections and the existence of natural infection in animals.

Several excellent reviews have been published on the disease, particularly by the Brazilian group of physicians and mycologists. Notable among these are the reviews of Lacaz[56] and Fonseca.[34] In 1971 the first Pan American Symposium on Paracoccidioidomycosis was held, and the proceedings of this meeting summarized the present state of our understanding of the disease and its etiology at that time.[85]

ETIOLOGY, ECOLOGY, AND DISTRIBUTION

The names of the disease and its etiologic agent give some indication of the confusion that has existed about these subjects. As mentioned above, the organism was initially thought to be a type of *Coccidioides* that exhibited cryptosporulation. The disease was also confused with lobomycosis. However, early in its history it was appreciated that almost all isolants were similar in morphology, and that probably only one agent was responsible for the specific disease entity, paracoccidioidomycosis. Unlike the etiologic agents of other fungal diseases, long lists of synonymous names were not forthcoming. The similarity of conidiation of the mycelial form and to some extent the budding of the yeast form have more recently been used to relate the agent of Lutz's disease to that of Gilchrist's disease. The inclusion of both species in the single genus *Blastomyces* had been proposed.[26] *B. dermatitidis* has a single, large, broad-based bud in the yeast phase, and *P. brasiliensis* has several small, narrow-necked buds. The work of Carbonell[20, 21, 22] has clearly demonstrated that the budding processes of the two species are fundamentally different. Careful mycological investigation of the mycelial phase indicates substantial differences between these two organisms.[91] Therefore, the accepted designation for the etiologic agent of paracoccidioidomycosis is *Paracoccidioides brasiliensis*.

The recent skin test studies of large populations and the recovery of the organism from soil have begun to clarify the ecology and epidemiology of the disease. In 1962 Chaves-Batista et al.[25] recovered from soil an organism that they considered to be the agent of paracoccidioidomycosis. This was later identified as *Aspergillus penicilloides*. Negroni claims to have isolated it once from soil in Argentina.[79, 80] More recently, in a study in a highly endemic region in Venezuela, Albornoz[3] recovered the organism several times from the area of Paracotos. This area has a population with a high rate of skin test sensitivity (up to 53 per cent).[4] The region is a coffee growing area and is characterized by "humid mountain forests" (Holdridge plant classification zone). The median temperature is 23° C, the altitude 648 M, and the rainfall is 1400 mm annually. Restrepo et al.[97, 101] have done retrospective studies on patients in Colombia. They found that at the time of diagnosis of the disease up to 89 per cent of patients had been born or lived in the subtropical sylvatic areas. These forests are included within a few of the Holdridge life zones characterized by particular climatic and biological associations.[66] The mean temperature is 18 to 23° C, and there is a rainfall of 800 to 2000 mm annually and an elevation of 500 to 1800 M. In an analysis by Rolon[106] on the distribution of the disease in Paraguay, similar findings were noted. Of the nearly 100 tabulated cases, most patients lived in the cooler (average temperature, 22 to 24° C), moister (1300 to 1500 mm average rainfall) regions of the southeast. Essentially no cases came from the hot, dry northwestern parts of the country. Most patients were men in the fifth decade of life. In the endemic zone the case rate was 3.54 per 100,000 for this population group. Laboratory studies by Restrepo et al.[96] had indicated that the fungus survives

longer in acid soil than in other soil types. Albornoz isolated the fungus from acidic surface soil, the "A horizon," an area of great biological activity and turnover.

The organism was reported to have been recovered from the intestinal contents of bats,[45] but there was no evidence of disease in the animals. Recent work by Greer et al.[43] indicates that neither phase of the organism can survive in the intestine of the bat and casts doubt on that report. The organism has been described in the tissue of a squirrel monkey,[48] but this report has not been confirmed. No valid isolations from natural infection in animals have yet been recorded. Borelli has suggested that the organism is a commensal or pathogen of poikilothermic animals.[15] At present, the saprophytic habitat and ecologic associations of the fungus remain to be elucidated.

Over 5000 clinically diagnosed cases of paracoccidioidomycosis have been recorded in the literature (Fig. 19–1).[16, 56] These represent only the serious, clinically apparent forms of the infection. The increased incidence of disease recorded in recent years probably reflects a greater degree of diagnostic accuracy. In Brazil, where over 3000 cases have been reported, the highly endemic areas are in the states of São Paulo, Rio de Janeiro, Rio Grande do Sul, Mato Grosso, and Minas Gerais. The disease is rare in the tropical rain areas of Manaus and Amazonas, and in the dry northeastern states of Brazil.[67] About 600 cases have been reported in Colombia, 600 in Venezuela, 100 in Argentina, 100 in Peru, 50 in Ecuador, 50 in Uruguay, 100 in Paraguay,[106] 20 in Central American countries, and 20 in Mexico.[16, 42] No cases have been reported as having originated in Chile, Guyana, Surinam, El Salvador, Nicaragua, or Panama. The disease is most prevalent in southern Brazil, certain areas of Colombia, Venezuela, and northeastern Argentina and southern Paraguay. There are also small areas of high endemicity in Ecuador, Peru, Mexico, and Guatemala.[67] It is estimated that 225 serious cases, or 0.8 per 100,000 population, occur annually.[106] This does not include the subclinical or mild forms of the disease. Imported and long-latent cases have been recorded in the United States and Europe.[1, 17, 53]

Using the paracoccidioidin skin test, Greer[44] studied the distribution of sensitivity in patients' families and in control groups in urban and rural environments. He found that in rural environments the number of reactors was the same as in families of pa-

Figure 19–1. Paracoccidioidomycosis. Approximate distribution of the disease in North, Central, and South America based on case reports. (After several authors, particularly the work of Dante Borelli.)

tients and control families. The reactivity rate, especially in the controls, was less in urban environments. Although Albornoz and Albornoz[4] had found a higher rate of skin sensitivity in males, Greer[44] and Restrepo[101] found no difference in sex distribution. Although members of a family would appear to have equal risk of exposure, no epidemics of clinical disease were noted, even when more than one person was skin test postive. These studies on skin test reactivity indicate that many subclinical infections go unrecognized.

The finding of 50:50 distribution between the sexes in skin test sensitivity contrasts sharply with the rate recorded for clinically significant disease. The sex distribution of cases in several series varies from 7:1 to 70:1, or from a general average of 90 per cent males to 10 per cent females. This is similar to the sex distribution noted in other systemic mycoses. Some of this unequal distribution of disease is attributable to exposure, but some is conjectured to be related to hormonal differences. Prepubertal children of either sex rarely have clinically apparent paracoccidioidomycosis. Athough the age range of patients is 10 to 70, the majority of clinically diagnosed cases is found in males between 20 and 50 years of age. In trying to associate the disease with other factors, it is noted that most patients are rural workers — many are tree cutters or work at similar occupations. In addition, many have other diseases, such as malaria, Chagas' disease, and tuberculosis, and essentially all are at least moderately malnourished. This suggests that physiologic, hormonal,[75] and immunologic factors play a role in the appearance and distribution of the clinically severe form of the disease. From time to time, authors have suggested that there can be human-to-human transmission of the disease, especially conjugal.[27, 44] This is a very difficult phenomenon to establish, as common-point source of exposure to the organism cannot be ruled out.

CLINICAL DISEASE

As noted earlier, the most commonly encountered form of clinical paracoccidioidomycosis reported in the literature was the presence of lesions on the oropharynx and gingivae. This, together with the less frequent occurrence of disease in the alimentary tract and anorectal region, gave the impression that primary infection occurred by implantation of the organism into the mucosa or by ingestion with food. It was suggested that cleaning the teeth with grass caused minor traumas that allowed invasion of the fungus. Recent careful observations by many investigators make it apparent that primary infection is pulmonary, with subsequent dissemination of the fungus to other regions. This is similar to the clinical course observed in essentially all other systemic mycoses. The disease falls into two clinical categories: benign and progressive. The latter category is subdivided into sections dependent on the distribution of the infection.

Clinical Types of Paracoccidioidomycosis[40, 61, 102, 109]

A. Primary benign disease
   1. Primary pulmonary paracoccidioidomycosis
   2. Pulmonary reinfection with allergic manifestations (?)
B. Acute and chronic progressive disease
   1. Acute and chronic progressive pulmonary disease
   2. Disseminated disease (latent or active) involvement
      a. Mucocutaneous lymphangitic
      b. Extracutaneous single organ involvement
      c. Generalized disease
   3. Acute juvenile paracoccidioidomycosis

Primary Benign Disease

Pulmonary Paracoccidioidomycosis

Benign pulmonary disease has only recently been recognized with regularity.[102] This is the inapparent subclinical infection that had previously been seen in Brazil in European immigrants residing in endemic areas.[64] Now a syndrome is developing consisting of minor lung changes and skin test reactivity in otherwise healthy individuals.[38] The lung changes appear commonly as bilateral lesions, with patchy or circumscribed infiltrates. The initial reaction is an alveolitis, with an influx of macrophages and a few giant cells, which is sometimes followed by caseation in small areas. At this stage of the disease, the macrophage is the probable mode of dissemination of the fungus to other areas of the body. A pyogenic reaction with numerous neutrophils then occurs, followed by granuloma formation.[65] Interstitial changes accompany the intraalveolar lesions. The lesions develop slowly over a period of time, losing their initial soft appearance and attaining the characteristics of interstitial fibrosis. A marked hilar adenopathy may be seen by x-ray examination (Fig. 19–2, E). In time, small calcifications may appear, but this finding is less common than in other mycoses or tuberculosis. While studying calcifications of the nodes of the hili pulmonis in routine autopsies, Angulo[8] found histologically identifiable yeast cells of P. brasiliensis in many patients who had died of other causes. Other patients are left with a primary lymph node complex similar to that in histoplasmosis.[8, 9]

In patients that live in highly endemic areas, there is some evidence for transient reinfection with allergic manifestations. This aspect of the disease is not as well defined as it is in histoplasmosis.[61]

X-ray. The radiographic presentations encountered in the various forms and stages of paracoccidioidomycosis are as varied as those seen in the other mycoses. There are a few patterns that recur with some frequency, however.[40, 61, 62, 102] As already noted, primary disease that is essentially subclinical shows minor bilateral infiltrates in the medullary regions. These resolve, leaving some interstitial fibrosis. In acute symptomatic disease, the presentation is exaggerated (Fig. 19–2, D), and resolution is notable for the enlarged hilum that remains (Fig. 19–2, E). This same picture of fibrosis and enlarged hilar regions is seen in patients without active pulmonary

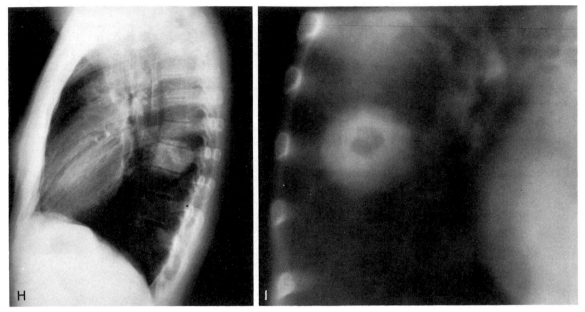

Figure 19–2 *Continued.* *I,* On tomogram focus, the lesion is seen to have a cavity that is slightly eccentric. (Courtesy of G. A. Agia.)

Acute and Chronic Progressive Disease

Acute and Chronic Pulmonary Paracoccidioidomycosis

Acute progressive disease in the adult is rare. When it occurs, rapid development of consolidation (Fig. 19–2, *F*) is seen in the lung fields. The course is often fulminant and fatal. Sometimes this occurs in immunosuppressed patients.[111]

Progressive chronic pulmonary disease occurs in about 50 per cent of cases.[40] Clinically it is accompanied by respiratory insufficiency, dyspnea, fever, rales, chest pain, and cough productive of sputum, which is sometimes blood-tinged. All lobes of the lung may be affected, but the involvement of the lower lobes seems to occur more frequently than of the upper (Fig. 19–2, *A*). Infiltration, striation, and the presence of small nodules are seen on x-ray. Such areas may become confluent. Chronic infections are associated with slight to moderate suppuration, but extensive fibrosis and deterioration of respiratory function occur, which may result in cor pulmonale. Cavities are not common and are seen in only about 15% of cases.

In the review series recently published by Giraldo et al.[40] and Londero et al.,[61, 62] the distribution of organ system involvement was almost evenly divided between tegumentary and pulmonary. Of the symptoms that prompted the patient to seek medical attention, 50 per cent were related to the skin or mucous membranes, 52 per cent were respiratory, and 11 per cent were related to the reticuloendothelial system; 5 per cent of patients presented with evidence of adrenal insufficiency. When these patients were examined, the finding of involvement in some other location was common. At least 90 per cent had evidence of pulmonary disease, about 60 per cent had at least some involvement of the skin or mucous membrane, and 25 per cent had lymph node involvement. From these and other studies,[107] it is apparent that the latent and residual disease may occur in several foci and remain quiescent for long periods of time.

Extensive mucocutaneous and lymphangitic disease may be present with only minor pulmonary changes. When the lungs are involved by secondary dissemination through hematogenous spread, numerous miliary granulomatous lesions are found throughout all areas.

Disseminated Disease

Mucocutaneous Lymphangitic Involvement. The primary pulmonary infection

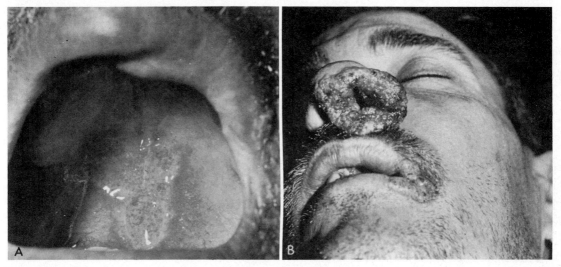

Figure 19–3. Mucocutaneous paracoccidioidomycosis. *A,* There is a papular lesion on the pharyngeal mucosa that has a blanched white exudative base, a rolled border, and small hemorrhagic dots in the center. *B,* Granulomatous lesions involving the nose. At this stage, the disease resembles secondary cutaneous leishmaniasis. (Courtesy of F. Pifano.)

with *P. brasiliensis* is often subclinical, and the secondary involvement of the oropharyngeal mucosa may be the most apparent presenting symptom. The lesions in the mouth begin as papules or vesicles, which then ulcerate. This condition is referred to as "moriform stomatitis." In the personally observed cases, the ulcerative lesions had a rolled border and a blanched white exudative base studded with small hemorrhagic dots (Fig. 19–3). Though they are initially painless, extensive lesions cause distress on ingestion and mastication and are painful. This may lead to cachexia. The lesions have a granulomatous appearance and spread slowly to sometimes form extensive vegetations. In advanced cases, the involvement becomes deeper and may be accompanied by destruction of the epiglottis and uvula, perforation of the hard palate, and extension to the lip and adjacent cutaneous areas of the nose (Figs. 19–4 and 19–5). Involvement of the lip may cause a hard edema and is followed by destruction of the area, with extensive scarring and fibrosis (Fig. 19–6). Gingival involvement, which is sometimes the presenting symptom, causes loss of teeth by spontaneous shedding.[34] Lesions may appear to occur in waves, so that many stages of their development may be seen in the same patient. Involvement of the ocular mucosa is infrequent but has been recorded.

Regional lymph nodes that drain the oronasal areas are involved soon after the appearance of initial lesions. Lymphadenopathy is a common and characteristic feature of this disease (Fig. 19–7). The cervical nodes are usually involved first and are often adherent to the overlying skin. Frequently, the nodes suppurate and drain, thereby forming sinus tracts. The drainage material contains abundant fungi. Massive nodes in the neck may be an early sign of the disease. Visceral node

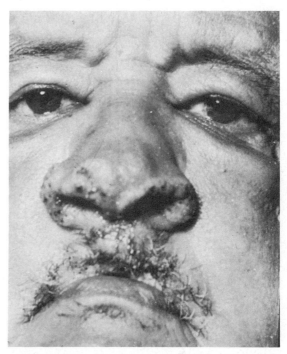

Figure 19–4. Mucocutaneous paracoccidioidomycosis. Chronic granulomatous lesions involving lips and nose.

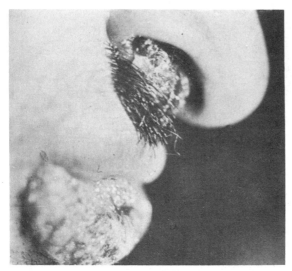

Figure 19–5. Mucocutaneous paracoccidioidomycosis. Granulomatous lesions of nose and lower lip. (Courtesy of A. Restrepo.)

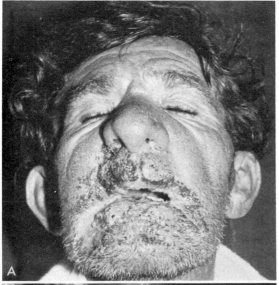

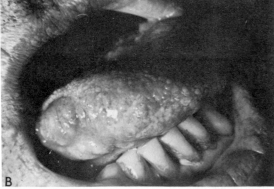

Figure 19–6. Mucocutaneous paracoccidioidomycosis. *A,* Extensive destruction of facial feature. *B,* Granulomatous lesions of the tongue. (Courtesy of F. Battistini.)

involvement may be so extensive as to suggest malignant neoplasia.[68]

Cutaneous lesions of various types occur in this disease.[19] Usually they begin at a mucocutaneous border or by extension from mucosal lesions. They are typically ulcerative and crusted granulomata and are described in descending frequency as papular dermoepidermal, papulopustular dermoepidermal, tuberous dermoepidermal, crusted and ulcerated dermoepidermal, vegetative dermoepidermal, tuberculoid dermoepidermal, and scrofulodermal lesions.

As a result of hematogenous or lymphatic dissemination, satellite lesions occur and in advanced cases not only are grouped on the face and neck but also develop on any area of the body (Figs. 19–8 and 19–9). Acneiform lesions are seen in fulminant disease of the adolescent. Chancriform lesions, which are an indication of primary inoculation disease, are extremely rare. The solitary skin lesion may be considered to be primary infection following subcutaneous implantation only when there is no evidence of lung involvement or disease elsewhere. The skin test and immunodiffusion test are negative in such cases, as they are in the early stages of the more classic infection.

The predilection of the disseminating organism for the mucocutaneous areas has been the subject of much conjecture and research. *In vitro* experiments indicate that growth of the organism occurs within a narrow range of temperature.[66] It had long been maintained by Mackinnon that the predilection for development of oropharyngeal and nasal lesions in man is owing to the cooling effect of air on these areas, and there is mounting experimental evidence[65, 116] to indicate this is so. With increased accommodation of the parasite to *in vivo* existence and diminution of host defenses, spread involves other areas.

Extracutaneous Disease. Foci of disease in other organ systems without evidence of involvement of the lungs or skin is uncommon but occurs. Probably the most common of the single organ system foci is the adrenals.[77] As is noted in the rare manifestations of blastomycosis, these patients present with frank Addison's disease or some degree of adrenal insufficiency. The proper diagnosis is often elusive; in retrospect, fibrotic areas are noted on chest films or firm cervical or inguinal lymph nodes reveal the organism on biopsy. Other single organ foci include posterior fossa tumor,[11] isolated lesion of the skull,[52] and cerebral tumor.[12]

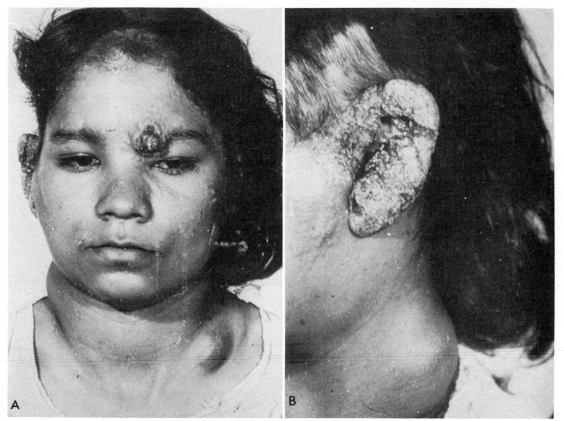

Figure 19–7. Cutaneous paracoccidioidomycosis. *A,* In this advanced case, lesions are found on the ear and forehead. Note the enlarged lymph nodes. *B,* Granulomatous lesion of the ear. (Courtesy of A. Restrepo.)

Generalized Disease (Disseminated Paracoccidioidomycosis). Disseminated visceral disease similar to that seen in other mycoses is infrequent. When dissemination occurs, the lymphatic system, spleen, intestine, adrenals, and liver are the common sites of systemic disease; other organs, such as the testes, brain, meninges, heart, large arteries, and bones, are rarely involved.[8, 76, 107]

Intestinal lesions are not infrequent and begin in the submucosal lymphoidal tissue. These erode into the lumen of the intestine, forming ragged ulcers. The conjectured occurrence of primary infection in the alimentary or anorectal areas has not been supported by experimental animal infection, and disease in such areas must be considered secondary. Radiographic findings in intestinal disease are nonspecific.[13] There is fixed narrowing adjacent to affected nodes and mucosal irregularities. The differential diagnosis includes regional enteritis, tuberculosis, and lymphoma. The stomach is seldom involved except by extension from abdominal lymph nodes. Symptoms of alimentary involvement are abdominal pain, anorexia, vomiting, and fever. Enlargement of the mesenteric lymph nodes is also a prominent feature. The liver,[115] spleen, and other organs are soon involved. In some series of reports, the spleen was found to be involved in 98 per cent of cases of disseminated disease. Ascites may be prominent. Invasion of the adrenals, with necrosis and destruction, is common and was present in 50 per cent of cases in one series.[77] Such patients often have Addison's disease as a further complication. The particular susceptibility of the adrenal glands to infection has been demonstrated in experimental animals.[65] Osseous lesions are relatively rare and, when found, are usually present in areas of bone that have a good blood supply. They usually consist of isolated osteolytic foci or, rarely, diffuse osteomyelitis.[102] Bone involvement may be more common than previously appreciated.[8, 9] The genitourinary system is rarely involved except in late state fungemia; even then, lesions in the kidney are small and few in number. Frias et al.[37] has noted an infection of the epididymis.

Central nervous system involvement has

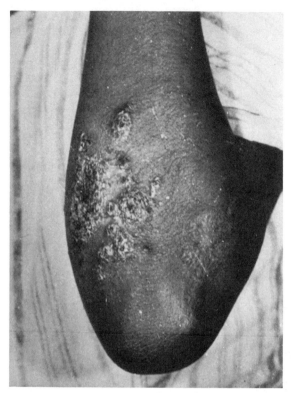

Figure 19–8. Cutaneous paracoccidioidomycosis. Crusted lesions on the forearm.

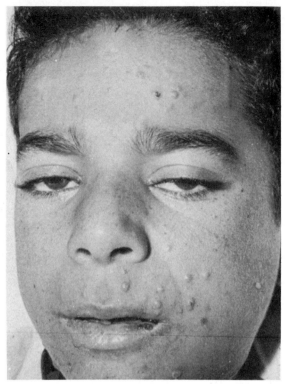

Figure 19–9. Cutaneous paracoccidioidomycosis. In this disseminated case, acneiform lesions are found on all body surfaces. (Courtesy of A. Restrepo.)

been recorded in about 80 cases (7 per cent) in several series tabulating the pathology of disseminated systemic disease. R. Negroni[82] recently described a tumor of the cerebellum that contained the fungus. The patient also had organisms in the sputum. A tumor at the level of the fourth ventricle was found in association with mucosal lesions of the mouth by Argollo et al.[12] Surgical removal of the tumor and treatment with sulfamethoxypyridazine, 1 g per day, brought about resolution of the disease. In a case reported by Krivoy et al.,[52] a frontotemporal mass caused severe headache. Surgical removal and sulfamethoxypyridazine again cured the infection. Double vision, headache, and ambulatory difficulty were associated with a tumor of the posterior fossa. In this case recorded by Araujo et al.,[11] the patient also had lesions in the lung and was successfully treated with amphotericin B. On rare occasion, meningitis is the presenting symptom. Papilledema and other evidence of intracranial pressure, as well as headache, vomiting, seizures, and hemiplegia, are present. Cerebrospinal fluid is within normal limits as far as pressure, color, glucose, and chlorides are concerned. The cell count may be increased, with a pre-

dominance of mononuclear leukocytes. The protein ranges from normal to 200 mg per 100 ml. The lesions are granulomatous and predominantly basilar in location. They may be single and massive or numerous and small. Tubercles composed of giant cells, macrophages, plasma cells, and phagocytized yeast cells are present and often show evidence of necrosis.

Acute Juvenile Paracoccidioidomycosis

About forty cases of acute "juvenile type" disease have been recorded.[61, 89, 99] They vary in severity, but almost all are fulminant in nature. In several, the clinically distinguishing characteristic was acneiform lesions of the skin (Fig. 19–9). The degree of fungemia in such patients is marked, since blood and marrow cultures are often positive. This leads to involvement of many organ systems, often with osteolytic lesions of the diverse skeletal areas. Most such fulminant cases are in children in puberty. Younger children[89] or older children usually have a more protracted course.[102]

DIFFERENTIAL DIAGNOSIS

Pulmonary disease emulates other mycoses and tuberculosis. Cases in which all three (tuberculosis, paracoccidioidomycosis, and histoplasmosis) were present have been recorded.[107] Mucocutaneous disease simulates cutaneous leishmaniasis. Both are endemic in the same geographic regions, and both cause secondary lesions leading to destruction of nasal septum and other areas of the oropharyngeal mucosa. Lymphatic enlargement simulates Hodgkin's disease and other malignant diseases, as well as scrofuloderma. The ulcerative or vegetative wartlike lesions suggest chromoblastomycosis and sporotrichosis. Advanced cases may simulate yaws, syphilis, systemic tuberculosis, and visceral leishmaniasis. With abdominal involvement, there are many features similar to abdominal actinomycosis, but no sinus tracts form.

As previously mentioned, paracoccidioidomycosis is frequently found in patients with other diseases. High on this list are malnutrition, ascariasis, and other helminth infections. Concomitant diseases also include Chagas' disease, tuberculosis, and schistosomiasis.

PROGNOSIS AND THERAPY

In a manner similar to histoplasmosis, paracoccidioidomycosis was formerly recognized only in its advanced, often fatal, form. A benign, self-limiting form of paracoccidioidomycosis also occurs, though probably not to such an extent as in the former disease.

Diagnosis of late-stage disease was universally fatal prior to the advent of the sulfas. In 1940, Oliveira Ribeiro demonstrated the efficacy of sulfonamides.[105] Blood levels of 5 mg per 100 ml of sulfadiazine or sulfamerazine (4 g per day) were able to control the infection, but with cessation of therapy clinical remission occurred. With the slowly excreted sulfas, such as sulfamethoxypyridazine and sulfadimethoxine, patients could be maintained on 0.5 g per day, following an initial dose of 1 g per day for a week. This schedule has resulted in many clinical cures, although cessation of therapy is sometimes followed by relapse of disease. From the combined experience of investigators using the sulfas for the treatment of paracoccidioidomycosis, Negroni[81] has recommended 3 to 5 years of 1 gram per day orally for most forms of the disease. Extensive pulmonary and severe disseminated forms of the infection require amphotericin B or some other agent. Even with supposed adequate levels and duration of sulfa therapy, the relapse rate is about 15 per cent, with a drug-induced resistance rate of clinical isolants of the fungus estimated to be 20 to 30 per cent in cases of treatment failure.

In 1959 amphotericin B was introduced by Sampaio and Lacaz[57] as a therapeutic regimen for paracoccidioidomycosis. The yeast phase of most strains is sensitive to 0.6 μg per ml of this drug. It is generally felt that a serum level twice the minimal inhibitory dose *in vitro* is necessary for effective treatment. Doses ranging from 0.25 to 1.2 mg per kg per day are given every day or on alternate days. The initial dose is 25 mg each day, increased to 50 mg, to 75 mg, and finally to the required regimen. It is given as the deoxycholate salt dissolved in 5 per cent glucose and administered as a slow infusion over several hours. The course of treatment is essentially the same as that used for other systemic mycoses. A total dose of 1 to 4 g is recommended, depending on the severity of disease and the degree of nephrotoxicity encountered.[81]

Sulfonamides can be effective in clearing the mildest forms of disease, but even in these cases it is recommended that a course of 1 g of amphotericin B be completed.[81] Dihydroxystilbamidine has usually failed when used alone. Iodides are contraindicated, as they tend to disseminate the infection.

Recently, two new drugs have received clinical trials. They are the imidazoles miconazole and ketoconazole. Both appear to be excellent drugs for the treatment of the disease. Miconazole has an *in vitro* efficacy of 0.0005 to 0.001 μg per ml (MIC). In a series using IV miconazole, Stevens et al.[113, 114] found good results in six patients with various forms of the disease. Three patients relapsed within six months of completion of therapy. A longer period of therapy and higher dose of the drug are recommended.[113] Lima et al.[58] used oral miconazole on twelve patients for 3 to 8 months. There were no relapses in a follow-up period of 3 to 15 months. The recommended dose is now ≥ 9 mg per kg body weight, given peak serum levels of > 1.0 μg per ml with a range of 200 to 1200 mg per day, depending on the severity of infection, continued for 2 to 16 weeks.

A clinical trial of ketoconazole was reported by Restrepo et al.[103] The fungus isolated from the five patients had a sensitivity of 0.008 μg per ml MIC. Plasma levels after 200 mg orally were found to be 6 μg per ml at two hours and > 1 μg per ml at four hours. All patients were initially treated with 200 mg per day. Depending on clinical response, this could be dropped to 100 mg per day. Therapy was continued for 12 months or until clinical or mycological cure. No toxicity was encountered.

PATHOLOGY

Gross Pathology. A picture similar to that seen in coccidioidomycosis and blastomy-cosis has been noted in autopsy examination of patients with disseminated paracoccidioi-domycosis. The main differences are the extensive involvement of lymphoid tissues in paracoccidioidomycosis and the occasional widespread involvement of the intestinal tract. Ulcers occur in the intestinal mucosa. The lesions appear to arise in the lymphoid tissue beneath the mucous membrane and then extend into it. This is also seen in nodes that drain from mucocutaneous lesions of other areas. The entire node may be involved, showing focal or diffuse necrosis. This predilection for lymphoid tissue also

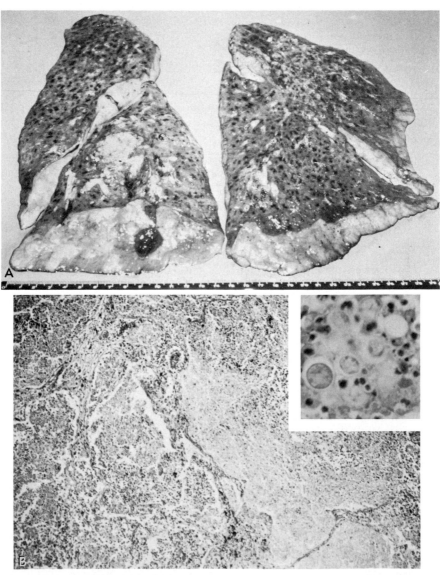

Figure 19–10. Acute pulmonary paracoccidioidomycosis. *A,* Cut surface of lung with abscess in right lower lobe and a pneumonic process in the left upper lobe. *B,* Interstitial and alveolar infiltrates in a section from the pneumonic process. ×63. *Inset:* Yeast cells. Hematoxylin and eosin stain. ×400.[89] (Courtesy of A. T. Londero.)

explains the almost universal presence of lesions in the spleen in disseminated cases. Nodules in the liver and, less frequently, the kidney are also found. Granulomatous lesions are also found in other organ systems with about the same frequency as is seen in blastomycosis. In contrast to the case in the latter disease, bone involvement is relatively rare in paracoccidioidomycosis.

In the several autospy series that have been published,[8, 9, 107] the pathology of tissues involved reflects the clinical forms of the disease already discussed. Again the preferential invasion of reticuloendothelial system and lymphoid tissue is noted. In almost 100 per cent of cases, lesions were found in the lungs, lymph nodes, and adrenals. In 90 per cent there were foci in spleen and liver, with granulomata also observed in skin, pharynx, larynx, and mouth. Seven per cent had lesions in the stomach and intestines; 5 per cent, in bone; about 5 per cent, in the central nervous system and trachea; and 3 per cent or less, in kidney, esophagus, and genitourinary and vascular systems.

The lesions in the lung are variable. In the cut section of the interstitial form there is thickening of the interlobular septa and a netlike appearance owing to dilated alveoli. The thickening is greatest in the hilar and parahilar regions and decreases as the septa approach the lung surface. Bullous emphysema is common and diffuse in the lung fields. The interstitial and alveolar forms represent the acute progressive disease and usually do not have a nodular or cavitary component. In older established disease, nodular forms are found. They vary from small miliary lesions to 1 to 2 mm, representing hematogenous dissemination, to large granules and granulomata several centimeters in diameter (Fig. 19–10).[89] Cavitary lesions are not common (Fig. 19–2, G, H, I). When they do occur, they are usually circular with an inner necrotic area surrounded by a granulomatous reaction with fibrosis. The condensed consolidated pneumonic process or confluent bronchopneumonic form represents an acute or subacute evolution and is not common[109] (Figs. 19–2,F, 19–10). Extensive interstitial and alveolar infiltrates that contain many organisms are noted (Figs. 19–10,B, 19–11). The solitary nodule or coin lesion so frequently found in histoplasmosis and coccidioidomycosis is very rare in paracoccidioidomycosis.

Lesions of other organs resemble those of tuberculosis but are without the tendency to

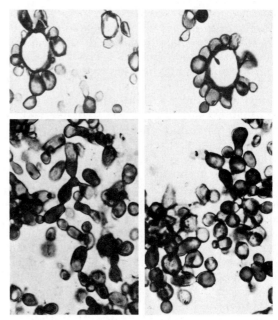

Figure 19–11. Varied morphology of yeasts. A, B, Multiple budding. C, D, Chains and elongated elements. GMS. ×400.[89] (Courtesy of A. T. Londero.)

calcification. Splenic involvement is very common, and most of the lesions are nodules of various sizes. These nodules most often have grayish necrotic centers and give the appearance of those described for Hodgkin's disease. In the central nervous system, involvement is of two types (1) pseudotumors called "blastomycomas," or "paracoccidioidomas," and (2) meningitic. The tumors are well circumscribed and are found (in descending order) in the cerebrum,[12] cerebellum,[82] or basal ganglia. The meningitic disease is a granulomatous process similar to tuberculosis. In the bone, osteolytic foci are seen with relative frequency. On occasion a diffuse osteomyelitis will accompany widespread dissemination. These frequently invade neighboring soft tissue and produce a subcutaneous abscess containing numerous yeast cells of the fungus.

Histopathology. The histopathologic reaction seen in cases of paracoccidioidomycosis is essentially identical to that of blastomycosis. There is a granulomatous reaction interspersed with pyogenic abscess formation. Langhans' and foreign body giant cells are numerous and may contain organisms. Focal necrotic and caseous areas are seen with zones of macrophages, histiocytes, lymphocytes, plasma cells, and fibroblasts.[116] Microabscesses with a polymorphonuclear reaction are interspersed between areas that are purely granulomatous. Epithelioid granulomas that are sharply outlined and without necrosis resemble Boeck's sarcoid. In older lesions, fibrosis may be very extensive.

Cutaneous lesions and those of the buccal mucosa characteristically show a pseudoepitheliomatous hyperplasia together with a marked pyogenic and granulomatous response. Intraepithelial microabscesses are frequent. Clusters of epithelioid cells, plasma cells, and lymphocytes are present. The same types of lesions are seen in coccidioidomycosis and particularly in blastomycosis. Differentiation from these diseases by examination of histopathologic material alone requires observation of the yeast cell of the fungus with its pathognomonic "pilot wheel" peripheral budding. The organism may be evident in hematoxylin-eosin stained material, but is more easily observed in methenamine silver stains. The typical budding cell varies from 12 to 40 μ in diameter. The periphery of the spherule is studded with small buds that communicate with the mother cell by small necks. The buds are visible when they are 2 μ in diameter, and they grow to be 5 μ or more. The parent cells may appear empty. Buds of many sizes or only a few large buds may occur on the same cell, giving a "Mickey Mouse cap" appearance to the yeast (Fig. 19–12). Young cells in hematoxylin and eosin preparations have chromatin dots representing nuclei in a centripetal or even distribution. The morphologic patterns seen in the liver have been tabulated by Teixeira et al.,[115] who found bile duct lesions to be prominent. A particular pattern was noted in which segments of bile ducts were disrupted and replaced by inflammatory reaction. It was felt that treatment (sulfas or amphotericin B) did not induce proliferation of bile ducts, fatty changes of liver cells, or cholestasis; rather, such changes were caused by the fungus. Less fibrosis was found in treated patients than in those who had not received drug therapy, however.

The diagnosis of paracoccidioidomycosis cannot be made histopathologically without the presence of the typical budding cell along with isolation and identification of the fungus in culture. Cells from 5 to 15 μ in size can be seen in many lesions that appear identical to young spherules of coccidioidomycosis. Single budding cells resemble *Blastomyces dermatitidis*, and small cells (2 to 3 μ) lying free or within macrophages or giant cells may be misidentified as the yeasts of *Histoplasma capsulatum*. In some necrotic areas, enormous numbers of cells that are sometimes crescent-shaped, distorted, and crumpled are seen (Fig. 19–11). So far there has been no report of mycelium in tissue sections, as is the case in the other mycoses. Coexistence of paracoccidioidomycosis and histoplasmosis in the same lesions is not uncommon.[107] Other concomitant infections such as cryptococcosis, aspergillosis, and, rarely, coccidioidomycosis have been noted. Paracoccidioidomycosis and tuberculosis are frequently encountered.

In disseminated disease, numerous epithelioid cell granulomas occur. These often have necrotic centers and resemble the lesions of miliary tuberculosis. Caseous necrosis of the adrenals is common and resembles that seen in histoplasmosis.[77] In most other organs in-

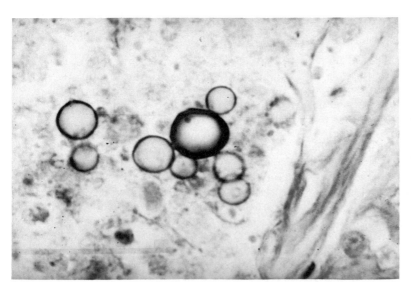

Figure 19–12. Paracoccidioidomycosis. Tissue section stained by the Gridley method. The buds, attached to the parent cell by a narrow neck, often enlarge to give a "Mickey Mouse type" cell appearance. (From Rippon, J. W. *In* Burrows, W. 1973. *Textbook of Microbiology.* 20th ed. Philadelphia, W. B. Saunders Company, p. 730. Courtesy of P. Graff.)

volved, the histopathology resembles that seen in blastomycosis and coccidioidomycosis. As in blastomycosis, the lesions show lack of calcification.

ANIMAL DISEASE

Natural infection in animals is unknown. In 1965 Grose and Tamsitt[45] reported on the isolation of *P. brasiliensis* from the intestinal content of three Colombian frugivorous bats (*Artibeus lituratus*). The authors speculated that the bats may be important in the dissemination of *P. brasiliensis*, filling a role similar to that of bats suspected of carrying *Histoplasma capsulatum*. Greer and Bolaños,[43] however, investigated the survival of *P. brasiliensis* in this species of bat and found that neither the yeast nor mycelial phase could survive passage through the alimentary tract. Recently, Johnson and Lang[48] reported on granulomatous lesions of a Squirrel monkey (*Saimiri sciureus*). They described caseating nodules of the liver and colon that contained fungal elements resembling *P. brasiliensis*. No cultures were performed.

In the laboratory *P. brasiliensis* is not highly virulent for most experimental animals.[60] Pus from lesions can be inoculated intratesticularly into guinea pigs, which then develop an orchitis. Examination of material from such lesions shows an abundance of yeast cells. The mountain rat (*Proechimys guayanensis*) is particularly susceptible to infection. Use of the animal has been suggested as a diagnostic aid when few organisms are present.[14] Systemic infections have been produced in mice, rabbits, guinea pigs, and hamsters, but with great difficulty. Mice inoculated by the pulmonary route developed an alveolitis initially. In some, this was followed by spread to the regional lymph nodes, with subsequent hematogenous dissemination.[65] If the mice were kept at an ambient temperature of 18° C, the infection ran a fulminant course, with fatal dissemination in all animals. When kept at an ambient temperature of 35° C, few animals developed disease.[116] Mackinnon[65] has demonstrated that growth of the organism in liquid media decreases sharply at temperatures above 35° C. Restrepo and Guzman[93] instilled conidia of *P. brasiliensis* into mice treated with cortisone. Only 38 per cent of animals developed pulmonary infections, and 14 per cent had lesions in the spleen and liver. The authors noted some difference in virulence between the strains used. In studies by Hay and Chandler,[46] mice injected IV with yeast cells of the fungus developed cranial lesions. These lesions invaded the nasal mucosa, with subsequent discharge of cells into the nasal cavity. It is suggested that person-to-person transmission can occur with organisms in nasal or oral mucosa. In studies by Montoya and Garcia-Moreno,[72] female mice were more resistant than male mice to experimental infections.

The course of experimental infection in hamsters was followed by Iabuki and Montenegro.[47] In their studies, dissemination occurred in 100 per cent of animals following intratesticular inoculation. In addition to finding granulomata containing fungi, they could also demonstrate antigen by immunofluorescence in nonspecific lesions of the liver (diffuse or nodular Kupffer cell hyperplasia). Chronic progressive paracoccidioidomycosis was repeatedly produced in hamsters during their investigations.

A self-limited infection with no tendency to disseminate or cure during a 5-month follow-up was produced in cattle by Costa et al.[30, 31] Lesions with granulomatous inflammatory foci with or without necrosis were found in testicles, spermatic cords, and testicular tunica. The animals were used to standardize the paracoccidioidin skin test. All animals became paracoccidioidin skin test positive by two to three weeks. They also developed histoplasmin skin test reaction. The latter faded by the fourth month, but the former persisted.

BIOLOGICAL STUDIES

As with other dimorphic fungi, the factors that affect mycelial to yeast transformation and the physiologic alterations that accompany this change have been the subject of much research. Yeast cells are seen to develop directly from mycelia by a budding process not unlike the production of conidia.[20] This process appears to be fundamentally different from that observed in *Histoplasma* or *Sporothrix*. As with other dimorphic fungi, the yeast form has a higher chitin content than the mycelium. The buds of yeast cells placed at 25° C elongated directly into hyphae to form a corona of many hyphal units from a single cell.[87] This transformation is accompanied by a change from α-glucan in the yeast cell wall to the β-glucan found in the mycelium.[23] From studies on composition and ultrastructure, Carbonell et al.[20, 21, 22, 23] were able to correlate biochemical properties with cell wall structure. They concluded that the M form has a single-layered wall containing interwoven chitin and β-glucan cemented together by disulfide containing protein. Enzymes capable of producing α-glucan were interspersed at random along the hyphal thread, and α-glucan was found in the areas forming conidia or chlamydoconidia. In contrast, the yeast (y) cell had a three-layered wall. The inner layer consisted of long fibers of chitin arranged in bundles. At regular intervals were islands of β-glucan. The outer layer was a mesh of α-glucan and/or a layer of galactomannan. If the yeast is grown at 37° C, β glucanase and protein disulfide reductase would cause a softening around the island β-1, 3 glucan and produce a bud. The round shape would be caused by equal wall growth on all sides due to the laying down of chitin and proteins. The proteins would not be rigid because of the high content of disulfide reductase in the yeast and its activity at 37° C. α-glucan is also favored at this temperature. At 25° C the synthesis of α-glucan would decrease in the budding site, and the β-glucan fibril would grow continuously in an apical fashion. The mycelial wall would have more rigid proteins, owing to the high content of disulfide bridges and low activity of protein disulfide reductase at room temperature. Thus, all yeasts can produce hyphae at the area of the β-glucan islands when the temperature is lowered. Hyphae, however, can produce yeast cells only in the areas along the wall where α-glucan is produced, i.e., areas that produce conidia or chlamydoconidia. San Blas and San Blas[108] reviewed this process (Fig. 19–13) and correlated virulence with the composition of the yeast cell wall.

Among the most interesting physiologic studies are those done by Muchmore et al.[75] They demonstrated that the yeast phase (Y) was inhibited by estrogen and related diols. Montoya et al.[72] have noted a greater resistance to infection in female mice when compaired with that of male mice. This may correlate with the preponderance of males compared to females with natural infections.

The effect of drying and temperature on the viability of the organism was studied by Conti-Diaz et al.[28] The fungus was not able to survive in dry soils, though it retained viabili-

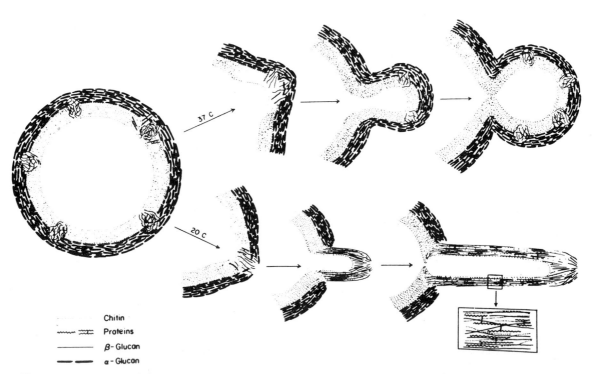

Figure 19–13. The budding process and mycelium process exhibited by the yeast phase of *P. brasiliensis*. The yeast phase has outer wall layers of glucan and galactomannan. The inner layer is chitin with "islands" of β glucan and protein. At 37°C, budding produces a "blow-out" round bud with wall growth equal on all sides and α glucan favored. At 25°C, β glucan is formed, and protein contains disulfide bonds; thus, expansion is linear and apical.[108] (Courtesy of G. San Blas and F. San Blas.)

ty in humid samples for long periods of time. Maintenance in humid or dry soils at 2° C was deleterious. Viability is retained for longer periods of time in acid soils than in neutral or alkaline soils.[96] The successful isolation of *P. brasiliensis* from nature was in soil of pH 5.3. These studies may explain the geographic limitations of the organism to the warm humid areas and acidic soils of South and Central America.

Growth curves of the M and Y phase were determined by Arango and Restrepo.[10] The survival time was 14 days for the yeast and 50 days for the mycelial phase. The mean growth of the Y in exponential phase was several times that of the M phase, as has been noted for other dimorphic fungi. With senescence both phases autolyzed. The deoxyribonucleic acid (DNA) content of the two growth phases was investigated by Garrison et al.[39] The guanine/cytosine (GC) content of DNA was found to be 45 per cent, similar to other fungi of the family Gymnoascaceae.

IMMUNOLOGY AND SEROLOGY

Immunity. The recent finding by surveys of skin test sensitivity that exposure to the fungus is much more prevalent than previously thought has modified several concepts of this disease.[44, 101] As has been established in histoplasmosis, it appears that a sizeable population encounters the etiologic agent of paracoccidioidomycosis, and a subclinical, self-limiting disease develops. This indicates that there is a high natural resistance to infection for both diseases. It appears that the patients who develop severe disease initially or from a latent infection long after exposure are physiologically or immunologically deficient, possibly only transiently. Patients with severe disease have an acquired deficiency of cell-mediated immunity. They are usually anergic and show no reactivity to a number of skin tests. In addition, there is a reduction of lymphocyte blast transformation after stimulation, and there is an increased tolerance to heterologous skin grafts.[69, 70] Restrepo and Velez,[104] however, found that neutrophils of patients with the disease had an increased capacity to kill yeast cells of *P. brasiliensis*. This was mediated by a specific component of serum. The effect of serum factors on killing and digestion of the fungus by leukocytes has been investigated by Goihman-Yahr et al.[41] and Calich et al.[18] Whether initial resolution of infection confers a strong immunity, as is

seen in coccidioidomycosis and histoplasmosis, is not known.

Serology. The first satisfactory and reasonably specific serologic test for paracoccidioidomycosis was developed by Fava-Netto in 1955.[32, 33] Using a polysaccharide antigen, he found that there were few cross-reactions with other mycotic diseases in either the complement fixation test or the precipitin test. Precipitins (by the tube technique) appear early in the disease and disappear after therapy. Complement-fixing antibodies appear later and are present for several months following clinical cure. The reappearance of positive serologic tests without clinical signs has been interpreted as indication of a smoldering infection. The complement fixation test is positive in 84 to 95 per cent of patients with active lesions. Restrepo has developed an immunodiffusion test reported in a series of papers.[90, 94, 100] She found that antigen prepared from the yeast phase was more reliable than that from the mycelial phase. There was a good correlation with the complement fixation test, but the immunodiffusion test was positive in 16 per cent of patients who were negative by complement fixation or whose sera were anticomplementary. In an extensive study, Kaufman[51] found low titer cross-reactions in the complement fixation test with sera from patients with aspergillosis and candidiasis. However, the immunodiffusion test of Restrepo was specific for the disease. This test usually becomes negative following clinical cure. Restrepo and Moncada[94] have also developed a latex slide agglutination test.

A number of common antigens between *P. brasiliensis* and *B. dermatitidis*, *H. capsulatum* and *H. capsulatum* var. *duboisii*, have been found, in that order of relatedness. By immunoelectrophoresis of cell fractions of *P. brasiliensis*, the specific antigen was determined to be a cathodic fraction identified as an alkaline phosphatase.[117, 118] A fluorescent antibody test for use with tissue sections and cultures has been described.[50]

Precipitin bands in the immunodiffusion (ID) test have been investigated by Restrepo and Moncada.[94] At time of diagnosis over 50 per cent of patients had 3 bands by ID test, and all had at least one. The highest number of bands correlated with the highest complement fixation titer. The bands disappeared with clinical cure of the infection. Band 3 was found to be an antigen in common with coccidioidin; bands 2 and 1 were specific. Band 3 disappeared first. Negroni et al.[83] found significant cross-reaction with histoplasmosis by the CF test; this was not true of the ID. The authors considered this test more specific but less sensitive. Specificity of the test was enhanced by modifications involving immunoelectroosmophoresis in investigations by Conti-Diaz et al.[29] The subject of serologic tests in paracoccidioidomycosis has been reviewed many times. One of the most informative is by Fava-Netto.[33]

Skin testing with antigens prepared from yeast cultures or tissue of infected guinea pigs has been used to evaluate the immune state of the patient. Patients recovering from disease have a strong delayed hypersensitivity exhibited by an area of induration of 10 mm or more. Patients in terminal stages of the disease are usually anergic. However, both types of patients will have a positive immunodiffusion test. Skin test reactivity along with a negative immunodiffusion test indicates the presence of a subclinical or mild, self-limited infection. A skin test antigen termed "paracoccidioidin" that was prepared by Mackinnon has been found to be specific for the disease in a 1:100 dilution. At a dilution of 1:10, however, there was cross-reaction with

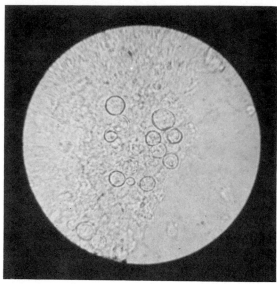

Figure 19–15. Paracoccidioidomycosis. Direct examination by KOH mount of scraping from lesion on oral mucosa. Most of the yeasts appear as single cells or with one bud. They may be mistaken for several other fungi. ×400. (Courtesy of E. Belfort.)

sera from cases of histoplasmosis and coccidioidomycosis. The antigen of Restrepo is even more specific,[94] but a few instances of cross-reaction in cases of histoplasmosis still occur.

LABORATORY IDENTIFICATION

Direct Examination. Sputum, biopsy material, crusts, material from the granulomatous bases or the outer edge of ulcers, and pus from suppurative draining lymph nodes contain numerous fungal elements. The material is placed on a slide with a drop of 10 per cent potassium hydroxide, then heated to clear the specimen (Figs. 19–14 and 19–15). The yeasts are readily observed. They vary from young, recently separated buds 2 to 10 μ in size to mature cells up to 30 μ or more. Some yeasts may be 60 μ in diameter. The cells have a double refractile wall (0.2 to 1 μ thick); they are spherical, oval, or elliptical and may occur in chains of four or more. From one to a dozen narrow-necked buds of uniform or variable size arise from the mother cells. Sometimes the yeasts grow in chains and have single buds, elongated distorted cells, and a number of other forms (Figs. 19–11 and 19–15). This can lead to confusion with other fungus infections.

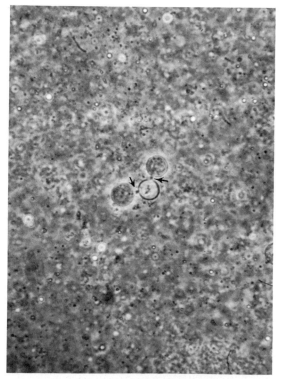

Figure 19–14. Paracoccidioidomycosis. Direct examination of sputum by KOH mount. Parent cell with large buds. Note the isthmus between mother and daughter cell *(arrow)*. (Courtesy A. Restrepo.)

Culture Methods. Material from pus, lesions, biopsies, and sputum can be plated on media containing antibacterial antibiotics and cycloheximide and incubated at 25° C. It can also be cultured at 37° C on blood agar with antibacterial antibiotics. The yeast phase of this organism, like that of *B. dermatitidis* and *H. capsulatum*, is sensitive to cycloheximide. At 25° C the fungus grows very slowly, and a colony usually does not appear for 15 to 25 days. It may require ten days more to reach a diameter of 1 cm. Growth of stock cultures is more rapid. Transfer to blood agar incubated at 37° C converts the organism to the yeast form. As no characteristic conidia are produced at 25° C, conversion to the typical morphology of the yeast form is essential for diagnosis. Restrepo and Corea[92] have found that yeast extract agar was superior to other media for the initial isolation of the fungus.

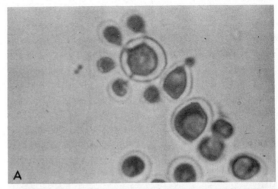
A

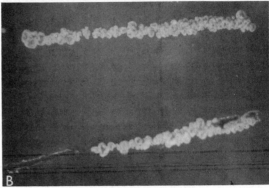

Figure 19–17. Paracoccidioidomycosis. *A,* Lactophenol mount of yeast form cells. Note the cytoplasmic connection from parent cell to bud. ×450. (Courtesy of A. Restrepo.) *B,* Yeast form growth at 37°C on Kelly's medium.

MYCOLOGY

Paracoccidioides brasiliensis (Splendore) Almeida 1930

Synonymy. *Zymonema brasiliense* Splendore 1912; *Zymonema histosporocellularis* Haberfield 1919; *Mycoderma brasiliensis* (Splendore) Brumpt 1912; *Mycoderma histosporocellularis* Neveu-Lemaire 1921; *Monilia brasiliensis* (Splendore) Vuillemin 1922; *Coccidioides brasiliensis* (Splendore) Almeida 1929; *Coccidioides histosporocellularis* Fonseca 1932; *Paracoccidioides cerebriformis* Moore 1935; *Proteomyces faverae* Dodge 1935; *Paracoccidioides tenuis* Moore 1935; *Lutziomyces histosporocellularis* Fonseca Filho 1939; *Blastomyces brasiliensis* (Splendore) Conant and Howell 1941; *Aleurisma brasiliensis* (Splendore) Neves and Boglioli 1951.

Colony Morphology. SGA, 25° C. The fungus grows slowly, producing a variety of colonial forms. These vary from a glabrous leathery, brownish, flat colony with a few tufts of aerial mycelium to a wrinkled, folded, floccose to velvety, white, pink, and later beige form. The colonial forms are indistinguishable from *B. dermatitidis* (Fig. 19–16,*A*). The

A
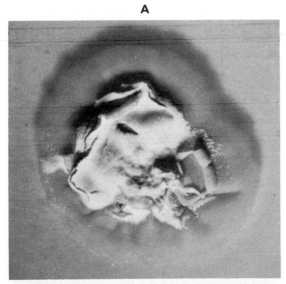

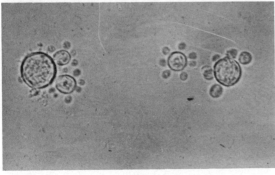
B

Figure 19–16. Paracoccidioidomycosis. *A,* Colony morphology at 25°C. The glabrous, leathery, cracked colony has some areas of aerial mycelium. *B,* Yeast form at 37°C. Note the "pilot wheel" and "Mickey Mouse type" forms. ×440.

color of the reverse is yellowish brown to brown.

SGA, Kelley's, BHI blood agar, 37° C. The mycelium converts to the yeast phase with ease at 37° C (Fig. 19–17,*B*). The organism grows slowly, producing a wrinkled, folded, glabrous, whitish colony.

Microscopic Morphology. SGA, 25° C. Microscopic examination of conidiating strains shows a variety of conidia, none of which is characteristic of the species (Fig. 19–18). Most strains grow for long periods of time (up to ten weeks) without the production of conidia. All cultures produce intercalary chlamydoconidia and coiled hyphae. In media deficient in glucose, some strains produce intercalary cells, which become rectangular, or triangular, thick-walled arthroconidia. They may be in an alternate pattern within the mycelium. Pear-shaped conidia and arthroconidia are also seen. Isolants vary in the proportion, timing, and variety of conidia produced. Conidiation is enhanced on yeast extract agar.[91]

When cultures are transferred to 37° C, distorted mycelial elements of varying lengths are seen to be intermixed with yeast cells. The yeasts are 2 to 30 μ or more, and are oval or irregular in shape (Figs. 19–16,*B* and 19–17,*A*). They have from one to several thin-necked, round buds of uniform or varying size, which develop from all areas of the parent cell. The walls of *P. brasiliensis* are thinner than those of *B. dermatitidis*, and the buds are easily dislodged. The bud of the latter species has a broad base. *P. brasiliensis* has no capsule. Biochemical properties are,

so far, not useful in its identification, and a perfect stage has not as yet been described.

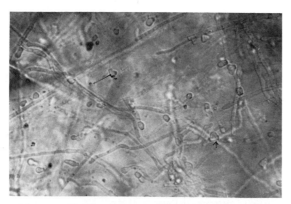

Figure 19–18. Paracoccidioidomycosis. Microscopic morphology of mycelial form at 25°C. There are aleurioconidia and intercalary chlamydoconidia, rectangular and bulging arthroconidia (at *arrows*), and arthroaleurioconidia. ×100. (From Restrepo, M. A. 1970. A reappraisal of the microscopical appearance of the mycelial phase of *Paracoccidioides brasiliensis*. Sabouraudia, *8*:141–144.)

REFERENCES

1. Agia, G. A., D. J. Hurst, et al. 1981. Paracoccidioidomycosis presenting as a cavitating pulmonary mass. Chest *78*:650–652.

2. Ajello, L. 1972. Paracoccidioidomycosis: a historical review. Pan American Health Organization. 1972. First Pan American Symposium on Paracoccidioidomycosis. Medellin Colombia, 25–27, October, 1971. Scientific Publication No. 254, Washington, D.C., World Health Organization, pp. 3–10.

3. Albornoz, M. B. 1971. Isolation of *Paracoccidioides* from rural soil in Venezuela. Sabouraudia, *9*:248–253.

4. Albornoz, M. B., and R. Albornoz. 1971. Estudio de la sensibilidad específica en residentes de un area endémica a la paracoccidioidomicosis en Venezuela. Mycopathologia, *45*:65–75.

5. Almeida, F. 1930. Estudos comparativos do granuloma coccidioidica nos Estados Unidos e no Brasil. Novo genero para o parasito brasiliero. An. Fac. Med. Univ. São Paulo, 5 3–19.

6. Almeida, F. 1938. Vacina comtra ogranuloma paracoccidioidico. Folia Clin. Biol., *10*:195–197.

7. Almeida, F., and C. S. Lacaz. 1941. Intrader morea cao com paracoccidioidina no diagnostico do granuloma paracoccidioidica II, A recao de Montenegrono granuloma paracoccidioidicoi. Folia Clin. Biol. (S. Paulo), *13*:177–182.

8. Angulo, A. 1971. Anatomo-clinical aspects of paracoccidioidomycosis. In reference no. 85, pp. 129–133.

9. Angulo, A., and L. Pollak. 1971. Paracoccidioidomycosis. The Pathologic Anatomy of Mycoses. *In* R. D. Baker (ed.). *Handbuch der Speziellen Pathologischen Anatomie und Histologie Dritter.* Band Funster Teil. Berlin, Springer-Verlag, 506–576.

10. Arango, M., and A. Restrepo M. 1976. Determination of growth curves of the mycelial and yeast forms of *Paracoccidioides brasiliensis*. Mycopathologia, *59*:163–170.

11. Araujo, J. C., L. Werneck, et al. 1978. South American blastomycosis presenting as a posterior fossa tumor. Case report. J. Neurosurg., *49*:425–428.

12. Argollo, A., V. L. Reis, et al. 1978. Central nervous system involvement in South American blastomycosis. Trans. R. Soc. Trop. Med. Hyg., *72*:37–39.

13. Avritchir, Y., and A. A. Perroni. 1978. Radiologic manifestations of small intestinal South American blastomycosis. Radiology, *127*:607–609.

14. Belfort, A. E. 1967. Paracoccidioidomycosis: Diagnostico mediante inoculacion a *Proechimys guayensis*. Rev. Dermatol. Venezol., *3*:91–97.

15. Borelli, D. 1961–1962. Hipótesis sobre ecologia de paracoccidioides. Derm. Venez., *3*:130–132.

16. Borelli, D. 1970. Prevalence of systemic mycoses in Latin America. *In International Symposium on Mycoses* (85). PAHO Scientific Publications No. 205. Washington, D.C., World Health Organization, pp. 28–38.

17. Bouza, E., D. J. Winston, et al. 1977. Paracoccidioidomycosis in the United States. Chest, *72*:100–102.

18. Calich, V. L. G., T. L. Kipnis, et al. 1979. The activation of the complement system by *Paracoccidioides brasiliensis* in vitro: its opsonic effect and possible significance for an in vivo model of infection. Clin. Immunol. Immunopathol., *12*:20–30.

19. Calle Velez, G. 1971. Dermatological aspects of paracoccidioidomycosis. Pan American Health Organization. 1972. First Pan American Symposium on Paracoccidioidomycosis. Medellin Colombia, 25–27 October, 1971. Scientific Publication No. 254, Washington, D.C., World Health Organization, pp. 118–121.

20. Carbonell, L. M. 1967. Cell wall changes during the budding process of *Paracoccidioides brasiliensis* and *Blastomyces dermatitidis*. An electron microscopic study. J. Bacteriol., *94*:213–223.

21. Carbonell, L. M., and J. Rodriguez. 1965. Transformation of mycelial and yeast forms of *Paracoccidioides brasiliensis* in cultures and in experimental inoculations. J. Bacteriol., *90*:504–510.

22. Carbonell, L. M., and J. Rodriguez. 1968. Mycelial phase of *Paracoccidioides brasiliensis* and *Blastomyces dermatitidis*. An electron microscopic study. J. Bacteriol., *96*:533–534.

23. Carbonell, L. M., F. Kanetsuna, et al. 1970. Chemical morphology of glucan and chitin in the cell wall of the yeast phase of *Paracoccidioides brasiliensis*. J. Bacteriol., *101*:636–642.

24. Carini, A. 1908. Un caso de Blastomycose, com lacalisa cao primaria na mucosa do bocca. Rev. do Soc. Scient. de São Paulo No. 10.

25. Chaves-Batista, A., S. K. Shome, et al. 1962. Pathogenicity of *Blastomyces brasiliensis* isolated from soil. Publicacao 373. Inst. de Micrologia, Universidade Recife, Brasil.

26. Conant, N. F., and A. Howell. 1942. The similarity of the fungi causing South American blastomycosis (paracoccidioidal granuloma) and North American blastomycosis (Gilchrist's disease). J. Invest. Dermatol., *5*:353–370.

27. Conti-Diaz, A., L. Calegari, et al. 1979. Paracoccidioidal infection in the wife of a patient with paracoccidioidomycosis. Sabouraudia, *17*:139–144.

28. Conti-Diaz, A., J. E. Mackinnon, et al. 1971. Effect of drying on *Paracoccidioides brasiliensis*. Sabouraudia, *9*:69–78.

29. Conti-Diaz, A., R. E. Somma-Moreira, et al. 1973. Immunoelectroosmophoresis-immunodiffusion in paracoccidioidomycosis. Sabouraudia, *11*:39–41.

30. Costa, E. O. da, C. Fara-Netto, et al. 1978. Bovine experimental paracoccidioidomycosis intradermic test standardization. Sabouraudia, *16*:103–113.

31. Costa, E. O. da. 1975. Paracoccidioidomicoses en animales domesticos. Thesis, Univ. São Paulo.

32. Fava-Netto, C. 1955. Estudos quantitativos sobre a fixacao do complement no blastomicose sulamericana, com antigeno polissa carido. Arquivos de Cirurgia Clinicae Experimental *18*:197–254.

33. Fava-Netto, C. 1976. Imunologia do Paracoccidioido micose. Rev. Inst. Med. Trop. São Paulo *18*:42–53.

34. Fonseca, J. B. 1957. Blastomycosis Sul-Americana Estudo das lesoes dentárias a paradentárias sob a ponto de vista clinico e histopathalogico. Thesis, Univ. São Paulo, pp. 1–182.

35. Fonseca Filho, O., and A. E. Area Leao. 1927. Réaction cutanée spécifique avec le filtrat de cellule du *Coccidioides immitis*. Compte Rendu des Séances de la Société de Biologie 97:1796–1797.

36. Fountain, F. F., and W. D. Sutliff. 1969. Paracoccidioidomycosis in the United States. Am. Rev. Respir. Dis., *99*:89–93.

37. Frias, F. A. S., S. Pasian, et al. 1979. South American blastomycosis of epididymis. Urology, *14*:85–87.

38. Furtado, T. 1975. Infection versus disease in South American Blastomycosis. Int. J. Dermatol., *14*:117–125.

39. Garrison, R. G., R. E. Bawdon, et al. 1974. Deoxyribonucleic acid composition of *Paracoccidioides brasiliensis*. Rev. Uruguaya Pat. Clin. y Microbiol., *54*:3–7.

40. Giraldo, R., A. Restrepo M. 1976. Pathogenesis of paracoccidioidomycosis: A model based on the study of 46 patients. Mycopathologia, *58*:68–70.

41. Goihman-Yahr, M., E. Essenfeld-Yahr, et al. 1979. New method for estimating digestion of *Paracoccidioides brasiliensis* by phagocytic cells in vitro. J. Clin. Microbiol., *10*:365–370.

42. González-Ochoa, A., and L. D. Soto. 1957. Blastomicosis sulamericana. Casos mexicanos. Rev. Inst. Salub. Enferm. Trop., *17*:97–101.

43. Greer, D. L., and B. Bolanos. 1977. Role of bats in the ecology of *Paracoccidioides brasiliensis:* the survival of *Paracoccidioides brasiliensis* in the intestinal tract of frugivorous bat. (*Artibeus literatus*). Sabouraudia, *15*:273–282.

44. Greer, D. L., D. D'Costa de Estrada, et al. 1971. Dermal reactions to paracoccidioidin among family members of patients with paracoccidioidomycosis. In reference no. 85, pp. 76–84.

45. Grose, E., and J. R. Tamsitt. 1965. *Paracoccidioides brasiliensis* recovered from intestinal tract of three bats (*Artibeus literatus*) in Colombia, S. A. Sabouraudia, *4*:121–125.

46. Hay, R. J., and F. W. Chandler. 1978. Experimental paracoccidioidomycosis: cranial and nasal localization in mice. Br. J. Exp. Pathol., *59*: 339–344.

47. Iabuki, K., and M. R. Montenegro. 1979. Experimental paracoccidioidomycosis in the Syrian hamster: morphology, ultrastructure, and correlation of lesions with presence of specific antigens and serum levels of antibodies. Mycopathologia, *67*:131–141.

48. Johnson, W. D., and C. M. Lang. 1977. Paracoccidioidomycosis (South American blastomycosis) in a squirrel monkey (*Saimiri sciureus*). Vet. Pathol., *14*:368–371.

49. Kanetsuna, F., L. M. Carbonell, et al. 1972. Biochemical studies on the thermal dimorphism of *Paracoccidioides brasiliensis*. J. Bacteriol., *110*:208–218.

50. Kaplan, W. 1971. Application of immunofluorescence to the diagnosis of paracoccidioidomycosis. In reference no. 85, pp. 224–226.

51. Kaufman, L. 1971. Evaluation of serological tests for paracoccidioidomycosis. Preliminary report. Pan American Health Organization. 1972. First Pan American Symposium on Paracoccidioidomycosis. Medellin Colombia, 25–27 October, 1971. Scientific Publication No. 254, Washington, D.C., World Health Organization, pp. 221–224.

52. Krivoy, O., E. A. Belfort, et al. 1978. Paracoccidioidomycosis of the skull. Case report. J. Neurosurg., *49*:429–433.

53. Kroll, J., and R. Walzes. 1972. Paracoccidioidomy-cosis in the United States. Arch. Dermatol., *106*:543–546.

54. Lacaz, C. S. 1940. O iodo no tratamento das micoses. An. Sao Paulo Med. Cirug., *39*:379.

55. Lacaz, C. S. 1967. Compendio de Micologia Medicas. Universidade de São Paulo. (Ed. Sataiva-Xavier).

56. Lacaz, C. S. 1955–1956. South American blastomycosis. An. Fac. Med. Univ. São Paulo, *29*:1–120.

57. Lacaz, C. S., and S. A. P. Sampaio. 1958. Tratamento da blastomicose sul-americana con anfotericino B. Resultados preliminases. Rev. Paulista de Medicina, *52*:443–450.

58. Lima, N. S., G. A. Teixeira, et al. 1978. Treatment of South American blastomycosis with orally administered miconazole. Rev. Inst. Med. Trop. São Paulo, *20*:347–352.

59. Limongelli, W. A., S. A. Rothstein, et al. 1978. Disseminated South American blastomycosis (paracoccidioidomycosis): report of a case. J. Oral Surg., *36*:625–630.

60. Linares, L., and L. Friedman. 1971. Pathogenesis of paracoccidioidomycosis in experimental animals. In reference no. 85, pp. 287–291.

61. Londero, A. T., C. D. Ramos, et al. 1976. Paracoccidioidomycosis: classificacao das formas clinicas. Rev. Urug. Pat. Clin. Microbiol., *14*:3–9.

62. Londero, A. T., C. D. Ramos, et al. 1978. Progressive pulmonary paracoccidioidomycosis. A study of 34 cases observed in Rio Grande do Sul (Brazil). Mycopathologia, *63*:53–56.

63. Lütz, A. 1908. Uma mycose pseudosoccidioidica localizada no boca e observada no Brasil. Contribuicao ao conhecimento das hyfoblastomycoses americanas. Brasil-Méd., *22*:121.

64. Machado, J., and S. L. Miranda. 1960. Concideraces relativas a blastomicose sul-americana. Da participacao pulmonar em 338 cases consecutivos. O. Hospital (Rio), *58*:431–449.

65. Mackinnon, J. E. 1959. Pathogenesis in South American blastomycosis. Trans. R. Soc. Trop. Med. Hyg., *53*:487–494.

66. Mackinnon, J. 1966. Algo mas sobre blastomicosis y temperatura ambiental. Tórax, *15*:127–129.

67. Mackinnon, J. 1970. On the importance of South American blastomycosis. Mycopathologia, *41*:187–193.

68. Martin, J. 1971. Blastomycose sul-americaine aigue evoluant sous le masque d'une lymphopathie maligne. Int. J. Dermatol., *10*:246–250.

69. Mendes, M., A. Raphael, et al. 1971. Impaired delayed hypersensitivity in patients with A. A. blastomycosis. J. Allergy, *47*:17–22.

70. Mendes, N. F., C. C. Musatti, et al. 1971. Lymphocyte cultures and skin allograft survival in patients with S. A. blastomycosis. J. Allergy Clin. Imm., *48*:40–45.

71. Mendez, G., G. Gonzalez, et al. 1979. Radiologic appearance of pulmonary South American blastomycosis. South. Med. J., *72*:1399–1401.

72. Montoya, F., and L. F. Garcia-Moreno. 1979. Effect of sex on delayed hypersensitivity responses in experimental mouse paracoccidioidomycosis. J. Reticuloendothelial Soc., *26*:467–478.

73. Moses, A. 1916. Fixacao do complemento no blastomicose. Memorias do Instituto Oswaldo Cruz, *8*:68–70.

74. Motta, L. C., and J. A. Pupo. 1936. Granulomatose paracoccidioidica (Blastomicose brasiliena). An. Faculdade Med. Univ. São Paulo, *12*:407–426.

75. Muchmore, H. G., B. A. McKovon, et al. 1972. Effect of steroid hormones on growth of *Paracoccidioides brasiliensis.* Pan American Health Organization. 1972. First Pan American Symposium on Paracoccidioidomycosis. Medellin Colombia, 25–27 October, 1971. Scientific Publication No. 254, Washington, D.C., World Health Organization, p. 300.

76. Murray, H. W., M. L. Littman, et al. 1974. Disseminated Paracoccidioidomycosis (South American Blastomycosis in the United States). Am. J. Med., *56*:209–220.

77. Negro, G. 1961. Localizacao supra-renal da blastomicose sul-americana. Thesis, Univ. São Paulo, pp. 1–144. São Paulo, Of. Graficas Saraiva, S. A.

78. Negroni, P. 1931. Estudio micologico sobre cincuenta casos de micosis observadas en Buenos Aires. Thesis, Buenos-Aires, Prensa Univ.

79. Negroni, P. 1966. El *"Paracoccidioides brasiliensis"* vive saprofiticamente en el suelo argentino. Prensa Med. Argent., *53*:2381–2382.

80. Negroni, P. 1967. Aislamiento del Paracoccidioides brasiliensis de una myestra de tierra de Chaco Argentino, Acad. Nac. Med. Buenos Aires, *45*:513–516.

81. Negroni, P. 1971. Prolonged therapy for paracoccidioidomycosis: Approaches, complications and risks. Pan American Health Organization. 1972. First Pan American Symposium on Paracoccidioidomycosis. Medellin Colombia, 25–27 October, 1971. Scientific Publication No. 254, Washington, D.C., World Health Organization, pp. 147–155.

82. Negroni, R. 1978. Paracoccidioidomicosis de localizacion cerebelosa. Rev. Argen. Micol., *1*:32–33.

83. Negroni, R., C. Lovannitti de Flores, et al. 1976. Estudio de las reacciones serologicas cruzatas entre antigenos de *Paracoccidioides brasiliensis* e *Histoplasma capsulatum.* Rev. Assoc. Argent. Microbiol., *8*:68–73.

84. Neves, J. S., and L. Bogliolo. 1951. Researches on the etiologic agents of American blastomycosis. I. morphology and systematic of Lütz' disease agent. Mycopathologia, *5*:133–146.

85. Pan American Health Organization. 1972. First Pan American Symposium on Paracoccidioidomycosis. Medellin Colombia, 25–27 October, 1971. Scientific Publication No. 254, Washington, D.C., World Health Organization.

86. Pollak, L., and A. Agulo-Octega. 1967. Las micosis broncopulmonares en Venezuela. Tórax, *16*:135–145.

87. Ramirez-Martinez, J. R. 1971. *Paracoccidioides brasiliensis.* Conversion of yeastlike forms into mycelia in submerged culture. J. Bacteriol., *105*:523–526.

88. Ramirez-Martinez, J. R. 1971. A comparative study of ribonucleic acid species on the yeastlike and mycelial forms of *Paracoccidioides brasiliensis.* Sabouraudia, *9*:157–163.

89. Ramos, C. D., A. T. Londero, et al. 1980. Pulmonary paracoccidioidomycosis in a nine-year-old girl. Mycopathologia, in press.

90. Restrepo M., A. 1966. La pruebra de immunodiffusion en el diagnostica de la paracoccidioidomicosis. Sabouraudia, *4*:223–230.

91. Restrepo M., A. 1970. A reappraisal of the mi-

croscopical appearance of the mycelial phase of *Paracoccidioides brasiliensis.* Sabouraudia, *8*:141–144.

92. Restrepo M., A., and I. Correa. 1973. Comparison of two culture media for primary isolation of Paracoccidioides brasiliensis from sputum. Sabouraudia, *10*:260–265.

93. Restrepo M. A., and G. Guzman de Espinosa. 1976. Paracoccidioidomicosis experimental del raton inducida por via aerogena. Sabouraudia, *14*:299–311.

94. Restrepo M., A., and L. H. Moncada. 1974. Characterization of the precipitin bands detected in the immunodiffusion test for paracoccidioidomycosis. Appl. Microbiol., *28*:138–144.

95. Restrepo M., A., and L. H. Moncada. 1978. Una prueba de latex en lamina para el diagnostico de la paracoccidioidomicosis. Bol. Ofic. Sanatar. Panam., *84*:520–522.

96. Restrepo M., A., L. H. Moncada, et al. 1969. Effect of hydrogen ion concentration and of temperature on the growth of *Paracoccidioides brasiliensis* in soil extracts. Sabouraudia, *7*:207–215.

97. Restrepo M., A., D. L. Greer, et al. 1971. Relationship between the environment and paracoccidioidomycosis. Pan American Health Organization. 1972. First Pan American Symposium on Paracoccidioidomycosis. Medellin Colombia, 25–27 October, 1971. Scientific Publication No. 254, Washington, D.C., World Health Organization, pp. 84–91.

98. Restrepo M., A., D. L. Greer, et al. 1973. Paracoccidioidomycosis: A review. Rev. Med. Vet. Mycol., *8*:97–121.

99. Restrepo M., A., F. Gutierrez, et al. 1970. Paracoccidioidomycosis (South American blastomycosis). A study of 39 cases observed in Medellin Colombia. Am. J. Trop. Med. Hyg., *19*:68–76.

100. Restrepo M., A., M. Restrepo, et al. 1978. Immune responses in paracoccidioidomycosis. A controlled study of 16 patients before and after treatment. Sabouraudia, *16*:151–163.

101. Restrepo M., A., M. Robledo, et al. 1968. Distribution of paracoccidioides sensitivity in Colombia. Am. J. Trop. Med. Hyg., *17*:25–37.

102. Restrepo M., A., M. Robledo, et al. 1976. The gamut of paracoccidioidomycosis. Am. J. Med., *61*:33–42.

103. Restrepo M., A., D. A. Stevens, et al. 1980. Ketoconazole in paracoccidioidomycosis: efficacy of prolonged oral therapy. Mycopathologia, *72*:35–46.

104. Restrepo M., A., and H. Velez A. 1975. Efectos de la fagocitosis in vitro sobre el *Paracoccidioides brasiliensis.* Sabouraudia, *13*:10–21.

105. Ribeiro, D. O. 1940. Nova terapentica para a blas-tomicose. Publicacoes Medicas, São Paulo, *12*:36–54.

106. Rolon, P. A. 1976. Paracoccidioidomicosis — epidemiologia en la republica del Paraguay centro de sud America. Mycopathologia, *59*:67–68.

107. Salfelder, K., G. Doehnert, et al. 1969. Paracoccidioidomycosis. Anatomic study with complete autopsies. Virchows Arch., *348*:51–76.

108. San Blas, G., and F. San-Blas. 1977. *Paracoccidioides brasiliensis:* Cell wall structure and virulence. A review. Mycopathologia, *62*:77–86.

109. Severo, L. C. 1979. Paracoccidioidomicose. Estudio clinico e radiologico das lesoes pulmonares e seu diagnostico. Thesis, Porto Alegre Universidade, Fed do Rio Grande do Sul.

110. Severo, L. C., G. R. Geyer, et al. 1979. The primary paracoccidioidomycosis in an immunosuppressed patient. Mycopathologia, *68*:171–174.

111. Severo, L. C., A. T. Londero, et al. 1979. Acute pulmonary paracoccidioidomycosis in an immunosuppressed patient. Mycopathologia, *68*:171–174.

112. Splendore, A. 1912. Una affezione micotica con localizazione nella mucosa della boca asservata in Brasile, determinate da funghi apparteneti alla tribú degli Exiascei (*Zymonema brasiliense* n. sp.). Estraido do volume in honore del Prof. Angello Celli nel 25° anno di inseqnamento. Roma, Tipografia Nazional di G. Bertero. and Bull. Soc. Pathol. Exot., *5*:313–319.

113. Stevens, D. A., and A. Restrepo M. 1979. Miconazole in paracoccidioidomycosis. Lancet, *1*:1301.

114. Stevens, D. A., A. Restrepo M., et al. 1978. Paracoccidioidomycosis (South American Blastomycosis). Treatment with micanazole. Am. J. Trop. Med. Hyg., *27*:801–807.

115. Teixeira, F., L. C. Gayotto, et al. 1978. Morphological patterns of liver in South American blastomycosis. Histopathology, *2*:231–237.

116. Yarzabal, L. A. 1971. Pathogenesis of paracoccidioidomycosis in man. In reference no. 85, pp. 261–270.

117. Yarzabal, L. A., and S. Andrew. 1976. Isolation of a specific antigen with alkalin phosphatase activity from soluble extracts of *Paracoccidioides brasiliensis.* Sabouraudia, *14*:275–280.

118. Yarzabal, L., J. M. Torres, et al. 1971. Antigenic mosaic of *Paracoccidioides brasiliensis.* Pan American Health Organization. 1972. First Pan American Symposium on Paracoccidioidomycosis. Medellin Colombia, 25–27 October, 1971. Scientific Publication No. 254, Washington, D.C., World Health Organization, pp. 239–244.

OPPORTUNISTIC INFECTIONS

20 Candidiasis and the Pathogenic Yeasts

THE YEASTS

The terms "yeast" and "yeastlike" are vernacular for unicellular fungal organisms that reproduce by budding and in their anamorph state are referred to as members of the class Blastomycetes. Such a definition is generally recognized as inadequate, in part because some yeasts reproduce by fission, in part because many yeasts produce mycelium or pseudomycelium under certain environmental and nutritional conditions, and in part because filamentous fungi (Hyphomycetes) may exist in a unicellular, yeastlike form that reproduces by budding, viz., oidia described in Chapter 6 and the parasitic stage of several pathogenic fungi. On the basis of their teleomorphic state, some yeasts are Ascomycota, others are Basidiomycota, and still others as yet have not been shown to have a sexual stage and are grouped together as Fungi Imperfecti (Deuteromycota). Clearly, then, the term "yeast" is of no taxonomic significance and is useful only to describe a morphologic form of a fungus.

YEASTS ASSOCIATED WITH HUMAN DISEASE

In the strictest sense of the word there are no inherently pathogenic yeasts. Those that are associated with human or animal disease are incapable of producing infection in the normal healthy individual. Some alteration of the host's cellular defenses, physiology, or normal flora must occur before colonization, infection, and disease production by yeasts can take place. The pathogenic potential of yeasts varies considerably, with the most virulent organism being *Candida albicans*. Slight changes in any of the three factors just mentioned may allow this normal human commensal to infect. As noted previously, the severity of the disease will depend on the seriousness of the alteration of the host rather than on any pathogenic properties exhibited by the fungus. Because of its rapid ability to colonize and take advantage of many types of host alterations, the clinical manifestations of *Candida* infection are protean. The disease may be cutaneous, mucocutaneous, subcutaneous, or systemic, or it may involve all of the aforementioned anatomic areas. *C. albicans* accounts for the vast majority of diseases caused by the yeasts. Other *Candida* spp., particularly *C. tropicalis*, also account for some infections, but the severity of the host's debilitation must be greater to allow these less virulent organisms to invade. Taken together, these infections are called candidiasis and are among the most common infectious diseases of man.

Diseases caused by *Cryptococcus* and *Rhodotorula* sp. are less frequently encountered than candidiasis. Although the former are usually systemic, cutaneous and mucocutaneous tissues may also be involved.

The methods of isolation, characterization, and identification of the yeasts are quite different from those of the mycelial fungi. Morphology is less important, and physiologic characteristics are of greater value for the identification of the yeasts. For their identification, emphasis is placed on carbohydrate fermentation and assimilation, nitrogen utilization, production of extracellular substances

484

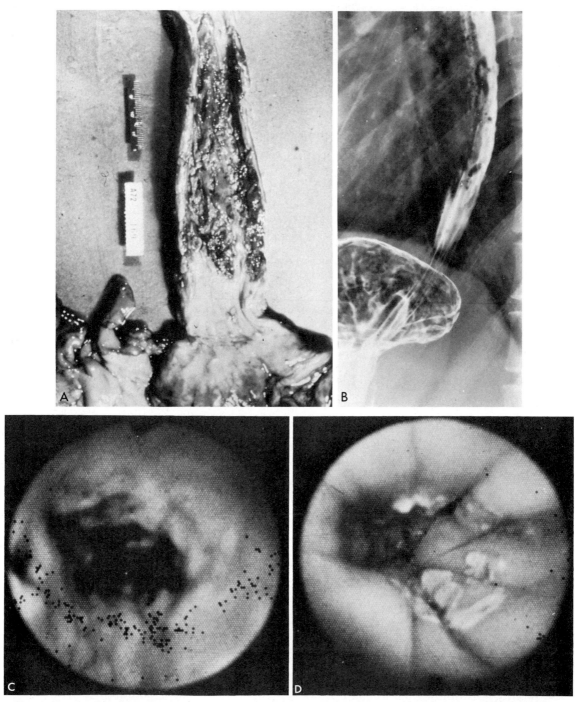

Figure 20–12. Candida esophagitis. *A*, Perforations and ragged ulcerations in late stage disease. *B*, Barium swallow of Candida esophagitis showing multiple plaquelike irregularities with ragged edges. *C*, Fiberoptic view of patient with confluent white plaques of lower esophagus. *D*, Same patient as in *C*. Fiberoptic esophagoscope after low dose amphotericin B treatment. Few lesions are seen and they are discrete. The patient had received nystatin suspension swallows but had failed to respond. The patient had hyperparathyroidism. (Courtesy C. Winans.)

such as capsules, and production of enzymes. The techniques used in enzymology, therefore, are similar to those used for the identification of bacteria.

The yeasts belong to three phyla of fungi: (1) Basidiomycota, which includes the basidiospore-forming yeasts of the families Filobasidiaceae and Sporobolomycetaceae of the order Ustilaginales (the medically important yeasts are included in three genera: *Filobasidiella, Leucosporidium,* and *Rhodosporidium*); (2) the ascospore-forming yeasts in the family Endomycetaceae of the phylum Ascomycota; and (3) the asporogenous yeasts in the form-family Cryptococcaceae in the form-phylum Deuteromycota (Fungi Imperfecti). The last group contains most of the species involved in human disease. Lodder in 1970[140] recognized 350 species in 22 genera of ascosporogenous yeasts: *Citeromyces, Coccidiascus, Debaryomyces, Dekkera, Endomycopsis, Hanseniaspora, Hansenula, Kluyveromyces, Lipomyces, Lodderomyces, Metschnikowia, Nadsonia, Nematospora, Pachysolen, Pichia, Saccharomyces, Saccharomycoides, Saccharomycopsis, Schizosaccharomyces, Schwanniomyces, Wickerhamia,* and *Wingea* and 12 genera of asporogenous yeasts: *Brettanomyces, Candida, Cryptococcus, Kloeckera, Oosporidium, Pityrosporum* (now *Malasezzia*), *Rhodotorula, Schizoblastosporion, Sterigmatomyces, Torulopsis,* Trichosporon,* and *Trigonopsis.* Excluded were the ballistosporogenous genera of yeasts from the phylum Basidiomycota *(Sporobolomyces, Bullera, Sporidiobolus, Itersonilia* and *Tilletiopsis)* and the genus *Geotrichum.* Recently about 200 more species have been described.[12] In all, there exist close to a thousand species of yeasts or yeastlike organisms. The vast majority of these have airborne spores or conidia and thus can be isolated as contaminants from skin, sputum, feces, or other clinical specimens. This compounds the difficulty of assessing the clinical importance of a yeast isolant. Only a few species in a few genera have regularly been associated with production of disease in man or animals. In a severely compromised host, however, many other species may be found as opportunists.

The industrial yeasts are, perhaps, the most familiar of the organisms. *Saccharomyces cerevisiae,* an Ascomycota, is the common brewers' and bakers' yeast. It occurs as two types: top yeasts, which cause a vigorous evolution of carbon dioxide and are found in the froth on the surface of the fermenting mixture, and bottom yeasts, which sink to the bottom. Beer and ale are produced from either type, depending on the variety of beverage. Another species of the genus, *S. ellipsoideus,* is the common wine yeast, occurring naturally on grapes and in the soil of vineyards. Its varieties are named for the various categories of wine which they produce. All these organisms are "perfect" or teleomorph yeasts, the cell body becoming an ascus after sexual union. Other yeasts are lactose fermenters, and are associated with the preparation of fermented milk beverages, such as Kefir and Koumiss, food staples commonly used in southern Europe and Asia. The commonest yeasts encountered as contaminants of bacterial cultures, growing on foods and air borne, or as transient flora of the human skin are members of the asporogenous genera *Rhodotorula, Trichosporon, Candida,* and *Cryptococcus.* Besides the frequently encountered species of the latter two genera, the saprophytic species, including several *Rhodotorula,* and *Trichosporon* spp. are occasionally isolated from human disease and represent opportunistic infectious agents. Even beer yeast has been encountered in lung infections.

In view of the ubiquitous distribution of yeasts, not only in air, dust, and soil but also on the surface of the body and in the mouth, intestinal tract, and vagina, it is not surprising that these forms have been found in a variety of pathologic processes. A great number of species have been described in this connection, most of them inadequately. In many instances, the yeast had no etiologic relation to the disease, and in others, the same yeast was repeatedly described as a new species, subsequently giving rise to a long list of synonymous names. For these reasons, a very long list of "pathogenic" yeasts has accumulated. Critical examination and consideration has now made it clear that only a few species of yeasts are actually regularly encountered as disease-producing organisms for man and animals. The yeasts of medical importance are listed below along with the generally accepted classification and basic differential characteristics (Table 20–1).

Though most of the above genera form a fairly well-defined group, the genera *Torulopsis* and *Candida* are extremely heterogeneous. Even the separation of the two genera is artificial and tenuous and recently the two genera have been combined,[256] though there

---

*In 1978 Yarnow and Memer[256] amended the genus *Candida* Berkhout *nomen conservadem* to include the species of the genus *Torulopsis.*

Table 20-1 *Classification of Medically Important Yeasts of the Form-Phylum Deuteromycota*

Form-Class: Blastomycetes

  Form-Family: Cryptococcaceae

    Genus 1: *Cryptococcus.* Unicellular budding cells only; reproduce by blastoconidia pinched off the mother cell. Most are urease-positive. Cell surrounded by a heteropolysaccharide capsule and produces starchlike compounds; carotenoid pigments are usually lacking. Inositol is assimilated; sugars are not fermented.
      Example: *Cryptococcus neoformans* (cryptococcal meningitis).

    Genus 2: *Malassezia.* Mostly unicellular budding cells which reproduce by blastoconidia that develop from a reduced phialide. Cells may adhere, forming short hyphal strands. Growth stimulated by lipids. There is no fermentative ability.
      Example: *Malassezia furfur* (pityriasis versicolor)

    Genus 3: *Rhodotorula.* Unicellular budding forms that rarely produce pseudomycelium, are generally encapsulated, but do not produce starchlike substance. They do not assimilate inositol or ferment sugars. Carotenoid pigments are produced.
      Example: *Rhodotorula rubra* (rare pulmonary and systemic infections)

    Genus 4: *Candida.* Reproduction is by pinched blastoconidia. They may form pseudomycelium or true mycelium; urease is generally negative; capsules are not formed; starch or carotenoid pigments are not produced; inositol is not assimilated.
      Example: *Candida albicans* (candidiasis)

    Genus 5: *Trichosporon.* Reproduction is by blastoconidia and arthroconidia. Mycelium and pseudomycelium are formed.
      Example: *Trichosporon beigelii* (white piedra and systemic infections)

    Genus 6: *Geotrichum.* Reproduction is by arthroconidia only. A true mycelium is formed.
      Example: *Geotrichum candidum* (rare pulmonary geotrichosis)

are many mycologists who oppose the "marriage."[109] A number of species of both genera have been shown to be Ascomycota, belonging in several teleomorphic genera, namely *Candida guilliermondii* (*Pichia guilliermondi*) and *Candida globosa* (*Citromyces metritensis*), among others. Some species have a teleomorphic stage in the Basidiomycota, namely *Candida scottii* (*Leucosporidium scottii*). There is also evidence that *Candida albicans* is a Basidiomycete and, like some other *Candida* species, has *Leucosporidium* as a teleomorph state. The genus *Rhodotorula* has a number of species with a teleomorph state in the Basidiomycete genus, *Rhodosporidium*. Certain of the Cryptococcaceae including *Cryptococcus neoformans* produce clamp connections and basidiospores, which ally them to the Basidiomycota in the genus *Filobasidiella.* Finally, electron micrographs of *Trichosporon beigelii* have demonstrated Basidiomycota-like septa with dolipores and parenthesomes. The yeasts are a very diverse group of organisms.

The relationship of *C. albicans* to other members of the genus has been investigated by many authors using a variety of techniques. It appears to be the only member of this genus regularly able to evoke fatal disease in man and animals.[161, 170] *C. stellatoidea* and *C. tropicalis,* though less pathogenic than *C. albicans,* are significantly more virulent than other *Candida* species. *C. albicans* is antigenically divided into two groups: A, which

it shares with *C. tropicalis,* and B, which is shared by *C. stellatoidea.*[238] Nucleic acid-base composition studies (G-C ratios) indicate that *C. albicans, C. tropicalis, C. clausenii,* and *C. stellatoidea* are related.[231] However, by DNA homology studies, *C. albicans* has close relationship to *C. clausenii* and *C. stellatoidea* but not *C. tropicalis.*[15]

YEASTS AS NORMAL FLORA

Hyphomycetes in the form of conidia or short mycelial strands can occur on the body surfaces or in the alimentary tract of man. Some dermatophytes, particularly anthropophilic strains of *Trichophyton mentagrophytes,* may reside in the interdigital spaces of the foot as occult tinea pedis and *Mucor* sp. in a rounded, budding yeastlike form and are known to colonize the vaginal surfaces and the large intestine. But in general mycelial fungi do not establish a permanent "normal flora" on human body surfaces. In a prospective study of skin colonization, Henney et al.[93] isolated 69 species of filamentous fungi over a 175-day period, but 71 per cent were isolated no more than twice. In contrast, many Blastomycetes (yeasts and yeastlike fungi) constitute a resident population regularly and universally part of the normal flora of skin surfaces, buccal mucosa, the intestinal tract, and vaginal mucosa. The species involved and the numbers present vary with the body surfaces but make up a balanced population in a particular ecological niche on the normal healthy individual. Some species are obligate animal saprotrophs[35] and do not form a stable population away from the animal body. These include *Malassezia furfur, M. ovalis,* and *M. pachydermatis* in the lipid-rich secretions of scalp and nose; *Candida albicans* and *C. glabrata* (formerly *Torulopsis*)[256] in the throat, buccal mucosa, and intestine; and *C. stellatoidea* in the vagina. Other common components of normal human flora are facultative animal saprotrophs, because they are often isolated from sources other than animal. These include *C. guilliermondii, C. parapsilosis, C. krusei* on the skin and *C. tropicalis* in the intestines. Other incidental facultative saprotrophs that are regularly encountered as normal flora but in small numbers are *C. pseudotropicalis* and *Rhodotorula* sp. on the skin; *Trichosporon beigelii* in throat and rectum and sometimes on the skin; and *Geotrichum candidum* in feces and occasionally in the throat and on the skin. In addition, *Saccharomyces, Endomycopsis, Leucosporidium,* other species of *Candida* and *Trichosporon,* the colorless alga *Prototheca, Cryptococcus sp.,* and even *C. neoformans* have been recovered from normal body surfaces.[64, 139, 149, 153, 205, 232, 233, 244, 247] In most surveys on the incidence of yeasts from normal skin, *Candida parapsilosis* and *C. guilliermondii* lead the list[139, 205, 233]; *C. krusei* is also fairly common. *C. tropicalis* and *C. albicans* are not regularly found on normal skin but can be recovered from the anogenital region, the area around the mouth, and the fingers.[170] Yeasts isolated from the mouth[64, 170, 225] are predominantly *C. albicans* (75 per cent), followed by *C. tropicalis* (8 per cent), *C. krusei* (3 to 6 per cent), and *C. glabrata* (2 to 6 per cent). As mentioned previously *Trichosporon beigelii, Geotrichum,* and other *Candida* species are also regularly recovered.

In swabs from the anorectal area or feces, 50 per cent of isolants were *C. albicans* and up to 20 per cent were *C. glabrata* and *C. tropicalis.*[35, 135, 170] *Geotrichum candidum* and *Trichosporon beigelii* are also present but in small numbers. In surveys of the normal vagina, between 11 and 30 per cent of healthy women harbor *C. albicans,*[135, 170] and up to 85 per cent of gynecological patients have positive cultures. The other common isolants in both groups were *C. glabrata* (9 to 30 per cent), *C. parapsilosis,* and *C. tropicalis.* In a survey of the external ear, Martini et al.[153] found *C. robusta, C. rhagii, C. parapsilosis,* and *C. guilliermondii* in normal ears and also associated with a variety of eczematous conditions. Occlusive dressings and use of topical antibiotics often lead to overgrowths of normal yeasts on various body surfaces.

A variety of other sources may contain yeasts associated with human disease. These are almost always associated with human activity, especially fecal contamination. In large enough concentration these organisms may elicit a variety of conditions, from eczematous rash to infections of the skin and body surfaces.[31] *C. albicans* is frequently isolated from thermal baths,[8] heavily utilized beaches,[7, 31, 224] and occasionally from swimming pools. It is sensitive to chlorination (10^5 cells inactivated by 4 ppm Cl at 30 min[116]), however, and will die rapidly in sea water.[31] *C. albicans* can survive for fairly long periods in beach sand,[7] dairy products, and other kinds of food[170] but not in soil, on plants, or in the atmosphere. The most thermotolerant species of disease associated yeast is *Trichosporon beigelii,* which is reported to have survived in sand held at 45° C for six months.[7]

Candidiasis

DEFINITION

Candidiasis is a primary or secondary infection involving a member of the genus *Candida.* Essentially, however, the disease is an infection caused by *Candida albicans.* The clinical manifestations of disease are extremely varied, ranging from acute, subacute, and chronic to episodic. Involvement may be localized to the mouth, throat, skin, scalp, vagina, fingers, nails, bronchi, lungs, or the gastrointestinal tract, or become systemic as in septicemia, endocarditis, and meningitis.

The pathologic processes evoked are also diverse and vary from irritation and inflammation to chronic and acute suppuration or granulomatous response. Since *C. albicans* is an endogenous species, the disease represents an opportunistic infection.

Candidiasis and its etiology are so common that it has received a number of names in the past. "Monilia" is perhaps the most popular of the older terms, but in reality the name refers to the anamorph stage of *Neurospora,* and the organism is not even a yeast. *Monilia* may cause brown spot of peach but not vaginitis. The term "moniliasis" still persists in some quarters, even though equally archaic designations such as "phlegms," "ptomaines," and "humors" have long since been abandoned.

Synonymy. Candidosis, thrush, dermato-candidiasis, bronchomycosis, mycotic vulvo-vaginitis, muguet, moniliasis.

HISTORY

Hippocrates[96] in his "Epidemics" describes aphthae or thrush (white patches) in debilitated patients, and the presence of this clinical condition has been recognized for centuries. The term "thrush" is probably derived from ancient Scandinavian or Anglo-Saxon. "Torsk" is the Swedish equivalent of this word. The French word for the condition is "le muguet," which means "lily of the vally." Galen described it as of common occurrence in children, particularly sickly children,[68] and the disease was noted in Pepys' diary of 1665.[253] It was recognized early as a condition of the newborn in textbooks on pediatrics by Rosen von Rosenstein in 1771[206] and Underwood in 1784.[243] The disease was so prevalent in France that the French Société Royale de Médecine offered an award at £1299 for its study in 1786.[207] Veron in 1835[245] postulated that it was acquired during passage through the womb. He also described the first cases of esophageal candidiasis. Berg (1846)[20] considered the fungus to be transmitted by unhygienic conditions and communal feeding bottles. It was also known to occur in patients with debilitating diseases, confirming Hippocrates' observation. Debilitation was proposed by Bennett in 1844[19] and Robin in 1853[201] as the most important prelude to candidal infection. Opposing views were held by Berg (1851) and Valleix (1838), who considered it an unfortunate disease of wide occurrence not related to other diseases or conditions.

Credit for the discovery of the "cryptogamic plant" in a lesion of thrush goes to Lagenbeck.[133] In 1839 he described a fungus in aphthae, one of the earliest accounts of a parasite in a human disease. He did not consider the fungus to be related to the disease on the tongue but thought it was the cause of typhus. However, Berg in 1841[20] and Bennett in 1844[19] conclusively demonstrated the fungal etiology of thrush. Bennett's work is notable for the excellent and accurate illustrations of the fungus in the lung and sputum of a patient with pneumothorax due to tuberculosis. Berg reproduced the disease in healthy babies by inoculating them with aphthous membrane material. One of them died of candidial bronchitis and pneumonia. His colleague David Gruby, who had earlier described the fungus of favus and tinea capitis, studied the thrush fungus and noted its similarities to other fungi. In 1842 he described this fungus as "le vrai Muguet des enfants."[82] He placed Lagenbeck's "cryptogramic plant" in the genus *Sporotrichum.* By 1853 Robin, who had great influence on later generations of physicians, recognized that the thrush fungus could become systemic as a terminal event of other illnesses. He renamed the organism *Oidum albicans,* the first use of the specific epithet. A bewildering array of names for the fungus began to accumulate. *Syringospora robinii* was proposed by Quinquand in 1868,[188] and Reess in 1877[194] redefined it as *Saccharomyces albicans.* The dimorphic nature of the fungus was noted by Grawitz in 1877,[79] and yeast-mycelial growth as a response to environment by a single species of fungus was conclusively demonstrated by Audrey in 1887.[14]

The first description of vaginal candidiasis was by Wilkinson in 1849. His patient, a 77-year-old woman, had a profuse vaginal discharge. Wilkinson was cautious when he quoted Voget by saying the "epiphytes" (*Candida* yeast cells) could not grow unless a "favorable soil was prepared for them." Previous to that time, and for the next forty years, vaginal discharge as well as thrush were defined in textbooks of medicine as the result of morbid secretions. The important association of vaginal candidiasis and thrush in the newborn was made by Haussmann in 1875.[91] He demonstrated the transmission of the fungus to the mouths of babies from lesions in the vagina. He also produced vaginitis in a healthy gravid female by inoculating *Candida* into her vagina. The relationship of dermatocandidiasis to thrush was recognized as early as 1771 by Rosen von Rosenstein,[206] and its relationship to subcutaneous infection by Virchow in the 1850's. Systemic disease by hematogenous spread was described by Zenker in 1861.[257] His patient was debilitated and had oral thrush, an association noted many times previously. The patient succumbed to a brain infection, however, and *Candida* was demonstrated in the lesion, indicating the fungus could spread to other organs. Intestinal disease was described by Parrot in 1870, and in 1877 he also noted the first pulmonary infection.[180] Disseminated infection involving many organ systems was recorded by Schmorl in 1890.[216] The first descriptions of the other major forms of *Candida* infections were onychomycosis by Dubendorfer in 1904,[54] and cutaneous disease by Jacobi in 1907.[112] Chronic mucocutaneous disease was probably first described by Forbes in 1909[63]; the distinction of it as a separate entity of familial inheritance was made by Schultz in 1925[218]; and its association with endocrine dysfunctions was pointed out by Sutphin in 1943.[235] *Candida* cystitis was first recorded by Rafin in 1910[189]; osteomyelitis, by Conner in 1928[45]; endocarditis as a separate entity by Joachim and Polayes in 1940[114]; and endogenous dissemination resulting in endophthalmitis in 1943, by Miale.[156] By the early 1940's it thus became evident that candidiasis was the most protean of fungus infections. Statistically it was a common infecting agent of the skin, mucosa, and vagina but was rarely a cause of serious systemic disease.

A revival of interest in systemic candidiasis and candidal endocarditis took place after 1940. The occurrence of candidiasis as a sequel to the use of antibacterial antibiotics, particularly broad-spectrum antibiotics, evoked a great surge of research. The results have demonstrated the delicate ecosystem of which *Candida* is a member. Many fatal cases of candidiasis occurred following abrogation of this balance. In 1940, Joachim and Polayes[44] described candidal endocarditis as a hazard of heroin injection. About the same time, the association of candidiasis and steroid therapy, immunosuppressive drugs, cytotoxic agents, and immune defects became apparent.[253] Presently, *Candida* is recognized as one of the most frequently encountered fungal opportunists and is now regarded as the commonest cause of serious fungal disease.

As noted previously, the classification of the organism *Candida albicans* has been the subject of controversy since its first association with human disease. The term "monilia," with which *Candida* is often confused, was first used by Hill in 1751 to describe fungi from rotting vegetation. This proved to be an invalid description, as the organisms were in reality *Aspergillus* sp. The genus *Monilia* was erected by Gmelin[72] in 1791 to include some species of *Mucor* and *Aspergillus*. Person in 1794 repeated this description, restricting it to *Aspergillus*-like organisms. The genus *Monilia,* as it is understood today, was utilized by A. Saccardo in 1886[208] to encompass certain species of fungi isolated from rotting fruit. These are now known to be the imperfect stage of certain ascomycetous genera, for example, *Neurospora* and *Sclerotinia*, i.e. peach mummies, and are in no way related to the yeasts of the genus *Candida*.

The original organism isolated by Lagenbeck in 1839[133] was restudied by Gruby.[82] He considered it a *Sporotrichum*. However, the dimorphic nature of the yeast was recognized as early as 1844 by Bennett,[19] and in 1847 Robin[201] placed it in the genus *Oidium* under the name *Oidium albicans*. This genus name was derived from the morphology of the egg-shaped, oval yeast cell. Robin described in detail the yeast and mycelial forms of the organism. Later, another fungus from rotting wood was termed *Monilia candida* by Bonorden[27] in 1851. Hansen in 1868 described it as a yeast and mycelial fungus, but he accepted the name of *Monilia candida* as valid. The identity of this organism is unclear. Plaut (1887)[185] found a wood-rotting fungus that produced experimental lesions in the throats of chickens. He concluded that it was the same as Hansen's fungus and, therefore, the etiologic agent of human thrush. This was accepted by Zopf, who in 1890 described it as *Monilia albicans*.[258] The name became popular in medical literature, though the genus *Monilia* (sensu Persoon) exists as a valid genus for plant pathogens unrelated to *C. albicans*. Much of this confusion is due to the influence of the eminent physician and mycologist, Castellani. He adamantly retained the name *Monilia* and described numerous varieties as new species that have since been shown to be identical to *Candida albicans*. At that time, it was a common practice to publish as new species organisms showing minor variations in morphology on clinical presentation. Lodder lists one hundred synonyms for *C. albicans* in the 1970 edition of *The Yeasts*.[140] To end the confusion, Berkhout in 1923 erected the genus *Candida* to encompass asporogenous yeasts that have "few hyphae, lying flat, falling apart into longer or shorter pieces. Conidia arise by budding on hyphae or each other. Small and colourless."[22] This name was accepted as a *nomen conservandum* by the Eighth Botanical Congress at Paris in 1954.

The terminology for the clinical disease has also been controversial. "Moniliasis" is obviously untenable, since the genus that contained the etiologic agents of the disease is invalid. "Candidiasis" came into common use in the United States, but in Canada, England, France, and Italy the term "candidosis" had been accepted as the more desirable descriptive term. Since the other mycotic diseases end with -osis, namely histoplasmosis, crypto-coccosis, blastomycosis, sporotrichosis, and so forth. For consistency, the term "candidosis" is preferable, but candidiasis is used presently in most of the world literature on the disease and will be utilized in the present text.

ETIOLOGY, ECOLOGY, AND DISTRIBUTION

Although the etiologic agent *Candida albicans* is usually encountered in most of the clinical forms of candidiasis, in some of the less common clinical conditions, such as endocarditis, other species are more frequently isolated. These other species represent normal flora of the cutaneous and mucocutaneous areas and are of very limited pathogenicity. All species may be involved in any form of candidiasis, but some are regularly encountered in one particular type. These include *C. parapsilosis* from paronychias, endocarditis, and otitis externa; *C. tropicalis* from vaginitis, intestinal disease,[252] bronchopulmonary and systemic infections, and onychomycosis; *C. stellatoidea* from vaginitis; *C. guilliermondii* from endocarditis, cutaneous candidiasis, and onychomycosis; *C. pseudotrophicalis* from vaginitis; *C. glabrata* from esophageal and vaginal lesions; *C. krusei* very rarely from endocarditis and vaginitis; and *C. zeylanoides* from onychomycosis.

Other species of *Candida* that have been unequivocally demonstrated in human infections include *C. viswanathii*,[2, 11, 212] *C. lusitaniae*,[99] *C. claussenii*, *C. intermedia*, *C. lambica*, *C. macedoniensis*, *C. robusta*,[85] *C. norvegensis*, *C. zeylanoides*, *C. catenula*,[234] *C. ravautii*,[48] and *C. lipolytica*.[168] The various clinical forms of candidiasis can also be produced by other yeast and yeastlike organisms. These include *Rhodatorula glutinis*, *Rh. rubra*, *C. famata (Torulopsis candida)*, *Trichosporon beigelii*, *Geotrichum candidum*, and others. Since most of the organisms will have the same general appearance in tissue sections, the specific diagnosis of the disease depends on isolation and identification of the organism in culture.

Candida albicans is a normal inhabitant of the alimentary tract and the mucocutaneous regions.[151, 152] It is regularly present in small numbers in the mouths of normal healthy adults. Poor oral hygiene or even small amounts of antibiotics promote an increase in the number of organisms, though usually without untoward results. In the newborn, however, before an oral ecology is estab-lished, even a few organisms presage clinical thrush. The incidence of oral candidiasis of the newborn varies in different surveys. Case rates as high as 18 per cent have been record-ed, but the average appears to be 4 per cent. It is well established that candidal vaginitis during pregnancy contributes to thrush of the newborn. A small but significant percent-age of cases are due to cross-contamination from other infants, mothers, and personnel. This is especially true in nursery epidemics. As noted in the section on normal flora, *C. albicans* is frequently found on fingers, and these are the probable vectors for intraper-sonal as well as person-to-person dissemina-tion.

The incidence of *C. albicans* in the normal vagina of healthy, nonpregnant women is about 5 per cent[151] and can be as high as 30 per cent in pregnant women or women on oral contraceptives.[139] Other studies showed no increase in the presence of *Candida* asso-ciated with oral contraceptives.[73] There is a distinct increase in clinical vaginitis in gravid females. Most studies indicate a rate of can-didiasis of about 18 per cent for nonpregnant women with vaginal discharge but an average rate of 30 per cent for gravid women and women on contraceptives.[73] These studies have been carried out in many geographic locations under varied climatic conditions, and such factors do not appear to contribute to the incidence or severity of disease.

The normal alimentary tract has a small but constant population of *C. albicans*. Under normal conditions this is probably influenced by foods, since diet markedly affects the total number of organisms present. In the young, before a balanced flora is established, initial colonization of the intestine is frequently as-sociated with clinical symptoms.[98] Perianal colonization may also occur, followed by diaper rash.[137] In the adult, two extrinsic factors alter the number of *C. albicans* in the intestine. First, it has been established that other members of the intestinal flora exert a control on the population density of the yeasts. Studies have implicated a variety of antimicrobial factors, and probably no single mechanism is totally responsible. However, lactic acid appears to be quite inhibitory to *C. albicans,* and a correlation has been found between the numbers of lactobacilli and other lactic acid-producing organisms present and the number of yeasts. Preparations of lac-tobacilli have been utilized in recalcitrant vaginitis.[213] Secreted inhibitory factors, oxidation-reduction potentials, and competition for available nutrients have also been implicated in yeast population control. It was observed quite early that a change in the intestinal flora following orally administered antibiotics greatly influenced the number of *C. albicans*. The overgrowth of organisms may manifest itself only as an irritating pruritis ani or progress to colonization of the intestinal tissue and eventual fatal systemic candidiasis. The second factor influencing the population of *C. albicans* is diet. A high-fruit diet appears to favor a rapid increase in the number of intestinal yeasts and probably explains the former postulated association of *Candida* and tropical sprue. In normal adults who maintain this diet, there does not appear to be any clinical symptomatology after acclimatization of the host to the presence of the organism.

As noted in the section on normal flora, the normal skin appears to have a resident yeast flora, but this does not include *C. albicans*. In 2444 scrapings from the skin of 118 normal people, *C. albicans* was recovered only three times.[151, 152] Many other surveys have been carried out which indicate that several factors influence colonization of the skin by *C. albicans*. Whereas normal skin does not harbor a resident flora of *C. albicans*, almost any damage to skin or environmental change leads to rapid colonization. For this reason, *Candida* is not infrequently isolated from a variety of dermatologic conditions. Most of the lesions are situated in moister areas, such as the inframammary folds, the perianal skin, and other intertriginous regions. Fruit pickers, canners, dishwashers, and so forth, are particularly prone to candidal infections of the fingers, since constant contact with a moist environment leads to maceration of skin. In some studies, living in a tropical environment alone or having contact with infected patients increases the recovery of *C. albicans* from the skin.[152] Endocrine balance, the administration of steroids, and other physiologic factors also influence the rapidity and extent of *C. albicans* colonization. Although this organism may not have initiated a particular lesion, once it is established, it contributes to the pathology of the disease.

C. albicans is common to the alimentary tract of almost all mammals and birds,[170] and all such species are susceptible to invasion by this opportunistic fungus. The organism has been recovered from soil rarely, but in these instances it probably represents fecal contamination. It appears unlikely that *C. albicans* normally survives and multiplies in a nonanimal environment. A bacterial parasite of *C. albicans* has been reported[4] and may be important in eliminating it from soil. Unsanitary conditions often lead to contamination of the environment or food with the organism.[71] The other species of *Candida* are regularly isolated from normal skin, particularly between the toes, under the toenails, and from the umbilicus. They also appear to be indigenous to many other animal species as well as to vegetable material and the soil.

Age and sex distribution as well as clinical manifestations of candidiasis are markedly affected by varying predisposing factors and the underlying disease of the patient. There are five general conditions in which the normal equilibrium between *Candida* and the host may be sufficiently upset to lead to a pathologic state:

1. Extreme youth. During the normal process of establishing a resident flora, the restricting factors for *Candida* may be absent, and a clinical condition is produced (thrush, diaper rash, etc.) In normal children this condition resolves rapidly, often without treatment.

2. Physiologic change. Pregnancy appears to affect the carbohydrate content of the vagina and leads to an increase in the population of *Candida*. This overgrowth may be sufficient to cause a clinically apparent vaginitis. The administration of steroids to males or females also leads to proliferation of *Candida*.[145] This would also include patients with endocrine dysfunctions, particularly diabetes. The "normal flora" carrier rate on the skin of diabetics is quite high, and the number of organisms in the usual colonized areas (buccal mucosa, intestine, vagina) is also increased. All of these factors contribute to the commonness of opportunistic infections in diabetics.[171]

3. Prolonged administration of antibiotics. Much evidence has accumulated associating clinical disease with the use of antibacterial antibiotics. The most important effect is the elimination and alteration of the bacterial flora that holds the population of *Candida* in check. Evidence also suggests that there is some effect of the antibiotic on the host tissue that predisposes it to invasion by the organism, and the antibiotic itself may stimulate the growth of the *Candida*. The occurrence of the latter effect has been controversial, and most of the current investigations tend to discount it.[253]

4. General debility and the constitutionally inadequate patient. The list of disease syndromes associated with candidiasis is long

and varied. The extent and severity of the disease usually correlates with the severity of the underlying illness. The term "debility" encompasses such things as the slight avitaminosis of the aged which leads to perlèche or thrush; diabetes and its associated cutaneous candidiasis; candidal vegetations of diseased heart valves; and pulmonary or generalized systemic candidiasis occurring as a sequela to chronic disease or as a terminal event in the various neoplasias. Debility may be iatrogenic also. Immunosuppressive agents, cytotoxins, and other drugs abrogate the normal defenses of the host and predispose them to candidiasis or invasion by other opportunistic organisms. Constitutionally inadequate patients include those with various immune defects and defects associated with abnormal leukocytic function. There are many genetic deficiencies for which *Candida, Aspergillus,* and a long list of other fungal, bacterial, parasitic, and viral opportunistic organisms exploit the defective host as culture milieu.

5. Iatrogenic and barrier-break candidiasis can result from a wide variety of insults. In the first type, colonization can occur in association with indwelling catheters, hyperalimentation, peritoneal dialysis, and surgical procedures or simply through the injection of material into the skin, muscle, or circulatory or central nervous system. Drug abusers often develop *Candida* infections at the site of injection or on heart valves. *Candida* infections may also result from accidental barrier breaks such as trauma, burns, or gun or knife wounds.

CLINICAL DISEASE

Candida albicans is perhaps the most protean infectious agent that afflicts man. Only syphilis presents with such a diversity of clinical pictures as are encountered with the various candidiases. All of the tissue and organ systems are subject to invasion, and the pathology evoked is as variable as are the clinical syndromes. In addition to active infection, *Candida albicans* is also involved in several allergic conditions. These various clinical manifestations will be discussed according to the primary organ system involved.

I. Infectious Diseases
  A. Mucocutaneous involvement
    1. Oral: thrush, glossitis, stomatitis, cheilitis, perlèche

    2. Vaginitis and balanitis
    3. Bronchial and pulmonary
    4. Alimentary: esophagitis, enteric, and perianal disease
    5. Chronic mucocutaneous candidiasis

  B. Cutaneous involvement
    1. Intertriginous and generalized candidiasis
    2. Paronychia and onychomycosis
    3. Diaper disease (napkin candidiasis)
    4. Candidal granuloma

  C. Systemic involvement
    1. Urinary tract
    2. Endocarditis
    3. Meningitis
    4. Septicemia
    5. Iatrogenic candidaemia (barrier-break candidaemia)
II. Allergic Diseases
  A. Candidids
  B. Eczema
  C. Asthma
  D. Gastritis

Infectious Diseases

Mucocutaneous Involvement

Oral Candidiasis — "Thrush." This is the commonest form of disease produced by overgrowth (colonization) of *Candida albicans.* The mouth of the newborn, similar to the vagina of gravid females, has a low pH, which may promote the proliferation of *C. albicans.* It is now well established that any *C. albicans* in the mouth of the newborn presages the clinical disease until a balanced flora has been established. A cream-white to gray pseudomembrane covers the tongue (Fig. 20–1), soft palate, buccal mucosa, and other oral surfaces. The distribution is discrete, confluent, or patchy. These well-marked signs are usually not evident until the child is one week of age. If *Candida* is absent on the third day of life, the condition rarely develops.[217] The membrane seen on the mucosa is composed of masses of fungi in both the mycelial and yeast growth form. The membranous patches often crumble and have the appearance of milk curds, and, in fact, the latter are sometimes mistaken for signs of thrush. The lesions begin as small focal areas of colonization which enlarge to become a patch. The mem-

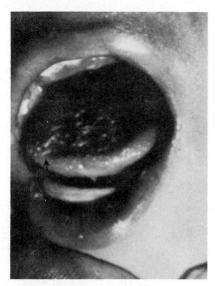

Figure 20–1. Thrush of newborn. Curdlike, patchy areas on tongue (*arrow*) and oral surfaces. (Courtesy of A. Padilha-Gonçalves.)

brane is rather closely adherent to the underlying mucosa, and its removal reveals a red, oozing base. The fungus is usually restricted to the cornea of the mucosa, as is seen in histologic sections. In severe disease, there may be ulceration and necrosis of the mucous membrane. The pseudomembrane of oral candidiasis is neither as firm nor as extensive as a diphtheritic membrane, but it often becomes large enough and the tissue involved sufficiently swollen to impede swallowing and, occasionally, breathing. Extensive involvement may also include the trachea, esophagus, and the angles of the mouth. Thrush of the newborn is most commonly associated with mothers having vaginal *Candida* infestations. Antepartum treatment with clotrimazole dramatically lowers both vaginal *Candida* contamination and thrush of newborn infants.[217]

Oral candidiasis also occurs in a condition known as "black hairy tongue." This disease is characterized by hypertrophy of the papillae of the tongue. *Candida* has no etiologic role in this disease but grows freely in this environment. Many such cases occur following antibiotic therapy and are sometimes called glossodynia (acute atrophic candidiasis) or antibiotic tongue. The etiology is obscure. The same is true in cases of chronic glossitis; the *Candida* proliferate in abundance because of favorable conditions for growth. In both cases, however, the *Candida* contributes to the overall pathology but not to its initiation.

One form of chronic *Candida* involvement of the tongue is called median rhomboid glossitis.[254] It appears to develop in hyperplastic conditions and can be associated with commensal leukoplakia and palatine "kissing" lesions.[81] *Candida* frequently invades leukoplakia,[86] causing exaggeration of the symptoms. It does not invade the related lesions of leukoedema, lichen planus, squamous papilloma, or carcinoma.

Oral thrush in older children and adults is clinically identical to that described for the newborn. In older children, chronic thrush usually indicates anatomic defect, polyendocrine disturbances (Fig. 20–2) or an underlying defect in natural[240] defenses. In adult patients it may be the result of mild avitaminosis, particularly riboflavin deficiency or a complication of diabetes (Fig. 20–3), polyendocrine disturbance, advanced neoplasia, or the administration of steroids, antibiotics, or other drugs (Fig. 20–4).

Adult stomatitis which is not related to other diseases is usually the result of badly fitting dentures or poor oral hygiene[32] (Fig. 20–5). Again, *Candida* grows in abundance without having initiated the disease. The condition occurs in about 25 per cent of denture wearers and up to 60 per cent of those over 65.[32] Lesions are erosive and painful. Good oral hygiene and cleansing of dentures often is sufficient to bring about clinical cure.

Deepening and exaggeration of the com-

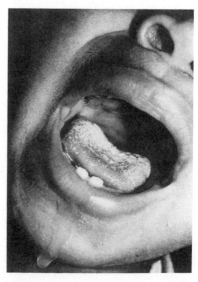

Figure 20–2. Oral candidiasis — "thrush." Cream white, curdy patches of pseudomembrane cover the back portion of the tongue. This is in an older child with polyendocrine disturbances.

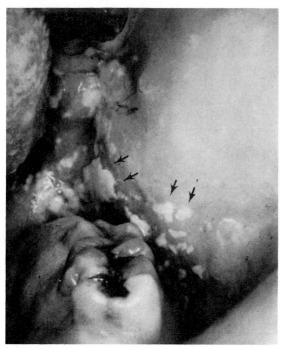

Figure 20–3. Candida stomatitis in an adult diabetic. (Courtesy of R. Goepp.) Patchy lesions indicated by arrows.

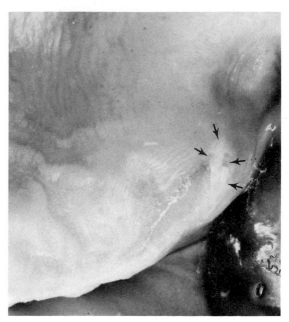

Figure 20–5. Denture stomatitis. Erosive erythematous lesions associated with the wearing of dentures (*arrows*). (Courtesy of R. Goepp.)

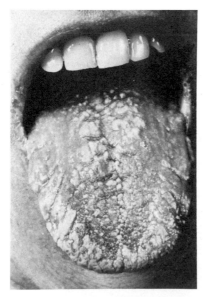

Figure 20–4. *Candida* stomatitis in the adult. The condition was associated with polyendocrine disturbance. Punctate, raised patches are seen involving most of the surface of the tongue.

missural folds in older patients often leads to a chronic chelitis, the perlèche syndrome (Fig. 20–6). *C. albicans* can often be isolated from the lesions, but it probably has little to do with the pathogenesis of the condition. In true candidal chelitis, there are scattered groups of satellite erosions over the lip, and *Candida* can be isolated from these in abundance.

Chronic oropharyngeal candidiasis has occurred as a complication of inhaling steroids as therapy for respiratory diseases. In a study by Milne and Crompton[157] up to 5.5 per cent of patients inhaling beclomethasone dipropionate had clinical disease and 41 per cent had positive cultures, as compared with 27 per cent positive cultures for patients not receiving inhaled steroids and only 0.7 per cent with clinically apparent disease. Sahay et al.[209] studied patients receiving aerosols of betamethasone valerate (800 μg per day). In their patient group, 10 per cent developed *Candida* infection, but it was not considered significant. Mouthwash prophylaxis was not beneficial.

Vaginitis and Balanitis. Diabetes, antibiotic therapy, oral contraceptives, and pregnancy may predispose to vaginal candidiasis. The disease is characterized by the presence of a thick yellow, milky discharge, and patches of gray-white pseudomembranes are seen on the vaginal mucosa. These curdlike patches are similar in appearance to those seen in thrush, esophageal candidiasis, and bronchial and intestinal disease. The lesions

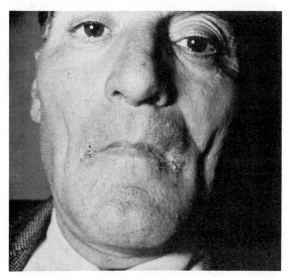

Figure 20–6. *Candida* chelitis — *perlèche.* Exaggeration of commissural folds, satellite lesions on lips, and chronic inflammation of commissures. Angular stomatitis.

ence symptoms after a warm bath or while in bed. The condition can be mimicked or may coexist with *Trichomonas* vaginitis. In the later disease, low level pruritus is constant and the curdlike patches are lacking.[58] Bacterial vaginitis such as *Corynebacterium vaginale* also mimic *Candida* disease.

Infection and reinfection occur by way of contamination from the digestive tract.[35, 244] *C. albicans* accounts for most cases of vaginitis, with some cases being caused by *C. glabrata, C. tropicalis, C. stellatoidea*, and other species.

Candida balanitis or balanoposthitis is an uncommon condition of males and was first described by Engman in 1920.[56] Often there is a history of vaginitis in the spouse, and the condition is probably a conjugal infection. There are superficial red erosions and thin-walled pustules over the glans and the sulcus corona (Fig. 20–8). Not infrequently, no yeasts are demonstrable in the lesions, suggesting an allergic rather than an infective disease. Some of these patients have an exaggerated response to the oidiomycin skin test. In more advanced disease, the white, curdlike appearance of thrush can be seen involving

vary from a slight eczematoid reaction with minimal erythema to a severe disease process with pustules, excoriations, and ulcers. The whole area is greatly inflamed, and pruritis is usually intense. Papular and rarely ulcerative lesions may occur, and the condition may extend to involve the perineum, the vulva, and the entire inguinal area (Fig. 20–7). *Candida* infections have long been associated with pregnancy. The rate is highest during the third trimester, when vaginal pH is lowest. In nonpregnant women the discomfort of vulvovaginitis may be particularly intense just prior to menstruation. Pruritus and pain in the introitus and labia minor can be aggravated by urination, sexual intercourse, or gynecological examination. Many patients experi-

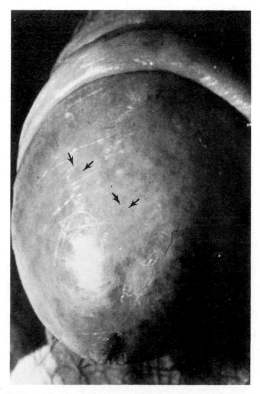

Figure 20–8. Balanitis. Superficial erosions with erythematous base and edges *(arrows)* and edema of the corona in the toxic, allergic stage. The fungus has not yet colonized the lesions.

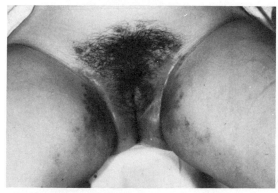

Figure 20–7. *Candida* vaginitis. Extension to involve perineum and inguinal area. The lesion is extremely pruritic and inflamed.

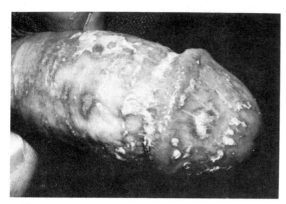

Figure 20–9. *Candida* balanitis. Superficial erosion and pustules over glans, sulcus, and shaft. (Courtesy of S. Lamberg.)

the glans, the sulcus, and sometimes the shaft (Fig. 20–9). In some individuals the disease spreads to the scrotum, the thighs, and the entire inguinal area. In severe cases lesions occur along the epithelium of the urethra. *C. albicans* is the most common agent of the infection, but a variety of other yeasts have also been recovered.[37]

The condition clears with cure of the vaginitis in the conjugal partner. *Candida* balanitis is also associated with diabetes and, usually, a tinea cruris–like syndrome called cutaneous candidiasis of the groin.

Bronchial and Pulmonary Candidiasis. Bronchial candidiasis is a chronic bronchitis with cough, production of sputum, and medium to coarse basilar rales and with linear fibrosis or peribronchial thickening seen on radiologic examination. The etiologic significance of *C. albicans* in this disease is difficult to determine. Many surveys have shown that the organism occurs, sometimes in considerable numbers, in essentially all chronic lung conditions.[170] These may be present in considerable numbers in direct preparations or in culture of sputum (Fig. 20–30). Bronchoscopy may not assess the extent or degree of candidal colonization of the bronchial tree. In advanced disease, small, white, curdlike patches similar to those observed in thrush or oral or vaginal infestations are seen. Ulcerative lesions may occur in advanced cases but are very rarely described.[186] Whether these colonizations are significant has yet to be determined. Many patients with bronchoscopically visible lesions go on for many years with no more symptoms than those of chronic obstructive lung disease. *Candida* may not be related at all to the disease process, or it may be a minor colonizer or an allergen. Recently,

more attention has been paid to the role of *Candida* in inciting allergic-type diseases. The second type of disease is a progressively fatal pulmonary one. It is a valid candidal infection, and the contribution of the yeast to the pathology is readily established.

Pulmonary candidiasis as a primary disease is extremely rare. The diagnosis of this disease must rest on unassailable evidence that rules out all other etiologies of the pathologic process.[44] *Candida* readily colonizes preexisting pathologic conditions attributable to other infectious agents, neoplasms, or chronic functional disorders. In these cases, the organisms may be present in numbers to 10^6 per ml and are easily visible on sputum smears. If their multiplication continues, the *Candida* may contribute to the pathology and cause severe distress and death. Often the patients have some underlying condition such as leukemia, lymphoma or severe respiratory compromise. Newborn infants[53] and children with cystic fibrosis have been reported to be more subject to primary pulmonary candidiasis than are normal sib-

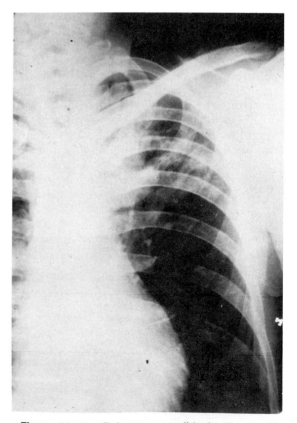

Figure 20–10. Pulmonary candidosis. Nonspecific infiltrate in left apex. There is density immediately above the left hilum. (Courtesy of J. Fennessey.)

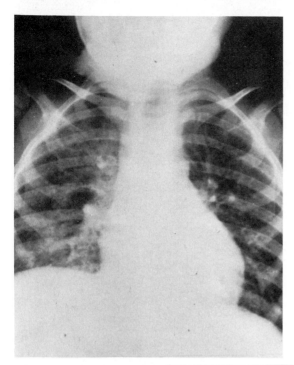

Figure 20–11. Pulmonary candidiasis in a child with leukemia. There is diffuse pulmonary infiltrate in many areas of the lung. The lung is underinflated. (Courtesy of J. Fennessey.)

lings.[113] Predisposing factors are prolonged antibiotic therapy, use of steroids, and use of intravenous fluids.

Primary pulmonary candidiasis presents with a cough, low-grade fever, night sweats, dyspnea, weight loss, and production of mucoid gelatinous sputum which is often blood-tinged. Roentgenographically there is hilar and peribronchial thickening, and lesions resembling miliary tuberculosis are seen (Fig. 20–10). In severe and extensive infections, the lesions are dense and smooth and often involve an entire lobe which may show complete consolidation (Fig. 20–11). Usually there are scattered patchy bronchopneumonic lesions, but in severe cases a lobar pneumonia develops. The lesions tend to be labile, and serial x-rays show clearing in some areas with development in others. Medium moist rales occur, but dullness and changes in breath sounds are detectable only in extensive disease. Cavitation of the type seen in bronchiectasis occurs. When two or more lobes are severely involved, death follows from respiratory insufficiency.

Pulmonary disease is frequently secondary to septicemia and dissemination of the organism from other loci. In such cases it is usually widespread and miliary (Fig. 20–22,*B*). These infections usually occur in the last stages of systemic disease, and the fungaemia usually heralds a rapidly fatal terminus. Such lesions in the lung are small microabscesses or necrotic areas containing yeasts and mycelium. On x-ray examination the chest shows "cotton-ball" nodular opacities or rarely, cavitation or a pneumonic-like presentation.[242]

Candida involvement of the larynx is extremely rare. On occasion laryngeal swabs will reveal *Candida* by culture in children with chronic sinusitis or tonsillitis. Other patients have had hoarseness attributed to *Candida*, and laryngitis as a terminal event in fatal disseminated disease[204] has been recorded.

Alimentary Candidiasis. Involvement of the esophagus and intestine occurs in several separate disease syndromes. Esophageal candidiasis is often an extension of lesions from the oral cavity, especially in thrush of the newborn. Chronic infection in older children is associated with genetic defects and polyendocrine deficiencies and will be discussed in a separate section, Chronic Mucocutaneous Candidiasis. In adults, *Candida* infection of the esophagus is associated with antibiotic therapy, corticosteroids, diabetes, and irradiation, as well as with various neoplasias, blood dyscrasias, endocrinopathies, and other debilitating conditions.[170] The significance of finding *Candida* in disease of the esophagus is often difficult to assess, especially in children. In adults, however, the diagnosis can be made by culture of material from esophagoscopy and the rather characteristic radiologic findings[83] (Fig. 20–12). Evidence of destruction of the mucous membrane, a ragged and irregular esophageal outline without loss of the longitudinal folds, is seen. The tract is distensible and has an identifiable pattern following barium swallow.[121, 122] In the absence of varices or history of other pathology of the esophagus, a diagnosis of candidiasis should be seriously entertained.

The clinical symptoms most often seen are dysphagia with retrosternal pain, gastrointestinal bleeding, nausea, and vomiting. Esophageal strictures are common.[177] These may remain following treatment with amphotericin B or nystatin and require surgical intervention and a gastrostomy feeding tube.[122] Other patients have been treated with esophageal dilation or have needed substernal colonic bypass because of perforations and many stenoses of the esophagus.[175]

Gastritis. *Candida* invasion of the stom-

ach wall is a very rare event and usually occurs during the final stages of disseminated disease. Secondary involvement of the stomach by *Candida* following surgery has been reported with increasing frequency, however. Katzenstein and Maksem found *Candida* superinfection common in surgical patients with gastric ulcers.[120] Though not the etiology of the disease, the fungus may contribute to the development and pathology. These patients had a high postoperative mortality. Strom et al.[234] also found a significant percentage of *Candida* infections as a postoperative septic complication. Radiologically there is a bezoar-like filling defect or multiple filling defects. Bezoar-like involvement of the jejunum also occurs after gastrectomy.[47] There is some interest in the possibility that use of cimetidine for ulcer patients raises gastric pH and encourages colonization by *Candida*.[166] Perforated duodenal ulcers have led to *Candida* peritonitis. These were not related to the use of cimetidine.[89]

Peritonitis. *Candida* peritonitis can result from perforated ulcers,[66, 181] surgery,[89] colonization of indwelling catheters for peritoneal dialysis,[182] or intraabdominal neoplasm that allows the fungus to traverse the intestinal wall.[128] Such patients usually respond to systemic antifungal agents[181, 182]; if untreated, the disease is fatal.[108]

Enteric Candidiasis. This is one of the most controversial clinical diseases attributed to *Candida*. The diagnosis is very difficult to establish, and the implication of *Candida* as the primary agent of disease is at best tenuous. The clinical condition that follows administration of tetracycline and its congeners is called candidal enteritis and occurs also when the yeast population is supposedly suppressed by the concomitant use of antifungal antibiotics. On the other hand, there have been many well-documented cases of fatal candidal invasion of the intestine following antibiotic therapy. The final diagnosis of this syndrome requires demonstration of candidal invasion of the intestinal mucosa or repeated isolation of the organism from ulcerative lesions. In biopsy specimens, colonization and invasion of the tissue is accompanied by the downward growth of mycelial elements. These mycelial elements are also seen in fecal smears and indicate invasion of the intestinal wall by *Candida*.[252] At autopsy numerous vegetations and a shaggy mucosal membrane are seen (Fig. 20–25). A quantitative increase in the population of *C. albicans* following antibiotic therapy or change in diet may also cause

a chronic "irritable colon" syndrome.[98] This is ascribed to hypersensitivity to the organism or its metabolic products rather than actual infection.

Perianal involvement is common in infants with oral thrush. The condition may exist with or without clinical disease of the intestine. The lesions are initially sharply defined, dull red patches which coalesce and spread with an irregular border. Vesicles form that later break to leave a ragged edge. Extension of the lesions to include the buttocks and inner thigh is common, and the genitalia, lower abdomen, and entire diaper area are sometimes involved. There is minimal pruritus, and in healthy children the condition resolves quickly following therapy. In adults the pruritus ani which may follow antibiotic therapy has been associated with *Candida* overgrowth (Fig. 20–13). The itching is severe, and the area presents as an intensely inflamed dermatitis. Various studies have implicated other organisms, and *Candida* colonization may be only incidental. Water-soluble toxins from *Candida* applied to the skin are able to elicit a similar clinical picture,[146] but the question of toxin in the pathogenesis of candidiasis is still unresolved.[25, 104, 111, 137, 190]

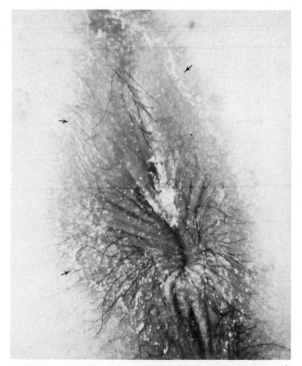

Figure 20–13. Pruritus ani from colonization of *Candida*. The patient was on broad spectrum antibiotics for one week as treatment for enteritis. Erosive erythematous lesions with ragged borders are indicated by arrows.

Chronic Mucocutaneous Candidiasis (CMCC). This is a general category describing a clinical manifestation that occurs in persons with various genetic defects. The patients are usually children and *Candida* is only one of several possible pathogens or opportunistic organisms that may establish disease.[42] The classification of these genetic diseases changes as new defects in leukocyte function or the endocrine systems are discovered.[9, 95] The first major group is associated with dysgenesis of the thymus and subsequent inability to elicit cellular immunity.[24] This category includes (a) dysplasia of the thymus with agammaglobulinemia (Swiss type of agammaglobulinemia); (b) dysplasia of the thymus without agammaglobulinemia (Nezelof-Allibone syndrome); and (c) congenital absence of the thymus and parathyroid glands (the third and fourth brachial pouch syndrome, DiGeorge syndrome). All the latter conditions are associated with an increased incidence of candidiasis. In contrast, when defects of humoral immunity alone are present (Bruton's hypogammaglobulinemia), there is no increased incidence of candidiasis.

The second group, which is associated with

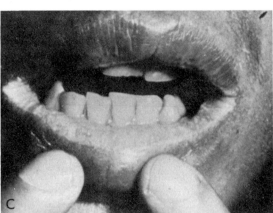

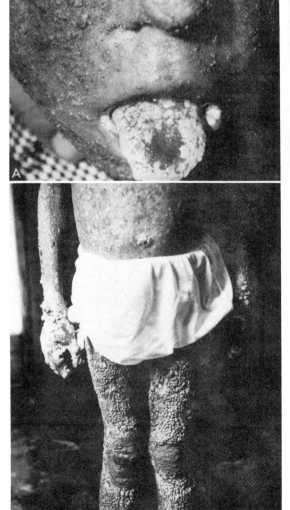

Figure 20–14. *A,* Chronic mucocutaneous candidosis involving face and tongue in a child with chronic granulomatous disease. (Courtesy of D. Windhorst.) *B,* Chronic generalized granulomatous candidosis in a child with an unknown type of leukocyte dysfunction. (Courtesy of S. Lamberg.) *C,* Chronic mucocutaneous candidiasis in juvenile polyendocrine disease. Erosive chelitis. (Courtesy of A. Padilha-Gonçalves.)

chronic mucocutaneous candidiasis, includes polyendocrine deficiencies, such as familial juvenile hypoparathyroidism (Fig. 20–14,*C*) and hypoadrenocorticism, and thymomas[101, 126] in adults. Patients with defects in leukocyte function (Fig. 20–14,*A*,*B*) are also prone to this type of infection, e.g., children with chronic granulomatous disease whose defective myeloperoxidase system precludes killing of *Candida* following phagocytosis.[9, 10] Idiopathic chronic mucocutaneous candidiasis is an additional category in which the underlying defect is as yet undefined.[131] Pathologically and clinically, it resembles the infection seen in patients with polyendocrine defects.

An autosomal recessive genetic defect is ascribed to patients with acrodermatitis enteropathica. These patients, usually children, have the triad of dermatitis, diarrhea, and alopecia. In addition they frequently have chronic mucocutaneous candidiasis. The disease has recently been found to be a disorder of zinc metabolism[162] and responds to zinc

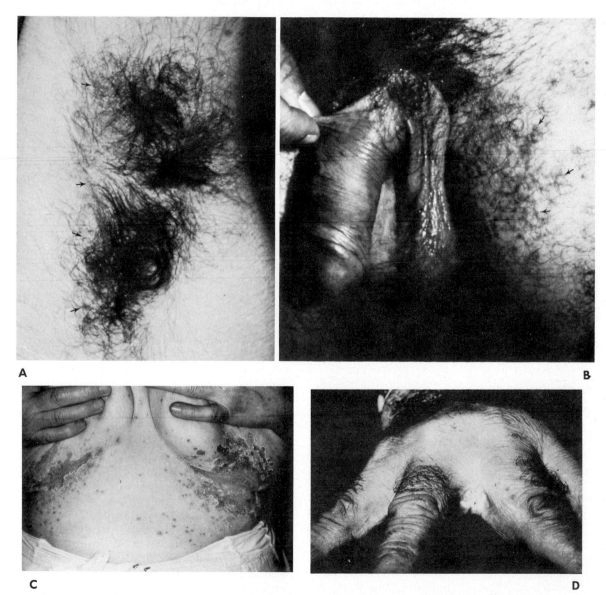

Figure 20–15. Cutaneous candidiasis of intertriginous areas. The lesions have an erythematous base and scalloped border (arrows), often with punctate satellite areas. *A,* Axilla. *B,* Groin. Note the involvement of scrotum and shaft along with satellite lesions. This usually differentiates candidiasis from dermatophytosis. (Courtesy of A. Padilha-Gonçalves.) *C,* Intertriginous candidiasis in a diabetic. Note "scalded skin" areas and satellite eruptions. *D,* Interdigital erosion in a bartender.

therapy.[34] With treatment of the underlying disease, the *Candida* infection resolves spontaneously. The role of iron deficiency, if any, in cases of chronic or recurrent candidiasis has not been established[9, 57] but has been speculated on by several authors.

Two final categories involve patients who are unresponsive solely to the *Candida* antigen,[40, 95, 210] or whose sera contain specific IgG which inhibits the clumping effect of normal serum for *Candida*.[9, 41] Skin test reactivity and the ability to produce leukocyte MIF return after treatment with transfer factor.[101] This is a transitory phenomenon, however. In a study by Sams et al.,[210] CMCC was studied in members of three generations in one family. The age of onset of the condition can be delayed considerably, as demonstrated by the 51-year-old patient of Aronson et al.[10]

Candida granuloma is the final category of disease. This entity is distinguishable clinically and pathologically from the mucocutaneous disease, although they both may exist in the same patient. The granulomata involve the skin or its appendages and are discussed in the following section (Fig. 20–14,*B* and Fig. 20–18).

Cutaneous Involvement

Intertriginous Candidiasis. Cutaneous candidiasis involves the intriginous areas of the glabrous skin directly or may occur as a colonization secondary to preexisting lesions on any part of the body caused by varying etiologies. Intertrigo most commonly is seen in the axillae (Fig. 20–15,*A*), groin (Fig. 20–15,*B*), intermammary folds (Fig. 20–15,*C*), intergluteal folds, interdigital spaces (Fig. 20–15,*D*) glans penis, and umbilicus. The lesions are quite characteristic and well-defined as weeping, "scalded skin" areas with an erythematous base and a scalloped border (Fig. 20–15,*C*). The lesion is surrounded by "satellite" eruptions, which develop as discrete vesicles, pustules, or bullae that break to leave a raw surface with eroded ragged edges. These develop and emulate the initial lesions. Clinical variants do occur, and the lesions may appear dry and scaly or, in some cases, papulopustular.[5] Candidiasis of the skin is associated with two types of patients. The first of these have metabolic disorders which predispose them to candidal colonization, such as diabetes,[5] obesity, or the sequelae of chronic alcoholism. In the second group, the skin is predisposed to infection by various environmental conditions, including moisture, occlusion, and maceration of the skin under dressings, caused by the wearing of boots and tight clothing in tropical climates and by frequent and continual immersion in water.[190] The latter is common in such occupations as housewife, dishwasher, barmaid, fruit canner, and so forth (Fig. 20–15,*D*).[192] Colonization of the skin under occlusive conditions without invasion of the epidermis produces a contact dermatitis of the primary irritant type.[146] Congenital anatomic defects and poor peripheral circulation also predispose to this type of disease.

Generalized cutaneous candidiasis with widely disseminated lesions occurs in diabetics[5] and in those with a wide variety of ectodermal defects. In the latter group, *Candida* colonizes damaged tissue but contributes little to the total pathology seen.[223] Another form of candidiasis results in the development of purulent follicular papulopustules in intertriginous areas. The pustules have sodden, annular fringes, and folliculitis can occur in the absence of frank intertrigo. The lesions may be potentiated by the use of zinc oxide. Widespread cutaneous lesions are sometimes seen in severely debilitated patients. These are often ulcerative, necrotic, or exfoliative. Usually they constitute a terminal event in disseminated disease, but some respond to therapy.[60]

Cutaneous *Candida* colonization is a frequent complication of patients with extensive body surface burns.[69] The colonization often leads to candidaemia (diagnosed by positive blood culture), with focal infections developing in many organ systems. Such infections often lead to fatal systemic candidiasis if initiation of therapy is delayed. In the series by Gauto et al.,[69] about 70 per cent responded to treatment with amphotericin B. All patients had had at least one episode of infection other than that of *Candida* and had received antibacterial antibiotics. In addition, most had long periods of urinary tract catheterization.

Paronychia and Onychomycosis. These are the commonest forms of cutaneous candidiasis. The paronychial folds are readily colonized, particularly in people whose occupations require frequent immersion of appendages in water.[195] The lesions are characterized by the development of painful, reddened swellings extending as far as one centimeter from the paronychial edge. They resemble pyogenic lesions, especially those

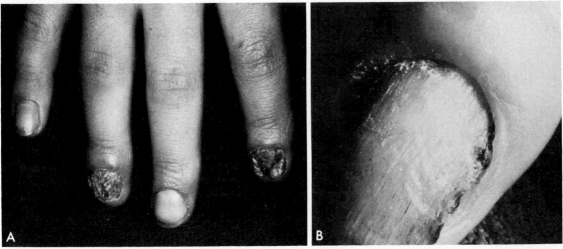

Figure 20–16. *A,* Chronic paronychia and onychomycosis in an adult diabetic. The same patient had chronic stomatitis (see Figure 20–3). (Courtesy of R. Goepp.) *B,* Chronic disease in patient without known underlying disease. The swollen paronychial area is evident and the dark, discolored nail plate is striated and grooved.

caused by staphylococci, in the same area. A mixture of bacteria and *Candida* is often present. In chronic paronychia the nail becomes invaded. The resulting onychomycosis appears as a hardened, thickened, brownish-discolored nail plate that is striated or grooved[174] (Fig. 20–16). The nail does not become friable, as it does in tinea unguium of dermatophyte etiology. The nail tissue is destroyed in chronic untreated cases. Similar nail disease can be produced by species of *Geotrichum* and *Trichosporon*[197] and by *Scopulariopsis brevicaulis* and other fungi.[174]

Diaper Rash. This is not an uncommon sequela of oral and perianal candidiasis of the newborn. It also occurs in infants under unhygienic conditions of chronic dampness and irregularly changed, unclean diapers. The initial colonization evokes a primary irritant dermatitis (Fig. 20–17). Invasion of the epidermis by the fungus may ensue, and the condition becomes severe, often spreading to the axillae, face, conjunctiva, and other areas.[137]

Candidal Granuloma. This very rare condition of children was described in detail by Hauser and Rothman, who reviewed 13 cases from the literature.[90] The lesions are quite distinct from those of other forms of cutaneous and mucocutaneous candidiasis. They are described as primary vascularized papules covered with a thick, adherent, yellow-brown crust. These may develop into horns or protrusions up to 2 cm in length (Fig. 20–18). On histologic examina-

tion, the lesions appear as poorly organized granulomatous tissue, with giant cells and a chronic inflammatory reaction (Fig. 20–19). The face is most commonly involved, but lesions are also found on the scalp, fingernails, trunk, legs, and pharynx. Defects of the immune system which may predispose to this disease are delayed cutaneous anergy of all types and a lymphopenia.[110] Specific immunologic unresponsiveness to *Candida* antigens has also been postulated.[164] The patients die of the underlying disease in time, often with infection as a terminal event.

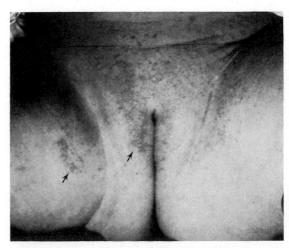

Figure 20–17. Diaper rash. Primary irritant dermatitis with a few colonized lesions. (Courtesy of A. Padilha-Gonçalves.)

Figure 20–18. Candidal granuloma. Many horns and protrusions are seen over the face of the child.

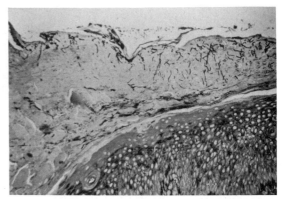

Figure 20–19. Candidal granuloma. Biopsy of hyperkeratic area. The fungus is seen to grow in the stratum corneum but does not invade the epidermis. Gomori-Grocott chromic acid methenamine silver stain (GMS). ×100. (Courtesy of H. Sommers.)

Systemic Involvement

This is a relatively rare condition except as the terminal event of a debilitating illness. In some cases, it is the result of chronic insult and continued seeding of yeasts into the body because of repeated injections by drug addicts, indwelling catheters, or long-term antibiotic or steroid therapy. The prognosis in these conditions is poor, and the death rate high. The conditions predisposing to severe candidal infections and aspects of therapy have been reviewed by Edwards et al.[55]

Urinary Tract. Clinical involvement of the urinary tract is reported in association with disseminated candidiasis, diabetes, pregnancy, administration of antibiotics, and use of unclean catheters. The condition is more common in women than men by a ratio of four to one.[74] Though bacterial cystitis is not uncommon, candidal invasion of the bladder is infrequent. In the normal patient, the condition clears readily under treatment.

The diagnosis is difficult to make, as *Candida* is frequently cultured from the urine. Counts in excess of 1000 colonies per milliliter are considered indicative of an active urinary tract infection.[74] In such cases the numbers can reach upwards of 24,000 or more per milliliter. On cystoscopy, candidal plaques can be seen on the mucosa of the bladder. These resemble the patches observed in thrush or vaginitis. In some patients, urinary tract "fungus balls" composed of *Candida* hypha and yeasts have been found.[62] These generally occur in the upper collection areas. They respond to irrigation with 5 FC or amphotericin B. The appearance of resistant organisms during treatment with miconazole is reported.[97] Pyelonephritis involving *Candida* occurs in adults, children, and the newborn as an ascending infection. Clumps of *Candida* may mechanically occlude the collecting tubules leading to obstruction and anuria.[219] In heavy urinary tract infestations, particularly when a pyelonephitis is present, serial blood cultures will be positive. In the absence of indwelling catheters, hyperalimentation, or other barrier breaks, recovery of *Candida* in blood cultures is frequently associated with urinary tract infection, less so with oropharyngeal or gastrointestinal involvement, and rarely with *Candida* endocarditis. Sometimes cutaneous lesions, infected surgical scars, and wounds can also lead to positive blood cultures. As a terminal event, a candidaemia may arise from a variety of foci.

Involvement of the kidney usually is the result of generalized dissemination from another focus of infection. Lesions are seen in the medulla and cortex or in the medulla alone. The latter is the result of an ascending infection. This is in contrast with the disease in experimental animals, which almost always involves the cortex only. In humans when a sudden rapid seeding occurs, numerous lesions of the cortex are seen (Fig. 20–20). Closely examined, these lesions contain more yeastlike than mycelial elements, indicating recent involvement. Invasion of the renal parenchyma[204] in candidal pyelonephritis is reflected by altered renal function.

Endocarditis. This is a rather special form of candidiasis, as the etiologic agents are usually species of *Candida* other than *C. albicans*. The clinical symptoms are similar to those of bacterial endocarditis and include fever, murmur, congestive heart failure, anemia, and splenomegaly. Large vegetations are seen on the valves, and there is a high incidence of embolization to the large arteries, both findings being uncommon in bacterial endocarditis. Echocardiography is an important noninvasive diagnostic procedure,[39] as is serology. Most of the patients with mycotic endocarditis are not immunosuppressed, so that serology by immunodiffusion (ID) using *Candida* antigen is positive. Many other fungi also cause mycotic endocarditis, particularly the cases associated with valve replacement or open heart surgery. A battery or fungal antigens in the ID test may help to sort out the true etiology.

Three groups of patients are susceptible to *Candida* endocarditis. The first includes those with preexisting valvular disease who also have had treatment with antibacterial antibiotics and an opportunity for entry of normal flora of the skin or intestinal tract into the blood stream. The latter is usually a result of indwelling catheters or prolonged intravenous infusions which predispose to candidemia and subsequent colonization of the valves by skin-inhabiting yeasts. A second group of patients includes drug addicts. Contaminated lots of heroin or needles have been implicated in some cases of candidal endocarditis, and repeated punctures of the uncleansed skin introduces yeasts of the normal flora (Fig. 20–21). The third group falls into a newly recognized category — endocarditis following heart surgery.[29, 117] In one hospital survey, fully 15 per cent of patients receiving a valve prosthesis developed *Candida* endocarditis. In many cases there was a long period between surgery and the onset of symptoms. Often the first and frequently the most serious complication is septic emboli to the central nervous system leading to cerebral abscess (Figs. 20–23 and 20–24).

Patients without valvular disease are sometimes infected by contaminated blood from oxygenators or by repeated cardiac catheterizations. Some of these infections were cleared with amphotericin B therapy and the patient received another replacement valve. Most, however, had a fatal outcome.[29, 241] This is true of other forms of candidal endocarditis, but despite therapy the immune state of the patient determines the outcome of the disease to a large extent.[184, 250] *C. parapsilosis* and *C. guilliermondii* are more common than *C. albicans* as etiologic agents of this syndrome. *C. stellatoidea* endocarditis has also

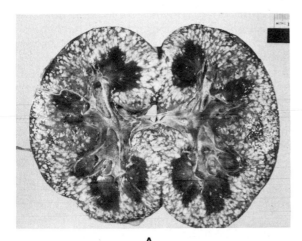

A

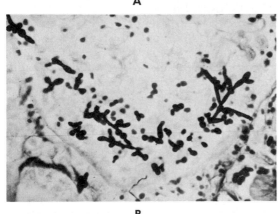

B

Figure 20–20. *A*, Renal candidiasis. Cut section of kidney. Miliary solitary and confluent punctate abscesses throughout the cortex and, to a lesser degree, the pyramides. The medulla in general is spared. The patient had acute leukemia. The massive involvement of the cortex indicates a recent, rapid seeding from another focus of infection. In the histologic section, most of the fungus was in the yeast form of growth, again indicating a recent infection. *B, Candida* in the glomerulus from hematogenous dissemination. Note mixture of yeast and mycelial forms. GMS. (Courtesy of R. Diamond.)

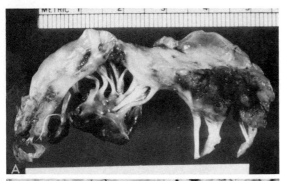

Figure 20–21. *A, Candida* endocarditis. Mitral valve. Acute and chronic endocarditis with prominent vegetations in a long-term drug addict. *B,* Histologic section of vegetation showing mycelial and yeast forms of *C. parapsilosis.* GMS. ×400. (Courtesy of S. Thomsen.)

been recorded, as has *Trichosporon beigelii,*[241] *Aspergillus,*[39] and a variety of other fungi.

Meningitis. Candidiasis of the central nervous system is relatively rare. About 100 cases of this form of the disease have been reported. *Candida* organisms appear to reach the brain by dissemination from foci in the gastrointestinal tract and respiratory system or as septic emboli from infected heart valves or they are introduced during intravenous therapy.[59] Most of the patients with this condition have had an underlying disease or were being treated with antibiotics, cytotoxins, or corticosteroids.[179] A few have had no detectable abnormality. The clinical symptoms are those of meningitis, including pain, nuchal rigidity, positive Kernig's and Brudzinski's signs, along with focal neurologic signs, such as aphasia and hemiparesis. Papilledema and increased intracranial pressure are rarely observed, but diplopia, tinnitis, vertigo, stupor, and coma are often present. There appears to be a lack of correlation between the clinical and laboratory signs in candidal meningitis. The spinal fluid is clear, with a low grade pleocytosis composed of mononuclear or polymorphonuclear leukocytes. The protein is generally very high, and the sugar low or normal. These signs may be present when there are no clinical symptoms of meningitis, and the spinal fluid may be normal when clinical symptoms are present. CNS lesions vary from large, solitary abscesses or widespread microabscesses to granuloma-like lesions.[203] Invasion of the blood vessel walls initiates thrombosis and secondary infarction. In rare cases, *Candida* meningitis has as fulminant a course as bacterial meningitis. Usually, however, it is more indolent. *C. albicans, C. guilliermondii C. tropicalis* and *C. viswanathii*[211, 212] have been isolated from this disease, although most cases are caused by *C. albicans.*

Septicemia. *Candida* septicemia is an increasingly encountered infection, particularly as a terminal event to an underlying disease. Most patients reported have been on antibiotic therapy, and about half were receiving corticosteroids. Approximately three-quarters of the patients had leukemia or other neoplasias. Where no barrier breaks or indwelling catheters have been involved, these cases are thought to be due to the lowering of cellular defenses that leads to translocation of *Candida* from the gastrointestinal or respiratory tract. *C. albicans* and *C. tropicalis* are most often involved.[252] The organisms are then found in microabscesses in all organ systems. As noted in the section on urinary tract infections, *Candida* fungemias can also occur in cases of heavy infestations of this organ system and also rarely with endocarditis. In the former, the *Candida* is usually cleared from the blood stream and does not induce focal microabscesses, with the exception of some cases of endophthalmitis. In the latter, septic emboli may be released, with the most serious complication being *Candida* cerebritis.

A transient fungemia may occur following indwelling catheterization, continuous intravenous infusions, or surgery, or it may be associated with Foley catheters, or hemodialysis, or other abrogations to natural barriers.[127] In most patients, the organisms are rapidly cleared, but in debilitated persons a septicemia may be followed by a systemic disease. In normal volunteers positive blood and urine cultures and transient toxemia were observed after oral ingestion of large numbers of *C. albicans*, but the yeasts were rapidly cleared. This demonstrates that overgrowth with large numbers of yeasts can result in some *Candida* passing through the intestinal wall.[130] A positive blood culture with *Candida* should always be considered of serious concern. Klein[127] noted a spontaneous resolution rate of 42 per cent, but two cases of endophthalmitis occurred after apparent resolution. In all, 36 per cent of patients required antifungal treatment because of serious disease.

The clinical signs of septicemia include fever, chills, and impaired renal function. The mortality rate with systemic colonization is about 56 per cent, even with treatment. *C. albicans* is the species involved in most cases, with *C. tropicalis* being recovered about 25 per cent of the time. *C. parapsilosis* and *C. glabrata* account for about 10 per cent of infections; other *Candida* species are infrequently encountered.

Iatrogenic Candidemia. A new category of *Candida* septicemia has only recently been recognized. This is fungemia as a complication of parenteral hyperalimentation. In one series, 6 of 15 infants and children developed *Candida* septicemia while on this regimen.[26] Meningitis,[154] endophthalmitis,[61] osteomyelitis[169] or other focal infestations may occur in such children or in adults on similar regimens.

In another study covering an eight-month period, 33 adults with fungal septicemia were seen.[49] In 22 (67 per cent) of these there was a history of receiving hyperalimentation for severe gastrointestinal disease. In no instance was there an association with steroid therapy or immunologic deficiency. If patients were on hyperalimentation for 20 days or more, fully 55 per cent developed candidemia. Ashcraft and Leape[13] advise constant monitoring by blood cultures of patients receiving this procedure. In immunologically competent patients, removal of the catheter alone or in conjunction with low-dosage amphotericin B therapy usually resulted in clearing of the infection. In children the drug can be given in as low a dose as 11 mg over an 11-day period.

Other Organ Systems. Dissemination to involve other organ systems is an infrequent occurrence. As noted before, it is most frequently associated with fungemia prior to death in a severely debilitated patient. Transient fungemias usually resolve spontaneously or with treatment; however, as a result of transient fungemias, isolated foci of *Candida* infection may involve a variety of body sites. The two most frequently occurring clinical forms of hematogenously disseminated candidiasis are multiple subcutaneous and cutaneous lesions and endophthalmitis. Disseminated cutaneous candidiasis usually occurs in the severely immunosuppressed patient, the leukemia or lymphoma patient in blast crisis, or the organ transplant patient during severe rejection syndrome. The lesions on the skin take a variety of forms. They are usually deep abscesses or pustular or bullous lesions located in the papillary dermis. On biopsy the nidus of the *Candida* organism is quite evident. *Candida albicans* as well as many other *Candida* species has been involved, as have *Trichosporon beigelii* and *Geotrichum* and *Rhodotorula* species.

Endophthalmitis is a serious and frequent sequela of candidaemia.[127] It is often associated with surgery, indwelling catheters,[226] or drug abuse,[1] and the clinical signs may not be evident until long after the candidaemia has cleared. These lesions can be treated with 5 FC and amphotericin B.

Septic thrombosis has occurred in patients with severe burns, at the site of indwelling catheters in leukemia and lymphoma patients, in patients with endocarditis, and in children with chronic granulomatous disease. These infections often respond to intensive amphotericin B therapy. Fungus balls in the biliary tract as a sequela of candidaemia have been reported by Magnussen[147] and by Marcucci.[148] The balls were surgically removed and the patients treated with amphotericin B. Svirsky-Fein[236] reported a case of osteomyelitis in two children and reviewed the pertinent literature. Reiser[196] reviewed the radiological aspects of *Candida* bone infections. Arthritis and osteomyelitis have also been recorded by Noble and Lynn[169] and Ide,[109] and therapy of such infections has been outlined by Firkin.[61] In the patient described by Ide there was a documented transient candidaemia prior to the development of the arthritis.

Allergic Diseases Involving Candida

Allergy to the metabolites of *Candida* is a well-established phenomenon, and the clinical condition is called candidids. This reaction is similar to the dermatophytids of ringworm infection in both appearance and genesis. Less well documented is the role of *Candida* in other syndromes involving hypersensitivity. These include eczemas, asthma, gastritis, and a condition of the eye similar to the so-called "histoplasma uveitis." A *Candida* gastritis syndrome has been the subject of many conflicting clinical and experimental reports.[170, 253] Extracellular products of *Candida* can induce both delayed and immediate types of hypersensitivity. Most adults without clinical symptoms of infection or allergy react to the intradermal injection of a culture filtrate of *C. albicans* termed "Oidiomycin" (Hollister-Stier). This indicates that antigenic material from the organism or the organism itself can pass through the mucosal wall in sufficient quantity to evoke an immunologic response. It is probable, therefore, that colonization of the bronchi or previously damaged cutaneous tissue by *Candida* or overgrowth of the organism in its normal habitat would be accompanied by manifestation of toxicity and allergy.

Candidids. The lesions of candidids are similar in clinical appearance, morphology, and distribution to those of dermatophytids. They are sterile, grouped, vesicular lesions (pompholyx type) which may occur in the interdigital spaces of the hands or on any part of the body. As with dermatophytids, they disappear following resolution of a *Candida* infection or after desensitization.

Eczema, Asthma, and Gastritis. Cutaneous allergy in the form of urticaria and eczema[221, 222] has been described following candidiasis. A skin test may be used to diagnose this condition; however, it must be appreciated that a high percentage of the normal population will give a positive reaction to the test. True allergy is usually indicated if there is an immediate reaction following intradermal injection of the test antigen. In such cases, resolution of the "allergic" diseases has followed desensitization of the patient by the intradermal or subcutaneous injection of extracts from *C. albicans.*[221]

Allergy to *Candida* in the form of urethritis, gastritis, balanitis, rhinosinusitis, and headache has also been described.[222] Hypersensitivity resulting from the overgrowth of *Candida* has been implicated in a condition termed "irritable colon syndrome" and also in skin lesions of *erythema annulare centrifugum* type.[98, 223] Much more study and clinical evaluation are necessary to substantiate the role of *Candida* in these various allergic conditions. The role of *Candida* in allergic uveitis is discussed under mycotic diseases of the eye.

DIFFERENTIAL DIAGNOSIS

In all cases of suspected candidiasis, complete cultural confirmation is necessary. A negative mycologic examination is significant, but a positive one is not unassailable proof of candidal involvement. Consideration of the numbers and morphology of the organisms present, as well as exclusion of other etiologies, are necessary before *Candida* is implicated as the inciting agent of any pathologic process. Isolation of *Candida* from the blood is usually significant and warrants attention. Thrush in the newborn is pathognomonic in its appearance and presents no diagnostic problem. Diagnosis depends on demonstration of organisms in direct smears, however, In other patients, leukoplakia, lichen planus, tertiary syphilis, and other lesions resemble cutaneous candidiasis. Vaginal candidiasis is similar to trichomonas vaginitis and requires laboratory studies for differentiation. Neither are as purulent or severe as acute gonococcal disease. The differential diagnosis of systemic infection must include other mycoses, tuberculosis, neoplasms, or chronic bacterial infections. Since *Candida* may colonize any preexisting cutaneous, mucocutaneous, or respiratory condition, it is very difficult to assess the contribution, if any, of isolated yeasts to the observed pathology.

PROGNOSIS AND THERAPY

As candidiasis is primarily an opportunistic infection, prognosis depends almost entirely on the type and severity of the predisposing conditions or diseases. Oral thrush in the newborn healthy child may clear uneventfully, but other forms of *Candida* infection are much more difficult to treat and usually do not clear spontaneously. Control of cutaneous candidiasis in the diabetic depends on proper hygiene and regulation of the diabetes. In candidiasis associated with macerating conditions, prolonged exposure to moisture, and so forth, elimination of these factors will cause resolution of the disease even without treatment. Chronic disease in the constitutionally inadequate patient can be controlled with therapy, but the condition will return with cessation of therapy.[101] In advanced systemic diseases, candidiasis is usually a terminal event which contributes to the ultimate demise of the patient.

Oral, Mucosal, and Cutaneous Candidiasis. Thrush of the newborn is almost always associated with passage through a birth canal contaminated with *Candida*. If the vagina is treated with clotrimazole six days prior to birth, the infection rate is markedly reduced. If clinical thrush does develop, it is often seen

Figure 20–22. *A*, P
tion of lung. There are
in some areas by a nar
is indented at this si
Pulmonary candidiasi
disease resulting from
tion from some othe
microabscesses (arrc
vessels and contain
neutrophils, and *Canc*
stain. ×400. (Courtes

reports of involve
been tabulated.[170]

Histopathology

Thrush. In or
branous lesion is
necrotic material
food material alo
hyphae, hyphae
epithelial cells. T
stratum corneun
reaction is that o
tis, with edema a
the subepiderma
found in other
(Figs. 20–28 and
stomatitis. In th
toxic reaction ar
organisms are se

to resolve spontaneously, but it should be treated because of the possibility for extensive overgrowth, interference with feeding, and sometimes aspiration pneumonia.

Treatment of thrush and cutaneous candidiasis was established many years ago and remains relatively unchanged. When compared with other modalities, 1 per cent crystal violet is as effective or better than such drugs as 1 per cent nystatin.[178] Treatment with crystal violet may cause necrosis of the mucosa, and it is recommended that application be limited to twice daily for three days.[115] Older topical agents, such as sodium caprylate and sodium propionate, are still as effective as Trichomycin or other new drugs. Nystatin can be applied to resistant mucosal lesions as a suspension containing 200,000 units per ml. This is applied every two to three hours for several days until clinical improvement is attained. For denture stomatitis, a 14-day course is recommended.[21] As a viscous suspension, nystatin has been successful in the treatment of chronic esophageal candidiasis,[83, 118] as has amphotericin B suspensions.[122] Some cases of chronic esophageal candidiasis are resistant to both drugs as suspensions. Such patients usually respond to low doses of IV amphotericin B. Nystatin is used in cutaneous, mucocutaneous, intertrigo, crural, pruritus ani, and axillary infections as an ointment or as suppositories. As tablets of 500,000 units administered three to four times daily, it is effective in gastrointestinal infection. In combination with steroids in ointment or cream bases, nystatin is effective in treating cutaneous lesions, and paronychias. It has been used without much success, as an aerosol in pulmonary candidiasis.[11] Natural or laboratory-induced resistance to nystatin apparently is rare.[28]

Oropharyngeal candidiasis in the debilitated or immunocompromised patient often does not respond well to nystatin or other older modalities. Yap and Bodey[255] report a clinical cure rate of 96 per cent using clotrimazole as a troche in patients with *C. albicans* or *C. tropicalis* infection. Other topical agents include hibitane (0.2 to 2 per cent) in the treatment of denture stomatitis[32] and ketoconazole in a variety of lesions. Ketoconazole has been found inferior to clotrimazole, however.[65] In other studies recalcitrant lesions in chronic mucocutaneous candidiasis have dramatically improved on 200 mg per day oral ketoconazole.

Vulvovaginitis and Balanitis. There are many preparations available for the treatment of vulvovaginitis, and they come in a variety of vehicles, such as creams, lotions, pessaries, and foaming pessaries. The form or vehicle appears to make little difference as far as efficacy is concerned and is a matter of personal preference. The two major groups of pharmaceuticals now in vogue are the imidazoles and the polyenes. Four polyenes—pimaricin, candicidin, amphotericin B, and nystatin — have been used topically, nystatin being the most popular. Most recent studies, however, indicate the imidazoles give higher cure rates than the polyenes. Clotrimazole, miconazole, ketoconazole, and econazole have all had clinical trials. At present, clotrimazole topically for two weeks appears to be the favored modality.[51] Resistant infections[103] and reinfections are common. Lactobacilli have been suggested in recalcitrant disease.[213] Balanitis usually resolves with cure of the infection in the sexual partner. Most cases are toxic or allergic eruptions and benefit from steroids. Extensive disease in which colonization by the fungus has occurred is treated with the same medications used in vulvovaginitis.

Systemic Disease. Amphotericin B (2 or 3 per cent) is sometimes used as a topical agent, though this drug is generally more irritating than nystatin. However, it is the only available effective agent for the treatment of systemic candidiasis. The dosage schedule is that recommended for other systemic mycoses (0.6 mg per kg body weight, total course: 1 to 3 g). Patients should be started on a low dosage of 5.0 mg given on alternate days and increased as can be tolerated. Clinical cures in all forms of systemic disease, endocarditis,[184, 250] meningitis,[52] peritonitis,[181] *Candida* granuloma, and chronic mucocutaneous candidiasis have been recorded. In the latter diseases relapse is the rule. In *Candida* septicemia resulting from indwelling catheters and hyperalimentation, removal of the needle alone is sometimes sufficient to clear the infection. In immunologically competent patients, low doses of amphotericin B may also be administered. Medoff et al.[153a] found that 10 to 355 mg given over a period of 4 to 18 days produced satisfactory blood levels of the drug. No side effects were noted. The patients had both mucocutaneous and systemic disease. The latter was the result of an indwelling catheter.

Candida albicans has a sensitivity to 5-fluorocytosine that ranges from 0.23 to 3.9 μg per ml, although some other species of *Candida* are resistant to 1000 μg per ml. (See

Chapter 29). Th
cessfully in treat
candidiasis.[182] Th
from 8 g daily (o
per kg per day
reports, treatme
as well as bone n
death.[193] Many st
tant to the drug
resistance. It ap
this drug in sy
ed.[138]

At the present
receiving clinica
systemic candid
has exhibited t
didiasis in canc
sponse), in vari
break infection:
(92 per cent res
cutaneous candi
The drug is usu
to 1200 mg p
depending on
appears to be no
combination wi
lapses are comn
courses may be
tive effects of
zole are often n
grids. Efficacy
still being evalu

Katz and Cas
had acute gran
oped systemic
with 400 mg mi
1.2 g per day) f
and no side eff
are investigatin
following initia
conazole. Keto
are all being u

PATHOLOGY

There has a
concerning the
mycelial and t
cans. It was fel
cell to form m
escape macrop
essary for inv
(mycelial) pha
the pathogeni
(yeast) stage tl
contrast to the
phic pathoger

Colony Morphology. SGA 25° C, three days — creamy white, smooth; one month — white to cream colored, dull, soft, smooth, reticulated or wrinkled, often with overgrowth of mycelium. Old stocks become hairy and tough.

Cell Morphology. 1. SGB 25° C, three days — globose-ovoid or short-ovoid cells (4 to 8 × 5 to 11 μ). A thin film forms over the broth that may have entrapped gas bubbles. 2. CM or N PTB 25° C, three days—abundant pseudomycelium composed of elongate cells with much branching, blastoconidia singly, along mycelium, or in clusters. True mycelium formed (Fig. 20–39). Some strains of *C. tropicalis* produce chlamydoconidia, especially on initial isolation. These differ somewhat from those of *C. albicans* in that they do not usually have a supporting cell. Production of these conidia ceases on subculture, whereas it is a constant feature of *C. albicans*. *C. tropicalis* also may appear to produce germ tubes in the RB test. The germ tube shows a narrowing or constriction at the emergence from the yeast cell. Such constrictions are not seen in germ tubes of *C. albicans*.

Fermentation. Glucose +, galactose +/w, sucrose +, maltose +, cellobiose −, trehalose +/w, lactose −, melibiose −, raffinose − melezitose +/−, inulin −.

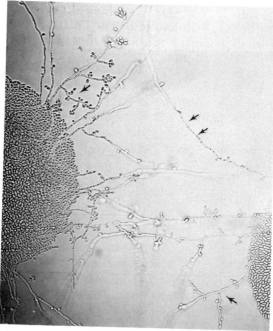

Figure 20–39. *Candida tropicalis.* Short, branched pseudohypha *(arrow)*, with single blastoconidia arising from edge of colony *(single arrows)*. Long, meandering true hypha is also seen *(double arrow)*.

Assimilation. Glucose +, galactose +, D-ribose −, L-rhamnose −, L-sorbose +/−, sucrose +, maltose +, cellobiose +/−/w, trehalose +, lactose −, melibiose −, raffinose −, inulin −, starch +, D-xylose +, L-arabinose +/w/−, D-arabinose −, ethanol +, glycerol +/w/−, D-mannitol −, ribitol +, salicin +/−/w, inositol −, melezitose +, potassium nitrate −.

This organism appears to be closely related to *C. albicans,* and there are reports of conversion between the two. It has been isolated from feces, shrimp, kefir and soil.

Candida pseudotropicalis (Castellani) Basgal 1931

Synonymy. *Endomyces pseudotropicalis* Castellani 1911; *Monilia pseudotropicalis* Castellani et Chalmers 1913; *Atelosaccharomyces pseudotropicalis* de Mello, Gonzagi, et Fernandez 1918; *Myzeblastanon pseudotropicalis* Ota 1928; *Mycocandida pseudotropicalis* Ciferri et Redaelli 1935; *Castellania pseudotropicalis* Dodge 1935; *Mycotorula pseudotropicalis* Redaelli et Ciferri 1947; *Torula cremoris* Hammer et Cordes 1920; *Candida mortifera* Redaelli 1925; *Mycocandida mortifera* Langeron et Talice 1932; *Monilia mortifera* Martin, Jones, Yao, et Lee 1937; *Blastodendrion procerum* Zach 1934.

Colony Morphology. SGA 25° C, three days — creamy smooth; one month — cream to yellowish, somewhat dull, soft, smooth or slightly reticulated.

Cell Morphology. SGB 25° C, three days — short-ovoid with a few elongate cells (2.5 to 5 × 5 to 10 μ). Pseudomycelium is abundant in most strains; in rare ioslants none is formed. The cells are very elongate; they fall apart and lie parallel like "logs in a stream." Blastoconidia not abundant (Fig. 20–40). When present, they are elongated and in verticils.

Fermentation. Glucose +, galactose +, sucrose +, maltose −, cellobiose −, trehalose −, lactose +, melibiose −, raffinose +/w, inulin +, melizitose −.

Assimilation. Glucose +, galactose +, L-sorbose −, sucrose +, sorbose −, maltose −, cellobiose +, trehalose −, lactose +, melibiose −, melezitose −, inulin +, D-xylose +, L-arabinose +, D-arabinose −, L-rhamnose −, ethanol +, glycerol +/w, ribitol −, D-mannitol +/w/−, salicin +, inositol −, raffinose +, starch −, D-ribose +/w/−, potassium nitrate −.

Teleomorph Stage. *Kluyveromyces fragilis.* Commonly isolated from nails and lung in-

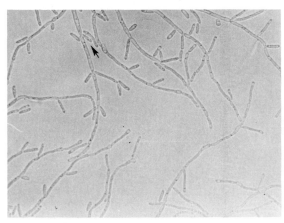

Figure 20–40. *Candida pseudotropicalis.* Fragile pseudohyphae of long cells. These fall apart to form "palisades" and "log jams" *(arrow).* Few blastoconidia. ×900.

fections. Found in cheese and dairy products.

Candida viswanathii Sandhu et Randhawa 1959

Colony Morphology. SGA 25° C, three days — cream-colored, soft, glistening; one month — creamy, soft to membranous, wrinkled, semi-dull.

Cell Morphology. 1. SGB 25° C, three days — cells globose, ovoid to cylindrical (2.5 to 7 × 4 to 12 μ). 2. CM or N PTB, long, wavy mycelium with irregular branches at angles up to 90 degrees, globose to ovoid blastoconidia in chains, verticillately arranged.

Fermentation. Glucose +, galactose +/w, sucrose −, maltose +, cellobiose −, trehalose +, lactose −, melibiose −, raffinose −, melezitose −, inulin −.

Assimilation. Glucose +, galactose +, L-sorbose +/−, sucrose +, maltose +, cellobiose +, trehalose +, lactose −, melibiose −, raffinose −, melezitose +, inulin −, starch +, D-xylose +, L-arabinose +/−, D-arabinose −, D-ribose −, L-rhamnose −, ethanol +, glycerol +, erythritol −, ribitol +, D-mannitol +, galactitol −, salicin +, inositol −, potassium nitrate −.

Isolated from spinal fluid and sputum.[211, 212]

REFERENCES

1. Aguilar, G. L., M. S. Blumen-Krantz, et al. 1979. Candida endophthalmitis after intravenous drug abuse. Arch. Ophthalmol. 97:96–100.
2. Ahern, D. G. 1974. Identification and ecology of yeasts of medical importance. *In* J. E. Prier and H. Friedman (ed.). *Opportunistic Pathogens.* Baltimore, University Park Press, pp. 129–146.
3. Ahern, D. G., J. R. Jannach, et al. 1966. Speciation and densities of yeasts in human urine specimens. Sabouraudia, 5:110–119.
4. Akiba, T., and K. Iwata. 1954. On the destructive invasion of a new species *Bacterium candidostruens* into *Candida* cells. Jap. J. Exp. Med., 24:159–166.
5. Alteras, I., E. J. Feuerman, et al. 1979. Widely disseminated cutaneous candidosis in adults. Sabouraudia, 17:383–388.
6. Anderson, A., and J. Yardley. 1972. Demonstration of *Candida* in blood smears. New Engl. J. Med., 286:108.
7. Anderson, J. H. 1979. *In vitro* survival of human pathogenic fungi in Hawaiian beach sand. Sabouraudia, 17:13–22.
8. Andrussy, K., and J. Horvath. 1979. Mycotic contamination of the public baths in Hajdu county. Egeszegtudomany, 23:46–49.
9. Aronson, I. K., and K. Soltani. 1976. Chronic mucocutaneous candidosis: A review. Mycopathologia, 60:17–25.
10. Aronson, I. K., G. H. Rieger, et al. 1979. Late onset mucocutaneous candidiasis with lymphoma and specific serum inhibitory factor. Cancer, 43:101–108.
11. Arthur, L. J. H. 1969. Pulmonary candidosis. Proc. R. Soc. Med., 62:906–907.
12. Arx, J. A., et al. 1977. The genera of yeasts and yeastlike fungi. Studies in Mycology No. 14. Centraalbureux voor Schimmelculture.
13. Ashcraft, K., and L. Leape. 1970. *Candida* sepsis complicating parenteral feeding. J.A.M.A., 217:454–456.
14. Audrey, C. 1887. Sur l'evolution du champignon du muguet. Rev. Med., 7:586–595.
15. Bak, A. L., and A. Stendcup. 1969. Deoxyribonucleic acid homology in yeasts. Genetic relatedness within the genus *Candida.* J. Gen. Microbiol., 59:21–30.
16. Balandren, L., H. Rothschild, et al. 1973. A cutaneous manifestation of systemic candidiasis. Ann. Intern. Med., 78:400–403.
17. Ballmann, G. E., and W. L. Caffin. 1979. Lipid synthesis during reinitiation of growth from stationary phase cultures of *Candida albicans.* Mycopathologia, 67:39–44.
18. Bedell, G. W., and D. R. Soll. 1979. Effects of low concentrations of zinc on the growth and dimorphism of *Candida albicans:* evidence for zinc resistant and sensitive pathways for mycelium formation. Infect. Immun., 26:348–354.
19. Bennett, J. H. 1844. On the parasite vegetable structures found growing in living animals. Trans. R. Soc., Edin., 15:277–294.
20. Berg, F. T. 1846. *Om torsk hos Barn.* Stockholm. L. J. Hjerta.
21. Bergental, T., K. Holinberg, et al. 1979. Yeast colonization in the oral cavity and feces in patients with denture stomatitis. Acta Odontol. Scand., 37:37–45.
22. Berkhout, C. M. 1923. De Schimmelgeschlachten *Monilia, Oidium, Oospora,* en *Torula.* Dissertation, Univ. of Utrecht.
23. Berye, T., and W. Kaplan. 1967. Systemic candidiasis with asteroid body formation. Sabouraudia, 5:310–314.
24. Block, M. B., L. M. Pachman, et al. 1971. Immu-

nological findings in familial juvenile endocrine deficiency syndrome associated with mucocutaneous candidiasis. Am. J. Med. Sci., *261*:213–218.

25. Blyth, W., and G. E. Stewart. 1978. Systemic candidiasis in mice treated with prednisolone and amphotericin I. Morbidity, mortality and inflammatory reaction. Mycopathologia, *66*:41–50. II. Ultrastructure and evidence for fungal toxin. Ibid. 51–57.

26. Boeckman, C. R., and C. E. Krill. 1970. Bacterial and fungal infections complicating parenteral alimentation in infants and children. J. Pediatr. Surg., *5*:117–126.

27. Bonorden. 1851. *Handbuch der Micologie.*

28. Boudru, I. 1969. De la resistance des *Candida albicans* à la nystatine. J. Pharm. Belg., *1969*: 162–185.

29. Boyd, A. D., F. Spencer, et al. 1977. Infective endocarditis. An analysis of 54 surgically treated patients. J. Thorac. Cardiovasc. Surg., *73*:23–30.

30. Brown-Thompson, J. 1966. Reverse variations between *Candida albicans* and *Candida tropicalis.* Acta Pathol. Microbiol. Scand., *66*:143–144.

31. Buck, J. D. 1977/78. Comparison of *in situ* and *in vitro* survival of *Candida albicans* in sea water. Microbiol. Ecol., *4*:291–302.

32. Budtz-Jorgensen, E. J. 1977. Hibitane in the treatment of oral candidiasis. J. Clin. Periodontol., *4*:117–128.

33. Buesching, W. J., K. Kurek, et al. 1979. Evaluation of modified API 20 C strip system for identification of clinically important yeasts. J. Clin. Microbiol., *9*:565–569.

34. Campo, A. C., and C. J. McDonald. 1976. Treatment of acrodermatitis enteropathica with zinc sulfate. Arch. Dermatol., *112*:687–689.

35. Carmo-Sousa, L. do. 1969. Distribution of yeasts in nature. *In* A. H. Rose and J. S. Harrison (eds.). *The Yeasts* Vol. 1. London, Academic Press, pp. 79–105.

36. Caroline, L., F. Rosner, et al. 1969. Elevated serum iron, low unbound transferrin and candidiasis in acute leukemia. Blood, *34*:441–451.

37. Chapel, T., W. J. Brown, et al. 1978. The microbiological flora of penile ulcerations. J. Inf. Dis., *137*:50–56.

38. Chattaway, F. W., F. C. Odds, et al. 1971. An examination of the production of hydrolytic enzymes and toxins by pathogenic strains of *Candida albicans.* J. Gen. Microbiol., *67*:255–263.

39. Child, J. S., and J. D. Shanley. 1979. Noninvasive detection of fungal endocarditis. Chest, *75*:539–540.

40. Chilgren, R. A., P. G. Quie, H. J. Meuwissen, et al. 1969. The cellular immune defect in chronic mucocutaneous candidiasis. Lancet, *1*:1286–1288.

41. Chilgren, R. A., R. Hong, et al. 1968. Human serum interactions with *Candida albicans.* J. Immunol., *101*:128–132.

42. Chipps, B. E., F. T. Saulsbury, et al. 1979. Noncandidad infections in children with chronic mucocutaneous candidiasis. Johns Hopkins Med. J., *144*:175–179.

43. Cobb, S. J., and D. Parratt. 1978. Determination of antibody levels to *Candida albicans* in healthy and hospitalized adults using a radioimmunoassay. J. Clin. Pathol., *31*:1161–1166.

44. Cohen, A. C. 1953. Pulmonary moniliasis. Am. J. Med. Sci., *226*:16–23.

45. Conner, C. L. 1928. Monilia from osteomyelitis. J. Infect. Dis., *43*:108–116.

46. Cooper, B. H., J. B. Johnson, et al. 1970. Clinical evaluation of the Uni-yeast-tek system for rapid presumptive identification of medically important yeasts. J. Clin. Microbiol., *7*:349–355.

47. Cross, R., and J. L. Clements. 1979. Jejunal monilial bezoar following total gastrectomy. Gastrointest. Radiol., *4*:29–31.

48. Crozier, W. J., and H. Coats. 1977. A case of onychomycosis due to *Candida ravautii.* Australas. J. Dermatol., *18*:139–140.

49. Curry, C. R., and P. G. Quie. 1971. Fungal septicemia in patients receiving parenteral hyperalimentation. New Engl. J. Med., *285*:1221–1225.

50. Demierre, G., and L. R. Friedman. 1979. Experimental endocarditis: prophylaxis of *Candida albicans* infection 5-fluorocytosine in rabbits. Antimicrob. Agents Chemother., *16*:252–254.

51. Dennerstein, G. 1979. Effective treatment of vaginitis. Curr. Ther., *20*:27–35.

52. De Vita, V. T., J. Utz, et al. 1966. *Candida* meningitis. Arch. Intern. Med., *117*:527–535.

53. Dixon, B. L., and C. S. Houston. 1978. Fatal neonatal pulmonary candidiasis. Radiology, *129*:132.

54. Dubendorfer, E. 1904. Ein Fall von Onychomycosis blastomycetica. Dermatol. Zentralblat., *7*:290–302.

55. Edwards, J. E., R. I. Lehrer, et al. 1978. Severe candidal infections. Clinical perspective, immune defence mechanisms and current concepts of therapy. Ann. Int. Med., *89*:91–106.

56. Engman, M. F. 1920. A peculiar fungus infection of the skin (soorpilze?). Arch. Dermatol. Syphilol., *1*:730.

57. Esterly, N. B., S. R. Brammer, et al. 1971. The relationship of transferrin and iron to serum inhibition of *Candida albicans.* J. Invest. Dermatol., *49*:437–442.

58. Felman, Y. M., and J. A. Nikitas. 1979. Trichomoniasis, candidiasis and *Corynebacterium vaginale* vaginitis. N.Y. State J. Med., *79*:1563–1566.

59. Fetter, B. F., G. K. Klintworth, et al. 1967. *Mycotic Diseases of the Central Nervous System.* Baltimore, Williams and Wilkins.

60. File, T. M., O. A. Marina, et al. 1979. Necrotic skin lesions associated with disseminated candidiasis. Arch. Dermatol., *115*:214–215.

61. Firkin, F. C. 1974. Therapy of deep seated infection with 5-fluorocytosine. Aust. N.Z. J. Med., *4*:462–467.

62. Fisher, R. J., G. B. Mayhall, et al. 1979. Fungus balls of the urinary tract. South Med. J., *721*:1281–84.

63. Forbes, J. G. 1909. A case of mycosis of the tongue and nails in a female child aged 3½ years. Br. J. Dermatol., *21*:221–223.

64. Fragner, P., and O. Mericka. 1977. Presence of yeasts in the sputum and their significance. Phtiseologica Cechoslovaca, *34*:364–372.

65. Fredricksson, J. 1979. Treatment of dermatomycoses with topical econazole and clotrimazole. Curr. Ther. Res., *25*:590–594.

66. Freund, U., and Z. Gimmon, et al. 1979. *Candida* infected ascites caused by perforated ulcer. Mycopathologia, *66*:191–192.

67. Fujiwara, A., J. W. Landau, et al. 1970. Responses of thymectomized and/or bursectomized chickens to sensitization with *Candida albicans*. Sabouraudia, *8*:9–17.

68. Galen, C. 130–200 A.D. De remediis parabilibus: I-III, from opera omnia ed. C. G. Kuhn. 1965. Mildesheim, George Olms.

69. Gauto, A., E. J. Law, et al. 1977. Experience with amphotericin B in the treatment of systemic candidiasis in burn patients. Am. J. Surg., *133*:175–178.

70. Gaylord, S. F. 1979. Gastric candidiasis. J.A.M.A., *241*:791.

71. Ghoniem, N. A., and M. Retai. 1968. Incidence of *Candida* species in Damieth cheese. Mykosen, *11*:295–298.

72. Gmelin. 1791. Syst. Nat., *2*:1487.

73. Goldacre, M. J., B. Watt, et al. 1979. Vaginal microbial flora in normal young women. Br. Med. J., *1*:1450–1453.

74. Goldberg, P. K., P. J. Kozinn, et al. 1979. Incidence and significance of candiduria. J.A.M.A., *241*:582–584.

75. Goldman, G. A., M. L. Littman, et al. 1960. Monilial cystitis effective treatment with instillations of amphotericin B. J.A.M.A., *174*:359.

76. Goldstein, E., M. H. Grieco, et al. 1965. Studies on the pathogenesis of experimental *Candida parapsilosis* and *Candida guilliermondii* infections in mice. J. Infect. Dis., *115*:293–302.

77. Gordon, M. A., J. C. Elliott, et al. 1967. Identification of *Candida albicans*, other *Candida* species and *Torulopsis glabrata* by means of immunofluorescence. Sabouraudia, *5*:323–328.

78. Goto, S., J. Susiyama, et al. 1969. A taxonomic study of Antarctic yeasts. Mycologia, *61*:748–774.

79. Grawitz, P. 1887. Zur Botanik des Soors und der Dermatomycosen. Dtsch. Z. Prakt. Med., pp. 209–211.

80. Grimley, P. M., L. D. Wright, et al. 1955. *Torulopsis glabrata* infection in man. Am. J. Clin. Pathol., *43*:216–223.

81. Grosshans, E., L. Dossman, et al, 1979. Glossite losangique mediane ou candidose lingual mediane? Ann. Dermatol. Venereol., *106*:259–264.

82. Gruby, D. 1842. Recherches anatomiques sur une plante eryptogame qui constituae le vrai muguet des enfants. C. R. Acad. Sci. (Paris), *14*:634–636.

83. Guyer, P. B., J. Brunton, et al. 1971. Candidiasis of the oesophagus. Br. J. Radiol., *44*:131–136.

84. Hache, J. 1966. Les candidoses aviaires. Influence de l'oxytetracycline sur la sensibilité du poulet à l'infection par *Candida albicans*. Ecole Nat. Vet. Alfort., Thesis. 79:49 pp.

85. Hantschke, D., and M. Zobel. 1979. The reaction of the physiological vaginal flora to topical antimycotics. Mykosen, *22*:267–273.

86. Hartshorne, J. 1977. *Candida* mycelia in leukoplaki, leukoedema, lichen planus, squamous papilloma, squamous cell carcinoma and chewing lesions of oral mucosa: a clinicopathological study. J. Dentol. Assoc. S. Africa, *32*:465–469.

87. Hasenclever, H. F., and W. O. Mitchell. 1961. Pathogenicity of *C. albicans* and *C. tropicalis*. Sabouraudia, *1*:16–21.

88. Hasenclever, H. F., and W. O. Mitchell. 1962. Pathogenesis of *Torulopsis glabrata* in physiologically altered mice. Sabouraudia, *2*:87–95.

89. Hassen, K. E., and M. K. Brown. 1978. *Candida* peritonitis and cimetidine. Lancet, *2*:1054.

90. Hauser, F. V., and S. Rothman. 1950. Monilial granuloma: Report of a case and review of the literature. Arch. Dermatol., *61*:297–310.

91. Haussmann, D. 1975. Parasites des organes sexuals femeilles de l'homme et de quelques animaux avec une notice sur developpement d'*oidium albicans*. Robin. Paris, J. B. Bailliere.

92. Heeres, J., I. J. Backx, et al. 1979. Antimycotic imidazoles. Part 4. Synthesis and antifungal activity of ketoconazole, a new potent orally active broad spectrum antifungal agent. J. Med. Chem., *22*:1003–1005.

92a. Heffner, D. K., and W. A. Franklin. 1978. Endocarditis caused by *Torulopsis glabrata*. Am. J. Clin. Pathol., *70*:420–423.

93. Henney, M. R., G. R. Raylor, et al. 1978. Mycological profile of crew during 56-day simulated orbital flight. Mycopathologia, *63*:131–144.

94. Herceg, M., B. Marzan, et al. 1977. Pathomorphological observations of spontaneous encephalitis in a monkey. Vet. Arhiv., *47*:183–187.

95. Hermens, P. E., J. A. Ulrich, et al. 1962. Chronic mucocutaneous candidiasis: a surface expression of deep seated abnormalities. Am. J. Med., *47*:503–519.

96. Hippocrates, circa 460–377 B.C. Epidemics. Book 3 translated by F. Adams 1939. Baltimore, Williams and Wilkins.

97. Holt, R. J., and A. Azmi. 1978. The emergence of *Candida albicans* resistant to miconazole during the treatment of urinary tract candidosis. Infection, *6*:198–199.

98. Holti, G. 1966. *Symposium on Candida Infections*. Edinburgh, Scotland, Livingstone, p. 73.

99. Holzschu, D. L., H. L. Presley, et al. 1979. Identification of *Candida lusitaniae* as an opportunistic yeast in humans. J. Clin. Microbiol., *10*:202–205.

100. Hopfer, R. L., and D. Gröschel. 1979. Detection by counterimmunoelectrophoresis of anti-*Candida* antibodies in sera of cancer patients. Am. J. Clin. Path., *72*:215–218.

101. Horsmanheimo, M., K. Krohn, et al. 1979. Immunologic features of chronic granulomatous mucocutaneous candidiasis before and after treatment with transfer factor. Arch. Dermatol., *115*:180–184.

102. Howard, D. H., and V. Otto. 1967. The intracellular behavior of *Toruplosis glabrata*. Sabouraudia, *5*:235–39.

103. Howat, R. C. L., B. W. Elito, et al. 1979. Cure and relapse of candidal vaginitis following treatment with miconazole clotrimazole and nystatin in Royal Society of Medicine Int. Congress and Symposium Series No. 7D. Gough (ED), pp. 3–8.

104. Howlett, J. A., and C. A. Squier. 1980. *Candida albicans* ultrastructure: colonization and invasion of oral epithelium. Infect. Immun., *29*:252–260.

105. Huang, S. Y., C. W. Berry, et al. 1979. A radioimmunoassay method for the rapid detection of *Candida* antibodies in experimental systemic candidiasis. Mycopathologia, *67*:55–58.

106. Hurley, R. 1966. Experimental infection with *Candida albicans* in modified hosts. J. Pathol. Bact., *92*:57–67.

107. Hurley, R., and V. C. Stanley. 1969. Cytopathic effects of pathogenic and nonpathogenic species of *Candida* on cultured mouse epithelium

cells: Relation to growth rate and morphology of the fungi. J. Med. Microbiol., *2*:63–74.

108. Hurwich, B. J. 1966. Monilial peritonitis. Report of a case and review of the literature. Arch. Int. Med., *117*:405–408.

109. Ide, A. S., Jacobelli, et al. 1978. Arthritis por *Candida albicans*. Rev. Medica Chile, *106*:884–886.

110. Imperator, P. J., C. E. Buckley, et al. 1968. *Candida* granuloma. A clinical and immunologic study. Arch. Dermatol., *97*:139–146.

111. Iwata, K., and Y. Yamamoto. 1978. Glycoprotein toxins produced by *Candida albicans*. In *The Black Yeasts and the White Yeasts*. Proceedings of the IV International Conference on the Mycoses. Pan American Health Association Washington Sci. Pub. No. 356, pp. 246–257.

112. Jacobi, E. 1907. Ein besondere Form der Trichophytie als Folgeerscheinung des permanenten Bades. Arch. Dermatol. Syphilogr. (Wien), *84*:289–300.

113. Jenner, B. M., and L. I. Landau, et al. 1979. Pulmonary candidiasis in cystic fibrosis. Arch. Dis. Child., *54*:555–556.

114. Joachim, H., and S. Paloyes. 1940. Subacute endocarditis and systemic mycosis (Monilia). J.A.M.A., *115*:205–208.

115. John, R. W. 1968. Necrosis of oral mucosa after local application of crystal violet. Br. Med. J., *1*:157–158.

116. Jones, J., and J. A. Schmitt. 1978. The effect of chlorination on the survival of cells of *Candida albicans*. Mycologia, *70*:684–689.

117. Juffe, A., A. L. Miranda, et al. 1977. Prosthetic valve endocarditis by opportunistic pathogens. Arch. Surg., *112*:151–153.

118. Kantrowitz, P. A., D. J. Fletschli, et al. 1969. Successful treatment of chronic esophageal moniliasis with a viscous suspension of nystatin. Gastroenterology, *57*:424–430.

119. Kapica, L., and A. Clifford. 1968. Quantitative determination of yeasts in sputum. Mycopathologia, *34*:27–32.

119a. Katz, M. E., and P. A. Cassileth. 1977. Disseminated candidiasis in a patient with acute leukemia. Successful treatment with miconazole, J.A.M.A., *237*:1124.

120. Katzenstein, A. L., and J. Maksem. 1979. Candidal infection of gastric ulcers. Histology, incidence and clinical significance. Am. J. Clin. Path., *71*:137–141.

121. Kaufman, S. A., S. Scheff, et al. 1960. Esophageal moniliasis. Radiology, *75*:726–732.

122. Kelvin, F. M., W. M. Clark, et al. 1978. Chronic esophageal stricture due to moniliasis. Br. J. Radiol., *51*:826–828.

123. Kiehn, T. E., E. M. Bernhard, et al. 1979. Candidiasis detection by gas-liquid chromatography of D-arabinitol, a fungal metabolite in human serum. Science USA, *206*:577–580.

124. Kim, M. H., G. E. Rodey, et al. 1969. Defective candidacidal capacity of polymorphonuclear leukocytes in chronic granulomatous disease of child. J. Pediatr., *75*:300–303.

125. King, D. S., and S. C. Jony. 1977. A contribution to the genus *Trichosporon*. Mycotaxon, *6*:391–417.

126. Kirkpatrick, C. H., and D. Windhorst. 1979. Mucocutaneous candidiasis and thymoma. Am. J. Med., *66*:939–945.

127. Klein, J. J., and C. Watanakunakorn. 1979. Hos-

128. Kopelson, G., M. Silva-Hutner, et al. 1979. Fungal peritonitis and malignancy: report of two patients and review of the literature. Med. Pediatr. Oncol., *6*:15–22.

129. Kovaks, E., B. Bucz, et al. 1969. Propagation of mammalian viruses in protista IV. Experimental infection of *C. albicans* and *S. cerevisiae* with polyoma virus. Proc. Soc. Exptl. Biol. Med., *132*:971–977.

130. Krause, W., H. Matheis, et al. 1969. Fungaemia and funguria after oral administration of *Candida albicans*. Lancet, *1*:598–599.

131. Kroll, J., J. M. Einbender, et al. 1973. Mucocutaneous candidiasis in a mother and son. Arch. Dermatol., *108*:259–262.

132. Land, G. A., B. A. Harrison, et al. 1979. Evaluation of the new API 20 C strip for yeast identification against a conventional method. J. Clin. Microbiol., *10*:357–364.

133. Langenbeck, B. 1839. Auffingung von Pelzen aus der Schleimhaut der speiseröhre einer Typhus Leiche. Neue Nat. Gebr. Natur-u-Heik (Froriep) *12*:145–147.

133a. Larson, P. A., R. L. Lindhstrum, et al. 1978. *Torulopsis glabrata* endophthalmitis after keratoplasty. Arch. Ophthalmol., *96*:1019–1022.

134. Lees, A. W., S. S. Rau, et al. 1971. Endocarditis due to *Torulopsis glabrata*. Lancet, *1*:943–944.

135. Lehrer, R. I., and M. J. Cline. 1969. Interaction of *Candida albicans* with human leukocytes and serum. J. Bact., *98*:996–1004.

136. Leuk, R. D., and R. J. Moon. 1979. Physiological and metabolic alterations accompanying systemic candidiasis in mice. Infect. Immun., *26*:1035–1041.

137. Leyden, J. J., and A. M. Kligman. 1978. The role of microorganisms in diaper dermatitis. Arch. Dermatol., *114*:56–59.

138. Lindquist, J. A., S. Robinovich, et al. 1973. 5-Flourocytosine in the treatment of experimental candidiasis in immunosuppressed mice. Antimicrob. Agents Chemoth., *4*:58–61.

139. Linares de, L. M., and C. Marin. 1978. Frequency of yeasts of the genus *Candida* in humans as pathogens and as part of normal flora. *In, The Black Yeasts and the White Yeasts*. Proceedings of the IV International Conference on the Mycoses. Pan American Health Association Washington Sci. Pub. No. 356, pp. 124–136.

140. Lodder, J. 1970. *The Yeasts: A Taxonomic Study*. 2nd ed. Amsterdam, North Holland.

141. Logan, D. A., J. M. Becker, et al. 1979. Peptide transport in *Candida albicans*. J. Gen. Microbiol., *114*:179–86.

142. Lopez Fernandez, J. F. 1952. Accionpatogena experimental de la levadura *Torulopsis glabrata* (Anderson 1917). Lodder y De Vries 1938, productora de lesions histo-patologicas semijantes a las de la histoplasmosis. *Ann. Fac. Med. Montevideo, 37*:470–483.

143. Louria, D. B. 1977. Experimental infections with fungi and yeasts. Contrib. Microbiol Immunol., *3*:31–47.

144. Louria, D. B., and R. G. Brayton. 1964. The behavior of *Candida* cells within leucocytes. Proc. Soc. Exp. Biol. Med., *115*:93–98.

145. Lynch, P. J., W. Minkin, et al. 1969. Ecology of *Candida albicans* in candidiasis of groin. Arch. Dermatol., *99*:154–160.

pital-acquired fungemia. Its natural course and clinical significance. Am. J. Med., *67*:51–58.

146. Maibach, H. I., and A. M. Kligman. 1962. The biology of experimental human cutaneous moniliasis. Arch. Dermatol., *85*:233–257.

147. Magnussen, C. R., J. P. Olson, et al. 1979. *Candida* fungus balls in the common bile duct. Unusual manifestations of disseminated candidiasis. Arch. Int. Med., *139*:821–822.

148. Marcucci, R. A., H. Whitely, et al. 1978. Common bile duct obstruction secondary to infection with *Candida*. J. Clin. Microbiol., *7*:490–492.

149. Marks, M. I. 1975. Yeast colonization in hospitalized and nonhospitalized children. J. Pediatr., *87*:524–27.

150. Marks, M. I., and E. O'Toole. 1970. Laboratory identification of *Torulopsis glabrata* typical appearance on routine bacteriological media. Appl. Microbiol., *19*:184–185.

151. Marples, M. J. 1965. *The Ecology of the Human Skin*. Springfield, Ill., Charles C Thomas.

152. Marples, M., and D. A. Somerville. 1968. The oral and cutaneous distribution of *Candida albicans* and other yeasts in Raratonga, Cook Islands. Trans. R. Soc. Trop. Med. Hyg., *62*:256–262.

153. Martini, A., G. Croce, et al. 1976. The yeast flora of the external ear of man. Boll. Soc. Ital. Biol. Sper., *52*:1943–1947.

153a. Medoff, G., W. E. Dismulces, et al. 1970. Therapeutic program for *Candida*. Antimicrob. Agents Chemother., 286–290.

154. Mercer, H. P., and J. M. Gupta. 1978. Candida meningitis causing aqueductal stenosis following parenteral nutrition in an infant with meconium peritonitis. Aust. Pediatr. J., *14*:286–287.

155. Merz, W. G., G. L. Evans, et al. 1977. Laboratory evaluation of serological tests for systemic candidiasis. A cooperative study. J. Clin. Microbiol., *5*:596–603.

156. Miale, J. B. 1943. *Candida albicans* infection confused with tuberculosis. Arch. Pathol., *35*:427–437.

157. Milne, L. J. R., and G. K. Crompton. 1974. Beclomethasone dipropionate and oropharyngeal candidiasis. Br. Med. J., *3*:797–798.

158. Minkowitz, S., D. Koffler, et al. 1963. *Torulopsis glabrata* septicemia. Am. J. Med., *34*:252–255.

159. Montes, L., and W. H. Wilborn. 1968. Ultrastructural features of host parasite relationship in oral candidiasis. J. Bacteriol., *96*:1349–1356.

160. Montes, L. F., R. Ceballos, et al. 1972. Chronic mucocutaneous candidiasis, myositis and thymoma. J.A.M.A., *222*:1619.

161. Mourad, S., and L. Friedman. 1961. Pathogenicity of *Candida*. J. Bacteriol., *81*:550–556.

162. Moyahan, F. J. 1974. Acrodermatitis enteropathica: a lethal inherited human zinc deficiency disorder. Lancet, *11*:399–400.

163. Murata, H. 1979. Experimental *Candida* induced arteritis in mice. Relation to arteritis and mucocutaneous lymph node syndrome. Microbiol. Immunol., *23*:825–831.

164. Newcomer, V. D., J. W. Landau, et al. 1966. *Candida* granuloma. Studies of host parasite relationships. Arch. Dermatol., *93*:149–161.

165. Newman, D. M., and J. M. Hogg. 1969. *Torulopsis glabrata* pyelonephritis. J. Urol., *102*:547–548.

166. Nicholls, P. E., and K. Henry. 1978. Gastritis and cimetidine: a possible explanation. Lancet, *1*:1095–1096.

167. Nickerson, W. J. 1954. An enzymatic locus participating in cellular division of yeast. J. Gen. Physiol., *37*:483–494.

168. Nitzulescu, V., and M. Niculescu. 1976. Three cases of ocular candidiasis caused by *Candida lypolitica*. Arch. Roum. Pathol. Exp. Microbiol., *35*:269–272.

169. Noble, H. B., and E. D. Lynn. 1974. *Candida* osteomyelitis and arthritis from hyperalimentation therapy. J. Bone Joint Surg., *56*:825–829.

170. Odds, F. C. 1979. *Candida and Candidosis*. Baltimore, University Park Press.

171. Odds, F. C., E. C. Evans, et al. 1978. Prevalence of pathogenic yeasts and humoral antibodies to *Candida* in diabetic patients. J. Clin. Pathol., *31*:840–844.

172. Oldfield, F. S. J., L. Kapica, et al. 1968. Pulmonary infection due to *Torulopsis glabrata*. Can. Med. Assoc. J., *98*:165–168.

173. Olsen, I., and O. Bondevik. 1978. Experimental *Candida*-induced denture stomatitis in the Wistar rat. Scan. J. Dent. Res., *86*:392–398.

174. Onsberg, P. 1977. The fungal flora of normal and diseased nails. Curr. Ther. Res., *22*:20–23.

175. Orringer, M. B., and H. Sloan. 1978. Monilial esophagitis: an increasingly frequent cause of esophageal stenosis. Ann. Thorac. Surg., *26*:364–374.

176. Otero, R. B., N. L. Goodman, et al. 1978. Predisposing factors in systemic and central nervous system candidiasis: histological and cultural observations in the rat. Mycopathologia, *64*:113–120.

177. Ott, D. J., and D. W. Gelfand. 1978. Esophageal stricture secondary to candidiasis. Gastrointest. Radiol., *2*:323–325.

178. Padhye, A. A., and M. J. Thirumalacher. 1967. Incidence of oral thrush in newborn infants 1960–1962 at Sassoon Hospitals, Poona, and a comparative evaluation of treatment with some antifungal agents. Hindustan Antibiot. Bull., *91*:143–151.

179. Parker, J. C., J. J. McCloskey, et al. 1978. The emergence of candidosis. The dominant postmortem cerebral mycosis. Am. J. Clin. Pathol., *70*:31–36.

180. Parrot, J. 1877. *Clinique des nouvaan-nes*. L'athrepsie Lecons recuellies par le Dr. Troisier. Paris, G. Massoon et cie.

181. Peterson, L. R., R. H. Ketty, et al. 1978. Therapy of *Candida* peritonitis: penetration of amphotericin B. into peritoneal fluid. Postgrad. Med. J., *54*:340–342.

182. Phillips, I., S. Eykyn, et al. 1973. *Candida* peritonitis treated with 5-fluorocytosine in a patient receiving hemodialysis. Clin. Nephrol., *1*:271–272.

183. Piken, E., R. Dwyer, et al. 1978. Gastric candidiasis. J.A.M.A., *240*:2181–2182.

184. Pinsloo, J. G., and P. J. Pretorius. 1966. *Candida albicans* endocarditis. Case successfully treated with amphotericin B. Am. J. Dis. Child., *111*:446–447.

185. Plaut, H. C. 1887. *Neue Beiträge zur systematischen Stellung des Soor pilzes in der Botanik*. Leipzig, H. Voigt.

186. Plummer, N. S. 1966. *Candida* infections of the lung. *In* Winner, H. I., and R. Hurley (eds.), *Symposium on Candida Infection*. London, Livingstone, pp. 214–220.

187. Pope, L. M., G. T. Cole, et al. 1979. Systemic and

gastrointestinal candidiasis of infant mice after intragastric challenge. Infect. Immun., 25:702–707.

188. Quinquaud, M. 1868. Nouvelles recherches sur le muguet. Arch. Physiol Norm. Pathol., 1:290–305.

189. Rafin, M. 1927. Muguet et calcul de la vessie. J. Urol., 23:32.

190. Ray, T. L., and K. O. Wuepper. 1978. Recent advances in cutaneous candidiasis, Int. J. Dermatol., 17:603–690.

191. Rebora, A., R. R. Marples, et al. 1973. Experimental infection with Candida albicans. Arch. Dermatol., 108:69–73.

192. Rebora, A., R. R. Marples, et al. 1973. Erosio interdigitales blastomycetica. Arch. Dermatol., 108:66–68.

193. Record, C. O., J. M. Skinner, et al. 1971. Candida endocarditis treated with 5-fluorocytosine. Brit. Med. J., 1:262–264.

194. Reess, M. 1877. Ueber den Soorpilz. S. B. Phys. Med. Soz. Erlangen, 9:190–193.

195. Refei, M., and M. Amer. 1974. Chronic paronychia in Egypt. Ernst Rodenwaldt Archiv., 1:55–63.

196. Reiser, M., N. Rupp, et al. 1978. Radiological findings in septic Candida arthritis. R.O.E.F.O., 129:335–339.

197. Restrepo M. A., and L. de Uribe. 1976. Isolation of fungi belonging to the genera Geotrichum and Trichosporum from human dermal lesions. Mycopathologia, 59:3–9.

198. Reynolds, I. M., P. W. Miner, et al. 1968. Cutaneous candidiasis in swine. J. Am. Vet. Med. Assoc., 152:182–186.

199. Reynolds, R., and A. I. Braude. 1956. The filament-inducing property of blood for Candida albicans; its nature and significance. Clin. Res. Proc., 4:40.

200. Rippon, J. W., and D. H. Anderson. 1978. Experimental mycosis in immunosuppressed rabbits. Acute and chronic candidosis. Mycopathologia, 64:91–96.

201. Robin, C. P. 1853. Histoire Naturelle des Vegetaux. Parasites qui croissent — sur l'homme et sur les animaux vivants. Paris. Bailliere.

202. Rodriguez, R., H. Shinya, et al. 1971. Torulopsis glabrata fungemia during prolonged intravenous alimentation therapy. New Eng. J. Med., 284:540–541.

203. Roessman, U., and R. L. Friede. 1967. Candidal infection of the brain. Arch. Pathol., 84:495–98.

204. Rogglis, V. L. 1978. Disseminated candidiasis with secondary involvement in the larynx. Arch. Otolaryngol., 104:546–547.

205. Rose, H. D., and V. P. Kurup. 1977. Colonization of hospitalized patients with yeastlike organisms. Sabouraudia, 15:251–256.

206. Rosen von Rosenstein, N. 1771. Underrättelse om Barms Sjukdomar och Deras Bote medal. Stockholm, Wenneberg and Nordstrom.

207. Roux, G., and G. Linoscier. 1890. Recherches morphologiques sur le champignon du muguet. Arch. Med. Exp. Anat. Pathol., 2:62–87.

208. Saccardo, P. A. 1886. Sylloge Fungorum.

209. Sahay, N. J., S. S. Chatterjee, et al. 1979. Inhaled corticosteroid aerosols and candidiasis. Br. J. Dis. Chest, 73:164–168.

210. Sams, W. M., J. L. Jorizzo, et al. 1979. Chronic mucocutaneous candidiasis. Immunologic studies of three generations of a single family. Am. J. Med., 67:948–959.

211. Sandhu, R. S., and H. S. Rundhawa. 1962. On the reisolation and taxonomic study of Candida viswanathii Viswanathan and Randhawa 1959. Mycopathologia, 18:179–183.

212. Sandhu, D. K., R. S. Sandhu, et al. 1976. Isolation of Candida viswanathii from cerebral spinal fluid. Sabouraudia, 14:251–254.

213. Sandler, B. 1979. Lactobacillus for vulvovaginitis. Lancet, 2:791–792.

214. Scherr, G. H., and R. H. Weaver. 1953. The dimorphism phenomenon in yeasts. Bacteriol. Rev., 17:51–92.

215. Scherr, G. H. 1955. The effect of hormones on experimental moniliasis in mice. Mycopathologia, 7:68–82.

216. Schmorl, G. 1890. Ein Fall von Soormetastase in der Nier. Zentralb I. Bakteriol., 7:329–335.

217. Schnell, D. 1977. The epidermiology and prophylaxis of mycoses in perinatology. Contrib. Microbiol. Immunol., 4:40–45.

218. Schultz, F. W. 1925. Two cases of thrush with unusual symptoms and skin manifestations. Am. J. Dis. Child., 29:283–285.

219. Schümann, R., and K. M. Meyer. 1978. Mechanische Anurie eines Säuglings bei Candidapyelonephritis. Urologe, 17:324–325.

220. Schwartzman, R. M., M. Denbler, and P. E. Dice. 1965. Experimentally induced moniliasis (Candida albicans) in the dog. J. Small Anim. Pract., 6:327–332.

221. Sclafer, J. 1956. Mycologie Medicale. Expansion Scientifique Française, Paris, p. 191.

222. Sclafer, J., and S. Hewitt. 1960. Valeur des tests cutane's dans l'allergie a Candida albicans. Pathol. Biol., 8:323–327.

223. Shelley, W. B. 1965. Erythema annulare centrifugum due to Candida albicans. Br. J. Dermatol., 77:383–384.

224. Sherry, J. P., S. R. Kuchma, et al. 1979. The occurrence of Candida albicans in Lake Ontario bathing beaches. Can. J. Microbiol., 25:1036–1044.

225. Sievers, I. 1976. Occurrence isolation and differentiation of the imperfect yeasts in material of 2492 mouth swabs of 468 men. Ernst Rodenwaldt Archiv., 3:133–141.

226. Sixbey, J. W., and Caplan E. S. 1978. Candida parapsilosis endophthalmitis. Ann. Intern. Med., 89:1010–1011.

227. Smiten, S. M. B. 1967. Candidiasis in animals in New Zealand. Sabouraudia, 5:220–225.

228. Solomkin, J. S., R. L. Simmons, et al. 1977. Serum opsonin for early diagnosis of disseminated candidasis. Surg. Forum, 28:46–48.

229. Sousa, H. M. de, and N. Van Uden. 1960. The mode of infection and reinfection in yeast vulvovaginitis. Am. J. Obstet. Gynecol., 80:1096–1100.

230. Stanel-Borowski, K. 1976. Experimental Candida vaginitis in the mouse. Arch. Gynecol., 221:313–322.

231. Stenderup, A., and A. L. Bak. 1968. Deoxyribonucleic acid base composition of some species within genus candida. J. Gen. Microbiol., 52:231–236.

232. Stenderup, A., and G. T. Pederson. 1962. Yeasts of human origin. Acta Pathol. Microbiol. Scand., 54:462–472.

233. Stenderup, A., E. Karuru, et al. 1978. Yeasts in Fijians. Fiji Med. J., *6*:334–337.

234. Strom, B. G., R. Beaudry, et al. 1978. Yeast overgrowth in operated stomachs. J. Canad. Assoc. Radiol., *29*:161–164.

235. Sutphin, A., F. Albright, et al. 1943. Five cases (three in siblings) of idiopathic hypoparathyroidism associated with moniliasis. J. Clin. Endocrinol., *3*:625–634.

236. Svirsky-Fein, S., L., Langer, et al. 1979. Neonatal osteomyelitis caused by *Candida tropicalis.* Report of two cases and review of literature. J. Bone Joint Surg., *61*:455–459.

237. Sweeney, J. C., G. Migaki, et al. 1976. Systemic mycoses in marine animals. J. Am. Vet. Med. Assoc., *169*:946–948.

238. Sweet, C. E., and L. Kaufman. 1970. Application of agglutinins for the rapid and accurate identification of medically important *Candida* species. Appl. Microbiol., *19*:830–836.

239. Taschdjian, C., M. B. Cuesta, et al. 1969. A modified antigen for sera diagnosis of systemic candidiasis. Am. J. Clin. Pathol., *52*:464–472.

240. Thorpe, E., and H. Handley. 1929. Chronic tetany and chronic mycelial stomatitis in a child aged four and one-half years. Am. J. Dis. Child., *36*:328–338.

241. Torok, I., and E. Moravcsik. 1979. Screening for candidosis in patients with open heart surgery. Mykosen, *22*:345–351.

242. Tsao, T. C., C. Wang, et al. 1979. Analysis of x-ray findings of 22 cases of acute pulmonary mycosis. Chinese Med. J., *92*:119–124.

243. Underwood, M. 1784. *A Treatise on the Diseases of Children.* London, J. Mathews.

244. Van Uden, N. 1960. The occurrence of *Candida* and other yeasts in the intestinal tract of animals. Ann. N.Y. Acad. Sci., *89*:59–68.

245. Veron. 1835. Memoire sur le Muguet. Arch. Gen. Med., *8*:466.

246. Vidal-Leiria, M., H. Buckley, et al. 1979. Distribution of maximum temperature for growth among yeasts. Mycologia, *71*:493–501.

247. Vieu, M., and G. Segretain. 1959. Contribution a l'etude de *Geotrichum* et *Trichosporum* d'origine humaine. Ann. Inst. Past., *96*:421–433.

248. Warnock, D. W., D. C. E-Speller, et al. 1979. Antibodies to *Candida* species after operations on the large intestine: observations on association with oral and faecal yeast colonization. Sabouraudia, *17*:405–414.

249. Warren, R. C., M. O. Richardson, et al. 1978. Enzyme-linked immunosorbent assay of antigens from *Candida albicans* circulating in infected mice and rabbits: the role of mannan. Mycopathologia, *66*:179–182.

250. Watanakunakorn, C., J. Carleton, et al. 1968. *Candida* endocarditis surrounding a Starr Edwards prosthetic valve. Recovery of Candida in hypertonic medium during treatment. Arch. Int. Med., *121*:243–245.

251. Wickerham, L. J. 1957. Apparent increase in frequency of infection involving *Torulopsis glabrata.* J.A.M.A., *165*:47–48.

252. Wingard, J. R., W. G., Merz, et al. 1979. *Candida tropicalis:* a major pathogen in immunocompromised patients. Ann. Intern. Med., *91*:539–543.

253. Winner, H. I., and R. Hurley. 1964. *Candida albicans.* Boston, Little, Brown and Co.

254. Wright, B. A. 1978. Median rhomboid glossitis: not a misnomer. Review of the literature and histologic study of twenty eight cases. Oral Surg., *46*:806–814.

255. Yap, B. S., and G. P. Bodey. 1979. Oropharyngeal candidiasis treated with troche form of chlortrimazole. Arch. Intern. Med., *139*:656–657.

256. Yarrow, D., and S. A. Meyer. 1978. Proposal for the amendment of the diagnosis of the genus *Candida* Berkhout nom. cons. Int. J. System. Bacteriol., *28*:611–615.

257. Zenker, W. 1862. Encephalitis mit Pilzentwicklung im Gehrin. Jahrb. Ges. Natur-u-Heik (Dresden), 51–52.

258. Zopf, W. 1890. *Die Pilz im Morphologischer, Physiologischer, Biologischer Beziehung.* Breslau, Trewendt.

21 Cryptococcosis

DEFINITION

Cryptococcosis is a chronic, subacute, or (rarely) acute pulmonary, systemic, or meningitic infection caused by the yeast *Cryptococcus neoformans.* The primary infection in man is almost always pulmonary following inhalation of the yeast; in animals it may follow implantation or ingestion. Pulmonary infection in man is usually subclinical and transitory; however, it may arise as a complication of other diseases in debilitated patients and become rapidly systemic or even fulminant. In Europe it is known as the signal disease (malade signal), as it signals an underlying debilitating disease. It has a predilection for the central nervous system. The tissue reaction is sparse, with few macrophages appearing in active infection, but the focus of infection may evolve into tuberculoid granulomas in chronic and healing disease. Suppuration and caseation necrosis are infrequent. The etiologic agent is unique among pathogenic fungi because of its production of a mucinous capsule in tissue and culture.

Synonymy. Torulosis, Busse-Buschke's disease, European blastomycosis, malade signal.

HISTORY

In medical literature the term "blastomycosis" has been used very loosely to designate any infection in which a budding yeast cell was found. Cryptococcosis in particular has been confused with blastomycosis. The cutaneous form of cryptococcosis, which was more often reported from Europe, was thought to be caused by a different species and came to be known as European blastomycosis. These questions were resolved in 1935 when Rhoda Benham[10] clearly differentiated blastomycosis from cryptococcosis. She showed that the cutaneous European type of cryptococcosis was caused by the same organism that produced the meningitic form more commonly reported in America.

Busse, a pathologist,[21] and Buschke, a surgeon,[20] reported separately the isolation of a yeast from the tibia of a 31-year-old woman. In an 1894 paper to the Greifswald Medical Society, Busse described the organisms as fungal in nature and resistant to treatment with sodium hydroxide. He isolated them on prune rice medium, then named the organisms "Hefe" (*Saccharomyces*) and coined the term Saccharomycosis hominis for the disease. The lesions were described as "gumma-like" or "sarcoma-like," and the patient had lymphadenopathy and cutaneous ulcers. Busse considered this yeast to be the cause of her malady; later authors used the name *neoformans* (cancer- or tumor-causing) as an epithet for the organism. A year later, Buschke[20] also described the case and the organism. He was aware of Busse's work, so he does not share in the discovery of the organism. About the same time, Sanfelice recovered an encapsulated yeast from peach juice which caused lesions in experimentally infected animals.[116] He named the organism *Saccharomyces neoformans,* thus giving taxonomic priority to this specific epithet. In 1896, Curtis described a case similar to that of Busse and Buschke. The patient had a yeast in a myxomatous tumor of the hip. He called the organism *Saccharomyces subcutaneous tumefaciens.* However, when the eminent French mycologist P. Vuillemin[137] examined the sev-

532

eral cultures, he did not find ascospores, which are characteristic of the genus *Saccharomyces;* therefore, in 1901 he transferred the yeasts to the genus *Cryptococcus* as *C. hominis* for the isolants of Busse and Curtis and *C. neoformans* for the isolant of Sanfelice. To confuse the taxonomy even more, the genus *Cryptococcus* had been used earlier by Kützing to describe algaelike white organisms. Sanfelice (1895) in Italy isolated another organism similar to Buschke's yeast, this time from a lymph node of an ox.[117, 118] He named it *Saccharomyces lithogenes.* Frothingham in 1902 recovered the same organism from a myxoma-like lung lesion in a horse in Massachusetts.[43] These latter findings established the fact that the disease occurs in both animals and man. The organisms from Sanfelice, one isolated by Klein in milk[73] and one isolated by Plimmer from a patient with cancer, were reported to be culturally identical by Weis in 1902.[140] The proper genus and specific epithet for these isolants had yet to be established.

The association of the encapsulated yeast with cancers, especially myxoma-like disease, was noted early and soon led to the postulation that it was an inciter of neoplasias. Von Hansemann in 1905[60] appeared to be the first to see this fungus in a case of meningitis. He described the disease as "tuberculous," with yeast present in "gelatinous cysts." It was first recognized ante mortem in 1914 by Verse in a case of leptomeningitis in a 29-year-old woman.[134]

Stoddard and Cutler (1916)[131] delineated the clinical and pathologic differences between cryptococcosis, blastomycosis, and other mycoses. They erroneously assumed that the capsules of the organisms were cysts in the tissue caused by digestive action of the fungus ("histolytic") and named the organism *Torula histolytica.* This name has no validity or priority, and it was based on an erroneous observation. Unfortunately it became popular, and the disease was known for a long time as torulosis or *Torula* meningitis. In a detailed analysis of 26 cases Baker and Haugen[6] discredited the "enzymic" action of the "torula" and demonstrated the compressive effect on tissue of expanding capsular masses.

From the time of its initial description and in many case reports that followed, cryptococcosis was considered a highly malignant infection that generally resulted in death. Most of these reports were of patients with generalized disease associated either with cancer or some other severe underlying con-

dition, or with the meningitic type of infection that, before the advent of amphotericin B, had a fatality rate of 80 to 90 per cent. Beginning in 1950, a self-limiting form of infection began to be recognized. Baker and Haugen[6] in 1954 described healing pulmonary foci in healthy persons. Since then, a healed and sometimes calcified cryptococcoma has come to be an incidental finding in routine autopsies.[113]

The original cultures of Busse, Buschke, and Curtis were included in a study of 27 isolants of pathogenic cryptococci by Benham.[10] She concluded that there was only one species and suggested retaining the name *C. hominis.* Lodder and Kreger-Van Rij (1952)[90] determined the priority of the name *Cryptococcus neoformans,* and this is now the accepted name. Benham also reported four serotypes of *C. neoformans:* A, B, C, and D. These serotypes are not constant, however, and can vary within a strain on repeated subculture.

Recently the organism has been shown to have a perfect or teleomorphic state. As defined by Kwon-Chung,[77, 79, 81] *Filobasidiella neoformans* represents the teleomorph of *Cryptococcus neoformans.*

Although the organism had been reported isolated from milk and fruit juices (including an erroneous report of isolation from the flux of mesquite), a saprobic habitat was not fully established until the work of Emmons.[35] He reported the isolation of virulent organisms from pigeon nests, roosts, and feces. Its constant association with such habitats has been amply verified and point-source infection recorded.[107]

A complete bibliography and review of the disease was published in a scholarly monograph by Littman and Zimmerman in 1956.[87]

ETIOLOGY, ECOLOGY, AND DISTRIBUTION

The pigeon (*Columba livia*), long a symbol of urban decay, appears to be the chief vector for the distribution and maintenance of *Cryptococcus neoformans.* The organism is recovered in large numbers from the accumulated filth and debris of pigeon roosts, such as attics of old buildings, cupolas, and cornices. In this desiccated environment, it is often the predominant organism present, sharing the alkaline, high-nitrogen, high-salt substrate with *Geotrichum candidum* and a few *Candida* and *Rhodotorula* species. Its prevalence in this particular ecologic niche is a situation similar to the restricted environmental habitat required by *Coccidioides immitis.* Both organisms appear unable to survive or compete with other species in other situations. Cryptococci disappear from infected debris when it is mixed with soil. This may be due to their being phagocytized and digested by amoebas.[19] Similarly *Coccidioides* cannot compete when irrigation and cultivation of its habitat allows other fungi to proliferate. In rural areas also, *Cryptococcus* is found in sites inhabited by pigeons, such as barn lofts and hay mows. The dust in such environments contains viable virulent organisms.[130] In this desiccated state the organism may be no more than 1 μ in diameter, a condition which allows it to be inhaled into alveolar spaces.

C. neoformans does not appear to infect the pigeon, whose average body temperature is 42°C, but survives passage through its gut.[86] The organisms survive but do not grow at 44°C. In moist or desiccated pigeon excreta, they remain viable for two years or more. In sheltered sites that are not in contact with the soil, the organism is found in large numbers,[31] its survival enhanced by increased relative humidity.[66] Direct exposure to sunlight, especially in summer months, soon sterilizes the habitats of *Cryptococcus;* however, it is maintained in similar locations during winter. In saprobic environments, the organisms are essentially nonencapsulated,[66] but such strains and unencapsulated mutants do become virulent following acquisition of the polysaccharide capsule.[18] The presence of, or the potential for, encapsulation rather than the degree of encapsulation appears to be the significant virulence factor.

Cryptococcus neoformans is essentially the only etiologic agent or cryptococcosis. In some severely immunosuppressed patients, *C. albidus* and a few other species have been isolated on rare occasions. The significance of these isolations remains to be determined. *C. albidus* was isolated in a case reported by Krumholz.[76] This appears to be a valid infection caused by the organism. *C. neoformans,* is worldwide in distribution and is constantly associated with avian habitats.[13] There is no significant difference in incidence of infection which can be related to age, race, or occupation. Pigeon breeders[40] have a higher than normal occurrence of antibody but no greater rate of frank infection. This mycosis is more often reported in males than in females. This may be a function of exposure or, as has been suggested for other mycoses, a matter of hormonal differences.[100] In summary, the epidemiologic data indicate that cryptococcosis is an opportunistic infection, since the organism requires some abrogation of patient defenses in order to establish infection.

The *C. neoformans* recovered from bird's nests and guano is almost always serotype A or D. These serotypes have also been found in ripened fruits, rotting vegetables, dairy products, and even a bagpipe.[27] As mentioned earlier, they are rapidly cleared from soil because they are a "gastronomic delight" of *Acanthamoeba* species.[19] Thus, the etiologic niche inhabited by the organism is limited. The confusion concerning the ecology of habitats invaded by cryptococci is augmented by the recent findings of Bennett et al.[13] In their survey, serotype A is the most common type from a clinical or natural habitat in the United States and accounts for over 95 per cent of isolants. Serotype D is common in clinical and natural isolations in Denmark and Italy. In southern California, 25 of 45 isolants from patients were serotypes B or C; however, these serotypes have never been recovered from any natural habitat. Thus, we have yet to find natural sources for these two serotypes. One dismayed worker suggested that the B and C serotypes are "sports" of the other serotypes that occur during infections,[65] and interconversion of types is noted, but rarely. This is only one of the many mysteries surrounding this organism that remain to be solved.

There are essentially two types of cryptococcal disease, but their manifestation depends on host response rather than on the strain of organism. In the normal patient, infection following inhalation of the organism is usually rapidly resolved with minimal symptoms, and the disease, if any, is subclinical. Growth-inhibiting substances are normally present in body fluids.[65] If the number of organisms inhaled is considerable, an infection may be initiated in the lungs, with occasional transitory foci of infection in other anatomic sites. Even though the tissue reaction and cellular defenses are evoked slowly, they usually are adequate to contain the infection. Infrequently in normal patients, a chronic infection is established which gives rise to occasional flares of systemic, cutaneous, or meningitic involvement. Such patients require treatment. A single pulmonary lesion in a relatively healthy individual frequently resolves spontaneously or, if persistent, often is amenable to surgical intervention followed by adequate drug therapy.

The second type of disease is associated

with neoplasias, debilitating diseases, and compromised hosts, usually as a result of drug therapy. In these cases the host defenses are minimal and inadequate, and the disease readily spreads to involve almost all organs, particularly the central nervous system. The course of the infection may be protracted, with slow spread from one organ system to another, or it may be fulminant and rapidly overwhelming. This clinical type was the first recorded and for many years was the only form of the disease recognized. Presently, cryptococcosis is particularly associated with the so-called "collagen diseases," such as lupus erythematosus,[28] very frequently with sarcoidosis,[129] as well as neoplasias[70, 91, 92] and endogenous and exogenous Cushing's syndrome.[39] Association of the disease with pregnancy[127] and alveolar proteinosis[132] has also been noted.

As is the case with other systemic mycoses, the subclinical or mild form of cryptococcosis in the uncompromised patient has only recently been recognized. Unlike the other fungal diseases, however, it has been impossible to estimate the frequency of occurrence of subclinical infection, for as yet no sensitive and reliable immunologic tests are available. That such a mild, subclinical form occurs quite commonly is attested to by the frequent incidental finding of old, healed, cryptococcal granulomata in routine autopsies. *Cryptococcus neoformans* can frequently be recovered from sputum, skin, or other areas open to the air as incidental flora or as a transient colonizer. Hammermann et al.[59] recovered the organism in the sputa of 80 patients. Actual disease could be verified in only 28. Current estimates of the incidence of cryptococcal disease range from 200 to 300 cases of cerebral meningitis per year in the United States[72] to 15,000 subclinical respiratory infections in New York City alone.[3, 86] It has been called the "awakening giant" of the fungus diseases.[72]

CLINICAL DISEASE

With rare exception, the portal of entry of the cryptococci is the lungs. The primary pulmonary infection may remain localized or disseminate to other organs. Involvement of other tissues and organ systems sometimes occurs in spite of resolution of lung lesions.[44] For this reason, it is often difficult to assess the site of initial infection. Although experimental disease has been produced by feeding large numbers of yeast cells to animals, it is doubtful that human infection via the alimentary tract occurs with any frequency. Entrance of the organism through the skin and nasopharyngeal mucosa is possible but is also considered extremely rare. The clinical types of cryptococcal disease include pulmonary, central nervous system, cutaneous and mucocutaneous, osseous, and visceral.

Pulmonary Cryptococcosis

Primary infection of the lungs has no diagnostic symptoms, and most cases are probably asymptomatic. When present, the symptoms include cough, low-grade fever, pleuritic pain, malaise, and weight loss, but these signs are seldom prominent.[22] Scanty mucoid sputum also is produced, and rarely there may be hemoptysis.[110] Following rupture of an eroding cryptococcal focus into a bronchial branch, there is a heavy discharge of mucoid sputum containing numerous organisms. Night sweats, which are prominent in tuberculosis, seldom occur in cryptococcosis. In the rare, fulminant cases of disease there is pulmonary consolidation and pronounced fever. Lesions may develop in any part of the lung. They are frequently bilateral, but they may appear in only one lobe. Diminished breath sounds and dullness to percussion are often present, along with a pleural friction rub. Cases of lobar pneumonia have also been reported.[61] These can be bilateral or may involve a single lobe. They mimic pneumococcal diseases rather than the necrotising gram-negative type of pneumonia. The course may be fulminant with a rapid fatal outcome, or the lung fields may clear with single or multiple focal areas of disease as a residuum. In time these also may resolve without treatment. In miliary disease, moist rales may be present over either the apex or the base. Asymptomatic infections are detected only by x-ray examination. Many of the lesions are small and often go unnoticed. In addition, many lesions heal without forming granulomas (cryptococcomas) and leave no residual history of infection. Pulmonary foci sometimes completely resolve, but infection may persist as a mediastinal mass. These have been shown to infiltrate the airways and great vessels and give the appearance of an advancing neoplasm.[128] Other pulmonary lesions resolved and calcify with time.[122]

X-ray. Roentgenographic examination for pulmonary cryptococcosis reveals a vari-

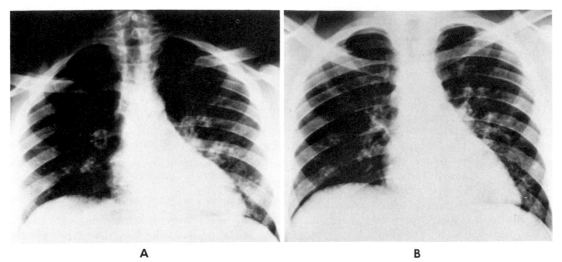

A   **B**

Figure 21–1. Acute pulmonary cryptococcosis. *A*, Ill-defined interstitial infiltrate of left lung. *B*, Similar changes ten days later with some progression. The disease healed with amphotericin B treatment, and an x-ray of one year later showed a normal chest with some residual fibrosis. This patient had been catching pigeons in an attic. Later he presented to the emergency room as a cryptococcal hepatitis.[107] (Courtesy of J. Fennessy.)

able picture, but lung changes generally fall into four categories. These depend on the recentness of the infection, the severity of the disease, and the status of host defense mechanisms.[92, 122]

The first and most common type is the discrete, solitary, moderately dense area of infiltration appearing in the upper and sometimes lower lung fields.[52] These lesions are usually from 2 to 7 cm wide and may gradually extend peripherally to simulate carcinoma, abscess, or hydatid cyst (Fig. 21–1). They are characterized by locular masses of yeasts mechanically displacing host tissue, which may rupture the bronchial wall and result in the production of mucoid, blood-tinged sputum; little or no hilar lymphadenopathy is seen. This is particularly associated with "gelatinous" cryptococcal pneumonia (see Pathology). If they heal, such lesions disappear altogether or form small granulomata which infrequently calcify.

In the second type of lesion there is a broader, more diffuse infiltrate involving the upper or lower lung areas. Vessel markings and nodular shadows are accentuated. In some patients, particularly those with chronic obstructive lung disease or some degree of immunosuppression, the infection can flare and spread rapidly to all lung fields. These become bilateral interstitial pneumonias,[41] often resulting in a relentless downhill course. Proper diagnosis and prompt appropriate therapy, however, can result in resolution of even severe, widespread disease. A

residual fibrosis remains following healing (Fig. 21–2). Other patients may develop a variety of clinical types following the initial diffuse infiltrative disease. These include a consolidating pneumonia,[39] (often segmental or cavitary), lymphadenopathy, pleural effusion, and, rarely, miliary disease.[52] Cryptococcal involvement of the lung may coexist with a variety of other fungal, neoplastic, and

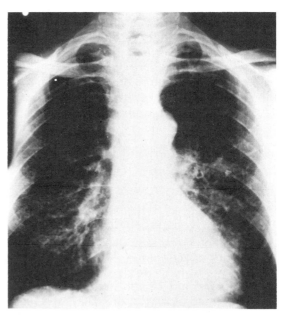

Figure 21–2. Cryptococcosis. Bilateral, interstitial infiltrate adjacent to the hilum; extension to lingula of left upper lobe.

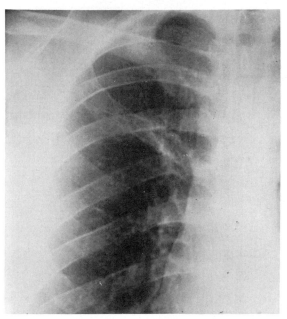

Figure 21–3. Cryptococcosis. Healing peribronchial disease. Ill-defined scarlike process superior to hilum of right lung.

bacterial diseases. Cryptococcal pneumonia is not infrequently a sequelae of viral or pneumococcal disease and has been encountered incidental to tuberculosis.

Extensive peribronchial infiltrates occur in the third type (Fig. 21–2). These extend widely to form woolly shadows which resemble active tuberculosis. However, fibrosis is minimal (Fig. 21–3), and caseation necrosis and cavitation are exceptional. Again calcification is rare.[122]

Any of the above types of lesions may occur in primary infections, the degree of involvement being related to the general health of the individual and the dose of the infectious agent inhaled. Dissemination to the central nervous system may occur in any of these types, apparently with equal frequency, or in infection with no evident pulmonary diseases.

The fourth form of cryptococcal lung disease resembles miliary tuberculosis and usually occurs in lymphoma and leukemia patients. Lesions consisting of small gelatinous granules are present in all lung fields, but little tissue response is elicited. In time, small granulomata may be formed, or the disease becomes relentlessly progressive and disseminates.

Differential Diagnosis. Several points are considered important in differentiating cryptococcal disease from other mycotic and bacterial infections.[87] In cryptococcosis there is marked predilection for disease to become established in the lower lung fields. Cavitation, fibrosis or calcification, hilar lymphadenopathy, and pulmonary collapse are rare occurrences. Coin lesions, which are so common in spent histoplasmosis or coccidioidomycosis, are uncommon in cryptococcosis (Fig. 21–4). Thin-walled cavities usually do not develop. "Crab claw" shadows, so-called because they extend from a focal lesion and emulate carcinoma, are found in blastomycosis but not in cryptococcosis. Definitive diagnosis of the disease depends on the isolation of the organism. Although *C. neoformans* is not part of the regular resident human flora, it may be a transient colonizer. It is necessary to identify specifically any encapsulated yeast isolated from sputum to establish the diagnosis of cryptococcosis. In some infections, particularly low-grade infections, however, the capsule may be absent or very small.[37, 61]

Central Nervous System Cryptococcosis

This is the most frequently diagnosed form of cryptococcosis, though certainly not the most common site of the disease. The reason for the predilection of *Cryptococcus* for the central nervous system has not been explained. The organism probably encounters less cellular (phagocytic) response there, and it has been theorized that selective nutritional factors for the yeasts are present. Simple sources of nitrogen, such as asparagine and creatinine, are found in the spinal fluid and may stimulate growth. The absence of inhibitory factors reputed to be in serum may also play a role here, but conclusive evidence is lacking. Before drug therapy became available, this form of the disease was almost always fatal, but the mortality has been cut to about 6 per cent by the use of amphotericin B. Retreatment is often necessary, as the treatment failure rate is about 25 per cent with this drug.

Cryptococcosis of the central nervous system is usually one of three clinical types. These are meningitis, meningoencephalitis, or expanding cryptococcoma.[141] The first is the most common. This disease is generally a leptomeningitis, since involvement of the dura mater is uncommon. The latter is subject to tumor-like cryptococcal masses, however. The only presenting symptom found in almost all patients with leptomeningeal disease

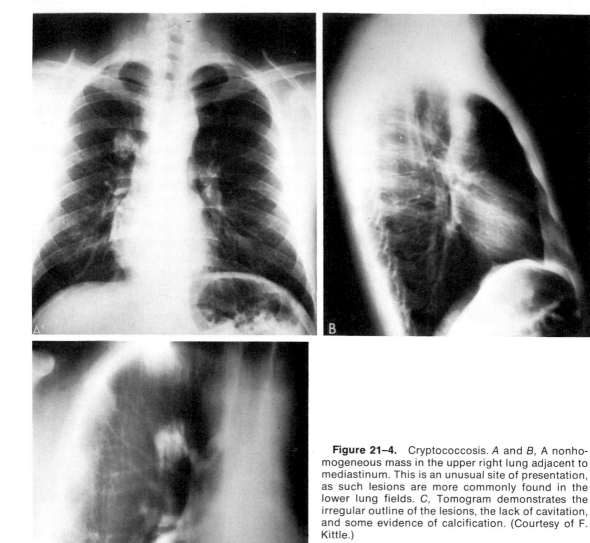

Figure 21–4. Cryptococcosis. *A* and *B,* A nonhomogeneous mass in the upper right lung adjacent to mediastinum. This is an unusual site of presentation, as such lesions are more commonly found in the lower lung fields. *C,* Tomogram demonstrates the irregular outline of the lesions, the lack of cavitation, and some evidence of calcification. (Courtesy of F. Kittle.)

is headache. It is usually frontal, temporal, or retro-orbital, intermittent, and of increasing frequency and severity. Fever is also frequently present. In time, most patients show the remaining signs of meningitis: nuchal rigidity, tenderness of the neck, and positive Kernig's and Brudzinski's signs.[49] The spinal fluid is usually clear, and none to few to many cryptococci can be seen in India ink preparations. In addition there is some cellularity (800 per ml) that consists primarily of lymphocytes. The opening pressure is elevated, and

glucose is usually low. It must be remembered that most patients with cryptococcal meningitis have only symptoms of headache and/or mental change for long periods before any other manifestations occur. Kornfeld et al.[74] described cases of cryptococcal disease in which the picture of a purulent meningitis was seen on spinal tap. All patients had very high cell counts that were predominantly neutrophils, thus implying a pyogenic meningitis; however, encapsulated yeasts were seen in wet mounts. All patients had some underlying abnormality, viz., chronic active hepatitis, polycystic kidney disease, or Hodgkin's disease. Meningoencephalitis, an uncommon form, is a fulminant, rapidly developing infection,[36] often leading to coma and death within a short time. Prompt diagnosis and appropriate treatment have averted this in some cases.[142]

A single, localized, cryptococcal granuloma of the brain (cerebral cryptococcoma) may produce signs only of an expanding intracranial mass. Nausea, vomiting, mental changes, coma, paralysis, and hemiparesis frequently develop. In other patients, headache develops first, followed slowly by intermittent drowsiness, slurred speech, double vision, and unsteadiness of gait.[133] These symptoms may appear over a period of several months, often steadily increasing in severity and resulting in prostration and coma.

Ocular manifestations such as blurring, vertigo, diplopia, photophobia, ophthalmoplegia, nystagmus, amblyopia, and papilledema occur.[89] The latter may require multiple lumbar punctures for control. The mental disturbances may be quite marked and include irritability, agitation, apathy, loquaciousness, confusion, defective memory, hallucinations,[89] or even frank psychosis. Some patients present with epileptic seizures. Hemiplegias and hemiparesis may develop as the result of expanding granulomatous masses. Urinary incontinence and loss of patellar and Achilles tendon reflexes also occur, but some patients become hyperreflexive. Symptoms of acute infection and toxemia are absent or rare. A spinal tap usually reveals increased pressure, often up to 700 mm of water. Release of pressure during a tap may restore comatose patients to consciousness, but later they may sink back into delirium and death. The spinal fluid is clear, rarely cloudy; the cell count is elevated but usually not over 800 per ml. Lymphocytes are more numerous, but occasionally neutrophiles predominate. The protein varies from 40 to 600 mg per 100 ml. Chlorides are low (183 mEq per L or lower) and sugar is quite low (often 10 mg per 100 ml). The colloidal gold curve is "paretic" or "tabetic." Organisms may be sparse or numerous, and a spun specimen examined with India ink usually confirms the clinical diagnosis (Fig. 21–5). Frequently no organisms are seen, particularly if the lesions are restricted to the brain itself. Cryptococcomas can now be envisioned with the aid of computed tomography.[133] No pattern is characteristic, however, and some indication

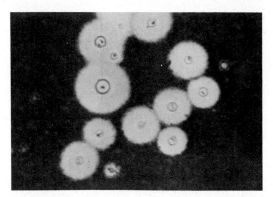

Figure 21–5. Cryptococcosis. India ink preparation of spinal fluid showing yeast cells surrounded by large capsule. From Rippon, J. W. *In* Freeman, B. A. 1979. *Textbook of Microbiology.* Philadelphia, W. B. Saunders Company, p. 773.

of the presence of the organism must be found by other procedures. Cryptococcal polysaccharides are usually present in the CSF in active infection,[41] and antibodies are detectable in healing or resolved disease. The clinical and spinal fluid pictures are indistinguishable from those of tuberculous meningitis, and double infections have been reported.[47] The latter disease usually has a more rapid course and is more frequently seen in children. Differentiation of cryptococcosis from other mycoses can be made only by isolation and identification of the organism.

The course of cryptococcosis extends from a few months to 20 or more years. Aside from the resolved subclinical disease, however, the usual outcome for immunocompromised patients or those with other deficiencies is a rapidly progressive deterioration. Untreated cryptococcal meningitis is fatal, with very rare exceptions.

Cutaneous and Mucocutaneous Cryptococcosis

The first human case of cryptococcosis described by Busse and Buschke had manifestations of cutaneous lesions, bone involvement, and systemic disease, but there was no meningeal disease. Subsequently most of the cases recognized in the United States were meningitis, but in Europe cutaneous and mucocutaneous disease was considered more common. No satisfactory explanation has been given for this difference, even though extrameningitic disease is now recognized more often, while cutaneous involvement is still considered uncommon. Cutaneous and mucocutaneous disease in man is usually a manifestation of disseminated disease and occurs in about 10 to 15 per cent of cases. Transient cutaneous lesions occur in experimental infections of monkeys.[7] Such lesions are not regularly encountered when other experimental animals are used. There is seldom convincing documentation that such lesions represent the primary sites of infection.[103] Data from animal infections suggest that traumatic implantation is an important means of entry of the yeast, but the same is not true in human disease. Lesions developing at sites of injury in man may represent seeding from a small internal focus rather than an exogenous inoculation. Absence of regional lymphadenopathy particularly casts doubt on an external origin of such cutaneous lesions.

The lesions of the skin present as papules, acneform pustules, or abscesses that ulcerate with time. In the few well-documented cases of primary cutaneous inoculation in immunologically normal patients, the lesions were chancreform and limited. Usually they resolve spontaneously, or, as in the case reported by Abdel-Fattah et al.,[1] surgical excision will suffice. Mucocutaneous lesions occur about one third less frequently than skin disease. They present as nodules, as granulomas, or as deep or superficial ulcers and are secondary to other foci of infection.

Primary and secondary cutaneous cryptococcosis is a regularly encountered clinical manifestation in the immunosuppressed patient. It is often difficult to determine if a lesion is primary, but Noble and Fajardo[103] and Iacobellis et al.[65a] indicate that primary lesions are superficial and ulcerative rather than deep and necrotic. Primary lesions often resolve spontaneously or with administration of intravenous amphotericin B.[65a] This obviates the possibility of dissemination of the disease in a setting of immunosuppression.

Secondary cutaneous cryptococcosis in a patient with underlying disease is generally a manifestation of wider dissemination. The lesions that develop are variable. They may appear as superficial ulcerations with evidence of necrosis and eschar formation (Fig. 21–6,*A*), but they also occur as deep thrombotic lesions[121] or with the appearance of a cellulitis.[46] Often very destructive, such lesions start as a cold abscess that develops a deep, wide area of necrosis and eschar formation and ends in sloughing (Fig. 21–6).[121] Schupbach et al. emphasize that these lesions most often are regarded as a bacterial cellulitis or an area of herpes vesiculation. An aspirate of the lesion or Tzanck preparation may show the true nature of the etiology.[121]

Osseous Cryptococcosis

Bone is involved in about 5 to 10 per cent of reported cases, which is considerably less than the frequency recorded in blastomycosis, coccidioidomycosis, and actinomycosis. As in other diseases caused by fungi, the cryptococci have a predilection for bony prominences, cranial bones, and vertebrae.[17, 104] The joints are usually spared except by direct extension, but isolated lesions have been recorded.[23] There is no characteristic x-ray appearance, but osseous lesions are similar to those of blastomycosis and coccidioidomycosis, i.e., they are usually multiple, discrete, widely disseminated, destructive, chronic, and slowly changing (Fig. 21–7). Swelling and pain may be present. For diagnosis, the relative stationary appearance of the lesion and the lack of periosteal proliferation suggest cryptococcosis rather than other mycoses.[30] Periosteal proliferation is common in actinomycosis and regularly seen in blastomycosis and coccidioidomycosis. Synovitis without bone involvement is quite rare but has been reported. Not infrequently, a solitary osseous lesion is the only evidence of cryptococcal disease.[83a, 104] Differential diagnosis of osseous cryptococcosis must also include other diseases of the bone. Unless special stains are

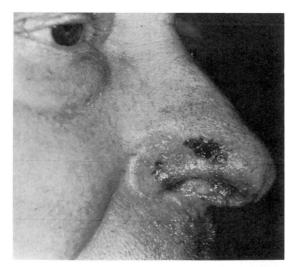

Figure 21–6. Cryptococcosis. Superficial ulcers involving mucocutaneous areas of nose in a case of disseminated disease. Note the necrotic area with the black eschar.

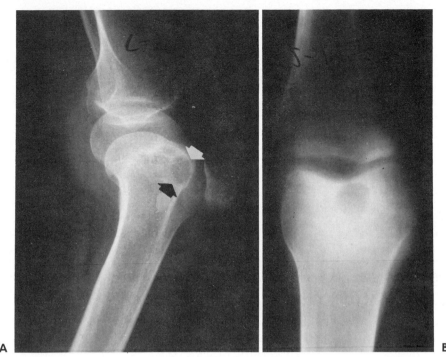

Figure 21–7. Cryptococcosis. *A,* Lateral film of knee. Spherical, well-defined lytic area in the distal end of femur (intercondylar fossa), adjacent to articular surface. Suggestive evidence of intra-articular effusion. *B,* Tomogram demonstration of sclerotic reaction about lytic lesion. The patient presented with a swollen knee from which nonencapsulated cryptococci were grown. The lung fields had areas of fibrosis. There was no other evidence of systemic disease.[83a]

used, examination of biopsy or excised material may lead to a diagnosis of osteogenic sarcoma or Hodgkin's disease, because the histologic picture is quite similar (Fig. 21–10,*D*). Chleboun and Nade[26] have recently reviewed the subject of skeletal cryptococcosis.

Visceral Cryptococcosis

In disseminated disease, any organ or tissue of the body may have foci of infection. Granulomatous gelatinous lesions are produced which symptomatically and even histologically resemble malignant neoplasias, particularly those of the myxomatous type. Heart, testis, prostate, and eye[113] are frequently involved, whereas kidney, adrenal, liver, spleen, and lymph nodes are usually spared. In rare cases, patients have presented with an acute abdomen with only liver involvement.[107] Rarely, massive involvement of the adrenals may give rise to signs of Addison's disease, as is sometimes seen in blastomycosis and paracoccidioidomycosis.[115] More frequently present when the adrenals

are involved are small areas of obliteration due to cryptococcal granulomata, in contrast with the massive toxic or caseating necrosis found in other fungal infections. Thus, signs of adrenal insufficiency are uncommon in cryptococcosis. Lesions in the gastrointestinal and genitourinary tract mimic tuberculosis, but they are much less frequently encountered in cryptococcosis. They may be of the gelatinous or granulomatous type but are rarely of sufficient size to manifest symptoms ante mortem.

In late-stage disease, the yeasts are spread hematogenously (cryptococcaemia),[111] and the organisms are seen in the glomerular loops of the kidneys as well as in the capillaries of many organ systems. If sufficient time has elapsed before death, lesions will have developed in some of these organs. Most commonly encountered are small gelatinous cysts in the spleen and small organizing granulomas of the lymph nodes. In the latter, necrotic foci and healed fibrotic granulomas, characteristic of histoplasmosis, are not found. Small lesions are also found, but much less frequently in the myocardium, the liver, and the pancreas. They occur as granulomata or cryptococcomas and are almost never symptomatic. Endocarditis lenta[29] is a rare form of involvement, as are cryptococcal mycotic aneurysms of the aorta.[109] Expanding granulomata in the prostate have occasionally been noted. These are sometimes the cause of symptomatic prostatitis. In the case illustrated in Figure 21–10,*D*, the cryptococcal lesion was old and sclerosing, but a superinfection with bacteria had caused a symptomatic prostatitis. In the published reports of extensive autopsy series, cryptococcal lesions have also been very rarely encountered in the testes, thymus, and thyroid but not in the pituitary gland. The skeletal muscle, larynx, trachea, esophagus, and stomach have not been recorded as having cryptococcal lesions.[87] Lesions in the human breast, a common site of bovine disease, have been reported only twice, and no reports of invasive lesions in the female genitalia have yet been published.

The eye is involved either by direct extension from the subarachnoid space or by hematogenous spread from other loci, resulting in the formation of multiple lesions in the uveal tract, retina, and vitreous (cryptococcal chorioretinitis).[5, 89] Such involvement has occurred as the only presenting symptom in patients on immunosuppressive drug therapy[5] and prednisone,[89] and subsequent to corneal transplantation.[14]

Nosocomial infections are very rarely encountered. Colonization of indwelling catheters by yeasts is common, but these are usually members of the genus *Candida* or, rarely, *Rhodotorula* or *Trichosporon*. Nosocomial septicemia due to *Cryptococcus neoformans* followed by establishment of infection and death has been documented, however.[71]

DIFFERENTIAL DIAGNOSIS

In general, cryptococcosis is most often mistaken for malignant neoplasias. This is true particularly of pulmonary and central nervous system disease, wherein there are signs and symptoms of an expanding mass. All forms of cryptococcosis mimic tuberculosis or other fungal diseases, but the lesions are obstructive and obliterative rather than necrotic. Meningitis of cryptococcal origin is more protracted than tuberculous disease but may resemble encephalitis, dementia psychosis, dementia paralytica, or bacterial infections, especially those caused by *Brucella* and *Listeria*.

PROGNOSIS AND THERAPY

Local lesions of the lungs in normal patients have a good prognosis. They heal slowly without treatment and disappear or leave a residual scar. Rarely there is hematogenous spread to the central nervous system. When this happens the prognosis becomes very grave unless treatment is instituted. In systemic infections, the outcome is usually fatal, unless treated, particularly if the patient is debilitated. Primary cutaneous or mucocutaneous lesions generally resolve spontaneously. Chronic disease of any organ system is marked by alternating intervals of remission and exacerbation, but usually the disease terminates fatally. Such periods of quiescence and recrudescence have posed great difficulties in the past in the evaluation of therapeutic modalities. At present there are several drugs available with clinically proven efficacy. Whereas minor pulmonary disease or primary cutaneous involvement may cure spontaneously or by surgical excision, meningeal or systemic disease requires adequate treatment with one of these drugs. The standard drug for therapy is amphotericin B. The dose regimen is similar to that for other systemic mycoses: 0.6 mg per kg per day to a total of 3 to 5 g intravenously ("piggyback" with 5 per cent glucose given as a one- to six-hour perfusion). If the test dose of the drug is not well tolerated or if the patient's BUN or creatinine level rises, the dosage can be lowered to a few milligrams per day, administered in conjunction with 50 mg diphenhydramine (Benadryl) and 10 grains aspirin.[12] Alternate-day therapy increases patient tolerance. The drug levels of sera and spinal fluid can be accurately gauged using the medium effective dose (ID_{50}) determination of Bennett.[11] Patients with underlying disease, such as lymphomas and leukemias, or in immunosuppressed states may require more than 3 g total dose. In the experience of most clinicians, intravenous therapy is adequate for meningeal as well as other forms of disease. Several studies have recently established the efficacy of combined drug therapy using amphotericin B and 5-fluorocytosine.[12, 68] In the study by Bennett et al.,[12] the combination cured or improved more patients, produced fewer failures or relapses, and resulted in more rapid sterilization of cerebrospinal fluid and less nephrotoxicity than amphotericin B alone. This appears to be the treatment regimen of choice for most forms of cryptococcal disease not amenable to other procedures. An appropriate dose was 0.35 mg per kg per day amphotericin B intravenously and 150 mg per kg per day 5-fluorocytosine orally. Failures on this drug regimen do occur and have been successfully treated subsequently with miconazole.[102]

Selected patients may require intrathecal administration, but arachnoiditis with severe residual damage is a not infrequent sequela to the use of amphotericin B by this method. Chronic pulmonary lesions as well as some osteal lesions have been cured by surgical excision.[22, 30] However, meningeal infections following surgical procedures are not uncommon, and prophylactic amphotericin B is recommended.[22]

Five-fluorocytosine (5FC) appears to have some efficacy in the treatment of cryptococcal meningitis.[124] Since it can be given orally and is relatively nontoxic, it can be instituted immediately at full dosage levels (8 to 10 g daily to a total of 370 g). This drug alone is often associated with treatment failure, so that combined therapy with amphotericin B is the mode of therapy currently used.[12] The failure rate of 5FC is about 60 per cent in the experience of some clinicians, whereas it is about 25 per cent for amphotericin B. Some strains of *C. neoformans* rapidly develop resistance to 5FC.

As noted above, even with the use of combined therapy, some patients relapse or fail to respond. There are records of response to intravenous miconazole following failure of combined amphotericin B and fluorocytosine,[102] and the drug has been used intraventricularly in meningitis when the disease was refractory to amphotericin B.[55] Miconazole alone as a first choice of treatment has been disappointing.[12] Garlic has been noted to be active against *Cryptococcus* as well as several other fungi.[42] An extract has been used as an oral and intramuscular agent in patients with cryptococcal meningitis.[4] Ketoconazole is being evaluated.

PATHOLOGY

Gross Pathology. Cryptococcal infection is accompanied by minimal inflammatory reaction; therefore, even in fatal cases of meningitis the gross appearance of the meninges and brain is almost normal. The lesions are of two principal types which correlate with the duration of the disease.[6, 113] Early lesions are gelatinous (mucinous), whereas older lesions become granulomatous. The early lesions characteristically contain aggregates of encapsulated budding cells intermixed with a fine reticulum of connective tissue. These lesions enlarge and mechanically compress the surrounding tissue. Fibrovascular proliferation is minimal in most lesions but tends to be pronounced in healing primary infection sites, such as discrete, subpleural granulomas. Granulation tissue is seen in localized cutaneous lesions, but the pseudoepitheliomatous hyperplasia frequently observed in blastomycosis and coccidioidomycosis is absent. Fibrous encapsulation which characterizes histoplasmomas and tuberculomas is infrequently encountered in cryptococcosis. Even in well-formed granulomata containing lymphocytes and epithelioid cells, caseation necrosis and cavitation are rarely observed. Significant calcification is unusual, but the spheroidal concretions which are present in granulomata of other etiology are sometimes mistaken for the yeast

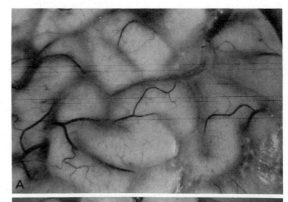

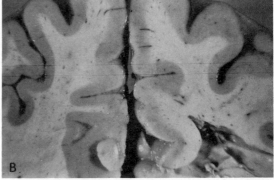

Figure 21–8. Cryptococcosis. *A*, Glistening surface of brain from fatal meningeal disease. The convoluted cortex is easily seen in some areas and obscured in others. There is cloudiness throughout the pia arachnoid with some small, cottony, white patches in some areas. *B*, In the sulci the pia is edematous, mucoid, and cloudy. The material within the space is gelatinous, glistening, and mucoid. There is some vascular proliferation, but no involvement of the cerebrum itself.

cells of cryptococci. The degree of the granulomatous response is influenced by the innate resistance of the patient to infection. In cases of Hodgkin's disease, lymphoma, or leukemia in which the resistance of the patient is lowered, the inflammatory response is essentially absent.

Pulmonary Cryptococcosis. Gross examination of the lung may reveal a localized process or extensive involvement. Miliary tubercles similar to those seen in tuberculosis are present in disseminated cases. Solitary cryptococcal granulomata of the lung from uncompromised patients vary in size from 2 to 7 cm in diameter. They most frequently are observed at the periphery of the lobes, at the hilum, or in the center of a lobe, in that

order. The lesions may be solid, firm, rubbery, nonspecific granulomata resembling those of sarcoid, or gelatinous, mucoid masses resembling myxomatous neoplasms. Cryptococcal lesions frequently coexist with tuberculosis, other mycoses, Hodgkin's disease, sarcoid, and lymphomas, thus confusing the diagnosis. The yeasts are sometimes seen infecting lymphoma lesions.

Cryptococcal Meningitis. As noted previously, the gross changes in the meninges, even in fatal cases of disease, may be so minimal that cryptococcal infection is not suspected. The organisms are dispersed so evenly over the subarachnoid spaces that a cloudiness is not perceived in x-ray examination (Fig. 21–8). Meningeal reaction is more

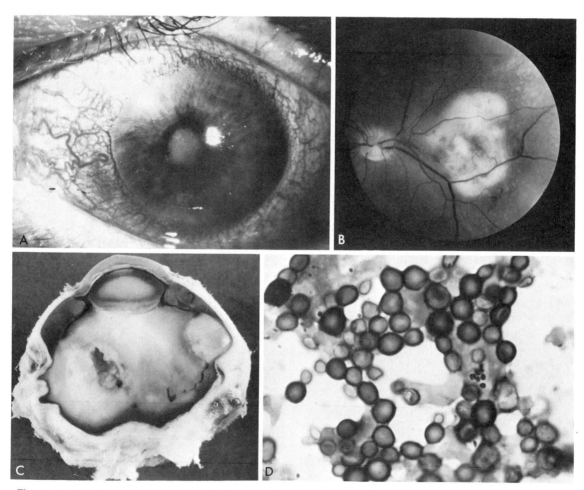

Figure 21–9. Cryptococcal chorioretinitis. A patient had obstructive hydrocephalus of obscure etiology. Bilateral inflammatory eye lesions developed. *A,* Blind left eye with epibulbar injection and cataract. Rubeosis iridis is present. *B,* Fundus photograph of right eye with 45-degree lens. Lesion has extended across large vein. *C,* Blind left eye was enucleated. Sectioning demonstrates dense white material filling globe. *D,* Meyer mucicarmine stain of material showing many budding yeasts. (From Shields, J. A. et al. 1980. Am. J. Ophthal. *89:*210–218. Used with permission.) In this patient, the process in the right eye was checked by use of amphotericin B and 5-fluorocytosine.[125] (Courtesy of J. A. Shields.)

pronounced at the base of the brain and in the dorsal area of the cerebellum. In these areas thickening and opacity of the membranes are seen. Although pachymeningitis has been observed, leptomeningitis is seen more frequently. The meninges are hyperemic, and there is slight flattening of the convolutions of the cerebrum. The subarachnoid space is distended by a grayish, mucinous exudate, and the brain surface appears to be covered by "soap bubbles."[113] Small tubercles sometimes develop along the small vessels of the brain and meninges. The surface of the brain may appear dimpled because of the presence of cystoid lesions. When the inflammatory reaction is patchy and granulomatous, the membranes may adhere to the cortex. If the meningeal disease progresses to encephalitis, the cut surface may appear normal or contain numerous cystic spaces. Brain involvement usually results from embolization from other foci of infection. Embolic lesions are usually deep and consist of large, mucoid cysts. These cysts are found beneath the ependyma and in the periventricular and periaqueductal gray matter, the basal ganglia, the cerebral white matter, and the cerebellar dentate nucleus. Single lesions are sometimes found in the thalamus. Enlarging cerebral lesions (cryptococcomas) have the radiologic appearance of a growing neoplasm or granuloma. As a consequence nearly one fourth of patients with central nervous system cryptococcosis have had a neurosurgical procedure performed before the correct diagnosis was known.

Involvement of Other Organs. In disseminated cryptococcosis, virtually any tissue may be colonized; however, localization of the infection outside of the lung and brain is infrequent. Cutaneous lesions are characterized by stretching and attenuation of the epidermis, but the hyperplasia seen in blastomycosis is absent. These lesions usually ulcerate. In the soft tissues, large cystic masses are formed that clinically and grossly mimic lipomas or myxomas. Upon excision these masses are seen to contain bright, gelatinous, shining material. Masses of organisms are present inside. In contrast, in small, hard,

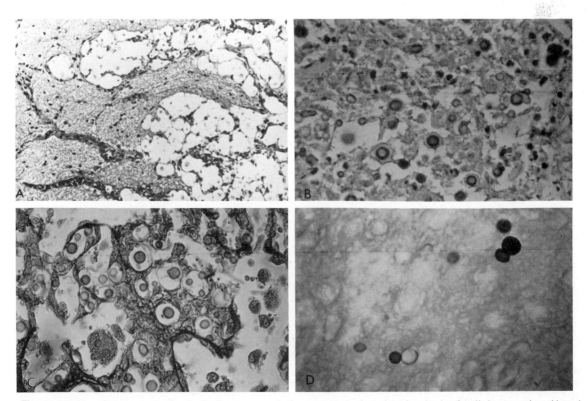

Figure 21–10. Cryptococcosis. *A,* Optic nerve, gelatinous-type lesion showing lack of cellular reaction. H and E. ×400. *B,* Mucicarmine stain of tissue from cryptococcosis of lung. Note "car tire" appearance of cells due to deep stain of capsule around yeast cell body. *C,* Gridley stain of lung tissue. The yeast cell body is stained, and the capsule appears as an unstained halo around it. *D,* Methenamine silver stain of synovial biopsy (same case as in Fig. 9–7), showing rare case of noncapsulated organism. ×400.[83a] (From Rippon, J. W. *In* Freeman, B. A. 1979. *Textbook of Microbiology.* 20th ed. Philadelphia, W. B. Saunders Company, p. 772.)

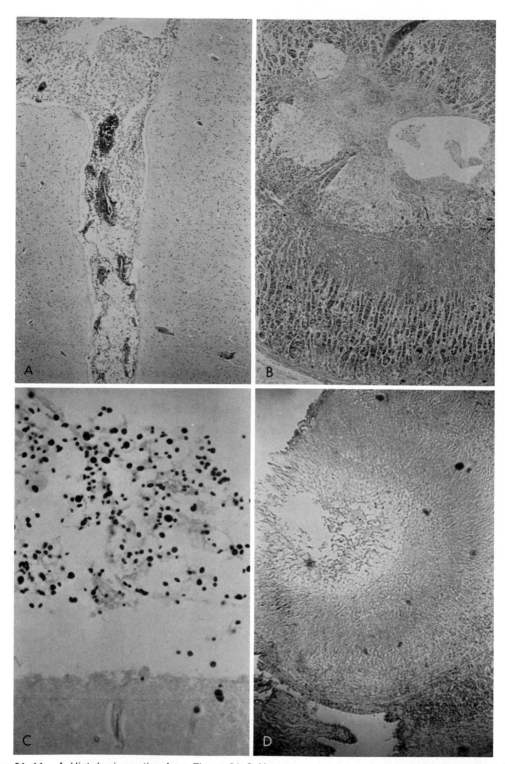

Figure 21–11. *A,* Histologic section from Figure 21–8. Numerous cryptococci are seen in pia arachnoid space, with focal lymphocytic aggregates around blood vessels. *B,* Adrenal gland from disseminated disease. Infiltrate and many cryptococci are seen in inner cortex, medulla, and vein walls. Note minimal cellular response, which consists of a few lymphocytes. H and E. ×100. *C,* Histologic section from Figure 21–8, *A* (methenamine silver, ×400). Numerous cryptococci in pia arachnoid space showing variation in size. *D,* Old inactive lesion in prostate. There is a central area of necrosis with margination of epithelial cells and a scattering of giant cells peripherally. Crytococci were demonstrable by Meyer mucicarmine stain. ×100.

granulomatous lesions, the organisms are not present in large numbers, and the diagnosis may be difficult to establish. In chorioretinitis of the eye (Fig. 21–9) and in solitary granulomas in bone, the diagnosis of cryptococcosis may be difficult to establish histopathologically. Visceral involvement usually does not cause clinically apparent enlargement of the involved organ, nor is distortion of the organ obvious upon gross examination at autopsy. The lesions are usually very small and are detected only on histologic examination.

Histopathology. Although the tissue reaction in cryptococcosis varies considerably from organ to organ and from case to case, in general the response is minimal. Frequently only tissue macrophages are found, but in rare cases there may be many giant cells and dense infiltration of plasma cells and lymphocytes. Two basic histologic patterns may occur: gelatinous and granulomatous.[6, 113] Both types may be present in the same lesion. In the gelatinous type, masses of organisms occur, and there is mucoid degeneration of the invaded tissue (Fig. 21–10,*A*; Fig. 21–11,*A–C*). Granulomatous lesions consist of histiocytes, giant cells, and lymphocytes, along with some fibroblastic activity (Fig. 21–9,*B,D*). The number of yeasts in granulomatous lesions is much fewer than in gelatinous type lesions. In the latter, the yeasts are mostly free in the tissue, whereas in the former they are almost all within giant cells and histiocytes. The average size of the yeasts ranges from 5 to 10 μ, and a large, gelatinous capsule is usually but not always present. The latter does not stain in hematoxylin and eosin (H and E) preparations but can be made visible by the Meyer mucicarmine stain (Fig. 21–10,*B*). The cells can be stained by the periodic acid–Schiff (Gridley) (Fig. 21–10,*C*) or methenamine silver (Grocott-Gomori) stain (Fig. 21–10,*D*). Very rarely the capsule is entirely absent or so small that the mucicarmine stain is negative.[37, 83a] In H and E slides the cells may appear as faint purple bodies surrounded by a blank space which was formerly occupied by capsular substance. The capsule is essentially immunologically inert, which accounts for the paucity of inflammatory cellular response. The yeast cells vary considerably in size, particularly in old healed lesions (Fig. 21–12). They are frequently as small as *Histoplasma capsulatum* and may be as large and thick-walled as *Blastomyces dermatitidis*. Unless budding is very apparent the yeasts also may resemble developing spherules of *Coccidioides immitis*.

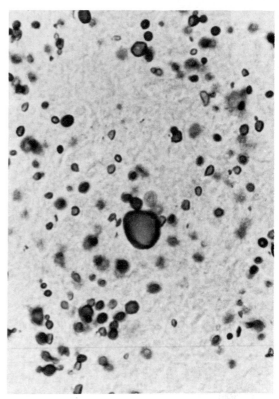

Figure 21–12. Cryptococcosis. Variation in size of yeasts present in healed lesion. This is from an old pulmonary granuloma of 1 cm in size. The lesion was calcified and fibrosed, and the organisms were dead. Note small bodies, giant yeasts, and crescent-shaped forms. Some cells retained enough capsular material to be Meyer mucicarmine-positive. Methenamine silver stain. ×400.

Candida yeast cells usually are mixed with mycelial units; however, hyphae can occur rarely in cryptococcal lesions[28] (Fig. 21–13). The capsule may be small or absent in old healed lesions, thus tending to confuse the diagnosis. As noted in the section on clinical disease, a cryptococcaemia often occurs late in the infection before death; thus, cryptococci are found in capillaries in many organ systems and in the glomeruli of the kidneys. These findings are not necessarily associated with the establishment of an invasive lesion in that organ system, however.

In the solitary, healed, fibrosed granuloma, the yeast are usually dead, and the capsules have mostly disintegrated. These granulomata are often encapsulated and partially calcified. They contain clefts of cholesterol and reveal considerable fibrosis. Such lesions therefore resemble healed tuberculomas or histoplasmomas, both of which may also be incidental findings in a routine autopsy. *Cryptococcus* will not be suspected unless a

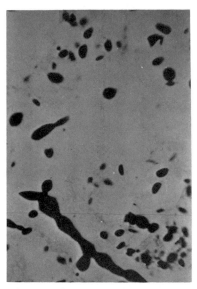

Figure 21–13. Cryptococcosis. Rare finding of mycelial-like elements in tissue of lung. Methenamine silver stain. ×600. (Courtesy of C. T. Dolan.)

methenamine silver or periodic acid–Schiff stain is done. The yeasts are essentially invisible when stained with H and E but may be extremely numerous on the specially stained preparations. The diagnosis of cryptococcosis may be verified by the following: the presence of yeasts varying considerably in size from 5 to 20 μ; no evidence of broad-based buds, but when budding is present it is thin-necked; no mature spherules with endospores; no mycelium; and a few cells with evidence of a capsule. A Meyer mucicarmine stain will demonstrate those few cells with capsules. A fluorescent antibody stain will confirm the diagnosis. It should be noted at this point that *C. neoformans*, like the other agents of systemic mycoses, can remain moribund but viable for long periods of time (decades) in lesion material. Thus, in old healed granulomata, using special techniques (hypertonic broth), cryptococci can sometimes be recovered.

ANIMAL DISEASE

Natural Infection. One of the earliest recordings of cryptococcal infection was that of a myxomatous lesion of an ox.[117] Since then, spontaneous infection has been reported in horse,[8] dog,[34] fox, cat,[95, 139] dolphin,[99] cheetah, civet, monkey, guinea pig, ferret, and dairy cattle.[105] As with human infection, in animals the central nervous system,[8] the

eye,[34] and the lung are most frequently involved; however, small lesions are also found on the skin. Although the lungs are probably the site of initial infection, in several reports of the disease, particularly in the horse, cat, and dog, prominent lesions were found in the facial region, nose, and hard palate. These could be interpreted as cases of direct inoculation. Massive granulomata in the stomach and intestine have been described, suggesting that the organism was swallowed, and the portal of entry was the alimentary tract. The pathology, tissue reaction, and histopathology of animal cryptococcal infection do not differ markedly from those of human disease.

Cryptococcal mastitis in dairy cows is not an uncommon disease and is worldwide in distribution. Cryptococcus was early associated with milk and dairy products. In one reported enzootic, 106 cases occurred in a herd of 235 Holstein-Friesian cows.[105] The infection was acquired through the teats from contaminated milking machines. The organisms proliferated in the udder, causing distension of the mammary, decrease in milk volume, and production of viscid secretions. In the more severe cases of infection, the regional and deeper lymph nodes were involved. Cryptococcal disease may be differentiated from bacterial mastitis by the slight elevation of temperature and absence of toxemia seen in the latter.

Circumscribed cryptococcal lesions are often amenable to surgical excision. Such manipulation can result in dissemination, however, so prophylactic drug therapy should be used as a cover. More extensive disease requires administration of antimycotic agents. At present, combined amphotericin B and 5-fluorocytosine therapy is the modality of choice. Weir et al.[139] recently reviewed therapeutic procedures that have been useful in small animals and reported the successful treatment of a cat with combined drug therapy.

Experimental Infection. This can be regularly produced in laboratory animals and is an important diagnostic procedure. Rabbits are more resistant than mice to infection. This may be because of the higher body temperature (39.6° C) of the rabbit.[38] Mice are the animals of choice for most experimental work. For infection, cryptococcal organisms are obtained from a 2- to 4-day-old culture and suspended in saline. Most strains are virulent within the range of 1×10^4 to $5 \times$

10^6 cells. This count is contained in 0.02 to 0.03 ml saline and injected into the tail vein of mice. A rapid, predictably progressive, and fatal disease is produced. Mice may begin to die within a week to ten days after infection. In most animals the skull is swollen, indicating meningitis. All viable mice should be autopsied at the end of two weeks. This technique of infection is so reproducible and accurate it can be used to evaluate the efficacy of chemotherapeutic agents.[56, 124] It is also the most critical test used in differentiating *C. neoformans* from other *Cryptococcus* species. Intracerebral as well as intravenous inoculation is used to test the virulence of clinical isolants. Diamond[32] has developed an experimental model using guinea pigs with the infection established by peritoneal inoculation. The results are noted in the section on biological studies.

IMMUNOLOGY AND SEROLOGY

The serodiagnosis of cryptococcal infection has received much attention and is only now becoming a standardized and useful procedure. Because of the immunologically inert nature of the capsular polysaccharide, little humoral or allergic response is elicited by infection with *Cryptococcus neoformans*. For this reason no satisfactory skin test or unequivocal complement fixation or hemagglutination tests have been developed. At present, diagnostic and prognostic tests can be carried out to measure cryptococcal capsular antigen in serum or spinal fluid.[41] The indirect fluorescent antibody test and the complement fixation test are used to detect the appearance of antibodies.[135] The presence of antigen is determined by the latex particle agglutination test (LCAT). In this procedure, hyperimmune rabbit antiserum is coated onto a suspension of latex particles,[16] which is then mixed with dilutions of serum or spinal fluid on a welled slide. The clumping of the latex particles is considered to be a positive reaction. Standardized test kits for carrying out this procedure are now available commercially and are recommended. This test appears to be very reliable for the detection of cryptococcal polysaccharide in body fluids, but it must be well controlled. For evaluating the progress of a patient during therapy, the latex agglutination test for antigen is combined with a method for detecting the appearance of antibody. A good prognosis is indicated by decrease in the titer of antigen concomitant with the appearance of antibody.[9] Correlation of the titers of antigen in the spinal fluid and serum has been a significant problem. Often the serum antigen titers remain positive, while those in the spinal fluid fall to undetectable levels. This is thought to indicate the presence of an extracerebral focus of infection. Persistence of either titer following a course of therapy denotes a grave prognosis and the probability of relapse.

There are several reproducible and accurate tests using immunofluorescent techniques (IFA) for the detection of antibody to cryptococci.[53, 135] In these, antigen is fixed to a slide, which is then covered with patient's serum or spinal fluid. The fixation of antibody to the antigen is detected by the subsequent addition of fluorescein-labeled, rabbit antihuman globulin to make an immunofluorescence "sandwich." In addition to these procedures, Gordon and Lapa[51] have developed a charcoal particle test to detect the presence of antibody. The value of serologic testing for following patients during treatment has been reviewed by Bardana[9] and Fisher and Armstrong.[41]

In summary, the serologic tests for cryptococcosis are useful, but it should be noted that both false-positive and false-negative results occur. The complement fixation test (CF Ag) for antigen appears to be most specific of all, since few false-positive results have been recorded. However, false-negative results run as high as 40 per cent. Latex agglutination (LCAT) for antigen is quite specific when used with proper controls,[16, 41, 106] but false-negative reactions still occur. This is particularly true when the disease is restricted to an isolated lesion of the bone or some other organ. In such cases the IFA or TA may be positive. False-positive latex tests occur even when controls for rheumatoid factor have been used. Such false-positive results are sometimes found in *Klebsiella* infections, but a dual infection must be ruled out. The indirect fluorescent test (IFA) for antibody is positive in 1 per cent of normal spinal fluids and sera and in up to 30 per cent of patients with blastomycosis. A tube agglutinin test (TA) is also used for the detection of antibody. In the experience of some investigators, utilization of a combination of the IFA, LCAT, and CF Ag tests still results in a false-negative result rate of up to 15 per cent.

BIOLOGICAL STUDIES

There have been numerous studies concerning acquired resistance to cryptococcosis,[93] and it appears that the injection of mice with capsular material or live unencapsulated yeasts affords some protection against infection.[45] By modifying the experiments which Hasenclever[62] performed with *Candida,* Abrahams has demonstrated that pertussis vaccine enhances acquired resistance of mice to cryptococcosis.[2] The exact mechanisms of this phenomenon have yet to be fully elucidated.

Growth inhibiting substances have been described that occur naturally in serum, saliva, and cerebral spinal fluid.[65] This may account in part for the high degree of resistance to infection by the fungus. Phagocytosis of capsulated and noncapsulated strains has been studied by Bulmer et al.[18, 18a] The presence of rather than the degree of capsulization appears to be significant.

The effect of antifungal agents on experimental murine infections has been reviewed by Graybill et al.,[54] as has the influence of cyclophosphamide.[56] Diamond,[32] using a guinea pig model, found that females survived longer than males (see also the study by Mohr et al.[100]) This could be negated by a brief course of cortisone. The cortisol effect on susceptibility to infection remained long after withdrawal of the drug. He confirmed the observations of others that cryptococci given in Freund's adjuvant increased survival when such animals were challenged by live organisms. A model of experimental cutaneous disease has been developed by Dykstra and Friedman.[33] Mice inoculated in the skin developed some degree of resistance when challenged by other routes. Green and Bulmer[58] found that mice fed cryptococci were positive by fecal culture for up to 12 months. None of these mice had clinical evidence of disease, but a few others (3 per cent) died of disseminated infection shortly after being fed the cryptococci. In other studies it has been shown that in mice that are B-cell deficient the mortality pattern is no different than that of control groups;[101] in T-cell–deficient or thymectomized mice, however, mortality is increased. Kozel and McGaw[75] have indicated the importance of opsonins in experimental infections and demonstrated their presence in human serum.

The antigenic and chemical structure of *Cryptococcus neoformans,* particularly the capsule, has been the subject of much investiga-

tion. Evans in 1949 divided the pathogenic strains of *C. neoformans* into three serologic types, A, B, and C, based on the capsular material. These types comprise group III *(C. neoformans)* of the serologic scheme devised in 1935 by Rhoda Benham. Her groupings included both saprophytic and pathogenic cryptococci. The capsule of *C. neoformans* is a polysaccharide composed of xylose, mannose, and a uronic acid, probably glucuronic acid.[15] It does not contain starch, glycogen, amino acid, protein, nucleic acids, amino sugars, or mucoitin sulfate. As noted in the section on epidemiology, several recent studies of the distribution of the serologic groups indicate that essentially all organisms from avian habitats and human infections are type A or D and that this varies geographically.[13, 82] In one study, most of the B and C types were from California,[82] and these were all from clinical material. Neither type has been isolated from natural sources. Some investigators have suggested that the serotypes can change in culture or during infection;[31, 138] however Bennett et al.[13] concluded that the serotypes are valid and that biochemical differences exist between them. Recently Bennett has modified his view to allow for natural variation of serotypes.

There have been several investigations into some of the biochemical and physiologic attributes of *Cryptococcus neoformans* and other cryptococci. Chaskes and Tyndall studied the aminophenol oxidases of the several species in the genus.[24] They found that the various species could be grouped into pigment patterns. One was shared by *C. neoformans* and *C. albidus.* "Killer" strains among the *Cryptococcus* sp. have been described by Kandel and Stern.[69] These yeasts have killer activity against strains of *Saccharomyces* sp. and *Torulopsis (Candida)* sp. Lysosomes in cryptococci have been identified by Mason and Wilson.[97]

LABORATORY IDENTIFICATION

Direct Examination. The yeasts of *Cryptococcus neoformans* are quite fragile and collapse or become crescentic in dried, fixed, or stained films. Although they are readily demonstrable in stained histopathologic slide preparations, direct examination of a wet mount should be carried out. The capsule is so distinctive that, in infected material mixed with a drop of India ink, nigrosin, or any colored colloidal mounting medium, the encapsulated organisms are outlined by negative contrast. This technique may be used to visualize the organisms in macerated biopsy material, centrifuged sediment of spinal fluid, cisternal fluid,[48] or touch slides of autopsy material. Viewed microscopically with diminished light, the budding yeast cells can be seen within the capsule and the organisms thereby differentiated from leukocytes, myelin globules, fat droplets, and tissue cells. India ink is often contaminated with a variety of artifacts and such microorganisms as diphtheroids, spirillae, other motile bacteria, and even encapsulated yeasts. In addition, the carbon particles sometimes spontaneously agglutinate. For this reason, it is advisable to run a saline control for comparison and regularly check the quality of the India ink. Sputum or pus can be digested in a potassium hydroxide mount, which eliminates most cells and other artifacts that may be misread as *Cryptococcus*. The capsule and organism resist this treatment and can be viewed by proper low-intensity lighting. Brain material can be crushed on a slide and examined directly or after mixing with India ink. The organism is also readily visible on Papanicolaou preparations of cerebrospinal fluid or other material.[120]

Culture Methods. Pathogenic strains of *C. neoformans* are not fastidious and grow well at 37° C. They are, however, variably sensitive to cycloheximide (Actidione), which is incorporated in most selective media for pathogenic fungi. This antibiotic was at one time used in the treatment of cryptococcal meningitis.[87] Material from patient sources can be streaked on SGA medium or SGA with antibacterial antibiotics. Several plates or a flask of broth (if the specimen is spinal fluid) should be used. These are incubated at 37° C.

Some reports of labile forms of the yeasts (L forms?)[94] have indicated that additional procedures may be valuable, particularly if the initial cultures are negative or the patient is being treated with amphotericin B. In some cases[94] incubation of material in hypertonic media containing salt and 0.3 M sucrose followed by subculture onto blood agar or SGA has been used to successfully recover aberrant cryptococci. For specimens such as sputum in which large numbers of contaminants may be present, a selective medium has been devised by Vogel.[136] It contains antibacterial antibiotics and potato glucose and can detect urease production. *C. neoformans* produces a white colony with a pink halo, whereas *Candida albicans,* which is also white, does not produce a halo. A rapid selective urease test for *C. neoformans* has recently been published by Zimmer and Roberts.[143] Among 286 isolants of *C. neoformans,* 99.6 per cent were positive within 15 minutes.

Growth from patient material on primary isolation media occurs in 24 to 48 hours; however, cultures should be retained for four to six weeks. Following initial isolation, colonies are streaked on heart infusion agar so that single clones can be picked and utilized for physiologic studies to confirm the identification.

Isolation of *Cryptococcus* from heavily contaminated material, such as pigeon nests and droppings, can be carried out using another selective medium. Staib[130] determined that creatinine is assimilated by *C. neoformans* but not by other members of its genus or by most species from the other common genera of the yeasts (including *Candida*). He observed also that *C. neoformans* selectively absorbs a brown pigment from the seeds of a common weed, *Guizotia abyssinica.* Shields and Ajello[126] then

Table 21–1 *Physiologic Differentiation of Common Cryptococcus Species*

	Nitrate Assimilation	*Growth at 37°C*	*Animal Pathology*	*Carbon Assimilation*						
				Gal	*Ma*	*La*	*Me*	*Er*	*Galtol*	*Su*
C. neoformans	−	+,v	+	+	+	−	−	v	+	+
C. laurentii and var.*	−	v	−	+	+	+	+/−	+/−	v	+
C. albidus and var.	+	v	−	+/w	+	+/−	+/v	−/w	v	+
C. terreus	+	−	−	+	v	+/w	−	−	v	−
C. luteolus	−	−	−	+	+	−/w	+/w	+	+	+
C. melibiosum	−	−	−	+	−	w	+	−	−	+
C. flavus	−	−	−	+	+	+	+	+	w	+
C. lactativorus	−	+	−	−	−	−	−	−	−	−

*Var. of species variable in some characteristics—see Lodder.[90]

Gal = galactose, Ma = maltose, La = lactose, Me = melibiose, Er = erythritol, Galtol = galactitol, Su = sucrose, w = weak, v = variable.

devised a selective medium for *C. neoformans* by combining purine and *Guizotia* seeds with diphenyl and chloramphenicol to act as mold and bacterial inhibitors. When incubated at 37° C on this medium, a few strains of *Cryptococcus laurentii* will also grow, but neither this species nor *Candida albicans* becomes pigmented, whereas almost all strains of *C. neoformans* grow well and develop a brown color. Such typical colonies can then be subcultured for further testing. Salkin[114] simplified the media by using pulverized seeds of *Guizotia*. Only *C. neoformans* and the var. *gatti* were positive.

Some isolants will have very small capsules on primary isolation. Capsule production is enhanced by growing the organisms on chocolate agar at 37° C in a CO_2 incubator.

The specific identification of *C. neoformans* requires the study of a combination of factors, including physiologic characteristics, temperature tolerance, and animal pathogenicity.[25, 67, 114] The characteristics that define the genus *Cryptococcus* are the assimilation of inositol, production of urease, and lack of mycelium on cornmeal agar. Carbon and nitrate assimilation profiles can also be carried out (see Table 21–1). The most reliable tests for the identification of *C. neoformans* are growth in culture at 37° C and pathogenicity for mice (see section on experimental animal disease).

MYCOLOGY

Cryptococcus neoformans (Sanfelice) Vuillemin 1901

Synonymy. *Saccharomyces neoformans* Sanfelice 1895; *Torula neoformans* Weiss 1902; *Blastomyces neoformans* Arzt 1924; *Torulopsis neoformans* Redaelli et Ciferri 1931; *Debaromyces neoformans* Redaelli, Ciferri, and Giordano 1937; *Saccharomyces lithogenes* Sanfelice 1895; *Cryptococcus hominis* Vuillemin 1901; *Atelosaccharomyces hominis* Verdun 1912; *Debaromyces hominis* Todd et Hermann 1936; *Torula histolytica* Stoddard et Cutler 1916; *Cryptococcus bacillisporus* Kwon-Chung, Bennett et al. 1978; Lodder[90] lists 39 synonyms.

Teleomorph[83] **State.** *Filobasidiella neoformans* Kwon-Chung 1975.

Synonymy. *Filobasidiella bacillispora* Kwon-Chung 1976

There were two varieties of the species proposed. They were *F. neoformans* var. *neoformans* for serotypes A and D that have a particular biochemical pattern (lack of malate assimilation and yellow color on creatine glucose bromthymol blue agar and temperature tolerant) and *F. neoformans* var. *gattii*, first described by Vanbreuseghem. The latter variety assimilates L-malate, is blue on CGB agar, and is temperature-sensitive. It also tends to be small, distorted, and sometimes mycelial in tissue. *Filobasidiella bacillispora*[77, 83] and its anamorph[83] were then considered in synonymy with *F. neoformans* var. *gattii*. However, more recent studies by Schmeding et al.[120a] indicate infertility between serotypes and biotypes in otherwise sexually competent strains. They conclude that there is no need to recognize *C. neoformans var. gattii*, *F. bacillispora* or *C. bacillispora*, and that there is only one species *F. neoformans* and its anamorph *C. neoformans*.

Morphology. SGB or malt extract broth: three days, 25° C. Cells are spherical or globose, occurring singly or in pairs, sometimes in groups. Budding is single or double anywhere on the cell; sometimes several buds appear. Size: 3.5 to 7.0 × 3.7 to 8 μ.

SGA or malt extract agar: three days, 25° C. Cells are globose to spheroidal, 2.5 to 8 μ in diameter. Encapsulation varies with the strain and the medium used. After one week, sectoring of the colony occurs in almost all species of the genus *Cryptococcus* (Fig. 21–14). This is because of changes in the composition and structure of the capsules. At one month, the colonies are cream-colored to yellowish or slightly pink. The texture is mucoid, and the colony may flow to the bottom of the slant. The edges are entire and without pseudomycelium. Occasional isolants produce little or no capsular material.[37] Colonies of such organisms appear dry or glabrous. Very rare strains produce mycelium and under particular conditions are seen to form clamp connections, basidia, and basidiospores. Organisms grown in the presence of deoxycholate tend to elongate and retain this morphology in experimental animals.[50]

Cornmeal agar: no pseudomycelium.

Fermentation. None.

Nitrogen Assimilation. Nitrate-negative; tryptone-positive.

Carbon Assimilation. Glucose +, galactose +, L-sorbose +/−, sucrose +, maltose +, cellobiose +/w, trehalose +, lactose −, melibiose −, raffinose +/w, melezitose +, inulin −, L- and D-arabinose +/w, D-ribose +/w

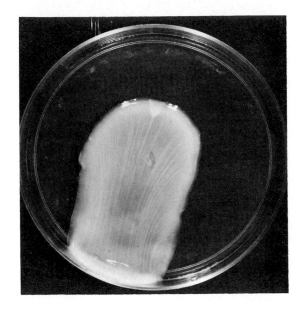

Figure 21–14. Cryptococcosis. Colony of *C. neoformans* on Sabouraud's glucose agar for one week, showing mucoid colony flowing over medium and evident sectoring.

(rare), L-rhamnose +, ethanol +/w (rare), ribitol +, galactitol +, D-mannitol +, inositol +, starch formation +/w, gelatin liquefaction −, D-xylose +.

Filobasidiella neoformans Kwon-Chung 1975.[77]

Synonymy. *Filobasidiella bacillispora* Kwon-Chung 1976.[79]

The basidia are produced on short hyphal units with clamp connections. They are single or clumped (none septate) and slender, with subglobase to clavate apices 20 to 60 μ long by 2 to 3.5 μ wide at base and up to 8 μ at the apex.[144] Continuous budding from four spots on the apex produces small 1.0 to 1.3 × 3 μ basidiospores. These enlarge and become budding yeasts with capsules. The initial basidiospores of the var. *gattii* are 1.0 to 1.5 × 3 to 8 μ. They then become round, budding encapsulated yeasts identical in size and morphology to the yeasts of *F. neoformans* var. *neoformans* (Figs. 21–15 and 21–16). The status of the anamorph and teleomorph states of *Cryptococcus neoformans* is uncertain at this time.

Kwon-Chung has assigned the teleomorph state of *Cryptococcus uniguttulatus* to the genus *Filobasidium*,[81] rather than to the genus *Filobasidiella*. Some investigators (Olive, personal communication and McGinnis[98a]) consider the two genera in synonymy with priority

Figure 21–15. *Filobasidiella neoformans.* The teleomorph of *Cryptococcus neoformans.* Large globose basidia giving rise to small basidiospores. Short mycelial units are seen and a clamp connection is evident (arrow). (Courtesy of D. Ahern.)

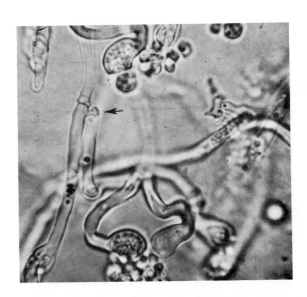

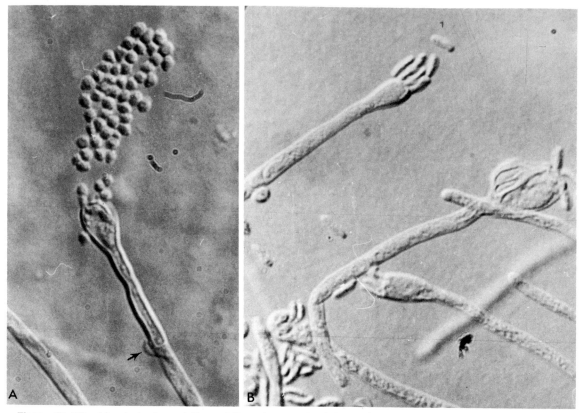

Figure 21–16. Morphologic variation of the teleomorph of *C. neoformans. A, F. neoformans.* Small round basidiospores emerging from a basidia. Note clamp connection *(arrow). B,* Elongate, slightly curved basidiospores. This latter type was called *F. bacillispora.* (Courtesy of M. McGinnis.)

going to *Filobasidium.* Malloch and Kane[96] describe a member of the genus *Filobasidiella* as *arachnophila* from spiders. It has no yeast-like state. It appears that the concepts of these several genera are unclear at this time. Further investigations should resolve these points of disagreement.

REFERENCES

1. Abdel-Fattah, A., M. S. Abuzeid, et al. 1976. Primary cutaneous cryptococcosis in Egypt. Int. J. Dermatol., *14*:606–608.
2. Abrahams, I. 1966. Further studies on acquired resistance to murine cryptococcosis: enhancing effect of Bordetella pertussis. J. Immunol., *98*:914–922.
3. Ajello, L. 1969. A comparative study of the pulmonary mucoses of Canada and the United States. Public Health Reports, *84*:869–877.
4. Anon. 1980. Garlic in cryptococcal meningitis. A preliminary report of 21 cases. Chinese Med. J., *93*:123–126.
5. Avedano, J., T. Tanishima, et al. 1978. Ocular cryptococcosis. Am. J. Opthalmol., *86*:110–113.
6. Baker, R. D., and R. K. Haugen. 1955. Tissue changes and tissue diagnosis in cryptococcosis

a study of 26 cases. Am. J. Clin. Pathol., *25*:4–24.
7. Baker, R. D., and G. Linares. 1971. Cryptococcal dermatropism in Rhesus monkeys. Bacteriol. Proc., *71*:118.
8. Barclay, W. P., and A. deLahunta. 1979. Cryptococcal meningitis in a horse. J. Am. Vet. Med. Assoc., *174*:1236–1238.
9. Bardana, E. J., L. Kaufman, et al. 1968. Amphotericin B and cryptococcal infection. Arch. Intern. Med., *122*:517–520.
10. Benham, R. W. 1950. Cryptococcosis and Blastomycosis. Ann. N.Y. Acad. Sci., *50*:1299–1314.
11. Bennett, J. E. 1966. Susceptibility of *Cryptococcus neoformans* to amphotericin B. *In Antimicrobial Agents and Chemotherapy,* 1966. Philadelphia. Am. Soc. for Microbiol., Oct. pp. 405–410.
12. Bennett, J. E., W. E. Dismukes, et al. 1979. A comparison of amphotericin B alone and combined with flucytocine in the treatment of cryptococcal meningitis. New Engl. J. Med., *301*:126–131.
13. Bennett, J. E., K. J. Kwon-Chung, et al. 1977. Epidemiologic differences among serotypes of *Cryptococcus neoformans.* Am. J. Epidemiol., *105*:582–586.
14. Beyt, B. E., and S. R. Waltman. 1978. Cryptococcal endophthalmitis after corneal transplantation. New Engl. J. Med., *298*:825–826.
15. Bhattacharjee, A. K., K. J. Kwon-Chung, et al. 1978. On the structure of the capsular polysac-

charide from *Cryptococcus neoformans* serotype C. Immunochem., *15*:673–679.

16. Bloomfield, N., M. A. Gordon, et al. 1963. Detection of *Cryptococcus neoformans* antigen in body fluids by latex particle agglutination. Proc. Soc. Exp. Biol. Med., *114*:64–67.

17. Bryan, C. S. 1977. Vertebral osteomyelitis due to *Cryptococcus neoformans.* J. Bone Joint Surg., *59*:275–276.

18. Bulmer, G. S., M. D. Sans, et al. 1967. *Cryptococcus neoformans.* I. Nonencapsulated mutants. J. Bacteriol., *94*:1475–1479.

18a. Bulmer, G. S., and M. D. Sans. 1967. *Cryptococcus neoformans.* II. Phagocytosis by human leukocytes. J. Bacteriol., *94*:1480–1483.

19. Bunting, L. A., L. B. Neilson, et al. 1979. *Cryptococcus neoformans:* a gastronomic delight of a soil ameba. Sabouraudia, *17*:225–232.

20. Buschke, A. 1895. Ueber eine durch Coccidien Hervergerufene Krankheit des menschen. Deutsche med. Wochenschr. 21 (No. 3):14.

21. Busse, O. 1894. Ueber parasitare zelleinschlüsse und ihre züchtung. Zentralbl. Bakterial., *16*:175–180.

22. Campbell, G. D. 1966. Primary pulmonary cryptococcosis. Am. Rev. Respir. Dis., *94*:236–243.

23. Chand, K., and K. S. Lall. 1976. Cryptococcosis (torulosis, European blastomycosis) of the knee joint. A case report with review of the literature. Acta Orthopaed. Scand. *47*:432–435.

24. Chaskes, S., and R. L. Tyndall. 1978. Pigment production by *Cryptococcus neoformans* and other *Cryptococcus* species from amino phenols and diamino benzenes. J. Clin. Microbiol., 7:146–152.

25. Chaskes, S., and R. L. Tyndall. 1978. Pigmentation and autofluorescence of *Cryptococcus* species after growth on tryptophan and anthranilic acid media. Mycopathologia, *64*:105–112.

26. Chleboun, J., and S. Nade. 1977. Skeletal cryptococcosis. J. Bone Joint Surg., *59*:509–514.

27. Cobcroft, R., H. Kronenber, et al. 1978. *Cryptococcus* in bagpipes. Lancet, *1*:1368–1369.

28. Collins, D. N., I. A. Oppenheim, et al. 1971. Cryptococcosis associated with systemic lupus erythematosus. Arch. Pathol., *91*:78–88.

29. Colmers, R. A., W. Irniger, et al. 1967. *Cryptococcus neoformans* endocarditis cured by amphotericin B. J.A.M.A., *199*:762–764.

30. Daveny, J. K., and M.D. Ross. 1969. Cryptococcosis of bone. Cent. Afr. J. Med., *15*:78–79.

31. Denton, J. F., and A. F. DiSalvo. 1968. The prevalence of *Cryptococcus* in various natural habitats. Sabouraudia, *6*:213–217.

32. Diamond, R. D. 1977. Effects of stimulation and suppression of cell-mediated immunity on experimental cryptococcosis. Infect. Immun., *17*:187–194.

33. Dykstra, M. A., and L. Friedman. 1978. Pathogenesis. Lethality and immunizing effect of experimental cutaneous cryptococcosis. Infect. Immun., *20*:446–455.

34. Edwards, N. J., and W. C. Rebhun. 1979. Generalized cryptococcosis: a case report. J. Am. Anim. Hosp. Assoc., *14*:439–445.

35. Emmons, C. W. 1955. Saprophytic sources of *Cryptococcus neoformans* associated with the pigeon (*Columba livia*). Am. J. Hyg., *62*:227–232.

36. Everett, B. A., J. A. Kusse, et al. 1978. Cryptococcal infection of the central nervous system. Surg. Neurol., *9*:157–163.

37. Farmer, S. G., and R. A. Komorowski. 1973. Histologic response to capsule-deficient *Cryptococcus* neoformans. Arch. Pathol., *96*:383–386.

38. Felton, F. G., W. E. Maldonado, et al. 1966. Experimental cryptococcal infection in rabbits. Am. Rev. Respir. Dis., *94*:589–594.

39. Ferguson, R. P., and J. K. Smith. 1977. Cryptococcosis and Cushing's syndrome. Ann. Intern. Med., *87*:65–66.

40. Fink, J. N., J. J. Barboriak, et al. 1968. Cryptococcal antibodies in pigeon breeders' disease. J. Allergy, *41*:297–301.

41. Fisher, B. D., and D. Armstrong. 1977. Cryptococcal interstitial pneumonia. New Engl. J. Med., *29*:1440–1441.

42. Fromtling, R. A., and G. S. Bulmer. 1978. *In vitro* effect of aqueous extract of garlic (*Allium sativum*) on the growth and viability of *Cryptococcus neoformans.* Mycologia, *70*:397–405.

43. Frothingham, L. 1902. A tumorlike lesion in the lung of a horse caused by a blastomycete (*Torula*) J. Med. Res., *3*:31–43.

44. Fusner, J. E., and K. L. McClain. 1979. Disseminated lymphonodular cryptococcosis treated with 5-fluorocytosine. J. Ped., *94*:599–601.

45. Gadebusch, H. H., and A. G. Johnson. 1966. Natural host resistance to infection with *Cryptococcus neoformans.* J. Infect. Dis., *116*:551–572.

46. Gauder, J. P. 1077. Cryptococcal cellulitis. J.A.M.A., *237*:672–673.

47. Gentry, R. H., W. E. Farrar, et al. 1977. Simultaneous infection of the central nervous system with *Cryptococcus neoformans* and *Mycobacterium intracellulare.* South Med. J., *70*:865–866.

48. Gonyea, E. F. 1973. Cisternal puncture and cryptococcal meningitis. Arch. Neurol., *28*:200–201.

49. Goodman, J. L., L. Kaufman, et al. 1971. Diagnosis of cryptococcal meningitis. New Engl. J. Med., *285*:434–436.

50. Gordon, M. A., and J. Devine. 1970. Filamentation and endogenous sporulation in *Cryptococcus neoformans.* Sabouraudia, *8*:227–234.

51. Gordon, M. A., and E. Lapa. 1971. Charcoal particle agglutination test for detection of antibody to *Cryptococcus neoformans.* Am. J. Clin. Pathol., *56*:354–359.

52. Gordonson, J., W. Birnbaum, et al. 1974. Pulmonary cryptococcosis. Diag. Radiol., *112*:557–561.

53. Goren, M. B., and J. Warren. 1968. Immunofluorescence studies of reactions at the cryptococcal capsule. J. Infect. Dis., *118*:216–229.

54. Graybill, J. R. 1979. Host defense in cryptococcosis III. *In vivo* alteration of immunity. Mycopathologia, *69*:171–178.

55. Graybill, J. R., and H. B. Levine. 1978. Successful treatment of cryptococcal meningitis with intraventricular miconazole. Arch. Intern. Med., *138*:814–816.

56. Graybill, J. R., and L. Mitchell. 1978. Cyclophosphamide effects on murine cryptococcosis. Infect. Immun., *21*:674–677.

57. Graybill, J. R., L. Mitchell, et al. 1978. Treatment of experimental murine cryptococcosis: a comparison of miconazole and amphotericin B. Antimicrob. Agents Chemother., *13*:277–283.

58. Green, J. R., and G. S. Bulmer. 1979. Gastrointestinal inoculation of *Cryptococcus neoformans* in mice. Sabouraudia, *17*:233–240.

59. Hammerman, K. J., K. E. Powell, et al. 1975. Pulmonary cryptococcosis: clinical forms and treatment. A center for disease control cooper-

ative mycoses study. Am. Rev. Respir. Dis., *108*:1116–1123.

60. Hansemann, D. von. 1905. Über eine bisher nicht beobachtete Gehirnerkrankung durch Hefen. Verh. Dtsch. Ges. Path., *9*:21–24.

61. Harding, S. A., W. M. Scheld, et al. 1979. Pulmonary infection with capsule deficient *Cryptococcus neoformans*. Virchows Arch., *362*:113–118.

62. Hasenclever, H. F., and E. J. Corley. 1968. Enhancement of acquired resistance in murine candidiasis by *Bordetella pertussis* vaccine. Sabouraudia, *6*:289–295.

63. Hay, R. J., and E. Reiss. 1978. Delayed type hypersensitivity responses in infected mice elicited by cryptoplasmic fractions of *Cryptococcus neoformans*. Infect. Immun. *22*:72–79.

64. Howard, D. H. 1973. The commensalism of *Cryptococcus neoformans*. Sabouraudia, *11*:171–174.

65. Howard, J. I., and R. P. Bolande. 1966. Humoral defense mechanisms in cryptococcosis: Substances in normal human serum, saliva and cerebral spinal fluid affecting the growth of *Cryptococcus neoformans*. J. Infect. Dis., *116*:75–83.

65a. Iacobellis, F. W., M. I. Jacobs, et al. 1979. Primary cutaneous cryptococcosis. Arch. Dermatol., *115*:984–985.

66. Ishaq, C. M., G. S. Bulmer, et al. 1968. An evaluation of various environmental factors affecting the propagation of *Cryptococcus neoformans*. Mycopathologia, *35*:81–90.

67. Jennings, A., J. E. Bennett, et al. 1968. Identification of *Cryptococcus neoformans* in routine clinical laboratory. Mycopathologia, *35*:256–264.

68. Jimbow, T., Y. Tejima, et al. 1978. Comparison between 5-fluorocytosine, amphotericin B. and the combined administration of these agents in therapeutic effectiveness for cryptococcal meningitis. Chemotherapy, *24*:374–389.

69. Kandel, J. S., and T. A. Stern. 1979. Killer phenomenon in pathogenic yeast. Antimicrob. Agents Chemother., *15*:568–571.

70. Kaplan, M. H., P. P. Rosen, et al. 1977. Cryptococcosis in a cancer hospital: clinical and pathological correlates in forty-six patients. Cancer, *39*:2265–2274.

71. Kauffman, C. A., and P. J. Severence. 1980. Nosocomial cryptococcal infection. South. Med. J., *73*:267.

72. Kaufman, L., and S. Blumer. 1978. Cryptococcosis the awakening giant. *In The Black and White Yeasts*. Pan American Health Org. Sci. Pub., 356. Proceedings of IV Int. Conf. Mycoses. Washington, D.C., pp. 176–184.

73. Klein E. 1901. Pathogenic microbes in milk. J. Hyg. (London), *1*:78–95.

74. Kornfeld, S. J., C. P. Wormser, et al. 1979. Purulent cryptococcal meningitis. Mt. Sinai J. Med., *46*:326–327.

75. Kozel, T. R., and T. G. McGaw. 1979. Opsonization of *Cryptococcal neoformans* by human immunoglobulin G: role of immunoglobulin G in phagocytosis by macrophages. Infect. Immun., *25*:255–261.

76. Krumholz, R. A. 1972. Pulmonary cryptococcosis. A case due to *Cryptococcus albidus*. Am. Rev. Respir. Dis., *105*:421–424.

77. Kwon-Chung, K. J. 1975. A new genus, *Filobasidiella* the perfect state of *Cryptococcus neoformans*. Mycologia, *67*:1197–1200.

78. Kwon-Chung, K. J. 1976. Morphogenesis of *Filobasidiella neoformans.* The sexual state of *Cryptococcus neoformans.* Mycologia, *68*:821–833.

79. Kwon-Chung, K. J. 1976. A new species of *Filobasidiella*, the sexual state of *Cryptococcus neoformans* B and C serotypes. Mycologia, *68*:942–946.

80. Kwon-Chung, K. J. 1978. Heterothallism vs. self-fertile isolates of *Filobasidiella neoformans* (*Cryptococcus neoformans*). *In The Black and White Yeasts*. Pan. Am. Health Org. Sci. Pub., 356. Proceedings of IV Int. Conf. Mycoses. Washington, D.C., pp. 204–213.

81. Kwon-Chung, K. J. 1977. Perfect state of *Cryptococcus uniguttulatus*. Int. J. Syst. Bacteriol., 27:293–299.

82. Kwon-Chung, K. J., and J. E. Bennett. 1978. Distribution of alpha and alpha mating types of *Cryptococcus neoformans* among natural and clinical isolates. Am. J. Epidemiol., *108*:337–340.

83. Kwon-Chung, K. J., J. E. Bennett, et al. 1978. *Cryptococcus bacillisporus* sp. nov. serotype B-C of *Cryptococcus neoformans*. Int. J. Syst. Bacteriol., *28*:616–620.

83a. Levinson, D. J., D. C. Silcox, et al. 1974. Septic arthritis due to noncapsulated *Cryptococcus neoformans* with coexisting sarcoid. Arthritis Rheum. *17*:1037–1047.

84. Lim, T. S., J. W. Murphy, et al. 1980. Host-etiological agent interactions in intranasally and intraperitoneally induced cryptococcosis in mice. Infect. Immun., *29*:633–641.

85. Littman, M. L., and R. Borok. 1968. Relation of the pigeon to cryptococcosis: natural carrier state, heat resistance and survival of *Cryptococcus neoformans*. Mycopathologia, *35*:329–345.

86. Littman, M. L., and J. E. Walker. 1968. Cryptococcosis: current issues. Am. J. Med., *45*:922–933.

87. Littman, M. L., and L. E. Zimmerman. 1956. *Cryptococcosis.* New York, Grune and Stratton.

88. Lacobellis, F. A., M. I. Jacobs, et al. 1979. Primary cutaneous cryptococcosis. Arch. Dermatol., *115*:984–985.

89. Lesser, R. L., R. M. Simon, et al. 1979. Cryptococcal meningitis and internal ophthalmoplegia. Am. J. Ophthal., *87*:682–687.

90. Lodder, J. (ed.) 1970. *The Yeasts*. 2nd ed. Amsterdam, North-Holland Publishing Company.

91. Lomvardia, S., and H. I. Lurie. 1972. Epipleural cryptococcosis in a patient with Hodgkin's disease: A case report. Sabouraudia, *10*:256–259.

92. Long, R. F., S. V. Berens, et al. 1972. An unusual manifestation of pulmonary cryptococcosis. Br. J. Radiol., *45*:757–759.

93. Louria, D. B., and T. Kaminski. 1965. Passively acquired immunity in experimental cryptococcosis. Sabouraudia, *4*:80–84.

94. Louria, D. B., T. Kaminski, et al. 1969. Aberrant forms of bacteria and fungi found in blood on cerebrospinal fluid. Arch. Intern. Med., *124*:39–48.

95. Madewell, B. R., C. A. Holmberg, et al. 1979. Lymphosarcoma and cryptococcosis in a cat. J. Am. Vet. Med. Assoc., *175*:65–68.

96. Malloch, D., J. Kane, et al. 1978. *Filobasidiella arachnophila* sp. nov. Can. J. Bot., *56*:1823–1826.

97. Mason, D. L., and C. L. Wilson. 1979. Cytochemical and biochemical identification of lysosomes in *Cryptococcus neoformans*. Mycopathologia, *68*:183–190.

98. McGaw, T. G., and T. R. Kozel. 1979. Opsonization of *Cryptococcus neoformans* by human immunoglobulin G: masking of immunoglobulin G by cryptococcal polysaccharide. Infect. Immun., *25*:262–267.

98a. McGinnis, M. R. 1980. Recent taxonomic developments and changes in medical mycology. Ann. Rev. Microbiol., *34*:109–135.

99. Migaki, G., R. D. Gunnels, et al. 1978. Pulmonary cryptococcosis in an Atlantic bottlenose dolphin *(Tursiops truncatus)*. Lab. Anim. Sci., *28*:603–606.

100. Mohr, J. A., H. Long, et al. 1972. *In vitro* susceptibility of *Cryptococcus neoformans* to steroids. Sabouraudia, *10*:171–172.

101. Monga, D. P., R. Kumar, et al. 1979. Experimental cryptococcosis in normal and B-cell-deficient mice. Infect. Immun., *26*:1–3.

102. Morgans, M. E., M. E. M. Thomas, et al. 1979. Successful treatment of systemic cryptococcosis with miconazole. Br. Med. J., *2*:100–101.

103. Noble, R. C., and L. F. Fajardo. 1972. Primary cutaneous cryptococcosis. Am. J. Clin. Pathol., *57*:13–22.

104. Poliner, J. R., E. B. Wilkins, et al. 1979. Localized osseous cryptococcosis. J. Pediatr., *94*:597–599.

105. Pounden, W. D., J. M. Amberson, et al. 1952. A severe mastitis problem associated with *Cryptococcus neoformans* in a large dairy herd. Am. J. Vet. Res., *13*:121–128.

106. Prevost, E., and R. Newell. 1978. Commercial cryptococcal latex kit. Clinical evaluation in a medical center. J. Clin. Microbiol., *8*:529–533.

107. Procknow, J. J., J. R. Benfield, et al. 1965. Cryptococcal hepatitis presenting as a surgical emergency. J.A.M.A., *191*:269–278.

108. Rhodes, J. C., L. S. Wicker, et al. 1980. Genetic control of susceptibility to *Cryptococcus neoformans* in mice. Infect. Immun., *29*:595, 494–499.

109. Rigdon, R. H., and O. Kirksey. 1952. Mycotic aneurysm cryptococcosis of the abdominal aorta. Am. J. Surg., *84*:486–491.

110. Riley, D. J., and N. H. Edelman. 1978. Hemoptysis in pulmonary cryptococcosis. J. Med. Soc. New Jersey, *75*:553–555.

111. Roberts, N. J., and R. G. Douglas. 1978. Cryptococcal meningitis. Cure despite cryptococcemia. Arch. Neurol., *35*:179–180.

112. Rogers, A. L., K. J. Kwon-Chung, et al. 1980. A scanning electron microscope comparison of basidial structures in *Filobasidiella neoformans* and *Filobasidiella bacillispora*. Sabouraudia, *18*:85–89.

113. Salfelder, K. 1971. Cryptococcosis. *In* Baker, R. D. (ed.). *The Pathologic Anatomy of the Mycoses.* Dritter Band, Fünfter Teil. Berlin, Springer-Verlag, pp. 383–464.

114. Salkin, I. F. 1979. Further simplification of the *Guiziotia abyssinica* seed medium for identification of *Cryptococcus neoformans* and *Cryptococcus bacillisporus.* Can. J. Microbiol., *25*:1116–1118.

115. Salyer, W. R., C. L. Moravec, et al. 1973. Adrenal involvement in cryptococcosis. Am. J. Clin. Pathol., *60*:559–561.

116. Sanfelice, F. 1894. Contributo alla morfologia e biologia dei blastomiceti che si sviluppano nei succhi di alcuni frutti. Ann. Igiene, *5*:239–262.

117. Sanfelice, F. 1895. Über einen neuen pathogenen Blastomyceten, welcher innerhalb de Gewebe unter Bildung Kalkartig aussenhender Masseи degereriert. Zbl. Bakt. I. Abt. Orig., *18*:521–526.

118. Sanfelice, F. 1895. Sull'azione patogena de blastomiceti como contributo alla etiologia dei tumori maligni. Nota preliminare II Policlinico Sez. Chir., *2*:204–211.

119. Sanfelice, F. 1898. Ein weiterer Beitrag zur Ätiologie der bösartigen Geschwulste. Zbl. Bakt. I. Abt. Orig., *24*:155–158.

120. Saigo, P., P. P. Rosen, et al. 1977. Identification of compatibility between serotypes of *Filobasidiella neoformans (Cryptococcus neoformans).* Curr. Microbiol., *5*:133–138.

120a. Schmeding, K. A., S. C. Jong, et al. 1981. Sexual compatibility between serotypes of *Filobasidiella neoformans (Cryptococcus neoformans).* Curr. Microbiol., *5*:133–138.

121. Schupbach, C. W., C. E. Wheeler, et al. 1976. Cutaneous manifestations of disseminated cryptococcosis. Arch. Dermatol., *112*:1734–1740.

122. Schwarz, J., and G. L. Baum. 1970. Cryptococcosis. Semin. Roentgenol., *5*:49–54.

123. Shadomy, H. J. 1970. Clamp connections in two strains of *Cryptococcus neoformans.* Spectrum Monograph series. Arts and Sci. Georgia State Univ., Atlanta. *1*:67–72.

124. Shadomy, S., H. J. Shadomy, et al. 1970. *In vivo* susceptibility of *Cryptococcus neoformans* to hemycin, Amphotericin B and 5-fluorocytosine. Infect. Immun., *1*:128–134.

125. Shields, J. A., D. M. Wright, et al. 1980. Cryptococcal chorioretinitis. Am. J. Ophthalmol., *89*:210–218.

126. Shields, A. B., and L. Ajello. 1966. Medium for selective isolation of *Cryptococcus neoformans.* Science, *151*:208–209.

127. Silberfarb, P. M., G. A. Saros, et al. 1972. Cryptococcosis and pregnancy. Am. J. Obstet. Gynecol., *112*:714–720.

128. Sinha, P., K. G. Maik, et al. 1978. Mediastinal cryptococcoma. Thorax, *33*:657–659.

129. Sokolowski, J. W., R. F. Schilaci, et al. 1969. Disseminated cryptococcosis complicating sarcoidosis. Am. Rev. Respir. Dis., *100*:717–722.

130. Staib, F., and G. Bethauser. 1968. Zum Nachweis von *Cryptococcus neoformans* in Stab von einem Taubenschlag. Mykosen, *11*:619–624.

131. Stoddard, J. L., and E. G. Cutler. 1916. *Torula* infection in man. Rockefeller Iиst. Med. Res. Monograph No. 62, pp. 1–98. (Jan. 31).

132. Sunderland, W. A., R. A. Campbell, et al. 1972. Pulmonary alveolar proteinosis and pulmonary cryptococcosis in an adolescent boy. J. Pediatr., *80*:450–456.

133. Tress, B., and S. Davis. 1979. Computed tomography of intracerebral toruloma. Neuroradiology, *17*:223–226.

134. Verse, M. 1914. Über einen Fall von general isolierter Blastomykose beim Menschen. Verh. Dtsch. Pathol. Ges., *17*:275–278.

135. Vogel, R. A. 1966. The indirect fluorescent antibody test for the detection of antibody in

human cryptococcal disease. J. Infect. Dis., *116*:573–580.

136. Vogel, R. A. 1969. Primary isolation medium for *Cryptococcus neoformans.* Appl. Microbiol., *18*:1100.

137. Vuillemin, P. 1901. Le Blastomycates pathogenes. Rev. Gen. Sci., *12*:732–751.

138. Walter, J. E., and E. G. Coffee. 1968. Distribution and epidemiological significance of serotypes of *Cryptococcus neoformans.* Am. J. Epidemiol., *87*:167–172.

139. Weir, E. C., A. Schwartz, et al. 1979. Short-term combination chemotherapy for treatment of feline cryptococcosis. J. Am. Vet. Med. Assoc., *174*:507–510.

140. Weis, C. 1902. Four pathogenic torulae (Blastomycetes). J. Med. Res., *2*:280–311.

141. Yoshikawa, T. T., N. Fujita, et al. 1980. Management of central nervous system cryptococcosis. Western J. Med., *132*:123–133.

142. Zelenger, K. S. 1978. Cryptococcal meningoencephalitis: an unusual clinical presentation. South. Med. J., *71*:212–213.

143. Zimmer, B. L., and G. D. Roberts. 1979. Rapid selective urease test for presumptive identification of *Cryptococcus neoformans.* J. Clin. Microbiol., *10*:380–381.

144. Zimmer, B. L., H. O. Hempel, et al. 1981. Technique for the purification of the hyphae of *Filobasidiella neoformans.* Mycopathologia *73*: 171–176.

22 Miscellaneous Yeast Infections

INTRODUCTION

As discussed in Chapter 20, the human integument has a resident yeast population, just as it has a normal bacterial flora.[1, 11, 24] There are, in addition, some species that are regular inhabitants of the gastrointestinal tract.[3, 28] All of these species have an established equilibrium with the animal host. Under certain circumstances, however, the members of this resident flora proliferate in great numbers and may incite a variety of clinical conditions. Some of these conditions are described in other chapters, viz., *Malassezia furfur* in pityriasis versicolor and *Trichosporon beigelii* in connection with white piedra (Chapter 7). In this chapter, other types of disease that are produced by these and related species of yeasts will be summarized. The infections involved are almost all opportunistic and occur as the result of a barrier break or in patients with some degree of immunodeficiency, physiologic imbalance, or disruption of the normal resident fungal and bacterial flora. Invasions due to deterioration or destruction of protective barriers constitute the most common predisposing factor in this type of yeast infection. Yeast colonization of indwelling catheters, intravenous lines, hemodialysis units, and other such devices leads to a septicemia by the colonizing organism.[7] Often such infections resolve spontaneously with removal of the catheter. The organisms themselves are of very low potential virulence, being second- or third-line opportunists and requiring a major alteration of host before an invasive disease is produced. When they are involved in an invasive process in a severely compromised patient, however, they can produce fatal fulminant infections.

Yeasts are such a constant and common part of the normal human[12, 25] and animal flora,[26, 28] it is not surprising that many reports of their association with disease have been made over the years. Most of these are unsubstantiated and probably represent isolation of normal flora as an incidental finding in sputum, urine and blood cultures,[7] or skin swabs.[24] Indeed, it is quite difficult to establish an unassailable diagnosis of such infections. Continued isolation of large numbers of the same organism is needed, together with demonstrable clinical symptoms and pathologic evidence. The last is probably the most important factor and depends on demonstration of the organism in tissue, with evidence of colonization and invasion. Some of the miscellaneous yeast infections that have been recently recorded and in which such evidence of infection appears valid will be enumerated in this chapter.

Rhodotorulosis

Rhodotorula is a common airborne contaminant of skin,[11, 21, 24] lungs, urine,[1] and feces.[22] Its isolation from sputum is usually without significance. Its occurrence in blood cultures when contamination has been ruled out is of greater significance.[7] It grows readily on almost all culture media, and its coral red, mucoid colonies are distinctive (Fig. 22–1). It is encapsulated and rarely forms mycelium. The organism has also been isolated from a

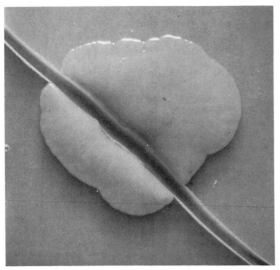

Figure 22–1. Rhodotorulopsis. Mucoid colony of *Rhodotorula rubra.*

number of food sources, cheese and milk products, air, soil, and water.[31]

Rhodotorula ruba and a few other *Rhodotorula* species have on rare occasion been isolated from patients in the terminal stages of debilitating diseases. Of these, *R. rubra,* formerly called *R. mucilaginosa,* is the species most frequently involved, and its presence has been documented in several fatal infections of the lung, kidney, and central nervous system.[18]

Rhodotorula fungaemia[20] is most often found to be due to colonized catheters, contaminated intravenous solutions, blood bank apparatus, and heart-lung and dialysis machines. The affected patients present with symptoms of endotoxic shock,[8] and cultures of their blood may be positive. Removal of the source of contamination usually leads to clearing of the symptoms. Some cases require short courses of amphotericin B therapy.

Rhodotorula has been involved in a case of a fatal endocarditis in which the organism was isolated from blood during life and from the aortic valve at autopsy.[9] Attempts to infect mice intracerebrally and intravenously with 5 × 10[6] cells were unsuccessful. Naveh et al.[15] described an endocarditis caused by *R. pilimanae* in a 7-year-old boy successfully treated with 5-fluorocytosine. Pore and Chen[18] reported a meningitis caused by *R. rubra* in a

21-year-old male with acute lymphoblastic leukemia. In the six months prior to death, the patient received 2.05 g of amphotericin B intravenously and 25 mg intrathecally.

A dermatitis in humans caused by *Rhodotorula* species has been reported but not documented; it has been verified in domestic animals, however. Page et al.[16] describe a dermatitis in chickens by *R. glutinis* that could be reproduced experimentally. Such infections are also recorded in dogs, cats, cattle,[17] and other animals.

MYCOLOGY

Rhodotorula rubra (Demme) Lodder 1934

Synonymy. *Saccharomyces ruber* Demme 1889; *Rhodotorula mucilaginosa* (Jörgensen) Harrison 1928; *Torulopsis mena* Dodge 1935; *Torulopsis sanguinea* (Ciferri et Redaelli) 1925; *Rhodotorula sanniei* Lodder 1934; Lodder, 1970, lists 34 names in synonymy.

Teleomorph State. The teleomorph state has not been described, but other species of *Rhodotorula* have a sexual stage in the phylum Basidiomycota, genus *Rhodosporidium.* The chief characteristics of the genus *Rhodotorula* are the presence of carotenoid pigments, the inability to assimilate inositol, the rudimentary or absent pseudomycelium, and the lack of fermentation.

Morphology. SGB or malt extract broth: three days, 25°C. Cells are short-ovoid to elongate, single, or in short chains or clusters, 2 to 6.5 μ in diameter, with elongate cells up to 14 μ. A ring may form.

SGA or malt extract agar: morphology is similar to above. The colony is moist, smooth to mucoid, glistening, and coral-red to pink or salmon in color (Fig. 22–1). Sectoring may be evident.

Cornmeal agar: rarely, rudimentary pseudomycelium is formed.

Fermentation. None.

Assimilation. Potassium nitrate. Glucose +, galactose +, w or −, sucrose +, maltose +, cellobiose + or −, trehalose +, raffinose +, inulin +, inositol −, soluble starch −, D-xylose +, ethanol + or −, D-mannitol + or −.

Vitamin Requirement. Thiamine.

Table 22–1 Reported Cases of Invasive Trichosporon beigelii Infection

Author	Age & Sex	Underlying Disease	Site	Treatment	Clinical Course
Watson & Kallichurum, 1970[32]	39/F	Bronchial adenocarcinoma of the lung with metastasis to the brain	Brain	None	Expired in 4 weeks
Sheik, Mahgoub, & Bedi, 1974[23]	70/M	Right cataract extraction	Right eye	Amphotericin-B	Vision lost in 6 months
	65/F	Right cataract extraction	Right eye	Amphotericin-B	Enucleation of the right eye
Rivera & Cangir, 1975[19]	12/M	Acute lymphocytic leukemia	Dissemination	Amphotericin-B 5-fluorocytosine	Expired
Madhaven et al., 1976[13]	55/F	Mitral valve replacement	Endocarditis with positive blood culture	Amphotericin-B	Recovered
	43/F	Kidney transplant	Septicemia, pulmonary edema	Amphotericin-B 5-fluorocytosine	Expired
Marier et al., 1978[14]	61/M	Rheumatic heart disease with aortic valve replacement	Endocarditis with positive blood culture	Amphotericin B 5-fluorocytosine	Expired
Terashima et al., 1978[27]	43/M	Acute leukemia	Dissemination	None	Expired
Evans et al., 1980[5]	4/M	Acute lymphocytic leukemia	Dissemination with cutaneous lesions	Amphotericin-B	Recovered
	57/F	Acute granulocytic leukemia	Dissemination	Amphotericin-B Miconazole	Expired
Yung et al., 1981[34]	70/F	Acute myeloblastic leukemia	Dissemination with cutaneous lesions	Amphotericin-B 5-fluorocytosine	Recovered. On long-term oral ketoconazole.
	63/M	Chronic myelogenous leukemia	Dissemination	None	Expired
Gold et al. 1981[6a]	58/M	Acute leukemia	Disseminated with cutaneous lesions	Amphotericin B	Expired

Trichosporonosis

Trichosporon species, particularly *T. beigelii* (Kuchenmetster et Rabenhorst) Vuillemin 1902 and *T. capitatum* Didden et Lodder 1942, are minor components of normal skin flora[11,23] and are widely distributed in nature.[29] *T. beigelii* is regularly associated with the soft nodules of white piedra (Chapter 7), but this species and, much less commonly, *T. capitatum* have been authentically involved in a variety of opportunistic infections. What appears to be the first case of a disseminated *T. beigelii* infection was reported by Watson and Kallichurum.[32] The patient had bronchial carcinoma and later developed a brain abscess. This and eleven other cases of invasive disease are summarized in Table 22–1. Sheik, Mahgoub, and Bedi described two cases of en-

dophthalmitis following a surgical procedure.[23] Endocarditis has been reported by Marier et al.[14] and Madhavan et al.[13] In the first case, septicemia developed following aortic valve replacement and the patient expired in spite of treatment. An endocarditis developed after mitral valve replacement in the report by Madhavan et al. Amphotericin B therapy was successful. These authors also reported pulmonary edema and *T. beigelii* septicemia in a kidney transplant patient. Treatment was unsuccessful. Almost all other cases of invasive disease were associated with immunosuppressive therapy for leukemia or other neoplasm (Table 22–1).

Cutaneous lesions secondary to septicemia were described in reports by Evans et al.,[5] Gold

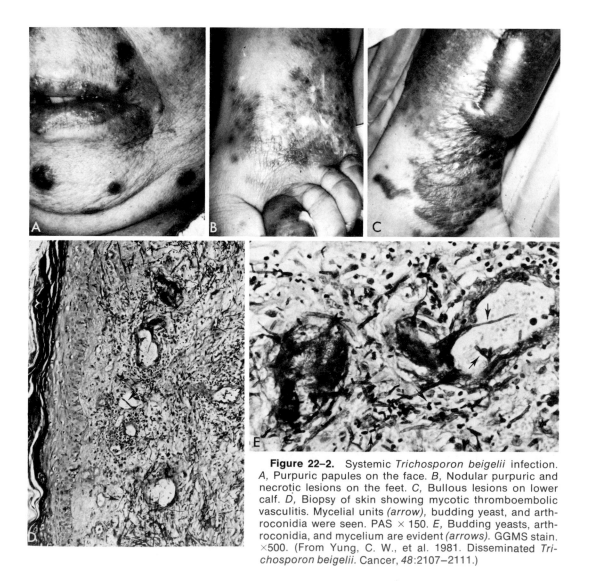

Figure 22–2. Systemic *Trichosporon beigelii* infection. *A,* Purpuric papules on the face. *B,* Nodular purpuric and necrotic lesions on the feet. *C,* Bullous lesions on lower calf. *D,* Biopsy of skin showing mycotic thromboembolic vasculitis. Mycelial units *(arrow),* budding yeast, and arthroconidia were seen. PAS × 150. *E,* Budding yeasts, arthroconidia, and mycelium are evident *(arrows).* GGMS stain. ×500. (From Yung, C. W., et al. 1981. Disseminated *Trichosporon beigelii.* Cancer, *48:*2107–2111.)

et al[6a] and Yung et al.[34] They were noted as multiple purpuric papules (Fig. 22–2) and nodules with central necrotic centers, along with large, tense hemorrhagic bullae. *T. beigelii* was isolated from biopsy of the lesions, blood cultures, and throat swabs. Yung's patient was cleared with administration of 1.6 g amphotericin B. She relapsed four weeks later and was treated with amphotericin B and 5-fluorocytosine. She was discharged on oral ketoconazole. The fungus was susceptible to amphotericin B (MIC 0.6 g per ml). In biopsy material of this case and autopsy material of other cases, budding yeasts, mycelium, and arthroconidia are seen in histologic section.

T. capitatum was described in lung lesions by Gemeinhardt.[6] The patient had chronic lung disease and the isolant was pathogenic for white mice. Winston et al.[33] report on endocarditis and disseminated disease caused by *T. capitatum* in a patient with idiopathic aplastic anemia. The patient expired.

Miscellaneous Yeasts

There are hundreds of other species of sporogenous (those with a known telemorph state) and asporogenous (anamorph state only) yeast species. On rare occasion these are implicated in and validated as causing human or animal infection. *Kluyveromyces fragilis* is a species widely distributed in nature, particularly in decaying plant material and rotting fruits. Lutwick et al.[10] described a case of opportunistic infection by this organism. A 51-year-old heart transplant patient developed a pneumonia in which large numbers of the yeast were isolated. It was marginally susceptible to amphotericin B but inhibited by 5-fluorocytosine and miconazole. The patient was successfully treated with amphotericin B.

Ascosporagenous yeasts, such as *Hansenula anomala*, *Pichia membranaefaciens*, *Saccharomyces carlsbergensis*, *S. cerevisiae*, and *S. fragilis*, may be a small part of the normal flora or transient flora of the throat and alimentary tract.[12] In Europe, *S. cerevisiae* has been implicated in several cases of pulmonary disease, especially in brewers. A well-documented case from the Mayo Clinic was reported by Dolan (Fig. 22–3). The organism was seen colonizing the lung parenchyma and was repeatedly isolated from sputum. It also has been reported to cause disease in rabbits.[4] *S. carlsbergensis* and *S. pastorianus* have been implicated in mycosis of the stomach.[2] *Hansenula polymorpha* was recovered from mediastinal lymph nodes by McGinnis et al.[11a] The patient had chronic granulomatous disease of childhood. The infection resolved with amphotericin B therapy.

Undoubtedly as the number of debilitated patients whose lives are extended by medical means increases, the occurrence and list of opportunistic yeast infections will also increase.

REFERENCES

1. Ahern, D. G., J. R. Jannach, et al. 1966. Speciation and densities of yeasts in human urine specimens. Sabouraudia, *5*:110–119.
2. Ahnlund, H. O., B. Pallin, et al. 1969. Mycosis of the stomach. Acta Chir. Scand., *133*:555–562.
3. Anderson, H. W. 1917. Yeastlike fungi of human intestinal tract. J. Infect. Dis., *21*:341–385.
4. Devos, A. 1969. Postmortem findings in rabbits. Vlaans Diergeneeskd, Tijdschr., *38*:275–280.
5. Evans, H. L., M. Kletzel, et al. 1980. Systemic mycosis due to *Trichosporon cutaneum.* Cancer, *45*: 367–371.
6. Gemeinhardt, H. 1965. Lugen pathogenitat von *Trichosporon capitatum* beim Menschen. Zentralbt. Bakteriol. (Orig.), *196*:121–133.
6a. Gold, J., W. Poston et al. 1981. Systemic infection with *Trichosporon cutaneum* in a patient with acute leukemia. Cancer *48*:2163–2167.
7. Kares, L., M. F. Biava. 1979. Levures isolées dans les hemocultures de 1962 à Mar 1979 in laboratoire de mycologie du CHU de Nancy. Bull. Soc. Fr. Mycol. Med., *8*:153–155.
8. Leeber, D. A., and I. Scher. 1969. *Rhodotorula* fungemia presenting as "endotoxic" shock. Arch. Intern. Med., *123*:78–81.

Figure 22–3. Saccharomycosis. Bronchial biopsy of a *Saccharomyces cerevisiae* infection of lung. Methenamine silver stain. × 425. (Courtesy of C. T. Dolan.)

9. Louria, D. B., S. M. Greenberg, et al. 1960. Fungemia caused by certain nonpathogenic strains of the family Cryptococcacea. New Engl. J. Med., *263*:1281–1284.

10. Lutwick, L. I., and H. J. Phatt. 1980. *Kluyveromyces fragilis* as an opportunistic fungal pathogen in man. Sabouraudia, *18*:69–73.

11. McGinnis, M. R., M. G. Rinalde, et al. 1975. Mycotic flora of the interdigital spaces of the human foot: a preliminary investigation. Mycopathologia, *55*:47–52.

11a. McGinnis, M. R., D. H. Walker et al. 1980. *Hansenula polymorpha* infection in a child with chronic granulomatous disease. Arch. Pathol. Lab. Med., *104*:290–292.

12. Mackenzie, D. W. H. 1961. Yeasts from human sources. Sabouraudia, *1*:8–14.

13. Madhavan, T., J. Eisses, et al. 1976. Infection due to *Trichosporon cutaneum:* an uncommon systemic pathogen. Henry Ford Hosp. Med. J., *24*:27–30.

14. Marier, R., B. Zakhireh, et al. 1978. *Trichosporon cutaneum* endocarditis. Scand. J. Infec. Dis., *10*:255–256.

15. Naveh, Y., A. Friedman, et al. 1975. Endocarditis caused by *Rhodotorula* successfully treated with 5-flurocytosine. Br. Heart J., *37*:101–104.

16. Page, R. K., O. J. Fletcher, et al. 1976. Dermatitis produced by *Rhodotorula glutinis* in broiler age chickens. Avian Diseases, *20*:416–421.

17. Pirlea, F., M. Garoiu, et al. 1976. Investigatii privind prezenta germenilor micotici in mastite in vaci. Rev. Cresterea Animal., *26*:33–37.

18. Pore, R. S., and J. Chen. 1976. Meningitis caused by *Rhodotorula*. Sabouraudia, *14*:331–335.

19. Rivera, R., and A. Cangir. 1975. *Trichosporon* sepsis and leukemia. Cancer, *36*:1106–1110.

20. Robineau, M., Y. Boussougart, et al. 1974. Fongemie a *Rhodotorula* et septicemie à colibacilles. Bull. Soc. Fr. Mycol. Med., *3*:195–198.

21. Rose, H. D., and V. P. Kurup. 1977. Colonization of hospitalized patients with yeastlike organisms. Sabouraudia, *15*:251–256.

22. Saez, H. 1979. Étude ecologique sur les *Rhodotorula* des homotherms. Rev. Med. Vet., *130*:903–908.

23. Sheik, E. S., S. Mahgoub, et al. 1972. Postoperative endophthalmitis due to *Trichosporon cutaneum*. Br. Ophthalmol., *58*:591–594.

24. Sonck, C. E. 1980. On the incidence of yeasts species from human sources in Finland. IV Yeasts from toe webs and nails. Mykosen, *23*:107.

25. Stenderup, A., and G. T. Pederson. 1962. Yeasts of human origin. Acta Path. Microbiol. Scand., *54*:462–472.

26. Taguchi, M., M. Tsukiji, et al. 1979. Rapid identification of yeasts by serological methods, a combined serological and biological method. Sabouraudia, *17*:185–191.

27. Terashima, E., T. Tsuchiya, et al. 1978. Recent causal organisms of bacteremia and a case of septicemia caused by *Trichosporon cutaneum*. Kansenshogakua Zasshi, *52*:102–104.

28. Van Uden, N., L. Do Carmo Sousa, et al. 1958. On the intestinal yeast flora of horses, sheep, goats and swine. J. Gen. Microbiol., *19*:435–445.

29. Verona, O. 1972. Yeasts in poultry feces. Ann. Microbiol., *22*:11–15.

30. Vieu, M., and G. Segretain. 1959. Contribution a l'étude de *Geotrichum* and *Trichosporon* d'origine humaire. Ann. Inst. Pasteur (Paris), *96*:421–433.

31. Volz, P. A., D. E. Jerger, et al. 1974. A preliminary study of yeasts isolated from marine habitats at Abaco Island. The Bahamas. Mycopathologia, *54*:313–316.

32. Watson, K. C., and K. Kallichurum. 1970. Brain abscess due to *Trichosporon cutaneum*. J. Med. Microbiol., *3*:191–193.

33. Winston, D. J., G. E. Balsley, et al. 1977. Disseminated *Trichosporon capitation* infection in an immunosuppressed host. Arch. Intern. Med., *137*:1192–1195.

34. Yung, C. W., S. B. Hanauer, et al. 1981. Disseminated *Trichosporon beigelii*. Cancer, *48*:2107–2111.

23 Aspergillosis

DEFINITION

Aspergillosis broadly defined is a group of diseases in which members of the genus *Aspergillus* are involved. The gamut of pathologic processes include (1) toxicity due to ingestion of contaminated foods; (2) allergy and sequelae to the presence of conidia or transient growth of the organism in body orifices; (3) colonization without extension in preformed cavities and debilitated tissues; (4) invasive, inflammatory, granulomatous, necrotizing disease of lungs and other organs; and rarely (5) systemic and fatal disseminated disease. The type of disease evoked depends on the local or general physiologic state of the host, as the etiologic agents are ubiquitous and opportunistic. The same agents are involved in many important animal diseases, such as mycotic abortion of sheep and cattle, pulmonary infections of birds, and toxicosis due to ingestion of grain containing fungal by-products. The spectrum of human infection also includes otomycosis, mycotic keratitis, and rarely mycetoma. The diversity of involvement attests to the adaptive capacity of the organisms, so that "aspergillosis" in reality is a spectrum of diseases.

HISTORY

Aspergillosis was one of the first fungal diseases of man or animals recognized. Mayer and Emmert[73] in 1815 described an infection in the lungs of a jay *(Corvus glandarius)*. The name *Aspergillus* had been coined much earlier by the Florentine botanist Micheli[76] in his "Nova Plantarum Genera" of 1729. The "aspergillum" (Latin: to sprinkle) referred to a perforated globe used to sprinkle holy water during religious ceremonies *("aspergus te")*. Micheli noted nine species, each of which he gave a sentence-long latin name (the binomial system of nomenclature developed by Linnaeus did not appear until 1753). Later, in 1809, Link[63] delineated several distinct types that he had isolated from decaying vegetation. Among these were *A. flavus, A. candidus,* and *A. glaucus.*

The term "aspergillosis" was first utilized by Fresenius in 1850 in his work on a fungus infection of the air sac of a bird — the bustard *(Otis tardaga)* — from the Frankfort Zoo. He named the isolant *Aspergillus fumigatus.*[40] Human infection was recognized by Bennett in 1842[13] and *Aspergillus* pneumomycosis was described by Sluyter in 1847.[126] By analyzing his description, the fungus in Bennett's case does not appear to be an *Aspergillus,* so that the report of Sluyter probably represents the first case of human aspergillosis.

A report of bronchial and pulmonary disease, together with such an accurate description of the etiologic agents as to allow them to be recognizable as *A. funigatus,* was published by Virchow in 1856.[138] This presentation of four cases is a classic paper of descriptive pathology. Except for the association of the pulmonary form with "pigeon feeder's" disease by Virchow and later by Dieulafoy, Changemesse, and Widal in 1897,[34] in the following years very few cases of the disease were reported. In 1897 Renon[101] published a book containing an excellent review of the field and also of the association of the disease with certain occupations in addition to pigeon handlers, such as wig cleaner, and concluded that moldy grain was the source of the conidia. By 1910 almost all forms of aspergillosis in animals had been delineated, including mycotic abortion and fatal respiratory disease.[8]

The association of *Aspergillus* infection with other diseases or with debilitation has occurred relatively recently. Cleland in 1924 found aspergillosis in necrotic lung tissue that resulted from a gunshot. Lapham in 1926 described aspergillosis as an infection secondary to tuberculosis. "Bronchiectasiant aspergilloma" (fungus ball) was defined by Deve in 1938.[33] However, it was not until 1952 that the diversity and importance of aspergillosis was realized. Monod et al.[81] expanded the work of Deve on aspergilloma,[33] and Hinson et al.[50] illustrated various clinical manifestations of the disease. The allergic form of aspergillosis and the colonizing type of aspergillosis are now realized to be common and frequently encountered diseases.[42, 46–48] Since that time a gamut of other disease processes has been delineated.

The importance of *Aspergillus* as an agent of opportunistic infections has only recently been recognized. Systemic disseminated disease is essentially a product of the antibiotic era and may be termed a "disease of medical progress." Fortunately this form of aspergillosis is still relatively rare.[131] The debilitated patient, however, offers a special milieu for opportunistic fungi. In one large series of 454 leukemia cases, 14 per cent of deaths were attributable to fungous infections.[16] Aspergillosis was second only to candidiasis in this series. Several recent reviews have emphasized the increasing occurrence of aspergillosis in patients with leukemias.[1, 79] In one series there was a greater increase of aspergillosis complicating leukemias as compared with lymphomatous disease.[74] Aspergillosis is also a significant problem in organ transplant recipients. In the review by Bach et al.[10] there was a significant rise in the incidence of aspergillosis associated with the extent of antirejection therapy necessary, and particularly with the use of antilymphocyte serum. The increased use of cytotoxin, steroids, and transplantation for prolongation of life will doubtless result in an increased frequency of disseminated aspergillosis.

More recently the toxic metabolites of *Aspergillus* sp. in foodstuffs have assumed great significance. Fungal toxins and other such agents had been known for some time. In 1902 Ceni and Besta in Italy described metabolites from *Aspergillus* sp. as being detrimental to animals,[26] as had Bodin and Gautier in France in 1906[17] and Henrici in the United States in 1939.[49] Henrici had described two endotoxins from *Aspergillus fumigatus*. One was hemolytic and the other a potent pyrogen, among the first "toxic" agents isolated from fungi. These observations were later extended by Tilden[134] and correlated with disease in animals. The great importance of these substances was not fully appreciated, however, until the famous "turkey X" epidemic of 1960–1961. In England, turkeys and, later, swine, poults, and cattle were afflicted with a variety of malformations and organ system dysfunctions. These were, in time, associated with consumption of particular foodstuffs. The active ingredient was soon found to be a metabolite of *Aspergillus flavus* produced when the fungus grows on particular substrates. Aflatoxins are now a favorite topic for study, although their lasting significance in human health has yet to be determined.

ETIOLOGY, ECOLOGY, AND DISTRIBUTION

There are about 600 species in the genus *Aspergillus*. They are among the most common fungi of all environments. They are common in the soil and decaying vegetation throughout the world,[37] but are also found on all types of organic debris: spilled food, wet paint, cracked dialysis bags, opened medications, refrigerator walls, "sanitizing" fluids, used dressings, and so forth. They are the "bottle imp" of fluffy growth that is found in chemical reagents and distilled water bottles that have remained undisturbed for some time. *Aspergillus* organisms have been recovered from sulfuric acid-copper sulfate plating baths and formalinized pathology museum specimens. *Aspergillus* conidia are airborne and constantly inhaled. After exposure to a cloud of *Aspergillus fumigatus* conidia, the fungus can be recovered from sputum for many days afterward. Conidia have been recovered by weather balloons in the upper atmosphere, from snow in the Antarctic, and from winds over the Sahara. The omnipresence and ubiquity of *Aspergillus* organisms are a major factor in designating this planet as "our moldy earth."[31]

Aspergilli, particularly the common etiologic agents of human disease, have sometimes been involved in nosocomical infections. Petheram and Seal[94] investigated the surgical theater utilized after a cluster of seven cases of *A. fumigatus* endocarditis had occurred following implantation of prosthetic valves. They found profuse growth of *A. fumigatus* in dust in the ventilating system. Growth of the fungus on fireproofing material was discovered by Aisner.[2] This investigation was instigated by the occurrence of eight cases in cancer patients following transfer to a new facility. A ventilating system again was implicated as the probable source of heavy inoculum, which may have resulted in three cases of endocarditis following heart surgery. In this instance, Gage et al.[41] found the air intake opening to be located just above a mound of pigeon excreta. Air inlets were also the source of heavy growth of *A. fumigatus* investigated by Burton et al.[20] following four cases of aspergillosis in kidney transplant patients. Rose[110] also found the ventilating system as a probable factor in eleven cases of *Aspergillus* pneumonia. Following a move to new facilities, no more cases occurred in the ensuing five years. Contaminated dust filtering through acoustic tile may have been significant in the case cluster studied by Arnow et al.[4] This occurred during hospital renovation. Many other investigators, however, have failed to find point sources of heavy inoculum following a cluster of cases within hospital facilities.[54] The association of such clusters with point sources of heavy fungal growth has raised concern about the public health hazards of activities that would enrich the environment with quantities of conidia. Such activities would include thermal enrichment of water effluent by nuclear power reactors leading to massive growth of thermophilic species such as *A. fumigatus,* and enrichment of aerosols from large-scale composting of sewage.[72] Careful investigation of such potential hazards[109, 132] has found the air within a few yards of these facilities to contain no more colony-forming units (cfu) of *A. fumigatus* than in the ambient air of an English apple orchard in spring[51] or the library of a midwestern university at any season.[19]

Although aspergilli are constant in the human environment, only about eight species have been consistently and authentically involved in human infectious disease. It is important to appreciate that any species could be involved in asthmatic allergy caused by inhaled conidia, and many species transiently germinate and grow on bronchial secretions,[114] wounds, and diseased and normal skin. These are the common "contaminants" found in cultures of such material, but in the past were often mistaken as etiologic agents for numerous diseases and found their way into medical literature. However, the species that have been recovered from authentic cases of aspergillosis are few in number. *Aspergillus fumigatus* accounts for almost all diseases, both allergic and invasive. Allergic

aspergillosis has also been caused by *A. flavus, A. ochraceus,*[84] *A. nidulans, A. niger, A. terreus,* and *A. clavatus.* Aspergillomas are frequently reported to be caused by *A. niger* in the Americas[121] and more commonly by *A. fumigatus* in England and Europe. Aspergillosis of the lungs may be caused by *A. fumigatus, A. terreus, A. flavus, A. nidulans,* and *A. niveus.* Three different species in pure culture (*A. flavus, A. fumigatus,* and *A. nidulans*) have been isolated from different lesions in the lungs and heart of a single case.[144] Disseminated aspergillosis has involved *A. fumigatus, A. nidulans,*[99] *A. flavus,* and *A. restrictus.*[131, 150] *A. niger* is frequently encountered in otomycosis and has been the cause of primary cutaneous disease.[21] *A. amsteloidami,* among others, has been recovered from cases of mycetoma. *A. flavus* is the cause of many primary infections of the nasal sinuses,[78] as is *A. fumigatus,*[117] and *A. terreus* is a rare etiology of primary cerebral aspergillosis.[107] All these species have thermotolerance in common, and many soil thermotolerant isolants have been shown to be pathogenic for experimental animals.[39, 95] Fungi, especially aspergilli, are remarkably adaptable and can be "trained" to become pathogenic.[108]

The first necropsy of a case of human aspergillosis was described by Virchow in 1856.[138] Thereafter, most cases of the disease were recorded from Europe and it was only much later that the infection was recorded in America. The infection has now been recognized in all age groups, both sexes, all races, and in all parts of the world.[131] The rare primary disease is more common in adult males than in children or females. The mean age is about 40. Secondary aspergillosis is a terminal event in the debilitated patient and depends on the availability of such subjects, not on sex or age. Small foci of primary infection are not infrequently found in routine lung necropsies, but these are incidental and not involved with the demise of the patient.

In the early French literature, the disease was associated with moldy barns, grain, pigeon fanciers (*graveurs des pigeons*), wig dusters, and so forth.[50, 138] These patients were exposed to clouds of conidia, and many were usually malnourished or tubercular. These could be described as opportunistic infections.[50] As discussed in a previous section, a point source of heavy contamination with conidia is always possible in primary disease. This was evident in the case of two sisters, both of whom died with acute pulmonary aspergillosis,[129] and in several case reports of farmers who inhaled masses of conidia.[137] Disease has also resulted from accidental injection of contaminated dialysis fluid, antibiotics, and so forth.[112] Primary disseminated aspergillosis not associated with underlying disease or heavy exposure is quite rare but does occur.[24, 111, 129] Secondary aspergillosis has been associated with antibiotics, steroids, radiation,[131] cytotoxins and neoplasias,[74, 79, 82] hematologic disorders,[65, 79, 91, 125] renal transplants,[20, 82] enterocolitis,[70] pneumonia,[38] chronic lung disease,[111] alcoholism,[15, 30] the neonatal period,[44] tuberculosis,[86] eclampsia and dialysis,[112] "barrier breaks" during surgery,[25, 41, 140] aspergillomas with tuberculosis cavities,[18] congenital heart disease,[90] sarcoid,[145] and histoplasmosis.[96]

There have been several good reviews of aspergillosis that have contributed to the emerging understanding of the disease: disseminated aspergillosis by Tan,[131] pulmonary aspergillosis by Pennington[92] allergic aspergillosis by Henderson,[46, 48] and Patterson[89, 141, 142] the French literature by Hinson et al.,[50] the English literature by Pepys et al.,[93] and the current European literature by Orie et al.[86]

CLINICAL DISEASE

It is now recognized that there are basically three categories of disease involving aspergilli. These are (1) allergic aspergillosis, (2) colonizing aspergillosis, and (3) invasive diseases. In addition, aspergilli can exist as saprobes in bronchi or on body surfaces without eliciting any pathology.

It is now appreciated that invasive or disseminated disease can be the result of one or a combination of three factors: (1) lowered resistance due to debilitating disease or drugs, (2) local point of entry for the fungus (barrier break) and (3) disruption of normal flora and inflammatory response by antibiotics and steroids. It must also be remembered that aspergillosis can be both primary and secondary. Allergic and colonizing aspergillosis are often primary conditions, whereas invasive and especially disseminated aspergillosis are secondary to other diseases. This division of primary and secondary is often overemphasized.[92] Obvious underlying disease presents no problem in classification, but to call aspergillosis primary when nothing else is found speaks more of our lack of diagnostic ability than the inherent pathogenicity of the fungus. Im-

paired defenses can be generalized, regional, local, or transient. The ubiquitous opportunist, *Aspergillus* would take advantage of any such situation.

The pathogenetic spectrum of aspergillosis is complicated, confusing, and, as stated earlier, represents a whole group of diseases. The clinical diseases will be discussed as categories that, most simply, allow separation and characterization of the particular syndrome. They are:

A. Pulmonary aspergillosis
  1. Allergic aspergillosis
    a. Extrinsic asthma
    b. Extrinsic allergic alveolitis
    c. Allergic bronchopulmonary aspergillosis
  2. Colonizing aspergillosis (aspergilloma)
  3. Invasive aspergillosis
B. Disseminated aspergillosis
C. Central nervous system aspergillosis
D. Cutaneous aspergillosis
E. Nasal orbital aspergillosis
F. Iatrogenic aspergillosis

Allergic aspergillosis is a disease that has only recently been defined, and its frequency appreciated.[89, 93, 116, 141, 142] The other forms of disease are still relatively rare. *Aspergillus* infections of the ear and eye are discussed in Chapter 27 and toxicosis due to ingestion of metabolic products in Chapter 28.

Pulmonary Disease

Allergic Aspergillosis

Extrinsic Asthma. Asthmatic allergy to conidia of the various *Aspergillus* species is a well-known and defined disease. It does not vary significantly from allergy to other types of dander and pollen or conidia. The conidia seldom germinate in the bronchial passages. The clinical picture is that of segmental shadowing, most frequently in the upper lobes. It has a tendency to recur in the same segment.[46, 93] Transient infiltrates are sometimes seen during the immediate reaction, and linear fibrosis is a rare sequela found in chronic disease. Infiltrates are more often associated with the other types of allergic pulmonary manifestations of aspergillosis. Cough, wheezing, chills, malaise, aches, pains, and peripheral blood and sputum eosinophilia are seen. Fever is rarely encountered in asthma but is a common symptom of extrinsic allergic alveolitis and allergic bronchopulmonary aspergillosis. Pulmonary function studies demonstrate a restrictive defect and impairment of gas exchange. Reaginic antibodies (IgE) are present and can be transferred into donors or monkeys.[42] These patients usually do not have precipitating antibodies or demonstrate an Arthus type reaction to injected antigen.[35, 89]

Extrinsic Allergic Alveolitis. This is a manifestation of allergy to *Aspergillus* conidia in the nonatopic individual. It usually occurs in individuals who have repeated exposure to organic dust that is heavily laden with conidia and mycelial debris.[93] One well known example of this is the so-called "malt worker's lung," which occurs in brewery workers and is associated with moldy (usually *A. clavatus*) barley.[105] Grain mill workers, silo workers, grain elevator workers, and farmers may be exposed to fungal conidia and debris from moldy oats, corn, wheat, and hay. It is to be noted that *Aspergillus* sp. is only one of many types of organic material that can lead to allergic alveolitis of varying severity, viz., thermophilic actinomycetes in "farmer's lung" and animal proteins in "bird fancier's lung." Extrinsic allergic alveolitis is a condition involving the lung parenchyma, whereas involvement of airways is seen in asthma (reversible) and allergic bronchopulmonary aspergillosis (irreversible).

Patients with extrinsic allergic alveolitis are seen to have symptoms some six hours after inhalation of dusts. The symptoms include cough, dyspnea, and sometimes fever and chills. Ronchi, but not wheezing, are present, and diffuse interstitial infiltrates are seen in all fields. Restrictive rather than obstructive deficits are noted in pulmonary function tests. No peripheral or pulmonary eosinophilia are present in lung biopsy. In the presensitized individual, continued exposure often leads to granulomatous disease and irreversible pulmonary fibrosis.

Considerable debate has surrounded the pathogenic mechanism or mechanisms of extrinsic allergic alveolitis. In general the patients have an elevated IgG level to the specific agent without IgE, and they have an Arthus reaction to skin testing. It has been assumed, therefore, that this was an immune complex disease (type III) of the lung. Recent evidence indicates a lymphocyte-directed hypersensitivity (Type IV) is also involved.

Allergic Bronchopulmonary Aspergillosis. Also called mucomembranous *Aspergillus* bronchitis, this form of aspergillosis may develop as an exaggeration of the above disease, in chronic or extrinsic asthma, late-onset asthma, or as some other manifestation of atopic diathesis. The clinical symptoms are similar to those of asthma but are more chronic and severe. Bronchial plugging occurs. The peripheral shadows are consistent, with areas of collapse distal to bronchial occlusion, thus resembling pulmonary infiltrate and eosinophilia (PIE) syndrome. Bronchoscopy reveals the fungus to be growing in patches of various sizes perpendicularly to the bronchi. Fruiting heads and conidia are sometimes produced. The underlying mucus membrane is red and congested but not invaded. Coughing produces sputum in which the fungus may be seen by direct smear (Fig. 23–1) and from which it may be cultured. In the more than 200 patients reviewed, positive sputum cultures were a constant finding, and they remained so during the length of the disease.[141, 142] A delayed type skin reaction, as well as immediate skin reactions, is demonstrated by these patients. In most of these cases, a type-III or Arthus hypersensitivity can be elicited. The pathology of the disease is thought to be owing to both the immediate and delayed types of hypersensitivity. The

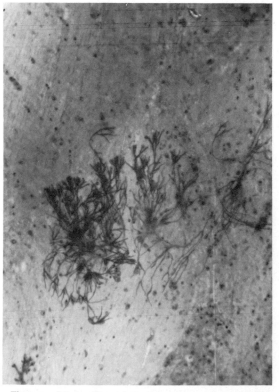

Figure 23–1. Bronchopulmonary aspergillosis. Direct smear of sputum. Clumps of aspergilli are seen that demonstrate fingerlike dichotomous branching. ×100.

most frequently isolated organism, *Aspergillus fumigatus,* is known to produce powerful endotoxins[134] and a C substance.[66] The former may also contribute to the disease. An absolute eosinophilia is present. Precipitins are demonstrable in the sera of almost all patients[46, 140, 141]; thus, both IgG and IgE antibodies are present in this disease. The IgE may be specific for the *Aspergillus* sp. involved or may be nonspecific.[89] It is now considered that allergic aspergillosis is one of the more common causes of pulmonary eosinophilia.[46] The other diagnostic criteria are episodic airway obstruction, rather than a restriction deficit; a one-second forced expiratory volume less than 70 per cent of vital capacity; and the finding of abundant fungi in culture. This latter criterion consists of culturing the fungus from more than one specimen or growing more than two colonies from one specimen. The sputum produced is gelatinous and sometimes bloody. The radiologic picture is that of segmental shadowing and a migrating type of recurrent pneumonitis, usually in the upper lobes and especially in the apical segments. The other lobes and lingula are not frequently involved. The shadows of mucoid impaction are continuous, with the hilum proximally and distally blunted. Dilated bronchi sometimes appear opaque and at other times lucent. Cylindrical bronchiectasis of the upper lobes is seen on bronchograms (Fig. 23–2).[67] A striking feature of this disease is the marked change in x-ray patterns associated with relatively mild clinical symptoms.

Although most of the patients with ABA are adults in whom the disease developed as a sequela of asthma, considerable numbers of pediatric cases are now being described.[141] Although the fungi grow in the lumen of the airways, systemic antifungal agents are of no benefit. Glucocorticosteroids[141] are effective and can prevent the serious complications of this condition.[102]

Colonizing Aspergillosis (Aspergilloma)

This condition, with the development of "fungus balls," may result from chronic aller-

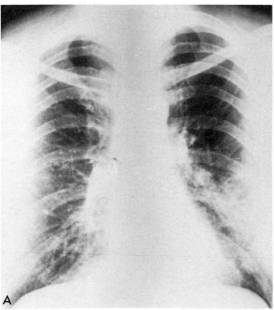

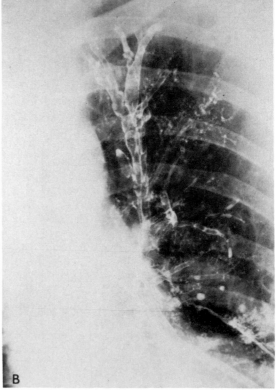

Figure 23–2. Allergic bronchopulmonary aspergillosis. *A,* Plane film of chest. Right lung is normal. In the left lung there are nonspecific, ill-defined infiltrates in the lingula of the upper lobe. *B,* Bronchogram of same patient. There is a fusiform or cylindrical dilation of the subsegmental bronchi of the apical posterior segment of the left upper lobe. There is tapering of posterior bronchi to close to the normal diameter. Elsewhere in the lung, chronic bronchitic changes are seen. (Courtesy of J. Fennessy.)

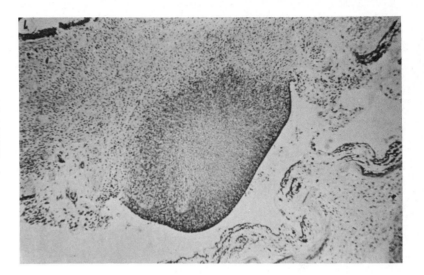

Figure 23–3. Aspergilloma. Small aspergilloma formed in an ectatic bronchus. There is erosion of underlying epithelium and in this case (sarcoidosis treated with steroids) the beginning of invasion of the bronchial wall. ×100.

gic aspergillosis[33] or from colonization of preformed cavities caused by other diseases. The former condition is associated with "eosinophilic" pneumonia and bronchiectasis. The patients usually have a history consistent with chronic allergic aspergillosis, and an aspergilloma forms in an ectatic bronchus. Clinically the symptoms are those of the allergic disease, but bouts of severe hemoptysis are more regularly seen. These patients have transferable reaginic (IgE) antibodies and precipitins (IgG), and demonstrate immediate and late skin sensitivity.

Another type of primary aspergilloma is seen in patients without allergic disease. Over a period of time, localized areas of infiltrate are seen to develop fuzzy edges that gradually become rounded and form cavities. These may show internal opacity with a radiolucent crest (Monod's sign).[32] If resected at this time, masses of live fungal elements are found. Some cavities may excavate or increase in size, and others may disappear with residual fibrosis. Resection of old unexcavated lesions show deliquescent masses of dead hyphae.

Aspergilloma secondary to other cavitary disease is now considered to be a frequently encountered disease (Fig. 23–3). According to a large survey conducted by the British Tuberculosis Association,[18] it occurred in 12 per cent of healed tuberculosis patients on radiologic evidence alone, and an additional 5 per cent had precipitating antibodies suggesting colonization. These patients lack reaginic antibodies, but IgG antibodies are universally present (Fig. 23–4). The immediate and delayed skin reactions are usually lacking. The second most common predisposing cavity is that resulting from healed lesions of sarcoido-

sis.[35, 53, 145] It is also found on bronchogenic cysts, histoplasmosis cavities,[96] and in areas of old radiation fibrosis.[55]

Aspergillomas or fungus balls have devel-

Figure 23–4. Aspergilloma. Immunodiffusion test of a patient with aspergilloma. The patient's serum is in the center well, and an *Aspergillus fumigatus* antigen is in the peripheral well. At least 18 bands were detected. More bands are present in cases of aspergilloma than in any other form of aspergillosis.

oped in other areas of the body as well. The most commonly involved extrapulmonary sites are the various nasal sinuses (see section on orbital nasal aspergillosis). In rare instances aspergillomas have developed in the urinary bladder. In the case of Torrington et al.,[135] this represented dissemination of pulmonary disease in a patient with acute myelocytic leukemia. The patient passed large fungus balls in the urine and was successfully treated with amphotericin B. Such cases have also been found with no underlying disease and no evidence of disseminated or renal aspergillosis.[113] Fungus balls in the gall bladder and in the bile ducts have also been described.[94]

The fungus grows as one or more brownish balls within the cavity. These consist of tangled masses of mycelium, which are often found conidiating if there is an air space above them. The cavity is usually not filled, and both cavity and fungus ball may be seen to enlarge over a period of time.[81] The smooth wall of the cavity is lined by metaplastic epithelium (Fig. 23–7,B). Inflammatory

reaction around the cavity may be severe, and fibrosis occurs. Two or more bronchi may be connected to the same cavity. Where they converge into the cavity they are both dilated, giving an x-ray picture essentially pathognomonic for the disease.[64]

The characteristic x-ray picture is that of a uniform opacity, not dense, which is round or oval in shape and has a radiolucent crescent (Monod's sign or *grelot* of the French authors[33, 81]) over the upper portion of an internal mass (Fig. 23–5). Sometimes a complete circle of air is seen around the mass.[81] Changing the position of the patient causes the density to shift. The most frequent location of aspergillomas is the apex of the lungs. Both right and left sides may be affected,[52] but there appears to be some predilection for the right.

The principal clinical features of aspergilloma is recurrent hemoptysis (Fig. 23–6). This varies from small episodes of blood-tinged sputa to fatal exsanguinating hemorrhage (Fig. 23–7). In a recent series, the mortality rate due to hemoptysis was considerably

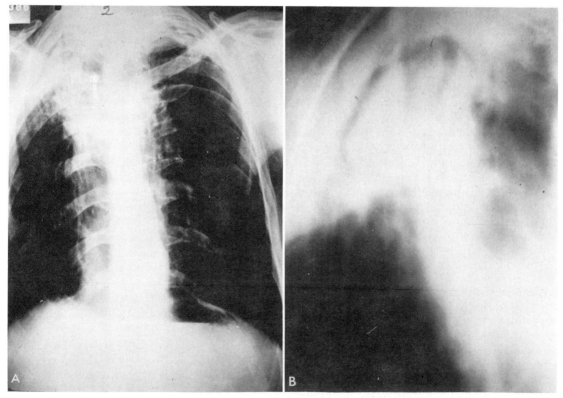

Figure 23–5. Aspergilloma. *A*, Plane film of chest. The lungs are over-inflated. There is scarring in upper lobes and bilateral, apical, pleural thickening, which is most marked on the right. There is opacification of the right apex. A density partly surrounded by air is seen in the center. There is also reduction in volume of the upper lobe. *B*, Tomogram of same patient. There is a large cavity containing a large solid density partly surrounded by a crescent of air (Monod's sign). (Courtesy of J. Kasik.)

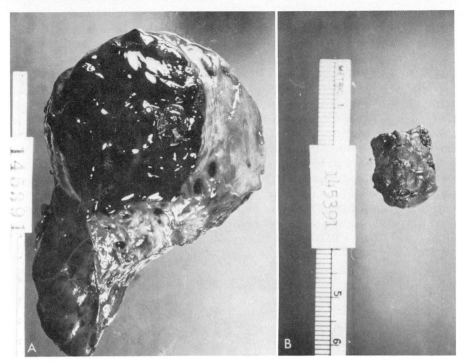

Figure 23–6. Aspergilloma. *A,* Aspergilloma in an old sarcoid cavity. The patient had recurrent hemoptysis. In the excised lobe, a large clot fills the cavity. The aspergilloma is in the upper central area. *B,* Fungus ball removed from cavity. (Courtesy of S. Thomsen.)

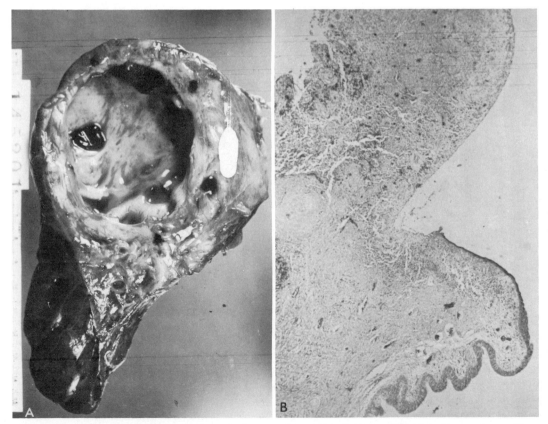

Figure 23–7. Aspergilloma. *A,* Cavity cleared of clot and fungus ball. The wall is thick, fibrous tissue. There are several expanded bronchi leading from the central cavity. *B,* Section of cavity wall. The epithelial lining has been eroded in one area, exposing granulation tissue. This necrosis and erosion may be due to endotoxins released by the organism. ×100. (Courtesy of S. Thomsen.)

573

higher than had previously been noted, indicating that aspergilloma should not be considered a benign condition.[35] Confirmation of the diagnosis by culture is often lacking, because the fungus ball may not be connected with a patent bronchus. In such cases, the etiology may not be established until after surgical removal and culture of the lesion (Fig. 23–8), but the disease can be unequivocally diagnosed by serologic methods. The serum of these patients contains antibodies to several aspergillus antigens. These various antigen-antibody systems can be demonstrated by the immunodiffusion procedure (see Immunology and Serology). Presence of four or more precipitin bands is pathognomonic for an aspergilloma in some body cavity. It should be remembered that fungus balls have also been produced by *Pseudallescheria boydii, Sporothrix schenckii, Coccidioides immitis,* and many other fungi. Serology in such cases would be negative if *Aspergillus* antigen is used, but by using a battery of antigens with the patient's serum, the true etiology can be determined.

Increase in size of the lesions has been noted in many series.[33] It has been postulated that the fungus ball acts as a valve, permitting entry of air during inspiration and preventing its escape at exhalation. This necessitates that the open end be downward and may infer the reason for the frequency of apical location. Lesions may show lobation of the fungal mass, with excavation of the surrounding tissue, indicating that an invasive process is possible. Zonation of the mycelial elements with layers of pyknotic host cellular debris verifies a dynamic process in the evolution of the lesion (Fig. 23–9). It has been postulated that some of the pathologic disease seen in the tissue is due to endotoxin released by the fungus (Fig. 23–7,*B*).

Resolution or regression of cavities has been noted,[45, 52] but this appears to be an uncommon occurrence. When the epithelial lining is eroded, the fungal mass comes into

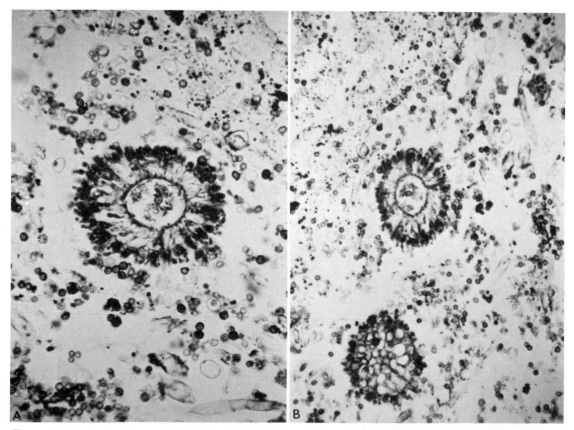

Figure 23–8. Aspergilloma. *A,* Unstained section from a case of aspergilloma caused by *Aspergillus niger.* Since there is an air space above the ball, fruiting heads of the aspergillus are formed. This does not occur in any other form of aspergillosis. *B,* By the color and architecture of the conidia and the presence of long primary and short secondary series of phialides, one can essentially identify the organism as *A. niger* on histologic section. ×440. (Courtesy of S. Thomsen.)

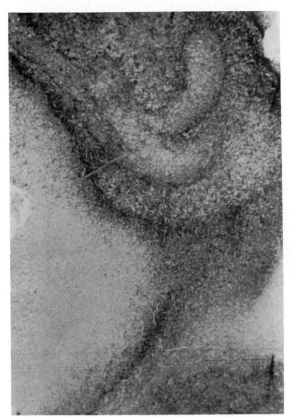

Figure 23–9. Aspergilloma. Cut section of aspergilloma, showing zonation of hyphal growth and deliquescence of central hyphal strands. GMS. ×100. (Courtesy of S. Thomsen.)

contact with formed granulation tissue, and invasion of the cavity by granulation tissue may occur. Old cavities may contain deliquescing masses of dead hyphal elements and show resorption of the debris by giant cells.[52, 53]

Although aspergillomas are associated with old tuberculosis cavities, the coexistence of the two diseases in the same area is controversial. *A. fumigatus* produces antibiotic substances, and tubercle bacilli are usually absent from lesions in which the fungus is present. Rarely, however, acid-fast bacilli have been found among hyphal masses in an aspergilloma.

The saprophytic infestation of the bronchial tree by *Aspergillus* is also known.[35] These patients show no evidence of hypersensitivity or clinically significant pathology, but aspergilli, such as *A. fumigatus,* are constantly excreted and can be regularly cultured from their sputum. Sometimes these develop into a pneumonitis or lung abscess, so that patients with several positive sputum cultures should be followed radiographically. Such patients may have a self-limited infection, or the

pneumonitis may be a precursor of more chronic *Aspergillus* lung disease.[5]

Invasive Aspergillosis

In this uncommon form of the disease, the mycelium is present in the lung tissue. This condition can be seen to develop from any of the preceding diseases (Fig. 23–3) or *de novo.* It is chronic or acute, sometimes with a rapid fulminating, fatal outcome. The usual predisposing factors for opportunistic infection are present, but the disease may occur in an apparently normal patient as well.[64] Rarely, the disease has developed after severe coccal or viral pneumonia.[38] Clinically the disease presents as a pneumonia with fever, cough, leukocytosis, and other signs of respiratory distress. Pleuritic pain is not common, but careful examination often reveals a pleuritic friction rub. Sometimes the pulmonary symptoms are mild and the patient may first present with other organ system involvement including pericarditis, cardiac tamponade,[66a] personality change,[82] superior vena cava syndrome,[41a] or some other obscure manifestation.

The x-ray findings may show diffuse involvement or consolidation with a single mass, resembling the picture of lung cancer.[87] Sometimes the lesions round up, excavate, and become aspergillomas or infarctions[125] (Figs. 23–10 and 23–11). The most common picture, however, is that of a bronchopneumonia with multiple patchy infiltrates that tend to be peripheral in distribution and are focal rather than diffuse.[87]

Several recent reviews have emphasized the increasing frequency of invasive aspergillosis among patients with leukemias and lymphomas.[74, 79] The former group seem to be particularly prone to developing this complication.[74] In one review of patients with renal transplants, aspergillosis was more frequently observed in those receiving antilymphocyte serum therapy.[10] These observations indicate the relative importance of the various host defense mechanisms in preventing or containing infections by *Aspergillus.* As noted in Chapter 20, the work of Diamond and others demonstrated the effectiveness of the human neutrophil in phagocytizing conidia or enveloping mycelial strands of *Aspergillus* and destroying them. This may account for the 20 times greater incidence of pulmonary aspergillosis in acute leukemia patients compared with that in lymphoma or transplant patients. On the other hand, glucocorticosteroids and lymphocyte-directed drugs

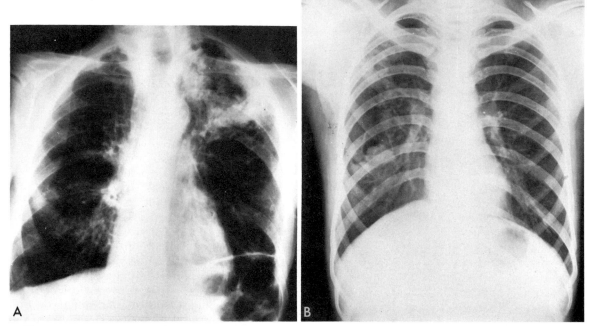

Figure 23–10. Invasive aspergillosis. *A,* A large, irregular cavity may be seen in the left upper lobe, surrounded by an infiltration that extends toward the hilum. Elsewhere there is evidence of obstructive lung disease. The patient had leukemia. *B,* Numerous thin-walled cavities are seen in the lungs. The one in the right lower lobe contains an air-fluid level. Irregular alveolar infiltrates containing small cavities are seen in the right upper lobe. The patient had leukemia.

such as azothioprine also predispose to invasive disease. This would argue for a role of lymphocyte-directed cell-mediated immunity.[127] As is the case in other mycotic infec-

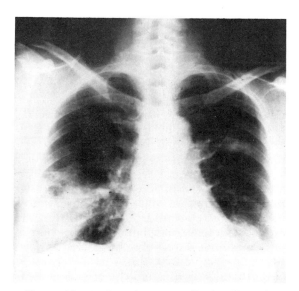

Figure 23–11. Invasive aspergillosis. There is a dense infiltrate in the right lower lobe, containing an irregularly outlined cavity. Nonspecific zones of alveolar infiltrate are present in the opposite lung. Subsequently these zones increased in size, and cavities appeared. The patient had hepatitis and was treated with steroids. (Courtesy of J. Fennessy.)

tions, humoral antibodies play an obscure role, if any, in combating fungal invasion. Immunosuppressed patients usually have no demonstrable anti-*Aspergillus* antibodies.[149] This adds to the difficulty of making a definitive diagnosis. The lack of significance of humoral antibodies in containing infection by aspergilli is attested to by patients with the various B-cell anomalies. In cases of dysgammaglobulinemia or hypogammaglobulinemia, aspergillosis is a rarely encountered infection. Thus, the primary defense against invasion by aspergilli as well as other opportunistic fungi is properly functioning phagocytic cells in adequate numbers, along with contributions from the cell-mediated immune system. Patients at particular risk are those with abrogations of these systems. Although aspergilli are omnipresent and ubiquitously distributed, patients at risk should be protected from exposure to high densities of fungal conidia. The important of exposure is dramatically demonstrated by case clusters in epidemiologic investigations of aspergillosis.[2, 4, 110]

The sinuous hyphae of the invading *Aspergillus* radiate from a central focus, branching freely and giving a characteristic picture histologically (Fig. 23–12). The surrounding tissue necrosis may be quite extensive. This has

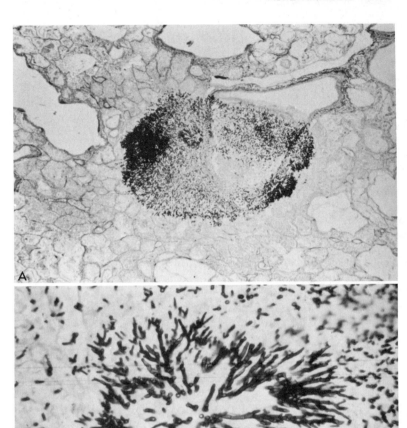

Figure 23–12. Invasive aspergillosis. *A,* Characteristic radiating pattern of invasive *Aspergillus.* An alveolar infiltrate is noted and the fungus has invaded a vessel. An embolus is beginning to form. ×100. *B,* Enlarged view of central area. GMS. ×440. (Courtesy of S. Thomsen.)

suggested to some that part of the pathologic disease is owing to an endotoxin derived from the fungus.[36] The characteristic radiating fungal masses are sometimes seen in routine autopsy studies. They apparently have no causal relationship to the demise of the patients, but represent an incidental or terminal colonization. This situation is especially common in farming areas, where frequent contact with conidia is possible. That these restricted invading colonies are common in healthy domestic animals has been demonstrated by Austwick.[8]

The clinical course of pulmonary aspergillosis is characterized as a rapidly advancing necrotizing pneumonia. The organisms soon invade the vasculature, causing thrombosis and ischemia of surrounding tissue. Septic emboli break off from such areas and are distributed to many parts of the body. Thus it is common to find *Aspergillus* abscesses in the brain, heart, liver, kidney, and thyroid[146] at autopsy when death resulted from a primary pulmonary focus. Pulmonary aspergillosis in a setting of immunosuppression is often a fulminant disease, with death occurring in one to two weeks.[79] Several reports imply, however, that if the diagnosis can be made early in the course of the infection, a successful resolution of the disease can be accomplished with proper and adequate treatment.[3, 29, 30, 91, 92, 125]

The major problem in *Aspergillus* pneumonia is the diagnosis. In immunosuppressed or myelosuppressed patients, the appearance of a pulmonary infiltrate and fever is a harbinger of many types of infectious complications. High on the list is aspergillosis, particularly if the lesions are focal. Confirming the diagnosis in such a setting is very difficult, however. Sputum is usually unavailable for cultures, and the significance of an isolated colony on such culture is negligible. Serologic tests are usually negative, even in widely disseminated disease due to the immunosuppression. Invasive procedures are often countermanded because of the thrombocytopenic and hypoxemic condition of the patient. *Aspergillus* pneumonitis is truly a clinical dilemma.

Some progress has been made in the development of diagnostic procedures for the compromised patient. Extensive experience has demonstrated the inadequacy of the usual serologic methods in invasive aspergillosis.[149] Recently two other approaches have received attention. The first, reported by Shoeffer et al.,[123] relies on three-fold concentration of serum and detection of antibody by immunodiffusion. There is still a significant false negative rate with this method, however. The second approach is the detection of *Aspergillus* antigen in serum of the patient. This method has not been rewarding in candidiasis, but some studies of aspergillosis look promising. Antigenemia in aspergillosis has been detected by Reiss and Lehman with CIE[100] and by Weiner[143] with radioimmunoassay. The procedures are positive in over 80 per cent of experimentally infected animals and now await clinical trials. Very recently, an ELISA system for the detection of minute amounts of antibody in immunosuppressed patients was reported.[119]

Several invasive procedures have been developed over the years to obtain biopsies, brushings, washings, and aspirates in patients with obscure pneumonias. Bronchial brushing, percutaneous transthoracic needle aspiration,[11] transbronchial biopsy,[91] open lung biopsy (thoracotomy), and more recently, fiberoptic bronchoscopic biopsy[29] have also been used to obtain material for diagnostic tests. The choice is dictated by the status of the patient. Material obtained is planted on culture media without actidione (inhibits fungi) or antibiotics (inhibits *Nocardia*), and preparations are stained with GMS (for fungi, pneumocystis, and Actinomycetes) and PAS (for fungi). Colonization of the nose sometimes precedes lung disease, so that nasal culture can sometimes predict future pulmonary involvement.[1]

Disseminated Aspergillosis

A relatively recent and as yet rare disease, this appears to be a product of the antibiotic-steroid-cytotoxin era. The first case of disseminated disease was reported by Linck[62] in 1939. The lung, brain and leptomeninges were involved. No predisposing condition could be assessed. Since that time, almost all subsequent cases have implicated either antibiotics, cytotoxins, or steroids.[1] Many had underlying disease or associated disease, including pneumonia, diarrhea, leukemia, eclampsias, and hepatitis.[112] A few patients did not have any apparent abnormality, however.

The presenting symptoms are those of an acutely ill patient. Dissemination to various organs causes associated symptoms, such as neurologic signs,[112] the Budd-Chiari syndrome,[148] and so forth. In the review by Tan et al.,[131] the lungs were the most frequently involved organ, followed by the brain and kidney. In widely disseminated disease, almost all organ systems are involved.[146] Endocarditis[77] is usually secondary to pulmonary disease[140] but may occur subsequent to surgery.[6, 28] Only rarely has dissemination occurred that does not involve the lung.[82] Other portals of entry aside from the pulmonary are possible, however. A "mycotic aneurysm" of the aorta was reported by Meyerwitz et al.[75] In this immunosuppressed patient, the primary focus appeared to be the lung. Involvement of bone is rare, but has been recorded.[6, 99, 118, 120] The lesions were osteolytic and not discernibly different from tuberculous or other bacterial osteomyelitis (Fig. 23–14).

Central Nervous System Aspergillosis

This disorder has been found as the result of iatrogenic procedures, in drug addicts following injection of contaminated material intravenously, and sometimes without known portal of entry or primary focus.[122] The greatest number of cases, however, have been associated with pulmonary aspergillosis in an immunosuppressed patient.[82] The presenting symptoms are those of acute meningitis, and the disease is rapidly fatal. Extensive necrotic lesions of brain, meninges, and vessels are seen at autopsy. Very few clinical cures have been recorded. Following intrathecal treatment with amphotericin B, the arachnoiditis resulting from therapy has often left the surviving patient paraplegic.[107] Some cases of meningeal disease have presented as spinal cord compression[120] or internal carotid stenosis.[7]

In the cases of cerebral aspergillosis described by Shapiro and Tabaddor[122] there were a variety of predisposing conditions. These included immunodeficiency, neoplasia, and narcotic addiction. In one case, no underlying condition was present and the patient presented in the emergency room as a generalized seizure. Essentially all cases of cerebral aspergillosis are mycotic abscesses, usually multiple (Fig. 23–16), and are fatal in spite of surgical or medical procedures. Shapiro and Tabaddor[122] note that the spinal fluid in cerebral aspergillosis usually shows moderate protein elevation, often above 100 mg per cent, a normal glucose, and a variable pleocytosis. The cellular response is usually neutrophilic, and marked elevations are seen if there is a meningitis. Meningitis is less common than cerebral aspergillosis and is usually the result of a barrier break,[107] whereas cerebral disease is almost always the result of septic embroli from some other focus of infection.

Cutaneous Aspergillosis

Although it has been reported, this is a rare disease. It is usually secondary to dissemination[137] but may be primary.[23] Essentially all cases of cutaneous aspergillosis, whether primary or secondary, occur in a setting of immunosuppression (Fig. 23–13). Cases have been described, however, in which no underlying factors could be discerned and in which the condition is a primary infection. In the latter, the lesions consist of multiple nodules, and the skin is thick, edematous, and has a purplish discoloration. One such case was mistaken for lepromatous leprosy and was so treated for over a decade. The lesions consisted of multiple granulomata containing *Aspergillus* in the superficial middle and deep dermis. Numerous giant cells were present, and there was central necrosis. The etiologic agent was *A. niger*. The disease responded to topical nystatin.[21]

The skin lesions in disseminated cases start as small red discrete papules which become pustular. Biopsy shows them to consist of small microabscesses and granulomata with central necrosis. Small radiating colonies of *Aspergillus* are seen.[137]

In a case of primary cutaneous disease in a leukemic child, the lesions were characterized by erythematous macules and papules associated with itching and pain. They progressed rapidly to ulcers with central necrosis and black eschars and a raised erythematous border. Carlile[23] observed a positive response to topical nystatin.

Aspergillus sp. are also known to colonize burned areas of skin. Panke et al.[88] reported an *A. niger* infection of a burn wound that resembled ecthyma gangrenosa or a *Pseudomonas* superinfection. Excision of the area was done and lesions did not recur.

Nasal-Orbital Aspergillosis

This disorder consists of an aspergilloma in the nasal sinuses that may eventually involve the orbit of the eye.[61] Throughout the world this is a rare disease, but it is not uncommon

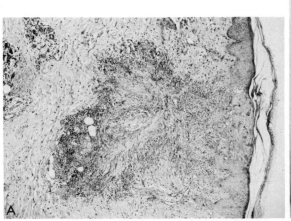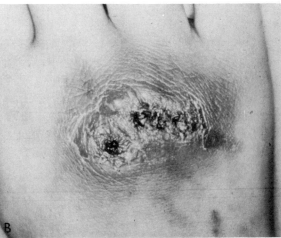

Figure 23–13. Cutaneous aspergillosis. *A,* Aspergillosis secondary to a pulmonary infection. There is thromboembolic occlusion of a vessel in the deep dermis, with hyphae radiating from the vessel into the surrounding tissue. A mixed cellular infiltrate is found in the periphery of the lesion. The epidermis is edematous, and loss of architecture is noted in areas that have begun to necrose. PAS stain. ×100. *B,* Primary aspergillosis of the skin in an immunosuppressed patient at the site of an IV line. Note the necrotic areas and eschar formation. *A, fumigatus.* (Courtesy of R. Diamond.)

in the Sudan. Seventeen cases have been reviewed by Milosev et al.[78] The lesions most often involved the ethmoid sinuses, although disease in all cavities was recorded. The most common presenting sign was unilateral proptosis and swelling of surrounding tissue of varying severity. The etiologic agent in all cases was *A. flavus*. Other species have also been reported.[68] Treatment consisted of surgical removal of granulomata and draining of cavities. An association of cure with the relief of partial anaerobiosis was hypothesized by the authors.

The maxillary sinus was involved in the four cases of Axelsson et al.[9] All had symptoms of chronic bacterial sinusitis and the x-ray picture was also consistent with that diagnosis. No underlying condition was noted. All were treated surgically with a Caldwell-Luc operation and there were no recurrences. A few cases have been associated with predisposing conditions, particularly steroids,[69] but in general sinusitis is independent of other diseases or conditions.[117] Very rarely primary aspergillus sinusitis is the portal of entry for disseminated disease in an immunosuppressed patient.[1] *Pseudallescheria boydii* also produces the same syndromes and is morphologically identical in histologic sections.

Iatrogenic Aspergillosis

Aspergilli are constantly present in the environment and grow on all types of organic debris. For this reason they may often contaminate hospital rooms and supplies.[2, 4, 6, 41, 54] They may thus inadvertently gain entrance to susceptible patients by many portals. *A. fumigatus* was growing in the procedure room and may have been growing on leaking dialysis bags in a case of fatal disseminated disease following dialysis.[112] Endocarditis has occurred after cardiac surgery,[41, 94] as has osteomyelitis[6, 118] (Fig. 23–14). Meningeal involvement has occurred following intrathecal penicillin therapy. In a personally observed case, severe meningitis occurred in an otherwise healthy man three weeks following a disk fusion. In this case *Aspergillus terreus* grew out from many spinal fluid cultures. Treatment with amphotericin B was successful in arresting the infection, but the patient is at present paraplegic due to the severe arachnoiditis that accompanied therapy.[107]

DIFFERENTIAL DIAGNOSIS

The diagnosis of pulmonary aspergillosis can be made only with demonstration of fungal elements in pathologic material and repeated isolation of the fungus in culture — the latter in the so-called "abundant fungus" method (see Laboratory Identification).

In allergic bronchopulmonary aspergillosis, hyphal strands are easily demonstrated in sputum by direct examination, and the organisms are easily grown on standard laboratory medium. In a typical case, the diagnosis is readily made. The mere presence of *Asper-*

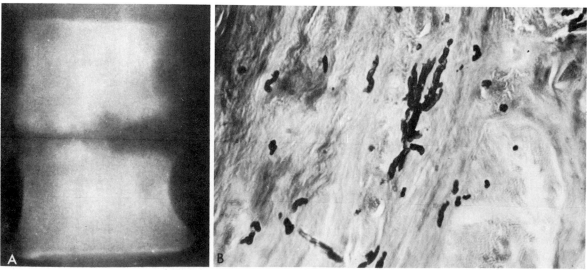

Figure 23–14. Iatrogenic aspergillosis of the vertebrae. *A,* X-ray of the spine at L1–L2. At the site of injection of a steroid, a lytic lesion has developed. There was soft tissue swelling and pain. *B,* GMS stain of bone biopsy. In some areas, aleurioconidia could be seen attached to hyphae. The culture was *Aspergillus terreus.*[119]

gillus conidia or a colony in culture is not sufficient evidence for this diagnosis. In some aspergillomas and invasive pulmonary aspergillosis, fungal elements are not present in sputum. The former may be diagnosed by typical radiologic picture; however, a recent blood clot within a cavity can give a similiar x-ray picture. Thus, serology is very important in confirming the clinical impression of an aspergilloma. In invasive disease, the diagnosis may not be made until surgical removal or at autopsy.

Aspergillosis is so frequently a secondary infection in tuberculosis, carcinoma, sarcoid, bronchiectasis and other fungal diseases that an underlying malady should be sought. Mucormycosis and aspergillosis as well as other opportunistic infectious agents are not uncommonly seen in the same patient. Both are fungal opportunists of the immunosuppressed patient, and both produce primary necrotizing pneumonia in the lung with invasion of vessels, embolization, and ischemia and necrosis of the lung parenchyma. In contrast, *Candida* involvement of the lung is a late-stage manifestation of hematogenous dissemination from some other source. *Candida* lesions tend to be focal rather than widespread and without consolidation.

PROGNOSIS AND THERAPY

The prognosis of aspergillosis depends almost entirely on the type of disease evoked and the physiologic state of the patient. Allergic aspergillosis may be benign[83] or, in some patients, may tend to become more severe and the duration of attacks longer and more chronic. The resulting pathology in these cases is that of increasing respiratory distress with bronchiectasis, segmental collapse, and fibrosis. Colonizing aspergillosis (aspergillomas) may remain chronic for years without much distress except for bouts of hemoptysis and associated lung pathology. Such colonization (fungus balls) are known to become invasive, and a high mortality is noted in some series.[18, 35]

Invasive aspergillosis is an extremely serious disease requiring intensive therapy, the success of which depends in large part on the patient's underlying illness and immune competence. In previous reviews, few patients were found who survived *Aspergillus* pneumonia;[16, 79] however, now with early diagnosis and appropriate therapy, invasive aspergillosis can often be cured.[125]

In terms of treatment, allergic aspergillosis is approached both as a hypersensitivity disease and as an infection. In some cases of short duration, prednisone has cleared the condition without relapse.[42] A dose of 25 mg daily was initiated, which led to improvement in one week, and the drug was gradually decreased and discontinued over the next few months.[142] Some investigators caution against the use of prolonged steroids in chronic cases.[35] Inhaled corticosteroids such as beclomethasone have also been used with success.[102] The main object of therapy is to minimize the periods of bronchial plugging by mucus containing the fungus. Vigorous physiotherapy, bronchoscopic aspiration, and lavage are recommended. Steroids are used only during the acute phase. Disodium chromoglycate, an antiallergic agent without the side effects of steroids, has been suggested.[35] Inhalation of nystatin or natamycin (pimaricin) has been used to clear the bronchial walls of the organism by some investigators,[47] but this form of therapy is now seldom used.

Aspergillomas have been considered benign conditions which sometimes resolve spontaneously.[45] However, several recent series have stressed the danger of death from massive exsanguinating hemoptysis during the development and enlargement of the cavity.[35, 55] Resumption of steroid therapy in patients with sarcoid has sometimes led to invasive disease (Fig. 23–3).[53] Therefore, therapy of aspergilloma is advised. The lesions are large and avascular, so that systemically administered antifungal agents are not usually helpful. Medical management has been used, especially in cases in which surgical resection was not possible. Ramirez introduced the method of infusing the lesion with amphotericin B through an indwelling catheter.[97] Ikemoto has used intrabronchial instillation of amphotericin B. This led to cessation of growth and expectoration of the ball. The drug (200 mg dissolved in 10 ml of water) was instilled by tracheal puncture using a lumbar puncture (7 cm, 22 gauge) needle.[52] In a six-year follow-up, there had been no recurrence of the disease.[52] Intracavitary needling for the instillation of a paste containing amphotericin B or nystatin has been described by Krakowka et al.[57] Nystatin had a small edge as far as effectiveness was concerned. Aerosol amphotericin B was not useful in one series.[52]

The treatment of choice for aspergilloma is surgical resection whenever it is feasible. In a review of 70 cases, with 14 new cases, Kilman et al.[55] concluded that lobectomy was followed by the least number of complications. Local surgical evacuation followed by irrigation with natamycin (pimaricin) has been used by Henderson and Pearson.[47] Natamycin (pimaricin) given as an aerosol, 0.25 per cent suspension in Alevaire, 1 ml four times daily, has also been used.[35] The old standard treatment for all fungous infections was potassium iodide. It has also been used in aspergillosis with some success.

Invasive aspergillosis is often rapidly fatal and its treatment difficult. Chronic aspergillosis in an immunologically competent patient has been successfully treated, however. The aspergilli are not particularly sensitive to amphotericin B by *in vitro* testing, but experience with treatment of early invasive disease has demonstrated its efficacy *in vivo*.[3, 117, 125] Combined with other drugs, amphotericin B appears to be particularly effective. Codish et al.[30] suggest a combination of amphotericin B and 5-fluorocytosine. The latter drug alone is not effective, but in combination allows a lower dose of amphotericin B.[125] Nystatin areosols and amphotericin B combined with rifampin have also been used. No single or combined modality has yet superceded amphotericin B and 5-fluorocytosine. Clinical evaluation of the newer imidazoles (ketoconazole and econazole) is now taking place. Clotrimazole is effective but is associated with serious side effects. Miconazole is considered of little value in invasive aspergillosis.

Surgical resection of the involved lung has been used when feasible. This procedure has varied from segmental lobectomy to pneumonectomy. Systemic and meningitic aspergillosis are generally fatal in spite of treatment.

PATHOLOGY

Gross Pathology. In routine autopsies of patients who have died of other diseases, it is not uncommon to find a colony of *Aspergillus* in the bronchi, in ectatic cavities, and in other stagnant air spaces. The mycelial mat growing on such areas may be quite

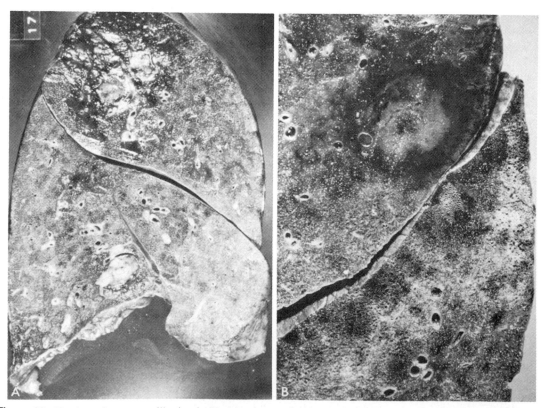

Figure 23–15. Invasive aspergillosis. *A*, Massive hemorrhagic consolidation and focal abscess formation of apical area of lungs. There is abscess formation in the lower lobe. *B*, Infarcted area of lung, showing hemorrhage and necrosis due to thrombosis in a mycotic abscess.

extensive, but there is usually little or no evidence of inflammation in the tissue.

The gross pathology of the several forms of aspergillosis is quite variable. In allergic aspergillosis, there is abundant hyphal growth in mucus plugs and a prominent inflammatory reaction in bronchi. In acute invasive disease, hyphae can be seen invading the wall of the bronchi and the small and large vessels but particularly the surrounding parenchyma. The pathologic picture here is that of an acute, necrotizing, pyogenic, pneumonitic process. The pulmonary lesions are usually necrotizing abscesses or infarcts (Fig. 23–15). Similar to the etiologic agents of mucormycosis, aspergilli have a marked tendency to invade blood vessels, causing thrombosis. Such invasion may also result in dissemination to other organs. The pulmonary abscesses may be as massive as those seen in mucormycosis, or they may be small, multiple, yellowish lesions. In the rare chronic granulomatous aspergillosis, consolidation of the entire lung may occur. Oxalosis has been described in several forms of aspergillosis.

Calcium oxalate crystals are found in sputum, pleural fluids, bronchial washings, and biopsy material in acute inflammatory exudate from pleural empyema[103] and in aspergilloma cavities,[58] particularly when *Aspergillus niger* is the etiologic agent. The presence of such crystals in the absence of other oxalosis-related conditions can be a diagnostic clue in *Aspergillus* infections.[103]

In contrast to invasive disease, there is little tissue destruction resulting from aspergillomas, as these develop in preformed stationary cavities. Cut sections of such cavities show brownish hyphal masses in the center and a fibrous or epithelial lining of the cavity wall (Fig. 23–6)

When dissemination to brain, kidney, and myocardium occurs, the metastatic lesions are acute necrotizing pyogenic abscesses. Infarcts surrounded by extravasated blood and ischemic tissue are also commonly seen (Fig. 23–16). Such reddish brown areas seen in cut sections of brain, heart, thyroid, or kidney are almost pathognomonic for disseminated aspergillosis.

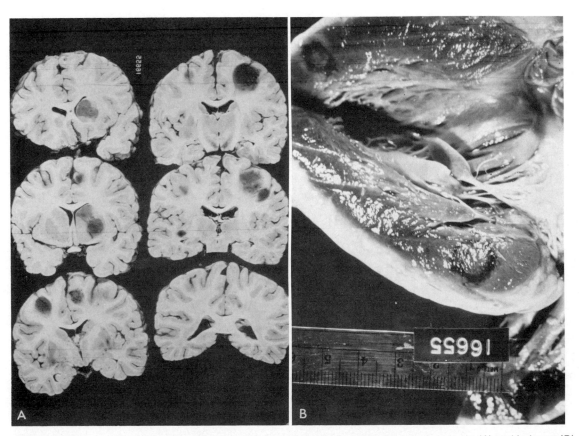

Figure 23–16. Invasive aspergillosis. Infarcted areas in disseminated aspergillosis in brain *(A)* and in heart *(B)*. Aspergilli are one of the few fungi that invade the myocardium. (Courtesy of M. Warnock.)

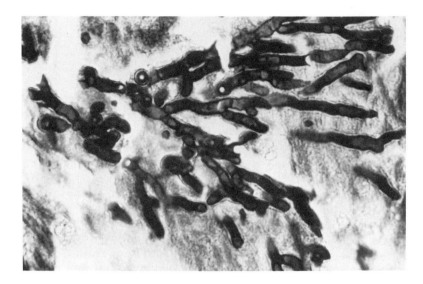

Figure 23–17. Invasive aspergillosis. Histologic section showing dichotomous branching, radiating, septate hyphae. GMA. ×500. (Courtesy of S. Thomsen.)

Histopathology. If the *Aspergillus* is growing in a cavity with an airspace (fungus ball), conidiophores and conidia are frequently observed (Fig. 23–8). This does not occur if the fungus is invading tissues. The hyphal units of aspergilli are stained reasonably well with hematoxylin and eosin, but unless searched for they will be missed. The Gridley and Grocott-Gomori methenamine silver stains readily stain the fungus (Fig. 23–9). The PAS stain has sometimes been disappointing when used to demonstrate the presence of aspergilli in pathologic specimens.

The diameter of the hyphae ranges from 2.5 to 4.5 μ, with an average of 3 μ. Hyphal strands characteristically demonstrate repeated dichotomous branching, with several hyphae oriented in the same direction (Fig. 23–17). This gives a brushlike or extended fingers-like appearance (Fig. 23–18). The hyphae are often seen to radiate from a focal point. The branches arise regularly at about a 45 degree angle. Multiple septation is usually present and demonstrated by the special fungal stains. If septae are not readily apparent, the hypha may resemble *Mucor* species. Conidia are not seen in tissue sections, but cross sections of hyphae may look like conidia or yeasts. In the center of hyphal aggregates there may be swollen cells and cystlike and distorted hyphae. These may measure up to 8μ in diameter. Members of the terreus-flavipes group produce aleurioconidia in culture and in tissue (Fig. 23–14).[95, 119] Eosinophilic halos are sometimes seen around hyphae[85] (Fig. 23–19). Aspergillosis is dif-

Figure 23–18. Invasive aspergillosis. Dichotomous branching and septation of hyphae. Sometimes there is an indentation of the hyphal wall at the point of septal formation. GMS. ×500.

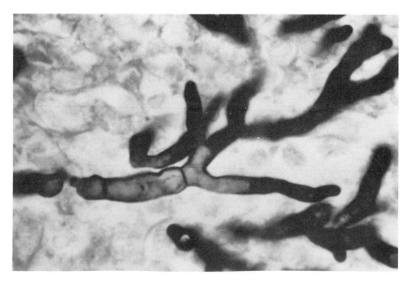

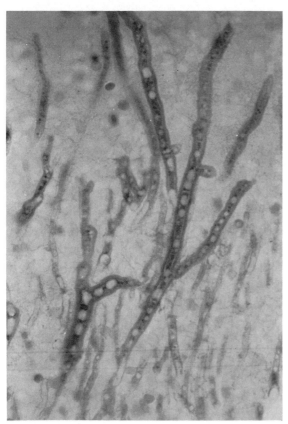

Figure 23–19. Invasive aspergillosis. This was from a case of chronic disease. There are eosinophilic sheaths around the hyphae, the Splendore-Hoeppli phenomenon. Hematoxylin and eosin stain. ×400.

ferentiated from candidiasis histologically by the following: (1) dichotomously branching hyphae and lack of budding yeasts are found in the former; (2) in candidiasis there is always a mixture of budding yeasts, pseudomycelium, and irregularly branching septate mycelium. The morphology of *Aspergillus* in tissue, however, is identical to several other agents of opportunistic fungal infections. These include *Pseudallescheria boydii* (see Chapter 24), *Fusarium* sp. (Chapter 26), and less commonly encountered fungi. *P. boydii* is now commonly found in settings that mimic *Aspergillus* in its various clinical manifestations, allergic bronchopulmonary pseudallescheriasis, *Pseudallescheria* fungus ball, and invasive pseudallescheriasis.

The histologic reaction evoked by the invading *Aspergillus* is generally that of the acute pyogenic type accompanied by necrosis. The leukocytic debris and perifocal necrosis may be extensive and is sometimes referred to as "diffusion necrosis." It is postulated that some of this necrosis is owing to the release of endotoxins from the fungus.[36]

Since chronic aspergillosis is an uncommon disease, granulomatous lesions of the lungs, nasal area, or other organs are rarely observed. Granulomata from nasal sinus aspergillosis are very hard and fibrosed.[78] Histologically they show aggregates of foreign body giant cells and well-formed fibrous tissue. The hyphae are usually sparse and distorted, and do not show the regular branching pattern seen in acute invasive aspergillosis.

ANIMAL DISEASE

Natural infection with aspergilli in birds has been known for many years and was the first form of the disease described. It is the most important fungal disease and one of the most important infectious agents of birds. Pleural, air sac, alimentary, meningeal, and systemic infections of chickens, turkeys, penguins,[22] and other birds have been described. Such infections sometimes occur in epidemics, resulting in severe economic loss. Many epidemics have been traced to moldy grain. Acute aspergillosis is particularly common in young chickens, and it accounts for 10 per cent of all deaths in this age group. Acute illness and diarrhea are followed by convulsions and death in 24 to 48 hours. The pathology seen in infected tissue is essentially the same as that noted in human disease. *Aspergillus* is also known to cause epidemics of cutaneous disease in chickens and other birds.

Aspergillosis in mammals is not as commonly encountered as in birds. It has been reported sporadically in the veterinary literature. The most important and well-defined form of animal aspergillosis is called "mycotic abortion." In this condition, there is an infection of the placenta of the cow, with subsequent loss of the calf. Radiating mycelial elements are seen in tissue section. Invasive aspergillosis has been reported in horses, sheep, pigs, and several other species. Foci of radiating hyphae of aspergilli have also been found in the tissues of normal healthy cattle.[8] Mastitis can also be caused by *Aspergillus* sp.[115] Purulent galactophoritis was seen in some cases, suggesting an ascending infection through the lactiferous ducts. A hematogenous source was implied in other cases in which no changes in ducts were seen, but hyphal units were cultured from milk.

Experimental Disease. Injection of the conidia of *A. fumigatus* or other thermotolerant species into mice regularly results in the production of disease. Strains vary markedly in virulence, however; Sidransky and Friedman obtained only a transient pneumonitis when mice treated with antibiotics were exposed to *Aspergillus* conidia.[124] A fulminating disseminated aspergillosis developed if the mice received steroids also. Chronic and acute aspergillosis in rabbits was produced by Rippon and Anderson.[106] Various modes of immunosuppression including steroids or cyclophosphamide greatly altered the course of infection and the serologic evaluation. Exposure of rabbits to aerosols of conidia of *Aspergillus* and *Pencillium* sp. was studied by Thurston et al.[133] Viable conidia were found throughout the intestinal tract within one hour of exposure. Viable fungi were present in lungs for two to three weeks after inhalation. Chick observed that experimental leukemia of chickens produced by injection of myoblastosis virus rendered the birds more susceptible to aspergillosis than normal controls.[27] In comparing the relative virulence of several *Aspergillus* species, Ford and Friedman[39] found that members of the *A. flavus* group consistently killed mice in doses as low as 10^4 conidia. In studies by Pore and Larsh,[95] torticollis, lateral and truncal ataxia, and multiple acute lesions of the cerebellum and cerebrum were produced by members of the terreus-flavipes group (Fig. 23–20).[107] Aleurioconida that are formed by the species of this group in culture were also found in histologic sections of cerebral lesions. Rippon et al.,[108] by manipulation of oxidation-reduction potentials and temperature, induced *A. sydowi* to grow as a budding yeastlike form. These cells caused a systemic mycosis when injected into mice. Smith[127] has found some improved re-

sistance in experimental infection in mice when the animals had been given a live vaccine of conidia prior to challenge. A new species, *A. bisporus*, is pathogenic for mice but has not as yet been recovered from human infection.[59]

A disease similar to allergic aspergillosis has been produced in rhesus monkeys by Golbert and Patterson.[42] Both the immediate (due to reaginic antibody) and toxic complex (due to precipitating antibody) types of reactivity were transferred to the animals. This was accomplished by infusion of the patient's serum into the animals (passive systemic transfer), followed by aerosol challenge with *Aspergillus* antigen. Cutaneous reactivity and pulmonary lesions consistent with the donor's illness were found.

BIOLOGICAL INVESTIGATIONS

The effect of steroids, amphotericin B, and nystatin on the growth of *Aspergillus* species in culture was studied by Mohr et al.[80] Of the steroids tested, estradiol in concentrations of 0.05 μg per ml, inhibited growth, but testosterone, progesterone, and potassium iodide (1 part per 100) did not. Amphotericin B at 5 μg inhibited one strain but not the others. Nystatin in a concentration of 100 units per ml inhibited all strains. A propensity for *A. fumigatus* to grow in microaerobic cavities and necrotic tissue was suggested by Okudaira and Schwartz.[85] However, they found the growth rates equal under aerobic or microaerobic environments, but greater at 37°C than at 25°C. Rippon et al.[107] studied the comparative virulence, metabolic rate, growth rate, and ubiquinone content of soil and human isolants of *Aspergillus terreus*. The isolant came from a case of human meningitis. In all parameters examined, including growth rate, ubiquinone content, tolerance of reduced oxidation potentials, and metabolic rate, this isolant showed pronounced increases over soil insolants at a 37°C incubation. This work suggests that occasional strains of soil organisms are particularly virulent in natural or experimental infections.

Endotoxins from *A. fumigatus* were first described by Henrici.[49] Acute hemorrhagic necrosis was produced when endotoxin derived from *Aspergillus* was inoculated into rabbits. This may explain the acute toxic reaction produced by inhalation of masses of conidia. Tilden[134] described and characterized a hemolysin and several endotoxins from

Figure 23–20. Aspergillosis. Torticollis of the mouse produced by infection of the central nervous system by *A. terreus*.[107] (From Rippon, J. W. and D. N. Anderson. 1974. Sabouraudia *13*:157–161.)

cultures of *A. fumigatus* and *A. flavus*. The endotoxins were extremely potent when tested in rabbits. A neurotoxin was isolated by Ceni and Besta in 1902,[26] but its identity is unknown. *A. flavus* produces a potent hepatotoxin and carcinogenic agent, aflatoxin, when growing on certain grains. It has been suggested that some human disease may be owing to this agent following the ingestion of contaminated food.[36] The production of toxins by pathogenic fungi is, in general, rare.

IMMUNOLOGY AND SEROLOGY

Aspergillosis is an opportunistic infection, and the normal individual is not susceptible to the disease. However, inhalation of large masses of conidia of some species, particularly *A. fumigatus*, can cause an acute toxic reaction or fulminating infection or both. There is no clinical or experimental evidence that an immunity develops if the patient survives the disease. Hypersensitivity reactions are developed by some atopic patients as a result of continued exposure to conidia, mycelium, or antigen. Reaginic antibodies are present in patients with asthma, and both reaginic and precipitating antibodies are found in cases of bronchopulmonary aspergillosis.

Serologic evaluation of the various forms of aspergillosis has seen significant advances in the past several years. It had long been known that atopic patients had a positive skin test to *Aspergillus* extracts and that some patients with the invasive disease had either an immediate and delayed reaction or only a delayed reaction. The studies by Pepys, Longbottom, and others have correlated the various clinical types of disease with skin reactions and immunodiffusion (ID) tests.[35, 46, 66] Serum precipitins demonstrated by the immunodiffusion test are present in a few (9 per cent) patients who have asthma but who do not have pulmonary eosinophilia. Precipitin lines occur in patients with pulmonary eosinophilia, allergic aspergillosis, and aspergillomas. In some studies, a stronger reaction and a larger number of bands occurred in patients with allergic aspergillosis who also had fungi growing in the lumen of the bronchial tree.[48] Kim et al. investigated the efficacy of various antigen preparations.[56] They concluded that highly refined antigen preparations had no diagnostic value over crude culture filtrates.

Precipitins are always present in aspergilloma patients. This is especially useful for diagnosis when sputa is culturally negative due to lack of connection of the cavity with a patent bronchiole. Since other fungi such as *Pseudallescheria boydii* also may form fungus balls, serology can define the etiology in the absence of culture. Antibodies to *P. boydii* do not cross-react with *Aspergillus* antigens and *vice versa*. The precipitin bands disappear after surgical removal of the lesion.

Few bands or none are found in cases of invasive aspergillosis. The latter patients are usually anergic. Other procedures may be useful, however. Marier et al.[71] described a solid phase radioimmunoassay (SP-RIA) for detecting minute amounts of antibody. SP-RIA was positive in 80 per cent of patients with invasive disease, whereas ID was positive in only 20 per cent. Richardson et al.[104] have a sensitive ELISA system for detection of antigen. Reiss and Lehman[100] detect antigen by CIE (countercurrent immunoelectrophoresis) using highly specific antibody. Sepulveda et al.[119] and Holmberg et al.[50a] have systems that incorporate ELISA for detection of antibodies in patient serum, and Weiner[143] has a RIA for antigen in rabbits. None of these procedures is a practical laboratory test as yet, but it is hoped that one will be developed in the near future.

In our experience using the immunodiffusion test, certain patterns of bands are present which correspond to the clinical type of disease. A single weak band may be found in asthmatic patients. Four bands or more are present in allergic aspergillosis with colonization of bronchi, and up to 18 bands have been found in cases of aspergilloma. Invasive disease, if chronic, may result in the detection of a few precipitin lines. However, the usual acute, fulminating, invasive aspergillosis occurs in patients with a defective or abrogated immune system who are unable to produce antibodies and whose ID test is therefore negative. It is important to note that sera may be negative to a single *Aspergillus* species antigen, so that a battery of antigens prepared from various strains and species should be used. Sometimes concentrating the sera is also necessary.[123] Complement fixation tests generally agree with immunodiffusion results, and may be positive earlier in the disease. They are quite nonspecific and their usefulness is debatable. Ky et al.[60] and others have shown that some of the antigenic components are enzymes of fungal origin, and that patients' sera contain antibodies specific for them. Bardana et al.[12] have studied the antigens and the types of antibody elicited during infection.

LABORATORY IDENTIFICATION

Direct Examination. The procedures and appearance of material for direct examination will vary somewhat, depending on the form of the disease. Sputum from allergic aspergillosis is usually gelatinous and thick. Therefore, the use of mucolytic agents is sometimes required before examination can be made. Long, branching, hyphal strands may be seen either on direct examination or, more easily, after digestion with 10 per cent potassium hydroxide tinted with ink. In aspergillomas in which the cavity is connected to a patent bronchiole, numerous tangled masses of hyphae may be present in bloody sputum. Such preparations often show the typical conidiophores topped by an expanded conidia-bearing vesicle. This finding indicates that the *Aspergillus* is colonizing the surface of a bronchus, cavity, or surface in contact with air. The vesicle and conidiophore of *A. fumigatus* are so distinctive as to make possible specific identification from such preparations. If the "fungus ball" of the aspergilloma is in a cavity that does not connect with a patent bronchiole, both culture and direct examination will be negative.

Biopsy material, surgical specimens, and tissue debris from cases of invasive aspergillosis can be digested with potassium hydroxide to render visible the mycelium of fungus. In tissue, conidiophores are usually absent, but characteristically dichotomously branching, septate hyphae are found. Many other fungi have the same appearance as aspergilli in tissue or clinical specimens.

Aspergillus species are common airborne contaminants of all surfaces, including skin, mouth, lungs, wounds, and so forth. For this reason it is necessary to be cautious in evaluating the isolation of an *Aspergillus* organism from patient materials. For the diagnosis of aspergillosis from cultures of sputa, Henderson[48] has defined as necessary what he terms the presence of "abundant fungus." This consists of growing the *Aspergillus* from several specimens or several colonies from a single specimen. The presence of mycelial elements on direct examination, in addition to positive culture, is a more reliable guide to correct diagnosis than simply the isolation of an *Aspergillus* organism. Similarly, a positive culture from surgically removed tissue is considered more significant if the organism can be demonstrated in histologic sections also.

The etiologic agents of aspergillosis grow readily on almost all laboratory media. They are usually sensitive to cycloheximide, so that only antibacterial antibiotics should be present in the isolation medium. If the specimen is a surgical biopsy from an immunosuppressed patient it should be planted on media without any antibiotics. This allows the isolation of *Nocardia* sp. as well as *Cryptococcus*, *Aspergillus*, *Pseudallescheria*, and other fungal opportunists. Most pathogenic aspergilli have an optimum growth temperature at 37°C, which inhibits some fungal contaminants. *A. fumigatus*, a very thermotolerant species, grows to a temperature of 45°C.[109, 132] Conidiophores characteristic of the organism are present within 48 hours when plated on agar and incubated at 37°C. Species identification is aided by preparation of slide cultures for the examination of conidiophores, conidia, vesicles, and conidia mass *in situ*, in addition to the observation of colony morphology, pigmentation, and growth rate at 25°C and 37°C. There are about 600 recognized species and varieties of *Aspergillus*. For identification of unusual species, reference should be made to Raper and Fennell's *The Genus Aspergillus*.[98] The commonly encountered species from pathologic material, such as *A. fumigatus*, *A. terreus*, *A. flavus*, and *A. niger*, are fairly easy to identify and are discussed in the following section.

MYCOLOGY

Aspergillus fumigatus Fresenius 1850[40]

Synonymy. *Aspergillus aviarius* Peck 1891; *Aspergillus bronchialis* Blumentritt 1901; *Aspergillus calyptratus* Gudemans 1901; *Aspergillus cellulosae* Hopffe 1919; *Aspergillus glaucoides* Spring 1852; *Aspergillus pulmonumhominis* Welcker 1855; *Aspergillus ramosus* Hallier 1870; *Aspergillus nigrescens* Robin 1853.

Colony Morphology. The organism grows rapidly on SGA or Czapek solution agar (25°C or 37°C), producing a flat, white colony that quickly becomes gray-green with the production of conidia (Fig. 23–21*A*). The texture may vary from strictly velvety to deep felt, floccose, or somewhat folded. The reverse is generally colorless. The conidia mass of the conidial heads are columnar, compact, and often crowded. They range in size from 200 to 400 μ by 50 μ. An albino variant of this species has been recovered from lesion material.[68]

Microscopic Morphology. The conidiophore is short, smooth, and up to 300 μ in length and 5 to 8 μ in diameter (Fig. 23–22). It may have a slightly green or brownish coloration, especially toward the upper part near the vesicle. The conidiophore gradually enlarges, passing imperceptibly to form the expanded flask-shaped vesicle. The vesicle is 20 to 30 μ in diameter, producing, on the upper half only, a single series of phialides (6 to 8 μ long). The phialides bend upward, paralleling the axis of the conidiophore. The conidia are green in mass, echinulate, globose to subglobose, and 2 to 3 μ in diameter. Elliptical conidia are found in the var. *ellipticus*. Sclerotia or cleistothecia are not found. *A. fisheri* is morphologically similar to *A. fumigatus*, but is homothallic, producing numerous cleistothecia. It has never been associated with human disease, although it is extremely thermotolerant.

Apergillus flavus Link 1809

Synonymy. *A. fasciculatum* Batista et Maia 1957; *A. humus* Abbott 1926; *A. luteus* (Van Teigh) Dodge 1935; *A. nolting* Hallier 1870; *A. pollinis* Howard 1896; *A. sojae* Sakaguchi et Yamada 1933; *A. wehmeri* Costantin et Lucet 1905.

Colony Morphology. SGA or Czapek's at 25°C. Growth is rapid (6 to 7 cm in ten days) or slow (3 to 4 cm), and consists of a close-textured basal mycelium, which is flat or radially furrowed or wrinkled (Fig. 23–21,*B*). Conidial heads are abundant and of intense yellow to yellow-green in color. The reverse is colorless to pinkish drab or darker. Sclerotia are prominent in some isolants. They become red-brown and may be 400 to 700 μ in diameter. Conidial heads are radiate, splitting to form loose columns. They have an

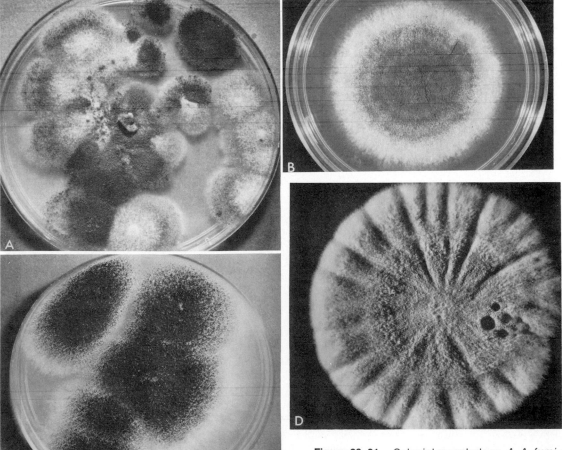

Figure 23–21. Colonial morphology. *A, A. fumigatus. B, A flavus. C, A. niger. D, A. terreus.*

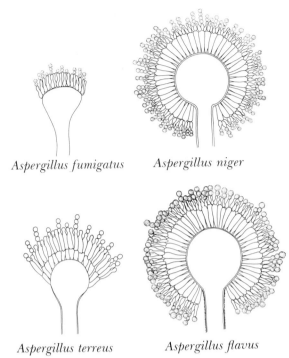

Aspergillus fumigatus   *Aspergillus niger*

Aspergillus terreus   *Aspergillus flavus*

Figure 23–22. Conidiophores of *A. fumigatus*, *A. niger*, *A. flavus*, and *A. terreus*.

average diameter of 300 to 400 μ. In the var. *columnaris*, the conidial heads are columnar, and phialides are uniserate (in one series).

Microscopic Morphology. The conidiophores are thick-walled, unpigmented, coarsely roughened, long (up to 1 mm in length or more), and 10 to 20 μ in diameter below the vesicles (Fig. 23–22). Vesicles are globose to subglobose, 10 to 65 μ in diameter, and fertile (produce phialides) over almost the entire area. Phialides are biserate or uniserate. The primary branches are up to 10 μ in length, and the secondary phialides are about 5 μ in length. The conidia are elliptical at first, but later they are mostly globose (3.5 to 4.5 μ in diameter) and conspicuously echinulate.

Aspergillus niger Van Tieghem 1867

Synonymy. *Sterigmatocystis antacustica* Cramer 1859; *Aspergillus fuliginosis* Peck 1934; *Aspergillus fumaricus* (Wehmer) Thom et Church 1926; *Aspergillus longobesidia* Bainer 1922; *Aspergillus nigricans* Wreden 1867; *Aspergillus nigraceps* Berk 1888; *Aspergillus pyri* English 1940; *Aspergillus welwitschiae* (Bresadola) Hennings 1907.

Colony Morphology. On SGA or Czapek solution agar (25°C) it grows to form a re-

stricted colony with a diamater of 2.5 to 3.0 cm in 10 days (Fig. 23–21,*C*). The compact basal mycelium is white to yellow and soon bears abundant conidial structures which are black. The conidial heads are large, black, and globose at first, becoming radiate or splitting to loosely columnar. There is a distinct "moldy" odor produced by the fungus.

Microscopic Morphology. Conidiophores are 1.5 to 3.0 mm by 15 to 20 μ, smooth, and colorless or turning dark towards the vesicle (Fig. 23–22). The vesicle is globose, about 60 μ in diameter, and bears phialides over all the surface. The phialides are in two series (biserate). The primaries are long (30 μ or more by 6 μ), brownish in color, and may be septate. The secondaries are short, 8 μ by 3 μ, and bud off conidia. These are globose, 4 to 5 μ in diameter, brown to black, and very rough.

A. niger is one of the most easily identifiable aspergilli, with its white to yellow mat later bearing black conidia. This species is very commonly found in aspergillomas and is the most frequently encountered agent of otomycosis.

Aspergillus terreus Thom 1918

Synonymy. *Sterigmatocystis hortai* Langeron 1922; *Aspergillus hortai* Dodge 1935; *Aspergillus galeritus* Blochwitz 1929; *Aspergillus terreus* var. *boedijni* Thom et Raper 1945.

Colony Morphology. SGA or Czapek's at 25°C. The growth of the colony is rapid, reaching a diameter of 3 to 5 cm by 10 days (Fig. 23–21,*D*). The consistency is floccose to velvety, sometimes furrowed or tufted, and the colony conidiates profusely. The massed conidia heads are columnar and range in color from cinnamon-buff to wood brown. The heads are 30 to 50 μ in diameter and 150 to 500 μ or more in length.

Microscopic Morphology. The conidiophores are long and slender (100 to 250 μ by 5 to 6 μ), smooth, and of uniform diameter throughout (Fig. 23–22). The vesicles are hemispherical or domelike, 10 to 16 μ in diameter, and merge imperceptibly with the conidiophore. The phialides are in two series. The primaries are 5 to 7 μ in diameter. The secondaries are 5 to 8 μ in length and 1 to 2 μ in diameter. The conidia are elliptical, smooth, and 2 to 2.5 μ in diameter. Aleurioconidia are produced by the mycelial mat submerged in the agar. They are 6 to 7 μ in diameter. Such cells are also produced in tissue during infection, which aids in the identification of the agent.[95, 119] *A. terreus* is another thermotolerant species that has been isolated from several cases of invasive aspergillosis, particularly meningitis.

REFERENCES

1. Aisner, J. J. Murillo, et al. 1979. Invasive aspergillosis in acute leukemia: correlation with nose cultures and antibiotic use. Ann. Intern. Med., *90*:4–9.

2. Aisner, J. A., S. C. Schimpff, et al. 1976. *Aspergillus* infection in cancer patients: association with fireproofing materials in a new hospital. J.A.M.A., *235*:411–412.

3. Aisner, J. A., S. C. Schimpff, et al. 1977. Treatment of invasive aspergillosis relation of early diagnosis and treatment to response. Ann. Intern. Med., *86*:539–543.

4. Arnow, P. M., R. L. Andersen, et al. 1978. Pulmonary aspergillosis during hospital renovation. Am. Rev. Respir. Dis., *118*:49–53.

5. Atkinson, G. S. and H. L. Israel. 1973. 5-Fluorocytosine treatment of meningeal and pulmonary aspergillosis. Am. J. Med., *55*:594.

6. Attah, C. A., and M. M. Cerruti. 1979. *Aspergillus* osteomyelitis of sternum after cardiac surgery. N.Y. State J. Med., *79*:1420–1421.

7. Aung, U. K., U. K. Lin, et al. 1979. Leptomeningeal aspergillosis causing internal carotid artery stenosis. Br. J. Radiol., *52*:328–339.

8. Austwick, P. K. 1962. The presence of *Aspergillus fumigatus* in the lungs of dairy cows. Lab. Invest., *11*:1065–1072.

9. Axelsson, H., B. Carlöö, et al. 1979. Aspergillosis of maxillary sinus. Acta Otolaryngol. *86*:303–308.

10. Bach, M. C., J. L. Adler, et al. 1973. Influence of rejection therapy on fungal and nocardial infections in renal transplant recipients. Lancet, *27*:180–184.

11. Band, P. D., N. Blank, et al. 1972. Needle diagnosis of pneumonitis value in high risk patients. J.A.M.A., *220*:1578.

12. Bardana, E. J., J. K. McClatchy, et al. 1972. The primary interaction of antibody to components of aspergilli. I. Immunological and chemical characterization of nonprecipitating antigen. J. Allergy Clin. Immunol., *50*:208–221.

13. Bennett, S. H. 1844. On the parasitic vegetable structures found growing in living animals. Trans. R. Soc., Edin., *15*:277.

14. Blount, W. P. 1961. Turkey "X" disease. Turkeys, *9*:52.

15. Blum, J., J. C. Reed, et al. 1978. Miliary aspergillosis as associated with alcoholism. A. J. R., *131*:707–709.

16. Bodey, G. P. 1966. Fungal infections complicating acute leukemia. J. Chronic Dis., *19*:667–687.

17. Bodin, E., and L. Gautier. 1906. Note sur toxine produite par *l'Aspergillus fumigatus*. Ann. Inst. Pasteur, *20*:209.

18. British Thoracic and Tuberculosis Association Report. 1970. Aspergilloma and residual tuberculous cavities — The results of a resurvey. Tubercle, *51*:227–245.

19. Burge, H. P., J. R. Boise, et al. 1978. Fungi in libraries an aerometric survey. Mycopathologia, *64*:67–72.

20. Burton, J. R., J. B. Zachery, et al. 1972. Aspergillosis in four renal transplant recipients. Ann. Intern. Med., *77*:385–388.

21. Cahill, K. M., A. M. El Mofty, et al. 1967. Primary cutaneous aspergillosis. Arch. Dermatol., *96*:545–546.

22. Campbell, G. 1972. Aspergillosis in captive penguins. Vet. Serv. Bull. Dept. Agricul. and Fisher. Dublin *2*:39–41.

23. Carlile, J. R., R. E. Millet, et al. 1978. Primary cutaneous aspergillosis in a leukemic child. Arch. Dermatol., *114*:78–80.

24. Case Records of the Massachusetts General Hospital (Case 44). 1971. New Engl. J. Med., *285*:337–346.

25. Case Records of the Massachusetts General Hospital (Case 24). 1973. New Engl. J. Med., *288*:1290–1296.

26. Ceni, C., and C. Besta. 1902. Über die Toxine von *Aspergillus fumigatus* and *A. flavescens* und deren Beziehungen zur Pellagra. Zentralbl. Allg. Pathol., *13*:930.

27. Chick, E. W., and N. C. Durham. 1963. Enhancement of aspergillosis in leukemic chickens. Arch. Pathol., *75*:81–84.

28. Child, J. S., J. D. Stanley. 1979. Noninvasive detection of fungal endocarditis. Chest, *75*:539–540.

29. Chung, C., and P. H. Lord. 1978. Diagnosis of invasive pulmonary aspergillosis by fiberoptic transbronchial lung biopsy. J.A.M.A., *239*:749–750.

30. Codish, S. O., J. S. Tobras, et al. 1979. Combined amphotericin B. flucytosine therapy in *Aspergillus* pneumonia. J.A.M.A., *241*:2418–2419.

31. Cooke, W. B. 1971. *Our Moldy Earth*. Cincinnati, Ohio, U.S. Dept. Interior. Fed. Water Pollution Control. Administ. Advanced Waste Treatment Research Laboratory.

32. Curtis, A., G. J. W. Smith, et al. 1979. Air crescent sign of invasive aspergillosis. Radiology, *133*:17–21.

33. Deve, F. 1938. Une nouvelle forme anatomoradiologique de mycose pulmonaire primitive. Le méga-mycétome intrabronchetasique. Arch. Med. Chir. App. Resp., *13*:337–361.

34. Dieulafoy, Chantemesse, Widal 1890. Une pseudo-tuberculose mycosique. Gas. Hopl., Paris, *63*:821.

35. Edge, J. R., D. Stansfeld, et al. 1971. Pulmonary aspergillosis in an unselected hospital population. Chest, *59*:407–413.

36. Eisenberg, H. W. 1970. Aspergillosis with aflatoxicosis. New Engl. J. Med., *283*:1348.

37. Emmons, C. W. 1962. Natural occurrence of opportunistic fungi. Lab. Invest., *11*:1026–1032.

38. Fischer, J. J., and D. H. Walker. 1979. Invasive aspergillosis associated with influenza. J.A.M.A., *241*:1493–1494.

39. Ford, S., and L. Friedman. 1967. Experimental study of the pathogenicity of aspergilli for mice. J. Bacteriol., *94*:928–933.

40. Fresenius, G. 1850–1863. *In Beitrage zur Mycologie*. Frankfort, H. L. Bronner, p. 81.

41. Gage, A. A., D. C. Dean, et al. 1970. *Aspergillus* infections after cardiac surgery. Arch. Surg., *101*:384–387.

41a. Gartenberg, G., K. Einstein, et al. 1978. Superior

vena cava syndrome caused by invasive aspergillosis. Chest, *74*:671–672.

42. Golbert, T. M., and R. Patterson. 1970. Pulmonary allergic aspergillosis. Ann. Intern. Med., *72*:395–403.

43. Goldberg, B. 1962. Radiological appearance in pulmonary aspergillosis. Clin. Radiol., *13*:106–114.

44. Gonzalez-Crussi, F., L. D. Mirkin, et al. 1979. Acute disseminated aspergillosis during neonatal period. Report of an instance in a 14-day-old infant. Clin. Pediatr., *18*:137–138.

45. Hammerman, K. J., 1973. Spontaneous lysis of aspergilloma. Chest, *644*:697–699.

46. Henderson, A. H. 1968. Allergic aspergillosis. Thorax, *23*:501–512.

47. Henderson, A. H., and J. E. G. Pearson, 1968. Treatment of bronchopulmonary aspergillosis with observations on the use of natamycin. Thorax, *23*:519–523.

48. Henderson, A. H., M. P. English, et al. 1968. Pulmonary aspergillosis. Thorax, *23*:513–518.

49. Henrici, A. T. 1939. An endotoxin from *Aspergillus fumigatus*. J. Immunol., *36*:319–338.

50. Hinson, K. E. W., A. J. Moon, et al. 1952. Bronchopulmonary aspergillosis: a review and report of eight new cases. Thorax, *7*:317–333.

50a. Holmberg, K., M. Berdischewsky, et al. 1980. Serologic immunodiagnosis of invasive aspergillosis. J. Inf. Dis. *141*:656–664.

51. Hudson, H. J. 1973. Thermophilous and thermotolerant fungi in the aerospora at Cambridge. Trans. Br. Mycol. Soc., *60*:596–598.

52. Ikemoto, H., K. Watanabe, et al. 1971. Pulmonary aspergilloma. Sabouraudia, *9*:30–35.

53. Israel, H. L., and A. Ostrow. 1969. Sarcoidosis and aspergilloma. Am. J. Med., *47*:243–250.

54. Kallenback, J., J. Dusheiko, et al. 1977. *Aspergillus* pneumonia — cluster of four cases in an intensive care unit. South. Med. J., *52*:919–923.

55. Kilman, J. W., N. C. Andrews, et al. 1969. Surgery for pulmonary aspergillosis. J. Thorac. Cardiovasc. Surg., *57*:642–647.

56. Kim, S. J., S. D. Chaparas, et al. 1979. Characterization of antigens from *Aspergillus fumigatus*. IV. Evaluation of commercial and experimental preparations and fractions in the detection of antibody to aspergillosis. Ann. Rev. Respir. Dis., *120*:1305–1311.

57. Krakowka, P., K. Traczyk, et al. 1970. Local treatment of aspergilloma of the lung with a paste containing nystatin or amphotericin B. Tubercle, *51*:184–191.

58. Kurrein, F., G. H. Green, et al. 1975. Localized deposition of calcium oxalate around a pulmonary *Aspergillus niger* fungus ball. Am. J. Clin. Pathol., *64*:556.

59. Kwon-Chung, K. J. 1971. A new pathogenic species of *Aspergillus,* Mycologia, *63*:478–489.

60. Ky, P. T. V., J. Biguet, et al. 1971. Immunoelectrophoretic analysis and characterization of the enzymic activities of *Aspergillus flavus* antigenic extracts. The significance for the differential diagnosis of human aspergillosis. Sabouraudia, *9*:210–220.

61. Lin, C. H., and C. H. Yuan. 1979. Mycotic abscess of paranasal sinuses and orbit. Plast. Reconstr. Surg., *63*:735–738.

62. Linck, K. 1939. Tödliche Meningitis aspergilloma beim Menschen, Virchows Arch. Path. Anat., *304*:408.

63. Link, H. F. 1809. Observations in ordines plantarum natureles. Ges. Natur f. Freunde Berlin Magazin, *3*:1.

64. Lipinski, J. K., G. L. Weisbrod, et al. Unusual manifestations of pulmonary aspergillosis. J. Canad. Assoc. Radiol., *29*:216–220.

65. Levine, A. S. 1974. Hematologic malignancies and other marrow failure states: progress in management of complicating infections. Semin. Hematol., *11*:141–202.

66. Longbottom, J. L., and J. Pepys. 1964. Pulmonary aspergillosis. Diagnostic and immunological significance of antigens and C. reactive substance in *Aspergillus fumigatus.* J. Pathol. Bacteriol., *88*:141–151.

66a. Luce, L. M., R. Ostenson, et al. 1979. Invasive aspergillosis presenting as pericarditis and cardiac tamponade. Chest, *76*:703–705.

67. McCarthy, D. S., G. Simon, et al. 1970. The radiological appearances in allergic bronchopulmonary aspergillosis. Clin. Radiol., *21*:366–375.

68. McGinnis, M. R., D. L. Buck, et al. 1977. Paranasal aspergilloma caused by an albino variant of *Aspergillus fumigatus.* South. Med. J., *20*:886–888.

69. McGuirt, W. F., and J. A. Harrill. 1979. Paranasal sinus aspergillosis. Laryngoscope, *89*:1563–1568.

70. Mangurten, H. M., and B. Fernandez. 1979. Neonatal aspergillosis accompanying fulminant necrotising enterocolitis. Arch. Dis. Child., *54*:559–562.

71. Marier, R., W. Smith, et al. 1979. A solid phase radioimmunoassay for the measurement of antibody to *Aspergillus* in invasive aspergillosis. J. Inf. Dis., *140*:771–779.

72. Marsh, P., P. Millner, et al. 1979. A guide to recent literature on aspergillosis as caused by *Aspergillus fumigatus*, a fungus frequently found in self-heating organic matter. Mycopathologia, *69*:67–81.

73. Mayer, A., and C. Emmert. 1815. Vershimmelung (Mucedo) in lebenden Korper. Dtsch. Arch. Anat. Physiol. (Meckl.) *1*:310–318. (cited by Urbain and Guillet. 1938. Rev. Pathol. Comp., *38*:929–955).

74. Meyer, R. D., L. S. Young, et al. 1973. Aspergillosis complicating neoplastic disease. Am. J. Med., *54*:6–15.

75. Meyerwitz, R. L., R. Friedman, et al. 1971. Mycotic "mycotic aneurysm" of the aorta due to *Aspergillus fumigatus*. Am. J. Clin. Pathol., *55*:241–246.

76. Micheli, P. H. 1729. Nova plantarum genera juxta tournefortii methodum disposita. Florence, p. 234.

77. Mikulski, S. M., L. J. Love, et al. 1979. *Aspergillus* vegetative endocarditis and complete heart block in a patient with acute leukemia. Chest, *76*:473–476.

78. Milosev, B., S. Mahgoub, et al. 1969. Primary aspergilloma of paranasal sinuses in the Sudan. Br. J. Surg., *56*:132–137.

79. Mirsky, H. S., and J. Cuttner. 1972. Fungal infection in acute leukemia. Cancer, *30*:1348–1352.

80. Mohr, J. A., B. A. McKown, et al. 1971. Susceptibility of *Aspergillus* to steroids, amphotericin B and nystatin. Am. Rev. Respir. Dis., *103*:283–284.

81. Monod, O., G. Pesle, et al. 1957. L'aspergillome

brochestasiant. Sem. Hop. (Paris), *33*:3588–3602.

82. Murray, H. W., J. O. Moore, et al. 1975. Disseminated aspergillosis in a renal transplant patient: diagnostic difficulties re-emphasized. Johns Hopkins Med. J., *137*:235–237.

83. Nichols, D., G. A. Pico, et al. 1979. Acute and chronic pulmonary function changes in allergic bronchopulmonary aspergillosis. Am. J. Med., *67*:631–637.

84. Novey, H. S., and I. D. Wells. 1978. Allergic bronchopulmonary aspergillosis caused by *Aspergillus ochraceus*. Am. J. Clin. Pathol., *70*:840–843.

85. Okudaira, M., and J. Schwartz, 1962. Tracheobronchopulmonary mycoses caused by opportunistic fungi. Lab. Invest., *11*:1053–1064.

86. Orie, N. G. M., G. A. deVries, et al. 1960. Growth of *Aspergillus* in the human lung. Am. Rev. Respir. Dis., *82*:649–662.

87. Orr, D. P., R. L. Meyerwitz, et al. 1978. Pathoradiologic correlation of invasive pulmonary aspergillosis in the compromised host. Cancer, *11*:2028–2039.

88. Panke, T. W., A. T. McManus, et al. 1979. Infection of a burn wound by *Aspergillus niger*. Gross appearance simulating ecthyma gangrenosa. Am. J. Clin. Pathol., *72*:230–232.

89. Patterson, R., M. Rosenberg, et al. 1977. Evidence that *Aspergillus fumigatus* growing in the airway of man can be a potent stimulus of specific and nonspecific IgE formation. Am. J. Med., *63*:257–262.

90. Paulk, E. A., R. C. Schlant, et al. 1965. Aspergilloma associated with congenital heart disease. Dis. Chest, *47*:113–117.

91. Pennington, J. E., and N. T. Feldman. 1977. Pulmonary infiltrates and fever in patients with hematologic malignancy: assessment of transbronchial biopsy. Am. J. Med., *62*:581–584.

92. Pennington, J. E. 1980. *Aspergillus* lung disease. Med. Clin. North Am., *64*:475–490.

93. Pepys, J. 1969. Hypersensitivity disease of the lungs due to fungi and organic dusts. Monographs in allergy. Vol. 4 Basal, S. Karger.

94. Petheram, E. S., and R. M. E. Seal. 1976. *Aspergillus* prosthetic valve endocarditis. Thorax, *31*:380–390.

95. Pore, R. S., and H. W. Larsh. 1968. Experimental pathogenicity of *Aspergillus terreus-flavipes* group species. Sabouraudia, *6*:89–93.

96. Procknow, J., and D. Loewen. 1960. Pulmonary aspergillosis with cavitation secondary to histoplasmosis. Am. Rev. Respir. Dis., *82*:101–111.

97. Ramirez, J. 1964. Pulmonary aspergilloma: Endobronchial treatment. New Engl. J. Med., *271*:1281–1285.

98. Raper, K. B., and D. I. Fennel. 1965. *The Genus Aspergillus*. Baltimore, Williams and Wilkins Company.

99. Redmond, A., I. J. Carré, et al., 1965. Aspergillosis (*Aspergillus nidulans*) involving bone. J. Pathol. Bact., *89*:391–395.

100. Reiss, E., and P. F. Lehman. 1979. Galatomannan antigenemia in invasive aspergillosis. Infect. Immun., *25*:357–365.

101. Renon, L. 1897. *Etude sur les Aspergilloses Chez les Animaux et Chex l'Homme*. Paris, Masson et Cie.

102. Research Committee of the British Thoracic Association. 1979. Inhaled beclomethasone dipropionate in allergic bronchopulmonary aspergillosis. Br. J. Dis. Chest, *73*:349–356.

103. Reyes, C. V., S. Katharia, et al. 1979. Diagnostic value of calcium oxalate crystals in respiratory and pleural fluid cytology. A case report. Acta Cytol., *23*:65–68.

104. Richardson, M. D., L. O. White, et al. 1979. Detection of circulating antigen of *Aspergillus fumigatus* sera of mice and rabbits by enzyme-linked-immunosorbent-assay. Mycopathologia, *67*:83–88.

105. Riddle, H. F. V., S. Channell, et al. 1968. Allergic alveolitis in a malt worker. Thorax, *23*:271.

106. Rippon, J. W., and D. N. Anderson. 1978. Experimental mycosis in immunosuppressed rabbits. II. Acute and chronic aspergillosis. Mycopathologia, *64*:97–100.

107. Rippon, J. W., D. N. Anderson, et al. 1974. Aspergillosis. Comparative virulence, metabolic rate, growth rate and ubiquinone content of soil and human isolates of *Aspergillus terreus*. Sabouraudia, *13*:157–161.

108. Rippon, J. W., T. P. Conway, et al. 1965. Pathogenic potential of *Aspergillus* and *Penicillium* species. J. Infect. Dis., *115*:27–32.

109. Rippon, J. W., R. Gerhold, et al. 1980. Thermophilic and thermotolerant fungi isolated from the thermal effluent of nuclear generating reactors: dispersal of human opportunistic and veterinary pathogenic fungi. Mycopathologia, *70*:169–179.

110. Rose, H. D. 1972. Mechanical control of hospital ventilation and *Aspergillus* infections. Am. Rev. Respir. Dis., *105*:306–307.

111. Roselle, G. A., and C. A. Kaufman. 1978. Invasive aspergillosis in a nonimmunosuppressed patient. Am. J. Med. Sci., *276*:357–361.

112. Ross, D. A., M. C. MacNaughton, et al. 1968. Fulminating disseminated aspergillosis complicating peritoneal dialysis in eclampsia. Arch. Intern. Med., *121*:183–188.

113. Sakamoto, S., J. Ogata, et al. 1978. Fungus ball formation of *Aspergillus* in the bladder: an unusual report. Eur. Urol., *8*:388–389.

114. Sandhu, D. K., and R. S. Sandhu. 1973. Survey of *Aspergillus* species associated with human respiratory tract. Mycopathologia, *49*:77–87.

115. Schallibaum, M., N. Nicolet, et al. 1980. *Aspergillus nidulans* and *Aspergillus fumigatus* as causal agents of bovine mastitis. Sabouraudia, *18*:33–38.

116. Schwartz, H. J., K. M. Citron, et al. 1978. A comparison of the prevalence of sensitization to *Aspergillus* antigens among asthmatics in Cleveland and London. J. Allergy Clin. Immunol., *62*:9.

117. Sekhon, L. N., M. Dujovny, et al. 1980. Carotid-cavernous sinus thrombosis caused by *Aspergillus fumigatus*. Case Report. J. Neurosurg., *52*:120–125.

118. Seligson, R., J. W. Rippon, et al. 1977. *Aspergillus terreus* osteomyelitis. Arch. Intern. Med., *137*:918–920.

119. Sepulveda, R., J. L. Longbottom, et al. 1979. Enzyme linked immunosorbent assay for IgG and IgE antibodies to protein and polysaccharide antigens of *Aspergillus fumigatus*. Clin. Allergy *9*:359–371.

120. Seres, J. L., H. Ono, et al. 1972. Aspergillosis presenting as spinal cord compression. Case report. J. Neurosurg., *36*:221–224.

121. Severo, L. C., J. L. Hetzel, et al. 1978. Aspergilloma pulmonar por *Aspergillus niger*. Apresentacao de um caso. J. Pneumologia, *4*:9–11.

122. Shapiro, K., and K. Tabaddor. 1975. Cerebral aspergillosis. Surg. Neurol., *4*:465–71.

123. Shoeffer, J. C., B. Yu, et al. 1976. An *Aspergillus* immunodiffusion test in the early diagnosis of aspergillosis in adult leukemia patients. Am. Rev. Respir. Dis., *113*:325–329.

124. Sidransky, H., and L. Friedman. 1959. The effect of cortisone and antibiotic agents on experimental pulmonary aspergillosis. Am. J. Pathol., *35*:169–183.

125. Sinclair, A. J., A. H. Rossof, et al. 1978. Recognition and successful management in pulmonary aspergillosis in leukemia. Cancer, *42*:2019–2024.

126. Sluyter, T., 1847. De vegetalibus organisimic animals paraśitis. Diss. Inaug. Berolini, p. 14.

127. Smith, G. R. 1972. experimental aspergillosis in mice: aspects of resistance. J. Hygiene., *70*:741–754.

128. Stinson, E. B., C. C. Bieber, et al. 1971. Infectious complications after cardiac transplantation. Ann. Intern. Med., *74*:22.

129. Strelling, M. K., K. Rhaney, et al. 1968. Fatal acute pulmonary aspergillosis in two children of one family. Arch. Dis. Child., *41*:34–43.

130. Symmers, W. S. C. 1962. Histopathologic aspects of the pathogenesis of some opportunistic fungal infections as exemplified in the pathology of aspergillosis and the phycomycoses. Lab. Invest., *11*:1073–1090.

131. Tan, K. K., K. Sugai, et al. 1966. Disseminated aspergillosis. Case report and review of world literature. Am. J. Clin. Pathol., *45*:697–703.

132. Tansey, M. R., C. B. Fiermans, et al. 1979. Aerosol dissemination of veterinary pathogenic and human opportunistic thermophilic and thermotolerant fungi from thermal effluents of nuclear power reactors. Mycopathologia, *69*:91–116.

133. Thurston, J. R., S. J. Cysewsk, et al. 1979. Exposure of rabbits to spores of *Aspergillus fumigatus* or *Penicillium* sp.: survival of fungi and microscopic changes in respiratory and gastrointestinal tracts. Am. J. Vet. Res., *40*:1443–1449.

134. Tilden, E. B., E. H. Hatton, et al. 1961. Preparation and properties óf the endotoxins of *Aspergillus fumigatus* and *A. flavus*. Mycopathologia, *14*:325–346.

135. Torrington, K. G., C. W. Old, et al. 1979. Transu-rethral passage of *Aspergillus* fungus balls in acute myelocytic leukemia. South. Med. J., *72*:361–363.

136. Upadhyay, M. P., E. P. West. 1980. Keratitis due to *Aspergillus flavus* successfully treated with thiabendazole. Br. J. of Ophthalmol., *64*:30–32.

137. Vedder, J. S., and W. F. Schorr. 1969. Primary disseminated pulmonary aspergillosis with metastatic skin nodules. J.A.M.A., *209*:1191–1195.

138. Virchow, R. 1856. Beiträge zur Lehre von den beim Menschen vorkommenden pflanzlichen Parasiten. Virchows Arch. Pathol. Anat., *9*:557–593.

139. Walker, J. E., and R. D. Jones. 1968. Serologic tests in diagnosis of aspergillosis. Dis. Chest., *53*:729–735.

140. Walsh, T. J., and G. M. Hutchins. 1979. *Aspergillus* mural endocarditis. Am. J. Clin. Pathol. *71*:640–644.

141. Wang, J. L., R. Patterson, et al. 1979. Allergic bronchopulmonary aspergillosis in pediatric practice. J. Pediatr., *94*:376–381.

142. Wang, J. L., R. Patterson, et al. 1979. The management of allergic bronchopulmonary aspergillosis. Am. Rev. Respir. Dis., *120*:87–92.

143. Weiner, M. H. 1980. Antigenemia detected by radioimmunoassay in systemic aspergillosis. Ann. Intern. Med., *92*:793–796.

144. Welsh, R. A., and J. M. Buchness. 1955. *Aspergillus* endocarditis, myocarditis and lung abscesses. Report of a case. Am. J. Clin. Pathol., *25*:782–786.

145. Winterbauer, R. H., and K. G. Kraemer. 1976. The infectious complications of sarcoidosis. Arch. Intern. Med., *136*:1356–1362.

146. Winzelberg, G. G., J. Gore, et al. 1979. *Aspergillus flavus* as a cause of thyroiditis in an immunosuppressed host. Johns Hopkins Med. J., *144*:90–93.

147. Yocum, M. W., A. R. Saltzman, et al. 1976. Extrinsic allergic alveolitis after *Aspergillus fumigatus* inhalation. evidence of a type IV immunologic pathogenesis. Am. J. Med., *61*:939–945.

148. Young, R. C. 1969. The Budd-Chiari syndrome caused by *Aspergillus*. Arch. Intern. Med., *124*:754–757.

149. Young, R. C., and J. E. Bennett. 1971. Invasive aspergillosis. Absence of detectable antibody response. Am. Rev. Respir. Dis., *104*:710–716.

150. Young, R. C., A. Jennings, et al. 1972. Species identification of invasive aspergillosis in man. Am. J. Clin. Pathol., *58*:554–557.

24 Pseudallescheriasis

DEFINITION

Pseudallescheriasis is a spectrum of clinical diseases involving *Pseudallescheria boydii,* a soil-inhabiting fungus. The organism is of low inherent virulence but is a fungal opportunist able to elicit infections similar in terms of variety and predisposing factors to those of *Aspergillus* species. These infections include pulmonary colonization, fungoma, and invasive pneumonitis in addition to mycotic keratitis, endophthalmitis, endocarditis, meningitis, primary cutaneous and subcutaneous infections, and disseminated systemic disease. Mycotic mycetoma, the disease historically associated with *P. boydii,* is described in detail in Chapter 5. In this chapter the other types of infections associated with this fungus will be examined, along with a summary of *Pseudallescheria* mycetoma.

Synonymy. Allescheriosis, allescheriasis, monosporiosis, petriellidiosis.

INTRODUCTION AND HISTORY

Allescheria boydii was described by C. L. Shear[112] in 1922 as the etiologic agent of pedal mycetoma. The isolant came from a black farmer in western Texas, a patient of Dr. Mark Boyd.[22] The agent defined by Shear was a homothallic Ascomycete that produced cleistothecia containing ascospores. In addition, he noted that it also produced two types of asexual conidiophores, byssoid (simple) and coremias. Several years earlier, however, Saccardo[108] had delineated an asexual hyphomycete isolated from a mycetoma in a patient of Dr. Tarozzi[117] in Italy. The fungus had been isolated in 1909, and Saccardo's diagnosis[108] and a complete description of the pathology by Radaeli[101] appeared in 1911. For the next 30 years, the anamorph (imperfect) phase name, *Monosporium apiospermum,* appeared in the literature as a frequently isolated agent of mycetoma in temperate countries. Interestingly, few cases were described as being caused by *A. boydii* during this period.

The connection between the two fungi was unequivocally demonstrated by Emmons in 1944.[42] He worked with a culture of *M. apiospermum* from the first Canadian case of mycetoma.[36] Emmons noted that in continuous transfer sclerotia began to appear, followed by fertile ascocarps with ascospores. Thus the two fungi were appreciated to be two phases of one species.

The taxonomy of the fungus has been particularly confusing. Recently the names for both phases have been changed. Malloch[79] had moved the teleomorph (perfect) stage to the genus *Petriellidium* as *Petriellidium boydii.* This is now considered in error and the organism has been placed in a previously described valid genus *Pseudallescheria.*[83a] Also, the genus *Monosporium* was deemed invalid by Hughes,[57] leaving the species *M. apiospermum* without a taxonomic home. However, an adequate description of the anamorph (imperfect) phase had been made by Castellani in 1927[25] as *Scedosporium apiospermum.* This is now considered the correct epithet. A more detailed discussion of the taxonomy will be found in the section on mycology.

What appears to be historically the first case of human infection, and indeed the first isolation of this fungus, was published in 1889. A fungus was isolated from the ear of a child with chronic otitis by Harz and Bezold and described in a volume by Siebenmann[113] on fungus infections of the ear. The description given is of the coremial or synnematous state of *P. boydii;* it was called *Verticillium graphii.* Apparently Bezold had isolated it three times between 1870 and 1879 from the ears of children. In modern literature, what apparently is the first case of an infection other than a mycetoma is again a case of chronic otitis. In 1935, Belding and Umanzio[13] described as a new species of *Monosporium* an isolant from an ear. Reinterpretation of their work indicates that the organism falls within the limits of the anamorph state of *P. boydii.* In 1938 an isolant from a patient with septicemia was reported by Zaffiro.[129] The authenticity of this case is questionable. Benham and Georg[18] in 1948 described a granulomatous meningitis due to *P. boydii.* There was no underlying disease in this patient, but the patient had recived a spinal anesthetic a month before the onset of symptoms.

The first reported cases of pulmonary disease appear to be those of Creitz and Harris[33] and Drouhet[37] in 1955. In the former, the patient had chronic bronchitis and emphysema that did not respond to months of multiple antibiotic therapy; in the latter, no clinical details were given. At present, about 50 cases of pulmonary infection have been recorded, a few cases each of meningitis, systemic mycosis, mycotic keratitis, endophthalmitis, prostatitis, endocarditis, and chronic otitis. All cases of chronic otitis have been in children. In the past few years an increasing number of cases of fatal disseminated disease have been encountered in severely compromised hosts, in addition to the other forms of infection. It is less commonly involved in these disease types than are *Aspergillus* species, and it may require a greater degree of debilitation of the host to initiate infection. *P. boydii* is now a frequently encountered agent of clinical mycoses, however, and may be considered a "second string" opportunist.

ETIOLOGY, ECOLOGY AND DISTRIBUTION

As *Monosporium apiospermum,* the species described by Saccardo has been isolated numerous times from cases of mycetoma in various parts of the world. The first case was a patient of Tarozzi[117] but was described fully by Radaeli.[101] An organism from chronic otitis called *Verticillium graphii* may in fact be the first isolation. Since that time colonial variants have given rise to additional species as etiologic agents. One described by Pepere[97] appears to be invalid, as the grains were black, whereas all the grains from cases of mycetoma due to *S. apiospermum* are white. Isolants were described as *Aleurisma apiospermum* by Montpellier and Gowillon in 1921,[90] *Indiella americana* by Delamare and Gatti in 1929, *Glenospora clapieri* by Catanei in 1927, and *Acromoniella lutei* by Leao and Lobo in 1940. The organism isolated from otitis by Belding and Unmanzio[13] was described as *Monosporium.* These all appear to be variants of the same species, and at present there is only one recognized etiologic agent, *P. boydii* and its anamorph *S. apiospermum.*

P. boydii appears to have a worldwide distribution and has been isolated from a variety of saprophytic situations. The first isolation from the soil was recorded by Ajello.[2] This was accomplished by injection of soil into mice in search for *Histoplasma capsulatum.* Since then it has been found in Ohio by direct spraying of soil on plates,[65] and from polluted streams and sewage sludge,[31] from the tide-washed ocean coast in California,[34, 62] from marine soil in Bombay,[100] from farm soils in Maharastra,[95] from heated soils in India,[70] and from manure of poultry and cattle.[15, 104] In these latter environments it was isolated in huge quantities. This may be relevant to the large number of cases of pulmonary colonization that involve farmers or rural inhabitants.

To these records of the saprophytic occurrence of *P. boydii* can be added the much more numerous publications of its isolation in clinical situations. The first records of infection due to the fungus concern ear infections in Switzerland[113] and mycetoma in Italy.[101] It is by far the most commonly isolated agent of mycetoma in the United States,[50, 125] Canada,[36] Roumania,[78] and the temperate regions of the world in general. Mahgoub[78] points out that it is most common in regions receiving between 1000 and 2000 mm of rainfall. He feels that rainfall is a more important climatic factor than temperature. Perhaps this accounts for its isolation from subjects in mountainous Venezuela, arable areas of Argentina, and coastal lands of Mexico and for its absence in arid regions of Mexico, Sudan, Somalia, and Senegal. It is also apparent that the presence of *P. boydii* in soil may not correlate with its isolation from pathologic material. It is frequently isolated from a variety of saprophytic environments in India,[70, 95] but, as pointed out by Padhye and Thirumalachar,[95] it is very rarely isolated from clinical material or from mycetoma. This discrepancy may be due to differences in pathogenic potentials of different strains and geographic distribution of those strains. Some experimental evidence of this was found by Lupan and Cazin,[72] who reported a wide variation in the pathogenicity of strains of *P. boydii,* all of which had been isolated from clinical material.

The age and sex distribution in human infections is about the same as for other causes of mycetoma. The occurrence is more frequent in men than in women (3:1 to 5:1), probably because of greater outdoor activity, and is commoner in rural than in urban areas, which may reflect the habit of not wearing shoes and exposure to soil. The greatest number of cases occur in the group between 20 and 40 years of age. Other types of infections caused by *P. boydii* are too uncommon for patterns to be discerned. They are usually associated with predisposing factors such as immunosuppressive drugs, underlying disease, trauma, barrier breaks, and more recently aspiration of soil or swamp water or working in sewers (Chapter 5). Almost all recorded cases of otitis associated with *P. boydii* have been in children. No predisposing factors have been noted except previous otitis of bacterial etiology.

CLINICAL DISEASE

The clinical manifestations of infection by *P. boydii* are quite varied. Though at least 99 per cent of such infections are mycetomas, the remainder include infections of the eye, ear, central nervous system, internal organs, and more commonly the lungs. The clinical presentations will be discussed in reference to the organ system involved.

Mycetoma

The most common clinical condition involving *P. boydii* is mycetoma. The general topic of mycetoma is discussed in Chapter 5. In this chapter, some of the unique aspects of *P. boydii* mycetoma will be emphasized. The symptomatology for this organism and for other agents of mycotic mycetoma is essentially the same. The usual history is that of a trauma or puncture wound to the feet, legs, arms, or hands.[51] There is some evidence that after initial implantation the organism remains quiescent for some time, possibly re-

quiring another insult to the area or time to adapt to the host before the symptoms become apparent. The primary lesion, when it does develop, is a locally invasive, indolent, tumorlike process or a small, painless subcutaneous swelling. This gradually enlarges and softens to become phlegmonous. It ruptures to the surface, forming sinus tracts, then burrows into the deeper tissue to produce swelling and distortion of the tissues. *P. boydii* invades primarily subcutaneous tissues and ligaments; tendons, muscle, and bone are usually spared. The burrowing follows fascial planes and there incites suppuration and the formation of abscesses that drain through the sinus tracts. The tracts may remain open for long periods of time, heal, and then reopen. When expressed, the drainage contains numerous small particles or grains representing microcolonies of the organism.[94]

The grains formed by *P. boydii* are large (up to 2 mm), white to yellowish, soft to firm, and round to lobulated. The hyphae are broad (up to 5 μ), septate, and intertwined and show numerous swollen cells 15 to 20 μ in diameter. These are called intercalary chlamydoconidia. There is no cementing substance between the hyphae composing the grain. In tissue sections the grains are seen to be in the lumen of the sinus tract. This is often filled with a pyogenic infiltrate, and large accumulations of neutrophils in all stages of degeneration are seen to be clinging to the grain. Immediately around this abscess formation is an area of dense fibrosis and granulation tissue that is rich in capillaries, epithelial cells, macrophages, and multinucleated giant cells.

Prognosis and Therapy. In uncomplicated mycetoma there are usually no other constitutional disturbances. Pyrexia may develop if there is secondary bacterial infection; otherwise, there are no other symptoms. The lesions of mycetoma that are associated with *P. boydii* seldom, if ever, heal spontaneously. The disease may be chronic and continue for 40 or 50 years. Unlike some agents of mycetoma, dissemination of *P. boydii* to other parts of the body is essentially unknown; however, in 1961 Oyston[94] reported isolation of the fungus in inguinal lymph glands in a case of far advanced mycetoma.

The medical management of mycotic mycetomas in general has been unrewarding. *P. boydii* is essentially resistant to all systemically useful agents, including 2-hydroxystilbamidine, potassium iodide, 5-

fluorocytosine, and amphotericin B. Nielsen[91] found that only three strains of *P. boydii* were inhibited by 2 µg per ml of amphotericin B at 48 hours. An isolant of Ernest and Rippon was resistant to 50 µg per ml at 48 hours.[43] Miconazole has been found to inhibit *P. boydii* at concentrations of 0.08 to 0.25 µg per ml. With a blood level of 1.0 µg per ml achievable following an intravenous dose of 9 mg per kg, the medical treatment of such infections in the future looks promising. Lutwick et al.[74, 75] reported in 1979 on the successful management of a mycetoma in a 32-year-old male. Four months of intravenous miconazole resulted in healing and elimination of swelling and induration. A personal communication from Dr. David Stevens indicates that a higher dose regimen is now recommended. From 600 to 1000 mg is given t.i.d. (3 gm per day). With this regimen no abnormal liver function tests or blood dyscrasias were noted. The one significant side effect was pruritus, and this was thought to be due to the vehicle. This occurred in 25 per cent of patients and was severe enough to require adjusting or interrupting therapy. In a personally observed case of pedal mycetoma, miconazole is now being administered. The patient is a sewer department worker with a four-year history of advancing disease. Ketoconazole appears to be ineffective in clinical trials.[20]

Up to the present time the most successful approach to the control of mycetoma caused by *P. boydii* was surgical. This is not difficult if the lesion is small and localized, but a relapse rate as high as 80% has been reported following inadequate surgical removal.[78] The margin of tumor in mycotic mycetoma is encapsulated; this aids in the isolation and removal of the lesion. In far advanced cases amputation is the only successful treatment. In most cases of mycetoma of the foot the heel is uninvolved and the calcaneum and talus unaffected; thus, the fore part of the foot can be removed with complete arthrodesis and union of tibia, fibula and calcaneum, leaving the stump capable of bearing weight. More advanced cases may require below-the-knee amputation. Lesions in areas where radical surgery is countermanded are more difficult to treat. Simple curettage of as much of the tissue as possible has some value. In the future, surgical intervention combined with miconazole therapy should lead to more satisfying results in the treatment of this disease.

Mycetoma caused by *P. boydii* in areas other than the feet is quite rare.[51] The definition of mycetoma includes tumefaction, sinus tract formation, and grains, so that the so-called mycetomas of the brain, lung, and other organs are erroneously defined. In one documented case described by Rippon and Carmichael[104] there was a mixed infection of *Actinomyces israelii* and *P. boydii* in a mycetoma involving the parotid. Serologic, cultural, and histopathologic confirmation were obtained (Fig. 24–1). There is also an incomplete description of "plectenchyma or sclerotia" in a fatal pulmonary case reported by Alture-Werber et al.[4] Whether this was a true grain or a fungoma (fungus ball) cannot be discerned.

Pulmonary Pseudallescheriasis

The most commonly encountered site of infection by *P. boydii* aside from pedal mycetoma is the lung and upper respiratory tract. Some 35 to 50 cases that have been reported of such infections fall into several well-defined categories: bronchial colonization, pulmonary colonization, fungus ball formation (fungoma), and invasive pseudallescheriasis (*Pseudallescheria* pneumonia) (Tables 24–1, 24–2, and 24–3).

Bronchial colonization has been reported twice. In the first case, reported by Reddy et al.,[102] a 77-year-old woman with some chronic pulmonary problems had repeated positive sputum cultures for *P. boydii*. Tests for tuberculosis and other fungi were noncontributory. No follow-up was given. In a second case, by Rippon and Carmichael,[104] strands of the fungus were seen on direct examination of sputum. The patient was on prednisone, 5 mg per day, for rheumatoid arthritis and had coughing, wheezing, and pulmonary congestion. The prednisone was stopped and the condition cleared. Steroids were started again and the condition reappeared. The condition cleared again following cessation of the medication. The patient had clinical symptoms similar to those of allergic bronchopulmonary aspergillosis with serum IgG antibodies to *P. boydii* and was skin-test positive.

Pulmonary colonization is by far the commonest manifestation of pseudallescheriasis of the lung. The report of Creitz and Harris[33] was the first to describe this condition. Their patient had a cavity subsequent to a pyogenic abscess that was secondarily invaded by *P. boydii*. The organism was recovered from sputum and in long-term follow-up the fungus was again recovered from bilateral upper lobe cavities at autopsy as strands and

Text continued on page 603

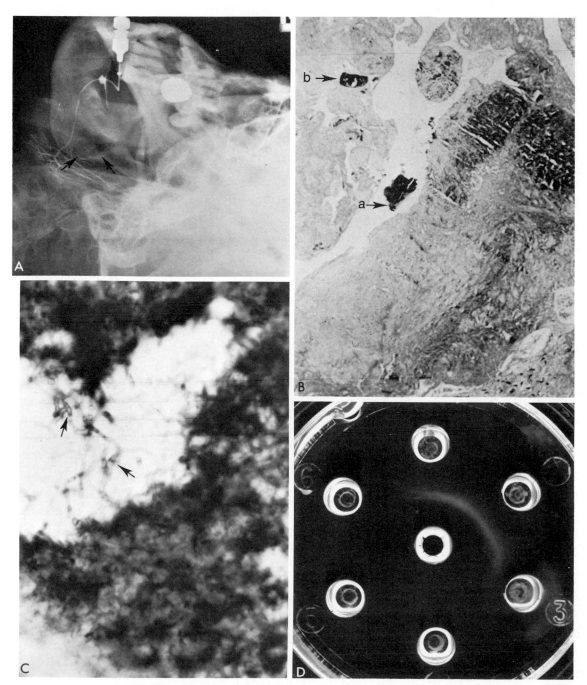

Figure 24–1. Mycetoma of *P. boydii* presenting as a mixed infection of the parotid with *Actinomyces israelii*.[104] *A,* Sialogram demonstrated a growing mass *(arrows)* distorting the channels of the parotid. (Courtesy of R. Goepp.) *B,* Section from surgical excision. Two grains are in the sinus tract: *A, A. israelii; B, P. boydii.* GMS. ×100. *C,* Close-up of grain. Note branching septate mycelium *(arrows).* ×950. *D,* Immunodiffusion plate from same patient. The center well contains patient serum. The three wells with precipitin bands contain types of antigen preparations of *P. boydii.*

Table 24–1 *Pulmonary Pseudallescheriasis 1955–1970. Transient Colonization, Grains, and Pseudallescheriomas*

Author	Year	Ref. No.	Age	Sex	Underlying Disease	Fungal Morphology	Culture Sputum	Culture Tissue	Treatment	Outcome
Drouhet	1955	37					+			No details; transient colonization?
Creitz	1955	33	56*	M	T.B. cavity		+			Died
Tong	1958	120	Same patient			Strands, "clumps"	+	+	2-hydroxy-stilbamidine	
Scharyj	1960	111	26*	F	Cavities	Fungoma		+	Surgery	Recovered
Stoeckel	1960	116	53*	M	T.B. cavity	Strands		+	Surgery	Recovered
Travis	1961	121	30*	M	Sarcoid	Fungoma	+		Steroid	Died
Dabrowa	1964	34					+			No details; transient colonization?
Louria	1966	71	43	F	Bronchogenic cyst; T.B.	Strands	–	+	Surgery	Recovered
Mirsa	1966	86	35*	F	Bacterial pneumonia	Strands	+		?	?
Ariewitsch	1968	5	38	F	Bronchitis	Strands	+		Antibodies	Recovered
Oury	1968	93	55	M	T.B. cavity	Strands	+		Surgery	Died
Adelson	1968	1	49	M	Rheumatoid spondylitis	1-cm Fungomas	+		Steroids, endocavitary iodides	Recovered, later died; *A. fumigatus*
Rosen, P.	1969	107	61	M	Ankylosing spondylitis	Fungomas			Endocavitary iodides	Died; *A. fumigatus*
McCarthy	1969	83	60	M	T.B. cavity?	Strands, fungomas	–[D]	+	Surgery[M]	*A. versicolor*, recovered
Reddy	1969	102	62*	M	T.B. cavity Sac. bronch.	Fungomas	+[S]		Surgery	Recovered
			52*	M	T.B. cavity	Strands		+	Surgery	Recovered
			77*	F	None	Strands	+		None	Transient colonization

*Farmers or farm family

D = Immunodiffusion positive; S = Positive skin test; M = Squamous metaplasia noted.

Table 24-2 *Pulmonary Pseudallescheriasis 1970–1979.*
Transient Colonization, Grains, and Pseudallescheriomas

Author	Year	Ref. No.	Age	Sex	Underlying Disease	Fungal Morphology	Culture Sputum	Culture Tissue	Treatment	Outcome
Thirumala-char	1973	118			T.B. cavity	No details	+			No details
El-Ani	1974	40	62	M	T.B. cavity	Fungoma	+		?	?
Belitsos	1974	14	24	M	Sarcoid, T.B. cavity	Fungoma	+	+	Antibiotic	Died
Häiner	1974	52	52*	F	T.B. cavity	Strands; fungoma	+D	+	Surgery	Recovered
Arnett	1975	6	59	F	Bronchial cyst	Fungoma		+	Surgery	Recovered
Rippon	1976	104	56*	M	T.B. cavity	Fungoma	+D	+	Surgery	Recovered
				M	Steroid for arthritis	Strands	+D	+	Dx Steroid	Recovered
Alture-Werber	1976	4	66	F	T.B. cavity	Grains?; strands; fungomas	+	+	None	Died
Jung	1977	60	64*	M	Ca. lung		+	–	Surgery	Died
			73*	M	T.B. cavity		+	–	Amphotericin B	Cured, relapse, died
			60*	F	T.B. infection			+	Surgery	Cured
			67*	F	Bronchiectasis		+	–	None	Resolved
			60*	M	T.B. cavity		+	+	Surgery	Sputum still +
			81*	M	Chronic obstruction		+	–	Amphotericin B	Cured
Bakerspigel	1977	10	53	F	Bronchitis; T.B.?	Fungoma		+	Amphotericin B; surgery	Recovered
Bousley	1977	21	39	M	Recurrent Resp. inf.	Strands in pleural fluid	+		Surgery	Recovered
Deloach	1979	35	52	F	Bronchiectasis	Strands	+		Surgery	Recovered
Carles	1979	23	71	M	T.B. cavity	Hyphal strands	+		Amphotericin B	Not effective
Kaplan	1979	61	?		Sarcoid	?			None	+ Culture

*Farmers or farm family D = Immunodiffusion positive

Table 24–3 *Pulmonary Pseudallescheriasis.*
Pseudallescheria Pneumonia and Invasive Disease

Author	Year	Ref. No.	Age	Sex	Underlying Condition/Therapy	Lung Disease	Antemortem Culture	Treatment	Postmortem Positive Fungi
Rosen, F.	1965	106	19	F	Nephritis/steroids	? Tonsilitis	–		Brain, thyroid
Forno	1972	44	56	M	Systemic lupus, renal/steriods	Rt. lower infiltrate	+		Brain, thyroid
Lutwick	1976	74	32	F	Systemic lupus/steroids.	Lt. lower infiltrate	+ (eye)	Miconazole	Recovered
			66	F	Lymphoma/steroids; cytotoxins	Rt. lower infiltrate	+	Amphotericin B	Brain, lung, heart, renal *Aspergillus*
Winston	1977	126	37	M	Leukemia/cytotoxins	Rt./Lt. upper consolidation, cavities	–	Amphotericin B	Lung, arm vessels and nerves
Bell	1978	17	3	M	Leukemia, rt. frontal cerebritis	None detected	+ staph. (brain)	Surgery; amphotericin B	Recovered
Walker	1978	124	37	M	Renal transplant/steroids	Rt./Lt. cavity infiltrates	+ (brain)	Amphotericin B	*Aspergillus*: Lung, brain, thyroid
Lutwick	1979	75	66	M	Lymphosarcoma/steroids	+ (also skin)	+ (sputum)	Miconazole	Recovered
Meadow	1980	84	15	F	Aspiration pneumonitis/steroids	Pneumothorax	+	Amphotericin B; miconazole	Eye, brain, liver, kidney, tissue

"clumps."[120] In the reviews by Lutwick et al.,[75] Reddy,[102] Jung,[60] and others, the authors note preformed cavities in all but one of 14 recorded cases (Tables 24–1 and 24–2). Louria and Liebermann[71] found mycelial strands within a cyst in the lung. Similar findings were noted by Travis[121] in a patient with sarcoidosis. Ropes of mycelium (plectenchyma) and conidia were found in large residual cavities in a resolved case of tubercu-

losis discussed by Alture-Werber et al.[4] They also describe "sclerotia similar to *Madurella mycetoma*" being present in some pulmonary tissue at autopsy. This may represent a true pulmonary mycetoma. Fungoma (fungus ball formation) has been reported many times but documented in only a few cases. In a case reported by Rippon and Carmichael[104] a fungoma developed in an old tuberculosis cavity (Fig. 24–2). On resection it was found to

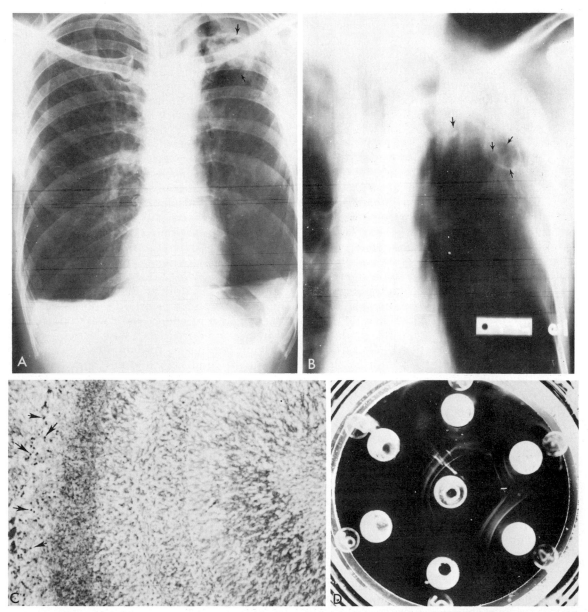

Figure 24–2. Pseudallescherioma. Fungus balls developed in old tuberculosis cavities. *A,* Several cavities are present in the apices of the lungs. Some opacifications are evident in two cavities located in the apex of the left upper lobe *(arrows). B,* Tomogram demonstrates balls within the cavities *(arrows). C,* Histologic section of a resected area of the pseudallescherioma. Concentric rings are evident. At the outer edge of the ball, where it was exposed to air, some conidia have been produced *(arrows).* GMS. ×450. *D,* Immunodiffusion plate from this patient. The center well is serum from the patient; the outer wells contain different *P. boydii* antigen preparations.[104]

consist of concentric rings of septate mycelium indistinguishable from an *Aspergillus* fungoma. On the periphery of the "oma," however, conidia typical of *P. boydii* were found. The fungus in such cases of pulmonary colonization consists of single hyphal units, loose mycelial strands that may sporulate occasionally, fungoma formation with concentric rings of mycelium and peripheral sporulation, and rarely true grain formation of the type found in mycetoma. The most commonly noted predisposing conditions are sarcoid, tuberculosis, and previous bacterial infections that resulted in cysts or cavities. In a report by Larsh,[67] some 20 patients with chronic lung problems were culture positive for *P. boydii*, although some only transiently. No clinical details were given.

The histologic findings in the reported cases of pulmonary colonization were similar to those seen in secondary invasion by *Aspergillus* spp. There are fungal abscesses partially walled off by granulation tissue, marked fibrosis obliterating the normal alveolar pattern, and small epithelioid tubercles with chronic inflammatory infiltrate and multinucleated giant cells. The surrounding parenchyma is often necrotic. In bronchial cysts colonized by *P. boydii* the squamous epithelial lining may show some degree of nuclear atypia.[71, 83] Unless conidiophores and conidia are seen in the specimen, it is not possible to distinguish the mycelium of *Aspergillus* sp. from that of *P. boydii*. Both may coexist in the same infection.[107] Culture and serologic analysis are necessary for confirmation of the etiology.

In pulmonary colonization the predisposing condition is usually some preformed cavity or cyst. Thus the patient is not severely debilitated and treatment is often successful. As noted in Tables 24–1 and 24–2, surgical resection of the areas has often led to complete recovery of the patient. Jung et al.[60] reviewed nine cases from the literature and added nine from the Missouri medical center, six previously unpublished. Ten of the 18 underwent resectional operations, with two operational deaths; the rest had uneventful recoveries. Amphotericin B was used in some cases,[10, 60] as were endocavitary iodides.[1, 107]

Transient colonization of the lower respiratory tract may give rise to *Pseudallescheria* pneumonia. This was recorded during the course of a bacterial pneumonia.[86] Although in this case the patient recovered, most such infections occur in a compromised host, and the resulting disease is often a fatal invasive systemic mycosis (Table 24–3). Probably the first such case was recorded by F. Rosen et al.[106] A young woman was being treated for glomerulonephritis with steroids. A fulminant disease occurred and at autopsy *P. boydii* was present in the thyroid and brain. Though no portal of entry was documented, the lungs were thought to have been the site of primary infection. Dissemination to the brain or to the brain and thyroid is recorded in the majority of cases of systemic pseudallescheriasis.

In the more recently reported cases the patients have also been severely immunosuppressed by receiving steroids or steroids and cytotoxins. In most, sputum or culture of some other site was positive ante mortem. The first case of visceral pseudallescheriasis in which the patient recovered was reported in 1976 by Lutwick et al.[74] The patient had systemic lupus erythematosus and was being treated with steroids. A lower lobe infiltrate was followed by an endophthalmitis in the right eye. *P. boydii* was grown from vitreous aspirate. The organism was sensitive to 0.125 mg per ml of miconazole. Intravenous miconazole therapy was started at a dose of up to 1000 mg three times a day until clinical improvement occurred. Aspiration of muddy water led to pneumonia and endophthalmitis, as reported by Meadow et al.[84] Disseminated disease was found at necropsy (details are given in the section on endophthalmitis). A second case of clinical cure was reported by Lutwick et al. in 1979.[75] The patient was receiving prednisone for lymphosarcoma. Tenosynovitis and skin nodules appeared, and sputum became positive for *P. boydii*. Miconazole therapy brought about resolution of the infection. Surgery with amphotericin B therapy was successful in a case of cerebral disease noted by Bell and Myers.[17] In this case and that of Rosen,[106] the lung was not definitely established as the portal of entry. Introduction of the organism at the site of repeated steroid injections was considered in the latter case. Some cases of invasive pulmonary disease occur in the absence of known predisposing factors. The patient in Figure 24–3 has slowly progressive pulmonary dysfunction of unknown etiology.

Pseudallescheria Sinusitis

That the sinuses and nasal septum may also act as a portal of entry for infections by *P. boydii* was demonstrated by Gluckman et al.[46]

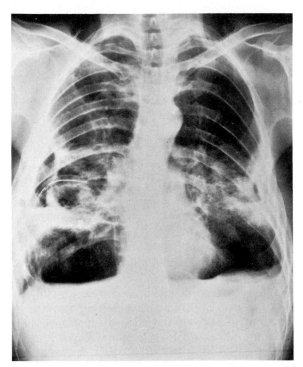

Figure 24–3. A 31-year-old male with slowly developing pulmonary dysfunction of unknown predisposing factors. Sputum, biopsy, bronchial wash, bronchial brush, and needle aspirate material all grew *P. boydii*. Mycelial elements were seen on surgical biopsy. The patient is being treated with miconazole. Chest film shows widespread destructive disease in both lung fields.

In this case the maxillary sinus was involved in a patient who was a diabetic in renal failure. He developed a syndrome resembling rhinocerebral mucormycosis. A second case also involved a diabetic with right antrum sinusitis. She was treated with sinus irrigation, with no response.[11] In a case reported by Mader,[77] no underlying condition was found. The patient complained of frontoparietal headaches. Transseptal sphenoidectomy followed by a five-week course of intravenous (IV) miconazole resulted in resolution of the disease. The nasal septum was thought to be the portal of entry in a case of fatal disseminated disease occurring in a leukemic patient. Winston et al.[126] described massive lesions containing fungi in the brain but in no other organ. The left side of the cerebrum and cerebellum were destroyed by fungal abscesses. Treatment with amphotericin B and nystatin irrigations together with systemic estrogens were unsuccessful in a case recorded by Hecht and Montgomerie.[56] The patient had lymphoblastic leukemia and was being treated with cytotoxins. No autopsy was performed. The use of estrogens was based on a case of pedal mycetoma in a woman reported by Mohr and Muchmore.[87] The lesions would regress during pregnancy (Table 24–4).

Pseudallescheria Meningitis

Meningitis, as distinct from dissemination to the central nervous system from some other focus of infection, is usually the result of a barrier break or trauma. In the first case, described separately by Wolf,[127] Benham,[18] and Aronson,[7] a middle-aged woman received spinal anesthesia for a hernia operation performed in Trinidad. About one month later she began to have headaches, malaise, and gradual deterioration. In a New York hospital, *P. boydii* was grown from 11 of 14 spinal taps taken before death. Selby[110] reported on a 49-year-old woman from Malaysia who developed an extradural lesion resulting in a pachymeningitis with progressive paraplegia. The lesion was interscapular at T-10 and thought to be a neuroma. The patient had fallen on a stair, striking the lower thoracic spine some time before. At surgery the mass involved the dura, and the

Table 24–4 Pseudallescheria *Sinusitis*

Author	Year	Ref. No.	Age	Sex	Site	Culture	Underlying Disease	Treatment/Outcome
Gluckman	1977	46	58	M	Maxillary sinus	+	Diabetic renal failure	Amphotericin B/cure
Winston	1977	126	57	M	Nasal septum	−	Myelomenocytic leukemia; cytotoxins	Died, massive lesions in brain
Hecht	1978	56	20	M	Maxillary sinus	+	Leukemia; cytotoxins	Amphotericin B, 5 FC/ died, no autopsy
Mader	1978	77	33	F	Sphenoidal sinus	+	None	Miconazole/recovered
Bark	1978	11	28	F	Rt. antrum sinusitis	+	Diabetic	Irrigation/no response

Table 24–5 Pseudallescheria *Meningitis*

Author	Ref. No.	Year	Age	Sex	Symptoms	Treatment	Outcome	Culture
Benham	18	1948	45	F	Meningoencephalitis after anesthesia	None	Death	+
Wolf	127	1948						
Aronson	7	1953						
Selby	110	1972	49	F	Extradural pachymeningitis after abrasion	Surgery	Resolved	−
Lytton	76	1975	16	F	Extradural paraplegia; skin abrasion	Surgery	Resolved	+
Peters	99	1976	25	M	Ventriculoperitoneal shunt	No details		+

arachnoid was quite thick and opaque. After surgery the patient recovered. A similar extradural lesion and subsequent pachymeningitis was reported by Lytton.[76] No history of injury was noted, but a skin abrasion was possible. Surgical removal resulted in recovery (Table 24–5).

Pseudallescheria Arthritis

Pseudallescheria arthritis also is most often associated with barrier break or trauma. The injuries range from puncture wounds with a rusty nail,[55] laceration while playing soccer,[75] and association with podiatric surgery[75] to a fall in a bathtub against a shower curtain. In the last situation, the fungus was growing on the shower curtain.[115] In most cases the lesions were limited and responded to surgery or treatment with miconazole. In a case recorded by Lutwick[75] there was no history of trauma. The lesions involved the fingers,

wrist, and hand of a farmer. *P. boydii* was grown from synovial aspirate. Treatment with a 30-day course of 600 mg IV of miconazole every eight hours was successful, but a relapse occurred after one year. Then a course of 65.2 grams of miconazole was given over 52 days, combined with intralesional irrigation with a 100 mg per ml solution of miconazole. Six weeks following termination of therapy the patient relapsed again. He is receiving a third course of therapy (Table 24–6).

Pseudallescheria Endocarditis

This is a rarely reported form of pseudallescheriasis and usually involves heart valve prosthesis. In the first case, recorded by Roberts et al.,[105] the patient (male, 48 years old) received a porcine valve replacement for mitral stenosis. Six weeks later petechial and other signs of infectious endocarditis began.

Table 24–6 Pseudallescheria *Arthritis*

Author	Year	Ref. No.	Age	Sex	Trauma	Involved Joint	Treatment	Outcome
Negroni	1943	90a	6	M	Fall	Knee, right	?	?
Nielsen	1967	91				Knee	No details given	
Stevens	1968	115		F	Bathroom fall	Knee	Surgery	Resolved
Hayden	1977	55	6	M	Puncture wound	Knee	Intra-articular amphotericin B	Resolved
Halperin	1977	54	46	M	Surgery; steroids	Hand	Miconazole	Resolved
Lichtman	1978	69	33	M	Shunt kidney transplant	Finger, hand	Surgery—amputation; miconazole	Resolved
Lutwick	1979	75	76	M	None known	4th & 5th fingers, wrist, hand	Miconazole	Resolved, relapsed
			32	M	Soccer injury	Knee	Miconazole	Resolved
			43	M	Podiatric surgery	Foot, synovium	Miconazole	Resolved

He did not respond to antibiotics. At necropsy mycotic lesions were present in the brain, kidneys, and myocardium. Culture of the porcine valve and one antemortem blood culture resulted in pure growth of *P. boydii*. In the case of Gordon and Axelrod,[47] a 52-year-old male with rheumatic valvular disease underwent a neurectomy and aortic valve replacement with a No. 24 duramater valve. About two months later, petechiae and other signs of infectious endocarditis appeared. *Clostridium limosum* was recovered from repeated blood cultures. Appropriate antibiotics for the bacterium were begun. A replacement valve (No. 29 Bjork-Shiley) was instituted, but the patient failed to respond and died. Postmortem examination was refused. Cultures of the duramater valve yielded *C. limosum* and *P. boydii*.

In the preparation of tissue-source replacement valves (porcine and duramater) the material is often soaked in amphotericin B and nystatin in addition to antibacterial antibiotics. Neither of these two antifungal agents is effective against most isolants of *P. boydii*.[91] However, most clinical isolants of *P. boydii* are sensitive to low concentrations (0.032 to 0.16 μg per ml) of miconazole.[74] Others have reported *P. boydii* endocarditis associated with homograph valve replacement.[123]

Cutaneous and Subcutaneous Pseudallescheriasis

Trauma with introduction of *Pseudallescheria boydii* into the dermal or subcutaneous tissue usually results in the clinical syndrome of mycetoma. A few other types of clinical disease are recorded, however. Zaffiro[129] in 1938 reported on a medical officer who received a puncture wound of the hand. Although a mycetoma did not develop, the patient had spiking fevers and multiple subcutaneous abscesses over all parts of the body. *P. boydii* was reported as having been isolated from these abscesses, although the cultural data are incomplete. Conti-Diaz recently described a syndrome occurring in a 50-year-old male farm worker that mimics the lymphocutaneous form of sporotrichosis[30] (Fig. 24–4). A bubo formed at the site of trauma; later, subcutaneous swelling occurred in the pattern of lymph drainage from the original site. These eroded and festered. A course of iodides failed to bring about resolution of the infection; it responded to sulfadiazine and levamisol, however.

Skin nodules in addition to tenosynovitis occurred in a patient with *Pseudallescheria* pneumonia. The patient had lymphosarcoma and was being treated with prednisone.[74] Intravenous miconazole for 33 days induced resolution of the infection.

Mycotic Keratitis

Colonization of an injured cornea by *Pseudallescheria boydii* is not as common as that by other agents of mycotic keratitis (see Chapter 27). Far more cases of infections caused by *Aspergillus* spp. or *Fusarium* spp. are recorded. The reported cases of mycotic keratitis have

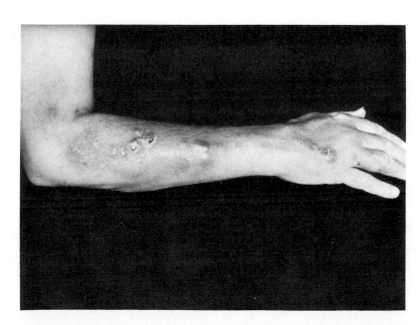

Figure 24–4. Lymphocutaneous form of *Pseudallescheria* infection mimicking sporotrichosis.[30] (Courtesy of A. Conti-Dias.)

been recently been reviewed by Zapater and Albesi.[128] They noted 14 cases, of which 12 were sufficiently documented to be considered as valid (Table 24–7). The predisposing conditions leading to infection, pathogenesis of the lesions, and response to therapeutic efforts are not markedly different from those seen in other types of mycotic keratitis (Chapter 27). In this section the salient features of *Pseudallescheria* keratitis will be reviewed.

The first isolation of *P. boydii* was recorded by Paulter et al. in 1955.[96] The patient was a farmer with a corneal ulcer, hypopyon, and iridocyclitis. Direct examination of the material revealed hypha and ovoid conidia. Treatment was unsuccessful, and the eye was enucleated. In the second case, noted by Gordon et al.[48] in 1959, a fish scale abraded the eye of a worker in a fish factory. Twenty-three days later a corneal ulcer and hypopyon were found. *P. boydii* was cultured from the eye and has been cultured from marine detritus. Lavage with amphotericin B and nystatin averted enucleation. In this case, as in most of the other cases, there was some previous injury or trauma to the eye. Several times this included abrasion by plant materials, twigs, splinters, or other soil-contaminated material. In a few cases no known injury or exposure to quantities of conidia was reported. Treatment has generally not been successful. Nystatin and amphotericin B lavage effected a cure in two cases; in one case, however,

scarring was so great that enucleation was elected.[43] The conjunctival flap exacerbated this infection and resulted in clinical cure in another,[122] although the latter patient was left with a leukoma. In the report by Zapater and Albesi,[128] lavage with pimaricin and amphotericin B was successful. Reports on the use of miconazole for keratitis have not as yet been noted. Hairstone and DeVoe[53] carried out ultrastructural studies on the development of the disease in an animal model.

A clinical variant of this disease was recorded by Persaud and Holroyd.[98] A 44-year-old black hotel gardener presented with a unilateral symblepharon. The tumor was extirpated surgically and the patient recovered. Within the tumor, grains comparable to those seen in mycetoma were found. Ten years prior to this, the patient had had "bumps" surgically removed from the forearm and ankle that also were due to *P. boydii*.

Endophthalmitis

Although fungal endophthalmitis is an increasingly recorded disease entity, few of the cases have involved *P. boydii*. Most such infections are in compromised hosts and *Candida* species or *C. albicans* is the most prevalent fungal agent. As in other forms of endophthalmitis, the fungi gain access to the inner aspects of the eye by two routes: exogenous, as the result of penetrating trauma or surgery,

Table 24–7 Pseudallescheria *Keratitis**

Author	Year	Ref. No.	Age	Sex	Trauma	Treatment/Outcome
Paulter	1955	96	39	M	None – rural inhabitant	Evisceration
Gordon	1959	48	27	M	Fish scale	Nystatin, amphotericin B/cure
Casero	1962	24	28	F	None	Evisceration
Ernest	1966	43	43	M	None	Nystatin, amphotericin B., flap leukoma, evisceration
Persaud	1968	98	44	M	None – gardener	Mycetoma – conjunctiva (excision)/cure
Matsuzaki	1968	81	41	F	Stick injury	Evisceration
Matsuzaki	1968	81	46	F	Rice leaf injury	Amphotericin B/cure
Ouchi	1969	92				
Bakerspigel	1971	9	41	M	Wood splinter	Evisceration
Levitt	1971	68	40	M	Foreign body	Evisceration
Elliot	1977	41	26	M	Burn – molten aluminum	Enucleation
Upadhayay	1978	122	16	M	None	Flap/cure
Zapater	1979	128	19	M	Foreign body	Pimaricin, amphotericin B/cure
Liesegang	1980	69a	?	M	No clinical details given	

*Modified from Zapater and Albesi.[128]

and endogenous, as the result of dissemination from some other focus of infection.

To date, three such cases of *Pseudallescheria* endophthalmitis have been recorded. The first case was described by Glassman et al.[45] A 49-year-old woman wtih a 15-year history of diabetes was subjected to cataract surgery with extraction and sector iridectomy. By the 17th postoperative day, many white spherules were noted about the iris and anterior vitreous. Aspirated material was cultured and yielded *P. boydii.* Amphotericin B instillation was begun and continued over a three-month period. The infection resolved, and the cornea and anterior chambers cleared. Visual acuity at eight months was 20/30. The anterior location of the infection probably aided the success of treatment.

The other two cases of endophthalmitis were of endogenous origin in immunosuppressed patients. In the first a 32-year-old woman with lupus was undergoing steroid therapy and developed pneumonia and, subsequently, a fungal endophthalmitis (see Table 24–3). *P. boydii* was isolated from posterior vitreous aspirate.[74] Treatment with miconazole was successful. The second case was reported by Meadow et al.[84] The patient, a 15-year-old girl, suffered a grand mal seizure while riding her bicycle. She fell into a puddle and aspirated some muddy water. After being hospitalized, multiple particles of mud and glass were removed from the bronchial tree. She developed aspiration pneumonia and was treated with antibiotics and methyl prednisolone (1 g per day). By the seventh day, depigmented areas appeared in the right and left fundi. Aspiration of these and pleural fluid both revealed *P. boydii* by culture. She was placed on IV miconazole but died within two days. At necropsy, *P. boydii* was found in the lung, the liver, the kidney, subcutaneous tissue, and the brain.

Otomycosis

Probably the first authentic isolations of *P. boydii* were from chronic infections of the ear. These were described by Harz and Bezold in a book on human otitic mycoses by Siebenmann in 1889.[113] The symptoms described then and in the published reports since[104] are the same. Chronic discharge from the ear canal is culture positive for various bacterial species, and the diagnosis of otitis externa is made. Some of the lesions became secondarily infected by fungi. The usual fungal species involved are *Aspergillus niger* and *Scopulariopsis brevicaulis*, with a few reports of *P. boydii.* Many persist, but some infections respond to nystatin lavage.[104]

Isolated reports of infections caused by *P. boydii* occurring in other body sites have appeared. One of these, by Meyer and Herrold,[85] involved a 57-year-old male with chronic prostatitis. He had recurrent episodes of cystitis as well. Urine and prostatic fluid yielded cultures of an enterococcus. Subsequently, many colonies of *P. boydii* were isolated in addition to the enterococcus. No follow-up was given.

ANIMAL DISEASE

Natural Infection

Considering the ubiquity of *Pseudallescheria boydii*, particularly in the farmyard environment,[15, 104] reports of natural disease in domestic or wild animals are rare. *P. boydii*, along with various *Candida* and *Aspergillus* spp., has been isolated from the throats of normal donkeys.[39] It can be recovered regularly from the teats of cattle[119] and the hooves of horses contaminated with mud, and disease resulting from penetrating trauma has been reported. Retrobulbar swelling caused severe exophthalmos in a horse that had received a kick on the eye one year previously.[58] At necropsy, mycelial strains and granules 1 to 5 mm in diameter were found in the orbit. Hyphae with chlamydoconidia and conidiophores were also found.

Bovine mycotic abortion is usually associated with *Aspergillus* or Zygomycetes. *P. boydii* is an uncommon agent of such infection. A serologic test developed by Knudtson is considered a diagnostic aid,[63] and a few cases have been uncovered. *P. boydii* has been isolated from milk of the infected quarter in a case of mastitis.[119]

A few cases of uterine infection in the horse have been noted. Persistent intrauterine bacterial infections were followed by recovery of *P. boydii* in culture.[103] Therapy was ineffective, and the animal was euthanized, whereupon *P. boydii* was isolated from the uterine lumen. In a second case, also reported by Reid.[103] persistent uterine infection was unresponsive to conventional antibiotic therapy. Intrauterine instillation of amphotericin B was effective, however. Natural infection in wild animals is very rarely reported.

Experimental Disease

Experimental disease in animals has been investigated by several authors. Lupan and Cazin[72] found wide strain variation in the ability to cause disease in mice; disease could regularly be produced in animals treated with cortisone, however. All strains used in their experiments were isolated from human infections. Bell[16] compared the virulence of human isolants to those found in soil or feedlot manure. Conidia contained in a 5 per cent mucin solution kill 95 per cent of inoculated mice in 21 days if the strains were from human infections. Eight times as many conidia from feedlot isolants killed only 28 per cent of mice. At autopsy of the mice, hyphae, conidia, and conidiophores were found in most internal organs. Torticollis was one of the commonest symptoms in infected mice. This was thought to be due to damage to the 11th cranial or spinal accessory nerve. Such damage is probably due to the lodging of conidia in the choroid and has been reported in *Aspergillus terreus* infections of mice. Antisera prepared from rabbits demonstrated common antigens in isolants from both natural infections and soil isolants. This discrepancy in virulence between soil isolants and isolants from natural infections has been noted in *Sporothrix schenckii* as well as other disease-producing fungi.

The humoral antibody response to experimental infection of mice was also studied by Lupan and Cazin.[73] Corbel and Eades[32] concluded that the T lymphocyte–dependent immune process played a major role in resistance to infections by *Candida* and *Cryptococcus* but not to those by *P. boydii* and other agents. They used NZB and CBA mice. Fungal arthritis has been produced in rabbits and its natural history has been studied by Edwards et al.[38]

BIOLOGICAL STUDIES

Few biological studies involving *Pseudallescheria boydii* have thus far been reported. Most such studies have concerned susceptibility to antibiotics. Nielsen,[91] Mohr and Muchmore,[87, 88, 89] Hecht,[56] Bell,[17] Lutwick,[74, 75] and others have documented the resistance of *P. boydii* to pharmacological doses of amphotericin B. There is, however, ample documentation of its sensitivity to miconazole.[17, 74]

Few nutritional studies have been done. Cazin and Decker[26] found strains of the organism able to assimilate a wide variety of carbohydrate sources. Only raffinose and insulin were not assimilated. The organism grows in media containing up to 8 mg per ml of cycloheximide[2, 7] and is completely unaffected by antibacterial antibiotics such as chloramphenicol as high as 1000 mg per liter. All six strains used in work by Mohr and Muchmore were inhibited by 0.01 and 0.003 mg per ml of estradiol.[87, 88, 89] Progesterone, testosterone, mestranol, and norethynodrel had no effect on growth of the fungus, nor did estrogen used in therapy of *P. boydii* infections.[56]

Serologic procedures have become standard in the diagnosis and prognostic evaluation of the various forms of pseudallescheriasis. They are particularly useful when cultural confirmation is lacking. Attempts to develop a complement fixation test have been unsuccessful, but immunodiffusion appears to be a reliable procedure.[73] Precipitin bands were found in cases of mycetoma,[12] bronchial colonization,[104] fungoma,[104] mixed mycetoma,[104] and bovine abortion.[64] These bands disappeared following resolution of the disease.[64]

In the work of Lupan and Cazin[73] an immunologically specific carbohydrate antigen (antigen I) was isolated from culture filtrates of several strains of *P. boydii*. Such antigens can be used in cases of possible fungus colonization when the etiology is unknown. Crude culture filtrates also appear to be specific and do not cross-react with *Aspergillus* or *Candida* antigens. These are adequate for simple serologic screening procedures.[104]

MYCOLOGY

The taxonomy of both the perfect and imperfect stages of this fungus is complicated. A summary will be given here. For a complete description of the growth characteristics and conidiation of the fungus, see Chapter 5. The type species of the perfect or teleomorph phase was described by Shear in 1922.[112] He thought it similar to an organism called *Eurotiopsis gayoni* described by Constantin in 1899.[29] This particular organism was reclassified as *Allescheria* by Saccardo and Sydow[109] because the name *Eurotiopsis* had already been used. Shear considered his isolant to be the second species of that genus. More recently, Malloch,[79] in reviewing Costantin's drawing (his type specimen had been lost), considered *E. gayoni* to be a *Monascus* and quite different from the fungus described by Shear. He therefore erected the genus *Petriellidium*, with *P. boydii* as the type species. There were six species within that genus. Malloch now considers this genus a "still birth," and *P. boydii* has been placed in *Pseudallescheria*, a previously described valid genus erected by Negroni and Fisher in 1943.[90a]

The imperfect or anamorph phase is more of a problem. As Shear noted, *P. boydii* had two asexual methods of conidiation. The simple, unbranched conidiophore (byssoid) ending in a single large conidiospore he called *Cephalosporium*. The other method consisted of conidiophores cemented together (coremia or synnema) and ending in a conidiospore, which he called *Dendrostibella*. Neither of these names is considered correct. Different strains of the organism produce these conidial types in varying quantities, sometimes one to the exclusion of the others. Thus a long list of names has evolved over the years. The name *Monosporium apiospermum* of Saccardo[108] has been the one most widely used, but the genus is now considered invalid.[57] Saccardo had proposed the name *Scedosporium* but never fully described it. Recently, an adequate description of that genus has been found, and the name is now considered valid.[25] The etiologic agent of *pseudallescheriasis* is *Pseudallescheria boydii* (Shear) McGinnis, Padhye et Ajello 1982.[83a] It is in the class Ascomycetes, order Microascales, family Microascaceae. The imperfect form is *Scedosporium apiospermum* (Saccardo) Castellani 1927.

The fungus grows readily and rapidly on most laboratory media. On rich media such as Sabouraud's agar the cleistothecia may not be formed. It sometimes requires repeated transfer on deficient media (such as oatmeal, potato glucose, or water agar) to induce ascocarp formation.

Pseudallescheria boydii (Shear 1922) McGinnis, Padhye et Ajello 1982 anamorph state Scedosporium apiospermum (Saccardo 1911) Castellani 1927

Colonies on oatmeal agar have a growth rate of 4 to 5 mm per day at 25° C. The colony is white initially and becomes a housemouse gray; it may sometimes have a rosy hue in some areas. Ascocarps (cleistothecia) are spherical, nonostiolate (closed), 140 to 200 μ in diameter, and usually submerged. The cells consist of yellow to brown thickwalled hyphae that are 2 to 3 μ in diameter. The wall of the fruiting body is 4 to 6 μ thick

and is composed of 2 or 3 layers of meanderingly interwoven filaments that give a polygonal appearance when viewed from above. The asci are nearly spherical, 12 to 18 × 9 to 13 μ in diameter, and evanescent. They contain eight ascospores that are straw-colored and ellipsoidal, symmetrical, or somewhat flattened. The spores have two germ pores and are 6 to 6.5 × 3.5 to 4 μ in diameter.

Two types of asexual conidia are produced. The first, *Scedosporium* type, consists of conidia that are broadly clavate or ovoid, rounded above and attenuated or truncate at the base, with a distinct brown wall. They are 6 to 12 × 3.5 to 6 μ in diameter. They are single, borne terminally or laterally on solitary conidiophores. The second type of conidiation (*Graphium* type) consists of smaller, hyaline conidia (5 to 7 × 2 to 3 μ) that are clavate or cylindrical, with a truncate base. These are borne on short annelides or elongating conidiophores, which are cemented together to form a synnemata or coremia. The conidia may form a globular mass after abstriction. This latter form of conidiation is infrequently seen in most cultures, but in a few it is the dominant conidia type. In order to verify the identification one must induce and examine the cleistothecia of the teleomorph state. This is because the conidial forms of almost all species of the genus *Pseudallescheria* and even some members of the genus *Petriella* as well as other members of the Microascaceae have almost the same morphology. Thus, *P. boydii* cannot be identified by examining the conidial form alone.

REFERENCES

1. Adelson, H. T., and J. A. Malcolm. 1968. Endocavitary treatment of pulmonary mycetomas. Am. Rev. Respir. Dis., *98*:87–92.
2. Ajello, L. 1952. The isolation of *Allescheria boydii* Shear, an etiologic agent of mycetomas from soil. Am. J. Trop. Med. Hyg., *1*:227–238.
3. Ajello, L., and L. D. Ziedberg. 1951. Isolation of *Histoplasma capsulatum* and *Allescheria boydii* from soil. Science, *113*:662.
4. Alture-Werber, E., S. C. Edberg, and J. M. Singer. 1976. Pulmonary infection with *Allescheria boydii*. Am. J. Clin. Pathol., *66*:1019–1024.
5. Ariewitsch, A. M., S. G. Stepaniszewa, and O. W. Tiufilna. 1968. Ein Fall des durch *Monosporium apiospermum* herorgerufenen lungenmyzetoms. Mycopathologica, *37*:171–177.
6. Arnett, J. C., and H. B. Hatch. 1975. Pulmonary allescheriasis. Report of a case and review of the literature. Arch. Intern. Med., *135*:1250–1253.
7. Aronson S, and R. Benham. 1953. Maduromycosis of central nervous system. J. Neuropathol. Exp. Neurol., *12*:158–168.

8. Avram, A., and G. Nicolau. 1969. Attempts to demonstrate complement-fixing antibodies in mycetoma. Mycopathologica, *39*:367–370.

9. Bakerspigel, A. 1971. Fungi isolated from keratomycosis in Ontario, Canada. I. *Monosporium apiospermum (Allescheria boydii)*. Sabouraudia, *9*:109–112.

10. Bakerspigel, A., T. Wood, and S. Burke. 1977. Pulmonary allescheriasis. Report of a case from Ontario, Canada. Am. J. Clin. Pathol., *68*:299–303.

11. Bark, C. J., L. J. Zaino, et al. 1978. *Petriellidium* sinusitis. J.A.M.A., *240*:1339–1340.

12. Baxter, M., I. G. Murray, and J. J. Taylor. 1966. A case of mycetoma with serologic diagnosis of *Allescheria boydii*. Sabouraudia, *5*:138–140.

13. Belding, D. L., and B. Umanzio. 1935. A new species of the genus *Monosporium* associated with chronic otomycosis. Am. J. Pathol., *11*:856–857.

14. Belitsos, N. J., W. G. Merz, et al. 1974. *Allescheria boydii* mycetoma complicating pulmonary sarcoid. Johns Hopkins Med. J., *135*:259–267.

15. Bell, R. G. 1976. The development in beef cattle manure of *Petriellidium boydii*, a potential pathogen for man and cattle. Can. J. Microbiol., *22*:552–556.

16. Bell, R. G. 1978. Comparative virulence and immunodiffusion analysis of *Petriellidium boydii* (Shear) Malloch strains isolated from feedlot manure and a human mycetoma. Can. J. Microbiol., *24*:856–886.

17. Bell, W. E., and M. C. Myers. 1978. *Allescheria (Petriellidium) boydii* brain abscess in a child with leukemia. Arch. Neurol., *35*:386–388.

18. Benham, R. W., and L. K. Georg. 1948. *Allescheria boydii*, causative agent in a case of meningitis. J. Invest. Dermatol., *10*:99–110.

19. Blank, F., and E. A. Stuart. 1955. *Monosporium apiospermum* (Sacc.) 1911 associated with otomycosis. Can. Med. Assoc. J., *72*:601.

20. Borelli, D., L. J. Bran, et al. 1979. Ketoconazole, an oral antifungal: laboratory and clinical assessment of imidazole drugs. *In* Cartwright, R. Y. (ed.). *Antifungal Therapy*. Postgrad. Med., *55*:657–661.

21. Bousley, P. H. 1977. Isolation of *Allescheria boydii* from pleural fluid. J. Clin. Microbiol., *5*:244.

22. Boyd, M. F., and E. D. Crutchfield. 1921. A contribution to the study of mycetoma in North America. Am. J. Trop. Med., *1*:215–289.

23. Carles, P., P. Recco, et al. 1979. Alleschériose pulmonaire. Poumon. Coeur, *35*:101–104.

24. Casero, L. 1962. Queratomicosis, su aumento en los ultimos anos y sutratamiento. Archiv. Soc. Oftal. Hisp.-Am., *22*:293.

25. Castellani, A. 1927. Fungi and fungous diseases. Arch. Dermatol., *16*:423.

26. Cazin, J., and D. W. Decker. 1964. Carbohydrate nutrition and sporulation of *Allescheria boydii*. J. Bacteriol., *88*:1624–1628.

27. Cazin, J., and D. W. Decker. 1965. Growth of *Allescheria boydii* in antibiotic containing media. J. Bacteriol., *90*:1308–1313.

28. Chupan, D. M., and J. Cazin. 1979. Humoral response to experimental petriellidiosis. Infect. Immun., *24*:843–850.

29. Costantin, M. J. 1893. *Eurotiopsis:* nouveau genre d'Ascomycetes. Bull. Soc. Bot. Fr., *40*:236.

30. Conti-Diaz, A. 1980. Micetomas y procesos premicetomatoses en el Uruguay. Mycopathologia, *72*:59–64.

31. Cooke, W. B., and P. W. Kahler. 1955. Isolation of potentially pathogenic fungi from polluted water and sewage. Public Health Rep., *70*:689–694.

32. Corbel, M. J., and S. M. Eades. 1982. T lymphocyte-dependent immune processes in resistance to experimental candida and cryptococcus infections. Mycopathologia, in press.

33. Creitz, S., and H. W. Harris, 1955. Isolation of *Allescheria boydii* from sputum. Am. Rev. Tuberc., *71*:126–130.

34. Dabrowa, N., S. W. Landau, and V. D. Newcomer. 1964. A survey of tide-washed coastal areas of Southern California for fungi potentially pathogenic to man. Mycopathologia, *24*:137–141.

35. Deloach, E. D., R. J. DiBenedetto, et al. 1979. Pulmonary infection with *Petriellidium boydii*. South. Med. J., *72*:479–481.

36. Dowding, E. S. 1935. *Monosporium apiospermum*, a fungus causing Madura foot in Canada. Can. Med. Assoc. J., *33*:128–132.

37. Drouhet, E., 1955. The status of fungus diseases in France. *In* Sternberg, T. H., and V. D. Newcomber (eds.): *Therapy of Fungus Diseases: An International Symposium*. Boston, Little, Brown & Co., pp. 43–53.

38. Edwards, C. C., and T. A. Hill. 1977. Natural history of fungal arthritis. Surg. Forum, *28*:497–499.

39. El-Allawy, T., M. Atia, and A. Amer. 1977. Mycoflora of the pharyngeo-tonsillar portion of clinically healthy donkeys in Assiut. Vet. Med. J., *4*:63–69.

40. El-Ani, A. S. 1974. *Allescheria boydii:* wild type and a variant from human pulmonary allescheriasis. Mycologia, *66*:661–667.

41. Elliot, D., C. Halde, and J. Shapiro. 1977. Keratitis and endophthalmitis caused by *Petriellidium boydii*. Am. J. Ophthalmol., *83*:16–18.

42. Emmons, C. W. 1944. *Allescheria boydii* and *Monosporium apiospermum*. Mycologia, *36*:188–193.

43. Ernest, J. T., and J. W. Rippon. 1966. Keratitis due to *Allescheria boydii*. Am. J. Ophthalmol., *62*:1202–1204.

44. Forno, L. S., and M. E. Billingham. 1962. *Allescheria boydii* infection of the brain. J. Pathol., *106*:195–198.

45. Glassman, M. I., P. Henkind, and E. Alture-Werber. 1973. *Monosporium apiospermum* endophthalmitis. Am. J. Ophthalmol., *76*:821–824.

46. Gluckman, S. J., K. Ries, and E. Abrutyn. 1977. *Allescheria (Petriellidium) boydii* sinusitis in a compromised host. J. Clin. Microbiol., *5*:481–484.

47. Gordon, G., and J. Axelrod. 1981. Combined *Petriellidium boydii* and *Clostridium limosum* endocarditis: a case report. (In press.)

48. Gordon, M. A., W. W. Vallotton, and G. S. Groffead. 1958. Corneal allescheriasis. Arch. Ophthalmol., *62*:758–763.

49. Green, W. R., J. E. Bennett, and R. D. Goos. 1955. Ocular penetration of amphotericin B. Arch. Ophthalmol., *73*:769–773.

50. Green, W. O., and T. E. Adams. 1964. Mycetoma in the United States. Am. J. Clin. Pathol., *42*:75.

51. Greither, A., and Z. Itani. 1974. Mycètome provoqué par *Allescheria boydii*. Bull. Soc. Fr. Dermatol. Syph., *81*:263.

52. Hainer, J. W., J. H. Ostrow, and D. W. R. Mackenzie. 1974. Pulmonary monosporosis. Chest, *66*:601–603.

53. Hairstone, M. A., and A. G. DeVoe. 1966. Keratomycosis: An ultrastructural study. Ophthalmologica, *152*:197–206.

54. Halperin, A. A., D. A. Nagel, and D. J. Schurman. 1977. *Allescheria boydii* osteomyelitis following multiple steroid injections and surgery. Clin. Orthop., *126*:232–234.

55. Hayden, G., C. Lapp, and F. Loda. 1977. Arthritis caused by *Monosporium apiospermum* treated with intraarticular amphotericin B. Am. J. Dis. Child., *131*:927.

56. Hecht, R., and J. Z. Montgomerie. 1978. Maxillary sinus infection with *Allescheria boydii (Petriellidium boydii)*. Johns Hopkins Med. J., *142*:107–109.

57. Hughes, S. J. 1958. Revisiones hyphomycetum aliquot cum appendice de nominilus rejiciendisi. Can. J. Bot., *36*:727–836.

58. Johnson, G. R., B. Schiefer, and J. F. Pantekoek. 1975. Maduromycosis in a horse in western Canada. Can. Vet. J., *16*:341–344.

59. Jones, J. W., and H. S. Alden, 1931. Maduromycotic mycetoma (Madura foot). Report of a case occurring in an American Negro. J. Am. Med. Assoc., *96*:256–260.

60. Jung, J. Y., R. Salas, et al., 1977. The role of surgery in the management of pulmonary monosporiosis. A collective review. J. Thorac. Cardiovasc. Surg., *73*:139–144.

61. Kaplan, J., and C. J. Johns. 1972. Mycetomas in pulmonary sarcoidosis: nonsurgical management. Johns Hopkins Med. J., *145*:157–161.

62. Kird, P. W. 1967. A comparison of saline tolerance and sporulation in marine and clinical isolated strains of *Allescheria boydii* Shear. Mycopathologia, *33*:65–75.

63. Knudtson, W. U., K. Wohlgemuth, C. A. Kirkbride, et al., 1974. Mycologic, serologic and histologic findings in bovine abortion associated with *Allescheria boydii*. Sabouraudia, *12*:81–86.

64. Knudtson, W. U., C. A. Kirkbride, et al. 1975. Bovine mycotic abortion: serology as an aid in diagnosis. Proc. Am. Assoc. Vet. Lab. Diag., *17*:299–303.

65. Kurup, P. V., and J. A. Schmitt. 1970. Human pathogenic fungi in the soils of Central Ohio. Ohio J. Sci., *70*:291–295.

66. Lacaz, C. da S., M. de S. Melhem, and L. C. Cuce. 1977. Maduramicose podal por *Petriellidium boydii* registro de um caso. Rev. do Hosp. das Clinicas Facultade de Medicina de Universidade de São Paulo, *32*:244–247.

67. Larsh, H. W. 1977. Opportunistic fungi in chronic diseases other than cancer and related problems. *In* Iwata, K. (ed.). *Recent Advances in Medical and Veterinary Mycology*. Baltimore, University Park Press, pp. 221–229.

68. Levitt, J. M., and J. Goldstein. 1971. Keratomycosis due to *Allescheria boydii*. Am. J. Ophthalmol., *71*:1190–1191.

69. Lichtman, D. M., D. C. Hohnson, et al. 1978. Maduromycosis *(Allescheria boydii)* infection of the hand. A case report. J. Bone Joint Surg., *60*:546–548.

69a. Liesegang, T. J., and R. K. Forster. 1980. Spectrum of microbial keratitis in south Florida. Am. J. Ophthalmol. *90*:38–47.

70. Lingappa, B. T., and Y. Lingappa. 1962. Isolation of *Allescheria boydii* Shear, from heated Indian soil. Curr. Sci. (India), *31*:70–72.

71. Louria, D. B., P. H. Lieberman, H. S. Collins, et al., 1966. Pulmonary mycetoma due to *Allescheria boydii*. Arch. Intern. Med., *117*:748–751.

72. Lupan, D. J., and J. Czain. 1973. Pathogenicity of *Allescheria boydii* for mice. Infec. Immun., *8*:743–751.

73. Lupan, D. M., and J. Cazin. 1976. Serological diagnosis of petriellidiosis (allescheriosis). I. Isolation and characterization of soluble antigens from *Allescheria boydii* and *Monosporium apiospermum*. Mycopathologia, *58*:31–38.

74. Lutwick, L. I., J. N. Galgiani, et al., 1976. Visceral fungal infections due to *Petriellidium boydii (Allescheria boydii)*. *In vitro* drug sensitivity studies. Am. J. Med., *61*:632–640.

75. Lutwick, L. I., M. W. Rytel, et al. 1979. Deep infections from *Petriellidium boydii* treated with miconazole. J.A.M.A., *241*:272–273.

76. Lytton, D. G., and D. R. Hamilton, 1975. Mycetoma in PNG with special reference to a case of extradura mycetoma. Papua New Guinea Med. J., *18*:61–65.

77. Mader, J. T., R. S. Ream, and P. W. Heath. 1978. *Petriellidium boydii (Allescheria boydii)* sphenoidal sinusitis. J.A.M.A., *239*:2368–2369.

78. Mahgoub, E. S., L. Sheik, and I. G. Murray. 1973. *Mycetoma*. London, W. Heinemann Medical Books Limited, p. 10.

79. Malloch, D. 1970 New concepts in the Microascaceae illustrated by two new species. Mycologia, *62*:727–739.

80. Matsuzaki, O. 1968. (In Japanese). Ocular infection with a fungus from a stick. Jap. J. Med. Mycol., *9*:218.

81. Matsuzaki, O. 1969. (In Japanese). Ocular infection with a fungus from rice leaf. Jap. J. Med. Mycol., *10*:239.

82. May, L. K., R. A. Knight, and H. W. Harris. 1966. *Allescheria boydii* and *Aspergillus fumigatus* skin test antigens. J. Bacteriol., *91*:2155–2159.

83. McCarthy, D. S., J. L. Longbottom, et al. 1969. Pulmonary mycetoma due to *Allescheria boydii*. Am. Rev. Respir. Dis., *100*:213–216.

83a. McGinnis, M. R., A. A. Padhye, et al. 1982. *Pseudallescheria* Negroni et Fischer, 1943 and its later synonym *Petriellidium* Mallock, 1970. Mycotaxon, *14*:94–103.

84. Meadow, W., M. Tipple, and J. Rippon. 1981. Endophthalmitis caused by *Petriellidium boydii* First report of a pediatric case. Am. J. Dis. Child., *135*:378–380.

85. Meyer, E., and R. D. Herrold. 1961. *Allescheria boydii* isolated from a patient with chronic prostatitis. Am. J. Clin. Pathol., *35*:155–159.

86. Mirsa, S. P., G. Y. Shende, S. N. Yerwade Kar, et al. 1966. *Allescheria boydii* and *Emmonsia ciferrina* isolated from patients with chronic pulmonary infections. Hindustan Antibiot. Bull., *9*:99–103.

87. Mohr, J. A., and H. G. Muchmore. 1968. Susceptibility of *Allescheria boydii* to amphotericin B. Antimicrob. Agents Chemother., 429–430.

88. Mohr, J. A., and H. G. Muchmore. 1968. Maduromycosis due to *Allescheria boydii*. J.A.M.A., *204*:335–336.

89. Mohr, J. A., and H. G. Muchmore. 1969. Inhibition of *Allescheria boydii* by naturally occurring estrogen. Antimicrob. Agents Chemother., 423–424.

90. Montpellier, J., and P. Gowillon. 1921. Mycetome du pied (type pied de Madura) due a *l'Aleurisma apiospermum*. Séance, *14*:285–290.

90a. Negroni, P., H. Herrmann, and I. Fisher. 1943.

Artritis aguda purulenta producida por el Ascomycete *Pseudallescheria sheari* n.g. n.sp. Prensa Med. Argentina. *30*:2389–2399.

91. Nielsen, H. S. 1967. Effects of amphotericin B in vitro on perfect and imperfect strains of *Allescheria boydii.* Appl. Microbiol., *15*:86–91.

92. Ouchi, E., E. Katsushita, and R. Inaba. 1969. A case of keratomycosis. Jap. Rev. Clin. Ophthalmol., *63*:33–35.

93. Oury, M., P. Hocquet, and P. Simard. 1968. Allescheriase pulmonaire (Mycetoma *Allescheria boydii*). J. Fr. Med. Chir. Thorac., *22*:425–437.

94. Oyston, J. K. 1961. Madura foot, a study of twenty cases. J. Bone Joint Surg. (Brit.), *43B*:259–267.

95. Padhye, A. A., and M. J. Thirumalachar. 1968. Distribution of *Allescheria boydii* Shear in soil of Maharashtra State. Hindustan Antibiot. Bull., *10*:200–201.

96. Paulter, E. R., W. Roberts, and P. R. Beamer. 1955. Mycotic infection of the eye. *Monosporium apiospermum* associated with corneal ulcer. Arch. Ophthalmol., *53*:385–389.

97. Pepere, A. 1914. Sul fungo parassita di un micetoma a grani neri del piede (Carter) nostrano. Lo Sperimentale, *68*:531–608.

98. Persaud, V., and J. B. Holroyd. 1968. Mycetoma of the palpebral conjunctiva caused by *Allescheria boydii.* Br. J. Ophthalmol., *52*:857–859.

99. Peters, B., T. Faggett, and C. C. Sampson. 1976. Isolation of *Monosporium apiospermum* from cerebrospinal fluid. J. Nat. Med. Assoc., *68*:512–513.

100. Pewar, V. A., A. A. Padhye, and M. J. Thirumalachar. 1963. Isolation of *Monosporium apiospermum* from marine soil of Bombay. Hindustan Antibiot. Bull., *6*:50–53.

101. Radaeli, F. R. 1911. Micosi del piede da *Monosporium apiospermum.* Lo Spermimentale, *65*:74.

102. Reddy, P. C., C. S. Christianson, and D. F. Gorelick. 1969. Pulmonary monosporiosis: an uncommon pulmonary mycotic infection. Thorax, *24*:722–728.

103. Reid, M. M., I. Frock, et al. 1977. Successful treatment of a maduramycotic infection of the equine uterus with amphotericin B. Vet. Med. Small Anim. Clin., *72*:1194–1196.

104. Rippon, J. W., and J. W. Carmichael. 1976. Petriellidiosis (allescheriasis): Four unusual cases and review of literature. Mycopathologia, *58*:117–124.

105. Roberts, F. J., P. Allen, and T. K. Maybee. 1977. *Petriellidium boydii (Allescheria boydii)* endocarditis associated with porcine valve replacement. Can. Med. Assoc. J., *117*:1250.

106. Rosen, F., J. H. Deck, et al. 1965. *Allescheria boydii*: unique systemic dissemination to thyroid and brain. Can. Med. Assoc. J., *93*:1125–1127.

107. Rosen, P. H., T. Adelson, and E. Burleigh. 1969. Bronchiectasis complicated by the presence of *Monosporium apiospermum* and *Aspergillus fumigatus.* Am. J. Clin. Pathol., *52*:182–187.

108. Saccardo, P. A. 1911. In Notae Mycologicae, Series XIII. Ann. Mycol., *9*:254–255.

109. Saccardo, P. A., and P. Sydow. 1899. Sylloge Fungorum, *14*:464.

110. Selby, E. 1972. Pachymeningitis secondary to *Allescheria boydii.* J. Neurosurg., *36*:225–227.

111. Scharyj, M., N. Levene, and H. Gordon. 1960. Pulmonary infection with *Monosporium apiospermum.* J. Infect. Dis., *106*:141–148.

112. Shear, C. L. 1922. Life history of an undescribed ascomycete isolated from granular mycetoma of man. Mycologia, *14*:239–243.

113. Siebenmann, F. 1899. *Die Schimmelmykosen des Menschlichen Ohnes.* Wiesbaden, J. F. Bergmann, p. 95.

114. Stevens, D. A. 1977. Miconazole in the treatment of fungal infections. Am. Rev. Respir. Dis., *116*:801–806.

115. Stevens, J. 1964. Personal communication.

116. Stoeckel, H., and C. Ermer. 1960. Ein fall von Monosporium mycetom der lunge. Beitr. Klin. Tuberk., *122*:30–38.

117. Tarozzi, G. 1909. Ric. anatomopat. bact. e sperim sopra un caso di actinomicosi del piede. Arch. perle. Sc. Med. *33*: n. 25.

118. Thirumalachar, M. J., and G. Y. Shende. 1973. Hamycin in pulmonary mycosis-complicated tuberculosis. Hindustan Antibiot. Bull., *15*:141–144.

119. Thompson, K. C., M. E. Menna, et al. 1978. Mycotic mastitis in two cows. N. Z. Vet. J., *26*:176–177.

120. Tong, J. L., E. H. Valentine, J. R. Durand, Jr., et al. 1958. Pulmonary infection with *Allescheria boydii*: report of a fatal case. Am. Rev. Tuberc., *78*:604–609.

121. Travis, R. E., E. W. Urich, and S. Phillips. 1961. Pulmonary allescheriasis. Ann. Intern. Med., *54*:141–152.

122. Upadhayay, M. P.: Personal communication quoted in Zapater and Albesil, 1979. Ophthalmologia, *178*:142–147.

123. Wain, W., M. Ahmed et al. 1979. The role of chemotherapy in the management of fungal carditis following homograph valve replacement. *In* Cartwright, R. Y. (ed.). *Antifungal Therapy.* Postgrad. Med., *55*:629–631.

124. Walker, D. H., T. Adamec, and M. Krigman. 1978. Disseminated petriellidiosis (allescheriosis) Arch. Pathol. Lab. Med., *102*:158–160.

125. Wintarwe, A., W. G. Winter, and N. L. Goodman. 1975. Maduromycosis (Madura foot) in Kentucky. South. Med. J., *68*:1570–1575.

126. Winston, D. J., M. C. Jordan, and J. Rhodes. 1977. *Allescheria boydii* infections in the immunosuppressed host. Am. J. Med., *63*:830–835.

127. Wolf, A., R. Benham, and L. Mount. 1948. Maduromycotic meningitis. J. Neuropath. Exp. Neurol., *7*:113.

128. Zapater, R. C., and E. J. Albesi. 1979. Corneal Monosporiosis. Opthalmologica (Basel), *178*:142–147.

129. Zaffiro, A. 1938. Forma sigolare di mycosi cutanea de *Monosporium apiospermum* a sviluppo clinicamente setticemico. Gior. Med. Mil., *86*:636–640.

25 Mucormycosis

DEFINITION

Mucormycosis is generally an acute and rapidly developing, less commonly chronic, infection of a debilitated patient caused by various species of the order Mucorales. Depending on the portal of entry, the disease involves the rhino-facial-cranial area, lungs, gastrointestinal tract, skin, or less commonly other organ systems. The infecting fungi have a predilection for invading vessels of the arterial system, causing embolization and subsequent necrosis of surrounding tissue. A suppurative, pyogenic reaction is elicited; granuloma formation is not frequently encountered. The infection is most commonly associated with the acidotic diabetic, in whom the disease runs a rapid, fatal course, usually within ten days, or with malnourished children, severely burned patients, or those with leukemia, lymphomas, and other debilitating diseases. Infection also occurs as a sequela to immunosuppressive therapy, use of cytotoxins and corticosteroids, or any procedure such as surgery or trauma that results in a "barrier break." Nosocomial infections have resulted from injections, intravenous therapy, hemodialysis, and contaminated elastic bandages. Several species of the genera *Rhizopus*, *Rhizomucor*, *Mucor* and *Absidia* are involved, but the clinical syndrome and pathologic findings are similar regardless of the specific etiology. The organisms are members of the order Mucorales of the Zygomycota, quite distinct taxonomically and in the type of disease evoked from other pathogenic zygomycetous fungi. For this reason the term "mucormycosis" is preferred to others that have been suggested to describe this infection. The disease is also encountered with some frequency in domestic animals and less commonly in wild animals. In addition to the aforementioned organisms, other genera, such as *Mortierella* and *Hyphomyces*, have been isolated from such infections. The latter is now tentatively identified as a *Pythium* sp. of the Oomycetes and is discussed in Chapter 26. In animals, mucormycosis has not been associated with any specific predisposing factors.

Synonymy. Phycomycosis, hyphomycosis, zygomycosis.

HISTORY

Even though the first reports of this disease are 100 to 150 years old, mucormycosis has been a rare finding in pathologic processes until recently. Most early records of infection are from animals, such as dogs, pigs, horses, and cows, with only rare reports of the disease occurring in man. The first published account of a zygomycetous disease was that of Mayer[98] in 1815. He described hyphae (Mucedo) in the lungs of a blue jay, but scrutiny of his drawings reveals that the fungi were, in fact, aspergilli. Furthermore, systemic avian mucormycosis has not been reported since. The earliest record of human involvement appears to be that described by Sluyter in 1847.[133] He described *Mucor* in a pulmonary cavity, but Virchow examined the material and considered it to also be *Aspergillus*.[141] Kurchenmeister in 1855 described a *Mucor* in a cancerous lung. His figures show a sporangium and nonseptate mycelium.[85] Thus his is probably the first authentic case of human mucormycosis. Fürbringer[56] in 1876 presented excellent data on the pulmonary disease–producing capacity of *Mucor* and described two cases of the infection. He saw nonseptate hyphae and sporangia in pathologic specimens, so that his report is probably valid,[9] although some authors have questioned it.[45] Paltauf in 1885[114] recorded the first disseminated case. The detail and accuracy of his description leave no question as to the acceptability of this as a case of mucormycosis. He called it an infection by *Mucor corymbifer*, although no cultures were taken. It appears to have started in the larynx and pharynx, so this probably was a case of rhinocerebral mucormycosis with dissemination. Experimental infections were produced by Lichtheim,[90] who described two species, *Mucor (Absidia) corymbifer* and *Mucor (Rhizopus) rhizopodiformis*. These species, when injected into rabbits, evoked widespread abscesses, and death occurred. In 1886 Lindt[91] described two more species, *Mucor pusillus* and *Mucor ramosus*, from human infections which were also pathogenic by animal inoculation. From this time on until well into the 20th century almost all cases of mucormycosis were pulmonary and associated with some type of cancer. In 1895 Herla[69] described a *Mucor* sp. in the pulmonary cavity of a woman who died of cancer of the liver. This is considered an authentic description. The cases described by Bostroem (1886) in tuberculous lungs and Obraszoe and Petrov (1890) in actinomycotic lungs probably represent contaminants. A review of all cases and a detailed description of the pathology caused by the organisms in the lung was published by Podack in 1899.[115] He described fungal filaments in alveoli, interstitial tissue, and the walls of veins. He named the fungus he isolated *Mucor corymbifera*. In 1901 Lucet and Costantin[92] described a woman with respiratory distress who coughed up strands of fungal mycelium. The organism was cultured and named *Rhizomucor parasiticus*. Arsenic and iodides were used in treatment, with apparent cure of the malady. Finally, in 1903 Barthelat[10] collected all the known reports of *Mucor* infections of animals and man along with descriptions of the pathogenic species then known. He produced accurate line drawings of nonseptate mycelium in tissue, sporangia, sporangiophores, and some species that produced rhizoids in culture preparations.

The most commonly seen form of the disease in the United States is the rhino-facial-cranial syndrome.[2, 107] Yet until 1943 this syndrome had only been recorded once, in 1885, by Paltauf.[114] A report of the disease

in 1929 by Christensen and Nielson is considered dubious, but the three cases recorded by Gregory et al.[63] in 1943 delineated the syndrome, its history, symptomatology, and development so accurately that this paper is considered a classic in descriptive medicine. About 400 case reports have been recorded since their paper,[2, 61, 87, 107] and this form of the disease is now regularly encountered.

Until recently there was a similar paucity of case reports describing mucormycosis of the digestive tract. In Paltauf's review of disseminated mucormycosis, a few patients had involvement of the gastric mucosa, and Gregory[63] found only five cases in the literature up to the year 1943. Only six case reports were accepted as valid up to 1960 by Neame and Rayner,[112] who tabulated 20 more. Gastrointestinal mucormycosis appears to be an infrequent form of the disease and is associated most frequently with malnutrition.[1] Infection or dissemination to other organs or body areas has been documented in a few scattered reports in the literature.[21, 50, 52, 65, 88, 94]

Pulmonary disease was a very rare medical curiosity following the first few case reports. After the review of 1903 published by Barthelat, pulmonary disease disappeared from the medical literature on mucormycosis and almost all reported cases of mucormycosis until 1960 were rhinocerebral.[8, 9] At present, pulmonary disease is becoming increasingly common as a complication of lymphoma and leukemia and occasionally diabetes.[110]

ETIOLOGY, ECOLOGY, AND DISTRIBUTION

The early case reports in the 1880's of Lindt, Paltauf, and others contained descriptions of a few species of the Mucorales as the etiologic agents of this disease. Since then many additional species and genera have been reported, but most of these have been placed in synonymy with the originally recorded species. The most frequently encountered agents of human mucormycosis are *Rhizopus arrhizus*, *R. oryzae*, and *Absidia corymbifera*, with some cases caused by *A. ramosa*, *R. rhizopodoformis* and *Rhizomucor pusillus*. It is very difficult to assess the variety of species that may evoke the infection, as cultural information is present in less than 10 per cent of all recorded cases of mucormycosis. Historically *A. corymbifera* under its various synonyms was the most frequently recorded agent in the early literature. However, in more recent reports essentially all human infections have been caused by the two species of *Rhizopus*. *A. corymbifera* is still commonly encountered in animal disease, however. *Saksenaea vasiformis* was recovered from a case of cerebral disease,[37] *Cunninghamella bertholletiae* has been isolated in about eight cases of human infection,[81, 86, 144] and *Mucor circinelloides* has been reported as having colonized (in its yeast form) human vagina, bladder, and intestine[25, 122] without evidence of disease production. Various other species of *Mucor, Rhizopus, Absidia, Mortierella, Cumminghamella,* and *Syncephalastrum*,[80, 82] among others, have been sporadically reported, but the authenticity of most of these is in doubt.[45] The first isolation of *Mucor ramosissimus* has recently been recorded from a chronic destructive lesion quite different from the usual forms of mucormycosis.[140] In this case the patient had no detectable underlying disease. In experimental infections, other species of the genus *Rhizopus* (all of which were thermotolerant) have been shown to have pathogenic potential,[119, 120] but so far have not been recorded from natural infection.

The spectrum of disease and etiologic agents of mucormycosis in animals is somewhat different from that associated with human infections. No set of predisposing factors has been established, and chronic granulomatous disease and cutaneous lesions are more commonly encountered. *Absidia corymbifera* is the most frequently recorded agent from systemic infections, and a species unique to animal disease, *Hyphomyces destruens*, is found in cutaneous and subcutaneous lesions. This is now considered to be a *Pythium* sp. and is discussed in Chapter 26.

The species of the Mucorales isolated from cases of mucormycosis are ubiquitous; a constant component of decaying organic debris; and thermotolerant.[124] The human infecting species grow rapidly on any carbohydrate substrate and usually produce numerous sporangia and sporangiospores which are airborne. This, together with their thermotolerance, rapid growth, and saccharolytic ability, may account for the predominance of the *Rhizopus* species causing human infections, especially rhinocerebral and pulmonary disease in diabetics. It has been postulated that some animal and human gastric infections probably are the result of ingestion of contaminated food, however.[112]

The distribution of the various clinical forms of mucormycosis is not based on age, sex, geography, or race, but on factors predisposing the patient to infection. Rhinocerebral disease is most often associated with acidosis of the uncontrolled diabetic;[87, 107] pulmonary infection is primarily a disease of patients with leukemia and lymphoma;[62, 68, 81] and gastric involvement is encountered most frequently in malnourished individuals, particularly children with kwashiorkor[1, 112] or in the immunosuppressed patient.[94, 134] In malnourished children it is surmised that the primary gastric and intestinal disease results from ingestion of moldy foodstuffs combined with lowered resistance. Focal or general tissue debility, such as burns or surgical procedures, also accounts for some infections.[52, 118, 138] Hepatic disease, renal failure, tuberculosis, corticosteroids,[132] cytotoxins, antibiotics, and antineoplastic chemotherapeutic agents have also been cited as predisposing conditions and agents, and some infections have followed injections[17, 59, 75] or use of intravenous lines or surgical bandages.[48, 65] The recent increase in mucormycosis is attributable in part to the use of new drugs that, although life-saving at the time, prolong the existence of the patient without curing the basic malady. This allows opportunistic infectious agents to proliferate. A comprehensive bibliography of mucormycosis was compiled and published by Ader and Dodd in 1979.[4]

CLINICAL DISEASE

Mucormycosis is the most acute and fulminant fungus infection known. Rhinocerebral disease in the acidotic patient terminates in death, often within seven days or less. The particular predilection of the organism for the patient with acidosis, whether due to diabetes,[2, 107] diarrhea,[38] uremia,[87] or aspirin intake, has been the subject of much investigation. Although significant, delayed cellular migration and diminished phagocytic function of the ketoacidotic patient alone cannot account for the rapid growth of the organism in tissue. Growth *in vitro* of *Rhizopus* is favored by the metabolic conditions encountered in the ketoacidotic diabetic. The organism has an optimum growth rate at an acid pH and a temperature of 39°C; it thrives in a medium of high glucose content and has an active ketoreductase.[116] Fortunately the disease is comparatively rare, even in such predisposed patients. This indicates the probable occurrence of other, as yet undefined, determinants in the host and in the prospective parasite.

Mucormycosis presents a spectrum of disease types dependent upon the type of previous debility of the patient and the portal of entry of the organism. The ensuing pathology following infection is similar, regardless of the anatomic site or species involved. The marked predilection of the organism for invading major vessels and the resultant emboli cause ischemia and necrosis of adjacent tissue. Landau and Newcomber,[87] Abramson et al.,[2] and others have classified the clinical forms of human mucormycosis according to the major anatomic areas involved. These are rhinocerebral, thoracic, abdominal-pelvic and gastric, and cutaneous mucormycosis.

Rhinocerebral Mucormycosis

This is the acute, rapidly progressive infection most commonly encountered in the North American and European literature. The patient is generally an uncontrolled diabetic who is acidotic. This form of mucormycosis is the exclusive domain of the genus *Rhizopus*. When cultures have been done, essentially all cases were caused by *R. oryzae* or

R. arrhizus. The age range of patients is from 16 days to 75 years, though about half are patients under 19 years old.[9] The infection begins in the upper turbinates or paranasal sinuses,[38, 49, 107] or less commonly in the palate[138] and pharynx, causing severe cellulitis. The acidotic patient may respond to treatment of the diabetes, but if an infection has been established, the course may be an insidious and relentless development of the mucormycosis. Sometimes, however, the disease may regress and remain quiescent when the diabetes is controlled, only to be reactivated by another onset of acidosis. The usual presenting symptoms involve first the nose and then the eye, brain, and possibly the meninges. The nasal area has a thick discharge that is dark and blood-tinged, and there are reddish black necrotic areas in the turbinates and septum. A scraping taken from the nasal area and examined as a direct KOH mount will often reveal hyphal strands and confirm the diagnosis (Fig. 25–10). The material should also be cultured. A cloudy sinus with a discernible fluid level is seen on x-ray (Fig. 25–1). Also present are paranasal sinusitus or pansinusitus, fistula, and sloughs of the hard palate (Fig. 25–2), loss of function of the fifth and seventh cranial nerves, and

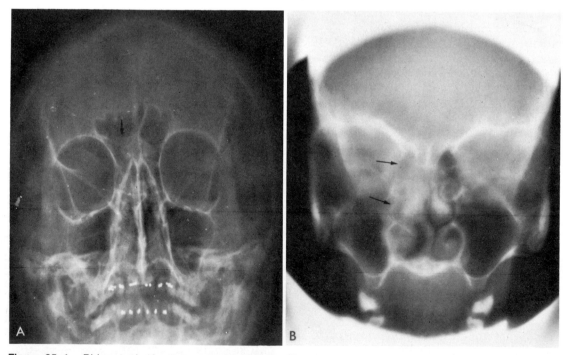

Figure 25–1. Rhinocerebral mucormycosis. *A,* Plane film sinus view. There is a well-defined air-fluid level in right ethmoid *(arrow). B,* Tomogram of same patient. There is total opacification of right ethmoid sinuses with demineralization of their lateral walls *(arrows)* and medial wall. Soft tissue of right nasal turbinates is enlarged, and the air passage is almost completely occluded. (Courtesy of J. Fennessy.)

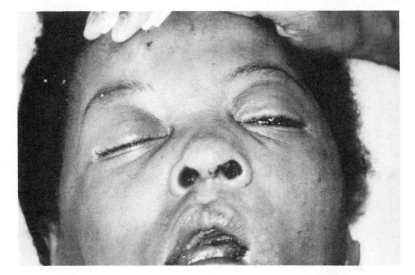

Figure 25–2. Rhinocerebral mucormycosis. The patient is an acidotic diabetic. There is a dark discharge in the nares, necrosis of the hard palate, a palpable swelling extending from the nasal bridge, facial nerve palsy and ptosis.

occasionally a hard, somewhat palpable, demarcated swelling on either side of the nasal bridge. Severe orbital cellulitis at this stage denotes a poor prognosis, as it heralds impending or actual invasion by the fungus of the orbit and central nervous system. The findings in the eye, when it is involved, include orbital pain, with ophthalmoplegia, ptosis, localized anesthesia, proptosis, limitation of movement of the bulbus oculi, fixation of the pupil, and loss of vision[51] (Fig. 25–3). Funduscopic examination may reveal normal findings, dilatation of the retinal veins, or thrombosis of retinal arteries, and hyphae may be seen coursing through the vitreous. Anesthesia of the areas supplied by the first and second branches of the trigeminal nerve occurs, along with homolateral facial nerve palsy due to involvement of the seventh cranial nerve (Fig. 25–4).

As the disease progresses, the fungus shows a predilection for invading large blood vessels. There is subsequent infarction with necrotic sequelae in the brain, which may be accompanied by softening of the frontal lobes. Lethargy progressing to coma ensues, and the patient succumbs in seven to ten days. Demise as early as two days and as late as two months following onset of the disease has been recorded. Additional symptoms are retro-orbital headaches and intracranial pressure, which may give rise to seizures and convulsions. Fever is not prominent. Laboratory findings include a white count of 10,000 to 30,000 cells per ml and up to 41,000, a high initial blood sugar, and a CO_2 of about 13 mEq per liter. Bone marrow necrosis with peripheral blood cytopenia[21] and potent hemolytic toxin[55] have been described in some cases of mucormycosis. Spi-

nal fluid findings are variable and nondiagnostic. Radiologic patterns are also variable, and application of computed tomography has been disappointing.[105] The infection can spread to the lungs and disseminate to other organ systems. The overall mortality rate was 80 to 90 per cent, but many survivors have been reported following treatment with amphotericin B,[42, 68, 104, 107] and at present the prognosis for recovery is quite good when appropriate medical and surgical management is applied.

There are rare records of chronic nasal mucormycosis in otherwise healthy patients.[140] These cases have consisted of tumors containing hyphae, surrounded by a granulomatous reaction. One such condition

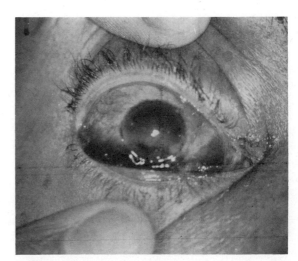

Figure 25–3. Rhinocerebral mucormycosis. The eye shows ophthalmoplegia, fixation of pupil, and haziness in the anterior chamber.

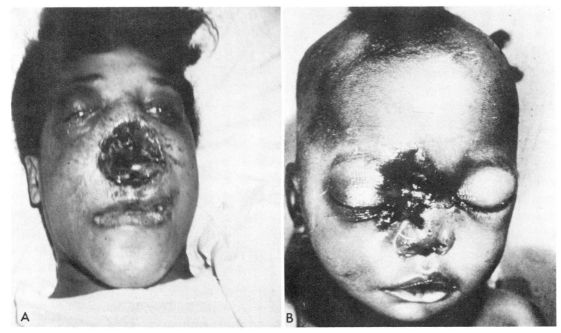

Figure 25–4. Rhinocerebral mucormycosis in acidotic diabetics. *A*, Adult with sloughing of nasal area. *B*, Child with sloughing of nasal area and frontal area of cranial bones. (Courtesy of L. Calkins.)

was successfully treated with amphotericin B.[15] There are also reports of a single granulomatous lesion of the brain that contained hyphae consistent with mucormycosis in an otherwise healthy individual. The lesion was excised owing to increased intracranial pressure, and the patient recovered.[108] Sometimes cerebral mucormycosis has occurred in the absence of clinically evident nasal involvement. In a report by Wilson et al.,[149] the patient had headaches and eye findings and was not diabetic. At autopsy, thrombosed vessels and necrosis were found to be limited to the brain and adjacent sinuses. Cranial infection has also occurred after trauma. Dean et al.[37] describe a 19-year-old male who sustained a head injury in an automobile accident. During the course of treatment, which included surgery, he developed neurological signs. *Saksenaea vasiformis* was recovered, and the patient responded to amphotericin B therapy. In addition to the nose, the ears may serve as the portal of entry for mucormycosis in the diabetic (Fig. 25–5).

Thoracic Mucormycosis

Pulmonary involvement may be primary, particularly in leukemia and lymphoma patients,[19, 66, 110, 111] following inhalation of

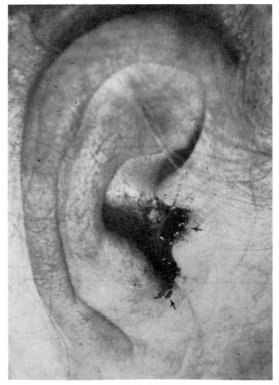

Figure 25–5. The ear as the portal of entry for rhinocerebral mucormycosis. A brown-flecked discharge is evident. Hyphal units were seen in direct mount of this material, and *Rhizopus arrhizus* was cultured. (Courtesy of L. Ajello.)

spores, or secondary to aspiration of infectious material from rhinofacial disease. Primary infection is rarely reported in the absence of some predisposing factors.[62] The presenting symptoms of pulmonary mucormycosis are those of a progressive nonspecific bronchitis and pneumonia, with superimposed signs of thrombosis and infarction.[8, 33] A severe and sudden onset may be accompanied by pain, a pleural friction rub, and bloody sputum. X-ray examination demonstrates signs of nonspecific pneumonia and infarction.[13] Massive cavitation can occur (Fig. 25–6). The disease is generally progressive to a fatal outcome in as few as three or as many as 30 days. Some patients, even with acute disease, have survived with amphotericin B therapy.[66, 99] It has been pointed out that pulmonary mucormycosis is more prevalent among patients with leukemia or lymphoma than among those with other types of cancer,[102] but primary pulmonary mucormycosis may occur in diabetics as well. Recently there were several cases recorded in organ transplant recipients,[18, 64] patients in intensive care,[5] and patients who had had surgery.[17] Cardiac involvement has been reported, and myocardial infarction due to direct invasion of coronary blood vessels has occurred,[24] as has massive fatal hemoptysis due to involvement of pulmonary arteries.[110] In a case reported by Connor et al.,[24] the patient presented with mediastinitis and cardiac friction rub.

Localized chronic pulmonary lesions have been recorded. Some have been successfully treated by lobectomy alone or with pre- and postoperative amphotericin B.[42, 135] Pulmonary lesions have also been incidental findings at autopsy when death was due to other causes.

On rare occasions a Mucorales species has been found as a fungus ball in ectatic bronchi or preformed pulmonary cavities. The radiologic picture is similar to that of an aspergilloma. One such case of fungoma was first described as being caused by a *Syncephalastrum* sp.[82] but was later attributed to *Aspergillus niger*. Authentic "mucoromas" have been encountered, however (Fig. 25–7).

Abdominal-Pelvic and Gastric Disease

Mucormycosis of the gastrointestinal tract as a primary disease is most often associated with undernourished patients, particularly children.[1, 36] In a review of twenty cases, Neame[112] found severe underlying disorders of the gastrointestinal tract to be predisposing factors. These included kwashiorkor,

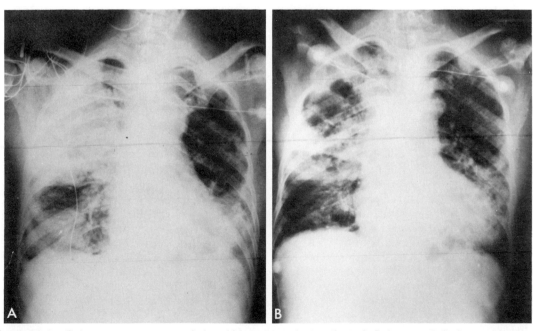

Figure 25–6. Pulmonary mucormycosis in a kidney transplant patient. *A,* Pulmonary infusion can be seen in both lung fields. It is particularly dense in the right lung, which shows consolidation. *B,* Two days later, massive cavitation of the right lung is evident. The patient died on the day this was taken.

Illustration continued on following page

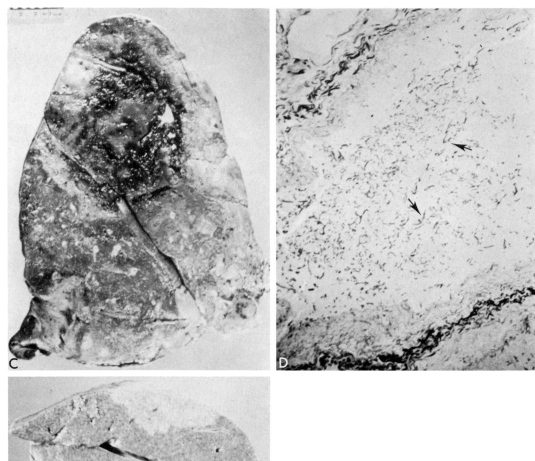

Figure 25–6 *Continued.* *C,* Gross section of lung showing infarctions, multiple cavities, and massive necrosis. *D,* Low-power view of central cavity showing masses of mucoraceous hyphae. Hematoxylin and eosin stain. ×100. *E,* Subcapsular infarction of the liver.

amebic colitis, typhoid, and in one case pellagra. All patients were undernourished or suffering from severe malnutrition with the associated physiologic imbalances. Primary gastric disease has been reported in diabetics[79] and leukemia patients,[134] and occasionally where no predisposing illness could be found.[94] In a report by Whiteway and Virata,[147] esophageal mucormycosis occurred subsequent to perforation caused by choking on a piece of ham.

The symptoms of abdominal mucormycosis vary considerably and depend on the site and extent of involvement. Nonspecific ab-

dominal pain, atypical peptic ulcer, pain, diarrhea, "coffee ground" hematemesis, and bloody stools are recorded. Ulceration of the gastric mucosa with thrombosis of associated vessels was most commonly observed at autopsy in one series.[112] Some patients have presented with symptoms of obstructive ileitis[134] or generalized abdominal distention.[94] No particular area is favored, as reports list involvement of the colon, stomach, esophagus,[147] ileum, and, by extension, gall bladder, liver, pancreas, and spleen.[9] Signs of peritonitis may become evident, as lesions commonly perforate the gastrointestinal wall. The

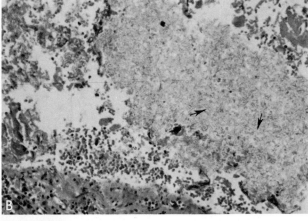

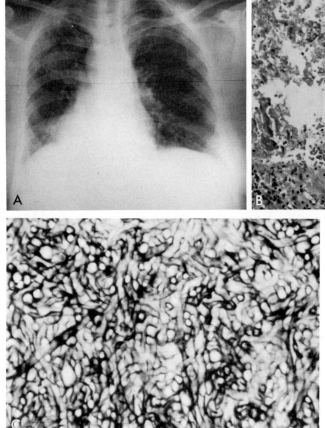

Figure 25–7. Pulmonary fungoma involving a Mucorales. *A,* Fungus ball on x-ray of chest. *B,* At surgical resection the fungal mass was seen to be lysing and in several pieces. Hematoxylin and eosin stain. ×100. *C,* GMS shows sparsely septate mycelium of the fungal mass. The patient was a diabetic and had had cavitary tuberculosis. The fungoma had been present for many years, causing occasional bouts of hemoptysis.

course of the disease is usually 70 days. The immediate cause of death is shock from hemorrhage of bowel, resulting in peritonitis and bowel infarction.[50]

Primary mucormycosis of other organs of the pelvic-abdominal area is quite rare. Unilateral renal involvement has been recorded. This patient presented with symptoms of renal infarction and was cured following nephrectomy and amphotericin B treatment.[117] Histologic examination showed focal necrosis of parenchyma and adipose tissue. Other cases of renal involvement have recently been reported.[21,88] In a report by Gartenberg et al.,[59] mucormycosis occurred subsequent to renal biopsy. Osteomyelitis involving the femur in an 18-year-old female leukemic was reported by Echols et al.[41] The patient was successfully treated with amphotericin B. The authors also reviewed the literature on bone involvement in mucormycosis. Diabetics may also be found to have isolated bony lesions of mucormycosis. In one personally observed case, disease developed in a heel following a trauma caused by a supermarket grocery cart. Repeated surgical and medical

procedures have failed to contain the disease. Infection limited solely to the bladder wall, endometrium of the uterus, arteriosclerotic thrombus of the abdominal aorta,[50] and saphenous vein graft[139] have also been recorded.

Cutaneous Lesions

Involvement of skin, particularly otitis externa, was frequently reported in the older literature, but more rigid criteria for establishing this diagnosis have invalidated most of these reports. Species of the Mucorales are quite frequently isolated from normal as well as diseased skin, ear canals, and so forth. That the ears may serve as a portal of entry for mucormycosis in the diabetic has also been documented (Fig. 25–5). Transient colonization of injured areas is not infrequent, and colonization with rapidly fatal disease is a particular danger to burn patients.[52] Such patients may be protected from bacterial disease by the use of antibiotics, only to succumb to fulminant mucormycosis.

Cutaneous and subcutaneous mucormycosis can be a primary manifestation of disease or the result of dissemination from some other sources.[121, 150] When the latter occurs it is usually subsequent to a pulmonary process. The lesions of secondary cutaneous mucormycosis begin as progressively enlarging, painful, nodular ecchymotic areas of cutaneous infarction. They may measure several centimeters across and have a pale periphery surrounded by a thin reddish ring. Necrosis, eschar formation, central ulceration, and sloughing occur. Most such patients are leukemics and in many the etiologic agent is *Rhizomucor pusillus*.[83] Primary cutaneous and subcutaneous disease is often the result of "barrier breaks." These include a variety of invasive procedures.[17, 59, 75] A spate of such cases was associated with contaminated bandages and surgical dressings.[65] The organism in all cases was *Rhizopus rhizopodoformis*. The lesions varied considerably in morphology but included plaques, pustules, ulcerations, deep abscesses, and ragged necrotic patches. Most healed with little treatment (debridement and amphotericin B), and they were not associated with dissemination.

A few cases of cutaneous mucormycosis based on histologic evidence alone are recorded. A lesion on a chest wall of a 6-year-old boy revealed nonseptate hyphae. The lesion consisted of a reticulated, atrophic, central area with a slowly progressive border of lichenified papules.[46] In another case, indolent chronic ulcers yielded a *Mortierella* species on culture.[45] Kamalan[80] described a diabetic with lesions of the thumb and a mycotic keratitis. *Syncephalastrum* was isolated. Granulomatous and abscessed areas contained nonseptate hyphae. In both cases the disease appeared to be similar to *Hyphomyces* infection of horses. Cutaneous lesions over the malleolus of a diabetic have been reported,[87] and another cutaneous lesion is reported to have developed following excision of a carbuncle.[9] Mycetoma-like lesions have also been reported from which *Rhizopus* and *Mucor* have been isolated.[46, 77] Disease of the oral mucosa is also known.[138] Cutaneous disease is most often a manifestation of disseminated infection as has been documented many times.[101, 150]

Colonization by Mucor Spp.

Under particular environmental conditions, members of the genus *Mucor* exhibit morphogenesis to a variety of cellular forms: branching mycelium, budding yeasts, and spherule-like chlamydoconidia. This morphologic transformation was first noted by Pasteur, who deduced that it involved semianaerobic conditions, the carbon dioxide content of the media, and an aqueous environment. The process has been studied in detail by Bartnicki-Garcia,[12] who published a simple method to demonstrate such morphogenesis as a laboratory exercise.[11] The first natural occurrence of the morphologically altered growth form associated with an animal species was published by Frank et al.[54] They noted budding spherules in the internal organs of dead frogs that had been housed with a moribund toad (*Bufo bufo*). The authors considered the fungus as the cause of the infection and could reproduce it in other amphibians but not in mice or guinea pigs. The isolant was identified as *Mucor circinelloides*. The first observation of this in material of human origin was made by Morris Gordon (personal communication). He noted spherules and yeasts in intestinal contents (Fig. 25–15,*A*). It was not associated with specific illness and was considered an incidental finding. An isolation was made and identified as a *Mucor* sp. A second observation was made by C. T. Dolan.[122, 123] In this case spherules, budding yeastlike cells, grouped chlamydoconidia, and short hyphal units were seen on a routine Papanicolau preparation from a vagina. The woman had no symptoms or distress and later gave birth to a normal healthy baby. The isolant was identified as a *Mucor* sp. (Fig. 25–15,*B,C,D*). The same morphologic forms were noted by Copper[25] in a routine examination of a urine specimen (Fig. 25–15,*E,F*). Again the fungus could not be associated with a disease process and was found only once but in large numbers in that specimen. The isolant was identified as *Mucor circinelloides*.

The normal human fungal flora found on the integument and within the alimentary tract and genitourinary tracts are almost exclusively yeasts. Few hyphal fungi appear to colonize these areas successfully, and their isolation represents an accidental or transient occurrence. Formation of yeasts and spherules by the *Mucor* spp. may permit a transient colonization. As yet no disease process has been defined, and such colonizations may be considered a commensal association. The work of Bartnicki-Garcia[12] indicates that this transformation ability is found only in the genus *Mucor* among the Mucorales.

DIFFERENTIAL DIAGNOSIS

In a diabetic patient with acute and rapidly spreading sinusitis, cellulitis of orbital tissues, bronchitis, or bronchial or lobar pneumonia, a diagnosis of mucormycosis should be seriously entertained. Demonstration of the fungus by direct examination and isolation in culture should be diligently sought. Because of the fulminant course of the disease, early diagnosis is imperative for administration of appropriate therapy. Other opportunistic mycoses may simulate the condition, but usually are not so rapidly overwhelming. Fulminating bacterial or viral infections also must be considered in the differential diagnosis.

In tissue the hyphal strands seen in mucormycosis resemble those present in entomophthoromycosis morphologically, but the pronounced eosinophilia and deposition of Splendore-Hoeppli material around the organism found in the latter disease are usually missing. *Mortierella* and *Pythium* (see Chapter 26), however, incite a chronic eosinophilic reaction similar to that seen in infections caused by *Basidiobolus* and *Conidiobolus* (see Chapter 11).

PROGNOSIS AND THERAPY

The overall mortality of rhinocerebral mucormycosis in the diabetic formerly was 80 to 90 per cent. The prime consideration is control of the diabetes, but even when antifungal therapy is instituted, the prognosis is grave. In recent series in which the disease is diagnosed ante mortem and adequate therapeutic[42] measures instituted, the survival rate is 50 per cent or better.[2, 14, 61] Based upon the experimental results reported by Chick et al.[22, 23] concerning the efficacy of amphotericin B in experimental mucormycosis in rabbits, the drug has been used in the successful treatment of the disease. Battock et al.[14] reported survival following institution of the following measures: alternate-day IV amphotericin B administration at a rate of 1.2 mg per kg, control of diabetes, and local surgical debridement of involved nasal tissue. Mean 48-hour serum levels of the drug were 0.34 to 0.45 μg per ml. This does not exceed the maximal tolerated levels in humans found by other investigators (0.5 to 1.5 μg per ml). The *in vitro* sensitivity of the isolated strains varies from 0.03 μg to more than 1000 μg per ml. In one of Battock's cases,[14] the minimal inhibitory concentration was 1000 μg per ml. This emphasizes that the drug may be of some efficacy *in vivo* in spite of lack of action *in vitro*. This discrepancy may be partly due to the testing procedure. Medoff and Kobayashi suggest use of germinated spores in broth with increments of the drug as being more physiologically relevant.[99] Abramson et al.[2] suggest a cumulative dose of 3 grams. The more recent experience, as published by Hauch[66] and Meyers et al.,[104] indicates that amphotericin B is effective and the drug of choice. It has been used in combination with a variety of other drugs, including 5-fluorocytosine, clotrimazole and miconazole.[107, 150] Winn et al.,[150] using the Saubolle[126] susceptibility testing technique, found two isolants of *Rhizopus arrhizus* to be sensitive to amphotericin B (0.05 and 0.025 μg per ml), resistant to 5-fluorocytosine, and moderately sensitive to miconazole (0.62 and 5.0 μg per ml). The isolants tested by Stevens[136] were completely resistant to miconazole. thus it appears that amphotericin B alone is the preferred treatment for mucormycosis at this time. No good evidence exists for the efficacy of iodides or desensitization. Both regimens have now been abandoned.

Surgical management varies with the extent and location of lesions. In rhinocerebral disease surgical debridement of necrosed tissue is mandatory, and often enucleation of the eye is required (see Chapter 27).

Most cases of gastric and pelvic disease are diagnosed at autopsy, so that there is not sufficient accumulated experience to comment on efficacy of treatment.[36] In renal disease and in the case of saphenous graft infection, amphotericin B has been used with clinical cure of the disease.[117, 139] Though most cases of pulmonary mucormycosis have been fatal or required much surgical manipulation, a few patients have survived. Medoff and Kobayashi report on a leukemic patient with acute pulmonary mucormycosis successfully treated with amphotericin B.[53] Alternate-day therapy and surveillance of blood levels of the drug and sensitivity of the organism to the drug were employed.

PATHOLOGY

Gross Pathology. Infarction and necrosis of tissue are the main findings in autopsy specimens from mucormycosis.[9] The findings are similar in all anatomic sites: brain, lung, stomach, intestine, and liver (Fig. 25–6). Gangrenous degeneration is a common appellation of such involvement. The infected areas of brain show softening and sloughing of adjacent tissue with punctate hemorrhage. There is usually gross consolidation in the lungs when involved. Cut surfaces show massive hemorrhage with recent emboli and gray "puttylike" material in old infected areas. In gastric and intestinal involvement, engorged vessels and ulcers up to 3 to 4 cm that have black necrotic central areas are seen. Perforations and subsequent evidence of peritonitis are common. In old established chronic lesions, some granulomatous changes may be found.

Histopathology. The tissue reaction to infection by species of the Mucorales is variable. Sections with marked invasion of hyphal elements sometimes show little or no cellular response.[9, 89, 137] More commonly, there are varying degrees of edema, necrosis, and accumulations of neutrophils, plasma cells, and sometimes giant cells. Eosinophils are not as frequently seen as in entomophthoromycosis. Often the tissue reaction is suppurative, but it may show some granulomatous changes.[89]

The most characteristic feature of mucormycosis is involvement of blood vessels. The fungi show a marked predilection to invade directly the walls of large and small arteries (Fig. 25–8), and less frequently veins. This invasion is followed by thromboses, which in turn cause infarction and necrosis of adjacent tissue. This sequence of events is the major contributing factor to symptomatology of the patient and the gross and histopathologic findings. In most series of rhinocerebral disease, thrombosis of the internal carotid artery was present in 33 per cent of patients, whereas cavernous sinus thrombosis was infrequent.[9, 87] The fungus also invades bones, nerves, and fatty tissues; muscles are usually spared. In one personally observed case, all major cranial vessels were affected, and the

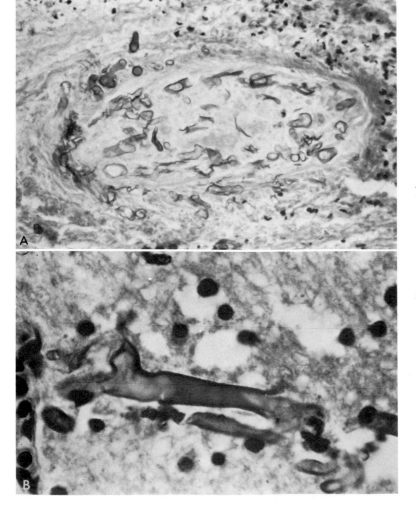

Figure 25–8. Mucormycosis. *A*, Invasion of arterial wall and thrombus formation which resulted in necrosis of adjacent tissue. Note the broad, irregular, nonseptate hyphae. Hematoxylin and eosin stain. ×150. *B*, Broad, irregular, nonseptate hyphae in necrotic debris. ×800. (From Rippon, J. W. *In* Burrows, W. 1979. *Textbook of Microbiology.* 20th ed. Philadelphia, W. B. Saunders Company, p. 777.)

sella turcica and pituitary were obliterated. The fungus invaded new areas, traveling through the vessels and nerve trunks.

In contrast to most fungi, the etiologic agents of mucormycosis are readily seen in hematoxylin and eosin preparations of tissue. The special fungus stains, such as PAS and Gridley, do not demonstrate the organisms well, although the Grocott-Gomori methenamine silver stain is usually quite adequate (Fig. 25–9). The hyphae are typically broad (10 to 20 μ), sparsely septate, and haphazardly branched. Widths varying from 6 to 50 μ have been recorded. The mycelium is sometimes quite distorted, and the walls vary considerably in thickness. Cross sections of hyphae have sometimes been mistaken for nematodes or empty spherules of *Coccidioides immitis*. Thick-walled hyphal strands may also resemble sclerosed capillaries or arterioles. Mucormycosis is easily differentiated from *Aspergillus* infection, in which the thinner width, regular septation, and brushlike or fingerlike branching are characteristic. True septa are rarely seen in mucormycosis. When they occur, they are complete. Hyphal folds may resemble a septum, but examination under oil immersion reveals them to be patent. The eosinophilic halo (Splendore-Hoeppli phenomenon) is usually absent or rare in human infection (Fig. 25–9), although "asteroid" or "club" formation has been reported in animal disease.[34]

ANIMAL DISEASE

Natural infection with the Mucorales is widespread among wild animals and is a significant veterinary problem in domesticated livestock.[3, 20, 142] Among wild animals, disease has been reported in monkeys,[72, 97] a

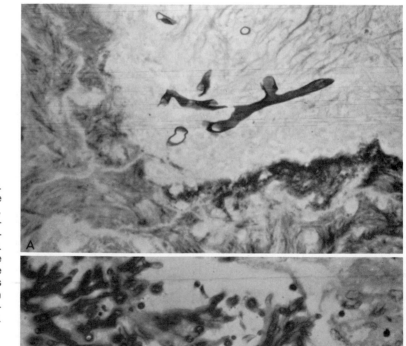

Figure 25–9. Mucormycosis. *A*, Irregular, broad, nonseptate hyphae in GMS stain. ×400. *B*, Unusual deposition of eosinophilic halo around hyphae in lesion from chronic lung disease. The hyphae are thinner, and there are more septations than are usually found. The organism was not cultured. This may have been entomophthoromycosis. Hematoxylin and eosin stain. ×400. (Courtesy of S. McMillen.)

whale,[16] a rattlesnake,[148] voles,[31] mice,[6] minks, okapi,[73] and others.[6] Both gastric[72] and rhinofacial types of the disease are found in monkeys. In the rhinofacial form,[97] areas involved included the paranasal sinuses, orbit, facial skin, and lymph nodes; in addition, osteomyelitis, cellulitis, vasculitis, lymphadenitis, and sialadenitis were also observed in this diabetic animal. An unidentified *Mucor* was isolated. The okapi infection was in a neonatal animal and involved the gastrointestinal tract. As is usual in animal infection, *Absidia corymbifera* was isolated and was thought to have been the result of ingestion of moldy grain.

Mucormycosis of domestic animals is chiefly known as a granulomatous disease involving lymph nodes. The disease falls into two categories. The commoner form consists of an acute or chronic involvement of mesenteric, bronchial, mediastinal, or submaxillary lymph nodes and sometimes liver, lung, and kidney. The granulomata are usually not discovered until the animal wastes away, or in routine inspection of abattoirs.[34, 142] Cattle and pigs are most frequently affected.[6, 39, 95, 106, 129] Fatal infection in adult animals has occurred following abortion[39] and in neonatal calves both as terminal systemic infection and incidental to slaughter.[32, 73a] Almost all internal organs are involved. Culture is usually not obtained, but *A. corymbifera* is the most commonly isolated species when fungal isolation has been done. In the two cattle and one pig described by Watanabe et al.,[142] the organism was *Rhizopus rhizopodoformis*. In one series, *Mucor circinelloides,* a rarely encountered species, was repeatedly cultured from diseased cattle[106] and from moribund and dead toads and frogs.[54] In the latter instance, the organisms were in the form of budding yeasts and spherules. Parrots fed millet infected with the fungus died of "toxic manifestations of the brain"; however, the organism did not elicit a disease when experimental infection of animals or birds was attempted. *Mortierella* spp. have also caused disease in cattle.[100]

The second common form of animal disease is that of ulceration of the stomach and intestinal tract. Any age may be affected, but most rapid fatal infections occur in young animals. Scouring of calves is the first symptom, and the disease progresses rapidly. This form accounts for practically all cases of abdominal ulcerations. The early lesions are small, raised, inflamed foci which become ulcers with raised hemorrhagic margins and gray, depressed, central necrotic areas.[6, 34] *Absidia corymbifera*, *Rhizopus microsporus*, and various other Mucorales have been reported. Ingestion of moldy grain is the probable cause. Bovine mycotic abortion, though usually caused by *Aspergillus* spp., can also be the result of invasion by members of the Mucorales.

Horses and dogs are infrequently affected.[3] Lesions involving the nose only, the nose and other areas, cutaneous lesions,[109] and fatal systemic infection have been recorded.[35, 47, 67, 93] In one case, a sinus tract formed at the site of a bite by another dog, and systemic involvement followed.[47] The histopathology is similar to that in human infection. *Rhizomucor pusillus* and other Mucorales have been identified.

Mucormycosis of birds is very rare. This may be due to their having a higher body temperature than mammals. In a report by Marjankova et al.,[96] *Mucor janssenii* was isolated from necrotic inflammatory lesions of the penis in ganders. It is similar to a disease process caused by *Candida* and reaches epidemic proportions in large flocks.

A special form of mucormycosis in horses is caused by *Hyphomyces destruens*. This disease is called "swamp cancer." Recently, *Hyphomyces* was found to be the tissue form of an Oomycete, *Pythium* spp. It is discussed in Chapter 26.

Experimental Disease. Mucormycosis has been the subject of experimental investigation for some time. Many of these studies have been concerned with the particular propensity of acidotic diabetics to acquire infection. One of the earliest records of experimental disease is that of Lichtheim, who produced fatal infections in rabbits with various species.[90] He also noted that the species that were pathogenic were able to grow at elevated temperatures. This work was the basis of Vuillemin's separation, in 1903, of the *Mucor* and *Absidia* species into a new genus, *Lichtheimia,* which was to contain all species demonstrated to be pathogenic. This artificial separation is no longer recognized.

Duffy[40] and others[119, 120] showed that alloxan diabetes in rabbits was similar to human diabetes and had an acute acidotic form, usually fatal, and a chronic form. Later a number of investigators demonstrated that intranasal or intravenous injection of various species of the Mucorales could produce cerebral or pulmonary mucormycosis. Elder and Baker[43] found that intrathecal injection of spores into rabbits with acute acidotic diabetes produced extensive or fatal infections, whereas normal rabbits or those with chronic diabetes did not succumb to disease. In normal rabbits, the spores did not germinate. These results have been duplicated in mice and rats similarly treated. Sheldon et al.[131] demonstrated that in experimental infection most mast cells in normal rats quickly degranulated, and a rapid inflammatory reaction occurred, with limitation and abortion of the disease process. Previous degranulation of mast cells by the histamine liberator 48/80 only slightly delayed the reaction, but in acute alloxan diabetic animals, degranulation was impaired. The inflammatory reaction was greatly delayed, and severe, sometimes fatal infection occurred. Gale and Welch[58] demonstrated that normal human serum markedly inhibited the growth of *Rhizopus oryzae,* whereas sera from an acidotic patient did not. Later, when the diabetic patient had recovered clinically, inhibition of growth by his sera was similar to that of normal controls. The serum factor inhibitory to the growth of *Mucor* is apparently independent from the anti-*Candida* and anti-dermatophyte complexes described in previous chapters.

Josefiak and Smith-Foushee[78] have shown that a limited disease could be produced in a Selye pouch (pneumoderm) in normal rats. Chick et al.,[22, 23] using this technique, demonstrated the efficacy of amphotericin B in alloxan diabetic rabbits. They noted that the drug inhibited hyphal formation when spores were introduced into the pneumoderm. This important finding led to the use of this drug in therapy of the disease.

BIOLOGICAL STUDIES

The extraordinary and extensive tissue damage, necrosis, and thrombosis (even in the presence of thrombocytopenia) that is seen in cases of mucormycosis has stimulated a search for toxins and thromboplastins.[84, 125] This is particularly relevant, since extensive marrow necrosis,[21] hemolysis, and other toxic effects have been described in human infections. The presence of thromboplastic substances or a substance of the host activated by the invading organism has been hypothesized, but as yet there is no laboratory evidence. An endotoxin has been described, but little research has been done on it.[125] Potent endotoxins, such as those of *Aspergillus fumigatus,* have been found in some fungi, and it is possible a similar substance exists in the

pathogenic Mucorales. A hemolysin, which was shown to be a lipid, has been recovered from some pathogenic Mucorales.[84] The biological significance remains to be determined, as hemolysis is not commonly a prominent feature of the natural disease in man or animals, although it sometimes occurs. Corbel and Eades[26-30] have published a series of studies on various aspects of mucormycosis in various animal models. They have addressed the role of acquired immunity,[27] the importance of T-cell function,[28] and factors affecting localization of lesions in experimental disease. Acquired resistance appeared to be of minor significance in repeated sublethal infections. Athymic mice had a propensity equal to that of heterozygous normals toward containing an infective challenge. This indicated that thymus-dependent processes do not play an essential role in primary resistance. Work by Diamond (Chapter 14) has documented the importance of the primary phagocytes (neutrophils) in killing hyphae of *Rhizopus.* Corbel and Eades[29] found the highest concentration of spores following inoculation to be in liver and spleen, but active lesions were confined to brain, kidneys, and myocardium. In the wood vole (*Clethrionomys glariolus*), natural and experimental infections were also confined to the central nervous system and the kidney, with few other organ systems involved.[31]

SEROLOGY AND IMMUNOLOGY

The rarity of mucormycosis indicates a high natural resistance to the disease. This has been amply verified by laboratory investigation. Mucormycosis is an almost perfect example of opportunism — a host with a very specific set of predisposing factors and a parasite, essentially avirulent normally, which has the potential for growth in that particular tissue environment.

There are no serologic procedures of consequence for the diagnosis of mucormycosis. Patients with the disease, as well as many normal patients, will react to intradermal injections of culture extracts. Such preparations are of no practical value and are remnants of the time when desensitization and autologous vaccines were in vogue. The Mucorales are ubiquitous members of the environment, and their airborne spores are omnipresent. The spores apparently are potent allergens, as are the conidia of *Aspergillus* spp. Up to 60 per cent of patients with allergic bronchitis or asthma react positively to scratch tests or interdermal injections of culture extracts of various Mucorales. White et al.[146] devised a system for determining chitin antigenemia during experimental infection in mice. Further developments of a practical laboratory procedure are awaited.

LABORATORY IDENTIFICATION

Direct Examination. Since mucormycosis is the most acutely fulminant fungal disease of man, rapid diagnosis is extremely important if management and therapy are to be successful. Unfortunately fungal elements are usually not numerous in discharges, so that diagnosis may have to rest on clinical evidence alone. For the "classic" disease in the "classic" patient, this is sufficient. Since spores of the Mucorales are common contaminants of the environment and sputa, direct examination with demonstration of the organism is more meaningful for diagnosis than is culture. Fungal elements can be found in scrapings from the upper turbinates or in aspirated material from sinuses in rhinocerebral disease, and in sputum in pulmonary disease. Broad, sparsely septate, branching hyphae are seen in well-prepared potassium hydroxide mounts. They are thick-walled and refractile, with an average diameter of 10 to 15 μ (Fig. 25–10). The size may range from 3 to 30μ. Swollen cells (up to 50μ) and distorted hyphae are sometimes seen.[89, 137] Biopsy material, whenever practical, is usually an excellent source for observing and culturing the organism.

Culture Methods. The Mucorales that

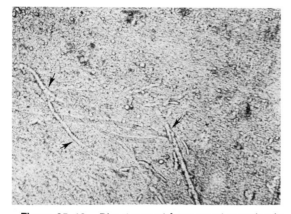

Figure 25–10. Direct mount from nasal scraping in a case of rhinocerebral mucormycosis. Distorted, branched, sparsely septate hyphae are seen in this KOH preparation. ×400.

are involved in human or animal disease all grow on standard laboratory media without cycloheximide. The growth is rapid and usually noticeable 12 to 18 hours after planting of the specimen. Establishing a diagnosis on cultural evidence alone is difficult. The pathogenic species of the Mucorales are constant inhabitants of the environment, contaminants of skin, discharges, and sputa, and grow on almost all moist organic substrates. Bronchiectatic patients as well as normal individuals may cough up spores of these organisms for days following exposure to a spore-ridden environment. Since the organisms overgrow the Petri dish so rapidly, colony counts that help in establishing the diagnosis of aspergillosis are difficult to carry out. An uncritical diagnosis of mucormycosis cannot be given on cultural evidence alone, and conversely the isolation of a Mucorales from a patient cannot be discarded as transient flora or a contaminant. All forms and sources of evidence must be marshaled and critically judged for an accurate diagnosis.

Discharges, scrapings, and biopsy material can be planted on malt agar, potato glucose agar, or Sabouraud's agar and incubated at 37° C or 25° C. Almost all the pathogenic Mucorales are easily isolated from such material. The media may contain antibacterial antibiotics, but most isolants are sensitive to cycloheximide. Organisms from animal disease and from gastric mucormycosis are often difficult to culture. Sometimes the pronounced saccharolytic abilities of the organisms are used to advantage by placing the specimen on a section of sterile bread which is on the surface of an agar plate.[112]

Once isolated, the specific identification of the organism is often difficult. There are many species of Mucorales in the environment, and many of these are thermotolerant. Even to an expert, species separation is a difficult task. Physiologic reactions as a basis for identification have not been established, although some have been reported.[128] In the heterothallic species, identification can be made by sexual crosses with known testor strains.

Sensitivity to amphotericin B varies considerably among isolants of the Mucorales.[136, 143, 150] For this reason, it is important to determine the sensitivity of the organism isolated to the drug. The usual procedure is to disperse known concentrations of amphotericin B into tubes and add known amounts of melted Sabouraud's agar (see Part Five). The tubes are then planted with some mycelial strands or spores of the isolant. Sensitivities of isolants may vary from 0.1 μg per ml to 1000 μg per ml. The drug appears to be fungistatic, as spores washed free of the drug from tubes showing inhibition are still viable. *In vivo* efficacy is not always able to be correlated with results obtained by *in vitro* testing. It has been observed that the drug is efficacious in treating some cases of mucormycosis in spite of the high concentration necessary for inhibition of growth determined by *in vitro* tests. Medoff and Kobayashi[99] devised a more "physiologic" method for testing. In their procedure, the spores are allowed to germinate in a liquid medium, and then amphotericin B is added. Growth is determined turbidometrically. Most laboratories now employ the antibiotic testing system of Saubolle[126] as modified by Winn et al.[150]

Figure 25–11. Characteristics of four of the genera of the Mucorales associated with human infection. *A, Rhizopus.* (1) Sporangiophores arise in groups directly above the rhizoids.(2) A columella extends into the sporangium. (3) Naked, round, crenated, or roughened zygospores formed by perfect species. *B, Absidia.* (1) Sporangiophores arise between nodes; rhizoids present. (2) Columella extends into sporangium. (3) Zygospores are enveloped by circinate hyphae. *C, Mucor.* (1) Branched sporangiophores arise randomly along aerial mycelium. There are no rhizoids. (2) Columella extends into sporangium. (3) Zygospores in perfect species resemble those of the genus *Rhizopus.* The genus *Rhizomucor* is similar except that it produces some rhizoids along the vegetative mycelium. *D, Mortierella.* (1) Sporangiophores are branched. (2) Sporangiophores taper toward the attachment of sporangium and lack columellae. (3) Zygospore covered by thick case of mycelium.

MYCOLOGY

The commonly encountered genera of the Mucorales can usually be identified by using Gilman's key.[60] For species identification of *Rhizopus* and other genera, the Zycha monograph should be consulted.[151] The key personally preferred by the author is the thesis on the Mucorales of C. W. Hesseltine.[70] Unfortunately, copies of this are hard to obtain. A modern approach to taxonomy is being devised by Schipper.[127] Ellis and Hesseltine[44] have extensively reviewed the genus *Absidia*.

Sporulation in the Mucorales is often difficult to observe in culture material. Ellis[44] described media other than Sabouraud's that augment fruiting in clinical isolants. These include potato glucose, malt agar, Czapek solution, agar, and hay infusion agar (see Part Five). The latter is one of the best for induction of sporulation.

Rhizopus Ehrenberg ex Corda 1838

This genus is characterized by simple or branched brown sporangiophores that arise singly or in groups from nodes directly above the rhizoids (hold fasts) (Fig. 25–11, *A*). The nodes are connected by a stolon. An evagination of the sporangiophore called a columella extends into the sporangium. The mycelium is tenacious, woolly, and coarse. Sporangia are dark-walled, spherical, and filled with hyaline or colored spores. Of the many reported species from human disease, *R. oryzae* and *R. arrhizus* are the most often implicated.

Rhizopus oryzae Went et Prinsen Geerlings 1895

Synonymy. *Rhizopus achlamydosporus* Takeda (fide Hesseltine) 1965; *Rhizopus formosaensis* Nakazawa 1913.

This is the most frequently reported isolant from mucormycosis.

Colony Morphology. The colony is fast growing, white at first, becoming yellowish brown with age (Fig. 25–12, *a, b*). Sporangia are dark brown.

Microscopic Morphology. The sporangiophores are simple or branched, up to 4 mm in length. Rhizoids are yellow-brown (Fig. 25–13). Sporangia are 100 to 350 μ in diameter. The sporangiospores are light brown, striated, irregularly shaped, and 6 to 8 μ by 7 to 9 μ in size. Numerous gemmae may be present. Zygospores are not known. The organism grows at a temperature of 40° C. This species is used in Japan for making koji.

Rhizopus arrhizus Fisher 1892

Synonymy. *Rhizopus nodosis* Namyslowski 1906; *Rhizopus ramosis* Moreau 1913; *Rhizopus maydis* Brudeslein 1917; *Mucor arrhizus*

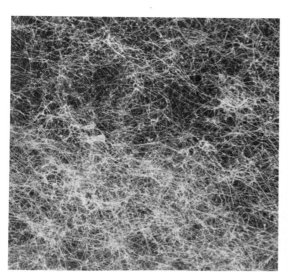

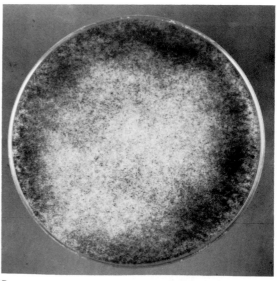

A **B**

Figure 25–12. *A, Rhizopus oryzae.* Loose, cottony colony. The light yellow sporangia are visible above the mycelium. × 5. *B,* Petri dish completely filled by mycelium of *R. oryzae.* Note the large dark sporangia.

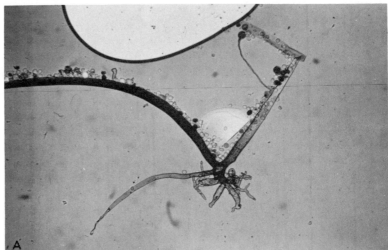

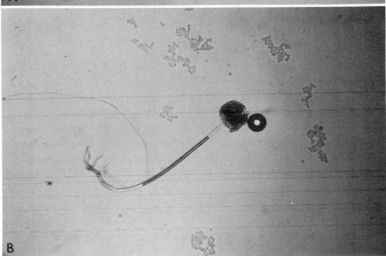

Figure 25–13. *Rhizopus sp. A,* Light yellow-brown rhizoids. ×400. *B,* Rhizoids and a small sporangiophore and sporangium. ×100.

Hagem 1908; *Mucor nurveqicus* Hagem 1908; *Rhizopus bovinus* van Beyma 1931; *Rhizopus chinensis* Nakazawa 1913.

This species is also frequently isolated from mucormycosis.

Colony Morphology. The general morphology is similar to that of *R. oryzae,* but the colony less often becomes brown.

Microscopic Morphology. Sporangiophores are similar to those of *R. oryzae,* but the spores are oval to flattened and angular with a ridged surface. They are 4.5 to 9 μ by 4.5 to 5.5 μ in size. Rhizoids are not as well developed in this species as in *R. stolonifer.* Growth at temperatures of 40° C is recorded. Gemmae are rarely seen.

Rhizopus rhizopodoformis (Cohn apud Lichtheim) Zopf 1890

Synonymy. *Mucor rhizopodoformis* Cohn apud Lichtheim 1884; *Rhizopus cohnii* (Cohn) Berlese et de Toni 1888; *Rhizopus equinus* Costantin et Lucet 1903; *Rhizopus pucillus* Naumov 1939.

Colony Morphology. Colonies are first white, then black-plumbeous; the obverse is nearly as black as the reverse, with few sterile mycelia, which are less than 3 mm in diameter. An odor is often present. Rhizoids are poorly developed. Sporangiophores are branched and erect.

Microscopic Morphology. Sporangiophores are up to 800 μ in height, usually 500 μ. Sporangia are spherical and white, then glistening black. They average 90 to 120 μ in size. Columellae are hyaline or colored, pyriform to globose, and less than 72 μ in diameter. Sporangiospores are not striate but are smooth, heavy-walled, spherical, and between 4 and 5 μ in size. Chlamydoconidia are cylindrical and up to 25 μ in size; zygospores are unknown. Good growth is achieved at 37° C. It has been isolated from a variety of clinical conditions.[17]

Rhizopus stolonifer (Ehrenberg et Fries) Vuillemin 1902

Synonymy. *Mucor stolonifer* Ehrenberg 1818; *Rhizopus nigricans* Ehrenberg 1820; *Rhizopus niger* Ciaglinski et Hewelke 1896; *Rhizopus artocarpi* Raciborski 1900; *Mucor niger* Gedoelst 1902.

Zygospores are often present (160 to 220 μ in diameter) (Fig. 25–14). The species is heterothallic. Sporangiophores are not branched. This is a very common contaminant and has not been isolated from true infections.[119] The general morphology is similar to *R. arrhizus*, but the rhizoids are well developed. The latter species is commonly encountered in nature but seldom as a laboratory contaminant. Spores of *R. stolonifer* are large, 7 to 15 μ in diameter. Poor growth is shown at 37° C.

R. microsporus Van Tieghem 1875 has also been reported from infections. *R. oligosporus* Saito 1905 has been shown to be pathogenic for animals, but so far it has not been isolated from natural disease.[119, 120]

Mucor Micheli ex Saint-Amans 1821

Members of this genus bear simple or branched sporangiophores which arise randomly along the aerial mycelium (Fig. 25–11). Stolons and rhizoids are absent. The colony is rapidly growing and generally gray or yellowish gray. The sporangium is large and spherical, and the wall may dissolve (deliquesce) at maturity. Few species have been recovered from well-documented cases of mucormycosis, and infection due to members of this genus is rare.

Mucor circinelloides van Tiegham 1875

Synonymy. *Calyptromyces circinelloides* (van Tiegham) Sumstine 1910; *Mucor paromychia* Sutherland et Plunckett 1934.

Colonies are fast growing, smoky gray, up to 2 cm in height, yellow to cream on reverse, and odoriferous. Sporangiophores are of two types: elongate sporangiophores bear large sporangia and are brown, filled with granules, and sympodially branched, with tapering branches. The second type of sporangiophore is short and grows haphazardly near the substrate with small sporangia (25 μ). The large sporangia are at first white, then greenish brown, globose or dorsoventrally compressed, and 40 to 84 μ (and up to 108 μ) in size; many are circinately borne and characterized by "bobbing heads." (See illustration in Part Five, Laboratory Mycology.) These sporangia deliquesce at maturity. Smaller sporangia have persistent walls that are smooth and hyaline. Fine-grained, hyaline columellae are variable in shape, with well-developed collars up to 35 μ in size. Spores are short, oval, and 4 to 7 μ in size; when free, they are often seen as budding yeasts. Chlamydoconidia are few and are found in sporangiophores. They are heavy-walled and are the diameter of the sporangiophore, usually less than 21 μ in size. No zygospores are present. Good growth is achieved at 37° C (Fig. 25–15).[25, 54, 123]

Mucor ramosissimus Sanutsevitsch 1927 was isolated from a chronic destructive facial le-

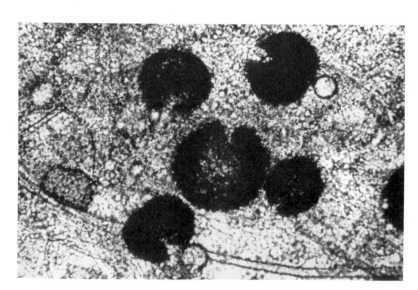

Figure 25–14. Zygospores of *Rhizopus stolonifer*. Some have been broken open. One developing zygote shows attachment of suspensor cell. ×450. (Courtesy of S. McMillen.)

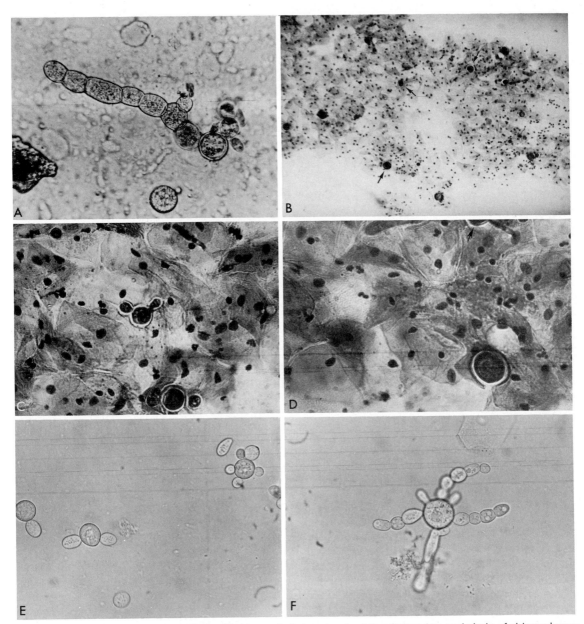

Figure 25–15. Human colonization by *Mucor* sp. *A,* Spherules, budding spherules, and chain of chlamydoconidia of *Mucor* sp. in fecal sample. (Courtesy of Morris Gordon.) *B,* Vaginal colonization by *Mucor* sp. Low power (× 100) view of Pap smear. Numerous spherules (*arrows*), budding spherules, and short hyphal units are noted. *C,* Budding cells in smear. × 450. *D,* Thick-walled spherules and elongate cell (*top*).[122, 123] (B, C. D. Courtesy of C. T. Dolan,) *E,* Spherules and budding cells in urine sample. There is a resemblance to the yeast cells of *Paracoccidioides brasiliensis. F,* Various forms of *M. circinelloides* in urine sample. (Courtesy of B. Cooper.)

sion of 24 years' duration.[140] The colony is cinnamon-buff to gray-olive, with sporangia 15 to 70 μ in diameter. Small sporangia lack a columella. Spores are ovoid to globose, 3.3 to 5.5 μ by 3.5 to 8 μ.[44]

M. javanicus Wehmer 1900, *M. racemosus* Fresenius 1850, and *M. spinosus* Van Tieghem 1876 have also been recorded as being isolated from disease processes, but the authenticity of these cases is in doubt. *M. janssenii*[96] causes penis infections in geese.

Rhizomucor (Lucet and Costantin) Wehmer ex Vuillemin 1931

This genus is similar to *Mucor* but has some poorly developed rhizoids. *R. pusillus* and *R. miehei* are thermophilic and common in heated environments.[124] Both have been involved in human and animal infections. *R. pusillus* Lindt Schipper 1978[127] has been reported from disseminated infections.[53, 83] Colonies are gray. Sporangiophores are short and less than 3 mm in size and they arise from stolons. Few rhizoids are present, and they are irregularly branched. Sporangia are black and spherical, and have spines; they are 60 to 70 μ in size, but many are smaller. Columellae are smooth and variable in shape, and have a collar. The colony becomes dark brown with age. Zygospores are unknown.

Absidia Van Tieghem 1876

This genus is characterized by the presence of rhizoids and by branching sporangiophores which arise between the nodes (Fig. 25–11,*B*). This latter characteristic is in contrast to the location of sporangiophores in the genus *Rhizopus*. The sporangia are pyriform. Two species of this genus are frequently isolated from animal infections and sometimes from human disease. These are *A. corymbifera* (Cohn) Saccardo and Trotter 1912 and *A. ramosa* (Lindt) Lendner 1909.[44] Scholer[128] could not distinguish them biochemically and indicates the species may in fact be identical.

Absidia corymbifera (Cohn) Saccardo et Trotter 1912

Synonymy. *Mucor corymbifer* Cohn 1884. Ellis and Hesseltine[44] list 16 other synonyms.

Colony Morphology. Growth is rapid and produces a floccose, light olive-gray colony. The color fades in time. Sporangiophores are very long, up to 450 μ, and 4 to 8 μ in diameter. They arise from stolons and are light gray in color. Numerous smaller and irregularly shaped sporangiophores are also produced. The sporangiophores branch repeatedly to form corymbs. Sporangia are 20 to 35 μ in diameter and grayish. Columellae are 16 to 27 μ in diameter, ovoid to spatulate, and may have a few spines. The spores are formed by internal cleavage in a manner similar to that found in *Coccidioides*. Mature spores are globose to oval, 2 to 3 × 3 to 4 μ. Rhizoids arise from swollen areas (nodes) along a stolon. They are 12 μ in diameter, up to 370 μ in length, and hyaline. Zygospores are produced. They are 40 to 80 μ in diameter, thick-walled, and roughened with equitorial ridges. The brown suspensor cells develop circinate filaments that almost completely encircle the zygospores. Giant cells are numerous. The fungus grows well at 37°C, and growth is obtained up to 42°C.

Absidia ramosa (Lindt) Lendner 1908 differs from *A. corymbifera* chiefly in having uniformly oval spores, and when mated it produces fewer zygospores. Scholer[128] has developed a carbohydrate utilization pattern to distinguish *A. corymbifera* from several other species of *Absidia*. However, the physiologic pattern of *A. ramosa* is identical to that of *A. corymbifera*. Therefore the two are probably conspecific.

Mortierella Coemans 1863

This genus is characterized by a sporangium lacking a columella. The sporangiophore tapers to the attachment of the sporangium (Fig. 25–11,*D*). The spores are small, 1 to 2 μ by 2 to 4 μ. Small one-celled conidia called stylospores are also present. They are spiny or echinulate. The colonies are gray to yellowish gray and do not develop much aerial mycelium. Some species, particularly *Mortierella wolfii* Mehrotra et Baijal, have been isolated from cases of animal disease such as mycotic abortion and fatal pneumonia of cattle.[100]

Other Mucorales have also been encountered, although rarely, in human disease. *Cunninghamella bertholletiae* Lendner 1908[86, 144, 145] has been isolated from documented cases. *Saksenaea vasiformis* (Fig. 25–16) was involved in a cerebral mucormycosis,[7, 37] and *Syncephalastrum* spp. have been recovered from a few human infections.[80, 82]

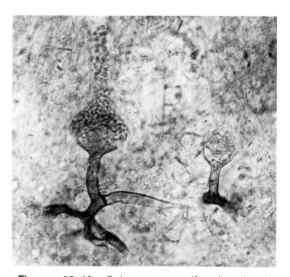

Figure 25–16. *Saksenaea vasiformis* showing characteristic flask-shaped sporangium, one full of sporangiospores, the other broken and empty. The sporangiophore is short and rhizoids are evident. Fruiting was accomplished by floating an agar block containing mycelium in distilled water. ×560. (Courtesy of L. Ajello.)

REFERENCES

1. Abramowitz, I. 1964. Fatal perforations of the stomach due to mucormycosis of the gastrointestinal tract. South Afr. Med. J., *38*:93–94.
2. Abramson, E., D. Wilson, et al. 1967. Rhinocerebral phycomycosis in association with diabetic ketoacidosis. Report of two cases and a review of clinical and experimental experience with amphotericin B therapy. Ann. Intern. Med., *66*:735–742.
3. Ader, P. L. 1979. Phycomycosis in fifteen dogs and two cats. J. Am. Vet. Med. Assoc., *174*:1216–1223.
4. Ader, P. L., and J. K. Dodd. 1979. Mucormycosis and entomophthoromycosis. A bibliography. Mycopathologia, *68*:67–99.
5. Agger, W. A., D. G. Maki, et al. 1978. Mucormycosis: a complication of critical care. Arch. Intern. Med., *138*:925–927.
6. Ainsworth, G. C., and P. K. C. Austwick. 1973. *Fungal Diseases of Animals*. 2nd ed. Bucks, England, Commonwealth Agricultural Bureau.
7. Ajello, L., D. F. Dean, et al. 1976. The zygomycete *Saksenaea vasiformis* as a pathogen of humans with a critical review of the etiologies of zygomycosis. Mycologia, *68*:52–62.
8. Baker, R. D. 1956. Pulmonary mucormycosis. Am. J. Clin. Pathol., *26*:1235.
9. Baker, R. D. 1971. Mucormycosis. *In* Baker, R. D., et al. *The Pathologic Anatomy of Mycoses*. Dritterband, Funster teil. Berlin, Springer-Verlag, Ch. 21.
10. Barthelat, G. J. 1903. Les mucorinées pathogènes et les mucormycoses chez les animaux et chez l'homme. Arch. parasitol., *7*:1–116.
11. Bartnicki-Garcia, S., 1972. The dimorphic gradient of *Mucor rouxii*: A laboratory exercise. ASM News, *38*:486–488.
12. Bartnicki-Garcia, S., and W. J. Nickerson. 1962. Induction of yeastlike development in *Mucor* by carbon dioxide. J. Bacteriol., *84*:829–840.
13. Bartram, R. J., M. Watnick, et al. 1973. Roentgenographic findings in pulmonary mucormycosis. Am. J. Roentgenol., *117*:810–815.
14. Battock, D. J., H. Grausz, et al. 1968. Alternate day amphotericin B therapy in the treatment of rhinocerebral phycomycosis (mucormycosis). Ann. Intern. Med., *68*:122–137.
15. Baum, J. L. 1967. Rhinoorbital mucormycosis occurring in an otherwise apparently healthy individual. Am. J. Ophthalmol., *63*:335–339.
16. Best, P. B., and R. M. McCully. 1979. Zygomycosis (phycomycosis) in a right whale (*Enbalaena autralis*). J. Comp. Pathol., *89*:341–348.
17. Bottone, E. J., I. Weitzman, et al. 1979. *Rhizopus rhizopodoformis:* emerging etiological agent of mucormycosis. J. Clin. Microbiol., *9*:530–537.
18. Braf, Z., et al. 1976. Fungal infections after renal transplantation. Israel J. Med. Sci., *12*:674–677.
19. Brown, J. F., et al. 1977. Pulmonary and rhinocerebral mucormycosis: Arch. Intern. Med., *137*:936–939.
20. Campos-Nieto, E. 1978. A case of bovine cerebral absidiomycosis. Bol. Soc. Mex. Micol., *12*:115–116.
21. Caraveo, J., A. A. Trowbridge, et al. 1977. Bone marrow necrosis associated with *Mucor* infection. Am. J. Med., *62*:404–408.
22. Chick, E. W., J. Evans, et al. 1958. The inhibitory effect of amphotericin B on localized *Rhizopus oryzae* (mucormycosis) utilizing the pneumoderm pouch of rats. Antibiot. Chemother., *8*:506–510.
23. Chick, E. W., J. Evans, et al. 1958. Treatment of experimental mucormycosis *(Rhizopus oryzae)* infection in rabbits with amphotericin B. Antibiot. Chemother., *8*:394–399.
24. Connor, P. A., R. J. Anderson, et al. 1979. *Mucor* mediastinitis. Chest, *75*:524–526.
25. Cooper, B. H. 1981. A case of pseudococcidioidomycosis, yeast phase of *Mucor circinelloides* in a clinical specimen. J. Clin. Microbiol. in press.
26. Corbel, M. J., and S. M. Eades. 1975. Factors determining the susceptibility of mice to experimental mucormycosis. J. Med. Microbiol., *8*:551–564.
27. Corbel, M. J., and S. M. Eades. 1976. Experimental phycomycosis in mice; examination of the role of acquired immunity in resistance to *Absidia ramosa*. J. Hyg., *77*:221–231.
28. Corbel, M. J., and S. M. Eades. 1977. Experimental mucormycosis in cogenitally athymic (nude) mice. Mycopathologia, *62*:117–120.
29. Corbel, M. J., and S. M. Eades. 1978. Observations on localization of *Absidia corymbifera in vivo*. Sabouraudia, *16*:125–132.
30. Corbel, M. J., and S. M. Eades. 1981. Cerebral mucormycosis following experimental inoculation with *Mortierella wolfii*. Mycopathologia, in press.
31. Corbel, M. J., D. W. Redwood, et al. 1980. Infection with *Absidia corymbifera* in bank voles (*Clethriohomys glariolus*). Lab. Anim., *14*:25–30.
32. Cores, D. O., W. A. Royal, et al. 1967. Systemic mycosis in neonatal calves. N. Z. Vet. J., *15*:143–149.
33. Darja, M., and M. Davy. 1963. Pulmonary mucormycosis with cultural identification. Can. Med. Assoc. J., *89*:1235–1238.
34. Davis, C. L., W. A. Anderson, et al. 1955. Mucormycosis in food producing animals. J. Am. Vet. Med. Assoc., *126*:261–267.
35. Dawson, C., N. G. Wright, et al. 1969. Canine phycomycosis. A case report. Vet. Rec., *84*:633–634.
36. Deal, W. B., and J. E. Johnson. 1962. Gastric phycomycosis. Report of a case and review of the literature. Gastroenterology, *57*:579–586.
37. Dean, D. F., L. Ajello, et al. 1977. Cranial zygomycosis caused by *Saksenaea vasiformis*. J. Neurosurg., *46*:97–103.
38. Deweese, D. D., A. J. Schleuning, et al. 1965. Mucormycosis of the nose and paranasal sinuses. Laryngoscope, *75*:1398–1407.
39. Donnelly, W. J. C. 1967. Systemic bovine phycomysosis: A report of two cases. Irish Vet. J., *21*:82–87.
40. Duffy, E. 1945. Alloxan diabetes in rabbits. J. Pathol. Bacteriol., *57*:199–212.
41. Echols, R. M., D. S. Selinger, et al. 1979. *Rhizopus* osteomyelitis: a case report and review. Am. J. Med., *66*:141–145.
42. Eden, O. B., and J. Santos. 1979. Effective treatment for rhinopulmonary mucormycosis in a boy with leukemia. Arch. Dis. Child., *54*:557–559.
43. Elder, T. D., and R. D. Baker. 1956. Pulmonary mucormycosis in rabbits with alloxan diabetes. Arch. Pathol., *61*:159–168.

44. Ellis, J. J., and C. W. Hesseltine. 1966. Species of *Absidia* with ovoid sporangiospores. Sabouraudia, *5*:59–77.

45. Emmons, C. W. 1964. Phycomycosis in man and animals. Rev. Pathol. Veg., *4*:329–337.

46. Englander, G. S. 1953. Mycetoma caused by *Rhizopus.* Arch. Dermatol., *68*:741.

47. English, M. P., and N. M. Lucke. 1970. Phycomycosis in a dog caused by unusual strains of *Absidia corymbifera.* Sabouraudia, *8*:126–132.

48. Everett, E. D., S. Pearson, et al. 1979. *Rhizopus* surgical wound infection with elasticized adhesive tape dressing. Arch. Surg., *114*:738–739.

49. Faillo, P. S., H. P. Sube, et al. 1959. Mucormycosis of the paranasal sinuses and the maxilla. Oral. Surg., *12*:304–309.

50. Feinberg, R., and T. S. Risley. 1959. Mucormycotic infection of an arteriosclerotic thrombus of the abdominal aorta. New Engl. J. Med., *260*:626–629.

51. Fleckner, R. A., and J. H. Goldstein. 1969. Mucormycosis. Br. J. Ophthalmol., *53*:542–548.

52. Foley, F. D., and J. M. Shuck. 1968. Burn wound infection with phycomycetes requiring amputation of the hand. J.A.M.A., *208*:596.

53. Fragner, P., J. Vitovec, et al. 1975. *Mucor pucillus* jako purodee uzlinove mukormykozy bycka. Cesk. Mykol., *29*:59–60.

54. Frank, W., U. Roester, et al. 1974. Sphaerule-Bildung bei einer *Mucor*-Spezies in inneren Organen von Amphibien. Vorläufige Mitteilung. Zentralbl. Bakteriol. *226*:405–417.

55. Fujiwara, A., J. W. Landau, et al. 1970. Preliminary characterization of the tremolysin of *Rhizopus nigricans.* Mycopathologia, *40*:139–144.

56. Fürbringer, P. 1876. Beobachtungen über Lungenmycose beim Menschen. Virchows Arch. Pathol. Anat., *66*:330.

57. Gale, A. M., and W. P. Kleitsch. 1972. Solitary pulmonary nodule due to phycomycosis (mucormycosis). Chest, *62*:752–755.

58. Gale, G. R., and A. M. Welch. 1961. Studies on opportunistic fungi. I. Inhibition of *Rhizopus oryzae* by human serum. Am. J. Med. Sci., *241*:604–612.

59. Gartenberg, G., E. J. Bottone, et al. 1978. Hospital acquired mucormycosis. *Rhizopus rhizopodoformis* of skin and subcutaneous tissue. New Engl. J. Med., *299*:1115–1118.

60. Gilman, J. C. 1957. *A Manual of Soil Fungi.* 2nd ed. Ames, The Iowa State College Press.

61. Ginsberg, J., A. G. Spaulding, et al. 1966. Cerebral phycomycosis (mucormycosis) with ocular involvement. Am. J. Ophthalmol., *62*:900–906.

62. Gong, N. C., K. Prathap, et al. 1978. Primary pulmonary mucormycosis. Singapore Med. J., *19*:109–111.

63. Gregory, J. E., A. Golden, et al. 1943. Mucormycosis of the central nervous system. Report of three cases. Bull. Johns Hopkins Hosp., *73*:405–419.

64. Hammer, G. S., E. J. Bottone, et al. 1975. Mucormycosis in a transplant recipient. Am. J. Clin. Pathol., *64*:389–398.

65. Hammond, D. E., and R. K. Winkelman. 1979. Cutaneous phycomycosis. Report of three cases with identification of *Rhizopus.* Arch. Dermatol., *115*:990–992.

66. Hauch, T. W. 1977. Pulmonary mucormycosis: another cure. Chest, *72*:92–93.

67. Heller, R. A., H. P. Hobson, et al. 1971. Three cases of phycomycosis in dogs. Vet. Med. Small Anim. Clin., *66*:472–476.

68. Henriquez, M., A. Levy, et al. 1979. Mucormycosis in a renal transplant recipient with successful outcome. J.A.M.A., *242*:1397–1399.

69. Herla, V. 1895. Note sur un cas de pneumomycose chez l'homme. Bull. Acad. R. Med. Belg., IV *9*:1021–1031.

70. Hesseltine, C. W. 1950. A revision of the mucorales based especially on a study of the representatives of the order in Wisconsin. Thesis, Madison, Wis. Univ. of Wisconsin.

71. Hesseltine, C. W., and J. J. Ellis. 1964. An interesting species of *Mucor, M. ramosissimus.* Sabouraudia, *3*:151–154.

72. Hessler, J. R., J. C. Woodard, et al. 1967. Mucormycosis in a rhesus monkey. J. Am. Vet. Med. Assoc., *151*:909–913.

73. Hewer, T. F., H. Pearson, et al. 1968. Aspergillosis and mucormycosis in two newborn *Okapi johnstoni* (Sclater). Br. Vet. J., *124*:282–286.

73a. Hogben, B. R. 1967. Systemic mycotic lesions in apparently normal bobby calves. N.Z. Vet. J., *15*:30–32.

74. Hutchins, D. R., and K. G. Johnson. 1972. Phycomycosis in a horse. Aust. Vet. J., *48*:269–278.

75. Jain, J. K., A. Markowitz, et al. 1978. Localized mucormycosis following intramuscular corticosteroid. Case report and review of literature. Am. J. Med. Sci., *275*:209–216.

76. Jones, K. O., and L. Kaufman. 1978. Development and evaluation of an immunodiffusion test for diagnosis of systemic zygomycosis (mucormycosis). Preliminary report. J. Clin. Microbiol., *7*:97–103.

77. Jonquieres, E. D., C. A. Castello, et al. 1972. Seudomicetoma cutaneo ficomicotico. Int. J. Dermatol., *11*:89–95.

78. Josefiak, E. J., and J. H. Smith-Foushee. 1958. Experimental mucormycosis in the healthy rat. Science, *127*:1442.

79. Kahn, L. B. 1963. Gastric mucormycosis: Report of a case with a review of the literature. South Afr. Med. J., *37*:1265–1269.

80. Kamalan, A., and A. S. Thambials. 1980. Cutaneous infection by *Syncephalastrum.* Sabouraudia, *18*:19–20.

81. Kiehn, T. E., F. Edwards, et al. 1979. Pneumonia caused by *Cunninghamella bertholletiae* complicating chronic lymphatic leukemia. J. Clin. Microbiol., *10*:374–379.

82. Kirkpatrick, M. B., H. M. Pollock, et al. An intracavitary fungus ball composed of *Syncephalastrum.* Am. Rev. Respir. Dis., *120*:943–947.

83. Kramer, B. S., A. D. Hernandez, et al. 1977. Cutaneous infarction. Arch. Dermatol., *113*:1075–1076.

84. Kujiwara, A., J. W. Landau, et al. 1970. Hemolytic activity of *Rhizopus nigricans* and *Rhizopus arrhizus* (et. seq.).Mycopathologia, *40*:131–139, 139–144.

85. Kurchenmeister, G. F. H. 1855. Die in undan dem Kärper des lebenden Menschen vor kommenden Parasiten. Ein Lehr und Handbuch der Diagnose und Behandlung der tierschen und pflanzlichen Parasiten des Menschen. Leipzig (Germany), B. G. Teubiner.

86. Kwon-Chung, K. J., R. C. Young, et al. 1975. Pulmonary mucormycosis caused by *Cunninghamella elegans* in a patient with chronic myelo-

genous leukemia. Am. J. Clin. Pathol., *64*:544–548.

87. Landau, J. W., and V. D. Newcomber. 1962. Acute cerebral phycomycosis (mucormycosis). J. Pediatr., *61*:363–383.

88. Langston, C., D. A. Roberts, et al. 1973. Renal phycomycosis. J. Urol., *109*:941–944.

89. LaTouche, C. J., T. W. Sutherland, et al. 1964. Histopathological and mycological features of a case of rhinocerebral mucormycosis (phycomycosis) in Britain. Sabouraudia, *3*:148–150.

90. Lichtheim, L. 1884. Über pathogene Mucorineen und die durch sie erzeugten Mykosen der Kaninchens. Z. Klin. Med., 7:140.

91. Lindt, W. 1886. Mitteilungen über einige neue pathogene Schimmelpilze. Arch. F. Exp. Pathol. Pharm., *21*:269.

92. Lucet, A., and J. Costantin. 1901. Contribution a l'étude des mucorinées pathogènes. Arch. Parasitol., *4*:362–408.

93. Lucke, V. M., D. G. Morgan, et al. 1969. Phycomycosis in a dog. Vet. Res., *84*:645–646.

94. Lyon, D. T., T. T. Schubert, et al. 1979. Phycomycosis of the gastrointestinal tract. Am. J. Gastroent., *72*:379–394.

95. MacKenzie, A. 1969. The pathology of respiratory infections in pigs. Br. Vet. J., *125*:294–303.

96. Marjánková, K., K. Krivanec, et al. 1978. Mass occurrence of necrotic inflammation of the penis in ganders caused by phycomyces. Mycopathologia, *66*:21–26.

97. Martin, J. E., D. J. Kroe, et al. 1969. Rhino-orbital phycomycosis in a rhesus monkey *(Macaca mulatta)*. J. Am. Vet. Med. Assoc., *155*:1253–1257.

98. Mayer, A. F., and J. L. Emmert. 1815. Verschimmelung (Mucedo) in leben der Körper. Dtsch. Arch. Anat. Physiol. (Meckel), *1*:310–318.

99. Medoff, G., and G. S. Kobayashi. 1972. Pulmonary mucormycosis. New Engl. J. Med., *286*:86–87.

100. Menna. M. E. D., M. E. Carter, et al. 1972. The identification of *Mortierella wolfii* isolated from cases of abortion and pneumonia in cattle and a search for its infection source. Res. Vet. Sci., *13*:439–442.

101. Meyer, R. D., and M. H. Kaplan. 1978. Cutaneous lesions in disseminated mucormycosis. J.A.M.A., *225*:737–738.

102. Meyer, R. D., P. Rosen, et al. 1972. Phycomycosis complicating leukemia and lymphoma. Ann. Intern. Med., *77*:871–879.

103. Meyer, R. D., M. H. Kaplan, et al. 1973. Cutaneous lesions in disseminated mucormycosis. J.A.M.A., *225*:737–738.

104. Meyers, B. R., G. Wormser, et al. 1979. Rhinocerebral mucormycosis. Premortem diagnosis and therapy. Arch. Intern. Med., *139*:557–560.

105. Mikhael, M. A. 1979. Cerebral phycomycosis. J. Comput. Assist. Tomogr., *3*:417–420.

106. Morquer, R., C. Lombard, et al. 1965. Pathologénie de quelques Mucorales pour les animaux. Une nouvelle mucormycose chez les Bovides. Bull. Trim. Soc. Mycol. Fr., *81*:421–449.

107. Moye, T. D., Jr., and R. J. Candill. 1980. Rhinomaxillary phycomycosis: report of a case. J. Oral Surg., *38*:132–134.

108. Muresan, A. 1960. A case of cerebral mucormycosis diagnosed in life with eventual recovery. J. Clin. Pathol., *13*:34–36.

109. Murray, D. R., P. W. Ladds, et al. 1978. Granulomatous and neoplastic diseases of the skin of horses. Aust. Vet. J., *54*:338–341.

110. Murray, H. W. 1975. Pulmonary mucormycosis with massive fatal hemoptysis. Chest, *68*:65–68.

111. Murray, H. W. 1977. Pulmonary mucormycosis: one hundred years later. (Editorial) Chest, *72*:1–2.

112. Neame, P., and D. Rayner. 1960. Mucormycosis. A report of twenty-two cases. Arch. Pathol., *70*:261–268.

113. O'Hara, K. 1978. Review of the rare cases of skin and nail mucormycosis. Jap. J. Med. Mycol., *19*:129–132.

114. Paltauf, A. 1885. Mycosis mucorina: ein Beitrag zur Kenntnis der menschlichen Fadenpilzer-Krankungen. Virchows Arch. Pathol. Anat., *102*:543.

115. Podack, M. 1899. Zur kenntnis des sog. Endothelkrebes der Pleura und die Mucormycosen im menschlichen Respirations apparat. Dtsch. Arch. Klin. Med., *63*:1.

116. Polli, C. 1965. On the distribution of ketoreductase in microorganisms. Pathol. Microbiol. (Basel), *28*:93–98.

117. Prout, G. R., and A. R. Goddard. 1960. Renal mucormycosis. Survival after nephrectomy amphotericin B therapy. New Engl. J. Med., *263*:1246.

118. Rabin, E. R., G. D. Lundberg, et al. 1961. Mucormycosis in severely burned patients. New Engl. J. Med., *264*:1286–1288.

119. Reinhardt, D. J., W. Kaplan, et al. 1970. Experimental cerebral zygomycosis in alloxan diabetic rabbits. I. Relationship of temperature tolerance of selected zygomycetes of pathogenicity. Infect. Immun., *2*:404–413.

120. Reinhardt, D. J., and J. J. Licota. 1974. *Rhizopus, Mucor* and *Absidia* as agents of cerebral zygomycosis. Abstracts Am. Soc. Microbiol., Mm 21.

121. Rippon, J. W. 1980. Cutaneous and subcutaneous mucormycosis. Curr. Concepts Skin Disord., *1*:8–11.

122. Rippon, J. W., and C. T. Dolon. 1979. Colonization of the vagina by fungi of the genus *Mucor*. Clin. Microbiol. Newsletter, *11*:4–5.

123. Rippon, J. W., and C. T. Dolon. 1980. Mucorales: colonization of human vagina and colon, pp. 39–42. *In* E. S. Kuttin and G. L. Baum (eds.). *Human and Animal Mycology, Proceedings of VII Congress ISHAM 1979.* Amsterdam, Excerpta Medica.

124. Rippon, J. W., R. Gerhold, et al. 1980. Thermophilic and thermotolerant fungi from thermal effluent of nuclear power generating reactions: dispersal of human opportunistic and veterinary pathogenic fungi. Mycopathologia, *70*:169–179.

125. Salvin, S. B. 1952. Endotoxin in pathogenic fungi. J. Immunol., *69*:89–99.

126. Saubolle, M. 1976. *The Antimicrobic Susceptibility Test: Principles and Practice.* Philadelphia, Lea and Febiger.

127. Schipper, M. A. A. 1978. I. On certain species of *Mucor* with a key to all accepted species. 2. On the genera *Rhizomucor* and *Parasitella.* Studies in Mycology No. 17, p. 17, Baarn, Netherland, Centralbureau voor Schimmelculture.

128. Scholer, H. J., and E. Müller. 1966. Beziehungen

zwischen biochemischer Leistung und Morphologie bei Pilzen aus der Familie der Mucoraceen. Pathol. Microbiol., *29*:730–741.

129. Scholz, H. D., and L. Meyer. 1965. *Mortierella polycephala* as a cause of pulmonary mycosis in cattle. Berl. Münch. Tieräztl. Wochenschr., *78*:27–30.

130. Sheldon, D. L., and W. C. Johnson. 1979. Cutaneous mucormycosis. Two documented cases of suspected nosocomial cause. J.A.M.A., *241*:1032–1034.

131. Sheldon, W. H., and H. Bauer. 1960. Tissue mast cells and acute inflammation in experimental cutaneous mucormycosis of normal, 48/80 treated and diabetic rats. J. Exp. Med., *112*:1069–1083.

132. Simon, R., G. G. Hoffman, et al. 1964. Phycomycosis. Aerospace Med., *35*:668–675.

133. Sluyter, 1847. De vegetabilibus organismi animalis parasitis ac de novo epiphyto in pityriasi versicolore obvio. Lnang. Diss. Berlin, Germany.

134. Sone, S., M. Tamura, et al. 1977. Phycomycosis in intestine complicated with acute leukemia: report of a case. Jap. J. Med. Mycol., *18*:87–91.

135. Souza, R. de, S. MacKinnon, et al. 1979. Treatment of localized pulmonary phycomycosis. South Med. J., *72*:609–612.

136. Stevens, D. A. 1977. Miconazole in the treatment of systemic fungal infections. Am. Rev. Respir. Dis., *116*:801–806.

137. Symmers, W. St. C. 1962. Histopathological aspects of pathogenesis of some opportunistic fungal infections, as exemplified in the pathology of aspergillosis and the phycomycetoses. Lab. Invest., *11*:1073–1090.

138. Taylor, R., G. Shklar, et al. 1964. Mucormycosis of oral mucosa. Arch. Dermatol., *89*:419–425.

139. Thomford, N. R., T. H. Dee, et al. 1970. Mucormycosis of a saphenous vein autograph. Arch. Surg., *101*:518–519.

140. Vignale, R., J. E. MacKinnon, et al. 1964. Chronic destructive mucocutaneous phycomycosis in man. Sabouraudia, *3*:143–147.

141. Virchow, R. 1856. Beitrage zür Lehre von den beim Menschen. Vorkommen der pflanzlichen Parasiten Virchows. Arch. Pathol. Anat., *9*:557.

142. Watanabe, K., K. Tabuchi, et al. 1929. Mucormycosis due to *Rhizopus rhizopodoformis:* a report of 3 cases. Bull. Azabu Vet. College, *4*:25–32.

143. Watson, K. C., and P. B. Neame. 1960. *In vitro* sensitivity of amphotericin B on strains of Mucoraceae pathogenic to man. J. Lab. Clin. Med., *56*:251–257.

144. Weitzman, I., and M. Y. Crist. 1979. Studies with clinical isolates of *Cunninghamella.* I. Mating behavior. Mycologia, *71*:1024–1033.

145. Weitzman, I., and M. Y. Crist. 1980. Studies with clinical isolates of *Cunninghamella.* II. Physiological and morphological studies. Mycologia, *72*:661–669.

146. White, L. O., H. C. Newham, et al. 1978. Estimation of *Absidia ramosa* infection of the brain and kidneys of cortisone treated mice by chitin assay. Mycopathologia, *3*:177–179.

147. Whiteway, D. E., and R. L. Virata. 1979. Mucormycosis. Arch. Intern. Med., *139*:944.

148. Williams, L. W., E. Jacobson, et al. 1979. Phycomycosis in a western massasauga rattlesnake *(Sisturus catenatus)* with infection of the telencephalon, orbit and facial structures. Vet. Med. Small Anim. Clin., *74*:1181–1184.

149. Wilson, B., J. C. Grotta, et al. 1979. Cerebral mucormycosis. Arch. Neurol., *36*:725–726.

150. Winn, R. E., A. I. Hartstein, et al. 1981. Cutaneous mucormycosis: report of two cases and review of the literature, in press.

151. Zycha, H. 1935. *Mucorineae* Kryptogamen: Flora, Leipzig Mark Brandenberg. *6A*:1–264. Berlin, Gebrueder Borntraeger.

26 Hyphomycosis, Phaeohyphomycosis, Miscellaneous and Rare Mycoses, Algosis, and Pneumocystosis

INTRODUCTION

From time to time numerous soil and air-borne fungi have been reported as being responsible for a variety of disease processes. The majority of these reports are invalid or at least questionable. Pathogenic fungi are often difficult to grow, whereas contaminants are frequently isolated and described as the source of a particular disease process. However, there are a few organisms not normally considered virulent that may adapt to and take advantage of a situation and establish an infection. This is particularly true of the so-called "opportunistic infections" — candidiasis, aspergillosis, mucormycosis, and so forth, covered in previous chapters. Such organisms as *Acremonium, Trichosporon, Penicillium, Rhodotorula*, and so forth have been isolated from such situations also and their etiologic roles substantiated in some cases. The circumstances usually involve a barrier break, such as by surgery or indwelling catheter, steroid therapy, or the use of immunosuppressive drugs or cytotoxins. The increased use of such modalities will vastly increase the opportunities for such superinfections, and the list of organisms involved will also increase.

The use of steroids, antibiotics, and cytotoxins has also brought to light the presence of a peculiar organism, *Pneumocystis carinii*, which appears to be of common occurrence in rodents and man; until recently, however, it was not recognized. It is now known to be of major importance as an opportunistic infectious agent. The taxonomic affinities of this organism to other microorganisms are as yet unsettled. It had previously been grouped with the protozoans, but its cell wall contains chitin, which, along with certain ultramicroscopic characteristics, indicates a relation to the fungi.

In addition to the true fungi and actinomycetes, there are a few valid cases of infection caused by the colorless alga *Prototheca*, and even reports of disease due to blue-green and green algae. About the only groups of organisms not yet incriminated in human disease are the mosses and the ferns.

In this chapter some of the well-established but infrequent mycoses will be discussed, together with a résumé of unique or very rare infections. Infrequently encountered diseases are given separate headings. Those that are even more rarely encountered are grouped into two categories. Hyphomycosis includes infections caused by fungi with hyaline (colorless) mycelium; phaeohyphomycosis refers to diseases in which the fungal hyphae have pigmented cell walls in tissue and/or culture. The rarely encountered yeasts are discussed in Chapter 22. Reviews of the rare mycoses have recently been published by English[78] and Parker and Klintworth.[185]

Geotrichosis

DEFINITION

Geotrichosis is an infection caused by the ubiquitous fungus *Geotrichum candidum.* Lesions are bronchopulmonary, bronchial, oral, cutaneous, and, very rarely, alimentary.

HISTORY

Geotrichum was first isolated from decaying leaves by Link in 1809. Bennett in 1842 described the organism as causing a superinfection in an old tuberculous cavity.[20] He differentiated it from *"Monilia" (Candida)*, and if this report is valid, it represents the first case of geotrichosis. Confusion with infections caused by *Candida* or isolation of the organism from normal flora invalidates most of the early records of *Geotrichum* as a fungal pathogen. The reports of pulmonary geotrichosis made by Linossier in 1916 and Martin in 1928 are probably authentic.[167] The description of the organism in eczematous dermatitis in 1935 by Ciferri and Redaelli, however, probably represents isolation of normal flora from a pathologic process of other etiology. Geotrichosis, when substantiated as valid, is usually secondary to some other debilitation, such as tuberculosis, or is a complication of steroid therapy.

ETIOLOGY AND DISTRIBUTION

Though several species of the genus have been described, only *Geotrichum candidum* has been associated with human infections. Schnoor[211] isolated the fungus from 29 per cent of 314 stool specimens obtained from healthy people. In a recent survey of sputum, feces, urine, and vaginal secretions from 2643 subjects, both healthy and with some other disease, the organism was present in 18 to 31 per cent of specimens.[188] However, it was not associated with any specific illness. *G. candidum* has also been recovered from cottage cheese, dairy products, plant material, and healthy skin. It is therefore endogenous as normal flora in man and appears to be ubiquitous in nature.

CLINICAL DISEASE

Pulmonary involvement is the most frequently reported form of the disease, but bronchial, oral, cutaneous, and alimentary infections have also been noted.

Pulmonary Geotrichosis

Pulmonary disease simulates tuberculosis and is usually secondary to it. Clinically there is a light gray, thick, and mucoid sputum which in some cases is purulent and rarely blood-tinged. Fine to medium rales are heard. The condition is most often chronic, and there is little debilitation or fever present. On x-ray examination, smooth, dense, patchy infiltrations and some cavities are seen. These are present in the areas of the lung commonly associated with tuberculosis, such as the hilar and apical regions. About 24 cases of pulmonary geotrichosis are recognized in the literature.[167] Although usually chronic, the disease may run a fulminant, sometimes fatal, course if patients are put on steroids.[199] To establish the diagnosis of geotrichosis, it is necessary to see the organisms in quantity in sputum and to obtain multiple positive cultures.[180] Other etiologies of lung disease must be excluded. Specific treatment for the disease includes iodides, aerosol nystatin, and amphotericin B.[167,185]

Bronchopulmonary geotrichosis, which is similar to the allergic type of aspergillosis, is another form of the disease. *Geotrichum* is seen growing in the lumen of the bronchi, and the patient has symptoms of severe asthma.[207] Colistin, methanesulfonate, desensitization, and prednisolone have relieved the symptoms, and patients recover without specific antifungal therapy.

Bronchial Geotrichosis

In this form, the lung is not involved, and the disease consists of an endobronchial infection. Symptoms include a prominent chronic cough, gelatinous sputum, lack of fever, and medium to coarse rales. On x-ray, diffuse peribronchial thickenings are seen. A fine mottling may be present on the middle and basilar pulmonary fields. About nine cases have been recorded.[167] In the early literature the allergic form of chronic bronchopulmonary geotrichosis was not differentiated from the infectious or invasive forms,[188, 244] and was sometimes called "tea taster's cough." Colonization of the bronchi by *Candida albicans* and *Geotrichum* can occur in these patients and those with chronic obstructive lung disease. On bronchoscopy, fine white patches are seen in the bronchial tree. They are similar in appearance to the lesions of thrush. Such lesions remain stable for years, and their clinical significance is questionable. The sputum contains large numbers of the organism. Specific antifungal therapy is usually not necessary for control of this form of geotrichosis.

Oral Geotrichosis

This disorder is identical in appearance to thrush and formerly was often confused with it. The two can be differentiated only by direct microscopic examination and culture. Few authentic cases of this disease exist.[167]

Gastrointestinal Geotrichosis

This disorder has been recorded a few times. The first case was that of Almeida and Lacaz.[9] More recently, Neagoe reported on enterocolitis associated with glutamic acid therapy.[172] Symptoms disappeared when the therapy was discontinued. The establishment of an etiologic relationship of the organism to the disease is very difficult, as *Geotrichum*, like *Candida*, is part of the normal flora.[45, 180] Intestinal geotrichosis has been described in animals.[148]

Cutaneous Geotrichosis

A few cases of cutaneous infection have been recorded. In a recent case the lesion, a cystic mass, appeared in the soft tissue of the hand following a skin grafting procedure. Drainage was followed by healing.[95]

PATHOLOGY

A case of disseminated geotrichosis was reported by Chang and Buerger in 1964.[45] The patient had carcinoma of the ascending colon and received fluorouracil as a treatment. Lesions containing *Geotrichum* were extensive and generalized throughout the body. Infection of the colonic mucosa adjacent to the carcinoma was seen. This was the probable point of entry for the organism. The gross lesions appeared as necrotic foci and were observed in heart, lungs, and spleen. Microscopic examination revealed suppuration and

necrosis. The fungus was seen as a mixture of yeastlike cells and septate mycelium with oval and spherical arthroconidia. Some areas contained the rectangular arthroconidia characteristic of *Geotrichum*. In addition, some of the lesions had *Candida*-like organisms in them. If this truly was a case of geotrichosis, it represents an opportunistic infection. Unlike the chronic bronchial forms, systemic infection is rapidly fatal.

Establishing a diagnosis of geotrichosis is very difficult. All other etiologies must be eliminated. Similar to *Candida*, the organism is part of the normal flora of the mouth and intestine but is a much less aggressive opportunist. The two cannot be differentiated on histologic appearance alone; only identification by culture is adequate. In the terminal stages of debilitating disease, it is possible that both organisms may invade and set up focal lesions.

A critical reading of past literature indicates that very few authentic cases of infection by *Geotrichum candidum* have occurred. Because of its ubiquity and its rapid colonization in numerous environmental situations, it heads the list of organisms erroneously associated with infectious diseases.

ANIMAL DISEASE

Natural disease in animals is also quite rare. The organism has caused an adenitis in pigs, intestinal disease in the ocelot,[41] and disseminated disease in the dog.[148] Early reports of mastitis in cows and infection in other domestic animals probably reflect confusion with *Candida* spp. In the case of the disseminated disease in the dog, lesions consisted of areas of coagulative necrosis and macrophage invasion in the lung, lymph nodes, and kidney. Small granulomas were seen in liver, spleen, bone marrow, and eye.

Experimental Infection. In laboratory animals, this is very difficult to establish.[168] In most of the recorded literature, the attempts have been unsuccessful or the identity of the organism questionable. The fungus is easily confused with other yeastlike organisms, especially *Trichosporon capitatum*. This latter yeast, when injected into rabbits, regularly produces abscesses.

LABORATORY IDENTIFICATION

There are no serologic procedures useful for the diagnosis of the disease, and the skin test is of no value. The latter and the oidio-mycin skin test only indicate host sensitization by normal flora. Direct and histologic examination may strongly suggest the presence of *Geotrichum*, but cultural confirmation is always necessary.

Direct Examination. Sputum, pus, or lesion material is examined as a potassium hydroxide mount. As in examination for other fungi, sputum may be digested first (by the methods used for isolation of tubercle bacilli), the specimen concentrated and spun, and the sediment examined. Gram stains are also useful. The characteristic morphology is an oblong or rectangular arthroconidium, 4 to 8 μ in size. The ends are square or rounded. Spherical cells, 4 to 10 μ in size, are also seen.

Culture Methods. Material to be cultured can be spread over media containing antibiotics and incubated at 25°C.

MYCOLOGY

Geotrichum candidum Link ex Persoon 1822

Synonymy. *Oidium pulmoneum* Bennett 1842.

Teleomorph State. *Endomyces geotrichum* Butler and Petersen 1972. Carmichael[40] lists 100 synonyms of *Geotrichum.* They are usually in the genera *Oidium, Oospora,* and *Mycoderma.*

Morphology. SGA, 25°C. The organism grows as a dry, mealy, white to cream colony (Fig. 26–1). At 37°C growth is very slow and subsurface. Microscopically the hyphae are seen to fragment into arthroconidia, which are quite variable in size. The arthroconidia are seen to germinate at one end, giving the appearance of a bud. The latter develops into

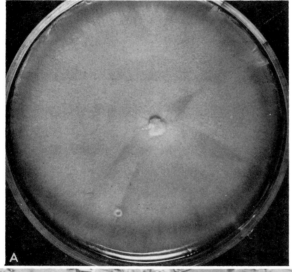

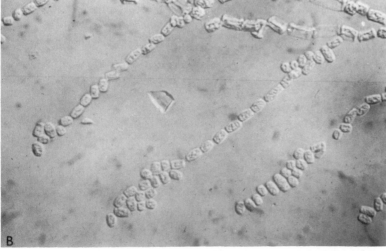

Figure 26–1. Geotrichosis. *A,* Colony of *Geotrichum candidum.* It is off-white, mealy, and glabrous. *B,* Fragmentation of mycelium to form typical arthroconidia. ×400. (Courtesy of S. McMillen.)

a septate mycelium, however. True blastoconidia production is not found in the genus. Carbohydrates are not fermented. Auxanograms usually show assimilation of glucose and galactose. Lactose, maltose, erythritol, cellibiose, melibiose, and sucrose are not assimilated. It is urease-negative. The identification of G. *candidum* is very difficult, especially its differentiation from *Trichosporon* species.

Recently Butler and Petersen[33] described the teleomorph state of the organism as *Endomyces geotrichum* in the class Ascomycetes. However, Von Arx considers the species to be in the genus *Dipodascas* as *D. geotrichum* Von Arx 1977.

Adiaspiromycosis

DEFINITION

Adiaspiromycosis refers to the *in vivo* development, without replication, of adiaconidia from inhaled conidia of the fungal genus *Chrysosporium (Emmonsia)*. The condition is commonly found in the lungs of rodents and other small animals that live or burrow in the soil. The environment inhabited by the animals and the fungus ranges from hot, dry desert and semi-arid habitats to banks of streams and ponds in tropical jungles or semiarctic, coniferous forests. The distribution is worldwide.[122, 123, 233] The finding of adiaconidia in human tissue is quite rare. Such conidia in sufficient numbers in human or animal tissue to cause distress or disease are also uncommon. The name "adiaconidia" signifies the enlargement, without replication, of a fungal conidium under the influence of elevated temperature.[77] Adiaspiromycosis is the term used when such development occurs within an animal body.

Synonymy. Haplomycosis, adiasporosis.

HISTORY

Many of the human and animal mycoses were initially described as protozoan diseases. This was usually done on pathologic evidence alone. It is only chance observation and appreciation that fungi may have different morphologic forms under different environmental conditions of growth that has led to the discovery of the true nature of the etiologic agent. Adiaspiromycosis was discovered in 1942 by Emmons while examining rodents from Arizona for infection by *Coccidioides immitis*.[76] Since that time the former disease has been found to be extremely common in rodents and small animals from all parts of the world. The disease is not new, as it now has been recognized in specimens of preserved lungs from animals trapped in Sweden in 1845. It is likely that the organism had been seen in tissue previously but described as a protozoan. The fungal nature of the condition was appreciated by Kirschenblatt[134] in 1939. He recorded cysts in the lungs of several rodents from Georgia *(Transcaucasia)*. He described the cause as a fungus, *Rhinosporidium pulmonale*, but without culture or a Latin description.

In a search for an animal reservoir of coccidioidomycosis, Emmons and Ashburn[76] in 1942 examined 303 rodents trapped in southern Arizona. *Coccidioides immitis* was isolated from 25 of these, but a new fungus was present in 101 animals. Both fungi were found in eight. Besides the spherules of *C. immitis*, another type of fungal entity was found. This had a pale-staining cytoplasm and very thick laminated walls, was uninucleate, and did not form endospores. Later work showed that the conidia from the cultures of the new fungus grew to these large bodies ($<40\ \mu$) when injected into mice or grown at 37°C. The new fungus was named *Haplosporangium parvum* because of the purported resemblance to the zygomycete *Haplosporangium bisporale*. Basing their taxonomy on critical mycologic observation, Ciferri and Montemartini[50] considered the organism to represent a new hyphomycete genus and named it *Emmonsia parvum*. Isolants of *E. parvum* from the desert areas of the southwest United States produced adiaconidia 20 to $40\ \mu$ in diameter. However, larger adiaconidia were being discovered in material from Canada and the northern United States. The strains that produced adiaconidia with a diameter up to 400 μ were designated a new species, *Emmonsia crescens*, in 1960.[77] Other workers have concluded that there is not sufficient evidence to separate the two species, and the latter may be regarded as a variety of the former.[39, 183] They conclude that the fungus is best regarded as a species of the genus *Chrysosporium.*

ETIOLOGY, ECOLOGY, AND DISTRIBUTION

The only agents so far identified as producing adiaconidia in the lungs of animals are *Chrysosporium parvum* and *C. parvum* var. *crescens.* The organisms identified as *E. brasiliensis* and *E. ciferrina* were later shown to be *Chrysosporium pruinosum.*[183] This latter species produces large conidia in culture. When these are injected into animals, they remain viable and can be recultured, but they do not enlarge to form adiaconidia.

Adiaspiromycosis has been recorded in a number of animal species. The disease is very common in mice, moles, rats, rabbits, ground squirrels, and other rodents.[115] It has also been found in carnivores, such as skunks, weasels, martens, and minks, and in other animals, such as armadillos, wallabies, and opposums.[122, 123] What was probably the first human case of the disease was described by Doby-Dubois[67] in 1964. A single adiaconidium was found in a nodule in the parenchyma near a bronchus in a patient who had pulmonary aspergillosis and cystic disease of the lung. Another case has been discovered in Honduras, again represented by a single adiaconidium in a nodule.[56] A few cases have been recorded in Czechoslovakia,[138] France,[193] and Venezuela.[209] In a review by Salfelder et al.,[209] the authors accept ten human infections as valid. In one report, enormous numbers of adiaconidia of the var. *crescens* were found in the lung biopsy of an 11-year-old boy. It appeared that functional impairment of the lung was present.[138] When large numbers of conidia are present in natural or experimentally produced disease, severe respiratory distress may ensue. Sometimes the infection occurs as an epidemic among animals and can cause severe pneumonia and death.[154] Skin infection has been claimed.[193]

A spring peak in the incidence of the disease in animals has been reported by Dvorák et al.[71] These authors also speculate that adiaconidia may serve in dissemination of the species to other locales through the mediation of the lungs of infected animals.[72] Following death of the animals, the mycelial stage of the organism grows in the burrows of the rodents or in surrounding soil.

CLINICAL DISEASE AND PATHOLOGY

Naturally occurring adiaspiromycosis is usually restricted to the lungs. The conidia do not migrate, replicate, or disseminate, so that lesions are found only in the endobronchial or alveolar spaces. When few conidia are present, there is little cellular response. A few mononuclear cells may surround the developing conidia. In cases of heavy exposure, granulomas may develop. In some of the original animals examined by Emmons,[76] adiaconidia and the spherules of *C. immitis* were found in the same granuloma. Very rarely conidia may be found in other organs.[138]

The size of the adiaconidia produced by *C. parvum* is quite variable. In natural disease, the average size is 10 to 13 μ. In experimental disease using white mice, the adiaconidia reach a diameter of 40 μ. This is an increase in volume of 10^4. *C. parvum* var. *crescens* is much more common throughout the world. The adiaconidia produced average 200 to 400 μ in diameter. Such conidia are seen surrounded by a few histiocytes.[24] Injection of the fungus into white mice may be followed by the production of adiaconidia up to 600 μ in diameter and of a wall with a thickness of 70 μ. This is a volume increase of 10^7.

Natural disease is limited to the lung, but systemic disease can be produced if aleurioconidia from a mycelial culture are injected into mice, dogs, rabbits, and rats. The conidia enlarge and produce adiaconidia in any organ, but they do not replicate. If conidia are given in sufficient numbers, death may ensue from mechanical obstruction. Recently Kamalan and Thambiah[125] claim to have found two cases of skin infection caused by *C. parvum* var. *crescens.*

On staining, the fungus shows a laminate wall and often a vacuolate cytoplasm. The inner layers of the wall take the PAS stain, but the outer layers are variable in reaction (Fig. 26–2). The cytoplasm is pale-staining and in *C. parvum* has one nucleus. *C. parvum* var. *crescens* is multinucleate.

Culture. The organism grows on most laboratory media at 25°C as a mycelial colony. At 37°C, adiaconidia are produced, but they do not proliferate at this temperature.

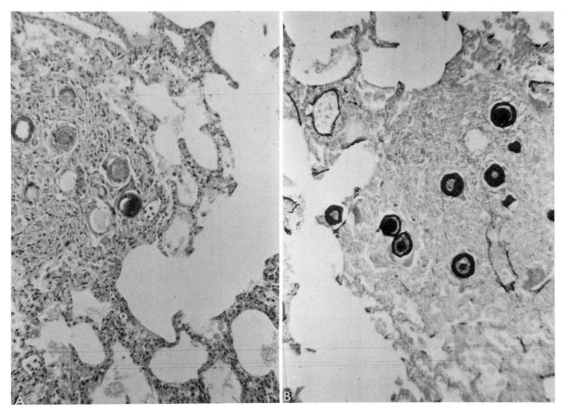

Figure 26–2. Adiaspiromycosis. *A*, Section of rat lung showing large vacuolated adiaconidia and granuloma formation. Hematoxylin and eosin stain. ×60. *B*, Section of lung stained by Gridley method showing thick laminate walls of adiaconidia. ×60.

MYCOLOGY

The taxonomy of this organism is still unsettled. The original epithet used by Emmons, *Haplosporangium*, cannot accommodate the organism, and a new monotypic genus was created to include it. Later Emmons divided the known strains into two species. *E. parvum* produced adiaconidia that were uninucleate and reached a diameter of 40 μ. This was the type commonly isolated from the southwestern United States and perhaps once in Kenya. In the thousands of lesions and isolants found in other parts of the world, the adiaconidia were larger and multinucleate. Emmons[77] considered this sufficient evidence to create a new species, *E. crescens*, for them. However, in culture the colonial characteristics, conidiation, and mycelial morphology of the two organisms are essentially identical. For this reason other workers consider them to be varieties of the same species. Furthermore, Carmichael[39] and Padhye and Carmichael[183] discuss the similar-

ity of the *Chrysosporium, Emmonsia, Blastomyces,* and *Paracoccidioides* genera. Critical mycologic investigation is needed before natural relationships among these organisms can be made.

Chrysosporium parvum (Emmons et Ashburn) Carmichael 1962

Synonymy. *Haplosporangium parvum* Emmons et Ashburn 1942; *Emmonsia crescens* Emmons et Jellison 1960; *Emmonsia parva* Ciferri et Montemartini 1959.

The organism grows initially as a glabrous, colorless colony, which in time produces white aerial mycelia. It reaches a diameter of 5 cm in two weeks. The colony often has alternate areas of tufted mycelium or mycelium in coremia, with areas of a glabrous consistency.

On microscopic examination the hyphae are seen to be septate and branching, with a diameter of 0.5 to 2 μ. Numerous conidia are produced on conidiophores that branch at right angles from the vegetative hyphae. These conidia are spherical, though slightly flattened in the vertical axis, 3 to 3.5 μ in diameter, and may have fine to coarse spinulation. Secondary conidia may be formed from the primary conidia. The var. *crescens* is reported to have slightly larger conidia. These may reach a diameter of 4.5 μ and are more ovoid in shape.

When the conidia and hyphae of *C. parvum* are placed on media at 40°C, the mycelium degenerates but the conidia enlarge. The size reached by the enlarging conidia will depend on how crowded they are. They average 15 to 25 μ, but some will grow to 40 μ or more, with a wall 2 μ thick. They remain uninucleate and do not replicate. Emmons cautions that a crowded inoculum when transferred will show apparent multiplication, but this is owing only to the further enlarging of conidia in a less crowded environment. The conidia of the var. *crescens* when grown at 40°C develop into adiaconidia from 200 to 700 μ in diameter. These may have walls up to 70 μ thick and contain several hundred nuclei. If adiaconidia are placed at 25°C, they sprout, producing numerous mycelial strands. A type of "budding" of the adiaconidia can be produced that resembles *Paracoccidioides brasiliensis.*[74] If conidia of var. *crescens* are incubated at 40°C for 12 days (attaining a diameter of 200 μ) and then incubated at a temperature of 25°C, they will sprout, with the formation of many hyphae. If the culture is reincubated at 40°C within 4 to 8 hours, the hyphae initially will form secondary adiaconidia, giving the appearance of the multiple budding yeast cells of *P. brasiliensis*. In some culture preparations, several secondary adiaconidia may be formed within the parent conidium, giving the impression of endosporulation. The process is quite different, however, and no true multiplication occurs.

Histoplasmosis farciminosum (Epizootic Lymphangitis)

Epizootic lymphangitis is an infection of horses and mules caused by *Histoplasma farciminosum*. The disease was first described by Rivolta in 1873.[5] He noted the budding yeast cells within lesions from horses and named the organism *Cryptococcus farciminosum*. In 1934, Redaelli and Ciferri transferred it to the genus *Histoplasma* because of the similarity in life cycle and the appearance of the macroconidia. Ajello[5] has reviewed the genus *Histoplasma* and contends that the etiologic agent of epizootic lymphangitis should not be included in it.

The disease is widespread throughout Scandinavia, Russia, central and southern Europe, northern Africa, India, and southern Asia.[2] It is particularly prevalent in the northern sections of Egypt and in India.[164, 198] The horse is most commonly involved and constituted 89 per cent of the cases in one survey.[198] Mules and sometimes donkeys are also involved. Male animals have a higher rate of infection (61.6 per cent) than females. The incidence of infection is highest in January in Egypt, and some 724 cases were reported between 1960 and 1970.

The type of clinical disease most commonly seen involves subcutaneous and ulcerated lesions of the skin. The local lymphatics are involved, and thus this is a primary inoculation mycosis. Frequently both front and hind legs are involved, as is the neck, other areas rubbed by a harness, or an area which is the site of repeated injury. A mixed necrotic, pyogenic, and granulomatous process is seen in tissue section. The yeast cells are seen free and intracellularly within giant cells. The yeasts are 3 to 5 × 2.5 to 3.5 μ in diameter.

Fawi[80, 81] has recently reported on the primary pulmonary form of the disease, which in time caused death of the animals. The horses had multiple soft gray granulomata throughout the parenchyma of the lung, which varied in size from 0.5 to 20 cm in diameter. The lymph nodes were not involved, nor were other organs of the body. Previously Bennett had recorded "cryptococcal" pneumonia in 1931.[21] Disseminated disease involving all organ systems has been recorded sporadically.[221] Experimental disease has been produced by intradermal or intraperitoneal injection of mice and rabbits.[220]

Treatment of the infection has included amphotericin B, hamycin, and several other modalities. The yeast form of the organism is quite sensitive to hamycin *in vitro.*[182] Recently Richer[204] reviewed the subject and reported on 12 cases of the disease in Senegal. Treatment of these cases included neoarsphenamine, thiabendazole, miconazole, econazole, potassium iodide, and mercuric iodide.

Diagnosis of the disease depends on isolation of the organism. Lesion material can be planted on Sabouraud's agar plates containing antibiotics and grown at 25°C. The yeast form is maintained by incubation at 37°C, and growth is enhanced if Hartley horse blood agar is used in an atmosphere of 20 per cent CO_2. Fawi[81] has devised a fluorescent antibody test. Sera of the horse are placed on fixed smears of the organisms. Forty-seven of 50 sera from proven cases were positive and controls negative. *H. farciminosum* produces H and M histoplasmin antigens, as do other members of the genus. Standard and Kaufman[228] report they are specific for the species.

Histoplasma farciminosum (Rivolta) Redaelli et Ciferri 1934

Synonymy. *Cryptococcus farciminosum* Rivolta 1873.

Morphology. SGA, 25°C. The fungus grows slowly and produces a grayish white colony. Microscopically septate mycelia are seen. A variety of conidia are found, including arthroconidia, chlamydoconidia, and some blastoconidia, but most isolants fail to conidiate. Round, smooth macroaleurioconidia are produced that somewhat resemble the macroconidia of *H. capsulatum* but lack the tubercles. At 37°C a yeast resembling *H. capsulatum* is seen.

Basidiomycosis

Basidiomycota, though very common in nature, are very rarely associated with human disease. Allergic reactions to inhalation of the spores have been frequently cited, but actual infection is very rare. The exception to this is the recent description of the teleomorphic stage of *Cryptococcus*, which is in the basidiomycota genus *Filobasidiella* (see Chapter 21).

Emmons in 1954[75] reported on the repeated isolation of the mushroom *Coprinus micaceus* from the sputa of one patient. Mycelial strands were found on direct examination. Ciferri et al.[51] reported that *Schizophyllum commune* was isolated and mycelial elements were seen in sputum from another patient with chronic lung disease.

Meningitis was attributed to an *Ustilago* sp. by Moore et al.[165] Lesions contained multinucleate giant cells and macrophages. Within these were seen structures that resembled the sprout mycelium and echinulate spores of *Ustilago zeae*. No cultures were obtained, however. In another case of meningitis, the organism *Schizophyllum commune* was isolated on 16 cultures from the first spinal tap and on 19 cultures from a second tap obtained four months later. The patient recovered uneventfully.[46] Restrepo et al.[201] also recovered *Schizophyllum commune* from human disease. Their patient, a 4-year-old girl, had ulcerations in the mouth and a perforated palate (Fig. 26–3). No underlying disease was noted, and the lesion responded to amphotericin B therapy. A nail infection has been described by Kligman.[135]

Probably the only well-documented case of a systemic Basidiomycota infection is that of Speller and MacIver in 1971.[226] In a patient who had had a mitral valve prosthesis, endocarditis of the aortic valve developed. When this was removed, it was found to contain much septate mycelium in association with an acute inflammatory process. On culture, an oidium producing fungus was isolated. At autopsy, involvement of other organs was not noted. It was determined by mating experiments that the organism was the haploid stage of *Coprinus cinereus,* one of the inky cap mushrooms. The mycelium of several *Coprinus* spp. grows on plaster, and this may have been the source of exposure during the original surgical procedure. Greer[98] has critically reviewed the reported cases of Basidiomycota infections of man.

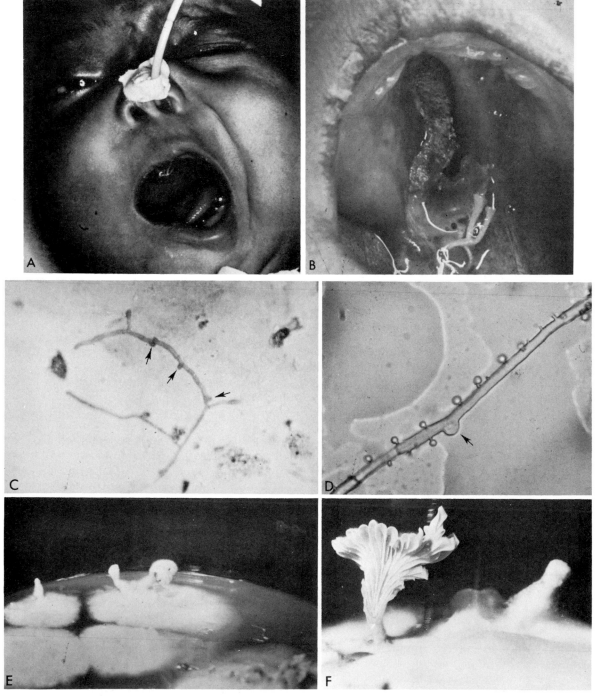

Figure 26–3. Basidiomycosis caused by *Schizophyllum commune*. *A*, Child severely dehydrated by viral diarrhea. *B*, Necrosed area of soft palate on roof of mouth. *C*, Scraping from roof of mouth. Mycelium is observed and clamp connections are evident. Clamp connections indicate this to be a Basidiomycota (*arrows*). *D*, Mycelium in culture showing clamp connection (*arrow*) and series of spicules. The latter are characteristic of *Schizophyllum*. *E*, Culture from lesion. The mycelium is beginning to organize into a basidiocarp. *F*, Basidiocarp of *Schizophyllum commune* from culture of lesion.[201] (Courtesy of A. Restrepo-Moreno and D. Greer.)

Protothecosis and Other Algoses

DEFINITION

Protothecosis is an infection caused by members of the genus *Prototheca.* These organisms are generally considered to be achloric algae. The disease has been recorded in man and several domestic and wild animals. The lesions have varied from cutaneous and subcutaneous infections to systemic invasion involving several internal organs. Most of such infections probably represent traumatic implantation into subcutaneous tissues; however, a few reports indicate disease as an opportunistic infection as well. The organisms are ubiquitous in nature and can be isolated from human and animal skin, feces, and sputum in the absence of disease.[53, 224] Infections by blue-green (Cyanobacteria) and green (Chlorophyta) algae have also been recorded.

HISTORY

Prototheca was a genus erected by Krüger in 1894 to encompass some nonpigmented unicellular organisms found in the mucinous flux of trees. They were considered to be yeasts and therefore included in the fungi. West in 1916 reclassified them as algae because their spores are produced internally in a manner identical to that of the green alga *Chlorella.* As late as 1957, however, Ciferri[49] reclassified them as saccharomycetes. They have the general appearance of achlorophyllic mutants of *Chlorella,*[32] but there are differences in cell wall composition,[58, 59] physiology, and ability to survive environmental stress. It is now generally considered that the genus developed from *Chlorella* at some point in evolution.

ETIOLOGY, ECOLOGY, AND DISTRIBUTION

Algae are autotrophic organisms. Their importance in medicine has generally been restricted to toxicity to animals and man following ingestion of contaminated water or to imparting an unpleasant taste and odor to drinking water.[212] Recently, however, infection caused by chlorophyll-containing algae has been substantiated.[54, 126] The achloric *Prototheca* are heterotrophic and require external sources of organic carbon and nitrogen. There are now several (about 65) authentic records of various types of infection being caused by species of this genus. Members of the genus are commonly isolated from soil, detritus, and aqueous environments throughout the world; they also are part of the aerospora and are transient colonizers of human and animal integument and alimentary tract. Early reports of the association of *Prototheca* and disease are for the great part invalid. *Prototheca ciferri* (*P. zopfii*) was isolated from feces in patients with tropical sprue, and it was suggested that there was a relationship.[12] Almeida et al.[9] found *Prototheca* in cases of actinomycosis and paracoccidioidomycosis. These were probably contaminants.

Prototheca zopfii and *P. wickerhamii* account for all infections in animals and man in which identification was made. The subject of human infection by algae was reviewed in 1940 by Redaelli[195] and more recently in 1978 by Kaplan.[126] He tabulated animal infections and infection caused by green algae as well.

CLINICAL DISEASE

Human Infection. About 65 valid cases of human infection by *Prototheca* have been recorded in Kaplan's review of the subject.[126] He divided the report cases of human protothecosis into three categories. The first form, **simple cutaneous,** is characterized by the development of single or multiple lesions of the skin or underlying tissue. These generally are on an exposed area. About half of the 20 or so recorded cases of human disease fit into this category. They are very slow to develop and usually show no tendency to spontaneously resolve.[68] The lesions commonly appear as papulonodular areas, crusted papules, ulcerations, and rarely as an extensive granulomatous eruption (Fig. 26–4). Antecedent trauma or surgery has been noted in several cases, the first by Davies et al. in 1964.[59] The patient was a rice farmer in Sierra Leone. The lesion began on the inner side of the right foot as a depigmented area that had been injured several times by the patient's walking barefoot. Within three years it had become a rugose papule with a raised edge covering two thirds of the foot. In tissue sections and culture the grouped, rounded bodies of *Prototheca* were observed. A skin biopsy showed hyperkeratosis and pseudo-

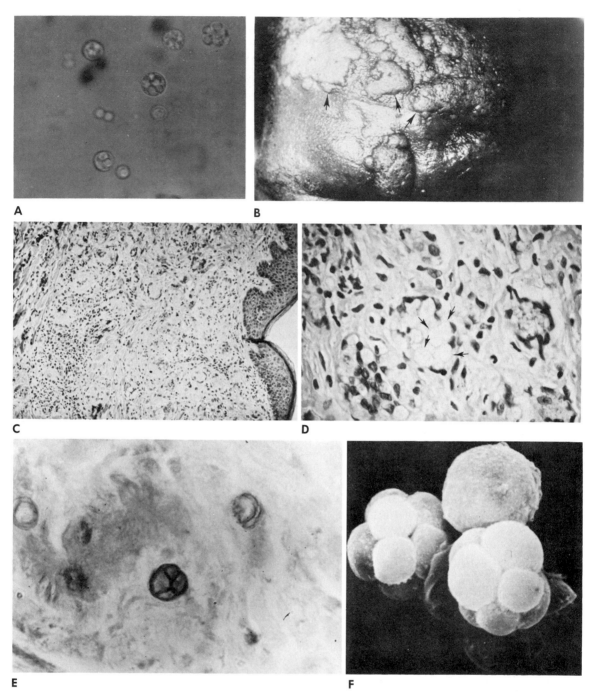

Figure 26–4. Protothecosis. *A, Prototheca zopfii.* Single cells are thecae with autospores. ×60. (Courtesy of M. Feo.) *B,* Nodular papules (*arrows*) on forehead of patient with no underlying disease. The infection was chronic and had gradually progressed for 10 years. *C,* Low power view of biopsy. Chronic inflammatory infiltrate, granulomatous reaction, and giant cells. Hematoxylin and eosin stain. ×100. *D,* Higher power view. Giant cells are seen to contain large bodies (*arrows*), colorless and indistinct with this stain. Hematoxylin and eosin stain. × 450. *E,* Same magnification, PAS stain. The theca and autospores are clearly visible as a "wheel-and-spoke" cell. *F,* Scanning electron microscope view of *Prototheca* autospores in a broken theca. (Courtesy of J. Williams.)

epitheliomatous hyperplasia. The organism was seen in the epidermis and in the papillary and reticular dermis. *Prototheca* had been noted to be sensitive to pentamidine. A total of 4.9 g was given to the patient but without

benefit. At last report,[58] the lesion was advancing, and organisms were found in the lower femoral lymph nodes. This isolant was named *P. segbwema* after the location of the hospital in which the organism was isolated, but the name has been reduced to synonymy with *P. zopfii*.

The second form discussed by Kaplan can be termed **prothecal olecranon bursitis.** About half the recorded cases of infection had involvement of this anatomic area. Most of these resulted from trauma to the site, such as an automobile accident, a fall, surgery, or some other injury.[176] The signs and symptoms were those of a persistent olecranon bursitis, with pain and soft tissue swelling. Most human infections, when recorded, are noted as having been caused by *P. wickerhamii*.[151, 229]

What may be termed an **opportunistic infection** by *Prototheca* was reported by Klintworth et al.[136] in 1968. The patient was a diabetic and had widespread metastases of breast cancer. *Prototheca wickerhamii* was isolated from several ulcerating papulopustular lesions on the leg. Neutrophils and macrophages were present, but the cellular response was minimal. The patient died of the carcinoma, but no autopsy was performed. The authors mentioned other cases, as yet unreported, of algosis. A clear case of opportunistic infection in an immunodeficient patient was recorded by Cox et al.[55a] The patient had an unknown defect in cellular immunity. Multiple lesions were found in the peritoneal cavity, nose, and subcutaneous areas of the face. Other opportunistic infections have been described.[247]

ANIMAL INFECTION

Over 40 cases of infection in wild and domestic animals have been reported. Ainsworth and Austwick recovered the organism from the inflamed udder of a cow.[3] *Prototheca* was cultured and seen in intestinal tissues and several organs in a case of enterocolitis of a dog reported by Van Kruiningen in 1970.[236] Three cases of the disease in cows have been recorded by Migaki.[162] Frese, Gedek, and Shiefer[86, 217] found the organism in the lymph nodes, bone, and subcutaneous tissue of the limbs of a two year old deer in Germany. Primary cutaneous disease in the dog[231] and cat[127] has been reported. Disseminated infections have been described several times in cows and dogs.[126] These involved all organ systems: heart, intestine, liver, kidney, brain, eyes, adrenals, and lymph nodes. Most animal infections are caused by *P. zopfii*.

Some seven cases of green alga infection have been reported so far. All of these were in animals.[54, 126] Four involved a cow or steer, two a sheep, and one the tail of a beaver (*Castor canadensis*). In these cases lesions containing organisms were found in skin, lymph nodes, liver, lung, and peritoneum. The identity of the organisms is still uncertain, but morphologically those that have been examined resemble *Chlorella* spp.

PATHOLOGY

Direct examination, when recorded, has shown that the organism is easily mistaken for cells of various fungi, particularly the etiologic agents of chromoblastomycosis and histoplasmosis duboisii. In *Prototheca* one can observe a number of daughter cells (autospores) forming within the theca. No budding (blastconidia formation) is found.

In tissue sections containing cells of *Prototheca*, little cellular reaction is observed. Depending on the species, the size of the cells varies from 8 to 20 μ in diameter. The morphology of the cells also varies among the several species described.

Identification in Tissue. *Prototheca* is a large nonbudding cell readily seen in tissue. It is spheroid, ovoid, or elliptical, with a prominent wall. The round cell (theca) contains several thick-walled autospores. The organisms are not readily apparent in hematoxylin and eosin stained smears, but stain well with Grocott's modification of the Gomori methenamine silver or the PAS stain. The PAS stain is particularly useful for observing starch grains within the cells of algae. Sudman and Kaplan[230] have developed an immunofluorescence stain for the identification of *Prototheca* in tissue.

PROGNOSIS AND THERAPY

Davies et al.[58] have shown that *Prototheca* are moderately sensitive to several therapeutic agents, particularly pentamidine. This latter drug was used in the treatment of one case but had little effect. The organism was inhibited by 10 μg of amphotericin B and 100 units per ml of nystatin. Segal[213] found both nystatin and amphotericin B inhibitory. Amphotericin B has been used alone and in combination with transfer factor.[55a] In general the clinical response to medical therapy, except for surgical excision,[176] has been disappointing. Experimental disease in animals has been tried by several authors, but the results have been equivocal.

ORGANISMS INVOLVED

The classification of the genus *Prototheca* has been examined by several authors.[11, 44, 53, 191] Arnold and Ahearn[1] investigated the carbohydrate and alcohol assimilation patterns of several isolants in 1972. They concluded that there are five species in the genus: *Prototheca filamenta*, *P. stagnora*, *P. zopfii*, *P. moriformis*, and *P. wickerhamii*. Since then the consensus is that there are only three species in the genus *Prototheca: P. wickerhamii, P. zopfii*, and *P. stagnora* (Table 26–1). Based on their extensive investigation, Kaplan and Sudman concluded[230] that *P. chlorelloides, P. pastoriensis, P. trispora, P. ubrisyi*, and *P. moriformis* are in synonymy with *P. zopfii. P. filamenta* was considered in synonymy with *Trichosporon inkin* and placed in a new genus, *Sarcinosporon*, as *S. inkin* by King and Jong.[132]

Table 26–1 *Schema to the Genus* Prototheca*

| Species | Assimilation of | | |
	Sucrose	Trehalose	Propanol
P. stagnora	+†	−	−
P. wickerhamii	−	+‡	−
P. zopfii	−	−	+

*From Kaplan[126]
†After 14 days
‡After 7 days

Pore et al.[191] reinvestigated the problem and proposed yet another genus for the organism: *Fissuricella filamenta*. Confusion reigns.

What the author knows as *S. inkin* is an incidental isolant as flora of the genitourinary region of the skin and mucous membrane. It is sometimes found in urine, from scrotal smegma, and from bladder serosa when catheters have been used. It appears as packets of yeast cells in fours, eights, and so forth, and the morphology resembles that of the bac-

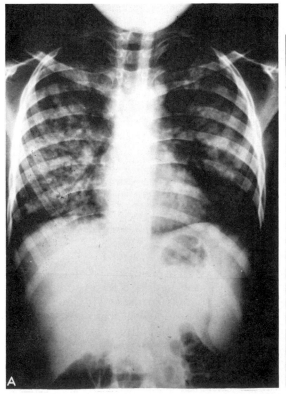
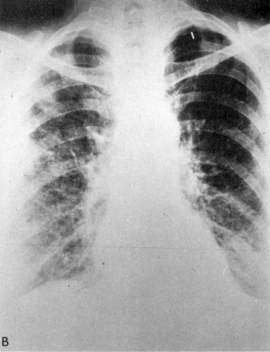

Figure 26–5. Pneumocystosis. The two most frequently seen x-ray patterns. *A*, Diffuse alveolar infiltrate throughout both lungs. Air bronchograms are seen. The liver and spleen are enlarged. Child with leukemia. *B*, Diffuse interstitial infiltrates in both lungs, but more evident in right lung. There are small areas of alveolar infiltrate as well. (Courtesy of J. Fennessy.)

terium *Micrococcus (Sarcina) tetragena* but the organism is much larger.

So far only *P. zopfii* and *P. wickerhamii* have been involved in human or animal infection. Pore[190] has devised a selective medium for the isolation of *Prototheca*.

PHYCOLOGY

Prototheca zopfii Kruger 1894

Synonymy. *P. moriformis* Kruger 1894; *P. chlorelloides* Beijerinck 1904; *P. pastoricensis* Ashford et al. 1930; *P. pastoricensis* var. *trispora* Ashford, Ciferri, et Salman 1930; *P. ciferii* Negroni et Blaisten 1941; *P. segbwema* Davies, Spencer, et Waklein 1964; *P. ubrisyi* Zsolt et Novak 1968.

Morphology. SGA, 25° C. The colony is dull white and yeastlike in consistency. It does not grow on medium containing cyclohex-imide. The cells are variable in size and shape, being 8.1 to 24.2 × 10.8 to 26.9 μ in diameter. The autospores are spherical and 9 to 11 μ in diameter (Fig. 26–5). The morphol-ogy varies, depending on the medium em-ployed for growth.

Assimilation. Glucose +, sucrose −, mal-tose −, lactose −, galactose +, cellobiose −, xylose −, raffinose −, trehalose −, methan-ol−, ethanol −, *n*-propanol +, *n*-butanol +, glycerol +, sorbitol −, inositol −, dulcitol −, erythritol −.

Prototheca wickerhamii Tubaki et Soneda 1959

Morphology. SGA, 25° C. The colony is moist and cream-colored. Growth is optimal at 30° C. The cells are similar to those of *P. zopfii* in shape but somewhat smaller. The cells vary from 8.1 to 13.4 × 10.8 to 16.1 μ when grown on glucose-containing media. The autospores are smaller (4 to 5 μ in diameter) and more numerous (up to 50 per theca).

Assimilation. Same as for *P. zopfii*, with the following exceptions: trehalose +, and *n*-propanol−.

P. stagnora Cooke 1968 is mucoid owing to the capsular material it produces. Unlike *P. zopfii* and *P. wickerhamii*, it does not grow well at 37° C.

Pneumocystosis

DEFINITION

Pneumocystosis is an infection of the al-veolar areas of the lung, leading to an inter-stitial pneumonia. The etiologic agent is *Pneumocystis carinii*. This organism is general-ly considered to be a protozoan but is proba-bly allied to the fungi. It is of worldwide distribution and appears to be part of the normal flora of rodents and perhaps other animals, including man. The disease is most often encountered in infants, in whom it may reach epidemic proportions, and in patients debilitated by a number of disease states or iatrogenic procedures. Pneumocystosis is be-coming increasingly more significant and prevalent in this latter group.

Synonymy. Chronic interstitial plasma cell pneumonia.

HISTORY

Pneumocystosis was first observed in the lungs of guinea pigs in 1909, by Chagas,[43] who was working on experimental infection of *Trypanosoma cruzi*. He considered the struc-tures to be a part of the life cycle of that organism. Carini[37, 38] in 1910 found the or-ganism in rats that had not been infected with other parasites. After studying the various structures seen in tissue sections, Delanöe and Delanöe[61] in 1912 named the organism *Pneumocystis carinii*. Following careful exami-nation and comparison with other parasites, Wenyon in 1926[245] classified the organism as a protozoan in the class Sporozoa, subclass Coccidiomorpha.

Many fungus diseases were first consid-ered to have protozoan etiologies because the initial descriptions were made on histologic evidence alone. Only later by use of several lines of investigation did their true biological affinity become clear. Several enzymes, in-cluding chitinase, attack the cell wall of *Pneu-mocystis*,[223] as well as the walls of true fungi. This, in addition to observations on its ultra-structure, suggests a relation of the organism to the fungi. This evidence is at most tenuous, so that the phylogenetic affinities of the or-ganism are as yet still in doubt. *Pneumocystis* infection is included in this chapter because the disease is assuming greater importance and must be considered in the differential diagnosis of several fungal and actinomycotic diseases, particularly those termed "opportunistic infections."

Rossle in 1923[208] described a chronic interstitial pneumonia of infants. Rossle's work evoked a great deal of interest and led to the discovery of the disease in several parts of central Europe. This type of pneumonia was not yet associated with the *Pneumocystis* organism, but in retrospect these were probably cases of pneumocystosis. Van der Meer and Burg[237] in 1942 were probably the first investigators to associate *Pneumocystis* with the form of chronic interstitial pneumonia that was characterized by large numbers of plasma cells. Rossle's work and that of Van der Meer stimulated investigation of this form of pneumonia, and it was realized that the disease was much more common than previously appreciated. Between 1941 and 1949 some 700 cases were reported in Switzerland alone, and the disease was found in many other countries thereafter. Essentially all these cases involved infections of prematurely born infants or "sickly" neonates.

The first cases in adults were noted by McMilland and Hamperl in Europe,[103] and by the mid 1960's some 19 cases had been tabulated from the United States.[79, 109] Of these, 14 were complications of neoplastic disease, including leukemias, Hodgkin's disease, lymphosarcoma, and multiple myeloma. In a retrospective study, Esterly and Warner reviewed 12 cases at the University of Chicago Medical Center.[79] Six were from children with lymphoreticular malignancies, and six were from adults with similar diseases. All but one had had steroid therapy, and all had received antibiotics. At various times and in various reviews, the disease has been associated with cytomegalic inclusion pneumonia, renal transplants, steroid therapy, particularly after cessation of the regimen, and hyaline membrane disease.[79] It now appears that any debilitation, particularly when complicated by cytotoxin or steroid therapy, may result in *Pneumocystis* pneumonia.

CLINICAL DISEASE

The clinical course of the pulmonary disease is insidious.[205] There is gradually progressive dyspnea, tachypnea, tachycardia, and cyanosis, with little or no fever unless there is complication by other infections. Chest x-ray reveals gradual spreading of perihilar haziness, with some granular components or formation of indistinct nodules.[29] There is an alveolar or interstitial infiltrate but no pulmonary infiltrate (Fig. 26–5). Very frequently the course is complicated by superinfection with bacterial or other fungal opportunists. Only careful analysis of the several possible causes of the infection results in adequate therapy and resolution of the disease. Walzer et al. in 1974 reviewed 194 cases from the United States.[242]

PATHOLOGY

Gross Pathology. At autopsy, grayish yellow or pink-tinged consolidations of the lung are seen[227] (Fig. 26–6). In outward appearance and consistency, these areas often resemble lobular pancreatic tissue. Septal tissue is prominent, but the pleura is usually normal. In most of the cases of Esterly and Warner,[79] prominent septal cells and hyaline membranes were found.

Histopathology. The most characteristic finding in severely involved tissue is pink, honeycombed, foamy, vacuolated, intraalveolar material (Fig. 26–7C). This material is PAS-positive, usually separated from the alveolar wall, and the cysts of *P. carinii* are

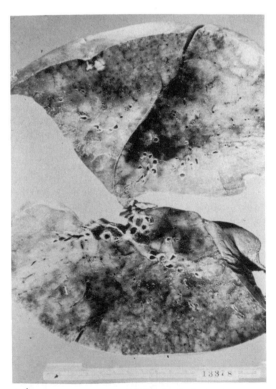

Figure 26–6. Pneumocystosis. Lobular "pancreas-like" consolidation of lung.

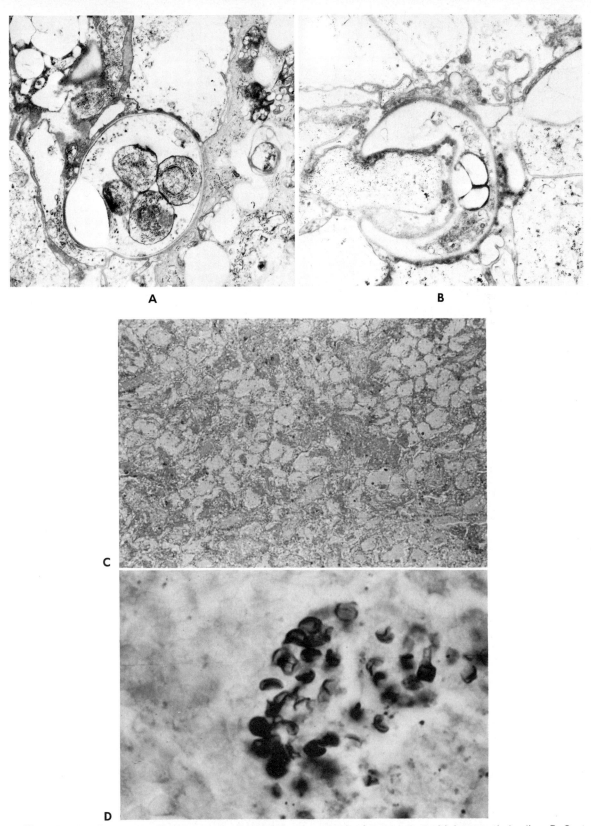

Figure 26–7. *Pneumocystis* pneumonia. *A,* Electron micrograph of mature cyst with intracystic bodies. *B,* Cyst being invaded by macrophage. ×44,000. (Courtesy of L. Vogel.) Pneumocystosis. *C,* Section of lung showing pink, honeycombed, foamy, vacuolated, intra-alveolar material. There are plasma cells, lymphocytes, histiocytes, and some giant cells. ×60. *D,* Wrinkled, crescent-shaped, cup-shaped, "parentheses" and round cells are seen within a giant cell. ×400. (Courtesy of S. Thomsen.)

Illustration continued on following page

657

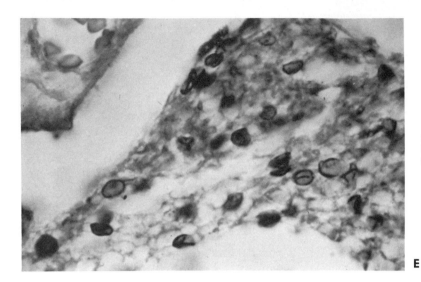

Figure 26–7 *Continued* Pneumocystosis. *E,* Free-living cells in foamy material within alveolar space. GMS. × 400.

E

easily seen within it. Plasma cells and macrophages may or may not be present,[79] probably reflecting the immune competency of the patient. The organism lives freely in the intra-alveolar airspace and within the alveolar wall. This accounts for the accompanying cyanosis resulting from interference with blood gas exchange.

In a study of lung biopsies from 36 patients with *P. carinii* pneumonia, Weber et al.[243] found atypical microscopic findings in 69 per cent. These included absence of exudate in 47 per cent, interstitial fibrosis in 33 per cent, and epithelioid granuloma in 6 per cent. In addition, focal multinucleated giant cells, dense interstitial infiltrates, massive infiltration of alveolar macrophages, and calcifications were found in some cases.

The most useful stain for observing the organisms is the Grocott-Gomori methenamine silver method.[79, 222] The cysts of *P. carinii* measure about 4 μ in diameter. Many show "parentheses" within the round cysts. Wrinkled, crescent-shaped and cup-shaped cells are also observed (Figs. 26–7,*B* and 26–8). Campbell,[34] from electron micrograph studies, has suggested a life cycle. Mature cysts (up to 5 μ in diameter) contain intracystic bodies (Fig. 26–7*A*). The crescent-shaped cells represent cysts from which small cells (1.2 to 2 μ in diameter) have escaped. These are at first thin-walled, but grow and possibly fuse to form thick-walled, mature cysts again containing intracystic bodies. In lung material studied after pentamidine therapy, the cysts were found to be abnormal, and did not contain intracystic bodies. Such "ghosts" were seen within macrophages, a finding not associated with active disease. In experimental

work by Masur and Jones,[152] it was determined that antibodies directed against *Pneumocystis* were necessary for interiorization of the organism by macrophages. The *Pneumocystis* was subsequently killed by the cell.

PROGNOSIS AND THERAPY

The disease is still too new to assess clinical response to a variety of therapeutic modalities. Previously, the most commonly used agent in successfully treated cases was pentamidine isothionate.[120, 186] The dosage recommended was 4 mg per kg for 14 days. Recurrences were noted in some series[118] and perhaps can be related to inadequate therapy. Kirby et al.[133] have successfully treated the disease with pyrimethamine and sulfadiazine. Trimethoprim and sulfamethoxazole have also been found effective and presently constitute the treatment of choice. Hughes et al.[116, 117] found the combined dosage of 150 mg trimethoprim and 750 mg of sulfamethoxazole per square meter per day to be an effective prophylaxis in high-risk patients. Seventeen of 80 patients at risk were given a placebo and developed *P. carinii* pneumonitis, whereas none of 80 patients given the drug combination developed disease.

DIAGNOSIS

The etiologic agent of *Pneumocystis* pneumonia has not as yet been cultured. Therefore, diagnosis depends on demonstration of the organism in tissue. Using the bronchial brush technique of Fennessy,[82] Repsher et al.[200] were able to confirm the presumptive diagnosis in 11 of 19 patients. They conclude that the procedure is well tolerated, has few serious complications, and is the most sensitive method presently available. The Gomori method was used to stain the smears. Smith et al.[222] have modified this stain using DMSO and increased its sensitivity for the demonstration of organisms. Other objects may closely resemble *Pneumocystis* in histologic preparations,[202, 222] such as conidia of various fungi.

An indirect fluorescent stain for the detection of antibodies in sera is being developed.[175] So far the test is positive in 50 per cent of provisionally diagnosed cases of pneumocystic pneumonia. Antibodies are present in 14 per cent of healthy contacts of such patients, but none were detected in 50 sera of a healthy control population. The former may represent transient experience and sensitization to the organism. A direct fluorescent antibody test has also been developed.[147]

Barton and Campbell[17] have reproduced the disease in rats by simply giving them cortisone. This indicates that the organism is probably widespread in nature and perhaps in small numbers is a part of the normal flora. The organism has been reported in many other animals, including rodents, horses, goats and monkeys. Hendley[108] was able to effect activation of the disease in rats and demonstrated transmission of the disease among cage mates.

Hyphomycosis

There are thousands of hyphomycetes in the terrestrial and aquatic biosphere of the earth. Millions upon millions of their conidia are present in the air above and in the water flowing on the earth's surface. On rare occasion a few of these organisms may be involved in human and animal infections, particularly opportunistic disease. The list of genera associated with these infections has now grown so long that it is necessary to categorize them into accessible groups. Most of them have the same tissue morphology and elicit the same pathologic response. Often, however, they do not respond to the same treatment regimen, so that their identification is of more than just academic interest.

For the sake of convenience, the diseases are assigned to two major divisions: hyphomycosis and pheohyphomycosis. The division is based on the one readily evident differentiating characteristic, the presence or absence of brown pigment in their mycelium in tissue and/or in culture. This is an artificial system, because completely unrelated organisms fall into the same category, but it is believed that the division will be useful in sorting out case reports. If an organism causes infections with regularity or if there is some other aspect that is particularly distinctive, it can be assigned to a separate category; otherwise it is included in one of two categories: hyphomycosis or phaeohyphomycosis. In the present usage, case reports are assigned to the genus of the organism involved and the various types of infection the organism elicits are summarized.

PENICILLIUM

The common blue-green molds, *Penicillium* spp., are ubiquitous in nature and constantly present in the human environment. Whereas species of the related genus *Aspergillus* are regularly associated with invasive human disease, documented infection by members of the genus *Penicillium* is very rare. The fungus is regularly isolated from sputum, bronchial secretions, and other body surfaces,[65] but even when repeatedly isolated and in quantity, its etiologic role is equivocal unless verified by histopathologic examination. There are well over 800 species of the genus, and some are known to produce toxins while growing on decaying vegetation. Sometimes these products produce severe and fatal toxicosis in animals and gastrointestinal disturbances in man. *Penicillium* spp. are also well established as agents of external ear infections. The toxic diseases are discussed in the chapter on mycotoxicoses (Chapter 28) and the otic disease in that on infections of the eye and ear (Chapter 27).

The ubiquity of *Penicillium* species and their ease of isolation from a variety of human sources have led to a long list of diseases in which these organisms have been the purported etiologic agents. In the vast majority of cases, the mold isolated was a contaminant and not associated with the pathology. One of the earliest records of human infection by *Penicillium* was that of Chute in 1911.[48] The organism *P. glaucum* was isolated from urine and said to be the etiologic agent of a bladder infection. Details are lacking. In 1915 Mantrelli and Negri[150] isolated a *Penicillium* described as *P. mycetogenum* from black-grained mycetoma. It has not been found since.

Bronchopulmonary penicilliosis has been the most frequently reported form of the disease. There are ten records in the literature, starting with Castellani in 1918, and the disease has been reported sporadically since. Delore reported a case in 1955,[62] and Huang and Harris in 1962.[114] The latter was a systemic infection. In most of these cases, adequate substantiation of the etiologic role of *Penicillium* in the disease process is lacking. In some, a matted mycelial mass was seen in a pulmonary cavity or cyst. There was no cellular reaction. It is probable that the fungus was growing saprophytically in a preformed cavity, possibly just before or immediately after the death of the patient. *P. crustaceum* is the organism most often cited. Recently, Liebler et al.[146] described a pulmonary "coin" lesion removed by surgical excision in an otherwise healthy man. The lesion was 2 cm in diameter, partially calcified, and interspersed with hyphae. *Penicillium* was grown, but it was not further identified and was subsequently discarded. What might be termed an allergic bronchopulmonary disease was described by Delore et al.[62] The patient improved on cortisone therapy. Although a mycotic septicemia was alleged, owing to "finding mycelium in a calf muscle biopsy," the evidence is inadequate. An authenticated case of *Penicillium* infection of the urinary tract was reported by Gillium and Vest[94] in 1951. The patient had sporadic attacks of fever and right flank pain. Several boli of pink-tinged mycelial mats were voided in the urine. By catheterization, it was found that these came only from the right side calyx. The organism was identified as *P. citrinum*. No treatment was instituted.

The only verified case of systemic penicilliosis was reported in 1963 by Huang and Harris.[114] The patient had acute leukemia and was treated with antibiotics and steroids. At autopsy, the patient was shown to have had disseminated cerebral and pulmonary penicilliosis and gastrointestinal candidiasis. Vascular invasion, thrombosis, and infarction were the basic pathologic features. Masses of mycelium were seen in the thrombi. Thus the lesions were identical to those found in invasive aspergillosis. The organism was identified as *P. commune*, an infrequently encountered soil saprophyte. Patients have also received intravenous fluids, dialysis fluid, and blood bank material contaminated with *Penicillium*[57] and other fungi. A pyrogenic reaction may occur owing to toxic products, but infection is extremely rare. Kinare et al.[131] claim *Penicillium* spp. to have caused endocarditis following valve replacement. A documented case was published by Hall.[102]

An unusual form of penicilliosis in rodents and in man has been described. The disease was first reported in 1956 in the bamboo rat, which is found throughout Southeast Asia in the high plateaus. Several animals were noted as having a disease characterized by ascites, splenomegaly, hepatomegaly, and lesions in the Peyer's patches. The organism *P. marneffei* was found as small bodies (4.5 μ), somewhat resembling *Histoplasma* yeasts, within macrophages. These bodies reproduced by planate division, however, and not by blastoconidia formation. Mycelial elements (15 to 20 μ) have also been found.[36, 214, 215] The human case reported by Di Salvo et al.[65] was in a 67-year-old male with Hodgkin's disease. The organism was demonstrated in and cultured from the spleen (Figs. 26–8, 26–9, 26–10). The case was recorded from South Carolina, indicating that this organism may have a worldwide distribution. A few other cases of opportunistic infection occurring in the United States have been verbally reported to the author. Publication is forthcoming.

BEAUVERIA

The organism *Beauveria bassiana* is of great historical interest in microbiology. Bassi in 1835[18] showed it to be responsible for the disease known as muscardine of silkworms. This was the first demonstration of a microorganism specifically evoking a disease process, and it preceded the classic works of Koch and Pasteur by many years. *Beauveria* has since been found to be a commonly encountered soil fungus pathogenic for many insects, particularly beetles.

Georg et al.[93] isolated the organisms from pulmonary disease in the giant tortoise, and fatal pulmonary infection in alligators has also been noted.[87] A human infection of the lung was recorded by Lahourcade in 1966.[85] The patient was a female (22 years old) who had lived in North Africa. For many years she had manifested ulcerative cervical lymphadenopathy and was diagnosed as having tuberculosis. A chest x-ray revealed a thick-walled pulmonary cavity plus other small pulmonary

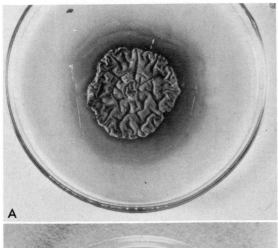

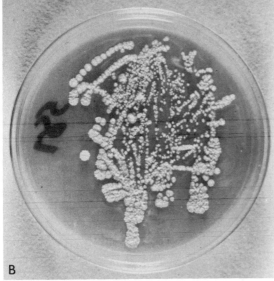

Figure 26–8. *Penicillium marneffei. A,* Colony at 25°C in Sabouraud's glucose agar. A defusable red pigment is produced. The colony is wrinkled, folded, and has red-stained mycelium and greenish conidia. *B,* At 37°C the organism grows as a yeastlike colony. (Courtesy of A. DiSalvo.)

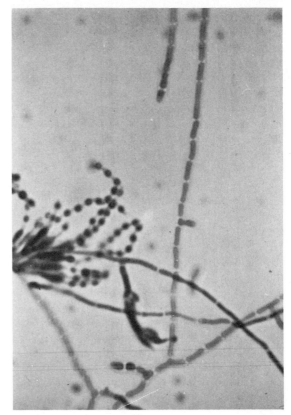

Figure 26–9. Culture mount of *Penicillium marneffei.* ×400. (Courtesy of A. DiSalvo.)

lesions. *B. bassiana* was isolated from surgically excised material and bronchial aspirates. Hyphae were visible in tissue section. She was treated with amphotericin B, and at last report her condition was stable.

ACREMONIUM

Acremonium spp. have been recovered from many cases of mycetoma (Chapter 5), onychomycosis (Chapter 8), and mycotic keratitis (Chapter 27). *Acremonium* spp. are abundant in soil, sewage, and so forth, and represent the anamorphic stage of several teleomorph fungi. Only rarely have they been unquestionably related to other types of human infection. In one case a midline granuloma with eroding destruction of the hard palate and involvement of the maxilla and mandible was described. In biopsy material from this case, hyphae were seen in tissue sections, and the organism was cultured on several occasions.[55] An authenticated case of meningitis was reported by Drouhet et al.[70] In this case, a 33-year-old woman developed sciatica following a spinal anesthetic given for a cesarean operation. She was treated with steroids. In time, after a second pregnancy, she developed neurologic and psychiatric problems. A surgical exploration revealed a granulomatous meningitis from which *Acremonium* sp. was cultured, and hyphal units were seen in tissue section. The patient died 15 months after onset of symptoms in spite of amphotericin B therapy.

Yeast bezoars following gastric surgery sometimes contain *Acremonium.*[187] Onorato et al.[181] reported on two renal transplant patients with *Acremonium* infections of their renal grafts; these were treated successfully with amphotericin B. The authors reviewed the literature on infection with *Acremonium* spp.

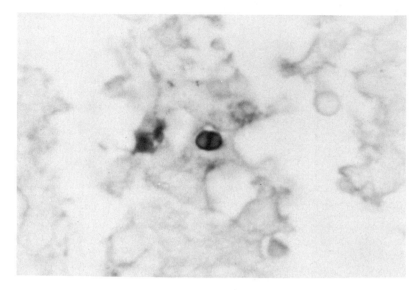

Figure 26–10. Planate dividing yeast cell of *P. marneffei* from liver biopsy of patient. ×440. (From DiSalvo, A. F., A. Fickling, et al. 1973. *Penicillium marneffei* infection in man. Description of first natural infection. Am. J. Clin. Pathol., 59:259–263.)

There are many reports in the literature of *Acremonium* being isolated from skin, nails,[197] bile, blood, gastric juice, pleural fluid, and so forth. Most of these lack verification.

FUSARIUM

Fusarium species are very common soil organisms and plant pathogens. A few are important agents of mycotic keratitis (Chapter 27) and onychomycosis.[64] Until recently the only authentic cases of human infection caused by *Fusarium* were those of colonization of burned skin. Holzegel and Kempf[113] reported such a case in 1964. Previously Peterson and Baker in 1959 had isolated *Fusarium roseum* from several cultures of burned skin.[189] There have been a few recent reports

of this form of the disease.[225, 245a] In one case observed personally, *Fusarium oxysporum* was isolated in large numbers from 21 consecutive cultures of a severely burned man. At autopsy, some mycelium was seen in the crusts and debris of the burned cutaneous tissue. Invasion of surrounding tissue was not observed, however. What, if any, contribution to the pathology was attributable to the fungus is unknown. Disseminated systemic infections have occurred in burn patients, however.[1]

Several other types of systemic infection have now been reported. Almost always these represent opportunistic infections in severely debilitated hosts.[139] Young and Myers[248] reported on a renal transplant patient; painful erythematous nodules had formed on a livedo

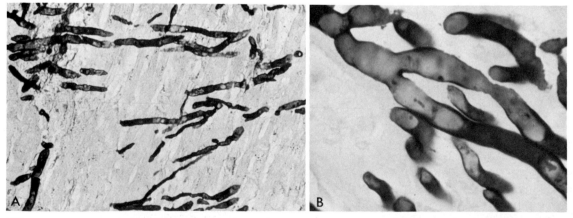

Figure 26–11. *Fusarium* in disseminated infection. *A,* Section of lung showing dichotomous branching septate hyphae. GMS. ×500. *B,* Higher magnification showing indentation at septa. The morphology is identical to that of invasive aspergillosis (compare with Figs. 23–15 and 23–16). The patient had malignant lymphoma. The isolant was *F. moniliforme.*[249] ×600. (Courtesy of K. J. Kwon-Chung.)

Table 26-2 *Cutaneous and Systemic Infections by Fusarium spp. Not Associated With Burns or Keratitis*

Year	Reference Number	Predisposing Factor	Sex	Age	Site Infected	Species	Treatment	Outcome
1948	143	?	F(7)	9–69	Urinary bladder	*Fusarium* spp.	None	Resolved spontaneously
1970	19	Chronic granulomatous disease	M	14	Facial granuloma	*Fusarium* spp.	Surgical	Cure
1973	47	Acute lympholeukemia	M	2.5	Skin, eye, brain (?)	*F. solani*	Amphotericin B	Cure
1975	99	Aplastic anemia	F	66	Subcutaneous granulomatosis	*F. oxysporium*	Amphotericin B	Cure
1976	28	Puncture wound	M	7	Osteomyelitis	*Fusarium* spp.	Amphotericin B	Cure
1977	52	None known	M	32	Pustules on hand	*F. moniliforme*	None	Resolved spontaneously
1978	249	Malignant lymphoma	M	32	Lung, heart, skin, kidney, pancreas, blood	*F. moniliforme* and *C. albicans*	None	Died
1978	104	Adenocarcinoma	M	36	Skin of foot	*Fusarium* spp.	Surgical	Cure
1979	248	Renal transplant	F	25	Subcutaneous livedo reticularis	*F. oxysporum*	Surgical	Cure

reticularis and had ulcerated. *Candida* and *Fusarium oxysporum* were recovered. Systemic granulomatous disease caused by *F. oxysporum* in a patient with aplastic anemia and myasthenic syndrome was described by Gutman et al.[99] Infections have resolved spontaneously[143] or with amphotericin B[47, 99] or surgical excision.[19, 248] Some disease resulted in death.[249] In histologic examination, one sees septate branching hyphae that are indistinguishable from infections caused by *Aspergillus, Penicillium, Pseudallescheria boydii, Acremonium,* and so forth (Fig. 26–11). A summary of cutaneous and systemic disease is found in Table 26–2, and the differential characteristics of species are listed in Table 26–3. Infection in the crocodile has been reported,[140] and *Fusarium* infections can occur as epidemics in sea turtle hatcheries, among fish, poults, shrimp, and calves. It has also caused mycotic abortion of cattle.[2, 3] *Fusarium* spp. grow as yeasts under certain conditions.[130]

PAECILOMYCES

Paecilomyces is seldom associated with human infections. Most of the cases recorded have been of a mycotic keratitis (Chapter 27). Several cases are now on record of endocarditis following valve replacement[101, 153, 219, 235]

Table 26–3 Fusarium *spp. Involved in Human Infection***

Species	Some Characteristic Features†
F. moniliforme	Fusoid macroconidia with catenulate (in chains) microconidia. Macroconidia few.
F. dimerum‡	Conidia are salmon-colored in mass micro- and macroconidia indistinguishable by size; almost all have a central septum.
F. solani§	Not as above; microconidiophores are elongate, macroconidia thick-walled, septate acentrally, upper part longest.
F. oxysporum	Short microconidiophores; walls of macroconidia are thin.

*Modified from Young, N. A., K. J. Kwon-Chung, et al. 1978. Disseminated infection by *Fusarium moniliforme* during treatment for malignant lymphoma. J. Clin. Microbiol., 7:589–594.

†For identification of species, consult Booth, C. 1971. *The Genus Fusarium.* Kew, Surrey, England, Commonwealth Mycological Institute.

‡Zapater, R. C. 1972. Queratomicosis por *Fusarium dimerum.* Sabouraudia, 10:274–275.

§*F. solani* is the most common cause of mycotic keratitis. See Chapter 27.

and endophthalmitis following lens implantation.[169, 178] This common airborne contaminant is resistant to most sterilizing techniques. Several nosocomial infections have occurred in which the organism was growing in the disinfectant used for surgical instruments.[163]

The first case of endocarditis was reported by Uys[235] in 1963, and it is typical of these cases. *Paecilomyces* was isolated from the blood, from a thrombus overlying the mitral valve, and from an embolus in the iliac artery following the death of a patient who had had a valve replacement. Following the surgery the patient had signs of cardiac failure and embolization along with pyrexia and emaciation. Fungal elements (1.5 to 3 μ) were demonstrated within thrombi of the mitral valve and in the iliac embolus. Rounded bodies were also seen. A tuberculoma-like lesion was found in the vessel wall near the iliac embolus. Caseous necrosis, epithelioid cells, and giant cells were present. *Paecilomyces variotii* was cultured.[235]

More recently, *Paecilomyces* has been encountered in infections in immunocompromised patients. Harris et al.[105] described a deep cutaneous cellulitis of the leg of a renal transplant patient. The lesion was a 1-cm erythematous area on the anterior aspect of the left leg. The patient noted that the lesion enlarged to 5 × 3 cm and became necrosed and ulcerated. Biopsy demonstrated hyphae and rounded budding cellular forms on tissue section, and *Paecilomyces* was grown. The MIC of the isolant for amphotericin B was 25 μg per ml. Miconazole therapy (10 mg per kg every 8 hours) was successful. Intraocular infections have been cured with miconazole and thiabendazole.[163] Most strains are resistant to amphotericin B. *P. lilacinus* is claimed to have caused scaly patches on the skin of an otherwise healthy woman. It was treated with griseofulvin[232] (Fig. 26–12).

George et al.[93] isolated *Paecilomyces fumosoroseus* from pulmonary lesions in a giant tortoise. Mycelium was present in many small abscesses throughout the lung. Infection in a monkey is also reported.[89] Subcutaneous granulomata were found at autopsy.

SCOPULARIOPSIS

Scopulariopsis brevicaulis is a common soil saprophyte of wide distribution. González-Ochoa and Dallal y Castillo[96] have isolated it from the soil of caves and mines inhabited by *Histoplasma.* They demonstrated pathogenic-

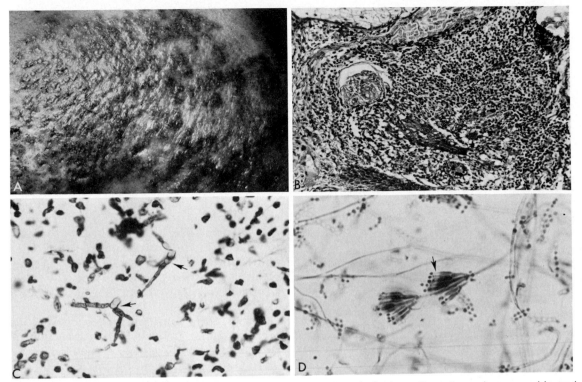

Figure 26–12. *Paecilomyces lilacinus* deep cutaneous infection. *A,* Scaly erythematous plaque on chin and cheek of 20-year-old female persistent for 15 years. No known underlying abnormality. *B,* Biopsy of skin shows chronic granulomatous changes composed of lymphocytes, giant cells, epithelioid cells in upper dermis. Hematoxylin and eosin stain. ×100. *C,* PAS stain at high power. Distorted mycelial units *(arrows).* ×800. *D,* Slide culture shows compactly arranged phialides with long narrow neck *(arrow)* with small young conidium at tip of phialide, which is attached to chain of older conidia ×600.[232] (From Takayasu, S., M. Agagi, et al. 1977. Arch. Dermatol., *113*:1687–1698.)

ity for mice of several strains. *In vivo* yeastlike forms were noted. Sekhon et al.[216] described a subcutaneous infection involving tendon sheath and muscle of the right ankle. The fungus isolated was sensitive to griseofulvin and hamycin. Scopulariopsis has been reported to form fungus balls in preformed pulmonary cavities.[142]

Phaeohyphomycosis

Phaeohyphomycosis is a clinical entity erected by Ajello[4] to include various invasive infectious processes for which the etiologic agent is a brown-pigmented, dematiaceous fungus in culture and the form found in tissue is mycelial. This separates it from other clinical types of disease involving brown-pigmented fungi when the tissue morphology of the organism is a grain (mycotic mycetoma) or a sclerotic body (chromoblastomycosis), and from the soil of caves and mines inhabited by *Histoplasma.* They demonstrated pathogenic-stratum corneum (tinea nigra), or to a subcutaneous cyst (phaeomycotic cyst). In hyphomycosis, as we have seen, the form found in tissue is also mycelial, but brown pigmenta-tion is absent in tissue or in culture. In phaeohyphomycosis, brown pigmentation is always seen in the fungus in culture but may or may not be discernible in tissue sections. The etiologic agents are an extremely diverse group of soil saprophytes and plant pathogens (Table 26–4,*A,B*). Many of these organisms are common laboratory contaminants, transiently and incidentally found on body surfaces and components of the aerospora; consequently, it is often difficult to establish a valid diagnosis. These organisms are also very difficult to correctly identify. The two-volume work on dematiaceous fungi by M. Ellis[73] should be consulted, and in cases of uncertainty the isolant should be sent to a

Table 26–4, A *Etiologic Agents of Phaeohyphomycosis in Man and Animals**

Alternaria	*Exophiala*
A. alternata	*E. jeanselmei*
A. state of *Pleospora infectoria*	*E. moniliae*
	E. spinifera
Aureobasidium	
A. pullulans	*Lasidiplodia*
	L. theobromae
Chaetoconidium	
	Mycocentrospora
Chaetomium	*M. acerina*
C. funicolum	
	Oidiodendron
Cladosporium	*O. cerealis*
C. bantianum	
C. cladosporoides	*Phialophora*
	P. hoffmannii
Curvularia	*P. parasitica*
C. geniculata	*P. repens*
C. lunata	*P. richardsiae*
C. pallescens	
C. senegalensis	*Phoma*
	P. cruris-hominis
Cylindrocarpon	*P. eupyrena*
C. tonkinesis	*P. hibernica*
	P. oculo-hominis
Drechslera	
D. hawaiiensis	*Phyllosticta*
D. rostrata	*P. citricarpa*
D. spicifera	
	Pseudomicrodochium
	P. suttonii
	Pyrenochaeta
	P. unguis-hominis
	Rhizoctonia
	Wangiella
	W. dermatitidis

*Modified from Ajello, L. 1980. The gamut of human infections caused by dematiaceous fungi. First Spanish and Italian Conf. Mycoses.

Table 26–4, B *Etiologic Agents of Phaeohyphomycosis Found So Far Only in Animals*

Dactylaria
  D. gallopava

Exophiala
  E. pisciphila
  E. salmonis

Scolecobasidium
  S. humicola
  S. tshawytschae

found in the vegetations, and at autopsy mycotic emboli were found in the left renal and internal iliac arteries. Mycotic infarcts were seen in the heart, spleen, left kidney, and right cerebral hemisphere, as were mycotic microabscesses in the thyroid. *Curvularia geniculata* was grown from all infected material.[129]

Harris and Downham[104] described a disseminated case involving *Curvularia* that occurred in a football player (male, 27 years old). The lesions began on the leg following injury and were noted as being hyperpigmented scaly patches. Later, swellings occurred on the submandibular lymph nodes. On surgical excision they were found to contain brown mycelium. A right hemithorax developed, and *C. geniculata* was isolated from the pleural fluid. T-8 and T-9 vertebrae were also involved, and T and B cell hypofunction were noted. Amphotericin B and miconazole were given. *C. geniculata* has also been isolated from cases of keratitis (Chapter 27) and mycetoma (Chapter 5).

In another case associated with a football injury, Rohwedder et al.[206] record a 25-year-old immunocompetent man with a ten-year history of progressive disseminated infection caused by *C. lunata*. Lesions started on the legs following abrasions. Deep soft tissue abscesses, pulmonary suppuration, a paravertebral abscess, and a cerebral abscess developed. There was also a paravertebral-mediastinal pleurocutaneous fistula (mycetoma?). Miconazole alleviated the cerebral involvement. Amphotericin B was given and the lesions began to resolve. After 5.4 g, the patient refused further treatment and is now bed-ridden with vertebral involvement. The isolant was sensitive to miconazole and amphotericin B but developed resistance to 5-fluorocytosine.

Curvularia pallescens was isolated from an apparently immunocompetent boy in a case reported by Lampert et al.[141] Lesions were

specialist. In this section some genera involved in phaeohyphomycosis will be described, along with representative case reports.

CURVULARIA

Curvularia geniculata is another common soil saprophyte that has recently been isolated from a case of endocarditis. The patient had received a Starr-Edwards aortic valve prosthesis. Four months later, the patient was readmitted to the hospital because of elevated temperature and signs of endocarditis. The chest was reopened and vegetations were found adhering to the valve. A replacement was made, but the patient died of renal failure shortly thereafter. Septal hyphae were found in the lung, and there was cerebral metastasis. Amphotericin B was started, but azotemia necessitated withdrawal. Miconazole therapy was successful.

In experimental infections, Whitcomb et al. found that *C. lunata* was capable of producing infection in mice,[246] but *C. spicifera* and *C. pallescens* were not. After radiation therapy (200 or 400 rads) alone or cortisone therapy (5.0 or 10.0 mg) alone, still only *C. lunata* caused infections. Combinations of the two allowed *C. pallescens* and *C. spicifera* also to cause infections. *C. spicifera* has been isolated from natural animal infections. In mice infected with *C. lunata*, lesions were found in liver and spleen (Fig. 26–13).

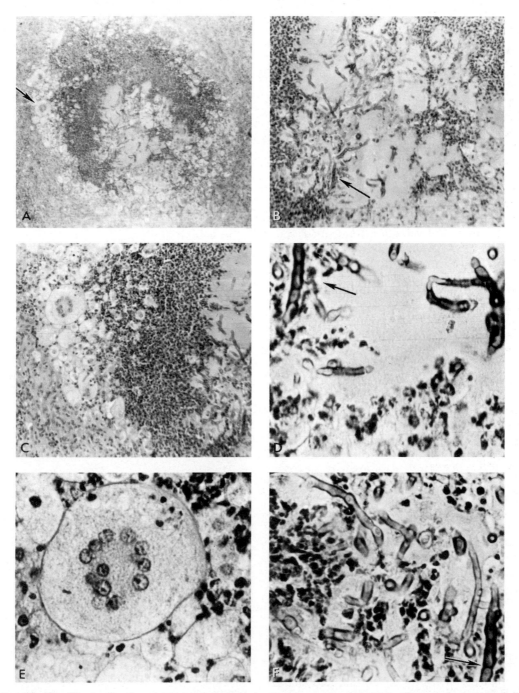

Figure 26–13. Tissue section from granulomata in the liver of mice given *Curvularia lunata* following x-irradiation (400 rads) and 5 mg cortisone. *A,* Low power view of granuloma *(arrow)* on balloon cell. *B* and *C,* Higher magnification of balloon cell. *D, E, F,* Mycelium and dematiaceous filaments *(arrows)* in granuloma. (From Whitcomb, M. P., C. D. Jefferies, et al. 1981. Mycopathologia, *75*:81–88.)

ALTERNARIA

Alternaria is a very common soil saprophyte and plant pathogen, and its conidia are frequently involved in human asthma. It is seldom as yet an established cause of human infectious disease. It may, however, colonize denuded, macerated, or previously injured skin. It has been isolated repeatedly from such areas. Whether the organism is contributing to the pathology is doubtful.[27, 110]

Some authentic infections caused by *Alternaria* spp. have been recorded. Moragas et al.[166] describe a 40-year-old male farmer who had chronic, slowly spreading inflammatory disease on the right knee. Biopsy showed rounded and elongate fungal units. The culture was identified as an *Alternaria* sp. The fungus was sensitive to amphotericin B and miconazole. The lesion did not respond to IV miconazole but resolved following infiltration of the lesion with the drug. In a case of osteomyelitis described by Garau et al.,[89] the infection of the turbinates occurred in a 23-year-old female following a submucous resection. The organism was *A. alternata.* Infection was not controlled by amphotericin B. Surgical debridement has not cured the infection either. The nose was involved in a case described to the author by G. Schertz. The patient was a 54-year-old male with acute myelogenous leukemia. A black crust formed on the nares during a blast crisis. Mycelium was seen on biopsy, and *A. alternata* was isolated. It resolved with remission of the leukemia. In the report of Lobritz et al.,[148a] a localized pulmonary nodule in a 36-year-old male was seen to have mycelium on tissue section and *Altenaria* species by culture. Two cases of cutaneous disease were reported by Pederson et al.[186a] Both patients had multiple ulcerated lesions covered by dry crusts on the backs of their hands. On isolation, one patient had *A. alternata* and the other an unknown *Alternaria* species. The authors found ten cases of *Alternaria* causing cutaneous infections prior to 1976. More recently, Nishimoto has reported another. The patient was a 6-year-old female, and the lesion regressed spontaneously.[174]

MYCOCENTROSPORA ACERINA

Among the stranger cases of human fungus disease is the one recorded by Lie-Kian-Joe et al.[145] and Emmons.[76a] The patient was a 12-year-old boy from Indonesia. The disease had appeared many years previously as a small nodule on the cheek. At the time of examination, there were extensive indurated, verrucous, and ulcerated cutaneous and subcutaneous lesions of the face. The nasal mucosa was also involved. There was no lymphadenopathy or fever, and no underlying disease was detected (Fig. 26–14).

Eleven biopsies were taken during a period of several years. Epithelial hypertrophy was seen, along with hyperkeratosis and intraepithelial abscesses. Granulomas were found in subcutaneous tissue. Brown septate mycelia, 4 to 8 μ in diameter, were seen in all areas affected. Cultures grew at 25° C but not at 37° C. The same fungus grew from all specimens. It produced an olive-gray colony. Conidiophores grew irregularly and could be seen to have terminal scars from which conidia were released. These were unicellular when young (4 to 6 μ) and multicellular when mature (26 to 120 μ). Experimental disease in animals was attempted but unsuccessful; however, lesions were produced in many plants. The organism was identified as *Cercospora apii* by Emmons.[76a] It was later reidentified as *Mycocentrospora acerina.*[60] The patient

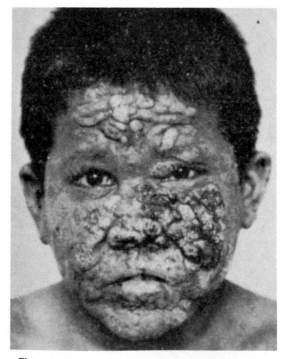

Figure 26–14. *Mycocentrospora acerina.* Numerous granulomatous lesions around face. (From Lie-Kien-Joe, et al. 1957. Arch. Dermatol., 75:864–870.)

died years later with extensive involvement of the face, but an autopsy was not obtained.

AUREOBASIDIUM

Aureobasidium pullulans has been isolated from a skin lesion by Vermeil et al.[239] The plaques were on the inferior aspect of the thigh, feet, and calves. The lesions were verrucous and had a black, studded appearance. There was no edema or induration. On histologic examination, the lesion resembled a keloid. There was extensive fibrosis and a granulomatous reaction, and numerous black, yeastlike bodies were seen free and within macrophages.

An infection on the skin of a porcupine (*Erethizon dorsatum*) was reported by Salkin et al.[210] Dematiaceous hyphal fragments and yeastlike budding cells were seen in tissue sections.

PHOMA

Phoma hibernica has been found to colonize previously injured skin. Bakerspigel's report[14] notes that the patient had had recurring lesions on the lower leg that had been treated with steroids, coal tar, and grenz rays. A granuloma developed in the treated area, from which *Phoma hibernica* was isolated. On histologic examination, hyphal elements were seen in tissue.

Bakerspigel et al.[16] have isolated *Phoma eupyrena* from facial lesions in an 18-month-old boy. There was no anticedent history of significance. Skin scrapings showed numerous hyphae (Fig. 26–15) and *Phoma eupyrena* was cultured. It responded to topical clotrimazole. An ear infection and cutaneous lesions on a white-tailed deer caused by *P. cava* were described by Gordon et al.[97] Sphaeropsidales in human disease was reviewed by Punithalingam.[192]

CLADOSPORIUM

C. bantianum has been recorded as cerebral phaeohyphomycosis from about two dozen cases of human disease of the brain[160] since its first modern description by Binford et al. in 1952.[22] The organism isolated by Binford et al. was named *C. trichoides* Emmons, sp. nov., but most investigators claim it is identical to

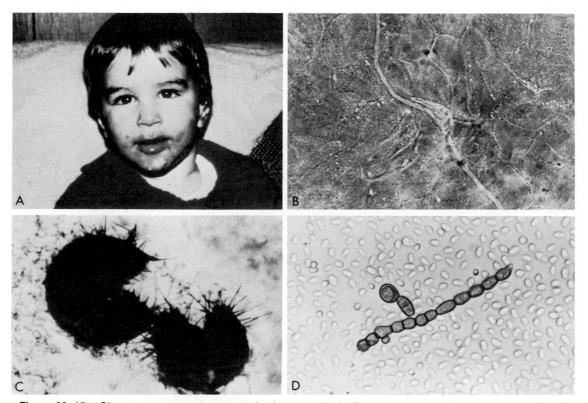

Figure 26–15. *Phoma eupyrena* cutaneous infection reported by Bakerspiegel et al.[16] *A,* Erythematous crusted lesions around the mouth. The boy was salivating excessively. *B,* KOH mount from skin lesion. Hyphae are seen coursing through squames. × 320. *C,* Mature pycnidia of *P. eupyrena.* × 1320. *D,* Large, brown, thick-walled chlamydoconidia and smaller hyaline pycnioconidia. × 480. (Courtesy of A. Bakerspiegel.)

one that had been isolated by Banti from brain disease many years previously. Although the same dematiaceous fungi have been recovered from chromoblastomycosis, from phaeomycotic cysts, and from cerebral phaeohyphomycosis (Chapter 9), *C. bantianum* has to date been found almost exclusively in cerebral disease. In addition to human infections, *C. bantianum* was recovered from lesions of the brain in two cats.[121, 196] The report of its isolation from chromoblastomycosis by Nsanzumuhire et al.[177] is considered in error, and the fungus from their patient has been named *Cladophialophora ajelloi*.[26] The infection of the foot recorded by Amma[10] has been confirmed as *C. bantianum*. It has been recovered in soil in Panama,[137] and from a tree stump[66] and from rural soil[160] in Virginia (Fig. 26–16).

Dixon and Shadomy[66] point out that the fungus cannot be identified on morphology alone. *C. bantianum* is neurotropic in animal pathogenicity studies, grows at 43° C, and does not liquefy gelatin. These characteristics separate it from other *Cladosporium* spp. However, McGinnis and Borelli[158a] maintain the morphology of *C. bantianum* is so distinct that it is not easily confused with other species.

In the cases of human infection, the lesions are single or multiple, the latter suggesting hematogenous spread from a focal lesion elsewhere. They consist of encapsulated abscesses, with masses of brown hyphal strands contained in the central area. Concurrently, small lesions have been found in the lungs and on the skin of the abdomen or the ear; these possibly represent primary sites of infection. Often the patients are debilitated or on steroids.

The presenting symptoms of cerebral phaeohyphomycosis include headache, weakness or paralysis, coma, diplopia, confusion, ataxia, incoordination, and seizures. The diversity of lesions tends to reflect the location of the lesion in the brain.[63] Except for elevation of protein and a cell count up to 1400 per cu mm, the spinal fluid is not grossly abnormal. In no case has the organism been recovered from spinal tap. Most patients are not diagnosed until autopsy, and in none of the cases have patients survived beyond a few months, even after surgical intervention or medical management. The differential diagnosis includes abscess of various other etiologies, particularly mycotic. Biopsy and culture establish the diagnosis. The mycology of *C. bantianum* is discussed in Chapter 9.

HELMINTHOSPORIUM

Helminthosporium is another common dematiaceous fungus of the soil. It is frequently isolated from sputum and skin, but it is rarely recorded as responsible for infectious disease. Dolan, Weed, and Dines reported two such cases. Both patients had chronic pulmonary disease and presented with purulent sputum, hemoptysis, and fever. Multiple cavities and abscesses were found in the lung. These contained histiocytes, lymphocytes, plasma cells, and granulocytes. Light brown, septate, branching mycelium was seen in the lesion tissue and found in bronchi, bronchioles, and alveoli. The fungus grown was called a *Helminthosporium* sp.[69] This may have been a *Drechslera* sp.[156]

CHAETOCONIDIUM

Chaetoconidium sp. was isolated from a lesion on an ankle and later from subcutaneous nodules on the knee. The patient was a 16-year-old male on immunosuppressive therapy following a renal allograph. The lesions developed at the site of an injury. He also had cryptococcosis. The fungus *Chaetoconidium* is a common soil saprophyte.[149] *Chaetoconidium* infection of the brain is reported by Huppert et al.[119]

DRECHSLERA

Drechslera hawaiiensis is a soil dematiaceous fungus of wide geographic distribution. It has been isolated from a fatal case of meningoencephalitis in a patient who had unsuspected lymphocytic lymphosarcoma.[88] Examination of autopsy material revealed a severe granulomatous and suppurative leptomeningitis and vasculitis. Brownish-colored mycelial fragments were seen in tissue sections from the base of the brain. The involved areas showed massive infiltration of lymphocytes and plasma cells with granulomatous reaction. In areas of vasculitis there were numerous giant cells, and the mycelial elements were greatly distorted, swollen, and bizarre. The fungus grew as a cottony blackish brown mold. The conidia resembled those of *Helminthosporium*, but it was classified by Ellis[73] as a *Drechslera* because of the acropetalous succession of porospores produced on a sympodially extending conidiophore. Dermal le-

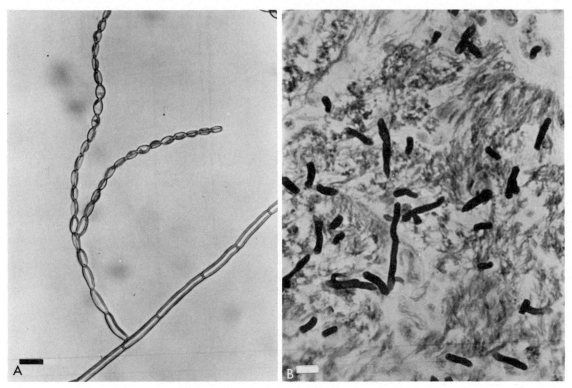

Figure 26–16. *Cladosporium bantianum. A, C. bantianum* from the stump of a juniper. Cornmeal, 30°C, one week. Bar = 10 μ. (From Dixon, D. M., and H. J. Shadomy. 1980. Mycopathologia, *70*:139–144.) *B, C. bantianum* in mouse brain. Bar = 10μ. (Courtesy of D. M. Dixon.[66])

sions in horses caused by *D. spicifera* have been reported by Kaplan et al.,[128] and subcutaneous infection in a cat by Muller and Kaplan.[170]

PHIALOPHORA

Phialophora parasitica was isolated from the subcutaneous tissue of a kidney transplant patient on immunosuppressive therapy.[6] This is another example of the ever-increasing list of soil fungi that can be isolated from opportunistic infections. The common denominator in all these cases is a patient with abrogated defenses. The fungus grows as mycelium in tissue as it does in soil, and there is no transformation to a "parasitic" form, as seen in the *Phialophora* species isolated from chromoblastomycosis (Fig. 26–17,*A*).

The fungus isolated in this case was *P. parasitica.* In culture it is characterized by formation of phialides with cylindrical collarettes that extrude ovate to elliptical phialoconidia. In tissue section the fungus appears as septate, branching, brown-colored

mycelial elements of varied morphology. *P. hoffmanii* has been isolated from cases of endocarditis.[189a, 221a] The mycology of these organisms is discussed in Chapter 9, as is that of *P. richardsiae.*

MISCELLANEOUS DEMATIACEAE

Exophiala spp. are commonly associated with mycetoma *(E. jeanselmei)*, tinea nigra *(E. werneckii)*, and phaeomycotic cyst. *E. jeanselmei*[106] and *E. moniliae,*[159] in addition to several other *Exophiala* spp., also cause phaeohyphomycosis in man and animals (Fig. 26–17 *B,C*). *E. pisciphila*[23, 157] was isolated from diseased clown fish, cod, seahorse, channel catfish, scup, and trigger fish; *E. salmonis,*[35, 203] from disease in trout and Atlantic salmon; and *E. spinifera,*[173] from a granulomatous infection in man (see Chapter 9). Other agents of phaeohyphomycosis in animals include *Scolecobasidium humicola* (Fig. 26–18)[8] in trout and *S. tshawytschae* in various fish.[158] *Dactylaria gallopara* was first isolated from the brains of

A

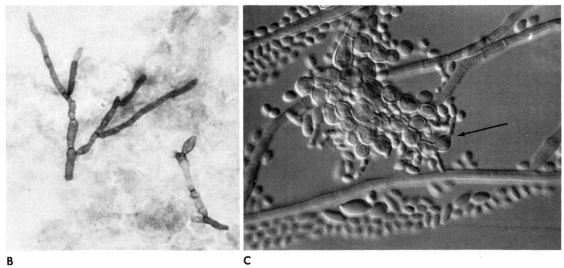

B   C

Figure 26–17. Phaeohyphomycosis. *A,* Tissue section showing irregular, branching hyphae. *Phialophora parasitica.* GMS. ×500. (Courtesy of L. Ajello.) *B, Exophiala moniliae* in pus.[159] (From McGinnis, M. R., D. F. Sorell, et al. 1981. Mycopathologia *73*:69–72.) *C, E. moniliae.* Swollen annellide with necklike extension demonstrating ringlike annellations *(arrow).* Note the conidia at end of neck. (Courtesy of M. McGinnis.)

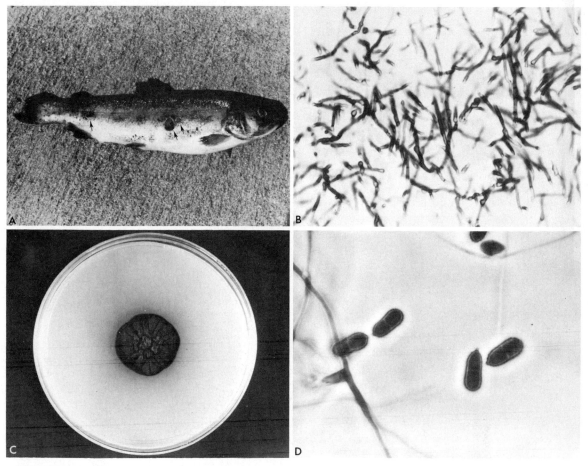

Figure 26–18. Phaeohyphomycosis of fish. *A,* Lateral lesions *(arrows)* on rainbow trout caused by *Scolecobasidium humicola.*[8] *B,* Phaeomycotic hyphae in kidney of trout. GMS. ×800. *C, Scolecobasidium humicola.* SGA, 3 weeks, 25°C. *D, S. humicola* showing finely echinulate walls of two-celled conidia. (Courtesy of L. Ajello and M. McGinnis.)

turkeys in 1964.[92] It has since become a major disease of chickens and turkeys and has occurred in epidemics.[194] It is a thermophilic fungus found in heated environments such as sloughs, ponds, and hot springs. It is also found in compost heaps and broiler litter.[241] *Wangiella dermatitidis* is a common agent of phaeomycotic cyst (Chapter 9). It has also been isolated from disseminated disease and opportunistic infections (Fig. 26–19). In one case described by Hohl et al.,[112] *W. dermatitidis* came from a septic arthritis in a diabetic, but the patient also had mucormycosis *(Rhizopus arrhizus)* and aspergillosis *(Aspergillus flavus)*. Amphotericin B therapy cleared the latter two mycoses, but control of *W. dermatitidis* was not achieved until 5-fluorocytosine was added to the regimen. Most agents of phaeohyphomycosis are resistant to amphotericin B and the imidazoles but usually sensitive to 5-FC.

SWAMP CANCER (OÖMYCOSIS)

A special form of mycosis of horses had the name of *Hyphomycosis destruens equi*. The disease was first noted in India in 1884 under the name bursautee and later described in France in 1896. In 1903, it was recorded in Indonesia by de Haan and Hooglamer[100] under the name *Hyphomycosis destruens*. Since then it has frequently been confused with infection by a nematode larvae under such names as Florida horse leech, swamp cancer, and bursatti. A disease with identical symptomatology is caused by larvae of *Habronema*. Only histologic examination will differentiate them.[124] About 40 cases of mycotic swamp cancer have been recorded in Texas, and several have been reported in Florida, New Guinea, and Australia.[171, 124] Infection usually starts at the site of a cut on the hoof, hock, or

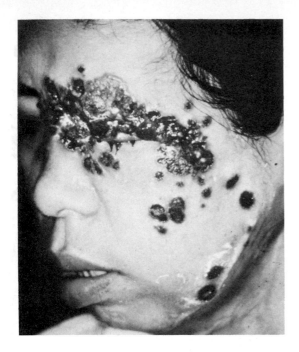

Figure 26–19. Phaeohyphomycosis as an opportunistic infection caused by *Wangiella dermatitidis*. The patient had leukemia. (Courtesy of L. Ajello.)

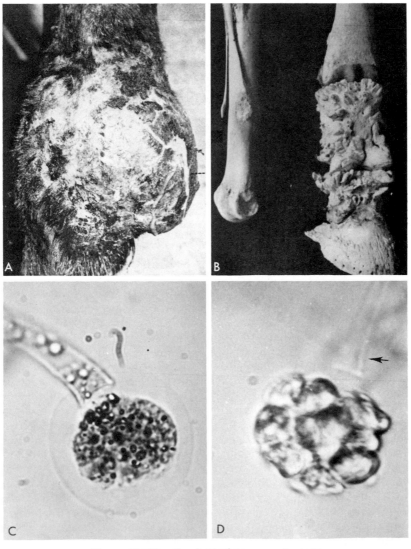

Figure 26–20. *See legend on opposite page*

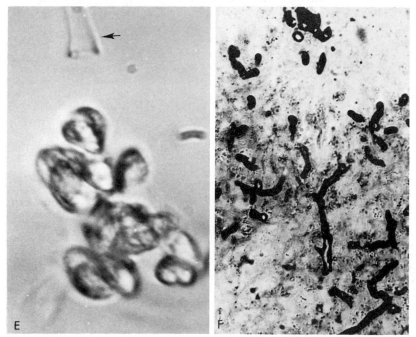

Figure 26–20. *Continued*   Swamp cancer, an Oömycete infection caused by *Pythium* sp. *A,* Mycotic tumor on foot of horse. *B,* Bone damage. (From thesis of J. Witkamp, University of Utrecht, 1924, p. 137.) *C,* Cytoplasmic vesicle at top of sporangium. *D,* Formation of zoospores within the vesicle at end of sporangiophore (*arrow*). *E,* Discharge of zoospores. Note trumpetlike opening of sporangiophore (*arrow*). *F,* Section from fetlock of horse stained with GMS showing mostly aseptate hyphae embedded in an eosinophilic matrix. (From J. W. Copland, Papua, New Guinea. Courtesy of P. Austwick.)

fetlock (Fig. 26–20). Necrosis, granulation tissue, and formation of fistulous tracts ensues. This is followed by formation of yellow necrotic lesions that contain masses of fungus. Histopathologic examination reveals the yellow masses to be foci of coagulative necrosis containing numerous neutrophils, eosinophils, and, occasionally, giant cells. Masses of branching, sparsely septate hyphae are seen. The fungi vary in width from 5 to 10 μ. Infections of the mucosa, abdomen, mammary gland, neck, head, and lips have also been recorded. The disease is chronic and progressive; spontaneous remission is unknown.[124] Fatal disseminated cases have, as yet, not been reported. Radical surgery or euthanasia is the usual management. The disease may also occur in dogs.[107]

For some time the fungus was considered a *Mortierella* sp.,[30] even though attempts to induce sporulation among the isolants were unsuccessful. The solution to the problem and the probable true identity of the organism were provided by Austwick and Copland.[13] They took a piece of the colony growing on Sabouraud's agar and floated it in sterile distilled water (see also *Saksenaea* in Mucormycosis), to which sterilized rotten maize silage was added. After two days at 25° C, biflagellate zoospores (Fig. 26–20) appeared that were 9 to 10 μ in diameter. They came from sporangia. The authors conclude the organism is a *Pythium* sp. in the family Pithiaceae of the order Peronosporales in the kingdom Protista, phylum Oömycetes. Other Oömycetes, such as *Saprolegnia parasitica,* also cause infectious disease in fish.

REFERENCES

1. Abramowski, C. R., D. Quin, et al. 1974. Systemic infection by *Fusarium* in a burned child. J. Pediatr., *84*:561–564.
2. Ainsworth, G. C., and P. K. C. Austwick. 1959. *Fungal Diseases of Animals.* Kew, Surrey, England, Commonwealth Mycological Institute, p. 22.
3. Ainsworth, G. C., and P. K. C. Austwick. 1955. A survey of animal mycoses in Britain. Trans. Br. Mycol. Soc., *38*:369–386.
4. Ajello, L. 1975. Phaeohyphomycosis: definition and etiology. *In Mycoses.* Sci. Publ. No. 304. Washington, D.C. 20037, Pan American Health Organization, pp. 126–130.
5. Ajello, L. 1968. Comparative morphology and immunology of members of the genus *Histoplasma.* Mycosen, *11*:507–514.
6. Ajello, L., L. K. Georg, et al. 1973. A case of phaeohyphomycosis caused by a new species of *Phialophora.* Abstracts. Am. Soc. Microbiol.

7. Ajello, L., M. Iger, et al. 1980. *Drechslera rostrata* as an agent of phaeohyphomycosis. Mycologia, *82*:1094–1102.

8. Ajello, L., M. R. McGinnis, et al. 1977. An outbreak of phaeohyphomycosis in rainbow trout caused by *Scolecobadium humicola*. Mycopathologia, *61*:15–22.

9. Almeida, F., C. S. Lacaz, et al. 1946. Consideracoes sobre tres cajos di mucoses humanas. Ann. Fac. Med. Univ. São Paulo, *22*:295–299.

10. Amma, S. M., C. K. J. Paniker, et al. 1979. Phaeohyphomycosis caused by *Cladosporium bantianum* in kerala (India). Sabouraudia, *17*:419–423.

11. Arnold, P., and D. G. Ahearn. 1972. The systemics of the genus *Prototheca* with a description of a new species, *P. filamenta*. Mycologia, *64*:265–276.

12. Ashforth, B. K., R. Ciferri, et al. 1930. A new species of *Prototheca* and a variety of the same isolated from the human intestine. Arch. Protisteuk., *70*:619–638.

13. Austwick, P. K. C., and J. W. Copland. 1974. Swamp cancer. Nature, *250*:84.

14. Bakerspigel, A. 1970. The isolation of *Phoma hibernica* from a lesion on a leg. Sabouraudia, 7:261–264.

15. Bakerspigel, A. 1968. Canadian species of *Sorex, Microtus, Peromyscus* infected with *Emmonsia*. Mycopathologia, *37*:273–279.

16. Bakerspigel, A., D. Howe, et al. 1981. The isolation of *Phoma eupyrena* from a human lesion. Arch. Dermatol., *117*:362–383.

17. Barton, E. G., Jr., and W. G. Campbell, Jr. 1969. *Pneumocystis carinii* in lungs of rats treated with cortisone acetate: Ultrastructural observations relating to the life cycle. Am. J. Pathol., *54*:209–236.

18. Bassi, A. 1835. Del male del segno calcinacio o muscordino malattia che antigge: Bachi da set a. Patel Patel. Teorica Tip. Orcesi. Lodi.

19. Benjamin, R. P., J. L. Callaway, et al. 1970. Facial granuloma associated with *Fusarium* infection. Arch. Dermatol., *101*:598–600.

20. Bennett, J. H. 1842. On the parasitic fungi found growing in living animals. Trans. R. Soc. Edin., *15*:277–294.

21. Bennett, S. C. J. 1931. *Cryptococcus* pneumonia in Equidae. J. Comp. Pathol., *44*:85–105.

22. Binford, C. H., K. Thompson, et al. 1952. Mycotic brain abscess due to *Cladosporium trichoides*, a new species. Am. J. Clin. Pathol., *22*:535–542.

23. Blazer, V. S., and R. E. Wolke. 1979. An *Exophiala*-like fungus as the cause of systemic mycosis of marine fish. J. Fish Dis., *2*:145–152.

24. Boisseau-Lebreuil, M. T. 1972. Evolution d'*Emmonsia crescens* agent de l'adiaspiromycose en adiaspores dans le poumon de souris de laboratoire expérimentalement infestées avec la phase saprophytique du champignon. Mycopathol. Mycol. Appl., *46*:267–281.

25. Booth, C. 1971. *The Genus* Fusarium. Kew, Surrey, England, Commonwealth Mycological Institute.

26. Borelli, D. 1980. Chromomycosis agents. *In Superficial Cutaneous and Deep Mycoses*. Sci. Publ. No. 396. Washington, D.C., Pan American Health Organization.

27. Botticher, W. W. 1966. *Alternaria* as a possible human pathogen. Sabouraudia, *4*:256–258.

28. Bouequiqnon, R. L., A. F. Walsh, et al. 1976. *Fusarium* species osteomyelitis: a case report. J. Bone Joint Surg., *58a*:722–723.

29. Bragg, D. C., and B. Janis. 1973. The roentgenographic manifestations of pulmonary opportunistic infections. Am. J. Roentgenol., *117*:798–801.

30. Bridges, C. H., and C. W. Emmons. 1961. A phycomycosis of horses caused by *Hyphomyces destruens*. J. Am. Vet. Med. Assoc., *138*:579–589.

31. Bullen, J. J. 1949. The yeastlike form of *Cryptococcus farciminosum*. J. Pathol. Bacteriol., *6*:117–120.

32. Butler, E. E. 1954. Radiation induced chlorophylless mutants of *Chlorella*. Science, *120*:274–275.

33. Butler, E. E., and L. J. Peterson. 1972. *Endomyces geotrichum*, a perfect state of *Geotrichum candidum*. Mycologia, *64*:365–375.

34. Campbell, W. G. 1972. Ultrastructure of *Pneumocystis* in human lung. Life cycle in human pneumocystosis. Arch. Pathol., *93*:312–324.

35. Carmichael, J. W. 1966. Cerebral mycetoma of trout due to a *Phialophora*-like fungus. Sabouraudia, *5*:120–123.

36. Capponi, M., P. Sureau, et al. 1956. Penicillose de *Rhizomys sinensis*. Bull. Soc. Pathol. Exot., *49*:418.

37. Carini, A., and J. Maciel. 1916. Ueber *Pneumocystis carinii*. Zentralblt. Bakteriol. (Orig.), *77*:46–50.

38. Carini, A. 1910. On protozoan parasites found in rodents. Bol. Soc. Med. Cirug. São Paulo, *18*:204.

39. Carmichael, J. W. 1962. *Chrysosporium* and some other aleuriosporic hyphomycetes. Can. J. Bot., *40*:1137–1173.

40. Carmichael, J. W. 1957. *Geotrichum candidum*. Mycologia, *49*:820–829.

41. Carroll, J. M., et al. 1968. Intestinal geotrichosis (*Geotrichum candidum*) in the ocelot (*Relis pardalis*). Am. J. Vet. Clin. Pathol., *2*:257–261.

42. Castellani, A. 1920. The higher fungi in relation to human pathology. Lancet, *1*:895–901.

43. Chagas, C. 1909. Nova tripanozomiaze cidoevolutiro do schizo trypanum n. gen. n. sp. agente etiolojio de nova entidade morbica do homen. Mem. Inst. Osw. Cruz, *1*:159–218.

44. Chandler, F. W., C. S. Callaway, et al. 1978. Differentiation of *Prototheca* species from morphologically similar green algae in tissue. Arch. Pathol. Lab. Med., *102*:353–356.

45. Chang, W. W. L., and L. Buerger. 1964. Disseminated geotrichosis. Case report. Arch. Intern. Med., *113*:356–360.

46. Chavez Batista, A., J. A. Maia, et al. 1955. Basidioneuromycosis in man. Soc. Biol. Pernambuco Anals., *13*:52–60.

47. Cho, C. T., T. S. Vast, et al. 1973. *Fusarium solani* infection during treatment for acute leukemia. J. Pediatr., *83*:1028–1031.

48. Chute, A. L. 1911. An infection of the bladder with *Penicillium glaucum*. Boston Med. Surg. J., *164*:420–422.

49. Ciferri, O. 1957. Metabolismo comparativo delle Protothecae e delle mutanti achloriche di *Chlorella*. G. Microbiol., *3*:97–108.

50. Ciferri, R., and A. Montemartini. 1959. Taxonomy of *Haplosporangium parvum*. Mycopathologia, *10*:303–316.

51. Ciferri, R., A. Chavez Batista, et al. 1956. Isolation of *Schizophyllum commune* from sputum. Atti Inst. Botanicó Laboratorio Crittogamico Univ. di Pavia, *14*(1–3):118–120.

52. Collins, M. S., and M. G. Rinaldi. 1977. Cutaneous infection in man caused by *Fusarium moniliforme*. Sabouraudia, *15*:151–160.

53. Cooke, W. B. 1968. Studies in the genus *Prototheca*. II. Taxonomy. J. Elisha Mitchell Sci. Soc., *84*:217–220.

54. Cordy, D. R. 1973. Chlorellosis in a lamb. Vet. Pathol., *10*:171–176.

55. Cowen, D. E., D. E. Dines, et al. 1965. *Cephalosporium* midline granuloma. Ann. Intern. Med., *62*:791–795.

55a. Cox, E. G. 1974. Protothecosis: a case of disseminated algal infection. Lancet, *1*:379–382.

56. Cueva, J. A., and M. D. Little. 1971. *Emmonsia crescens* infection adiaspiromycosis in man in Honduras. Am. J. Trop. Med., *20*:282–287.

57. Daisy, J. A., E. A. Abrutyn, et al. 1979. Inadvertent administration of intravenous fluids contaminated with fungi. Ann. Intern. Med., *91*:563–565.

58. Davies, R. R., and J. L. Wilkinson. 1967. Human protothecosis. Supplementary studies. Ann. Trop. Med. Parasitol., *61*:112–115.

59. Davies, R. R., H. Spencer, et al. 1964. A case of human protothecosis. Trans. R. Soc. Trop. Med. Hyg., *58*:448–451.

60. Deighton, F. C., and J. L. Mulder. 1977. *Mycocentrospora acerina* as a human pathogen. Trans. Br. Mycol. Soc., *64*:326–327.

61. Delanöe, P., and Mme. P. Delanöe. 1912. Sur les rapports de cystes de carini du poumon des rats avec le *Trypanosoma lewisi*. C. R. Acad. Sci. (Paris), *155*:658–660.

62. Delore, P., J. Candent, et al. 1955. Un cas de mycose bronchique avec localisations musculaires septicémiques. Prensa Med., *63*:1580–1582.

63. Desai, S. C., L. M. Bhatikar, et al. 1966. Cerebral chromoblastomycosis due to *Cladosporium trichoides* (*Bantianum*). II. Neurol. India, *14*:6–17.

64. DiSalvo, A., and A. M. Fickling. 1980. A case of nondermatophytic toe onychomycosis caused by *Fusarium oxysporum*. Arch. Dermatol., *116*:699–700.

65. DiSalvo, A. F., A. M. Fickling, et al. 1973. *Penicillium marneffei* infection in man. Description of first natural infection. Am. J. Clin. Pathol., *59*:259–263.

66. Dixon, D. M., and H. J. Shadomy. 1980. Taxonomy and morphology of dermatiaceous fungi isolated from nature. Mycopathologia, *70*:139–144.

67. Doby-Dubois, M., M. L. Cherrel, et al. 1964. Premier cas humain d'adiaspiromycose par *Emmonsia crescens*. Bull. Soc. Pathol. Exot., *57*:240–244.

68. Dogliotti, M. 1975. Cutaneous protothecosis. Br. J. Dermatol., *93*:473–474.

69. Dolan, C. T., L. A. Weed, et al. 1970. Bronchopulmonary helminthosporiosis. Am. J. Clin. Pathol., *53*:235–242.

70. Drouhet, E., L. Martin, et al. 1965. Mycose méningocérébrale a *Cephalosporium*. Presse Med., *31*:1809–1814.

71. Dvořák, J., M. Otcenasek, et al. 1969. The spring peak of adiaspiromycosis due to *Emmonsia crescens* Emmons et Jellison 1960. Sabouraudia, *7*:12–14.

72. Dvořák, J., M. Otcenasek, et al. 1966. Conceptions on the circulation of *Emmonsia crescens* Emmon et Jellison 1960 in nature. Folia Microbiol. (Praha), *13*:150–157.

73. Ellis, M. B. 1971. Dermatiaceous Hyphomycetes, Kew, Surrey, England, Commonwealth Mycological Institute.

74. Emmons, C. W. 1964. Budding in *Emmonsia crescens*. Mycologia, *56*:415–419.

75. Emmons, C. W. 1954. Isolation of *Myxotrichum* and *Gymnoascus* from lungs of animals. Mycologia, *46*:334–338.

76. Emmons, C. W., and L. L. Ashburn. 1942. The isolation of *Haplosporangium parvum* n. sp. and *Coccidioides immitis* from wild rodents. Public Health Rep., *57*:1715–1727.

76a. Emmons, C. W., Lie-Kien-Joe, et al. 1957. *Basidiobolus* and *Cercospora* from human infections. Mycologia, *49*:1–10.

77. Emmons, C. W., and W. L. Jellison. 1960. *Emmonsia crescens* sp. n. and adiaspiromycosis (haplomycosis in mammals). Ann. N.Y. Acad. Sci., *89*:91–101.

78. English, M. P. 1967. Some unusual mycoses. Rev. Med. Vet. Mycol., *6*:103–108.

79. Esterly, J. A., and N. E. Warner. 1965. *Pneumocystis carinii* pneumonia. Arch. Pathol., *80*:433–441.

80. Fawi, M. T. 1971. *Histoplasma farciminosum*, the etiologic agent of equine cryptococcal pneumonia. Sabouraudia, *9*:123–125.

81. Fawi, M. T. 1969. Fluorescent antibody test for the sero diagnosis of *Histoplasma farciminosum* infections in Equidae. Br. Vet. J., *125*:231–234.

82. Fennessy, J. J. 1966. A technique for the selective catheterization of segmental bronchi using arterial catheters. Am. J. Roentgenol. Radium Ther. Nucl. Med., *96*:936–943.

83. Fetter, B. F., G. K. Klintworth, et al. 1971. Protothecosis — Algal infection. *In* Baker, R. D. (ed.) *Handbuch der Speziellen Pathologischen Anatomie und Histologie*. Berlin, Springer-Verlag.

84. Fleischman, R. W., and D. McCracke. 1977. Paecilomycosis in a nonhuman primate (*Macaca mulatta*). Vet. Pathol., *14*:387–391.

85. Freour, P., M. Lahourcade, et al. 1966. Une mycose nouvelle. Étude clinique et mycologique d'une localisation pulmonaire de "*Beauveria*." Bull. Soc. Med. Hop. Paris, *117*:197–206.

86. Frese, K., and B. Gedek. 1968. Ein Fall von Prothecosis beim Reh. Berlin Muchen Tieraerztl. Wschr., *81*:174–178.

87. Fromtling, R. A., J. M. Jensen, et al. 1979. Fatal mycotic pulmonary disease of captive alligators. Vet. Pathol., *16*:428–431.

88. Fuste, F. J., L. Ajello, et al. 1973. *Drechslera hawaiiensis*: Causative agent of a fatal fungal meningoencephalitis. Sabouraudia, *11*:59–63.

89. Garau, J., R. D. Diamond, et al. 1977. *Alternaria* osteomyelitis. Ann. Intern. Med., *86*:747–748.

90. Garcia, N. P., and E. Ascanietol. 1972. Queatomycosis por *Fusarium dimerium*. Arch. Oftalmol. (Buenos Aires), *47*:332–334.

91. Garrison, R. G., and K. S. Boyd. 1973. Dimorphism of *Penicillium marneffei* as observed by electron microscopy. Can. J. Microbiol., *19*:1305–1309.

92. Georg, L. K., B. W. Bierer, et al. 1964. Encephalitis in turkey poults due to a new fungus species. Sabouraudia, *3*:239–244.

93. Georg, L. K., W. M. Williamson, et al. 1962. Mycotic pulmonary disease of captive giant tortoises due to *Beauveria bassiana and Paecilomyces fumosoroseus.* Sabouraudia, *2*:80–86.

94. Gillium, J., and S. A. Vest. 1951. *Penicillium* infection of the urinary tract. J. Urol., *65*:484–489.

95. Goldman, S., R. R. Lipscomb, et al. 1969. *Geotrichum* tumefaction of the hand. J. Bone Joint Surg., *51*:587–590.

96. González-Ochoa, A., and E. Dallal y Castillo. 1960. Frequencia de *Scopulariopsis brencicaulis* en muestras de suelos en cuevas y minas del pais. Rev. Inst. Salub. Enferm. Trop. (Mexico)., *20*:247–252.

97. Gordon, M. A., I. F. Salkin, et al. 1975. *Phoma* as a zoopathogen. Sabouraudia, *13*:329–333.

98. Greer, D. L. 1978. Basidiomycetes as agents of human infections: a review. Mycopathologia, *65*:133–139.

99. Gutman, L., S. M. Chou, et al. 1975. Fusariosis, myasthenic syndrome and aplastic anemia. Neurology, *25*:922–926.

100. Haan, J. de, and L. J. Hooglamer. 1903. Arch. wissi prakt. tierheilk, *29*:395.

101. Haldane, E. V., J. L. MacDonald, et al. 1974. Prosthetic valvular endocarditis due to the fungus *Paecilomyces.* Can. Med. Assoc. J., *111*:963–965.

102. Hall, W. J. 1974. *Penicillium* endocarditis following open heart surgery and prosthetic valve insertion. Am. Heart J., *87*:501–506.

103. Hamperl, H. 1957. Variants of *Pneumocystis* pneumonia. J. Pathol. Bact., *74*:353–356.

104. Harris, J. J., and T. F. Downham. 1978. Unusual fungal infections associated with immunological hyporeactivity. Int. J. Dermatol., *17*:323–330.

105. Harris, L. F., B. M. Dan et al. 1979. *Paecilomyces* cellulitis in a renal transplant patient — successful treatment with intravenous miconazole Southern Med. J. *72*:897–898.

106. Haschek, W. M., and U. B. Kasali. 1977. A case of cutaneous feline phaeohyphomycosis caused by *Philalophora gougerottii.* Cornell Vet., *67*:467–471.

107. Heller, R. A., H. P. Hobson, et al. 1971. Three cases of phycomycosis in dogs. Vet. Med. Small Anim. Clin., *66*:472–476.

108. Hendley, J. D. 1971. Activation and transmission in rats of infection with *Pneumocystis.* Proc. Soc. Exp. Biol. Med., *137*:1401–1404.

109. Hendry, W. S., and R. Patrick. 1962. Observations on thirteen cases of *Pneumocystis carinii* pneumonia. Am. J. Clin. Pathol., *38*:401–405.

110. Higashi, N., and Y. Asada. 1973. Cutaneous alternariosis with mixed infections of *Candida albicans.* Arch. Dermatol., *108*:558–560.

111. Hogben, B. R. 1967. Systemic mycotic lesions in apparently normal bobby calves: Incidence and appearance. N. Z. Vet. J., *15*:30–32.

112. Hohl, P., and A. E. Prevost, 1980. Ineffectiveness of amphotericin B in a human case of phaeohyphomycosis due to *Wangiella dermatitidis.* Am. Soc. Microbiol. Abst.

113. Holzegel, K., and H. F. Kempf. 1964. *Fusarium* mykose an der Haut eines Verbrannten. *(Fusarium* mycosis of the skin of a burned patient.) Dermatol. Wochenschr., *150*:651–658.

114. Huang, S. N., and L. S. Harris. 1963. Acute disseminated penicillosis. Am. J. Clin. Pathol., *39*:167–174.

115. Hubalek, Z., and W. Sixl. 1979. Adiaspiromycosis of wild small animals in Austria. Folia Parasitol., *26*:159–164.

116. Hughes, W. T., S. Kuhn, et al. 1977. Successful chemoprophylaxis for *Pneumocystis carinii* pneumonitis. New Engl. J. Med., *297*:1419–1426.

117. Hughes, W. T., P. C. McNabb, et al. 1974. Efficacy of trimethoprim and sulfamethoxazole in the prevention and treatment of *Pneumocystis carinii* pneumonitis. Antimicrob. Agents Chemother., *5*:289–293.

118. Hughes, W. T., W. W. Johnson, et al. 1971. Recurrent *Pneumocystis carinii* pneumonia following apparent recovery. J. Pediatr., *79*:751–759.

119. Huppert, M., D. J. Oliver, et al. 1978. Combined methenamine silver nitrate and hematoxylin and eosin stain for fungi in tissues. J. Clin. Microbiol., *8*:598–603.

120. Ivady, G. 1962. Further experience in the treatment of interstitial plasma cell pneumonia with pentamidine. Wochenschr. Kinderbeilkd., *111*:297–299.

121. Jang, S. S., E. L. Biberstein, et al. 1977. Feline brain abscess due to *Cladosporium trichoides.* Sabouraudia, *15*:115–123.

122. Jellison, W. L. 1969. *Adiaspiromycosis: Haplamycosis.* Missoula, Montana, Mountain Press.

123. Jellison, W. L., and J. W. Vinson. 1961. The distribution of *Emmonsia crescens* in Europe. Mycologia, *53*:524–535.

124. Johnston, K. G., and D. R. Hutchins. 1972. Phycomycosis in the horse. Aust. Vet. J., *48*:269–278.

125. Kamalam, A., and A. S. Thambiah. 1979. Adiaspiromycosis of human skin caused by *Emmonsia crescens.* Sabouraudia, *17*:377–381.

126. Kaplan, W. 1978. Prototothecosis and infections caused by morphologically similar green algae, pp. 218–232. *In The Black and White Yeasts.* Pan Am. Hlth. Org. Publ. 256, Proceedings of IV Int. Conf. Mycoses. Washington, D.C., World Health Organization.

127. Kaplan, W., et al. 1976. Prototothecosis in a cat: first recorded case. Sabouraudia, *14*:281–286.

128. Kaplan, W., F. W. Chandler, et al. 1975. Equine phaeohyphomycosis caused by *Drechslera spicifera.* Can. Vet. J., *16*:205–207.

129. Kaufman, S. M. 1971. *Curvularia* endocarditis following cardiac surgery. Am. J. Clin. Pathol., *56*:466–470.

130. Kidd, G. H., and F. T. Wolf. 1973. Dimorphism in a pathogenic *Fusarium.* Mycologia, *65*:1371–1375.

131. Kinare, S. G., A. P. Chaukar, et al. 1978. Fungal endocarditis after cardiac valve surgery. J. Postgrad. Med., *24*:164–170.

132. King, D. S., and S. C. Jong. 1975. *Sarcinosporon:* a new genus to accommodate *Trichosporon inkin* and *Prototheca filamentosa.* Mycotaxon, *3*:84–94.

133. Kirby, H. B., B. Kenameare, et al. 1971. *Pneumocystis carinii* pneumonia treated with pyrimethamine and sulfadiazine. Ann. Intern. Med., *75*:505–509.

134. Kirschenblatt, J. D. 1939. A new parasite of the

lungs in rodents. C. R. (Doklady) Acad. Sci. (USSR), *23*:406–408.

135. Kligman, A. M. 1950. A basidiomycete probably causing onychomycosis. J. Invest. Dermatol. *14*:67–70.

136. Klintworth, G. K., B. F. Fetter, et al. 1968. Protothecosis, an algal infection. J. Med. Microbiol., *1*:211–216.

137. Klite, P. D., H. B. Kelley, et al. 1965. A new soil sampling technique for pathogenic fungi. Am. J. Epidemiol., *81*:124–130.

138. Kodousek, R., V. Vojtek, et al. 1970. Systemic pulmonary adiaspiromycosis (caused by *E. crescens*). Cas. Lek. Cesk., *109*:923–924.

139. Krick, J. A., and J. S. Remington. 1976. Opportunistic invasive fungal infections in patients with leukemia and lymphoma. Clin. Haematol., *5*:249–310.

140. Kuttin, E. S., J. Mullar, et al. 1978. Mykosen bei krokodilen. Mykosen, *21*:39–48.

141. Lampert, R. P., J. H. Hutto, et al. 1977. Pulmonary and cerebral mycetoma caused by *Curvularia pallescens*. J. Pediatr., *91*:603–605.

142. Larsh, H. W. 1977. Opportunistic fungi in chronic disease other than cancer and related problems. *In* K. Iwata (ed.). *Recent Advances in Medical and Veterinary Mycology*. Proceedings 6th Congress ISHAM. Baltimore, University Park Press, pp. 221–229.

143. Lazarus, J. A., and L. H. Schwarz. 1948. Infestation of urinary bladder with an unusual fungus strain, *Fusarium*. Urol. Cutan. Rev., *52*:185–189.

144. Lee, W. 1975. Wound infection by *Prototheca wickerhamii:* a saprophytic alga pathogenic for man. J. Clin. Microbiol., *2*:62–66.

145. Lie-Kien-Joe, Njo-Injo Tjoei, et al. 1957. A new verrucous mycosis caused by *Cercospora apii*. Arch. Dermatol., *75*:864–870.

146. Liebler, G. A., G. J. Magovern, et al. 1977. *Penicillium* granuloma of the lung presenting as a solitary pulmonary nodule. J.A.M.A., *234*:671.

147. Lim, S. K., W. C. Eveland, et al. 1973. Development and evaluation of a direct fluorescent antibody method for the diagnosis of *Pneumocystis carinii* infections in experimental animals. Appl. Microbiol., *26*:666–671.

148. Lincoln, S. D., and J. L. Adcock. 1968. Disseminated geotrichosis in a dog. Pathol. Vet., *5*:282–289.

148a. Lobritz, R. W., T. H. Roberts, et al. 1979. Granulomatous pulmonary disease secondary to *Alternaria*. J.A.M.A., *241*:596–597.

149. Lomvardias, S., and G. E. Madge. 1972. *Chaetocenidium* and atypical acid fast bacilli in skin ulcers. Arch. Dermatol., *106*:875.

150. Mantrelli, C., and G. Negri. 1955. Richerche sperimentali sull' agente exiologico di' un micetoma a grani negri *(Penicillium mycetogenum)* n.f. G. Acad. Med. Torino, *21*:161–167.

151. Mars, P. W., A. R. Robson, et al. 1971. Cutaneous protothecosis. Br. J. Dermatol., *85*:76–84.

152. Masur, H., and T. C. Jones. 1978. The interaction *in vitro* of *Pneumocystis carinii* with macrophages and L-cells. J. Exp. Med., *147*:157–169.

153. McClellan, J. R., J. D. Hamilton, et al. 1976. *Paecilomyces varioti* endocarditis, on a prosthetic valve. J. Thorac. Cardiovasc. Surg., *71*:472–475.

154. McDiarmid, A., and P. K. C. Austwick. 1954. Occurrence of *Haplosporangium parvum* in the lungs of the mole *Talpa europea*. Nature (Lond.), *174*:843–844.

155. McGinnis, M. R. 1978. Human pathogenic species of *Exophiala, Phialophora* and *Wangiella. In The Black and White Yeasts*. Sci. Publ. No. 356. Washington, D.C., Pan American Health Organization, pp. 37–59.

156. McGinnis, M. R. 1978. *Helminthosporium* corneal ulcers. Am. J. Ophthalmol., *86*:853.

157. McGinnis, M. R., and L. Ajello. 1974. A new species of *Exophiala* isolated from channel catfish. Mycologia, *66*:518–520.

158. McGinnis, M. R., and C. Ajello. 1974. *Scolecobasidium tshawytschae*. Trans. Br. Mycol. Soc., *63*:202–203.

158a. McGinnis, M. R., and D. Borelli. 1981. *Cladosporium bantianum* and its synonym, *Cladosporium trichoides*. Mycotaxon, *13*:127–136.

159. McGinnis, M. R., D. F. Sorell, et al. 1981. Subcutaneous phaeohyphomycosis caused by *Exophiala moniliae*. Mycopathologia, *73*:69–72.

160. Middleton, F. G., P. F. Jurgenson, et al. 1976. Isolation of *Cladosporium trichoides* from nature. Mycopathologia, *62*:125–127.

161. Migaki, G. 1969. Bovine protothecosis. A report of three cases. Pathol. Vet., *6*:444–453.

162. Migaki, G., F. M. Garner, et al. 1969. Bovine protothecosis. Pathol. Vet., *6*:444–453.

163. Miller, G. P., G. Rebell, et al. 1978. Intravitreal antimycotic therapy and the cure of mycotic endophthalmitis caused by a *Paecilomyces lilacinus* contaminated pseudophakos. Ophthal. Surg., *9*:54–63.

164. Mohan, R. N., K. N. Sharma, et al. 1966. A note on an outbreak of epizootic lymphangitis in equines. Indian Vet. J., *43*:338–339.

165. Moore, M., W. O. Russel, et al. 1946. Chronic lepromeningitis and ependymitis caused by *Ustilago*, probably *U. zea:* Ustilagomycosis, the second reported instance of human infection. Am. J. Pathol., *22*:761–773.

166. Moragas, J. M., G. Prats, et al. Cutaneous alternariosis treated with miconazole. In press.

167. Morenz, J. 1971. Geotrichosis. *In* Baker, R. D. (ed.). *Handbuch der Speziellen Pathologischen Anatomie*. Vol. 3. Berlin, Springer-Verlag.

168. Morquer, R., C. Lombard, et al. 1955. Pouvoir pathogène de quelques éspèces de *Geotrichum*. C. R. Acad. Sci. (Paris), *240*:378–380.

169. Mosier, N. A., B. Lusk, et al. 1977. Fungal endophthalmitis following intraocular lens implantation. Am. J. Ophthalmol., *83*:1–8.

170. Muller, G. H., and W. Kaplan. 1975. Phaeohyphomycosis caused by *Drechslera spicifera* in a cat. J. Am. Vet. Med. Assoc., *166*:150–154.

171. Murray, D. R., P. W. Landis, et al. 1978. Granulomatous and neoplastic diseases of the skin of horses. Aust. Vet. J., *54*:338–341.

172. Neagoe, G., and M. Neagoe. 1967. Enterocolitis durch *Geotrichum candidum* nach Therapie mit Glutaminsaure. Dtsch. Z. Verdau. Stoffwechselkr., *27*:205–208.

173. Nielsen, H. S., and N. F. Conant. 1968. A new human pathogenic *Phialophora*. Sabouraudia, *6*:228–231.

174. Nishimoto, K. 1979. A case of cutaneous alternariosis. Jap. J. Med. Mycol., *20*:148–152.

175. Norman, L., and I. G. Kagan. 1972. A preliminary report of an indirect fluorescent antibody test

for detecting antibodies to cysts of *Pneumocystis carinii* in human sera. Am. J. Clin. Pathol., *58*:170–176.

176. Nosanchuk, J. S., and R. D. Greenberg. 1973. Prototheosis of the olecranon bursa caused by achloric algae. Am. J. Clin. Pathol., *59*:567–573.

177. Nsanzumuhire, H., D. Vollum, et al. 1974. Chromomycosis due to *Cladosporium trichoides* treated with 5-fluorocytosine. Am. J. Clin. Pathol., *61*:257–263.

178. O'Day, D. M. 1977. Fungal endophthalmitis caused by *Paecilomyces filacinus* after intraocular lens implantation. Am. J. Ophthalmol., *83*:130–131.

179. Ohash, Y. 1960. On a rare disease due to *Alternaria tenuis* Nees (alternariasis). Tohoku J. Expt. Med., *72*:78–82.

180. Olin, R. 1971. Pulmonary geotrichosis and candidiasis. Minnesota Med., *54*:881–886.

131. Onorato, I. M., J. L. Axelrod, et al. 1979. Fungal infections of dialysis fluid. Ann. Intern. Med., *91*:50–52.

182. Padhye, A. A. 1969. *In vitro* antifungal activity of hamycin against *Histoplasma farciminosum*. Mykosen, *12*:203–205.

183. Padhye, A. A., and J. W. Carmichael. 1968. *Emmonsia brasiliensis* and *Emmonsia ciferrina* are *Chrysosporium pruinosum*. Mycologia, *60*:445–447.

184. Padhye, A. A., J. G. Baker, et al. 1979. Rapid identification of *Prototheca* species by the API 20c system. J. Clin. Microbiol., *10*:579–582.

185. Parker, J. C., and G. K. Klintworth. 1971. Miscellaneous and uncommon disease attributed to fungi and actinomycetes. *In* Baker, R. D. (ed.). *Handbuch der Speziellen Pathologischen Anatomie*. Vol. 3. Berlin, Springer-Verlag.

186. Patterson, J. H. 1966. *Pneumocystis carinii* pneumonia and altered host resistance: Treatment of one patient with pentamidine isothionate. Pediatrics, *38*:388–397.

186a. Pederson, N. B., and P. A. Mordh. 1976. Cutaneous alternariosis. Br. J. Dermatol., *94*:201–209.

187. Perttala, Y., P. Peltokallio, et al. 1975. Yeast bezoar formation following gastric surgery. Am. J. Roentgenol. Radium Ther. Nuc. Med., *125*:365–373.

188. Peter, M., C. Horváth, et al. 1967. Date referitoare la frecventa genului *Geotrichum* in diferite produse biologice umane. Med. Intern. (Bucur), *19*:875–878.

189. Peterson, J. D., and J. J. Baker. 1959. An isolate of *Fusarium roseum* from human burns. Mycologia, *51*:453–456.

189a. Pierarch, C. A., G. Gulman, et al. 1973. *Phialophora mutabilis* endocarditis. Ann. Int. Med., *79*:900–901.

190. Pore, R. S. 1973. Selective medium for the isolation of *Prototheca*. Appl. Microbiol., *29*:648–649.

191. Pore, R. S., et al. 1977. *Fissuricella* gen. nov.: a new taxon for *Prototheca filamenta*. Sabouraudia, *15*:69–78.

192. Punithalingam, E. 1979. Sphaeropsidales in culture from humans. Nova Hedwigia, *31*:119–158.

193. Quilici, M., A. Orsini, et al. 1977. Adiaspiromycose pulmonaire disséminée. Arch. Anat. Cytol. Pathol., *25*:227–234.

194. Ranck, F. M., L. K. Georg, et al. 1974. Dactylariosis: a newly recognized fungus disease of chickens. Avian Dis., *18*:4–20.

195. Redaelli, P. 1940. As algoses Resenha. Clin. Cient. São Paulo, *9*:443–447.

196. Reed, C. J., G. Fos, et al. 1974. Leukemia in a cat with concurrent *Cladosporium* infection. J. Small Anim. Pract., *15*:55–62.

197. Refai, M., and M. Amer. 1974. Chronic paronychia in Egypt. Ernst Rodenwalt Archiv., *1*:55–63.

198. Refai, M., and A. Loot. 1970. Incidence of epizootic lymphangitis in Egypt. Mykosen, *13*:247–252.

199. Reninou-Castaing, S., et al. 1964. Geotrichose pulmonaire d'évolution fatale associée à une tuberculose pulmonaire. Poumon, *20*:287–296.

200. Repsher, I. H., G. Schröter, et al. 1972. Diagnosis of *Pneumocystis carinii* pneumonitis by means of endobronchial brush biopsy. New Engl. J. Med., *287*:340–341.

201. Restrepo, A., D. L. Greer, et al. 1971. Ulceration of the palate caused by a Basidiomycete, *Schizophyllum commune*. Sabouraudia, *9*:201–204.

202. Rhinehardt, D. J., W. Kaplan, et al. 1977. Morphologic resemblance of zygomycete spore to *Pneumocystis carinii* cysts in tissue. Am. Rev. Respir. Dis., *115*:170–192.

203. Richards, R. H., A. Holliman, et al. 1978. *Exophiala salmonis* infection in Atlantic salmon (*Salmo. salar*) L. J. Fish Dis., *1*:357–368.

204. Richer, F. J. C. 1977. La lymphangite epizootique C. *In Revue Générale de la Maladie et Observations Cliniques en Republique du Senegal*. Alfort, France, École Nationale Vétérinaire.

205. Rifkind, D., T. D. Faris, et al. 1966. *Pneumocystis carinii* pneumonia: Studies on the diagnosis and treatment. Ann. Intern. Med., *65*:943–956.

206. Rohwedder, J. J., J. L. Simmons, et al. 1979. Disseminated *Curvularia lunata* infection in a football player. Arch. Intern. Med., *139*:940–941.

207. Ross, J. D., K. D. G. Reid, et al. 1966. Bronchopulmonary geotrichosis with severe asthma. Br. Med. J., *1*:1400–1402.

208. Rossle, R. 1923. Referat über Ehtzundung. Verh. Dtsch. Pathol. Ces., *9*:18–68.

209. Salfelder, K., A. Fingerland, et al. 1973. Two cases of adiaspiromycosis. Beitr. Pathol., *145*:94–160.

210. Salkin, I. F., M. Gordon, et al. 1976. Cutaneous infection of a porcupine (*Erethizon dorsatum*) by *Aureobasidium pullulans*. Sabouraudia, *14*:47–49.

211. Schnoor, T. G. 1939. The occurrence of *Monilia* in normal stools. Am. J. Med., *19*:163–169.

212. Schwimmer, M., and D. Schwimmer. 1955. *The Role of Algae and Plankton in Medicine*. New York, Grune & Stratton.

213. Segal, E. 1976, Susceptibility of *Prototheca* species to antifungal agents. Antimicrob. Agents Chemother., *10*:75–79.

214. Segretain, G. 1959. *Penicillium marneffei* n. sp. agent d'une mycose du système reticuloendothelial. Mycopathologia, *11*:327–353.

215. Segretain, G. 1962. Some new or infrequent fungous pathogens. *In* Dalldorf, G. (ed.). *Fungi and Fungous Diseases*. Springfield, Ill., Charles C Thomas.

216. Sekhon, A. S., D. J. Williams, et al. 1974. Deep scopulariopsosis: A case and sensitivity studies. Abs. Am. Soc. Microbiol., *74*:18.

217. Shiefer, B., and B. Gedek. 1968. Zum Verhalten von *Prototheca* species im Gewebe von Sauge-

tieren. Berlin, Munchen Tierarztl. Wochenscht., *81*:485–490.

218. Sileo, L., and N. C. Palmer. 1973. Probable cutaneous protothecosis in a beaver. J. Wildlife Dis., *9*:320–322.

219. Silver, M. D., P. G. Tuffnell, et al. 1971. Endocarditis caused by *Paecilomyces varioti* affecting an aortic valve allograft. I. Thorac. Cardiovasc. Surg., *61*:278–281.

220. Singh, T., and B. M. L. Varmani. 1966. Studies on epizootic lymphangitis. A note on pathogenicity of *Histoplasma farciminosum* (Rivolta) for laboratory animals. Indian J. Vet. Sci., *36*:164–167.

221. Singh, T. 1966. Studies on epizootic lymphangitis. Study of clinical cases and experimental transmission. Indian J. Vet. Sci., *36*:45–49.

221a. Slifkin, M., and H. M. Bowers. 1975. *Phialophora mutabilis* endocarditis. Am. J. Clin. Pathol., *63*:120–130.

222. Smith, J. W., and W. T. Hughes. 1972. A rapid staining method for *Pneumocystis carinii*. J. Clin. Pathol., *25*:269–271.

223. Smith, J. W., W. T. Hughes, et al. 1971. Characterization of *Pneumocystis carinii* by biophysical and enzymatic methods. Bacteriol. Proc., *71*:121.

224. Sonck, C. E., and Y. Koch. 1971. Vertreter der Gattung *Prototheca* als Schmarotzer auf der Haut. Mykosen, *14*:475–482.

225. Spebar, M. F., and R. J. Lindberg. 1979. Fungal infection of the burn wound. Am. J. Surg., *138*:879–882.

226. Speller, D. C. E., and A. C. MacIver. 1971. Endocarditis caused by a *Caprinus* species. A fungus of the toadstool group. J. Med. Microbiol., *4*:370–374.

227. Spencer, H. 1968. *Pathology of the Lung*. Oxford, Pergamon Press, pp. 449–453.

228. Standard, P. C., and L. Kaufman. 1976. Specific immunological test for rapid identification of members of the genus *Histoplasma*. J. Clin. Microbiol., *3*:191–199.

229. Sudman, M. S. 1974. Protothecosis. A critical review. Am. J. Clin. Pathol., *61*:10–19.

230. Sudman, M. S., and W. Kaplan. 1973. Identification of the *Prototheca* species by immunofluorescence. Appl. Microbiol., *25*:981–990.

231. Sudman, M. S., J. A. Majka, et al. 1973. Primary mucocutaneous protothecosis in a dog. J. Am. Vet. Med. Assoc., *163*:1372–1374.

232. Takayasu, S., M. Agagi, et al. 1977. Cutaneous mycosis caused by *Paecilomyces lilacinus*. Arch. Dermatol., *113*:1687–1698.

233. Taylor, R. L., D. C. Cavanaugh, et al. 1968. Adia-

spiromycosis in small mammals of Viet Nam. Mycologia, *60*:450–451.

234. Travis, R. E., E. W. Ulrich, et al. 1961. Pulmonary allescheriosis. Ann. Intern. Med., *54*:141–152.

235. Uys, C. J., P. A. Don, et al. 1963. Endocarditis following cardiac surgery due to the fungus *Paecilomyces*. South Afr. Med. J., *37*:1276–1280.

236. Van Kruiningen, H. J. 1970. Protothecal enterocolitis in a dog. J. Am. Vet. Med. Assoc., *157*:56–63.

237. Van der Meer, G., and S. L. Burg. 1942. Infection à *Pneumocystis* chez l'homme et chez les animaux. Ann. Soc. Belge Med. Trop., *22*:301–307.

238. Vanek, J., O. Jirovec, et al. 1953. Interstitial plasma cell pneumonia in infants. Ann. Pediatr., *180*:1–21.

239. Vermeil, C., A. Gordeff, et al. 1971. Blastomycose cheloidiïnne à *Aureobasidium pullulans*. Mycopathologia, *43*:35–39.

240. Vlassopoulos, K., and S. Bartsakas. 1972. Intestinal geatrickasis — case report. Acta Microbiol. Hellenica, *17*:197–206.

241. Waldrip, D. A., A. A. Padhye, et al. 1974. Isolation of *Dactylaria gallopava* from boiler-house litter. Avian Dis., *18*:446–451.

242. Walzer, P. D., D. P. Perl, et al. 1974. *Pneumocystis carinii* pneumonia in the United States. Ann. Intern. Med., *80*:83–93.

243. Weber, W. R., F. B. Askin, et al. 1977. Lung biopsy in *Pneumocystis carinii* pneumonia. Am. J. Clin. Pathol., *67*:11–19.

244. Webster, B. H. 1959. Bronchopulmonary geotrichosis: A review of four cases. Dis. Chest, *35*:273–281.

245. Wenyon, C. M. 1926. *Protozoology*. Vol. 2. London, Balliere, Tindall and Cox.

245a. Wheeler, M. S., M. R. McGinnis, et al. 1981. *Fusarium* infection in burned patients. Am. J. Clin. Pathol., *75*:304–311.

246. Whitcomb, M. P., C. D. Jeffries, et al. 1981. *Curvularia lunata* in experimental phaeohyphomycosis. Mycopathologia. *75*:81–88.

247. Wolfe, I. D. 1976. Cutaneous protothecosis in a patient receiving immunosuppressive therapy. Arch. Dermatol., *112*:829–832.

248. Young, C. N., and A. M. Meyers. 1979. Opportunistic fungal infection by *Fusarium oxysporum* in renal transplant patient. Sabouraudia, *17*:219–223.

249. Young, N. A., K. J. Kwon-Chung, et al. 1978. Disseminated infection by *Fusarium moniliforme* during treatment for malignant lymphoma. J. Clin. Microbiol., *7*:589–594.

27 Mycotic Infections of the Eye and Ear

INTRODUCTION

The eye and the ear are continually subjected to challenge by a variety of fungi, bacteria, yeasts, and other microorganisms that are present in the external environment. Through evolution these organs have developed several means of preventing colonization and infection from airborne organisms. Any breach in the normal defenses preventing infection may lead to colonization and disease by the omnipresent potential invaders. Injury to the cornea of the eye is the most common predisposing factor leading to infection of this organ, whereas the accumulation of debris, particularly in damp tropical environments, allows colonization and infection of the external ear. The majority of the fungal organisms involved are soil saprophytes whose airborne conidia find the injured tissue a suitable environment for growth. Therefore, the mycology involved is quite different from that encountered in systemic or cutaneous infections. The list of fungal species associated with such infections is very long and may include essentially every fungus that inhabits the earth. However, a few species seem to be more aggressive opportunists and account for the majority of infections recorded. *Fusarium* sp., *Aspergillus fumigatus*, and *A. niger* are most frequently involved in external mycotic keratitis, whereas *A. niger* and *A. fumigatus* most commonly cause infections of the external ear. The characteristics of these particular organisms that enhance their ability to colonize these organs are as yet undetermined.

MYCOTIC INFECTIONS OF THE EYE

Mycotic infections of the eye have assumed increasing importance in ophthalmology. The several reviews that have appeared on this subject since 1960[16, 31, 38, 64, 86, 97, 98] attest to the increased incidence and awareness of this disease. These reviews also emphasize the difficulties involved in predicting its occurrence, defining its morbidity, and instituting effective treatment.

There are three categories of eye disease in which fungi are involved. Each has a particular set of predisposing factors, type of pathology, group of etiologic agents, and specific therapeutic procedures. These categories are (1) mycotic keratitis, which is an infection of the cornea following trauma or superficial disease by fungi from an external source; (2) endogenous oculomycosis, which is ocular infection by dissemination (usually hematogenous) of systemic fungal disease; and (3) extension oculomycosis, which is the extension into the orbit of fungal disease from adjacent tissue. In addition, there are miscellaneous infections of the adnexa, tear ducts, eye lids, and conjunctiva.

History

Recorded cases of fungal infections of the eye have appeared sporadically in the literature. What was probably canaliculitis was described by Cesoni in 1670,[81] and the diagnosis of the disease and its etiologic agent (an

682

actinomycete) was made in 1854 by Graefe.[35] The first report of mycotic keratitis was made by Leber in 1879.[51] He described an infection of the cornea .by an *Aspergillus*. However, mycotic keratitis was considered to be a very rare entity, and by 1951 only 63 cases were recorded.[16] In a review of the pathology of fungus infections of the eye made in 1958, only 31 cases had been reported in which the organism had been demonstrated in tissue.[64] Other reviews emphasized the rarity of this type of infection until recently. The first instance of keratomycosis recorded at the Armed Forces Institute of Pathology was in 1933, and only two other cases had been noted by 1952. In the next four years, however, twelve more patients were recorded.[99] By 1963 a total of 150 reports of mycotic infections of the eye had appeared in the literature, and of these 85 had occurred since 1951. This increase coincided in time with the general use of corticosteroids in ophthalmology. As a result of the general awareness that the use of steroids predisposes to mycotic keratitis, the use of this drug has abated, as has the incidence of the sequelae. Thus, at present mycotic keratitis occurs as a sequela of injury or surgery to the eye.[78]

The endogenous dissemination of systemic mycosis to involve the orbit, retina, optic nerve, sclera, conjunctiva, and adjacent tissues has also been reported sporadically in the literature.[27, 30, 40] In most of these reports, ocular involvement has been incidental to primary pathology elsewhere in the body, and the eye does not appear to be a target organ following dissemination of these diseases. Endophthalmitis is now considered a hazard associated with drug abuse[22] and has occurred without apparent injury to the eye or known infection in another organ system.[54]

The third category of disease is the special case of rhinocerebral mucormycosis. This disease also was considered rare,[12] and by 1966 only 55 cases were recorded in the world literature. Of these, most had occurred between 1960 and 1966. This reflects both a real increase in the number of cases and improved methods of diagnosis for the recognition of the disease. Rhinocerebral mucormycosis is now a well-defined syndrome that has a particular set of predisposing factors (see Chapter 25).

Mycotic Keratitis

Clinical Features. Mycotic ulcers of the eye usually occur subsequent to trauma to the cornea by vegetative matter, soil, or surgery. Colonization of debilitated tissue may occur as a sequela to exposure keratitis, congenital defects, or ulcers initiated by other causes. Usually there is a severe inflammatory reaction, with vascularization, ciliary flush, flare of the anterior chamber, and folding of Descemet's membrane. The initial tissue reaction may be minimal if masked by corticosteroid treatment. However, the ulcer itself is characterized by a raised epithelium and often a white shaggy border (Fig. 27–1). Frequently there is a distinct radiating margin with penetration of fungal elements into the corneal stroma. This fuzzy, hyphate border extends beyond the ulcer edge and may form

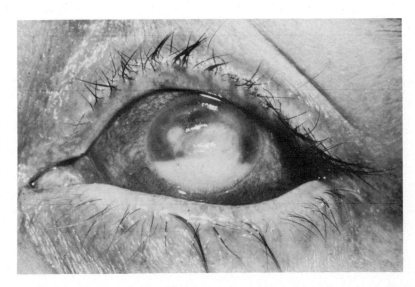

Figure 27–1. Mycotic keratitis. Central, shaggy-edged ulcer and satellite lesion with marked hypopyon. *Pseudallescheria boydii.* (From Ernest, J. T., and J. W. Rippon. 1966. Am. J. Ophthalmol., *62*:1202–1204.)

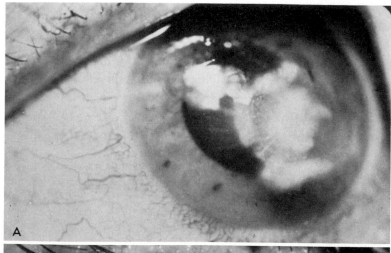

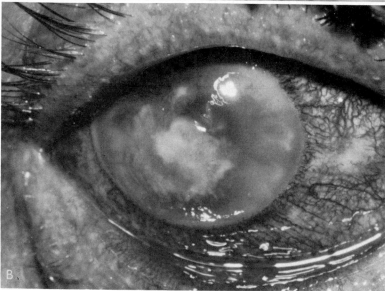

Figure 27–2. Mycotic keratitis. *A,* Fuzzy, hyphate border and satellite lesions. There is little inflammatory reaction owing to use of steroids. *Acremonium* sp. *B,* Typical fungal ulcer with necrotic, hyphate infiltrate within cornea underlying ulcer. Satellite lesions are present, along with a small hypopyon. This is a case from southern Florida caused by *Fusarium solani.* The infection was cured by use of pimaricin. (Courtesy of R. K. Forster.)

satellite lesions (Fig. 27–2). This picture, seen in conjunction with white endothelial plaques in the center of the cornea, strongly suggests fungal infection.[46, 99] Eventually a sterile hypopyon develops (Fig. 27–3), and there may be a persistent "corneal ring" beyond the edge of the ulcer, composed of neutrophils, eosinophils, and plasma cells.

Another clinical syndrome, endophthalmitis, may be a complication of surgery[30, 63] (Fig. 27–6), the result of deep penetrating injury or progression into the globe of the above described ulcerative lesions. Francois[31] has tabulated 74 cases up to 1968 in which the infection followed surgery. All but two cases followed cataract operations. Starch from glove powder has been implicated as the vector carrying the fungi.[74] The lesion may not be apparent for several days or up to six months following surgery. A developing

fuzzy-edged white mass forms behind the iris and ciliary body in the vitreous, and a grayish infiltrate occurs which gradually becomes more extensive. Vision is not appreciably impaired for some time, however. Redness and pain may develop, along with a hypopyon. The latter clinical features may recede in time, but the infection itself runs a relentless progressive course. Almost all cases have been resistant to treatment and have required enucleation.

Mycotic keratitis is frequently encountered in animals also. As in human cases, the most common predisposing condition is a penetrating injury. The disease has been reported in snakes,[18] horses,[11] dogs, and many other animals. The pathology, etiology, and treatment are the same as for human mycotic keratitis.

Pathology. Unless eradicated rapidly, the

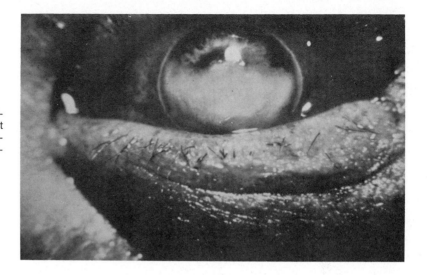

Figure 27–3. Mycotic keratitis. Deep central ulcer without hyphate border; extensive hypopyon. *Candida albicans.* (Courtesy of A. Maumenee.)

lesions of mycotic keratitis advance into the deep stroma of the cornea. It has been repeatedly emphasized that hyphae were present throughout the thickness of the cornea but frequently only in the middle and deeper layers. Histologically, an acute suppurative inflammatory process is seen early in the disease, which may be accompanied by coagulative necrosis. This response often subsides, and the reaction becomes minimal. Fungal hyphae are usually aligned parallel to the lamellae of the cornea, and Descemet's membrane appears to act as a barrier to penetration of the globe proper (Fig. 27–4). However, in some patients, especially if steroids were used, perpendicular penetration through the membrane leading to perfora-tion of the cornea and invasion of the internal orbit has been reported[46, 64] (Figs. 27–5 and 27–6).

Predisposing Factors. As has been pointed out by several authors, the rise in the frequency of keratomycosis corresponded to the general use of corticosteroids in the treatment of inflammatory eye disease. In experimental fungal infections of the cornea in rabbits, infection was noted in 80 per cent of test animals when cortisone was instilled along with fungal conidia. By comparison, there was only a 20 per cent incidence of infection in controls that had had conidia instilled but no steroids.[53] In addition, external corneal trauma (especially caused by foreign bodies containing vegetative materi-

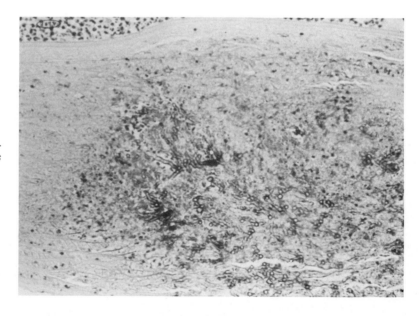

Figure 27–4. Mycotic keratitis. Hyphal strands of *Aspergillus* in cornea. ×100.

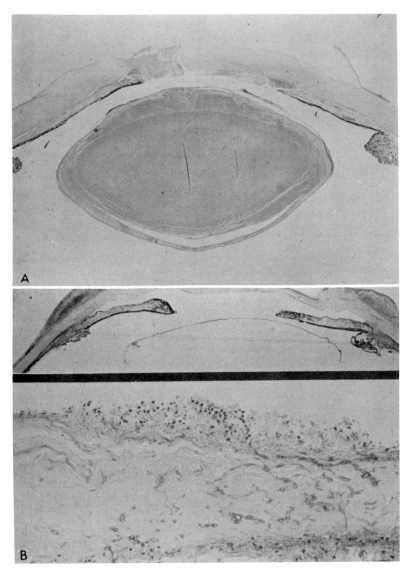

Figure 27–5. Mycotic keratitis. *A*, Ulceration and perforation of cornea. *B*, Ulceration of cornea without perforation. Mycelial elements are seen in high-power view. The fungus was not cultured. ×100.

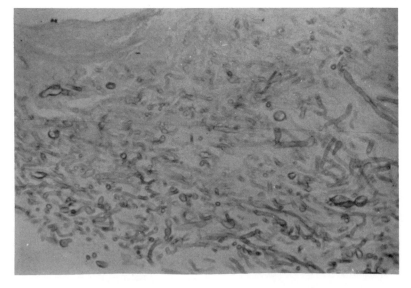

Figure 27–6. Mycotic keratitis. Mycelium of *Acremonium* sp. in a sloughed corneal transplant. ×400.

al) and many other conditions predispose the eye to colonization by fungi.[29, 88] These other conditions include previous corneal disease, exposure keratitis, radiation keratitis, herpetic lesions, serpiginous lesions, and functional disorders, such as facial palsy and bullous keratopathy.

Jones et al.,[42-44] in reviewing 39 cases that had occurred in southern Florida, detected a seasonal distribution. The majority of cases occurred in the rainy winter months in that geographic area. Twenty-nine of the cases were caused by *Fusarium solani*, and almost all patients were fruit pickers or agricultural workers. In a review by Chick and Conant,[16] trauma was noted in 97 per cent and previous use of steroids in 37 per cent and of antibiotics in 39 per cent of cases. As noted previously, there has been a decrease in the use of steroids as a general medicament. Currently most cases of mycotic keratitis have a deep penetrating injury as their antecedent.

The second most common predisposing factor in cases of mycotic infection of the eye is ophthalmologic surgery. Many cases of mycotic keratitis and endopthalmitis resulting from corneal transplants and other surgical procedures are now on record. Most often it has been found that fungi were contaminating the solution used for washing instruments, a fluid used for the lens, powder for gloves, or any of numerous other materials used in the procedure.[63] A variety of fungi have been involved. In a recent study associated with lens transplants, amphotericin B–resistant *Paecilomyces lilacinus* was recovered.[63]

Etiology. The list of fungi isolated from patients with keratitis is long and varied. There is a considerable fungal flora in the normal eye[2, 31, 72, 73] and an even greater flora in the diseased eye. For this reason many of the reports of oculomycosis in the literature are invalid, as the organisms isolated probably represent normal fungal flora. In patients in whom infection is substantiated by demonstration of fungal elements in debrided material or on histologic examination of tissue, organisms belonging to at least 30 genera have been isolated.[31, 77, 88] Essentially all of these are saprophytic soil fungi not usually associated with infections in man. Fungi are notorious opportunists. It is not surprising, therefore, to find fungal colonization when natural defenses of the eye are abrogated. Even in cases of keratitis in which the etiologic agent was one of the pathogenic

fungi, the organism was observed in its saprophytic phase in the diseased cornea.[25]

Depending on the series,[38, 99] in up to one half the cases, members of the genus *Aspergillus* were involved, particularly *A. fumigatus, A. flavus,* and *A. niger*. This is not surprising, as these are particularly aggressive opportunists and are frequently involved in pulmonary and systemic disease in patients with a variety of debilitations. As might be expected, another common opportunistic organism, *Candida albicans,* and related species also account for a large percentage of cases.[30, 55, 61] The third most commonly isolated agent, *Fusarium solani,* is uniquely associated with mycotic keratitis, as it does not cause any other form of human disease.[6, 43, 44, 68] The factors that account for its particular affinity for colonizing injured corneal tissue are not known, but there is some suggestive evidence. It appears to be collagenolytic and grows rapidly in corneal tissue. It also is somewhat heat-tolerant. *Fusarium oxysporum*[78] and *F. moniliforme* have also been recovered from several cases of mycotic keratitis and endophthalmitis.

Some of the other reported causes of keratitis include phaeomycotic agents such as *Aureobasidium pullulans, Alternaria alternata, Cladosporium oxysporum, Cylindrocarpon tonkinensis,*[56] *Curvularia lunata,*[94] *C. geniculata,*[67] *C. pallescens, C. senegalensis, C. verruculosa, Cladorrhinum,*[97] *Drechslera* sp.,[1] *D. rostrata, D. spiciferum, Lasiodiplodia theobromae, Phialophora verrucosa,*[92] *Phoma oculo-hominis, Pleospora infectoria, Botryodiplodia,*[50] *Tetraploa,*[65] *Rhizoctonia,*[86] and *Macrophoma.* Hyphomycetes and yeasts reported from mycotic keratitis include *Rhodotorula,*[32] *Trichosporon, Ustilago, Scopulariopsis,*[31] *Pseudallescheria* (about 12 cases are discussed in Chapter 24), *Sporothrix schenckii, Sterigmatocystis nigra, Paecilomyces lilacinus,*[63] *Periconia keratitidis, Verticillium, Acremonium*[14] (Fig. 27–7, *A* and 27–8, *A*), *Fusidium, Neurospora, Volutella, Glenospora, Penicillium, Graphium,*[5] and so forth. In addition, a few actinomycetes such as *Actinomyces bovis (israelii)*[33] have also been reported. How many of these are valid infections is difficult to determine. Unequivocal evidence for a valid infection includes demonstration of the fungus or yeast in lesion material or scrapings, isolation and identification of the organism from that material, and elimination of other causes for the disease.

One of the most notorious fungi involved in recent nosocomial infections following surgical procedures is *Paecilomyces lilacinus.*[63] It

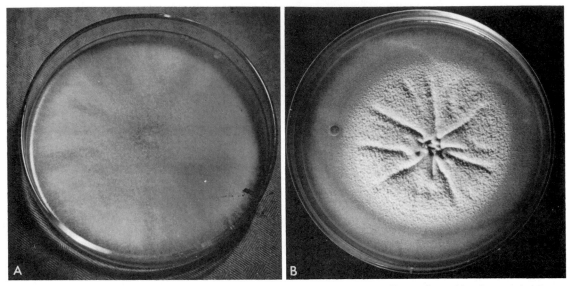

Figure 27–7. *A,* Colony of *Fusarium* sp. It is rapidly growing, cottony, and flat. At first white, it soon develops a rose-mauve or other color, depending on the species. *B,* Colony of *Acremonium* species. It is often glabrous and rusty brick-colored at first, but becomes fluffy (often folded) and cream-white with age.

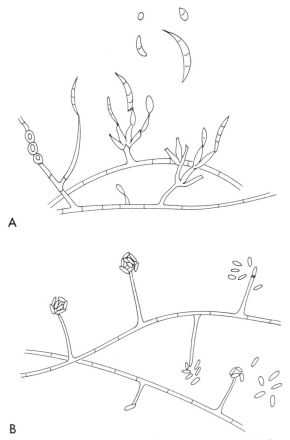

Figure 27–8. Microscopic morphology of (*A*) *Fusarium,* with characteristic crescent-shaped, septate macroconidia and of (*B*) *Acremonium,* with the conidia in a spherical cluster atop a slender, tapering conidiophore.

is resistant to many sterilization procedures and to most antimycotic agents (see Chapter 26).

Diagnosis.[63, 77, 93] In making the diagnosis of mycotic keratitis, it is important to remember two factors. Fungal elements are difficult to find, and the etiologic agents are saprophytic soil organisms. In keratitis, the fungi are usually found deep within the corneal structure and are often absent on its surface. Therefore, a superficial swabbing of the affected area may not be sufficient for demonstration of the etiologic agent. It is often necessary to use extensive debridement in order to obtain viable fungal material. A direct smear is valuable for rapid diagnosis. In potassium hydroxide mount, fungal elements are easily seen (Fig. 27–9). If the organisms are present in small numbers, a Gram stain is useful, especially if a yeast is the etiologic agent. The Gridley stain may also be employed.

The second point of importance in the mycology of eye disease is that the fungi are saprophytic soil organisms. Therefore, they require cultural procedures different from those used for systemic or cutaneous fungal pathogens. Multiple cultures should be taken. A single colony on one culture may represent a contaminant. However, multiple colonies of the same organism usually indicate fungal etiology of the disease, especially if there is histologic evidence as well. The preferred medium for culture is Sabouraud's agar slants or plates that do not contain antibiotics.

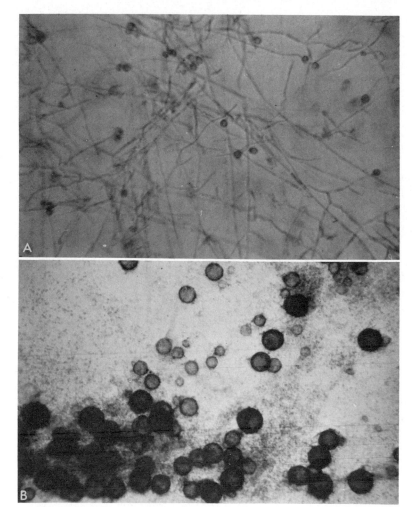

Figure 27–9. Mycotic keratitis due to *Pseudallescheria boydii*. *A*, KOH mount of corneal scraping showing mycelium and conidia characteristic of *P. boydii*. This fungus is associated with mycetoma and demonstrates a type of tissue-induced morphologic dimorphism which forms granules in tissue. In this eye infection (Fig. 27–1), it is growing in its "saprophytic" form. ×300. *B,* In culture, the fungus produced cleistothecia filled with ascospores in addition to the conidia and mycelium. ×5. (From Ernest, J. T., and J. W. Rippon. 1966. Am. J. Ophthalmol., *62*:1202–1204.)

The culture media usually used for the isolation of pathogenic fungi contain cyclohexamide (Actidione), which inhibits growth of saprophytic fungi. For this reason their use is contraindicated for the isolation of the etiologic agents of mycotic kerititis. Fungi in general are not fastidious and thus will grow on almost all antibiotic-free laboratory media. Most of the organisms involved in mycotic keratitis are inhibited by temperatures of 37°C, so that the cultures should be kept at room temperature. Growth is usually apparent in about three days, sometimes within 24 hours. Cultures should be kept for several weeks before being discarded, however, as some species grow very slowly. If growth occurs, slide cultures can be made. The fungus is identified by the morphology and color of its colony, and by the production type and arrangement of its conidia. Most of the organisms are species rarely encountered in a medical mycology laboratory, so that identification is often difficult. Such cultures should be sent to reference laboratories and other medical mycologists for verification of identification.

Therapy. In the past, many agents have been used for the treatment of mycotic keratitis, and essentially all have failed. There are few specific drugs for the treatment of any form of mycotic disease, and this is particularly true of mycotic keratitis. Of those available griseofulvin is without benefit, nystatin is of limited effective range, and amphotericin B is quite irritating to the involved tissue. A relatively new drug, pimaricin (natamycin) appears to be very efficacious. Fortunately it has recently been released for general usage[4] and is the drug of choice for most cases of mycotic keratitis. Yeasts may be less sensitive than hyphomycetes to this drug. Sensitivity testing of the isolated organism should be done, because the response of this group of fungi to available drugs is unpredictable. Several studies have been done on the efficacy of these drugs in experimental mycotic keratitis.[29, 76]

Amphotericin B. Many of the organisms isolated from eye infections are resistant to this drug at pharmacologic levels. However, higher concentrations may be used with benefit as drops or a wash. The standard regimen is instillation of one or two drops in the conjunctival sac at intervals of one-half to two hours. The concentration is usually 2 to 4 mg per ml aqueous solution. If the infection is severe, deep, or systemic, intravenous therapy is used as an adjunct. Administered either as drops or by subconjunctival injection, the drug is severely irritating, and treatment may be painful. An exaggerated inflammatory response, scarring, and permanent damage may be associated with the use of this drug. How much of this is caused by the drug itself

and how much is due to the bile salts surfactant included in the preparation is difficult to assess. Unfortunately the drug is unavailable without additives. Concerning this latter point, studies have shown that, as an orally administered preparation used in conjunction with antibacterial antibiotics and without surfactants, it does not seem to cause untoward distress. It is necessary to conclude, therefore, that if other treatments are available, use of amphotericin B is contraindicated.

In the review of experience with amphotericin B, Jones et al.[42-44] noted treatment failures in 13 of 20 cases when the drug was used alone. These were all infections caused by *Fusarium*. Some of their patients required conjunctival flaps or penetrating grafts in addition to drug therapy. They had greater success with the drug when the disease was

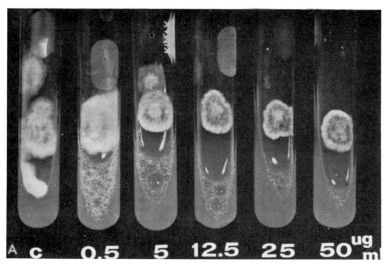

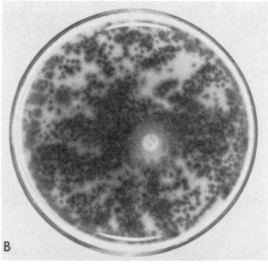

Figure 27–10. Antibiotic testing in a case of mycotic keratitis. *A,* The etiologic agent *(P. boydii)* was resistant to amphotericin B, but there was a zone of inhibition around a disk containing 100 units of nystatin *(B).* (From Ernest, J. T., and J. W. Rippon. 1966. Am. J. Ophthalmol., *62*:1202–1204.)

caused by fungi other than *Fusarium*. This was also noted in the review of cases by Anderson and Chick.[3] These authors stressed extensive mechanical debridement as a part of the therapeutic procedure.

Nystatin. This agent has been used successfully, even in some cases in which the fungus was resistant to amphotericin B[25, 59] (Fig. 27–10). Treatment consists of instillation of one drop of the antibiotic every hour in a concentration of 8000 to 20,000 (and sometimes 100,000) units per ml in saline. It is somewhat irritating, but much less so than amphotericin B. Again it is necessary to emphasize the necessity for determining the sensitivity of the etiologic agent to the drug used.

Pimaricin (Natamycin). This drug was derived from the actinomycete *Streptomyces nataliense* in 1955. It is another polyene antibiotic in the group of the tetraenes. This group also includes nystatin and Rimocidin. Its clinical usefulness has been studied since 1958, but it has not gained wide acceptance in the treatment of systemic infections. Although pimaricin is insoluble and moderately toxic when administered systemically, it is essentially without irritation or discomfort when used topically.[43, 44] Several series have recently been published in which this drug was used successfully in mycotic keratitis. It is particularly effective in *Fusarium* infections.[43, 44, 66, 73] The treatment schedule has now been standardized.[4] The drug is administered as a 5 per cent solution. This is prepared by dissolving the dry antibiotic in distilled water with 4 N sodium hydroxide. This is then quickly neutralized to pH 7 with 4 N hydrochloric acid. It is stirred for two hours until it becomes a lotionlike suspension. After adjusting the pH to 6.5 or 7.0, it is autoclaved for 20 minutes at 110° C, 15 psi, and then dispensed in plastic dropper bottles. It should be stored in the dark at 4° C. The suspension is given as drops on the lower tarsal conjunctiva every one to three hours until the infection subsides. There were only two treatment failures noted by Jones et al.[43] in 18 consecutive cases. In previous studies there had been more failures when a 1 per cent suspension was used.

Following the resolution of the infection, a variety of surgical procedures may be necessary, depending on the extent of damage to the eye. A conjunctival flap or penetrating keratoplasty is contraindicated as a treatment of the infection without previous medical management.[42] Prior to the establishment of specific drug therapy, these procedures were the only form of treatment available, but their success rate was very low.

The drug 5-fluorocytosine (5-FC) was used in one case in which there was endogenous disease and mycotic keratitis due to *Candida albicans*. Treatment was successful.[76] This patient had had several courses of intravenous and intraocular amphotericin B and intraocular 1 per cent pimaricin without resolution of the disease. The patient was subsequently given 5-fluorocytosine, 150 to 200 mg per kg per day orally, and hourly drops of a 1 per cent solution, and the infection was brought under control. There was a relapse, however, that required more treatment. 5-FC has been shown to have *in vivo* activity against very few fungi other than a few yeasts, such as *Candida* and *Cryptococcus*. Fishman,[30] however, has found that some strains of *Candida* are resistant to the drug. In addition to *Candida* and *Cryptococcus*, many phaeomycotic agents are also sensitive to 5-FC.

Miconazole has been used successfully in some cases of *Paecilomyces*, *Pseudallescheria* and other organisms that are resistant to amphotericin B[63] and 5-FC.

Endogenous Oculomycosis

This type of disease results from dissemination of fungus from other loci in the body. In almost all the systemic mycoses, eye involvement has been reported but is considered a rare complication of such diseases. Often involvement of the eye is a terminal event in which there is widespread dissemination of the disease to all organ systems. The diseases noted in order of frequency are candidiasis, cryptococcosis, coccidioidomycosis, blastomycosis, sporotrichosis, paracoccidioidomycosis, and histoplasmosis. About 25 cases of ocular involvement as a sequela of cryptococcosis have been recorded.[31, 82] Most of these involved a chorioretinitis. Approximately nine cases of eye disease in systemic coccidioidomycosis have been found,[27] and about six cases of blastomycosis have been recorded.[15] Eye involvement is much more common in blastomycosis of the dog than in human disease. Almost 50 per cent of cases of this disease in dogs involve the eye as well as some other organ system. The latter oculomycosis was first described in 1914 by Churchill and Stober.[17] In another case described by Font,[30a] this was the only manifestation, and the diagnosis of blastomycosis was made only after finding the organisms in the enucleated eye. There were retrospective chest findings.

The most common cause of endogenous oculomycosis is *Candida albicans.* More than 30 cases of ocular involvement in disseminated candidiasis have been described in the recent literature.[30, 61, 76] In a series of six, Fishman[30] described *Candida* endophthalmitis as a complication of candidemia. All patients had received antibiotics, and all had had indwelling catheters for prolonged periods of time. Only one patient had a neoplasm, and none had received steroid or immunosuppressive therapy. Symptoms of ocular distress occurred 3 to 15 days after positive blood cultures were obtained. The symptoms included painful eye, pain around the eye, blurring of vision, and spots before the eye. However, in two patients there were no symptoms, though lesions of the eye were found on examination. Intravitreous extension occurred in three cases, and anterior uveitis was found in the other three. The treatment consisted of intravenous amphotericin B. The dosage varied, depending on response. As little as 300 mg was used with success in one case, but others required up to 1.4 g. The minimum inhibitory concentration of amphotericin B for the *Candida* isolated from these cases varied between 0.04 and 0.078 μg per ml. Two isolants were resistant to 5-fluorocytosine, and one had a minimum inhibitory concentration (MIC) of 0.05.

The usual presenting symptoms in all types of disseminated mycosis involving the eye are a progressive granulomatous uveitis (Fig. 27–

11), diffuse retinitis, and deep vitreous abscess. Glaucoma may also be present. The inflammatory response is noted to be minimal in ocular cryptococcosis. The diagnosis of endophthalmic disease usually depends on isolation of the organism from other loci of infection, and treatment involves systemically administered amphotericin B. Mycotic endophthalmitis has occurred in the absence of trauma to the cornea or known fungus infection in another organ system. In a case reported by Lieberman et al.[54] no discernible portal of entry was found in a *Fusarium* endophthalmitis. The patient had no underlying disease, but she did have many potted plants that were diseased by *Fusarium* in her home. The authors speculate that the conidia were inhaled and subsequently carried to the eye hematogenously.

Presumed *Histoplasma* choroiditis[48, 49, 80] (benign histoplasmosis, *Histoplasma* granulomatous uveitis, or multifocal choroiditis) appears to be an allergic disease rather than a direct infection. Its association with histoplasmosis is tenuous, and substantiated evidence of a causal relationship is lacking. In true disseminated histoplasmosis, the eye is rarely involved. In such cases the pathology resembles that of tuberculosis and other granulomatous infections and does not resemble the clinical or pathologic features seen in multifocal choroiditis. In the latter disease, the primary lesion in the acute stage is a yellowish white, slightly elevated choroidal infiltrate that is usually multifocal. Fluid leakage may occur from these macular lesions. During periods of chronic serous fluid leak-

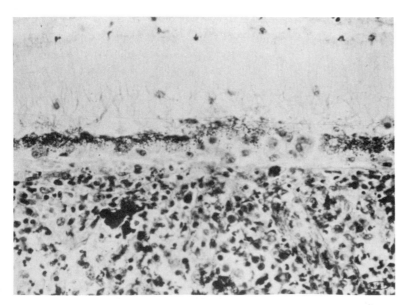

Figure 27–11. Endogenous oculomycosis. Granulomatous uveitis in a case of disseminated blastomycosis. Some yeast cells can be seen among the cellular infiltrate within choroid. × 400.

age, new vessels from the choroid may be formed. Their rupture leads to hemorrhage.[48] Eventually, hypertrophic scars form in such areas, with severe visual loss.

Presumed *Histoplasma* uveitis was first described by Woods and Wahlen in 1959.[95] They reported that a preexisting uveitis worsened following a histoplasmin skin test and concluded there was a causal relationship. Since that time several hundred cases of a similar (though possibly not identical) disease have been diagnosed.[48, 49, 80] In no case has the organism been identified in tissue or has specific fluorescent staining been positive for debris of the *Histoplasma* yeast cell. Furthermore, there is no well-documented evidence that treatment with amphotericin B improved the patient's condition, although such has been reported.[26] However, steroids and photocoagulation have been efficacious in controlling the disease.[48]

Although it has been reported that 90 per cent or more of patients with multifocal choroiditis have a positive histoplasmin skin test,[48, 49] it is difficult to assess or correlate the association of skin test reactivity with the disease. In some areas of the United States, between 80 and 90 per cent of the general population will demonstrate a skin test reactivity, and cross-reactions with other fungal antigens are common.[20] It has also been shown that patients with presumed histoplasmin choroiditis are hyperergic to several antigens other than histoplasmin.[90] In this same study the curve expressing frequency of uveitis at different ages paralleled the appearance of skin testing reactivity in different age groups. In recent studies concerned with the disease, serum antibodies to double stranded RNA and DNA have been found in patients who had the idiopathic and secondary types of uveitis.[13, 24] In France a type of uveitis similar to multifocal choroiditis was associated with sensitivity to *Candida* antigen.[89]

Extension Oculomycosis

This disease is usually associated with acute rhinocerebral mucormycosis of the diabetic, but it is also seen in cases of rhinosporidiosis and meningeal cryptococcosis. These latter infections are discussed in other chapters. In acute mucormycosis the infection starts in the upper portion of the nasal septum and extends into the orbit of the eye, the frontal sinuses, the major cerebral vessels, and subsequently into the central nervous system. On occasion the infection may begin in the ear and progress internally from there. A complete discussion of this disease is given in Chapter 25. In this section a summary of the associated ocular manifestations will be presented.

Clinical Symptoms. The presenting symptoms of mucormycosis involve first the nose and then the eye. The nasal area shows a thick discharge, dark and blood-tinged, as well as reddish-black necrotic areas of the turbinates and septum. A cloudy sinus, sometimes with a discernible fluid level, can be seen on x-ray examination. In addition, there may be paranasal inflammation or pansinusitis; fistula formation and sloughs of the hard palate; loss of function of fifth and seventh cranial nerves; and often a hard, somewhat palpable, demarcated swelling on either side of the nasal bridge. Severe orbital cellulitis at this stage denotes a poor prognosis, as it heralds impending or actual invasion by the fungus of the eye and usually the central nervous system. The findings of the eye include orbital pain, with ophthalmoplegia, ptosis, localized anesthesia, proptosis, limitation of movement of the eyeball, fixation of the pupil, and loss of vision. Hyphae may be seen coursing through the vitreous. The disease is unilateral or bilateral.

As the infection progresses, the fungus shows a predilection for blood vessels, with subsequent infarction and necrotic sequelae in the brain and softening of the frontal bones. Lethargy progressing to coma ensues, and the patient succumbs in seven to ten days. In some personally observed cases of fulminating disease, death occurred as early as two days after initial symptoms of infection. The overall mortality rate of this disease was between 80 and 90 per cent. It is now greatly lessened because of prompt medical and surgical management and the use of antifungal drugs. Most of the patients who survive have received antifungal therapy in addition to control measures for their diabetes.

Predisposing Factors. Acute mucormycosis has classically been associated with diabetes. There is experimental evidence, however, indicating that the acidotic state is the prime factor predisposing to infection. Rats or rabbits made diabetic by alloxan treatment were resistant to *Mucor* infection, but when made acidotic as well, they rapidly succumbed to infection by the organism. Acute mucormycosis has been recorded in patients who were acidotic as a result of other conditions, such as diarrhea or salicylate ingestion.

Treatment. Control of diabetes appears to be the most important factor in halting the infection. In the few reports of survivors before antifungal therapy became available, the infection receded when the acidotic state of the patient was corrected. However, the disease sometimes relapsed during another bout of uncontrolled diabetes with acidosis. At present, use of systemic and local amphotericin B has controlled the infection in many patients. It is important to do sensitivity studies on the fungus isolated from the infection, as the organisms vary considerably in their response to the drug.

Etiology and Diagnosis. The fungus most commonly isolated from this form of the disease is *Rhizopus oryzae* and *R. arrhizus*. A few other species in the genera *Rhizomucor* and *Rhizopus* have also been reported as etiologic agents of the infection. All organisms are common saprophytic soil fungi, often termed the "black bread molds." In general the organisms are readily cultivated from patient material. A scraping of the upper nares is used for preparation of a direct mount (KOH mount is preferred), and this same material can be used for culture. The fungi grow on any media without antibiotics. On direct mount, the fungus appears as wide (4 to 15 μ), nonseptate, branching hyphae. The culture is incubated at room temperature, and growth usually occurs within a few days or as early as 18 hours.

Infection of the Lacrymal Ducts and Conjunctiva

Cases of canaliculitis (dacryocanaliculitis) have been described for many years, and the disorder is a well-recognized syndrome. The patients present with a swelling at the opening of the duct, tearing, pruritus, and inflammation of the conjunctiva. A hard, knobby concretion is felt within the duct, causing stasis and pressure. Most often only one duct is involved, and it is usually the inferior canal. In a review of 146 cases, the superior duct was involved in only 41.[31] The treatment is simple and effective and consists of removal of the plug. This is generally accompanied by administration of antibiotics, such as sulfas and penicillin.

Examination of the excised material usually demonstrates intertwined hyphal elements and bacilliform bodies that are 1 μ in diameter, Gram-positive, and usually acid-fast–negative. The etiologic agent was called *Streptothrix* in the older literature and the disease named streptotricosis. Culture of material from canaliculitis has yielded a variety of organisms, including *Actinomyces israelii*,[72] *Nocardia asteroides*[70] (acid-fast–positive), and various *Corynebacterium*.

Another form of the disease is dacryocystitis. This is an inflammatory disease of the lacrymal duct caused by a variety of bacteria and fungi. Among these are *Actinomyces, Candida, Aspergillus, Blastomyces (?), Paecilomyces*,[41] *Sporothrix, Rhinosporidium, Acremonium*, and *Trichophyton*. Probably only a few such reports are valid cases of disease in which there was direct involvement of a fungal agent.

Although the conjunctiva may be involved in a number of systemic mycoses, conjunctivitis is rarely a presenting symptom of a mycotic disease. Infection of the conjunctiva in the absence of other foci of disease is known in rhinosporidiosis but in essentially no other fungus infection. Infection of the eyelid alone, however, has been recorded on several occasions. This is usually due to dermatophytes[69] or *Malassezia* sp.[31]

Otomycosis

Otomycosis, or otitis externa of fungal etiology, is a chronic or subacute infection of the pinna, the external auditory meatus, and the ear canal. Various fungi may be involved, along with several bacteria.[36, 37] The presenting symptoms include scaling, pruritus, and pain. The canal is seen to be crusted, edematous, and erythematous, and there is a collection of cerumen. A feeling of fullness of the ear is often present, along with impairment of hearing. Suppuration and a foul odor may also be found and are caused by bacterial invasion of the subepithelial layers. In disease in which there is little bacterial involvement, the lesions are dry and eczematized. In severe acute infection, the cartilagenous structures of the ear may be involved. In almost all cases, the tympanic membrane is spared, however. In 80 to 90 per cent of cases of otitis externa, a variety of bacteria are found, including *Pseudomonas, Proteus, Micrococcus, Streptococcus, Escherichia,* and *Corynebacterium.*[83] Frequently the condition responds to antibacterial therapy alone. A fungal etiology in such cases is doubtful, even if it is positive by culture. The establishment of a true fungal etiology for otitis externa requires demonstration of mycelial elements in scrapings, as well as a positive culture.[37, 47, 60]

External ear infection due to invasion by fungi is a well-established syndrome which is frequently encountered. The fungi most commonly involved are *Aspergillus fumigatus* and *A. niger.*[87] *A. niger* alone accounts for more than 90 per cent of valid infections.[21] Otoscopic examination sometimes reveals the fruiting heads of the organisms lining the wall of the canal (Fig. 27–12). Other species of fungi have also been documented as causing this type of infection, but only rarely. These include *Scopulariopsis,*[96] *Aspergillus*

flavus, other *Aspergillus* species, *Penicillium, Rhizomucor,* and *Candida* sp. *Peyronellaea* has been reported in animal infections.[19] External infection of the ears of dogs is a common veterinary problem. It is usually caused by *Malassezia* sp., which responds to topical therapy such as 1 per cent clotrimazole drops. Numerous other species have been listed, but few have been unequivocally documented as the etiologic agent. Dermatophytosis has been reported in infections of the pinna proper. Such infections usually are extensions from disease elsewhere on the skin.

Treatment of the disease has included many agents and modalities. In a recent series of six cases in which *A. niger* was the etiologic agent, application of nystatin ointment for three to four weeks cleared the infection,[9] and this drug has also been successfully used in infections caused by a variety of other fungi.[30] In reviewing therapy, El

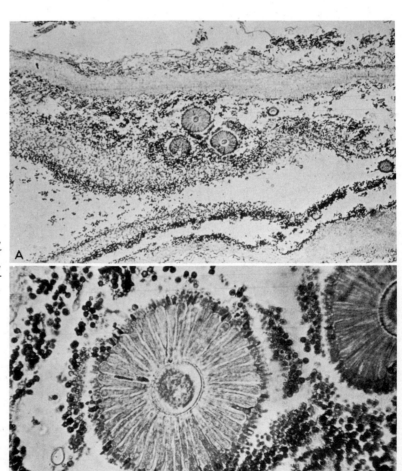

Figure 27–12. Otomycosis. *A,* Fruiting heads of *Aspergillus niger* seen in the ear canal. ×100. *B,* High-power view of the vesicle metulae and conidia of *A. niger.* ×400.

Gothamy[21] indicated success with 1 per cent clotrimazole, amphotericin B, or nystatin. Phenylmercuric borate ointment has been used with good results in a series of 34 cases.[37]

The usual treatment involves cleaning the ear and removing the macerated debris. Swabs with Burow's solution, 5 per cent aluminum acetate solution, or urea-acetic acid solution accomplish debridement of this material. Often this is sufficient to effect a clinical cure. As there is usually a bacterial component, antibacterial therapy is often necessary. The preparations most generally utilized are chloramphenicol, bacitracin, polymyxin, neomycin, and Aureomycin. In chronic ear infection in which the bacterial component is minimal, metacresol acetate (Cresatin), thymol (1 per cent) in metacresyl acetate, and phenylmercuric acetate (0.02 per cent) in water[96] or as a borate ointment (0.005 and 0.04 per cent) have effected clinical cure. Amphotericin B ointment also has been tried, but the results have been equivocal.

The fungi responsible for the disease are generally airborne saprophytes, so that cultures should be made on Sabouraud's agar without cycloheximide. Optimal growth is obtained at room temperature. Isolation of several colonies of the same fungus, in addition to evidence by direct examination, is necessary for a specific diagnosis. Otomycosis has been reproduced experimentally in animals.[85]

Mycotic infections of the inner ear can result from an extension of an organism colonized in the external auditory canal, by extension of a process in surrounding tissue, or from hematogenous dissemination from some other organ system. The ear has been the portal of entry for meningoencephalitic mucormycosis.[62] Most cases of inner ear involvement resulting from hematogenous dissemination were caused by *Candida* sp.[62]

REFERENCES

1. Ajello, L. 1981. The gamut of human infections caused by dematiaceous fungi. 1st Conf. of Italian and Spanish Mycology.
2. Albesi, E. J., and R. Zapater. 1972. Flora fungica de la conjunctiva en ojos sanos. Arch. Oftalmol. B. Aires, 47:329–334.
3. Anderson, B., and E. W. Chick. 1963. Mycokeratitis: Treatment of fungal corneal ulcers with amphotericin B and mechanical debridement. South. Med. J., 56:270–273.
4. Anon, 1979. Natamycin approved — first U.S. drug for fungal keratitis. FDA Drug Bulletin, 8:37–38.
5. Apostol, J. C., and S. L. Meyer. 1972. *Graphium* endophthalmitis. Am. J. Ophthalmol., 73:566–569.
6. Arrechea, A., R. Zapater, et al. 1971. Queratomicosis por *Fusarium solani*. Arch. Oftalmol. B. Aires, 46:123–127.
7. Bakerspigel, A. 1971. Fungi isolated from keratomycosis in Ontario, Canada. I. *Monosporium apiospermum*. Sabouraudia, 9:109–112.
8. Beaney, G. P. E., and A. Broughton. 1967. Tropical otomycosis. J. Laryngol. Otol., 81:987–997.
9. Bezjak, V., and O. P. Arya. 1970. Otomycosis due to *Aspergillus niger*. East Afr. Med. J., 47:247–253.
10. Birge, H. L. 1962. Reclassification of mycotic disease in clinical ophthalmology. Am. J. Ophthalmol., 53:630–635.
11. Bisttrer, S. L., and R. C. Riis. 1970. Clinical aspects of mycotic keratitis in the horse. Cornell Vet., 69:364–374.
12. Borland, D. S. 1959. Mucormycosis of central nervous system. Am. J. Dis. Child., 97:852–856.
13. Burns, R. M., M. S. Rheins, et al. 1967. Anti-DNA in the sera of patients with uveitis. Arch. Ophthalmol., 77:777–779.
14. Byers, L., M. G. Holland, et al. 1960. *Cephalosporium* keratitis. Am. J. Ophthalmol., 49:247–269.
15. Cassady, J. V., 1946. Uveal blastomycosis. Arch. Ophthalmol., 35:84.
16. Chick, E. W., and N. F. Conant. 1962. Mycotic ulcerative keratitis. A review of 148 cases from the literature (abstract). Invest. Ophthalmol., 1:419.
17. Churchill, T., and A. M. Stober. 1914. A case of systemic blastomycosis. Arch. Intern. Med., 13:568–574.
18. Collette, B. E., and O. H. Curry. 1978. Mycotic keratitis in a reticulated python. J. Am. Vet. Med. Assoc., 173:1117–1118.
19. Dawson, C. D., and A. W. D. Lopper. 1970. *Peyronellaea glomerata* infection of the ear pinna in goats. Sabouraudia, 8:145–148.
20. Edwards, P. W., and J. H. Klaer. 1956. Worldwide geographic distribution of histoplasmosis and histoplasmim sensitivity. Am. J. Trop. Med., 5:235–257.
21. El Gothamy, M. A. B., and Z. El Gothamy. 1977. Otomycosis — a new line of treatment. Castellania, 5:215–216.
22. Elliot, J. H., D. M. O'Day, et al. 1979. Mycotic endophthalmitis in drug abuses. Am. J. Ophthal., 88:66–72.
23. English, M. P. 1957. Otomycosis caused by a ringworm fungus. J. Laryngol. Otol., 71:207–208.
24. Epstein, W. V., M. Tan, et al. 1971. Serum antibody to double stranded RNA and DNA in patients with idiopathic and secondary uveitis. New Engl. J. Med., 285:1502–1506.
25. Ernest, J. T., and J. W. Rippon. 1966. Corneal ulcer due to *Allescheria boydii*. Am. J. Ophthalmol., 62:1202–1204.
26. Falls, H. F., and C. L. Giles. 1960. The use of amphotericin B in selected cases of chorioretinitis. Am. J. Ophthalmol., 49:1288–1298.
27. Faulkner, R. F. 1962. Ocular coccidioidomycosis. Am. J. Ophthalmol., 53:822–827.
28. Fazakas, A. 1955. Mycotic infections of the eye and their treatment. Klin. Monatsbl. Augenheilkd., 127:701–721.

29. Fine, B. S., and L. E. Zimmerman. 1960. Therapy of experimental intraocular *Aspergillus* infection. Arch. Ophthalmol., *64*:849–861.

30. Fishman, L. S., J. R. Griffin, et al. 1972. Hematogenous *Candida* endophthalmitis. A complication of candidemia. New Engl. J. Med., *286*:675–681.

30a. Font, R. L., A. G. Spaulding, et al. 1967. Endogenous mycotic panophthalmitis caused by *Blastomyces dermatitidis*. Arch. Ophthal., *77*:217–222.

31. François, J. 1968. *Les Mycoses Oculaires*. Paris, Masson and Cie.

32. François, J., and M. Rijsselaere. 1979. Corneal ulcer infections by *Rhodotorula*. Ophthalmologica, *178*:241–249.

33. Gingrich, W. D., and M. E. Pinkerton. 1952. Anaerobic *Actinomyces bovis* corneal ulcer. Arch. Ophthalmol., *67*:549–553.

34. Gordon, M. A., W. W. Vallotton, et al. 1959. Corneal allescheriosis. Arch. Ophthalmol., *62*:758–763.

35. Graefe, A., von. 1854. Polypen des Tränenschlauchs. Albrecht v. Graefes Arch. Ophthalmol., *1*:283–284.

36. Gregson, A. E. W., and C. J. La Touche. 1961. The significance of mycotic infection in the aetiology of otitis externa. J. Laryngol. Otol., *75*:167–170.

37. Grigoriu, D., and N. Font. 1970. Les otomycoses. Dermatologica, *141*:138–142.

38. Halde, C., and J. Okumoto. 1966. Ocular mycoses. A study of 82 cases. Amsterdam Excerpta Medical International Congress, 1966, pp. 705–712.

39. Hammeke, J. C., and P. P. Ellis. 1960. Mycotic flora of the conjunctiva. Am. J. Ophthalmol., *49*:1174–1178.

40. Harley, R. D., and S. E. Mishler. 1959. Endogenous intraocular fungus infection. Trans. Am. Acad. Ophthalmol. Otolaryngol., *63*:264–268.

41. Henig, F. E., N. Lehrer, et al. 1973. Paecilomycosis of the lacrimal sac. Mykosen, *16*:25–28.

42. Jones, B. R., D. B. Jones, et al. 1970. Surgery in the management of keratomycosis. Trans. Ophthalmol. Soc., U.K., *89*:887–897.

43. Jones, D. B., R. K. Forster, et al. 1972. *Fusarium solani* keratitis treated with natamycin (pimaricin). Arch. Ophthalmol., *88*:147–154.

44. Jones, D. B., R. Sexton, et al. 1970. Mycotic keratitis in South Florida. A review of 39 cases. Trans. Ophthalmol. Soc. U.K., *89*:781–787.

45. Kaufman, L., P. T. Terry, et al. 1967. Effect of a single histoplasmim skin test on the serologic diagnosis of histoplasmosis. J. Bacteriol., *94*:798–803.

46. Kaufman, H. E., and R. M. Wood. 1965. Mycotic keratitis. Am. J. Ophthalmol., *59*:993–1000.

47. Kingery, F. A., 1965. The myth of otomycosis. J.A.M.A., *191*:129.

48. Krill, A. E., M. Chreshti, et al. 1969. Multifocal inner choroiditis. Trans. Am. Acad. Ophthalmol. Otolaryng., *73*:722–742.

49. Krill, A. E., and D. Archer. 1970. Choroidal neovascularization in multifocal (presumed histoplasmin) choroiditis. Arch. Ophthalmol., *84*:595–604.

50. Laverde, S., L. H. Moncada, et al. 1973. Mycotic keratitis; five cases caused by unusual fungi. Sabouraudia, *11*:119–123.

51. Leber, T. 1879. Keratomykosis aspergillina als Ursache von Hypopion keratitis. Albrecht v. Graefes Arch. Ophthalmol., *25*:285–301.

52. Levitt, J. M., and J. Goldstein. 1971. Keratomycosis due to *Allescheria boydii*. Am. J. Ophthalmol., *71*:1190–1191.

53. Ley, A. P. 1956. Experimental fungus infections of the cornea. Am. J. Ophthalmol., *42*:59–71.

54. Lieberman, T. W., A. P. Ferry, et al. 1979. *Fusarium solani* endophthalmitis without primary corneal involvement. Am. J. Ophthal., *88*:764–767.

55. Manchester, P. T., and L. K. Georg. 1959. Corneal ulcer due to *Candida parapsilosis (C. parakrusei)*. J.A.M.A., *171*:1339–1341.

56. Matsumoto, T., J. Masaki, et al. 1979. *Cylindrocarpon tonkinensis* as a cause of keratomycosis. Trans. Br. Mycol. Soc., *72*:503–504.

57. Maumenee, A. E., and A. M. Silverstein (eds.). 1964. *Immunopathology of Uveitis*. Baltimore, Williams and Wilkins Co.

58. McGinnis, M. R. 1978. *Helminthosporium* corneal ulcer. Am. J. Ophthal., *86*:853.

59. McGrand, J. C. 1970. Symposium on direct fungal infection of the eye. Keratomycosis due to *Aspergillus fumigatus* cured by nystatin. Trans. Ophthalmol. Soc., U.K., *89*:799–802.

60. McGonigla, J. J., and O. Jilson. 1967. Otomycosis, an entity. Arch. Dermatol., *95*:45–46.

61. Mendelblatt, D. L. 1953. A review and a report of the first case demonstrating the *Candida albicans* in the cornea. Am. J. Ophthalmol., *36*:379.

62. Meyerhoff, W. L., M. M. Parella, et al. 1978. Mycotic infections of the inner ear. Laryngoscope, *89*:1725–1734.

63. Miller, G. R., G. Rebell, et al. 1978. Intravitreal antimycotic therapy and cure of mycotic endophthalmitis caused by a *Paecilomyces lilacinus* contaminated pseudophakos. Ophthalmic Surg., *9*:54–63.

64. Naumann, G., W. R. Green, et al. 1967. Mycotic keratitis. A histopathologic study of 73 cases. Am. J. Ophthalmol., *64*:668–682.

65. Newmark, E., and F. M. Polack. 1970. *Tetraploa* keratomycosis. Am. J. Ophthalmol., *70*:1013–1015.

66. Newmark, E., A. C. Ellison, et al. 1970. Pimaricin therapy of *Cephalosporium* and *Fusarium* keratitis. Am. J. Ophthalmol., *69*:458–466.

67. Nityananda, K., P. Sivasubramaniam, et al. 1964. A case of mycotic keratitis caused by *Curvularia geniculata*. Arch. Ophthalmol., *71*:456–458.

68. O'Day, D. M., P. L. Akrabawi, et al. 1979. An animal model of *Fusarium solani* endophthalmitis. Br. J. Ophthalmol., *63*:277–280.

69. Ostler, H. B., M. Okumoto, et al. 1971. Dermatophytosis affecting the periorbital region. Am. J. Ophthalmol., *72*:934–938.

70. Penikett, E. J. K., and D. L. Rees. 1962. *Nocardia asteroides* infection of the nasal lacrimal system. Am. J. Ophthalmol., *53*:1006–1008.

71. Pine, L. H., Hardin, et al. 1960. Actinomycotic lacrimal canaliculitis. A report of two cases with a review of characteristics which identify the causal organism *Actinomyces israelii*. Am. J. Ophthalmol., *49*:1278–1288.

72. Pine, L., W. A. Shearin, et al. 1961. Mycotic flora of the lacrimal duct. Am. J. Ophthalmol., *52*:619–625.

73. Polack, F. M., H. E. Kaufman, et al. 1971. Keratomycosis, medical and surgical treatment. Arch. Ophthalmol., *85*:410–416.

74. Posner, A. 1960. The role of starch derivative glove powder in hospital infection. Eye, Ear, Nose, Throat Monthly, *39*:175.

75. Prabhakar, H., N. L. Chitakara, et al. 1969. Mycotic and bacterial flora of the conjunctival sacs in healthy and diseased eyes. Indian J. Pathol. Bacteriol., *12*:158–161.

76. Richards, A. B., B. R. Jones, et al. 1970. Corneal and intraocular infection by *Candida albicans* treated with 5-fluorocytosine. Trans. Ophthalmol. Soc., U.K., *89*:867–885.

77. Rippon, J. W., 1972. Mycotic infections of the eye. Diagnosis and treatment. Ophthalmol. Dig., *34*:18–25.

78. Rowsey, J. J., T. E. Acers, et al. 1979. *Fusarium oxysporum* endophthalmitis. Arch. Ophthalmol., *97*:103–105.

79. Savir, H., E. Henig, et al. 1978. Exogenous mycotic infections of the eye and adnexa. Ann. Ophthalmol., *10*:1013–1018.

80. Schlaegel, T. F., J. C. Weber, et al. 1967. Presumed histoplasmic choroiditis. Am. J. Ophthalmol., *69*:919–925.

81. Segelken, Von. 1902. Ein kasuistischer Beitrag zur Aeriologic der konkremente in den Tränenröhrchen. Klin. Monatsbl. Augenheilk., *40*(11): 134–143.

82. Shields, J. A., D. M. Wright, et al. 1980. Cryptococcal choricoretinitis. Am. J. Ophthalmol., *89*:210–218.

83. Singer, D. E., et al. 1952. Otitis externa; bacteriological and mycological studies. Ann. Otol. Rhin. & Laryng., *61*:317–330.

84. Snyder, W. C., and T. A. Tousson. 1965. Current status of taxonomy of *Fusarium* species and their perfect stages. Phytopathology, *55*:833–837.

85. Sood, V. P., A. Sinha, et al. 1967. Otomycosis: A clinical entity. Clinical and experimental study. J. Laryngol. Otol., *81*:999–1012.

86. Srivastava, O. P., B. Lal, et al. 1977. Mycotic keratitis due to *Rhizoctonia* sp. Sabouraudia, *15*:125–131.

87. Stuart, E. A., and F. Blank. 1955. Aspergillosis of the ear. A report of twenty-nine cases. Can. Med. Assoc. J., *72*:334–337.

88. Theodore, F. H. 1962. The role of so-called saprophytic fungi in eye infection. *In* Daldorf, F. (ed.) *Fungi and Fungus Diseases.* Springfield Ill., Charles C Thomas, Publisher, pp. 22–32.

89. Vallery-Radot, C. 1966. Uvéite et allergie á *Candida albicans*. Rev. Fr. Allerg., *6*:27–32.

90. Van Metre, T. E., and A. E. Maunseuee. 1964. Specific ocular uveal lesions in patients with evidence of histoplasmosis. Arch. Opthalmol., *71*:314–324.

91. Weber, J. C., and T. F. Schlaegel. 1969. Delayed skin test reactivity of uveitis patients. Am. J. Ophthalmol., *62*:732–744.

92. Wilson, L. A., R. R. Sexton, et al. 1966. Keratomycosis. Arch. Ophthalmol., *76*:811–816.

93. Wilson, L. A., and R. R. Sexton. 1968. Laboratory diagnosis in fungal keratitis. Am. J. Ophthalmol., *66*:647–653.

94. Wind, C. A., and F. M. Polack. 1970. Keratomycosis due to *Curvularia lunata*. Arch. Ophthalmol., *84*:694–696.

95. Woods, A. C., and H. E. Wahlen. 1960. The probable role of benign histoplasmosis in the etiology of granulomatous uveitis. Am. J. Ophthalmol., *49*:205–220.

96. Yamashita, K., and T. Yamashita. 1972. *Polypaecilum insolitum (Scopulariopsis divaricata)* isolated from cases of otomycosis. Sabouraudia, *10*:128–131.

97. Zapater, R. C., and F. Scattini. 1979. Mycotic keratitis by *Cladorrhinum*. Sabouraudia, *17*:65–69.

98. Zapater, R. C., A. de Arrochea, et al. 1972. Queratomicosis por *Fusarium dimerum*. Sabouraudia, *10*:274–275.

99. Zimmerman, L. E. 1963. Keratomycosis. Survey Ophthalmol., *8*:1–25.

Allergic Diseases, Mycetismus, and Mycotoxicosis

28 Allergic and Toxic Diseases Associated with Fungi

INTRODUCTION

In addition to actual infection by a fungus, there are several diseases that are caused by the inhalation of fungal debris, eliciting an allergic reaction; by the ingestion of organisms, usually mushrooms, containing toxins (mycetismus); or by the ingestion of products that have been altered by previous growth of fungi that contain toxic substances (mycotoxicosis). An overview of these several types of disease will be given in this section. Each of these disease types has been the subject of a great amount of investigation, and several books have been devoted to them. The present chapter is intended to be an introduction to these specialized subjects and a guide to the major reference sources.

Fungal toxins and allergies have been known to affect man and animals since antiquity. Fatal mushroom poisoning was the subject of much Greek, Roman, and Hindu literature, and interest in the subject has continued through the middle ages and renaissance to the present. Formulas for the practical or political use of such poisons, as well as means of alleviating the distress if one, perchance, is the victim, are extant in these sources. On a wider scale of economic importance, the entire population and probably the survival of western civilization has been dramatically affected by such intoxications as ergot poisoning. Most of these accounts were relegated to curious sidelights of history until recently. In 1960 hundreds of turkey poults were killed by a strange disease in England termed "turkey X disease." This disease, caused by a toxin produced by *Aspergillus flavus* growing on grain, sparked a renewed interest in the mycotoxicoses. It is surmised that many previously unexplained diseases of man and animals may be caused by as yet unknown toxins that result from the growth of fungi on foodstuffs. At present, aflatoxicosis, as well as other similar syndromes, is being investigated, and a voluminous literature is evolving.

Distress and death mixed with much superstition and religous folklore have been associated with the ingestion of mushrooms and other fleshy fungi during all recorded history. Even in eastern European countries where mycophagy has been practiced for centuries, there are still significant numbers of fatalities each year due to misidentification of mushrooms used for human consumption. The intoxication and hallucination following ingestion of certain species in certain Indian tribal rites is the subject of much faddist, anthropologic, and scientific "lore." Many mushrooms are superior items on a gourmet's menu; others, however, may lead to a variety of gastrointestinal, neurologic, or other types of disturbances or may ultimately be fatal. Thus the study of the various effects of ingesting fungi on man has popular as well as scientific and medical interest.

Although asthma and general allergy to the inhalation of fungal conidia has long been recognized as a clinical entity, only recently have specific occupational diseases, some causing irreversible damage or death, been recognized. Many cases of "idiopathic" chronic lung disease are now being recognized as manifestations of hypersensitivity to fungal conidia, fungal products, or fungal debris present in a particular environment.

New syndromes of all three types of mycotic disease are continually being recognized.

701

These diseases will be grouped for convenience into three major subheadings: (1) mycetismus, or disease caused by the ingestion of fungi (mushroom poisoning); (2) mycotoxicosis, or ingestion of preformed fungal products or products formed under the influence of fungi; and (3) allergies or hypersensitivities manifest by particular individuals to the presence of fungal allergens in the environment.

Mycetismus (Mushroom Poisoning)

Mycophagy, the consumption of fleshy fungi, has been practiced by man since before recorded history. Experience over the centuries has implicated certain species as having undesirable or prized side effects. These range from minor gastrointestinal distress to hallucination, delirium, coma, and death. Needless to say, such a variety of responses has held a measure of fascination for the "mushroom hunter" equal to or surpassing

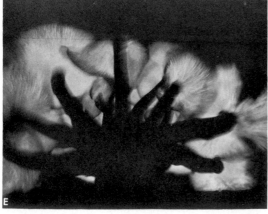

Figure 28–1. *A, Amanita virosa* showing white volva and annulus. The cap is off-white and without scales. *B, A. muscaria* showing white gills, volva, and annulus. This is the yellow-capped variety. Note the white scales on cap. *C, Omphalotus olearius,* the "jack-o-lantern." This brilliant orange mushroom glows in the dark. *D, Inocybe fastigiata.* The gills are purplish brown. The cap is smooth, brownish, and split. *E,* A group of *Omphalotus olearius* growing from a stump and demonstrating bioluminescence of the gills. It is this trait that led it to be called "jack-o-lantern," or "ghost of the forest." Its bright orange color arouses curiosity in children, who often eat it. They are sick for 72 hours and then recover spontaneously and dramatically.

the "hunter" of other forms of dangerous game. Despite these inherent dangers, the quest of a determined mycophagist or the naive amateur continues in experimentation with new forms, hopefully without unpleasant sequelae.

The author is a devoted mycophagist and has practiced this study for some years. One learns, for example, that *Hericium coralloides,* found in the fall, is sweet and has a definite shrimplike flavor when sautéed. *Collybia radicata* has a chickenlike quality but may give the impression of a lingering pepperlike aftertaste, and *Clavicorona pyxidata* (a coral fungus), slightly sautéed in butter, is nutlike and superb as an appetizer or dessert. The gourmet discerns that white truffles are more delicate in flavor than the black, and in spring the advent of morels is to be savored with much joy. On the other hand, folklore and scientific investigation delineate certain species as forbidden fruit. There is no way to determine which is an edible and which is a poisonous species except by careful and expert identification. Peeling of the cap (pileus) to avoid possible toxins, turning silver dark when the fungus is placed in water, or changing color when the mushroom is bruised do not constitute accurate delineation of a "poisonous" from a "safe" mushroom.[9, 14, 60] Even the most experienced mycophagist must be cautious in picking the prey. Many books are available on the subject, but it must be appreciated that there are thousands of species of mushrooms, many difficult to identify and quite variable in morphologic characteristics, and that idiosyncrasies of the consumer may determine the pleasure or distress of the feast.[16, 67, 79, 80] A "toadstool" is the vernacular for a toxic mushroom. This has no basis on scientific grounds. In the final analysis, a toadstool is a mushroom on which a toad is sitting.

Mushroom poisoning is not as common among inhabitants of the United States as it is among those of eastern Europe. In the latter area, the gathering and cooking of fungi is a centuries-old practice. Even though tradition and folklore have determined what are good mushrooms and what are "toadstools" (poisonous forms), the highest rate of mycetismus is still among peoples who regularly collect and consume known "safe" species. Each year there are many cases of poisoning and death, especially among children. In the western hemisphere, the popularity of mushroom gathering is not so great, so that there is less

experience in these forms of toxicity. However, there are a significant number of cases each year, so that attention and interest should be maintained. In Canada there are approximately 150 cases per year of mycetismus, with the most occurring in Ontario (75 per cent).[64] Similar to the situation in Europe, the majority (>80 per cent) are in children. In the United States the rate of cases of intoxication is low but constant from year to year. There are about 350 cases of mushroom poisoning annually in the United States; about 70 per cent of these occur in children under five years of age. Fatalities are rare. (Center for Disease Control: Foodborne and Waterborne Disease Outbreaks Annual Summary, 1978, Atlanta CDC, Nov. 1979.) Inexperience in identification is the main reason for the hazards, and thankfully most cases are not serious.

Probably the first report of fatal mycetismus in the United States was recorded in 1838. This is noted on a tombstone in Trinity Episcopal Church, Fishkill, New York. The inscription reads: William Gould, wife, and son were "poisoned by eating fungi (toadstools)." Since that time numbers of small, often family-centered, epidemics have occurred.[5, 31, 51, 60]

There have been many attempts to classify the particular distresses associated with the various fungi involved. The one proposed by Lincoff and Mitchel[60] in their book on mushroom poisoning is among the most useful. A modified schema from their work and that of others will be presented here. Mycetismus can be conveniently divided into eight categories based on the major active ingredients: (1) Cyclopeptides, the amatoxins and phallotoxins found in poisonous *Amanita* and *Galerina* spp.; (2) Monomethyl hydrazine, the gyromitrin of *Gyromitra esculata;* (3) Coprine, the disulfiramlike material of *Coprinus atrementarius;* (4) Muscarine, the parasympathomimetic agent of *Inocybe fastigiata* and other species; (5) Ibotenic acid and muscimol, from *Amanita muscaria,* the "old time religion" mushroom of old world priests, prophets, and cultists; (6) Psilocybin, the "flesh of God" from *Psilocybe* sp. mushrooms of new world religions and faddists; (7) Gastrointestinal irritants from *Rhodophyllus lividum, Lepiota morgani,* and the jack-o-lantern mushroom, *Omphalotus olearius;* and (8) Miscellaneous components that produce exotic effects, including the dreaded orellana syndrome of *Cortinarius orellana* and related species.

CYCLOPEPTIDES

For humans the most toxic substances derived from mushrooms are the cyclopeptides of the poisonous *Amanita* sp. These species are aptly called "death angels" and "destroying angels," for a single cap of these deadly organisms is capable of killing a healthy adult. Chemically there are two groups of cyclopeptides: the amatoxins and the phallotoxins. One of the amatoxins appears to be important in the symptomology and pathology engendered by consumption of the mushrooms; the importance of the phallotoxins has not been elucidated.

The symptoms associated with ingestion of poisonous *Amanita* sp. are developed in stages. Stage 1 is the prodomal period of 6 to 24 hours that elapses after the eating of the mushrooms and until overt symptoms begin. Although the patient has no discomfort, the amatoxins are being absorbed, excreted, and reabsorbed, first from the intestine and later from the intestine and kidneys. In stage 2 the patient experiences violent diarrhea, cramps, and abdominal pain. If blood chemistries are done at this time they are often normal, but liver enzymes and creatinine levels may begin to rise. Dizziness and other neurologic manifestations occur. This stage lasts about a day. During the following day (stage 3) there is remission of gastrointestinal symptoms and the patient feels better. Sometimes if a patient has been admitted to a hospital, he is released. In stage 4 the effects of the amatoxins on liver and kidney become apparent. The liver transaminases (SGOT, SGPT) skyrocket (Fig. 28–2), bilirubin and other parameters indicate serious hepatotoxicity, and other tests reflect renal failure. If large quantities of the fungus were consumed, kidney dysfunction may occur on the third day and hepatic

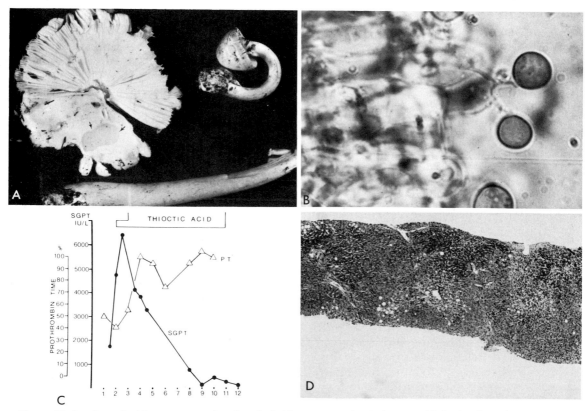

Figure 28–2. *Amanita bisporogera* poisoning. *A, A. bisporogera* aborted egg and fully developed cap and stem. Remnants of the annulus are visible on the right side of the stem and bits of the volva on the left. KOH was applied to the gill area and caused it to turn yellow (chrysocystidia test) see *arrows*; this, together with the round, amyloid-positive (with Melzer's reagent) spores, and two spores to a basidium, identify the specimen as *A. bisporogera. B,* Two spores on a basidium. ×1000. (Photo by Linnea Gillman.) *C,* Response of a patient to therapy with thioctic acid. *SGPT,* serum glutamic-pyruvic transaminase; *P.T.,* prothrombin time. Abscissa is days after hospitalization. *D,* Needle biopsy of liver with acute *Amanita* poisoning. Note centralobular necrosis and steatosis with sparing of other areas. (Courtesy of D. M. Jensen.)

coma on the fourth. Central nervous system and cardiac problems may be major or not apparent. There is still no reliable antidote for amanitin poisoning. Many drugs and regimens have been proposed, but symptomatic treatment and intensive care are mandatory and in the majority of cases are successful in saving the life of the patient.

The remedies utilized in the treatment of amatoxin (amanitin) poisoning have varied over the years. In the 1820's Roques used quantities of sugared or honeyed water. This at least helped in the hypoglycemia that occurs during the poisoning. Later, salts were added to this. At present, intravenous fluid therapy for electrolyte balance and hypoglycemia is standard practice in such cases. The Pasteur Institute in Paris developed a horse serum antitoxin that was widely hailed at first as a cure but is now generally disregarded. There is, however, some basis for this therapy. Dr. Pierre Bastien has given himself small inoculations of amatoxin and has built up a sufficient serum titer that he can eat 50 g (a lethal amount) of *A. phalloides*. His treatment includes antibiotics and intravenous vitamin C. In nature, antibodies to the amatoxins are found in squirrels and other rodents that regularly eat mushrooms. More recently hemodialysis and hemoperfusion using activated charcoal have been developed, and some dramatic recoveries have been reported. Large quantities of penicillin, penicillamine, and silybin[86] have also been used with varying success. The most controversial drug therapy presently being evaluated is thioctic acid. Some clinicians who regularly treat mushroom poisoning are convinced of its efficacy; others contend that it is useless. In one personally observed case, the results were impressive, although the response may have been coincidental (Fig. 28–2). Thioctic acid (α lipoic acid) is a coenzyme in the electron transport system of man, plants, animals, and fungi. In large quantities, it has been found useful by some investigators in the treatment of liver disease in man and animals, in heavy metal poisoning in man, and recently in *A. phalloides* poisoning.

Several toxic components have been described over the years as being present in the poisonous *Amanita* sp. In 1891 Kobert isolated a hemolytic principle now called phallolysin. This substance is heat labile, however, and is destroyed in the intestine. From 1937 to 1947 the Wielands[95] described two groups of cyclopeptides derived from the *Amanita* sp. These are the phalloidins (phallotoxins) and the amanitins (amatoxins). The phalloidins, of which there are at least five, are composed of seven amino acids (heptapeptides) and differ from one another mainly by substitution of H for OH on particular amino acids (Fig. 28–3,*A*). If injected into mice, hemorrhagic necrosis of the liver ensues immediately, providing the dose is large enough (LD_{50} = 2 mg per kg). The effect is on a specific receptor on the cell membrane and lysis is rapid. This does not occur if pretreatment by silybin[86] is used. The phallotoxins are active only if injected; they cause no effect on isolated liver slices or if given by the oral route. Their role in *Amanita* poisoning by ingestion is doubtful. In the mushroom, they may occur as a polymer attached to a carbohydrate moiety.

The toxicity of the amanitins (amatoxins) has been fully established. They are fully active by whatever route of administration and are thermostabile. The amanitins are octapeptides and exist as at least six chemical species. Again, the chemical species differ by substitutions of H for OH on some of the amino acids. Alpha amanitin inhibits RNA polymerase (Fig. 28–3,*B*) (77 per cent diminution in 24 hours); thus, it interrupts RNA synthesis and, indirectly, DNA transcription. The next effects are termination of protein synthesis and cell death, and the process is slow and cumulative. Even at high injected levels, death of laboratory animals requires 15 hours (LD_{50} = 0.1 mg per kg). The α amanitins are filtered out of the blood by the kidneys but are reabsorbed and recirculated into the blood stream. The proximal tubules are affected and show necrosis. This reabsorption occurs in humans and mice but not in rats, squirrels, and some other rodents. Animal species differ greatly in their susceptibility to mushroom poisoning. Other toxic components exist in the *Amanita* species and may or may not be important in immediate or delayed symptoms.[10] The Meixner test is a crude method of determining the presence of amatoxins in a given mushroom[76]; it depends on the acid-catalyzed reaction of amatoxin with lignin. A drop of mushroom juice is squeezed onto cheap bulk paper (newspaper) and dried. When a drop of concentrated HCl is added, a blue color develops if amatoxins are present. A negative test does not assure the absence of toxins in the mushroom. Quantitative techniques for recovery of amatoxin involve paper chromatography. Toxic amounts of α amanitin have been found in the death angel, *A. phalloides*, and in

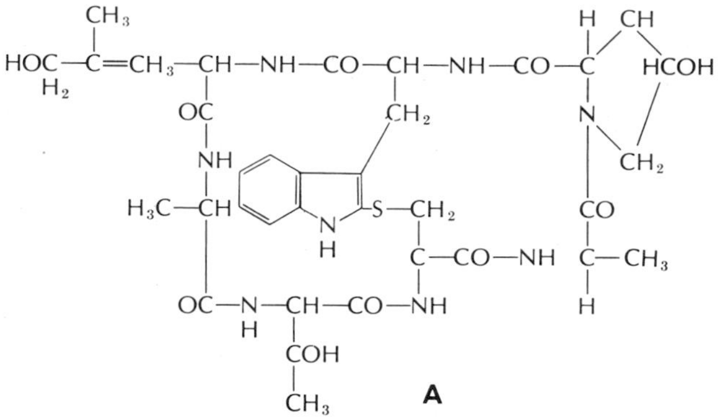

Figure 28–3. A, Chemical structure of a phallotoxin. B, Chemical structure of α amatoxin. (From Wieland, T., and H. Fulstick. 1978. CRC Crit. Rev. Biochem., 5:185–260.)

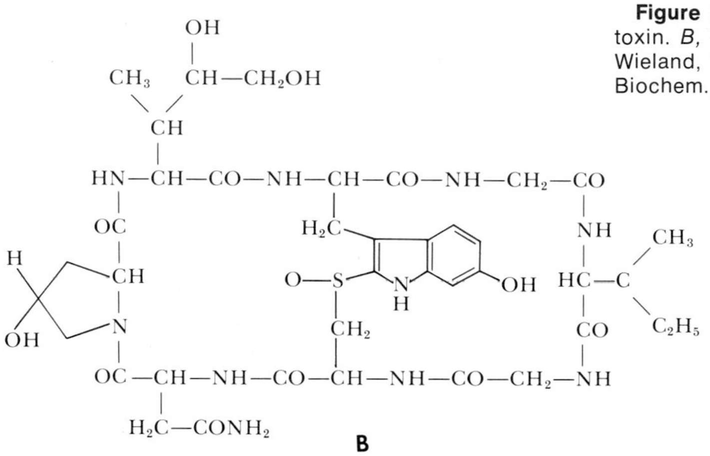

the destroying angels, A. verna, A. virosa, and A. bisporigera (possibly also A. ocreata). Three species of Galerina — G. autumnalis, G. marginata, and G. venenata — also contain α amanitin, which has caused fatal poisoning. Other species are reported to contain minute amounts of amatoxins but have not as yet been involved in recorded poisonings. The concentrations of the toxins vary considerably from specimen to specimen. An average was determined by Weiland and Fulstick.[95] One hundred grams of fresh Amanita phalloides (3.2 oz) contains 10 mg phalloidin, 8 mg α amanitin, 5 mg β amanitin, and 0.5 γ amanitin. A lethal dose for a 150-lb. man ($LD_{50} = 0.1$ mg per kg) would be contained in 50 grams or about 2 ounces of mushroom.

There is some evidence that the amatoxins exist as a polymer attached to a carbohydrate moiety in the mushroom. A current hypothesis holds that when the amatoxins attached to this carbohydrate moiety are ingested the peptides separate from the carrier in the acid milieu of the stomach. When passed to the intestine they reattach and are absorbed through the intestinal wall. They then slowly detach from the carbohydrate carrier and are taken up by the kidney and liver. In animal species that are unaffected by the toxins, the carbohydrate carrier is retained in the wall of the stomach. The naked peptides in the intestine are susceptible to degradation and, without being reattached to the large carbohydrate molecule, are not readily absorbed through the intestinal wall. The molecules that are reattached are neutralized by antibodies. Furthermore, the molecules that are excreted through the kidneys are not reabsorbed. More work on all of these points is now in progress.

MONOMETHYL HYDRAZINE

The Gyromitra species are Ascomycota that resemble the Morels. Eight to ten species exist in the North American continent and about two or three in Europe. They are much sought after in Europe as an edible species (Gyromitra esculata), although 2 to 4 per cent of all mushroom fatalities are associated with them. The active ingredient is called gyromitrin (N-methyl-N-formylhydrazine). Eaten raw, most of the *Gyromitra* sp. are quite poisonous. Some are parboiled to eliminate the chocolate odor of gyromitrin, but it can be absorbed through the nose, and enough is left in the fungus to cause illness. It is a hemolytic toxin in man, other primates, and dogs. It is toxic to the central nervous system and damages the liver and gastrointestinal tract. It may act by interfering with transaminases, particularly those having a pyridoxal phosphate cofactor. Vitamin B_6 is used in the treatment. As in cyclopeptide poisoning, a long latent period ensues (6 to 12 hours) between ingestion and symptoms. The symptoms include nausea, vomiting, diarrhea, cramps, distention, weakness, lassitude, and headache; if the condition is severe, these may develop into jaundice, convulsions, coma, and death. Methemoglobinuria and very low blood sugar are found in laboratory tests.

Recently Toth et al.[83] have found *N*-methyl-*N*-formylhydrazine (Fig. 28–4,*A*) to be carcinogenic (Fig. 28–5). When it was administered to mice at concentrations of 0.0039 per cent in drinking water, tumors were found in lungs, liver, blood vessels, and bile ducts. The tumors included adenomas, adenocarcinomas of lungs, benign hepatomas and liver cell carcinomas, angiomas and angiosarcomas, and other types. Gyromitrin is found in all *Gyromitra* species in significant amounts. Small amounts have been detected in related ascomycetous fungi and even in *Agaricus brunessens,* the cultivated mushroom. Carcinogenesis of this compound for humans has not been determined.

COPRINE POISONING

Coprinus atrementarius, the inky cap, when collected young and prepared as a cream butter sauce is a gastronomic delight, but if alcohol is consumed sometime after ingestion of the mushroom, an antabuse-like reaction occurs in some people. Coprine, the compound responsible, is an alcohol synergist (N^5-(1-hydroxycyclopropyl) L glutamine) and contains a cyclopropanone equivalent. It chelates molybdenum in the same way as does disulfiram, blocking acetaldehyde dehydrogenase and arresting the metabolism of ethanol at acetaldehyde. The symptoms include an unpleasant, hot, flushed feeling due to dilation of blood vessels. There is also a metallic taste, palpitation, chest pain, headache, and nausea, and vertigo, vomiting, weakness, and confusion may develop. Recovery is spontaneous. In susceptible people, the symptoms can occur for up to five days after eating the mushroom. Only *C. atrementarius* has been documented in this type of poisoning (Fig. 28–4,*B*).

MUSCARINE POISONING

Muscarine is a small molecule (Fig. 28–4,*C*) having a cholinergic effect. It is a quaternary compound and does not cross the blood-brain barrier. Its effects are parasympathomimetic and peripheral. They are negated by large doses of atropine, but this is rarely needed. Some of the compounds found in

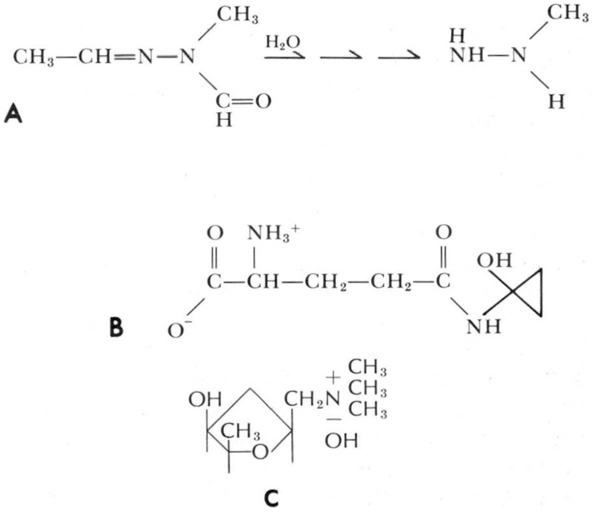

Figure 28–4. *A,* Chemical structure of gyromitrin and monomethyl hydrazine. *B,* Structure of coprine. *C,* Structure of muscarine.

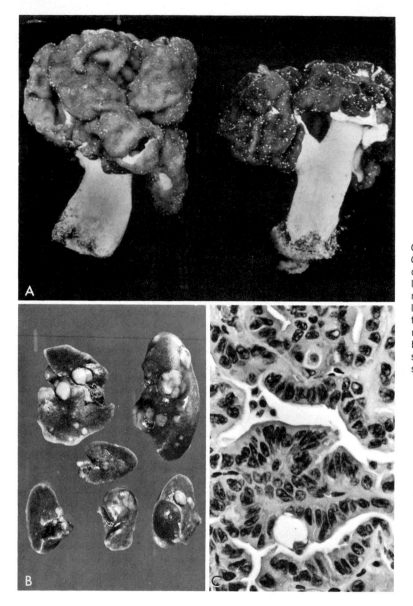

Figure 28–5. Carcinogenesis of *Gyromitra esculata* for mice. *A, Gyromitra esculata. B,* Liver cell carcinoma showing multiple nodular tumor growths. 58-week-old female. ×2. *C,* Adenocarcinoma of lung. Irregular acinotubular formations are lined by malignant columnar or cuboidal cells supported by a poorly developed fibrous stroma. Hematoxylin and eosin stain. × 400.[83] (Courtesy of H. Toth.)

mushrooms in addition to muscarine are histaminic. Muscarine effects are notable within two hours, and usually within 30 minutes, of ingestion. The major symptoms are perspiration, lacrimation, and salivation (PLS). This triad is unique to this type of mushroom poisoning. Cramps, diarrhea, headache, and blurred vision may follow, and contraction of pupils, hypotension, and slow pulse sometimes occur. Supportive therapy is usually sufficient; atropine, though commonly used previously, is seldom necessary. *Inocybe fastigiata, I. napipes,* and other *Inocybe* species (Fig. 28–1,*D*) contain large quantities of muscarine. *Clitocybe dealbata* and other related species also contain it. Whether it is present in jack-o-lantern (Fig. 28–1,*C*), *Omphalotus*

olearius (Cylitocybe illudens), is debatable. *Amanita muscaria* and *A. pantherina* do not cause muscarine poisoning. They contain only minute amounts of the compound.

IBOTENIC ACID, MUSCIMOL, MUSCAZONE

For many years the type of poisoning attributed to *Amanita muscaria* was muscarine. The muscarine effect (atropine negated) was first described in *A. muscaria,* but there are such small quantities of the substance that it is of no importance. The effect of the consumption of *A. muscaria* is now known to be due to the psychoactive compounds ibotenic acid, muscimol, and muscazone. Neither *A. muscaria* nor *A. pantherina* ingestion causes symptoms typical of muscarine poisoning (PSL) and neither contains toxins related to the toxins of the death angel or destroying angel (*A. phalloides* and *A. bisporogera*). Instead, the symptoms following ingestion include inebriation, derangement of senses, manic behavior, delirium, and deep sleep. In time these resolve and almost all people affected recover spontaneously. The fly agaric *(Amanita muscaria)* (Fig. 28–1,*B*) has figured prominently in ethnic, religious, faddist and anthropologic literature.[91, 92] Its use in Sweden and Siberia in tribal days is well known. It was what made the berserkers berserk. It appears to have been the Soma of the Vedic priests, and many hymns and sacred writings refer to it. The priest class of the near eastern religions knew of its use, and some of the writings found in the canon of judeo-christian sacred works may have been done under its influence.[2] Although the fly agaric was widely utilized for both recreational and religious activity in the old world, it does not appear to have been employed extensively in the new world. Mesoamerican tribes did develop a religious cult based on mushrooms, but this was the *Psilocybe cubensis* and related species.

Following ingestion, symptoms appear between 30 minutes and two hours. These include incoordination (alcoholic stagger), dizziness, twitching, and sometimes jerking. There is usually no vomiting. The subject has a "creative set," meaning the feeling that perception and understanding are expanded; the subject wants to do things. The psychologic effects are usually influenced by environment and the companions present. Speech and articulation are acute so that visions can be related in great detail and clarity. These visions are often religious in nature, and this effect is known as the "old-time religion" set. An initial vision-filled sleep is followed by a long period of physical and mental activity. The toxic effects, if present, are anticholinergic: tachycardia, dilated pupils, dry mouth, and decreased bowel sounds. Usually all symptoms are neurologic: the soporific period is followed by visions and voices, then by macropsia (small things appear large) and spatiotemporal expansion. Following this the subject sleeps again and all effects gradually wear off.

The active ingredients are chemically isoxazole derivatives (Fig. 28–6). Ibotenic acid is reportedly concentrated in the epidermis of the cap, but it degrades on drying. The

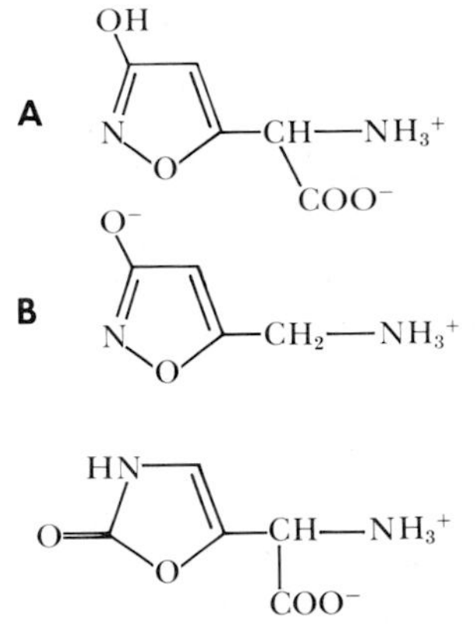

Figure 28–6. Chemical structures of *(A)* ibotenic acid, *(B)* muscimol, and *(C)* muscazone.

degradation product, muscimol, is five to ten times more potent. A derivative of muscimol is used as an insecticide. Muscazone is found only in European specimens of *Amanita muscaria*. In addition to *A. muscaria* and *A. pantherina*, ibotenic acid–like compounds have been reported in *A. cothurnata*, *A. gemmata* (an ill-defined species), and a few *Paneolus* species.

PSILOCYBIN-PSILOCIN INTOXICATION

As "teonanacatl" (flesh of the gods), *Psilocybe cubensis* was an established part of the religious experience of Mesoamerica.[92] In recent years it has been popularized by Carlos Castaneda in his "The Teachings of Don Juan: a Yaqui Way of Knowledge," and it has been the subject of experimentation as a "magic mushroom." The active ingredients (Fig. 28–7) are the hallucinogens psilocybin and psilocin. They are indole derivatives related to baeocystin and lysergic acid diethylamide (LSD). Almost all of the effects are on the central nervous system, but some peripheral action (probably by a norepinephrine pathway) akin to bufotenin has also been observed. Within a half hour to an hour after ingestion, a hallucinogenic dysphoric experience begins. The experience can be a "good trip" or a "bad trip," depending on a number of factors. In contrast to ibotenic acid–muscimol, psilocybe intoxication is a "passive set." The patient recedes into a trance and does not usually wish to exert himself physically. He may experience compulsive movements and laughter along

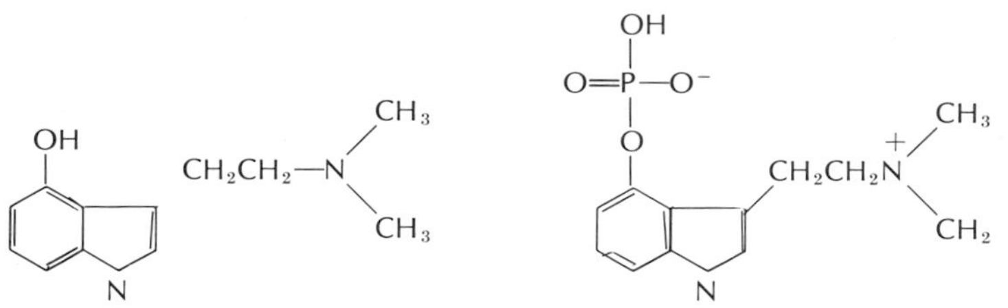

Figure 28–7. Chemical structure of psilocin (left) and psilocybin (right).

with physical symptoms of mydriasis, vertigo, ataxia, and parasthesias. Eventually the patient drops into a dreamless sleep. The effects of a 30-gram intake of fresh mushroom lasts for six hours. Fatalities are very rare, and most of these have been in children. Psilocybin is found in various species of *Psilocybe, Paneolus,* and *Conocybe.* Notable amounts are in *Psilocybe cubensis, P. mexicana, P. caerulescens, P. caerulipes,* and other "blue staining" stalked *Psilocybe* sp. *Paneolus foenisecii* and numerous other LBD mushrooms (little brown devils) have been cited as also being hallucinogenic, although most are simply gastrointestinal irritants. Intoxication with these species is usually deliberate;[18] on occasion, however, they have been inadvertently used in cooking, such as in the famous pizza parlor incident in Norway.[32] Psilocybin, $C_{12}H_{17}N_2O_4P$ (*O*-phosphoryl-4-hydroxy-*N,N*-dimethyltryptamine), is the form found in the fungus. Following ingestion this is converted to psilocin (Fig. 28–7) by dephosphorylation. The ingestion of as little as 20 to 25 μg produces a hallucinogenic effect equal to 2.5 μg of LSD, to which it is structually related.

GASTROINTESTINAL IRRITANTS

There are a large number of fleshy fungi that induce a degree of discomfort following ingestion in some if not all people. In this type of intoxication there is no hepatic damage, as found in cyclopeptide and gyromitrin poisoning; no central or peripheral nervous system manifestations, as in coprine or muscarine poisoning; and no psychic aberrations, as in ibotenic acid, muscimol, or psilocybin poisoning. Symptoms are almost entirely restricted to the gastrointestinal tract and include nausea, vomiting, cramps, and diarrhea. The severity of any and all of these varies, depending on quantity consumed, individual idiosyncrasies, and species of mushroom involved. Symptoms manifest

themselves anywhere from 30 minutes to two hours following consumption. In severe cases the patients may complain of weakness, faintness, and chills. Hospitalization is usually not necessary and symptoms subside in three to four hours; recovery is complete in one to two days. The list of involved species is long. Commonly encountered types are *Lepiota morgani (Chlorophyllum molybdites), Boletus luridus, B. satanas, Entoloma lividum, E. griseus, Gomphus (Coantharellus) floccosus, Hebeloma* sp., *Naematoloma fasciculare, Paxillus involutus,* and so forth. The active ingredient in only a few has been isolated. In *Gomphus* sp. the chemical is norcaperatic acid (α tetradecycleitric acid), which is structurally similar to citric acid and inhibits the enzyme aconitase in the citric acid cycle.

MISCELLANEOUS TOXINS

There are some mushroom species that cause poisonings, either mild or serious and fatal, that do not appear to fit in the previously described categories. *Cortinarius orellanus* (associated with the dreaded orellana syndrome) is probably the most famous of these. Until about 1950, the *Cortinarius* species were considered edible. However, Grzymala in Poland[60] reported on almost 100 cases of fatal disease attributed to this fungus. The symptoms appear as soon as three days to two weeks and as long as 150 days post ingestion. The symptoms are variable, but a relentless necrosis of kidney and liver ensues. The onset was always accompanied by extreme thirst. The active ingredients appear to be a group of polypeptides referred to as orellanines.

Galerina sulciceps causes a very rapid fatal toxemia following ingestion. The patients experience spasms and nausea but not vomiting or diarrhea. Death occurs as early as seven hours after ingestion. The fungus is found in southeast Asia and is not known to grow in North America. Some poisoning is recorded in England[60] and the United States.

Mycotoxicosis

Mycotoxicosis is an intoxication due to the ingestion of preformed substances produced by the action of certain molds on particular foodstuffs. It is analogous to *Clostridium botulinum* or staphylococcal food poisoning in which the organism has produced the offending agent in the food before consumption. In the case of mycotoxins the evidence indicates that the toxins are produced only in particular environments, on particular substrates, under certain conditions, and only by particular strains of the species involved. Examples include the following:

1. The common soil contaminant *Cladosporium cladosporoides*, growing on tall fescue (grass), produces a skin erythema–causing toxin that is not formed when the fungus grows on other substrates.[52]

2. *Claviceps purpurea* elaborates a great variety of toxins (ergot alkaloids) while growing on the fruiting heads of rye, but few or none when growing on other substrates.[8]

3. Only a few strains of *Fusarium moniliforme* or other molds produce the toxin responsible for human and bovine toxic alimentary aleukia, and these only under conditions of cold.[48]

4. Some strains of *Aspergillus flavus* produce the hepatoma-inducing toxin, aflatoxin, when growing on peanuts or wheat, but not during the formation of fermented soya sauce from soy beans.[41] This indicates that particular chemical moieties must be present in the substrate in order for the particular toxic metabolites to be formed.

Throughout the ages epidemics of poisoning due to the ingestion of contaminated foods have been recorded. Probably the most famous of these were the waves of ergot poisoning that occurred during the Middle Ages. The symptoms of this toxicosis — hallucination, black leg (due to vascular necrosis), and other distresses — were well known and the etiology established long before the science of mycology was founded.[8] Picking only the rye heads free of diseased grain served to prevent the disease. After this, interest in mycotoxins diminished, and little was made of such diseases up to and including the first half of the twentieth century. Ergot is now grown commercially for extraction of pharmacologically active alkaloids. Some of these are used to control bleeding, induce uterine contractions, and treat some vascular diseases. Mycotoxicosis, in general, remained of some interest to veterinary medicine due to the economic loss that occurred when domestic animals ate certain moldy forage, but in human medicine it was largely forgotten.

All this was changed in 1960 with the investigations surrounding turkey X disease.[6, 58] The symptoms in turkey poults, ducklings, calves, and pigs were loss of weight, followed by ataxia, convulsions, and death, and the animals often showed signs of jaundice. Hepatic necrosis, fibrosis, and sometimes neoplasia were found in affected animals, particularly in ducklings. The common factor was found to be Brazilian ground nut (peanut meal) in the feed. Further investigation revealed that a fluorescent toxic substance was present in the peanuts and was attributable to growth on the nuts of *Aspergillus flavus*. A large amount of investigation has occurred since, and it has been realized that mycotoxins are present in many foodstuffs contaminated by a number of different organisms and can cause a wide variety of diseases.[1, 53, 88, 100]

Aflatoxins

Aflatoxin was isolated from grains infected by *Aspergillus flavus* and by experimentation defined as the cause of the strange manifestations of turkey X disease. Later investigations indicated that there were a series of compounds with similar activity. These were designated aflatoxin B_1, B_2, B_{2a}, G_1, G_2 and G_{2a}. The best described is aflatoxin B, which has a molecular weight of 312 and a formula of $C_{17}H_{12}O_6$ (Fig. 28–8). It has a highly unsaturated structure and is chemically related to the coumarins. In quantities of 0.3 ppm it regularly produces hepatomas and hepatic necrosis in ducklings, trout, and other animals. After a simple procedure for its identification had been devised,[19, 49] it was found to contaminate many human foods. In one case it was found in grain used in the making of beer in a Chicago brewery. Since the entire

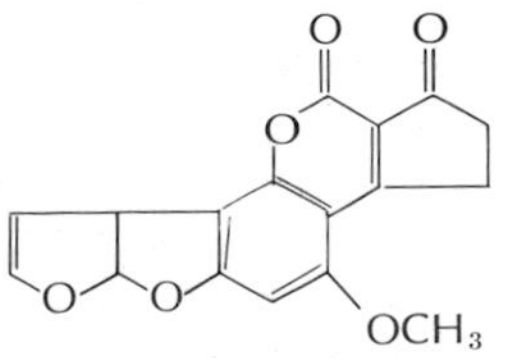

Figure 28–8. Chemical structure of aflatoxin B.

supply of its grain was affected, the company went bankrupt. Aflatoxin has been found in the ground nuts (peanuts) used as a major staple of food of some African tribes and has been associated with the high rate of liver and renal disease (including cancer) of these people,[50] although other investigators have associated the maladies of the tribes with the use of copper- and lead-lined cooking utensils. It was determined, however, that strains used for the production of koji and shoyu in the Orient do not produce the active ingredients.[41] At present, such products as peanut butter, beer, stored grains, nuts, meal, and animal feed can be monitored to prevent intoxication by aflatoxin.[25] Aflatoxin sensitivity varies considerably from species to species. The duckling and trout fry are exquisitely sensitive, rats and mice less so. The importance of aflatoxin in human disease is still being assessed. Human cell lines are not very sensitive to the effects of the toxin. Incidental exposure of adults to contaminated foods appears to carry very little risk. Continued exposure of adults and especially children, as is the case in southeast Asia, may lead to demonstrable abnormalities and neoplasias. The toxin is produced by *Aspergillus flavus, A. parasitius*, and *Penicillium puberulum*. Interestingly, cancer production in animals given aflatoxin is prevented by hypophysectomy.[28]

Toxic Alimentary Aleukia

Toxic alimentary aleukia in animals is a well-established syndrome of forage herds, particularly in Russia. During World War II, conditions necessitated the storage of grain for human consumption in the open field during the winter. Thereafter there were epidemics of disease and death among the population that consumed the grain. The disease was characterized by necrotic rashes on the skin, leukopenia, agranulocytosis, necrotic and hemorrhagic diathesis, vertigo, and ulcerative and gangrenous lesions of the pharynx that led to aphasia and death. It was found that particular strains of *Fusarium sporotrichoides, Cladosporium, Alternaria, Mucor*, and *Penicillium* produced toxins responsible for the condition. When stored under normal conditions, the infestation of grain and production of toxin were not observed. Three compounds with toxic properties have been isolated and identified. They are fusariogenin, epicladosporic acid, and fagicladosporic acid. Chemically, they belong to a group known as the trichothecene toxins. This

group includes nivalenol, deoxynivalenol, and T_2. The use of these agents in biological warfare is under investigation. Nivalenol and T_2 are reputed to have been used in some limited military engagements ("yellow rain"). These toxins are formed by many strains of *Fusarium* spp., particularly under conditions of cold (8° C). Humans may absorb them percutaneously. They cause severe blistering and later internal bleeding. Another fungus, *Fusarium roseum*, produces a mildly intoxicating substance when growing on grain used for flour. When bread or other staples are made from it, a syndrome known as "drunken bread eater" is produced.

Ipomeanols

One recent epidemic of atypical interstitial pneumonia of cattle was found to be the result of feeding on moldy sweet potatoes. Several toxins termed "ipomeanols" were found to be responsible.[21, 97] They were produced by the sweet potato in response to infection by *Fusarium javanicum*. After extensive investigation into the problem, Wilson et al.[97] found that there were several toxins elaborated. Two were the hepatotoxins ipomeamarone (Fig. 28–9) and ipomeamaronal and a third, referred to as "lung oedema (LO) factor," was shown to be structurally related to the others and called 4-ipomeanol. It has an empirical formula of $C_9H_{12}O_3$ and a molecular weight of 168. A dose of 1 mg IV in mice causes death in five to eight hours owing to severe pulmonary interstitial edema, but it is without effect on the liver.

Leukoencephalomalacia

Fusarium moniliforme when growing on corn or other grain produces a potent toxin. This toxin causes a disease called leukoencephalomalacia in donkeys and horses that consume the infected grain. The affected animals show increasing ataxia, and at autopsy large areas of necrosis in the brain are seen. This disease

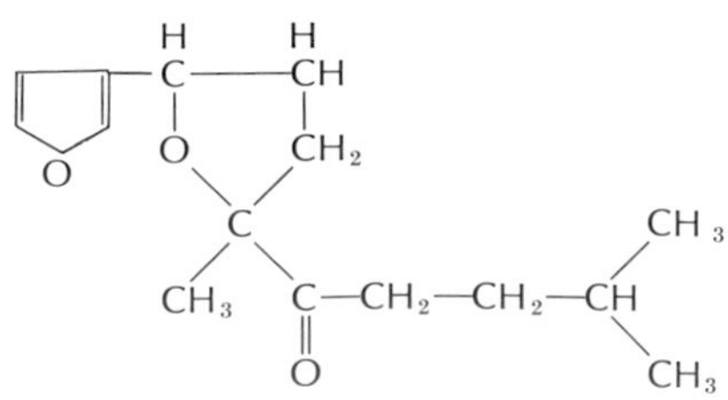

Figure 28–9. Ipomeamarone.

has been known since the nineteenth century.[96] Other *Fusarium* species produce a variety of toxins when growing on fodder, forage, or grain.

Stachybotrytoxicosis

Stachybotryotoxicosis has been extensively studied in the Soviet Union. The disease syndrome was first described in 1931. It was known as MZ *(massavie zabolivanie,* massive illness), because it was enzootic in that country, causing the death of thousands of horses, cattle, sheep, and swine. The first manifestations are epithelial desquamation and swelling of lymph nodes. A leukopenia develops, and death follows within five days of onset of symptoms. At autopsy profuse hemorrhage is found, along with necrosis of many organs. The organism responsible is *Stachybotrys alterans (altra)* which grows on hay and forage.[24, 26, 88, 100] Farm workers inhaling the toxin or absorbing it percutaneously have been affected to a milder degree. They demonstrate cutaneous rashes, pharyngitis, and leukopenia. The active ingredients are trichothecene derivatives. These are also being investigated for use in biological warfare.

Facial Eczema

A rather strange disease, first described in New Zealand, is called facial eczema. It appears in sheep and cattle who eat forage infected with *Pithomyces chartarum.* When growing on grass, this organism elaborates a group of powerful toxins called "sporidesmins," or "sporidesmolides," which have the empirical formula of $C_{33}H_{58}N_4O_8$ (Fig. 28–10). The toxin causes liver damage characterized by acute cholangitis, which in time evolves into fibrotic obstruction of the involved portals. In addition, vasculitis and lymphangitis occur. Severely affected animals also become photosensitized and develop a severe sloughing eczema on exposed skin sites. The face and nose desquamate in sheep, as does the udder in cows. The photosensitization is due to retention of phylloerythrin, a normal breakdown product of chlorophyll usually excreted in bile. *Pithomyces* is now known to be of worldwide distribution.[59]

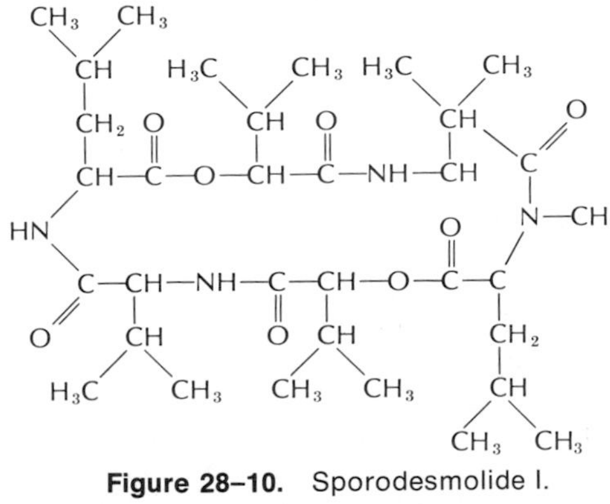

Other Toxins

The endotoxins of *Aspergillus fumigatus* have been known since the pioneering work of Henrici (see Chapter 23). As early as 1902 Ceni and Besta correlated certain pellagra-like syndromes with the eating of corn infected by this fungus. Red clover infested by *Rhizoctonia leguminacala* causes excessive salivation of foraging animals; the active principle is a parasympathomimetic alkaloid,[20] but the chemical structure has not yet been elucidated.

A toxin called sterigmatocystin produced in quantity by *Aspergillus versicolor* is another carcinogenic fungal product. It produces hepatomas in rats, and long-term administration to mice results in adenomas and adenocarcinomas in bronchial and alveolar epithelium. No natural human intoxication has yet been recorded. Sterigmatocystin shares with aflatoxin the presence of a bis-dihydrofuran structure in its molecule. *A. parsiticus* converts sterigmatocystin to aflatoxin.[88]

Some other examples of mycotoxicosis include the following: *Penicillium cyclopium* when growing on hay, produces a tremogenic toxin that kills cattle, horses, and sheep. The animals suffer severe convulsions and death. A chronic and eventually fatal respiratory disease of certain natives of New Guinea is associated with moldy sweet potatoes. The etiology is as yet unknown, but this may be similar to the *Fusarium javanicum* toxicosis described by Wilson et al.[96, 97] (see Ipomeanols (opposite page). Vomitoxin[90] is produced by the growth of *Fusarium graminearum (Gibberella zeae)* on corn. In small concentrations it induces vomiting in domestic animals; at concentrations of 1.4 ppm or more, however, feed is rejected by animals because of the bad taste.[98]

Aspergillus ochraceus produces ochratoxins A and B when growing on feed grains (Fig.

Figure 28–10. Sporodesmolide I.

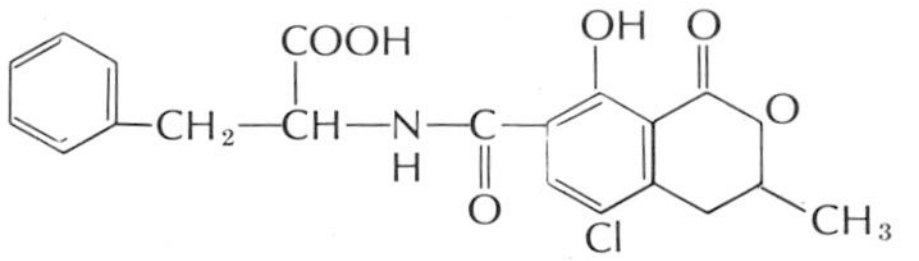

Figure 28–11. Ochratoxin A.

28–11). The disease produced in chickens and other animals is similar to several other mycotoxicoses. There is suppression of hemopoiesis, acute nephrosis, hepatic necrosis, and enteritis.[71, 85] Nephropathy and edema are evoked in swine fed corn on which *Penicillium viridicatum* has grown.[15] The edema produced in these animals is so massive that ascites, hydrothorax, and hydropericardium develop, in addition to subcutaneous edema and mesenteric edema. Affected animals die within a few hours. Lupinosis is a disease of sheep in South Africa and other countries that is characterized by hepatic necrosis. The disease was thought to be caused by eating the stems and pods of various legumes, particularly *Lupinus albus*. It has now been shown to be the result of a toxin produced by the fungus *Phomopsis leptostromiformus* while growing on the plants.[89] *Penicillium rubrum* also produces several potent agents called rubritoxins.[36]

It is probable that in the future many more diseases of animals and man, both toxic and neoplastic, will be found to be caused by the chronic ingestion of toxins elaborated by fungi growing on articles of consumption.[17, 66, 73] An interesting history of the mycotoxicoses before 1960 and subsequent to turkey X disease is given by Uraguchi and Yamazaki in the text on the toxicology of mycotoxins.[88]

Allergic Diseases

Hypersensitivity to the inhalation of fungal conidia or fungal products manifests itself in many ways. In certain atopic individuals, the sensitivity presents as bronchial asthma, and the allergens are usually the airborne conidia of common soil fungi. Sensitivity may also develop in normal individuals who are chronically exposed to the conidia of certain fungi in the course of their occupation. These include farmer's lung, teapicker's disease, maple bark stripper's disease, bagassosis, and so forth. In the first group, asthma, the disease is due to an idiosyncrasy of immunologic response, and a long list of fungal conidia have been involved along with dust, animal dander, pollens, and so forth. In the second group, the allergies are produced in normal people, and the disease is a consequence of chronic contact with quantities of specific conidia that are particularly allergenic. A few of the well-documented allergic diseases will be briefly discussed.

ASTHMA[30]

Fungal conidia have long been incriminated as a common allergen in provoking bronchial asthma. *Alternaria, Helminthosporium, Drechslera, Cladosporium, Penicillium,* and *Aspergillus* conidia are most frequently involved. There is a definite seasonal incidence, with a high count of conidia in the fall and a lesser peak in the spring. This corresponds to the growth peaks of these fungi.[30, 69]

There are two general types of asthma. The childhood form reaches a peak in early adolescence and gradually subsides, and the second form begins after forty and may become more severe with increasing age. The latter runs a more rapid and fatal course if unremitting asthma (status asthmaticus) develops. The patients have reaginic antibodies (IgE) to the offending substances. On challenge, the antigen-antibody complex causes activation of mast cells, release of histamine, and there is subsequent bronchospasm and formation of mucus plugs in bronchioles. A delayed Arthus-like reaction (IgG-mediated) may complicate the disease. In fatal cases, the lungs are overaerated and show extensive bronchial plugging; however, there is little evidence of emphysematous changes. Both large and small bronchi are thickened and filled with semisolid yellowish plugs of mucus. The mucus contains large numbers of layered eosinophils, Charcot-Leyden crystals, and clusters of epithelial cells (the epithelial *zellballen* described by Vierordt in 1883). Histologically the bronchial mucus glands are thickened and in greater than normal numbers. The lumen of the glands contains the thick mucinous material. Edema of the epithelium is pronounced, and areas denuded of epithelium are present. The basement membrane is thickened and is sometimes ruptured, thus allowing extravasations of underlying tissue into the bronchial lumen. There is a large invasion of eosinophils and some neutrophils. The bronchi as well as bronchioles down to 0.2 cm are involved and often completely obstructed. A proteinaceous edema fluid is present in many alveolar spaces. Fungal conidia (unsprouted) are sometimes observed in the plugs. Patients with bronchial asthma usually do not have precipitating antibodies in their sera. They have an immediate wheal and flare to the intracutaneous injection of the allergen, but do not have a delayed reaction. A delayed reaction in addition to the immediate reaction indicates the asthma is complicated by IgG-mediated Arthus-like reactions in the lung as well. This occurs in bronchopulmonary aspergillosis (see Chapter 23). In this disease there are sprouted conidia and mycelium in the bronchial lumen, and the patients have IgG antibodies as well as IgE.

FARMER'S LUNG

This is an acute or chronic, sometimes fatal disease of persons continually exposed to moldy hay or similar material.[34, 72, 75] It is a manifestation of sensitization to several allergens, principally the spores of the actinomycetes *Micromonospora vulgaris*, *Micromonospora (Micropolyspora) faeni*, and *Thermopolyspora polyspora*. Several fungi also cause the identical syndrome.[44] *Aspergillus*, especially *A. niger*, *A. fumigatus*, and *A. flavus* and *Penicillium*, most commonly *P. simplicissimum*, *P. herquei*, *P. rubrum*, *P. italicum*, and *P. caseicolum*, are frequently involved. In particular areas or in particular occupations, other genera and species may be the predominant allergens.

The syndrome in which farm workers showed acute symptoms and chronic lung changes was first delineated in Canada by Cadham in 1924.[12] Subsequently Campbell[13] described the "farmer's lung" in England, and it has now been found to be a frequent and serious occupational disease of worldwide distribution.[34, 72] It occurs more often in the temperate than tropical zones, and is more frequent in areas of high rainfall where the conditions for mold growth on hay are optimum. Although most patients are in the third or fourth decade of life and are males, the disease has been described in women and in children as young as five years of age.[11]

Office workers have suffered the same disease owing to the growth of the organism in air conditioning vents.[3, 33]

Three stages of clinical disease are described in the several series on farmer's lung.[55] They are (1) the acute stage, found most frequently in harvesters and thrashers, in whom there is an overwhelming initial exposure to the spores; (2) the subacute stage, found sometimes in harvesters and animal keepers who have long-spaced exposure periods, but this form occurs more frequently in silo and grain mill workers; and (3) the chronic form, found in silo and mill workers who have low-level but constant exposure to the allergens. A similar disease occurs in horses (the "heaves"), in cows ("fog fever"), and among bird fanciers.[35]

On x-ray examination, the acute stage shows diffuse mottling of the middle and lower lobes. The patients have chills, fever, and general malaise. This form usually lasts only a few days and may leave no residual, or only slight fibrotic, changes. The subacute or subchronic form shows areas of infiltrated foci which are variable in size. Small tubercle-like granulomas are sometimes seen, and there is segmental atelectasis. The condition may resolve without major consequence if exposure challenge is prevented.

Airway obstruction is the major feature of chronic farmer's lung, but gas exchange studies are usually normal. The patients suffer from acute dyspnea following exertion. They have livid lips, shallow breathing, chronic cough, and some weakness. Radiographically fine to gross fibrotic changes and interstitial infiltrates are seen, and sometimes scattered miliary nodules (Fig. 28–12). Chronic disease is debilitating in varying degrees and often progressive to a fatal terminus.

The histopathology of the disease has been described by several investigators.[43, 75] In the acute phase there is an interstitial edema and some lymphocytic infiltration. This may resolve but leave some intralobular fibrosis. In progressive forms, granulomatous lesions develop. These resemble sarcoid. The lesions are in the parenchyma near the bronchioles, in the septa, or in subpleural areas. These contain histocytes which develop into layers of epithelioid cells, which are in turn surrounded by a ring of lymphocytes and plasma cells. Large foreign-body giant cells are interspersed in the lesions. These may contain nonbirefringent fibrous material. There are also bronchiolitis and vasculitis.[75] As the disease becomes chronic, more severe and per-

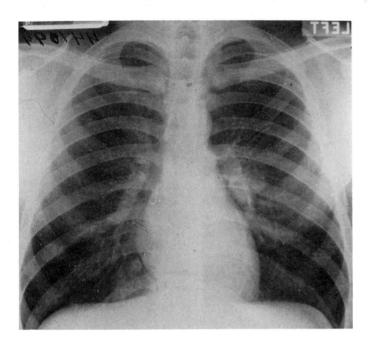

Figure 28–12. Farmer's lung of several years' duration. There are diffuse, reticulated, interstitial infiltrates in the bases of both lungs. (Courtesy of N. Gross.)

manent changes occur. The granulomas may calcify and sometimes enlarge. Pulmonary fibrosis becomes massive; there are cystic changes and changes associated with pulmonary hypertension.

Whereas asthma is associated with an immediate type 1 reaction mediated by IgE, the changes in farmer's lung are compatible with an Arthus (type III) reaction, and IgG precipitating antibodies have been demonstrated in sera. Later in the disease, there are numerous epithelioid tubercles in the lung, indicating that a delayed hypersensitivity (type IV) reaction of the tuberculin type is also a part of the syndrome. In addition, the presence of foreign-body giant cells within the granulomas suggests a nonallergic foreign-body, inflammatory component to the disease. It can be appreciated, therefore, that farmer's lung is a complicated disease syndrome involving several allergic and nonallergic components.

A serologic procedure for diagnosis has been devised by Kobayashi et al.[54] and Pepys and Jenkins.[72] This is an immunodiffusion test in which patients' sera is placed in one well, and either a decoction of moldy hay or specific actinomycete antigen is placed in the other. Precipitin lines appear if specific antibody is present. Jameson[46] has described a rapid immuno-osmophoresis technique to detect the same antibodies. Positive tests for the detection of precipitins have been found to correlate well with clinical disease, and this test together with a positive delayed reaction to the skin test is considered sufficient for diagnosis of disease.[22, 94]

The organisms responsible for the disease are thermophilic fungi and actinomycetes that grow on rotting hay or other material. The hay does not appear to contribute to the pathology. The hay flora thrives at temperatures of 40 to 60°C. The usual etiology is a mixture of *Thermopolyspora, Micromonospora, (Micropolyspora)*, and other actinomycetes and various fungi. Exposure to *Aspergillus fumigatus* alone has caused acute fatal disease.[43]

MAPLE BARK STRIPPER'S DISEASE (CONIOSPOROSIS OR CRYPTOSTROMOSIS)

This disease was first described by Towey in 1932.[84] It occurred in 35 bark peelers in a lumbering community in northern Michigan. The logs had been stored over the winter. The symptoms included cough, dyspnea, fever, night sweats, and substernal pain. The disease resolved uneventfully. More recently it has been described in Wisconsin and other places.[23, 84] In the series by Emmanuel, lung biopsy showed small cream-colored granulomas. Microscopically there were chronic granulomatous changes that consisted of histiocytes, fibroblasts, giant cells, lymphocytes, and a few neutrophils. Small spores 4 to 5 μ in diameter were sometimes present within the phagocytic cells. These were brownish red in color, and no budding was evident.

Initially there was confusion with the yeast cells of *Histoplasma capsulatum*, but the spores were shown to be from the fungus *Cryptostroma corticale*, which grows on the bark of maple trees. The inhaled spores do not grow in the lung but remain dormant. When lung tissue containing spores is planted on Sabouraud's agar and incubated at 25°C, the fungus will grow. A white pleomorphic colony develops. The hyphae are septate and profusely branched. The spores are oval, 4 to 5 μ, and brownish red in color. Patients with the disease have both positive skin tests and precipitins by the immunodiffusion test.[82] An allergic pneumonitis caused by *Alternaria* sp. growing on logs has recently been described by Schlueter et al.[74] and called wood pulp workers disease.

ALLERGIC SCOPULARIOPSOSIS

A disease similar to cryptostromosis was brought to my attention by Dr. Hans Grieble[29] of the Veterans Administration Hospital, Hines, Illinois. The patient had been in Viet Nam, where he had injected himself over a period of a month with a brown fluid sold to him by the natives as a crude opium preparation. In time he developed severe debilitating respiratory distress and was shipped back to the United States. On chest x-ray, several nodular lesions could be seen in the lower lung fields (Fig. 28–13,*A*). One of these was removed surgically for diagnosis. Histologic examination revealed a forming granulomatous reaction (Fig. 28–13,*B*) and the presence of numerous brown bodies that

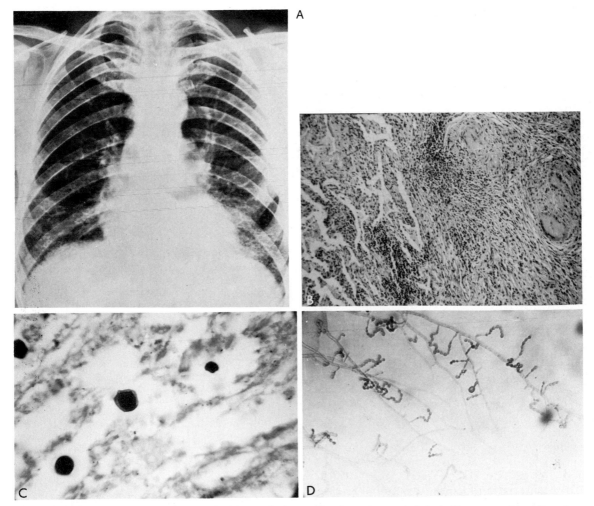

Figure 28–13. Scopulariopsosis. *A,* X-ray of chest. The lung is overinflated. There are two discrete, well-defined nodules in the lower lobe of the right lung. In the left lung there is a solitary, discrete, well-defined nodule in the superior segment of the lower lobe. (Courtesy of H. Grieble.) *B,* Well-formed granuloma with foreign body giant cells. ×100. *C,* High-power magnification showing various size conidia of fungus. Note lack of budding. ×500. *D,* Culture mount of *Scopulariopsis brumptii* (Salvanet-Duval, 1935) isolated from lesion. ×100.[29]

were 4 to 7 μ in diameter (Fig. 28–13,C). The material was cultured on a variety of media. Numerous colonies of a grayish white fungus grew on all media (Fig. 28–13,D). The organism was identified as *Scopulariopsis brumptii*. Antibodies to an antigen prepared from the fungus were present in the patient's sera. The disease in the patient resolved uneventfully and without treatment. The disease was reproduced in experimental animals.

BAGASSOSIS

This disease is the result of inhaling dust from bagasse, the material left after the juice has been expressed from sugar cane. Until 1921 the material was burned, but it is now used in making wallboard and similar products. In addition to fiber, silica, and quartz, the dust contains numerous fungal conidia.[45] The fungi represent many species and genera. Actinomycetes are also present and may be involved in the disease.[57] The syndrome was first described in 1941 by Jamison and Hopkins.[47] The symptoms are cough, dyspnea, some fever, and production of blood-streaked sputum. Mottling of the lung fields is seen on x-ray examination. Histologically there is an interstitial inflammatory reaction, with giant cells and intra-alveolar foam cells present in the early stages of the disease. The giant cells are seen to contain birefringent material. As the disease progresses, interstitial fibrosis occurs, along with chronic non-specific changes in the lung. As is seen in farmer's lung, continued exposure may result in fatal pulmonary disease.[7] Patients have precipitating antibodies in their sera.[37]

BYSSINOSIS

The disease is similar to bagassosis but involves inhalation of dust from cotton processing plants. The disease was first described by Kay in 1831 among cotton workers, carders, and sorters in Lancashire cotton mills. Nicholls et al.[68] demonstrated that the pericarp of the cotton seed pods was the active material and caused histamine release. Bronchoconstriction results, and, in cases of chronic exposure, inflammatory, fibrotic, and nonspecific changes occur. Fungal conidia are found in the dust, but their contribution to this particular allergy is probably not great.

LYCOPERDONOSIS

This disease is caused by the inhalation of the conidia of the fleshy basidiomycete, *Lycoperdon* (puffball). In folk medicine this material is used to treat epistaxis. Inhalation of large quantities of conidia results in an allergic pulmonary disease. The symptoms include rales, fever, and cough. Conidia can be seen in sputum and will sprout if placed on culture media. Chest x-ray reveals fine miliary nodules throughout several lung fields.[81]

REFERENCES

1. Allcroft, R. 1965. Aspects of aflatoxicosis in farm animals. *In* Wogan, G. N. 1965. *Mycotoxins in Foodstuffs*. Cambridge, MIT Press, pp. 153–162.
2. Allegro, J. 1971. *The Sacred Mushroom and the Cross*. New York, Bantam.
3. Banaszak, E. F., W. H. Thiede, et al. 1970. Hypersensitivity pneumonitis due to contamination of an air conditioner. New Engl. J. Med., *283*:271–276.
4. Bell, D. K., and B. Doupnik. 1971. Infection of ground nut pods by isolates of *Aspergillus flavus* with different aflatoxin producing potentials. Trans. Br. Mycol. Soc., *57*:166–169.
5. Blayney, D., E. Rosen Kranz, et al. 1980. Mushroom poisoning from *Chlorophyllum molybdites*. West. J. Med., *132*:74–77.
6. Blount, W. P. 1961. Turkey x disease. Turkey (J. Br. Turkey Fed.), *9*(2):52, 55–58, 61, 77.
7. Boonpucknavig, V., N. Bhamarapravati, et al. 1973. Bagassosis. Am. J. Clin. Pathol., *59*:461–474.
8. Bove, F. J. 1970. *The Story of Ergot*. Basel, S. Karger.
9. Buck, R. W. 1969. Mycetism. New Engl. J. Med., *280*:1363.
10. Buku, A., T. Wieland, et al. 1980. Amaninamide, a new toxin of *Amanita virosa* mushroom. Experientia, *36*:33–34.
11. Bureau, M. A., C. Fecteau, et al. 1979. Farmer's lung in early childhood. Am. Rev. Respir. Dis., *119*:671–675.
12. Cadham, F. T. 1924. Asthma due to grain dust. J.A.M.A., *83*:27.
13. Campbell, J. M. 1932. Acute symptoms following work with hay. Br. Med. J., *2*:1143–1144.
14. Cann, H. M., and H. L. Verhulst. 1961. Mushroom poisoning. Am. J. Dis. Child., *101*:128–131.
15. Carlton, W. W., and J. Tuite. 1970. Nephropathy and edema syndrome induced in miniature swine by corn cultures of *Penicillium viridicatum*. Pathol. Vet., *7*:68–80.
16. Christensen, C. M. 1965. *Common Fleshy Fungi*. Minneapolis, Burgess Publishing Co.
17. Christensen, C. M. 1975. *Molds, Mushrooms and Mycotoxins*. Minneapolis, University of Minnesota Press.
18. Cooles, P. 1980. Abuse of mushrooms *Paneolus foenisecii*. Br. Med. J., *280*:446–447.
19. Coomes, T. J., and J. C. Sander. 1963. The detec-

tion and estimation of aflatoxin in ground nuts and ground nut materials. Part I. Paper chromatographic procedure. Analyst, 88:209–213.

20. Crump, M. H., E. B. Smalley, et al. 1963. Mycotoxicosis in animals fed legume hay infested with *Rhizoctonia leguminocola*. J. Am. Vet. Med. Assoc., 143:996–997.

21. Doupnik, B., O. H. Jones, et al. 1971. Toxic fusaria isolated from mouldy sweet potatoes involved in an epizootic of atypical interstitial pneumonia in cattle. Phytopathology, 6:890.

22. Edwards, J. H. 1972. The isolation of antigens associated with farmer's lung. Clin. Exp. Immunol., 11:341–355.

23. Emmanuel, D. A., F. J. Wenzel, et al. 1966. Pneumonitis due to *Cryptostroma corticales* (maple bark disease). New Engl. J. Med., 244:1413–1418.

24. Eppley, R. M., and W. J. Bailey. 1973. 12,13 Epoxy-Δ9-trichothecenes as the probable mycotoxins responsible for stachybotryotoxicosis. Science, 181:758–760.

25. Food and Agriculture Organization of the United Nations. 1979. Mycotoxin surveillance. A guideline. FAO Food Control Series No. 4, pp. 68. ISBN-92-5-100315-7.

26. Forgacs, J., and W. T. Carll. 1962. Mycotoxicoses. Adv. Vet. Sci., 7:273–282.

27. Frankland, A. W. 1971. Seasonal allergic rhinitis. Proc. R. Soc. Med., 64:447–450.

28. Goodall, C. M., and W. H. Butler. 1969. Aflatoxin carcinogenesis: Inhibition of liver cancer induction in hypophysectomized rats. Int. J. Cancer, 4:422–429.

29. Grieble, H., J. W. Rippon, et al. 1975. Scopluriopsosis and hypersensitivity pneumonitis in an addict. Ann. Intern. Med., 83:326–329.

30. Gross, N. 1974. *Bronchial asthma: Current Immunologic, Pathophysiologic, and Management Concepts*. New York, Harper, Row and Co.

31. Grossman, C. M., and B. Malbin. 1954. Mushroom poisoning: A review of the literature and report of 2 cases caused by a previously undescribed species. Ann. Intern. Med., 40:249–259.

32. Gundersen, T. 1979. Pizza med fleinsopp. Narkotika i maten. Tidsskr. Nor. Laegeforen., 99:424–434.

33. Hales, C. A., and R. H. Rubin. 1979. A 34-year old man with recurrent pulmonary infiltrates. New Engl. J. Med., 301:1168–1174.

34. Hapke, E. J., M. E. Seal, et al. 1968. Farmer's lung. Thorax, 23:451–468.

35. Hargreave, F. E., J. Pepys, et al. 1966. Bird breeder's (fancier's) lung. Lancet, 1:445–449.

36. Hayes, A. W., and B. J. Wilson. 1970. Effects of rubratoxin B on liver composition and metabolism in the mouse. Toxic. Appl. Pharmacol., 17:481–493.

37. Hearn. C. E. D., and V. Halford-Strevens. 1968. Immunological aspects of bagassosis. Br. J. Ind. Med., 25:283–292.

38. Heim, R. 1963. *Les Champignons Toxiques et Hallucinogenes*. Paris, Editions N., Boubee et Cie.

39. Heim, R., and R. G. Wasson. 1958. *Les Champignons Hallucinogenes du Mexique*. Paris, Mus. Natl. Histoire Natural.

40. Herzberg, M. (ed.). 1969. Proc of UNJR Conference on Toxic Microorganisms (Honolulu), Washington, U.S. Dept. of Agriculture.

41. Hesseltine, C. W., O. L. Shotwell, et al. 1966. Aflatoxin formation by *Aspergillus flavus*. Bacteriol. Rev., 30:795–805.

42. Hodges, F. A., J. R. Zust, et al. 1964. Mycotoxins: Aflatoxins isolated from *Penicillium puberulum*. Science, 145:1439.

43. Höer, P. W., L. Horbach, et al. 1964. Das Krankeitsbild der Farmer lunge und seine Beziehung zu den Pilzinfektionen. Z. Klin. Med., 158:1–21.

44. Hořejší, N., J. Šach, et al. 1960. A syndrome resembling farmer's lung in workers inhaling spores of *Aspergillus and Penicillium* moulds. Thorax, 15:212–217.

45. Hunter, D., and K. M. A. Perry. 1946. Bronchiolitis resulting from handling of bagasse. Br. J. Ind. Med., 3:64–74.

46. Jameson, J. E. 1968. Rapid and sensitive precipitin test for diagnosis of farmer's lung using immunoosmophoresis. J. Clin. Pathol., 21:376–382.

47. Jamison, S. C., and J. Hopkins. 1941. Bagassosis: Fungus disease of the lung. New Orleans Med. Surg., J., 93:580–582.

48. Joffe, A. 1965. Toxin production in cereal fungi causing toxic alimentary aleukia in man. *In* Wogan, G. N. 1965. *Mycotoxins in Foodstuffs*. Cambridge, MIT Press.

49. Jones, B. D. 1972. Methods of aflatoxin analysis. Trop. Prod. Inst. Reports No. 670, 58 pp.

50. Keen, P., and P. Martin. 1970. Is aflatoxin carcinogenic in man? The evidence in Swaziland. Trop. Geogr. Med., 23:44–53.

51. Kempton, P. E., and V. L. Wells. 1969. Mushroom poisoning in Alaska: Helvella. Alaska Med., 10:24–32.

52. Keyl, A. C., J. C. Lewis, et al. 1966. Toxic fungi isolated from tall fescue. Mycopathologia, 31:327–331.

53. Kirksey, J. W., and R. J. Cole. 1973. New Toxin from *Aspergillus flavus*. Appl. Microbiol., 26:827–828.

54. Kobayashi, M., M. A. Stakmann, et al. 1963. Antigens in mouldy hay as the cause of farmer's lung. Proc. Soc. Exp. Biol. Med., 113:472–476.

55. Kovats, F., and B. Bufy. 1968. *Occupational Mycotic Diseases of the Lung*. Budapest, Akademiai Kiada.

56. Kuschinsky, G. (ed.) 1970. *Taschenbuch der moderne Arzneimittel behandlung*. 5th ed. Stuttgart, Georg Thieme Verlag, p. 664.

57. Lacey, J. 1971. *Thermoactinomyces sacchari* sp. nov. A thermophilic actinomycete causing bagassosis. J. Gen. Microbiol., 66:327–338.

58. Lancaster, M. C., F. P. Jenkins, et al. 1961. Toxicity associated with certain samples of ground nuts. Nature (Lond.), 192:1095–1096.

59. Leach, C. M., and M. Tulloch. 1971. *Pithomyces chartarum*, a mycotoxin producing fungus, isolated from seed and fruit in Oregon. Mycologia, 63:1086–1089.

60. Lincoff, G., and D. H. Mitchel. 1977. *Toxic and Hallucinogenic Mushroom Poisoning*. New York, Van Nostrand Rheinhold.

61. List, P. H., and H. Hackenberg. 1969. Velleral und iso Velleral, scharf schmeckende Stoffe aus *Lacterius vellereus*. Fries. Arch. Pharm. (Weinheim), 302:125–145.

62. List, P. H., and P. Luft. 1967. Gyromitrin das Gift der Früjahrslorchel *Helvela (Gyromitra) esculata* Pers. ex Fr. Z. Pilzkunds, 34:1–8.

63. List, P. H., and H. Reith. 1960. Der faltentintling *Coprinus atramentarius* Bull., und seine dem tetraäthylthiuram disulfid ähnliche Wirkung. Arzneimittelforsch, *10*:34–40.

64. Lough, J., and D. G. Kinnear. 1970. Mushroom poisoning in Canada: A report of a fatal case. Can. Med. Assoc. J., *102*:858–860.

65. Louria, D. B. 1967. Deep seated mycotic infections, allergy to fungi and mycotoxins. New Engl. J. Med. *227*:1065–1071; 1126–1134.

66. Louria, D. B., J. K. Smith, et al. 1970. Mycotoxins other than aflatoxins: Tumor producing potential and possible relation to human disease. Ann. N.Y. Acad. Sci., *174*:583–591.

67. Miller, O. K. 1972. *Mushrooms of North America*. New York, E. P. Dutton & Co.

68. Nicholls, P. J., and G. R. Nicholls. 1967. *In* Davies, C. N. (ed.). *Inhaled Particles and Vapors*. 2nd ed. Oxford, Pergamon Press, pp. 93–100.

69. Ordman, D. 1970. Seasonal respiratory allergy in Windhook: The pollen and fungus factors. South Afr. Med. J., *44*:250–253.

70. Ott, J., and J. Bigwood. 1978. Teonancatyl. *Hallucinogenic Mushrooms of North America.* Seattle, Madrona Pub. Inc.

71. Peckham, J. C., B. Doupnik, et al. 1971. Acute toxicity of ochratoxins A and B in chicks. Appl. Microbiol., *21*:492–494.

72. Pepys, J., and P. A. Jenkins. 1965. Precipitin (F.L.H.) test in farmer's lung. Thorax, *20*:21–35.

73. Purchase, I. F. H. 1971. *Mycotoxins in Human Health*. London, MacMillan.

74. Schlueter, D. P., J. N. Fink, et al. 1972. Wood pulp worker's disease: a hypersensitivity pneumonitis caused by *Alternaria*. Ann. Intern. Med., *77*:907–914.

75. Seal, R. M. E., E. J. Hapke, et al. 1968. The pathology of acute and chronic stages of farmer's lung. Thorax, *23*:469–489.

76. Simons, D. M. 1979. The Meixner test for amatoxins in mushrooms. McIlvainea, *4*:1.

77. Singer, R. 1958. Observations on agarics causing cerebral mycetismus. Mycopathologia, *9*:261–284.

78. Slifkin, M., and J. Spaulding. 1970. Studies of the toxicity of *Alternaria mali*. Toxic. Appl. Pharmacol., *17*:375–386.

79. Smith, A. H., H. V. Smith, et al. 1979. How to know the gilled mushrooms. Dubuque, Wm. C. Brown.

80. Smith, H. V., A. H. Smith, et al. 1979. How to know the non-gilled fleshy fungi. Dubuque, Wm. C. Brown.

81. Strand, R. D., E. D. Neuhauser, et al. 1967. Lycoperdonosis. New Engl. J. Med., *277*:89–91.

82. Tewksbury, D. A., F. J. Wenzel, et al. 1968. An immunological study of maple bark disease. Clin. Exp. Immunol., *3*:857–863.

83. Toth, B., K. Patil, et al. 1979. False morel mushroom *Gyromitra esculata* toxin: *N*-methyl *N-N*-formylhydrazine carcinogenesis in mice. Mycopathologia, *68*:121–128.

84. Towey, J. W., H. C. Sweany, et al. 1932. Severe bronchial asthma apparently due to fungus spores found in maple bark. J.A.M.A., *99*:453–459.

85. Trenk, H. L., M. E. Butz, et al. 1971. Production of ochratoxins in different cereal products by *Aspergillus ochraceus*. Appl. Microbiol., *21*:1032–1035.

86. Tuchweber, B., R. Sieck, et al. 1979. Prevention by silybin of phalloidion-induced acute hepatotoxicity. Toxicol. ed. Appl. Pharmacol, *51*: 265–275.

87. Tyler, V. E. 1963. Poisonous mushrooms. Progr. Chem. Toxicol., *1*:339–384.

88. Uraguchi, K., and M. Yamazaki. 1978. *Toxicology, Biochemistry and Pathology of Mycotoxins*. New York, John Wiley and Sons.

89. Van Warmelo, W., F. O. Marases, et al. 1970. Experimental evidence that lupinosis of sheep is a mycotoxicosis caused by the fungus *Phomopsis leptostromiformis* (Kuhn) Bubak. J.S. Afr. Vet. Med. Assoc., *41*:235–247.

90. Vesonder, R. F., and A. Ciegler. 1979. Natural occurrence of vomitoxin in Australian and Canadian Corn. Euro. J. Appl. Microbiol. Biotech., *8*:237–240.

91. Wasson, R. G. 1971. *Soma: Divine Mushrooms of Immortality*. New York, Harcourt Brace Jovanovich.

92. Wasson, R. G. 1961. The hallucinogenic fungi of Mexico. Bot. Mus. Leaflets (Harvard Univ.)., *19*:137–162.

93. Watling, R. 1980. Orellanin poisoning in Britain. Bull. Br. Mycol. Soc., *14*:63.

94. Wenzel, F. J., D. A. Emmanuel, et al. 1972. A simplified hemagglutination test for farmer's lung. Am. J. Clin. Pathol., *57*:206–208.

95. Wieland, T., and H. Fulstick. 1978. Amatoxins, phallotoxins, phallolysin and antamanide: the biologically active components of poisonous *Amanita* mushrooms. CRC Crit. Rev. Biochem. *5*:185–260.

96. Wilson, B. J., and P. R. Manonpot. 1971. Causative fungus agent of leucoencephalomalacia in equine animals. Vet. Rec., *88*:484–486.

97. Wilson, B. J., M. R. Boyd, et al. 1971. A lung oedema factor from mouldy sweet potatoes *(Ipomoea batatas)*. Nature (Lond.), *231*:52–53.

98. Wright, D. E. 1968. Toxins produced by fungi. Ann. Rev. Microbiol., *22*:269–282.

99. Wogan, G. N. 1965. *Mycotoxins in Foodstuffs*. Cambridge, MIT Press.

100. Wogan, G. N., and R. S. Pony. 1970. Aflatoxins. Ann. N.Y. acad. Sci., *174*:623–635.

Pharmacology———IV
of Actinomycotic
Drugs

29 Antimycotic Agents

INTRODUCTION

There is a wide range of antibiotics available for the treatment of bacterial infections. This is in part due to vast differences between the physiology of the prokaryotic bacteria and the eukaryotic host. Because of these differences certain substances (antibiotics) will interfere with the metabolism of the infecting agent but will have little or no effect on the host. Though some of the antibiotics used in the treatment of bacterial infections may have moderate to serious side effects, one has the prerogative of changing from one to another antibiotic agent if the response to treatment is not adequate or if side effects occur. This is not true in infections caused by fungi. As a consequence of an eukaryotic organism parasitizing a eukaryotic host, the range of physiologic differences between host and parasite is much smaller.

So far the search for antifungal agents has been disappointing. We have at this point three well-established drugs, one for each of the three major categories of infection, together with a few other drugs that are of limited use and only in particular types of infections. The polyene antibiotic amphotericin B, a very toxic agent, is the standard drug for the treatment of systemic fungus infections. Nystatin, another polyene antibiotic and essentially insoluble, is restricted to the topical treatment of yeast infections. Griseofulvin is the most efficacious agent for treating all forms of dermatophyte infections, but it is not useful for other types of fungus diseases. In addition to these, there are some other agents being investigated that have been found useful in some types of disease. These include tolnaftate and haloprogin for some forms of dermatophyte infections, and hydroxystilbamidine and 5-fluorocytosine in a few subcutaneous and systemic infections. A new group of compounds, the imidazoles, are now being investigated and hold promise. Already such drugs as clotrimazole and miconazole have found wide application as topical agents in dermatophyte and candida infections, and micronazole and ketoconazole in paracoccidioidomycosis and systemic candidiasis. Other uses for these drugs will doubtless be found, and newer derivatives are currently under investigation.

The most satisfactory presentation of the history, development, and present application of antifungal drugs is found in David Speller's book, *Antifungal Chemotherapy*[74] and treatment modalities for systemic mycoses in Medoff and Kobayashi[58a] *Strategies of Treatment for Systemic Mycoses.*

POLYENE ANTIBIOTICS

The first antifungal antibiotic to be isolated was nystatin in 1949.[36] The compound was found to contain a series of aliphatic unsaturated chains ($-CH=CH-CH=CH-$) also present in carotene or vitamin A and was designated a "polyene." Since that time numerous polyene antibiotics have been described.[46] Of these, only a few have received much attention as useful antifungal agents owing to the high degree of toxicity inherent in all of them.[47]

The number of conjugated double bonds varies in the various polyene antibiotics and allows them to be classified according to that number. The basic structure of all of them is a macrolide ring containing a number of

723

conjugated double bonds. Those that have been investigated as antifungal agents are listed in Table 29–1. The macrolide ring may also contain a number of substitutes, such as an amino sugar, carboxyl groups, and so forth. The addition of some of these confers amphoteric properties to the molecule. Thus amphotericin A and B, nystatin, pimaricin, and etruscomycin are amphoteric, whereas filipin, lacking an ionizable radical, is a neutral polyene. The presence or absence of such qualities does not appear to affect the mode of action of the antibiotic or its toxicity.

The biological effects of the polyene antibiotics have been the subject of much investigation.[28, 30, 32] It appears that in low concentrations all the polyenes affect only those cells that contain sterols in their cell membrane.[48] This is also the basis for their selective toxicity and application as therapeutic agents. In cells that contain rigid cell walls, such as fungi and helminths, they interfere with selective membrane functions, which may result in stasis or death of the organism. In cells without rigid external walls (fungal protoplasts, protozoa, mammalian erythrocytes), the alteration of permeability of the membrane may result in lysis of the cell. In organisms that do not contain sterols in their cell membranes and thus are not sensitive to polyenes (i.e., bacteria), it has been possible to confer sensitivity by growing them on media containing cholesterol.[84] Mutants of yeasts that do not contain cholesterol in their membranes are also resistant to nystatin.[28, 75] Fungi can be protected from the effect of polyene antibiotics by including sterols in the growth medium.[32] The amount of damage to a given membrane appears to be a function of the number of carbon atoms in the antibiotic molecule. The larger antibiotics, nystatin and amphotericin B, are less damaging than the smaller molecules, filipin and etruscomycin.

The major effect of polyene antibiotics appears to be on the selective permeability of biological membranes. It has been demonstrated that certain substances (hydrogen ions) normally excluded gain entrance, and that certain essential cytoplasmic constituents are lost. The effect and degree of interference with normal membrane function are then reflected by other cellular abnormalities, such as cessation of growth, abnormal growth, decrease in metabolic rate, ballooning of cells, lysis, and so forth. In some cases the effect may be reversed by removal of the antibiotic from the environment. Thus *Coccidioides* spherules inhibited by the presence of amphotericin B will resume activity and growth if transferred within a certain amount of time to antibiotic-free media.

The extent of damage to the membrane and the nature of the cytoplasmic constituents released are influenced by several factors. Among these are the particular polyene molecule investigated, the concentration of the substance in the environment, and the time of exposure. One of the first effects noted in experimental studies on the mode of action of polyene antibiotics is the loss of intracellular potassium. This is followed by the appearance in the media of other low-molecular-weight cellular constituents. As time of exposure increases, larger compounds, including proteins and nucleic acids, leak from the cell. Also the toxicity of the polyene antibiotics for mammalian tissues may be more pronounced on particular membrane systems. Thus amphotericin B seems to have a special affinity or toxicity for cells of the renal tubules.

The two polyene antibiotics most commonly employed in antimycotic therapy, amphotericin B and nystatin, will be discussed in detail. In addition, some compounds having promise as useful agents of the future or established as effective treatments for certain specific diseases will be included.

An excellent review of the subject of antifungal antibiotics up to 1968 has been published by Hildick-Smith.[37]

Table 29–1 *Polyene Antibiotics of Interest as Antimycotic Agents and Date of Discovery*

Tetraenes	
Nystatin	Hazen and Brown, 1949
Amphotericin A	Vanderputte et al., 1956; Gold et al., 1956
Pimaricin (Tennecetin)	Struyk et al., 1958; Divekar et al., 1961
Etruscomycin	
Pentaenes	
Filipin	Gottlieb, 1955
Fungichromin	Tytell, 1955
Hexaenes	
Flavacidin	Albert, 1947
Mediocidin	Utahara et al., 1954
Heptaenes	
Amphotericin B	Vanderputte et al., 1956; Gold et al., 1956
Trichomycin	Nakano, 1960
Candicidin	Kligman and Lewis, 1953
Candidin	Taber et al., 1954

Nystatin

Nystatin (Mycostatin, Moronal, Nystan, Fungicidin) $C_{47}H_{75}NO_{17}$ yellow powder, m.p.

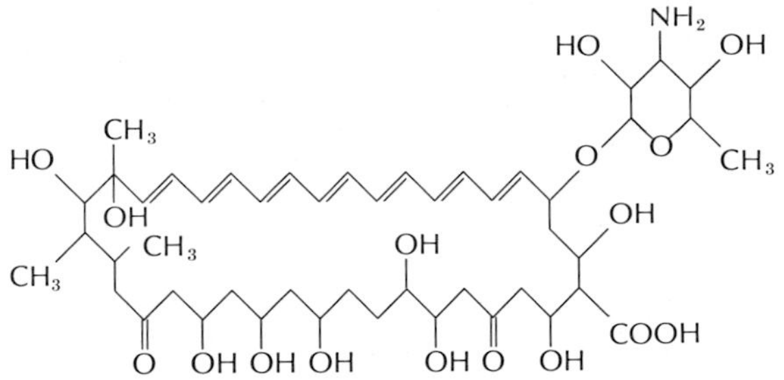

Figure 29–1. Nystatin.

dec. above 160°C. No definite m.p., $[\alpha]D^{25}$ − 10° (glacial acetic acid). An amphoteric tetraene with a sugar moiety mycosamine (Fig. 29–1). U.V. max., 290, 307, 322. Essentially insoluble in water 4.0 mg per ml, methanol (11.2 mg per ml), or ethanol (1.2 mg per ml). Quite soluble in dimethylformamide, propylene glycol, dimethylacetamide, glacial acetic acid, 0.5 N methanolic hydrochloric acid, and sodium hydroxide. It is inactivated rapidly in the latter three. All solutions begin to lose activity after preparation. Sensitive to heat, light, and oxygen, $LD_{50}IP$ in mice, 200 mg per kg; IV, 3 mg per kg. Activity not diminished by blood or serum. Extracted from mycelium of *Streptomyces noursei* by MeOH.

History. A search for antimycotic agents by Hazen and Brown led to the discovery of nystatin in 1949.[36] The first of the polyene antibiotics was isolated from an actinomycete recovered from a pasture of the Nourse dairy farm in Fouquier County, Virginia. The organism was named *Streptomyces noursei* and was shown to produce two antifungal antibiotics. One was soluble and identified as Actidione (cycloheximide). The other was insoluble and was later named "nystatin" after the New York State laboratory where the developers were employed. Most of the common fungal pathogens are inhibited *in vitro* by 1.5 to 13 μg per ml of nystatin. However, the lack of solubility of the compound in water is paralleled by its lack of absorption following ingestion. This together with its significant toxicity when administered parenterally prevents the use of this drug in the treatment of systemic fungus infection.[36] It has been found to be of great value in the topical treatment of cutaneous and mucocutaneous infections, especially those caused by the *Candida* species.

Range of Antimicrobial Activity. When first isolated, a milligram of nystatin was arbitrarily assigned an activity value of 1000 units. In bioassays, the activity of the drug in dilution series was expressed as "units," depending on the fraction of a milligram per milliliter needed to inhibit growth. In time, purification procedures allowed the produc-

tion of the drug with a range of up to 6000 units per mg. The antibiotic now generally available for clinical use has a range of 2500 to 3500 units per mg. For bioassay, nystatin is dissolved in dimethylformamide to give 1000 units per ml, then diluted with phosphate buffer pH 6 to 30 units per ml. Aliquots are placed in liquid or solid medium and test organisms inoculated.[16] The compound is more active between pH 4.4 to 6.6 on solid media, but in liquid media the optimum pH is 7. The inhibitory range of nystatin is given in Table 29–2. A simple laboratory procedure for sensitivity testing uses a disk containing 100 units. This is done by inoculating a plate with organism and placing a disk of antibiotic on it. Zones of inhibition can be read within 24 to 72 hours.

The antibiotic has no effect on bacteria, and in concentrations of 100,000 units per ml it is not toxic to ECHO, coxsackie, or polio virus. In tissue culture it is used in a concentration of 30 units per ml to control fungal contaminants. In addition to antifungal activity, nystatin inhibits *Trichomonas vaginalis* at

Table 29–2 *Antifungal Range of Nystatin**

Fungi	Least Amount Inhibiting Growth (units/ml agar)†
Candida albicans, C. stellatoidea C. parapsiosis, C. krusei* C. tropicalis* *Histoplasma capsulatum* (yeast) *Paracoccidioides brasiliensis*	7.8–7.9
Cryptococcus neoformans	3.5–3.9
Blastomyces dermatitidis (yeast)	1.9
Sporothrix schenckii (yeast)	31.2
Rhodotorula sp., *Saccharomyces* sp.	7.9
Penicillium sp.	7.9
Fusarium sp.	31.0
Pseudallescheria boydii	500.0
Trichophyton tonsurans, Microsporum audouinii, M. canis	16.0

*Derived from several authors and personal experience.

†Average of several experiments. It must be noted that strains of species vary greatly in sensitivity, particularly among myceliate fungi. Most of the yeasts, particularly *Candida,* are rather constant in the range of sensitivity.

dilutions of 1:60,000. It also has leishmanicidal activity.[29] Nystatin resistance in *Candida* sp. has been produced in the laboratory.[28, 52] Mutants of yeasts with low or no ergosterol in their cell membranes are also resistant to nystatin.[86] Such *in vitro*–produced resistant strains are nonpathogenic in experimental infection, and there is usually cross-resistance to other polyenes.[87] Back mutation to ergosterol- and cholesterol-containing membrane systems has been accomplished. Such cells are again pathogenic and sensitive to nystatin. Significant natural resistance to the drug in organisms recovered from patients following treatment has not been substantiated or is quite rare.[49]

Pharmacology. Even after massive oral doses, levels of no more than 1 to 2.5 μg per ml can be attained in serum. This indicates the compound is not absorbed to any degree from the intestinal tract. The use of intravenous therapy is precluded by the high toxicity, as demonstrated in several clinical trials.

Indications and Dosage. Nystatin is available as oral tablets containing 500,000 units. These can be given as two tablets t.i.d. in cases of intestinal overgrowth of *Candida* and continued until stool or oral cultures show normal numbers. As a prophylactic measure, it has been given at a dose of one tablet t.i.d. for seven to ten days. There is no evidence that this is efficacious in preventing colonization by *Candida* following abdominal surgery.

A nystatin suspension is used in the treatment of oral candidiasis of children. It is available commercially in the form of a powder. After addition of water, each ml contains 100,000 units. It is stable for seven days. For children with recalcitrant thrush, 1 to 2 ml of the suspension is placed directly in the mouth q.i.d. Treatment is continued for 7 to 14 days or until clinical response. In adults this suspension may be used for oral candidiasis and as a lavage both for esophageal candidiasis and in the treatment of mycotic keratitis. For treatment of oral lesions, patients can be given nystatin suppositories to suck on. This prolongs contact with the drug. The efficacy of nystatin in most cases of esophageal candidiasis is equivocal. Systemic administration of amphotericin B is usually met with greater success. For oral disease, the suspension is given at a dose of 4 ml q.i.d. until clinical response is achieved. As a lavage in the treatment of mycotic keratitis, it is given as conjunctival drops every one to two hours or as necessary. This is done only after demonstration of sensitivity to the antibiotic by the etiologic agent. It is particularly useful in *Candida* infections.

Cutaneous Candidiasis. Nystatin ointment at a concentration of 100,000 units per g is applied to the affected site twice daily. It is commonly combined with antibacterial antibiotics and a steroid, and contained in a cream or ointment base. One popular preparation includes neomycin, 2.5 mg, gramicidin, 0.25 mg, and triamcinolone acetonide, 1 mg, in addition to 100,000 units per g of nystatin.

For diaper disease, a nystatin topical powder is available. Each gram of talc contains 100,000 units of nystatin. It is applied liberally two or three times a day. *Candida* vaginitis is treated with vaginal suppositories containing 100,000 units of nystatin dispersed in lactose, ethyl cellulose, stearic acid, and starch. One to two tablets are inserted daily for 14 days. Recalcitrant infections may require some oral nystatin to lower the anal contamination.

Nystatin has been used as an aerosol in pulmonary candidiasis (Chapter 20) and aspergillosis (Chapter 23). It was administered in a nebulizer containing 25,000 units suspended in 5 ml of saline twice daily for six weeks.[57]

Side Effects. Rare but occasional hypersensitivity has been reported in the topical use of nystatin. Usually, as a cutaneous, vaginal, or oral preparation, it generally produces no serious side effects or irritations. Some records of diarrhea, nausea, and vomiting exist, and allergic contact dermatitis has been recorded.[18, 83]

Amphotericin B

Amphotericin B (Fungizone) $C_{46}H_{73}NO_{20}$, MW 960.10; deep yellow prisms or needles from dimethylformamide; m.p. gradually dec. above 170°C:$[\alpha]_D^{24}$ + 333° (in acidic dimethylformamide); $[\alpha]_D^{24}$ − 33.6° (in 0.1 N methanolic HCl). Amphoteric heptaene with mycosamine moiety. Molecular structure as in Figure 29–2. U.V. max., 364, 383, 408. Insoluble in water at pH 7, pH 2, or pH 11 ~ 0.1 mg per ml. Solubility in water increased by sodium deoxycholate. Solubility in dimethylformamide, 2 to 4 mg per ml; dimethylformamide + HCl, 60 to 80 mg per ml; dimethyl sulfoxide, 30 to 40 mg per ml. Solid and solutions stable between pH 4 and 10 for long periods at room temperature if protected from air. LD_{50} IP mice, 280 mg per kg; IV, 4.5 mg per kg; LD_2 IV, 2.2 mg per kg.

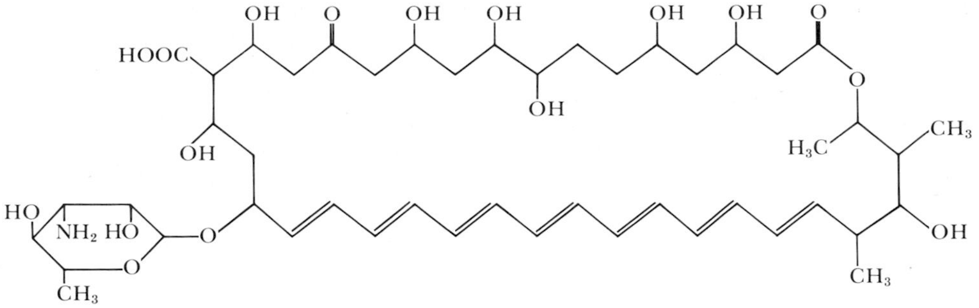

Figure 29–2. Amphotericin B.

Blood level of human tolerance is 0.5 to 1.5 μ per ml. Extracted from broth cultures of *Streptomyces nodosus* by *iso*-PrOH at pH 10.5.

History. Although the first specific antifungal agent isolated was the polyene nystatin, the great toxicity of the substance when used intravenously and the lack of absorption from the gut precluded its use for treating systemic fungus infection. Gold et al.[30] in 1956 described a culture of streptomyces (known as M-4575) that was inhibitory to the *in vitro* growth of *Saccharomyces cerevisiae, Rhodotorula glutinis, Candida albicans,* and *Aspergillus niger.* Vandeputte et al.[81] isolated and characterized two antifungal substances from this streptomycete now known as amphotericin A and B. Amphotericin B is the more effective antifungal agent. Commercial amphotericin B preparations contain about 1 to 2 per cent amphotericin A. The streptomycete is now called *Streptomyces nodosus* and was isolated from a soil sample collected from near Tembladoro on the Orinoco River in Venezuela.[79] The usefulness of the antibiotic was at first restricted by its complete insolubility in water and its unpredictable absorption from the gut. It was found, however, that the wetting agent deoxycholate greatly increased the solubility of the drug. Currently, the marketed powder contains 50 mg of amphotericin B, 41 mg of sodium deoxycholate, and a phosphate buffer. When needed, it is dissolved in 10 ml of sterile water. It can then be added to more sterile water or to 5 per cent glucose up to an optimum dilution of 1 mg per 10 ml water. Though it appears to be a solution, the mixture is a fine colloidal dispersion. Any contamination with electrolytes, viz., sodium chloride, will cause the formation of flocs, making the mixture unsuitable for clinical use. A solution of amphotericin B in water or 5 per cent glucose is stable for about 24 hours.

The advent of a soluble form of the drug allowed its investigation as an antifungal agent for systemic infections.[51] Even though it has numerous serious side effects, particularly nephrotoxicity, it is the only proven standardized agent for the treatment of most systemic mycoses. In addition, it has been used in the treatment of Chagas' disease, cutaneous leishmaniasis[67] and leprosy,[27] and schistosomiasis.[31] The drug is particularly efficacious in advanced mucocutaneous leishmaniasis and has been effective in the treatment of visceral leishmaniasis and Chagas' disease.

Range of Antimicrobial Activity. Laboratory tests have shown that the growth of a wide variety of fungi is inhibited by amphotericin B. Most of the fungi causing systemic infections in man are sensitive to the drug and at fairly constant levels.[30, 37, 38, 78] Tests on many strains of several species by a number of investigators show little variation in level of drug causing inhibition. Therefore, the drug is always useful in treating disease caused by these organisms. These fungi are included in Group I of Table 29–3. There are some fungi that are quite variable in sensitivity, and strains may differ significantly in *in vivo* and *in vitro* response to the drug. Group II of Table 29–3 includes some important pathogens and opportunists, such as *Mucor* and *Rhizopus* sp. isolated from mucormycosis, *Pseudallescheria boydii* from both eye infections and mycetoma, and some of the *Candida* species, such as *C. parapsilosis* and *C. tropicalis.* The drug may or may not be of benefit in diseases caused by these organisms. It has been the author's experience that results of sensitivity testing of organisms in Group I vary little from the figures given, and routine sensitivity testing is not necessary. The organisms in Group II are totally unpredictable, and each isolant must be tested for its level of sensitivity. It should be

Table 29–3 *Minimal Inhibitory Concentrations (µg/ml) of Amphotericin B for Various Fungi* and Protozoans*

Blood Level of Human Tolerance: 0.5–1.5 µg/ml

Group	Organism	Average MIC	Range MIC
I	Candida albicans	1.9	0.05–3.25
	Histoplasma capsulatum (Y)	0.04	
	Blastomyces dermatitidis (Y)	0.1	
	Coccidioides immitis	0.5	
	Paracoccidioides brasiliensis	0.1	
	Sporothrix schenckii (Y)	0.07	
	Cryptococcus neoformans	0.2	
	Leishmania brasiliensis	0.01	
	Candida glabrata	0.25	
	Rhodotorula mucilaginosa	0.5	
II	Candida parapsilosis		3.0–>40
	Candida tropicalis	3.5	0.2–25
	Rhizomucor heimalis		1.9–>40
	Rhizopus oryzae		0.9–>40
	Pseudallescheria boydii	>40	
	Aspergillus fumigatus		1.0–>40
III	Microsporum audouinii	1.0	
	Microsporum canis	3.7	
	Trichophyton mentagrophytes	2.5	
	Trichophyton rubrum	7.3	
	Sporothrix schenckii (M)	>40	
	Geotricum candidum		3.0–14.0
	Phialophora verrucosa	>40	
	Fonsecaea pedrosoi	>40	
	Fonsecaea compaetum	>40	
	Acremonium kiliense	>40	
	Fusarium sp.		3–>40

*Results read at 24°C 72 hours post inoculation either broth or agar.

cautioned, however, that organisms relatively resistant by *in vitro* testing have been isolated from disease in which the drug was at least somewhat efficacious. The reverse is also true and probably involves host factors as well as inherent sensitivity of the fungus to the drug. In general, the testing for sensitivity of a fungus to a drug should be viewed only as a plus or minus reaction: the organism is sensitive or it is not. The range and levels of sensitivity, while useful in bacteriologic procedures, are of limited value when applied to medical mycology. Details of the sensitivity test are given in Section Five. Group III includes the *in vitro* sensitivities of various other fungi with which the drug is ineffective, of questionable value, or not used. Resistance to amphotericin B has been produced *in vitro*[52, 73] but not authenticated *in vivo*. Cross-resistance with nystatin may or may not develop in *in vitro* studies.[87]

Absorption, Distribution, and Excre- tion.[53] Amphotericin B is poorly and unpredictably absorbed from the gastrointestinal tract following oral administration. Intake of 3 g per day produces plasma levels of about 0.1 to 0.5 µg per ml. In another study, 5 g was given during one day. Following this, 100 to 300 µg were detected in the 24-hour urine specimen. A daily dose of 1.6 to 5.6 g was followed by blood levels of up to 0.18 µg per ml. Some studies have shown no detectable blood level following ingestion of as much as 7.2 g.[51]

Oral amphotericin B was reported as being very useful in the treatment of chronic cutaneous candidiasis. In the protocol outlined by Montes et al.,[60] patients received 1 to 1.8 g per day for six months. This was followed by resolution of the disease in most cases.

The intravenous test dose injection of 1 to 5 mg of amphotericin B is done initially (1 mg in 250 ml D_5W infused over 30 minutes). The dose is then gradually or rapidly increased to the desired level of 0.6 mg per kg per day, which will lead to a peak serum level between 0.5 to 3.5 µg per ml, with an average of 1 µg per ml. The escalation of dosage depends on the tolerance of the patient. The test dose serves to determine the response of the patient to the drug. If immediate adverse reactions occur, these can be controlled with appropriate added drugs (50 mg Solucortef 1000 units heparin added to amphotericin B solution and infused over 4 to 6 hours). Some patients are acutely sensitive to even small amounts of amphotericin B. These patients often require some other antifungal agent or combination of agents. There are cases, however, in which amphotericin B is the only effective agent and must be given in spite of severe adverse reactions. Most all patients will become tolerant of the drug in time. The peak of 1 µg per ml is maintained for about six to eight hours, with the half-life usually 24 hours. It is estimated that only 2 to 5 per cent of the drug given is excreted in a biologically active form in the urine. There are several molecular alterations that occur, particularly in the liver. Other studies on blood levels have shown a single 1 mg per kg dose was followed by levels of 0.36 µg per ml at 24 hours, 0.3 to 5.5 mg per ml at 48 hours, and demonstrable levels persisting at 72 hours.[15] For this reason alternate-day therapy is the most frequently used modality today.[5, 6] Cerebrospinal fluid levels are about 1/40 that of blood levels.

The drug is slowly excreted in the urine.

Detectable levels are frequently present at 72 and 96 hours after cessation of treatment. It has been detected in the kidney one year after the last intravenous dose.[63] For this reason, manifestation of nephrotoxicity may develop long after cessation of the therapy. Renal insufficiency does not appear to influence excretion or plasma levels.[26] There is some evidence that the drug is initially bound to lipoproteins and then gradually released over a period of time. Several studies have shown that the drug is able to penetrate cells such as erythrocytes and leukocytes. In the latter case, it inhibited the growth of intracellular yeasts of *Histoplasma capsulatum*.

Side Effects.[38] Essentially no one receiving a full dose of amphotericin B therapy escapes without some side effects.[59] These vary from mild headache, chills, and fever to severe hemolytic anemia and acute nephritis. The list of reported side effects following intravenous administration includes fever, chills, headache, nausea, vomiting, anorexia, anaphylaxis, malaise, severe pain, phlebitis, hemolytic anemia, hypokalemia, hypotension, arrhythmias, diarrhea, melena, myalgia, arthralgia, and various manifestations of nephrotoxicity. The latter is the most constant and serious problem. Intrathecal administration is complicated by arachnoiditis, vertigo, diplopia, peripheral neuropathy, and convulsions. Arachnoiditis is the most serious and may be followed by paraplegia or other residua.

In most series almost all patients suffer some degree of chills and fever; anorexia is found in about half, headache in 40 per cent, and nausea and vomiting in about 20 per cent. Usually patients develop a degree of tolerance to some of these effects, and the dosage may be increased. However, chills and fever accompany almost every injection of the drug. If administration of the drug has been interrupted by seven days or more, it is necessary to gradually build up tolerance again.

Liver. Most reports have indicated there is little demonstrable damage to the liver in the majority of patients. In a few patients, particularly infants with widely disseminated disease, hepatic failure and jaundice have been seen. Focal necrosis was observed at autopsy.

Kidney. Renal toxicity is the most important single side effect of amphotericin B therapy. Some degree of azotemia is almost always seen during a course of therapy. No detectable azotemia occurs if dose is below 20 to 25 mg per day.

The severity of this side effect may necessitate interruption of treatment. Blood urea nitrogen and particularly creatinine clearance must be monitored as an index of this complication. It is therefore necessary to have a complete renal function profile to act as a baseline before beginning treatment.

From studies in experimental animals, it appears that renal blood flow is reduced to about 10 per cent following injection of the drug.[14a] This vasoconstriction is almost immediate, and no tolerance develops. At first there is a tubular leak, manifested by a few cells in the urine. Later, distal tubular necrosis occurs. Hypokalemia and hyposthenuria precede the development of azotemia. Nephrocalcinosis and hypokalemia are attributed to tubular acidosis, and alkali therapy during during a prolonged treatment regimen may be necessary.[56] The urine from patients with moderate to severe renal damage contains the usual signs of nephritis: red and white blood cells, albumin, and granular and cellular casts. Though the response of patients varies, most of the damage is reversible. Permanent impairment or progressive fatal degeneration may occur, however. It is generally held that no permanent damage occurs when the regimen is less than 3 g and little if less than 5 g, but permanent damage is almost universally found if more than 5 g has been administered. The high doses (10 to 20 g) formerly used in treatment protocols led to a number of deaths due to progressive renal failure.[38] The methyl ester of amphotericin B is about 1.8 times less toxic,[41] but at this writing it has received disappointing clinical evaluations.

Hematopoietic System. When high blood levels of amphotericin B are reached (0.5 to 5 μg per ml), there may develop a moderate to severe hemolytic anemia.[42, 43] After long-term treatment at lower doses, some depression of the erythropoietic system may occur, but without effect on the production of leukocytes or thrombocytes. The drug causes a normochromic or normocytic anemia accompanied by hypoferremia. The decrease in hemoglobin may be as much as 2.0 mg per 100 ml. This is almost always reversible.

Cardiac Effects. The cardiac effects of drug therapy are rare and usually not severe. However, increase in heart size[17] and decrease in cardiac force and frequency have been noted. These symptoms usually disappear after therapy. However, irreversible cardiac arrest has occurred.[38] Tissue culture studies indicate cytotoxic effects occur at concentrations of 0.5 to 1 μg per ml.

Impairment of prostate and testicular function has been found in animal studies,[77] but so far it has not been documented in human studies. Many patients have a transient hypocholesterolemia during treatment.[68] This reverses with the termination of therapy.

Therapeutic Uses. The drug was formerly available as oral tablets containing 250 mg and used for enteric candidiasis and chronic mucocutaneous candidiasis.[60] The latter disease responds to oral ketoconazole, which has largely displaced the oral preparations of amphotericin B. A 3 per cent lotion is available for topical application. The most commonly used form for intravenous or intrathecal therapy is a lyophilized powder that contains 50 mg of amphotericin B, 41 mg sodium deoxycholate, buffers, and diluent. The contents are dissolved in 10 ml of sterile water and added to 5 per cent dextrose to give a concentration of 0.1 mg per ml. It is given as a slow infusion over six hours.[5]

Several treatment schedules are available.[3, 6, 15, 74] One involves the administration of 0.25 mg per kg the first day after a test dose, followed by increments of 0.25 mg per kg each day until the 0.6 mg per kg level is reached. Alternate-day double dose therapy is recommended to increase patient tolerance.[6] Intrathecal amphotericin B has been used in cases of meningitis, but it is of questionable benefit to the patient owing to the high percentage of severe complications,[2, 22] unless it is a life-threatening situation. The usual method is to dissolve 0.5 mg in 5 ml of spinal fluid and to reinject the spinal fluid. This is done two to three times a week until a total dose of 15 mg is reached. This is considered adequate by most investigators (see chapters on specific diseases). Intrathecal steroids are often used. Some of the side effects engendered by intravenous therapy can be ameliorated by additives. Commonly aspirin, cortisone, and heparin are used (Chapter 16). The total course required varies with the severity of the disease, the immune state of the patient, and the etiologic agent.

The total experience with amphotericin B has shown it to be efficacious in most systemic fungal infections. The series vary, but in general the following data have been accumulated. Essentially all cases of pulmonary and disseminated blastomycosis have been cured with adequate treatment. About 50 per cent of acute pulmonary or disseminated coccidioidomycosis responded to therapy. Many of the treatment failures were associated with pregnancy. Whereas the mortality rate of disseminated cryptococcosis formerly was 83 per cent and that of cryptococcal meningitis 100 per cent, amphotericin B has decreased this to about 30 per cent.[74] About 75 per cent of patients with disseminated or chronic cavitary histoplasmosis are improved or cured following therapy. Pulmonary and disseminated candidiasis usually respond to a schedule of 1 to 3 g. Iatrogenic candidemia (from catheters and such) has responded to as little as 10 to 300 mg total dose. In cases of *Candida* cystitis, the urinary bladder was irrigated with 15 mg per day, which cleared the infection. *Candida* endocarditis does not usually respond to amphotericin B therapy. American cutaneous leishmaniasis has been cured following a total dose of 200 to 1450 mg of the drug.

The one area in which amphotericin B is of limited value is in some of the rarer forms of opportunistic infections. From the experience recorded in the literature, amphotericin B is of little or no value in infections caused by *Pseudallescheria boydii*, and disseminated infections caused by dematiaceous fungi. Other agents such as miconazole or 5-fluorocytosine are often efficacious in these infections. Laboratory studies have indicated some synergistic effects with other antibiotics, such as rifampicin[44, 45, 58] and 5-fluorocytosine. It may be possible in the future to augment the efficacy of amphotericin B by using combinations of drugs. This may allow the administration of the drug at lower levels, hopefully lessening the toxic side effects.

Amphotericin B preferentially attaches to ergosterol over cholesterol in cell membranes. This appears to be the basis of its selective affect on fungi.[32] Its effects also appear to be antagonized by ketoconazole, as this compound interferes with ergosterol synthesis.

Pimaricin

Pimaricin (Tennecetin, Natamycin, Pimifucin, Myprozine) $C_{33}H_{47}O_{13}N$ (Fig. 29–3), MW 665.75, amphoteric colorless crystal insoluble in H_2O, acetone, alcohol, chloroform; soluble in dilute acid or alkali; m.p. ca 200°C (dec), $[\alpha]_D^{25} + 250°$ (MeOH), U.V. max. (in MeOH), 279, 290, 303, 318 mμ, similar to other tetraenes (nystatin, amphotericin A, and Rimocidin). Extracted from mycelium of *Streptomyces*

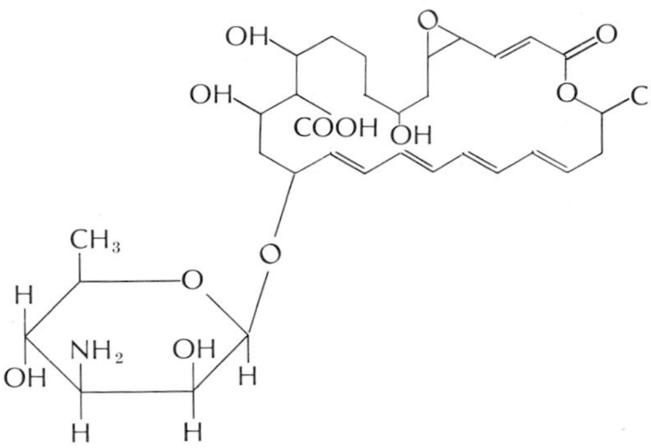

Figure 29-3. Pimaricin.

natalensis by lower alcohols, glycols, or forma-mides. Original isolant from Pietermaritz-burg in Natal, South Africa. LD in mice, 650 mg per kg IP, 1500 mg per kg per os.

This drug has been under investigation since 1959.[1] It has been advocated for use as an aerosol in pulmonary candidiasis and aspergillosis.[36a] and appears to be popular as a topical treatment for various cutaneous fungus infections in European countries and some others.[62] To date, its most important and almost unique place as an antifungal agent is in mycotic keratitis (Chapter 27).

Griseofulvin

Griseofulvin (Fulcin, Grifulvin, Grisovin, Lamoryl, Grisactin, Sporostatin, Spirofulvin, Likuden, Poncyl) $C_{17}H_{17}ClO_6$, MW 352.77 (Fig. 29-4). Antibiotic from *Penicillium griseo-fulvum*[13, 61] and *P. janczewskii* Zal. [= *P. nigri-cans* (Banier) Thom], octrahedra from ben-zene, m.p. 220°C, $[\alpha]D^{17} + 370°$ absorption max., 286,325. Practically insoluble in water; slightly soluble in ethanol, methanol, acetone, benzene, chloroform, ethyl acetate, and acetic acid; soluble in dimethylformamide (25°C) 12 to 14 g per 100 ml. Also soluble in dimethyl sulfoxide. It is thermostable and unaltered by autoclaving. Several spectrophotometric pro-cedures for its determination have been devised.[4, 25, 85]

History. In a series of works on the bio-chemistry of microorganisms, Oxford, Rai-strick, et al. discovered griseofulvin.[61] In 1939 they reported on its isolation from *Penicillium griseofulvum* (Dierk.) and its chemical and physical properties, but did not study its biological effects. Sometime later (1956) a compound was described by Brian, Curtis, and Hemming from a culture of *P. janczew-skii*.[13] This substance caused the mycelium of *Botrytis allii* to become stunted and form spirals, and the compound became known as curling factor. Later Brian recognized that the "curling factor" was identical to the gris-eofulvin of Oxford et al. It was proposed by Brian that griseofulvin would be useful as an antimycotic agent of plants.[12] It was not until the work of Gentles in 1958 that it was shown to be useful in the treatment of animal ring-worm infections. Since that time, it has be-come the standard treatment for all forms of human and animal dermatophytosis.

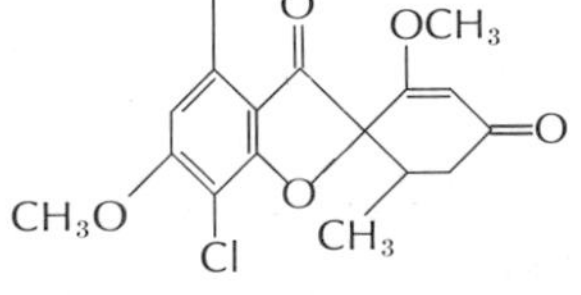

It is now known that griseofulvin is pro-duced by a number of *Penicillium* sp. The antibiotic is extracted from the mycelial mat following growth on broth. Several chemical alterations have been tried. Dechlorogriseo-fulvin and bromogriseofulvin are less active on assay (curling of *B. allii* test) than the parent substance. However, the *n*-butyl and *n*-propyl homologues are twenty times more active. The biosynthesis of griseofulvin was described by Rhodes et al.[64]

Range of Antimicrobial Activity. *In vitro* griseofulvin is active against almost all the dermatophytes at concentrations less than 1 μg per ml when tests are read at 72 hours (Table 29-4). In higher concentrations, it inhibits many other fungi to varying de-grees.[37] Therapeutically it is useful against dermatophyte infections, but has little or no efficacy in other types of fungus infections. It is not active against bacteria, most yeasts, Zygomycetes, viruses, or rickettsiae.[12, 37, 66]

In vitro resistance can be produced by growing some strains of *Trichophyton, Epider-mophyton,* and *Microsporum* on increasing con-centrations of the drug. This resistance does not disappear with subsequent transfer to antibiotic-free media. When the fungus is used to infect animals, the disease produced is the same as that elicited by parent antibiotic sensitive strains. Natural resistance of strains from human infections in which the drug has been used in therapy appears to occur and with increasing frequency (W. Artis, personal communication). The antibiotic is fungistatic and not fungicidal. Actively growing mycelia are killed by the drug, but it does not affect

Figure 29-4. Griseofulvin.

Table 29–4 *Minimal Inhibitory Concentrations of Griseofulvin for Common Dermatophytes**

Organism	Average MIC
Microsporum audouinii	0.4 –0.6
M. canis	0.2 –0.24
M. gypseum	0.4 –0.46
Trichophyton mentagrophytes	0.38–0.43
T. rubrum	0.14–0.18
T. tonsurans	0.30–0.34
T. gallinae	0.40–0.44
T. schoenleinii	0.34–0.38
T. violaceum	0.36–0.40
T. megninii	0.30–0.34
T. verrucosum	0.28–0.30
T. concentricum	0.26–0.30
Epidermophyton floccosum	0.38–0.42

*In vitro agar plates 25°C read at 72 hours; at 160 hours the MIC are double, on the average.

From Roth, F. J., B. Sallman, et al. *In vitro* studies of the antifungal antibiotic griseofulvin. J. Invest. Dermatol., *33*:403–418. © 1959 The Williams & Wilkins Co., Baltimore.

dormant hyphae or arthroconidia. In time, the mycelium destroys griseofulvin by demethylation.[11]

Mode of Action. Large quantities of the antibiotic are taken up by actively growing mycelium. There appears to be a two-stage binding of the drug. The first step is an immediate reaction not requiring energy, whereas the second stage is prolonged and requires active metabolism by the organism. Griseofulvin is bound to lipids within the cell but not to RNA or DNA. DNA metabolism is enhanced, and there is a greater content of this substance in drug-treated cells than in normal cells.[85]

The drug appears to have a colchicine-like action and has been used in the treatment of gout. It also has some anti-inflammatory activity[20] and has been used in a number of diseases, such as gout, angina pectoris,[21] and shoulder-hand syndrome.[37]

Therapeutic Uses.[88] The drug is useful in all forms of dermatophyte infection (Chapter 8). The usual oral daily dose is 10 mg per kg for children between 30 and 50 pounds, and 125 to 250 mg for children over 50 pounds. The usual adult dose is 0.5 to 1.0 g per day. In severe infection, up to 2 g per day can be given. Absorption is enhanced if the drug is taken after a fatty meal.[19] It is rapidly deposited in the epidermis and subsequently in the stratum corneum.[25]

Side Effects. Usually there are no serious side effects to therapy. About 55 per cent of patients have headache at the onset of thera-

py. This usually disappears as therapy is continued. Some patients complain of gastrointestinal distress, nausea, vomiting, and diarrhea. Rare neurologic symptoms include neuritis, lethargy, mental confusion, syncope, vertigo, blurred vision, and transient macular edema. Other rare side effects include urticaria, stomatitis, glossodynia, albuminuria, cylinduria, and renal insufficiency. All these resolve after cessation of treatment. Estrogen-like effects have been noted in some children. In most studies, topical application of griseofulvin is without significant benefit.

A serious consequence of griseofulvin therapy has been porphyria cutanea tarda,[88] as well as other porphyrias.[65] Protoporphyrin deposition in the liver, liver cirrhosis, and hepatomas have been produced in the mouse[40] but not in rats, guinea pigs, or rabbits. Rats have received up to 2000 mg per kg for three months without notable effects, and guinea pigs maintained on 500 mg per kg for 12 months have shown no ill effects. No case of neoplasia in humans has been attributed to the use of griseofulvin.

Tolnaftate

Tolnaftate (Focusan, Hi-Alazin, Sporiline, Tinactin, Tinaderm, Tonoftal) $C_{19}H_{17}NOS$, *m,N*-dimethylthiocarbanilic acid *O*-2 naphthyl ester (Fig. 29–5). Crystals from alc., m.p. 110.5–111.5°. Soluble in chloroform; sparingly soluble in ether; slightly soluble in alcohol; essentially insoluble in water.

This compound was one of the first synthetic chemical agents to have topical antifungal activity. *In vitro* studies indicate that it inhibits the growth of *T. mentagrophytes* at concentrations of 0.0075 to 0.075 μ per ml. The compound is without effect on yeasts (*C. albicans*) or bacteria. As a 1 per cent compound in polyethylene glycol or vanishing cream base, it is effective against infections of the cutaneous regions caused by most dermatophytes. It is not effective in tinea capitis or tinea unguium. The relapse rate for this drug

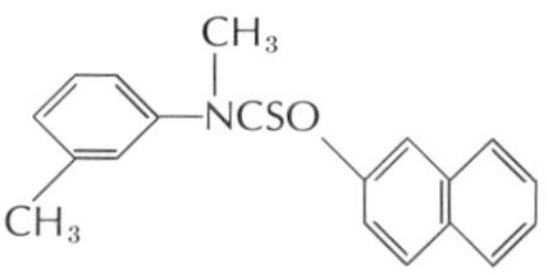

Figure 29–5. Tolnaftate.

in cutaneous infections is about the same as that for griseofulvin. The main advantage of this preparation is that it can be applied topically. Toxic or allergic reactions have not yet been recorded with frequency, although skin sensitivity occurs in some patients. Tolnaftate USP (Tinactin) is available as a cream, powder, or 1 per cent solution. It is usually applied twice daily until symptoms disappear. Pruritus is usually relieved within 72 hours, but treatment should be continued for 14 to 21 days to prevent relapse of infection.[33]

Haloprogin

Haloprogin (Halotex, Polik) $C_9H_4Cl_3IO$, MW 361.41, 3-iodo-2-propynl-2,4,5-trichlorphenyl ether (Fig. 29–6). Crystals with m.p. at 113–114°C.

This compound under the name of Halotex has been marketed as a topical antifungal agent. In *in vitro* studies,[70] it has a recorded MIC of 0.78 μg per ml for *T. rubrum* and 0.48 μg per ml for *T. tonsurans*. It has approximately equal efficacy[35] against most of the etiologic agents of dermatophyte infection. In *in vivo* studies of its metabolism, the compound was rapidly absorbed percutaneously (in rats) and appeared unchanged in urine. No tissue storage was detected. As a 1 per cent compound per gram in various bases, it appears to be well tolerated and efficacious in most dermatophyte infections and has an activity about equal to that of tolnaftate as currently used. It is not effective in tinea capitis and tinea unguium. There have been many reports of significant irritation or contact allergy. For this reason many practitioners have discontinued its use. It has the advantage of being effective against *Candida* sp. also.

5-Fluorocytosine

5-Fluorocytosine (Ancobon, Flucytosine) has recently been introduced as an antifungal agent[5, 34, 69, 78, 80, 82] (Fig. 29–7). It was first shown to have antifungal activity in 1964 and has since been demonstrated to have efficacy in cryptococcal meningitis and *Candida* sepsis.[76] Unfortunately, relapses

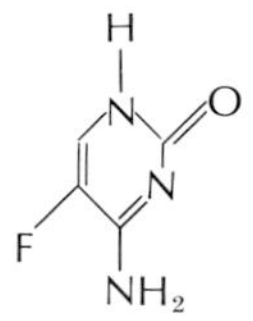

Figure 29–7. 5-Fluorocytosine.

have occurred in both. It has the advantage of being relatively nontoxic by oral dose[80] of 150 mg per kg per day. Treatment must be continued until clinical cure is assured. *C. neoformans* appears to develop resistance only after treatment with lower doses.[8] However, almost 50 per cent of *Candida albicans* strains are resistant without previous exposure at the prescribed human tolerance levels.[69, 71, 72] Because it can be given orally and appears to have few side effects, it may be considered as a first choice in cryptococcal meningitis or as an adjunct to IV amphotericin B. It has also been used successfuly in infections due to *Candida glabrata* and the rare infections caused by *Saccharomyces cerevisiae*. Most strains of the latter organism are sensitive to 0.05 to 0.1 μg per ml. Some cases of pancytopenia after its use have been recorded (Chapter 21). Hepatic insufficiency does not influence serum levels.[7] However, renal failure does affect excretion. The dosage should be adjusted in such cases and serum levels monitored. Investigators in Mexico and Latin America have been using the drug at maximal levels for the treatment of mycotic mycetoma, sporotrichosis, chromoblastomycosis, and other fungal diseases. In most of these diseases it was found to be of little efficacy. However, it is now considered the treatment of choice in chromoblastomycosis and other infections caused by dematiaceous fungi. In experimental infection using *Cladosporium bantianum* it was superior to other drugs available.[9]

It should be noted that *in vitro* studies must be carried out in media free of cytosine or uridine. Thus media containing beef extract, peptone, or yeast extract will negate the effect of the compound. It appears that the range of activity of 5-fluorocytosine is restricted to a few yeasts. It has, however, been demonstrated to potentiate the effect of amphotericin B on yeasts *in vitro*.[58]

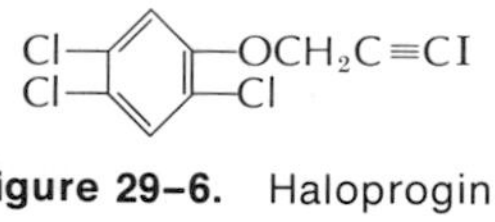

Miconazole

R 14889 Janssen Lab. MW 479.16, $C_{18}H_{14}Cl_4N_2O$, 2,4-dichloro-β(2,4-dichlorobenzyl oxylphenethyl) imidazole nitrate (Fig.

Figure 29–6. Haloprogin.

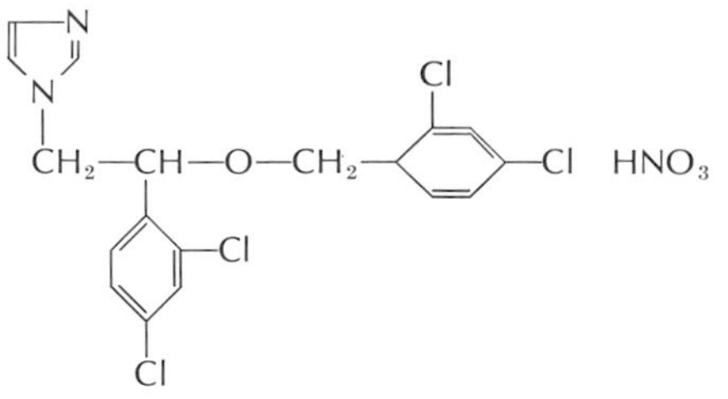

Figure 29–8. Miconazole.

29–8). Slightly soluble in water (0.03 per cent) or organic solvents. Pharmacology: Administration of 40 mg per kg per day orally in rats produced no signs of toxicity after three months. Necropsy normal. Acute toxicity orally 600 mg per kg in rats. *In vitro* range of activity is similar to that of tolnaftate. Most dermatophytes are sensitive in a range of 1–10 μg per ml. *Nocardia* sp., *Streptomyces* sp., *Histoplasma capsulatum* (Y), and *Blastomyces dermatitidis* (Y) are sensitive in a range of 0.1–1 μg per ml. Other agents of deep mycoses vary from 1–100 μg per ml.

This product is new, and at this writing has been found superior to other drugs in at least two types of infection: certain forms of candidiasis and *Pseudallescheria boydii* infections. In other systemic infections (coccidioidomycosis, blastomycosis, paracoccidioidomycosis, aspergillosis, histoplasmosis, and cryptococcosis) it has been disappointing.

Most other studies have involved its use as a topical agent. As a 2 per cent cream applied one or two times per day for ten days, it is about as effective as tolnaftate for the treatment of dermatophytosis and more effective than topical nystatin for cutaneous candidiasis. It is also very effective in vaginal infections caused by *Candida albicans.* Infections with *Trichophyton verrucosum* were found resistant. One investigator has reported complete cure of tinea unguium after eight months of therapy.

Clotrimazole

Clotrimazole (Bay b 5097) bis-phenyl(2-chlorophenyl)-1-imidazole methane (Fig. 29–9) (Delbay Pharmaceuticals) is a chlorinated trityl imidazolyl, which has been shown to have *in vitro* activity against a number of pathogenic fungi at a concentration of 4 μg per ml or less.[71] *Histoplasma capsulatum, Blastomyces dermatitidis, Sporothrix schenckii, Cryptococcus neoformans,* and *Coccidioides immitis* were found to be inhibited in a range of 0.20 to

3.13 μg per ml. This is comparable to the range of amphotericin B, which is 0.10 to 6.25 μg per ml. The activity against *Candida albicans, Aspergillus,* and *Pseudallescheria* was quite variable. Against dermatophytes, it was found that 0.78 μg per ml would suppress the growth of most species. At present it is being marketed as a cream and an ointment for the treatment of dermatophyte and *Candida* infections of the skin and as suppositories for the treatment of vaginal candidiasis. As a topical agent its efficacy is generally good, and at present it is widely used in cutaneous infections. It is not used for systemic infections because of adverse side effects.

Ketoconazole

Ketoconazole (Nizoral) (Janssen Pharmaceutica) is cis-1-acetyl 4[4]2-(2-4-dichlorophenyl)-2(1H imidazol-lylmethyl)-1-3-dioxolan-4-methoxy]phenyl] piperazine. Its mode of action is impairment of ergosterol synthesis in fungal cell membranes. It is active *in vitro* against dermatophytes, yeasts, dimorphic fungi, zygomycetes, and various other fungi. It is now established in the treatment of various *Candida* infections (particularly chronic mucocutaneous disease), paracoccidioidomycosis, and some forms of histoplasmosis and blastomycosis. It is being investigated for use in other fungal infections.

The drug is supplied as a 200 mg tablet, usually given once a day. Mean peak plasma levels of about 3.5 μg per ml are achieved within 1 to 2 hours. Adult dose is one tablet a day; the dosage for children is modified based on weight. The drug is well tolerated, and serious side effects at this writing have not been reported. Relapse in coccidioidomycosis, disseminated histoplasmosis, and some other infections has been documented.

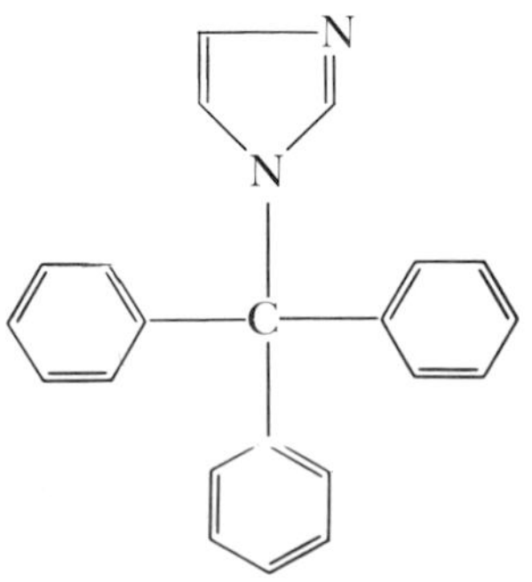

Figure 29–9. Clotrimazole.

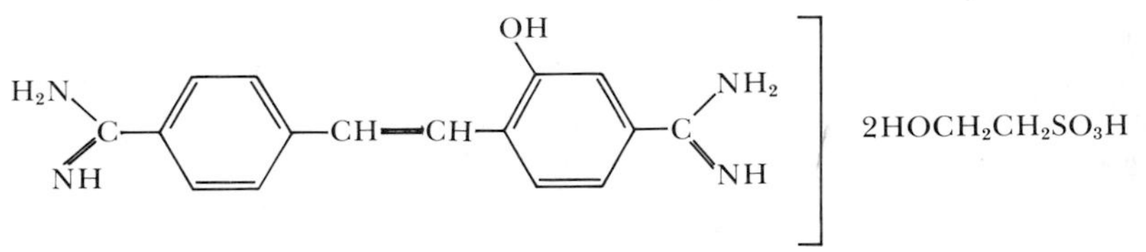

Figure 29–10. Hydroxystil-bamidine.

$$\left[\begin{array}{c} H_2N \\ \diagdown \\ C \\ \diagup \\ NH \end{array} \right. \text{—} \bigcirc \text{—} CH \text{—} CH \text{—} \bigcirc^{OH} \left. \begin{array}{c} NH_2 \\ \diagup \\ C \\ \diagdown \\ NH \end{array} \right] \quad 2HOCH_2CH_2SO_3H$$

Antagonism appears to occur when used with amphotericin B in therapy.

In dermatophyte infection, ketoconazole has achieved a wide application. It appears to be useful in most all types of disease, including dermatophyte or other fungal infection of the nails. Moncada[59a] found that three months of one tablet a day combined with 40 per cent urea cream resulted in cure of 13 of 16 patients. Ketoconazole appears to be most useful in cases wherein topical therapy is inadequate or infection involves yeasts as well.

Econazole is currently receiving clinical evaluation.

Hydroxystilbamidine Isothionate

This drug is an aromatic diamidine that was first described many years ago (Fig. 29–10). It was utilized in various systemic mycoses but was found efficacious only in one, blastomycosis. *Blastomyces dermatitidis* is sensitive to it *in vitro* and *in vivo*. At present its only use in antifungal therapy remains in certain forms of blastomycosis (see Chapter 18). It is given intravenously and it is known to accumulate in the liver and skin. The intravenous dose is a vial containing 225 mg that is added to 200 ml of saline or glucose. It is administered over a period of about an hour. It can be given daily. The usual therapeutic regimen calls for a total dose of 8 grams. The common side effects are malaise, nausea, anorexia, and moderate to severe headache. The parent compound (stilbamidine) was associated with peripheral neuropathies such as paresthesias, anesthesias, and palsy of the fifth cranial nerve. These are not associated with the hydroxy derivative. Liver function tests may be altered during the course of therapy, and use of the drug in patients with pre-existing liver dysfunction is countermanded.

REFERENCES

1. Alteras, I., and I. Cojocaru. 1969. Aperçu critique sur l'action antifongique de la pimaricine. Mykosen, *121*:139–149.

2. Alazraki, N. P., J. Fierer, et al. 1974. Use of hyperbaric solution for administration of intrathecal amphotericin B. New Engl. J. Med., *290*:641–646.

3. Andriole, V. T., and H. M. Kavetz. 1962. The use of amphotericin B in man. J.A.M.A., *180*:269–272.

4. Bedford, C., D. Busfield, et al. 1959. Spectrophotofluorometric assay of griseofulvin. Nature (Lond.), *184*:364–365.

5. Bennett, J. E. 1974. Chemotherapy of systemic mycoses. New Engl. J. Med., *289*:30–32 and *290*:320–324.

6. Bindschadler, D. D., and J. E. Bennett. 1969. A pharmacologic guide to the clinical use of amphotericin B. J. Infect. Dis., *20*:427–436.

7. Block, E. R. 1973. Effect of hepatic insufficiency on 5-fluorocytosine concentrations in serum. Antimicrob. Agents Chemother., *3*:141–142.

8. Block, E. R., A. E. Jennings, et al. 1973. 5-Fluorocytosine resistance in *Cryptococcus neoformans*. Antimicrob. Agents Chemother., *3*:649–656.

9. Block, E. R., A. E. Jennings, et al. 1973. Experimental therapy of cladosporiosis and sporotrichosis with 5-fluorocytosine. Antimicrob. Agents Chemother., *3*:95–98.

10. Blum, S. F., S. B. Shalet, et al. 1969. The effect of amphotericin B on erythrocyte membrane cation permeability; its relation to in vivo erythrocyte survival. J. Lab. Clin. Med., *73*:980–987.

11. Boothroyd, B., E. J. Napier, et al. 1961. The demethylation of griseofulvin by fungi. Biochem. J., *80*:34–37.

12. Brian, P. W. 1960. Control of fungal diseases of plants with griseofulvin. Trans. St. John's Hosp. Dermatol. Soc., *45*:4–6.

13. Brian, P. W., P. J. Curtis, et al. 1946. A substance causing abnormal development of fungal hyphae produced by *Penicillium janczewskii* Zal. I. Biological assay, production and isolation of curling factor. Trans. Br. Mycol. Soc., *29*:173–187.

14. Brian, P. W., P. J. Curtis, et al. 1955. Production of griseofulvin by *Penicillium raistrickii*. Trans. Br. Mycol. Soc., *38*:305–308.

14a. Butler, W. T., G. J. Hill, et al. 1964. Amphotericin B renal toxicity in dogs. J. Pharm. Exp. Ther., *143*:47–56.

15. Campbell, G. D., H. E., Einstein, et al. 1969. Indications for chemotherapy in the pulmonary mycoses. A report of the committee on fungus diseases, subcommittee on therapy — American College of Chest Physicians. Dis. Chest. *55*:160–162.

16. Carlson, J. R., and J. W. Synder. 1959. *Candida albicans* plate assay of nystatin. Antibiot. Chemother. (Basel). *9*:139–144.

17. Chung, D. K., and M. G. Koenig. 1971. Reversible cardiac enlargement during treatment with amphotericin B. and hydrocortisone. Report of three cases. Am. Rev. Respir. Dis., *103*:831–841.

18. Cosey, R. J. 1971. Contact dermatitis due to nystatin. Arch. Dermatol., *103*:228.

19. Crounse, R. G., 1961. Human pharmacology of griseofulvin. The effect of fat intake on gastrointestinal absorption. J. Invest. Dermatol., *37*:529–533.

20. D'Ary, F., E. M. Howard, et al. 1960. The anti-inflammatory action of griseofulvin in experimental animals. J. Pharm. Pharmacol., *12*:659–665.

21. DePasquale, N. P., J. W. Burks, et al. 1963. The treatment of angina pectoris with griseofulvin. J.A.M.A., *184*:421–422.

22. Diamond, R. D., and J. E. Bennett. 1973. A subcutaneous reservoir for intrathecal therapy of fungal meningitis. New Engl. J. Med., *288*:186–188.

23. Douglas, J. B., and J. K. Healy. 1969. Nephrotoxic effects of amphotericin B including renal tubular acidosis. Am. J. Med., *46*:154–162.

24. Eisenberger, R. S., and W. H. Oatway. 1971. Nebulization of amphotericin B. Am. Rev. Respir. Dis., *103*:289–292.

25. Epstein, W. L., V. P. Shah, et al. 1972. Griseofulvin levels in stratum corneum. Arch. Dermatol., *106*:344–348.

26. Feldman, H. A., J. D. Hamilton, et al. 1973. Amphotericin B therapy in an anephric patient. Antimicrob. Agents Chemother., *4*:302–305.

27. Fustardo, T. A., E. Cisalpiro, et al. 1962. Amphotericin B in the treatment of lepromatous leprosy. O. Hospital (Rio), *62*:1099–1109.

28. Gale, G. R. 1963. Cytology of *Candida albicans* as influenced by drugs acting on the cytoplasmic membrane. J. Bacteriol., *86*:151–157.

29. Ghosh, B. K., D. Haldar, et al. 1961. Leishmanicidal property of nystatin and its clinical application. Ann. Biochem. Exp. Med., *21*:25–28.

30. Gold, W., H. A. Stout, et al. 1956. Amphotericin A and B: Antifungal antibiotics produced by a streptomycete. I. *In vitro* studies of A *Antibiotics Annual*, 1955–1956. New York, Medical Encyclopedia, pp. 579–586.

31. Gordon, B. L., P. A. S. St. John, et al. 1963. Chemotherapy of schistosomiasis in mice with amphotericin B. Bacteriol. Proc., p. 95.

32. Gottlieb, D., H. E. Carter, et al. 1958. Protection of fungi against polyene antibiotics by sterols. Science, *128*:361.

33. Gould, A. H. 1964. Tolnaphtate in dermatology. A new potent topical fungicide. Dermatol. Trop., *3*:255–259.

34. Hamilton-Miller, J. M. T. 1972. A comparative *in vitro* study of amphotericin B clotrimazole and 5-fluorocytosine against clinically isolated yeasts. Sabouraudia, *10*:276–283.

35. Harrison, E. F., P. Zwadyk, et al. 1970. A topical antifungal agent. Appl. Microbiol., *19*:746–750.

36. Hazen, E., and R. Brown. 1950. Two antifungal agents produced by a soil actinomycete. Science, *112*:423.

36a. Henderson, A. H., and S. E. G. Pearson. 1968. Treatment of bronchopulmonary aspergillosis with observations of the use of natamycin. Thorax, *25*:519–523.

37. Hildick-Smith, G. 1968. Antifungal antibiotics. Pediatr. Clin. North Am., *15*:107–118.

38. Holeman, C. W., and H. Einstein. 1963. The toxic effects of amphotericin B in man. California Med., *99*:90–93.

39. Holt, R. J., and L. Newman. 1972. Laboratory assessment of the antimycotic drug clotrimazole. J. Clin. Pathol., *25*:1089–1097.

40. Hurst, E. M., and G. E. Paget. 1963. Protoporphyrin cirrhosis and hepatoma in the livers of mice given griseofulvin. Br. J. Dermatol., *75*:105–112.

41. Keim, G. R., J. W. Poutsiaka, et al. 1973. Amphotericin B methyl ester hydrochloride and amphotericin B: Comparative acute toxicity. Science, *179*:584–585.

42. Kinsky, S. C. 1970. Antibiotic interaction with model membranes. Ann. Rev. Pharmacol., Palo Alto, Annual Reviews, pp. 119–142.

43. Kinsky, S. C., J. Avruch, et al. 1962. The lytic effect of polyene antifungal antibiotics on mammalian erythrocytes. Biochem. Biophys. Res. Commun., *9*:503–509.

44. Kobayashi, G., G. Medoff, et al. 1972. Amphotericin B potentiation of rifampicin on antifungal agent against the yeast phase of Histoplasma capsulatum. Science, *177*:709–710.

45. Kobayashi, G., C. Sze-Schuen, et al. 1974. Effects of rifamycin derivatives, alone and in combination with amphotericin B against *Histoplasma capsulatum*. Antimicrobiol. Agents Chemother., *5*:16–18.

46. Korzybski, T., Z. Kowszk-Gindifer, et al. 1967. Antifungal antibiotics. *In Antibiotics*. Vol. I Oxford, Pergamon Press, pp. 769–846.

47. Lampen, J. O. 1969. Amphotericin B and other polyenic antifungal antibiotics. Am. J. Clin. Pathol., *52*:138–146.

48. Lane, J. W., R. G. Garrison, et al. 1972. Drug-induced alterations in the ultrastructure of *Histoplasma capsulatum* and *Blastomyces dermatitidis*. Mycopathologia, *48*:289–296.

49. Larsh, H. W. 1962. The prevalence of the drug resistance of *Histoplasma capsulatum* and *Candida albicans* in patients with pulmonary histoplasmosis. Lab. Invest., *11*:1140–1145.

50. Lepper, M. H., S. Lockwood, et al. 1959. Studies on the epidemiology of strains of *Candida* in hospital wards. *Antibiotics Annual*, 1958–1959. New York, Medical Encyclopedia, pp. 666–671.

51. Littman, M. L., and P. L. Horowitz. 1958. Coccidioidomycosis and its treatment with amphotericin B in man. Am. J. Med., *24*:568–592.

52. Littman, M. L., M. A. Pisano, et al. 1958. Induced resistance of *Candida* species to nystatin and amphotericin B. *Antibiotics Annual,* 1957–1958. New York, Medical Encyclopedia, pp. 981–987.

53. Louria, D. B. 1958. Some aspects ot absorption, distribution and excretion of amphotericin B. Antibiot. Med. Clin. Ther., *5*:295–301.

54. Mahgoub, E. S. 1972. Laboratory and clinical experience with clotrimazole (Bay b 5097) Sabouraudia, *10*:210–217.

55. McCoy, E., and J. S. Kiser. 1959. An evaluation of nystatin and candicidin against a standardized systemic *Candida albicans* infection in mice. *Antibiotics Annual*, 1958–1959. New York, Medical Encyclopedia, pp. 903–909.

56. McCurdy, D. K., M. Frederic, et al. 1968. Renal tubular acidosis due to amphotericin B. New Engl. J. Med., *278*:124–130.

57. McKendrick, G. D. W., and J. M. Medlock. 1958. Pulmonary moniliasis treated with nystatin aerosol. Lancet, *1*:631.

58. Medoff, G., G. S. Kobayashi, et al. 1972. Potentiation of rifampicin and 5-fluorocytosine as antifungal antibiotics by amphotericin B. Proc. Natl. Acad. Sci., USA, *69*:196–199.

58a. Medoff, G., and G. S. Kobayashi. 1980. Strategies in the treatment of systemic fungal infections. N. Engl. J. Med., *302*:145–155.

59. Miller, R. P., and J. H. Bates. 1969. Amphotericin B toxicity. A follow-up report of 53 patients. Ann. Intern. Med., *71*:1089–1095.

59a. Moncada, B. 1982. Onychomycosis. Successful therapy using ketoconazole and nonsurgical evulsion of affected nails. Cutis, in press.

60. Montes, L., M. D. Cooper, et al. 1971. Prolonged oral treatment of chronic mucocutaneous candidiasis with amphotericin B. Arch. Dermatol., *104*:45–55.

61. Oxford, A. E., H. Raistrick, et al. 1939. XXIX. Studies on the biochemistry of microorganisms. LX. Griseofulvin, $C_{18}H_{17}O_6Cl$, a metabolic product of *Penicillium griseofulvum.* Dierck. Biochem. J., *33*:240–248.

62. Polay, A. 1969. Erfahrungen mit einem neuen antimykotikum dem Pimaricin. Mykosen, *12*:677–686.

63. Renolds, B. S., Z. M. Tomkiewicz, et al. 1963. The renal lesions related to amphotericin B treatment of coccidioidomycosis. Med. Clin. North Amer., *47*:1149–1154.

64. Rhodes, A., G. A. Sommerfield, et al. 1963. Biosynthesis of griseofulvin. Biochem. J., *88*:349–356.

65. Rimington, C., P. N. Morgan, et al. 1963. Griseofulvin administration and porphyrin metabolism. A survey. Lancet, *2*:318–322.

66. Roth, F. J., B. Sallman, et al. 1959. *In vitro* studies of the antifungal antibiotic griseofulvin. J. Invest. Dermatol., *33*:403–418.

67. Sampaio, S. A., J. T. Godoy, et al. 1960. Treatment of American mucocutaneous leishmaniasis with amphotericin B. Arch. Dermatol., *82*:627–635.

68. Schaffner, C. P., and H. W. Gordon. 1968. The hypocholesterolemic activity of orally administered polyene macrolides. Proc. Natl. Acad. Sci. USA, *61*:36–41.

69. Schönbeck, J., and S. Ansehu. 1973. 5-Fluorocytosine resistance in *Candida* sp. and *Torulopsis glabrata.* Sabouraudia, *11*:10–20.

70. Seki, S., B. Nomiya, et al. 1963. Laboratory evaluation of M-1028 (2,4 5-trichlorophenyl iodoporparyl ether) a new antimicrobial agent. In *Antimicrobial Agents and Chemotherapy.* Sylvester, J. C. (ed.), Ann Arbor, American Society of Microbiology, pp. 569–572.

71. Shadomy, S. 1971. *In vitro* and fungal activity of clotrimazole (Bay G5097). Infect. Immun., *4*:143–148.

72. Shadomy, S., C. B. Kirchoff, et al. 1973. *In vitro* activity of 5-fluorocytosine against *Candida* and *Torulopsis* species. Antimicrob. Agents Chemother., *3*:9–14.

73. Sorsensen, L. J., E. G. McNall, et al. 1959. The development of strains of *Candida albicans* and *Coccidioides immitis* which are resistant to amphotericin B. *Antibiotics Annual,* 1958–1959. New York, Medical Encyclopedia, pp. 920–923.

74. Speller, D. (ed.) 1980. *Antifungal Chemotherapy.* New York, John Wiley.

75. Struyk, A. P., I. Hoette, et al. 1958. Pimaricin, a new antifungal antibiotic. *Antibiotics Annual,* 1957–1958. New York, Medical Encyclopedia, pp. 878–885.

76. Tassel, D., and M. A. Medoff. 1968. Treatment of *Candida* sepsis and *Cryptococcus* meningitis with 5-fluorocytosine: A new antifungal agent. J.A.M.A., *206*:830–832.

77. Texter, J. H., and D. X. Coffey. 1969. The effects of amphotericin B on prostatic and testicular function in the dog. Trans. Am. Assoc. Genitourin. Surg., *61*:101–114.

78. Titsworth, E., and E. Grunberg. 1973. Chemotherapeutic activity of 5-fluorocytosine and amphotericin B against *Candida albicans* in mice. Antimicrob. Agents Chemother., *4*:306–308.

79. Trejo, W. H., and R. E. Bennett. 1963. *Streptomyces nodosus* sp. nov. The amphotericin producing organism. J. Bacteriol., *85*:436–439.

80. Utz, J. 1972. Flucytosine. Editorial. New Engl. J. Med., *286*:777–778.

81. Vandeputte, J., J. L. Wachtel, et al. 1956. Amphotericin A and B, antifungal antibiotics produced by a streptomycete. The isolation and properties of crystalline.amphotericins. *Antibiotics Annual,* 1955–1956. New York, Medical Encyclopedia, pp. 587–595.

82. Warner, J. F., A. F. McGee, et al. 1971. 5-Fluorocytosine in human candidiasis. *Antimicrobial Agents* and *Chemotherapy.* II. Bethesda, American Society of Microbiology, pp. 473–475.

83. Wasilewski, C. 1970. Allergic contact dermatitis from nystatin. Arch. Dermatol., *102*:216–217.

84. Weber, M. M., and S. C. Kinsky. 1965. Effect of cholesterol on the sensitivity of *Mycoplasma laidlawii* to the polyene antibiotic filipin. J. Bacteriol., *89*:306–312.

85. Weinstein, G. D., and H. Blank. 1960. Quantitative determination of griseofulvin by a spectrophotometric assay. Arch. Dermatol., *81*:746–749.

86. Woods, R. A. 1971. Nystatin resistant mutants of yeast: alterations in sterol content. J. Bacteriol., *108*:69–73.

87. Woods, R. A., and K. A. Ahmed. 1968. Genetically controlled cross resistance to polyene antibiotics in *Saccharomyces cerevsiae.* Nature (Lond.), *218*:1080–1084.

88. Zaias, N., D. Taplin, et al. 1966. Evaluation of microcrystallin griseofulvin in tinea capitis. J.A.M.A., *198*:805–807.

89. Ziprkowski, L., A. Zeinberg, et al. 1966. The effect of griseofulvin in hereditary porphyria cutanea tarda. Investigations of porphyrins and blood lipids. Arch. Dermatol., *93*:21–22.

Laboratory ———— V
Mycology

Section I COMMON LABORATORY CONTAMINANTS

To the general mycologist the term "contaminant" is abhorrent, because it conveys an unpleasant connotation about a perfectly respectable fungal organism. It would be more appropriate to distinguish "incidental" isolants from "etiologic" isolants. The latter represent organisms involved in a particular human or animal fungal infection; the former, organisms that are present in the specimen, the container, the medium, or the air and that have been isolated incidentally. Sometimes it is difficult to separate the two and to be certain that what has been isolated was or was not involved in an infection. This is particularly true with the ever-lengthening list of opportunistic infections discussed in previous chapters. The criteria necessary for designating and demonstrating the involvement of an organism in a mycosis have also been discussed in previous sections. In this section, some of the incidentally isolated fungi that are most frequently encountered in the clinical mycology laboratory will be illustrated, and the characteristics of the colony and microscopic morphology noted briefly. It should be emphasized that almost any fungus may be involved in an infectious process, so that all isolants from clinical material should be considered suspect.

The taxonomy utilized in this section is largely drawn from the *Genera of the Hyphomy-cetes* of J. W. Carmichael, W. Bryce Kendrick, I. L. Conners, and Lynne Sigler (1980, University of Alberta Press). Three other guides that are indispensable in laboratory mycology: *A Manual of Soil Fungi,* by J. C. Gilman (1965, Iowa State College Press); *The Genera of Hyphomycetes from Soil,* by G. L. Barron (1968, The Williams and Wilkins Co.); *Illustrated Genera of Imperfect Fungi,* by H. L. Barnett and Barry B. Hunter (1972, Burgess Publishing Company), and *Laboratory Handbook of Medical Mycology* by M. R. McGinnis (1980, Academic Press).

The fungi included here are considered in three groups for convenience of discussion: (1) Zygomycetous fungi; (2) Hyaline hyphomycetes and (3) Dematiaceous hyphomycetes. In general, the zygomycetous fungi have broad, sparsely septate mycelium; the hyaline hyphomycetes have septate mycelium without discernible melanin pigment in their cell walls either in culture or in tissue section; and the dematiaceous hyphomycetes are those that generally have discernible melanin in their cell wall in culture and usually in tissue section when the organisms are involved in an infectious process.

Most of the photographs in the section were taken by Janet Gallup. I am deeply indebted to her for permission to use these fine pictures. Other photographs are from a variety of sources.

Group 1: Zygomycetous Fungi

Mucor Micheli ex Saint-Amans 1821

Rapidly growing fungus that fills the Petri dish within a few days. Turf short and floccose, velvety or elongate, and "cotton-candy"–like. Light mouse gray to dull yellow, with branching aerial and substrate mycelium. Hyphae are broad, sparsely septate, but producing complete septa on aging.

Sporangiophores are numerous, highly branched, irregular in length with a columella. Sporangia walls usually evanescent. Sporangia are 60 to 300 μ in diameter and filled with sporangiospores. Fragments of the wall are retained (collarettes) at base of columellae. The columella is an extension of the sporangiophore into the sporangium and is seen only when the sporangial walls are broken as spores are dispersed (Figs. A–1, A–2).

Gemmae (chlamyoconidia) are thick- to thin-walled units that are intercalary and terminal on mycelium, have a variety of forms, and are not found in all species. They are smooth and colorless and are sometimes noted to become yeastlike and to show single

741

Figure A–1. Floccose turf of a *Mucor* sp. filling a tall Petri dish. Sporangia are visible on right side.

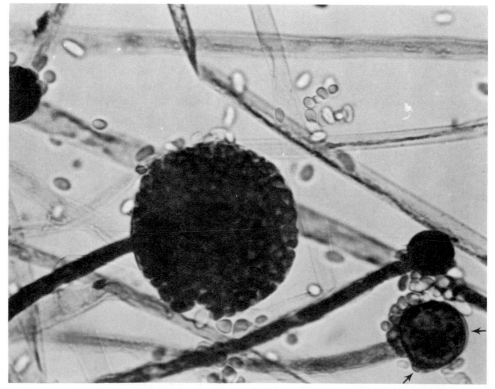

Figure A–2. *Mucor* sp. Mature sporangium deliquescing and releasing sporangiospores. Note columella and collarette of emptied sporangium *(arrows).*

and multiple budding. Some species produce gemmae only under particular environmental conditions.

Rhizopus Ehrenberg ex Corda 1838

Mycelium of two types: one submerged in substrate with an appearance of rootlike branchings (rhizoids); the other aerial with arching filaments (stolons) (Fig. A–3). The sporangiophores arise from nodes on the stolons above the point at which the rhizoids enter the substrate. The sporangiophores occur singly or in groups of two, three, or more. The top of the sporangiophore gradually enlarges to form a sporangium. The columella arises above the juncture of the sporangiophore filament and the sporangium (see Fig. A–4). The sporangium thus looks like a mushroom cap. The sporangia are white at first, becoming bluish-black at maturity. Walls are thin, delicate (not cuticularized), often encrusted with crystals, and entirely diffluent, leaving no collarette. Columella is oval, pyriform, and easily collapsed and broken off; looks like the pileus of a mushroom. Spores round, oval, angular, cuticularized, and smooth or striate. Some species form zygospores. Gemmae (chlamydoconidia) formed in some species.

Syncephalastrum Schröter 1886

Rapidly growing mycelia are feltlike (Fig. A–5), similar to those of *Mucor*. Mycelia are arranged sympodially with many branches; the side branches are often curved toward the axis of the main branch. Cross walls are sparse in young cultures and appear with age. Sporangiophores arise irregularly in turf and develop a globose to ovoid vesicle at the apex. This bears many fingerlike merosporangia radially (Fig. A–6). The merosporangia have a complete wall that encloses many spores stacked like marbles in a tube. There is a cross wall separating the merosporangium from the vesicle, and there is no basal cell. At first glance the sporangiophore and merosporangia may be mistaken for those of *Aspergillus* sp.

Absidia van Tieghen 1876

The many-branched, arching stolons produced by *Absidia* sp. are similar to those of

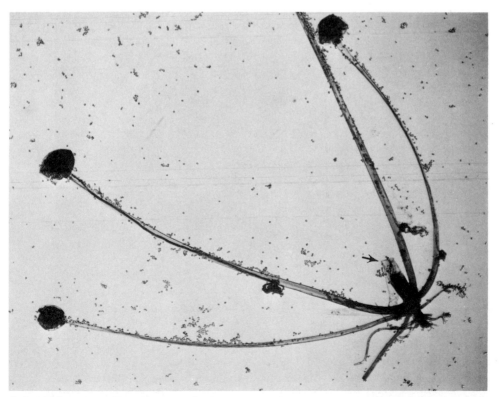

Figure A–3. *Rhizopus* sp. Several sporangiophores arising from a node above the rhizoids. Note broken stolon *(arrow)*.

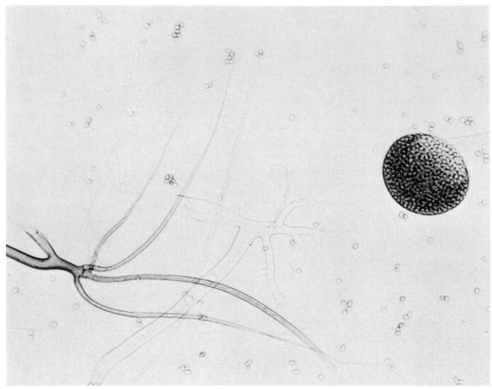

Figure A–4. *Rhizopus* sp. Sporangium filled with spores and elongate branching rhizoids. ×1000.

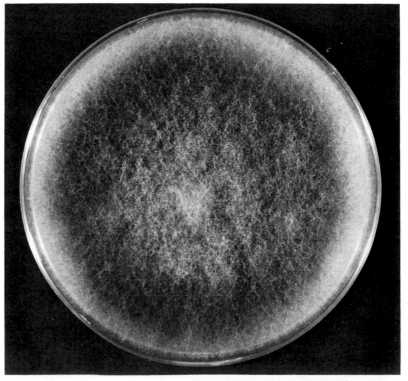

Figure A–5. *Syncephalastrum* sp. Turf is white in the periphery, becoming gray in center, about 6 mm high. Pseudo-holdfasts may be seen on lid of Petri dish.

Figure A–6. *Syncephalastrum* sp. Merosporangia radiating from vesicle. One has become detached *(arrow)*. Note stack of sporangiospores within the merosporangium. ×800.

Rhizopus sp. They form rhizoids at the nodes, and the rhizoids penetrate the substrate. Sporangiophores occur at the curve of the stolons (internodal), in contrast to those of *Rhizopus* sp., which occur at the nodes (Fig. A–7). The sporangiophores gradually expand, forming an apophysis topped by a columella. Membrane of sporangium is thin (not cuticularized), not encrusted, and is diffluent, leaving a short collarette at the juncture of the columella and apophysis. The columellae are hemispheric or elongate, sometimes topped by a long spine. They are cuticularized and often have a definite color. A cross wall occurs at a distance below the sporangium on the sporangiophore.

Circinella van Tieghem and Le Monnier 1873

Mycelium strongly branched; and at first coencytic, later becoming septate. Sporangiophores erect and branching sympodially. The sporangiophore tip grows indefinitely and does not end in a sporangium. Sporangia arise as lateral branches either singly or as whorls. They curve downward and inward toward the main axis (adaxial). The sporangium is many-spored, spherical, encrusted with calcium oxalate crystals, and nondiffluent but breaking into an irregular collarette. Columella is large, spherical, and concrescent at base (Fig. A–8).

Cunninghamella Matruchot 1903

White, floccose mycelium is 3 to 6μ in diameter and coencytic when young, later becoming septate. Sometimes primitive rhizoids are formed in substrate. Sporangiophores are straight-branched, ending in a swollen spherical head covered with spinelike projections (Fig. A–9). These projections are the point of attachment of the sporangiola. The sporangiola walls are often encrusted with crystals. Globose chlamydoconidia are often seen.

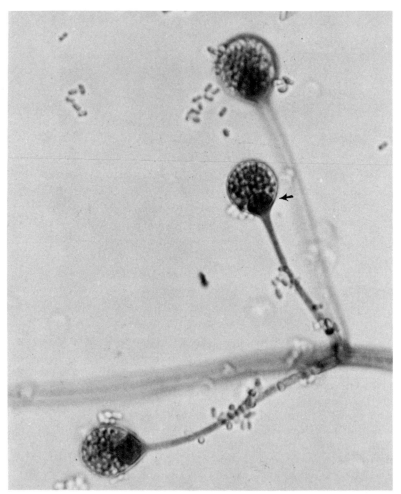

Figure A–7. *Absidia* sp. Several sporangia arising internodally on stolon. Note junctures of hemispheric columella and apophysis *(arrow).* The sporangium looks like a mushroom cap. ×800.

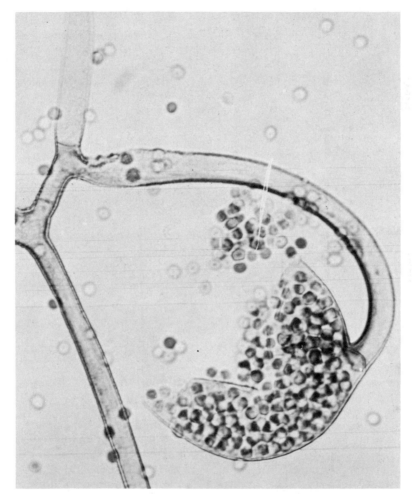

Figure A–8. *Circinella* sp. Broken sporangium curved inward toward main axis. × 1200.

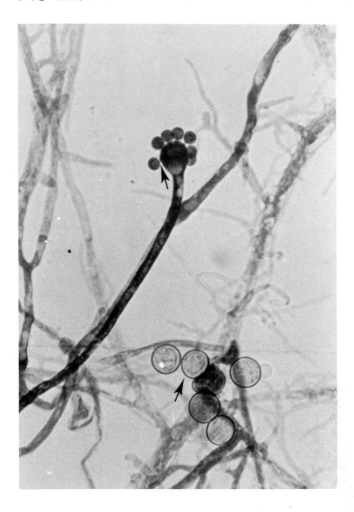

Figure A–9. *Cunninghamella* sp. Sporangiola atop a sporangiophore. The sporangiola are attached by a short stipe *(upper arrow)* or denticle. Chlamydoconidia (gemmae) are evident in lower field *(lower arrow).* × 1000.

Group 2: Hyaline Hyphomycetes

Aspergillus Micheli ex Link 1821

Mycelium grows rapidly but not indefinitely. Conidiophores arise from a foot cell and terminate in a vesicle. The vesicle produces phialides in one or two series, giving rise to unicellular conidia (ameroconidia) that are arranged basocatenulately (in a chain, with the youngest conidia at the base or proximal end where they arise from the phialide) (Fig. A–10). The conidia are colorless (hyaline) or darkly pigmented (phaeoconidia) in colors of yellow, green, blue-green, gray, black, tan, brown, and so forth. Teleomorph states are various, including *Eurotium, Dichlaena, Emericella,* and *Sartorya.*

Penicillium Link ex Gray 1821

Mycelium white, septate, and multibranched. Growth is rapid to a finite colony. Conidiophores arise in various forms, producing phialides singly or in groups or from branched metulae, giving a brushlike appearance (Fig. A–11). Conidia single celled (ameroconidia) and in chains (basocatenulate) with youngest at base. Conidia are colorless or have dark green and blue-green pigmentation. Sometimes conidiophores aggregated into stalks (synnemata) called coremia. Various teleomorph states, such as *Eupenicillium, Talaromyces, Hamigera, Penicilliopsis,* and *Trichocoma.*

Paecilomyces Bainier 1907

Similar to *Penicillium* sp., but the latter are always green or blue-green. *Paecilomyces* are gold, green gold, lilac, or tan, but never blue or green. The phialides are single or verticillate on well-developed conidiophores or directly on vegetative hyphae (Fig. A–12).

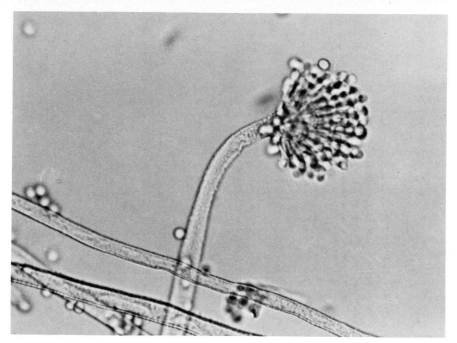

Figure A–10. *Aspergillus* sp. Phialides in two series arise from a terminal vesicle. ×600.

Figure A–11. Brushlike conidiophores showing primary branches and metulae and ending in phialides that bear chains of conidia (basocatenulate). ×800.

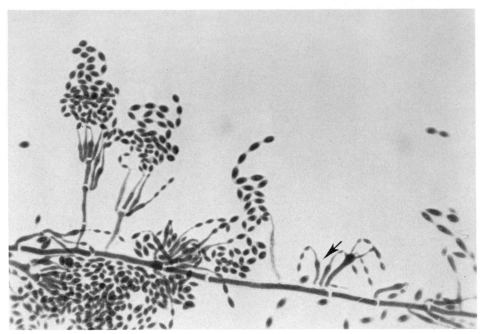

Figure A–12. *Paecilomyces* sp. Note well-developed conidiophores (*left*) with phialides, which are the long, tapering tubes developing on the terminus of conidiophore. They also arise directly from main hypha (*arrow*). ×1200.

Figure A–13. *Gliocladium* sp. Metulae arise from primary branch of conidiophore *(arrow).* The metulae are topped by elongate phialides with massed conidia in slimy balls. ×800.

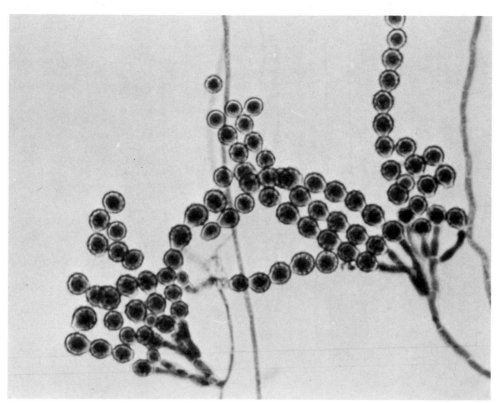

Figure A–14. *Scopulariopsis* sp. Annellophores variously arranged, with conidia arising from top of annelide; the youngest conidium is at the bottom (basocatenulate). ×1000.

The phialides end in a long tapered tube and bear long chains of lemon-shaped or nearly cylindrical conidia. Chlamydoconidia and aleurioconidia are also produced. Teleomorph states various: *Byssochlamys*, *Talaromyces*, and *Thermoascus*.

Gliocladium Corda 1840

Colonies broadly spreading, floccose, and white to pale cream, pink, salmon, green, or olive. Conidiophores arise irregularly, consisting of three parts: primary branches, metulae, and phialides. Single conidia (ameroconidia) held together in slimy ball (gloiospora) (Fig. A–13); it is colorless (hyaline) but may have color in mass. Some species have a teleomorph state, such as *Nectria*.

Scopulariopsis Bainier 1907

Turf is white at first, thin or floccose, and then gray, tan, beige, or deep ivory in color, often resembling *Microsporum gypseum*. The conidial structures are penicillate. The conidiophore is well developed, or the conidia are produced by annellides singly or in groups

directly from the mycelium (Fig. A–14). The ameroconidia are thick walled, roughened, and in chains with the youngest at the base (basocatenulate). Annellides proliferate percurrently during the production of conidia and so become annellated (ringed) (see Chapter 6). Teleomorph states take various forms, including *Microascus* and *Chaetomium*.

Trichoderma Persoon ex Gray 1821

Colonies spread rapidly and are thin to floccose and white at first, becoming yellowish green to deep green often only in small areas. These areas are the site of conidiation. The conidia are unicellular and hyaline and are produced in a slimy ball (gloiospora) from ovate to flask-shaped phialides (Fig. A–15). These are borne singly or in groups. Teleomorph states include *Hypocrea* or *Podostroma*.

Verticillium Nees ex Stendel 1821

Rapidly growing, thin, velvety to floccose, and white at first, becoming powdery and shades of rose, red, pink, green, or yellow. Conidiophores erect, septate, and branched.

Figure A–15. *Trichoderma* sp. Phialides in groups topped by mass of conidia *(arrow).* ×550.

Branches of first order whorled, opposite or alternate; secondary branches whorled, dichotomous, or trichotomous; third order same, ending in verticillate phialides with distinctly pointed apices. The ameroconidia are produced in balls (gloiosporae), are hyaline, grouped at the end of the phialides (Fig. A–16). Teleomorph states various, including *Nectria, Cordyceps, Torrubiella, Ephemeroascus, Hypocrea,* and so forth. Other anamorph states: *Humicola, Volutella, Mycogone,* and so forth.

Fusarium Link ex Gray 1821

Colony fluffy to cottony owing to extensive mycelium. Sometimes diffusible pigment produced on reverse. Conidiophores single or grouped into sporodochia (a compact mass of interwoven conidiophores). In colors of red, pink, purple, green, and various other shades. Conidia produced singly or in conidia balls (gloiosporae), are hyaline and unicellular or transversely septate (phragmoconidia) (Fig. A–17). Microconidia are one-celled and often numerous and in chains or balls. Macroconidia are elongate and cylindrical but more often crescent or falciform. Intercalary or terminal chlamydoconidia also produced. Teleomorphs: *Gibberella, Nectria,* and so forth. (See page 688.)

Beauveria Vuillemin 1912

This is moderately fast-growing, white to tan colony that becomes powdery in time. The colony resembles *Histoplasma* or a dermatophyte. The conidiogenous cell is inflated at the base and sympodially proliferates in a zig-zag (rachiform or geniculate) fashion (Fig. A–18). The conidiogenous cells are often aggregated into sporodochia or synnemata. *Beauveria* sp. are pathogens of insects.

Botrytis Micheli ex Saint-Amans 1821

A large genus with numerous forms that have little in common. Turf moderately fast growing, white or gray. Conidiophores simple or with many branches. Branches thin or thick, and narrowing to a point, truncate (see Fig. A–19) or with swollen warts on tips. The conidiogenous cell swells to form an ampulla that bears many simultaneously produced conidia (arrow). Teleomorph as in *Sclerotinia,* and so forth.

Trichothecium Link ex Gray 1821

Colony grows moderately fast; white, thin, later becoming rosy, pink, fleshy or orange-pink. Conidiophores erect, unbranched, producing conidia at apex in groups. Conidia are

Figure A–16. *Verticillium* sp. Conidiophore *(double arrows)* branches several times, ending in verticillately arranged phialides *(single arrow).* ×1000.

Figure A–17. *Fusarium* sp. Phialides *(arrows)* give rise to macroconidia. Microconidia when present are also produced from phialides. ×1200.

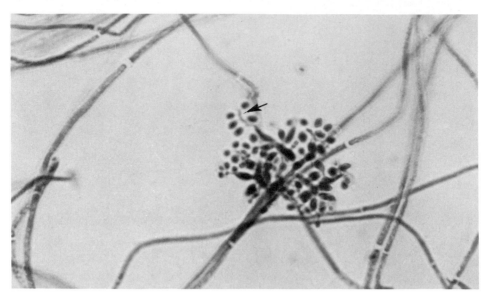

Figure A–18. *Beauveria* sp. Group of conidiogenous cells that produce zigzagging rachis *(arrow)* growth that is sympodial in development. A conidium is produced first on one side, then on the other. ×1000.

Figure A–19. *Botrytis* sp. Thin, thick, and truncate irregular branching of the conidiophore. These end in swollen ampulliform conidiogenous cells, which bear numerous conidia *(right arrow),* or they narrow to a point to form a single conidia *(left arrow).* ×800.

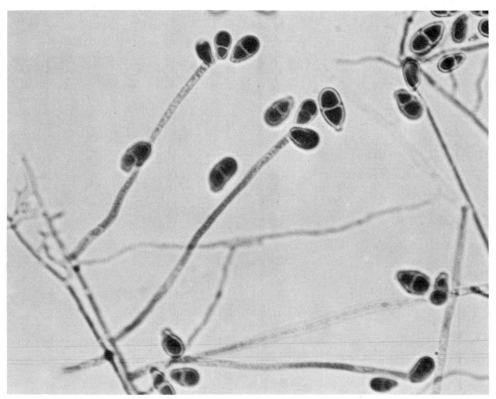

Figure A–20. *Trichothecium* sp. Typical microscopic view of alternating pyriform two-celled conidia (didymoconidia). The end conidium is the oldest (holoblastic), the penultimate (enteroblastic) is next, and so forth, using up the conidiogenous cell (retrogressive development). × 1000.

terminal single, two-celled, retrogressively produced (one after another) and forming a head or chain (Fig. A–20). They are pyriform, two-celled, with the apical cell larger. Teleomorph state: *Hypomyces.*

Sepedonium Link ex Greville 1824

This is a moderately fast-growing colony, usually white to golden yellow. The hyphae are multibranched, and conidiogenous cells are nonspecialized, resembling short branches of vegetative mycelium. Large, warty-covered macroconidia and small, thin-walled, eggshaped microconidia are produced (Fig. A–21). Microconidia and macroconidia are aleuric (requiring fracture or dissolution of the supporting cell for release). Microconidia also produced from phialides. The microscopic morphology resembles that of *Histoplasma. H. capsulatum* was once included in this genus. The teleomorph state includes *Hypomyces, Apiocrea, Thielavia, Corynascus* and so forth.

Chrysosporium Corda 1833

This is the anamorph state of several teleomorph genera, such as *Ctenomyces, Arthroder-* *ma,* and *Thielavia.* It is a moderately fast-growing fungus, producing a white to tan to beige colony, often powdery or granular. It closely resembles the dermatophytes *Histoplasma* and *Blastomyces.* Single-celled (amero) hyaline conidia are produced directly on vegetative hyphae by a nonspecialized conidiogenous cell (Fig. A–22). They are aleuric and require fracture of the supportive cell for release. Some are keratinophilic, thermophilic, or cellulolytic.

Acremonium Link ex Fries 1821

This genus includes the organism formerly called *Cephalosporium.* Colony is fast growing; in shades of white-gray, rose, or brick. Initial isolants often budding and yeastlike. The conidia are single-celled (amero) basocantenulate (in chains with the youngest at bottom) or gloiosporae (in a conidial mass) from short, unbranched, simple, tapered phialides (Fig. A–23). Teleomorphs: *Emericellopsis, Nectria, Wallrothiella, Mycocitrus, Ceratocystis, Hypocrea,* and so forth. Other conidial (anamorph) states include, among others, *Volutella.* (Some teleomorph fungi have several anamorph states, often in the same colony. See page 688.)

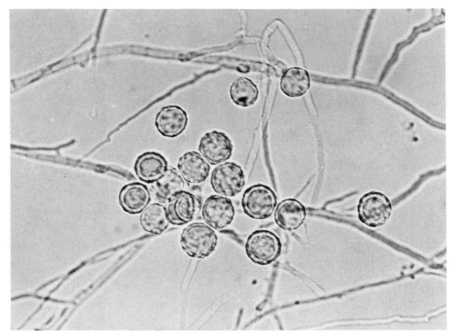

Figure A–21. *Sepedonium* sp. Warty macroconidia arising from nonspecialized short branches of vegetative mycelium. ×1000.

Epicoccum Link ex Stendel 1821

The turf of this fungus is yellow to orange because of pigment in the mycelium. Some species elaborate a deep purple-red diffusible pigment observable on the reverse. The colony develops dark areas as conidia mature. The conidiogenous cells are grouped in aggregates (sporodochia) (Fig. A–24). Conidiophores are short, dark, abstricted on all sides (cut off from other cells), and produce aleuric conidia. The conidia are dark (phaeo) and

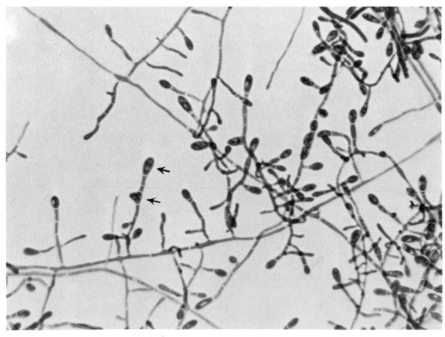

Figure A–22. *Chrysosporium* sp. Numerous laterally developing conidia *(arrows)* in conjunction with randomly dispersed arthroconidia can be seen in this microscopic view. ×650.

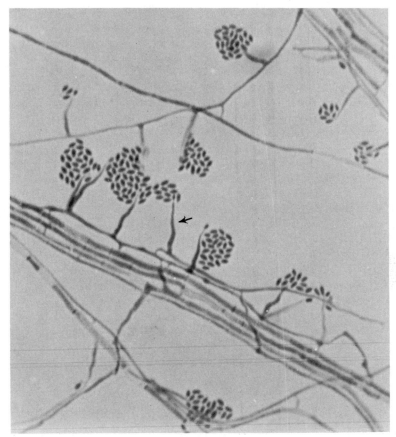

Figure A–23. *Acremonium* sp. Tapered conidiogenous cell (a phialide) giving rise to single-cell conidia. These accumulate as balls or heads (cephalosporium) and may also occur as chains. ×500.

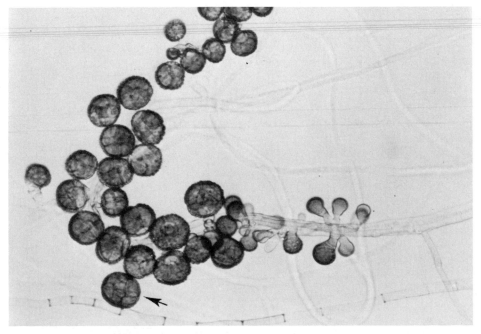

Figure A–24. *Epicoccum* sp. Stages in development of conidia from the young one-celled conidium to mature multiseptate dictyoconidium (muriform) *(arrow)*. ×1000.

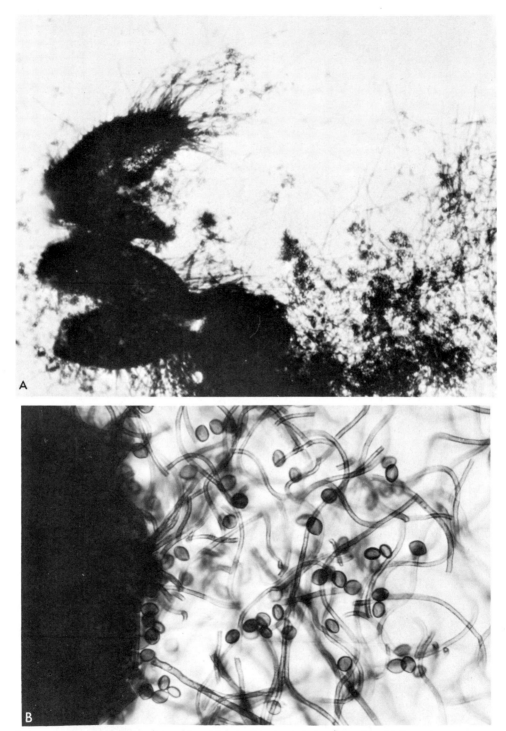

Figure A–25. *A, Chaetomium* sp. Several perithecia showing bristles (setae) and released ascospores. ×300. *B,* Dark ascospores among the bristles surrounding the ascoma. ×1200.

multiseptate both longitudinally and transversely netlike (dictyoconidium). Coelomycete state: *Phoma.*

Chaetomium Kunze ex Fries 1829

This is a common homothallic ascomycete that produces a dirty white mycelium at first. Later it develops dark areas where the peri-

thecia are produced. The perithecium has setae of various sizes and shapes on the body itself and around an opening at the top (the ostiole). Within the perithecium, evanescent (autolysing-disappearing) asci are formed and, within these, ascospores (Fig. A–25). The ascospores are dark and lemon-shaped (Fig. A–25). *Chaetomium* sp. may produce a variety of conidial (anamorph) states.

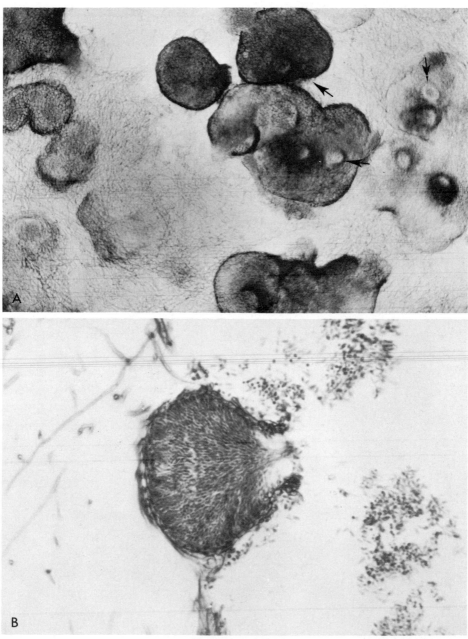

Figure A–26. *A, Phoma* sp. View of colony from above showing some mature pycnidia. This species has the appearance of a mountain range; the ostioles at top resemble craters of volcanos (arrows). At maturity the conidia ooze from the top as a mass embedded in sticky material. ×350. *B,* Cross section of a pycnidium showing masses of conidia within. ×650.

Phoma Saccardo 1880 nom. cons.

Turf is light gray to soft brown. Mycelium abundant in some species, usually submerged in others. Fruiting body is a pycnidium: a tough leathery to membranous structure that is globose to lens-shaped, with a small papilla at top. Conidia are produced in abundance within the pycnidia on threadlike conidiophores. At maturity conidia ooze out of the volcano-shaped pycnidia (Fig. A–26). A common plant pathogen.

Group 3: Dematiaceous (Phaeo) Hyphomycetes

Cephalotrichum Link ex Gray 1821

The species included in this genus were formerly called *Doratomyces* and *Stysanus*. The hyphae are dark. The conidiophores are dark, single, or compacted into synnemata with dense heads of conidiogenous cells at the top. There is a long stalk, thus giving it a Douglas fir tree–like appearance. The conidia are single-celled (amero-), darkly pigmented (phaeo), and produced as anneloconidia on annellophores in chains, with the youngest at bottom (basocatenulate). The compacted synnematous conidiophores form stalks that are 400 μ high and 30 μ wide (Fig. A–27). The stalk is black. Other conidial (anamorph) states: *Scopulariopsis*, and so forth.

Curvularia Boedijn 1933

Colony is fast-growing, floccose, brown with black reverse. Conidiophores simple, bearing conidia apically on new sympodial growth (rachiform or raduliform). The conidia are transversely septate (phragmo-) and cylindrical or slightly curved, with one of the central cells being larger and darker (Fig. A–28). In some species the conidiophores

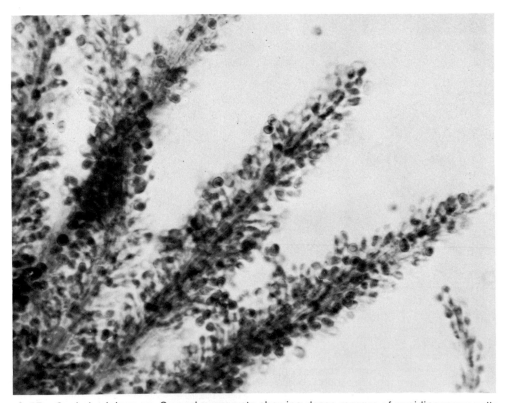

Figure A–27. *Cephalotrichum* sp. Several synnemata showing dense masses of conidiogenous cells. ×400.

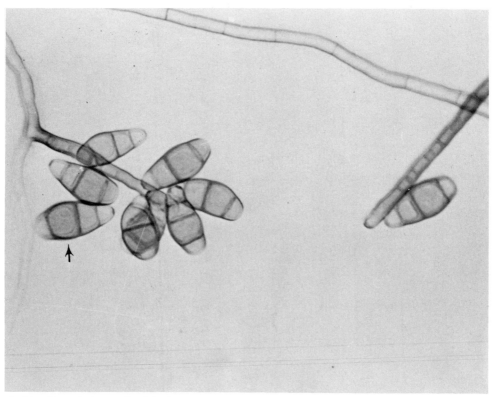

Figure A–28. *Curvularia* sp. Conidia (poroconidia) arise from a pore in the sympodially elongating conidiophore similar to *Drechslera*. One of the central cells is darker and larger *(arrow)* and some conidia are slightly curved. ×800.

arise from a stroma. Teleomorph state: *Cochliobolus*.

Drechslera Ito 1930

Dematiaceous hyphae; black reverse. Colony floccose to velvety. Conidiophores brown, mostly simple, producing conidia at apex of sympodially elongated conidiogenous cell. The conidia are produced from a pore at the apex, then new growth occurs at a point below apex. A conidia (poroconidia) is produced at the new apex, and new growth occurs just below this point (Fig. A–29). This gives a zigzag (rachiform) appearance. In some species this zigzag is replaced by clavate or an inflated area at point of new growth (raduliform). The conidia are transversely (phragmo-) septate. Teleomorphs: *Pyrenophora, Cochliobolus,* and so forth.

Helminthosporium Link ex Fries 1821 nom. cons.

Floccose dematiaceous hyphae on obverse; reverse black. Conidiophores erect, simple,

determinate (of a definite length, i.e., not elongating), often in clusters. Conidia produced from pores on the conidiophore (Fig. A–30). The pores and poroconidia often appear in whorls. The conidiophore axis may be swollen at the points of conidia production or may curve slightly (raduliform). Conidiophores often arise from a stroma. The conidia are transversely (phragmo-) septate. Teleomorph as *Pseudocochliobolus*, and so forth.

Alternaria Nees ex Wallroth 1833 nom. cons.

Floccose dematiaceous hyphae on obverse; reverse black. Conidiophore simple, sometimes branched, short or elongate, bearing a chain of conidia. The conidia are transversely or transversely and longitudinally (muriform) septate. This is dictyo- and phragmoseptation of conidia. The conidia form elongate chains with the youngest at top (blastocatenulate) (Fig. A–31). Conidiophores extend sympodially and often form zigzag structures (rachiform). Some conidia have elongate appendages (beaks) on them. Some species form synnemata with the conidio-

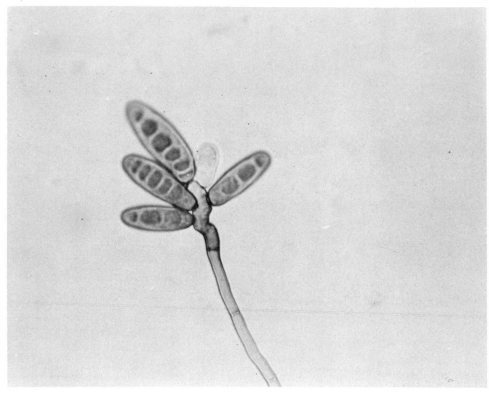

Figure A–29. *Drechslera* sp. Poroconidia arising from a zigzag sympodially elongating conidiophore. ×800.

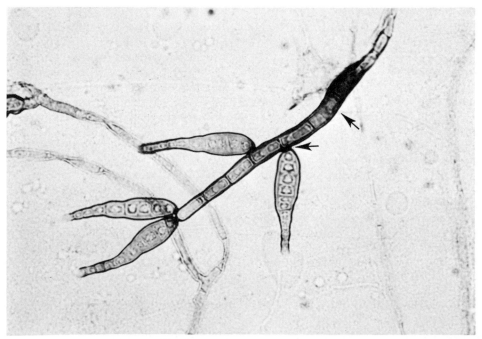

Figure A–30. *Helminthosporium* sp. Simple determinate conidiophore giving rise to poroconidia from a pore along axis. Note prominent pores *(arrows)*. These organisms are rarely encountered in the clinical laboratory. Most records of previous isolation were probably *Drechslera* sp. × 1000.

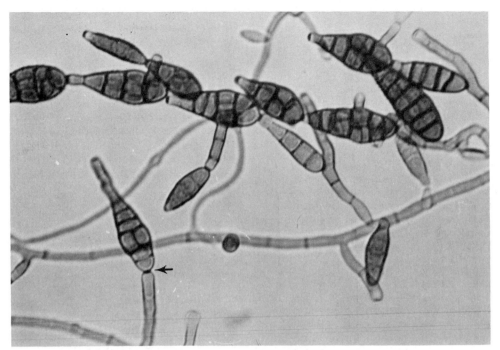

Figure A–31. Chains of dictyoconidia (also called muriform) of *Alternaria* sp. One conidium is shown arising from a pore at the apex of a conidiophore *(arrow)*. The long tube that can be seen at its apex will develop either into an appendage (a hypha-like structure) or into more conidia forming a chain of conidia. ×900.

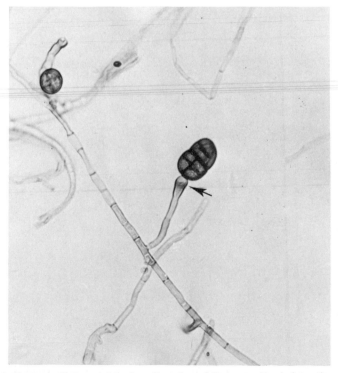

Figure A–32. *Stemphylium* sp. Dictyoseptate (muriform) conidia borne singly from the swollen end *(arrow)* of a percurrent conidiophore. ×900.

phores aggregated. Teleomorphs: *Pleospora, Clathrospora (Leptosphaeria);* sclerotial state: *Sclerotium.*

Stemphylium Wallroth 1833

Hyphae floccose dematiaceous; reverse black. Conidiophores dark, simple, few branching, with darker terminal swelling bearing a single terminal conidium or succession of conidia on new growing tip (percurrent) (Fig. A–32). Conidiophores as they become older will show a series of collars (annellides). Conidia septate transversely and longitudinally (dictyo-). Conidiophores unspecialized and appearing as short or long branches of vegetative hyphae. The teleomorph state is *Pleospora.*

Ulocladium Preuss 1851

Floccose dematiaceous hyphae on obverse; reverse black. Conidiophores simple with dark terminal swellings, unspecialized, producing conidia as a series from sympodially extending phore that is straight, curved, or bent and may have swelling at the point of conidia production (Fig. A–33). Often produces zigzag (rachiform) appearance. Teleomorph state: *Lasiobotrys.* Conidia are muriform, septate, netlike (dictyo-).

Pithomyces Berkeley et Broome 1873

Common on litter and soil. Floccose dematiaceous hyphae; black on reverse. Conidiophores unspecialized, with short, simple, peglike branches of vegetative hyphae (Fig. A–34). Conidia are single, aleuric (require fracture for release), transversely (phragmo-) or transversely and longitudinally (dictyo-) septate, broadly elliptical, pyriform, oblong, and commonly echinulate or verrucose. *P. chartarum* is involved with facial eczema of sheep.

Cladosporium Link ex Gray 1821

Floccose, dematiaceous hyphae on obverse; reverse black. Many common species are velvety or suedelike in primary culture. Conidiophores tall, branched variously near apex; branching and conidiation change axis and

Figure A–33. *Ulocladium* sp. Several conidia are produced on sympodially extending simple conidiophores. The conidia are smooth in some species, verrucose or echinulate in others.

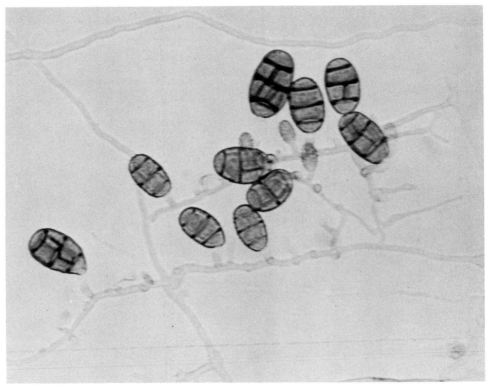

Figure A–34. *Pithomyces* sp. Multiseptate conidia arising from peglike conidiophores. ×800.

often have swellings (raduliform). The branches and conidia continue to extend (Fig. A–35), producing irregular chains with the youngest cell on top (blastocatenulate). Attachment scars are prominent, often with evident hilum (short projection of point of attachment). Includes species of *Hormodendrum, Heterosporium,* and others. Teleomorph states: *Mycosphaerella, Venturia, Apiosporina, Zopfia.*

Nigrospora Zimmerman 1902

Hyphae white at first and floccose, later dark. Reverse becomes dark. Conidiophores simple or rarely with short branches producing conidiogenous cells that are swollen and then narrow at point of conidial attachment (Fig. A–36). Conidia single-celled (amero-), aleuric (require fracture for release), black, solitary, subglobose, smooth. Teleomorph: *Apiospora.*

Humicola Traaen 1914

Plain, off-white creeping mycelium, which becomes fluffy to thin, gray or deep brown. The characteristic conidiophores and conidia are single-celled, large (6 to 16 μ), globose or pyriform, thick-walled, dark conidia borne on simple or branched nonspecialized conidiogenous cells (Fig. A–37). These conidia are aleuric (require fracture for release). The various species of *Humicola* produce, in addition, a variety of conidial states, such as *Acremonium* (from phialides and often in chains), *Chalara, Idriella, Scolecobasidium, Trichoconis, Phaeoisaria, Microdochium, Chloridium,* and *Verticillium.* Teleomorph as *Ceratocystis* or *Chaetormium.* Many species are thermophilic (*Thermomyces*).

Aureobasidium Vrala et Boyer 1891

Initial isolants often as a white to pink yeast turning black, wrinkled, with a fringe of submerged mycelium in time. Hyphae not extensive, becoming multiseptate, with thick walls containing pigment (Fig. A–38). Little wartlike pegs are formed and give rise to hyaline ameroconidia. These form from several places at once on the conidiogenous cell. This is referred to as the *Aureobasidium* anamorph. The conidia then may bud to form chains and darken. Several conidia may be seen aggregating in a mass (gloiospora) around the conidiogenous cell. Hyaline hyphae also produce hyaline one-celled conidia

Text continued on page 770

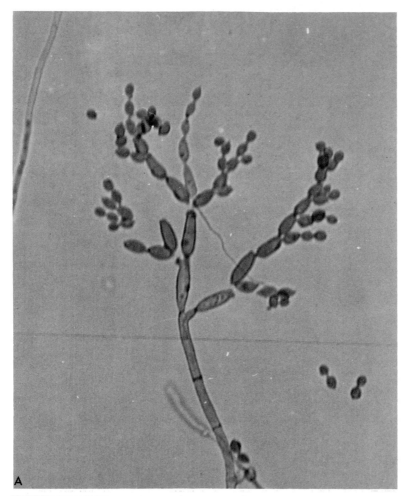

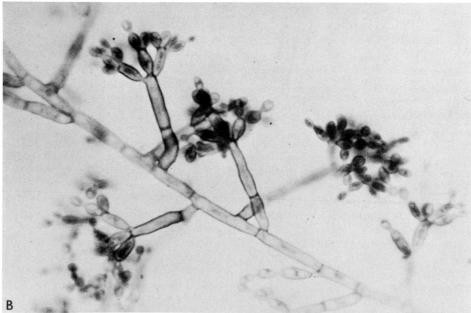

Figure A–35. *A, Cladosporium* sp. Simple erect conidiophores bearing branched (acropetal) chains of blastoconidia. These conidia have a dark hilum observable in this preparation. ×600. *B, Cladosporium* type conidiation in *Fonsecaeae pedrosoii* showing much more compacted fruiting structures. ×900.

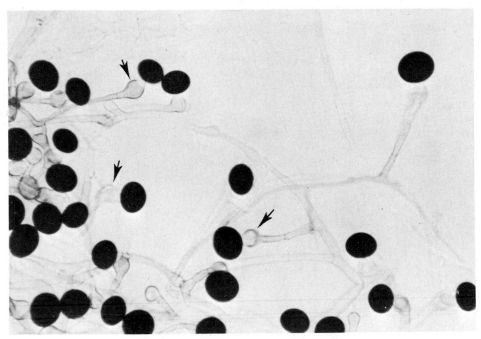

Figure A–36. *Nigrospora* sp. Single-cell black conidia produced from expanded then narrowed conidiogenous cells *(arrows).* ×800.

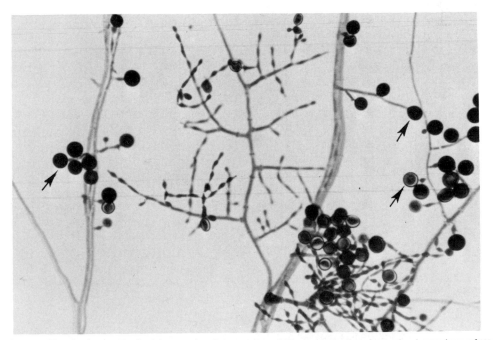

Figure A–37. *Humicola* sp. Typical large aleuric amerooonidia produced on irregular branches of vegetative hyphae *(arrow).* Also evident are other types of conidiation, including chains of conidia. Many conidiation types can be seen in these species. ×500.

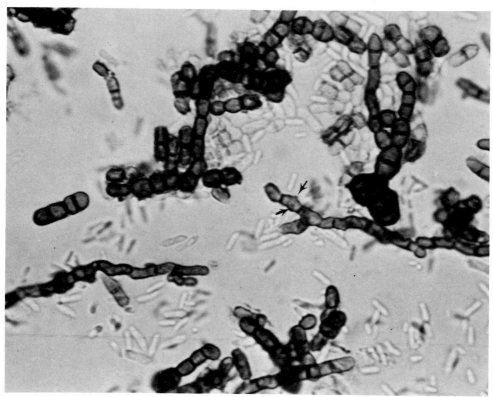

Figure A–38. *Aureobasidium* sp. Much septate mycelium giving rise to many hyaline ameroconidia. These conidia then bud. Note several peglike denticles *(arrows)* on a single conidiogenous cell that gives rise to conidia. ×800.

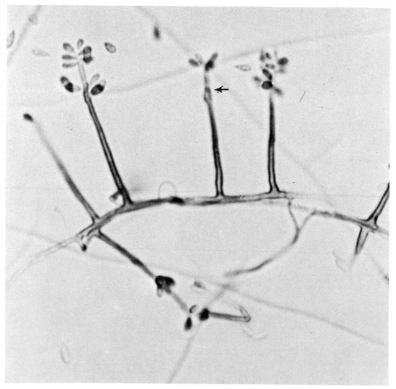

Figure A–39. *Rhinocladiella* sp. Short, simple conidiophores that extend sympodially bear ameroconidia along the side of the growing tip. When released, these leave a series of small, wartlike scars *(arrow).* ×500.

Figure A–40. *Stachybotrys* sp. *A,* The conidiophores on the left are young and hyaline. Those on the right are older and darkening. Elongate or elliptical, dark ameroconidia are evident and are produced from phialides. × 700. *B. Stachybotrys* sp. with basocatenulate chains. These species were formerly included in *Memnoniella.* Hohnel. ×800.

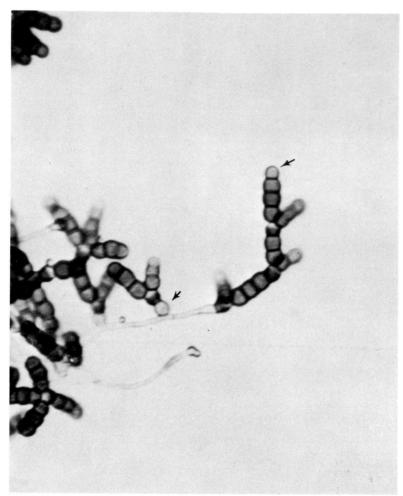

Figure A–41. *Torula* sp. Conidia are produced directly on vegetative hyphae or on short, sometimes bulbous conidiophores *(left arrow).* Conidia sometimes form blastocatenulate chains with youngest at top *(right arrow).*

synchronously. Other hyaline hyphae give rise to conidia from a common point non-synchronously. All these hyaline conidia may then bud (blastoconidia) and grow as a non-pigmented yeast resembling *Candida.* Colonies may also form dematiaceous arthroconidia, which is referred to as the *Scytalidium* anamorph. All of the types and morphologic forms are found in isolants in varying quantities; the amounts can be influenced by media composition.

This is the black yeast most commonly isolated in clinical laboratories. As noted previously, the morphology is extremely variable. Some isolants are white and remain so for long periods of time, causing confusion and consternation to the mycologist. Teleomorph states: *Guignardia, Potebniamyces, Dothidea,* and so forth.

Rhinocladiella Nannfeldt 1934

Hyphae variable, white to gray to brown to black, sparse or fluffy. The conidiophores extend sympodially; conidiogenous cells produce single-celled (amero-) conidia on short denticles (raduliform) along the extending rachis. The conidia are hyaline or phaeo (Fig. A–39). Other conidial states: Racodium, *Fonsecaea,* and so forth.

Stachybotrys Corda 1837

Colonies grow rapidly, producing abundant aerial mycelia that are white at first, later becoming dark. Conidiophores are erect, simple or branched, hyaline and smooth at first, becoming pigmented and often roughed with age. There are groups of phialides at the apex which bear ameroconidia that are usually dark, elliptical, smooth or roughed and aggregate in mass (gloiosporae) (Fig. A–40,*A*). Some species produce hyaline or phaeoconidia that are basocatenulate and may form synnemata (Fig. A–40,*B*). *S. alternas* is an important agent of mycotoxicosis. Teleomorph states: *Melanopsamma* and *Chaetosphaeria*.

Torula Persoon et Gray 1821

Restricted shiny or hairy black colony. Mycelium sparse, may be whitish at first. Conidiophores are lacking or are short bulbous pegs (left arrow) (Fig. A–41). Conidia are produced holoblastically from pores on this conidiogenous cell. The terminal cell of the conidium can then act as a conidiogenous cell, producing several more conidia. In this way, multiple branched conidia are formed.

Section II CULTURE METHODS, MEDIA, STAINS, AND SEROLOGIC PROCEDURES

INTRODUCTION

As one gains experience and, hopefully, expertise in medical mycology, one develops a set of procedures for the isolation and identification of fungi, for serodiagnosis, and for histopathologic methods. These are usually derived from the work of others and are modified for the particular laboratory situation in which one is working. In this section will be listed procedures and methods that have been found useful in the author's experience. As newer methods are described, the older ones are modified or dropped, so that any procedure listed here may be altered to fit a particular need.

The procedures listed herein have been derived from a number of sources, added to, and improved over a number of years. In general they are the ones used in our hospital laboratories, and they have been developed and codified by Ms. Daila Scheffner.

Culture Methods

As has been noted previously in the text, the identification of mycelial fungi depends primarily on morphologic criteria. Morphology, color, conidia production, and so forth are all affected by the particular medium on which the organism is growing, as well as by other environmental factors. A fungus growing on blood agar or an eosin methylene blue plate would hardly be recognized as the same organism when grown on Czapek-Dox agar. Since the identification of the fungus is based on observed morphology, there have evolved standard descriptions of organisms growing on particular media. The standard most universally used in medical mycology is the morphology of the organism growing on Sabouraud's agar. Raymond Sabouraud developed a peptone and honey or sugar medium in the late 1800's and described the morphology of many species of dermatophytes. Today the medium bearing his name, Sabouraud's agar, is composed of 4 per cent glucose and neopeptone at a pH of 5.6 and is still the stand-ard one used in the medical mycology laboratory. This is not because the organisms grow best on it or produce the most conidia; it is simply because of tradition. The standard description of the dermatophytes and most of the systemic and subcutaneous infecting organisms is based on the morphology of the organism when grown on this medium. The *Aspergillus, Penicillium,* and other opportunists or contaminants are identified by their morphology when growing on Czapek-Dox or similar "deficient" media. Morphologic description of the yeasts is often based on the growth observed on cornmeal or a similar medium. Thus the medical mycology laboratory requires only three media, basically: Sabouraud's agar, Czapek-Dox agar, and cornmeal agar. This is sufficient for the identification of essentially all the mycelial fungi one will encounter.

The approach used for the identification of yeasts is entirely different from that employed for mycelial fungi. Yeasts are identified primarily in terms of physiologic reactions, viz., fermentation of a carbohydrate, assimilation of a nitrogen source, and so forth. The methods and techniques utilized are discussed in Chapter 20 and in the section that follows.

To facilitate the isolation and identification of fungi, some modifications of the above media have been made. Sabouraud's formula plus antibiotics becomes a selective medium for the isolation of dermatophytes and most systemic pathogens, and is called Mycosel or Mycobiotic agar. In our experience and that of others, however, we find that some pulmonary pathogens are more easily isolated from blood agar than from Sabouraud's agar with antibiotics. Therefore, a blood agar plate is often added to the regimen of primary media used in respiratory infections. For the isolation of some pathogens, such as *Histoplasma*, from material that is heavily contaminated, the Smith medium is recommended. It suppresses the growth of *Candida* and allows the *Histoplasma* or other hyphomycete to grow. See Chapter 16 for other laboratory methods.

AN APPROACH TO THE IDENTIFICATION OF MYCELIAL FUNGI

The culture procedures for the isolation of fungi used in our laboratory are as follows: For dermatologic specimens a wide slant (25 × 100-mm tube) containing Mycosel or Mycobiotic agar is used. Skin scrapings or hairs are furrowed into the slant, which is then incubated at room temperature for four weeks. When mycelium is visible on the slant, a transfer is made to a slide culture. This consists of 1 per cent water agar in a Petri dish to act as a moist chamber, and a bent glass rod to act as support for the slide culture. The slide culture itself has a small amount (a glob) of Sabouraud's agar (without antibiotics) on one side and cornmeal 1 per cent glucose agar on the other. Both are inoculated with the growth to be identified, and a sterile cover slip is placed on top of each of the media. The slide culture is then placed in the moist chamber, covered, and allowed to grow for ten days to two weeks. The Sabouraud's agar is for observation of the morphology of conidia and hyphal structures of the organism. The cornmeal 1 per cent glucose agar stimulates pigment production in *T. rubrum* and generally suppresses it in other dermatophyte species (see Fig. 8–42). After growth has occurred, the cover slip is removed from the slide, and by using a razor blade the glob of medium is gently moved. This is done so that the strands of mycelium adherent to the slide remain undisturbed. A drop of lactophenol cotton blue is placed on the slide where the growth occurred, and the cover slip is replaced. It can now be examined with a microscope to observe the conidia morphology, size and arrangement, and mycelial appendages.

For the culture of material containing pathogens causing subcutaneous and systemic infections, two media are recommended. The first is a Mycosel or Mycobiotic agar (or the equivalent) plate, and the second is a brain-heart infusion (BHI) sheep blood (5 per cent) agar plate. If sputum is to be cultured, a viscous blood-tinged area is selected. A KOH mount and/or PAS stain of a sample from the area is made for direct examination, and an aliquot of the specimen is streaked thoroughly on the two agar plates. These are incubated at room temperature for four to six weeks. To prevent drying, plates can be stored in a plastic bag. The plates should be checked every day for five days for the appearance of any mycelial growth. If present, it should be transferred immediately to a Sabouraud's slant. After five days, the plates are checked biweekly until discarded. After the organism has grown on Sabouraud's agar, it can be processed for identification. For identification of the Hyphomycetes, we refer the reader to the several standard manuals and keys that are noted in the individual chapters and beginning on page 741.

Mites

Anyone who has worked extensively in a mycology laboratory has had the uncomfortable experience of looking through a microscope at a culture and seeing what appear to be eight-legged monsters browsing on the fungal mycelium. These are tarsonemid mites or other acarids. They are particularly common when plant debris or other such material is brought into the laboratory for testing. There are a few precautions that, if exercised regularly, will keep them under control. Containers for housing plates, racks for tubes, and other such items must be regularly sterilized. The work areas are regularly swabbed with 3 per cent phenol. Stock cultures are stored in closed cabinets with open plates of *p*-dichlorobenzene (we use the cakes sold as urinal deodorants). A culture already contaminated can be saved by the following procedure: Two drops of Kelthane (1.8 per cent di(p-chlorophenyl) trichloromethyl carbinol) are placed on an unoccupied area or agar plate at a slant. Set aside for two or three days. Adults, eggs, and larvae can be killed in this way. Pick a few conidia or some mycelium from the plate and transfer to fresh media. Isolate this on a tray of *p*-dichlorobenzene to prevent spread until you are sure no mites are present.

AN APPROACH TO THE IDENTIFICATION OF YEASTS AND YEASTLIKE ORGANISMS

The following is a flow-sheet of the methods currently used in our laboratories for the identification of yeasts. The main sources are the methodology of L. Haley and the procedures of J. Land, B. Cooper, and D. Ahern. (See key page 777; also see Chapters 20, 21, and 22 for additional keys.)

If an isolant from clinical material appears to be a yeast, the first thing to be done is to make a direct mount (e.g., using lactophenol) and examine it. Observe for (1) budding, as in blastoconidia or fission yeasts; (2) size and general shape of cells (round, globose, subglobose, cylindric, spindle-shaped, squared) or mycelia (the size of some small yeasts approaches that of some bacteria, such as *Micrococcus*); (3) internal structures such as ascospores in *Saccharomyces* and other ascomycetous yeasts, autospores in a theca in *Prototheca,* and refractile globules and oil droplets (helpful when using Lodder's key to the yeasts; see Chapter 20); and (4) presence of a capsule (use low intensity light or India ink preparation).

The next step is to determine if the organism is *Candida albicans* using the RB germ tube test. If it is not, *all* yeasts are to be put on morphology plates (cornmeal Tween agar). Morphology of the hyphae and conidia are useful in the identification of certain *Candida, Trichosporon, Geotrichum,* and *Rhodotorula* species. Some hyphomycetes, such as *Aureobasidium, Scopulariopsis, Acremonium, Mucor, Penicillium,* and *Aspergillus,* have also been recovered in primary isolation as budding yeastlike cells. On morphology plates, most hyphomycetes will rapidly convert to identifiable mycelium.

There are many automated procedures for yeast identification that are gaining popularity in the routine diagnostic laboratory. For most clinical isolants, these procedures are quite adequate and useful. It is necessary nonetheless to be familiar with the standard techniques used in yeast identification. These techniques must be mastered by the person responsible for identification so that the accuracy and efficiency of "machine done" or "assembly line" automated identification can be monitored. Yeast morphology plates are necessary both for standard identification methods and for the automated techniques.

Procedure for Identification of Yeastlike Fungi

Candida albicans *by RB Germ Tube Technique*

Suspend a loop full of inoculum in 0.5 ml serum. Incubate at 37° C. Read in two to three hours. If germ tubes are present, organism is reported as *Candida albicans.*

Yeast But Not Candida albicans

Carry out preceding test. If germ tubes are not present, inoculate a morphology plate

(cornmeal with Tween 80). Incubate at room temperature for from 24 to 48 hours. At the same time, streak Sabouraud's glucose agar to obtain pure cultures. If mycelia *are* present on the morphology plates, proceed to:

1. Nitrate (tryptone and/or KNO_3) assimilation in YC base agar. If tryptone is not assimilated, the culture is not growing. If KNO_3 is assimilated, the culture is not usually a pathogenic yeast.
2. Carbohydrate assimilation (glucose, maltose, sucrose, lactose, galactose, trehalose, inositol, melibiose, cellibiose, raffinose, xylose, and dulcitol).
3. Fermentation or gas production (glucose, maltose, sucrose, lactose, galactose, and trehalose). Acidification in the presence of carbohydrates is not reliable in identification of yeasts; gas production is the criterion used here.

Based on the information obtained, it should be possible to identify the following organisms:

Candida albicans
Candida guilliermondii
Candida krusei
Candida parapsilosis
Candida pseudotropicalis
Candida stellatoidea
Candida tropicalis
Candida vini
Geotrichum species (see Chapter 26)
Trichosporon species (see Chapter 22)

If mycelia are *not* present, the procedure outlined above should also be followed to identify the following organisms:

Candida famata
Candida glabrata
Cryptococcus species (see Chapter 21)
Rhodotorula species (see Chapter 22)

For a more definitive identification of *Cryptococcus,* urease production and bird seed agar are useful. (See Table 20–1 and flow charts in Chapters 20 and 21.)

Haley-Standard Auxanographic Method for Identifying Yeast Species

Carbon Assimilation

1. Melt a tube of H-S yeast nitrogen base medium. Allow to cool to 47 to 55° C.
2. Make a suspension of 24- to 72-hour-old yeast* culture in 3 to 4 ml sterile distilled H_2O.

---

*Many authors suggest growing on HI agar without sugar to starve yeast of fermentable carbohydrate.

3. Pour yeast suspension into a 150 mm × 15 mm Petri plate.
4. Pour yeast nitrogen medium into the above 150 mm × 15 mm plate. Mix well and allow to harden.
5. With either dispenser or tweezers, place the following Minitek* disks: dextrose (glucose), maltose, sucrose, lactose, galactose, trehalose, inositol, melibiose, D-xylose, raffinose, arabinose and cellibiose.
6. Incubate plate at room temperature for 18 to 24 hours, then examine it for presence or absence of growth around each disk. The size of the zone of growth of the density is not significant. Note: Ignore the yellowish pigment diffused through the media if Minitek disks are used. They contain a phenol red indicator which does not indicate presence or absence of growth.

Nitrate Assimilation

1. Melt tube of H-S yeast carbon base medium. Cool to 47 to 55° C.
2. Make an aqueous suspension of yeast with a density of about 1+ with the Wickerham card.
3. Pour yeast suspension into a 150 mm × 15 mm Petri dish.
4. Pour cooled yeast carbon medium into the above plate. Mix well and allow to harden.
5. Place disks that are saturated with 1 per cent KNO_3 and 1 per cent tryptone on the hardened plate at a distance from each other.
6. Incubate at room temperature for up to 96 hours.

Check for growth in the tryptone area. If growth is absent after 72 hours, the test is invalid. If growth is seen in the tryptone area, examine for growth in the KNO_3 area. Tryptone is the control, since all yeasts assimilate it.

Fermentation (Wickerham).[†]

These tests are not needed routinely but are recommended when identification is in doubt.

---

*BBL Laboratories, Baltimore, Md.
[†]Wickerham, L. J. 1951. *Taxonomy of Yeasts.* Technical Bulletin No. 1029. U.S. Department of Agriculture, Washington, D.C. This publication is no longer available.

1. Inoculate tubes containing fermentation broth and inverted Durham tubes from sterile suspensions of the unknown yeast used for either YC or YN assimilations.
2. Incubate test at 25 to 30°C for 10 to 14 days.
3. Agitate tubes daily, taking care not to get air bubbles in Durham tubes. Examine for gas production.

Media and Reagents for Nitrogen and Carbon Assimilation

Nitrogen Assimilation Test

Sterilize 1 per cent potassium nitrate solution and pour over No. 740-E-Schleicher and Schuell disks of absorbent paper 1/2 inch in diameter and allow disks to become impregnated with the solution.

One per cent tryptone: Use in same way as potassium nitrate.

Modified Yeast Carbon Base (H–S) for Nitrogen Assimilation (YC base)

Washed agar (Noble), Difco #0142-01	20 g
Yeast carbon base, Difco #0391-15	1.17 g
Distilled water	1000 ml

Heat until dissolved, dispense in 20 ml aliquots in screw-capped tubes, and sterilize by autoclaving.

Modified Yeast Nitrogen Base (H–S) for Carbon Assimilation (YN base)

Washed agar (Noble) Difco #0142-01	20 g
Yeast nitrogen base, Difco #0335-15	0.67 g
Distilled water	1000 ml

Heat until dissolved, dispense in 20 ml aliquots in screw-capped tubes, and sterilize by autoclaving.

Carbon Sources for Assimilation Tests

Carbohydrate-impregnated disks may be prepared by soaking #740-E-Schleicher and Schuell disks as described before in 20 per cent solutions. For routine identification, the following carbohydrates are used: glucose, maltose, sucrose, lactose, galactose, trehalose, inositol, melibiose, cellibiose, raffinose, arabinose, and xylose. These can also be obtained commercially (Minitek, BBL Laboratories).

Media for Fermentation Tests

Prepare phenol red broth base (Difco #0092-01 or equivalent) 16 g in 1000 ml distilled water as per instructions on the bottle. Dispense 7 ml per tube (narrow tube) and place a Durham's tube in each tube with the open end down. Autoclave. After the medium has cooled, check to see if there are any air bubbles trapped in the Durham's tubes. If there are, turn the capped tubes upside down and tap gently until the bubble disappears.

Next add 0.5 ml of the following 20 per cent carbohydrate solutions: glucose, maltose, sucrose, lactose, galactose, and trehalose. For identification, the medium is inoculated and incubated at 25 to 30° C for up to ten days and inspected for the presence of air bubbles. Ignore the color change.

Wickerham Card

Using India ink, draw 3 or 4 black lines measuring 3.75 mm on a white card. A yeast suspension that obliterates the lines is 3+; if lines are seen as diffuse bands it is 2+; if lines are distinct but edges are slightly fuzzy the suspension is 1+. Standardized suspensions of yeasts are used for automated (API, etc.) and other procedures.

Special Media for Other Tests Used in Identifying Candida, Cryptococcus, Trichosporon, Geotrichum and Rhodotorula

Morphology Plates

Prepare cornmeal agar (Difco #0386-01-Bacto-Cornmeal Agar), 17 g per liter of distilled water as per instructions on the bottle.

After autoclaving, 3 ml of Tween 80 is added and then the medium is poured into Falcon 60 × 15 mm No. 1007 Petri plates. Inoculation of the medium is accomplished using a straight inoculating needle with a small amount of inoculum. The agar is then slightly scarified in one direction at a 90-degree angle. A cover slip is placed over a portion of the agar, and the plate is incubated at room temperature from 24 to 48 hours.

If the cornmeal with Tween 80 (CM/TW80) is prepared in butts, it can be redissolved in hot water, and placed in Petri plates when needed. *Candida albicans* chlamydoconidia can sometimes be seen in 24 to 48 hours.

Malt Extract Broth

Malt extract broth can be used for enhancement of blastoconidia production by *Trichosoporon* species.

Malt extract	0.8 g
Peptone (Difco)	1.2 g
Glucose	2.5 g
Distilled water	250 ml

Place 50 ml aliquots in cotton-stoppered flasks and autoclave at 15 p.s.i. for 15 minutes. The inoculated broth is incubated at room temperature. Check after 72 to 96 hours. If blastoconidia production is not enhanced, check every two to three days for at least two weeks.

Urease Test Media

Christensen's urea agar is used for urease production.

Peptone	0.1 g
Glucose	0.1 g
NaCl	0.5 g
KH_2PO_4	0.2 g
Agar	1.5 g
H_2O	100 ml
Phenol red	0.0012 g (0.012 per liter)

Mix ingredients and dissolve by placing flask in boiling water. Adjust to pH 6.8. Place in tube in 4.5 ml aliquots and autoclave 15 p.s.i. for 15 minutes. When these are cool, add 0.5 ml of 20 per cent Seitz filtered urea solution to each tube. Slant and allow to harden.

When using for identifying an unknown yeast, streak a slant with an aqueous suspension of the yeast isolate, then incubate at room temperature and examine daily for no longer than four days. An uninoculated tube should be opened and used as a control for pH change due to atmospheric gases in the

```
┌─────────────────────────────────────────────────────────┐
│                                                         │
│                  API WICKERHAM CARD                     │
│                                                         │
├─────────────────────────────────────────────────────────┤
├─────────────────────────────────────────────────────────┤
├─────────────────────────────────────────────────────────┤
├─────────────────────────────────────────────────────────┤
├─────────────────────────────────────────────────────────┤
└─────────────────────────────────────────────────────────┘
```

room. A positive result is a color change from pink to red.

Cryptococcus sp., *Rhodotorula* sp., *Trichosporon beigelii, T. pullulans,* and some species of *Candida* produce urease. Most other *Candida* spp., *Geotrichum* spp., *G. penicillatum,* and *G. capitatum* do not.

Bird Seed Agar for Cryptococci (Stalb)

Guizotia abyssinica seeds (niger or thistle seeds)	70.00 g
Creatinine	0.78 g
Chloramphenicol	0.05 g
Glucose	10.00 g
Agar	20.00 g
Water	1000 ml
Diphenyl	100.00 mg

Preparation

1. Suspend seed powder in 350 ml of water and autoclave 10 minutes at 121° C.

2. Filter and bring volume to 1000 ml.

3. Dissolve other ingredients except diphenyl and autoclave 121° C for 15 minutes.

4. Cool to 45° C. Add diphenyl (dissolved

A Key to Asporogenous Clinical Yeast Isolates

1. Germ tubes positive 2
   Germ tubes negative 3
2. Sucrose positive .... *Candida albicans*
   Sucrose negative .... *Candida stellatoidea*
3. Pseudohyphae absent or sparse.. 4
   Pseudohyphae well developed 8
4. Inositol positive (*Cryptococcus*) 5
   Inositol negative (*Torulopsis* now included in *Candida*) 17
5. Potassium nitrate positive.. 6
   Potassium nitrate negative.. 7
6. Maltose and sucrose positive .... *Cryptococcus albidus*
   Maltose positive, sucrose negative .... *Cryptococcus terreus*
7. Maltose, sucrose, and dulcitol positive; lactose and melibiose negative .... *Cryptococcus neoformans*
   Maltose, sucrose, lactose, melibiose and dulcitol positive............................ *Cryptococcus laurentii*
   Maltose and sucrose positive; lactose, melibiose and dulcitol negative .... *Cryptococcus uniguttalatus*
8. Arthroconidia produced (*Trichosporon*) 9
   Arthroconidia not produced (*Candida*) 10
9. Potassium nitrate negative; lactose and melibiose positive .... *Trichosporon beigelii*
   Potassium nitrate positive; lactose positive; melibiose variable .... *Trichosporon pullulans*
10. Potassium nitrate positive .... some *Candida* species
    Potassium nitrate negative.. 11
11. Lactose positive and fermented .... *Candida pseudotropicalis*
    Lactose negative 12
12. Raffinose positive and melibiose positive... *Candida guilliermondii*
    Raffinose negative 13
13. Trehalose positive... 14
    Trehalose negative 15
14. Cellibiose positive; maltose fermented .... *Candida tropicalis*
    Cellibiose negative 16
15. Glucose positive; galactose, sucrose, maltose negative .... *Candida krusei* complex
16. Maltose and sucrose fermented .... *Candida tropicalis*
    Maltose fermentation negative; sucrose fermentation variable (generally negative) .... *Candida parapsilosis*
    Maltose fermented; sucrose negative (repeat germ tube test) .... *Candida albicans*
17. Only glucose and trehalose fermented...*Candida glabrata*

in 10 ml 95 per cent alcohol). Stir and pour in plates.

Use. *Cryptococcus neoformans* forms dark brown gelatinous colonies; other yeasts are colorless or light brown.

Quick Method Bird Seed Agar[7]

Prepare 1 liter *Guizotia abyssinica* seed extract in an electric blender. Pulverize seeds (70 g) first, then add powder to 1 liter of water and boil for 25 to 30 minutes. Filter through gauze and filter paper, q.s. 1 liter. Add 15 g agar and autoclave at 110° C for 25 minutes. When cooled add 50 mg chloramphenicol dissolved in 10 ml absolute alcohol. Pour as plates.

REFERENCES

1. Ahearn, D. G. 1970. Systematics of yeasts of medical interest. Pan American Health Organization International Symposium on Mycoses, *205*:64–70.
2. Haley, L., and C. S. Callaway. 1978. *Laboratory Methods In Medical Mycology.* 4th ed. Washington, D. C., HEW, 78-8361.
3. Kendrick, B. (ed.) 1971. *Taxonomy of Fungi Imperfecti.* Toronto, University of Toronto Press.
4. Lennette, E. H., J. Spaulding, and J. P. Traunt. (eds.) 1974. *Manual of Clinical Microbiology.* 2nd ed. Washington, D. C., American Society for Microbiology.
5. Lodder, J. (ed.) 1971. *The Yeasts: A Taxonomic Study.* Amsterdam, North-Holland Publishing Company.
6. McGinnis, M. R. 1980. Laboratory Handbook of Medical Mycology. New York, Academic Press.
7. Paliwal, D. K., and H. S. Rundhawa. 1978. A rapid identification test for *Cryptococcus neoformans.* Antonie van Leeuwenhoek, *44*:243–246.

MEDIA FOR THE GENERAL MEDICAL MYCOLOGY LABORATORY

Most media for general diagnostic mycology are available commercially. The general mycology laboratory requires few types of media when compared with the microbiology laboratory. For mycology, one requires Sabouraud's agar, cornmeal agar, potato-glucose agar, blood agar, and water agar. Selective media for primary isolation include Sabouraud's agar with antibiotics and the Smith medium. In the following section, the preparation and use of these media are described, along with some media for special purposes.

Water Agar

Agar	20 g
Water	1000 ml

Preparation. Heat to dissolve and then autoclave at 121°C for 15 minutes.

Use. This is used as the base of Petri plates for slide cultures. It is a deficient medium and is sometimes useful in suppressing mycelial growth and stimulating asexual conidiation or cleistothecium production of some fungi. *Pseudallescheria boydii* often produces asci and ascospores when grown on water agar.

Sabouraud's Glucose Agar

Glucose	40 g
Peptone	10 g
Agar	20 g
Water	1000 ml

Preparation. Heat to dissolve and then autoclave at 121°C for 15 minutes. Dispense in plates or tubes.

Use. This is the most commonly used medium in medical mycology and the basis of most morphologic descriptions. It is not good for maintaining stock cultures. For this, Emmons' modification (2 per cent glucose), sugar-free media, or cornmeal agar should be used. The last, however, does not stimulate typical pigment production, and thus growth on such agars cannot be used for comparison with the standard morphologic descriptions of most species.

Sabouraud's Cycloheximide-Chloramphenicol Agar*

Glucose	10.0 g
Peptone	10.0 g
Agar	15.5 g
Cycloheximide	0.4 g†
Chloramphenicol	0.05 g‡
Water	1000 ml

---

*Formulations available as Mycobiotic agar (Difco Laboratories, Detroit, Michigan) and Mycosel (Baltimore Biological Laboratory, Inc., Baltimore, Maryland).
†Actidione (Upjohn).
‡Chloromycetin (Parke, Davis).

Preparation. Heat to dissolve and then autoclave at 121°C for 15 minutes. Dispense in tubes or plates.

Use. This preparation is the selective medium most commonly used in medical mycology. The cycloheximide inhibits the growth of most saprophytic fungi. Some pathogenic fungi are also sensitive to it to some degree. Among these are *Cryptococcus neoformans; Candida tropicalis* and a few other *Candida* species; *Trichosporon beigelii;* and, at 37°C, the yeast forms of *Blastomyces dermatitidis* and *Histoplasma capsulatum*. The medium is used only as an isolation medium at room temperature. The chloramphenicol inhibits the growth of bacteria, so that *Actinomyces, Nocardia,* and *Streptomyces* will not grow.

Cornmeal Agar and Cornmeal Tween Agar

Cornmeal	125 g
Water	3000 ml

Preparation. Heat to 60°C with stirring for one hour. Filter. Then add sufficient water to make 3 liters and add 50 g agar. Heat to dissolve, divide to manageable portions, autoclave at 121°C for 15 minutes, and dispense in tubes or plates.

Use. This is a deficient medium useful for stimulating mycelium production in *Candida* and other yeasts. For this purpose Tween 80 at the rate of 3 ml per liter is added. Without Tween it suppresses vegetative growth of many fungi and stimulates conidiation. It can also be used as a stock culture medium.

When made up with glucose (1 per cent), it is used as a stimulant for pigment production of *T. rubrum*.

Casein Medium

Prepare separately:
1. Skim milk (dehydrated
    nonfat milk)   10 g
    Water   100 ml
2. Water   100 ml
    Agar   2 g
Autoclave both at 121°C for 15 minutes.

Preparation. Cool broth to approximately 45°C, mix, and pour in small Petri dishes.

Use. *N. asteroides* does not hydrolyze casein; *N. brasiliensis* and *Streptomyces* do. The medium is also used for differentiation of some agents of mycotic mycetoma.[6]

Gelatin Test Medium

Gelatin	4.0 g
Water	1000 ml

Preparation. Dissolve and adjust to pH 7. Dispense in tubes; autoclave at 121°C for five minutes.

Use. Inoculate with small fragments of suspected organisms. *N. asteroides* grows poorly; *Streptomyces* produces poor to good growth (stringy or flaky); *N. brasiliensis* produces compact, round colonies.

Xanthine or Tyrosine Agar

Nutrient agar	23 g
Tyrosine	5 g
or	
Xanthine	4 g
Water	1000 ml

Preparation. Dissolve nutrient agar in 1000 ml of water. Add tyrosine or xanthine crystals and shake. Autoclave at 121°C for 15 minutes. Shake to disperse crystals evenly and dispense into Petri plates.

Use. For the differentiation of *Nocardia* and *Streptomyces*. See Chapters 2 and 4.

Starch Hydrolysis Test Medium

Heart infusion broth	25 g
Casitone	4 g
Yeast extract	5 g
Soluble starch	5 g
Agar	15 g
Water	1000 ml

Preparation. Heat to dissolve and adjust to pH 7. Autoclave at 121°C for 15 minutes. Dispense in tubes for *Actinomyces* or on plates for *Nocardia* and fungi.

Use. To test for starch hydrolysis. Inoculate and incubate for several days to two weeks. Add Gram's iodine to plate. If the starch has been hydrolyzed, there will be a clear halo around the colony.

Starch Production Agar

$(NH_4)_2SO_4$	1 g
$MgSO_4$	0.5 g
KH_2PO_4	1 g
Glucose	10 g
Thiamine	200 μg
Agar	25 g
Water	1000 ml

Preparation. Dissolve and adjust to pH 4.5. Autoclave at 121°C for 15 minutes. Dispense in plates.

Use. For detecting extracellular starch-like substances of the cryptococci. Inoculate plate and incubate at room temperature. After three weeks, pour Lugol's iodine over plate. If starch is present, there will be a blue halo around the colony.

Media for the Stimulation of Ascospore Formation of Perfect Yeasts and Fungi

Alphacel Agar

Alphacel*	20 g
$MgSO_4 \cdot 7H_2O$	1 g
KH_2PO_4	1 g
$NaNO_2$	1 g
Tomato paste (Hunt's)	10 g
Oatmeal, baby (Beechnut)	10 g
Agar	18 g
Water	1000 ml

Preparation. Dissolve and adjust to pH 5.6. Autoclave at 121°C for 15 minutes. Dispense in plates for molds or on slants for yeasts.

Use. For mating test cultures for ascospore formation of *Histoplasma, Blastomyces,* and other teleomorphic Ascomycetes (after K. J. Kwon-Chung, 1973).

Malt Agar for Yeast Ascospore Formation (Wickerham)

Malt extract	50 g
Agar	30 g
Water	1000 ml

---

*Nutritional Biochemical Co. Cleveland, Ohio.

Preparation. Dissolve in water, autoclave at 121°C for 15 minutes, and dispense in plates.

Use. Streak yeasts on agar slants. Incubate at room temperature for 24 to 48 hours. To observe ascospores of yeasts, make slides of growth and stain by Dohrn method.

Stain for Ascospores (Dohrn)
1. Prepare film on slide, dry, and heat fix.
2. Flood slide with carbolfuchsin; heat over flame for four minutes.
3. Cool. Wash with water; decolorize with acid alcohol (1 per cent HCl in ethanol).
4. Drip down 1 per cent nigrosin and allow to dry.

USE. Ascospores red, ascus colorless, yeast cells pale pink, background black.

Soil Extract Agar

Soil extract:		
	Garden loam	500 g
	Tap water	1200 ml

Autoclave for three hours at 121°C and filter through Watman No. 1 filter paper while still hot; q.s., 1 liter.

Soil extract	1 liter
Glucose	2 g
Yeast extract	19 mg
KH_2PO_4	0.5 g
Agar	15 g

Preparation. Adjust to pH 7.0, then autoclave at 121°C for 20 minutes. Make tube or plates.

Use. For mating of *Histoplasma* and *Blastomyces* for formation of gymnothecia. Also for conservation and maintenance of *Histoplasma* and *Blastomyces* isolants (after E. S. McDonough. See Chapter 18). This agar is also useful for inducing conidiation in many hypomycetes.

Oatmeal-Tomato Paste Agar (Weitzman and Silva Hutner, 1967)

Beechnut baby oatmeal	10 g
Hunt's tomato paste	10 g
$MgSO_4 - 7 H_2O$	1 g
KH_2PO_4	1 g
$NaNO_3$	1 g
Agar	18 g
Distilled water	1000 ml

Preparation. Adjust pH with NaOH to 5.6. Sterilize at 121°C for 20 minutes. Pour as plates.

Use. Ascospore production of *Nannizzia* and *Arthroderma* sp. Some clipped, sterile hair is often useful to get the mated pairs in a romantic mood. Beetle wings and iron filings sometimes help also.

Agar for Mating Fungi and Maintenance of Stock Cultures

Glucose	5.0 g
Yeast extract	0.9 g
Soil	20.0 g
Agar	20.0 g
Hair (human or horse)	As needed

Preparation. Dissolve and autoclave at 121°C for 15 minutes. Dispense into Petri plates.

Use. Mating of *Aspergillus heterothallicus*, *A. fennelli*, and many other fungi can be carried out by simply planting the two mating types on the agar near each other and allowing them to grow toward each other. At the boundary of the intermingling of the two thalli, a line of cleistothecia will be formed. For mating *Arthroderma* and *Nannizzia* species, it is often necessary to add keratin. This can be done by placing a few strands of hair in the center of the plate and then inoculating. Experience varies, but it appears that horse tail hair is the most satisfactory; however, sterile human hair can also be used. Some authors use chopped or ground hair. If mating is not successful, other additives may be utilized to enhance the romantic environment. These include beetle wings, oatmeal, tomato paste, chopped chicken feathers, and so forth. See oatmeal–tomato paste agar. For *Blastomyces* and *Histoplasma* mating, see soil extract procedure.

Homothallic organisms are often stimulated to produce their perfect stage by being grown in deficient media. Soil agar is excellent for most of these. Some will require salts, and some produce ascospores best on plain 2 per cent water agar. *Pseudallescheria* particularly is stimulated by this. Deficient medium is also the best for stock cultures. Various combinations of ingredients have been formulated and modified. These include potato-carrot, tomato-oatmeal, cornmeal, lima bean, and so forth. Many sound quite appetizing. This is probably because most mycologists of my acquaintance are also gourmets and excellent cooks.

Different formulations for the maintenance of stock culture have been devised over the years. Many species require special media or special care. The most useful and simplest techniques that have been successful in the author's experience will be outlined here.

Stock cultures of all organisms are kept on Sabouraud's agar (Emmons' formula) and cornmeal agar. Both media are available commercially. The slants are stored at room temperature. An additional set is flooded with sterile litmus milk (or reconstituted dry milk) and frozen. It is stored in the minus 20°C deep freeze. This combination of four tubes in two temperatures and one set protected by a colloid has been found to maintain most species satisfactorily, both alive and with cultural and conidia characteristics intact. The room temperature stocks are transferred every six to eight months, and the deep freeze cultures every two to three years. If available, the lyophilization of conidia in a colloid medium is the superior method of preserving fungi.

Other Media for Special Purposes

Potato-Carrot Media

Carrots	20 g
Potatoes	20 g

Add some water and purée in blender; q.s. water to 1000 ml. Add 20 g agar.

Preparation. Heat to dissolve agar and autoclave 15 minutes at 121°C. Dispense in tubes and slant.

Use. Stock culture medium for many hyphomycetes.

Hypertonic Media for Growth of L Forms of Fungi

Sabouraud's broth plus 30 per cent sucrose. Sterilize and pour over previously made Sabouraud's slants.

Asparagine Broth

l-Asparagine	7.00 g
K$_2$HPO$_4$ c.p. anhydrous	1.31 g
Ammonium chloride	7.00 g

Na₃C₆H₅O₇ · 5H₂O (sodium citrate, c.p.)	0.90 g

$Na_3C_6H_5O_7 \cdot 5H_2O$ (sodium citrate, c.p.)   0.90 g
$MgSO_4 \cdot 7H_2O$ 1.50 g
Ferric citrate USP   0.30 g
Glucose USP   10.00 g
Glycerine c.p.   25.00 g
Water   1000 ml

Preparation. The asparagine is dissolved in 300 ml of hot (50°C) distilled water. Each organic salt is dissolved in separate aliquots of 25 ml of distilled water. Hot distilled water is used for the ferric chloride. Each salt is added in the order given above to the hot asparagine, mixing well after each addition. The glucose and glycerine are added last. The solution is brought up to volume by addition of distilled water. It is then dispensed in a wide-bottom flask, such as the Fernback flask, to give a depth of 1 to 1½ inches. The flasks are autoclaved at 121°C for 15 minutes.

Use. For the preparation of histoplasmin, coccidioidin, or other antigens. Conidia are spread over the surface of the medium, and it is incubated (stationary) at room temperature for four weeks. A mycelial mat should cover the surface. At the end of four weeks the mat is separated from the culture broth by filtration. The antigens are prepared as described for the *Aspergillus* antigen preparations in a later section.

Medium for Production of Culture Spherules of Coccidioides Immitis From Counts and Crecelius (1970)*

Basic Medium (tenfold concentration):

KH_2PO_4	6.260 g
$ZnSO_4$	0.036 g
$MgSO_4$	3.944 g
$NaHCO_3$	1.176 g
$CaCl_2$	0.029 g
NH_4 acetate	12.334 g
Glucose	39.635 g
$KHPO_4$	6.846 g
NaCl	0.014 g
Casein hydrolysate	20.000 g
Distilled water	1000 ml

Preparation. The pH is adjusted to 6.0 and the medium autoclaved at 121°C for ten

*Roberts, J. A., J. M. Counts et al. 1970. Production in vitro of *Coccidioides immitis* spherules and endospores as a diagnostic aid. Am. Rev. Resp. Dis., *102*:811–813.

minutes. The medium is then diluted 1:10 with sterile distilled water and dispensed in 10-ml aliquots into a 50-ml sterile Erlenmeyer flask stoppered with a cotton plug and the cap covered with aluminum foil. The basic medium can be stored this way for some time without deterioration. Final medium is prepared when needed and cannot be stored.

Additives for Final Medium:

Biotin	1:10,000
Glutathione	0.005 M

These are prepared fresh and sterilized by filtration. The final medium is prepared by the addition of 0.5 ml of the glutathione solution and 0.5 ml of the biotin solution to each flask of basic medium to be used.

Use. Mycelium and arthroconidia of *C. immitis* are inoculated and the cultures incubated without shaking at 40°C. Within three weeks spherules are produced.

TESTING FUNGI FOR DRUG SENSITIVITIES

Testing fungal sensitivity to antimycotic agents is an arduous and often thankless and useless task. The differences between bacteriology and its techniques and mycology are nowhere more apparent than in testing drug sensitivities. Minimum inhibitory concentrations and minimum bactericidal concentrations of drugs for bacterial isolants are routinely done and have proven to be of utmost clinical significance. The same is not true of fungi. Yeast sensitivities are fairly straightforward, but those of mycelial fungi, particularly the nonconidiating isolants, are difficult and tedious to ascertain. In the following section some procedures used in the author's laboratory and some alternative methods are given.

Amphotericin B Sensitivity (Method 1)

Amphotericin B (Fungizone, intravenous), 50 mg, Sabouraud's agar or broth.

Preparation of Medium. Prepare 18 tubes of Sabouraud's broth or agar (the latter must be kept molten) containing 9.9 ml in each tube. Two control tubes containing 10 ml are also needed.

Preparation of Drug. Into the phial containing 50 mg of amphotericin B, place 10

Outline of Drug Dilution (Method 1)

Tube Number	1	2	3	4	5	6	7	8	9
Amphotericin B (5 mg/ml solution)	Undil soln.	1 ml of no. 1	1 ml of no. 2	0.5 ml of no. 1	0.5 ml of no. 4	1 ml of no. 5	1 ml of no. 5	1 ml of no. 7	1 ml of no. 8
Water		1 ml	1 ml	4.5 ml	4.5 ml	1 ml	4 ml	1 ml	1 ml
Concentration of drug in 0.1 ml	500 μg	250 μg	125 μg	50 μg	5 μg	2.5 μg	1 μg	0.5 μg	0.25 μg

ml of distilled water (sterile). This solution contains 5 mg per ml.

Preparation of Drug Dilutions. Set up eight empty tubes labeled 2 to 9 and handle as shown in the chart above.

Preparation of Drug Dilutions in Medium. Into the tube of medium labeled no. 1, place 0.1 ml of the original solution of amphotericin B. Into the tube of medium labeled no. 2, place 0.1 ml of dilution tube no. 2. Repeat with the remaining tubes 3 to 9. Mix thoroughly and, if agar is used, slant the tube.

Final Concentration of Amphotericin B in Medium. See chart at the bottom of this page.

When cool, the medium is inoculated with a standardized suspension of spores of the fungus or yeast cells. Control tubes containing no antibiotics are inoculated. The test should be done in duplicate. The test is incubated at 37°C and read at 48 hours.

Serum Levels of Amphotericin B

To test serum levels of amphotericin B, use the standard strain of *Candida tropicalis* ATCC#13803. Dilutions of serum are made in molten Sabouraud's agar. The medium is slanted and inoculated with a standard inoculum (usually 10^6) of *C. tropicalis*. It is incubated at 37°C. The test can also be done using conidia of *Paecilomyces variotii*. An amphotericin B sensitivity series is done at the same time to ascertain the sensitivity of the organism. All tests are read at 48 hours. By comparison of end points, the level of amphotericin B in the serum can be ascertained.

Haley Modification of Amphotericin B Sensitivity Test (Method 2)*

Modified Sabhi Agar (molten)	500 ml
Brain Heart Infusion (BHI)	500 ml in 12.5 ml aliquots in 14 ml bottles (12 for drug dilution, and one bottle with 24 ml, 2 for controls). Keep rest of broth in flask.

Preparation of Amphotericin B. *Stock Drug Solution No. 1 Concentration: 10,000 μg per ml).* Dissolve 50 mg of active amphotericin B in 5.0 ml of 100 per cent DMSO in 5 ml volumetric flask. Can keep frozen for several months at 10°C.

Stock Drug Solution No. 2 (Concentration: 5000 μg per ml). Make a 1:2 dilution of 1 ml of stock no. 1 in 100 per cent DMSO.

Stock Drug Solution No. 3 (Concentration: 200 μg per ml). Dilute 1.0 ml of stock no. 2 in 24 ml of BHI broth in medicine bottle.

Preparation of Final Drug Dilutions in Medium

1. Make 1:2 dilution of stock no. 3 (with 25-ml pipette) by taking 12.5 ml from the 25 ml of stock no. 3 and transferring it to one of the medicine bottles containing 12.5 ml BHI.

2. Continue in the same manner, transferring 12.5 ml from bottle to bottle, for dilutions of from 100 μg per ml to 0.1 μg per ml. Discard 12.5 ml of last solution.

*Adapted from Haley, L. D., and C. S. Callaway 1978. *Laboratory Methods in Medical Mycology.* HEW Publication #CDC 78-8361.

Concentration in Medium (Method 1)

Tube Number	1	2	3	4	5	6	7	8	9
Concentration in μg/ml	50	25	12.5	5.0	0.5	0.25	0.1	0.05	0.025

Chart of Concentrations for Haley Modification

Tube number	1	2	3	4	5	6	7	8	9	10	11	12
Concentration of drug in μg/ml	100	50	25	12.5	6.25	3.13	1.56	0.78	0.39	0.20	0.10	0.05

3. Add 12.5 ml of molten Sabhi medium to first drug solution (i.e., stock no. 3, 200 μg per ml), mix, and dispense in 2.0 ml volume in 12 tubes. Add 12.5 ml of molten Sabhi to second drug dilution; mix and dispense in 2.0 ml, as done previously. Continue doing the same for each drug dilution. See above chart for final concentration of drug.

Preparation of Control Media. As in Method 1.

Inoculation of Medium. As in Method 1.

Note: Semisolid media are in fact semisolid and cannot be slanted. Media must be inoculated by merely touching the top of the media. Growth will be indicated by a streak going down the tube.

Procedure for Determining 5-FC MIC (Method of E. R. Block*) (Method 1)

1. Each organism to be tested is subcultured on yeast-nitrogen base agar slants, incubated at 30°C for 48 hours, and then harvested in sterile normal saline.
2. The saline suspension of each organism is adjusted to obtain an optical density of 0.300 at a wavelength of 600 nm (using a spectrophotometer tube with an internal diameter of 13 mm). Normal saline is used as a blank.
3. This latter suspension (O.D. = 0.300) is then diluted 1:10,000 (10^{-4}) in sterile distilled water and is ready to be used in the test procedure.

*Modified by Block from Block, E. R., and J. E. Bennett. 1972. Pharmacological studies with 5-fluorocytosine. Antimicrob. Agents Chemother., *1*:476–482.

4. Preparation of the test media is as follows:
   a. Nine tubes are required for each organism to be tested.
   b. This includes a control tube which has organisms but no 5-fluorocytosine (FC) and eight tubes with twofold serial dilutions of 5-FC extending from 2.5 to 320 μg per ml.
   c. For each set of nine tubes, 35 ml of medium is prepared as below:

Yeast-nitrogen base (YNB)	5 ml	(each 100 ml contains 6.7 g YNB and 10.0 g dextrose in distilled H_2O); to use, dilute 1:10.
50% Glucose	2 ml	
Distilled water	28 ml	

   d. Add 5 ml of the final suspension of the test organism (from step 3) to 35 ml of medium to make a final volume of 40 ml. Mix well and then add 4 ml of this mixture to each of the nine test tubes. This will leave 4 ml extra. This step will be repeated with each organism to be tested.
   e. To prepare the 5-FC concentrations, 160 mg of powdered 5-FC (obtained as bulk powder from Hoffman–La Roche, Inc.) is dissolved in 100 ml of distilled water (with help of KOH) and filter sterilized. Serial twofold dilutions of this original stock concentration are made as below.
   f. If 1.0 ml of A is added to the first of the nine tubes containing 4 ml of medium and test organism, the final concentration of 5-FC will be 320 μg

Outline of Drug Dilution (5-FC)

A = 160 mg/100 ml 5-FC
B = 20 ml of 160 mg/100 ml stock + 20 ml distilled water = 80 mg/100 ml 5-FC
C = 20 ml of 80 mg/100 ml dilution + 20 ml distilled water = 40 mg/100 ml 5-FC
D = 20 ml of 40 mg/100 ml dilution + 20 ml distilled water = 20 mg/100 ml 5-FC
E = 20 ml of 20 mg/100 ml dilution + 20 ml distilled water = 10 mg/100 ml 5-FC
F = 20 ml of 10 mg/100 ml dilution + 20 ml distilled water = 5 mg/100 ml 5-FC
G = 20 ml of 5 mg/100 ml dilution + 20 ml distilled water = 2.5 mg/100 ml 5-FC
H = 20 ml of 2.5 mg/100 ml dilution + 20 ml distilled water = 1.25 mg/100 ml 5-FC

per ml. Similarly, if 1.0 ml of B through H is added to the next seven tubes, twofold dilutions of 5-FC, ranging from 160 to 2.5 μg per ml, will be obtained. In the ninth tube (control tube), 1.0 ml of sterile distilled water is added.

5. All tubes are then incubated on a rotating drum (2 rpm) at 32°C and read at 48 hours. The minimum inhibitory concentration is defined as the lowest concentration of drug in which no visible growth is observed.

Haley Modification of 5-Fluorocytosine Testing (Method 2)

Preparation of Medium. Prepare yeast nitrogen base (YNB) at 10 × strength. (Sterilize by filtration.) Prepare 100 ml of 1 × YNB in 90 ml sterile H_2O. Dispense 5 ml of 1 × YNB into each of 11 sterile cotton-plugged test tubes. Dispense 9 ml into one large, cotton-plugged test tube.

Preparation of Drug Dilutions

Solution No. 1 (Concentration: 10,000 μg per ml). Make 5 ml aqueous stock solution of 5-FC containing 0.05 g in 5 ml, 50 g in 5 ml. (Sterilize by filtration.)

Solution No. 2 (Concentration: 1000 μg per ml). Make 1:10 dilution of solution no. 1 in sterile 10 × YNB. (0.5 ml into 4.5 ml, 10 × YNB.)

Solution No. 3 (Concentration: 100 μg per ml). Make 1:10 dilution of solution no. 2 in sterile 1 × YNB. (One ml into 9 ml 1 × YNB.) Serially dilute 5 ml of solution no. 3 (100 μg per ml) through the 11 tubes with 5 ml of 1 × YNB. (Discard last 5 ml.) Dispense 1 ml of solution in first tube (50 μg per ml) to each of 4 cotton-plugged tubes. Do this with each drug solution.

Preparation of Control Media. As in Method 1.

Inoculation of Medium. As in Method 1.

Griseofulvin Sensitivity Test

The contents of one capsule containing 125 mg of griseofulvin are dissolved in 62.5 ml of acetone (concentration 2 mg per ml). Also prepare six tubes containing 9.9 ml of melted Sabouraud's agar. Six dilutions of griseofulvin are then prepared as follows:

1. Original dilution — griseofulvin in acetone

2. 3 ml of tube 1 and 1 ml H_2O
3. 1 ml of tube 1 and 1 ml H_2O
4. 1 ml of tube 1 and 3 ml H_2O
5. 1 ml of tube 2 and 4 ml H_2O
6. 1 ml of tube 3 and 9 ml H_2O

Add 0.1 ml of dilution 1 to agar tube A, then 0.1 ml of dilution 2 to tube B, and so forth.

Tube	Concentration of Drug in 0.1 ml	Final Concentration in Tube
A	200 μg	20 μg
B	150	15
C	100	10
D	50	5
E	30	3
F	10	1

Slant tubes and inoculate each and a control without the drug with standard inoculum of the organism to be tested.

The Smith Shadomy–Espinel-Ingraff Method for Antifungal Drug Sensitivities

This is a broth dilution technique that gives results comparable to the MIC and MBC levels used in clinical bacteriology. For details, the reader is referred to the third edition of the *Manual of Clinical Microbiology,*[*] Chapter 62. A brief summary will be given here.

In general, the selection of medium to be used in broth dilution testing depends upon the drug. Unbuffered yeast nitrogen base supplemented with glucose is used in tests for both imidazoles and 5-FC, and tests with the other three are usually performed in Difco antibiotic medium #3 (Difco #0243) or the equivalent. The inoculum used should be 24- to 48-hour cultures, and any one of the following controls is recommended:

Saccharomyces cerevisiae	ATCC	#36375
Saccharomyces cerevisiae	ATCC	#9763
Saccharomyces cerevisiae	ATCC	#2601
Candida albicans	ATCC	#10231
Candida tropicalis	ATCC	#13803

---

[*]Shadomy, S., and A. Espinel-Ingraff. 1980. Susceptibility testing with antifungal drugs. Chapter 62. *In* Lennette, E. H., A. Balows, et al. *Manual of Clinical Microbiology.* 3rd edition. Washington, D.C., American Society of Microbiology, pp. 647–653.

Range of Susceptibility

Drug		*Level of Concentration*
5-FC	MIC	12.5 mg/ml or less
	MFC	25.0 mg/ml or less
Amphotericin B	MIC	0.39 mg/ml or less
	MFC	0.39 mg/ml or less
Nystatin	MIC	3.13 μg/ml or less
	MFC	3.13 μg/ml or less
Pimaricin	MIC	1–10 μg/ml
Miconazole	MIC	0.5–10 μg/ml or less
	MFC	slightly higher
Clotrimazole	MIC	5 to 10 $\times$ more active than miconazole, depending upon media

Definition of Terms

MIC: The lowest concentration of drug that inhibits clearly visible growth, ignoring a faint haze or slight turbidity.

MFC: The lowest concentration of drug from which subcultures are negative or yield fewer than 3 colonies. This is determined by subculturing 0.01 ml from each negative tube and the first positive tube to drug-free Sabouraud, and incubating at room temperature for 48 hours.

Broth Dilution Method. For 5-FC, miconazole, and clotrimazole:

Yeast Nitrogen Base Broth and
10 × Yeast Nitrogen Base Broth

Yeast Nitrogen Base	
(Difco #0392)	6.7 g
L-asparagine	1.5 g
Dextrose	10.0 g
Distilled water	100 ml

Filter sterilize; unused portion can be stored at 4°C. One × solution is prepared by making a 1:10 dilution in sterile distilled water.

For amphotericin B, nystatin, and pimaricin:

Antibiotic medium #3

Antibiotic medium #3	17.5 g
Distilled water	1000 ml

Sterilize by autoclaving 15 minutes at 15 p.s.i.

Drug Solutions. 5-FC: 10,000 μg per ml solution of standard 5-FC powder is prepared in distilled water and filter sterilized. May be stored at −30°C indefinitely.

Amphotericin B, pimaricin, and *nystatin:* 5000 μg per ml solution to be made. First solution to be made in 100 per cent DMSO.

Imidazoles: 10,000 μg per ml solution. First solution to be made in 100 per cent DMSO.

If standards are not available, patient treatment preparations of the drug may be used but are not recommended.

Inoculum. This is prepared from 24- to 40-hour cultures. Suspensions are prepared in sterile saline or distilled water to a 1+ on Wickerham card 1.

Tube no.	1	2	3	4	5	6	7	8	9	10
Dilution (μg/ml)	100	50	25	12.5	6.25	3.12	1.56	.78	.39	.19

11	12	13	14
.01	.005	inoculum, no drug	no drug, no inoculum

Drug Dilutions and Performance of Test. Using either $1 \times$ YNB, or antibiotic medium no. 3, prepare 10 ml of 100 μg per ml solution of drug to be tested. For amphotericin B, start with 10 μg per ml.

Place 12 sterile tubes in a rack and number them 1 through 14. No. 13 contains an organism, but not a drug; no. 14 contains neither drug nor organism. Add 5.0 ml of appropriate media to tubes no. 2 through 14. Add 5.0 ml of 100 μg per ml solutions if imidazoles or 5-FC (or 10 μg per ml if amphotericin B or other polyene) to tubes no. 1 and 2. Mix contents of tube no. 2; serially dilute, using 5.0 ml volumes and fresh pipettes through to tube no. 12. Dilutions will be obtained as shown in table at bottom of page 786.

When amphotericin B or other polyene is tested, the concentrations will be a factor of 10 higher.

Transfer 1.0 ml from each dilution and to each of 4 sterile tubes. Inoculate each tube with 0.05 ml of the standardized suspensions of the test and control organisms. Incubate at room temperature for 48 hours or until growth is visible in the growth control tubes.

SEROLOGIC TECHNIQUES

Serologic techniques used in the diagnosis of fungus infections have been the subject of an immense amount of literature and controversy. A few of these techniques have proved to be of great value and are standard procedures used in the diagnosis and prognostic evaluation of fungus infections. These procedures include complement fixation tests, precipitin tests, agglutination tests, and various immunofluorescent procedures. The complement fixation test is technically difficult, in addition to being time consuming, and requires experience and skill in its methodology and interpretation. Many of the immunofluorescent procedures are very useful but require expensive equipment and special training. Therefore, these tests are usually restricted to the large facilities of state and federal laboratories.

For the small laboratory, the most promising serologic procedures which are both rapid and inexpensive are the several latex agglutination tests and the immunodiffusion techniques. The latex agglutination tests that are commercially available at this writing are screening tests for histoplasmosis and coccidioidomycosis and a test for the presence of antigen in cryptococcosis. These all appear to be quite specific and, if properly carried out, are of great value in the diagnosis of disease. When these screening tests are positive, sera can be sent to reference laboratories for evaluation by complement fixation and other serologic procedures. The immunodiffusion test is also simple to carry out and can be used as a screening procedure. Double diffusion plates are available commercially that require only small amounts of both serum and antigen. In our laboratory the routine test consists of the patient's serum in the center well and a battery of six antigens in the outer wells. These are antigens prepared from *Histoplasma, Blastomyces, Candida, Aspergillus, Coccidioides,* and *Sporothrix schenckii.* The tests are read at 48 and 72 hours after incubation at room temperature and then placed in the cold and read after five days. If lines appear, a drop of 1.5 per cent EDTA (ethylenediaminetetraacetic acid) is placed in the center well and allowed to diffuse for one hour. EDTA will cause the disappearance of soluble lines. These are generally due to C-reactive substance in the antigen preparation. If lines are present, they are considered a probable positive result, and the serum is tested with specialized serologic procedures, such as the complement fixation test. The interpretation of such lines is discussed in the various chapters concerned with specific diseases.

Production of good antigens is, at best, an art. So far, consistency of antigen preparation has not been achievable. In this section will be presented the methods that have been developed or found to be useful by S. McMillen, myself, and several other investigators. As a final note, it is hoped that better and more specific tests will be developed in the future, as most of these methods are quite crude. In some patients, especially those with severe debilitation and those suffering from candidiasis or aspergillosis, the tests are often negative owing to diminished or absent antibody production by the patient. A better test in these as in all cases would lead to the detection of antigen produced by the fungus in the infected patient. One such test is currently available and is quite useful: the antigen test in cryptococcosis. It is hoped that in the future more such procedures will be developed and made available.

Aspergillus *Antigen*

Medium. Sabouraud's broth.
Inoculum. *Aspergillus fumigatus* var. *ellipticus.* In our experience the antigen from this

strain is the most sensitive for all forms of aspergillosis caused by most species of *Aspergillus*. In our laboratory we have also prepared antigen from *A. fumigatus* (five strains), *A. niger* (three strains), *A. terreus*, and *A. flavus*. On rare occasions the test is negative using the var. *ellipticus* antigen and is picked up by one of the other antigens when it is the etiologic agent.

Method. Dispense broth to a depth of 1 to 1½ inches in a large flask in order that you have a wide surface for the growth of the mycelial mat. Autoclave. Inoculate with a quantity of conidia laid on the surface. Incubate (stationary) in the light at room temperature. A large surface mat will be produced.

At four weeks, aseptically remove 5 to 10 ml of the broth and test as is with positive control serum. If there is no activity, add 2 volumes cold acetone to the remainder of the sample and refrigerate for 24 to 48 hours. Centrifuge and collect the sediment. This is redissolved in 0.5 ml of distilled water. Retest the concentrate with known sera. If lines appear, challenge with a drop of 1.5 per cent EDTA or 5 per cent sodium citrate. Soluble lines representing C-reactive substance are commonly found in *Aspergillus* antigen preparations.

In general, most culture filtrates of *Aspergillus* have the greatest antigen content after four weeks' incubation. Those that grow more slowly may require six weeks. There is little increase in antigen content after that time.

To process the broth filtrate, decant from the flask and filter. Precipitate with 2 volumes cold acetone and refrigerate for 48 hours. The material is now stable and may be stored until convenient for further processing. Decant acetone and collect precipitate. Redissolve in water, pool the solutions from several flasks, and reprecipitate with equal amounts of (v/v) cold acetone. Decant and collect filtrate. Redissolve in water. Serial dilutions are necessary to establish potency and proper working dilution. In general, this is double the point of solubility of the precipitate. It can be stored frozen and does not lose potency. Most antigen is inactivated by heat.

Blastomyces *Cell Sap Antigen*

Medium. Brain-heart infusion broth. 100 ml in 250-ml cotton-stoppered flask.

Inoculum. *Blastomyces dermatitidis* yeast cells grown on BHI blood agar slants at 37°C for two weeks.

Method. Wash down two slants of *Blastomyces* yeast cells with 5 ml of sterile saline. Pipette 1 ml of saline suspension of yeast cells to each of five flasks of BHI broth. Incubate in a shaking water bath at 37°C for two weeks or until the cell mass is large. Harvest cells by filtration and wash with several changes of distilled water. Collect cells and freeze in liquid nitrogen. Break cells in a French press immersed in liquid nitrogen. It is important that the French press be immersed in liquid nitrogen at least one hour before using. Place the French press on a hydraulic press and bring diaphragm pressure to 2000 lb per in². Collect debris in stainless steel test tube. Check for per cent breakage of cells by examining with a microscope some debris suspended in water. If the breakage is less than 90 per cent, cells must be reprocessed. The debris is then suspended in sufficient saline to make an opalescent slurry. This is spun at 3000 rpm for 30 minutes to separate cell sap from cell walls. The cell sap may be frozen and stored at −95°C. When it is to be used, the sap is diluted with saline to give a reading of 2.5 g protein per 100 ml, using a Hitachi refractometer (see method for *Candida* cell sap). Culture filtrate and cell walls contain antigens that cross-react with many fungi. In our laboratory, the cell sap has been found to be the most specific antigen for use in *Blastomyces* immunodiffusion and complement fixation tests. Serum to be tested must be fresh, as many false-negatives have been found using stored sera. In our experience the French press and liquid nitrogen are the only way to obtain breakage of *Blastomyces* yeast cells.

Candida *Cell Sap Antigen*

Preparation. *Candida albicans* is grown in Sabouraud's broth for 24 to 48 hours at 37°C in a shaking water bath. The cells are harvested by centrifugation, washed three times in saline, and resuspended in sterile distilled water. The latter should be a very turbid suspension. To 20 ml of such a suspension, add 5 ml of a 5-micron glass bead suspension. A 10-ml lot is then sonicated in a stainless steel "cold shoulder" cell in a sonifier (Branson High Intensity). The exposure needed is 90 to 95 watts for 40 minutes. This should yield 80 to 90 per cent breakage of cells. The material is spun at 3000 rpm for 30 minutes in order to separate the cell walls from the cell sap. The cell sap is diluted to give a reading about 1.015 specific gravity on a Hitachi refractometer or equivalent apparatus. This is equivalent to 1.5 mg protein per 100 ml. This constitutes the "s" antigen of Taschdjian et al. Merthiolate, to a final concentration of 1:10,000, may be added as a preservative. The antigen remains stable for months in the refrigerator or indefinitely in the freezer. About equal results can be obtained using the commercially available Hollister-Stier *Monilia albicans* (Oidiomycin) antigen.

Sporothrix schenckii *Antigen*

There are several methods described for the preparation of this antigen. A good M form antigen useful as a screening test can be made using the technique described for *Aspergillus* antigen. A yeast form antigen (Y) which is more specific and comparable to yeast form histoplasmin antigen can also be prepared. Several media may be used, such as nutrient broth, Sabouraud's broth, or asparagine broth. Some authors prefer *Trichophyton* #3 (Difco) broth. The formula is as follows.

Bacto-Vitamin-free Casamino Acids	2.5 g
Bacto-Inositol	50 mg
Thiamin	200 μg
Glucose	40 g
KH_2PO_4	1.8 g
$MgSO_4\cdot 7H_2O$	0.1 g
Distilled water	1000 ml

Dissolve and dispense in 250-ml flasks. Autoclave at 121°C for 15 minutes. Inoculate with a heavy suspension of yeasts. Incubate at 37°C (rotating) for 30 days. Harvest cells and save both filtrate and yeasts. The filtrate antigen is prepared as described for *Aspergillus* antigen using cold acetone. A cell sap antigen "s" can be prepared by sonicating yeast cells or putting them through a French press (method for *Blastomyces* antigen). The cell sap antigen is the most specific in our hands.

An M form broth antigen can be prepared using *Trichophyton* #3 broth and the method for *Aspergillus* antigen.

Histoplasma capsulatum *Antigen*

There are many formulas for the preparation of this antigen. The most successful in our experience is the simple "filtrate" antigen. This is prepared by growing the organism as a mycelial mat in asparagine broth medium (stationary) at room temperature for four weeks. The procedure for concentration of the antigen by cold acetone precipitation is the same as noted above for *Aspergillus* antigen. The *Histoplasma* antigen prepared from the mycelial stage (M) has been found to be superior to that prepared from the yeast stage (Y) in the immunodiffusion test.

Coccidioides *Antigen*

A good *Coccidioides* antigen can be prepared from mycelial mat growth in asparagine broth. The culture is kept stationary at room temperature for four weeks. The antigen is concentrated by twofold acetone precipitation, as noted above in the section on *Aspergillus* antigen. This is the immunodiffusion antigen and complement fixation antigen (IDCF) of Huppert and Bailey.

Huppert and Bailey have prepared another antigen from the mycelial mat of *Coccidioides*. After the mat is separated from the culture filtrate, a measured volume of water is added to make a thick slurry. Toluene at a rate of 3 ml per 100 ml of water is added to the slurry. The flask is stoppered and incubated at 37°C for three days. The residue is filtered off through a Buchner filter. The lysate is then used as the antigen. This is termed the IDTP (immunodiffusion, tube precipitation) antigen, as it corresponds to that used in the old tube precipitin tests. (Hupper, M., and J. W. Bailey. 1965. The use of immunodiffusion tests in coccidioidomy cases. Am. J. Clin. Pathol., *44*:369–373.)

Complement Fixation Test Using Fungal Antigens (After C. E. Smith)

The serum is inactivated for 30 minutes at 56°C. Standard controls are run simultaneously. Serial dilutions are made of the serum with saline — 1:2, 1:4, 1:8, 1:16, and so forth, up to 1:512. To each 0.25 ml of diluted serum is added 0.25 ml of coccidioidin (or other antigen) and two units of complement contained in 0.5 ml. The mixture is incubated for two hours in a 37°C water bath. Then 0.5 ml of 2 per cent sheep red blood cells sensitized with two units of antisheep ambocepter is added. The mixture is agitated and incubated at 37°C in a water bath for one hour. Readings are made before and after overnight refrigeration. Complete hemolysis, representing no previous fixation of complement, is read as negative. No hemolysis (complete fixation of complement) is read at 4+. Intermediate degrees are recorded as +, ++, or +++. If complement is fixed, serum is retained for comparison with future samples. A modification of this procedure used in some laboratories is to mix sera, complement, and coccidioidin (or other antigen) and refrigerate (4°C) overnight. The sheep erythrocytes (sensitized) are then added and the mixture incubated at 37°C for two hours. This method is reputed to be more sensitive and to give a one-tube dilution higher than the method employing two-hour binding at 37°C.

PROCEDURES FOR DIRECT AND HISTOPATHOLOGIC EXAMINATION FOR FUNGI

Direct Examination

The Potassium Hydroxide Mount. The most useful technique for direct examination of all types of specimens for fungi is the potassium hydroxide mount. One drop of 10 per cent potassium hydroxide is placed on the slide and mixed with the specimen. It is cleared by warming the slide. Do not boil, as this produces KOH crystals. The walls of the fungi are tough and not harmed by the alkali. Most tissue cells and many artifacts are dissolved.

India Ink Mount. India ink slides are a traditional method used for the detection of cryptococci in spinal fluid. The spun sediment of spinal fluid is mixed with one small drop of India ink and examined. The preparation should be thin and hazy brown, not thick and black. It is examined with the high dry lens in the microscope. There are numerous artifacts that resemble cryptococcal yeasts. Most are red and white blood cells, glial cells, macrophages, and so forth. These are destroyed by KOH, so that this is a necessary control to rule out artifact. Only when the typical budding yeast cell with its halolike capsule is seen is the examination termed positive (see Fig. 21–5). Airborne cryptococci as well as other fungi have been found growing in bottles of India ink, so the reagent must be checked from time to time.

Cryptococci in tissue can be seen by using touch slide. The slide is touched to the suspected formalinized tissue or the tissue contents squeezed onto a slide and India ink

added. A KOH control must be performed. Macerated brain can also be examined as an India ink preparation.

Scrape Mounts. These are prepared by scraping the cultures of fungi to obtain mycelium and conidia. An inoculating needle bent to form a hook is useful. The mycelium and conidia are teased apart in a drop of lactophenol cotton blue on a slide. The teasing is performed with the corner of a cover slip. After sufficient separation of mycelial strands, the cover slip is placed on it and the preparation examined. The use of slide cultures is superior to scrape mounts for the identification of fungi. "Scotch" tape mounts are also useful for quick observations. A strip of tape is touched (adhesive side down) to the colony. The strip is then placed on a slide (adhesive side down), on which is a drop of LCB.

Lactophenol Cotton Blue (LCB)

Phenol crystals	20 g (melt in water bath before weighing)
Lactic acid (85%)	20 g
Glycerin	40 g
Cotton blue (C4B) (Poirrier's blue)	0.05 g
Distilled water	20 ml

Permanent mounts of slide cultures or scrape mounts can be made by sealing the edge of the cover slip with nail polish. A superior method is to dip the preparation in collodion before staining. After drying, the slide is stained with LCB for 10 to 15 minutes. Then the following dehydration procedure is followed: (From Rivalier, E., and S. Seydel. 1932. Cultures minces sur lames gelasees calorees et ecaminees in situ in preparatines definitines pour étude des cryptogámes microscopiques. C. R. Soc. Biol., *110*:181–184.)

1. Wash in 70% ethanol rapidly.
2. 95% ethanol 10 minutes.
3. Absolute alcohol 10 minutes or acetone.
4. Acetone-xylene 10 minutes.
5. Xylene.
6. Immediately (before drying) add mounting media (histology lab as Permount) and cover slip.

PAS Stained Direct Mounts. Direct mounts of clinical material can be examined immediately under the microscope. Clearing with KOH and heating will often aid in the observation of the fungus. It has become our practice to take most specimens one step further and stain the material with PAS and examine it for fungi. This is done on digested sputum that will be inoculated for AFB and fungi cultures and on a wide variety of clinical specimens. The PAS stain becomes a rapid procedure with practice, and we find that it greatly facilitates the search for fungi in clinical specimens.

Direct Mount Periodic Acid–Schiff (PAS).* Technique for staining:

1. Fix specimen on slide.†

2. Place fungal smear in absolute ethyl alcohol for 1 minute.

3. Drain alcohol and immediately place slide in 5 per cent periodic acid for 5 minutes.

4. Wash in running water for 2 minutes.

5. Place in basic fuchsin for 2 minutes.

6. Wash 2 minutes in running water.

7. Immerse slide in sodium metabisulfite solution for 3 to 5 minutes.

8. Following removal of slide from metabisulfite, wash 5 minutes.

9. Place slide in Light Green stain for 5 seconds.

10. Wash for 5 to 10 seconds.

11. Dip slide for 5 second intervals in 85 per cent, 95 per cent, and absolute alcohols.

12. Dip in xylene and mount with Hydromount.

This procedure is used to stain certain polysaccharides found in cell walls of fungi as detailed below.

Mayer's Egg Albumen.‡ Very satisfactory egg albumen can be procured from laboratory supply houses. For those who prefer to make their own, the formula follows:

Egg white	50 cc
Glycerin	50 cc

Beat the egg white and glycerin together and filter through coarse filter paper or through several thicknesses of gauze. Add a thymol crystal to preserve the albumen and prevent the growth of molds.

For fixing nails, skin scrape, or hair for staining.

Histopathologic Examination

Tissue specimens for examination of fungi are handled in the same manner as routine tissue. They are fixed in 10 per cent formalin solution and processed by the usual procedures for tissue sectioning. Few fungi are visible in routine hematoxylin and eosin stained preparations, so that special staining procedures are often necessary to visualize them. The various modifications of the periodic acid–Schiff stain are the most frequently employed or the Grocott variation of the Gomori methenamine silver stain. All have general acceptance and are excellent aids for diagnosis. They also have many pitfalls in the staining technique and the presence of artifacts. The author prefers the Gridley stain and the Grocott-Gomori methenamine stain. In this section will be described some of the more useful stains, the procedures used, and the difficulties encountered.

The Periodic Acid–Schiff Stains. The periodic acid–Schiff stains are based on the Feulgen reaction and have found wide use in biology. The Feulgen reaction depends on the formation of aldehyde groups in a polysaccharide chain and the *in situ* recoloration of the Schiff base (leukofuchsin). The Schiff base is produced by the action of sulfurous acid on the dye, basic fuchsin. This involves the addition of sulfur dioxide to the two amino groups of pararosaniline (basic fuchsin). The dye is then colorless.

The cell walls of fungi contain chitin and other complex polysaccharide chains. When these are treated with periodic or chromic acid, the carbon to carbon chain is broken in places, and pairs of aldehydes are formed. When Schiff's base comes in contact with the properly paired aldehydes, a reaction takes place in which the leukosulfonic acid–Schiff base is oxidized and the aldehyde reduced. The dye is attached to the polysaccharide molecule at the site of the reaction. This is termed the Feulgen reaction (Fig. A–42). The color produced is a magenta. The depth of the color produced depends on the number of available aldehyde groups produced by hydrolysis. Since the cell walls of fungi have a high concentration of complex polysaccharides, they stain an intense purple-red. Both the cell walls and the capsular material of *Cryptococcus neoformans* stain by the PAS technique. The cell wall will be dark, and a halo of deep, red-staining, concentrated capsular material will surround the cell. As the capsule becomes thinner, the color becomes progressively paler.

Advantages. This stain is very useful for observing fungi in tissue. The fungus is usu-

---

*Courtesy of Haley, L. D., and C. C. Callaway. Laboratory Methods in Medical Mycology, USDHEW. Publication #(cdc) 78-8361.

†For sputum, pleural fluid, CSF, and so forth, the usual methods of fixing tissues on a slide will do: air drying or drying over a flame. For nails, skin scrapes, or hair for fixing for staining, use Mayer's egg albumen.

‡Ann Preece, M. T. (ASCP) 1959. *A Manual for Histologic Technicians.* London, J. & A. Churchill, Ltd.

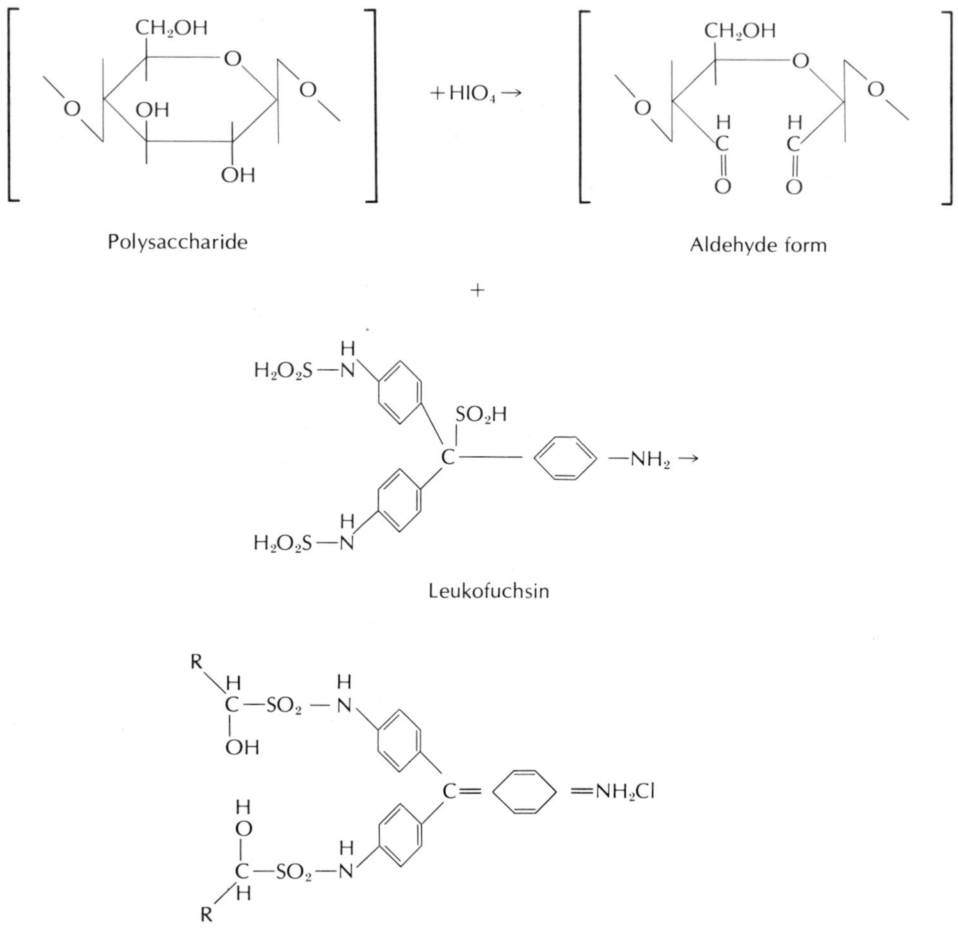

Figure A–42

ally stained much darker than the surrounding tissue. The morphology of the fungus is easily discernible, and even a few organisms in a field will be observable, especially if the background is stained by metanil yellow or fast green. Some pathologists prefer to have the slides stained by the hematoxylin-eosin method as a counterstain after the PAS has been completed. This makes individual fungus cells difficult to see, as there are many red or magenta objects present. The author prefers a slide stained by the Gridley method (metanil yellow), a second slide stained by the methenamine silver stain, and a third slide stained by hematoxylin and eosin to observe the tissue reaction and morphology.

Disadvantages. Many components of tissue that are carbohydrate in composition are stained by the PAS method. These include glycogen, starch, cellulose, glycolipids, mucin, fibrin threads, and elastic tissue. They

are not usually stained as intensely as the fungi, however. Bacteria such as *Nocardia* or *Actinomyces* are not stained by the PAS method but are by the methenamine stains.

Gridley Stain for Fungi[*]
Solutions
A. Chromic acid solution (this must be prepared fresh)
    Chromic acid   4.0 g
    Distilled water   100.0 ml
B. Feulgen reagent (Coleman)
Bring to boil 200 ml distilled water. Remove from flame and add 1 g basic fuchsin. When dissolved, cool and filter. To the solution add 2 g sodium metabisulfite (hypo — photographic) and 10 ml hydrochloric acid (normal). Permit the solution to bleach for 24 hours. Add 0.5 g activated charcoal powder

---

[*]Gridley, H. F. 1953. A stain for fungi in tissue sections. Am. J. Clin. Pathol., *23*:303–307.

(Norit or equivalent), shake for one minute, and filter using coarse filter paper. The solution should be colorless. The finished reagent may be stored in the refrigerator. It is usable until it becomes pale pink.

C. Normal hydrochloric acid
HCl, specific gravity 1.19   83.5 ml
Distilled water   916.5 ml

D. Sodium metabisulfite solution (10 per cent)
Sodium metabisulfite   10.0 g
Distilled water   100.0 ml

E. Sulfurous acid rinse
Sodium metabisulfite, 10 per cent   6.0 ml
Normal hydrochloric acid   5.0 ml
Distilled water   100.0 ml

F. Aldehyde-fuchsin solution
Basic fuchsin   1.0 g
Alcohol, 70 per cent   200.0 ml
Paraldehyde   2.0 ml
Hydrochloric acid (conc.)   2.0 ml

The solution is allowed to stand for three days at room temperature until it has turned a deep blue. It may be stored in the refrigerator. Before use, filter and bring to room temperature.

G. Metanil yellow solution
Metanil yellow   0.25 g
Distilled water   100 ml
Glacial acetic acid   2 drops

Procedure. Paraffin blocks are cut at 6 microns and fixed to the slide. The slides are passed through the following steps.

1. Deparaffinize through xylene, absolute alcohol, and 95 per cent alcohols to distilled water.
2. Place in 4 per cent chromic acid (oxidizer) for one hour.
3. Wash in running water for five minutes.
4. Place in Coleman's Feulgen reagent for 15 minutes.
5. Put slides through three changes of sulfurous acid rinse.
6. Wash in running tap water for 15 to 30 minutes.
7. Place slides in aldehyde-fuchsin solution for 15 to 30 minutes (the time allowed depends on personal experience and quality of reagents).
8. Rinse off excess stain in 95 per cent alcohol.
9. Rinse in water.
10. Lightly counterstain with metanil yellow for one minute.
11. Rinse in water.
12. Dehydrate through the usual series of alcohols to absolute alcohol.
13. Clear with xylene.

Results. Mycelium and yeasts, deep blue or deep rose; conidia, deep rose to purple; background, yellow; elastic tissue, mucin, and so forth, deep blue.

Grocott Modification of Gomori Methenamine Silver Stain*

Solutions

A. Chromic acid (5 per cent)
Chromic acid (CrO_3)   5.0 g
Distilled water   100.0 ml

B. Methenamine (3 per cent)
Hexamethylene tetramine USP $(CH_2)_6N_4$ 3.0 g
Distilled water   100.0 ml

C. Methenamine silver nitrate stock solution
Silver nitrate, 5 per cent solution   5.0 ml
Methenamine, 3 per cent solution   100.0 ml

A white precipitate will form, but it disappears on shaking. The solution is stable for months when refrigerated.

D. Sodium bisulfite (1 per cent)
Sodium bisulfite ($NaHSO_3$)   1.0 g
Distilled water   100.0 ml

E. Light green stock
Light green SF (yellow)   0.2 g
Distilled water   100.0 ml
Glacial acetic acid   0.2 ml

F. Working light green solution
Light green (stock)   10.0 ml
Distilled water   50.0 ml

G. Silver nitrate (5 per cent)
Silver nitrate ($AgNO_3$)   5.0 g
Distilled water   100.0 ml

H. Borax solution (5 per cent)
Borax (USP) $Na_2B_4O_7 \cdot 10H_2O$ 5.0 g
Distilled water   25.0 ml

I. Methenamine silver nitrate working solution
Borax, 5 per cent solution   2.0 ml
Distilled water   25.0 ml

Mix and add 25 ml of methenamine silver nitrate stock solution.

J. Gold chloride (0.1 per cent)
Gold chloride, 1 per cent solution ($AuCl_3 \cdot HCl \cdot 3H_2O$)   10 ml
Distilled water   90 ml

This solution may be used repeatedly.

---

*Grocott, R. G. 1955. A stain for tissue section and smears using Gomori's methenamine-silver nitrate technic. Am. J. Clin. Pathol., *25*(no. 7):975–979.

Procedure. Cut paraffin blocks at 6 microns, place on slides, and fix. Deparaffinize through xylene (two changes), absolute alcohol, and 95 per cent alcohols to distilled water. Previously stained sections may also be used, as chromic acid will remove previous stains.

1. Place slides in chromic acid solution for one hour.

2. Wash in running tap water for 15 seconds.

3. Rinse in 1 per cent sodium bisulfite for one minute to remove residual chromic acid.

4. Wash in tap water for five to ten minutes.

5. Wash in four changes of distilled water.

6. Place in working methenamine silver nitrate solution, and then place container in oven at 58 to 60°C for 30 to 60 minutes (until sections are a light yellowish brown). Most fungi require 60 minutes. To remove slides use paraffin-coated forceps. Dip slides in distilled water. Check control slide with microscope for silver impregnation. Fungi should be a dark brown at this stage.

7. Rinse in six changes of distilled water.

8. Place in 0.1 per cent gold chloride to tone for two to five minutes.

9. Rinse in distilled water.

10. Remove unreduced silver with 2 per cent sodium bisulfite for two to five minutes.

11. Wash thoroughly in running tap water.

12. Counterstain in light green for 30 to 45 seconds.

13. Dehydrate with two changes 95 per cent alcohol, absolute alcohol, and clear with two changes xylene. Mount slides in Permount.

Results. Fungi, sharply delineated in black; bacteria (*Nocardia, Actinomyces*), black; *Nocardia* may require up to 90 minutes staining in the methenamine silver solution; mucin, taupe to dark grey; inner parts of mycelium, old rose; background, pale green. In addition, depending on staining procedure, reticulum fibers, elastin fibers, and so forth are stained black. Red blood cells and some other cells are old rose to grayish rose.

Acid Retention Stain for *Nocardia*

1. Heat-fix slide.

2. Flood slide with Kinyoun carbolfuchsin and heat for three to five minutes. Wash thoroughly.

3. Decolorize with 1 per cent H_2SO_4 for $1\frac{1}{2}$ minutes.

4. Counterstain with methylene blue for five to ten seconds.

Results. *Nocardia*, red; background, blue.

Mayer's Mucicarmine Histologic Stain for *Cryptococcus neoformans* *
Solutions

A. Weigert's iron hematoxylin
1. Solution A

Hematoxylin	1.0 g
Alcohol, 95 per cent	100.0 ml

2. Solution B

Ferric chloride, 29 per cent aqueous solution	4.0 ml
Distilled water	95.0 ml
HCl (conc.)	1.0 ml

3. Working solution

Equal parts of solution A and B	Prepare fresh

B. Metanil yellow solution

Metanil yellow	0.25 g
Distilled water	100.0 ml
Glacial acetic acid	0.25 ml

C. Mucicarmine stain

Carmine	1.0 g
Aluminum chloride, anhydrous	0.5 g
Distilled water	2.0 ml

Mix stain in small test tube. Heat over small flame for two minutes. Liquid will become almost black and syrupy. Dilute with 100 ml of 50 per cent alcohol and allow to stand for 24 hours. Filter and dilute 1:4 with tap water for use.

Staining Procedure

1. Cut paraffin sections at 6 microns and fix to slide. Deparaffinize and rehydrate through xylene, absolute alcohol, and 95 per cent alcohols in the usual manner to distilled water. Remove mercury precipitates through iodine and hypo (photographic) solutions if necessary.

2. Stain for seven minutes in working solution of Weigert's hemotoxylin.

3. Wash in tap water for five to ten minutes.

4. Rinse in diluted mucicarmine solution for 30 to 60 minutes or longer. (Check control slide after 30 minutes.)

5. Rinse in distilled water.

6. Stain in metanil yellow solution for one minute.

---

*Mallory, F. B. 1942. *Pathologic Techniques.* Philadelphia, W. B. Saunders Company, p. 130.

7. Rinse quickly in distilled water.

8. Rinse quickly in 95 per cent alcohol.

9. Dehydrate in two changes of absolute alcohol; clear with two to three changes of xylene and mount in Permount.

Results. Mucin, deep red to rose; nuclei, black; background, yellow. Care should be used not to overstain with metanil yellow.

Brown and Brenn Modification of Gram's Stain for *Nocardia* and *Actinomyces* in Tissue*

Solutions

A. Crystal violet solution

Crystal violet	1.0 g
Distilled water	100.0 ml

B. Sodium bicarbonate solution

Sodium bicarbonate	5.0 ml
Distilled water	100.0 ml

C. Gram's iodine

Iodine	1.0 g
Potassium iodide	2.0 g
Distilled water	300.0 ml

D. Ether-acetone mixture

   Equal parts of acetone and ether (v/v)

E. Basic fuchsin solution

Saturated aqueous solution of basic fuchsin	0.1 ml
Distilled water	100.0 ml

F. Picric acid–acetone

---

*Brown, J. H., and L. Brenn. 1931. Methods for differentiational staining of gram-positive and gram-negative bacteria in tissue sections. Bull. Johns Hopkins Hosp., *48*:69–73.

Picric acid	0.1 g
Acetone	100.0 ml

G. Acetone-xylene mixture

   Equal parts of acetone and xylene (v/v)

Procedure

1. Cut paraffin sections 6 microns and fix to slides. Deparaffinize and rehydrate with xylene, absolute alcohol, and 95 per cent alcohol to distilled water in usual manner.

2. Place slides on rack and pour on approximately 1 ml of 1 per cent crystal violet and 5 drops of 5 per cent sodium bicarbonate. Allow to stand for one minute while agitating slowly.

3. Wash in water.

4. Flood slide with Gram's iodine solution for one minute.

5. Rinse in water. Gently blot dry with filter paper.

6. Decolorize in ether-acetone mixture by dropping on slide until no more color comes off.

7. Stain with basic fuchsin solution for one minute.

8. Wash in water. Blot dry.

9. Transfer slide to acetone.

10. Decolorize in picric acid–acetone solution until sections are yellowish pink.

11. Rinse in acetone.

12. Rinse in acetone-xylene mixture.

13. Clear in several changes of xylene.

14. Mount in Permount or equivalent.

Results. Gram-positive bacteria, blue; gram-negative bacteria, red; nuclei, red; background, yellow.

A Glossary of Terms
for Medical Mycology

Acicular: needle-shaped.

Acrocatenulate: a chain of conida having the youngest cell at the apex; also called blastocatenate.

Acrogenous: produced at the tip or apex of the conidiogenous cell.

Acropleurogenous: produced at the tip and along the sides of the conidiogenous cell.

Acuminate: tapering to a narrow tip.

Adiaspore: a conidium of a fungus that increases greatly in size when incubated at elevated temperature (*in vitro* or *in vivo*). It does not replicate during this process, thus becoming a vesicle.

Aerial mycelium: hypal units above the colony agar interface.

Aleurioconidium (pl. aleurioconidia): a conidium that develops as the expanded end of a conidiogenous cell or hyphal branch. It is released by the lysis or fracture of its supporting cell: aleuric

Amero-: single-cell

Anamorph: a somatic or reproductive structure that originates without nuclear recombination (asexual reproduction). Cf. *Teleomorph.*

Annellide: a percurrent conidiogenous cell that produces conidia in succession, each leaving a ringlike collar of cell wall material when released. The tip of an annellide increases in length and becomes narrower as each subsequent conidium is formed. The first is holoblastic; subsequent conidia are enteroblastic.

Annelloconidium: a conidium produced by an annellide.

Antheridium: the motile, smaller, sperm-producing or "male" gametagial structure in the teleomorph state of some fungi.

Anthropophilic: a fungus (dermatophyte) that preferentially grows on man rather than other animals or in the soil.

Apiculus: a short, sharp projection from a conidium or spore.

Arthroconidium (pl. arthroconidia): a thallic conidium released by the fragmentation or lysis of hypha. It is not notably larger than the hypha from which it was produced, and separation occurs at a septum.

Ascocarp: a structure bearing asci and ascospores; a fruiting body of the Ascomycetes also called an ascoma.

Ascogenous: ascus-bearing.

Ascogonium: the "female" gametangium, a large and/or stationary unit that is fertilized from an antheridium or spermagonium.

Ascoma: a fruiting body that gives rise to asci and ascospores.

Ascomycetous: referring to the Ascomycetes.

Ascospore: a spore, usually the result of nuclear recombination, formed in an ascus, the teleomorphic reproductive propagule of the Ascomycota. The spore is formed internally by "free cell" formation following meiosis. Cf. *Sporangiospore.*

Ascus (pl. asci): a saclike cell in which ascospores are formed by "free cell" formation following meiosis.

Asperate: roughened.

Asperulate: finely roughened.

Ballistoconidium: a conidium that is forcibly discharged from its conidiogenous cell.

Ballistospore: a spore that is forcibly discharged from its sporogenous cell, as in some Basidiomycota.

Basidiospore: a spore borne on a basidium that develops following meiosis in the Basidiomycota.

Basidium (pl. basidia): a club-shaped structure that gives rise to basidiospores.

Basocantenulate: a chain of conidia having the youngest conidia at the base of the chain above the apex of the conidiogenous cells.

Blastic: enlargement (thus new growth) of a conidium prior to being delimited by a septum.

Blastocatenate. See *Acrocatenate.*

Blastoconidium (pl. blastoconidia): a holoblastic conidium that detaches at a predetermined point at maturity and leaves a bud scar. Example: the budding process in yeast cells. Cf. *Phialoconidium.*

Capsule: a hyaline mucopolysaccharide covering aground the cell body of certain yeasts (*Cryptococcus, Rhodotorula*) and some spores and conidia.

Catenulate: in chains or end-to-end series.

Chlamydoconidium (pl. chlamydoconidia): a thallic conidium formed from a preexisting hyphal cell by modification. This usually involves increased thickness of the wall and sometimes enlargement of the cell, with the protoplasm becoming dense. The conidium is released by fracture or lysis of the adjacent hypha.

Clamp connection: a specialized hyphal bridge over a septum in the Basidiomyces. It is used for nuclear migration after mitosis during the formation of a new cell.

Clavate: club-shaped.

Cleistothecium (pl. cleistothecia): an ascoma that typically does not have a pore or opening and contains random asci and ascospores.

Coencytic: having many nuclei.

Collarette: the remnant of a cell wall present at the tip of a phialide. It is the result of the rupture of the tip of the conidiogenous cell during the release of the first phialoconidium. Also, the ring of cell wall material around a sporangiophore left by the peridium of a sporangium when it dissolves or is broken.

Columella (pl. columellae): a sterile invagination of the sporangiophore into the fertile area of the sporangium.

Conidiogenous cell: the cell that gives rise to a conidium.

Conidium (pl. conidia): a reproductive propagule produced in the absence of nuclear recombination, thus representing anamorphic or asexual reproduction.

Conidophore: a specialized hypha that gives rise to or bears a conidium.

Dehiscence: the opening of a pore at maturity or the breaking of a stem (at the septa) for release of a propagule.

Deliquescence: autolysing or becoming liquid at maturity.

Dematiaceous: a fungus having brown or black melanotic pigment in the cell wall, thus phaeo-.

Denticle: a small projection on which the conidium develops.

Denticulate: a conidiogenous cell bearing denticles.

Determinate conidiation: a type of conidiation in which the end conidium in a chain does not proliferate. Determinate conidiophore. The conidiogenous cell (conidiophore) does not extend or continue to grow after the formation of a conidium.

Dichotomous: a type of branching of hyphae that is repetitious without pattern; the branches are approximately equal in size and equal the stem from which they originated.

Dictyoconidium: a conidium having transverse and longitudinal septa. Same as *Muriform.*

Didymoconidium: a two-celled conidium (Septa transverse).

Dimorphic: having two forms.

Diploid: having the 2n chromosome number.

Echinulate: covered with delicate spines.

Ectoendothrix: arthroconidia formed on the outside and inside of a hair shaft.

Ectothrix: forming a sheath of arthroconidia on the outside of a hair shaft. The cuticle of the hair remains intact.

Endogenous: from within.

Endospore: a spore formed within some other unit, such as in a spherule.

Endothrix: arthroconidia formed inside a hair shaft. The cuticle of the hair is usually destroyed.

Enteroblastic: formed from the inner cell wall layers of a conidiogenous cell.

Epithet: the species name or second part of a Latin binominal.

Exogenous: from without.

Floccose: cottony.

Furfuraceous: branny, scaly.

Geniculate: bent like a knee.

Geophilic: soil-seeking, having a soil reservoir.

Germ tube: initial hypha from a sprouting conidia, spore, or yeast.

Glabrous: smooth.

Gloiospora: hyphomycete having slimy conidia, also called gloeoid balls.

Gymnothecium (pl. gymnothecia): an ascoma composed of loosely intertwined hyphae; this arrangement allows many open spaces for the release of ascospores formed within the fruiting structure. The asci are randomly distributed.

Halophilic: tolerant of high salt concentration.

Haploid: having the n or reduced numbers of chromosomes.

Heterokaryotic: said of an organism that has genetically different nuclei in the same protoplasm.

Heterothallic: in its restricted sense, the requirement for mating and karyogamy (production of teleomorph state) of two genetically dissimilar nuclei.

Holoblastic: said of a conidium in which the walls are formed from all walls of the conidiogeneous cells.

Holothallic: the walls of thallic conidia utilize all layers of the wall of the hypha in which it is formed.

Homothallic: production of the teleomorph state, i.e., karyogamy and nuclear recombination, from genetically similar nuclei of the same thallus.

Hyalo-: colorless; also hyaline.

Hypha (pl. hyphae): a vegetative filament of a fungus.

Hyphomycete: an anamorph fungus that produces mycelium with or without discernible dark pigment in the cell walls. If the hypha is pigmented, it is called dematiaceous.

Indeterminate: growth that may extend beyond a definite size.

Indurated: hard.

Intercalary: formed within a hyphal unit. Cf. *Terminal.*

Lanose: woolly.

Lateral: on the side.

Lenticular: shaped like a double convex lens or a lentil.

Macroconidium: the larger of two types of conidia produced in the same manner by the same fungus.

Macronematous: pertaining to a conidiophore that is markedly different from the vegetative hyphae. Also, a large, coarse or thick mycelial structure. Cf. *Micronematous.*

Meristematic: pertaining to an area from which new growth occurs.

Merosporangium: small, usually cylindric sporangium containing a few spores in a row.

Microconidium (pl. microconidia): the smaller of two types of conidia produced in the same manner by the same fungus.

Micronematous: pertaining to a conidiophore morphologically like vegetative hyphae.

Moniliform: having swellings that occur at regular intervals in a hypha.

Muriform: like a wall; multicellular, with traverse and longitudinal septations.

Mycelium: same as *Hyphae.*

Nodose: having intermittent thickenings.

Ostiole: an opening or pore.

Pectinate: like the teeth of a comb.

Penicillate: having a tuft of fine hairs, as on a brush.

Percurrent: continued growth of a conidiogenous cell after the production of the first conidium. The growth occurs through the open end of the conidiophore after the dehiscence of the conidium.

Peridium: the wall of the structure (ascoma) surrounding the asci; also, the wall of a sporangium and the wall around some basidiocarps.

Perithecium: an ascoma that has solid walls and a pore (ostiole) through which ascospores may escape. The asci are arranged on a basal bush or hymenial layer.

Phaeo-: darkly pigmented.

Phialide: a conidiogenous cell in which the meristematic region remains the same diameter and the cell does not increase in length with each conidium produced. The successive conidia produced within the cell are enteroblastic and extruded.

Phialoconidium (pl. phialoconidia): conidium produced by a phialide.

Phragmoconidium: a conidium with two or more transverse septa.

Pleurogenous: borne on the sides of a conidiophore or hypha.

Poroconidium (pl. poroconidia): a conidium produced through a small pore in a conidiogenous cell.

Progressive cleavage: sporulation in a sporangium, involving a series of cleavage planes that are produced in succession. The first result is a multinucleate *protospore*; more cleavages eventually result in numerous sporangiospores, each having one nucleus.

Protospore: the first product (multinucleate) of "progressive cleavage" spore formation.

Pseudohypha (pl. pseudohyphae): a fragile string of cells that result from the budding of blastoconidia that have remained attached to each other. The septa separating the cells are complete and there is no cytoplasmic connection, as is found in most true septate hypha.

Pseudoparenchyma: a mass of hyphae arranged together to form a tissuelike structure.

Rhizoid: a rootlike structure.

Sclerotium (pl. sclerotia): a hard mass composed of intertwined mycelium; usually with thick pigmented walls that are resistant to adverse environmental conditions. It will germinate to produce new hypha under favorable conditions or in response to a chemical stimulant from a prospective host.

Scutulum: the compact mass of hyphae, keratinous material, and hair found in favus.

Septum (pl. septa): a cross wall.

Spinose: covered with small spines.

Sporangiophore: a specialized hypha that gives rise to a sporangium.

Sporangiospore: a reproductive unit formed in a sporangium.

Sporangium: a cell that produces spores internally by a series of "progressive cleavages." The resulting spores are called sporangiospores. Cf. *Ascospore.*

Spore: a reproductive propagule produced internally by "free cell" formation, as in the ascomycete, i.e., complete spores formed all at once around the nuclei available or by "progressive cleavage," as in a sporangium.

Sporodochium (pl. sporodochia): a cushion-shaped stroma or mat of hyphae covered with conidiophores.

Sterigma (pl. sterigmata): a narrowed point that bears a basidiospore on a basidium.

Stolon: a runner; the aerial hypha that forms rhizoids when it contacts the agar (or substrate) surface. Often there is swelling (called a node) at this point.

Sympodula: a conidiogenous cell that extends beyond the point of the apex at which a conidium was produced, forming a new apex; thus a series of apices are formed on one side, then on the other.

Teleomorph: a reproductive structure of a fungus that is the result of plasmogamy and nuclear recombination; asexual state.

Thallic: said of conidia produced where there is no enlargement or new growth of the conidia initially prior to its delineation by a septa. The entire parent cell becomes the conidium.

Uncinate: bent like a hook.

Verticillate: branches or conidia radiating from a common point on a conidiogenous cell or a hypha; whorled.

Vesicle: a swollen or bladderlike cell.

Yeast: a single fungal cell, usually ovoid, that replicates by blastoconidia formation, planate division, or reduced phialide or annellide.

Zoophilic: having reservoir or preferential host in an animal other than man.

Zygote: the result of fusion of two haploid cells to form a diploid cell. Meiosis may occur immediately, as in most zygomycetes.

Author Index

801

Subject Index

Page numbers in *italics* indicate illustrations; t indicates table.